GRAPH STRUCTURE AND MONADIC SECOND-ORDER LOGIC

A Language-Theoretic Approach

The study of graph structure has advanced significantly in recent years: finite graphs can now be described algebraically, enabling them to be constructed out of more basic elements. One can obtain algebraic characterizations of tree-width and clique-width, two graph complexity measures that are important for the construction of polynomial-time graph algorithms. Separately the properties of graphs can be studied in a logical language called monadic second-order logic. In this book, these two features of graph structure are brought together for the first time in a presentation that unifies and synthesizes research over the last 25 years. The authors not only provide a thorough description of the theory, but also detail its applications, on the one hand to the construction of graph algorithms, and on the other to the extension of formal language theory to finite graphs. This extension combines algebraic notions (equational and recognizable sets) and logical ones (graph transformations specified by logical formulas). Applications of these tools to languages of words and terms are also presented.

Consequently the book will be of interest to graduate students and researchers in graph theory, finite model theory, formal language theory and complexity theory.

Encyclopedia of Mathematics and Its Applications

This series is devoted to significant topics or themes that have wide application in mathematics or mathematical science and for which a detailed development of the abstract theory is less important than a thorough and concrete exploration of the implications and applications.

Books in the **Encyclopedia of Mathematics and Its Applications** cover their subjects comprehensively. Less important results may be summarized as exercises at the ends of chapters. For technicalities, readers can be referred to the bibliography, which is expected to be comprehensive. As a result, volumes are encyclopedic references or manageable guides to major subjects.

ENCYCLOPEDIA OF MATHEMATICS AND ITS APPLICATIONS

All the titles listed below can be obtained from good booksellers or from Cambridge University Press.
For a complete series listing visit www.cambridge.org/mathematics.

ENCYCLOPEDIA OF MATHEMATICS AND ITS APPLICATIONS

Graph Structure and Monadic Second-Order Logic

A Language-Theoretic Approach

BRUNO COURCELLE

Université de Bordeaux

JOOST ENGELFRIET

Universiteit Leiden

CAMBRIDGE
UNIVERSITY PRESS

CAMBRIDGE UNIVERSITY PRESS
Cambridge, New York, Melbourne, Madrid, Cape Town,
Singapore, São Paulo, Delhi, Mexico City

Cambridge University Press
The Edinburgh Building, Cambridge CB2 8RU, UK

Published in the United States of America by Cambridge University Press, New York

www.cambridge.org
Information on this title: www.cambridge.org/9780521898331

First published 2012

Printed in the United Kingdom at the University Press, Cambridge

A catalogue record for this publication is available from the British Library

Library of Congress Cataloguing in Publication data
Courcelle, B.
Graph structure and monadic second-order logic : a language-theoretic approach / Bruno Courcelle,
Joost Engelfriet.
p.   cm. – (Encyclopedia of mathematics and its applications ; 138)
Includes bibliographical references and index.
ISBN 978-0-521-89833-1 (hardback)
1. Logic, Symbolic and mathematical – Graphic methods.   I. Engelfriet, Joost.   II. Title.
QA9.C748   2012
511.3–dc23   2012008159

ISBN 978-0-521-89833-1 Hardback

Additional resources for this publication at www.labri.fr/perso/courcell/TheBook.html

To Dominique and Louco

Contents

Foreword
by Maurice Nivat

The genesis of this great and beautiful book spans more than 20 years. It collects and unifies many theoretical notions and results published by Bruno Courcelle and others in a large number of articles.

The concept of a language to communicate with a computer, a machine or any kind of device performing operations is at the heart of Computer Science, a field that has truly thrived with the emergence of symbolic programming languages in the 1960s. Formalizing the algorithms that enable computers to calculate an intended result, to control a machine or a robot, to search and find the relevant information in response to a query, and even to imitate the human brain in actions such as measuring risk and making decisions, is the main activity of computer scientists as well as of ordinary computer users.

The languages designed for these tasks, which number by thousands, are defined in the first place by syntactic rules that construct sets of words and to which are then attached meanings. This understanding of a language was first conceived by structural linguists, in particular Nicolaï Troubetskoï, Roman Jacobson and Noam Chomsky, and has transformed Linguistics, the study of natural languages, by giving it new directions. It has also been extended to programming languages, which are artificial languages, and to the Lambda Calculus, one of many languages devised by logicians, among whom we can cite Kurt Gödel, Alonzo Church and Alan Turing, who aspired to standardize mathematical notation and to mechanize proofs. This same idea has inspired all research on computation theory and programming. Thanks to the results of this research, planes can fly with continuously monitored flight parameters, providing us with unprecedented reliability: this is so because millions of lines of code have been formally proved to be correct.

Words are strings of symbols taken from finite alphabets. They constitute the basic elements. They can represent all the information one might wish to capture, use, process, disseminate or share in a world that is fast becoming more and more "digital," as Gérard Berry emphasized recently in his lectures at the Collège de France.

Most information, though represented always by words, is nevertheless structured hierarchically and can thus be presented in a natural way as a tree or as a graph. Most of the countless electronic chips that make up computers but are also used in an ever-growing number of other machines as well, from washing machines to nuclear power plants, calculate on graphs: by connecting vertices, their edges can represent virtually any relationship of subordination, analogy, neighborhood or causality. From the early 1960s, the algorithms for graphs (and for trees, which are particular graphs) have been developed swiftly, and most of the current computing applications are based on these algorithms. Thousands of them have been designed by numerous researchers and engineers, and they fuel a burgeoning literature.

It was around 1980 that Bruno Courcelle, a former student of the prestigious Ecole Normale Supérieure (Rue d'Ulm in Paris), a logician by training, and a young but already established researcher, tackled a seemingly impossible task: to build a theory of tree languages that would classify all of these algorithms and present them in a unified and rational way. Bruno Courcelle is not a "problem solver" who happened to discover more-or-less elegant and clever answers to questions; he likes well-founded and harmonious theories, and is always looking for unifying concepts. Armed with a knowledge of logic and with a familiarity with Fundamental Computer Science, and in particular with Formal Language Theory which he gained during his years as a researcher at INRIA (Institut National de Recherche en Informatique et Automatique) while preparing his thesis, Bruno Courcelle got down to work with perseverance and determination.

Upon his arrival at Bordeaux-1 University in 1979 (LaBRI, the Laboratoire Borde-lais de Recherche en Informatique, was created in 1988), Bruno Courcelle found an excellent work environment. The concept of attribute grammar, which is important in compilation, provided the model that he has used to develop an algebraic approach to graph grammars and a logical approach to the proof of properties of the graphs they generate. The first published work he devoted to attribute grammars is the source of the theory presented here, based on Logic and Universal Algebra.

The impact of his work surpassed all expectations, even taking into consideration the remarkable qualities of method and rigor that characterize Bruno Courcelle. For when the first elements of his theory began to spread among those who work on design-ing and improving graph algorithms, these researchers realized that Bruno Courcelle had provided a convenient formal framework in which many problems could be solved. In particular, Bruno's theory brought a logical lightening to the profound works of Paul Seymour and his collaborators on graph minors. Other researchers have been inspired by his theory to study new problems and invent new algorithms. A daunting theory that was originally seen as arcane and abstract proved to be rich and fertile. In 2004, Bruno Courcelle was awarded by INIST (an institute depending on the Centre National pour la Recherche Scientifique) and the ISI-Web of Science (Thomson-Reuters) the prize of the "most cited researcher in Computer Science in France."

I am not going to analyze his work further; in any case Chapter 1 is a long overview that is perfectly readable, even by those who are well versed neither in algebra nor in mathematical logic. I would rather emphasize that the work of a computer scientist, as of any scientist, can be very diverse. The quest for new results is one endeavor, but bringing up to date the underlying structures, unifying concepts and simplifying the presentation of results, is quite another. The rapid, sometimes frantic, growth of publications in Computer Science has led most researchers to choose the former in pursuit of better results and more efficient algorithms. It gives me great pleasure to preface a work that is of the opposite nature: a long book produced by a comprehensive study, which leads to a very interesting result: the formation of a theory that brings order, that explains and simplifies a vast collection of results obtained by others and, at the same time, that proposes methods, yields results and raises new questions.

While still a student I was very struck when I first read André Lichnérowicz's book on linear algebra. I had already taken courses in linear algebra which, I confess, were not very helpful, and suddenly this book made everything clear. The mysterious operations which we were taught to perform on the square tables called "determinants" started making sense; the concepts of both vector space and the dimension of a vector subspace finally allowed me to understand what it meant to agonize over a determinant and, moreover, why this notion is important. Lichnérowicz's book is a classic that has enabled generations of students to learn linear algebra with ease, and it has become a mathematical tool widely used by engineers and technicians who are not professional mathematicians. I believe that this book will get the same reputation, quickly become a classic and provide an easy access to the burgeoning world of graph algorithms and its numerous applications throughout the sciences and beyond.

The comments above were written two years ago, when Bruno Courcelle's book was only 500 pages long, and I cannot change what I wrote then: it is a great and beautiful book that is going to take its place very soon on the library shelves of all the departments of Computer Science around the world. But now the book is 700 pages long and has two authors, Bruno and Joost Engelfriet. What happened is that Bruno sent the previous version to Joost to read and suggest corrections and improvements. Joost is a very old acquaintance of both Bruno and myself, and we have always known him as one of the most knowledgeable researchers in the field of grammars, automata and transducers on words and trees. And Joost had so many things to suggest that it is another book that I present today: thicker, with new results and a number of proofs that have been replaced by simpler and more elegant ones. Obviously the cooperation between Bruno and Joost was a very fruitful one indeed.

Knowing Joost as I do, this is not a surprise: when I asked him to referee papers submitted to the journal *Theoretical Computer Science*, in most cases Joost's report was longer and sometimes richer than the refereed article. His comments always led to a major improvement of the original text. Clearly Bruno's manuscript inspired Joost. And we all have to be grateful to him for, as usual, his comments and the work he did on the manuscript resulted in a major improvement and a sizable enlargement.

Thus today I am very happy to thank the two authors of this beautiful book, which I consider to be a wonderful source of knowledge in Computer Science. It is a theoretical book, and for that reason some people may find it hard to read, but reading it is worth the pain, because the formalism introduced and the methods presented have already led to many new algorithms on graphs (as the number of citations of Bruno's published papers show) and they will lead to many others in the future. To anyone interested in graph algorithms I can only recommend that they read this book first.

For indeed this book lies at the very heart of Computer Science, which is the expressiveness of the languages used to represent and manipulate information and information structures, graphs being among the most widely used information structures. Progress in the efficiency, liability and simplicity of algorithms comes mainly from the use of better representations, better structures and a better understanding of the different ways in which one can describe sets of data and express their properties. This book provides a huge number of conceptual tools to design and study graph algorithms that no one should ignore.

In the name of the young but fast-growing science that in French we call *Informatics*, in the name of all future researchers in this field, I just say to Bruno and Joost: Thanks, you have done a good job!

Introduction

This book contributes to several fields of Fundamental Computer Science. It extends to finite graphs several central concepts and results of Formal Language Theory and it establishes their relationship to results about Fixed-Parameter Tractability. These developments and results have applications in Structural Graph Theory. They make an essential use of logic for expressing graph problems in a formal way and for specifying graph classes and graph transformations. We will start by giving the historical background to these contributions.

Formal Language Theory

This theory has been developed with different motivations. Linguistics and compilation have been among the first ones, around 1960. In view of the applications to these fields, different types of *grammars*, *automata* and *transducers* have been defined to specify *formal languages*, i.e., sets of *words*, and transformations of words called *transductions*, in finitary ways. The formalization of the semantics of sequential and parallel programming languages, that uses respectively *program schemes* and *traces*,[1] the modeling of biological development and yet other applications have motivated the study of new objects, in particular of sets of *terms*.[2] These objects and their specifying devices have since been investigated from a mathematical point of view, independently of immediate applications. However, all these investigations have been guided by three main types of questions: comparison of descriptive power, closure properties (with effective constructions in case of positive answers) and decidability problems.

A *context-free grammar* generates words, hence specifies a formal language. However, each generated word has a *derivation tree* that represents its structure relative to the considered grammar. Such a tree, which can also be viewed as a term, is usually

---

[1] Traces are equivalence classes of words for congruences generated by commutations of letters; see the book [*DiekRoz]. For program schemes, see [*Cou90a]. The list of references is divided into two parts. The first part lists books, book chapters and survey articles: the * in, e.g., [*DiekRoz] indicates a reference of this kind. The second part lists research articles and dissertations.

[2] In Semantics, one is also interested in *infinite* words, traces and terms. In this book these will not be considered.

the support of further computation, typically a translation into a word of another language (this is the case in linguistics and in compilation). Hence, even for its initial applications, Formal Language Theory has had to deal with trees as well as with words. In Semantics, terms are even more important than words. Thus, sets of terms, usually called *tree languages*[3], and transductions of terms, called *tree transductions*, have become central notions in Formal Language Theory.

Together with context-free grammars, *finite* (also called *finite-state*) *automata* are among the basic notions of Language Theory, in particular for their applications to lexical analysis and pattern matching. They were also used early on (around 1960) for building algorithms to check the validity of certain logical formulas, especially those of *monadic second-order logic*, in certain relational structures. On the other hand, monadic second-order logic can be used to specify and to classify sets of words and terms.[4] There are deep relationships between monadic second-order formulas and finite automata that recognize words and terms (see [*Tho97a]). The *fundamental result* is that every language that is specified by a sentence of monadic second-order logic (expressing a property of words) can be recognized by a finite automaton, and vice-versa. Moreover, the finite automaton can be constructed effectively from the sentence. This means that monadic second-order logic can be viewed as a high-level specification language that can be compiled into "machine code": a finite automaton that recognizes the words that satisfy the specification. The same result holds for terms, with respect to finite automata on trees. As a consequence of this fundamental relationship, monadic second-order logic is now one of the basic tools used in Formal Language Theory and its applications, in addition to context-free grammars, finite automata and finite transducers (which are finite automata with output).

The extension of the basic concepts of Formal Language Theory to *graphs* is a natural step because graphs generalize trees. However, graphs have already been present from the beginnings in several of its fields. In compilation, one uses *attribute grammars* that are context-free grammars equipped with semantic rules ([*AhoLSU], [*Cre]). These rules associate graphs (called *dependency graphs*) with derivation trees. An attribute grammar is actually the paradigmatic example of a *context-free graph grammar* (based on *hyperedge replacement* rewriting rules, [*DreKH]). In the semantics of parallelism, traces are canonically represented by graphs, and an important concern is to specify them by finite automata ([*DiekRoz]).

One starting point of the research presented in this book has been the development of a robust theory of *context-free graph grammars*, of *recognizability of sets of graphs* (to be short, an algebraic formulation of finite automata) and of *graph transductions*. In order to use the theory of context-free grammars and recognizability in arbitrary algebras initiated by Mezei and Wright in [MezWri], we choose appropriate

---

[3] In addition to being words, terms have canonical representations as labeled, rooted and ordered trees. They are thus called "trees" but this terminology is inadequate.

[4] This logical language and the related one called μ-calculus ([*ArnNiw]) are also convenient for expressing properties of programs.

(and natural) operations on graphs. Thus, graphs become the value of terms that are built with these (infinitely many) operations. Roughly speaking, a *context-free graph grammar* is a finite set of rules of the form $A_0 \to f(A_1, \ldots, A_n)$, $n \geq 0$, where each A_i is a nonterminal of the grammar and f is one of the chosen graph operations. The rule means that if the graphs $G_1, \ldots, G_n$ are generated by respectively $A_1, \ldots, A_n$, then A_0 can generate the graph $f(G_1, \ldots, G_n)$. Such grammars have useful applications to Graph Theory: they can be used to describe many graph classes in uniform ways and to prove by inductive arguments certain properties of their graphs. Still roughly speaking, a set of graphs is *recognizable* if there is a finite automaton that recognizes all the terms that evaluate to a graph in the set. Thus, the automaton does not work directly on the given graph, but rather on any term that represents that graph. In a similar way one can define graph transductions through the use of tree transducers. Note that, to describe a set of graphs or a graph transduction in a finitary way, one can necessarily use only finitely many graph operations. As we will see, that is a rather severe, but natural restriction.

Our main goal will be to show that the fundamental use of monadic second-order logic as a high-level specification language carries over to graphs, not only for the specification of recognizable sets of graphs, but also for context-free sets of graphs and for certain types of graph transductions. This gives a new dimension to the above-mentioned fundamental result for words and terms, because the properties of graphs that can be specified in monadic second-order logic are more varied and useful than those of words and terms.

We will specify a set of graphs by a monadic second-order sentence, and a graph transduction by a tuple of monadic second-order formulas that define an "interpretation" of the output graph in the input graph. From such a specification we will show how one can construct a finite automaton on terms, or a tree transducer in the second case, that is related to the specification as explained above. Note that the logic "acts" directly on the graphs, whereas the automata and transducers work on the terms that denote these graphs. Thus, monadic second-order logic can be viewed as playing the role of "finite automata on graphs" and "finite transducers of graphs" in our Formal Language Theory for Graphs.

Graph algorithms

The above-mentioned developments have important applications for the construction of polynomial-time algorithms on graphs. In his 16th NP-completeness column, published in 1985 [John], Johnson reviews a number of NP-complete graph problems that become polynomial-time solvable if their inputs are restricted to particular classes of graphs such as those of trees, of series-parallel graphs, of planar graphs to name a few. For many of these classes, in particular for trees, *almost trees* (with parameter k), *partial k-trees*, *series-parallel graphs*, *outerplanar graphs* and *cographs*, the efficient algorithms take advantage of certain hierarchical structures of the input graphs. Because of these structures, these graphs are somehow close to trees.[5] The notion of a partial k-tree has emerged as a powerful one subsuming many other types of "tree-like graphs." (The cographs have a canonical hierarchical structure but they are not included in the class of partial k-trees for any fixed k.) Many articles have produced polynomial-time algorithms for NP-complete problems restricted to partial k-trees. In 1994, Hedetniemi has compiled a list of 238 references [*Hed] on partial k-trees and algorithms concerning them. The notion of a partial k-tree has also been used with a different terminology (*tree-width, tree-decomposition*) by Robertson and Seymour in their study of the structure of graph classes that exclude fixed graphs as minors. They formulate this notion in terms of particular decompositions of graphs, called tree-decompositions, that are at the basis of the construction of polynomial-time algorithms. Each tree-decomposition has a *width*, and a graph is a partial k-tree if and only if it has tree-width at most k, which means that it has a tree-decomposition of width at most k.

The recent theory of Fixed-Parameter Tractability (the founding book by Downey and Fellows [*DowFel] was published in 1999) now gives a conceptual framework to most of these results. The notion of a *fixed-parameter tractable algorithm* specifies how the multiplicative constant factor of the time-complexity of a polynomial-time algorithm depends on certain parts of the data. It happens that for most of the graph algorithms based on tree-decompositions, the exponent of the polynomial is 1: these algorithms are linear-time in the size of the input graphs, with multiplicative "constant" factors that depend exponentially (or more) on the widths of the input tree-decompositions.

The explanation for this fact is one of the main goals of this book. We will show that, for a certain natural choice of graph operations, tree-decompositions correspond to terms, and tree-decompositions of width at most k correspond to terms that are built from a finite subset of those operations. A general algorithmic result that encompasses many of the above-mentioned results, follows from the fundamental relationship between monadic second-order logic and finite automata discussed before: if the considered problem is specified by a monadic second-order sentence (and this is the case for many NP-complete graph problems not using numerical values in their inputs), then a finite automaton on the terms that encode the tree-decompositions of width at most k can be constructed (for each k) to give the answer to the considered question (for example, *Is the given graph 3-colorable?*) where the input graph is given by a tree-decomposition (or a term encoding it). The linearity result follows because finite automata can be implemented so as to work in linear time (and because a tree-decomposition of a graph can be found in linear time).

---

[5] These classes can actually be generated by certain context-free graph grammars and the corresponding hierarchical structures of the generated graphs are represented by their derivation trees. There is thus a close relationship between the algorithmic issues and the extensions of language theoretic concepts discussed above.

We will extend the case of tree-width bounded graphs (already discussed in [*DowFel]) to another type of graph decompositions, based on another natural choice of graph operations. This leads to the notion of *clique-width* of a graph. Clique-width is more powerful than tree-width in the sense that every set of graphs of bounded tree-width has bounded clique-width but not vice-versa, an example being the set of cographs. On the other hand, in the above general result, the monadic second-order sentences must be restricted to use quantifications on sets of vertices (instead of both vertices and edges), so fewer graph problems can be specified. The algorithms are cubic-time instead of linear-time because, for these graph operations, cubic time is needed to find a term for a given graph.

The theory that will be exposed in the nine chapters of this book has arisen from the confluence of the two main research directions presented above. The remainder of this introduction will present in a more detailed way, but still informally, the main concepts and results.

The role of logic

We will study and compare finitary descriptions of sets of finite graphs by using concepts from Logic, Universal Algebra and Formal Language Theory. We first explain the role of Logic. A graph[6] can be considered as a *logical structure* (also called *relational structure*) whose *domain* (also called its *universe*) consists of the vertices, and that is equipped with a binary relation that represents adjacency. Graph properties can thus be expressed by logical formulas of different languages and classified accordingly.

First-order formulas are rather weak in this respect because they can only express local properties such as that a graph has maximum degree or diameter bounded by a fixed integer. Most properties of interest in Graph Theory can be expressed by *second-order formulas*: these formulas can use quantifications on relations of arbitrary arity. Unfortunately, little can be obtained from the expression of a graph property in second-order logic. Our favorite logical language will be its restriction called *monadic second-order logic*. Its formulas are the second-order formulas that only use quantifications on unary relations, i.e., on sets. They can express many useful graph properties like *connectivity*, *p-colorability* (for fixed *p*) and *minor inclusion*, whence *planarity*. Such properties are said to be *monadic second-order expressible*, and the corresponding sets of graphs are *monadic second-order definable*.

These logical expressions have interesting algorithmic consequences as explained above, but only for graphs that are somehow "tree-like" (because 3-colorability is NP-complete and expressible by a monadic second-order sentence). Monadic second-order sentences are also used in Formal Language Theory to specify *languages*, i.e., sets of words or terms. The fundamental result establishes that monadic second-order sentences and finite automata have the same descriptive power. But

---

[6] In order to simplify the discussion, we only discuss simple graphs, i.e., graphs without parallel edges.

monadic second-order formulas are even more important for specifying sets of graphs than for specifying languages because there is no convenient notion of graph automaton. They replace finite automata, not only for specifying sets of graphs, but also for specifying graph transformations. Such transformations, called *monadic second-order transductions*, generalize the transductions of terms and words defined by finite automata with output called *finite transducers*.[7] Independently of these language theoretic applications, monadic second-order transductions are technically useful for constructing monadic second-order formulas because the inverse image of a monadic second-order definable set of relational structures under a monadic second-order transduction is monadic second-order definable.

However, monadic second-order logic alone yields no interesting results. In order to be useful for the construction of algorithms, the expression of a graph property by a monadic second-order sentence must be coupled with constraints on the graphs of interest such as having bounded tree-width or bounded clique-width. The language theoretical issues to be discussed below will also combine monadic second-order sentences and the very same constraints. Hence, we will study certain *hierarchical graph decompositions*, such as tree-decompositions, that fit with monadic second-order logic.

Graph algebras

Graph decompositions will be formalized algebraically by terms written with appropriate graph operations. Hence, we will use concepts from Universal Algebra in addition to ones from Logic.

For treating graphs as algebraic objects, i.e., as elements of appropriate algebras (words and traces are elements of monoids), we will define *graph operations* that generalize the concatenation of words. We will consider two natural ways to "concatenate" two graphs. One way is to "glue" them together, by identifying some of their vertices. The other way is to "bridge" them (or rather, "bridge the gap between them"), by adding edges between their vertices. Clearly, to obtain single valued operations, we have to specify which vertices must be "glued" or "bridged." By means of labels attached to vertices, we will specify that vertices with the same label must be identified, or that edges must be created between all vertices with certain labels. Hence, we will define "concatenation" operations on *labeled graphs*. To allow the flexible use of vertex labels, we also define (unary) operations that modify these labels. Terms written with these operations evaluate to finite (labeled) graphs. The value G of a term $t = f(t_1, t_2)$ is a certain combination, specified by f, of the values of its subterms t_1 and t_2. These values are, roughly speaking, subgraphs of G (only "roughly" because the labels of the vertices of the graphs defined by t_1 and t_2 may differ from their labels in the resulting graph G). The same holds for all subterms of t, hence, t represents a hierarchical decomposition of G.

---

[7] In particular, the *rational transductions* that are transductions of words defined either by finite(-state) transducers or, algebraically, in terms of homomorphisms and regular languages.

Based on the idea of "gluing" graphs (and using the numbers $1, \ldots, k+1$ as labels), we will define, for each k, a finite set of graph operations, $F^{\mathrm{HR}}_{[k+1]}$, that generates exactly the graphs of tree-width at most k. Hence, these operations formalize algebraically an existing combinatorial notion. They yield a graph algebra (that generalizes the monoid of words) having countably many operations. We will call it the *HR algebra* for reasons explained below. Another countable family of graph operations, also indexed by positive integers and based on the idea of "bridging" graphs, will yield a different graph algebra, called the *VR algebra*, and a graph complexity measure called *clique-width*. By definition, a graph has clique-width at most k if it is generated by the analogous finite set of graph operations $F^{\mathrm{VR}}_{[k]}$. As observed before, clique-width is more powerful than tree-width in the sense that every set of graphs of bounded tree-width has bounded clique-width but not vice-versa. Many definitions and results will be similar for these two graph algebras. We will explain below why both algebras are interesting.

The introduction of graph operations is essential for our project of extending to graphs the basic concepts of Formal Language Theory in a clean way. We will use for that the algebraic notions of an *equational* set and of a *recognizable* set. An equational set is a component of the least solution of an equation system written with set union and the operations of the considered algebra. Equation systems formalize context-free grammars that generate elements of the algebra: if such a context-free grammar has, e.g., three rules $A \to f(B,C)$, $A \to g(A)$ and $A \to a$ for the nonterminal A, where B and C are two other nonterminals, f and g are operations of the algebra and a is a constant of the algebra, then the corresponding equation system has the equation $A = f(B,C) \cup g(A) \cup \{a\}$ (where A, B and C now stand for sets of elements of the algebra). The context-free languages are actually the equational sets of the monoids of words over their terminal alphabets (due to the least fixed-point characterization of context-free grammars of [GinRic] and [ChoSch]). A recognizable set is a set saturated by a congruence having finitely many classes. The regular languages are thus the recognizable sets of the monoids of words. When all elements of the algebra can be denoted by a term (which is the case for the HR and VR algebra), a set is recognizable if and only if there exists a finite automaton on terms that recognizes all the terms that evaluate to an element of the set.

The chart of Figure I.1 shows some relationships between the notions defined above. An arrow means "used for a definition or a construction."

Two graph algebras

Since we will define two graph algebras, we will obtain two types of equational sets, called the HR- and the VR-equational sets. For each k, the set of graphs of tree-width at most k is HR-equational (because it is generated by the finite set of operations $F^{\mathrm{HR}}_{[k+1]}$), and similarly, the set of graphs of clique-width at most k is VR-equational. There are also two types of recognizable sets of graphs, the HR- and the VR-recognizable sets. Every HR-equational set is VR-equational and every VR-recognizable set is

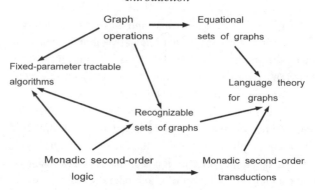

Figure I.1 The main notions.

HR-recognizable, but not vice-versa. The class of HR-equational sets is incomparable with the class of HR-recognizable sets, and similarly for the VR algebra.

These facts show some important differences with the case of words. For words, we have a unique algebraic structure based on a single operation, and the class of recognizable sets (the regular languages) is properly included in that of equational sets (the context-free languages). But graphs are intrinsically more complicated than words: this explains why we need countably many operations and not just one. We will explain next why we have *two* algebras and two (robust) classes of equational sets that both generalize the class of context-free languages.

The two graph algebras have been defined initially in such a way that their equational sets coincide with existing *context-free sets of graphs*: the HR-equational sets are actually (but not by definition) those generated by certain context-free graph grammars based on a rewriting mechanism called *hyperedge replacement* (that uses "gluing" of graphs) and we call the corresponding algebra the HR algebra to refer to this fact; the other algebra, called the VR algebra, has been designed similarly so that its equational sets are those generated by the context-free graph grammars based on *vertex replacement* (that uses "bridging" of graphs); see [*DreKH] and [*EngRoz] respectively for these two types of graph grammars.

Many properties of the equational and recognizable sets of graphs of both kinds are just particular instances of those of the equational and recognizable sets in arbitrary algebras. By using this algebraic approach, we generalize the context-free languages without having to define a graph rewriting mechanism and check that such rewriting is actually context-free (the general notion of context-free rewriting is defined in [Cou87]). Similarly, we generalize the regular languages without having to define any notion of graph automaton and to look for the closure properties of the class of sets of graphs that are recognized by such automata.

Monadic second-order logic and the VR graph algebra

We first discuss the equational sets and the recognizable sets of the VR algebra, and their relationships with monadic second-order logic.

Two main results of this book are the Recognizability Theorem and the Equationality Theorem. They relate two ways of handling graphs: the "logical way" by which graphs are characterized in terms of what they are made of and contain (vertices, edges, paths, minors, subgraphs with particular properties) and the "algebraic way" by which sets of graphs are characterized more globally by means of equation systems and congruences. In the latter approach, graphs are treated as elements of algebras and related with other elements that are not necessarily among their subgraphs.

The *Recognizability Theorem* says that if a set of graphs is monadic second-order definable, then it is VR-recognizable. The *Equationality Theorem* says that a set of graphs is VR-equational if and only if it is the image of the set of finite trees under a monadic second-order transduction[8]. We now describe some consequences of these two results.

The Recognizability Theorem entails that if a graph G is defined by a term t written with operations of the VR algebra belonging to any fixed finite set F, then one can check in time $O(|t|)$ whether or not G satisfies a fixed monadic second-order property. This fact, based on a compilation of monadic second-order formulas into finite automata over F, is one of the keys[9] to the construction of fixed-parameter tractable algorithms for the verification of monadic second-order properties of graphs of bounded clique-width (whence also of graphs of bounded tree-width since bounded tree-width implies bounded clique-width). Another consequence is the *Filtering Theorem*, which says that the graphs of a VR-equational set that satisfy a fixed monadic second-order property (for example planarity) form a VR-equational set. This is based on a Filtering Theorem that holds in all algebras and says that the intersection of an equational set and a recognizable one is equational (generalizing the corresponding fact for context-free and regular languages). Since the emptiness of an equational set is decidable, we get as another corollary that the *monadic second-order satisfiability problem* is decidable for each VR-equational set L. This means that one can decide whether or not a given monadic second-order sentence is satisfied by some graph in L.

The Equationality Theorem entails that the class of VR-equational sets of graphs is preserved under monadic second-order transductions, because the class of monadic second-order transductions is closed under composition. This corollary strengthens the Filtering Theorem. It is similar to the fact that the image of a context-free language under a rational transduction is context-free.

Monadic second-order logic and the HR graph algebra

The Recognizability and Equationality Theorems have versions relative to the HR algebra. To describe them, we must go back to the initial definition of monadic second-order formulas (MS formulas in the sequel) interpreted in graphs: they only

---

[8] This means, informally, that it is the set of graphs "defined inside finite trees" by a fixed finite tuple of monadic second-order formulas. These transductions are based on, and extend, the model-theoretical notion of "interpretation."

[9] The other one is a polynomial-time algorithm that finds a term evaluating to a given graph G if one exists.

use quantifications on vertices and sets of vertices. This is due to the chosen representation of a graph by a relational structure whose domain is its set of vertices. However, we can also express logically the properties of a graph G via its *incidence graph* $Inc(G)$. The vertices of this (bipartite) graph are the vertices and the edges of G, and its adjacency relation links a vertex and the edges incident with it. Thus, monadic second-order formulas to be interpreted in $Inc(G)$ (MS_2 formulas in the sequel) can also use quantifications on edges and sets of edges. A graph property is MS_2-*expressible* if it is expressible by an MS_2 formula, and the corresponding set of graphs is MS_2-*definable*. The notation MS_2 refers to this extension of the initially defined language (referred to as MS in the sequel). It is strictly more expressive. For example, the existence of a perfect matching is MS_2-expressible but not MS-expressible. However, MS_2 formulas are not more expressive than MS formulas for properties of words, of trees and of certain types of graphs such as planar graphs and, for each k, of graphs of degree at most k. These facts show the existence of deep links between structural graph properties (such as planarity) and the expressive power of MS_2 versus MS sentences.

The Recognizability Theorem for the HR algebra says that every MS_2-definable set of graphs is HR-recognizable, and the Equationality Theorem says that a set of graphs is HR-equational if and only if the set of its incidence graphs is the image of the set of finite trees under a monadic second-order transduction. We obtain an algorithmic consequence similar to the one we have discussed for MS-expressible problems and the VR algebra: if a graph is defined by a term t over $F_{[k]}^{HR}$ for some fixed k, then one can check in time $O(|t|)$ whether or not it satisfies a fixed MS_2 property. Since there exists a polynomial-time algorithm that decomposes appropriately the input graphs, we obtain, for each MS_2 property, a fixed-parameter tractable verification algorithm, tree-width being the parameter. The algorithm for MS properties applies to larger classes of graphs, because bounded tree-width implies bounded clique-width, but to less properties than this one, because not every MS_2-expressible property is MS-expressible. The notions of tree-width and clique-width are thus both useful, for solving different problems. We also have a Filtering Theorem for the HR-equational sets and MS_2-expressible properties, whence the decidability of the MS_2-satisfiability problem for each HR-equational set. The Equationality Theorem for the HR algebra entails that the class of HR-equational sets of graphs is preserved under the monadic second-order transductions that transform incidence graphs.

A graph is uniformly k-sparse if its number of edges is at most k times its number of vertices, and the same holds for all its subgraphs. Another main result of this book is the *Sparseness Theorem*: MS_2 formulas are not more expressive than MS formulas for properties of uniformly k-sparse graphs, for each fixed k. The above-mentioned types of graphs are uniformly k-sparse for some k.

Logical and language theoretical issues

Tree-width and clique-width are closely related with the decidability of monadic second-order satisfiability problems for particular sets of graphs. The satisfiability problem of MS_2 sentences for the set of graphs of tree-width at most some fixed k is decidable (because it is decidable for each HR-equational set), and the same holds for MS sentences and the set of graphs of clique-width at most k. Some converse results also hold: bounded tree-width is a necessary (but not sufficient) condition for a set of graphs to have a decidable MS_2-satisfiability problem, and a similar result holds for clique-width and MS sentences. Their proofs use monadic second-order transductions and deep results of Graph Theory.

The Recognizability and Equationality Theorems contribute to establishing the foundations of a sound and robust extension of the theory of formal languages to the description of sets of finite graphs. In this extension, monadic second-order logic plays a major role. From the above informal (and simplified) statements, this extension may seem to be straightforward. However, graphs are intrinsically more complex than words and terms, and some results do not extend as one could expect or hope. We give two examples. First, the set of all graphs is not equational in any of the two graph algebras, whereas the set of all words over a finite alphabet is (almost trivially) context-free. Second, there are uncountably many VR- and HR-recognizable sets of graphs, and this fact prevents any exact characterization of these sets in terms of graph automata or logical formulas. Such a characterization would generalize nicely the classical characterization of the recognizable (i.e., the regular) languages in terms of finite automata and monadic second-order sentences, but it cannot exist. These examples are related to the fact that the sets of operations of the two graph algebras are infinite, and that this infiniteness is somehow unavoidable.[10]

Graph structure

Graph structure is a flexible concept covering many different cases. Graph decompositions form an important type of structuring. We have already discussed those that yield the notions of tree-width and clique-width in connection with algorithmic applications. There exist other types of graph decomposition that are useful for algorithmic purposes or for proving results. Examples are the modular decomposition defined by Gallai [Gal], the decomposition in 3-connected components defined by Tutte [*Tut] and the clique-sum decomposition used by Robertson and Seymour ([*Gro], [RobSey03]). The existence of an embedding in a fixed surface, or of a homomorphism into a fixed graph (a proper vertex coloring with p colors of a loop-free graph can be defined as a homomorphism of this graph into the complete graph K_p) is a type of structure. Finally, the nonexistence in a graph of particular induced

---

[10] One can generate all finite graphs by a finite number of graph operations, but the Recognizability Theorem fails for the corresponding algebra. So this algebra is useless for our purposes.

subgraphs, minors or vertex-minors is also an important type of structural property. (See [*Die] for minors and [Oum05] for vertex-minors.) There exist nontrivial relations between these different notions, for example: the graphs without a fixed planar graph as a minor have tree-width bounded by a value computable from this graph and those embeddable in a fixed surface are characterized by finitely many excluded minors [*Die]; forbidding certain induced subgraphs implies bounded clique-width ([BraDLM], [BraELL]).

Monadic second-order sentences can express many such structural properties. The expression of p-colorability (for fixed p) is immediate. It is easy to construct a monadic second-order sentence expressing that a given graph has no minor or no induced subgraph isomorphic to a fixed graph. Hence the sets of planar graphs and of graphs of tree-width at most k (for all k) are MS-definable because each of them is characterized by finitely many excluded minors. A set of graphs defined by finitely many excluded induced subgraphs (this is the case of cographs) or by an infinite but MS-definable set of minimal excluded induced subgraphs is also MS-definable. The latter observation applies to *comparability graphs* ([Cou06a], [Gal]) and to *perfect graphs* ([*ChuRST], [ChuRST]). Their definitions are not directly expressible by monadic second-order sentences, and finding the minimal excluded induced subgraphs requires difficult proofs.

In many situations concerning graph structure, we need more than a yes or no answer. For example, that a graph does not contain K_5 or $K_{3,3}$ as a minor implies that it is planar, but this negative fact, when it is valid, does not help to find a planar embedding. In other words, we are not only interested in checking that a given graph "has some structure," e.g., has a planar embedding or a tree-decomposition of width bounded by a fixed integer, but we are also interested in having a monadic second-order transduction that constructs from the given graph some planar embedding or some tree-decomposition. Such transductions may be difficult to construct. Some constructions are given in [Cou96a], [Cou99], [Cou00], [Cou06b], [Cou08a], and challenging questions remain open in this area.

To conclude with this aspect, we can state that many constructions of monadic second-order formulas and transductions use in an essential way results of Graph Theory, and even very deep ones in some cases. Conversely, the methods developed in this book bring new results in Graph Theory apart from algorithmic applications. For example, the infinite set of minimal excluded induced subgraphs that characterizes the comparability graphs has a certain regularity that we can formalize by observing that this set is VR-equational. Such applications deserve further study.

The main contributions of this book

We will now summarize the main ideas and results to be developed in this book. We define two graph algebras called the HR algebra and the VR algebra, from which we get two classes of equational and two classes of recognizable sets of graphs. The terms of these algebras denote graphs and formalize certain hierarchical decompositions from which we get the graph complexity measures called tree-width and clique-width.

Monadic second-order logic in its two variants denoted by MS and MS_2 can be used to express formally graph properties and thus to specify sets of graphs. The Recognizability Theorem says that every MS-definable set of graphs is recognizable in the VR algebra and that every MS_2-definable set is recognizable in the HR algebra. We obtain from it fixed-parameter tractable algorithms for checking MS and MS_2 properties with, respectively, clique-width and tree-width as parameters. It entails that the corresponding monadic second-order satisfiability problems are decidable for the equational sets of the two algebras. The Sparseness Theorem says that MS and MS_2 logic have the same power for defining sets of uniformly k-sparse graphs.

Graph transformations called monadic second-order transductions can be specified by MS or by MS_2 formulas. The Equationality Theorem says that they generate from the set of trees, respectively, the equational sets of the VR and of the HR algebra. This shows the robustness of this theory that combines algebraic and logical notions. Its main definitions and results are actually formulated for *relational structures* which generalize graphs and incidence graphs, but several problems are open regarding this extension.

We will only consider finite graphs and finite relational structures. Another book would be necessary to cover the rich existing theory of countable structures that is important in Program Semantics.

Summary

The letters GT, UA, LT, L and A indicate that a chapter deals mainly with Graph Theory, Universal Algebra, (Formal) Language Theory, Logic and Algorithmic applications respectively.

Chapter 1 presents an overview of the main definitions and results in an informal way, with the help of examples.

Chapter 2 (GT, UA) defines two families of graph operations and the associated graph complexity measures of tree-width and clique-width. The two corresponding graph algebras, called the HR algebra and the VR algebra, are first defined as single sorted algebras and, later on, they are "refined" into many-sorted algebras.

Chapter 3 (UA, LT) defines and studies the equational and recognizable sets of many-sorted algebras in general. Its main result is the (algebraic version of the) Filtering Theorem.

Chapter 4 (GT, LT) applies the definitions and results of Chapter 3 to the graph algebras defined in Chapter 2 and establishes results which do not follow solely from the general algebraic definitions.

Chapter 5 (L, UA, GT) introduces monadic second-order logic and develops tools for expressing graph properties by monadic second-order formulas. Definitions and proofs are given for relational structures. In particular, the Recognizability Theorem is proved for a many-sorted algebra of finite relational structures. The particular cases of this theorem for the HR and the VR graph algebras follow as immediate corollaries.

Chapter 6 (L, LT, A) is devoted to algorithmic applications. It reviews the *parsing algorithms* that construct the necessary expressions of the input graphs by terms over the operations of the HR and of the VR algebra. It develops in detail the "compilation" of monadic second-order formulas into finite automata intended to run on the terms resulting from the parsing step. This construction is hopefully more usable than the one of Chapter 5. It yields alternative proofs of weaker versions of the Recognizability Theorem.

Chapter 7 (L, LT, GT) defines monadic second-order transductions and establishes their main properties: closure under composition and preservation of monadic second-order definability under inverse monadic second-order transductions; we call this latter result the *Backwards Translation Theorem*. The Equationality Theorem characterizes the VR- and the HR-equational sets as the images of the set of trees (equivalently of terms over any rich enough functional signature) under monadic second-order transductions of appropriate types. Four types of transductions come from the two possible representations of a graph by a relational structure for the input and the output. The Equationality Theorems characterize bounded clique-width and bounded tree-width in a way that does not depend on the graph operations chosen in Chapter 2. Hence, VR- and HR-equationality as well as the properties of having bounded clique-width and tree-width are robust in the sense that they are stable under the monadic second-order transductions that are respectively specified by MS and by MS_2 formulas.

Chapter 8 (L, LT) shows that the classical automata-theoretic characterization (recalled in Chapter 5) of the monadic second-order definable sets of terms (hence, also of words) extends to monadic second-order transductions. More precisely, these transductions are characterized in terms of *two-way finite-state transducers* on words, and of *tree-walking transducers* on terms, where "tree" refers to the representation of terms by labeled ordered trees. Characterizations of the VR-equational (equivalently of the HR-equational) languages of words and terms are obtained. Every (functional) monadic second-order transduction of graphs of bounded tree-width can be realized by a tree-walking transducer on the level of terms (of the HR algebra), i.e., it can be realized by parsing the input graph, applying the tree transducer to the resulting input term, and evaluating the output term of the transducer to produce the output graph. The same result holds for clique-width and the VR algebra. This can be viewed as a generalization of the Recognizability Theorem to graph transductions.

Chapter 9 (L, GT, UA, A) extends to finite relational structures the definitions and results of Chapters 2 to 7. It contains in particular an extension to relational structures of the Equationality Theorem. Although many results extend easily from graphs to relational structures, some seemingly difficult questions remain open. Additionally, this chapter proves the Sparseness Theorem, which establishes that, for expressing properties of uniformly k-sparse graphs (or relational structures[11]) by monadic second-order formulas, quantifications over sets of edges (or sets of tuples,

---

[11] Relational structures are used to prove the case of the theorem that concerns graphs. A relational structure is uniformly k-sparse if its number of tuples is at most k times the cardinality of its domain, and the same holds for all its substructures.

respectively) bring no additional power: every MS_2 formula can be translated into an equivalent MS formula.

A concluding section reviews some open problems and some results not presented in the previous chapters.

The references section is organized in two parts: the first part (with reference labels starting with *) lists books, book chapters and survey articles. The second lists research articles and dissertations.

All necessary definitions will be given, but the reader is expected to be familiar with the basic notions of Logic (mainly first-order logic), Universal Algebra (algebras, congruences), Formal Language Theory (context-free grammars, finite automata), and Graph Theory (basic notions). Chapters 2 to 9 present detailed proofs of results that have been published in articles. It was not an easy task to elaborate consistent definitions and notations for many different notions from various fields, namely Language Theory, Universal Algebra, Logic and Graph Theory. By giving precise definitions and carefully written proofs, our first aim is to give a robust foundation to the field described in this introduction. Our second aim is that these definitions and proofs can be adapted to related notions, implemented and improved by researchers without too much effort.

The web page www.labri.fr/perso/courcell/TheBook.html will maintain reference updates, new results answering the open questions and errata.

Acknowledgements

We thank M. Nivat for writing a foreword. The LaTeX typing of the first drafts of most chapters has been done by graduate and doctoral students of Bordeaux 1 University: M. Kanté, R. Chen, R. Synave, R. Li and S. Abbas. We thank them warmly for their care and patience. We also thank A. Arnold, A. Blumensath, S. Djelloul, I. Durand and S. Oum who have read and commented on some chapters.

We are grateful to the participants of the European workshops and projects on graph grammars and graph transformations, initiated by H. Ehrig and G. Rozenberg, for the many years of discussion and collaboration.

Without being a member of the Institut Universitaire de France (IUF), B. Courcelle could not have worked on this book. He thanks M. Nivat and W. Thomas who presented his application to IUF, and all those who supported it by writing recommendation letters. He dedicates his work to the memory of Ph. Flajolet (1948–2011), a 40-year friend with whom he began his career of researcher at INRIA in 1973.

J. Engelfriet wishes to thank the Leiden Institute of Advanced Computer Science (LIACS), and in particular the members of the Theoretical Computer Science group, headed by G. Rozenberg, for a stimulating research environment over the past 28 years.

1

Overview

This chapter presents the main definitions and results of this book and their significance, with the help of a few basic examples. It is written so as to be readable independently of the other chapters. Definitions are sometimes given informally, with simplified notation, and most proofs are omitted. All definitions will be repeated with the necessary technical details in subsequent chapters.

In Section 1.1, we present the notion of equational set of an algebra by using as examples a context-free language, the set of cographs and the set of series-parallel graphs. We also introduce our algebraic definition of derivation trees.

In Section 1.2, we introduce the notion of recognizability in a concrete way, in terms of properties that can be proved or refuted, for every element of the considered algebra, by an induction on any term that defines this element. We formulate a concrete version of the Filtering Theorem saying that the intersection of an equational set and a recognizable one is equational. It follows that one can decide if a property belonging to a finite inductive set of properties is valid for every element of a given equational set. We explain the relationship between recognizability and finite automata on terms.

In Section 1.3, we show with several key examples how monadic second-order sentences can express graph properties. We recall the fundamental equivalence of monadic second-order sentences and finite automata on words and terms.

In Section 1.4, we introduce two graph algebras. They are called the VR and the HR algebra because their equational sets are those that are generated by the context-free vertex replacement and hyperedge replacement graph grammars respectively. The cographs and the series-parallel graphs are respectively our current examples of a VR- and an HR-equational set. We state (a weak version of) the Recognizability Theorem which says, in short, that monadic second-order definability implies recognizability. From it we obtain a logical version of the Filtering Theorem where the recognizable sets are defined by monadic second-order sentences.

In Section 1.5, we review the basic definitions of Fixed-Parameter Tractability and state the algorithmic consequences of the (weak) Recognizability Theorem. This theorem actually has two versions, relative to the two graph algebras defined in Section 1.4, and yields two Fixed-Parameter Tractability Theorems.

In Section 1.6, we describe the consequences of the Recognizability and Filtering Theorems for the problem of deciding whether a given monadic second-order sentence is satisfied by some graph of tree-width at most a given k or, more generally, by some graph of an equational set.

In Section 1.7, we introduce the notion of a monadic second-order transduction by means of examples that have some graph theoretic content, and we state the Equationality Theorem for the VR algebra. It gives a characterization of the VR-equational sets, and in particular of the sets of graphs of bounded clique-width, that is formulated in purely logical terms.

In Section 1.8, we consider monadic second-order formulas interpreted in incidence graphs (as opposed to in graphs "directly"). These formulas can use edge set quantifications. We compare the corresponding four types of monadic second-order transduction and state the Equationality Theorem for the HR algebra: it is based on monadic second-order transductions that transform incidence graphs.

In Section 1.9, we define relational structures and extend to them some results relative to graphs represented by their incidence graphs. We introduce betweenness and cyclic ordering as examples of combinatorial notions that are based on linear orderings but are defined in a natural way as ternary relations.

1.1 Context-free grammars

By starting from the standard notion of a context-free grammar, we introduce the notion of an equational set and define two equational sets of graphs. We define the equational sets of a (one-sorted) algebra and the corresponding sets of derivation trees.

1.1.1 Context-free word grammars

By using *context-free grammars*, one can specify certain formal languages, namely the *context-free languages*, in a finitary way. Context-free grammars are usually defined as rewriting systems satisfying particular properties, conveyed by the term "context-free" and axiomatized in [Cou87]. However, the *Least Fixed-Point Characterization* of context-free languages due to Ginsburg and Rice [GinRic] and to Chomsky and Schützenberger [ChoSch] is formulated in terms of systems of recursive equations written with the operations of union and concatenation over languages. This algebraic view has been developed by Mezei and Wright [MezWri] and has many advantages. First, it is more synthetic in that it deals with languages rather than with words produced individually by derivation sequences. Second, it puts the study of context-free languages in the more general framework of recursive definitions handled as least solutions of systems of equations, and, last but not least, it is applicable to any algebra. This latter aspect is especially important for the extension to graphs.

We recall how context-free languages can be characterized as the components of the least solutions of certain systems of equations in languages. A context-free grammar G is a finite set of rewriting rules defined with two alphabets, a *terminal alphabet A* and a *nonterminal alphabet N*. For every S in N, the context-free language over A generated by G from S is denoted by $L(G,S)$.

Example 1.1 We consider for example the context-free grammar G with terminal alphabet $A = \{a,b,c\}$, nonterminal alphabet $N = \{S,T\}$ and rules named respectively $p,q,\dots,w$ (where ε denotes the empty word):

$$
\begin{aligned}
p: & \quad S \rightarrow aST, \\
q: & \quad S \rightarrow SS, \\
r: & \quad S \rightarrow a, \\
s: & \quad T \rightarrow bTST, \\
u: & \quad T \rightarrow a, \\
v: & \quad T \rightarrow c, \\
w: & \quad T \rightarrow \varepsilon.
\end{aligned}
$$

It defines two languages $L(G,S)$ and $L(G,T)$ over A, i.e., two sets of words in A^*. These languages satisfy the equations of the following system Σ_G:

$$
\Sigma_G \begin{cases} K = aKL \cup KK \cup \{a\}, \\ L = bLKL \cup \{a,c,\varepsilon\}, \end{cases}
$$

with $K = L(G,S)$ and $L = L(G,T)$. The pair $(L(G,S),L(G,T))$ is thus a solution of Σ_G. However, it is not the only one. The pair of languages $(A^*, bA^* \cup \{a,c,\varepsilon\})$ is another solution, as one can easily check.[1] The Least Fixed-Point Characterization of context-free languages establishes that the pair $(L(G,S),L(G,T))$ is the least solution of Σ_G for component-wise inclusion. □

1.1.2 Cographs

We give two examples of similar definitions of sets of graphs. We first consider as a ground set the set $\mathcal{G}^u$ of undirected simple graphs.[2] Two isomorphic graphs are considered as the same object. We will use $\oplus$ to denote the disjoint union of two graphs G and H. This means that $G \oplus H$ is the union of G and of a copy of H disjoint with G (hence $G \oplus G \neq G$). We will also use the *complete join*, $G \otimes H$, defined as $G \oplus H$ augmented with undirected edges linking every vertex of G and every vertex

---

[1] Since $L(G,S) \subseteq A^+ = AA^*$, the pair $(A^*, bA^* \cup \{a,c,\varepsilon\})$ is a solution of Σ_G that differs from $(L(G,S),L(G,T))$.

[2] In this book, all graphs are finite. A graph is *simple* if it has no two *parallel edges*, i.e., no two edges with the same ends, and the same directions in the case of directed graphs. Parallel edges are also called *multiple* edges. An edge with equal ends is a *loop*. The superscript "u" in $\mathcal{G}^u$ refers to undirected graphs.

of H. We let $\mathbf{1}$ denote any graph with one vertex and no edges. Note that both $\oplus$ and $\otimes$ are commutative and associative operations.

The *set of cographs* C can be defined as the least set of graphs satisfying the equation

$$C = (C \oplus C) \cup (C \otimes C) \cup \{\mathbf{1}\}. \tag{1.1}$$

This set (it is a proper subset of $\mathcal{G}^{\mathrm{u}}$) has alternative characterizations (see Section 1.3.1 below). From this equation, one can derive definitions of certain subsets of C. Consider for example the following system of two equations:

$$\begin{cases} C_0 = (C_0 \oplus C_0) \cup (C_1 \oplus C_1) \cup (C_0 \otimes C_0) \cup (C_1 \otimes C_1), \\ C_1 = (C_0 \oplus C_1) \cup (C_0 \otimes C_1) \cup \{\mathbf{1}\}. \end{cases} \tag{1.2}$$

Its least solution in $\mathcal{P}(\mathcal{G}^{\mathrm{u}}) \times \mathcal{P}(\mathcal{G}^{\mathrm{u}})$ is the pair of sets (C_0, C_1), where C_0 (resp. C_1) is the set of cographs having an even (resp. odd) number of vertices.[3] We will give general and effective methods for deriving from an equation or a system of equations that defines a set L, an equation or a system of equations defining $\{x \in L \mid P(x)\}$, where P is a property of the objects under consideration. This is possible if P has an appropriate "inductive behavior" relative to the operations with which the given equation or system of equations is written.

From the definition of cographs as elements of the least subset C of $\mathcal{G}^{\mathrm{u}}$ satisfying (1.1), it follows that each of them is *denoted by a term* or, more formally, is the *value of a term* in an *algebra* of graphs. Examples of terms denoting cographs are

$$\mathbf{1}, \ \mathbf{1} \oplus \mathbf{1}, \ (\mathbf{1} \oplus \mathbf{1}) \otimes \mathbf{1}, \ (\mathbf{1} \oplus \mathbf{1}) \otimes (\mathbf{1} \oplus \mathbf{1}).$$

The cograph of Figure 1.1 is the value of the term $t = (\mathbf{1} \otimes \mathbf{1} \otimes \mathbf{1}) \otimes (\mathbf{1} \oplus (\mathbf{1} \otimes \mathbf{1}))$. Since $\otimes$ is associative, we have written t by omitting some parentheses as usual, for readability. These terms belong to the set $T(\{\oplus, \otimes, \mathbf{1}\})$ of all terms written with the constant $\mathbf{1}$ and the two binary operations $\oplus$ and $\otimes$. Equation (1.1) can also be solved with ground set $T(\{\oplus, \otimes, \mathbf{1}\})$. For this interpretation of (1.1) the unknown C denotes subsets of $T(\{\oplus, \otimes, \mathbf{1}\})$. Clearly, the set of terms $T(\{\oplus, \otimes, \mathbf{1}\})$ itself is the least (in fact, the only) solution of (1.1) in $\mathcal{P}(T(\{\oplus, \otimes, \mathbf{1}\}))$.

A similar fact holds for System (1.2). Its least solution in $\mathcal{P}(T(\{\oplus, \otimes, \mathbf{1}\})) \times \mathcal{P}(T(\{\oplus, \otimes, \mathbf{1}\}))$ is a pair of sets (T_0, T_1) where T_0, $T_1 \subseteq T(\{\oplus, \otimes, \mathbf{1}\})$ and, for each $i = 0, 1$, the set C_i is the set of cographs which are the values of the terms in T_i.

This example shows that a *grammar*, i.e., a system of equations like (1.2), specifies not only a tuple of sets of objects, here graphs, but also denotations by terms of the specified objects. These objects can be words, terms, trees, graphs, as we will see. Each term is a formalization of the structure of the object it denotes, as specified by

---

[3] We denote by $\mathcal{P}(X)$ the powerset of a set X, i.e., its set of subsets.

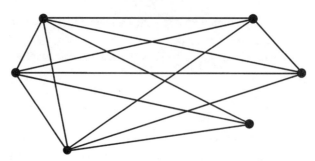

Figure 1.1 A cograph.

the grammar; it provides a hierarchical decomposition of that object. In many cases, an object can be denoted by several terms that are correct with respect to the grammar. In such a case, we say that the grammar is *ambiguous*. The grammars (1.1) and (1.2) are ambiguous: since $\oplus$ and $\otimes$ are commutative and associative, most cographs are denoted by more than one term.

As an example of structure, consider again the term $(\mathbf{1} \otimes \mathbf{1} \otimes \mathbf{1}) \otimes (\mathbf{1} \oplus (\mathbf{1} \otimes \mathbf{1}))$ that denotes the cograph of Figure 1.1. It provides a decomposition of that cograph, because the subterm $\mathbf{1} \otimes \mathbf{1} \otimes \mathbf{1}$ denotes the triangle at the left of Figure 1.1, whereas the subterm $\mathbf{1} \oplus (\mathbf{1} \otimes \mathbf{1})$ denotes the three vertices at the right of Figure 1.1 together with the edge between two of them.

1.1.3 Series-parallel graphs

The ground set of graphs is here the set $\mathcal{J}_2^{\mathrm{d}}$ of directed graphs G equipped with two distinct distinguished vertices marked 1 and 2 called its *sources*, denoted respectively by $src_G(1)$ and $src_G(2)$. These graphs may have multiple edges.[4] Let e be a constant denoting the graph with two vertices and only one edge from source 1 to source 2. The operations are the *parallel-composition*, denoted by $/\!/$, and the *series-composition*, denoted by $\bullet$. For G and H in $\mathcal{J}_2^{\mathrm{d}}$, the graph $G /\!/ H$ is the union of G and an isomorphic copy H' of H such that $src_G(1) = src_{H'}(1)$, $src_G(2) = src_{H'}(2)$, and G and H' have nothing else in common. We define $src_{G/\!/H}(1) := src_G(1)$ and $src_{G/\!/H}(2) := src_G(2)$. Note that $G /\!/ G$ has twice as many edges as G, hence $G \neq G /\!/ G$ in general.

Series-composition is defined similarly. For $G, H \in \mathcal{J}_2^{\mathrm{d}}$, we let $G \bullet H$ be the union of G and an isomorphic copy H' of H such that $src_G(2) = src_{H'}(1)$ and G and H' have nothing else in common. We let $src_{G \bullet H}(1) := src_G(1)$ and $src_{G \bullet H}(2) := src_{H'}(2)$. These operations are illustrated in Figure 1.2.

---

[4] The letter "J" in the notations $\mathcal{J}_2^{\mathrm{d}}$ and, below, in $\mathbb{J}_2^{\mathrm{d}}$, $\mathbb{JS}$ and related notions refers to graphs that can have multiple edges. By contrast, the letter "G" used in the notations $\mathcal{G}^{\mathrm{u}}$ and, below, $\mathbb{G}^{\mathrm{u}}$, $\mathbb{GP}$, etc. refers to simple graphs. The subscript "2" refers to the two sources, and the superscript "d" to directed graphs.

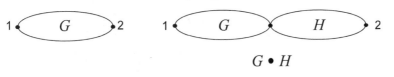

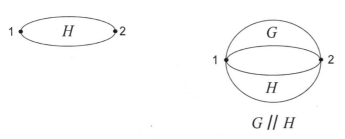

$$G \bullet H$$

$$G \mathbin{/\!/} H$$

Figure 1.2 Series- and parallel-compositions.

The set of *series-parallel graphs*[5] is defined by the equation

$$S = (S \mathbin{/\!/} S) \cup (S \bullet S) \cup \{e\}, \tag{1.3}$$

where by "defined" we mean that S is the least subset of $\mathcal{J}_2^{\mathrm{d}}$ satisfying (1.3). As for cographs, this definition gives a notation of series-parallel graphs by terms. The set of terms is here $T(\{\mathbin{/\!/}, \bullet, e\})$. Examples of terms are:

$$e, \; e \mathbin{/\!/} e, \; (e \mathbin{/\!/} e) \bullet (e \mathbin{/\!/} e), \; ((e \mathbin{/\!/} e) \bullet e) \mathbin{/\!/} (e \bullet e).$$

The graph denoted by the last of these terms is shown in Figure 1.3. Note that the subterm $(e \mathbin{/\!/} e) \bullet e$ denotes the three edges at the left of Figure 1.3, with their three incident vertices, whereas the subterm $e \bullet e$ denotes the two edges at the right, with their three incident vertices.

1.1.4 The general setting

Let F be a (functional) *signature*, that is, a set of *function symbols* such that each symbol f is given with a nonnegative integer intended to be the number of arguments of the corresponding function. This number is called its *arity* and is denoted by $\rho(f)$. A function symbol of arity 0 is also called a *constant symbol*.

An *F-algebra* $\mathbb{M}$ is a set M equipped with total functions $f_{\mathbb{M}} : M^{\rho(f)} \to M$ for all f in F. We write it $\mathbb{M} = \langle M, (f_{\mathbb{M}})_{f \in F} \rangle$. We call M the *domain* and $f_{\mathbb{M}}$ an *operation* of

---

[5] The term "series-parallel" is also used for partial orders ([*Möhr]) and, in a wider sense, for undirected graphs without K_4 as a minor ([*Die]). Our series-parallel graphs are called *two-terminal series-parallel digraphs* in [*Möhr].

Figure 1.3 A series-parallel graph.

$\mathbb{M}$; if f has arity 0, then $f_{\mathbb{M}}$ is also called a *constant* of $\mathbb{M}$. The F-algebra $\mathbb{M}$ is *finite* if M is finite.

Let $X = \{x_1, \ldots, x_n\}$ be a set of *unknowns* (or variables), intended to denote subsets of M. A *polynomial* is an expression of the form $p = m_1 \cup \cdots \cup m_k$, where each m_i is a *monomial*, i.e., a term written with the symbols of $F \cup X$ and well formed with respect to arities (the unknowns are of arity 0).

For each n-tuple $(L_1, \ldots, L_n)$ of subsets of M and each monomial m, the set $m(L_1, \ldots, L_n)$ is a subset of M. This subset is defined by taking $x_i = L_i$ and by interpreting each function symbol f as $f_{\mathbb{M}}$, where, for all $A_1, \ldots, A_{\rho(f)} \subseteq M$:

$$f_{\mathbb{M}}(A_1, \ldots, A_{\rho(f)}) := \{f_{\mathbb{M}}(a_1, \ldots, a_{\rho(f)}) \mid a_i \in A_i\}.$$

Hence $f_{\mathbb{M}}$ also denotes the extension to sets of the function $f_{\mathbb{M}} : M^{\rho(f)} \to M$. For a polynomial $p = m_1 \cup \cdots \cup m_k$ we define:

$$p(L_1, \ldots, L_n) := m_1(L_1, \ldots, L_n) \cup \cdots \cup m_k(L_1, \ldots, L_n).$$

A *system of polynomial equations* (an *equation system* for short) is a system of the form

$$S = \langle x_1 = p_1, \ldots, x_n = p_n \rangle, \tag{1.4}$$

where $p_1, \ldots, p_n$ are polynomials.

Example 1.2 In the particular case of the grammar G considered in Example 1.1, we let $F = \{\cdot, \varepsilon, a, b, c\}$ and $\mathbb{M} = \langle A^*, \cdot, \varepsilon, a, b, c \rangle$, where $A = \{a, b, c\}$ and $\cdot$ denotes concatenation; the equation system Σ_G can be written formally as follows:

$$\begin{cases} x_1 = a \cdot (x_1 \cdot x_2) \cup x_1 \cdot x_1 \cup a, \\ x_2 = b \cdot ((x_2 \cdot x_1) \cdot x_2) \cup a \cup c \cup \varepsilon, \end{cases}$$

where the associativity of concatenation is no longer taken for granted. Note that for the constant symbol a of F we have $a_{\mathbb{M}} = a$ and also, according to the above extension, $a_{\mathbb{M}} = \{a\}$; similarly, the constant symbol ε denotes both the empty word ε and the language $\{\varepsilon\}$. □

Going back to the general case, a *solution* of a system S as in (1.4) is an n-tuple $(L_1, \ldots, L_n)$ in $\mathcal{P}(M)^n$ that satisfies the equations of S, which means that $L_i = p_i(L_1, \ldots, L_n)$ for every $i = 1, \ldots, n$. Solutions are compared by component-wise inclusion and every system has a least solution. The components of the least solutions of such systems are called the *equational sets* of the F-algebra $\mathbb{M}$. We will denote by **Equat**($\mathbb{M}$) the family of equational sets of $\mathbb{M}$.

For a signature F, we denote by $T(F)$ the set of terms written with the symbols of F and well formed with respect to arities. The usual notation for terms is with the function symbols in leftmost position, their arguments are between parentheses and separated by commas. In this notation, the term denoting the graph of Figure 1.3 is written $/\!\!/(\bullet(/\!\!/(e,e),e),\bullet(e,e))$.[6] As is well known, terms can be represented by certain labeled, directed and rooted trees. This representation is the reason that terms are usually called trees in Formal Language Theory.

The set $T(F)$ is turned into an F-algebra, denoted by $\mathbb{T}(F)$, by defining the operation $f_{\mathbb{T}(F)}$ by

$$f_{\mathbb{T}(F)}(t_1, \ldots, t_{\rho(f)}) := f(t_1, \ldots, t_{\rho(f)}).$$

This operation performs no computation; it combines its arguments which are terms into a larger term.

For every F-algebra $\mathbb{M}$, a term $t \in T(F)$ has a *value* $t_{\mathbb{M}}$ in M that is formally defined as follows:

$$t_{\mathbb{M}} := f_{\mathbb{M}} \text{ if } t = f \text{ and } f \text{ has arity } 0 \text{ (it is a constant symbol)},$$

$$t_{\mathbb{M}} := f_{\mathbb{M}}(t_{1\mathbb{M}}, \ldots, t_{\rho(f)\mathbb{M}}) \text{ if } t = f(t_1, \ldots, t_{\rho(f)}).$$

Since every term can be written in a unique way as f or $f(t_1, \ldots, t_{\rho(f)})$ for terms $t_1, \ldots, t_{\rho(f)}$, the value $t_{\mathbb{M}}$ of t is well defined. The mapping $t \mapsto t_{\mathbb{M}}$, also denoted by

---

[6] For associative binary operations the more readable infix notation will be used, although it is ambiguous as already observed. The infix notation of this term is $((e /\!\!/ e) \bullet e) /\!\!/ (e \bullet e)$.

val$_\mathbb{M}$, is the unique F-algebra homomorphism from $\mathbb{T}(F)$ into $\mathbb{M}$.[7] An F-algebra $\mathbb{M}$ is *generated* by F if every element of M is the value of some term in $T(F)$.

An equation system S of the form (1.4) has a least solution in $\mathcal{P}(T(F))^n$ that is an n-tuple $(T_1,\ldots,T_n)$ of subsets of $T(F)$. The least solution $(L_1,\ldots,L_n)$ of S in $\mathcal{P}(M)^n$ is also characterized by $L_i = \{t_\mathbb{M} \mid t \in T_i\}$, for each $i = 1,\ldots,n$. This is an immediate consequence of a result of [MezWri] saying that the least fixed-point operator commutes with homomorphisms. A term t in T_i represents the structure of the element $t_\mathbb{M}$ of L_i as specified by the system S.

It follows in particular that for each i, $L_i = \emptyset$ if and only if $T_i = \emptyset$. Hence the least solutions of a system S in all algebras have the same empty components. The emptiness of each set T_i can be decided by the algorithm that decides the emptiness of a context-free language. Each set T_i is actually a context-free language over the alphabet consisting of F, parentheses and comma.

We will use these definitions for *algebras of graphs* $\mathbb{M}$ in the following way: M will be a class of graphs like $\mathcal{G}^u$ or $\mathcal{J}_2^d$ in the examples of cographs and series-parallel graphs, and F will be a set of total functions $f_\mathbb{M} : M^{\rho(f)} \to M$ that will be used to construct graphs. These functions, called the *operations* of $\mathbb{M}$, generalize the concatenation of words. The *constants* will be basic graphs. For each such graph algebra $\mathbb{M}$, the class of equational sets **Equat**$(\mathbb{M})$ generalizes the class of *context-free languages* since they are characterized as the components of the least solutions of equation systems as recalled in Section 1.1.1. There is thus no unique notion of a context-free set of graphs because this notion depends on the considered algebra.

However, even in the case of languages, several algebras can be considered, because one can enrich the monoid structure of A^* by new operations. This increases the class of equational sets, hence defines richer notions of context-free languages, if we take this term in the algebraic sense. The *squaring function* that associates with a word u the word uu, can be such an operation. Another one is the *shift* that associates with a word au the word ua, where a is a letter. The corresponding classes of equational sets have not received specific attention.

In the case of graphs, we will show that there are only two robust classes of equational sets, where *robust* means that they are closed under certain graph transformations definable by formulas of *monadic second-order logic*. These transformations, called *monadic second-order transductions*, play the role of rational transductions in the theory of formal languages.

Each of the two classes of context-free sets of graphs is the class of images of the set of finite binary trees under monadic second-order transductions of appropriate forms. Somewhat similarly, the class of context-free languages is the class of images of the language defined by the equation $L = aLbLc \cup \{d\}$, under all rational transductions. This language encodes binary trees. Hence trees play a major role in all three cases.

---

[7] In general, a *homomorphism* from $\mathbb{N}$ to $\mathbb{M}$, where $\mathbb{N} = \langle N, (f_\mathbb{N})_{f \in F} \rangle$ is another F-algebra, is a mapping $h : N \to M$ such that for every $f \in F$ and all $n_1,\ldots,n_{\rho(f)} \in N$, we have $h(f_\mathbb{N}(n_1,\ldots,n_{\rho(f)})) = f_\mathbb{M}(h(n_1),\ldots,h(n_{\rho(f)}))$.

1.1.5 Derivation trees

Context-free grammars specify languages. However the real importance of the notion of a context-free grammar is that, when a word is recognized as well-formed, the grammar specifies one or several *parse trees* for this word. These trees are obtained as results of the *syntactic analysis* (or *parsing*) of the considered word. They represent the syntactical structures of the considered word as generated by the grammar. In compiling applications, grammars are constructed so as to be unambiguous, and each recognized word has a unique parse tree. This tree is the support of further computation, in particular type checking and translation into intermediate code.

Similarly, an equation system specifies a set of objects and, as we have seen, it additionally specifies terms that denote those objects and represent their structure. Let S be an equation system and $\mathbb{M} = \langle M, (f_\mathbb{M})_{f \in F} \rangle$ an algebra, and consider an algorithm that, for each element m of M, computes a term t that denotes m as specified by S (if such a term exists). Due to the similarity with context-free grammars, we will say that this is a *parsing algorithm* for S.

However, if we view a context-free grammar G such as the one of Example 1.1 as an equation system $S = \Sigma_G$ over the signature $F = \{\cdot, \varepsilon, a, b, c\}$, as indicated in Example 1.2, then the terms in $T(F)$ specified by the system Σ_G are not the parse trees of G (because they do not show which rules of G are applied). Nevertheless, it is possible to view G as an equation system S' in a different way, such that the terms of S' *do* correspond to the parse trees of G, or rather a variant of parse trees called *derivation trees*. Let us illustrate this for the context-free grammar G of Example 1.1.

Example 1.3 We consider again the grammar G of Example 1.1. Its rules are named by symbols $p, q, \ldots, w$, which we will consider as function symbols with arities defined by ρ such that $\rho(s) = 3$, $\rho(p) = \rho(q) = 2$, $\rho(r) = \rho(u) = \rho(v) = \rho(w) = 0$; they form a signature P. The arity of a rule is the number of occurrences of nonterminals in the right-hand side of the rule.

Consider, for example, the word *baaac* generated from nonterminal T by the derivation sequence D:

$$T \Rightarrow bTST \Rightarrow bTaSTT \Rightarrow baaSTT \Rightarrow baaaTT \Rightarrow baaaT \Rightarrow baaac,$$

where the rules s, p, u, r, w, v are successively applied. (The arrow $\Rightarrow$ denotes the one-step derivation relation of G.) Assuming that rule w is applied to the leftmost T in *baaaTT*, we associate with D the term $d = s(u, p(r, w), v)$ of $T(P)$. This term contains more information than the sequence (s, p, u, r, w, v); from it one can find all derivation sequences of the word *baaac* that are equivalent to D by permutations of steps. In particular the leftmost derivation sequence uses successively rules s, u, p, r, w, v, and the rightmost one uses rules s, v, p, w, r, u. Figure 1.4 shows the parse tree of D and the corresponding term d.

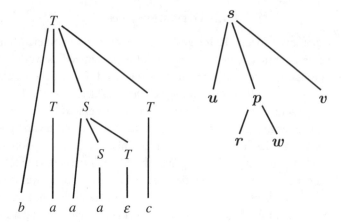

Figure 1.4 The parse tree of D and the term d.

Terms like d will be called *derivation trees*. We keep the name *parse tree* for the trees like the one of Figure 1.4 (left part) that are used in the theory of parsing. (Good textbooks exposing this theory are the "Dragon Book" by Aho *et al.* [*AhoLSU] and the book by Crespi-Reghizzi [*Cre]).

The equation system Σ_G of Example 1.1 can be rewritten into the following system:

$$\Sigma_G' \begin{cases} K = p(K,L) \cup q(K,K) \cup r, \\ L = s(L,K,L) \cup u \cup v \cup w. \end{cases}$$

Instead of solving this system for the F-algebra $\mathbb{M} = \langle \{a,b,c\}^*, \cdot, \varepsilon, a, b, c \rangle$ (which is the algebra for Σ_G in Example 1.2), we solve it for the P-algebra $\mathbb{M}'$ with the same domain $\{a,b,c\}^*$ but with the following interpretation of the symbols of P. If we interpret the symbols p,q,s by the following operations on $A^* = \{a,b,c\}^*$ (where x,y,z denote words in A^*):

$$p(x,y) := axy,$$

$$q(x,y) := xy,$$

$$s(x,y,z) := bxyz,$$

and the constant symbols r,u,v,w by the following words:

$$r := a,$$

$$u := a,$$

$$v := c,$$

$$w := \varepsilon,$$

then Σ'_G is just an alternative writing of Σ_G, and its least solution for the algebra $\mathbb{M}'$ is also $(L(G,S),L(G,T))$. But the system Σ'_G has also a least solution (K',L') in $\mathcal{P}(T(P)) \times \mathcal{P}(T(P))$, and the derivation tree d is an element of L'. More generally we define the sets of *derivation trees of G* respectively associated with S and T as the sets of terms K' and L'. For the above interpretation of the symbols of P, we can evaluate every term t of $T(P)$ into a word $t_{\mathbb{M}'}$ in A^*. In particular $d_{\mathbb{M}'} = baaac$. Clearly, $L(G,S) = \{t_{\mathbb{M}'} \mid t \in K'\}$ and $L(G,T) = \{t_{\mathbb{M}'} \mid t \in L'\}$. Thus, since a parsing algorithm for Σ'_G produces derivation trees of G, it corresponds to a classical parsing algorithm of the context-free grammar G.

The system Σ'_G and the derivation trees of G represent the *abstract syntax* of the grammar G, whereas the P-algebra $\mathbb{M}'$ represents its *concrete syntax*. It should be clear that the construction of Σ'_G and $\mathbb{M}'$ can be realized for every context-free grammar G. It should be noted, however, that the signature P and the algebra $\mathbb{M}'$ both depend on G.

A term that is associated with the word $baaac$ according to the equation system Σ_G in Example 1.2, is $b \cdot ((a \cdot (a \cdot (a \cdot \varepsilon))) \cdot c)$. That term can be obtained from derivation tree d by (re)interpreting the symbols p,q,s as the following operations on terms in $T(\{\cdot, \varepsilon, a, b, c\})$: $p(x,y) := a \cdot (x \cdot y)$, $q(x,y) := x \cdot y$, and $s(x,y,z) := b \cdot ((x \cdot y) \cdot z)$. Thus, a parsing algorithm for Σ'_G (producing derivation trees) can easily be transformed into one for Σ_G (producing terms). $\qquad\square$

In fact, derivation trees can be defined for the elements of general equational sets. The transformation of Σ_G into Σ'_G can be generalized into the transformation of an arbitrary equation system $S = \langle x_1 = p_1, \dots, x_n = p_n \rangle$ into a system $S' = \langle x_1 = p'_1, \dots, x_n = p'_n \rangle$ such that each polynomial p'_i is a union of monomials of the form $r(x_{i_1}, \dots, x_{i_{\rho(r)}})$ corresponding one-to-one to the monomials of p_i, where r belongs to a signature P associated with S. If $m \in T(F \cup \{x_1, \dots, x_n\})$ is the monomial of p_i to which $r(x_{i_1}, \dots, x_{i_{\rho(r)}})$ corresponds, then $x_{i_1}, \dots, x_{i_{\rho(r)}}$ is the sequence of unknowns that occur in m. Furthermore, we impose that each r has a unique occurrence in S'. The least solution of S' in $\mathcal{P}(T(P))^n$ defines the n-tuple of sets of *derivation trees* of S.

Let F be the signature over which S is written, and $\mathbb{M}$ the F-algebra for which S is to be solved. The function $M^{\rho(r)} \to M$ that interprets a symbol r in P is the one defined[8] (in the usual sense) by the unique term t_r in $T(F \cup \{y_1, \dots, y_{\rho(r)}\})$ such that (i) the variables $y_1, \dots, y_{\rho(r)}$ occur in t_r in that order and no variable y_j occurs more than once, and (ii) the monomial of S to which the monomial $r(x_{i_1}, \dots, x_{i_{\rho(r)}})$ of S' corresponds is obtained by substituting x_{i_j} for y_j in the term t_r, for every $j = 1, \dots, \rho(r)$.

We obtain thus a P-algebra $\mathbb{M}'$ (with the same domain as $\mathbb{M}$) and the system S' interpreted in $\mathbb{M}'$ is the same as the system S interpreted in $\mathbb{M}$.

The value mapping $t \mapsto t_{\mathbb{M}'}$ maps each set of derivation trees D_i (the i-th component of the least solution of S' in $\mathcal{P}(T(P))^n$) to the component L_i of the least solution of S

---

[8] A function $M^k \to M$ defined by a term in $T(F \cup \{y_1, \dots, y_k\})$ is a k-ary *derived operation* of $\mathbb{M}$.

in $\mathcal{P}(M)^n$. Taking $\mathbb{M} = \mathbb{T}(F)$, D_i is mapped to the set of terms T_i: the i-th component of the least solution of S in $\mathcal{P}(T(F))^n$. Thus, a parsing algorithm for S' can easily be transformed into one for S. Since a parsing algorithm for S' produces derivation trees that represent the syntactical structure of the elements of M as specified by the system S, it will also be called a parsing algorithm for S; thus, from now on, parsing algorithms produce derivation trees and/or terms.

Here is an example of the construction of S' from S.

Example 1.4 We let S be the following system:

$$\begin{cases} x_1 = x_2 \cup a \cup f(x_1, x_2, x_1), \\ x_2 = h(g(x_1, x_1), a) \cup f(x_1, x_2, x_1). \end{cases}$$

Then S' is:

$$\begin{cases} x_1 = r_{1,1}(x_2) \cup r_{1,2} \cup r_{1,3}(x_1, x_2, x_1), \\ x_2 = r_{2,1}(x_1, x_1) \cup r_{2,2}(x_1, x_2, x_1), \end{cases}$$

and the functions that interpret $r_{i,j}$ are defined by the following terms:

$$t_{r_{1,1}} := y_1,$$

$$t_{r_{1,2}} := a,$$

$$t_{r_{1,3}} := f(y_1, y_2, y_3),$$

$$t_{r_{2,1}} := h(g(y_1, y_2), a),$$

$$t_{r_{2,2}} := f(y_1, y_2, y_3).$$

Note that $r_{1,1}$ is interpreted as the identity function and that $r_{1,3}$ and $r_{2,2}$ are interpreted as the same function. The construction of S' from S does not depend on any F-algebra for which S is to be solved, and hence neither do the derivation trees of S. $\qquad\square$

1.2 Inductive sets of properties and recognizability

This section is a mild introduction to the algebraic notion of recognizability. This notion can be defined in several equivalent ways and we begin with its most concrete characterization, based on finite sets of properties that can be checked inductively.

1.2.1 Properties of the words of a context-free language

Let us consider the problem of proving an assertion of the form $\forall w \in L. P(w)$, where L is a context-free language, an equational set of graphs or, more generally, an equational set in an F-algebra with domain M, and where P is a property of the elements of M. Such an assertion expresses that P is universally valid on L.

Example 1.5 Let $X := \{f, x, y\}$ and $L \subseteq X^*$ be the language defined as the least solution (it is actually the unique solution) of the equation $L = fLL \cup \{x, y\}$. This language is the set of *Polish prefix notations* of the terms in $T(\{f, x, y\})$, where $\rho(f) = 2$ and $\rho(x) = \rho(y) = 0$. It satisfies the assertion $\forall w \in L. P(w)$, where property $P(w)$ is defined by

$$2|w|_f = |w| - 1 \land \forall u \in X^*(u < w \Rightarrow 2|u|_f \geq |u|).$$

Here $|w|$ denotes the length of a word w, $|w|_f$ the number of occurrences of f in w and $<$ the strict prefix order on words. Establishing the following facts is a routine exercise:

(i) $P(w)$ holds if $w = x$ or $w = y$;
(ii) $P(w)$ holds if $w = f w_1 w_2$ and $P(w_1)$ and $P(w_2)$ both hold.

Then the proof that $\forall w \in L. P(w)$ holds can be done by induction on the length of a *derivation sequence* of a word w in L, relative to the context-free grammar with rules $A \to f AA$, $A \to x$ and $A \to y$.

However this proof can also be formulated in terms of equation systems. By the Least Fixed-Point Theorem (Theorem 3.7), the least solution in $\mathcal{P}(M)^n$ of a system S of the form (1.4) is also the least solution of the corresponding system of inclusions:

$$\begin{cases} L_1 \supseteq p_1(L_1, \ldots, L_n) \\ \vdots \quad \vdots \\ L_n \supseteq p_n(L_1, \ldots, L_n), \end{cases} \tag{1.5}$$

where $L_i \subseteq M$. In the above example, Facts (i) and (ii) can be restated as the inclusion

$$K_P \supseteq f K_P K_P \cup \{x, y\}, \tag{1.6}$$

where $K_P := \{w \in X^* \mid P(w)\}$. Since the language L, defined by $L = fLL \cup \{x, y\}$, is also the least solution of (1.6), we have $L \subseteq K_P$, which yields the validity of $\forall w \in L. P(w)$.

We will say that an assertion of the form $\forall w \in L. P(w)$ is *provable by fixed-point induction* in order to express that this method applies, i.e., that property P satisfies Facts (i) and (ii). A property Q may be universally valid on the language L without this being provable by fixed-point induction. For example consider the property $Q(w)$ defined for $w \in X^*$ by $|w| = 1$ or $w \neq \tilde{w}$ (where $\tilde{w}$ denotes the mirror image of w) and $K_Q := \{w \in X^* \mid Q(w)\}$. It is not true that $f K_Q K_Q \subseteq K_Q$ (since x and f belong to K_Q and $fxf \notin K_Q$). However, $L \subseteq K_Q$. Hence the valid assertion $\forall w \in L. Q(w)$ is not provable by fixed-point induction.

In order to establish that Q is universally valid on L it suffices to find a property R such that:

(1) $\forall w \in X^*(R(w) \Rightarrow Q(w))$ is true; and

(2) $\forall w \in L.R(w)$ is provable by fixed-point induction.

We can take $R(w) :\Longleftrightarrow w \in \{x,y\} \cup f X^* \{x,y\}$. We prove in this way a stronger assertion than $\forall w \in L.Q(w)$, which was the initial goal.

Finding such R is always possible in a trivial way, by taking $R(w)$ to mean that w belongs to L, which does not yield any proof since (1) is just what is to be proved and (2) holds in a trivial way. Hence this observation is interesting if R can be found such that (1) and (2) are "easily provable," which is not a rigorous notion. A proof method can be defined by requiring that R is expressed in a particular language and/or that the proofs of (1) and (2) can be done in a particular proof system. We will give below an example of such a situation (Proposition 1.6). □

We generalize the notion of an assertion provable by fixed-point induction to systems of equations. Let S be an equation system of the general form (1.4) and let $(L_1,\ldots,L_n)$ be its least solution in $\mathcal{P}(M)^n$ for some F-algebra $\mathbb{M}$. Let $(P_i)_{1 \le i \le n}$ be an n-tuple of properties of elements of M, such that the assertion

$$\forall i \in [n], \forall d \in L_i.P_i(d) \tag{1.7}$$

is true. We say that this assertion is *provable by fixed-point induction* if

$$K_{P_i} \supseteq p_i(K_{P_1},\ldots,K_{P_n}), \tag{1.8}$$

for each $i = 1,\ldots,n$, where K_{P_i} denotes $\{d \in M \mid P_i(d)\}$. It follows from the Least Fixed-Point Theorem that the validity of (1.8) for all i implies that $L_i \subseteq K_{P_i}$ for all i, hence that the considered assertion (1.7) is true. The proof method consisting in proving (1.8) to establish (1.7) is thus sound.

Let us go back to context-free languages. Let the context-free language $L_1 \subseteq A^*$ be defined as the first component of the least solution $(L_1,\ldots,L_n)$ of an equation system $S = \langle x_1 = p_1,\ldots,x_n = p_n\rangle$.

Proposition 1.6 For every regular language K such that $L_1 \subseteq K$, there exists an n-tuple of regular languages $(K_1,\ldots,K_n)$ such that $K_1 \subseteq K$ and such that the assertion

$$\forall i \in [n], \forall d \in L_i.d \in K_i$$

is provable by fixed-point induction.

Proof sketch: We have $K = h^{-1}(N)$, where h is a homomorphism of A^* into a finite monoid[9] $\mathbb{Q} = \langle Q, \cdot_{\mathbb{Q}}, 1_{\mathbb{Q}}\rangle$ and $N \subseteq Q$ ([*Eil, *Sak]). We define $K_i := h^{-1}(h(L_i)) = \bigcup\{h^{-1}(q) \mid q \in Q, \ h^{-1}(q) \cap L_i \ne \emptyset\}$. Note that this definition does not depend on N.

---

[9] $\mathbb{Q} = \langle Q, \cdot_{\mathbb{Q}}, 1_{\mathbb{Q}}\rangle$ is a monoid if the binary operation $\cdot_{\mathbb{Q}}$ is associative with $1_{\mathbb{Q}}$ as a unit element.

We first prove that $K_1 \subseteq K$. We consider a word v in K_1. For some word v' in L_1, we have $h(v) = h(v')$. Since $L_1 \subseteq K$, we have $v' \in K$. Hence $v \in K$ since $h(v') \in N$ and $K = h^{-1}(N)$. This achieves the first goal.

In order to prove that for each i we have $p_i(K_1, \ldots, K_n) \subseteq K_i$, we need only consider a monomial α of p_i (i.e., the right-hand side of a rule $x_i \rightarrow \alpha$ of the context-free grammar S) and prove that

$$w_0 K_{i_1} w_1 \cdots K_{i_k} w_k \subseteq K_i, \tag{1.9}$$

where $\alpha = w_0 x_{i_1} w_1 x_{i_2} \cdots x_{i_k} w_k$ with $w_0, \ldots, w_k \in A^*$. Let $v_j \in K_{i_j}$ for $j = 1, \ldots, k$. There exist $v'_1, \ldots, v'_k$ such that $h(v'_j) = h(v_j)$ and $v'_j \in L_{i_j}$ for each j. Hence the word $v' = w_0 v'_1 w_1 \cdots v'_k w_k$ belongs to L_i, because $(L_1, \ldots, L_n)$ is a solution of S. Letting $v = w_0 v_1 w_1 \cdots v_k w_k$ we have $h(v) = h(v')$ hence $v \in K_i$. This proves inclusion (1.9) and completes the proof of the proposition. ∎

This result shows that assertions of the form $L \subseteq K$, where L is a context-free language and K is a regular one, can be proved by fixed-point induction. The proofs that a "guessed" n-tuple $(K'_1, \ldots, K'_n)$ satisfies $K'_1 \subseteq K$ and the inclusions $K'_i \supseteq p_i(K'_1, \ldots, K'_n)$ establish that $L_1 \subseteq K$ and use only algorithms on finite automata: Boolean operations, concatenation and emptiness test.

1.2.2 Some properties of series-parallel graphs

We now show that fixed-point induction can also be used for proving universal properties of equational sets of graphs. We use the example of the set of series-parallel graphs defined by Equation (1.3) considered in Section 1.1.3:

$$S = (S \,/\!/\, S) \cup (S \bullet S) \cup \{e\},$$

where $S \subseteq \mathcal{J}_2^{\mathrm{d}}$. We will prove the assertions $\forall G \in S. P_i(G)$, where the properties P_i are defined as follows:

$P_1(G) \; :\Longleftrightarrow \; G$ is connected,
$P_2(G) \; :\Longleftrightarrow \; G$ is bipolar,
$P_3(G) \; :\Longleftrightarrow \; G$ is planar,
$P_4(G) \; :\Longleftrightarrow \; G$ has no directed cycle.

A directed graph G with two sources denoted by $src_G(1)$ and $src_G(2)$ (cf. Section 1.1.3) is *bipolar* if it has no directed cycle and every vertex belongs to a directed path from $src_G(1)$ to $src_G(2)$.

Following the same method as for the language L of Example 1.5, we need only prove that $P_i(e)$ holds and that, for all graphs G, H in $\mathcal{J}_2^{\mathrm{d}}$: $P_i(G) \wedge P_i(H)$ implies $P_i(G \,/\!/\, H) \wedge P_i(G \bullet H)$.

These facts are clearly true for properties P_1 and P_2. The assertions that every graph in S satisfies P_1 on the one hand and P_2 on the other are thus provable by fixed-point induction, hence P_1 and P_2 are universally valid on S.

Property P_3 is not provable in this way because it is not true that, for all graphs G, H in $\mathcal{J}_2^d$, $P_3(G) \wedge P_3(H)$ implies $P_3(G \parallel H)$. For a counterexample, take H to be an edge, G to be K_5 minus one edge (K_5 is a complete simple undirected graph with five vertices) and equipped with sources in such a way that $G \parallel H$ is isomorphic to K_5, which is nonplanar. However, G and H are both planar, hence satisfy P_3.

For proving that every series-parallel graph is planar, we can use the property Q saying that a graph has a planar drawing with its two sources on the external face. (The books [*Die] and [*MohaTho] give formal definitions about graphs on surfaces.) This property is provable by fixed-point induction (with respect to the equation defining S), hence it is true for all graphs in S. Since $Q(G)$ implies $P_3(G)$ for all graphs G in $\mathcal{J}_2^d$, we obtain the announced result.

The case of property P_4 (the absence of directed cycles) is similar. The assertion that every graph in S satisfies P_4 is not provable by fixed-point induction; however, it is true. To prove it, one takes the stronger property P_2 considered above.

This proof technique can be applied to systems of equations (and not only to single equations) and to graph properties expressed in monadic second-order logic (see Section 1.3). More precisely, for every such graph property P and every equational set of graphs L:

(1) we can decide whether or not P is universally valid on L;
(2) if P is, then we can build a set of auxiliary properties, like the sets $K_1, \ldots, K_n$ in Proposition 1.6, that yields a proof by fixed-point induction of the universal validity of P on L.

These constructions and the verification of conditions like (1.8) can be done by algorithms.

1.2.3 Inductive sets of properties

We now consider properties that are not necessarily universally valid on the considered sets of graphs, so that they raise decision problems. We will consider a graph property as a function from graphs to $\{True, False\}$.

Example 1.7 (Cographs) The property $E(G)$ that a cograph G (cographs are defined in Section 1.1.2) has an even number of vertices is not universally valid. However, it satisfies the following rules:

$$E(\mathbf{1}) = False,$$

$$E(G \oplus H) = (E(G) \Leftrightarrow E(H)),$$

$$E(G \otimes H) = (E(G) \Leftrightarrow E(H)).$$

For Boolean values p and q, $p \Leftrightarrow q$ is defined as usual as $((p \wedge q) \vee (\neg p \wedge \neg q))$. It follows that if a cograph G is the value of a term t in $T(\{\oplus, \otimes, \mathbf{1}\})$ (we denote this by $G = val(t)$), then the validity of $E(G)$ can be determined by computing $E(val(t'))$ for all subterms t' of t, starting with the smallest ones. This type of computation can be done by automata on terms that we will present in Section 1.2.4.

The property $F(G)$, defined as "G has an even number of edges," can be checked in a similar way by computing simultaneously $E(val(t'))$ and $F(val(t'))$ for every subterm t' of t. This computation uses the following rules:

$$F(\mathbf{1}) = True,$$

$$F(G \oplus H) = (F(G) \Leftrightarrow F(H)),$$

$$F(G \otimes H) = \left(\left(F(G) \Leftrightarrow F(H) \right) \wedge \left(E(G) \vee E(H) \right) \right)$$
$$\vee \left(\left(F(G) \Leftrightarrow \neg F(H) \right) \wedge \left(\neg E(G) \wedge \neg E(H) \right) \right).$$

Hence, $F(G)$ can be checked with the help of $E(G)$ as additional information. □

We now generalize this computation method. We introduce a definition relative to an arbitrary F-algebra $\mathbb{M}$. Let $\mathcal{P}$ be a set of properties, i.e., of mappings: $M \to \{True, False\}$. We say that $\mathcal{P}$ is *F-inductive* if for every $P \in \mathcal{P}$, for every $f \in F$ of arity $n > 0$, and for every $m_1, \ldots, m_n$ in M, the Boolean value $P(f_{\mathbb{M}}(m_1, \ldots, m_n))$ can be computed by a fixed Boolean expression depending on P and f, in terms of finitely many Boolean values $Q(m_i)$ with Q in $\mathcal{P}$ and $i = 1, \ldots, n$.

In the previous example, the set of properties $\{E, F\}$ is $\{\oplus, \otimes\}$-inductive for the algebra of cographs, but the set $\{F\}$ is not. The computation of $F(G \otimes H)$ can be expressed by

$$F(G \otimes H) = B(E(G), F(G), E(H), F(H)),$$

where $B(p_1, p_2, p_3, p_4)$ is the Boolean expression

$$\left((p_2 \Leftrightarrow p_4) \wedge (p_1 \vee p_3) \right) \vee \left((p_2 \Leftrightarrow \neg p_4) \wedge (\neg p_1 \wedge \neg p_3) \right).$$

In order to have a uniform notation, if $\mathcal{P}$ is finite and enumerated as $\{P_1, \ldots, P_k\}$, we write

$$P_i(f_{\mathbb{M}}(m_1, \ldots, m_n))$$
$$= B_{i,f} \left(P_1(m_1), \ldots, P_k(m_1), P_1(m_2), \ldots, P_k(m_2), \ldots, P_1(m_n), \ldots, P_k(m_n) \right)$$

to formalize the way $P_i(f_{\mathbb{M}}(m_1,\ldots,m_n))$ can be computed from the Boolean values $P_i(m_j)$. In this writing, each $B_{i,f}$ is a Boolean expression in the propositional variables $p_{h,j}$, $1 \le h \le k$, $1 \le j \le n$, and $P_h(m_j)$ is substituted in $B_{i,f}$ for $p_{h,j}$.

An important theorem is the following one:

Theorem 1.8 (Filtering Theorem, concrete version)   Let $\mathbb{M}$ be an F-algebra and $\mathcal{P}$ be a finite F-inductive set of properties. For every equational set L of $\mathbb{M}$, for every P in $\mathcal{P}$, the set $L_P := \{x \in L \mid P(x)\}$ is equational. If the Boolean expressions involved in the definition of the inductivity of $\mathcal{P}$ are given, then the construction of a system of equations defining L_P from one defining L is effective, i.e., can be done by an algorithm.   □

The classical result saying that the intersection of a context-free language and a regular one is context-free is a special case of this theorem. Let us consider the language L of Example 1.5 defined by the equation $L = f\,LL \cup \{x,y\}$. From this language we want to keep only the words whose length is a multiple of 3. For $i \in \{0,1,2\}$, we let $L_i := \{w \in L \mid \mathrm{mod}_3(|w|) = i\}$. The triple (L_0, L_1, L_2) is the least solution (and actually also the unique one) of the following system:

$$\begin{cases} L_0 = f\,L_0 L_2 \cup f\,L_1 L_1 \cup f\,L_2 L_0, \\ L_1 = f\,L_0 L_0 \cup f\,L_1 L_2 \cup f\,L_2 L_1 \cup \{x,y\}, \\ L_2 = f\,L_0 L_1 \cup f\,L_1 L_0 \cup f\,L_2 L_2. \end{cases}$$

It follows that the language L_0 is context-free. A similar example for cographs is System (1.2) in Section 1.1.2 (cf. the discussion after (1.2)).

Corollary 1.9 Let $\mathbb{M}$ and $\mathcal{P}$ be as in Theorem 1.8. For every equational set L of $\mathbb{M}$, one can decide whether or not a property P in $\mathcal{P}$ is universally valid on L, and whether or not it is satisfied by some element of L.

Proof sketch: We assume that L is given by a system of equations S. By using Theorem 1.8 we can construct a system S' that defines L_P. As noted in Section 1.1.4, we can test the emptiness of the components of the least solution of S', hence in particular of L_P. We can thus decide if P is satisfied by some element of L.

Since $\mathcal{P}$ is inductive, so is $\mathcal{P} \cup \{\neg P\}$. We can apply the previous result to $\neg P$. Then P is universally valid on L if and only if $L_{\neg P} = \emptyset$, which is decidable.   ∎

Example 1.10 We again let L be the language of Example 1.5 defined by the equation $L = f\,LL \cup \{x,y\}$. We know from this example that every word of L has odd length (because $|w| = 2|w|_f + 1$ for every $w \in L$), but we will see how the algorithm of Corollary 1.9 "discovers" this fact. Let K_0 and K_1 be the sets of words in L of even and odd length respectively. These languages are defined by the two equations:

$$\begin{cases} K_0 = fK_0K_1 \cup fK_1K_0, \\ K_1 = fK_0K_0 \cup fK_1K_1 \cup \{x,y\}. \end{cases}$$

It is easy to see that K_0 is empty (just look at the corresponding context-free grammar). Hence $K_1 = L$ and every word of L has odd length. $\square$

It is useful to have a proof by fixed-point induction that a property is universally valid on an equational set although an algorithm can also give the answer, because a proof is more informative than the yes or no answer of an algorithm: it shows the properties of all components of the solution of the equation system that "contribute" to the validity of the proved property. This is clear in the case of Proposition 1.6.

We now consider one more example about graphs.

Example 1.11 (The 2-colorability of series-parallel graphs) We illustrate Theorem 1.8 with the 2-colorability of series-parallel graphs. A *proper vertex k-coloring* of a graph assigns to each vertex a *color*, i.e., an element of $\{1,\dots,k\}$ such that two adjacent vertices have different colors. A graph is *k-colorable* if it has a proper vertex *k*-coloring. We consider three properties of a series-parallel graph G defined as follows:

$\gamma_2(G) :\Longleftrightarrow G$ is 2-colorable,

$\sigma(G) :\Longleftrightarrow G$ is 2-colorable with the two sources of the same color,

$\delta(G) :\Longleftrightarrow G$ is 2-colorable with the two sources of different colors.

The set of series-parallel graphs is defined by Equation (1.3):

$$S = (S \mathbin{/\!/} S) \cup (S \bullet S) \cup \{e\},$$

in the algebra $\langle \mathcal{J}_2^{d}, /\!/, \bullet, e \rangle$.

Property γ_2 is not universally valid on S because $\gamma_2(e \mathbin{/\!/} (e \bullet e)) = \textit{False}$.[10] The set $\{\gamma_2\}$ is not inductive because $\gamma_2(e) = \textit{True}$, $\gamma_2(e \bullet e) = \textit{True}$, $\gamma_2(e \mathbin{/\!/} e) = \textit{True}$ and $\gamma_2(e \mathbin{/\!/} (e \bullet e)) = \textit{False}$. It follows that the validity of $\gamma_2(G \mathbin{/\!/} H)$ cannot be deduced from those of $\gamma_2(G)$ and $\gamma_2(H)$. Hence, as in the case of property F for cographs in Example 1.7, we need additional properties. They will be σ and δ. The set $\{\sigma, \delta\}$ is

---

[10] One can prove by fixed-point induction that every series-parallel graph is 3-colorable in such a way that its two sources have different colors.

inductive; this is clear from the following facts:

$$\sigma(G \bullet H) = (\sigma(G) \wedge \sigma(H)) \vee (\delta(G) \wedge \delta(H)),$$
$$\delta(G \bullet H) = (\delta(G) \wedge \sigma(H)) \vee (\sigma(G) \wedge \delta(H)),$$
$$\sigma(G \,/\!/\, H) = \sigma(G) \wedge \sigma(H),$$
$$\delta(G \,/\!/\, H) = \delta(G) \wedge \delta(H). \tag{1.10}$$

One can thus compute for every term t in $T(\{/\!/, \bullet, e\})$ the pair of Boolean values $(\sigma(val(t)), \delta(val(t)))$ by induction on the structure of t (where $val(t)$ is the graph in $\mathcal{J}_2^d$ defined by a term t). From the pair of Boolean values associated with a term t such that $val(t) = G$, one can decide whether $\gamma_2(G)$ holds or not, since for every graph G in $\mathcal{J}_2^d$, $\gamma_2(G)$ is equivalent to $\sigma(G) \vee \delta(G)$. This computation can be formalized as the run of a finite deterministic automaton on t because the considered inductive set of properties, here $\{\sigma, \delta\}$, is finite. The finiteness of a set of inductive properties is the essence of the notion of recognizability that we now introduce informally. $\qquad\square$

1.2.4 Recognizability

In Formal Language Theory, the *recognizability* (or *regularity*) of a set of finite or infinite words or terms means frequently that this set is defined by a finite automaton of some kind. Recognizable sets of finite words and terms are defined by finite *deterministic* automata and this fact yields algebraic characterizations of recognizability in terms of homomorphisms into finite algebras. In particular, a language $L \subseteq A^*$ is recognizable if and only if $L = h^{-1}(N)$ where $h : A^* \to \mathbb{Q}$ is a monoid homomorphism, $\mathbb{Q}$ is a finite monoid and $N \subseteq Q$. (We have used this fact in the proof of Proposition 1.6.)

This characterization has the advantage of being extendable in a meaningful way to any algebra, whereas the notion of automaton has no immediate generalization to arbitrary algebras. Furthermore, it fits very well with the notion of an equational set. The Filtering Theorem shows this, as we will explain in Section 1.2.5.

Following Mezei and Wright [MezWri], we say that a subset L of an F-algebra $\mathbb{M}$ (where F is finite) is *recognizable* if $L = h^{-1}(N)$ for some homomorphism of F-algebras $h : \mathbb{M} \to \mathbb{Q}$, where $\mathbb{Q}$ is a finite F-algebra and $N \subseteq Q$. We will denote by **Rec**($\mathbb{M}$) the family of recognizable subsets of $\mathbb{M}$.

The above definition of recognizability of L is equivalent to saying that the property P^L of the elements of M, such that $P^L(x)$ is *True* if and only if $x \in L$, belongs to a finite F-inductive set of properties. In fact, assume that L is recognizable and let $Q = \{q_1, \ldots, q_k\}$. Define for each $i \in [k]$ the property P_i by: $P_i(m) :\Longleftrightarrow h(m) = q_i$. Since $h(f_{\mathbb{M}}(m_1, \ldots, m_n)) = f_{\mathbb{Q}}(h(m_1), \ldots, h(m_n))$, the Boolean value $P_i(f_{\mathbb{M}}(m_1, \ldots, m_n))$ can be computed from the Boolean values $P_h(m_j)$, by a Boolean expression. Thus,

$\{P_1, \ldots, P_k\}$ is F-inductive in $\mathbb{M}$. Hence, also the set $\{P^L, P_1, \ldots, P_k\}$ is inductive, because $P^L(m) = \bigvee_{q_i \in N} P_i(m)$. The other direction of the equivalence is discussed in Section 1.2.5.

Proposition 1.6 also holds in general for recognizable sets instead of regular languages (with exactly the same proof, replacing A^* by $\mathbb{M}$). Thus, inclusions $L \subseteq K$ with $L \in \mathbf{Equat}(\mathbb{M})$ and $K \in \mathbf{Rec}(\mathbb{M})$ are provable by fixed-point induction, using auxiliary properties that correspond to recognizable subsets of M. Together with Corollary 1.9, this shows that the analogues of statements (1) and (2) at the end of Section 1.2.2 hold for properties that correspond to recognizable sets, in arbitrary algebras.

We now recall the link between recognizability for subsets of $T(F)$ and finite automata on terms (i.e., bottom-up finite tree automata). Let us consider a set $L \in \mathbf{Rec}(\mathbb{T}(F))$ defined as $L = h^{-1}(N)$, where h is the unique homomorphism: $\mathbb{T}(F) \to \mathbb{Q}$, $\mathbb{Q}$ is a finite F-algebra and $N \subseteq Q$. (Note that $h(t) = t_\mathbb{Q}$.) The pair $(\mathbb{Q}, N)$ corresponds to a finite deterministic and complete F-*automaton* $\mathscr{A}(\mathbb{Q}, N)$ (the general definitions can be found in the book [*Com+] and in the book chapter [*GecSte], Section 3.3) with set of states Q, set of accepting states N and transitions consisting of the tuples $(a_1, \ldots, a_{\rho(f)}, f, a)$ such that $f \in F$, $a_1, \ldots, a_{\rho(f)}, a \in Q$ and $a = f_\mathbb{Q}(a_1, \ldots, a_{\rho(f)})$.

On each term t in $T(F)$, the automaton $\mathscr{A} = \mathscr{A}(\mathbb{Q}, N)$ has a unique ("bottom-up") *run*, defined as a mapping $run_{\mathscr{A}, t} : Pos(t) \to Q$ such that $run_{\mathscr{A}, t}(u) = h(t/u)$ for every position[11] u of t. Hence t is accepted by $\mathscr{A}$ if and only if $h(t) = run_{\mathscr{A}, t}(root_t) \in N$.

Conversely, if L is the set of terms in $T(F)$ accepted by a finite, possibly not deterministic, automaton $\mathscr{B}$, then it is also accepted by a finite deterministic and complete automaton $\mathscr{A}$ (that one can construct from $\mathscr{B}$) and there is a unique pair $(\mathbb{Q}, N)$ such that $\mathscr{A}(\mathbb{Q}, N) = \mathscr{A}$. Hence, L is recognizable in $\mathbb{T}(F)$.

By a *recognizable set of graphs*, we will mean a subset of a graph algebra that is recognizable with respect to that algebra. No notion of "graph automaton" arises from this definition. However, we obtain finite automata accepting the sets of terms that denote the graphs of recognizable sets (this is true because the signature is finite). We will give a more precise statement in Theorem 1.12 below.

1.2.5 From inductive sets to automata

Let $\mathcal{P} = \{P_1, \ldots, P_k\}$ be a finite inductive set of properties on an F-algebra $\mathbb{M}$, where F is finite. We associate with $\mathcal{P}$ a finite deterministic and complete F-automaton $\mathscr{A} = \mathscr{A}(\mathbb{Q}, N)$ as follows. Its set of states is $Q = \{True, False\}^k$; its transitions, i.e., the operations of $\mathbb{Q}$, are defined in such a way that for every f in F of arity n, we have: $f_\mathbb{Q}(q_1, \ldots, q_n) = q$ if and only if $q_i = (a_{1,i}, \ldots, a_{k,i})$, $q = (b_1, \ldots, b_k)$ belong to

---

[11] The set $Pos(t)$ of positions of t is the set of occurrences of the symbols of F. We denote by t/u the subterm of t issued from u and by $root_t$ the first position of t. Formal definitions are in Chapter 2.

{*True,False*}^{k} and (we use the notation introduced in Section 1.2.3):

$$b_i = B_{i,f}(a_{1,1},\ldots,a_{k,1},a_{1,2},\ldots,a_{k,2},\ldots,a_{1,n},\ldots,a_{k,n}).$$

It follows that for every $t \in T(F), t_\mathbb{Q} = (P_1(t_\mathbb{M}),\ldots,P_k(t_\mathbb{M})) \in \{$*True,False*$\}^k$.

Hence if we want to specify by an automaton the set of objects $m \in M$ that satisfy $P_3(m)$ (to take an example), we take as set N of accepting states the set of Boolean vectors $(b_1,\ldots,b_k)$ such that $b_3 = $ *True*. More precisely, a term $t \in T(F)$ is accepted by $\mathscr{A}$ if and only if $t_\mathbb{M}$ has property P_3.

This proves the implication (2) $\Rightarrow$ (3) in the next result. The other direction also holds, provided $\mathbb{M}$ is generated by F (as defined in Section 1.1.4).

Theorem 1.12 Let $\mathbb{M}$ be an F-algebra generated by F, where F is finite, and let $L \subseteq M$. The following are equivalent:

(1) L is recognizable in $\mathbb{M}$;
(2) $L = \{m \in M \mid P(m)\}$, where P belongs to a finite F-inductive set of properties;
(3) the set of terms t in $T(F)$ such that $t_\mathbb{M}$ belongs to L is recognizable in $\mathbb{T}(F)$, equivalently, is the set accepted by a finite F-automaton. $\square$

The equivalence of (1) and (2) gives a concrete meaning to the algebraic notion of recognizability (and does not need $\mathbb{M}$ being generated by F). The implication (1) $\Rightarrow$ (2) was shown in Section 1.2.4. The other direction follows from the above discussion. If $h : M \to Q$ is defined by $h(m) := (P_1(m),\ldots,P_k(m))$, then h is a homomorphism from $\mathbb{M}$ to the finite F-algebra $\mathbb{Q}$, because $\mathcal{P}$ is inductive. And, e.g., $h^{-1}(N) = \{m \in M \mid P_3(m)\}$.

Let us now assume that L is an equational set of $\mathbb{M}$, defined by an equation system $S = \langle x_1 = p_1,\ldots,x_n = p_n \rangle$, and that $L \subseteq \{m \in M \mid P_3(m)\}$. The equivalence of (1) and (2) implies that such an inclusion is provable by fixed-point induction, using auxiliary properties $R_1,\ldots,R_n$ that belong to a finite inductive set of properties. In fact, considering the proof of Proposition 1.6, with the above definition of $\mathbb{Q}$, N and h, it can be seen that every R_i is a Boolean combination of $P_1,\ldots,P_k$ and hence $\mathcal{P} \cup \{R_1,\ldots,R_n\}$ is an inductive set of properties.

The equivalence of (1) and (3) implies that the membership in a recognizable set of an element of M, given as $t_\mathbb{M}$, for some term t in $T(F)$, or the validity of $P(t_\mathbb{M})$, where P belongs to a finite inductive set of properties, can be checked in time $O(|t|)$, i.e., in time linear in the size of t.

Let us go back to Example 1.11 about the 2-colorability of series-parallel graphs. For the inductive set $\mathcal{P} = \{\sigma,\delta\}$ we obtain the set of states

$$Q = \{\textit{True,False}\} \times \{\textit{True,False}\} = \{(\sigma,\delta),(\sigma,\overline{\delta}),(\overline{\sigma},\delta),(\overline{\sigma},\overline{\delta})\},$$

where, for readability, we use σ and δ to denote *True* and $\overline{\sigma}$ and $\overline{\delta}$ to denote *False*. For every state $q \in Q$ we let S_q be the set of series-parallel graphs G such that

$(\sigma(G), \delta(G)) = q$, i.e., $S_q = h^{-1}(q) \cap S$ with $h(G) = (\sigma(G), \delta(G))$ (cf. the proof of Proposition 1.6). Thus, $S_{\sigma,\delta}$ is the set of series-parallel graphs that satisfy σ and δ, $S_{\sigma,\bar{\delta}}$ the set of those that satisfy σ and not δ, $S_{\bar{\sigma},\delta}$ the set of those that satisfy δ and not σ, and $S_{\bar{\sigma},\bar{\delta}}$ the set of those that satisfy neither σ nor δ. From Properties (1.10) we obtain the operations $/\!/_\mathbb{Q}, \bullet_\mathbb{Q}$ and the constant $e_\mathbb{Q}$, which determine the transitions of the automaton $\mathscr{A}$. For example, since the graph defined by e satisfies δ and not σ, we have $e_\mathbb{Q} = (\bar{\sigma}, \delta)$. As another example, if $G = H \bullet K$ and both H and K satisfy δ and not σ, then G satisfies σ and not δ; hence $\bullet_\mathbb{Q}((\bar{\sigma}, \delta), (\bar{\sigma}, \delta)) = (\sigma, \bar{\delta})$.

From the defining equation $S = (S /\!/ S) \cup (S \bullet S) \cup e$ and the transitions of $\mathscr{A}$ we obtain the following system of equations that define the sets $S_{\sigma,\delta}, S_{\bar{\sigma},\delta}, S_{\sigma,\bar{\delta}}$ and $S_{\bar{\sigma},\bar{\delta}}$ (we omit parentheses around terms like $S_{\sigma,\delta} /\!/ S_{\sigma,\delta}$ for better readability):

(a) $\quad S_{\sigma,\delta} = S_{\sigma,\delta} /\!/ S_{\sigma,\delta}$

$\qquad \cup S_{\sigma,\delta} \bullet S_{\sigma,\delta} \ \cup \ S_{\sigma,\delta} \bullet S_{\sigma,\bar{\delta}} \ \cup \ S_{\sigma,\delta} \bullet S_{\bar{\sigma},\delta}$

$\qquad \cup S_{\sigma,\bar{\delta}} \bullet S_{\sigma,\delta} \ \cup \ S_{\bar{\sigma},\delta} \bullet S_{\sigma,\delta},$

(b) $\quad S_{\bar{\sigma},\delta} = e \ \cup \ S_{\bar{\sigma},\delta} /\!/ S_{\bar{\sigma},\delta} \ \cup \ S_{\sigma,\delta} /\!/ S_{\bar{\sigma},\delta} \ \cup \ S_{\bar{\sigma},\delta} /\!/ S_{\sigma,\delta}$

$\qquad \cup S_{\sigma,\bar{\delta}} \bullet S_{\bar{\sigma},\delta} \ \cup \ S_{\bar{\sigma},\delta} \bullet S_{\sigma,\bar{\delta}},$

(c) $\quad S_{\sigma,\bar{\delta}} = S_{\sigma,\delta} /\!/ S_{\sigma,\bar{\delta}} \ \cup \ S_{\sigma,\bar{\delta}} /\!/ S_{\sigma,\delta} \ \cup \ S_{\sigma,\bar{\delta}} /\!/ S_{\sigma,\bar{\delta}}$

$\qquad \cup S_{\sigma,\bar{\delta}} \bullet S_{\sigma,\bar{\delta}} \ \cup \ S_{\bar{\sigma},\delta} \bullet S_{\bar{\sigma},\delta},$

(d) $\quad S_{\bar{\sigma},\bar{\delta}} = S_{\bar{\sigma},\bar{\delta}} /\!/ S_{\bar{\sigma},\bar{\delta}} \ \cup \ S_{\bar{\sigma},\bar{\delta}} /\!/ S_{\sigma,\delta} \ \cup \ S_{\bar{\sigma},\bar{\delta}} /\!/ S_{\bar{\sigma},\delta} \ \cup \ S_{\bar{\sigma},\bar{\delta}} /\!/ S_{\sigma,\bar{\delta}}$

$\qquad \cup S_{\sigma,\delta} /\!/ S_{\bar{\sigma},\bar{\delta}} \ \cup \ S_{\bar{\sigma},\delta} /\!/ S_{\bar{\sigma},\bar{\delta}} \ \cup \ S_{\sigma,\bar{\delta}} /\!/ S_{\bar{\sigma},\bar{\delta}} \ \cup \ S_{\bar{\sigma},\delta} /\!/ S_{\sigma,\bar{\delta}}$

$\qquad \cup S_{\sigma,\bar{\delta}} /\!/ S_{\bar{\sigma},\delta} \ \cup \ S_{\bar{\sigma},\bar{\delta}} \bullet S_{\sigma,\delta} \ \cup \ S_{\bar{\sigma},\delta} \bullet S_{\bar{\sigma},\delta} \ \cup \ S_{\bar{\sigma},\delta} \bullet S_{\sigma,\delta}$

$\qquad \cup S_{\bar{\sigma},\bar{\delta}} \bullet S_{\bar{\sigma},\bar{\delta}} \ \cup \ S_{\sigma,\delta} \bullet S_{\bar{\sigma},\bar{\delta}} \ \cup \ S_{\bar{\sigma},\delta} \bullet S_{\bar{\sigma},\delta} \ \cup \ S_{\sigma,\bar{\delta}} \bullet S_{\bar{\sigma},\delta}.$

These equations are constructed as follows. Since $e_\mathbb{Q} = (\bar{\sigma}, \delta)$, the constant symbol e is put in the right-hand side of the equation that defines $S_{\bar{\sigma},\delta}$ and nowhere else. Moreover, for $f \in \{/\!/, \bullet\}$, if $f_\mathbb{Q}(q_1, q_2) = q$, then we put the monomial $f(S_{q_1}, S_{q_2})$ in the right-hand side of the equation that defines S_q. Thus, $S_{\bar{\sigma},\delta} \bullet S_{\bar{\sigma},\delta}$ is in the right-hand side of Equation (c).

Since we have e in the right-hand side of Equation (b), we have $S_{\bar{\sigma},\delta} \neq \emptyset$. Using this fact and since we have the term $S_{\bar{\sigma},\delta} \bullet S_{\bar{\sigma},\delta}$ in the right-hand side of Equation (c), we have $S_{\sigma,\bar{\delta}} \neq \emptyset$. And by these facts and since we have $S_{\bar{\sigma},\delta} /\!/ S_{\sigma,\bar{\delta}}$ in the right-hand side of Equation (d), we have $S_{\bar{\sigma},\bar{\delta}} \neq \emptyset$. Since every term in the right-hand side of Equation (a) contains $S_{\sigma,\delta}$, we have $S_{\sigma,\delta} = \emptyset$. This proves that no series-parallel graph has one coloring of type σ and another one of type δ. Moreover, according to the proof of Proposition 1.6, this property is provable by fixed-point induction, as the reader can easily check. Using this and the commutativity of $/\!/$, we can simplify the system into the following one:

(b') $S_{\overline{\sigma},\delta} = e \cup S_{\overline{\sigma},\delta} \parallel S_{\overline{\sigma},\delta} \cup S_{\sigma,\overline{\delta}} \bullet S_{\overline{\sigma},\delta} \cup S_{\overline{\sigma},\delta} \bullet S_{\sigma,\overline{\delta}},$

(c') $S_{\sigma,\overline{\delta}} = S_{\sigma,\overline{\delta}} \parallel S_{\sigma,\overline{\delta}} \cup S_{\sigma,\overline{\delta}} \bullet S_{\sigma,\overline{\delta}} \cup S_{\overline{\sigma},\delta} \bullet S_{\overline{\sigma},\delta},$

(d') $S_{\overline{\sigma},\overline{\delta}} = S_{\overline{\sigma},\overline{\delta}} \parallel S_{\overline{\sigma},\overline{\delta}} \cup S_{\overline{\sigma},\overline{\delta}} \parallel S_{\overline{\sigma},\delta} \cup S_{\overline{\sigma},\overline{\delta}} \parallel S_{\sigma,\overline{\delta}} \cup S_{\overline{\sigma},\delta} \parallel S_{\sigma,\overline{\delta}}$

$\qquad \cup S_{\overline{\sigma},\overline{\delta}} \bullet S_{\overline{\sigma},\delta} \cup S_{\overline{\sigma},\overline{\delta}} \bullet S_{\sigma,\overline{\delta}} \cup S_{\overline{\sigma},\overline{\delta}} \bullet S_{\overline{\sigma},\overline{\delta}}$

$\qquad \cup S_{\overline{\sigma},\delta} \bullet S_{\overline{\sigma},\overline{\delta}} \cup S_{\sigma,\overline{\delta}} \bullet S_{\overline{\sigma},\overline{\delta}}.$

Thus, this construction proves that every series-parallel graph either has no 2-coloring (it is then generated by $S_{\overline{\sigma},\overline{\delta}}$) or has one of type σ and none of type δ (it is generated by $S_{\sigma,\overline{\delta}}$) or has one of type δ and none of type σ (it is generated by $S_{\overline{\sigma},\delta}$). Let us for clarity replace $S_{\sigma,\overline{\delta}}$ by T_σ and $S_{\overline{\sigma},\delta}$ by T_δ. Then the set T of 2-colorable series-parallel graphs is defined by the equation system:

$$\begin{cases} T = T_\sigma \cup T_\delta, \\ T_\sigma = T_\sigma \parallel T_\sigma \cup T_\sigma \bullet T_\sigma \cup T_\delta \bullet T_\delta, \\ T_\delta = e \cup T_\delta \parallel T_\delta \cup T_\sigma \bullet T_\delta \cup T_\delta \bullet T_\sigma. \end{cases}$$

The construction of this system is based on Properties (1.10). A similar construction can be done for every equation system and every finite inductive set of properties, which proves the Filtering Theorem (Theorem 1.8).

1.3 Monadic second-order logic

We now introduce *monadic second-order logic*, a logical language with which we will specify finite inductive sets of properties. It is actually a favorite language among logicians because it is decidable for many sets of (finite or infinite) structures. Furthermore, it is suitable for expressing numerous graph properties.

1.3.1 Monadic second-order graph properties

We first explain how a graph can be made into a logical structure, hence can be a model of a sentence. For every graph G, we let $\lfloor G \rfloor$ be[12] the relational structure[13] $\langle V_G, edg_G \rangle$ with domain V_G, the set of vertices. Its second component is the binary relation $edg_G \subseteq V_G \times V_G$, such that $(x,y) \in edg_G$ if and only if there exists an edge from x to y if G is directed, and an edge between x and y if G is undirected.

The classical undirected graphs K_n and $K_{n,m}$ are represented by the following relational structures:[14]

---

[12] In some cases, we will write G instead of $\lfloor G \rfloor$.

[13] Relational structures are first-order logical structures without functions of positive arity. See Section 1.9 and Chapter 5 for detailed definitions.

[14] $[n]$ denotes $\{1,\dots,n\}$.

$$\lfloor K_n \rfloor \quad := \quad \langle [n], edg_n \rangle,$$

$$edg_n(x,y) :\Longleftrightarrow x,y \in [n] \text{ and } x \neq y,$$

$$\lfloor K_{n,m} \rfloor \quad := \quad \langle [n+m], edg_{n,m} \rangle$$

$$edg_{n,m}(x,y) :\Longleftrightarrow 1 \leq x \leq n \text{ and } n+1 \leq y \leq n+m, \text{ or}$$

$$1 \leq y \leq n \text{ and } n+1 \leq x \leq n+m.$$

Properties of a graph G can be expressed by *sentences*[15] of relevant logical languages, that are interpreted in $\lfloor G \rfloor$. For example, if G is a directed graph, then

$$\lfloor G \rfloor \models \forall x \exists y, z (edg(y,x) \wedge edg(x,z))$$

if and only if every vertex of G has at least one incoming edge and at least one outgoing edge (we may have $y = z = x$). If G is a simple undirected graph, then we have

$$\lfloor G \rfloor \models \forall x \Big(\neg edg(x,x) \Big) \wedge \neg \exists w,x,y,z \Big(edg(w,x) \wedge edg(x,y) \wedge edg(y,z)$$

$$\wedge \neg edg(w,y) \wedge \neg edg(w,z) \wedge \neg edg(x,z) \Big)$$

if and only if G has no loop and no induced subgraph isomorphic to P_4 (P_4 is the graph $\bullet - \bullet - \bullet - \bullet$). If G is assumed finite and nonempty, this property expresses that it is a cograph. This is an alternative characterization of cographs that has no immediate relation with the grammatical definition given in Section 1.1.2.

A simple graph G is completely defined by the relational structure $\lfloor G \rfloor$: we will say that the representation of G by $\lfloor G \rfloor$ is *faithful*. This representation is not faithful for graphs with multiple edges: the graphs e and $e \parallel e$ (we use here the notation of Section 1.1.3) are not the same but the structures $\lfloor e \rfloor$ and $\lfloor e \parallel e \rfloor$ are. The graph properties expressed by logical formulas via the structures $\lfloor G \rfloor$ are necessarily independent of the multiplicity of edges. We will present in Section 1.8 a representation of a graph G by a relational structure denoted $\lceil G \rceil$ that is faithful, where each edge of G is also an element of the domain of $\lceil G \rceil$. The incidence between edges and vertices is represented by two binary relations in $\lceil G \rceil$ if G is directed and by only one if G is undirected. By using this alternative representation, we will be able to express properties that distinguish multiple edges.

The above two examples use first-order formulas whose variables denote vertices. Monadic second-order formulas have a richer syntax and wider expressive power. They also use variables denoting sets of vertices. Uppercase variables will denote

---

[15] A sentence is a formula without free variables. The notation $S \models \varphi$ means that a sentence φ is true in the relational structure S; in that case S is said to be a model of φ.

sets of vertices, and lowercase variables will denote individual vertices. The property

$$\lfloor G \rfloor \models \exists X \Big(\exists x. x \in X \;\wedge\; \exists y. y \notin X \;\wedge\; \forall x, y (edg(x,y) \Rightarrow (x \in X \Leftrightarrow y \in X)) \Big)$$

holds if and only if G is not connected. (We consider the empty graph as connected.) In this sentence, X is a set variable. Let γ_3 be the sentence

$$\exists X, Y, Z \; \Big(Part(X,Y,Z)$$
$$\wedge \forall x, y \Big(edg(x,y) \;\wedge\; x \neq y \;\Rightarrow\; \neg (x \in X \;\wedge\; y \in X)$$
$$\wedge \neg (x \in Y \;\wedge\; y \in Y) \;\wedge\; \neg (x \in Z \;\wedge\; y \in Z) \Big) \Big)$$

where $Part(X,Y,Z)$ expresses that (X,Y,Z) is a partition[16] of the domain. The formula $Part(X,Y,Z)$ is written as follows:

$$\forall x \Big(\big(x \in X \vee x \in Y \vee x \in Z \big) \;\wedge\; \big(\neg (x \in X \;\wedge\; x \in Y)$$
$$\wedge \neg (x \in Y \;\wedge\; x \in Z) \;\wedge\; \neg (x \in X \;\wedge\; x \in Z) \big) \Big).$$

Then $\lfloor G \rfloor \models \gamma_3$ if and only if G is 3-colorable. That the ends of an edge are in different sets X, Y, Z means that they have different colors.

For each integer k, one can construct a similar sentence γ_k such that, for every graph G, we have $\lfloor G \rfloor \models \gamma_k$ if and only if G is k-colorable.

Many graph constructions can be expressed in terms of basic ones like choosing subsets and computing transitive closures of binary relations. The example of 3-colorability illustrates the first of these basic constructions. We have given a sentence expressing nonconnectivity. Its negation expresses connectivity, hence a property of the transitive closure of the relation $edg_G \cup edg_G^{-1}$. We now give an explicit construction of the transitive closure of an arbitrary binary relation.

Let R be a binary relation that is either a relation of the considered relational structure ($\lfloor G \rfloor$ or a more general one) or is defined by a formula $R(u,v)$ with free variables u and v. We say that a set X is *R-closed* if it satisfies the condition $\forall u, v \big(u \in X \wedge R(u,v) \Rightarrow v \in X \big)$. The formula $\varphi(x,y)$, defined as

$$\forall X (x \in X \wedge \text{“X is R-closed”} \Rightarrow y \in X),$$

where "X is R-closed" is to be replaced by the formula expressing this condition, expresses that (x,y) belongs to R^*, the *reflexive and transitive closure* of R, i.e., that there exists a finite sequence $z_1, \ldots, z_n$, such that $x = z_1, y = z_n$ and $(z_i, z_{i+1}) \in R$ for all $i = 1, \ldots, n-1$. We sketch the proof of this claim. If $x = z_1, y = z_n$ with $(z_i, z_{i+1}) \in R$ for all i, then for every R-closed set X such that x belongs to X, we have $z_i \in X$ for

---

[16] A partition may have empty components.

all $i = 1, \ldots, n$, hence $y \in X$ and $\varphi(x, y)$ holds. Conversely, if $\varphi(x, y)$ holds then one takes $X = \{z \mid (x, z) \in R^*\}$. It is R-closed, hence $y \in X$ and (x, y) belongs to R^*.

In order to have a uniform notation, we denote by $\mathrm{TC}[R; x, y]$ this formula $\varphi(x, y)$. We can use it to build a formula with free variable Y expressing that $G[Y]$, the induced subgraph of G with set of vertices Y, is connected. We let $\mathrm{CONN}(Y)$ be the formula

$$\forall x, y \big(x \in Y \wedge y \in Y \Rightarrow \mathrm{TC}[R; x, y] \big),$$

where R is the relation defined by the formula φ_R with free variables u, v and Y:

$$u \in Y \wedge v \in Y \wedge \big(edg(u, v) \vee edg(v, u) \big).$$

The variable Y is free in φ_R, hence it is also free in the monadic second-order formula $\mathrm{TC}[R; x, y]$. It is clear that the formula $\mathrm{CONN}(Y)$ expresses the desired property. This formula can be used for expressing further properties. The following sentence expresses that an undirected graph G has a cycle with at least three vertices:

$$\exists x, y, z \, \Big(x \neq y \wedge y \neq z \wedge x \neq z \wedge edg(x, z) \wedge edg(z, y)$$
$$\wedge \exists Y \big(z \notin Y \wedge x \in Y \wedge y \in Y \wedge \mathrm{CONN}(Y) \big) \Big).$$

Together with the expressibility of connectivity, we can thus express that a simple undirected graph is a tree.[17]

Aiming at the expression of planarity, we examine the monadic second-order expressibility of minor inclusion. We consider undirected graphs. We say that H is a *minor* of G, denoted by $H \unlhd G$ if and only if H is obtained from a subgraph G' of G by edge contractions. A graph G is planar if and only if it has no minor isomorphic to K_5 or to $K_{3,3}$. (This is a variant due to Wagner of a well-known result by Kuratowski; it is proved in the books [*Die] and [*MohaTho].)

Lemma 1.13 Let H be a simple, loop-free, undirected graph with set of vertices $[n]$. A graph G contains a minor isomorphic to H if and only if there are in G pairwise disjoint nonempty sets of vertices $Y_1, \ldots, Y_n$ such that each graph $G[Y_i]$ is connected and, for every edge of H between i and j, there exists an edge in G between u and v such that $u \in Y_i$ and $v \in Y_j$. $\qquad\square$

Corollary 1.14 For every graph H as in Lemma 1.13, there exists a monadic second-order sentence MINOR_H such that, for every undirected graph G, we have $\lfloor G \rfloor \models \mathrm{MINOR}_H$ if and only if G has a minor isomorphic to H.

---

[17] A tree is a nonempty connected undirected graph without cycles. This last condition implies that a tree has no loops and no multiple edges. The absence of loops is expressed by the sentence $\forall x(\neg edg(x, x))$, but the absence of multiple edges cannot be expressed by a sentence interpreted in $\lfloor G \rfloor$.

Proof: The construction follows from Lemma 1.13. One takes for MINOR$_H$ the following sentence:

$$\exists Y_1, \ldots, Y_n \Big(\bigwedge_{1 \leq i \leq n} \big((\exists y. y \in Y_i) \wedge \text{CONN}(Y_i) \big)$$

$$\wedge \bigwedge_{1 \leq i < j \leq n} \neg \exists y \big(y \in Y_i \wedge y \in Y_j \big)$$

$$\wedge \bigwedge_{(i,j) \in edg_H} \exists u, v \big(u \in Y_i \wedge v \in Y_j \wedge edg(u,v) \big) \Big).$$

■

Corollary 1.15 An undirected graph is planar if and only if it satisfies the sentence $\neg\text{MINOR}_{K_5} \wedge \neg\text{MINOR}_{K_{3,3}}$. □

With this collection of examples, the reader should have a good idea of how one can express graph properties in monadic second-order logic. However, not all graph properties can be expressed in this language. Here are some properties of a graph G and of subsets X, Y of its vertex set V_G that are not monadic second-order expressible.

P_1 : the cardinality $|X|$ of the set X is even;
P_2 : $|X|$ is a prime number;
P_3 : $|X| = |Y|$;
P_4 : G has a nontrivial automorphism;
P_5 : G has a Hamiltonian cycle.

There are some differences between these properties, however, and we have remedies in some cases. For Property P_1, the remedy consists of extending the language by adding a set predicate, $Even(X)$, expressing that the set X has even cardinality. All results that we will prove for monadic second-order logic hold for the extended language called *counting modulo 2 monadic second-order logic*. The notation $C_2\text{MS}$ will refer to it (and MS will refer to formulas written without cardinality predicates).

Property P_5 is actually expressible by a sentence of monadic second-order logic that additionally uses quantifications on sets of edges, and the incidence relations between edges and vertices. This language is based on the representation of a graph G by the richer relational structure than $\lfloor G \rfloor$ that we will define in Section 1.8 and denote by $\lceil G \rceil$. It can be viewed as another extension of monadic second-order logic that we will denote by MS_2, where the index 2 recalls that there are two types of elements in the domain of $\lceil G \rceil$, vertices and edges. There are some significant differences between the languages MS and MS_2, but our main results presented in the next sections and their applications to the construction of fixed-parameter tractable algorithms have formulations that apply to MS_2 as well as to MS.

Concerning the other three properties, there is nothing to do. Adding new set predicates, say $Card_{Prime}(X)$ expressing that $|X|$ is a prime number, or $EqCard(X, Y)$ expressing that $|X| = |Y|$, or $Auto(X)$ expressing that $G[X]$ has a nontrivial automorphism, yields extensions of monadic second-order logic for which the results to be presented in Sections 1.4, 1.5 and 1.6 fail.[18]

1.3.2 Monadic second-order logic and recognizability

Logical sentences express properties of relational structures of the appropriate type. They can also be viewed as finite specifications of sets of structures, namely, their sets of models. We first make precise the corresponding terminology. For a logical language $\mathcal{L}$ (such as MS, C_2MS or MS$_2$), we say that a property of relational structures over a fixed finite set of relation symbols is $\mathcal{L}$-*expressible* if it can be expressed by a sentence of $\mathcal{L}$. A set L of such structures is $\mathcal{L}$-*definable* if the membership of a structure in L is $\mathcal{L}$-expressible. These definitions are applicable to graphs represented by relational structures. Hence, with respect to a fixed representation, we will say that a graph property is $\mathcal{L}$-expressible. Examples have been given above. Let $\mathcal{C}$ be a set of graphs; an element of $\mathcal{C}$ will be called a $\mathcal{C}$-graph. We will say that a set of graphs $L \subseteq \mathcal{C}$ is an $\mathcal{L}$-*definable subset of* $\mathcal{C}$ (or, an $\mathcal{L}$-definable set of $\mathcal{C}$-graphs) if the membership of a graph in L is $\mathcal{L}$-expressible *and the considered representation is faithful for* $\mathcal{C}$-*graphs*. Hence, the connectedness of a graph G is MS-expressible with respect to its representation by $\lfloor G \rfloor$, but the set of connected graphs is not an MS-definable set of graphs, because this representation is not faithful for graphs with multiple edges. On the other hand, the set of connected simple graphs is an MS-definable set of simple graphs. In the first case $\mathcal{C}$ is the set of all graphs, and in the second case it is the set of all simple graphs.

Let F be a finite signature. There is a bijection between $T(F)$ and a set of labeled trees that are simple graphs. It follows that every term t in $T(F)$ can be faithfully represented by a relational structure $\lfloor t \rfloor$ over a finite set of relations (the binary edge relation of the tree, and a unary relation for each label). We say that a set $L \subseteq T(F)$ is *MS-definable* if there exists an MS sentence φ such that $L = \{t \in T(F) \mid \lfloor t \rfloor \models \varphi\}$. Since the property that a finite relational structure is isomorphic to $\lfloor t \rfloor$ for some term t in $T(F)$ is itself MS-expressible, the set $L \subseteq T(F)$ is MS-definable[19] if and only if the set of relational structures that are isomorphic to some structure $\lfloor t \rfloor$ for t in L is MS-definable.

The following fundamental theorem is due to Doner [Don] and to Thatcher and Wright [ThaWri] (see Section 1.10 for related references). We will prove it in Chapters 5 and 6.

---

[18] See the end of Section 7.5.
[19] Using MS$_2$ or C_2MS sentences for defining sets of terms would not yield a wider class of definable sets.

Theorem 1.16 A set of terms over a finite signature is MS-definable if and only if it is recognizable, i.e., accepted by a finite automaton. □

For the two graph algebras $\mathbb{G}^{\mathrm{u}} := \langle \mathcal{G}^{\mathrm{u}}, \oplus, \otimes, \mathbf{1} \rangle$ and $\mathbb{J}_2^{\mathrm{d}} := \langle \mathcal{J}_2^{\mathrm{d}}, /\!\!/, \bullet, e \rangle$ whose operations are defined respectively in Sections 1.1.2 and 1.1.3, we have the following results:

Proposition 1.17 Every MS-definable subset of $\mathcal{G}^{\mathrm{u}}$ is recognizable in $\mathbb{G}^{\mathrm{u}}$. Every MS_2-definable subset of $\mathcal{J}_2^{\mathrm{d}}$ is recognizable in $\mathbb{J}_2^{\mathrm{d}}$. □

This proposition is a corollary of the Recognizability Theorem (stated below in Section 1.4.3) that applies to algebras that extend $\mathbb{G}^{\mathrm{u}}$ and $\mathbb{J}_2^{\mathrm{d}}$.

1.4 Two graph algebras

Up to now, we have only given two examples of graph algebras, the algebra $\mathbb{G}^{\mathrm{u}}$ in which we have defined the cographs and the algebra $\mathbb{J}_2^{\mathrm{d}}$ in which we have defined the series-parallel graphs. These algebras are subalgebras of two larger algebras that we now define. They will differ by the way in which graphs will be composed: the first algebra has operations that "bridge" two disjoint graphs by creating edges between them (vertex labels determine how these edges are created) and the second one has operations that "glue" two disjoint graphs by fusing certain vertices specified by labels. The operations from which cographs and series-parallel graphs are defined illustrate these two types of graph composition.

1.4.1 The algebra of simple graphs with ports

Our first graph algebra, called the *VR algebra*,[20] generalizes the algebra $\mathbb{G}^{\mathrm{u}}$. In order to define more powerful edge creating operations than $\otimes$, we will use vertex-labeled graphs. We let $\mathcal{A}$ be a countable set of labels. In this overview chapter, we take it equal to $\mathcal{N}$, the set of nonnegative integers. This will simplify some statements; in Chapter 2 we will assume that $\mathcal{N} \subseteq \mathcal{A}$. We let $\mathcal{G}$ be the set of (abstract) simple directed graphs.[21] This set contains $\mathcal{G}^{\mathrm{u}}$ because for simple graphs we consider an undirected edge as a pair of directed opposite edges. This is coherent with the representation of a graph G by $\lfloor G \rfloor$ defined in Section 1.3.1. We let $\mathcal{GP}$ be the set of (abstract) *graphs with ports*, or *p-graphs* for short, defined as pairs $G = \langle G^\circ, port_G \rangle$, where $G^\circ \in \mathcal{G}$ and $port_G$ is a mapping $V_{G^\circ} \to \mathcal{A}$. If $port_G(u) = a \in \mathcal{A}$ we say that u is an *a-port* of G and that a is its *port label*. Every graph in $\mathcal{G}$ will be considered as the p-graph, all

---

[20] We give it this name because its equational sets, the VR-*equational sets of graphs*, are the sets of graphs generated by certain context-free graph grammars whose rewritings are based on *vertex replacement*. See [*EngRoz] or [*Eng97] for comprehensive surveys.

[21] "Abstract" means that two isomorphic graphs are considered as equal. This notion will be formalized in Chapter 2.

vertices of which are 1-ports; hence $\mathcal{G} \subseteq \mathcal{GP}$. The operations on $\mathcal{GP}$ are the following ones. First, the disjoint union:[22]

$$G \oplus H := \langle G° \oplus H°, port_G \cup port_H \rangle.$$

Then we define unary operations that manipulate port labels. For $a, b \in \mathcal{A}$ and $G = \langle G°, port_G \rangle$ we let

$$relab_{a \to b}(G) := \langle G°, port \rangle, \text{ where}$$
$$port(u) := \texttt{if } port_G(u) = a \texttt{ then } b \texttt{ else } port_G(u).$$

The next unary operations add directed edges.[23] For $a \neq b$, we define

$$\overrightarrow{add}_{a,b}(G) \text{ as } \langle G', port_G \rangle, \text{ where } V_{G'} = V_G \text{ and}$$
$$edg_{G'} \text{ is } edg_{G°} \cup \{(u, v) \mid (port_G(u), port_G(v)) = (a, b)\}.$$

This operation adds an edge linking u to v whenever u is an a-port and v is a b-port, unless there already exists one (we only consider simple graphs). It does not add loops.

We let **1** be a constant symbol denoting a single vertex that is a 1-port and we let $\mathbf{1}^\ell$ denote the same graph with a loop. The only way to define loops is by means of these constant symbols. Moreover, we let $\varnothing$ be a constant symbol denoting the empty graph, that we denote also by $\varnothing$. We denote empty sets[24] by the different symbol $\emptyset$.

We obtain the algebra $\mathbb{GP}$ of (abstract) p-graphs, also called the *VR algebra*. Its domain is $\mathcal{GP}$ and its signature, denoted by F^{VR}, is the countable set of operations introduced above. For every term $t \in T(F^{VR})$, we denote by $t_{\mathbb{GP}}$ its value, which is a p-graph, computed according to the definitions of the operations of F^{VR}. The equational sets of $\mathbb{GP}$ are called the VR-*equational sets*.

The complete join $G \otimes H$ can be defined in terms of the operations of F^{VR} as follows:

$$G \otimes H := relab_{2 \to 1}(\overrightarrow{add}_{2,1}(\overrightarrow{add}_{1,2}(G \oplus relab_{1 \to 2}(H)))),$$

hence by the term $relab_{2 \to 1}(\overrightarrow{add}_{2,1}(\overrightarrow{add}_{1,2}(x_1 \oplus relab_{1 \to 2}(x_2))))$. Such an operation, defined by a term with variables, is called a *derived operation* of the relevant algebra, here $\mathbb{GP}$.[25] Note that although this operation uses the port label 2, it transforms two

---

[22] We assume G and H are disjoint. If they are not, for instance if $H = G$, we replace H by an isomorphic copy disjoint from G. It follows that $\oplus$ is a well-defined binary function on isomorphism classes of graphs, i.e., on abstract graphs.

[23] To add undirected edges, we use $\overrightarrow{add}_{b,a}(\overrightarrow{add}_{a,b}(G))$. We will denote by $add_{a,b}$ the unary operation that transforms G into $\overrightarrow{add}_{b,a}(\overrightarrow{add}_{a,b}(G))$.

[24] We consider that an empty set of numbers is not equal to an empty set of words.

[25] The unary operation $add_{a,b}$ is also a derived operation of $\mathbb{GP}$, defined by the term $\overrightarrow{add}_{b,a}(\overrightarrow{add}_{a,b}(x_1))$. The operations r of the P-algebra $\mathbb{M}'$ in Section 1.1.5 are derived operations of the F-algebra $\mathbb{M}$, defined by terms t_r.

graphs G, H in $\mathcal{G}$ ($\mathcal{G}$ is the set of graphs, all vertices of which are 1-ports) into a graph in the same set.

Systems of equations (that define VR-equational sets) can frequently be written more clearly with the help of derived operations. For example, Equation (1.1) in Section 1.1.2 can be written as the following equation:

$$C = (C \oplus C) \cup relab_{2 \to 1}(\overrightarrow{add}_{2,1}(\overrightarrow{add}_{1,2}(C \oplus relab_{1 \to 2}(C)))) \cup \mathbf{1},$$

where C defines a subset of $\mathcal{GP}$, but it is more readable as in (1.1).

For each $k \in \mathcal{N}$ we denote by $F_{[k]}^{\mathrm{VR}}$ the finite subsignature of F^{VR} consisting of the operations written with port labels in $[k]$.[26] Hence

$$F_{[k]}^{\mathrm{VR}} := \{\oplus, relab_{a \to b}, \overrightarrow{add}_{a,b}, \mathbf{1}, \mathbf{1}^{\ell}, \varnothing \mid 1 \leq a, b \leq k\}.$$

It is not hard to see that every p-graph with n vertices and port labels in $[n]$ is the value of a term t in $F_{[n]}^{\mathrm{VR}}$. In many cases, however, much fewer labels suffice, and for algorithmic applications it is useful to use as few labels as possible.

In this perspective, we define the *clique-width* of a graph G, either directed or undirected, as the minimum k such that G is the value of a term in $T(F_{[k]}^{\mathrm{VR}})$. This number is denoted by $cwd(G)$. Trees have clique-width at most 3. Cographs have clique-width at most 2: this is clear because the above equation that defines cographs uses only two port labels.

Proposition 1.18 Every VR-equational set of graphs has bounded clique-width. For each k, the set of graphs of clique-width at most k is VR-equational.

Proof sketch: Let L be a VR-equational set of graphs. It is defined by an equation system written with port labels in some set $[k]$. It follows that all p-graphs belonging to the components of its least solution are the values of terms in $F_{[k]}^{\mathrm{VR}}$. In particular, the graphs in L have clique-width at most k.

Conversely consider the sets of p-graphs S and T defined by the following equation system:

$$\begin{cases} S = (S \oplus S) \cup \bigcup f(S) \cup \mathbf{1} \cup \mathbf{1}^{\ell} \cup \varnothing, \\ T = relab_{2 \to 1}(\cdots(relab_{k \to 1}(S)\cdots)) \end{cases} \tag{1.11}$$

where the union extends to all unary operations f belonging to $F_{[k]}^{\mathrm{VR}}$. In the equation that defines T, all port labels are transformed into 1. The set S consists of all p-graphs defined by terms in $T(F_{[k]}^{\mathrm{VR}})$ and the set T consists of those whose vertices are all 1-ports, hence of all graphs of clique-width at most k. ∎

---

[26] Recall that $[k] = \{1, \ldots, k\}$ for $k \in \mathcal{N}$ with $[0] = \emptyset$.

At the cost of some technicalities that we want to avoid here, the notion of clique-width and Proposition 1.18 can be extended to graphs with ports. However, port labels are just a tool to construct graphs. Hence Proposition 1.18 states the main facts about the relationship between VR-equational sets of graphs and clique-width.

The CLIQUE-WIDTH CHECKING problem consists of deciding whether or not $cwd(G) \leq k$ for given (G,k). It is NP-complete ([FelRRS]). It is not known whether this problem is polynomial for fixed k, when $k \geq 4$, but it is when $k \leq 3$ ([CorHLRR]).

1.4.2 The algebra of graphs with sources

We now define an algebra of graphs with multiple edges, called the *HR algebra*,[27] that extends the algebra $\mathbb{J}_2^d = \langle \mathcal{J}_2^d, /\!/, \bullet, e \rangle$ considered in Section 1.1.3.

We consider (abstract) directed or undirected graphs, possibly with multiple edges. They form the set $\mathcal{J}$. For a graph in $\mathcal{J}$, E_G denotes its set of edges (and V_G its set of vertices). We let $\mathcal{A}$ be a countable set of labels (as in Section 1.4.1 we take it equal to $\mathcal{N}$) that will be used to distinguish particular vertices. These distinguished vertices will be called *sources*, and $\mathcal{A}$ is the set of *source labels*. (This notion of source is unrelated with edge directions.)

A *graph with sources*, or *s-graph* for short, is a pair $G = \langle G^\circ, src_G \rangle$, where $G^\circ \in \mathcal{J}$ and src_G is a bijection from a finite subset $\tau(G)$ of $\mathcal{A}$ to a subset of V_{G°. We call $\tau(G)$ the *type* of G and $src_G(\tau(G))$ the set of its sources. The vertex $src_G(a)$ is called the *a-source* of G; its *source label*, also called its *source name*, is a.

We let $\mathcal{JS}$ denote the set of s-graphs; $\mathcal{J}$ is thus the set of s-graphs having an empty type. Clearly, $\mathcal{J}_2^d \subseteq \mathcal{JS}$ and the elements of $\mathcal{J}_2^d$ are s-graphs of type $\{1,2\}$. We define operations on $\mathcal{JS}$: first a binary operation called the *parallel-composition*, a particular case of which has been defined in Section 1.1.3. For $G, H \in \mathcal{JS}$ we let

$$G /\!/ H := \langle G \cup H', src_G \cup src_{H'} \rangle,$$

where H' is isomorphic to H[28] and is such that

$$E_{H'} \cap E_G = \emptyset,$$
$$src_G(a) = src_{H'}(a) \text{ if } a \in \tau(G) \cap \tau(H'),$$
$$V_G \cap V_{H'} = \{src_G(a) \mid a \in \tau(G) \cap \tau(H')\}.$$

This operation "glues" G and a disjoint copy of H by fusing their sources that have the same names. The s-graph $G /\!/ H$ is well defined up to isomorphism. Its type is $\tau(G) \cup \tau(H)$.

---

[27] We give it this name because its equational sets, the HR-*equational sets of graphs*, are the sets of graphs generated by certain context-free graph grammars whose rewritings are based on *hyperedge replacement*. We will define them in Chapter 4, Section 4.1.5. For a thorough study, see [*DreKH] or [*Hab].

[28] H' isomorphic to H implies that $\tau(H') = \tau(H)$.

We define unary operations that manipulate source labels. Let $G = \langle G^\circ, src_G \rangle$ be an s-graph. For $a \in \mathcal{A}$, we let $fg_a(G) := \langle G^\circ, src' \rangle$, where $src'(b)$ is $src_G(b)$ if $b \in \tau(G) - \{a\}$ and is undefined otherwise. We say that fg_a *forgets* the a-source: if G has an a-source, then this vertex is no longer distinguished as a source in $fg_a(G)$ and is turned into an "ordinary," or *internal*, vertex. Hence $\tau(fg_a(G)) = \tau(G) - \{a\}$.

The next operation modifies source names; it is called a *renaming*. For $a, b \in \mathcal{A}$, $a \neq b$, we let

$$ren_{a \leftrightarrow b}(G) := \langle G^\circ, src' \rangle, \text{ where}$$
$$src'(a) := src_G(b) \quad \text{if } b \in \tau(G),$$
$$src'(b) := src_G(a) \quad \text{if } a \in \tau(G),$$
$$src'(c) := src_G(c) \quad \text{if } c \in \tau(G) - \{a, b\},$$

and src' is undefined otherwise. Hence

$$\tau(ren_{a \leftrightarrow b}(G)) = \begin{cases} \tau(G) & \text{if } a, b \in \tau(G) \text{ or } \tau(G) \cap \{a, b\} = \emptyset, \\ (\tau(G) - \{a\}) \cup \{b\} & \text{if } a \in \tau(G),\ b \notin \tau(G), \\ (\tau(G) - \{b\}) \cup \{a\} & \text{if } b \in \tau(G),\ a \notin \tau(G). \end{cases}$$

We can write this more succinctly as $\tau(ren_{a \leftrightarrow b}(G)) = \tau(G)[b/a, a/b]$, where, for every set C, we denote by $C[c/a, d/b]$ the result of the simultaneous replacement in C of a by c and of b by d.

We also define constant symbols: $\mathbf{ab}$, $\overrightarrow{\mathbf{ab}}$, $\mathbf{a}$, $\mathbf{a}^\ell$, $\varnothing$ to denote respectively an undirected edge linking an a-source and a b-source, a directed edge from an a-source to a b-source, a single vertex that is an a-source, an a-source with a loop, and the empty graph. (Since we can change source names, it would suffice to use the constant symbols $\mathbf{12}$, $\overrightarrow{\mathbf{12}}$, $\mathbf{1}$, $\mathbf{1}^\ell$, $\varnothing$. However, using many renaming operations would make terms denoting graphs unreadable.)

We obtain a countable signature denoted by F^{HR} and an F^{HR}-algebra of graphs denoted by $\mathbb{JS}$ and called the *HR algebra*. It has domain $\mathcal{JS}$. As for $\mathbb{GP}$, for every term $t \in T(F^{\mathrm{HR}})$, we denote by $t_{\mathbb{JS}}$ the s-graph that is its value. The equational sets of $\mathbb{JS}$ are called the HR-*equational sets*.

As for VR-equational sets, the equations that define HR-equational sets can be shortened if they are written with derived operations. For example, the series-composition of graphs of type $\{1, 2\}$ is a derived operation that can be expressed by

$$G \bullet H := fg_3(ren_{2 \leftrightarrow 3}(G) \,\|\, ren_{1 \leftrightarrow 3}(H)).$$

We denote by $F^{\mathrm{HR}}_{[k]}$ the finite subsignature of F^{HR} consisting of the operations $\|, fg_a, ren_{a \leftrightarrow b}$, for $a, b \in [k]$ and of the constant symbols denoting graphs with source names in $[k]$.

Each s-graph G with n vertices and source names in $[n]$ is the value of a term in $T(F^{\mathrm{HR}}_{[n]})$. The least integer k such that a graph G (without sources) is the value of a term in $T(F^{\mathrm{HR}}_{[k+1]})$ is a well-known graph complexity measure called the *tree-width* of G and denoted by $twd(G)$. It has been defined previously in a combinatorial way by Robertson and Seymour [RobSey90] and by other authors using a different terminology. In the combinatorial definition, $twd(G)$ is the least integer k such that G has a so-called *tree-decomposition* of *width* k. A tree-decomposition of G is a decomposition of G into a tree of subgraphs, where each node of the tree corresponds to a subgraph of G; its width is the maximal number of vertices of those subgraphs, minus 1. The notion of tree-width is important for the construction of graph algorithms and for the study of graph minors: see the books [*Die], [*DowFel] and [*FluGro], and the survey articles [*Bod93] and [*Bod98]. We will study this combinatorial notion in Chapter 2 and prove the following result which is an algebraic characterization of it:

Proposition 1.19 A graph has tree-width at most k if and only if it is the value of a term in $T(F^{\mathrm{HR}}_{[k+1]})$. $\qquad\square$

By contrast clique-width, defined in terms of graph operations, has yet no alternative combinatorial definition. Using Proposition 1.19 the following result can be proved in the same way as Proposition 1.18.

Proposition 1.20 Every HR-equational set of graphs has bounded tree-width. For each k, the set of graphs of tree-width at most k is HR-equational. $\qquad\square$

1.4.3 A weak Recognizability Theorem

The VR algebra $\mathbb{GP}$ has a countable signature F^{VR} that generates it: this means that each element is the value of some term. For each k in $\mathcal{N}$, we let $\mathbb{GP}^{\mathrm{gen}}[k]$ be the subalgebra of $\mathbb{GP}$ that is generated by the finite subsignature $F^{\mathrm{VR}}_{[k]}$ of F^{VR}. Its domain $\mathcal{GP}^{\mathrm{gen}}[k]$ consists of the graphs with ports that are values of terms in $T(F^{\mathrm{VR}}_{[k]})$. We define similarly $\mathbb{JS}^{\mathrm{gen}}[k]$ as the subalgebra of the HR algebra $\mathbb{JS}$ that is generated by the finite subsignature $F^{\mathrm{HR}}_{[k]}$ of F^{HR}. Its domain is $\mathcal{JS}^{\mathrm{gen}}[k]$. Proposition 1.17 extends into the following theorem:

Theorem 1.21 (Weak Recognizability Theorem)
(1) Let L be a CMS-definable set of simple graphs. For every $k \in \mathcal{N}$, the set $L \cap \mathcal{GP}^{\mathrm{gen}}[k]$ is recognizable in the algebra $\mathbb{GP}^{\mathrm{gen}}[k]$.
(2) Let L be a CMS$_2$-definable set of graphs. For every $k \in \mathcal{N}$, the set $L \cap \mathcal{JS}^{\mathrm{gen}}[k]$ is recognizable in the algebra $\mathbb{JS}^{\mathrm{gen}}[k]$. $\qquad\square$

In this statement, CMS refers to *counting monadic second-order* logic, i.e., the use in monadic second-order formulas of set predicates $Card_p(X)$ expressing that $|X|$ is a multiple of p (*Even*(X) is thus $Card_2(X)$). Note that $L \cap \mathcal{GP}^{\mathrm{gen}}[k]$ is the set of graphs in L that have clique-width at most k, and similarly for $L \cap \mathcal{JS}^{\mathrm{gen}}[k]$ and tree-width at most $k - 1$.

This theorem is a consequence of the (more powerful) Recognizability Theorem to be stated and proved in Chapter 5. The latter theorem is more powerful in several respects. First it yields recognizability with respect to the algebras $\mathbb{GP}$ and $\mathbb{JS}$ that have infinite signatures and not only with respect to their finitely generated subalgebras $\mathbb{GP}^{\text{gen}}[k]$ and $\mathbb{JS}^{\text{gen}}[k]$. We postpone to Chapters 3 and 4 the detailed definitions. Second, the Recognizability Theorem can be stated and proved for an algebra of relational structures denoted by $\mathbb{STR}$, as a unique statement that entails the cases of $\mathbb{GP}$ and $\mathbb{JS}$.

However, Theorem 1.21 already leads to some interesting consequences. We first state the logical version of the Filtering Theorem and its applications to decidability results for monadic second-order sentences. Applications to fixed-parameter tractability will be considered in the next section. We will discuss decidability results in more detail in Section 1.6.

Theorem 1.22 (Filtering Theorem, logical version)
(1) For every VR-equational set of graphs L and every CMS-expressible graph property P, the set L_P consisting of the graphs of L that satisfy P is VR-equational.
(2) The analogous result holds for HR-equational sets and CMS_2-expressible properties. $\square$

This result is a direct consequence of Theorems 1.8, 1.12 and 1.21; note that each VR-equational set is equational in some algebra $\mathbb{GP}^{\text{gen}}[k]$, cf. the proof of Proposition 1.18. All constructions are effective: an equation system defining L_P can be constructed from one defining L and a sentence expressing P. Now consider Corollary 1.9 and its proof. Since the emptiness of an equational set is decidable, one can decide if L_P is nonempty, i.e., if P is satisfied by some graph in L: the CMS-*satisfiability problem* (resp. the CMS_2-*satisfiability problem*) is decidable for VR-equational sets (resp. for HR-equational sets) and, in particular, for the sets $CWD(\leq k)$ of graphs of clique-width at most k (resp. for the sets $TWD(\leq k)$ of graphs of tree-width at most k).

We now come back to statements (1) and (2) at the end of Section 1.2.2. The set $L_{\neg P}$ (we use the notation of Theorem 1.22) is also VR-equational (respectively, HR-equational in the second case). Its emptiness can be tested. That $L_{\neg P}$ is empty means that P is universally valid on L. From the equation system defining $L_{\neg P}$, one can also obtain a proof of this fact by fixed-point induction, according to the generalization of Proposition 1.6 to recognizable sets. As observed in Section 1.2.5, if P belongs to a finite inductive set $\mathcal{P}$ of properties, then the auxiliary properties used in that proof are Boolean combinations of the properties in $\mathcal{P}$. In Chapter 5, the proof of Theorem 1.21 shows that, more precisely, every CMS-expressible property P belongs to a finite inductive set $\mathcal{P}$ of *CMS-expressible* properties in the algebra $\mathbb{GP}^{\text{gen}}[k]$ (and similarly for CMS_2 and $\mathbb{JS}^{\text{gen}}[k]$). Hence, the auxiliary properties in the proof by fixed-point induction are CMS-expressible (respectively CMS_2-expressible).

1.5 Fixed-parameter tractability

In this section, we describe some algorithmic applications of the Weak Recognizability Theorem. We first recall the basic definition of *fixed-parameter tractability*. This notion has been introduced in order to describe in an abstract setting the frequently met situation where the computation time of an algorithm is bounded by an expression of the form $f(p(d)) \cdot |d|^c$, where the set of inputs is equipped with two computable integer valued functions, a *size* function $d \mapsto |d|$ (d is the generic input) and a *parameter* function $d \mapsto p(d)$, such that $0 \leq p(d) \leq |d|$, and where f is a fixed computable function of nonnegative integers and c is a fixed positive integer. If these conditions are satisfied, the considered algorithm is *fixed-parameter tractable with respect to the parameter p*. If $c = 1, 2$ or 3, we say that it is *fixed-parameter linear, quadratic* or *cubic*, respectively.

As size of an input graph G, we will use either the number of its vertices and edges, denoted by $\|G\|$, or, in particular if G is simple, its number of vertices. Parameters can be the degree of G, its tree-width or its clique-width. Our results will use these last two values; other examples can be found in the books [*DowFel] and [*FluGro] which present in detail the theory of fixed-parameter tractability.

The size $|t|$ of a term t or the size $|\varphi|$ of a formula φ is, roughly speaking, the number of symbols with which t or φ is written, i.e., the length of the corresponding word.[29]

Example 1.23 The *subgraph isomorphism problem* consists in deciding for a pair of simple graphs (G, H) whether H is isomorphic to a subgraph of G. It is NP-complete (Problem GT48 in [*GarJoh]). For each fixed graph H, this problem can be solved in time $O(n^m)$, where $n = |V_G|$ and $m = |V_H|$. However, it is fixed-parameter linear with respect to $Deg(G)$, the degree of G. This follows from a result by Seese [See96] (another proof is given in [DurGra]) saying that for every first-order sentence φ, one can decide in time bounded by $f(|\varphi| + Deg(G)) \cdot |V_G|$ whether $\lfloor G \rfloor$ satisfies φ, for some fixed computable function f. The property that G has a subgraph isomorphic to a fixed graph H is expressible by a first-order sentence φ_H to be interpreted in $\lfloor G \rfloor$. Hence the subgraph isomorphism problem can be solved in time at most $f(|\varphi_H| + Deg(G)) \cdot |V_G|$. $\square$

Fixed-parameter tractable algorithms that check monadic second-order graph properties can be derived from Theorems 1.16 and 1.21. The *model-checking problem* for a logical language $\mathcal{L}$ and a class of structures $\mathcal{C}$ consists of deciding whether $S \models \varphi$, for a given structure $S \in \mathcal{C}$ and a given sentence $\varphi \in \mathcal{L}$.

Theorem 1.24 For every finite signature F, every monadic second-order sentence φ and every term t in $T(F)$, one can decide in time at most $f(F, \varphi) \cdot |t|$ whether $\lfloor t \rfloor \models \varphi$, where f is a fixed computable function.

---

[29] Formal definitions of $|t|$ and $|\varphi|$ will be given in Chapters 2 and 5, respectively.

Proof: The proof of Theorem 1.16 being effective, it yields an algorithm that constructs from F and φ a finite deterministic F-automaton $\mathscr{A}$ accepting the set of terms s in $T(F)$ such that $\lfloor s \rfloor \models \varphi$. By running $\mathscr{A}$ on a given term t, one gets the answer in time proportional to $|t|$. The total computation time is thus $f_1(F,\varphi) + f_2(F,\varphi) \cdot |t|$, where $f_1(F,\varphi)$ is the time taken to compute $\mathscr{A}$ and $f_2(F,\varphi)$ is the maximum time taken by $\mathscr{A}$ to perform one transition.   ∎

In other words, the model-checking problem for monadic second-order sentences and the class of terms is fixed-parameter linear with respect to $\|F\| + |\varphi|$, where $\|F\|$ is the sum of arities of the symbols in F plus the number of constant symbols in F. This follows from Theorem 1.24 because, up to the names of the symbols in F and of the variables in φ, there are finitely many pairs (F,φ) such that $\|F\| + |\varphi| \le p$, so that $f(F,\varphi)$ can be bounded by $g(\|F\| + |\varphi|)$ for some computable function g.

We now consider the model-checking problem for monadic second-order sentences on graphs and its two possible parametrizations by tree-width and clique-width. For a graph G given by a term t in $T(F_{[k]}^{\mathrm{HR}})$ or in $T(F_{[k]}^{\mathrm{VR}})$, one gets a fixed-parameter linear algorithm as in Theorem 1.24 because, by Theorems 1.21 and 1.12, one can construct from k and φ a finite deterministic $F_{[k]}^{\mathrm{HR}}$-automaton $\mathscr{A}_{\varphi,k}$ that accepts the set of terms t such that $t_{\mathbb{JS}}$ satisfies φ, and similarly with $F_{[k]}^{\mathrm{VR}}$. This situation happens in particular if G belongs to an HR- or VR-equational set of graphs and is given by a corresponding term or derivation tree (cf. Section 1.1.5). However, if such a term or derivation tree is not given, it must be computed from the input graph by a parsing algorithm.

Theorem 1.25
(1) The model-checking problem for CMS sentences and the class of simple graphs is fixed-parameter cubic with respect to $cwd(G) + |\varphi|$. The input sentence is φ and the size of the input graph G is its number of vertices.
(2) The model-checking problem for CMS_2 sentences and the class of graphs is fixed-parameter linear with respect to $twd(G) + |\varphi|$. The input sentence is φ and the size of the input graph G is its number $\|G\|$ of vertices and edges.[30]

Proof: We first consider the parametrization by tree-width. We let n be the number of vertices and edges of the input graph G, i.e., $n := \|G\|$. There exists an algorithm (by Bodlaender [Bod96], see also [*DowFel]) that, for every graph G and integer k, decides if $twd(G) \le k$ in time at most $g(k) \cdot |V_G|$ (for some fixed computable function g), and that constructs if possible a tree-decomposition of width k of G; this tree-decomposition can be converted in linear time (in n) into a term t in $T(F_{[k+1]}^{\mathrm{HR}})$ that evaluates to G, cf. Proposition 1.19.

---

[30] In many cases, φ is fixed because one is interested in a particular graph property, and then the parameters are just tree-width and clique-width.

Step 1: For given G and by repeating this algorithm for $k = 1, 2, \ldots$ at most $twd(G)$ times one obtains an optimal tree-decomposition of G. This computation takes time at most $twd(G) \cdot g'(twd(G)) \cdot n$ for some fixed computable function g' and builds a term in $T(F)$ where $F := F^{HR}_{[twd(G)+1]}$, that evaluates to G. (The function g' takes into account the time needed to transform a tree-decomposition into a term.)

Step 2: By using Theorems 1.21 and 1.12, one constructs a finite deterministic F-automaton $\mathscr{A}_{\varphi,k}$, where $k = twd(G) + 1$, which accepts the set of terms $t \in T(F)$ that evaluate to a graph satisfying φ.

Step 3: By running this automaton on t, one obtains the answer.

We now consider the parametrization by clique-width. The main difference concerns Step 1. There exists an algorithm (that combines algorithms from [HliOum] and [OumSey], see Section 6.2.3 for details) that, for every simple graph G and every integer k, either reports (correctly) that $cwd(G) > k$ or constructs a term in $T(F^{VR}_{[h(k)]})$ that evaluates to G (and hence G has clique-width at most $h(k)$), where $h(k) = 2^{k+1} - 1$. This algorithm takes time $g''(k) \cdot n^3$ for some fixed computable function g'', where $n = |V_G|$. Note that it does not determine the exact clique-width of G.

Step 1 of the case of tree-width is replaced by the following: for given G and by repeating this algorithm for $k = 1, 2, \ldots$ at most $cwd(G)$ times, one obtains a term t in $T(F^{VR}_{[m]})$ that evaluates to G for some $m \le h(cwd(G))$. This computation takes time at most $cwd(G) \cdot f(cwd(G)) \cdot n^3$, where $f(k)$ is the maximum of $g''(i)$ for $i = 1, \ldots, k$. The computation continues then by constructing a finite $F^{VR}_{[m]}$-automaton (by Theorems 1.21 and 1.12) and running it on t like in Steps 2 and 3 above. ∎

The algorithms of Theorem 1.25 are actually not directly implementable because of the sizes of the automata to be constructed. Specifically, the automata $\mathscr{A}_{\varphi,k}$ in the two cases of Theorem 1.25 have a number of states that is not bounded by a function of the form $\exp \circ \exp \circ \cdots \circ \exp(|\varphi|)$ with a fixed number of iterated exponentiations ($\exp(n) = 2^n$ for every n). (This is also the case for the automaton constructed in the proof of Theorem 1.24.) This fact is not a weakness of the construction, but a consequence of the fact that complicated properties can be expressed by short formulas. This phenomenon occurs for first-order as well as for monadic second-order logic, as proved in [FriGro04]. Concrete constructions of $F^{VR}_{[k]}$-automata for small values of k and simple graph properties will be presented in Section 6.3.

A second reason that makes these algorithms difficult, if not impossible, to implement so as to run for arbitrary graphs is the time needed for parsing the input graphs, i.e., for building terms in $T(F^{HR}_{[k]})$ or in $T(F^{VR}_{[k]})$ for given values of k that evaluate to them. The linear algorithm by Bodlaender [Bod96] used in the proof of Theorem 1.25 takes time $2^{32k^3} \cdot n$. We will review other more efficient, even if not linear, algorithms in Chapter 6. The cubic algorithm of [HliOum] is not implementable either.

1.6 Decidability of monadic second-order logic

Apart from model-checking discussed in the previous section, another major problem in Logic consists in deciding whether a given sentence holds in some relational structure (or in a graph represented by a relational structure) of a fixed set L. In this case, the input of the problem is an arbitrary sentence from a logical language $\mathcal{L}$. This problem is called the $\mathcal{L}$-*satisfiability problem* for the set L. A related problem consists in deciding if a given sentence of $\mathcal{L}$ belongs to the $\mathcal{L}$-*theory* of L, that is, to the set of sentences of $\mathcal{L}$ that hold for all graphs (or structures) in L. We say that the $\mathcal{L}$-theory of L is *decidable* if this problem is decidable. As logical languages $\mathcal{L}$, we will consider fragments and extensions of monadic second-order logic that are closed under negation. For such languages, the $\mathcal{L}$-satisfiability problem for a set L is decidable if and only if the $\mathcal{L}$-theory of L is decidable.

The main motivation for the fundamental Theorem 1.16 was to prove the decidability of the MS-theory of the set of terms $T(F)$ (more precisely, the set of structures $\{\lfloor t \rfloor \mid t \in T(F)\}$), for every finite signature F. From Theorem 1.22 we obtain a similar result for graphs, as observed at the end of Section 1.4.3.

Theorem 1.26 The CMS-theory of the set $CWD(\leq k)$ of simple graphs of clique-width at most k, or of a VR-equational set of graphs is decidable. So is the CMS_2-theory of the set $TWD(\leq k)$ of graphs of tree-width at most k, or of an HR-equational set of graphs.   □

One obtains results which are quite powerful in that they apply to many different sets of graphs. One might hope to use them in order to obtain automatic proofs of conjectures or of difficult theorems in graph theory. However, the situation is not so favorable. Let us take the example of the 4-Color Theorem, stating that every planar graph is 4-colorable. In Section 1.3.1 we have shown the existence of two MS sentences, π and γ_4, expressing respectively that a graph is planar (Corollary 1.15) and that it is 4-colorable. The 4-Color Theorem[31] can thus be stated in the following logical form:

Theorem 1.27 We have $\lfloor G \rfloor \models \pi \Rightarrow \gamma_4$ for every graph G in $\mathcal{G}$.   □

This means that the MS sentence $\pi \wedge \neg\gamma_4$ is not satisfiable in $\mathcal{G}$.

Certain conjectures can be formulated in a similar way. For example, a conjecture by Hadwiger states that for every integer k, if a graph is not k-colorable, then it has a minor isomorphic to K_{k+1}. For each k, the corresponding instance of this conjecture is equivalent by Corollary 1.14 to the statement:

Conjecture 1.28 We have $\lfloor G \rfloor \models \neg\gamma_k \Rightarrow \text{MINOR}_{K_{k+1}}$ for every graph G in $\mathcal{G}$.   □

---

[31] Its proof by Robertson *et al.* ([RobSanST]) has been checked by computer by Gonthier [Gon] with the software Coq based on Type Theory.

Robertson *et al.* have proved it in [RobST] for $k = 5$. It is known to hold for smaller values and is otherwise open.

Could one prove the 4-Color Theorem or this Conjecture for some fixed $k \geq 6$ by an algorithm able to check the satisfiability of a monadic second-order sentence?

This is not possible without some further analysis that would limit the size, or the tree-width, or the clique-width of a minimal graph G that could contradict the considered properties.[32] The reason is that the MS-satisfiability problem for the set $\mathcal{G}$ of finite (simple directed) graphs is undecidable: there is no algorithm that would take as input an arbitrary monadic second-order sentence and tell whether this sentence is valid in the logical structure $\lfloor G \rfloor$ for some graph G in $\mathcal{G}$. This undecidability result actually holds for first-order logic (see Theorem 5.5 in Section 5.1.6 or the books [*EbbFlu] and [*Lib04]).

From Theorem 1.26, it follows that the *particular cases* of these theorems or conjectures obtained by restricting to the sets of graphs of clique-width at most k for fixed values of k can, at least in principle, be proved by machine. However, since we observed that the algorithms underlying Theorem 1.26 are not implementable, this possibility is presently purely theoretical. Furthermore, the difficult open questions of graph theory concern usually all graphs rather than graphs of bounded tree-width or clique-width. There are, however, some exceptions. Whether the oriented chromatic number of an oriented graph[33] is equal to k is expressible by a formula of monadic second-order logic (one formula for each k). Several articles, in particular by Sopena [Sop] and Fertin *et al.* [FerRR], determine the maximal value of the oriented chromatic number of outerplanar graphs, of 3-trees, and of the so-called "fat trees" and "fat fat trees." Since these four sets of graphs are HR-equational, the maximal values of the oriented chromatic numbers of their graphs can also be determined, in principle, by algorithms based on Theorem 1.26.

Seese raised in [See91] the question of understanding which conditions on a set of graphs L are necessary for its MS-satisfiability problem to be decidable. The two main results regarding this question are collected in the following theorem:[34]

Theorem 1.29 Let L be a set of finite, simple, undirected graphs.
(1) If L has a decidable MS_2-satisfiability problem, then it has bounded tree-width.
(2) If L has a decidable C_2MS-satisfiability problem, then it has bounded clique-width. □

---

[32] For Conjecture 1.28, Kawarabayashi [Kaw] has found such bounds for all k. It follows that each level of the conjecture is decidable, but by an intractable algorithm.

[33] An oriented graph G (i.e., a graph without loops and pairs of opposite directed edges) has *oriented chromatic number* at most k if there exists a tournament (a complete oriented graph) H with k vertices and a homomorphism $h : G \to H$ that maps V_G into V_H and every directed edge $u \to v$ of G to a directed edge $h(u) \to h(v)$ of H.

[34] We recall (from Section 1.3.1) that the acronym C_2MS refers to MS logic extended by the even cardinality set predicate.

Assertion (1) is proved by Seese in [See91]. He asks in this article whether every set L having a decidable MS-satisfiability problem (which is weaker than having a decidable C_2MS-satisfiability problem) is "tree-like," which is actually equivalent (by the Equationality Theorem for the VR algebra presented below in Section 1.7.3) to having bounded clique-width. Assertion (2) proved in [CouOum] is thus a partial answer to Seese's question.

1.7 Graph transductions

The theory of formal languages studies finite descriptions of languages (sets of words or terms) by grammars and automata, and also finite descriptions of transformations of these objects. Motivations come from the theories of coding, of compilation and of computational linguistics, to cite just a few. The finite devices that specify these transformations are called *transducers* and the corresponding transformations of words and terms are called *transductions*. Typical questions are the following:

> *Is a given class of transductions closed under composition? Is it closed under inverse?*
> *Does it preserve a given family of languages?*
> *Is this family the set of images of a particular language (called a generator) under the transductions of the class?*
> *Is the equality of the transductions defined by two transducers of a certain type decidable?*

In the study of sets of words, *rational transductions* play a prominent role originating from the work by Nivat [Niv]. The inverse of a rational transduction and the composition of two rational transductions are rational transductions. The families of regular and of context-free languages are preserved under rational transductions, and there exist context-free languages (like the one defined by the equation $L = f(L, L) \cup a$, where $L \subseteq A^*$ and A consists of a, f, parentheses and comma), whose images under all rational transductions are all context-free languages. Transductions of words are studied in the books [*Ber] and [*Sak]. Transductions of terms, usually called *tree transductions*, are studied in [*Com+] and [*GecSte].

Transducers are usually based on finite automata or on more complicated devices like macros (see, e.g., the *macro tree transducers* in [EngMan99]). Can one define similar notions for graphs? Since graphs can be denoted by terms, one can use transductions of terms to specify transductions of graphs (as, e.g., in [*Eng94, DreEng]). However, doing this requires the input to be processed in a *parsing* step that is algorithmically difficult (cf. Sections 1.5 and 6.2). For building the output, each term produced by such a transduction must be evaluated into a graph. The main drawback of the detour through terms over (necessarily) finite signatures is that it limits the input and output graphs of transductions to have bounded tree-width or clique-width. It is natural to try to avoid such a detour and to define transducers that work "directly" on graphs, by traversing them according to some rules and by starting from some specified vertex. However, no notion of finite graph automaton has been defined that would generalize conveniently finite automata on words and terms. Monadic second-order logic offers a powerful alternative. In this section, we define *monadic second-order transductions* through examples rather than formally. These transductions have good interactions with the HR- and VR-equational sets (our "context-free sets of graphs") that follow from the *Equationality Theorem* presented below in Section 1.7.3. It will be shown in Chapter 8 that every (functional) monadic second-order transduction of graphs of bounded tree-width or clique-width can be realized by a macro tree transducer on the level of terms.

1.7.1 Examples of monadic second-order transductions

Monadic second-order transductions are transformations of graphs specified by monadic second-order formulas. The basic notion is that of a monadic second-order transduction of relational structures over a fixed set of relation symbols. It applies to graphs faithfully represented by relational structures.

All examples and results presented in this section (Section 1.7) will concern simple graphs, for which $G \mapsto \lfloor G \rfloor$ is a faithful representation. Hence, we will consider mappings f from simple graphs to simple graphs such that, for every G, the structure $\lfloor f(G) \rfloor$ representing its image under f is defined from $\lfloor G \rfloor$ by monadic second-order formulas.

The simplest case is when

$$\lfloor f(G) \rfloor := \langle V_{f(G)}, edg_{f(G)} \rangle,$$

$$V_{f(G)} := \{u \in V_G \mid \lfloor G \rfloor \models \delta(u)\},$$

$$edg_{f(G)} := \{(u,v) \in V_G \times V_G \mid \lfloor G \rfloor \models \delta(u) \wedge \delta(v) \wedge \theta(u,v)\},$$

where δ and θ are monadic second-order formulas with free variables u, and u and v respectively. The formula δ defines the set of vertices of $f(G)$ as a subset of V_G and the formula θ defines the edge relation of $f(G)$ in terms of that of G. The mapping f is thus specified by a pair $\langle \delta, \theta \rangle$ of monadic second-order formulas. We will say that f is a *monadic second-order transduction* and that $\langle \delta, \theta \rangle$ is its *definition scheme*. If δ and θ are first-order formulas, we will say that f is a *first-order transduction*. The general definition is actually more complicated. We introduce it step by step by giving several examples.

Example 1.30 (Edge-complement) The *edge-complement* associates with a simple, undirected and loop-free graph G, the simple, undirected and loop-free graph $\overline{G}$ such that $V_{\overline{G}} = V_G$ and $u - v$ is an edge of $\overline{G}$ if and only if $u \neq v$ and $u - v$ is not an edge of G. This transformation is a first-order transduction with definition scheme $\langle \delta, \theta \rangle$, where

$\delta(x)$ is the Boolean constant *True*,

$\theta(x,y)$ is the formula $x \neq y \wedge \neg edg(x,y)$.

Example 1.31 (Elimination of loops and isolated vertices) The transformation that eliminates loops and then, isolated vertices is a first-order transduction with definition scheme $\langle \delta, \theta \rangle$, where

$\delta(x)$ is the formula $\exists y\big((edg(x,y) \vee edg(y,x)) \wedge x \neq y\big),$

$\theta(x,y)$ is the formula $edg(x,y) \wedge x \neq y$.

Example 1.32 (Transitive closure of a directed graph) For every directed graph G, we let G^+ be its transitive closure, i.e., the simple graph defined by

$$V_{G^+} := V_G,$$

$$edg_{G^+} := edg_G^+$$

$$:= \{(u,v) \mid \text{there is in } G \text{ a nonempty directed path}^{35} \text{ from } u \text{ to } v\}.$$

Note that u has a loop in G^+ if it belongs to a directed cycle in G. The mapping $G \mapsto G^+$ is defined by the definition scheme $\langle \delta, \theta \rangle$, where

$\delta(x)$ is the formula *True*,

$\theta(x,y)$ is the formula[36] $edg(x,y) \vee \exists z(edg(x,z) \wedge TC[edg;z,y])$.

We use here the definition of the reflexive and transitive closure of a binary relation by a monadic second-order formula presented in Section 1.3.1.

Example 1.33 (Transitive reduction of a directed acyclic graph) A *directed acyclic graph* (a *DAG*) is a simple directed graph without directed cycles. Every such finite graph G has a unique minimal subgraph H such that $H^+ = G^+$. It is called the *transitive reduction* of G and is denoted by $Red(G)$. (The Hasse diagram of a partial order $\langle D, \leq \rangle$ is a graphical representation of the transitive reduction of the directed acyclic graph $\langle D, < \rangle$.) The edge relation of $Red(G)$ is characterized by

$$edg_{Red(G)}(x,y) :\Longleftrightarrow edg(x,y) \wedge \neg \exists z(edg_{G^+}(x,z) \wedge edg_{G^+}(z,y)),$$

hence is defined in G by a monadic second-order formula θ' built with the formula θ of the previous example. It follows that the mapping Red from directed graphs

---

[35] That is a sequence of pairwise distinct vertices $w_1, w_2, \ldots, w_n$ such that $u = w_1$, $v = w_n$ and $w_i \rightarrow w_{i+1}$ in G for each $i = 1, \ldots, n-1$. A directed cycle is a sequence of this form with $w_1 = w_n$, $n \geq 2$, $w_i \neq w_j$ for $1 \leq i \leq j \leq n-1$.

[36] The atomic formula $edg(x,y)$ in $\theta(x,y)$ can be omitted, but putting it in makes the formula more clear.

to directed graphs is a partial function that is a monadic second-order transduction specified by a sentence χ and two formulas δ, θ' such that

$$\lfloor G \rfloor \models \chi \qquad \text{if and only if it is acyclic (hence}$$
$$\text{if and only if } Red(G) \text{ is well defined),}$$

$$\lfloor G \rfloor \models \delta(u) \qquad \text{for every } u \text{ in } V_G \text{ (so that } V_{Red(G)} = V_G),$$

$$\lfloor G \rfloor \models \theta'(u,v) \quad \text{if and only if } Red(G) \text{ has an edge } u \to v.$$

☐

It is frequently necessary to consider functions f that assign a graph to the pair of a graph G and a set of vertices X. In this case, the definition scheme consists of formulas χ, δ and θ with a free set variable X called a *parameter*. The formula χ expresses the conditions to be verified by G and X so that $f(G,X)$ be defined. Here is an example.

Example 1.34 (The largest connected subgraph of G containing X) The partial function f such that $f(G,X)$ is the largest connected induced subgraph of G containing a nonempty subset X of V_G is a monadic second-order transduction with parameter X. Its definition scheme is $\langle \chi, \delta, \theta \rangle$, where

$\chi(X)$ is the formula $(\exists x. x \in X) \wedge \exists Y (X \subseteq Y \wedge \mathrm{CONN}(Y))$,[37]
$\delta(X,x)$ is the formula $\exists Y (X \subseteq Y \wedge x \in Y \wedge \mathrm{CONN}(Y))$,
$\theta(X,x,y)$ is $\exists Y (X \subseteq Y \wedge x \in Y \wedge y \in Y \wedge edg(x,y) \wedge \mathrm{CONN}(Y))$. ☐

In the previous example, a parameter X is necessary because the function f to be defined does not depend only on G but also on a set of vertices. However, in some cases, parameters may be necessary even if the output graphs depend only, up to isomorphism, on the input graphs (and not on the values of the parameters). Here is an example of such a case.

Example 1.35 (The DAG of strongly connected components of a directed graph) Let G be a directed graph. Let $\approx$ be the equivalence relation on V_G defined by

$$u \approx v \quad \text{if and only if } u = v \text{ or there exists a directed}$$
$$\text{path from } u \text{ to } v \text{ and a directed path from } v \text{ to } u.$$

An induced subgraph $G[X]$, where X is an equivalence class of this relation, is called a *strongly connected component*. The DAG of strongly connected components of G is the *quotient graph* $G/\approx$ defined as follows: its vertices are the strongly connected components; there is in $G/\approx$ an edge $X \to Y$ if and only if $X \neq Y$ and $u \to v$ in G for some $u \in X$ and $v \in Y$. Our objective is to prove that there exists a monadic second-order transduction associating with every graph G a graph H isomorphic to

---

[37] The formula $\mathrm{CONN}(Y)$ is defined in Section 1.3.1. We use $X \subseteq Y$ as a shorthand for $\forall x (x \in X \Rightarrow x \in Y)$.

$G/\approx$. Its definition uses a parameter X for denoting sets $U \subseteq V_G$ required to contain one and only one vertex of each strongly connected component. Such sets are used as sets of vertices of H and its edges are defined accordingly. We use the following formulas:

$Eq(x,y)$ is the auxiliary formula $\mathrm{TC}[edg;x,y] \wedge \mathrm{TC}[edg;y,x]$,
$\chi(X)$ is defined as:

$$\forall x,y\Big(\big(x \in X \wedge y \in X \wedge Eq(x,y)\big) \Rightarrow x = y\Big) \wedge \forall x \exists y\Big(Eq(x,y) \wedge y \in X\Big),$$

$\delta(X,x)$ is defined as: $x \in X$,
$\theta(X,x,y)$ is defined as:

$$x \in X \wedge y \in X \wedge \neg Eq(x,y) \wedge \exists z,z'\Big(edg(z,z') \wedge Eq(x,z) \wedge Eq(y,z')\Big).$$

The definition scheme $\langle \chi, \delta, \theta \rangle$ specifies a monadic second-order transduction that associates with a directed graph G and a "well-chosen" subset U of V_G a graph $f(G,U)$ with vertex set U that is isomorphic to $G/\approx$ ($u \in U$ corresponds to an equivalence class, hence to a vertex of $G/\approx$). It is clear that for every graph G there exists a set U satisfying χ, and that, for any two sets U and U' satisfying χ, the graphs $f(G,U)$ and $f(G,U')$ are isomorphic (the corresponding bijection $h : U \to U'$ is defined by $h(u) = v$ if and only if $u \in U$, $v \in U'$ and $u \approx v$).

This construction extends actually to every equivalence relation definable by a monadic second-order formula in place of $\approx$ and proves that the mapping that associates with a graph G its quotient $G/\approx$ by a monadic second-order definable equivalence relation $\approx$ is a monadic second-order transduction. $\qquad\square$

In all the above examples, the set of vertices of the output graph is a subset of the set of vertices of the input graph. The general definition of a monadic second-order transduction includes the possibility of *enlarging the input graph*, by "copying it" a fixed number of times.

A k-*copying monadic second-order transduction* associates with a graph G a graph H such that

$$V_H := (V_1 \times \{1\}) \cup \cdots \cup (V_k \times \{k\}),$$

$$edg_H := \{((u,i),(v,j)) \mid 1 \le i,j \le k,\ (u,v) \in E_{i,j}\},$$

where the sets $V_1,\dots,V_k \subseteq V_G$ and the relations $E_{i,j} \subseteq V_G \times V_G$ are defined in G by monadic second-order formulas, respectively $\delta_1,\dots,\delta_k$ and $\theta_{i,j}$ for $1 \le i,j \le k$ and possibly written with parameters. Let us give an example.

Example 1.36 (Graph duplication) For every simple directed graph G, its *duplication* is the simple graph $H = dup(G)$ defined as follows:

$V_H := V_G \times \{1,2\},$

the edges of H are the edges of each copy of G, together with the edges $(u,1) \rightarrow (u,2)$ for all $u \in V_G$.

The mapping dup has the definition scheme $\langle \delta_1, \delta_2, \theta_{1,1}, \theta_{1,2}, \theta_{2,1}, \theta_{2,2} \rangle$ where

> $\delta_1(x)$ and $\delta_2(x)$ are both the Boolean constant *True*,
>
> $\theta_{1,1}(x,y)$ and $\theta_{2,2}(x,y)$ are both $edg(x,y)$,
>
> $\theta_{1,2}(x,y)$ is $x = y$,
>
> $\theta_{2,1}(x,y)$ is the Boolean constant *False*.

The formulas δ_1 and δ_2 define V_1 and V_2 as V_G; the formulas $\theta_{i,j}$ define $E_{1,1} := E_{2,2} := edg_G$, $E_{1,2} := \{(u,u) \mid u \in V_G\}$, and $E_{2,1} := \emptyset$. Informally, H consists of two disjoint copies of G (i.e., $G \oplus G$) together with an edge from any vertex of the first copy to the corresponding vertex of the second copy. $\qquad\square$

The general definition of a monadic second-order transduction combines the above presented features. To summarize, a monadic second-order transduction associates with a relational structure S and subsets $U_1, \ldots, U_m$ of its domain D_S that must satisfy a monadic second-order formula $\chi(X_1, \ldots, X_m)$, a relational structure $T = f(S, U_1, \ldots, U_m)$ defined as follows. Its domain is $D_T := (D_1 \times \{1\}) \cup \cdots \cup (D_k \times \{k\})$, where each set D_i is defined as $\{d \in D_S \mid S \models \delta_i(U_1, \ldots, U_m, d)\}$ for some monadic second-order formula δ_i. If R is an n-ary symbol of the relational signature of T, the corresponding n-ary relation on D_T is defined as

$$\bigcup_{i_1,\ldots,i_n \in [k]} \{((d_1,i_1),\ldots,(d_n,i_n)) \mid d_1 \in D_{i_1}, \ldots, d_n \in D_{i_n},$$

$$S \models \theta_{R,i_1,\ldots,i_n}(U_1,\ldots,U_m,d_1,\ldots,d_n)\},$$

where each $\theta_{R,i_1,\ldots,i_n}$ is a monadic second-order formula.

Such a transduction f is specified by a tuple of formulas called a *definition scheme* of the form $\langle \chi, \delta_1, \ldots, \delta_k, (\theta_w)_{w \in W} \rangle$, where W consists of all tuples $(R, i_1, \ldots, i_{\rho(R)})$ such that R is a relation symbol of T and $i_1, \ldots, i_{\rho(R)} \in [k]$. The following fact is clear from the definitions:

Fact 1.37 For every monadic second-order transduction f there exists an integer k such that, if f transforms a relational structure S into a relational structure T, then $|D_T| \leq k \cdot |D_S|$. $\qquad\square$

We have $k = 1$ in the first six examples and $k = 2$ in the last one. We now give another example, with $k = 1$ and without parameters.

Example 1.38 (The cograph denoted by a term) As a final example we consider the mapping that evaluates a term t in $T(\{\oplus, \otimes, \mathbf{1}\})$ into the cograph $val(t)$ (cf. Section 1.1.2 and Example 1.7). First of all we must explain how t is represented by a relational structure. We let $\lfloor t \rfloor = \langle N_t, son_t, lab_{\oplus t}, lab_{\otimes t}, lab_{\mathbf{1}t} \rangle$ be the relational structure such that:

- N_t is $Pos(t)$, the set of positions of t, i.e., of occurrences of symbols from $\{\oplus, \otimes, \mathbf{1}\}$; we will consider N_t as the set of nodes of a rooted labeled tree representing t and also denoted by t;
- son_t is the binary relation such that $son_t(u, v)$ holds if and only if v is a son of u in the tree[38] t;
- $lab_{\oplus t}$, $lab_{\otimes t}$ and $lab_{\mathbf{1}t}$ are the unary relations such that $lab_{\oplus t}(u)$ holds if and only if u is an occurrence of $\oplus$ in t (i.e., is labeled by $\oplus$ as a node of the tree t) and similarly for $lab_{\otimes t}$ and $lab_{\mathbf{1}t}$.

Since the operations $\oplus$ and $\otimes$ are commutative, we need not express, when v is a son of u, whether it is the left or the right son. Hence we can use a simpler representation than the general one to be defined in Section 5.1.1. Here is an example. We let

$$s := \left(\left(\mathbf{1}_1 \otimes_2 \mathbf{1}_3 \right) \oplus_4 \mathbf{1}_5 \right) \otimes_6 \left(\mathbf{1}_7 \oplus_8 \mathbf{1}_9 \right),$$

where we number from left to right the nine occurrences of $\mathbf{1}$, $\oplus$, $\otimes$ (this numbering is indicated by subscripts). The corresponding labeled tree is shown in Figure 1.5. Then

$$\lfloor s \rfloor = \langle [9], son, lab_\oplus, lab_\otimes, lab_\mathbf{1} \rangle, \text{ with}$$
$$son = \{(6,4), (6,8), (4,2), (4,5), (2,1), (2,3), (8,7), (8,9)\},$$
$$lab_\oplus = \{4, 8\},$$
$$lab_\otimes = \{2, 6\},$$
$$lab_\mathbf{1} = \{1, 3, 5, 7, 9\}.$$

In the general case, the cograph $G = val(t)$ that is the value of a term t in $T(\{\oplus, \otimes, \mathbf{1}\})$ can be defined from $\lfloor t \rfloor$ as follows:

V_G is the set of elements of N_t that are labeled by $\mathbf{1}$ (i.e., that are the leaves of the tree t); for distinct vertices u and v of G, there is an edge between u and v if and only if the least common ancestor of u and v in t is labeled by $\otimes$.

In the above example of s, the vertices of the cograph $val(s)$ are $1, 3, 5, 7, 9$. There is an edge between 3 and 9 because their least common ancestor in s is 6, which is

---

[38] We describe the relations between occurrences of symbols in t with the terminology of trees. Chapter 2 will detail the terminology about terms and the trees that represent them.

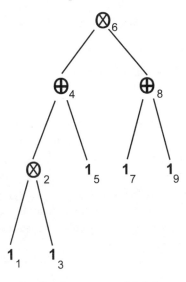

Figure 1.5 The term s as a labeled tree.

labeled by $\otimes$. There is no edge between 1 and 5 because their least common ancestor is 4, labeled by $\oplus$.

The monadic second-order transduction that associates with $\lfloor t \rfloor$, for any term t in $T(\{\oplus, \otimes, \mathbf{1}\})$, the relational structure $\lfloor val(t) \rfloor$ is specified by the definition scheme $\langle True, \delta, \theta \rangle$, where

$\delta(x)$ is the formula $lab_1(x)$,

$\theta(x,y)$ is the formula $x \neq y \wedge \exists z(\mathrm{LCA}(x,y,z) \wedge lab_\otimes(z))$.

In this writing, $\mathrm{LCA}(x,y,z)$ stands for the following formula expressing that z is the least common ancestor of x and y:

$$\mathrm{TC}[son;z,x] \wedge \mathrm{TC}[son;z,y] \wedge$$
$$\forall w\big((\mathrm{TC}[son;w,x] \wedge \mathrm{TC}[son;w,y]) \Rightarrow \mathrm{TC}[son;w,z]\big).$$

This formula is a straightforward translation of the definition of the least common ancestor. Note that if x is an ancestor of y or if $x = y$, then $\mathrm{LCA}(x,y,x)$ is valid. Furthermore, $\mathrm{LCA}(x,y,z)$ defines z in a unique way from x and y.

1.7.2 The main properties of monadic second-order transductions

A *monadic second-order transduction* is a subset f of $\mathcal{S} \times \mathcal{T}$, where $\mathcal{S}$ and $\mathcal{T}$ are classes of graphs or, more generally, of relational structures, that is specified as explained in the examples by monadic second-order formulas. If $L \subseteq \mathcal{S}$ and $f \subseteq \mathcal{S} \times \mathcal{T}$, we let $f(L) := \{T \mid (S,T) \in f,\ S \in L\}$ be the *image* of L under f. Hence, f can also be seen as the multivalued mapping $\widehat{f} : S \to \mathcal{P}(T)$ such that $\widehat{f}(G) := f(\{G\})$. In Examples 1.30, 1.31 and 1.32, $\widehat{f}(G)$ consists of a unique graph. In Example 1.33 it consists of at most one graph and in Example 1.34 of several graphs (the connected components). However, a monadic second-order transduction is always based on a (single-valued) partial function that transforms *parametrized relational structures* into relational structures. A parametrized relational structure is a tuple $(S, A_1, \ldots, A_n)$ consisting of a relational structure S and an n-tuple of subsets of its domain D_S. From a partial function $\overline{f}$ that transforms $(S, A_1, \ldots, A_n)$ into $\overline{f}(S, A_1, \ldots, A_n)$ in T for S in S and for certain subsets $A_1, \ldots, A_n$ of D_S, we define f as the transduction $\{(S, \overline{f}(S, A_1, \ldots, A_n)) \mid A_1, \ldots, A_n \subseteq D_S$ and $\overline{f}(S, A_1, \ldots, A_n)$ is defined$\})$. We also say that $\overline{f}$ and $\widehat{f}$ are monadic second-order transductions.

The *composition* of two monadic second-order transductions is the composition of the corresponding binary relations: if $f \subseteq S \times T$ and $g \subseteq T \times \mathcal{U}$, then $f \cdot g := \{(S, U) \mid (S, T) \in f$ and $(T, U) \in g$ for some $T \in T\}$. If f and g are functional, we will also use the notation $g \circ f$ for $f \cdot g$.

Theorem 1.39 The composition of two monadic second-order transductions is a monadic second-order transduction. $\square$

A more precise statement showing how parameters are handled in the composition of monadic second-order transductions will be given in Chapter 7 (see Theorem 7.14).

Theorem 1.40 (Backwards Translation Theorem) If the set $L \subseteq T$ is MS-definable and $f \subseteq S \times T$ is a monadic second-order transduction, then the set $f^{-1}(L) := \{S \in S \mid \widehat{f}(S) \cap L \neq \emptyset\}$ is MS-definable. $\square$

Theorem 1.40 is formally a consequence of Theorem 1.39 but it is actually used for proving Theorem 1.39 (see Chapter 7). We call Theorem 1.40 the Backwards Translation Theorem because its proof yields an algorithm that transforms a formula defining L into one defining $f^{-1}(L)$. We now give the idea of its proof. Let us go back to the edge-complement transformation (Example 1.30). It is a first-order transduction that transforms G into $\overline{G}$. If P is a first-order graph property, then the property Q defined by $Q(G)$ if and only if $P(\overline{G})$ is also first-order. The edge relation of $\overline{G}$ is defined by $edg_{\overline{G}}(u, v) :\Longleftrightarrow u \neq v \wedge \neg edg_G(u, v)$. Hence one obtains a first-order sentence expressing Q by replacing in the given first-order sentence expressing P each atomic formula of the form $edg(x, y)$ by $(x \neq y \wedge \neg edg(x, y))$. Here is an example. We let P be defined by the sentence

$$\exists x, y, z \Big(x \neq y \wedge y \neq z \wedge x \neq z \wedge edg(x, y) \wedge edg(y, z) \wedge \neg edg(x, z) \Big),$$

expressing that an undirected graph G has an induced path P_3 (of the form $\bullet - \bullet - \bullet$).
The sentence defining Q is then

$$\exists x,y,z\Big(x \neq y \wedge y \neq z \wedge x \neq z \wedge (x \neq y \wedge \neg edg(x,y))$$

$$\wedge \big(y \neq z \wedge \neg edg(y,z)\big) \wedge \neg\big(x \neq z \wedge \neg edg(x,z)\big)\Big).$$

It can actually be simplified into:

$$\exists x,y,z\Big(x \neq y \wedge y \neq z \wedge x \neq z \wedge \neg edg(x,y) \wedge \neg edg(y,z) \wedge edg(x,z)\Big),$$

expressing that G has an induced subgraph of the form $\bullet - \bullet \;\; \bullet$. This construction
extends easily to monadic second-order transductions without parameters, like those
of Examples 1.31, 1.32 and 1.33.

Let us now assume that f is defined by a definition scheme of the form
$\langle \chi(X), True, \theta(X,x,y)\rangle$ with one parameter X. Here $\chi(X)$ is a formula that imposes
some conditions on the parameter X. In Example 1.34, the condition "X denotes a
singleton" might be imposed on X: the corresponding transduction associates with
each vertex its connected component.

Let β be a sentence and let L be the set of graphs such that $\lfloor G \rfloor \models \beta$. Let then
$\beta^\#$ be the formula with free variable X obtained by replacing in β every atomic
formula $edg(x,y)$ by $\theta(X,x,y)$ (by using appropriate substitutions, and, if necessary,
renamings of bound variables in $\theta(X,x,y)$). For $A \subseteq V_G$, the graph $f(G,A)$ is well
defined if and only if $\lfloor G \rfloor \models \chi(A)$, and, then, $f(G,A) \models \beta$ if and only if $\lfloor G \rfloor \models \beta^\#(A)$.
It follows that $f^{-1}(L)$ is defined by the sentence $\exists X\big(\chi(X) \wedge \beta^\#(X)\big)$.

1.7.3 The Equationality Theorem

The Equationality Theorems for the VR and the HR algebras (and for an algebra
of relational structures that generalizes the VR algebra) are among the main results
established in this book.

As an introduction to the Equationality Theorem for the VR algebra, we recall
that the mapping that associates the cograph $val(t)$ with a term t in $T(\{\oplus, \otimes, \mathbf{1}\})$ is a
monadic second-order transduction (Example 1.38). More generally:

Theorem 1.41 For every k, the mapping that associates with a term t in $T(F^{VR}_{[k]})$ the
p-graph $t_{\mathbb{GP}}$ is a monadic second-order transduction. $\qquad\square$

This implies that the set of graphs of clique-width at most k, which is a VR-
equational set, is the image of the set of terms (over some finite signature) under a

monadic second-order transduction. The following includes a converse of this result and is very important for the theories of graph structuring and of graph grammars.

Theorem 1.42 (Equationality Theorem for the VR algebra) A set of simple graphs is VR-equational if and only if it is the image of the set of binary rooted trees under a monadic second-order transduction.   □

As an immediate consequence and by using Theorem 1.39, we obtain that the image of a VR-equational set under a monadic second-order transduction is again VR-equational. Note that this implies, as a special case, the logical version of the Filtering Theorem (Theorem 1.22): for a sentence χ, the transduction $f_\chi = \{(G,G) \mid \lfloor G \rfloor \models \chi\}$ is a monadic second-order transduction with definition scheme $\langle \chi, True, edg(x,y) \rangle$, and $f_\chi(L) = \{G \in L \mid \lfloor G \rfloor \models \chi\}$ for every set of graphs L. Another immediate consequence, using Proposition 1.18, is the following:

Corollary 1.43 A set of simple graphs has bounded clique-width if and only if it is included in the image of the set of binary rooted trees under a monadic second-order transduction.   □

Thus, again using Theorem 1.39, the image of a set of graphs of bounded clique-width under a monadic second-order transduction from graphs to graphs has bounded clique-width.

In Theorem 1.42 and Corollary 1.43, one can replace "the set of binary rooted trees"[39] by "the set of trees" or by "$T(F)$ where F is any finite signature with at least one constant symbol and at least one symbol of arity at least 2."

1.8 Monadic second-order logic with edge set quantifications

If a graph G is represented by a relational structure whose domain also contains the edges, instead of by $\lfloor G \rfloor$, then the expressive power of monadic second-order logic is increased, even for expressing properties of simple graphs. We will also compare the four types of monadic second-order transductions obtained by representing input and output graphs G either by $\lfloor G \rfloor$ or by the alternative structure denoted by $\lceil G \rceil$.

1.8.1 Expressing graph properties with edge set quantifications

For every undirected graph G, we let $\lceil G \rceil$ be the pair $\langle V_G \cup E_G, in_G \rangle$, where[40] $in_G = \{(e,u) \mid e \in E_G, u \in V_G, u$ is an end vertex of $e\}$. If G has several edges (called *multiple edges*) between two vertices, these edges are distinct elements of the domain of $\lceil G \rceil$.

---

[39] A rooted tree is directed in such a way that the root is the unique node of indegree 0 and every node is accessible from the root by a directed path. It is *binary* if each node has outdegree 0 or 2. The associated (undirected) tree has degree at most 3.

[40] Recall that V_G is the set of vertices and E_G is the set of edges.

The structure $\lceil G \rceil$ can be seen as $\lfloor Inc(G) \rfloor$, where $Inc(G)$ is a bipartite directed graph called the *incidence graph* of G, whose edge relation is denoted by the binary relation symbol *in*. In a relational structure $S = \langle D_S, in_S \rangle$ isomorphic to $\lceil G \rceil$ for some graph G, the elements of D_S corresponding to edges are those, say u, such that $(u, v) \in in_S$ for some v. The element u corresponds to a loop if and only if there is a single such v. It follows that G can be reconstructed from S in a unique way. Thus, the representation of G by $\lceil G \rceil$ is faithful for all undirected graphs.

For directed graphs, we will use $\lceil G \rceil := \langle V_G \cup E_G, in_{1G}, in_{2G} \rangle$, where $(e, u) \in in_{1G}$ (resp. $(e, u) \in in_{2G}$) if and only if u is the tail[41] (resp. the head) of e. Hence $\lceil G \rceil$ can be seen as a directed bipartite graph, also denoted by $Inc(G)$, with edges labeled either by 1 or by 2. Loops in G are multiple edges (with different labels) in $Inc(G)$.

In this setting, edges are considered, like vertices, as objects that form a graph and not as the pairs of some binary relation over vertices. In particular, we do not consider an undirected edge as a pair of opposite directed edges.

Graph properties can be expressed logically, either via the representation of a graph G by $\lceil G \rceil$, or via the initially defined representation $\lfloor G \rfloor := \langle V_G, edg_G \rangle$. The representation $\lfloor G \rfloor$ only allows quantification on vertices and on sets of vertices in monadic second-order formulas, whereas the representation $\lceil G \rceil$ also allows quantification on edges and sets of edges. A graph property is MS_2-*expressible* if it is expressible by a monadic second-order formula interpreted in $\lceil G \rceil$. The index 2 refers to the possibility of "two types of quantification," on sets of vertices and sets of edges. A property is MS_1-*expressible* if it is by a monadic second-order formula interpreted in $\lfloor G \rfloor$. Unless for emphasizing the contrast with MS_2, we will write MS instead of MS_1.[42]

Let us stress that we *do not modify* the logical language, but only the representation of graphs by relational structures. Since an incidence graph is a graph, we still deal with a single language that we use to express formally graph properties. We now compare the power of these two ways of expressing graph properties. It is clear that a property of a graph G like "for every two vertices u, v, there are no more than three edges from u to v" cannot be expressed by any sentence interpreted in $\lfloor G \rfloor$ because this relational structure cannot identify the existence of multiple edges, and thus no sentence can take into account their multiplicity. However, even for expressing properties of simple graphs, monadic second-order sentences with edge set quantifications are more powerful. The property that a loop-free undirected graph G has a perfect matching is equivalent to $\lceil G \rceil \models \exists X . \psi$, where ψ is a formula with free variable X expressing that X is a set of edges and that for every vertex u there exists a unique $e \in X$ such that $(e, u) \in in_G$. The formula ψ is easy to write with first-order quantifications only. However, there is no monadic second-order sentence φ expressing this property by $\lfloor G \rfloor \models \varphi$. The formal proof of this assertion (to be done

---

[41] If e is an edge directed from u to v, then we say that u is its *tail* and that v is its *head*.

[42] By an MS_2 *formula*, we mean a monadic second-order formula written with the binary relation symbols *in*, in_1, in_2, intended to be interpreted in logical structures of the form $\lceil G \rceil$. An MS_1 *formula* is written with the binary relation symbol *edg* and is to be interpreted in $\lfloor G \rfloor$.

in Chapter 5) is based on the observation that the complete bipartite graph $K_{n,m}$ has a perfect matching if and only if $n = m$, and on the theorem saying that no monadic second-order formula can express that two sets have the same cardinality. It is easy to express that a *given* set of edges, say X, is a perfect matching, and this is what formula ψ does. But one cannot replace "there exists a set of edges satisfying ψ" by an MS formula without edge set quantification. The property "G has a Hamiltonian cycle" can be expressed similarly by an MS formula interpreted in $\lceil G \rceil$. The graphs $K_{n,m}$ can be used as counter-examples, as above for perfect matchings, to prove that this is not possible by an MS formula over $\lfloor G \rfloor$.

It is clear that every MS_1-expressible graph property is MS_2-expressible. Although some properties of simple graphs are MS_2-expressible but not MS_1-expressible, MS_2 formulas are in many cases no more expressive than MS_1 formulas.

Theorem 1.44 (Sparseness Theorem) Let L be a set of simple graphs that are all, either planar, or of degree at most k, or of tree-width at most k for some fixed k. Every MS_2 sentence φ can be translated into an MS_1 sentence ψ such that for every graph G in L:

$$\lceil G \rceil \models \varphi \text{ if and only if } \lfloor G \rfloor \models \psi.$$

$\square$

This result actually extends to sets of *uniformly sparse* graphs, as we will prove in Section 9.4.

1.8.2 Monadic second-order transductions over incidence graphs

For expressing properties of graphs G, we can choose between the two representing relational structures $\lfloor G \rfloor$ and $\lceil G \rceil$. For defining monadic second-order graph transductions, we get thus four possibilities arising from two possible representations for the input as well as for the output. All examples of Section 1.7.1 use the first representation for the input and the output. We give below examples using $\lceil G \rceil$. We first fix some notation.

A graph transduction f is an $MS_{i,j}$*-transduction*, where $i,j \in \{1,2\}$, if there exists a monadic second-order transduction g such that $(G,H) \in f$ if and only if $(S,T) \in g$, where S is $\lfloor G \rfloor$ if $i = 1$ and $\lceil G \rceil$ if $i = 2$, and, similarly, T is $\lfloor H \rfloor$ if $j = 1$ and $\lceil H \rceil$ if $j = 2$. Hence, the indices i and j indicate which representations are used, respectively for the input and the output graphs.

It is clear that every $\mathrm{MS}_{1,j}$-transduction is also an $\mathrm{MS}_{2,j}$-transduction, because every MS_1 formula can be rewritten into an equivalent MS_2 formula. Hence, changing 1 into 2 in this way makes "easier" the task of writing formulas to specify a transduction. For the "output" side we get that every $\mathrm{MS}_{i,2}$-transduction is an $\mathrm{MS}_{i,1}$-transduction, and not vice-versa as one might think, because in the former case the transduction must define the edges (and not only the vertices) from elements of the input structure, either $\lfloor G \rfloor$ or $\lceil G \rceil$. We have the following inclusions of classes of monadic second-order transductions:

$$MS_{1,2} \subseteq MS_{1,1} \subseteq MS_{2,1},$$
$$MS_{1,2} \subseteq MS_{2,2} \subseteq MS_{2,1}.$$

We will give examples proving that these inclusions are proper, that $MS_{1,1}$ and $MS_{2,2}$ are incomparable, that $MS_{1,1} \cup MS_{2,2}$ is a proper subclass of $MS_{2,1}$ and that $MS_{1,2}$ is a proper subclass of $MS_{1,1} \cap MS_{2,2}$.

Example 1.45 (The line graph transduction) The *line graph Line(G)* of an undirected graph G is the loop-free undirected graph H such that $V_H = E_G$ and e, f are adjacent vertices of H if and only if they have at least one common vertex as edges of G. The mapping $\lceil G \rceil \to \lfloor H \rfloor$ is an $MS_{2,1}$-transduction with definition scheme $\langle \chi, \delta, \theta_{edg} \rangle$ such that:

$$\chi \qquad :\Longleftrightarrow \quad True,$$
$$\delta(x) \qquad :\Longleftrightarrow \quad \exists z. in(x,z),$$
$$\theta_{edg}(x,y) \quad :\Longleftrightarrow \quad x \neq y \wedge \exists z(in(x,z) \wedge in(y,z)).$$

Could one use $\lfloor G \rfloor$ instead of $\lceil G \rceil$ for the input? The answer is no by Fact 1.37 because for arbitrary graphs G, if $H = Line(G)$ then we do not have $|D_{\lfloor H \rfloor}| = O(|D_{\lfloor G \rfloor}|)$ since $D_{\lfloor H \rfloor} = V_H = E_G$ and $D_{\lfloor G \rfloor} = V_G$.

Could one use $\lceil H \rceil$ instead of $\lfloor H \rfloor$ for the output? Again the answer is no by a similar argument: if $G = K_{1,n}$, then $|D_{\lceil G \rceil}| = 2n + 1$, $|V_H| = n$, $|E_H| = n(n-1)/2$, hence $|D_{\lceil H \rceil}| = n(n+1)/2$ and is not $O(|D_{\lceil G \rceil}|)$.

Hence the line graph transduction is neither in $MS_{1,1}$ nor in $MS_{2,2}$.

Example 1.46 (Transitive closure) We have seen in Example 1.32 that the transitive closure $G \mapsto G^+$ on directed graphs is an $MS_{1,1}$-transduction. It is not an $MS_{2,2}$-transduction: consider Q_n, the directed path with n vertices. We have $|D_{\lceil Q_n \rceil}| = 2n - 1$ and $|D_{\lceil Q_n^+ \rceil}| = n(n+1)/2$, hence we do not have $|D_{\lceil Q_n^+ \rceil}| = O(|D_{\lceil Q_n \rceil}|)$. Consequently the transitive closure of directed graphs is not an $MS_{2,2}$-transduction.

Example 1.47 (Edge subdivision) For G simple, directed and loop-free, we let $Sub(G)$ be the graph of the same type such that

$$V_{Sub(G)} := V_G \cup E_G,$$
$$E_{Sub(G)} := \{(u,e), (e,v) \mid (e,u) \in in_{1G}, (e,v) \in in_{2G}\}.$$

This transformation, called *edge subdivision* consists in replacing directed edges by directed paths of length 2. We have $|V_{Sub(G)}| = |V_G| + |E_G|$ and $|E_{Sub(G)}| = 2|E_G|$. Edge subdivision is an $MS_{2,2}$-transduction because the transformation of $\lceil G \rceil$ into $\lceil Sub(G) \rceil$ is a 3-copying monadic second-order transduction: a vertex v of G is made

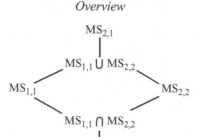

Figure 1.6 The different classes of monadic second-order transductions.

into a vertex $(v, 1)$ of $Sub(G)$, and an edge e linking u to v is made into a vertex $(e, 1)$ and into edges $(e, 2)$ and $(e, 3)$ of $Sub(G)$ that link respectively $(u, 1)$ to $(e, 1)$ and $(e, 1)$ to $(v, 1)$. Its definition scheme will be given in Chapter 7 (see Example 7.44). Since $|V_{Sub(G)}|$ is not $O(|V_G|)$, edge subdivision is not an $MS_{1,1}$-transduction.

Example 1.48 (Identity) The identity mapping is trivially an $MS_{1,1}$- and an $MS_{2,2}$-transduction, and hence also an $MS_{2,1}$-transduction. It is not an $MS_{1,2}$-transduction, as a clear consequence of Fact 1.37. However we have the following theorem which yields Theorem 1.44 with the help of Theorem 1.40.

Theorem 1.49 On each set of simple graphs that are planar, or of degree at most k, or of tree-width at most k for some fixed k, the identity is an $MS_{1,2}$-transduction. □

We conclude this discussion with a diagram (Figure 1.6) relating the different types of MS-transductions, where lines indicate strict inclusions from bottom-up. All inclusions are clear from the definitions and the above observations. That they are strict and that $MS_{1,1}$ and $MS_{2,2}$ are incomparable is proved by Examples 1.45, 1.46, 1.47 and 1.48.

The results in Section 1.7.3 are stated for $MS_{1,1}$-transductions and the VR algebra. There are analogous statements for $MS_{2,2}$-transductions and the HR algebra. We first observe a technical point: since the identity on trees or terms is an $MS_{1,2}$-transduction and since the composition of two MS-transductions is an MS-transduction, a transduction from trees or terms to graphs is an $MS_{1,j}$-transduction if and only if it is an $MS_{2,j}$-transduction. There are thus only two types of transductions taking trees or terms as input to consider and we get the following fully similar results:

Theorem 1.50 For every k, the mapping that associates with a term t in $T(F_{[k]}^{HR})$ the s-graph t_{JS} is an $MS_{1,2}$-transduction. □

Theorem 1.51 (Equationality Theorem for the HR algebra) A set of graphs is HR-equational if and only if it is the image of the set of binary rooted trees under an $MS_{1,2}$-transduction. □

As in Section 1.7.3, this implies, using the closure of monadic second-order transductions under composition (Theorem 1.39), that the image of an HR-equational set under an $MS_{2,2}$-transduction is again HR-equational (generalizing the logical version of the Filtering Theorem for HR). Moreover, using the Equationality Theorems for both the VR and the HR algebra, we obtain that the image of an HR-equational set under an $MS_{2,1}$-transduction is VR-equational, and the image of a VR-equational set under an $MS_{1,2}$-transduction is HR-equational. Note that if f is an $MS_{i,j}$-transduction and g is an $MS_{j,k}$-transduction $(i,j,k = 1,2)$, then their composition $f \cdot g$ is an $MS_{i,k}$-transduction.

Using Proposition 1.20, Theorem 1.51 implies the following corollary:

Corollary 1.52 A set of graphs has bounded tree-width if and only if it is included in the image of the set of binary rooted trees under an $MS_{1,2}$-transduction. $\square$

As for Theorem 1.42 and its corollary, both Theorem 1.51 and Corollary 1.52 hold with "the set of binary rooted trees" replaced by "the set of trees" or by "$T(F)$ where F is any finite signature with at least one constant symbol and at least one symbol of arity at least 2." In these results, graphs are without ports or sources in order to have simpler statements. Slightly more general results will be stated in Chapter 7.

From Corollaries 1.52 and 1.43 we get the following corollary from which quick proofs that certain sets of graphs have bounded or unbounded tree-width or clique-width can be obtained:

Corollary 1.53

(1) The image of a set of graphs of bounded tree-width under an $MS_{2,2}$-transduction (resp. under an $MS_{2,1}$-transduction) has bounded tree-width (resp. bounded clique-width).

(2) The image of a set of graphs of bounded clique-width under an $MS_{1,1}$-transduction (resp. under an $MS_{1,2}$-transduction) has bounded clique-width (resp. bounded tree-width).

(3) A set of simple planar graphs or of simple graphs of bounded degree has bounded tree-width if and only if it has bounded clique-width. $\square$

Statements (1) and (2) follow from Corollaries 1.52 and 1.43, by Theorem 1.39. Since all proofs are effective, the new bound can be computed from the given bound and the definition scheme of the MS-transduction. Using again Theorem 1.39, statement (3) follows from Theorem 1.49 and the fact that the identity is an $MS_{2,1}$-transduction (Example 1.48). We will give in Section 2.5.5 a proof of statement (3) that does not use transductions and gives a good estimate of the bound on tree-width.

To conclude this section, let us stress the nice parallelism between two groups of definitions:

(1) the VR algebra, clique-width and MS formulas;
(2) the HR algebra, tree-width and MS_2 formulas.

The Recognizability Theorem and the Equationality Theorem have fully analogous statements for both groups. Furthermore, the same graph theoretic conditions, i.e., those of Theorem 1.44 (and more general ones), ensure the equivalence of clique-width and tree-width, and simultaneously of MS and MS_2 formulas. The $MS_{1,2}$- and $MS_{2,1}$-transductions define "bridges" between the "world of bounded clique-width" and that of "bounded tree-width." The facts show how intimate are the relationships between logical and combinatorial notions.

1.9 Relational structures

Terms, graphs, labeled graphs, hypergraphs of different types are, or rather can be conveniently represented by *relational structures*. Up to now, we have only seen relational structures with unary and binary relations that correspond to vertex- and edge-labeled graphs. However, many of our results can be proved without any difficulty for general relational structures.

In order to illustrate the usefulness of relational structures in Discrete Mathematics, we will present the examples of *betweenness relations* and *cyclic orderings*, two combinatorial notions defined in a natural way as ternary relations. Furthermore, in a different domain, the theory of *relational databases* is based on the concept of relational structure (see the book by Abiteboul, Hull and Vianu [*AbiHV]). However, our theory will not bring much to this field for reasons that we will discuss briefly.

1.9.1 Relational signatures and structures

A *relational signature* (to be contrasted with the notion of a functional signature defined in Section 1.1.4) is a finite set $\mathcal{R}$ of *relation symbols* where each symbol R of $\mathcal{R}$ has an associated arity $\rho(R)$ in $\mathcal{N}_+ := \mathcal{N} - \{0\}$. A *relational structure* of type $\mathcal{R}$, called simply an *$\mathcal{R}$-structure*, is a tuple $S = \langle D_S, (R_S)_{R \in \mathcal{R}} \rangle$ consisting of a finite (possibly empty) *domain D_S* and of a $\rho(R)$-ary relation[43] R_S for each $R \in \mathcal{R}$. The set of $\mathcal{R}$-structures is denoted by $STR(\mathcal{R})$. We let $\rho(\mathcal{R}) := \max\{\rho(R) \mid R \in \mathcal{R}\}$. A signature $\mathcal{R}$ (resp. an $\mathcal{R}$-structure) is *binary* if $\rho(\mathcal{R}) \leq 2$.

Since every k-ary function can be considered as a $(k + 1)$-ary relation, there is no loss of generality in considering relational structures as opposed to more general logical structures also containing functions. Although a *constant*, i.e., a 0-ary function can be replaced by a (singleton) unary relation, it will be convenient (for instance for representing the sources of graphs) to allow constants. In this introductory section, however, we will only consider relational structures without constants.

---

[43] A k-ary relation can be defined as a subset of D_S^k or, equivalently, as a total function: $D_S^k \to \{True, False\}$.

Formulas are written with atomic formulas of the two forms $x_1 = x_2$ and $R(x_1, \ldots, x_{\rho(R)})$, where $x_1, \ldots, x_{\rho(R)}$ are individual variables. The notion of an *MS-expressible* property of $\mathcal{R}$-structures follows immediately. A subset L of $STR(\mathcal{R})$, the set of all $\mathcal{R}$-structures, is *MS-definable* if it is the set of finite models of a monadic second-order sentence φ, formally, if $L = \{S \in STR(\mathcal{R}) \mid S \models \varphi\}$.

For expressing graph properties by monadic second-order formulas, we have defined two relational structures associated with a graph G, denoted by $\lfloor G \rfloor$ and $\lceil G \rceil$. We have observed that certain graph properties are monadic second-order express-ible *via* the "rich" representation $\lceil G \rceil$, but not *via* the "natural" one $\lfloor G \rfloor$. The former properties are called MS_2-expressible. A similar extension of monadic second-order logic can be defined for relational structures. We let $\mathcal{R}^{Inc} := \mathcal{R} \cup \{in_i \mid 1 \leq i \leq \rho(R)\}$ with $\rho(R) = 1$ for $R \in \mathcal{R}$ and $\rho(in_i) = 2$ for $i = 1, \ldots, \rho(\mathcal{R})$. The *incidence structure* of $S = \langle D_S, (R_S)_{R \in \mathcal{R}} \rangle$ is the $\mathcal{R}^{Inc}$-structure $Inc(S)$ defined as

$$\langle D_S \cup T_S, (R_{Inc(S)})_{R \in \mathcal{R}}, in_{1\,Inc(S)}, \ldots, in_{k\,Inc(S)} \rangle,$$

where $k := \rho(\mathcal{R})$ and

$$T_S := \{(R, d_1, \ldots, d_{\rho(R)}) \mid R \in \mathcal{R}, (d_1, \ldots, d_{\rho(R)}) \in R_S\},$$
$$R_{Inc(S)}(d) :\Longleftrightarrow d = (R, d_1, \ldots, d_{\rho(R)}) \in T_S \text{ for some } d_1, \ldots, d_{\rho(R)} \in D_S,$$
$$in_{i\,Inc(S)}(d, d') :\Longleftrightarrow d \in T_S, d' \in D_S \text{ and } d = (R, d_1, \ldots, d_{\rho(R)}) \text{ for some } R \in \mathcal{R}$$
$$\text{and } d_1, \ldots, d_{\rho(R)} \text{ such that } d' = d_i.$$

It is clear that two $\mathcal{R}$-structures S and S' are isomorphic if and only if $Inc(S)$ and $Inc(S')$ are isomorphic. The incidence structure $Inc(S)$ of S is actually a vertex- and edge-labeled bipartite directed graph. Furthermore, each relation $in_{i\,Inc(S)}$ is func-tional. A set of $\mathcal{R}$-structures is MS_2-*definable* if it is $\{S \in STR(\mathcal{R}) \mid Inc(S) \models \varphi\}$ for a monadic second-order sentence φ over the signature $\mathcal{R}^{Inc}$. As for graphs, we obtain the notion of an MS_2-*expressible* property of $\mathcal{R}$-structures by replacing S by $Inc(S)$.

Since the incidence structure of a relational structure is a labeled graph, the results concerning MS_2 formulas and labeled graphs of bounded tree-width transfer easily to relational structures. It is not difficult to see that the identity on incidence structures (of $\mathcal{R}$-structures) is an $MS_{1,2}$-transduction, which implies that $Inc(S)$ has the same MS_2-expressible and MS_1-expressible properties (cf. Theorems 1.49 and 1.44).

We define $twd^{Inc}(S) := twd(Inc(S))$ to be used as parameter. For each k, we define $STR_k(\mathcal{R})$ as the class $\{S \in STR(\mathcal{R}) \mid twd^{Inc}(S) \leq k\}$ and we get the following result:

Theorem 1.54 Let $\mathcal{R}$ be a relational signature.

(1) The model-checking problem for CMS_2 sentences and the class of $\mathcal{R}$-structures is fixed-parameter linear with respect to $twd^{Inc}(S) + |\varphi|$ where S is the input structure and φ is the input sentence.

(2) For each $k \in \mathcal{N}$, the CMS_2-satisfiability problem for the class $STR_k(\mathcal{R})$ is decidable.

(3) If a subset of $STR(\mathcal{R})$ has a decidable MS_2-satisfiability problem, then it is contained in $STR_m(\mathcal{R})$ for some m. □

This theorem generalizes to $\mathcal{R}$-structures the parts of Theorems 1.25, 1.26 and 1.29 that concern tree-width and CMS_2-expressible graph properties. Establishing analogous results for CMS properties (as opposed to CMS_2 properties) raises difficult open problems.

1.9.2 Betweenness and cyclic ordering

We now present the two combinatorial notions of betweenness and cyclic ordering that are naturally defined as ternary relations. They raise open questions relative to monadic second-order expressibility. All results stated below will be proved in Section 9.1.

With a finite linear order $\langle D, \leq \rangle$ such that $|D| \geq 3$ we associate the following ternary relation, called its *betweenness relation*:

$$B(x,y,z) :\Longleftrightarrow (x < y < z) \vee (z < y < x),$$

where $x < y$ means "$x \leq y$ and $x \neq y$". We denote it by $B(\leq)$. This relation satisfies the following properties, for all $x,y,z,t \in D$:

(B1) $B(x,y,z) \Rightarrow x \neq y \wedge x \neq z \wedge y \neq z$;
(B2) $B(x,y,z) \Rightarrow B(z,y,x)$;
(B3) $B(x,y,z) \Rightarrow \neg B(y,z,x)$;
(B4) $B(x,y,z) \wedge B(y,z,t) \Rightarrow B(x,y,t) \wedge B(x,z,t)$;
(B5) $B(x,y,z) \wedge B(y,t,z) \Rightarrow B(x,y,t) \wedge B(x,t,z)$;
(B6) $x \neq y \wedge x \neq z \wedge y \neq z \Rightarrow B(x,y,z) \vee B(y,z,x) \vee B(z,x,y)$.

Conversely, if B is a ternary relation satisfying these properties, it is $B(\leq)$ for some linear order on D, hence it is a betweenness relation. A set $X \subseteq D^3$ is *consistent for betweenness* if $X \subseteq B$ for some betweenness relation B on D. The problem BETWEENNESS consisting in deciding whether a given set $X \subseteq D^3$ is consistent for betweenness is NP-complete ([*GarJoh]).

If X is consistent for betweenness, we define

$$\widehat{X} := \bigcap \{B \mid X \subseteq B, \ B \text{ is a betweenness relation}\}.$$

The set $\widehat{X}$ satisfies properties (B1)–(B5). We say that X is a *partial betweenness relation on D* if $\widehat{X} = X$. These definitions raise the following open questions, where a ternary relation $X \subseteq D^3$ is identified with the $\mathcal{R}$-structure $\langle D, X \rangle$ (for some fixed singleton $\mathcal{R}$):

Question 1.55

(1) Is the set of relational structures $\{\langle D,X \rangle \mid X$ is consistent for betweenness$\}$ MS-definable?
(2) Is the set of partial betweenness relations MS-definable?

For MS_2 the answers to these questions are positive.

Proposition 1.56 The set of partial betweenness relations and the set of ternary relations that are consistent for betweenness are MS_2-definable. □

For $X \subseteq D^3$, we define the *size* of $\langle D,X \rangle$ as $|D| + |X|$ and $twd^{Inc}(X)$ as the tree-width of the labeled graph $Inc(\langle D,X \rangle)$. From Proposition 1.56 and Theorem 1.54(1) we immediately obtain the next result.

Corollary 1.57 The problem BETWEENNESS is fixed-parameter linear with respect to twd^{Inc}. □

We now consider the similar notion of cyclic ordering. With a finite linear order $\langle D, \leq \rangle$ such that $|D| \geq 3$, we associate the ternary relation

$$C(x,y,z) :\Longleftrightarrow (x < y < z) \vee (y < z < x) \vee (z < x < y).$$

Let $D := \{d_1, \ldots, d_n\}$ with $d_1 < d_2 < \cdots < d_n$, where $d_1, \ldots, d_n$ are points on a circle such that, according to some orientation of the plane, d_{i+1} follows d_i and d_1 follows d_n. Then $C(x,y,z)$ expresses that, if one traverses the circle according to this orientation by starting at x, one meets y before z. We denote by $C(\leq)$ the ternary relation associated with $\leq$ in this way. A relation of this form is a *cyclic ordering*. A cyclic ordering C satisfies the following properties, for every x,y,z,t of its domain D:

(C1) $C(x,y,z) \Rightarrow x \neq y \wedge x \neq z \wedge y \neq z$;
(C2) $C(x,y,z) \Rightarrow C(y,z,x)$;
(C3) $C(x,y,z) \Rightarrow \neg C(x,z,y)$;
(C4) $C(x,y,z) \wedge C(y,t,z) \Rightarrow C(x,y,t) \wedge C(x,t,z)$;
(C5) $x \neq y \wedge x \neq z \wedge y \neq z \Rightarrow C(x,y,z) \vee C(x,z,y)$.

Every ternary relation satisfying (C1)–(C5) is a cyclic ordering. A subset X of D^3 is *consistent for cyclic ordering* if $X \subseteq C$ for some cyclic ordering on D. The problem CYCLIC ORDERING, which consists of deciding if a set $X \subseteq D^3$ is consistent for cyclic ordering, is NP-complete ([GalMeg]).

As for betweenness, for $X \subseteq D^3$, we let $\widehat{X}$ be the intersection of all cyclic orderings C on D such that $X \subseteq C$ (and $\widehat{X}$ is undefined if there is no cyclic ordering containing X). A *partial cyclic ordering* on a set D is defined as a subset of D^3 such that $\widehat{X} = X$. For cyclic ordering, we have the same results and open questions as for betweenness.

1.9.3 Relational databases

The theory of relational databases (exposed in the book [*AbiHV]) is based on relational structures. In this theory, a relational signature $\mathcal{R}$ is called a *database schema*, its elements are called *relation schemas*, and an $\mathcal{R}$-structure is called a *database instance*. A *query* is a syntactic or algorithmic description of a relation with specified arity written in some *query language* and defined in terms of the relations stored in the considered database instance. One concern is to compare the expressive powers of several such languages. Another one is to construct efficient algorithms for evaluating these relations, that is, to list their tuples or, sometimes, only to count them.

Theorem 1.54 yields linear-time algorithms for such computations (and for fixed queries) in cases where the input structures are constrained to belong to $STR_k(\mathcal{R})$ for some fixed k, or to satisfy some similar condition (e.g., for binary structures, to have bounded clique-width), and formulas are required to be monadic second-order. In the case of databases, there is usually no reason to assume that the relational structure modeling the database instance satisfies such constraints. Constraints are rather put on the formulas expressing queries in order to ensure the existence of efficient algorithms. These constraints are formulated in terms of tree-width and *hypertree-width* of certain graphs associated with formulas: we refer the reader to the comprehensive article by Gottlob *et al.* [GotLS]. Hence, the basic concepts of relational structures and logical formulas are the same as in the algorithms of Section 1.5, but the methods for constructing fixed-parameter tractable algorithms are not.

1.10 References

The collective book [*Com+] by Comon *et al.*, readable online, is a thorough study of finite automata on terms. Another reference is the book chapter [*GecSte].

Graph grammars defined in terms of graph rewritings are surveyed in two chapters ([*EngRoz], [*DreKH]) of the first volume of the handbook of graph grammars and graph transformations [*Roz] edited by Rozenberg. Another similar survey is the book chapter [*Eng97]. Most of the material referred to in Sections 1.1–1.8 is surveyed in [*Cou97], another chapter of [*Roz].

The books by Diestel [*Die] and by Mohar and Thomassen [*MohaTho] are our main references for general graph theory and for graphs embedded on surfaces respectively.

The books by Downey and Fellows [*DowFel] and by Flum and Grohe [*FluGro] present in detail the theory of Fixed-Parameter Tractability and contain important sections on tree-decompositions and their algorithmic applications. The surveys by Grohe [*Gro] and by Kreutzer [*Kre] focus on algorithms for problems expressed by first-order and monadic second-order sentences relative to graphs that are structured in various ways.

One of our objectives is to extend to finite graphs the algebraic view of Formal Language Theory initiated by Mezei and Wright [MezWri]. The least fixed-point characterization of context-free languages due to Ginsburg and Rice [GinRic] and to Chomsky and Schützenberger [ChoSch] has inspired the notion of equational sets, defined in [MezWri]. This article extends to general algebras the notion of recognizability studied for monoids by Eilenberg, Schützenberger and many others: see the books by Eilenberg [*Eil] and Sakarovitch [*Sak].

Monadic second-order logic on words, terms and trees, either finite or infinite, and its relationships with automata is a vast domain presented in the two book chapters by Thomas: [*Tho90] and [*Tho97a]. From this theory, we will only use Theorem 1.16 by Doner [Don] and Thatcher and Wright [ThaWri] that generalizes to terms the corresponding basic result established for words by Büchi [Büc], Elgot [Elg] and Trakhtenbrot [Tra].

We will not study countable graphs and structures. For this rich topic we refer the reader to the book chapters by Thomas ([*Tho90] and [*Tho97a]), and to the books [*GräTW] and [*FluGräW].

2

Graph algebras and widths of graphs

We will define graph operations that generalize the concatenation of words. Some of these operations have been used in Section 1.1 to define the cographs and the series-parallel graphs. We will define actually two signatures of graph operations, hence two graph algebras. Both algebras will be defined in such a way that their equational sets are exactly the sets defined by certain context-free graph grammars. (The notion of an equational set has been presented informally in Section 1.1.4 and will be studied in detail in Chapter 3.)

Two main types of *context-free sets of graphs* defined by certain graph rewriting mechanisms have emerged from the intense research conducted from around 1980 and synthesized in the handbook [*Roz] edited by Rozenberg.

We will first define the HR graph algebra that corresponds in this respect to the *hyperedge replacement grammars*. It turns out that the operations of this algebra yield an exact characterization (as equational sets) of the sets of graphs of *tree-width* bounded by fixed integers. The terms built with the operations of the HR algebra can be seen as algebraic expressions of tree-decompositions. (Tree-width and the corresponding tree-decompositions are important for the construction of efficient graph algorithms and also for the characterization of the graphs that exclude a fixed graph as a minor.)

The second algebra, called the VR algebra, is defined so that its equational sets are those generated by the (context-free) *vertex replacement grammars*. A new graph complexity measure called *clique-width* has arisen in a natural way from the definition of the VR algebra, without having (yet) any independent combinatorial characterization. Clique-width and tree-width are interesting parameters yielding *fixed-parameter tractable algorithms* for problems expressible in monadic second-order logic.

This chapter introduces by necessity many definitions and a lot of notation. We define algebras and terms in Section 2.1, graphs in Section 2.2, the HR algebra in Section 2.3, tree-decompositions and their relationships with the HR algebra in Section 2.4, the VR algebra and clique-width in Section 2.5. The HR and VR algebras are one-sorted, but for defining the notion of recognizability, we will need the many-sorted versions of these algebras that we define in Section 2.6.

2.1 Algebras and terms

For every set A, we denote by $Seq(A)$ the set of finite sequences of elements of A and by $s = (a_1, \ldots, a_n)$ the generic sequence. Its length is n, denoted by $|s|$. The empty sequence of length 0 is denoted by (). The concatenation of two sequences is denoted by $\cdot$, hence $(a_1, \ldots, a_n) \cdot (b_1, \ldots, b_m) = (a_1, \ldots, a_n, b_1, \ldots, b_m)$.

If A is an *alphabet*, i.e., a set[1] of letters (or symbols), an element of $Seq(A)$ is called a *word over A*. The notations A^*, $a_1 a_2 \cdots a_n$ and ε replace respectively $Seq(A)$, $(a_1, \ldots, a_n)$ and (). The $\cdot$ representing concatenation is frequently omitted. The i-th element of a sequence s (in particular of a word s) is denoted by $s[i]$. A *language over A* is subset of A^*. (A language in general is a set of words or a set of terms, see Definition 2.2 below.)

If k belongs to $\mathcal{N}$, the set of natural numbers (including 0), then $[k]$ denotes the set $\{1, \ldots, k\}$, so that $[0] = \emptyset$. The set $\mathcal{N} - \{0\}$ of positive integers is denoted by $\mathcal{N}_+$.

Definition 2.1 (Algebras) A *functional signature* F is a set of *function symbols*, each being given with a natural number called its *arity*. We denote by $\rho(f)$ the arity of f. We say that f is a *constant symbol* if it has arity 0. We let $F_i := \{f \in F \mid \rho(f) = i\}$, $F_+ := F - F_0$ and $\rho(F) := \max\{\rho(f) \mid f \in F\}$. A signature is *unary* (resp. *binary*) if $\rho(F) \leq 1$ (resp. $\rho(F) \leq 2$).

Let F be a (finite or infinite) functional signature. An *F-algebra* $\mathbb{M}$ is a set M equipped with total functions $f_\mathbb{M} : M^{\rho(f)} \to M$ for $f \in F$. If $f \in F_0$, then $f_\mathbb{M}$ is a function with no arguments, hence an element of M that we call a *constant*. We write $\mathbb{M} = \langle M, (f_\mathbb{M})_{f \in F} \rangle$. We call M the *domain* of $\mathbb{M}$ and the functions $f_\mathbb{M}$ (for f in F_+) its *operations*. The domain may be empty unless F_0 is nonempty.

The graph algebras $\mathbb{G}^{\mathrm{u}} = \langle \mathcal{G}^{\mathrm{u}}, \oplus, \otimes, \mathbf{1} \rangle$ and $\mathbb{J}_2^{\mathrm{d}} = \langle \mathcal{J}_2^{\mathrm{d}}, /\!/, \bullet, e \rangle$ have been defined in Section 1.1. In these cases and in many others we denote an operation $f_\mathbb{M}$ by f, that is, we do not distinguish an operation from the symbol denoting it. Algebras of terms and words are defined below (Definitions 2.2 and 2.7).

A signature H is a *subsignature* of F, which we denote by $H \subseteq F$, if H is a subset of F and the arity of f in H is the same with respect to H and to F. If $\mathbb{M}$ is an F-algebra and $\mathbb{N}$ is an H-algebra, then we say that $\mathbb{N}$ is a *subalgebra* of $\mathbb{M}$, which we denote by $\mathbb{N} \subseteq \mathbb{M}$, if H is a subsignature of F, $N \subseteq M$ and $f_\mathbb{N} = f_\mathbb{M} \upharpoonright N^{\rho(f)}$ for every $f \in H$ (in particular $f_\mathbb{N} = f_\mathbb{M}$ if $f \in F_0$).

A *homomorphism* $h : \mathbb{M} \to \mathbb{N}$, where $\mathbb{M}$ and $\mathbb{N}$ are two F-algebras, is a mapping $h : M \to N$ such that for every $f \in F$ and $m_1, \ldots, m_{\rho(f)} \in M$, we have

---

[1] Alphabets and signatures will always be finite or countably infinite.

$h(f_{\mathbb{M}}(m_1,\ldots,m_{\rho(f)})) = f_{\mathbb{N}}(h(m_1),\ldots,h(m_{\rho(f)}))$. In particular, $h(f_{\mathbb{M}}) = f_{\mathbb{N}}$ if $f \in F_0$. If h is a bijection, it is an *isomorphism* and $\mathbb{M}, \mathbb{N}$ are *isomorphic* algebras.

The *Cartesian product* of two F-algebras $\mathbb{M}$ and $\mathbb{N}$ is defined as the F-algebra $\mathbb{M} \times \mathbb{N} := \langle M \times N, (f_{\mathbb{M} \times \mathbb{N}})_{f \in F} \rangle$, where $f_{\mathbb{M} \times \mathbb{N}}((m_1, n_1), \ldots, (m_{\rho(f)}, n_{\rho(f)})) :=$ $(f_{\mathbb{M}}(m_1, \ldots, m_{\rho(f)}), f_{\mathbb{N}}(n_1, \ldots, n_{\rho(f)}))$ for all $m_1, \ldots, m_{\rho(f)}$ in M and $n_1, \ldots, n_{\rho(f)}$ in N.

A *monoid* is an F-algebra $\mathbb{M}$ such that $F = \{\cdot, 1\}$, where $\cdot$ is binary, 1 is a constant symbol, and the operation $\cdot_{\mathbb{M}}$ is associative with unit element $1_{\mathbb{M}}$. The symbol ϵ will also be used to denote the unit element.

Definition 2.2 (Terms)   Let F be a functional signature. The set $T(F)$ of *terms over* F is the unique subset L of F^* (here we consider F as an alphabet) such that:

$$L = \bigcup_{f \in F} fL \cdots L, \tag{2.1}$$

where, in each $fL \cdots L$, we have $\rho(f)$ occurrences of L.[2] This way of writing terms, due to Łukaciewicz, is called *Polish prefix notation*. This notation is *unambigous*, which means that every element of L has a unique expression of the form $ft_1 \cdots t_{\rho(f)}$ for $f \in F$ and $t_1, \ldots, t_{\rho(f)} \in L$. Inductive definitions and proofs will be based on this fact. The set of *subterms* of a term t is defined by the following induction:

$$Subterm(t) := \{t\} \cup Subterm(t_1) \cup \cdots \cup Subterm(t_{\rho(f)}),$$

if $t = ft_1 \cdots t_{\rho(f)}$ and $t_1, \ldots, t_{\rho(f)}$ are terms.

For signature F, a *language over* F is a subset of $T(F)$; since $T(F) \subseteq F^*$, it is also a language over the alphabet F.

Let $\mathbb{M}$ be an F-algebra. Every term $t \in T(F)$ *evaluates into an element* of M denoted by $t_{\mathbb{M}}$. Formally, we have:

$$t_{\mathbb{M}} := f_{\mathbb{M}}(t_{1\mathbb{M}}, \ldots, t_{\rho(f)\mathbb{M}}),$$

if $t = ft_1 \cdots t_{\rho(f)}$ and $t_1, \ldots, t_{\rho(f)} \in T(F)$; the value $t_{\mathbb{M}}$ is defined in a unique way because Polish prefix notation is unambigous. We will also denote by $val_{\mathbb{M}} : T(F) \rightarrow M$ the mapping that associates $t_{\mathbb{M}}$ with $t \in T(F)$. Let $F' \subseteq F$. A subset of M is *generated by* F' if it is a subset of $val_{\mathbb{M}}(T(F'))$. The subalgebra of $\mathbb{M}$ with signature F' and domain $val_{\mathbb{M}}(T(F'))$ is called the subalgebra of $\mathbb{M}$ generated by F'. In particular, we say that $\mathbb{M}$ is generated by F' if $M = val_{\mathbb{M}}(T(F'))$.

We let $\mathbb{T}(F)$ be the F-algebra with domain $T(F)$ such that:

$$f_{\mathbb{T}(F)}(t_1, \ldots, t_{\rho(f)}) := ft_1 \cdots t_{\rho(f)},$$

---

[2] There is a unique set $L \subseteq F^*$ satisfying Equality (2.1), see Proposition 3.15. If F is finite, then L is a context-free language.

for all $f \in F$ and $t_1, \ldots, t_{\rho(f)} \in T(F)$. It is called the *initial F-algebra* because, for every F-algebra $\mathbb{M}$, the mapping $val_{\mathbb{M}}$ is the unique homomorphism : $\mathbb{T}(F) \to \mathbb{M}$. This homomorphism is surjective if and only if $\mathbb{M}$ is generated by F. We also define $\mathbb{S}(F)$ as the F-algebra $\langle F^*, (f_{\mathbb{S}(F)})_{f \in F} \rangle$ such that each operation $f_{\mathbb{S}(F)}$ is defined like $f_{\mathbb{T}(F)}$, that is:

$$f_{\mathbb{S}(F)}(u_1, \ldots, u_{\rho(f)}) := fu_1 \cdots u_{\rho(f)},$$

for all $f \in F$ and $u_1, \ldots, u_{\rho(f)}$ in F^*. Then $\mathbb{T}(F)$ is the subalgebra of $\mathbb{S}(F)$ generated by F.

We now introduce some definitions relative to the internal structure of terms.

Definition 2.3 (Positions and occurrences of symbols) Since every term t over F is a word, it has a length $|t|$ in $\mathcal{N}$, also called its *size*. A *position* of t is an integer[3] i in the set $[|t|]$. It is an *occurrence* of the symbol $t[i]$. We denote by $Pos(t)$ the set of positions of t and by $Occ(t, f)$ the set of occurrences in t of a symbol f. Hence $Pos(t) = \bigcup \{Occ(t, f) \mid f \in F\}$. The *leading symbol of* t is $t[1]$. The *height* of t is defined inductively as follows:

$$ht(f) = 0 \text{ if } f \in F_0,$$
$$ht(ft_1 \cdots t_{\rho(f)}) = 1 + \max_{1 \leq i \leq \rho(f)} ht(t_i) \text{ otherwise.}$$

Definition 2.4 (Notational improvements) In order to improve readability, we will write terms with commas and parentheses. For example the term $fghabcggbcc$, where $\rho(f) = 3$, $\rho(g) = 2$, $\rho(h) = 1$, $\rho(a) = \rho(b) = \rho(c) = 0$, will be written $f(g(h(a),b),c,g(g(b,c),c))$. (We have used this notation in Chapter 1 in the definition of terms.)

With this notation the arities of the symbols occurring in a term need not be specified because they can be determined from the term (whereas for parsing Polish prefix notation, one needs to know the arities of symbols). In other words, we use in terms the same notation as for function application in the metalanguage.

Another notational variant will be used. If f is a binary function symbol intended to denote an associative operation, we will write $t_1 f t_2$ instead of $f(t_1, t_2)$. This notation is ambiguous because the two different terms $f(f(t_1, t_2), t_3)$ and $f(t_1, f(t_2, t_3))$ are denoted in the same way by $t_1 f t_2 f t_3$, but this ambiguity does not matter if we use this notation in cases where f denotes an associative operation. This notation can be used for a single associative binary operation. For writing terms with several such operations, we add parentheses. For example, we will write $(1 \oplus 1 \oplus 1) \otimes (1 \oplus 1 \oplus 1)$ the term $\otimes \oplus 1 \oplus 11 \oplus 1 \oplus 11$ that denotes the cograph $K_{3,3}$ (cf. Section 1.1.2). Another notation for this term is $\otimes(\oplus(1, \oplus(1,1)), \oplus(1, \oplus(1,1)))$. We might also use the notation $\oplus(1, \oplus(1,1)) \otimes \oplus(1, \oplus(1,1))$.

---

[3] Positions can also be designated by finite sequences of integers called *Dewey sequences*. This method is used in many articles and books, e.g., [*Cou83], [*Com+] and [*GecSte].

Definition 2.5 (Terms with variables) We will use terms written with *first-order variables*, i.e., variables that can denote arbitrary elements of the domains of algebras (and not functions or relations). If X is a set of variables (implicitly, of first-order variables), then $T(F,X)$ is the set of terms over $F \cup X$, where variables are symbols of arity 0. We always assume implicitly that $F \cap X = \emptyset$.

Let $X = \{x_1, \ldots, x_n\}$ be a set of variables listed in this order and t be a term in $T(F,X)$. In every F-algebra $\mathbb{M}$, the term t defines an n-ary total function $t_\mathbb{M} : M^n \to M$ such that for all $m_1, \ldots, m_n$ in M, $t_\mathbb{M}(m_1, \ldots, m_n)$ is the value of t in $\mathbb{M}$ computed by considering x_i as a constant symbol denoting m_i. This function will also be denoted by $\lambda m_1, \ldots, m_n \cdot t_\mathbb{M}(m_1, \ldots, m_n)$ or by $\lambda x_1, \ldots, x_n \cdot t$ if $\mathbb{M}$ is known from the context. (This notation has the advantage of specifying the list of argument variables.) Such a function is called a *derived operation* of $\mathbb{M}$. A term t with variables is *linear* if each variable has at most one occurrence in t. A derived operation is *linear* if it is defined by a linear term.

We denote by X_n the *standard set of variables* $\{x_1, \ldots, x_n\}$ (listed in this order) and similarly $Y_k := \{y_1, \ldots, y_k\}$.

Definition 2.6 (Substitutions, contexts and subterms) If $t \in T(F,X)$, if $v_1, \ldots, v_n$ are pairwise distinct variables in X and $t_1, \ldots, t_n \in T(F,X)$, then we denote by $t[t_1/v_1, \ldots, t_n/v_n]$ or equivalently by $t[t_i/v_i; i \in [n]]$ the result of the substitution in t of t_i for each occurrence of v_i. A variable v_i may have no or several occurrences in t, hence an argument t_i may disappear or be duplicated. For example, if $t = fgav_1v_1v_3$, then $t[t_1/v_1, t_2/v_2, t_3/v_3] = fgat_1t_1t_3$ for all terms t_1, t_2, t_3 in $T(F,X)$.

This operation has the following semantic meaning: if $s = t[t_1/v_1, \ldots, t_n/v_n]$ where $t, t_1, \ldots, t_n \in T(F, \{v_1, \ldots, v_n\})$, then for every F-algebra $\mathbb{M}$, the mapping $s_\mathbb{M}$ is the composition of the derived operations : $M^n \to M$ associated with $t, t_1, \ldots, t_n$, i.e., $s_\mathbb{M}(\overline{m}) = t_\mathbb{M}(t_{1\mathbb{M}}(\overline{m}), \ldots, t_{n\mathbb{M}}(\overline{m}))$ for every $\overline{m} \in M^n$.

A *context* over $F \cup X$ is a term t in $T(F, X \cup \{w\})$ where w is an auxiliary variable assumed not to belong to X and that has a unique occurrence in t. The term w is the *empty context*. We denote by $Ctxt(F,X)$ the set of contexts over $F \cup X$, and write $Ctxt(F)$ instead of $Ctxt(F, \emptyset)$. If $c \in Ctxt(F,X)$ and $t \in T(F,X)$ then $c[t]$ denotes the result of the substitution in c of t for w. The actual choice of w as auxiliary variable is irrelevant provided $w \notin X$. This variable does not appear in the notations $Ctxt(F,X)$ and $c[t]$.

If $t \in T(F,X)$ and $i \in Pos(t)$, then t can be written in a unique way as $t = \alpha t' \beta$, where $\alpha, \beta \in (F \cup X)^*$, $|\alpha| = i - 1$ and $t' \in T(F,X)$. We call t' the *subterm of t issued from position i*, and we denote it by t/i. (Hence $t/1 = t$.) Let w be a variable (not in X) to be used for defining contexts as terms. The term $c = \alpha w \beta$ belongs to $Ctxt(F,X)$ and we have $t = c[t']$. We will denote it by $t \uparrow i$ and call it the *context of t issued from i*. Hence, we have $t = (t \uparrow i)[t/i]$. For example, if $t = fgav_1v_1v_3$ (with $\rho(f) = 3, \rho(g) = 2, \rho(a) = 0$) then $t/2 = gav_1$ and $t \uparrow 2 = fwv_1v_3$.

If $t \in T(F,X)$, we denote by $ListVar(t)$ the word in X^* obtained by deleting from t all symbols from F. Every term t in $T(F,X_n)$ such that $ListVar(t) = x_{i_1} \cdots x_{i_k}$ is equal to $\tilde{t}[x_{i_1}/y_1,\ldots,x_{i_k}/y_k]$ for a unique term $\tilde{t}$ in $T(F,Y_k)$ such that $ListVar(\tilde{t}) = y_1 y_2 \cdots y_k$. For $t = fgax_3x_3x_1$ with f,g,a as above, we have $ListVar(t) = x_3x_3x_1$, $\tilde{t} = fgay_1y_2y_3$, $i_1 = i_2 = 3$ and $i_3 = 1$. It is clear that $t_{\mathbb{M}} = \lambda m_1,\ldots,m_n \cdot \tilde{t}_{\mathbb{M}}(m_{i_1},\ldots,m_{i_k})$.

Definition 2.7 (Three algebras of words) The set of words A^* over an alphabet A is an F_A-algebra where the signature F_A consists of a binary symbol $\cdot$ to be interpreted as the concatenation, a constant symbol ϵ to be interpreted as the empty word ε and a constant symbol a to be interpreted as the word a for each $a \in A$. We let $\mathbb{W}(A) := \langle A^*, \cdot, \varepsilon, (a)_{a \in A} \rangle$ be the corresponding algebra. It is generated by F_A (every word in A^* is the value of some term in $T(F_A)$). The algebra $\langle A^*, \cdot, \varepsilon \rangle$ is the free monoid generated by A.

However, two other algebraic structures on A^* will be relevant to our purposes. We let $U_A := A \cup \{\epsilon\}$ be the unary signature such that $\rho(\epsilon) = 0$ and $\rho(a) = 1$ for each a in A. We then let $\mathbb{W}_{right}(A) := \langle A^*, \varepsilon, (r_a)_{a \in A} \rangle$, where $r_a(w) := wa$ for every $a \in A$ and $w \in A^*$. Dually, $\mathbb{W}_{left}(A) := \langle A^*, \varepsilon, (l_a)_{a \in A} \rangle$, where $l_a(w) := aw$ for every $a \in A$ and $w \in A^*$.

These two algebras are generated by U_A and, for each of them, a word in A^* is the value of a unique term in $T(U_A)$. It is clear that if $a_1,\ldots,a_n \in A$ and $t = a_n(a_{n-1}(\cdots(a_1(\epsilon))\cdots))$ then $a_1 \cdots a_{n-1}a_n = val_{\mathbb{W}_{right}(A)}(t)$ and $a_na_{n-1}\cdots a_1 = val_{\mathbb{W}_{left}(A)}(t)$. Hence, we have two bijections from $T(U_A)$ to A^*, which are isomorphisms from $\mathbb{T}(U_A)$ to $\mathbb{W}_{right}(A)$ and $\mathbb{W}_{left}(A)$.

Definition 2.8 (Effectively given sets, signatures and algebras) We will use infinite signatures for defining graph algebras. We now discuss effectivity questions in order to obtain algorithms and not only abstract formalizations.

An *encoding* of a finite or countable set A is a triple $(|A|, enc_A, \xi_A)$ such that $|A|$ belonging to $\mathcal{N} \cup \{\omega\}$ is the cardinality of A, enc_A is a decidable subset[4] of $\mathcal{N}$ and ξ_A is a bijection : $A \to enc_A$. Since $enc_A = \xi_A(A)$, an encoding is uniquely determined by the bijection ξ_A, which will therefore also be called an encoding.

We fix *standard encodings* $(\omega, \mathcal{N}, \xi)$ of sets like $\mathcal{N}, \mathcal{N} \times \{0,1\}, \mathcal{N} \times \mathcal{N}$ and $Seq(\mathcal{N})$ where we assume that ξ and ξ^{-1} are computable. For instance, for the set $\mathcal{N}$, the mapping ξ is the identity on $\mathcal{N}$, and for the set $\mathcal{N} \times \{0,1\}$ we define $\xi(i,b) := 2i + b$. The set $\{True, False\}$ has the standard encoding $(2, \{0,1\}, \xi)$ with $\xi(True) := 1$ and $\xi(False) := 0$, and similarly for any fixed finite set. The set $\{x_1, x_2, \ldots\}$ of standard variables (see Definition 2.5) has the standard encoding $(\omega, \mathcal{N}, \xi)$ with $\xi(x_i) := i$. Let $A, A', A_1, \ldots, A_k$ be encoded sets. Standard encodings of sets like $A \cup A'$ (for A and A' disjoint), $A \times A'$ and $Seq(A)$ are derived in the obvious way from $\xi_A, \xi_{A'}$ and those of $\mathcal{N} \times \{0,1\}, \mathcal{N} \times \mathcal{N}$ and $Seq(\mathcal{N})$, respectively. To be precise, for $a, a_1, \ldots, a_m \in A$

---

[4] This means that the membership problem of enc_A is decidable, i.e., an algorithm can check whether or not an element of $\mathcal{N}$ belongs to enc_A.

and $a' \in A'$ we let $\xi_{A \cup A'}(a) := 2\xi_A(a)$ and $\xi_{A \cup A'}(a') := 2\xi_A(a') + 1$, $\xi_{A \times A'}(a,a') := \xi_{\mathcal{N} \times \mathcal{N}}(\xi_A(a), \xi_{A'}(a'))$, and $\xi_{Seq(A)}(a_1, \ldots, a_m) := \xi_{Seq(\mathcal{N})}(\xi_A(a_1), \ldots, \xi_A(a_m))$. A subset B of A is a *decidable subset* of A if $\xi_A(B) := \{\xi_A(b) \mid b \in B\}$ is a decidable subset of enc_A (and hence of $\mathcal{N}$). The standard encoding of B is the restriction of ξ_A to B, i.e., the triple $(|B|, \xi_A(B), \xi_A \upharpoonright B)$. A function $f : A_1 \times \cdots \times A_k \to A$ is *computable* (implicitly, via the encodings of $A_1, \ldots, A_k, A$) if the mapping

$$\widetilde{f} : enc_{A_1} \times \cdots \times enc_{A_k} \to enc_A,$$

such that $\widetilde{f}(x_1, \ldots, x_k) := \xi_A(f(\xi_{A_1}^{-1}(x_1), \ldots, \xi_{A_k}^{-1}(x_k)))$, is computable. For example, B is a decidable subset of A if and only if the mapping $g : A \to \{True, False\}$ is computable, where $g(a)$ is defined to be *True* if $a \in B$ and *False* otherwise.

Encoded sets will be given as input to algorithms and they will be produced as output by algorithms. An encoded set is *effectively given* if it is specified by its cardinality $|A| \in \mathcal{N} \cup \{\omega\}$ and an algorithm that decides the membership of an element of $\mathcal{N}$ in enc_A. It is *semi-effectively given* if it is specified by a membership algorithm only. If an encoded set A is finite and effectively given, then the set enc_A can be computed,[5] but if it is only semi-effectively given, then that is not possible in general. Conversely, if the finite set enc_A is given, then it is effectively given (i.e., its cardinality can be computed and a membership algorithm for enc_A can be constructed).

A signature F is *effectively given* if the set F is effectively given and the arity mapping $\rho : F \to \mathcal{N}$ is computable, i.e., the mapping $\widetilde{\rho} := \rho \circ \xi_F^{-1}$ is computable (and an algorithm computing $\widetilde{\rho}$ must be specified). It follows that the set of terms $T(F)$ is semi-effectively given by its standard encoding as a decidable subset of $Seq(F)$. If the integer i encodes the word $f_1 \cdots f_m$, with $f_1, \ldots, f_m \in F$, then the sequence $(\xi_F(f_1), \ldots, \xi_F(f_m))$ of integers that encode its function symbols can be computed from i, because $\xi_{Seq(\mathcal{N})}^{-1}$ is computable. It can then be checked if that sequence encodes a term t in $T(F)$, and if this is the case, the integers encoding its subterms can be computed because $\xi_{Seq(\mathcal{N})}$ is computable. In a similar way, the sets of terms $T(F, X_n)$ are effectively given, where X_n is the set of standard variables $\{x_1, \ldots, x_n\}$. With the standard encoding of $T(F)$, the algebra $\mathbb{T}(F)$ of terms over F is a semi-effectively given F-algebra, according to the next definition, as can easily be verified.

An F-algebra $\mathbb{M}$ is *effectively given* (respectively *semi-effectively given*) if its signature F is effectively given, its domain M is effectively given (respectively semi-effectively given) and its operations $f_{\mathbb{M}}$ are computable in the following uniform way: the mapping $\zeta_{\mathbb{M}} : Seq_{\mathbb{M}} \to M$ is computable (and an algorithm computing $\widetilde{\zeta}_{\mathbb{M}}$ must be specified), where $Seq_{\mathbb{M}}$ is the decidable subset of $F \times Seq(M)$ consisting[6] of all tuples $(f, x_1, \ldots, x_p)$ such that $f \in F$, $p = \rho(f)$ and $x_1, \ldots, x_p \in M$, and $\zeta_{\mathbb{M}}$ is defined

---

[5] Enumerate the elements i of $\mathcal{N}$, test whether i belongs to enc_A, and halt when $|A|$ elements of enc_A have been found.

[6] When an encoding of a set is not mentioned, as here for $Seq(M)$ and $Seq_{\mathbb{M}}$, we silently assume that its standard encoding is used.

by $\zeta_{\mathbb{M}}(f, x_1, \ldots, x_p) := f_{\mathbb{M}}(x_1, \ldots, x_p)$. It follows in particular that each operation is computable. Furthermore, the mapping $val_{\mathbb{M}} : T(F) \to M$ is also computable. The derived operations of $\mathbb{M}$ are computable in a similar way (with a term t in $T(F, X_p)$ instead of a function symbol f in F_p).

If $\mathbb{M}$ is semi-effectively given and generated by F, then there exists a computable "parsing" mapping $\pi : M \to T(F)$ that associates with every element m in M a term t such that $val_{\mathbb{M}}(t) = m$. To compute $\widetilde{\pi}(x)$ it suffices to enumerate the elements of $\mathcal{N}$ until some integer is found that encodes a term t such that $\xi_M(val_{\mathbb{M}}(t)) = x$, and this equality can be checked since $val_{\mathbb{M}}$ is computable. If M is finite and effectively given, such a mapping $\widetilde{\pi}$, or equivalently π itself, can be computed once and for all into a table.

2.2 Graphs

We assume that the basic terminology of graph theory is well known. In this section we mainly fix notation and give some definitions that are not standard. We pay particular attention to the distinction between a *concrete graph* with specified sets of vertices and edges, and an *abstract graph*, which is the isomorphism class of a concrete graph.

There are many types of graphs. They can be directed or not, with or without loops and/or multiple edges; they can be "decorated" with labels, colors or weights. We give formal definitions for a few basic cases and we indicate how they extend to "decorated" graphs.

Definition 2.9 (Graphs) A *directed graph* G consists of a set of *vertices* V_G, a set of *edges* E_G and a mapping $vert_G$ that associates with every edge e in E_G an ordered pair of vertices $vert_G(e) \in V_G \times V_G$. An *undirected* graph G is a triple $\langle V_G, E_G, vert_G \rangle$ as above except that $vert_G(e)$ is a set of one or two vertices for each edge e in E_G.[7]

In many cases, we leave $vert_G$ implicit and only use V_G and E_G. We always assume that $V_G \cap E_G = \emptyset$, and when we discuss several graphs G, H, we assume that $V_G \cap E_H = \emptyset$ except when specified otherwise (e.g., for *line graphs*; cf. Section 1.8.2, Example 1.45). If G is directed and $vert_G(e) = (u, v)$ (we may have $u = v$) we often write $e : u \to v$ and say that e *links* u to v or that e is *from* u to v. If G is undirected and $vert_G(e) = \{u, v\}$ (possibly $u = v$) we write $e : u - v$ and say that e links u *and* v or that e is *between* u and v. (We write $u \to_G v$ or $u -_G v$ if it is useful to specify the graph G.) If $e : u \to v$ or $e : u - v$ we say that e is *incident* with u and v, and that u and v are the *end vertices* (or *ends*) of e. In the former case, we say that u is the *tail* of e and v its *head*. In both cases we say that u and v are *adjacent* or that they are *neighbors*. The *degree* of a vertex is the number of its incident edges, and (in the directed case) its *indegree* (*outdegree*) is the number of its *ingoing* (*outgoing*)

---

[7] For a more precise set theoretical formalization, one could choose a countable set $\mathcal{U}$ containing $\mathcal{N}$ that is closed under pairing (i.e., is such that (x, y) and $\{x, y\}$ are in $\mathcal{U}$ if $x, y \in \mathcal{U}$), and require for all graphs G that V_G and E_G are contained in $\mathcal{U}$. In this way, the collection of all graphs forms a set.

edges, i.e., the edges of which it is the head (the tail). A vertex is *isolated* if it has degree 0. The degree of a graph is the maximal degree of its vertices, and similarly for the indegree and outdegree of a graph. Two (distinct) edges e and f are *parallel* if $vert_G(e) = vert_G(f)$. A graph is *simple* if it has no pair of parallel edges. Parallel edges are also called *multiple edges*. We say that e is a *loop* if $e : u \to u$ or $e : u - u$, and in this case say that u *has a loop*. A graph is *loop-free* if no vertex has a loop.

For a simple directed graph G, one can identify an edge e with the ordered pair $vert_G(e)$. Such a graph can be specified more simply as a pair $\langle V_G, edg_G \rangle$, where $edg_G \subseteq V_G \times V_G$ is the binary relation such that $(u, v) \in edg_G$ if and only if $e : u \to v$ for some edge e. Every binary relation on a set V defines a graph with set of vertices V.

A simple undirected graph G can be specified similarly as a pair $\langle V_G, edg_G \rangle$, where edg_G is the symmetric relation such that $(u, v) \in edg_G$ if and only if $e : u - v$ for some e in E_G. It follows that a simple undirected graph can be considered as a simple directed graph with a symmetric edge relation, i.e., such that every edge (u, v) that is not a loop has an *opposite edge* (v, u). However, this identification is not always convenient for our grammatical and logical treatments. In many cases, we will consider directed and undirected graphs as distinct species and use explicit codings between directed and undirected graphs whenever appropriate.

For $G = \langle V_G, E_G, vert_G \rangle$, directed, we denote by $und(G)$ the undirected graph $\langle V_G, E_G, vert_{und(G)} \rangle$ such that $vert_{und(G)}(e) = \{u\}$ if $e : u \to_G u$ and $vert_{und(G)}(e) = \{u, v\}$ if $e : u \to_G v$, $u \neq v$, and call it the undirected graph *underlying* G. Even if G is simple, the graph $und(G)$ may have pairs of parallel edges (coming from pairs of opposite directed edges). An *orientation* of an undirected graph is a directed graph obtained by choosing a direction for each edge (i.e., choosing a head and a tail). Hence, G is an orientation of H if and only if $und(G) = H$. The *core* of a graph G is the graph $core(G)$ obtained from $und(G)$, or from G if G is undirected, by deleting loops and by fusing parallel edges. Thus, $core(G)$ is simple, loop-free and undirected.

A *clique* is a simple, loop-free, undirected graph, such that any two distinct vertices are adjacent. We denote by K_n the isomorphism class of cliques with n vertices. (Graph isomorphisms are formally defined in Definition 2.10 below.) A *clique in a graph G* (directed or not) is a set of vertices of G such that any two vertices are adjacent.

Let G be an undirected graph, and $u, v \in V_G$. A *walk* from u to v (or, linking u to v) is a sequence of edges $(e_1, \dots, e_n)$ such that, for some sequence of vertices $(u_0, \dots, u_n)$ with $u = u_0$ and $u_n = v$, we have $vert_G(e_i) = \{u_{i-1}, u_i\}$ for each $i = 1, \dots, n$. This walk is a *path* if $u_i \neq u_j$ for every i and $j \neq i$. It will be convenient to view, for $n = 0$, the empty sequence () as an *empty path* from u to u for each vertex u. A *cycle* is a walk $(e_1, \dots, e_n)$ such that $n \geq 1$, $u_0 = u_n$, $e_i \neq e_j$ if $i \neq j$ and $u_i \neq u_j$ if $0 \leq i < j < n$. In cases where this does not create ambiguity, especially for simple graphs (such as trees), we will designate a path by the sequence $(u_0, u_1, \dots, u_n)$ of its vertices. This sequence is not empty, even if $n = 0$.

If G is a directed graph, the notions of walk, path and cycle are those relative to $und(G)$. A *directed walk* is a walk $(e_1,\ldots,e_n)$ as above such that $vert_G(e_i) = (u_{i-1},u_i)$ for each i. The notions of *directed path* and of *directed cycle* (also called a *circuit*) follow accordingly.

A graph G is *connected* if any two vertices are the ends of a path.

A graph G is *finite* if the set $V_G \cup E_G$ is finite. We will mostly consider finite graphs. However, certain expressions of graph properties by logical formulas will be valid for infinite graphs. We will indicate if finiteness is crucial for the validity of such formulas.

We denote by $\varnothing$ the *empty graph*, such that $V_\varnothing = E_\varnothing = \emptyset$ (where $\emptyset$ denotes the empty set). The empty graph is connected.

We define the *size* $\|G\|$ of a finite graph G as $|V_G| + |E_G|$. This notion will be used for analyzing algorithms. In many cases, and in particular for algorithms taking simple graphs as input, the appropriate notion of size will be $|G| := |V_G|$.

Definition 2.10 (Isomorphisms) Let G and H be graphs, either both directed or both undirected. An *isomorphism* $h : G \to H$ is a pair of bijections (h_V, h_E), where $h_V : V_G \to V_H$, $h_E : E_G \to E_H$, $vert_H(h_E(e)) = (h_V(u), h_V(v))$ if $e : u \to v$, and $vert_H(h_E(e)) = \{h_V(u), h_V(v)\}$ if $e : u - v$. If G and H are simple, it suffices to specify h_V because, if (h_V, h_E) is an isomorphism $: G \to H$, then h_E is determined in a unique way from h_V by the conditions of the definition. In this case, an isomorphism $:$ $G \to H$ can be more simply defined as a bijection $h_V : V_G \to V_H$ such that some bijection $h_E : E_G \to E_H$ exists that makes (h_V, h_E) into an isomorphism. We denote by $G \simeq H$ the existence of an isomorphism $: G \to H$ and we say that G and H are *isomorphic*.

All our results will concern *graphs up to isomorphism*. To establish them, we will need to perform constructions on graphs with specific sets of vertices and edges. Hence we refine our terminology as follows: a graph $G = \langle V_G, E_G, vert_G \rangle$ is called a *concrete graph*. By an *abstract graph*, we mean the isomorphism class, denoted by $[G]_{iso}$, of a concrete graph G. We say that a concrete graph G is *isomorphic to an abstract graph* $[H]_{iso}$ if $G \simeq H$, i.e., if G belongs to the isomorphism class $[H]_{iso}$; this is also written as $G \simeq [H]_{iso}$.

A *graph property* is a property of concrete graphs invariant under isomorphism, hence a property of abstract graphs. By a *class of graphs* we mean a set of concrete graphs closed under isomorphism, hence equivalently, a set of abstract graphs.[8] In most circumstances we will leave implicit the distinction between a concrete and an abstract graph. Roughly speaking, statements concern abstract graphs while proofs, usually starting with "Let $G = \langle V_G, E_G, vert_G \rangle$ be a graph," deal by necessity with concrete graphs.

---

[8] We will never use "class" in the set-theoretical sense of a collection of objects that may not form a set. By the footnote in Definition 2.9, every collection of graphs is a set.

Definition 2.11 (Concrete and abstract labeled graphs)   A *labeled* (or *colored*) *graph* is a graph equipped with additional information, formalized as follows. A *labeling function* (or *coloring function*) of a concrete graph G is a mapping γ from $V_G \cup E_G$ into a set Γ. In most cases, Γ will be finite.

For example, finite automata recognizing words can be seen as labeled graphs: a finite automaton is a directed graph such that certain vertices, called its *states*, are labeled as *initial* or *final* (or *accepting*). A state may be labeled as both initial and final. Furthermore, each *transition*, i.e., each edge of the corresponding graph, has a label which is a letter from a finite alphabet or the symbol ε denoting the empty word.

The following notion of labeled graph includes these graphs. Let K and Λ be two finite disjoint[9] sets of labels. A (K, Λ)-*labeled graph* is a graph, directed or not, such that:

- each edge has one and only one label, which belongs to Λ; and
- each vertex has a possibly empty set of labels belonging to K.

The corresponding labeling function maps $V_G \cup E_G$ into $\Gamma := \mathcal{P}(K) \cup \Lambda$. Such a graph G is *simple* if no two edges have the same pair (ordered if G is directed, unordered if G is undirected) of end vertices and the same label. We may have parallel edges with different labels. For example, a finite automaton over an alphabet A is a simple, directed, (K, Λ)-labeled graph where $K = \{initial, final\}$ and $\Lambda = A \cup \{\varepsilon\}$.

If $\mathcal{R}$ is a binary relational signature (cf. Section 1.9.1) then $\mathcal{R}$-structures correspond bijectively to simple, directed $(\mathcal{R}_1, \mathcal{R}_2)$-labeled graphs, where $\mathcal{R}_i$ is the set of symbols in $\mathcal{R}$ of arity i. The graph G corresponding to an $\mathcal{R}$-structure $S = \langle D_S, (R_S)_{R \in \mathcal{R}} \rangle$ has vertex set D_S, it has an edge from u to v labeled by R in $\mathcal{R}_2$ if and only if $(u, v) \in R_S$, and a vertex u has label R in $\mathcal{R}_1$ if and only if $u \in R_S$. Conversely, every simple, directed (K, Λ)-labeled graph G is associated in this way to a binary relational $\mathcal{R}$-structure such that $(\mathcal{R}_1, \mathcal{R}_2) = (K, \Lambda)$.

According to the definition, a $(K, \emptyset)$-labeled graph has no edges. However, we will call $(K, \emptyset)$-labeled a graph with unlabeled edges (equivalently, with all edges having the same label) and vertices with labels in K. In this way, we need not specify a default edge label.

We now define isomorphisms of labeled graphs, from which follows the notion of an *abstract labeled graph*. If G and H are concrete graphs with labeling functions γ_G and γ_H, we say that an isomorphism $h : G \to H$ is an *isomorphism of labeled graphs* if $\gamma_H(h_V(u)) = \gamma_G(u)$ for every $u \in V_G$, and similarly for h_E. Hence, we can define abstract labeled graphs as isomorphism classes of concrete labeled graphs. We use the same terminology as in the case of unlabeled graphs.

These definitions apply to *weighted graphs*, where a *weight function* is a total or partial function from $V_G \cup E_G$ to the set of real numbers or to another set usually equipped with an algebraic structure. Each time we will use labeled or weighted

---

[9] The assumption that K and Λ are disjoint simplifies some definitions and statements but is not essential.

graphs, we will specify the particular properties of the labeling or weight functions under consideration.

Definition 2.12 (Subgraphs) Let G and H be concrete graphs, either both directed or both undirected. We say that H is a *concrete subgraph of* G, written $H \subseteq G$, if $V_H \subseteq V_G, E_H \subseteq E_G$ and $vert_H(e) = vert_G(e)$ for all $e \in E_H$. It is an *induced subgraph*, written $H \subseteq_i G$, if E_H is the set of edges of G having their ends in V_H, and it is a *spanning subgraph* if $V_H = V_G$. If G and H are abstract graphs, saying that H is a *subgraph of* G means that H is isomorphic to a concrete subgraph of a concrete graph isomorphic to G.

If G is labeled and $H \subseteq G$, then the edges and vertices of H have the same labels as in G.

The induced subgraph of a concrete graph G with set of vertices $X \subseteq V_G$ is denoted by $G[X]$. We let $G - X := G[V_G - X]$ and $G - u := G - \{u\}$ if u is a vertex. If $F \subseteq E_G$, then the subgraph $G[F]$ of G is such that $E_{G[F]} = F$ and $V_{G[F]}$ is the set of vertices incident with an edge in F, and its subgraph $G - F$ is defined by $V_{G-F} = V_G$ and $E_{G-F} = E_G - F$; we let $G - e := G - \{e\}$ if e is an edge.

Definition 2.13 (Trees) A *tree* is a nonempty connected undirected graph without cycles. It is thus simple, loop-free and every two vertices are linked by a unique path. A *rooted tree* is a directed graph T such that $und(T)$ is a tree, all vertices have indegree 1 except one which has indegree 0; this vertex is called the *root* of T and is denoted by $root_T$. In a rooted tree, every vertex is accessible from the root by a unique directed path. The edge directions of a rooted tree T are defined in a unique way from $und(T)$ and $root_T$. Any vertex of a tree can be chosen as a root so as to turn the tree into a rooted tree. Hence, in many cases, we will specify a rooted tree as a pair (U,r) consisting of a tree U and a vertex r.

A *forest* (a *rooted forest*) is a (possibly empty) graph, each connected component of which is a tree (a rooted tree). A rooted forest has thus a (possibly empty) set of roots, and can be defined by the pair (F,R) of a forest F and a set of vertices R containing one and only one vertex from each connected component. Every subgraph of a (rooted) forest is a (rooted) forest, and every nonempty connected subgraph of a (rooted) tree is a (rooted) tree. The height $ht(H)$ of a rooted forest H is the maximal length (i.e., number of edges) of a directed path of H.

In many cases, we will discuss simultaneously a graph and a tree representing its structure. It will be convenient to call the vertices of the tree *nodes* in order to distinguish them from the vertices of the graph.

Since it is a simple graph, a tree or a forest H can be specified as a pair $H = \langle N_H, edg_H \rangle$, where N_H is the set of its nodes. If H is a rooted forest and $u \in N_H$, we denote by H/u the *subtree issued from* u. It is the rooted tree such that $N_{H/u}$ is the set of nodes reachable from u by a directed path (this path may be empty, so that u belongs to $N_{H/u}$) and $edg_{H/u} = edg_H \cap (N_{H/u} \times N_{H/u})$. If H is labeled or weighted, then H/u inherits the labels or weights of H in the obvious way. For a rooted forest H we will denote by son_H the relation edg_H, and will use the classical terminology for rooted trees: leaf, son, father, ancestor, descendant, least common ancestor of two nodes, etc. More precisely, for $u, v \in N_H$ we say that u is an *ancestor* of v, or that v is a *descendant* of u, which we denote by $v \leq_H u$, if there is a directed path in H from u to v, i.e., if $v \in N_{H/u}$; in order to exclude the equality[10] $u = v$, we say that u is a *proper ancestor* of v, or that v is a *proper descendant* of u, and write $v <_H u$. We also say that v is *below* u and that u is *above* v if $v \leq_H u$; to exclude equality, we say that v is *strictly below* u or that u is *strictly above* v. A *topological order on H* is a linear order $\leq$ on its set of nodes that contains $\geq_H$ (i.e., $v \leq_H u$ implies $u \leq v$).

A computation on H is a *bottom-up computation* if it processes sequentially the nodes of H, in such a way that the computations done at node u can use results of computations done at some nodes v below u. Hence, it can be done in any order on the nodes such that v is processed before u if $v <_H u$, that is, in the reverse order of any topological order.

A *spanning tree* of an undirected graph G is a spanning subgraph of G that is a tree. If $F \subseteq E_G$ and $G[F]$ is a spanning tree of G, then we say that F *forms* the spanning tree $G[F]$. A spanning tree of a directed graph G is, by definition, one of $und(G)$. A *rooted spanning tree* is a rooted tree T such that $und(T)$ is a spanning tree. The direction of an edge of a rooted spanning tree need not coincide with its direction in G when G is directed. A rooted spanning tree T is *normal* if every two vertices that are adjacent in G are comparable with respect to $\leq_T$. Every nonempty connected graph has a normal spanning tree that can be constructed by a depth-first traversal starting from any given vertex.[11]

Definition 2.14 (Syntactic trees of terms)   Let $t \in T(F)$. Its *syntactic tree* is the concrete labeled rooted tree T denoted by $Syn(t)$ such that $N_T = Pos(t)$ and:

(i) $root_T$ is 1, the first position of t;
(ii) a node u of T is labeled by $f \in F$ if and only if $u \in Occ(t, f)$, i.e., if and only if $t[u] = f$ (u is a positive integer and the symbol at occurrence u in t is the u-th element of t, since we use Polish prefix notation for terms);
(iii) there is in T an edge $u \to v$ labeled by i in $\mathcal{N}_+$ if and only if $t = \alpha f t_1 \cdots t_{\rho(f)} \beta$ for some $\alpha, \beta \in F^*$ and $f \in F$ such that $|\alpha| = u - 1$ (so that $u \in Occ(t, f)$), $t_1, \ldots, t_{\rho(f)} \in T(F)$ and $v = |\alpha f t_1 \cdots t_{i-1}| + 1$; hence the outdegree of u is $\rho(f)$. (We have $t/u = f t_1 \cdots t_{\rho(f)}$ and $t/v = t_i$.)

Hence $Syn(t)$ is an $(F', [\rho(F')])$-labeled graph for every finite subsignature F' of F such that $t \in T(F')$. Figure 2.1 shows the tree $Syn(t)$ for the term $t =$

---

[10] For every partial order $\leq$ we denote by $<$ the corresponding strict partial order, i.e., $x < y$ if and only if $x \leq y$ and $x \neq y$. This convention applies to all partial orders defined here or below: $\leq_H$, $\leq_t$, etc.

[11] Every *countable* connected graph has a normal spanning tree that is not necessarily obtained by depth-first search (Theorem 8.2.4 of [*Die]).

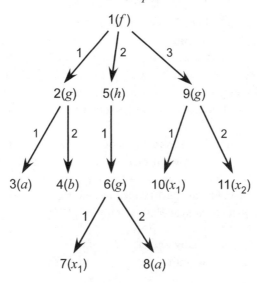

Figure 2.1 A syntactic tree.

$f(g(a,b),h(g(x_1,a)),g(x_1,x_2))$ that belongs to $T(\{f,g,h,a,b\},\{x_1,x_2\})$. The set of nodes of $Syn(t)$ is $[11]$. The symbol $t[i]$ at position i in t is indicated between parentheses.

For each $u \in Pos(t)$, the tree $Syn(t)/u$ is an induced subgraph of $Syn(t)$ and is isomorphic to $Syn(t/u)$. (Its set of vertices is the set of descendants of u in $Syn(t)$.) The corresponding bijection $h_u : N_{Syn(t/u)} \to N_{Syn(t)/u} \subseteq N_{Syn(t)}$ is defined by $h_u(i) = i + u - 1$. If for example $t = g(f(a,b),f(a,b))$ then we have $t/2 = t/5 = f(a,b)$. The two mappings h_2 and h_5 such that $h_2(1) = 2$, $h_2(2) = 3$, $h_2(3) = 4$, $h_5(1) = 5$, $h_5(2) = 6$, $h_5(3) = 7$ are isomorphisms from $Syn(f(a,b))$ to the two subtrees $Syn(t)/2$ and $Syn(t)/5$ of $Syn(t)$. More generally, two subterms t/u and t/v of a term t are equal if and only if $Syn(t)/u \simeq Syn(t)/v$.

Similarly, if $u \in Pos(t)$, then the syntactic tree of the context $t \uparrow u$ is isomorphic (up to the label of u) to the subgraph of $Syn(t)$ that is induced by the set of vertices consisting of u and those that are not descendants of u. We denote by $Syn(t) \uparrow u$ the labeled graph obtained from this subgraph by changing the label of u into the special variable w (or another special variable if this is more convenient). Then $Syn(t \uparrow u)$ is isomorphic to $Syn(t) \uparrow u$: a position i of $t \uparrow u$ corresponds to i in $Pos(t)$ if $1 \leq i \leq u$, and to $i + |t/u| - 1$ if $u < i \leq |t|$.

We will use for positions in terms the terminology relative to trees: leaf, son, ancestor, root. In particular, if u is an occurrence in t of $f \in F_k$, then it has a sequence of k sons, $(u_1,\dots,u_k)$, (u_i will be called the *i-th son* of u) and we have $t/u = f(t/u_1,\dots,t/u_k)$.

The syntactic tree $T = Syn(t)$ of $t \in T(F)$ can be specified by the tuple $\langle N_T, (son_{iT})_{1 \leq i \leq r}, lab_T \rangle$, where r is the maximal arity of a symbol that has an occurrence in t, $son_{iT}(u,v)$ holds if and only if v is the i-th son of u, and $lab_T(u) = t[u]$. If F is finite, this tuple can be replaced by a relational structure over a binary relational signature (see Chapter 5). We will usually denote N_T by N_t instead of by $N_{Syn(t)}$, $\leq_T$ by $\leq_t$, son_{iT} by son_{it}, etc., when T is the syntactic tree of t. Moreover, for a node $u \in N_t$, we will denote the set of nodes of $Syn(t)/u$ by N_t/u, to distinguish it from $N_{t/u}$ which is the set of nodes of $Syn(t/u)$; similarly, $N_t \uparrow u$ will denote the set of nodes of $Syn(t) \uparrow u$.

The term t can be recovered from any labeled rooted tree that is isomorphic to the syntactic tree $Syn(t)$. We will call such a tree a syntactic tree of t, as opposed to *its* syntactic tree. Thus, the mapping $t \mapsto [Syn(t)]_{iso}$ is injective, which shows that terms can be represented by abstract labeled rooted trees. A labeled rooted tree T is *a* syntactic tree of *some* term over F if and only if:

every node u of T has exactly one label f from F, its outdegree is $\rho(f)$ and its outgoing edges are numbered by $1, \ldots, \rho(f)$, such that each number in $[\rho(f)]$ labels a single edge.

For a finite signature F, a mapping $t \mapsto val(t)$ from $T(F)$ to some finite set M is *computable inductively on the structure of* t if $val(f(t_1, \ldots, t_k)) = v_f(val(t_1), \ldots, val(t_k))$ for all $f \in F$ and $t_1, \ldots, t_k \in T(F)$ where $k = \rho(f)$ and v_f is a fixed mapping : $M^k \to M$. Equivalently, the mapping val is equal to $val_{\mathbb{M}}$ for some F-algebra $\mathbb{M}$ with domain M (and $v_f = f_{\mathbb{M}}$). Such a mapping val can be computed bottom-up on the syntactic tree of t, and in linear time (in the size of t) if the mappings v_f are computable in constant time (that is in time depending only on "fixed" parameters).

A *parsing algorithm for* a finite signature F takes as input a word in F^* and either outputs the answer that this word is not a term in $T(F)$ or constructs its syntactic tree if it is (with the appropriately linked data structure that makes it easy to traverse this tree later on). Textbooks on compilation (e.g., [*AhoLSU] and [*Cre]) present such algorithms. Let us only note that the natural order on positions of a term t (defined as integers) is a topological order of its syntactic tree. By reading the word t from right to left and by using a pushdown to store the necessary information, one can evaluate t in a given algebra, compute the unique run on it of a deterministic automaton (see Section 3.3), build its syntactic tree or, more generally, perform a bottom-up computation.[12] Such a computation takes linear time on t given as a word if it takes linear time on the syntactic tree of t.

Definition 2.15 (Quotient graphs and minors) Let G be a concrete graph, directed or not. Let $\approx$ be an equivalence relation on V_G. The *quotient graph* of G by $\approx$ is obtained, informally, by fusing any two equivalent vertices into a single one. We do not fuse or delete edges. However, we wish to define this graph as a concrete

---

[12] In this book, we will not give any more details on data structures and implementations of algorithms.

graph by specifying its vertices. We denote by $[u]$ the equivalence class of $u \in V_G$. A *cross-section* of $\approx$ is a set $X \subseteq V_G$ that has one and only one element $\bar{u}$ in each equivalence class $[u]$.

For every cross-section X of $\approx$, we define the *concrete quotient graph* $(G/\approx)_X$ as the graph H such that:

$$V_H := X,$$
$$E_H := E_G,$$
$$vert_H(e) := (\bar{u},\bar{v}) \text{ or } \{\bar{u},\bar{v}\} \text{ or } \{\bar{u}\}$$
$$\quad \text{if } vert_G(e) \text{ is } (u,v) \text{ or } \{u,v\} \text{ or } \{u\} \text{ respectively.}$$

If the edges of G are labeled, then their labels are transferred to those of H in the obvious way. If G is a (K,Λ)-labeled graph, then a vertex of H inherits all labels of the vertices of G that are equivalent to it.

For any two cross-sections X and Y, the graphs $(G/\approx)_X$ and $(G/\approx)_Y$ are isomorphic in a canonical way (u in X is mapped to u' in Y such that $u \approx u'$, and the correspondence between the edge sets $E_{(G/\approx)_X}$ and $E_{(G/\approx)_Y}$ is the identity). Hence, we will write $G/\approx$ instead of $(G/\approx)_X$ if X need not be specified and we will consider it as the quotient graph of G by $\approx$. (This definition will facilitate the description of the quotient construction as a monadic second-order transduction to be done in Section 7.1.1, cf. Example 1.35 in Section 1.7.1.)

If G and G' are isomorphic by an isomorphism h that maps an equivalence relation $\approx$ on V_G to an equivalence relation $\approx'$ on $V_{G'}$ (i.e., $h(u) \approx' h(v)$ if and only if $u \approx v$), then $(G/\approx)_X$ and $(G'/\approx')_Y$ are isomorphic for any two cross-sections X and Y of $\approx$ and $\approx'$. Hence in such a case the quotient by an equivalence relation is well defined on abstract graphs.

Edge contraction is a particular type of quotient. Let G be a concrete graph and $F \subseteq E_G$. We let $\approx_F$ be the equivalence relation generated by the set of pairs $vert_G(e)$ for e in F. A cross-section of $\approx_F$ is thus a set of vertices X that has one and only one element in each connected component of the subgraph of G with vertex set V_G and F as set of edges. For a cross-section X of $\approx_F$, we define $(G/F)_X$ as the graph $((G-F)/\approx_F)_X = (G/\approx_F)_X - F$. Again $(G/F)_X$ and $(G/F)_Y$ are canonically isomorphic if X and Y are any two cross-sections. We will use the simplified notation G/F if we need not specify any cross-section. We let $G/e := G/\{e\}$ if e is an edge. Contracting a loop is the same thing as deleting it.

A concrete graph H is a *minor* of a concrete graph G if $H = (G'/F)_X$ for some concrete subgraph G' of G, some set $F \subseteq E_{G'}$ and some cross-section X of $\approx_F$ relative to G'. Note that $V_H = X \subseteq V_{G'} \subseteq V_G$ and $E_H = E_{G'} - F \subseteq E_{G'} \subseteq E_G$, but the incidences are not the same in H and in G. We denote this relation by $H \trianglelefteq G$. Here is some further notation: $H \trianglelefteq_c G$ means $H = (G/F)_X$ (hence $G' = G$ in the initial definition). We say that H is a *proper minor* of G if $F \neq \emptyset$ or $G' \neq G$ or both conditions hold.

Proposition 2.16 Let H be a minor of a concrete graph G. The following are equivalent:

(1) $H = G$;
(2) $V_H = V_G$ and $E_H = E_G$;
(3) $|V_H| = |V_G|$ and $|E_H| = |E_G|$.

Then H is a proper minor of G if and only if $H \lhd G$, i.e., $H \unlhd G$ and $H \neq G$.

Proof: Let $H \unlhd G$. By the definition we have $V_H \subseteq V_G$ and $E_H \subseteq E_G$.

(1) $\Longrightarrow$ (3) is clear, and (3) $\Longrightarrow$ (2) is immediate from the two above inclusions, since graphs are finite. We now prove (2) $\Longrightarrow$ (1). Let $H = (G'/F)_X$ with $V_H = V_G$ and $E_H = E_G$. As observed above, $V_H = X \subseteq V_{G'} \subseteq V_G$ and $E_H = E_{G'} - F \subseteq E_{G'} \subseteq E_G$. Hence $V_{G'} = V_G$, $F = \emptyset$, and $E_{G'} = E_G$. Since $G' \subseteq G$, we have $G' = G$ and $F = \emptyset$. Hence, by definition, $H = (G'/F)_X = G$. ∎

Proposition 2.17

(1) The relation $\unlhd$ is a partial order on concrete graphs.
(2) A concrete graph H is a proper minor of a concrete graph G if and only if H is obtained from G by a nonempty sequence of edge deletions $K \mapsto K - e$, vertex deletions $K \mapsto K - u$ and edge contractions $K \mapsto K/e$ (where K is a concrete graph, $e \in E_K$ and $u \in V_K$).

Proof: (1) It is clear that $G \unlhd G$. Let us assume $H \unlhd G \unlhd H$. We have $V_H \subseteq V_G \subseteq V_H$ and $E_H \subseteq E_G \subseteq E_H$ hence $H = G$ by (2) $\Longrightarrow$ (1) of Proposition 2.16.

Next we consider transitivity. We let $K \unlhd H \unlhd G$. By the definitions we have $K \unlhd_c H' \subseteq H \unlhd_c G' \subseteq G$. We first prove that if $K \unlhd_c H \unlhd_c G$ then $K \unlhd_c G$. We let $K = (H/L)_Y$ and $H = (G/F)_X$ with $Y \subseteq X$. We have $K = (G/(L \cup F))_Y$, which gives $K \unlhd_c G$.

To complete the proof we verify that, if $H' \subseteq H \unlhd_c G'$ with $H = (G'/F)_X$ then $H' \unlhd_c G'' \subseteq G'$ and $H' = (G''/F')_Y$, where G'', F' and Y are defined as follows. We construct G'' by removing from G':

(a) the vertices that yield vertices in $V_H - V_{H'}$ by the contraction of the edges of F;
(b) the edges incident with vertices removed in Step (a);
(c) the edges of $E_H - E_{H'}$ that have not been removed by Step (b). We let $F' = F \cap E_{G''}$ and $Y = X \cap V_{G''}$. The verification that $H' = (G''/F')_Y$ is easy.

Then going back to the proof of transitivity, we have $K \unlhd_c H' \subseteq H \unlhd_c G' \subseteq G$, hence $K \unlhd_c H' \unlhd_c G'' \subseteq G' \subseteq G$, hence $K \unlhd_c G'' \subseteq G$ and $K \unlhd G$ as was to be proved.

(2) "If" is clear from (1) because an edge deletion, a vertex deletion (we delete a vertex and all incident edges) and an edge contraction transform a graph G into a minor H of it, such that $|V_H| + |E_H| < |V_G| + |E_G|$, whence $H \lhd G$.

"Only if". A subgraph of G can be obtained from it by edge and vertex deletions. If $H = (G/F)_X$ and $F = \{f_1, \ldots, f_n\}$ then H can be obtained from G by the successive contractions of $f_1, \ldots, f_n$. We omit the details. ∎

On abstract graphs the minor relation $\trianglelefteq$ can be defined as follows:

$H \trianglelefteq G$ if and only if H and G are the isomorphism classes of concrete graphs H' and G' such that $H' \trianglelefteq G'$.

Proposition 2.18 The relation $\trianglelefteq$ is a partial order on abstract graphs.

Proof: Let G and H be abstract graphs. Clearly $G \trianglelefteq G$. Assume that $H \trianglelefteq G \trianglelefteq H$, which means that $H' \trianglelefteq G'$ and $G'' \trianglelefteq H''$ where $H = [H']_{iso} = [H'']_{iso}$, $G = [G']_{iso} = [G'']_{iso}$ and G', G'', H', H'' are concrete. By the definitions we have $|V_{H'}| \leq |V_{G'}| = |V_{G''}| \leq |V_{H''}| = |V_{H'}|$ hence $|V_{H'}| = |V_{G'}|$ and similarly for sets of edges. It follows from (3) $\Longrightarrow$ (1) of Proposition 2.16 that $H' = G'$. Hence $H = G$.

Transitivity follows in a similar way from Proposition 2.17(1) and the fact that if concrete graphs satisfy $G' \trianglelefteq H' \simeq H''$, then there is a concrete graph G'' such that $G' \simeq G'' \trianglelefteq H''$. ∎

Remark 2.19 The definitions of edge contraction and minor inclusion apply to infinite graphs. Proposition 2.16 is no longer valid because Assertion (3) does not imply Assertion (2) if G is infinite. However Proposition 2.17(1), the proof of which uses no cardinality argument, is valid for infinite graphs. Proposition 2.18 is not valid: the minor relation on infinite abstract graphs is only a quasi-order because we may have $H \trianglelefteq G \trianglelefteq H$ and $G \neq H$. As an example[13] take $H = K_\omega$, the complete undirected graph with countably many vertices, and G the union of two disjoint copies of K_ω. However, even on finite concrete graphs, the minor relation is usually referred to as a quasi-order. The reason is that it is usually defined on concrete graphs G, H as follows:

H is a minor of G if and only if H is isomorphic to a concrete graph H' such that $H' \trianglelefteq G$.

This definition is a mixture of concepts relative to concrete and abstract graphs. A cleaner situation occurs if we separate definitions relative to abstract graphs from those relative to concrete graphs. □

We now review basic definitions and facts concerning graph classes defined by excluded (or forbidden) minors.

Definition 2.20 (Well-ordered sets) Let $\langle \mathcal{D}, \leq \rangle$ be a partially ordered set. We say that $\leq$ is *Noetherian* if there is no strictly decreasing infinite sequence, i.e., no sequence $(d_i)_{i \in \mathcal{N}}$ such that $d_i \in \mathcal{D}$ and $d_0 > d_1 > \cdots > d_i > \cdots$. We say that $\mathcal{D}$ is *well-ordered*

---

[13] This example also works for the subgraph relation.

by $\le$ if this partial order is Noetherian and has no infinite *antichain*, i.e., no infinite subset of pairwise incomparable elements.

Let $\le$ be an arbitrary partial order on a set $\mathcal{D}$. For every $A \subseteq \mathcal{D}$, we define:[14]

$Forb(A) := \{d \in \mathcal{D} \mid a \le d \text{ for no } a \text{ in } A\},$

$Obst(A) := \{d \in \mathcal{D} \mid d \notin A \text{ and for all } d' \in \mathcal{D} \text{ if } d' < d \text{ then } d' \in A\}.$

Hence, $Obst(A)$ is the set of $\le$-minimal elements of $\mathcal{D} - A$. A subset A of $\mathcal{D}$ is $\le$-*closed*, or is an $\le$-*ideal*, if $d' \le d$ and $d \in A$ imply $d' \in A$.

Proposition 2.21 Let $\langle \mathcal{D}, \le \rangle$ be a partially ordered set.
(1) If $\mathcal{D}$ has no infinite antichain, then $Obst(A)$ is finite for every $A \subseteq \mathcal{D}$.
(2) If $\mathcal{D}$ is Noetherian and A is an $\le$-ideal, then $A = Forb(Obst(A))$.
(3) If $\mathcal{D}$ is well-ordered by $\le$, then every $\le$-ideal A is $Forb(B)$ for some finite set B, and in particular for $B = Obst(A)$.

Proof: (1) Clear since $Obst(A)$ is, by definition, an antichain.

(2) Let $\mathcal{D}$, $\le$ and A be as in the statement. Let $d \in \mathcal{D} - A$. Since there is no infinite strictly decreasing sequence in $\mathcal{D}$, the set $\{e \mid e \le d \text{ and } e \notin A\}$ contains a minimal element d'. Hence $d' \in Obst(A)$ and $d \notin Forb(Obst(A))$. This proves that $Forb(Obst(A)) \subseteq A$. In this proof we have not used the hypothesis that A is $\le$-closed. We will use it for proving the opposite inclusion.

We now assume that some d belongs to $A \cap (\mathcal{D} - Forb(Obst(A)))$. Then $d' \le d$ for some $d' \in Obst(A)$. But $d' \notin A$, and this contradicts the hypothesis that A is $\le$-closed. Hence $A \subseteq Forb(Obst(A))$, which completes the proof.

(3) Immediate consequence of (1) and (2). ∎

These definitions and results can be applied to graphs as follows. We let $\mathcal{D}$ be the set of abstract simple and loop-free undirected graphs, partially ordered by minor inclusion. This order is Noetherian because $H \lhd G$ implies $|V_H| + |E_H| < |V_G| + |E_G|$ by Proposition 2.16. The Graph Minor Theorem proved by Robertson and Seymour establishes that $\mathcal{D}$ is well-ordered by $\lhd$. (See the book [*DowFel], Section 7.6, for an overview of the proof and references.)

It follows that if a class $\mathcal{C}$ of simple, loop-free, undirected graphs is a $\lhd$-ideal, i.e., is closed under taking minors, then there exists a finite subset $\mathcal{K}$ of $\mathcal{D}$ such that $\mathcal{C}$ is the class of graphs in $\mathcal{D}$ that have no minor isomorphic to a graph in $\mathcal{K}$. This applies to the class $\mathcal{C}$ of planar graphs and the corresponding set $\mathcal{K}$ is $\{K_5, K_{3,3}\}$ (see [*Die] Section 4.4 or [*MohaTho] Section 2.3). The Graph Minor Theorem also holds for directed graphs [RobSey04] but this instance is seldom used.

Definition 2.22 (Union and intersection of graphs) Two concrete graphs H and K are *compatible* if the mappings $vert_H$ and $vert_K$ agree, i.e., $vert_H(e) = vert_K(e)$ for

---

[14] Here "Forb" stands for "Forbidden" and "Obst" for "Obstruction."

every $e \in E_H \cap E_K$. This implies in particular that the ends of an edge in $E_H \cap E_K$ are in $V_H \cap V_K$.

If H and K are compatible, we can define their *union* $M := H \cup K$ as the graph such that $V_M := V_H \cup V_K$, $E_M := E_H \cup E_K$ and $vert_M := vert_H \cup vert_K$. Clearly, H and K are subgraphs of M. We can also define their intersection as the graph $N := H \cap K$ such that $V_N := V_H \cap V_K$, $E_N := E_H \cap E_K$ and $vert_N := vert_H \upharpoonright E_N = vert_K \upharpoonright E_N$. Clearly $H \cap K$ is a subgraph of both H and K. If $H \subseteq G$ and $K \subseteq G$, then H and K are compatible and we have $H \cap K \subseteq H \subseteq H \cup K \subseteq G$.

Two concrete graphs G and H are *disjoint* if $V_G \cap V_H = \emptyset$ and $E_G \cap E_H = \emptyset$. This implies that they are compatible, that $G \cap H = \varnothing$ (the empty graph) and that $G \cup H$ is defined.

It is clear that the union and the intersection of two concrete graphs are not always defined. In the next section we will define the *disjoint union*, a binary operation that is well defined for any two abstract graphs.

2.3 The HR algebra of graphs with sources

In this section we define operations on abstract graphs that generalize the parallel-composition and series-composition used in Section 1.1.3 to define series-parallel graphs. Sources are distinguished vertices that allow to glue graphs in uniquely defined ways. Thus graphs are built from smaller graphs like words are built by concatenating shorter words. But graphs are intrinsically more complex than words, and so we cannot build them by means of a single concatenation operation. We need countably many operations. However, finite sets of these operations generate significant graph classes. In particular, there is for each k a finite subset of these operations that generates the graphs of tree-width at most k.

The acronym HR stands for hyperedge replacement and will be justified in Chapter 4.

2.3.1 The HR graph operations

We make the definitions of Section 1.4.2 more precise by paying attention to the distinction between abstract and concrete graphs.

Definition 2.23 (Disjoint union of abstract graphs) Let G, H, K be concrete graphs. We write $G = H \oplus K$ if H and K are disjoint and $G = H \cup K$. For abstract graphs G, H and K we write $G = H \oplus K$ if $G \simeq H' \oplus K'$, where $H' \simeq H$, $K' \simeq K$ and the graphs H' and K' are disjoint concrete graphs.[15] There is not a single such pair of concrete graphs (H', K'). However, G is uniquely defined as an abstract graph because any

---

[15] One can also take $H' = H$ and choose K' disjoint with H.

two pairs (H',K') and (H'',K'') satisfying these conditions yield isomorphic graphs $H' \oplus K'$ and $H'' \oplus K''$. The following facts are clear from the definitions: for every concrete graph G, we have $G = G \cup G$ but $G \oplus G$ is undefined unless G is empty; for every finite[16] nonempty abstract graph G the graph $G \oplus G$ is defined and different from G.

Disjoint union is not powerful enough to generate interesting classes of graphs from finitely many basic abstract graphs. In order to "glue" graphs, we introduce distinguished vertices that we call *sources*.

Definition 2.24 (Graphs with sources) Let $\mathcal{A}$ be a fixed countable set, the elements of which will be used as *names* (or *labels*) of distinguished vertices. We assume that $\mathcal{N} \subseteq \mathcal{A}$ and that $\mathcal{A}$ is ordered linearly by an ordering that extends the usual order on $\mathcal{N}$.[17] A *concrete graph with sources*, or a *concrete s-graph* in short, is a pair $G = \langle G^\circ, slab_G \rangle$ consisting of a concrete graph G° and a partial injective function $slab_G : V_{G^\circ} \to \mathcal{A}$. We will simplify the notation V_{G° into V_G, and similarly for other notations. The domain of $slab_G$ is a set of vertices denoted by $Src(G)$ and called the *set of sources* of G, and $\tau(G) := slab_G(Src(G))$ is the *type* of G, a finite subset of $\mathcal{A}$. If $slab_G(u) = a$, then we say that u is the *a-source* of G and that a is the *name* (or *label*) of source u. We will also use the bijection $src_G : \tau(G) \to Src(G)$ defined as the inverse of $slab_G$. A graph is an s-graph of empty type. The set of vertices that are not sources, called the *internal* vertices, is denoted by Int_G.

For example, a series-parallel graph as defined in Section 1.1.3 is an s-graph with two sources, a 1-source and a 2-source. A rooted tree (Definition 2.13) can be defined as a tree with one source, the vertex selected as root.

Isomorphisms of concrete s-graphs are defined in the obvious way, and still denoted by $\simeq$. (We mean by this that if $G \simeq H$, the bijection $h_V : V_G \to V_H$ maps $Src(G)$ onto $Src(H)$ and that $slab_H(h_V(u)) = slab_G(u)$ for every u in $Src(G)$.) Hence the above terminology and notation concerning sources apply to *abstract s-graphs*. The type of an abstract s-graph is well defined.

We denote by $\mathcal{JS}^u$ and $\mathcal{JS}^d$ the classes of undirected and directed abstract s-graphs.[18]

Definition 2.25 (Parallel-composition) Let G, H be concrete s-graphs. We say that H is a *subgraph of* G, written $H \subseteq G$, if $H^\circ \subseteq G^\circ$ and $slab_H = slab_G \upharpoonright V_H$ (so that $\tau(H) \subseteq \tau(G)$).

---

[16] We have $G = G \oplus G$ if G is the abstract graph isomorphic to the union of countably many disjoint copies of a concrete graph H.

[17] Additionally, we assume that $\mathcal{A}$ is effectively given (see Definition 2.8) and that the linear order $<$ on $\mathcal{A}$ is decidable. The latter means that the mapping $f : \mathcal{A} \times \mathcal{A} \to \{True, False\}$ such that $f(a,b) = True$ if $a < b$ and $False$ otherwise, is computable.

[18] As explained in Section 1.4, the letter J in notations like $\mathcal{JS}$ and $\mathbb{JS}$ is intended to remind the reader that the graphs of the corresponding algebras may have multiple edges. By contrast, the letter G in $\mathcal{GP}$ and $\mathbb{GP}$ refers to algebras of simple graphs.

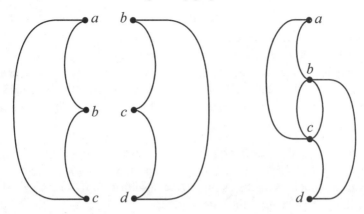

Figure 2.2 From left to right: H, K and $H \parallel K$.

Let G, H, K be concrete s-graphs. We write $G = H \parallel K$, and we say that G is the *parallel-composition* of H and K, if and only if:

(1) $H \subseteq G, K \subseteq G, G^\circ = H^\circ \cup K^\circ$;
(2) $V_H \cap V_K = Src(H) \cap Src(K)$;
(3) $E_H \cap E_K = \emptyset$;
(4) $slab_G = slab_H \cup slab_K$.

These conditions imply that $slab_H$ and $slab_K$ agree and that $slab_H(v) \neq slab_K(w)$ if $v \in Src(H) - V_K$ and $w \in Src(K) - V_H$. From the definitions, we get the following:

Proposition 2.26 For concrete s-graphs G, H, K, we have $G = H \parallel K$ if and only if the following conditions hold:
(1) $V_G = V_H \cup V_K$;
(2) $V_H \cap V_K = Src(H) \cap Src(K)$, and if $u \in V_H \cap V_K$ then $src_G(a) = src_H(a) = src_K(a) = u$ for some $a \in \mathcal{A}$;
(3) $u \in Src(G)$ if and only if $u \in Src(H) \cup Src(K)$; and its name (as a source) is the same with respect to G and to H and/or to K.
(4) $E_G = E_H \cup E_K, E_H \cap E_K = \emptyset$;
(5) $vert_G(e) = \begin{cases} vert_H(e) & \text{if } e \in E_H, \\ vert_K(e) & \text{if } e \in E_K. \end{cases}$ $\qquad\square$

We have $\tau(H \parallel K) = \tau(H) \cup \tau(K)$; if $\tau(H) \cap \tau(K) = \emptyset$, then $(H \parallel K)^\circ = H^\circ \oplus K^\circ$.

Parallel-composition is not defined for *all* pairs of concrete s-graphs. We will turn it into a total binary function on abstract s-graphs, more or less as we did for disjoint union in Definition 2.23.

Definition 2.27 (Parallel-composition of abstract s-graphs) Let H and K be abstract s-graphs. The abstract s-graph $H \parallel K$, called the *parallel-composition* of H and K, is the isomorphism class of $H' \parallel K'$ for any two concrete s-graphs H'

and K' such that $H' \simeq H$, $K' \simeq K$ and $H' \parallel K'$ is defined. For any two pairs (H',K') satisfying these conditions, the resulting s-graphs $H' \parallel K'$ are isomorphic. Hence $H \parallel K$ is well defined for any two s-graphs H and K. Informally we will say that H and K are "glued at their sources with the same name." See Figure 2.2 for an illustration. (In this figure and in Figure 2.3, the graphs H, K and G have three sources; their possible internal vertices and incident edges are not shown.)

It is clear that for all s-graphs G, H and K, we have $\tau(G \parallel H) = \tau(G) \cup \tau(H)$ and $(G \parallel H) \parallel K = G \parallel (H \parallel K)$. Since parallel-composition is associative, we will use infix notation without parentheses for it. Note also that, for G finite, $G = G \parallel G$ if and only if $E_G = \emptyset$ and all vertices are sources.

Definition 2.28 (Quotient s-graphs) The notion of a quotient graph has been defined for graphs in Definition 2.15. For a concrete s-graph G and an equivalence relation $\approx$ on V_G such that no two distinct sources are equivalent, we define its (concrete) quotient s-graph $G/\approx$ of the same type as G, as follows. For every cross-section X of $\approx$ we let $(G/\approx)_X$ be the concrete s-graph H such that $H^\circ := (G^\circ/\approx)_X$ and, for every $a \in \tau(G)$, $src_H(a) = u$ if $u \in X$ and $u \approx src_G(a)$. Any two cross-sections yield isomorphic quotients. We denote $(G/\approx)_X$ by $G/\approx$ when X need not be specified.

The following proposition is an immediate consequence of the definitions. It describes the "gluing of sources" by a quotient.

Proposition 2.29 If H and K are disjoint concrete s-graphs, then the abstract s-graph $[H]_{iso} \parallel [K]_{iso}$ is equal to $[L/\approx]_{iso}$, where L is the concrete s-graph such that $L^\circ = H^\circ \cup K^\circ$, $src_L(a) = src_H(a)$ if $a \in \tau(H)$, $src_L(a) = src_K(a)$ if $a \in \tau(K) - \tau(H)$, and $\approx$ is the equivalence relation generated by the pairs $(src_H(a), src_K(a))$ for all $a \in \tau(H) \cap \tau(K)$. $\qquad\square$

Definition 2.30 (Source manipulating operations) Let $h : \mathcal{A} \to \mathcal{A}$ be a partial injective function that is the identity outside of a finite subset C of $\mathcal{A}$ (hence $Dom(h) \supseteq \mathcal{A} - C$ and $h(C) \subseteq C$).[19] For example, $h(d) = d$ for $d \notin \{a,b,c\}$, $h(c)$ is undefined, $h(a) = b$ and $h(b) = a$ (or $h(b) = c$). For every concrete s-graph G, we let $app_h(G)$ be the s-graph H such that $H^\circ := G^\circ$ and $slab_H := h \circ slab_G$. It follows that $Src(app_h(G)) \subseteq Src(G)$, $\tau(app_h(G)) = h(\tau(G))$ and $|\tau(app_h(G))| \leq |\tau(G)|$. The mapping app_h commutes with isomorphisms hence applies to (abstract) s-graphs. It will be convenient actually to distinguish two main cases.

The first case is when $h(a)$ is either undefined or equal to a, for every $a \in \mathcal{A}$. Hence $B := \mathcal{A} - Dom(h)$ is finite and we will denote app_h by fg_B. The effect of fg_B is to make internal all a-sources for $a \in B$. We have $Src(fg_B(G)) = Src(G) - src_G(B)$

---

[19] $Dom(h)$ is the *definition domain* of h, i.e., the set of elements a such that $h(a)$ is defined.

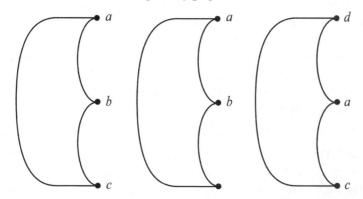

Figure 2.3 From left to right: $G, fg_c(G)$ and $ren_h(G)$.

and $\tau(fg_B(G)) = \tau(G) - B$. Clearly $fg_B(G) = G$ if $B \cap \tau(G) = \emptyset$. We call fg_B a *source forgetting operation*. It will be technically useful to allow $B = \emptyset$ in fg_B. The corresponding operation is the identity. We will write fg_b instead of $fg_{\{b\}}$.

In the second case, h is a bijection : $A \to A$ and $D := \{a \in A \mid h(a) \neq a\}$ is finite; thus, h permutes the finite subset D of A and is the identity outside of D. For $D \subseteq C \subseteq A$ (and C possibly infinite), h is called a *finite permutation of C*; if C is finite it is just called a permutation of C. We denote by $Perm_f(C)$ the set of finite permutations of C. If h is a finite permutation of A, we denote app_h by ren_h and call it a *source renaming operation*. We have $Src(ren_h(G)) = Src(G)$, $\tau(ren_h(G)) = h(\tau(G))$ and $|\tau(ren_h(G))| = |\tau(G)|$. We have $ren_h(G) = G$ if h is the identity on $\tau(G)$. The effect of ren_h is to turn an a-source into an $h(a)$-source for each a. Only finitely many source names are modified by ren_h. See Figure 2.3 for an example where $h(a) = d$, $h(b) = a$, $h(c) = c$ and $h(d) = b$.

In the general case, we have that $app_h(G) = ren_{h'}(fg_B(G))$ where B is the set $A - Dom(h)$ and h' is any bijection : $A \to A$ that extends h (the existence of which follows from Lemma 2.37 below). Thus, app_h is a derived operation.

Each operation fg_B is defined in a finitary way since B is required to be a finite set. Each operation ren_h can also be specified in a finitary way because $D = \{a \in A \mid h(a) \neq a\}$ is a finite set. Hence it can be written $ren_{\{b_1 \to c_1, b_2 \to c_2, ..., b_n \to c_n\}}$, where $D = \{b_1, ..., b_n\} = \{c_1, ..., c_n\}$ and $c_i = h(b_i)$ for all $i = 1, ..., n$. We will also abbreviate $ren_{\{a \to b, b \to a\}}$ into $ren_{a \leftrightarrow b}$, and for $n = 0$, $ren_{\{\}}$ is the identity. Since A is countable, the set of operations of the forms fg_B and ren_h is countable.

Definition 2.31 (Basic s-graphs) We now specify some basic (abstract) s-graphs that will be the values of the constant symbols $\mathbf{a}$ and $\mathbf{a}^\ell$ for $a \in A$. These symbols denote s-graphs consisting of a vertex that is the a-source, and, in the latter case, that has also a loop. (We need not distinguish directed and undirected loops.) If $a, b \in A$, $a \neq b$, the constant symbol $\mathbf{ab}$ denotes an s-graph consisting of two vertices, an a-source and a

b-source, linked by a single undirected edge, and $\overrightarrow{\mathbf{ab}}$ denotes the similar s-graph with an edge directed from the a-source to the b-source. We can consider **ab** and **ba** as the same symbol. Assuming fixed a linear order $<$ on $\mathcal{A}$, we will only use **ab** with $a < b$. It will be useful for some constructions to have also a constant symbol $\varnothing$ denoting the empty graph. Every term evaluating to a nonempty graph can be transformed into an equivalent one without $\varnothing$, by means of properties of the operations $/\!/$, ren_h and fg_B that will be stated in Proposition 2.48.

Definition 2.32 (The HR algebra of s-graphs)   We have defined a binary operation $/\!/$ that "glues" s-graphs, and unary operations fg_B and ren_h that transform them by modifying their source names. We have also defined constant symbols that define basic s-graphs, directed or not, having at most one edge. We obtain thus two algebras, one with domain $\mathcal{JS}^{\mathrm{u}}$, the set of (abstract) undirected s-graphs, and the other one with domain $\mathcal{JS}^{\mathrm{d}}$, the similar set of directed s-graphs. We let:[20]

$$F^{\mathrm{HR}}_+ := \{ /\!/, fg_B, ren_h \mid B \in \mathcal{P}_f(\mathcal{A}), \ h \in Perm_f(\mathcal{A}) \},$$

$$F^{\mathrm{HRu}}_0 := \{ \mathbf{a}, \mathbf{a}^\ell, \mathbf{ab}, \varnothing \mid a,b \in \mathcal{A}, \ a < b \},$$

$$F^{\mathrm{HRd}}_0 := \{ \mathbf{a}, \mathbf{a}^\ell, \overrightarrow{\mathbf{ab}}, \varnothing \mid a,b \in \mathcal{A}, \ a \neq b \},$$

$$F^{\mathrm{HRu}} := F^{\mathrm{HR}}_+ \cup F^{\mathrm{HRu}}_0,$$

$$F^{\mathrm{HRd}} := F^{\mathrm{HR}}_+ \cup F^{\mathrm{HRd}}_0.$$

These signatures are countably infinite. We let $\mathbb{JS}^{\mathrm{u}}$ denote the F^{HRu}-algebra with domain $\mathcal{JS}^{\mathrm{u}}$ and $\mathbb{JS}^{\mathrm{d}}$ denote the F^{HRd}-algebra with domain $\mathcal{JS}^{\mathrm{d}}$. In order to simplify notation, we will not distinguish the operations of these algebras from the corresponding symbols in their signatures. We call $\mathbb{JS}^{\mathrm{u}}$ and $\mathbb{JS}^{\mathrm{d}}$ the (undirected or directed) *HR algebra*.

Every term t in $T(F^{\mathrm{HRu}})$ has a value in $\mathcal{JS}^{\mathrm{u}}$ denoted by $t_{\mathbb{JS}^{\mathrm{u}}}$ or by $val_{\mathbb{JS}^{\mathrm{u}}}(t)$ according to the notation of Definition 2.2. If t is in $T(F^{\mathrm{HRd}})$, then its value $t_{\mathbb{JS}^{\mathrm{d}}}$ (or $val_{\mathbb{JS}^{\mathrm{d}}}(t)$) belongs to $\mathcal{JS}^{\mathrm{d}}$. In both cases and for other graph algebras, we will use the generic notation $val(t)$, the relevant algebra being understood from the context. We will say that the term t *evaluates to* the s-graph $val(t)$ or that t *denotes* or *defines* $val(t)$. Two terms t and t' are *equivalent* if $val(t) = val(t')$.

We now extend these definitions to (K, Λ)-labeled graphs (cf. Definition 2.11) equipped with sources, called (K, Λ)-*labeled s-graphs* (or *labeled s-graphs* if we need not specify K and Λ). We always assume that $(K \cup \Lambda) \cap \mathcal{A} = \emptyset$. Each vertex has a possibly empty set of labels that belong to K and each edge has a unique label in Λ. If G and H both have an a-source, then we take as set of labels of the a-source of $G /\!/ H$ the set $\gamma_G(src_G(a)) \cup \gamma_H(src_H(a))$, i.e., we take the labels from G and those from H. The operations that manipulate source names are the same.

---

[20] For a set X, $\mathcal{P}_f(X)$ denotes the set of finite subsets of X.

We will use the constant symbols $\overrightarrow{\mathbf{ab}}_\lambda$, $\mathbf{ab}_\lambda$ and $\mathbf{a}_\lambda^\ell$ to denote an edge labeled by $\lambda \in \Lambda$ that is, respectively, directed from the a-source to the b-source, undirected between the a-source and the b-source, and a loop incident with the a-source. For specifying vertex labels, we will use the constant symbol $\mathbf{a}_\kappa$ to denote the s-graph with a single vertex that is the a-source and is labeled by $\kappa \in K$. We denote by $F_{[K,\Lambda]}^{\mathrm{HRu}}$ and by $F_{[K,\Lambda]}^{\mathrm{HRd}}$ the corresponding modifications of the signatures F^{HRu} and F^{HRd}. (The modifications concern only the constant symbols.) We denote by $\mathcal{JS}_{[K,\Lambda]}^{\mathrm{u}}$ and $\mathcal{JS}_{[K,\Lambda]}^{\mathrm{d}}$ the sets of abstract directed and undirected (K,Λ)-labeled s-graphs, and by $\mathbb{JS}_{[K,\Lambda]}^{\mathrm{u}}$ and $\mathbb{JS}_{[K,\Lambda]}^{\mathrm{d}}$ the corresponding algebras.

Let us stress that we have defined algebras of abstract s-graphs. We have no algebraic structure on concrete s-graphs because the parallel-composition of two concrete s-graphs is not always defined. The terms "graph" and "s-graph" will mean "abstract graph" and "abstract s-graph" respectively. In many statements, we will refer to $\mathbb{JS}$, $\mathcal{JS}$, F^{HR} without specifying the variant, relative to directed or undirected s-graphs. Any such statement will cover two statements, one for directed s-graphs and one for undirected s-graphs. For example, the statement "every term $t \in T(F^{\mathrm{HR}})$ evaluates to an s-graph $val(t)$ in $\mathcal{JS}$" means that if t belongs to $T(F^{\mathrm{HRu}})$, then $val(t) \in \mathcal{JS}^{\mathrm{u}}$, and also that if t belongs to $T(F^{\mathrm{HRd}})$, then $val(t) \in \mathcal{JS}^{\mathrm{d}}$. The same convention applies to the algebras of labeled graphs.

Every term in $T(F^{\mathrm{HR}})$ is written with finitely many source names. We make this precise as follows. For $C \subseteq \mathcal{A}$, we let:

$$F_{C+}^{\mathrm{HR}} := \{ /\!/, fg_B, ren_h \mid B \in \mathcal{P}_f(C), \ h \in Perm_f(C) \},$$

$$F_{C0}^{\mathrm{HRu}} := \{ \mathbf{a}, \mathbf{a}^\ell, \mathbf{ab}, \varnothing \mid a,b \in C, \ a < b \},$$

$$F_{C0}^{\mathrm{HRd}} := \{ \mathbf{a}, \mathbf{a}^\ell, \overrightarrow{\mathbf{ab}}, \varnothing \mid a,b \in C, \ a \neq b \},$$

$$F_C^{\mathrm{HRu}} := F_{C+}^{\mathrm{HR}} \cup F_{C0}^{\mathrm{HRu}},$$

$$F_C^{\mathrm{HRd}} := F_{C+}^{\mathrm{HR}} \cup F_{C0}^{\mathrm{HRd}}.$$

The notation F_C^{HR} will mean ambiguously F_C^{HRu} as well as F_C^{HRd}. Since F^{HR} is the union of the signatures F_C^{HR} for all finite subsets C of $\mathcal{A}$, and $F_C^{\mathrm{HR}} \subseteq F_{C'}^{\mathrm{HR}}$ if $C \subseteq C'$, every finite subsignature of F^{HR} is a subsignature of F_C^{HR} for some finite set C. Hence, every term in $T(F^{\mathrm{HR}})$ belongs to $T(F_C^{\mathrm{HR}})$ for some finite set C. We denote by $\mathcal{JS}^{\mathrm{d}}[C]$ the set of s-graphs in $\mathcal{JS}^{\mathrm{d}}$ that have their type included in C. We obtain an F_C^{HRd}-algebra denoted by $\mathbb{JS}^{\mathrm{d}}[C]$, and similarly, $\mathcal{JS}^{\mathrm{u}}[C]$ and $\mathbb{JS}^{\mathrm{u}}[C]$ for undirected s-graphs. We use $\mathbb{JS}[C]$ and $\mathcal{JS}[C]$ for denoting simultaneously the F_C^{HR}-algebras of directed and undirected s-graphs and their domains. For (K,Λ)-labeled s-graphs, we will denote by $F_{C,[K,\Lambda]}^{\mathrm{HR}}$ and $\mathbb{JS}_{[K,\Lambda]}[C]$ the corresponding signatures and algebras, defined by restricting source names to belong to C.

The signature F^{HR} is *effectively given* (Definition 2.8). In fact, we can assume that the countable set $\mathcal{A}$ is effectively given with a fixed bijection $\xi_{\mathcal{A}}$ onto $\mathcal{N}$ (one could take $\mathcal{A} = \mathcal{N}$, but more flexibility will be convenient in some proofs). Each function symbol of F^{HR} can be formally written as a sequence in $Seq(\mathcal{N} \cup P)$, where P is a finite auxiliary set of symbols. For example, the renaming operation $ren_{\{a \to b, b \to c, c \to a\}}$ with $a, b, c \in \mathcal{A}$ can be written as $(ren, \bar{a}, \bar{b}, \bar{b}, \bar{c}, \bar{c}, \bar{a})$, where $ren \in P$ and $\bar{a}, \bar{b}, \bar{c}$ are integers that encode respectively a, b and c (i.e., $\bar{a} = \xi_{\mathcal{A}}(a)$ and similarly for b and c). These sequences form an infinite decidable subset of $Seq(\mathcal{N} \cup P)$ and can be encoded as integers by the standard encoding of that set, with a computable arity mapping.

The algebra $\mathbb{JS}$ is also effectively given. Although that should be intuitively clear, let us look at some details. Without loss of generality, we can describe an s-graph by a concrete s-graph G with vertex set $[n]$ and edge set $[n+1, p]$ for some $p \geq n$ (we have $p = n$ and $[n+1, p] = \emptyset$ if G has no edges). The mappings $vert_G$ and $slab_G$ can be written as sequences of triples and of pairs of integers. For example, the directed s-graph $G = a\bullet \to \bullet \leftarrow \bullet \to \bullet b$ can be written as the sequence $(4, 0, 5, 1, 2, 0, 6, 3, 2, 0, 7, 3, 4, 0, 0, 1, \xi_{\mathcal{A}}(a), 0, 4, \xi_{\mathcal{A}}(b))$, where the first occurrence of 4 indicates the number of vertices, the integers $5, 6, 7$ represent edges and are followed by the associated pairs of vertices, and 0 separates the different objects of the sequence. Several such sequences can describe the same s-graph. As *the* description of an s-graph we can take the first sequence that describes it, with respect to some standard linear order on $Seq(\mathcal{N})$. Hence, we obtain a bijection between s-graphs and a decidable subset[21] of $Seq(\mathcal{N})$, whence an encoding $\xi_{\mathcal{JS}}$ of $\mathcal{JS}$ in $\mathcal{N}$. The operations of $\mathbb{JS}$ are computable via this encoding (cf. Definition 2.8). The algebra $\mathbb{JS}$ is thus effectively given. It should, however, be stressed here that when discussing efficiency of algorithms we will not use this encoding to represent s-graphs, but rather the usual representation by a traditional data structure like an incidence matrix.[22] The notion of an effectively given algebra will only play a role when the efficiency of algorithms is not an issue.

These remarks clearly extend to labeled graphs.

Proposition 2.33 Let $C \subseteq \mathcal{A}$ and let G be an s-graph such that $\tau(G) \subseteq C$ and $|V_G| = |C|$. Then G is the value of a term t over F_C^{HR} of size $\Theta(|V_G| + |E_G|)$. Furthermore, if G is (K, Λ)-labeled, then it is the value of a term over $F_{C,[K,\Lambda]}^{\text{HR}}$.

Proof: We first consider an undirected G. We extend the mapping $src_G : \tau(G) \to V_G$ to a bijection $: C \to V_G$. Each vertex can be identified with an element of C. Now let $B = C - \tau(G)$. We let t be the term

$$fg_B(\mathbf{x_1 y_1} /\!/ \cdots /\!/ \mathbf{x_m y_m} /\!/ \mathbf{z_1} /\!/ \cdots /\!/ \mathbf{z_p} /\!/ \mathbf{w_1}^\ell /\!/ \cdots /\!/ \mathbf{w_q}^\ell),$$

---

[21] An algorithm can check if a sequence in $Seq(\mathcal{N})$ actually describes some s-graph and another one can check if two sequences describe the same s-graph; hence an algorithm can check if a sequence is *the* description of an s-graph.

[22] These representations are equivalent: an algorithm can find, for every concrete s-graph G given by an incidence matrix, the integer that encodes the s-graph $[G]_{iso}$, and vice versa. However, the algorithms that translate one representation into the other are not efficient.

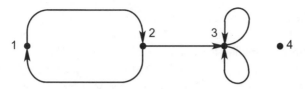

Figure 2.4 An s-graph G.

where the nonloop edges of G are $e_i : x_i - y_i$ (with $x_i < y_i$; $<$ is a fixed strict linear order on $\mathcal{A}$) for $i = 1,\dots,m$, its isolated vertices are $z_1,\dots,z_p$, and the loops are $e'_j : w_j - w_j$ for $j = 1,\dots,q$. Because we allow multiple edges and loops, we may have $\{x_i,y_i\} = \{x_j,y_j\}$ or $w_i = w_j$ for $i \neq j$; also, we may have $w_i \in \{x_1,\dots,x_m,y_1,\dots,y_m\}$. Note that $t = fg_C(\varnothing)$ if $G = \varnothing$.[23] It is clear from this construction and the definitions that $val(t) = G$.

For directed s-graphs, we use constant symbols $\overrightarrow{x_iy_i}$ instead of x_iy_i. The construction extends in an obvious way to labeled s-graphs; a vertex label κ can be attached to a vertex $\mathbf{a}$ by adding $/\!/\,\mathbf{a}_\kappa$ to the parallel-composition in t.

It is clear from these constructions that

$$|t| = 2(|E_G| + |Isol_G|) \leq 2(|E_G| + |V_G|),$$

where $Isol_G$ is the set of isolated vertices of G, whereas, for every term t evaluating to an s-graph G:

$$|E_G| + |V_G| \leq 3|t|.$$

For a (K,Λ)-labeled graph, with $K \neq \varnothing$, we have

$$|t| \leq 2(|E_G| + |K| \cdot |V_G|),$$

and the second inequality is also valid. ∎

This proposition entails that the algebra $\mathbb{JS}$ is generated by its signature F^{HR}. The construction used in its proof is equivalent to the specification of an s-graph as a list of vertices and edges, and of sources with their names. It yields no interesting structuring similar to that of series-parallel graphs described in Chapter 1.

Example 2.34 Let G be the s-graph shown in Figure 2.4, and of type $\tau(G) = \{1,2,3,4\}$. The construction of Proposition 2.33 gives:

$$G = val(\overrightarrow{12} /\!/ \overrightarrow{21} /\!/ \overrightarrow{23} /\!/ 4 /\!/ 3^\ell /\!/ 3^\ell) \text{ and}$$

$$und(G) = val(12 /\!/ 12 /\!/ 23 /\!/ 4 /\!/ 3^\ell /\!/ 3^\ell).$$

□

---

[23] Here and from now on, we use the convention that $t_1 /\!/ \cdots /\!/ t_n$ denotes $\varnothing$ if $n = 0$.

For the algorithmic applications that we will detail in Chapter 6, it is useful to denote graphs by terms in $T(F_C^{HR})$ such that C is of smallest possible cardinality. It is also necessary for finding for a given graph such a term for a set C of fixed size, as efficiently as possible (cf. Definition 2.46). Since there are close relationships between these terms and the well-developed theory of tree-decompositions, as we will see in Section 2.4, we will get answers to these algorithmic questions.

Definition 2.35 (The type and extended type of a term) The *type* of a term t in $T(F^{HR})$ is $\tau(t) := \tau(val(t))$. If $u \in Pos(t)$, then the type of u is $\tau(u) := \tau(t/u)$. Clearly $\tau(t) = \tau(root_t)$ (where $root_t$ denotes the first position of t, cf. Definition 2.14). Let us collect some observations already made above about types:

$$\tau(t_1 /\!/ t_2) = \tau(t_1) \cup \tau(t_2),$$
$$\tau(fg_B(t_1)) = \tau(t_1) - B,$$
$$\tau(ren_h(t_1)) = h(\tau(t_1)),$$
$$\tau(\mathbf{a}) = \tau(\mathbf{a}^\ell) = \{a\},$$
$$\tau(\mathbf{ab}) = \tau(\overrightarrow{\mathbf{ab}}) = \{a, b\},$$
$$\tau(\varnothing) = \emptyset.$$

These rules can be used to determine the types of all positions in a term t, by using a bottom-up computation on the syntactic tree of this term. We can also compute in this way the set $\hat{\tau}(t)$ defined as $\bigcup\{\tau(t/u) \mid u \in Pos(t)\}$. This is the set of all names of the sources of s-graphs defined by the subterms of t.

We have noted in Definition 2.32 that every term t in $T(F^{HR})$ belongs to $T(F_C^{HR})$ for some finite set $C \subseteq \mathcal{A}$. We let $\mu(t)$ denote the minimal such set, where minimality is relative to set inclusion. This set is well defined because $F_C^{HR} \cap F_D^{HR} \subseteq F_{C\cap D}^{HR}$ for all $C, D \subseteq \mathcal{A}$. It can be computed inductively by the following rules:

$$\mu(t_1 /\!/ t_2) = \mu(t_1) \cup \mu(t_2),$$
$$\mu(fg_B(t_1)) = \mu(t_1) \cup B,$$
$$\mu(ren_h(t_1)) = \mu(t_1) \cup \{a \in \mathcal{A} \mid h(a) \neq a\},$$
$$\mu(c) = \tau(c) \text{ for every constant symbol } c.$$

We call $\mu(t)$ the *extended type of* t and $|\mu(t)|$ the *width* of t denoted by $wd(t)$. We will establish in Section 2.4.5 its relationship with tree-width. Both $\tau(t)$ and $\hat{\tau}(t)$ are included in the extended type of t, as shown next.

Lemma 2.36 For every $t \in T(F^{HR})$, we have $\tau(t) \subseteq \hat{\tau}(t) \subseteq \mu(t)$.

Proof: We prove that if C is a finite subset of $\mathcal{A}$ and $t \in T(F_C^{HR})$, then $\tau(t) \subseteq C$. Then also $\hat{\tau}(t) \subseteq C$, because $t/u \in T(F_C^{HR})$ for every u. The proof is by induction on the structure of t and uses the above rules for τ.

(1) If t is a constant symbol, then $\tau(t) \subseteq C$ by definition of F_C^{HR}.
(2) If $t = t_1 \,/\!/\, t_2$, then $\tau(t) = \tau(t_1) \cup \tau(t_2) \subseteq C$ since, by induction, we have $\tau(t_1) \subseteq C$ and $\tau(t_2) \subseteq C$.
(3) If $t = fg_B(t_1)$, then $\tau(t) = \tau(t_1) - B \subseteq C$ by the induction hypothesis.
(4) If $t = ren_h(t_1)$, then $\tau(t) = h(\tau(t_1))$. We have $\tau(t_1) \subseteq C$ by the induction hypothesis. By the definition of F_C^{HR}, h is a permutation of C and so $h(\tau(t_1)) \subseteq h(C) = C$. ∎

However, the inclusion $\widehat{\tau}(t) \subseteq \mu(t)$ may be strict: take, for example, $t_1 = fg_{\{a,b\}}(\mathbf{a})$. Then $\widehat{\tau}(t_1) = \{a\} \subset \{a,b\} = \mu(t_1)$. As another example, take $t_2 = ren_{\{a \to b, b \to c, c \to a\}}(\mathbf{a})$. Then $\widehat{\tau}(t_2) = \{a,b\} \subset \{a,b,c\} = \mu(t_2)$. The term t_1 can actually be replaced by the equivalent term $t_1' = fg_a(\mathbf{a})$ for which $\mu(t_1') = \{a\}$. Similarly, the term t_2 can be replaced by the equivalent term $t_2' = ren_{a \leftrightarrow b}(\mathbf{a})$, for which $\mu(t_2') = \{a,b\}$. In order to generalize these observations, we say that a term t is *reduced* if for every subterm t' of t we have the following:

(a) if $t' = fg_B(t'')$ then $B \subseteq \tau(t'')$;
(b) if $t' = ren_h(t'')$ then h is permutation of $\tau(t'') \cup h(\tau(t''))$.

Lemma 2.37 Let $g : B \to B'$ be a bijection between finite sets B and B'. There exists a bijection $f : B \cup B' \to B \cup B'$ such that $f \upharpoonright B = g$.

Proof: Since g is a bijection, $|B - B'| = |B' - B|$. Let $B - B' = \{b_1, \ldots, b_n\}$ and $B' - B = \{b_1', \ldots, b_n'\}$. We define $f(b) = g(b)$ for every $b \in B$, and $f(b_i') = b_i$ for every $i \in [n]$. ∎

It is clear that the function f can be constructed in a canonical way from g and a fixed strict linear order $<$ on $\mathcal{A}$ (where B and B' are subsets of $\mathcal{A}$): in the above proof, we take $b_i < b_{i+1}$ and $b_i' < b_{i+1}'$ for every $i \in [n-1]$.

Proposition 2.38 Every term t in $T(F_C^{HR})$ is equivalent to a reduced term $\bar{t}$ in $T(F_{\widehat{\tau}(t)}^{HR})$; we have $\mu(\bar{t}) = \widehat{\tau}(\bar{t}) = \widehat{\tau}(t)$.

Proof: The proof is by induction on the structure of t. If t is a constant symbol, then $\bar{t} = t$. If $t = t_1 \,/\!/\, t_2$, then $\bar{t} = \bar{t}_1 \,/\!/\, \bar{t}_2$. If $t = fg_B(t_1)$, then $\bar{t} = fg_{B'}(\bar{t}_1)$, where $B' = B \cap \tau(t_1)$. If $t = ren_h(t_1)$, then $\bar{t} = ren_{h'}(\bar{t}_1)$, where h' is a permutation of $\tau(t_1) \cup h(\tau(t_1))$ that coincides with h on $\tau(t_1)$. This permutation exists by Lemma 2.37. The following assertions can be proved by this induction:

$$val(\bar{t}) = val(t), \quad \widehat{\tau}(\bar{t}) = \widehat{\tau}(t) \text{ and } \mu(\bar{t}) = \widehat{\tau}(\bar{t}).$$

To prove the last equation, it suffices by Lemma 2.36 to show that $\mu(\bar{t}) \subseteq \widehat{\tau}(\bar{t})$, i.e., that $\bar{t} \in T(F_{\widehat{\tau}(t)}^{HR})$. Hence $\bar{t}$ belongs to $T(F_{\widehat{\tau}(t)}^{HR})$, which completes the proof. ∎

Remark 2.39 (1) For each finite set C, F_C^{HR} is a finite subsignature of F^{HR}. Two F_C^{HR}-algebras that are subalgebras of $\mathbb{JS}$ can be defined. The first is $\mathbb{JS}[C]$ (cf. Definition 2.32). Its domain is the set $\mathcal{JS}[C]$ of s-graphs of type included in C. The second is the subalgebra $\mathbb{JS}^{gen}[C]$ of $\mathbb{JS}[C]$ that is generated by F_C^{HR}. Its domain $\mathcal{JS}^{gen}[C]$ consists of the s-graphs that are the values of terms over F_C^{HR}. We will prove in Section 2.4.5 (Theorem 2.83) that its s-graphs are those in $\mathcal{JS}[C]$ of tree-width at most $|C| - 1$.

(2) For constructing a term t over F^{HR} evaluating to a given s-graph G, we must use a set of labels $\mu(t)$ such that $\tau(G) \subseteq \mu(t)$. But the source names in $\mu(t) - \tau(G)$ can be chosen arbitrarily and changed like bound variables in logical formulas. In most cases, we are interested in defining graphs (without sources). If a graph G is defined by a term t in F_C^{HR}, and C' is in bijection with C, then $G = val(t')$ for some $t' \in T(F_{C'}^{HR})$. The term t' is obtained from t by replacing, in each operation symbol, every source name (it belongs to C) by the corresponding element of C'.

(3) We have not formulated Definition 2.35, Lemma 2.36 and Proposition 2.38 for labeled graphs because the extensions are obvious. In the sequel, we will leave implicit the extensions to labeled graphs when they are obvious.

2.3.2 Construction of the s-graph defined by a term

The mapping $val : T(F^{HR}) \to \mathcal{JS}$ is defined by induction on the structure of terms in $T(F^{HR})$ in a straightforward manner because the constant symbols of F^{HR} evaluate to fixed (abstract) s-graphs and its function symbols define the unary and binary operations of Definitions 2.27 and 2.30. However, this definition is not fully satisfactory for several reasons: it does not construct $val(t)$ as a precise concrete s-graph, it does not relate the subterms of t to the corresponding subgraphs of $val(t)$ and it gives no tool to verify that a given concrete s-graph G is isomorphic to $val(t)$ where t is a given term. For the latter verification, we must construct $val(t)$ as a concrete s-graph by applying Definitions 2.27, 2.30 and 2.31 and look for an isomorphism between G and this concrete s-graph.

The construction we will define remedies these drawbacks, yields a clear correspondence between terms and tree-decompositions (to be developed in Section 2.4.5) and is directly translatable into a monadic second-order transduction (see Chapter 7 and Section 1.7). We first explain the idea.

The edges of a concrete s-graph G defined by a term t in $T(F^{HR})$ are in bijection with the occurrences in t of constant symbols of the forms $\mathbf{ab}$, $\overrightarrow{\mathbf{ab}}$ or $\mathbf{a}^\ell$. We say that such an occurrence *describes* the corresponding edge. Hence, one part of the correspondence between G and t will be a bijection between E_G and these occurrences. The description of vertices is more complicated. We first take an example. Let t be the term $fg_a(\mathbf{ab}) /\!/ \mathbf{bc}$. The vertex x is common to the edges described by $\mathbf{ab}$ and $\mathbf{bc}$. It is described by the occurrences of these two symbols, and also by occurrences of others. Here is the definition: a vertex is *described by* u in $Pos(t)$ if it is a source of the value of t/u. In the example above, the vertex x is described by all occurrences of the symbols of t. The a-source of $val(\mathbf{ab})$ is described by the occurrence of $\mathbf{ab}$ only. The c-source of $val(\mathbf{bc})$ is described by the occurrences of $\mathbf{bc}$ and of $/\!/$.[24]

We will construct a concrete s-graph isomorphic to $val(t)$ by defining its vertices as equivalence classes of pairs (u,a) such that $u \in Pos(t)$ and $a \in \tau(val(t/u))$, hence, as a quotient of an s-graph $Exp(t)$ called the *expansion of* t. Furthermore, by selecting in each equivalence class a canonical representing element, we will get a canonical concrete s-graph $cval(t)$. We now give the formal definitions.

Definition 2.40 (The expansion of a term and its quotients) Let $t \in T(F^{\mathrm{HR}})$ be a term with its syntactic tree specified as $\langle N_t, son_{1t}, son_{2t}, lab_t \rangle$ (cf. Definition 2.14) where N_t is $Pos(t)$, the set of positions of t. Each $u \in N_t$ has a type $\tau(u) := \tau(val(t/u)) \subseteq \mu(t)$ (we use here Lemma 2.36). The types of all nodes can be computed bottom-up on the syntactic tree of t (cf. Definition 2.35), in linear time (provided $\mu(t) \subseteq C$ with C fixed). We let $Exp(t)$ denote the concrete s-graph called the *expansion of* t and defined as follows:[25]

$$V_{Exp(t)} := \{(u,a) \mid u \in N_t, a \in \tau(u)\},$$

$$E_{Exp(t)} := \{u \in N_t \mid lab_t(u) = \mathbf{ab}, \overrightarrow{\mathbf{ab}} \text{ or } \mathbf{a}^\ell \text{ for some } a,b \in \mathcal{A}\},$$

$$vert_{Exp(t)}(u) := \begin{cases} \{(u,a),(u,b)\} & \text{if } lab_t(u) = \mathbf{ab}, \\ ((u,a),(u,b)) & \text{if } lab_t(u) = \overrightarrow{\mathbf{ab}}, \\ \{(u,a)\} & \text{if } lab_t(u) = \mathbf{a}^\ell \text{ and } t \in T(F^{\mathrm{HRu}}), \\ ((u,a),(u,a)) & \text{if } lab_t(u) = \mathbf{a}^\ell \text{ and } t \in T(F^{\mathrm{HRd}}), \end{cases}$$

$$src_{Exp(t)}(a) := (root_t, a) \text{ for every } a \in \tau(t) = \tau(root_t).$$

We let R_t be the binary relation on $V_{Exp(t)}$ defined as follows:

$$R_t := \{((u,a),(v_i,a)) \mid lab_t(u) = /\!/, son_{it}(u,v_i), i \in \{1,2\}\}$$
$$\cup \{((u,a),(v,a)) \mid lab_t(u) = fg_B, son_{1t}(u,v)\}$$
$$\cup \{((u,h(a)),(v,a)) \mid lab_t(u) = ren_h, son_{1t}(u,v)\}.$$

We let $\approx_t$ be the equivalence relation generated by R_t. The vertex set of $val(t)$ will be defined as the quotient set of $V_{Exp(t)}$ by $\approx_t$. With these definitions we have the following proposition:

Proposition 2.41 For every term $t \in T(F^{\mathrm{HR}})$, we have

$$val(t) = [Exp(t)/\approx_t]_{iso}.$$

---

[24] We could also decide that the a-source of $val(\mathbf{ab})$ is described by the occurrence of fg_a. This idea works well for terms defining graphs (without sources) because each vertex corresponds to a unique occurrence of an operation fg_a. It will be used in Chapter 6.

[25] Clearly, $Exp(\varnothing)$ is the empty graph.

Proof: Let $t \in T(F^{\mathrm{HR}})$. For every u in N_t, we define $V_u := \{(w,a) \in V_{Exp(t)} \mid w \leq_t u\}$, $E_u := \{w \in E_{Exp(t)} \mid w \leq_t u\}$ and $R_{t,u} := R_t \cap (V_u \times V_u)$. We let also $\approx_{t,u}$ be the equivalence relation on V_u generated by $R_{t,u}$.

We let $Exp(t) \restriction u$ be the s-graph H of type $\tau(u)$ such that H° is the subgraph of $Exp(t)^\circ$ with vertex set V_u, edge set E_u, and such that $src_H(a) = (u,a)$ for every a in $\tau(u)$.

Claim 2.41.1 For every u in N_t, if $(u,a) \approx_{t,u} (u,b)$, then $a = b$.

Proof: For fixed t, by bottom-up induction[26] on u. $\square$

It follows that the quotient graph $(Exp(t) \restriction u)/\approx_{t,u}$ is well defined for every u in N_t.

Claim 2.41.2 For every u in N_t, we have

$$val(t/u) = [(Exp(t) \restriction u)/\approx_{t,u}]_{iso}.$$

Proof: For fixed t, by bottom-up induction on u. In the case where u is an occurrence of $/\!/$ we use Proposition 2.29 with $H := (Exp(t) \restriction v_1)/\approx_{t,v_1}$ and $K := (Exp(t) \restriction v_2)/\approx_{t,v_2}$, where $son_{it}(u,v_i)$ for $i \in \{1,2\}$. $\square$

We now complete the proof of the proposition. If $u = root_t$, then we have $Exp(t) \restriction u = Exp(t)$ and $\approx_{t,u}$ is $\approx_t$. Hence Claim 2.41.2 yields the result. $\blacksquare$

This proposition will entail (in Section 7.4) that the mapping *val*, from terms to graphs, is a monadic second-order transduction.

In the above proposition, we have shown how to construct from a term t a concrete s-graph isomorphic to $val(t)$. Actually, for each cross-section X of $\approx_t$ we have a particular concrete s-graph, defined as $(Exp(t)/\approx_t)_X$. However, we can define a canonical cross-section, whence, a canonical concrete s-graph isomorphic to $val(t)$.

Lemma 2.42 Each equivalence class of $\approx_t$ contains a unique pair (u,a) such that u is maximal for $\leq_t$.

Proof: It follows from the definitions that if $(u,a) \approx_t (u',b)$ then $(u,a) \approx_t (w,c) \approx_t (u',b)$ for some (w,c), where w is a common ancestor of u and u', and furthermore $(u,a) \approx_{t,w} (w,c) \approx_{t,w} (u',b)$. Hence, each equivalence class of $\approx_t$ contains pairs (u,a) for a unique maximal node u. If it contains (u,a) and (u,b), then $(u,a) \approx_{t,u} (u,b)$ hence $a = b$ by Claim 2.41.1. $\blacksquare$

We will denote by $max(u,a)$ the unique maximal pair equivalent to (u,a), that exists by the previous lemma, and by $X_{\max}$ the set of all such maximal pairs. This set is a cross-section of $\approx_t$ that we call the *canonical cross-section* and we let $cval(t)$ be the concrete s-graph $(Exp(t)/\approx_t)_{X_{\max}}$, called the *canonical* (concrete) *s-graph denoted by* t.

---

[26] The property for u is proved by using the validity of the same property for nodes strictly below u.

We now characterize what it means for an arbitrary concrete s-graph G to be isomorphic to $val(t)$.

Definition 2.43 (Witnesses) Let G be a concrete s-graph and t be a term over F^{HR}. A *witness* of $G \simeq val(t)$ is a mapping $w : V_G \cup E_G \to (N_t \times \mu(t)) \cup N_t$ satisfying the following conditions:

(1) w defines a bijection from E_G to the set $\{u \in N_t \mid lab_t(u) = \mathbf{ab}, \overrightarrow{\mathbf{ab}} \text{ or } \mathbf{a}^\ell \text{ for some } a, b \in \mathcal{A}\}$; it is a subset of the set of leaves of the syntactic tree of t, i.e., of occurrences of constant symbols;

(2) w defines a bijection from V_G to a cross-section X of $\approx_t$ (this equivalence relation is defined in Definition 2.40);

(3) if $e \in E_G$, $lab_t(w(e)) = \overrightarrow{\mathbf{ab}}$ and $vert_G(e) = (x,y)$, then $w(x) \approx_t (w(e), a)$ and $w(y) \approx_t (w(e), b)$; if $lab_t(w(e)) = \mathbf{ab}$ and $vert_G(e) = \{x, y\}$, then this condition or the one with a and b exchanged holds;

(4) if $e \in E_G$, $lab_t(w(e)) = \mathbf{a}^\ell$ and $vert_G(e)$ is either $\{x\}$ or (x,x), then $w(x) \approx_t (w(e), a)$;

(5) x is an a-source of G if and only if $w(x) \approx_t (root_t, a)$.

Proposition 2.44 Let G be a concrete s-graph and t a term over F^{HR}. Then:

(1) w is a witness of $G \simeq val(t)$ with cross-section X of $\approx_t$ if and only if $(w \restriction V_G, w \restriction E_G)$ is an isomorphism from G to $(Exp(t)/\approx_t)_X$; and

(2) $G \simeq val(t)$ if and only if there exists a witness of $G \simeq val(t)$.

Proof: Assertion (1) is easy to check from the definitions. Assertion (2) is then immediate from Proposition 2.41; in the "only-if" direction, if (h_V, h_E) is an isomorphism from G to $(Exp(t)/\approx_t)_X$, then $w := h_V \cup h_E$ is a witness of $G \simeq val(t)$. ∎

As an illustration, consider the term t of the proof of Proposition 2.33, where

$$t = fg_B(/\!/ \cdots /\!/ \; x_1 y_1 \cdots x_m y_m z_1 \cdots z_p w_1^\ell \cdots w_q^\ell)$$

is here written in Polish prefix notation (which was not the case in the proof for readability). We let G be the associated graph. A witness w of $G \simeq val(t)$ can be defined as follows. Note that the first symbol fg_B of t is followed by $m + p + q - 1$ occurrences of $/\!/$. We let $w(e_i) := m + p + q + i$, $w(z_k) := 2m + p + q + k$, $w(e'_j) := 2m + 2p + q + j$, and for every nonisolated vertex x, we let $w(x)$ be equal to $w(e_i)$ or $w(e'_j)$ for some edge e_i or e'_j that is incident with x.

A *canonical witness* of $G \simeq val(t)$ is one such that $X = X_{max}$ in Condition (2) of Definition 2.43. By Assertion (1) of Proposition 2.44 it corresponds to an isomorphism from G to $cval(t)$. Thus, the canonical witnesses of $cval(t) \simeq val(t)$ correspond to the automorphisms of $cval(t)$. A *ground witness* of $G \simeq val(t)$ is one such that u is a leaf of t for each pair $(u, a) \in X$. It is not hard to see that each equivalence class of $\approx_t$ contains at least one such pair. Hence, one can also construct concrete s-graphs isomorphic to $val(t)$ whose vertices are such pairs (u,a); the corresponding witnesses are ground witnesses. By selecting from an equivalence class of $\approx_t$ the pair (u,a) such that u is the leftmost leaf (according to the linear order of positions in t), we get a unique such concrete s-graph which we denote by $gval(t)$; a corresponding ground witness is called a *leftmost ground witness*. Thus, a leftmost ground witness of $G \simeq val(t)$ corresponds to an isomorphism from G to $gval(t)$.

Let us say more generally that two witnesses w_1 and w_2 of $G \simeq val(t)$ are *equivalent* if $w_1(e) = w_2(e)$ for every $e \in E_G$ and $w_1(x) \approx_t w_2(x)$ for every $x \in V_G$. Then, every witness of $G \simeq val(t)$ is equivalent to a unique canonical witness and to a unique leftmost ground witness of $G \simeq val(t)$.

Proposition 2.45 For each finite set C, the concrete s-graph $cval(t)$ can be constructed from $t \in T(F_C^{HR})$ in linear time, i.e., in time $O(|t|)$. One can also construct within the same time the concrete s-graph $gval(t)$ and a leftmost ground witness of $cval(t) \simeq val(t)$. These results extend to (K, Λ)-labeled s-graphs with the same computation times.

Proof: Let $t \in T(F_C^{HR})$. We know that we can compute the types of its nodes in linear time by a bottom-up computation. Hence, we can also compute in linear time the sources of the s-graph $Exp(t)$ and its sets of vertices and edges. It is clear that a pair $(u,a) \in V_{Exp(t)}$ belongs to X_{max} if and only if either u is the root of t or its father is an occurrence of fg_B such that $a \in B$. Hence X_{max}, the vertex set of $cval(t)$, can also be computed bottom-up in linear time. It remains to compute the mapping $vert_{cval(t)}$. The set R_t of Definition 2.40 is the set of edges of a rooted forest H, with set of roots X_{max}. It has at most $|C| \cdot |t|$ edges and can also be constructed in linear time by a bottom-up computation on t. Then, by using depth-first traversals of its trees (there is one such tree for each element of X_{max}), one can compute the mapping max that associates with each node (u,a) of H the root $max(u,a)$ of the tree to which it belongs. In particular, we obtain for every occurrence u (in t) of a constant symbol that defines an edge, say $\mathbf{ab}$, its end vertices $max(u,a)$ and $max(u,b)$ in $cval(t)$. These traversals can be done in linear time in the size of H, hence in time $O(|C| \cdot |t|)$. By similar depth-first traversals, one can associate with each element x of X_{max} an equivalent pair $left(x) := (u,a)$ such that u is a leftmost leaf. Hence, one can also construct $gval(t)$ and a leftmost ground witness w of $cval(t) \simeq val(t)$. In fact, $w(x) = left(x)$ for every $x \in V_{cval(t)}$ and $w(e) = e$ for every $e \in E_{cval(t)}$.

The extension of these definitions and results to labeled graphs is easy: edge labels are specified in the constant symbols that specify edges. The different labels of a vertex are specified by different occurrences of constant symbols. However, the depth-first traversal that defines the mapping max can also compute the set of labels of each vertex: whenever one reaches a leaf (u, a) of H such that u is an occurrence of $\mathbf{a}_\kappa$ in t, one adds κ to the current set of labels of $max(u, a)$. ∎

Definition 2.46 (Parsing) The *parsing problem* relative to the HR algebra is defined as follows:

    Input: A finite set C of source names and a concrete s-graph G of type included in C.

    Output: A term t in $T(F_C^{HR})$ and a witness of $G \simeq val(t)$ if they exist, or the answer that they do not exist.

It follows from the proof of the previous proposition that an arbitrary witness can be converted in linear time into an equivalent canonical or leftmost ground one. In fact, if $w : V_G \cup E_G \rightarrow V_{Exp(t)} \cup E_{Exp(t)}$ is a witness of $G \simeq val(t)$, then the mapping $\overline{w}$ such that $\overline{w}(x) = max(w(x))$ if $x \in V_G$ and $\overline{w}(e) = w(e)$ if $e \in E_G$ is the unique canonical witness of $G \simeq val(t)$ equivalent to w. The algorithm sketched in the proof of Proposition 2.45 can be adapted to compute $\overline{w}$ from w and t in linear time. Similarly, with *left* instead of *max*, w can be converted in linear time into the unique leftmost ground witness $\overline{w}$ of $G \simeq val(t)$ equivalent to w.

By the results of the next section, the parsing problem is equivalent to that of finding a tree-decomposition of G of width at most some given integer. We will review some algorithms that solve it in Section 6.2. Having parsed a graph G, it is often easy to parse a graph that is a modification of G obtained by, e.g., adding loops and multiple edges, adding labels, changing labels or changing the direction of edges, as shown by the following proposition. In particular, any algorithm solving the parsing problem can easily be extended into one solving the corresponding problem for labeled graphs.

We recall from Definition 2.9 that the core of a graph is obtained by removing edge directions and loops, and by fusing multiple edges. If G is a labeled s-graph, we let $core(G)$ be the unlabeled, simple, loop-free, undirected s-graph $\langle core(G^\circ), slab_G \rangle$. Thus, sources and their names are not modified.

Proposition 2.47

(1) Let be given a concrete s-graph G, a term t in $T(F_C^{HR})$ and a witness w of $core(G) \simeq val(t)$. We can compute in linear time a term $\overline{t}$ in $T(F_C^{HR})$ that evaluates to G, together with a witness of this fact.[27]

(2) Let be given a concrete s-graph G, a term t in $T(F_C^{HR})$ and a witness w of $G \simeq val(t)$. Let γ be a labeling function turning G into a (K, Λ)-labeled s-graph $\widehat{G}$. In linear time (for fixed C, K and Λ), we can compute $\widehat{t} \in T(F_{C,[K,\Lambda]}^{HR})$ that evaluates to $\widehat{G}$, together with a witness of this fact.

(3) Let G, t and w be as in (2) with G possibly (K, Λ)-labeled. Let G' be obtained from G by changing certain vertex and edge labels and by reversing certain edge

---

[27] Here, linear time means time $O(|V_G| + |E_G|)$.

directions (if G is directed). We can compute in linear time a term t' in $T(F_C^{HR})$ or $T(F_{C,[K,\Lambda]}^{HR})$ that evaluates to G', such that w is a witness of this fact.

Proof: Let us assume that G is undirected; the proof for the directed case is similar.

(1) By the above observations, we first transform the given witness into a ground one that we still denote by w. Moreover, since $\mathbf{ab} = \mathbf{ab} \mathbin{/\!/} \mathbf{a}$, we may assume that for every vertex x of G, if $w(x) = (u,a)$ then the leaf u is an occurrence of $\mathbf{a}$ in t. We construct $\bar{t}$ from t by the following simultaneous replacements. It is then easy to transform w into a ground witness of $G \simeq val(\bar{t})$.

For every vertex x of G, we do the following: we have $w(x) = (u,a)$ for some leaf u in t that is an occurrence of a constant symbol $\mathbf{a}$. If x has loops $e_1,\dots,e_m$ in G with $m > 0$, then we replace the occurrence at u by the term $\mathbf{a}^\ell \mathbin{/\!/} \cdots \mathbin{/\!/} \mathbf{a}^\ell$, the parallel-composition of m constant symbols $\mathbf{a}^\ell$.

For every edge $e : x - y$ of $core(G)$, we do the following: the witness w specifies a position $w(e)$ in t that is an occurrence of some $\mathbf{ab}$. If G has edges $e_1,\dots,e_n$ in G with ends x and y, then we replace the occurrence at $w(e)$ by the parallel-composition of n constant symbols $\mathbf{ab}$.[28]

(2) The proof is similar (using that also $\mathbf{a}^\ell = \mathbf{a}^\ell \mathbin{/\!/} \mathbf{a}$). For every vertex x of G, if $\{\kappa_1,\dots,\kappa_p\} = \gamma(x)$ with $p > 0$, i.e., if $\kappa_1,\dots,\kappa_p$ are the labels of x in $\widehat{G}$, then we replace the occurrence of $\mathbf{a}$ by the term $\mathbf{a}_{\kappa_1} \mathbin{/\!/} \cdots \mathbin{/\!/} \mathbf{a}_{\kappa_p}$.

For every edge e of G, $w(e)$ is an occurrence of $\mathbf{ab}$ or $\mathbf{a}^\ell$. If $\lambda = \gamma(e)$, i.e., if λ is the label of e in $\widehat{G}$, then we replace the constant symbol at $w(e)$ by $\mathbf{ab}_\lambda$ or $\mathbf{a}_\lambda^\ell$ respectively.

(3) The proof is similar but we need not assume that w is ground. At certain positions specified by w, it suffices to replace constant symbols of the form $\mathbf{a}_\lambda$ (or $\mathbf{ab}_\lambda$ or $\overrightarrow{\mathbf{ab}}_\lambda$) by $\mathbf{a}_\mu$ (or $\mathbf{ab}_\mu$ or $\overrightarrow{\mathbf{ab}}_\mu$), and to replace $\overrightarrow{\mathbf{ab}}_\lambda$ by $\overrightarrow{\mathbf{ba}}_\lambda$ (or $\overrightarrow{\mathbf{ab}}$ by $\overrightarrow{\mathbf{ba}}$ for unlabeled graphs). ∎

These transformations do not modify the source names that are used in t, hence they preserve the width of t.

2.3.3 Algebraic properties and derived operations defined by contexts

Three equational axioms characterize the monoid of words over any alphabet. The situation for graphs is more complicated.

Proposition 2.48 For all s-graphs G, H, J, all unary operations $fg_B, fg_{B'}, ren_h, ren_{h'}$ in F^{HR} and all $a, b \in \mathcal{A}$ we have:
(1) $G \mathbin{/\!/} H = H \mathbin{/\!/} G$;
(2) $G \mathbin{/\!/} (H \mathbin{/\!/} J) = (G \mathbin{/\!/} H) \mathbin{/\!/} J$;
(3) $G \mathbin{/\!/} \varnothing = G$;
(4) $fg_B(fg_{B'}(G)) = fg_{B \cup B'}(G)$;

---

[28] In the directed case, there are k edges $x \rightarrow_G y$ and n edges $y \rightarrow_G x$. If $w(x) = (u,a)$ and $w(y) = (v,b)$, then we replace the occurrence of $\mathbf{ab}$ by the parallel-composition of k symbols $\overrightarrow{\mathbf{ab}}$ and n symbols $\overrightarrow{\mathbf{ba}}$.

(5) $fg_B(\varnothing) = \varnothing$;

(6) $fg_\varnothing(G) = G$;

(7) $ren_h(ren_{h'}(G)) = ren_{hoh'}(G)$;

(8) $ren_h(G \parallel H) = ren_h(G) \parallel ren_h(H)$;

(9) $ren_h(\varnothing) = \varnothing$;

(10) $ren_{\{\}}(G) = G$;

(11) $ren_h(fg_B(G)) = fg_{h(B)}(ren_h(G))$;

(12) $ren_h(\mathbf{ab}) = \mathbf{cd}$, where $\{c,d\} = \{h(a), h(b)\}$, $a < b$ and $c < d$;

(13) $ren_h(\overrightarrow{\mathbf{ab}}) = \overrightarrow{\mathbf{cd}}$, where $c = h(a)$, $d = h(b)$;

(14) $ren_h(\mathbf{a}) = \mathbf{c}$, where $c = h(a)$;

(15) $ren_h(\mathbf{a}^\ell) = \mathbf{c}^\ell$, where $c = h(a)$;

(16) $\mathbf{a} \parallel \mathbf{a} = \mathbf{a}$ and $\mathbf{a} \parallel \mathbf{a}^\ell = \mathbf{a}^\ell$;

(17) $\mathbf{a} \parallel \mathbf{ab} = \mathbf{ab}$, $\mathbf{a} \parallel \overrightarrow{\mathbf{ab}} = \overrightarrow{\mathbf{ab}}$ and $\mathbf{a} \parallel \overrightarrow{\mathbf{ba}} = \overrightarrow{\mathbf{ba}}$.

Furthermore, for every edge label λ and every vertex label κ, we have:

(12') $ren_h(\mathbf{ab}_\lambda) = \mathbf{cd}_\lambda$; where $\{c,d\} = \{h(a), h(b)\}$, $a < b$ and $c < d$;

(13') $ren_h(\overrightarrow{\mathbf{ab}}_\lambda) = \overrightarrow{\mathbf{cd}}_\lambda$, where $c = h(a)$, $d = h(b)$;

(14') $ren_h(\mathbf{a}_\kappa) = \mathbf{c}_\kappa$, where $c = h(a)$;

(15') $ren_h(\mathbf{a}_\lambda^\ell) = \mathbf{c}_\lambda^\ell$, where $c = h(a)$;

(16') $\mathbf{a} \parallel \mathbf{a}_\kappa = \mathbf{a}_\kappa$, $\mathbf{a}_\kappa \parallel \mathbf{a}_\kappa = \mathbf{a}_\kappa$ and $\mathbf{a} \parallel \mathbf{a}_\lambda^\ell = \mathbf{a}_\lambda^\ell$;

(17') $\mathbf{a} \parallel \mathbf{ab}_\lambda = \mathbf{ab}_\lambda$, $\mathbf{a} \parallel \overrightarrow{\mathbf{ab}}_\lambda = \overrightarrow{\mathbf{ab}}_\lambda$ and $\mathbf{a} \parallel \overrightarrow{\mathbf{ba}}_\lambda = \overrightarrow{\mathbf{ba}}_\lambda$.

Proof: Easy to check from the definitions. ∎

Equalities (6) and (10) could be omitted from this list because they just say that $fg_\varnothing$ and $ren_{\{\}}$ are identity operations. Hence, they express properties of the syntax and not properties of nontrivial graph operations. However, we retain them here so that the reader will have a comprehensive list of equalities usable for proofs.

We have not listed conditional equalities, in particular the following one, which we will number for later reference:

(18) $fg_b(G \parallel H) = G \parallel fg_b(H)$ if $b \notin \tau(G)$.

Such conditional equalities are best expressed as equalities of typed terms of the many-sorted HR algebra that we will define in Section 2.6.2. However, we will not develop this aspect.

Let us ask the question whether the equalities of Proposition 2.48 form a complete description of the equational[29] properties of $\mathbb{JS}$?

We can formulate this question more precisely, as follows. Every term t in $T(F^{HR}, \{x_1, \ldots, x_n\})$, where $x_1, \ldots, x_n$ are variables (intended to denote

---

[29] The term "equational" refers here to equational logic and term rewriting systems [*BaaNip], not to the equational sets to be studied in Chapter 3.

s-graphs) defines a derived operation $t_{JS} : \mathcal{JS}^n \to \mathcal{JS}$. Two terms t and t' in $T(F^{HR}, \{x_1,\ldots,x_n\})$ are *equivalent* if $t_{JS} = t'_{JS}$.

Does there exist, for each finite set $C \subseteq \mathcal{A}$, a finite set C' such that $C \subseteq C' \subseteq \mathcal{A}$ and a finite set of equational axioms of the form $t = t'$ where $t, t' \in T(F_{C'}^{HR}, \{x_1,\ldots,x_m\})$ that characterizes in equational logic the equivalence of pairs of terms in $T(F_C^{HR}, \{x_1,\ldots,x_n\})$ for all $n \geq 0$?

The many-sorted framework of Section 2.6.2 is perhaps necessary, or at least convenient. For a quite similar algebra, an equational characterization of the equivalence of terms without variables, hence that evaluate to constant s-graphs,[30] is given in [BauCou]. However, the intermediate steps of the transformation of a term into another one may use functions of the signature that are not in a fixed finite superset $F_{C'}^{HR}$ of F_C^{HR}, as required. Furthermore, the (open) problem is to have C' as small as possible.

Proposition 2.49 Every term t over F_C^{HR} evaluating to a nonempty s-graph is equivalent to a term $\widehat{t}$ over F_C^{HR} written without $fg_\emptyset$, $\emptyset$ and the source renaming operations, and such that $|\widehat{t}| \leq |t|$.

Proof: It is clear that, by using Equalities (1), (3), (5), (6) and (9) of Proposition 2.48, we can transform a term t in $T(F_C^{HR})$ into an equivalent term t' in $T(F_C^{HR})$ with $|t'| \leq |t|$ such that either $t' = \emptyset$ or t' is without $\emptyset$ and $fg_\emptyset$. The first case is excluded since $val(t) \neq \emptyset$. We now eliminate the source renaming operations from t assumed to be already without $\emptyset$ and $fg_\emptyset$. The construction is by induction on the size of t.

(1) If t is a constant symbol then $\widehat{t} = t$.
(2) If $t = t_1 \,/\!/\, t_2$ then $\widehat{t} = \widehat{t_1} \,/\!/\, \widehat{t_2}$.
(3) If $t = fg_B(t_1)$ then $\widehat{t} = fg_B(\widehat{t_1})$.
(4) If $t = ren_h(t_1)$, then we distinguish several subcases:

    (4.1) If t_1 is a constant symbol, then $\widehat{t}$ is the constant symbol obtained by Equalities $(12)-(15)$ and $(12')-(15')$ of Proposition 2.48.

    (4.2) If $t_1 = t_2 \,/\!/\, t_3$, then $\widehat{t} = \widehat{ren_h(t_2)} \,/\!/\, \widehat{ren_h(t_3)}$ and the correctness follows from Equality (8).

    (4.3) If $t_1 = fg_B(t_2)$, then $\widehat{t} = fg_{h(B)}(\widehat{ren_h(t_2)})$ and the correctness follows from Equality (11).

    (4.4) If $t_1 = ren_{h'}(t_2)$, then $\widehat{t} = \widehat{ren_{h \circ h'}(t_2)}$ and the correctness follows from Equality (7).

The inductive construction is well defined because in each case, $\widehat{t}$ is defined either directly as in Cases (1) and (4.1) or in terms of $\widehat{t'}$ for terms t' of size strictly smaller

---

[30] Such terms are said to be *ground* in the theory of term rewriting systems.

than t. For Case (4.4), we observe that if $h, h' \in Perm_f(C)$ then $h \circ h' \in Perm_f(C)$. It is clear by this inductive construction that $\widehat{t}$ has no occurrences of renaming operations, that $\widehat{t}$ is equivalent to t and that $|\widehat{t}| = |t| - |t|_{ren}$, where $|t|_{ren}$ is the number of occurrences of renaming operations in t. ∎

The constant symbols **a** are actually useful only for defining isolated vertices. If a term t in $T(F^{HR})$ evaluates to an s-graph G without isolated vertices, then the term t' obtained from t by replacing each symbol **a** by $\varnothing$ evaluates to G, because all its vertices are defined as the ends of edges, hence by constant symbols of the forms **ab**, $\overrightarrow{\mathbf{ab}}$ and $\mathbf{a}^\ell$. The constant symbols $\mathbf{a}_\kappa$ are useful to define vertex labels. These labels are not otherwise specified because the constant symbols $\mathbf{ab}_\lambda$, $\overrightarrow{\mathbf{ab}}_\lambda$ and $\mathbf{a}^\ell_\lambda$ that define labeled edges do not specify the labels of the ends of these edges.

We will write a context over F^{HR} as a term in $T(F^{HR}, \{x_1\})$ where the variable x_1 has a single occurrence (see Definition 2.6). Such a term defines a unary derived operation of the algebra $\mathbb{JS}$. (For all other terms t in $T(F^{HR}, \{x_1\})$, such as the term $x_1 \,/\!/\, x_1$, the unary derived operation $t_{\mathbb{JS}}$ cannot be defined by a context.) We will prove that each such derived operation can be concretely described as the result of gluing a fixed s-graph to the input s-graph, followed by one forgetting and one renaming operation. The basic idea is that all renaming and forgetting operations of a term can be "moved outwards."

We define two contexts as *equivalent* if they define the same derived operation.

Proposition 2.50 Let C be a finite set of source names. For every context c in $Ctxt(F^{HR}_C)$, there exists a context $\overline{c}$ in $Ctxt(F^{HR}_C)$ written without renaming operations, and a permutation h of C such that the context $ren_h(\overline{c})$ is equivalent to c.

Proof: The proof is by induction on the structure of c.

(1) If $c = x_1$, then we let $\overline{c}$ be c and h be the identity.

(2) If $c = c_1 \,/\!/\, t$ (or $c = t \,/\!/\, c_1$), where $c_1 \in Ctxt(F^{HR}_C)$ and $t \in T(F^{HR}_C)$, then, by the induction hypothesis, c_1 is equivalent to $ren_{h_1}(\overline{c}_1)$ for some $\overline{c}_1$ in $Ctxt(F^{HR}_C)$ written without renaming operations, and some permutation h_1 of C. It follows that c is equivalent to $ren_{h_1}(\overline{c}_1) \,/\!/\, t$ and to $ren_{h_1}(\overline{c}_1 \,/\!/\, ren_{h_1^{-1}}(t))$ by Equality (8) of Proposition 2.48. Hence we can take $h := h_1$ and $\overline{c} := \overline{c}_1 \,/\!/\, t'$, where t' is a term without renaming operations that is equivalent to $ren_{h_1^{-1}}(t)$. Such a term can be constructed by Proposition 2.49.

(3) If $c = ren_{h'}(c_1)$ and $h_1, \overline{c}_1$ are obtained from c_1 by the induction hypothesis, we can take $\overline{c} := \overline{c}_1$ and $h := h' \circ h_1$ by Equality (7) of Proposition 2.48.

(4) If $c = fg_B(c_1)$ and $h_1, \overline{c}_1$ are as in the previous case, then c is equivalent to $fg_B(ren_{h_1}(\overline{c}_1))$ and to $ren_{h_1}(fg_{h_1^{-1}(B)}(\overline{c}_1))$ by Equality (11) of Proposition 2.48. Hence, we can take $h := h_1$ and $\overline{c} := fg_{h^{-1}(B)}(\overline{c}_1)$. ∎

Proposition 2.51 Let B be a finite set of source names, and let c be a context in $Ctxt(F^{HR}_B)$. For every $A \subseteq B$ there exists a set C of source names such that $C \cap B = \varnothing$

and $|C| \leq |A|$, a term t over $F_{B \cup C}^{\mathrm{HR}}$, a subset A' of A and a permutation h of $B \cup C$ such that for every s-graph G of type included in A we have:

$$c_{\mathbb{JS}}(G) = ren_h(fg_{A'}(G \parallel t_{\mathbb{JS}})).$$

In other words, the restriction of $c_{\mathbb{JS}}$ to s-graphs of type included in A is defined by the context $ren_h(fg_{A'}(x_1 \parallel t))$ that belongs to $Ctxt(F_{B \cup C}^{\mathrm{HR}})$.

Proof: We first use Proposition 2.50 to reduce the proof to the case of contexts written without renaming operations: a context c in $Ctxt(F_B^{\mathrm{HR}})$ is equivalent to $ren_k(\overline{c})$ for some permutation k of B and some context $\overline{c}$ without renaming operations. If $\overline{c}$ is equivalent on s-graphs of type included in A to a context of the form $ren_{h'}(fg_{A'}(x_1 \parallel t))$ with $A' \subseteq A$ and $h' \in Perm_f(B \cup C)$, then c is equivalent to $ren_{k \circ h'}(fg_{A'}(x_1 \parallel t))$ on the same s-graphs, because k is the identity on C (we use Equality (7) of Proposition 2.48).

We now prove the statement for a context c without renaming operations. Such a context is a term in $T(F_B^{\mathrm{HR}}, \{x_1\})$ and we let u be the occurrence in c of the variable x_1. Without loss of generality we assume that c is written with the elementary operations fg_a for a in B, and not with their compositions fg_D for $D \subseteq B$ (see Equalities (4) and (6) of Proposition 2.48). For convenience, the unary derived operation $c_{\mathbb{JS}}$ defined by c will also be denoted by c.

We prove the statement by induction on the structure of c, together with the additional statements that (i) $C = \{\overline{a} \mid a \in A'\}$, where $\{\overline{a} \mid a \in A\}$ is a set of source names in bijection with A and disjoint with B, and (ii) h is such that $h(\overline{a}) = a$, $h(a) = \overline{a}$ for a in A' and $h(b) = b$ for $b \notin A' \cup C$. Thus, C and h are determined uniquely by A'. We will also write t instead of $t_{\mathbb{JS}}$ for a term t over F^{HR}.

(1) If $c = x_1$, then we take $A' := \emptyset$ and $t := \emptyset$, because $c(G) = ren_{\{\}}(fg_\emptyset(G \parallel \emptyset)) = G$.

(2) If $c = c_1 \parallel s$, where $c_1 \in Ctxt(F_B^{\mathrm{HR}})$ and $s \in T(F_B^{\mathrm{HR}})$, then, by the induction hypothesis, there are A' and t_1 such that $c_1(G) = ren_h(fg_{A'}(G \parallel t_1))$ for every s-graph G with $\tau(G) \subseteq A$. For c we take the same set A' and the term $t := t_1 \parallel ren_h(s)$. Then $ren_h(fg_{A'}(G \parallel t)) = ren_h(fg_{A'}(G \parallel t_1) \parallel ren_h(s))$ by the fact that for all s-graphs H, K and every set of source names A', we have $fg_{A'}(H \parallel K) = fg_{A'}(H) \parallel K$ if $\tau(K) \cap A' = \emptyset$ (and note that $\tau(ren_h(s)) \subseteq h(B)$ and $h(B) \cap A' = \emptyset$), see Equality (18) after Proposition 2.48. By Equalities (8) and (7) of Proposition 2.48 this equals $ren_h(fg_{A'}(G \parallel t_1)) \parallel ren_{h \circ h}(s)$, which equals $c_1(G) \parallel s$ because $h \circ h$ is the identity.

(3) If $c = fg_b(c_1)$, with $b \notin A$ or $b \in A'$, and A', t_1 are obtained from the induction hypothesis, then we can take the same A' and $t := fg_{h(b)}(t_1)$ by Equality (18) (and note that $h(b) \notin \tau(G)$) and by Equalities (4) and (11) of Proposition 2.48.

(4) If $c = fg_a(c_1)$, with $a \in A - A'$, and A', t_1 are obtained from the induction hypothesis, then we can take for c the same term $t := t_1$ and the set $A'' := A' \cup \{a\}$

with the corresponding permutation h'. In fact, again by Equalities (10) and (4) of Proposition 2.48, we have that $c(G) = fg_a(ren_h(fg_{A'}(G /\!/ t_1))) = ren_h(fg_a(fg_{A'}(G /\!/ t_1))) = ren_h(fg_{A''}(G /\!/ t_1))$ because $h(a) = a$, and this equals $ren_{h'}(fg_{A''}(G /\!/ t_1))$ because $h'(b) = h(b)$ for every $b \in (B \cup C) - A''$. ∎

The extensions of Propositions 2.49, 2.50 and 2.51 to labeled graphs are straightforward.

Example 2.52 We let $A = \{a\}$ and c be the context

$$fg_a(fg_a(x_1 /\!/ \mathbf{ab}) /\!/ \mathbf{ab} /\!/ \mathbf{ac}) /\!/ \mathbf{ac}.$$

For $\tau(G) \subseteq \{a\}$ we have (with obvious simplifications):

$$
\begin{aligned}
G &= ren_{\{\}}(fg_\emptyset(G /\!/ \emptyset)), \\
G /\!/ \mathbf{ab} &= ren_{\{\}}(fg_\emptyset(G /\!/ \mathbf{ab}), \\
fg_a(G /\!/ \mathbf{ab}) &= ren_{a \leftrightarrow \bar{a}}(fg_a(G /\!/ \mathbf{ab})), \\
fg_a(G /\!/ \mathbf{ab}) /\!/ \mathbf{ab} /\!/ \mathbf{ac} &= ren_{a \leftrightarrow \bar{a}}(fg_a(G /\!/ \mathbf{ab} /\!/ \overline{\mathbf{ab}} /\!/ \overline{\mathbf{ac}})), \\
fg_a(fg_a(G /\!/ \mathbf{ab}) /\!/ \mathbf{ab} /\!/ \mathbf{ac}) &= ren_{a \leftrightarrow \bar{a}}(fg_a(G /\!/ fg_{\bar{a}}(\mathbf{ab} /\!/ \overline{\mathbf{ab}} /\!/ \overline{\mathbf{ac}}))), \\
c(G) &= ren_{a \leftrightarrow \bar{a}}(fg_a(G /\!/ fg_{\bar{a}}(\mathbf{ab} /\!/ \overline{\mathbf{ab}} /\!/ \overline{\mathbf{ac}}) /\!/ \overline{\mathbf{ac}})),
\end{aligned}
$$

by cases (1), (2), (4), (2), (3) and (2) respectively. Thus, $ren_h = ren_{a \leftrightarrow \bar{a}}$, $fg_{A'} = fg_a$, and $t = fg_{\bar{a}}(\mathbf{ab} /\!/ \overline{\mathbf{ab}} /\!/ \overline{\mathbf{ac}}) /\!/ \overline{\mathbf{ac}}$. □

2.4 Tree-decompositions

In this section, we define tree-decompositions and we establish that the terms over F^{HR} describe them. We get in this way a characterization of the associated graph complexity measure known as tree-width. For the numerous equivalent definitions and properties of tree-decompositions and tree-width, we refer the reader to the two comprehensive overview articles by Bodlaender [*Bod93, *Bod98] and to the books [*DowFel] and [*FluGro]. We only present and discuss the aspects of tree-decompositions that will be useful in the forthcoming chapters.

2.4.1 Tree- and path-decompositions

Our definition of tree-decomposition differs from the usual one on two minor technical points: it uses rooted trees and concerns graphs with sources. (It generalizes the definition given in Section 1.4.2.) Since tree- and path-decompositions do not depend on vertex or edge labels, we only define them for (unlabeled) s-graphs.

Definition 2.53 (Tree-decompositions)   Let G be a concrete s-graph. A *tree-decomposition* of G is a pair (T,f) such that T is a concrete rooted tree and f is a mapping from N_T to $\mathcal{P}(V_G)$ satisfying the following conditions, where a *box* of (T, f) is a set $f(u)$ for some u in N_T:

(1) every vertex of G belongs to some box;
(2) every edge of G that is not a loop has its end vertices in some box;
(3) for every vertex x of G, the set $f^{-1}(x)$ of all nodes u of T such that $x \in f(u)$, is connected (i.e., $T[f^{-1}(x)]$ is connected); this condition is called the *connectivity condition*;
(4) all sources of G belong to the box $f(root_T)$, called the *root box* of the decomposition.

An equivalent formulation of (3) is that if u and v are two nodes of T, then $f(u) \cap f(v) \subseteq f(w)$ for every node w on the path in T between u and v.[31] The *width* of (T,f) is defined as $wd(T,f) := \max\{|f(u)| \mid u \in N_T\} - 1$. The *tree-width* of G, denoted by $twd(G)$, is the smallest width of a tree-decomposition of G.

A tree-decomposition of a concrete s-graph G is *optimal* if its width is equal to $twd(G)$. Isomorphic concrete s-graphs have isomorphic tree-decompositions, hence the same tree-width. Tree-decompositions and tree-width are thus well defined for (abstract) s-graphs. We denote by $TWD(\leq k)$ the set of graphs of tree-width at most k, and by $TWD(\leq k, C)$ the set of s-graphs of tree-width at most k and of type included in a set of source names C. The *tree-width* $twd(L)$ of a set of s-graphs L is defined as the least upper-bound of the tree-widths of its elements. We say that L has *bounded tree-width* if $twd(L) \in \mathcal{N} \cup \{-1\}$ (otherwise $twd(L) = \omega$).

Definition 2.54 (Rich tree-decompositions) A *rich tree-decomposition* of an s-graph G is a pair (T,f) where T is as in Definition 2.53, f is a mapping : $N_T \to \mathcal{P}(V_G \cup E_G)$ and Condition (2) is replaced by the following one:

(2′) every edge of G belongs to a unique box and its end vertices belong to this box.

The width of (T, f), denoted by $wd(T, f)$, is defined as $\max\{|f(u) \cap V_G| \mid u \in N_T\} - 1$. It is clear that if (T, f) is a rich tree-decomposition of G, then (T, f') such that $f'(u) := f(u) \cap V_G$ for all u is a tree-decomposition of G, having the same width as (T, f). Conversely, every tree-decomposition (T, f') can be turned into a rich one, (T, f), such that $f'(u) = f(u) \cap V_G$ for every u. It suffices to define $f(u)$ as the set $f'(u) \cup \{e \in E_G \mid u$ is the least common ancestor of the nodes w such that $f'(w)$ contains the end vertices of $e\}$. It follows that the tree-width of an s-graph is the same whether we define it with respect to tree-decompositions or to rich tree-decompositions.

---

[31] Recall from Definition 2.13 that, in a rooted tree T, there is a unique (undirected) path between any two nodes u and v of T. This path contains the least common ancestor of u and v.

From a rich tree-decomposition (T,f) of an s-graph G, we will construct in Section 2.4.5 a term in $T(F^{HR})$ evaluating to G, and conversely, a rich tree-decomposition will be associated with each term evaluating to it.

Remark 2.55 (a) If (T,f) is a tree-decomposition or a rich tree-decomposition of G, then every vertex x of G belongs to a unique $\leq_T$-maximal box $f(u)$ of (T,f). This follows from Conditions (1) and (3) of the definitions and the fact that T is rooted (otherwise $\leq_T$ is not defined). In fact, u is the least common ancestor of all nodes in $f^{-1}(x)$, and it is the root of the rooted tree $T[f^{-1}(x)]$.

(b) For a graph, Condition (4) of Definition 2.53 is trivially true and any node of the tree T can be chosen as the root. Equivalently, the tree T of a tree-decomposition (T,f) of a graph can be defined as a tree without root.

(c) Let (T,f) be a tree-decomposition or a rich tree-decomposition of G with $wd(T,f) = k - 1$. By Conditions (1) and (2), we have that $|V_G| \leq k \cdot |N_T|$ and, if G is simple, that $|E_G| \leq k^2 \cdot |N_T|$. On the other hand, since (T,f) can consist of a unique box containing all vertices and since some box must contain all sources by Condition (4), we have $|\tau(G)| \leq k \leq |V_G|$. This implies:

$$|\tau(G)| - 1 \leq twd(G) \leq |V_G| - 1.$$

The tree-width of the empty graph is -1. A nonempty s-graph has tree-width 0 if and only if it has only loops and isolated vertices, and has at most one source.

(d) The notion of a tree-decomposition depends neither on edge directions nor on colors, labels or weights possibly attached to vertices and / or edges. Adding or deleting a loop, or fusing two edges with the same ends into a single one does not change the tree-width. Thus, $twd(G) = twd(core(G))$ for every s-graph G (for the core of G, see Definition 2.9 and Proposition 2.47). □

Example 2.56 The following examples concern graphs.

(1) A tree T with at least one edge has tree-width 1. To see this, we choose any node r as a root, which turns T into a rooted (hence directed) tree (cf. Definition 2.13). We let $f : N_T \to \mathcal{P}(N_T \cup E_T)$ be defined by:

(a) $f(r) := \{r\}$;
(b) $f(u) := \{u,v,e\}$ if $e : v \to u$ is the unique edge with head u (which implies $u \neq r$).

Clearly (T,f) is a rich tree-decomposition of T and its width is 1. For constructing a tree-decomposition of a forest F, we add edges to turn F into a tree to which we apply the above construction. We obtain a tree-decomposition of width 1 of F.

(2) Figures 2.5, 2.6 and 2.7 show a graph G and a rich tree-decomposition (T,f) of this graph. The root of T is the node a in Figure 2.7, the boxes of the decomposition are represented in Figure 2.6 as disjoint graphs with continuous lines representing edges. A dotted edge links two vertices of these disjoint graphs that correspond to the same vertex of G. We denote by $G(T,f)$ the graph representing (T,f) as it is

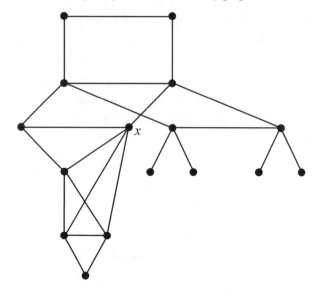

Figure 2.5 A graph G.

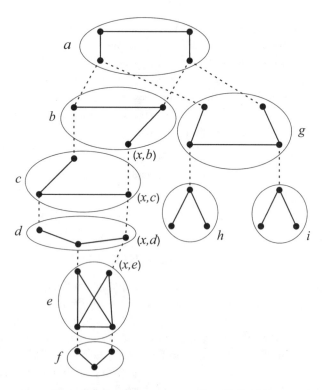

Figure 2.6 A rich tree-decomposition (T, f) of G.

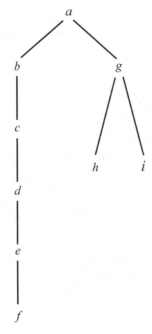

Figure 2.7 The tree T.

shown in Figure 2.6 with its two types of edges. Its dotted edges form a forest. The graph G is obtained from it by contracting all dotted edges (see Definition 2.15 for edge contractions). The vertex x of G belongs to boxes b,c,d,e. The four vertices (x,b), (x,c), (x,d), (x,e) of the graph $G(T,f)$ of Figure 2.6 get fused and yield the vertex x of G. Nodes b,c,d,e form a path in T, hence $T[f^{-1}(x)]$ is connected. The width of (T,f) is 3.

(3) Here are the tree-widths of some classical graphs. The graphs K_n (the *cliques*) and $K_{n,m}$ (the *complete bipartite graphs*) are defined in Section 1.3.1; C_n is a cycle with n vertices and $G_{n\times m}$ is the *rectangular $n \times m$ grid* defined as $\langle [n] \times [m], edg \rangle$ with $((i,j),(i',j')) \in edg$ if and only if either $i' = i+1$ and $j' = j$, or $i' = i$ and $j' = j+1$. For $n,m \geq 1$ we have $twd(K_n) = n-1$, $twd(K_{n,m}) = \min\{n,m\}$, $twd(C_n) = 2$ if $n \geq 3$, and $twd(G_{n\times m}) = \min\{n,m\}$. The inequality $twd(K_n) \leq n-1$ follows from Remark 2.55(c), and the inequalities $twd(K_{n,m}) \leq \min\{n,m\}$, $twd(C_n) \leq 2$, $twd(G_{n\times m}) \leq \min\{n,m\}$ follow from easy constructions (see also (6) below). Theorem 12.3.9 of [*Die] shows that $twd(G_{n\times n}) \geq n-1$ (with a proof technique that can be used for finding other lower-bounds). The lower-bound $twd(G_{n\times n}) \geq n$ is proved in [*Bod98], Corollary 89. Hence we get by Corollary 2.60(1) stated below that $twd(G_{n\times m}) = \min\{n,m\}$. The other opposite inequalities will be proved in Section 2.4.2.

(4) Rectangular grids, hence more generally, planar graphs have unbounded tree-width but outerplanar graphs have tree-width at most 2 (see [*Bod98]).

(5) The *incidence graph* $Inc(G)$ of an undirected graph G (already defined in Section 1.8.1) has vertex set $V_G \cup E_G$ and edges between e and x whenever $x \in V_G$, $e \in E_G$ and x is an end of e. The tree-width of $Inc(G)$ is at most $\max\{2, twd(G)\}$. To see this, take a rich tree-decomposition (T, f) of G. For each edge e of G, add to the node u of T such that $e \in f(u)$ a new son u_e and extend f in such a way that $f(u_e)$ consists of e and its ends. The resulting pair is a tree-decomposition of $Inc(G)$, as the reader will easily verify.

(6) Let us say that a rooted tree T is *normal for* a graph G if $N_T = V_G$ and every two adjacent vertices of G are comparable with respect to $\leq_T$ (like in a normal spanning tree, except that here we do not require that two adjacent nodes of T are adjacent in G, see Definition 2.13). A *normal tree-decomposition* of a graph G is a tree-decomposition (T, f) such that T is normal for G and f is defined as follows from G and T:

$$f(u) := \{u\} \cup \{w \mid u \leq_T w \text{ and } w \text{ is adjacent in } G \text{ to some } v \leq_T u\}.$$

This is a tree-decomposition as one checks easily: if $u \in f^{-1}(w)$, then $u \leq_T w$ and all vertices on the path from w to u in T are in $f^{-1}(w)$, and so $f^{-1}(w)$ is connected.

As an example, a normal spanning tree T of the cycle C_n is obtained by removing an edge and taking one of its ends as root (thus, T is a directed path). The corresponding normal tree-decomposition (T, f) has width 2. As another example, a normal tree for the complete bipartite graph $K_{n,m}$, with $n \leq m$, is the directed path with edges $(i, i+1)$ (it first enumerates the n nodes and then the m nodes of the bipartition). The corresponding normal tree-decomposition has width n. As a final example, a normal tree for the grid $G_{n \times m}$, with $n \leq m$, is the directed path with root $(1,1)$ and edges $((i,j),(i+1,j))$ and $((n,j),(1,j+1))$, and the normal tree-decomposition has width n.

We will show in Corollary 2.73(3) that every nonempty graph has an optimal normal tree-decomposition. □

Definition 2.57 (Path-decompositions and path-width)   A *path-decomposition* (resp. a *rich path-decomposition*) of an s-graph is a tree-decomposition (resp. a rich tree-decomposition) (T, f) such that T is a directed path (and hence its root is one end of this path). The *path-width* of an s-graph G is the smallest width of its path-decompositions. It is denoted by $pwd(G)$. We denote by $PWD(\leq k)$ the set of graphs of path-width at most k, and by $PWD(\leq k, C)$ the set of s-graphs of path-width at most k and of type included in a set of source names C.

A path-decomposition of an s-graph G can be specified as a sequence $D \in Seq(\mathcal{P}(V_G))$: it corresponds to (T, f), where $T = \langle [|D|], \{(i, i+1) \mid i \in [|D|-1]\}\rangle$ and $f(i) = D[i]$.

Since every path-decomposition is a tree-decomposition, we have

$$twd(G) \leq pwd(G) \leq |V_G| - 1.$$

The tree-width of an s-graph may be smaller than its path-width. In particular, trees have unbounded path-width, but tree-width at most 1. The graphs K_n, $K_{n,m}$, C_n, $G_{n \times m}$ considered in Examples 2.56(3) and 2.56(6) have equal path-width and tree-width. For every graph G we have, by a result of [BodGHK]:

$$pwd(G) \leq (twd(G) + 1) \cdot \log(|V_G|).$$

We now detail the case of the *complete binary tree* T_n of height n: its set of nodes is the set of words over $\{0, 1\}$ of length at most n and a node u is an ancestor of w if and only if u is a prefix of w. Hence ε is the root. For $n \geq 1$, we have $twd(T_n) = 1$ and $pwd(T_n) = \lceil n/2 \rceil$. A path-decomposition of T_2 of width 1 is

$$D_1 := (\{0, 00\}, \{0, 01\}, \{0, \varepsilon\}, \{\varepsilon, 1\}, \{1, 10\}, \{1, 11\}).$$

By induction on n, we define

$$D_{n+1} := (00D_n + 0) \cdot (01D_n + 0) \cdot (\{0, \varepsilon\}, \{\varepsilon, 1\}) \cdot (10D_n + 1) \cdot (11D_n + 1),$$

where we use the following notation: if D is a path-decomposition of a graph G with vertices in $\{0, 1\}^*$, then $D + u$ is the path-decomposition (of a graph with at most one more vertex than G) obtained by adding u to each box, and uD the path-decomposition of a graph isomorphic to G obtained by replacing each word w by uw. The reader will verify easily that D_n is, for each n, a path-decomposition of T_{2n} of width n. The path-width of T_{2n+1} is larger than n: this follows from the characterization given in [TakUK] of the acyclic excluded minors for graphs of path-width at most n.

2.4.2 Some properties of tree-decompositions

We review a few results that will be useful in the sequel. Proofs and references can be found in [*Bod98]. For two disjoint sets A and B, we denote by $A \otimes B$ the complete bipartite concrete graph with $V_{A \otimes B} = A \cup B$ and edges between each vertex of A and each vertex of B.

Proposition 2.58 Let (T, f) be a tree-decomposition of a concrete s-graph G and $A, B \subseteq V_G$. If A is a clique in the graph $core(G)$, then A is included in some box of (T, f). If $A \cap B = \emptyset$ and $A \otimes B$ is a subgraph of $core(G)$, then at least one of A and B is included in some box. $\qquad \square$

It follows that $twd(K_n) \geq n - 1$ and $twd(K_{n,m}) \geq n - 1$ if $n \leq m$. We know from Examples 2.56(3) and 2.56(6) that $twd(K_n) \leq n - 1$ and $twd(K_{n,m}) \leq n$. Thus, $twd(K_n) = n - 1$. Since K_{n+1} is a minor of $K_{n,n}$, we have (by Corollary 2.60(1) below) $twd(K_{n+1}) \leq twd(K_{n,m})$ and thus $twd(K_{n,m}) = n$ (if $n \leq m$). (To prove that $K_{n+1} \trianglelefteq K_{n,n}$ we consider the graph isomorphic to $K_{n,n}$ with vertices $u_1, \ldots, u_n, v_1, \ldots, v_n$ and edges $e_{i,j} : u_i - v_j$ for all $i, j \in [n]$; we contract the edges $e_{i,i}$ for $i = 2, \ldots, n$, which yields a graph H with vertices $u_1,\ldots,u_n,v_1$; it is clear that K_{n+1} is isomorphic to $core(H)$. Since $core(H) \subseteq H$, we obtain from Proposition 2.17 that $K_{n+1} \trianglelefteq K_{n,n}$.)

Similarly, $K_3 \trianglelefteq C_n$ for $n \geq 3$ and so $twd(C_n) \geq 2$ (and we know from Examples 2.56(3) and 2.56(6) that $twd(C_n) \leq 2$).

Proposition 2.59 If a graph H is obtained from a graph G by deleting or contracting an edge, or by deleting a vertex and the edges incident with this vertex, then $twd(G) - 1 \leq twd(H) \leq twd(G)$. $\qquad\square$

The proof is easy. We obtain thus from Proposition 2.17(2) the first assertion of the following corollary.

Corollary 2.60
(1) If H is a minor of a graph G, then $twd(H) \leq twd(G)$.
(2) For every k there is a finite set Ω_k of simple, loop-free, undirected graphs such that a graph G has tree-width at most k if and only if $core(G)$ has no minor isomorphic to a graph in Ω_k. $\qquad\square$

The second assertion follows from the first assertion, the fact that $twd(G) = twd(core(G))$ and a particular case of the Graph Minor Theorem recalled in Section 2.2 (after Proposition 2.21) that is proved in [RobSey90]. The sets Ω_k are known for k at most 3: $\Omega_1 = \{K_3\}$, $\Omega_2 = \{K_4\}$ and Ω_3 consists of four graphs (see [ArnPC]). Each set Ω_k contains K_{k+2}.

A result in [Lag] shows that graphs in Ω_k have size at most $\exp(\exp(O(k^5)))$, where for every integer n, $\exp(n) = 2^n$. It follows that they can be, in principle, determined. However, this upper-bound is too large to be usable in any implementable algorithm.

Proposition 2.59 and Corollary 2.60 hold for path-width instead of tree-width, with a finite set Θ_k instead of Ω_k for the graphs of path-width at most k. The trees in Θ_k have been determined in [TakUK]: there are more than $(k!)^2$ trees in Θ_k, all with $(5/2) \cdot (3^k - 1)$ vertices.

Proposition 2.61 For each k, the set Ω_k contains at least one planar graph. Conversely, for every planar graph P, the set of graphs not containing P as a minor has bounded tree-width.

Proof: Since $twd(G_{(k+1)\times(k+1)}) = k + 1$ (Example 2.56(3)), the undirected grid $core(G_{(k+1)\times(k+1)})$ has a minor P in Ω_k by Corollary 2.60(2). This graph P is planar since every minor of a planar graph is planar. The second assertion is due to Robertson and Seymour and is proved in Theorem 12.4.3 of [*Die]. The proof yields an upper-bound $\exp(O(n^5))$, where n is the number of vertices of P. $\qquad\blacksquare$

The set Ω_k characterizes the graphs in $TWD(\leq k)$. It can also be used to characterize the s-graphs in $TWD(\leq k, C)$, where C is a set of at most $k + 1$ source names. Let C be such a set and let K_C be the simple loop-free undirected s-graph of type C such that K_C° is the clique $K_{|C|}$. It follows from Proposition 2.58 and Remark 2.55(b) that an s-graph

G of type included in C has tree-width at most k if and only if the (undirected) graph $fg_C(und(G) \parallel K_C)$ has tree-width at most k (where $und(G) := \langle und(G^\circ), slab_G \rangle$), hence, if and only if the graph $core(fg_C(und(G) \parallel K_C))$ has no minor in Ω_k.

We conclude this review with easier results. Every graph can be decomposed into its 2-*connected components* (its maximal induced subgraphs without separating vertices), and these components form a tree. Furthermore, every 2-connected graph can also be decomposed into a tree of so-called 3-*blocks*. A 3-block is either a cycle, or a *bond* (a set of pairwise parallel edges) or a 3-connected graph (a 2-connected graph without separating set of two vertices). The 3-blocks of a graph, i.e., the 3-blocks of its 2-connected components, are minors of it.

Proposition 2.62 The tree-width of a nonempty graph G is the maximal tree-width of its connected components. It is also the maximal tree-width of its 2-connected components and the maximal tree-width of its 3-blocks. $\qquad\square$

The last assertion is proved in [Cou99]. The two others follow from straightforward constructions.

2.4.3 Transformations of tree- and path-decompositions

We consider transformations of tree- and path-decompositions of a given s-graph intended to reduce some parameters like the number of boxes without, hopefully, increasing the width, or by increasing it in a controlled way. By iterating certain of these elementary transformations, one can put tree-decompositions in certain normal forms.

Definition 2.63 (Edge contractions in a tree-decomposition) Let (T, f) be a (rich) tree-decomposition of an s-graph G, and let $e : u \to v$ be an edge of T. We let T/e be the result of the contraction of e. Its set of nodes is defined as $N_{T/e} = N_T - \{v\}$: from the two vertices u and v that are fused, we keep u as result of the fusion, and the edges from v to w of T are made into edges from u to w in T/e. We let $(T/e, f_e)$ be the pair such that:

$$
f_e(w) := \begin{cases} f(w) & \text{if } w \in N_T - \{u, v\} = N_{T/e} - \{u\}, \\ f(u) \cup f(v) & \text{if } w = u. \end{cases}
$$

With these notations and hypotheses:

Lemma 2.64 The pair $(T/e, f_e)$ is a (rich) tree-decomposition of G if (T, f) is a (rich) tree-decomposition of G. We have

$$
wd(T/e, f_e) = \max\{wd(T, f), |(f(u) \cup f(v)) \cap V_G| - 1\}
$$

where u and v are the end vertices of e. The same result holds for (rich) path-decompositions. $\qquad\square$

Remark 2.65 If E is a set of edges of a rooted tree T, then $(((T/e_1)/e_2)\cdots)/e_n$ is the same for every enumeration $e_1,\ldots,e_n$ of E. This is so because rooted trees are directed so that each node has indegree at most 1 and when we contract an edge, we take its tail as the vertex resulting from this contraction. We can thus define T/E, the result of the simultaneous contraction of all edges of E. The nodes in T that are in T/E are the roots of the rooted forest $T[E]$. Similarly, if (T, f) is a tree-decomposition of G, then $(T/E, f_E)$ is also one, where, for a root r of $T[E]$, $f_E(r)$ is the union of all $f(u)$ such that u is a descendant of r in $T[E]$. □

We consider some conditions that can be imposed on tree-decompositions and on path-decompositions without increasing the associated widths. In all cases, we give an algorithm that transforms an arbitrary decomposition into one of no larger width that satisfies the considered conditions.

Definition 2.66 (Downwards increasing decompositions) A tree-decomposition (T, f) of an s-graph G is *downwards increasing* if, for every edge $u \to v$ of T, we have $f(v) - f(u) \neq \emptyset$. This means that $f(v)$ is not a subset of $f(u)$. In this case $wd(T, f) \leq k$ implies $|f(u) \cap f(v)| \leq k$ for every edge $u \to v$. Assuming that the box $f(root_T)$ is nonempty, we also have $|N_T| \leq |V_G|$.

A tree-decomposition (T, f) is *1-downwards increasing* if $|f(v) - f(u)| = 1$ for every u, v as above. In this case, $|V_G| = |N_T| + |f(root_T)| - 1$.

Every normal tree-decomposition (Example 2.56(6)) is 1-downwards increasing. The tree-decomposition of Figure 2.6 is downwards increasing but not 1-downwards increasing.

Proposition 2.67

(1) Let (T, f) be a tree- or a path-decomposition of an s-graph G. By applying to T a series of edge contractions, one can transform it into a tree- or a path-decomposition of G of the same width, such that every two boxes are incomparable for inclusion, and that is thus downwards increasing.

(2) Every graph has an optimal tree-decomposition (path-decomposition) that is 1-downwards increasing.

Proof: (1) Let (T, f) be such that $f(v) \subseteq f(u)$ or $f(u) \subseteq f(v)$ for some edge $e : u \to v$. By contracting this edge, we get by Lemma 2.64 a tree- or a path-decomposition $(T/e, f_e)$ of G. Since $f(u) \cup f(v)$ is equal to $f(u)$ or to $f(v)$ we have $wd(T/e, f_e) = wd(T, f)$. By repeating this elementary step at most $|edg_T|$ times, we obtain a decomposition of the desired form.

(2) By (1), every graph has an optimal tree-decomposition (path-decomposition) that is downwards increasing. This decomposition can be made 1-downwards increasing by insertion of new nodes in the tree. We omit the easy formal details. ∎

Definition 2.68 (Binary tree-decompositions) Let (T, f) be a (rich) tree-decomposition. It is *binary* if every node of T has outdegree at most 2. For $u \in N_T$, we let $d(u) := \max\{0, \text{outdegree}(u) - 2\}$ and $d(T) := \sum_{u \in N_T} d(u)$. Hence (T, f) is binary if and only if $d(T) = 0$.

Proposition 2.69 Every (rich) tree-decomposition (T, f) of an s-graph G can be transformed into a binary (rich) tree-decomposition (T', f') of G of the same width, such that $ht(T') \leq (\max_{u \in N_T} \{\lceil \log(d(u) + 2) \rceil\}) \cdot ht(T)$. The reverse transformation of (T', f') into (T, f) can be done by edge contraction.

Proof: Let (T, f) and G be as in the statement. It obviously suffices to consider the nonrich case. If $d(T) = 0$ there is nothing to do. If $d(T) \neq 0$, we take any node u with outgoing edges $u \to v_1, \ldots, u \to v_p, p \geq 3$, we introduce a new node u' and we replace the edges $u \to v_{p-1}$ and $u \to v_p$ by the edges $u \to u'$, $u' \to v_{p-1}$ and $u' \to v_p$. We obtain a tree T_1 with $d(T_1) = d(T) - 1$. We turn it into a tree-decomposition (T_1, f_1) by defining $f_1(u') := f(u)$ and $f_1(w) := f(w)$ for $w \in N_T$. It is clear that (T_1, f_1) is a tree-decomposition of G that gives back (T, f) by contracting the new edge $u \to u'$.[32] Since $f_1(u') = f(u)$ we have $wd(T, f) = wd(T_1, f_1)$.

By repeating this elementary transformation $d(T)$ times, we obtain from (T, f) a binary tree-decomposition (T', f') of the s-graph G of the same width that gives back (T, f) by $d(T)$ edge contractions. The height of T' is at most $(\max_{u \in N_T} \{d(u) + 1\}) \cdot ht(T)$, which is not the desired bound.

We now modify the construction so as to obtain a binary tree-decomposition (T', f') of height $ht(T') \leq (\max_{u \in N_T} \{\lceil \log(d(u) + 2) \rceil\}) \cdot ht(T)$. In the case where $p \geq 4$, we add two new nodes u' and u'' and we replace the edges $u \to v_1, \ldots, u \to v_p$ by $u \to u'$, $u \to u''$, $u' \to v_1, \ldots, u' \to v_q$, $u'' \to v_{q+1}, \ldots, u'' \to v_p$ where $q = \lfloor p/2 \rfloor$. We obtain a tree T'' and we define f'' by $f''(u') := f(u)$, $f''(u'') := f(u)$, and $f''(w) := f(w)$ for $w \in N_T$. We obtain a tree-decomposition (T'', f'') of the s-graph G such that $wd(T'', f'') = wd(T, f)$, $d(T'') < d(T) - 1$ and that gives back (T, f) by contracting two edges. By repeating this step at most $d(T)$ times, we obtain a tree-decomposition having the desired properties. ∎

Remark 2.70 The transformations of Propositions 2.67 and 2.69 go in opposite directions. This suggests the following question:

Does every graph have an optimal tree-decomposition that is binary *and* downwards increasing?

To show that the answer is no, we give a counter-example. We let G be the simple undirected graph such that $V_G = \{1, 2, 3, 4, 1', 2', 3', 4'\}$, its edges are between i and j for $1 \leq i < j \leq 4$ and between i and j' for $1 \leq i, j \leq 4$, $i \neq j$. This graph is shown in Figure 2.8.

---

[32] In this contraction the "new node" u' disappears, by Definition 2.63.

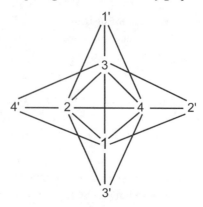

Figure 2.8 Graph G of Remark 2.70.

The graph G is the union of five cliques with sets of vertices $f(0) = \{1,2,3,4\}$, $f(1) = \{1',2,3,4\}, f(2) = \{1,2',3,4\}, f(3) = \{1,2,3',4\}$ and $f(4) = \{1,2,3,4'\}$. It has a tree-decomposition (T,f), where T consists of the edges $0 \to 1$, $0 \to 2$, $0 \to 3$, $0 \to 4$. Let (U,g) be a downwards increasing optimal tree-decomposition. Its width cannot be less than 3 (because by Proposition 2.58 each set $f(i)$ is included in a box $g(u)$ for some u in N_U). Hence, it is 3. Assuming that U is binary, we will get a contradiction.

Let $u \in N_U$ be a node such that $g(u) = f(0)$. Consider a son v of u. It must contain one of $1',2',3',4'$. Assume it contains $1'$. Necessarily $g(v) = f(1)$ and v is a leaf. Now u has at most two sons, v and v'. Without loss of generality we assume that $g(v) = f(1)$ and $g(v') = f(2)$. Hence (whether u has 2, 1, or 0 sons) there exist two nodes w and w' such that $g(w) = f(3)$ and $g(w') = f(4)$, and none of them is below u. The father y of u belongs to the (undirected) paths in U between u and w and between u and w'. (We may have $y = w$ or $y = w'$.) It follows that $g(y)$ includes $\{1,2,4\}$ (because y is on the path between u and w) and $g(y)$ includes $\{1,2,3\}$ (because y is on the path between u and w'). Hence $g(y)$ includes $\{1,2,3,4\} = g(u)$, which contradicts the fact that (U,g) is downwards increasing. Thus, the graph G has no optimal, binary and downwards increasing tree-decomposition.

2.4.4 Tree-decompositions and chordal graphs

We consider the graphs that have tree-decompositions, all boxes of which induce cliques. Efficient algorithms for chordal graphs, based on their tree-decompositions, are presented in [Gav] and [*Gol].

Definition 2.71 (Chordal graphs) Let $\mathcal{C}$ be the class of nonempty, connected, simple and loop-free undirected graphs. Let $k \in \mathcal{N}_+$. A graph G in $\mathcal{C}$ is *k-chordal* if it has a tree-decomposition, each box of which is a nonempty clique with at most k vertices.

(The graph of Figure 2.8 is 4-chordal.) A *k-perfect elimination order* for a graph G in C is an enumeration $\pi = (v_1, \ldots, v_n)$ of V_G, such that, for each $i = 2, \ldots, n$, the set $N_i := \{v_j \mid j < i$ and $v_j -_G v_i\}$ is a nonempty clique with at most $k - 1$ vertices.[33] A *k-perfect spanning tree* is a normal (rooted) spanning tree T of G (cf. Definition 2.13) such that for every $u \in V_G$, the set $M_u := \{w \in V_G \mid u -_G w$ and $u \leq_T w\}$ is a clique in G with at most $k - 1$ vertices. A graph is *chordal* if it is k-chordal for some k, and similarly *perfect* for a spanning tree or an elimination order means k-perfect for some k. The *clique number* of a graph G is $\omega(G)$, the maximal number of vertices of a clique in G.

A *simplicial orientation* of a graph $G \in C$ is an acyclic orientation H of G such that for every vertex u, the set $Adj_H^-(u)$ of vertices w such that $w \to_H u$ is a clique in H (hence, a clique in G). Its indegree is the maximal indegree of its vertices.

Proposition 2.72 For every graph G belonging to C the following properties are equivalent:

(1) G is k-chordal;
(2) G has a k-perfect elimination order;
(3) G has a simplicial orientation of indegree at most $k - 1$;
(4) G has a k-perfect spanning tree;
(5) G has no induced cycle with at least four vertices and its clique number is at most k.

Proof: (1) $\Longrightarrow$ (2). The proof is by induction on $n = |V_G|$. The case $n = 1$ requires no proof. Otherwise, let (T, f) be a tree-decomposition of G, each box of which is a nonempty clique with at most k vertices. We can assume that it is downwards increasing because the transformation of Proposition 2.67(1) preserves the property that each box is a nonempty clique with at most k vertices. If $|N_T| = 1$, then any linear order on $V_G = f(root_T)$ is a k-perfect elimination order. Otherwise, let u be any leaf of T. The box $f(u)$ contains a vertex v that belongs to no other box, hence the set $f(u) - \{v\}$ of its neighbors is a nonempty clique (because G is connected) with at most $k - 1$ vertices. The graph $G - v$ (i.e., G minus v and the incident edges) is also k-chordal (it is still connected) and, by the induction hypothesis, it has a k-perfect elimination order $(v_1, \ldots, v_{n-1})$. Then $(v_1, \ldots, v_{n-1}, v)$ is a k-perfect elimination order for G, which completes the proof.

(2) $\Longrightarrow$ (3). Let $G \in C$ have a k-perfect elimination order $(v_1, \ldots, v_n)$. We define an acyclic orientation H of G by letting $v_j \to_H v_i$ if and only if $j < i$ and $v_j -_G v_i$. It is simplicial and of indegree at most $k - 1$.

(3) $\Longrightarrow$ (4). Let $G \in C$ have a simplicial orientation H of indegree at most $k - 1$. Since H is acyclic (and finite), it has a vertex u_1 of indegree 0. Every vertex u of H is reachable from u_1 by a directed path, as one proves easily by induction on the length

---

[33] The elimination order is usually taken to be $(v_n, \ldots, v_1)$, but it is more convenient to reverse it while keeping the well-known terminology.

of an undirected path from u_1 to u (if u_{i-1}, u_i, u_{i+1} are three consecutive vertices on that path with $u_{i-1} \to_H u_i$ and $u_i \leftarrow_H u_{i+1}$, i.e., $u_{i+1} \to_H u_i$, then $u_{i-1} -_G u_{i+1}$ because $Adj_H^-(u_i)$ is a clique).

We let T be the transitive reduction of H, i.e., the graph T such that $V_T := V_H = V_G$ and E_T is the set of edges $w \to_H u$ such that there is no directed path in H of length at least 2 from w to u. The vertex u_1 has also indegree 0 in T. Every other vertex u has at least one incoming edge because it is reachable from u_1. If it had indegree at least 2 in T, we would have $w \to_T u$ and $w' \to_T u$ with $w \neq w'$, hence $w \to_H w'$ (or vice versa) because $Adj_H^-(u)$ is a clique, and then $w \to u$ could not be an edge of T. It follows that T is a rooted spanning tree of H and of G. If $w \to_H u$, then there is a directed path in T from w to u, i.e., $u \leq_T w$. Hence T is a normal spanning tree. It is k-perfect since $M_u = Adj_H^-(u)$ and $Adj_H^-(u)$ is a clique with at most $k - 1$ vertices.

(4) $\Longrightarrow$ (1). Let T be a k-perfect spanning tree of G and let (T, f) be the associated normal tree-decomposition (as defined in Example 2.56(6)). We have actually $f(u) = \{u\} \cup M_u$: from the definition of f, we have $\{u\} \cup M_u \subseteq f(u)$; for proving the opposite inclusion, we note that, since T is perfect, if $v <_T w$ and w is adjacent to v, then w is adjacent to every vertex u such that $v <_T u <_T w$ (if u is the father of v, then both w and u are in M_v). Since T is k-perfect, each set $f(u)$ is a nonempty clique with at most k vertices, hence G is k-chordal.

(2) $\Longrightarrow$ (5). Let G in $\mathcal{C}$ have a perfect elimination order $\pi = (v_1, \ldots, v_n)$. Assume that C is an induced cycle in G of length at least 4 with last element v_i. There are v_j and $v_{j'}$ in C that are adjacent to v_i, that are not adjacent and are such that $j, j' < i$. This contradicts the hypothesis that π is a perfect elimination order. Since G also satisfies (1), its clique number is at most k (by Proposition 2.58).

(5) $\Longrightarrow$ (1). This is proved as Proposition 5.5.1 in [*Die]. $\blacksquare$

Other characterizations and properties of chordal graphs can be found in the books [*BranLS], [*Die] and [*Gol].

Corollary 2.73

(1) Every chordal graph G is $\omega(G)$-chordal, and its tree-width is $\omega(G) - 1$.
(2) The tree-width of a graph G with at least one edge that is not a loop, is the minimal tree-width of a chordal graph H such that $V_G = V_H$ and $core(G) \subseteq H$.
(3) Every nonempty graph has an optimal normal tree-decomposition.[34]

Proof: (1) That $\omega(G) - 1 \leq twd(G)$ follows from Proposition 2.58. If G is chordal, it has a tree-decomposition (T, f), each box of which is a clique, hence G is $\omega(G)$-chordal and $twd(G) \leq \omega(G) - 1$.

(2) Let m be the minimal tree-width of a chordal graph H containing $core(G)$ as subgraph. Then $twd(G) = twd(core(G)) \leq m$ by Corollary 2.60(1).

---

[34] Recall the definition of normal tree-decomposition from Example 2.56(6).

Let conversely (T, f) be an optimal tree-decomposition of G. Since G has at least one edge that is not a loop, the width of (T, f) is at least 1. By Proposition 2.67(1) we may assume that (T, f) has no empty boxes. Moreover, we may assume that (T, f) has the special property that adjacent boxes are not disjoint. In fact, for every edge $e : u \to v$ of T with $f(u) \cap f(v) = \emptyset$, we introduce a new node w_e and we replace the edge e by the edges $u \to w_e$ and $w_e \to v$. We extend f by defining $f(w_e) = \{x, y\}$ for some $x \in f(u)$ and $y \in f(v)$. It is clear that this leads to a tree-decomposition of G of the same width.

Now let H be the graph such that $V_H = V_G$ and any two vertices in a same box of (T, f) are adjacent. Since (T, f) has the above special property, we obtain that H is connected, hence $H \in C$. Clearly $core(G) \subseteq H$, (T, f) is a tree-decomposition of H of width $twd(G)$, and H is chordal. This completes the proof.

(3) Let G be a nonempty graph. The result is obvious if all edges of G are loops. So, let G have at least one edge that is not a loop, and let $k := twd(G)$. By (2), there is a chordal graph H such that $V_G = V_H$, $core(G) \subseteq H$, and $twd(H) = k$. Then H is $(k+1)$-chordal by (1). Hence, by the proof of Proposition 2.72 (in particular (4) $\Longrightarrow$ (1)), H has a normal tree-decomposition (T, f) of width k. Since $core(G) \subseteq H$, the tree T is normal for G and if (T, f') is the corresponding normal tree-decomposition of G, then $f'(u) \subseteq f(u)$ for every u in V_G. Hence (T, f') has width k. $\blacksquare$

Normal tree-decompositions will be used in Example 5.2(4) for defining logical representations of tree-decompositions.

Corollary 2.74

(1) Every simple, loop-free, undirected graph G of tree-width at most k has at most $k \cdot |V_G| - k(k+1)/2$ edges if $|V_G| \geq k+1$.

(2) For all $k, n > 0$, there are less than $2^{kn \cdot (\log(n)+1)}$ simple, loop-free, undirected concrete graphs of tree-width at most k with vertex set $[n]$.

Proof: (1) Let G be as in the statement with at least one edge (the result is obvious for graphs without edges). Using Corollary 2.73(2), let H be a chordal graph of tree-width at most k such that $G \subseteq H$ and $V_H = V_G$. By Corollary 2.73(1), H is $(k+1)$-chordal. Thus, by (1) $\Longrightarrow$ (2) of Proposition 2.72, the graph H has a $(k+1)$-perfect elimination order $\pi = (v_1, \ldots, v_n)$. We let $G_i = G[\{v_1, \ldots, v_i\}]$. The property holds for G_{k+1}, and for each i we have $|E_{G_i}| \leq |E_{G_{i-1}}| + k$ (because $|N_i| \leq k$), which gives the result for each graph G_i, by induction on i, hence for G.

(2) We let G, H, π be as in (1) with $V_G = [n]$. It follows from the existence of π that H is the union of at most k edge-disjoint forests,[35] hence is a subgraph of the union of k trees with set of nodes V_G. The same holds for G, and the same holds, obviously, for graphs without edges.

---

[35] The *arboricity* of a loop-free graph is the least number of edge-disjoint forests of which it is the union.

There are n^{n-2} trees[36] with set of nodes $[n]$, at most $n^{k(n-2)}$ unions of k such trees, and at most $2^{k(n-1)}$ ways to delete edges from any set of at most $k(n-1)$ edges, so that there are at most $2^{k(n-1)} \cdot 2^{k(n-2) \cdot \log(n)} < 2^{kn \cdot (\log(n)+1)}$ graphs that are the union of k forests with vertex set $[n]$. This gives the claimed result.   ∎

Since there are n^{n-2} trees with set of nodes $[n]$, the upper-bound $2^{kn \cdot (\log(n)+1)}$ cannot be lowered to $2^{o(n \cdot \log(n))}$.

2.4.5 A syntax for tree-decompositions

We will establish correspondences between terms in $T(F^{\mathrm{HR}})$ and tree-decompositions of the s-graphs they define. These correspondences will yield the equality $twd(G) = wd(G) - 1$, where $wd(G)$ is the *width* of an s-graph G defined as the minimal cardinality of a set C such that G is the value of a term in $T(F_C^{\mathrm{HR}})$.

We first give an idea of the constructions. The expression of an s-graph G as $val(t)$ for $t \in T(F_C^{\mathrm{HR}})$ yields a tree-decomposition of G based on the syntactic tree T of t. The size of a box of this decomposition is at most $|C|$, which gives $twd(G) \leq wd(G) - 1$. For the other direction, we will transform a rich tree-decomposition (T, f) of an s-graph G into a term that evaluates to it and is written with at most $wd(T, f) + 1$ source names. Although neither of these two transformations is the inverse of the other, they are closely related. The specific features of a tree-decomposition, such as being binary or unary (i.e., being a path-decomposition), or its size and height, are reflected in the corresponding terms.

Definition 2.75 (The rich tree-decomposition associated with a term over F^{HR})
Let $t \in T(F^{\mathrm{HR}})$. We recall from Definition 2.35 that $\mu(t)$ is the set of source names in $\mathcal{A}$ that are used in the operations of t, and that $wd(t)$, the width of t, equals $|\mu(t)|$. We denote by $X_{\max}$ the canonical cross-section of $\approx_t$ (cf. Section 2.3.2) that consists of the pairs (u, a) such that u is maximal for $\leq_t$. The concrete s-graph $cval(t) = (Exp(t)/\approx_t)_{X_{\max}}$ with vertex set $X_{\max}$ is isomorphic to $val(t)$. We take the syntactic tree T of t (without its labels) as the rooted tree of the rich tree-decomposition (T, f) of $cval(t)$ to be constructed. We define f as follows.

Case 1: u is an occurrence of a constant symbol that describes an edge of the s-graph $cval(t)$. Then u is this edge in $Exp(t)$ as well as in $cval(t)$. We let

$$f(u) := \{u\} \cup \{w \in X_{\max} \mid w \approx_t (u, a) \text{ for some } a \text{ in } \tau(u)\}.$$

Case 2: u is not an occurrence of such a symbol. Then we let $f(u)$ be the set $\{w \in X_{\max} \mid w \approx_t (u, a) \text{ for some } a \text{ in } \tau(u)\}$.

In both cases, there are at most $wd(t)$ such pairs (u, a), because $\tau(u) \subseteq \mu(t)$; hence $|f(u) \cap X_{\max}| \leq wd(t)$.

---

[36] A result by Cayley: see Section II.5.1 of [*FlaSed].

For example, the decomposition associated with the term

$$t = \overrightarrow{\mathbf{cd}} \mathbin{/\mkern-5mu/} ren_{a \leftrightarrow c}(\overrightarrow{\mathbf{ab}} \mathbin{/\mkern-5mu/} fg_b(\overrightarrow{\mathbf{ab}} \mathbin{/\mkern-5mu/} \overrightarrow{\mathbf{bc}}))$$

is shown in Figure 2.9. In each box $f(u)$ the elements of $\tau(u)$ are indicated, i.e., the names of the sources of $val(t/u)$. The elements of X_{max} are circled.

With this notation we have:

Proposition 2.76

(1) For each finite set C, if $t \in T(F_C^{HR})$, then the pair (T, f) defined in Definition 2.75 is a rich tree-decomposition of the s-graph $cval(t)$ of width at most $|C| - 1$. It can be constructed from t in time $O(|t|)$.

(2) We have $twd(G) \leq wd(G) - 1$ for every (labeled) s-graph G.

Proof: (1) Let $t \in T(F_C^{HR})$ and (T, f) be as in Definition 2.75. All conditions of the definition of a tree-decomposition are clearly valid, except possibly the connectivity condition (Condition (3) of Definition 2.53). For each $w \in X_{max}$, the set $f^{-1}(w)$ defined as $\{u \in N_T \mid w \in f(u)\}$, is the set of nodes that are the first components of the pairs (u, a) that are $\approx_t$-equivalent to w. But the equivalence relation $\approx_t$ is generated by pairs $((u, a), (v, b))$ such that u and v are adjacent nodes of T. Hence the nodes in $f^{-1}(w)$ induce a connected subgraph of T. This proves the connectivity condition. It is clear that $|f(u) \cap V_{cval(t)}| \leq wd(t) \leq |C|$ for every $u \in N_T$, which gives $wd(T, f) \leq |C| - 1$. It is also clear, from the proof of Proposition 2.45, that (T, f) can be constructed from t in time $O(|C| \cdot |t|)$: $w \in f(u)$ if and only if $w = max(u, a)$ for some a in $\tau(u)$.

(2) is an immediate consequence of (1). ∎

Remark 2.77 (1) If (u, a) is a vertex of $cval(t)$, then it belongs to a unique maximal box of the tree-decomposition (see Remark 2.55(a)). This box is $f(u)$ because X_{max}, the vertex set of $cval(t)$, is the canonical cross-section and thus consists of the maximal pairs (u, a). See Figure 2.9 where the circled vertices are those of X_{max}.

(2) The rich tree-decompositions (T, f) produced by this construction have the particular property that only the boxes at the leaves of T contain edges, and they never contain more than one edge. Hence, not all rich tree-decompositions are associated with terms by the above construction.

Furthermore, these tree-decompositions are binary but not downwards increasing. The number of boxes is clearly not minimal; it can be reduced by edge contractions (cf. Section 2.4.3). In particular, if u is an occurrence of a renaming operation ren_h, and its son is v, then $f(u) = f(v)$. One can thus contract the edge $u \to v$ of T and one gets a smaller rich tree-decomposition of the same s-graph, having the same width.

(3) If t is a term that defines a (K, Λ)-labeled graph, then we perform the same construction by neglecting the labels belonging to $K \cup \Lambda$ that are attached to some constant symbols. □

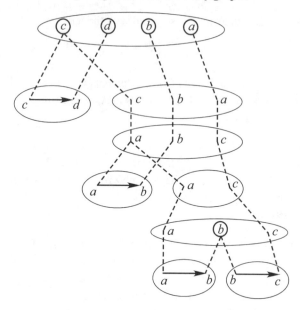

Figure 2.9 The rich tree-decomposition for the term t of Definition 2.75.

A converse construction

We need some notation and a lemma. Let (T, f) be a rich tree-decomposition of an s-graph G. For every node u of the rooted tree T, we recall that T/u denotes the subtree of T issued from u, with $N_{T/u} = \{w \in N_T \mid w \leq_T u\}$. We denote by $G(u)$ the s-graph defined as follows. The graph $G(u)^\circ$ is the subgraph of G° such that $V_{G(u)^\circ} \cup E_{G(u)^\circ} = \bigcup \{f(w) \mid w \in N_{T/u}\}$. The sources of $G(u)$ are those of G that belong to $V_{G(u)^\circ}$, i.e., $slab_{G(u)} = slab_G \restriction V_{G(u)^\circ}$. These sources belong to $f(u)$ by the connectivity condition. Clearly, $G(root_T) = G$.

Lemma 2.78 Let (T, f) be a tree-decomposition of width at most $k - 1$ of an s-graph G and let $C \subseteq \mathcal{A}$ be a set of cardinality k such that $\tau(G) \subseteq C$. There exists a coloring $\gamma : V_G \to C$ that extends $slab_G$ and is injective on each box of the decomposition. Such a coloring can be determined in linear time in the size of T, i.e., in time $O(|N_T|)$.

Proof: Let G, C, T, f be as in the statement. Obviously, there is an injective mapping $\delta_0 : f(root_T) \to C$ that extends $slab_G$. We will prove that for every $u \in N_T$ the following holds:

Every injective mapping $\delta : f(u) \to C$ can be extended into a mapping $\gamma : V_{G(u)} \to C$ that is injective on $f(w)$ for each w in $N_{T/u}$.

The proof is by bottom-up induction on u. If u is a leaf of T there is nothing to prove. Otherwise, let $u_1, \ldots, u_p$ be the sons of u. For each of them one can find

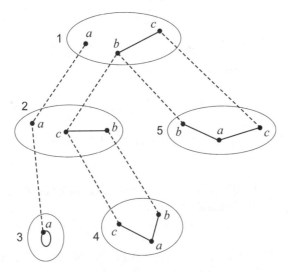

Figure 2.10 Illustration of Lemma 2.78.

an injective mapping $\delta_i : f(u_i) \rightarrow C$ that coincides with δ on $f(u_i) \cap f(u)$. By the induction hypothesis, it can be extended into γ_i defined on $V_{G(u_i)}$. Then, the common extension γ of these mappings γ_i and of the mapping δ is the desired coloring. This extension exists because if $x \in N_{T/u_i} \cap N_{T/u_j}$, $i \neq j$, then $x \in f(u_i) \cap f(u) \cap f(u_j)$ by the connectivity condition, and so $\gamma_i(x) = \gamma_j(x) = \delta(x)$.

For example, Figure 2.10 shows a rich tree-decomposition of width 2 that is colored with colors a, b, c. Boxes are numbered from 1 to 5, and 1 is the root.

It is routine work to construct a linear algorithm computing γ. ∎

Remark 2.79 Since the ends of any edge are in the same box, the mapping γ constructed by the proof of Lemma 2.78 is a *proper* (vertex) k-coloring, i.e., a coloring γ of V_G using k colors such that $\gamma(x) \neq \gamma(y)$ if $x \neq y$ and $x - y$ is an edge. However, this coloring is not optimal. The cycle C_6 is properly 2-colorable, however it has tree-width 2 and the above construction yields a coloring using three colors.

Definition 2.80 (The term representing a rich tree-decomposition) Let (T, f) be a rich tree-decomposition of width at most $k - 1$ of a concrete s-graph G. We let $C \subseteq \mathcal{A}$ be a set of cardinality k that contains $\tau(G)$. Let $\gamma : V_G \rightarrow C$ be a vertex coloring of G that is injective on each box and is such that $\gamma(x) = slab_G(x)$ for every source x of G (Lemma 2.78).

For each $u \in N_T$, we let $\overline{G}(u)$ be the s-graph $\langle G(u)^{\circ}, slab_{\overline{G}(u)} \rangle$ such that $slab_{\overline{G}(u)} := \gamma \restriction (f(u) \cap V_G)$. (This mapping $slab_{\overline{G}(u)}$ extends $slab_{G(u)}$ because all vertices of $f(u) \cap V_G$ are sources of $\overline{G}(u)$.) It follows then from the definitions that

$$G = fg_{C-\tau(G)}(\overline{G}(root_T)). \tag{2.2}$$

By bottom-up induction on u, we construct for each $u \in N_T$ a term $t(u) \in T(F_C^{HR})$ such that $\overline{G}(u) \simeq val(t(u))$ and a ground witness of this fact.

Let u have sons $u_1, \ldots, u_p, p \geq 0$; we can assume that we have already constructed $t(u_1), \ldots, t(u_p)$ and we have

$$\overline{G}(u) = fg_{B_1}(\overline{G}(u_1)) /\!/ \cdots /\!/ fg_{B_p}(\overline{G}(u_p)) /\!/ H(u) \tag{2.3}$$

where the sets B_i and the s-graph $H(u)$ are defined as follows:

(1) $B_i := \{\gamma(x) \mid x \in f(u_i) - f(u)\}$ for $i = 1, \ldots, p$.[37]
(2) $H(u) := (H, slab)$, where H is the subgraph of G° such that $V_H \cup E_H = f(u)$ and $slab := \gamma \restriction V_H$. To optimize the construction, the isolated vertices of $H(u)$ that are in $f(u_1) \cup \cdots \cup f(u_p)$ can be removed, including their source names.

The s-graph $H(u)$ has at most k vertices and a term $t'(u)$ in $T(F_C^{HR})$ evaluating to it can be constructed by Proposition 2.33; a ground witness w' of this fact is easy to determine (cf. the discussion after Proposition 2.44). Hence the term

$$t(u) := fg_{B_1}(t(u_1)) /\!/ \cdots /\!/ fg_{B_p}(t(u_p)) /\!/ t'(u) \tag{2.4}$$

has value $\overline{G}(u)$. A ground witness w of $\overline{G}(u) \simeq val(t(u))$ is easy to obtain from ground witnesses w_i of $\overline{G}(u_i) \simeq val(t(u_i))$ and the ground witness w'. Here are the details. Assuming that the term $t(u)$ officially starts with p symbols $/\!/$, we define the positions $v_1, \ldots, v_{p+1}$ of $t(u)$ by $v_1 := p + 2$ and $v_{i+1} := v_i + |t(u_i)| + 1$ for $i \in [p]$. Thus, $t(u)/v_i = t(u_i)$ for $i \in [p]$ and $t(u)/v_{p+1} = t'(u)$. For $i \in [p]$ let h_i be the isomorphism : $Syn(t(u_i)) \to Syn(t(u))/v_i$, and let h_{p+1} be the isomorphism : $Syn(t'(u)) \to Syn(t(u))/v_{p+1}$ (see Definition 2.14). Moreover, let $h_j((u,a)) := (h_j(u), a)$ for $a \in A$. Then we define the ground witness w as follows, for $x \in V_{\overline{G}(u)} \cup E_{\overline{G}(u)}$: $w(x) = h_{p+1}(w'(x))$ for $x \in f(u)$, and $w(x) = h_i(w_i(x))$ for $i \in [p]$ and $x \in (V_{\overline{G}(u_i)} - f(u)) \cup E_{\overline{G}(u_i)}$. For the optimized construction the definition of w is slightly more complicated.

By Equality 2.2, the term $t := fg_{C - \tau(G)}(t(root_T))$ has value G. A witness w of this fact is easy to obtain from a witness $\overline{w}$ of $\overline{G}(root_T) \simeq val(t(root_T))$: if $\overline{w}(x)$ is (u,a) or u, then $w(x)$ is defined as $(u+1,a)$ or $u+1$, respectively. As observed after Definition 2.46, w can be converted in linear time into a canonical witness of $G \simeq val(t)$.

This construction need not distinguish the cases of directed and undirected s-graphs because Proposition 2.33 covers both cases. The extension to labeled graphs is straightforward. It can also be obtained from Proposition 2.47(2).

---

[37] If $B_i = \emptyset$, we can omit fg_{B_i} because $fg_\emptyset$ defines the identity mapping. If (T, f) is downwards increasing, we have $B_i \neq \emptyset$ for all i. We can also take $B_i := C - f(u)$ for each $i = 1, \ldots, p$.

Example 2.81 Consider the rich tree-decomposition of Figure 2.10. The vertices are colored by a, b, c. The optimized construction of Definition 2.80 gives the term

$$t = fg_c\big(\mathbf{a}^\ell \,/\!/\, fg_a(\mathbf{ab} \,/\!/\, \mathbf{ac}) \,/\!/\, \mathbf{bc}\big) \,/\!/\, fg_a(\mathbf{ab} \,/\!/\, \mathbf{ac}) \,/\!/\, \mathbf{bc}.$$

The full construction has $\mathbf{a} \,/\!/\, \mathbf{bc}$ instead of $\mathbf{bc}$ (at its two occurrences). If the loop is in box $f(1)$ (and not in box $f(3)$ as shown in Figure 2.10), then the same (optimized) construction gives the term

$$fg_c(\mathbf{a} \,/\!/\, fg_a(\mathbf{ab} \,/\!/\, \mathbf{ac}) \,/\!/\, \mathbf{bc}) \,/\!/\, fg_a(\mathbf{ab} \,/\!/\, \mathbf{ac}) \,/\!/\, \mathbf{a}^\ell \,/\!/\, \mathbf{bc}.$$

Proposition 2.82

(1) Let C be a set of k source names. The construction of Definition 2.80 associates with every rich tree-decomposition (T, f) of width at most $k - 1$ of a concrete s-graph G of type included in C a term in $T(F_C^{\mathrm{HR}})$ (hence of width at most k) that evaluates to G, and a canonical witness of this fact. This construction can be done in linear time in the size of (T, f), i.e., in time $O(|E_G| + |N_T|)$.

(2) For every (labeled) s-graph G, we have $wd(G) \leq twd(G) + 1$.

Proof: The proof of (1) is straightforward by bottom-up induction on the node u of T, and (2) is an immediate consequence of (1). ∎

Propositions 2.76 and 2.82 give the following theorem:

Theorem 2.83

(1) Let $C \in \mathcal{P}_f(\mathcal{A})$. An s-graph G is the value of a term in $T(F_C^{\mathrm{HR}})$ if and only if $\tau(G) \subseteq C$ and $twd(G) \leq |C| - 1$. The corresponding terms (with witnesses) and tree-decompositions can be constructed from each other in linear time. The same result holds for (K, Λ)-labeled s-graphs and terms in $T(F_{C,[K,\Lambda]}^{\mathrm{HR}})$.

(2) For every $((K, \Lambda)$-labeled) s-graph G, we have $twd(G) = wd(G) - 1$.

Proof: If w is a witness of $G \simeq val(t)$ and (T, f) is the rich tree-decomposition of $cval(t)$ obtained by Proposition 2.76(1), then $(T, w^{-1} \circ f)$ is a rich tree-decomposition of G (cf. Proposition 2.44(1)). The total time is $O(|t|)$, assuming that w^{-1} can be applied in constant time.

If (T, f) is a tree-decomposition of G of width at most $|C| - 1$, then a rich tree-decomposition (T, f') of G of the same width can be obtained in time $O(|E_G| + |N_T|)$ by a depth-first search of T during which each edge is assigned to the first box to which both its ends belong. Then Proposition 2.82(1) can be applied. The total time is $O(|E_G| + |N_T|)$. Note that, in particular, if G is simple, then the total time is $O(|N_T|)$, cf. Remark 2.55(c). ∎

For a given s-graph G, the source names in $C - \tau(G)$ are "auxiliary" and can be chosen arbitrarily. In particular, if $\tau(G) = \emptyset$, then G has tree-width at most $k - 1$ if and only if $G = val(t)$ for some term t over $F_{[k]}^{\mathrm{HR}}$.

Remark 2.84 (1) The construction of Definition 2.80 does not use the source renaming operations ren_h. For the purpose of defining s-graphs by terms, they are dispensable: not using them does not result in the necessity of augmenting the number of source names. We have already observed this fact in Proposition 2.49. In other words, the tree-width of an s-graph can be characterized as in Theorem 2.83 in terms of $wd'(G)$, where this width wd' is relative to the subsignature of F_C^{HR} obtained by omitting the source renaming operations.

(2) In Definition 2.80, a term $t(u)$ in $T(F_C^{\mathrm{HR}})$ (for some finite set C) is associated with each node u of T; if u has sons $u_1,\ldots,u_p$ then Equality (2.3) can be rewritten $\overline{G}(u) = g(\overline{G}(u_1),\ldots,\overline{G}(u_p))$ where g is the linear derived operation $\lambda x_1,\ldots,x_p \cdot fg_{B_1}(x_1) \mathbin{/\!/} \cdots \mathbin{/\!/} fg_{B_p}(x_p) \mathbin{/\!/} t'(u)$. By Proposition 2.69 it is not a loss of generality to assume that the given tree-decomposition is binary, i.e., that $p \leq 2$. With this assumption and for defining simple graphs, the number of different terms $t'(u)$ is finite and consequently, so is the number of different derived operations g for each fixed set C. Without the restriction to simple s-graphs, the s-graphs $H(u)$ can have arbitrarily many edges and the set of derived operations g is countably infinite.

(3) Every term in $T(F_C^{\mathrm{HR}})$ written with the operations fg_a (and not with the operations fg_B) that evaluates to a nonempty s-graph G has size at least $|Int_G| + 2 \cdot (|E_G| + |Isol_G|) - 1$ as one can check easily (cf. the proof of Proposition 2.33). This lower-bound can be realized by such a term in $T(F_C^{\mathrm{HR}})$ with C of cardinality $twd(G) + 1$. This term can be obtained by the construction of Proposition 2.82(1) from an optimal downwards increasing tree-decomposition. We omit the details. As an example, consider the term t of Example 2.81. It is (also) obtained from the optimal downwards increasing tree-decomposition that one gets from Figure 2.10 by removing box 3 and putting the loop in box 2.

(4) A set of s-graphs of tree-width at most k and of type included in a finite set C is a subset of the domain of the F_C^{HR}-algebra $\mathbb{JS}[C]$ and also of that of the F_D^{HR}-algebra $\mathbb{JS}^{\mathrm{gen}}[D]$ (cf. Remark 2.39(1)), where D is any finite set that includes C and has at least $k+1$ elements.

(5) We recall from Section 1.4.2 that the series-composition $G \bullet H$ of two directed s-graphs of type [2] can be defined as a derived operation of $\mathbb{JS}$, as follows: $G \bullet H = fg_3(ren_{2\leftrightarrow3}(G) \mathbin{/\!/} ren_{1\leftrightarrow3}(H))$. It follows that every series-parallel graph, as defined in Section 1.1.3, is the value of a term over $F_{[3]}^{\mathrm{HRd}}$. Hence, by Theorem 2.83, series-parallel graphs have tree-width at most 2.

Conversely, the graphs of tree-width at most 2 and of type included in [2] are those generated by the signature $F_{[2]}^{\mathrm{HR}} \cup \{\bullet\}$. Moreover, by generalizing series-composition to a k-ary operation, we obtain a similar characterization for graphs of tree-width at most k. We let H_k be the signature $F_{[k]}^{\mathrm{HR}} \cup \{S_k\}$ where S_k is the *generalized series-composition* defined as follows:

$$S_k(G_1,\ldots,G_k) := fg_0(ren_{0\leftrightarrow1}(G_1) \mathbin{/\!/} \cdots \mathbin{/\!/} ren_{0\leftrightarrow k}(G_k)).$$

Hence, S_k is a derived operation defined with the help of the source label 0. It is intended to be used for s-graphs $G_1, \ldots, G_k$ of type included in $[k]$. It is clear that $G \bullet H = S_2(H, G)$ for G and H of type $[2]$.

The signature H_k generates the set of s-graphs of type included in $[k]$ and of tree-width at most k. (The simple loop-free undirected graphs in this set are called *partial k-trees*, see [*Bod98] for details.) This fact is used in [ArnCPS], [CorRot] and [CouOla]. Apparently, the signature H_k uses fewer source labels than $F_{[k+1]}^{HR}$ to generate the same s-graphs of type included in $[k]$; however, the definition of S_k uses the source label $0 \notin [k]$. $\square$

Our next objective is to characterize s-graphs of bounded path-width as values of particular terms over F^{HR}.

A term t over a signature F is *slim* if, for every subterm of t of the form $f(t_1, \ldots, t_k)$, at most one of $t_1, \ldots, t_k$ is not a constant symbol. We will denote by $Slim(F)$ the set of slim terms over F.

Proposition 2.85 Let $C \in \mathcal{P}_f(\mathcal{A})$. An s-graph G is the value of a term in $Slim(F_C^{HR})$ if and only if $\tau(G) \subseteq C$ and $pwd(G) \leq |C| - 1$.

Proof: Let G be defined by a slim term t in $T(F_C^{HR})$. This means that for every subterm of t of the form $t_1 /\!/ t_2$, the term t_1 or t_2 (or both) is of the form $\mathbf{ab}$, $\overrightarrow{\mathbf{ab}}$, $\mathbf{a}$, $\mathbf{a}^\ell$ or $\varnothing$. Let T be the syntactic tree of t and (T, f) be the rich tree-decomposition of G constructed by Definition 2.75. If no node has outdegree 2, then (T, f) is a path-decomposition and the result is proved.

Otherwise let u be a node of T with two sons u_1 and u_2. It is an occurrence of $/\!/$, and one of the two sons, say u_i, is an occurrence of a constant symbol. Note that $f(u_i) \cap V_G \subseteq f(u) \cap V_G$. By contracting the edge $u \to u_i$ of T we obtain a tree-decomposition of G having the same width as (T, f) and one less node of outdegree 2, cf. Remark 2.77(2). By repeating this step finitely many times, we obtain a path-decomposition of G of width at most $k - 1$.

Let us assume conversely that (T, f) is a path-decomposition of an s-graph G of width $k - 1$. We apply the construction of Definition 2.80. Since T is a path-decomposition, we have $p \leq 1$ at each occurrence u. Let $t'(u) = c_1 /\!/ \cdots /\!/ c_n$ for constant symbols $c_1, \ldots, c_n$ (the construction of Proposition 2.33 is applied with $B = \emptyset$). With the same notation as in Definition 2.80, we can write the term $t(u)$ as

$$t(u) = (\cdots(((s /\!/ c_1) /\!/ c_2) /\!/ c_3) /\!/ \cdots c_n),$$

where $s := \varnothing$ if $p = 0$ and $s := fg_{B_1}(t(u_1))$ if $p = 1$.

The term t for G is then defined as $t := fg_{C - \tau(G)}(t(root_T))$. It is a slim term (provided we write the parallel-compositions with parentheses as indicated). $\blacksquare$

For each finite set C, we let P_C^{HR} be the signature consisting of the constants $\varnothing$, $\mathbf{ab}$, $\overrightarrow{\mathbf{ab}}$, $\mathbf{a}$, $\mathbf{a}^\ell$ for $a, b \in C$, $a \neq b$, of the unary operations $ren_{a \leftrightarrow b}$ for $a, b \in C$, $a \neq b$, the unary operations fg_B for $B \subseteq C$, and the unary derived operations $\lambda x \cdot x \parallel c$ where c is any of the above constant symbols except $\varnothing$. It follows from Proposition 2.85 that an s-graph G is the value of a term in $T(P_C^{\mathrm{HR}})$ if and only if $\tau(G) \subseteq C$ and $pwd(G) \leq |C| - 1$. We omit a detailed proof. Needless to say that the extension of the previous proposition to labeled s-graphs is straightforward.

Tree-decompositions and their algebraic expressions will be used in Chapter 6 for some algorithmic applications.

2.5 The VR algebra of simple graphs with ports

We define graph operations that generalize the complete join of two graphs used in Section 1.1.2 for defining cographs. These operations will provide us with another algebra of graphs that we will call the VR algebra because its equational sets (cf. Section 1.1.4) are those generated by the context-free *vertex replacement* graph grammars (see [*EngRoz] or [*Eng97]). These equational sets will be studied in Chapter 4. A graph complexity measure called *clique-width* is associated in a natural way with the definition of a graph by a term over the signature of the VR algebra. Taken as parameter, it yields fixed-parameter tractable algorithms for problems expressible in monadic second-order logic. In this section, we only consider simple, possibly labeled graphs. This restriction is due to the way edges will be defined.

2.5.1 The VR graph operations

A simple (directed) graph will be handled as a binary relation on a finite set of vertices,[38] hence such a graph G will be defined as a pair $\langle V_G, edg_G \rangle$ consisting of a set of vertices V_G and a binary relation $edg_G \subseteq V_G \times V_G$. This relation is symmetric if the graph is undirected. There is a loop incident with a vertex x if and only if $(x,x) \in edg_G$. Undirected graphs are just particular (directed) graphs.

Definition 2.86 (Simple graphs with ports) Let $\mathcal{A}$ be a fixed countable set of *port labels* with the same properties as in Definition 2.24 (in particular, $\mathcal{N} \subseteq \mathcal{A}$). A *concrete graph with ports*, or a *concrete p-graph* in short, is a pair $G = \langle G^{\circ}, port_G \rangle$ consisting of a concrete simple (directed) graph G° and a mapping $port_G : V_{G^{\circ}} \to \mathcal{A}$. A vertex x is an *a-port* of G if $port_G(x) = a$. We denote by $\pi(G)$ the set $port_G(V_G)$ (we denote $V_{G^{\circ}}$ also by V_G, and similarly for other notations). The set of port labels $\pi(G)$ is called the *type* of G, like $\tau(G)$ in the HR algebra.

Let G and H be concrete p-graphs. We say that G is a *subgraph* of H if G° is a subgraph of H° and $port_G$ is the restriction of $port_H$ to V_G (so that $\pi(G) \subseteq \pi(H)$). An *isomorphism* : $G \to H$ is a bijection $h : V_G \to V_H$ such that for every u, v in V_G we have

---

[38] Vertex sets will be included in $\mathcal{U}$, the universal set of vertices mentioned in the footnote in Definition 2.9.

$(h(u), h(v)) \in edg_H$ if and only if $(u, v) \in edg_G$, and $port_H(h(u)) = port_G(u)$. Since these graphs are simple, isomorphisms between them can be specified by bijections between their vertex sets only (cf. Definition 2.10) and these bijections must preserve the vertex labelings defined by the port mappings (cf. Definition 2.11). We obtain thus the notion of an *abstract p-graph*.

Every simple graph will be considered as a p-graph, all vertices of which are a-ports for some fixed default label a, usually 1.

We now define operations on concrete and abstract p-graphs. We denote by $\mathcal{GP}$ the class of abstract p-graphs and by $\mathcal{GP}^u$ the class of undirected abstract p-graphs. In the sequel, the term "abstract" will be omitted, except for emphasis.

Definition 2.87 (Operations on p-graphs) We define a signature of graph operations on $\mathcal{GP}$ that we will denote by F^{VR}.

Disjoint union For disjoint concrete p-graphs G and H, we let $G \oplus H$ be the graph $G^\circ \cup H^\circ$ equipped with the port mapping $port_{G \oplus H} := port_G \cup port_H$. If G and H are not disjoint, we replace them by disjoint isomorphic copies. In this way, we obtain a well-defined binary operation on (abstract) p-graphs. Clearly

$$\pi(G \oplus H) = \pi(G) \cup \pi(H).$$

Edge addition Let $a, b \in \mathcal{A}$, with $a \neq b$. For every concrete p-graph G, we let $\overrightarrow{add}_{a,b}(G)$ be the p-graph G' such that $V_{G'} := V_G$, $edg_{G'} := edg_G \cup \{(x, y) \mid x, y \in V_G, port_G(x) = a, port_G(y) = b\}$ and $port_{G'} := port_G$. This operation is well defined for all concrete p-graphs, and obviously for p-graphs as well. Clearly

$$\pi(\overrightarrow{add}_{a,b}(G)) = \pi(G).$$

Note that $\overrightarrow{add}_{a,b}(G) = G$ if a or b does not belong to $\pi(G)$, or if there are already edges from every a-port to every b-port.

For adding undirected edges, we use $add_{a,b}(G)$ defined as $\overrightarrow{add}_{b,a}(\overrightarrow{add}_{a,b}(G))$. Hence $add_{a,b}$ is a derived operation of the algebra we are defining. However, to allow an easy description of undirected p-graphs, it is more convenient to consider $add_{a,b}$ as a basic operation. Note that $add_{a,b}$ and $add_{b,a}$ are two function symbols with the same associated operation. We also have $\pi(add_{a,b}(G)) = \pi(G)$. None of these operations creates loops. Loops can be specified by constant p-graphs defined below.

Port relabeling If $h : \mathcal{A} \to \mathcal{A}$ is a mapping that is the identity outside of a finite subset of $\mathcal{A}$, then we let $relab_h$ be the unary operation such that $G' = relab_h(G)$ if $V_{G'} := V_G$, $edg_{G'} := edg_G$ and $port_{G'} := h \circ port_G$.

This operation is well defined for concrete as well as for abstract p-graphs and

$$\pi(relab_h(G)) = h(\pi(G)).$$

A particular case deserves an easier notation. For $a, b \in \mathcal{A}$, $a \neq b$, we let $relab_{a \to b}$ denote $relab_h$ where $h : \mathcal{A} \to \mathcal{A}$ is defined by $h(a) = b$ and $h(c) = c$ for every $c \in \mathcal{A}$, $c \neq a$. We call this operation an *elementary relabeling*. We have $relab_{a \to b}(G) = G$ if $a \notin \pi(G)$. Furthermore, $relab_h \circ relab_{h'} = relab_{h \circ h'}$ for all mappings h, h'. We can thus express as a single operation $relab_h$ a composition of elementary relabelings $relab_{a_1 \to b_1} \circ relab_{a_2 \to b_2} \circ \cdots \circ relab_{a_k \to b_k}$, and vice versa.

For $C \subseteq \mathcal{A}$ and a mapping $h : C \to \mathcal{A}$ that is the identity outside of a finite subset of C (which holds in particular if C is finite), we also denote by $relab_h$ the operation $relab_{h'}$, where h' agrees with h on C and is the identity outside of C. For each set $C \subseteq \mathcal{A}$, we denote by $[C \to C]_f$ the set of mappings $h : C \to C$ such that h is the identity outside of a finite subset of C.

Basic graphs   We let the constant symbols $\mathbf{a}$ and $\mathbf{a}^\ell$ denote p-graphs with a single vertex that is an a-port, and, for the second, that also has an incident loop. It will be useful for some constructions to have also the symbol $\varnothing$ to denote the empty graph. We will see that every term denoting a nonempty p-graph can be replaced by an equivalent one not using $\varnothing$.

The VR algebra of p-graphs   We obtain thus a countably infinite (but effectively given) signature:

$$F^{VR} := \{\oplus, \overrightarrow{add}_{a,b}, add_{a,b}, relab_h, \mathbf{a}, \mathbf{a}^\ell, \varnothing \mid a, b \in \mathcal{A}, \ a \neq b, \ h \in [\mathcal{A} \to \mathcal{A}]_f\},$$

and, for dealing with undirected graphs, its subsignature:

$$F^{VRu} := \{\oplus, add_{a,b}, relab_h, \mathbf{a}, \mathbf{a}^\ell, \varnothing \mid a, b \in \mathcal{A}, \ a \neq b, \ h \in [\mathcal{A} \to \mathcal{A}]_f\}.$$

We let $\mathbb{GP}$ denote the F^{VR}-algebra with domain $\mathcal{GP}$. It is effectively given. We call $\mathbb{GP}$ the *VR algebra*, and we denote by $\mathbb{GP}^u$ its subalgebra with domain $\mathcal{GP}^u$ and the operations of F^{VRu}. Each term t over F^{VR} evaluates into a p-graph $t_{\mathbb{GP}}$, also denoted by $val(t)$. It t is a term over F^{VR} with variables, then $t_{\mathbb{GP}}$ denotes the corresponding derived operation.

The signatures F_C^{VR} and F_C^{VRu} for $C \subseteq \mathcal{A}$ are obtained by restricting a, b to C and h to $[C \to C]_f$ in the above definitions. It is easy to show (by induction on t) that $\pi(val(t)) \subseteq C$ for every $t \in F_C^{VR}$, cf. Lemma 2.36.

As for the HR algebra (cf. Definition 2.32 and Remark 2.39(1)), we can define two F_C^{VR}-algebras that are subalgebras of $\mathbb{GP}$. The first one is $\mathbb{GP}[C]$ of which the domain is the set $\mathcal{GP}[C]$ of p-graphs of type included in C, and the second one is the subalgebra $\mathbb{GP}^{gen}[C]$ of $\mathbb{GP}[C]$ that is generated by F_C^{VR}. Its domain $\mathcal{GP}^{gen}[C]$

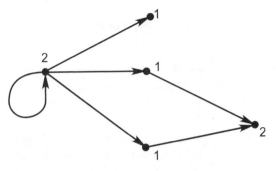

Figure 2.11 A graph with ports.

consists of the p-graphs that are the values of terms over F_C^{VR}. Similar notions can be defined for undirected p-graphs: $\mathbb{GP}^u[C]$ and $\mathbb{GP}^{gen,u}[C]$. We will shortly extend these definitions to labeled p-graphs.

Example 2.88 (1) The p-graph G of type $\pi(G) = \{1,2\}$ in Figure 2.11 is the value of the term $relab_{3\to 2}(\overrightarrow{add}_{3,1}(\overrightarrow{add}_{1,2}(\mathbf{1} \oplus \mathbf{1} \oplus \mathbf{2}) \oplus \mathbf{1} \oplus \mathbf{3}^\ell))$.

(2) For each $n \geq 1$, the clique K_n with all vertices being 1-ports is the value of the term t_n defined inductively by:

$$t_1 \quad := \quad \mathbf{1},$$
$$t_{n+1} \quad := \quad relab_{2\to 1}(add_{1,2}(t_n \oplus \mathbf{2})) \text{ for each } n \geq 1.$$

(3) The mapping that associates the complete join $G \otimes H$ with two graphs G and H (two p-graphs of type $\{1\}$) is the derived operation $t_{\mathbb{GP}}$ defined by the term $t := relab_{2\to 1}(add_{1,2}(x_1 \oplus relab_{1\to 2}(x_2)))$ that belongs to $T(F_{[2]}^{VRu}, \{x_1, x_2\})$.

It follows that cographs (see Section 1.1.2) can be defined by terms in $T(\{\oplus, add_{1,2}, relab_{1\to 2}, relab_{2\to 1}, \mathbf{1}\})$. The set of cographs C is thus defined by the equation

$$C = (C \oplus C) \cup t_{\mathbb{GP}}(C, C) \cup \{\mathbf{1}\}.$$

We have already presented such definitions in Sections 1.1.2 and 1.4.1, and we will consider them in more detail in Chapter 4.

(4) We now consider rooted trees. We let the root of a rooted tree T be a 1-port and all the other nodes be 2-ports. The type of a rooted tree T is $\{1\}$ or $\{1,2\}$. If T is the union of pairwise disjoint rooted trees $T_1, \ldots, T_p$ augmented with a new node r that is the root of T, and with edges from r to the roots of $T_1, \ldots, T_p$, then we have

$$T = relab_{3\to 1}(relab_{1\to 2}(\overrightarrow{add}_{3,1}(\mathbf{3} \oplus T_1 \oplus \cdots \oplus T_p))).$$

It follows that rooted trees are defined by the system of two equations:

$$\Sigma_{Trees} \begin{cases} S &= \{\mathbf{1}\} \cup relab_{3\to 1}(relab_{1\to 2}(\overrightarrow{add}_{3,1}(\mathbf{3} \oplus F))), \\ F &= S \cup (F \oplus S). \end{cases}$$

This system defines the set S of rooted trees and the set F of nonempty rooted forests.

Definition 2.89 (Clique-width) The *clique-width* of a p-graph G is the minimal cardinality of a set C such that G is the value of a term in $T(F_C^{VR})$. (We generalize the definition given for graphs in Section 1.4.1.) It is denoted by $cwd(G)$. A term $t \in T(F_C^{VR})$ with $|C| = cwd(G)$ is called an *optimal clique-width term* for G.

It is easy to check that the clique-width of an undirected p-graph is the minimal cardinality of a set C such that this graph is the value of a term in $T(F_C^{VRu})$. In fact, for a term t over F_C^{VR}, if $val(t) = G$ and t' is obtained from t by changing every $\overrightarrow{add}_{a,b}$ into $add_{a,b}$, then $val(t')$ is the undirected p-graph $\langle und(G^\circ), port_G \rangle$ with $und(G^\circ) = \langle V_G, edg_G \cup edg_G^{-1} \rangle$.

Since the type of $val(t)$ is included in C for every term t in $T(F_C^{VR})$, every p-graph G has clique-width at least $|\pi(G)|$. By a proof similar to the one of Proposition 2.33 (see Proposition 2.104(1) below), we obtain that every p-graph is the value of a term in $T(F_C^{VR})$ for every set C such that $\pi(G) \subseteq C$ and $|C| = |V_G|$. Hence the algebra $\mathbb{GP}$ is generated by its signature F^{VR}, and every p-graph G has clique-width at most $|V_G|$. More precise results will be given in Section 2.5.4.[39]

If a p-graph G is the value of a term t in F_C^{VR} and $h \in [\mathcal{A} \to \mathcal{A}]_f$ is injective on C, then $relab_h(G) = val(t')$ for some term $t' \in T(F_{h(C)}^{VR})$; this term is obtained from t by replacing every port label a of C by $h(a)$. If, in particular, h is the identity on $\pi(G)$, then $G = val(t')$. This can be used to prove the following easy analogue of Theorem 2.83(1): a p-graph G is the value of a term in $T(F_C^{VR})$ if and only if $\pi(G) \subseteq C$ and $cwd(G) \leq |C|$. (For the proof of this fact, the only-if direction is clear; for the if-direction, let $G = val(t)$ for some $t \in T(F_D^{VR})$ with $|D| \leq |C|$. Choose some $h \in [\mathcal{A} \to \mathcal{A}]_f$ with $h(D) \subseteq C$ that is injective on D and is the identity on $\pi(G) \subseteq D$. Then $G = val(t')$ and $t' \in T(F_C^{VR})$.)

In particular, if G is a graph (i.e., $\pi(G) = \{1\}$), then G has clique-width at most k if and only if $G = val(t)$ for some term t over $F_{[k]}^{VR}$.

The *clique-width* $cwd(L)$ of a set of p-graphs L is the least upper-bound (possibly ω) of the integers $cwd(G)$ for $G \in L$. We say that L has *bounded clique-width* if $cwd(L) \in \mathcal{N}$. The class of all graphs has unbounded clique-width: this fact follows from the exact characterization of the clique-width of square grids in Proposition 2.106(1), and also by a different proof based on Corollary 2.122 and using a counting argument.

The *linear clique-width* of a p-graph G is the minimal cardinality of a set C such that G is the value of a slim term in $T(F_C^{VR})$. It is denoted by $lcwd(G)$. The notion of linear clique-width is thus similar to that of path-width, characterized in Proposition 2.85 by means of slim terms over F_C^{HR}. The *linear clique-width* of a set of p-graphs is the least upper-bound of the linear clique-widths of its elements.

---

[39] Clique-width is defined in [CouOla] in terms of the operations $relab_{a \to b}$ and not in terms of the larger set of operations of the form $relab_h$. However the corresponding values are the same as we will prove in Section 2.5.6 (Proposition 2.118).

Linear clique-width is to clique-width what path-width is to tree-width. It plays a central role in the NP-completeness proof ([FelRRS]) of the CLIQUE-WIDTH CHECKING PROBLEM, i.e., of the problem of checking that $cwd(G) \leq k$ for given (G,k).

Definition 2.90 (Labeled graphs with ports) We let K and Λ be two finite, disjoint, sets of labels such that $(K \cup \Lambda) \cap \mathcal{A} = \emptyset$, and we let $\mathcal{GP}_{[K,\Lambda]}$ be the set of simple[40] directed $[K, \Lambda]$-labeled graphs equipped with port labels, called $[K, \Lambda]$-*labeled p-graphs*. In a $[K, \Lambda]$-labeled p-graph, each edge has one and only one label in Λ and each vertex has, in addition to its port label in $\mathcal{A}$, a possibly empty set of labels from K. Hence, we extend Definition 2.11 to graphs with ports (as we did in Definition 2.32 for graphs with sources). We will apply to $[K, \Lambda]$-labeled p-graphs the disjoint union $\oplus$ and the relabeling operations $relab_h$, generalized from Definition 2.87 to $\mathcal{GP}_{[K,\Lambda]}$ in the obvious way, and edge addition operations that depend on edge labels: for every $\lambda \in \Lambda$ and $a, b \in \mathcal{A}$ with $a \neq b$, we define an operation $\overrightarrow{add}_{a,b,\lambda}$ that adds λ-labeled edges from every a-port x to every b-port y (if a λ-labeled edge already exists from x to y, then no other edge is added between these vertices).

We also define new constant symbols (in addition to $\varnothing$). For $M \subseteq K \cup \Lambda$ and $a \in \mathcal{A}$, the constant symbol $\mathbf{a}_M$ denotes the p-graph with a single vertex that is an a-port, has label κ for each $\kappa \in K \cap M$ and an incident λ-labeled loop for each $\lambda \in \Lambda \cap M$. Hence, $\mathbf{a}_\emptyset$ defines the same p-graph as $\mathbf{a}$, and $\mathbf{a}_{\{\iota\}}$ defines the same p-graph as $\mathbf{a}^\ell$ if ι is taken as default edge label.

The operations $\oplus$, $relab_h$ and $\overrightarrow{add}_{a,b,\lambda}$ do not modify the labels from K and Λ once they have been specified for a vertex or a loop by some constant $\mathbf{a}_M$ or, for an edge that is not a loop, at the moment of its introduction by an operation $\overrightarrow{add}_{a,b,\lambda}$.

We denote the corresponding signature by $F^{\mathrm{VR}}_{[K,\Lambda]}$ and its restriction to operations using port labels in C by $F^{\mathrm{VR}}_{C,[K,\Lambda]}$. The *clique-width* of a $[K, \Lambda]$-labeled p-graph G is defined as the minimal cardinality of a set C of port labels such that G is the value of a term over $F^{\mathrm{VR}}_{C,[K,\Lambda]}$. Specialized definitions for undirected[41] p-graphs and for linear clique-width are straightforward to set up.

Example 2.91 Let G be the $(\emptyset, \{\lambda, \mu\})$-labeled graph with vertex set $\{1,2,3,4\}$, λ-labeled edges $2 \to 1, 2 \to 3, 4 \to 3$ and μ-labeled edge $4 \to 1$. It is the value of the term

$$relab_h(\overrightarrow{add}_{2,1,\lambda}(\overrightarrow{add}_{2,3,\lambda}(\overrightarrow{add}_{4,3,\lambda}(\overrightarrow{add}_{4,1,\mu}(\mathbf{1} \oplus \mathbf{2} \oplus \mathbf{3} \oplus \mathbf{4)))))},$$

where $h(i) = 1$ for every $i \in [4]$. Its clique-width is thus at most 4 and is actually 4, as one checks by trying all possibilities of constructing this graph by a term in

---

[40] Recall from Definition 2.11 that simple $[K, \Lambda]$-labeled graphs may have multiple edges with different labels.

[41] A $[K, \Lambda]$-labeled p-graph is undirected if for every edge from x to y there is an edge from y to x with the same label.

$T(F^{VR}_{[3],[\emptyset,\{\lambda,\mu\}]})$; in fact, only finitely many possibilities need to be tried, see the last paragraph of Section 2.5.3 below. If we make labels λ and μ identical then the clique-width of the resulting graph is only 2. It is the value of $relab_{2\to1}(\overrightarrow{add}_{1,2,\lambda}(2 \oplus 1 \oplus 2 \oplus 1))$.

2.5.2 Construction of the p-graph defined by a term

Our aim is to construct from a term $t \in T(F^{VR})$ a concrete p-graph isomorphic to $val(t)$, and to define a notion of witness of $G \simeq val(t)$, where G is a concrete graph, as we did in Definition 2.43 for terms in $T(F^{HR})$. Fortunately, the definitions and constructions will be much simpler because, when $G \simeq val(t)$, we have a bijection from V_G to the set $Occ_0(t)$ of occurrences in t of constant symbols different from $\emptyset$. A witness of $G \simeq val(t)$ will be such a bijection w, satisfying some conditions ensuring that w is an isomorphism between G and $val(t)$. No quotient graph will be needed.

Definition 2.92 (Terms denoting concrete p-graphs) For every $a \in \mathcal{A}$ (the countable set of port labels) and every $x \in \mathcal{U}$ (the universal set of vertices, see Definition 2.9) we let $\mathbf{a}(x)$ and $\mathbf{a}^\ell(x)$ denote the concrete p-graphs with vertex x that is an a-port, and also has an incident loop in the second case.

We let $F^{cVR}_0 := \{\mathbf{a}(x), \mathbf{a}^\ell(x), \emptyset \mid a \in \mathcal{A}, x \in \mathcal{U}\}$ and F^{cVR} be the signature obtained by replacing in F^{VR} the set of constant symbols F^{VR}_0 by F^{cVR}_0. A constant symbol $\mathbf{a}(x)$ or $\mathbf{a}^\ell(x)$ *describes* the vertex x. We obtain a new signature F^{cVR} (and its variants F^{cVR}_C, F^{cVRu}_C, etc.), where the superscript "c" stands for "concrete."

Not every term $T(F^{cVR})$ defines a concrete p-graph because for concrete p-graphs G and H, $G \oplus H$ is well defined only if $V_G \cap V_H = \emptyset$. (However, $f(G)$ is well defined for every unary operation f in F^{VR}.) We now formalize this observation. Let $t \in T(F^{cVR})$. Two occurrences of constant symbols are *conflicting* if they describe the same vertex. The *value* of a term $t \in T(F^{cVR})$ is the concrete p-graph denoted by $cval(t)$ defined as follows:

$$
\begin{aligned}
&cval(t) \text{ is } \mathbf{a}(x), \mathbf{a}^\ell(x) \text{ or } \emptyset &&\text{if } t \text{ is } \mathbf{a}(x), \mathbf{a}^\ell(x) \text{ or } \emptyset \text{ respectively,}^{42} \\
&cval(t) := f(cval(t_1)) &&\text{if } t = f(t_1), \\
&cval(t) := cval(t_1) \oplus cval(t_2) &&\text{if } t = t_1 \oplus t_2 \text{ and } V_{cval(t_1)} \cap V_{cval(t_2)} = \emptyset, \\
&\text{otherwise } cval(t) \text{ is undefined.}
\end{aligned}
$$

Hence, $V_{cval(t)}$ is the set of vertices x that are described by the constant symbols occurring in t. Note in particular that if $cval(t)$ is defined and u and v are two positions in t that are incomparable with respect to $\leq_t$, then the p-graphs $cval(t/u)$ and $cval(t/v)$ are disjoint.[43] The next lemma can be proved by induction on the structure of t.

---

[42] We denote in the same way a constant symbol in F^{cVR} and the corresponding graph.
[43] Even if $t'/u = t'/v$ where $t' \in T(F^{VR})$ is obtained from t by replacing each $\mathbf{a}(x)$ or $\mathbf{a}^\ell(x)$ respectively by $\mathbf{a}$ or $\mathbf{a}^\ell$.

Lemma 2.93 For every term t over F^{cVR}, the concrete p-graph $cval(t)$ is defined if and only if no two occurrences of constant symbols in t are conflicting. If $cval(t)$ is defined, its vertex set is the set of vertices described by the constant symbols. □

We define concrete p-graphs that are values of terms over F^{VR}. Let t be such a term and w be a bijection : $V \to Occ_0(t)$ where V is a finite subset of $\mathcal{U}$, the universal set of vertices, and as before, $Occ_0(t)$ is the set of occurrences in t of constant symbols different from $\varnothing$. We define the term t_w in $T(F^{cVR})$ by replacing in t at each occurrence u of a constant symbol $\mathbf{a}$ or $\mathbf{a}^\ell$ this symbol by $\mathbf{a}(w^{-1}(u))$ or $\mathbf{a}^\ell(w^{-1}(u))$ respectively. Clearly, this term has no conflicting occurrences, hence $cval(t_w)$ is a concrete p-graph G such that $V_G = V$ and $G \simeq val(t)$. (Here, $val(t)$ is the abstract p-graph that is the value of t.) In particular, we can take $V := Occ_0(t)$ and w equal to the identity function on $Occ_0(t)$, denoted by Id. Thus, the set of vertices of $cval(t_{Id})$ is $Occ_0(t)$. We will also denote $cval(t_{Id})$ by $cval(t)$; moreover, for every $u \in N_t$, we will also denote $cval(t_{Id}/u)$ by $cval(t)/u$.[44]

If a concrete p-graph G is isomorphic to $val(t)$ for a term t in $T(F_C^{VR})$, then $G = cval(t')$ for some term $t' \in T(F_C^{cVR})$: if w is an isomorphism from G to $cval(t_{Id})$, then $G = cval(t_w)$ as one checks easily. Hence, a concrete p-graph G is the value of a term in $T(F_C^{cVR})$ if and only if $[G]_{iso}$ is the value of a term in $T(F_C^{VR})$.

Our next aim is to describe the concrete p-graph $cval(t) = cval(t_{Id})$, for t in $T(F^{VR})$. We need auxiliary definitions.

Definition 2.94 (Ports of p-graphs defined by subterms) Let t belong to $T(F^{VR})$ and $Syn(t)$ be its syntactic tree. The set of nodes of $Syn(t)$, denoted by N_t, is $Pos(t)$. We recall that $lab_t(u) = f$ if and only if $u \in Occ(t,f)$. If $u \in N_t$, then $Syn(t)/u$, the subtree of $Syn(t)$ issued from u, is isomorphic to the syntactic tree of the subterm t/u of t. The vertex set of the (concrete) p-graph $cval(t)/u$ is $Occ_0(t) \cap N_t/u = \{x \in Occ_0(t) \mid x \leq_t u\}$. For $u \in N_t$ and $x \in Occ_0(t)$ such that $x \leq_t u$ (which means that x is below u in $Syn(t)$), we define

$$port_t(u,x) := port_{cval(t)/u}(x).$$

Hence, if $G = cval(t)$, then $V_G = Occ_0(t)$ and $port_G(x) = port_t(root_t,x)$ for every $x \in Occ_0(t)$. The following is clear from the definitions.

---

[44] Note that if u is not the root, then $t_{Id}/u \neq (t/u)_{Id}$ and $cval(t)/u \neq cval(t/u)$ (however, they are isomorphic). If u and v are incomparable in t, then $cval(t)/u$ and $cval(t)/v$ are disjoint concrete p-graphs.

Lemma 2.95 The partial function $port_t : N_t \times Occ_0(t) \to \mathcal{A}$ satisfies the following properties:

1. if $x = u$, then $port_t(u,x) = a$, where $lab_t(x)$ is $\mathbf{a}$ or $\mathbf{a}^\ell$;
2. if $x <_t u$ and u' is the son of u such that $x \leq_t u' <_t u$, then we have:

$$port_t(u,x) = \begin{cases} port_t(u',x) & \text{if } lab_t(u) \in \{\oplus, add_{a,b}, \overrightarrow{add}_{a,b}\}, \\ h(port_t(u',x)) & \text{if } lab_t(u) = relab_h. \end{cases}$$

$\square$

The second case of this lemma can be written as follows:

$$port_t(u,x) = (h_1 \circ h_2 \circ \cdots \circ h_n)(port_t(x,x)),$$

where $h_1, h_2, \ldots, h_n$ is the sequence of mappings that define relabelings on the path in $Syn(t)$ from u to x.

Lemma 2.96 Let $t \in T(F^{VR})$ and let G be the concrete p-graph $cval(t)$ with vertex set $Occ_0(t)$. The edges of G are the pairs (x,y) in $Occ_0(t) \times Occ_0(t)$ such that either $x = y$ and $lab_t(x) = \mathbf{a}^\ell$ (for some $a \in \mathcal{A}$) or $x \neq y$ and there exists u in N_t with $lab_t(u) \in \{add_{a,b}, add_{b,a}, \overrightarrow{add}_{a,b}\}$, $x \leq_t u$, $y \leq_t u$, $port_t(u,x) = a$ and $port_t(u,y) = b$ (for some $a, b \in \mathcal{A}$).

Proof: Let t and G be as in the statement. One can prove by bottom-up induction on u that, for every u in N_t, the relation $edg_{cval(t)/u}$ is the set of pairs (x,y) such that either $x = y$, $x \leq_t u$ and $lab_t(x) = \mathbf{a}^\ell$, or $x \neq y$ and there exists u' in N_t with $u' \leq_t u$ and $lab_t(u') \in \{add_{a,b}, add_{b,a}, \overrightarrow{add}_{a,b}\}$, $x \leq_t u'$, $y \leq_t u'$, $port_t(u',x) = a$ and $port_t(u',y) = b$. Taking $u = root_t$ the result is obtained. $\blacksquare$

In Section 7.2 we will use these two lemmas to prove that the mapping *val* from terms over F^{VR} to p-graphs is a monadic second-order transduction. (Proposition 2.41 will be used in a similar way in Section 7.4 for terms over F^{HR}.)

Definition 2.97 (Witnesses) Let G be a concrete p-graph and $t \in T(F^{VR})$. A *witness* of $G \simeq val(t)$ is a bijection $w : V_G \to Occ_0(t)$ satisfying the following conditions:

(1) edg_G is the set of pairs (x,y) in $V_G \times V_G$ such that:

either $x = y$ and $lab_t(w(x)) = \mathbf{a}^\ell$ (for some $a \in \mathcal{A}$),

or $x \neq y$ and there exists u in N_t with $lab_t(u) \in \{add_{a,b}, add_{b,a}, \overrightarrow{add}_{a,b}\}$, $w(x) \leq_t u$, $w(y) \leq_t u$, $port_t(u, w(x)) = a$ and $port_t(u, w(y)) = b$ (for some $a, b \in \mathcal{A}$), and

(2) $port_G(x) = port_t(root_t, w(x))$ for all $x \in V_G$.

Proposition 2.98 Let G be a concrete p-graph and t a term over F^{VR}. Then $G \simeq val(t)$ if and only if there exists a witness of $G \simeq val(t)$.

Proof: By Lemma 2.96, a witness of $G \simeq val(t)$ is an isomorphism from G to $cval(t)$.[45] Since $cval(t) \simeq val(t)$, the result follows. ∎

The above definitions and results on the construction of concrete p-graphs from terms extend to labeled p-graphs and to the terms that define them in a straightforward way.

Definition 2.99 (Parsing) The *parsing problem* relative to the VR algebra is defined as follows:

Input: Finite pairwise disjoint sets C of port labels, K of vertex labels, Λ of edge labels and a concrete (K, Λ)-labeled p-graph G of type included in C.

Output: A term t in $T(F^{VR}_{C,[K,\Lambda]})$ and a witness of $G \simeq val(t)$ if they exist, or the answer that they do not exist.

Efficient algorithms for this problem are more difficult to build than for parsing relative to the HR algebra. We will discuss them in Section 6.2.

The proof of the following proposition is similar to that of Proposition 2.47 (but easier: it suffices to replace constant symbols).

Proposition 2.100

(1) Let be given a concrete p-graph G, a term t in $T(F^{VR}_C)$ and a witness w of $G \simeq val(t)$. Let γ be a labeling function turning G into a $(K, \emptyset)$-labeled p-graph $\widehat{G}$. In linear time (for fixed C and K), we can compute $\widehat{t} \in T(F^{VR}_{C,[K,\emptyset]})$ that evaluates to $\widehat{G}$, such that w is a witness of this fact.

(2) A similar result holds for deleting or adding loops, or modifying their labels, and for modifying vertex labels. □

Proposition 2.47 is similar for s-graphs and the HR algebra, but here we cannot add edge labels or reverse edge directions. Examples showing that will be given in Section 2.5.4.

2.5.3 Algebraic properties and derived operations defined by contexts

The results of this section are analogous to those of Section 2.3.3 relative to the HR algebra.

We define some notation. If $(f_i)_{i \in I}$ is a finite family of unary total functions on some set that commute pairwise, i.e., are such that $f_i \circ f_j = f_j \circ f_i$ for every i, j in I, then we denote by $\bigcirc_{i \in I} f_i$ their composition in any order (if $I = \emptyset$, it denotes the identity on the set). The resulting operation is the same for every order of composition.

The following proposition is stated for labeled p-graphs; its proof is an easy verification from the definitions.

---

[45] This implies, by the way, that a bijection $w : V_G \to Occ_0(t)$ is a witness of $G \simeq val(t)$ if and only if $G = cval(t_w)$.

Proposition 2.101 For all (K, Λ)-labeled p-graphs G, H, J, all mappings h, h' in $[\mathcal{A} \to \mathcal{A}]_f$, all a, b, c, d in $\mathcal{A}$ with $a \neq b$, $c \neq d$, all λ, μ in Λ, and all subsets M of $K \cup \Lambda$, we have:

(1) $G \oplus H = H \oplus G$;

(2) $G \oplus (H \oplus J) = (G \oplus H) \oplus J$;

(3) $G \oplus \varnothing = G$;

(4) $relab_h(relab_{h'}(G)) = relab_{h \circ h'}(G)$;

(5) $relab_h(G \oplus H) = relab_h(G) \oplus relab_h(H)$;

(6) $relab_h(\varnothing) = \varnothing$;

(7) $relab_{Id}(G) = G$, where Id is the identity on $\mathcal{A}$;

(8) $relab_h(\mathbf{a}_M) = \mathbf{c}_M$ if $c = h(a)$;

(9) $\overrightarrow{add}_{a,b,\lambda}(\overrightarrow{add}_{c,d,\mu}(G)) = \overrightarrow{add}_{c,d,\mu}(\overrightarrow{add}_{a,b,\lambda}(G))$;

(10) $\overrightarrow{add}_{a,b,\lambda}(G \oplus H) = \overrightarrow{add}_{a,b,\lambda}(\overrightarrow{add}_{a,b,\lambda}(G) \oplus H)$;

(11) $\overrightarrow{add}_{a,b,\lambda}(\varnothing) = \varnothing$;

(12) $\overrightarrow{add}_{a,b,\lambda}(relab_h(G)) = relab_h((\bigcirc_{a' \in h^{-1}(a), b' \in h^{-1}(b)} \overrightarrow{add}_{a',b',\lambda})(G))$;

(13) $\overrightarrow{add}_{a,b,\lambda}(\mathbf{c}_M) = \mathbf{c}_M$.

Equalities (9)–(13) hold with *add* instead of $\overrightarrow{add}$. □

Some further equalities can be derived from those of this proposition. In particular, we have, for unlabeled p-graphs:

$$(10') \quad \overrightarrow{add}_{a,b}(\overrightarrow{add}_{a,b}(G)) = \overrightarrow{add}_{a,b}(G),$$

by letting $H = \varnothing$ in (10) and by using (3), and

$$(9') \quad add_{a,b}(G) = add_{b,a}(G),$$

by (9) since $add_{a,b}(G)$ is defined as $\overrightarrow{add}_{b,a}(\overrightarrow{add}_{a,b}(G))$. If h is injective, then (12) can be rewritten into:

$$(12') \quad relab_h(\overrightarrow{add}_{a,b}(G)) = \overrightarrow{add}_{h(a),h(b)}(relab_h(G)).$$

Equality (7) states that $relab_{Id}$ denotes the identity, and could have been omitted. (We made a similar remark after Proposition 2.48.)

Conditional equalities can be formalized as equalities in the many-sorted setting of Section 2.6.3. We can cite in particular:

(14) $relab_h(G) = relab_{h'}(G)$ if h and h' agree on $\pi(G)$;

(15) $\overrightarrow{add}_{a,b}(G) = G$ if $a \notin \pi(G)$ or $b \notin \pi(G)$;

(16) $\overrightarrow{add}_{a,b}(G \oplus H) = \overrightarrow{add}_{a,b}(G) \oplus H$ if $a \notin \pi(H)$ and $b \notin \pi(H)$.

If t, t' belong to $T(F^{VR}_{C,[K,\Lambda]}, X_n)$ and t' is obtained from t by using the rules of Proposition 2.101, then t and t' are *equivalent*, i.e., they define the same derived operation of $\mathbb{GP}$ or the same labeled p-graph if $n = 0$. For transforming, conversely, two equivalent terms in $T(F_{C,[K,\Lambda]}^{VR}, X_n)$ one into the other, conditional rewriting rules (or rules with terms in the many-sorted signature of Section 2.6.3) seem necessary. The questions raised after Proposition 2.48 for the signature F^{HR} can also be raised for F^{VR}.

Corollary 2.102 Every term in $T(F_{C,[K,\Lambda]}^{VR})$ that denotes a nonempty labeled p-graph is equivalent to a term in $T(F_{C,[K,\Lambda]}^{VR} - \{\varnothing\})$.

Proof: Let t be a term containing occurrences of $\varnothing$. By using rules (1), (3), (6) and (11) of Proposition 2.101, we can transform t into an equivalent term t' that is $\varnothing$ or has no occurrence of $\varnothing$. The first case is excluded since $val(t)$ is not the empty graph. These transformations do not introduce new port labels, hence $t' \in T(F_{C,[K,\Lambda]}^{VR})$ if $t \in T(F_{C,[K,\Lambda]}^{VR})$. ∎

As we did in Section 2.3.3 for the HR algebra, we will express in a concise way the unary derived operations defined by contexts. If C is a finite set of port labels and $R \subseteq C \times C$, we denote by ADD_R the derived operation $\bigcirc_{(a,b)\in R, a\neq b}\overrightarrow{add}_{a,b}$. In the (K,Λ)-labeled case we define $ADD_R := \bigcirc_{(a,b,\lambda)\in R, a\neq b}\overrightarrow{add}_{a,b,\lambda}$ for $R \subseteq C \times C \times \Lambda$. It is well defined because the result does not depend on the order of composition by Equality (9) of Proposition 2.101. Note that Equality (12) of that proposition can now be formulated as follows: $\overrightarrow{add}_{a,b,\lambda}(relab_h(G)) = relab_h(ADD_R(G))$ with $R = h^{-1}(a) \times h^{-1}(b) \times \{\lambda\}$. The next result is stated for the unlabeled case.

Proposition 2.103 Let B be a finite set of port labels and let $c \in Ctxt(F_B^{VR})$. There exist a set C of port labels with $C \cap B = \emptyset$ and $|C| \leq |B| \cdot 2^{2|B|}$, a term t over $F_{B\cup C}^{VR}$ with $\pi(t_{\mathbb{GP}}) \subseteq C$, a relation $R \subseteq (B \times B) \cup (C \times B) \cup (B \times C)$ and a mapping $h : B \cup C \to B$ such that, for every p-graph G with $\pi(G) \cap C = \emptyset$, we have $c_{\mathbb{GP}}(G) = relab_h(ADD_R(G \oplus t_{\mathbb{GP}}))$. If c is written without $\oplus$, then $C = \emptyset$ and $t = \varnothing$.

Proof: We will use x_1 as special variable in contexts. For convenience, we will write $c(G)$ instead of $c_{\mathbb{GP}}(G)$.

We first consider the $\oplus$-free case, i.e., the case where c is written without $\oplus$. By using Equalities (4), (9) and (12) of Proposition 2.101 and the derived equality $(10')$, we can transform c into an equivalent context of the form $relab_h(ADD_R(x_1))$. For doing this we need not introduce new port labels, hence $C = \emptyset$ and so $R \subseteq B \times B$ and $h : B \to B$. The transformation can be done in linear time for fixed B.

We now consider the case where c contains at least one $\oplus$. In this case we let $C := B \times \mathcal{P}(B) \times \mathcal{P}(B)$.[46] We define $R_C \subseteq (C \times B) \cup (B \times C)$ by

$$R_C := \{((b,A,A'),a) \in C \times B \mid a \in A\} \cup \{(a,(b,A,A')) \in B \times C \mid a \in A'\},$$

---

[46] We assume here that the elements of $B \times \mathcal{P}(B) \times \mathcal{P}(B)$ are appropriately encoded as natural numbers that are not in B, which is possible because B is finite.

and we define $h_C : C \to B$ such that $h_C(b,A,A') := b$ for every $(b,A,A') \in C$. Thus, if G and H are p-graphs with $\pi(G) \cap C = \emptyset$ and $\pi(H) \subseteq C$, then $ADD_{R_C}(G \oplus H)$ is obtained from $G \oplus H$ as follows: for every $x \in V_H$, with $port_H(x) = (b,A,A')$, edges are added from x to every $y \in V_G$ with $port_G(y) \in A$, and edges are added from every $y \in V_G$ with $port_G(y) \in A'$ to x; note that b is irrelevant.

By the $\oplus$-free case, it suffices to prove that there exist a context $c' \in Ctxt(F_B^{VR})$ that is written without $\oplus$ and a term $t \in T(F_{B \cup C}^{VR})$ with $\pi(t_{\mathbb{GP}}) \subseteq C$ such that

$$c(G) = relab_{h_C}(c'(ADD_{R_C}(G \oplus t_{\mathbb{GP}})))$$

for every p-graph G with $\pi(G) \cap C = \emptyset$. In fact, if $c' = relab_{h'}(ADD_{R'}(x_1))$, with $R' \subseteq B \times B$ and $h' : B \to B$, then $R := R' \cup R_C$ and $h := h' \cup h_C$ satisfy the requirements. (The operation $relab_h$ is equivalent to $relab_{h'} \circ relab_{h_C}$ by Equality (4) of Proposition 2.101.)

We prove the existence of c' and t by induction on the structure of c. We will also write s instead of $s_{\mathbb{GP}}$ for a term s over F^{VR}.

(1) If $c = x_1$, then we take $c' := x_1$ and $t := \emptyset$. This is correct because $relab_{h_C}(ADD_{R_C}(G)) = G$ if $\pi(G) \cap C = \emptyset$, cf. Equalities (14) and (15) after Proposition 2.101.

(2) If $c = c_1 \oplus s$ where $c_1 \in Ctxt(F_B^{VR})$ and $s \in T(F_B^{VR})$, then there are c_1' and t_1 such that $c_1(G) = relab_{h_C}(c_1'(ADD_{R_C}(G \oplus t_1)))$ for every p-graph G with $\pi(G) \cap C = \emptyset$, by the induction hypothesis. For c we take $c' := c_1'$ and $t := t_1 \oplus relab_j(s)$ where $j : B \to C$ maps each $a \in B$ to $(a,\emptyset,\emptyset)$ in C. Then $c_1'(ADD_{R_C}(G \oplus t_1 \oplus relab_j(s))) = c_1'(ADD_{R_C}(G \oplus t_1)) \oplus relab_j(s)$, by the fact that for all p-graphs J,K and every unary operation $f \in F_D^{VR}$, we have $f(J \oplus K) = f(J) \oplus K$ if $\pi(K) \cap D = \emptyset$, cf. Equalities (5), (14) and (16) of Proposition 2.101. Applying $relab_{h_C}$ to both sides of the equality, we obtain the required equality for $c(G)$ by Equality (5) of Proposition 2.101 and by the fact that $relab_{h_C}(relab_j(s)) = s$.

(3) If $c = relab_g(c_1)$ and c_1', t_1 are obtained from the induction hypothesis, then we can take $c' := relab_g(c_1')$ and $t := relab_{\overline{g}}(t_1)$, where $\overline{g} : C \to C$ applies g to the first component of the port labels in C, i.e., $\overline{g}(a,A,A') := (g(a),A,A')$. Indeed, $c(G) = relab_g(relab_{h_C}(c_1'(ADD_{R_C}(G \oplus t_1))))$ by the induction hypothesis and that clearly equals $relab_{h_C}(relab_g(c_1'(ADD_{R_C}(G \oplus relab_{\overline{g}}(t_1)))))$.

(4) If $c = \overrightarrow{add}_{a,b}(c_1)$ and c_1', t_1 are obtained from the induction hypothesis, then we can take $c' := \overrightarrow{add}_{a,b}(c_1')$ and define t as follows. First, we define ADD_S such that it simulates $\overrightarrow{add}_{a,b}$ on the first components of the port labels in C: the set $S \subseteq C \times C$ consists of all pairs $((a,A,A'),(b,D,D'))$ with $A,A',D,D' \subseteq B$. Second, assuming that $c_1' = relab_{g_1}(ADD_{R_1}(x_1))$, we define $k : C \to C$ by $k(a,A,A') := (a,A \cup g_1^{-1}(b),A')$, $k(b,A,A') := (b,A,A' \cup g_1^{-1}(a))$, and $k(d,A,A') := (d,A,A')$ for $d \neq a,b$. And finally, we define $t := relab_k(ADD_S(t_1))$. We have $c(G) =$

$\overrightarrow{add}_{a,b}(relab_{hc}(c'_1(ADD_{RC}(G \oplus t_1))))$ by the induction hypothesis and that clearly equals

$$relab_{hc}(\overrightarrow{add}_{a,b}(ADD_T(c'_1(ADD_{RC}(G \oplus ADD_S(t_1)))))),$$

where T is the set of all pairs $((a,A,A'),b)$ and $(a,(b,D,D'))$ with $A,A',D,D' \subseteq B$. By Equalities (9) and (12) of Proposition 2.101, this equals

$$relab_{hc}(\overrightarrow{add}_{a,b}(c'_1(ADD_{RC}(ADD_{T'}(G \oplus ADD_S(t_1)))))),$$

where T' consists of all pairs $((a,A,A'),d)$ with $d \in g_1^{-1}(b)$ and all pairs $(d,(b,D,D'))$ with $d \in g_1^{-1}(a)$. From this it should be clear that

$$ADD_{RC}(ADD_{T'}(G \oplus ADD_S(t_1))) = ADD_{RC}(G \oplus relab_k(ADD_S(t_1)))$$

and so we obtain that

$$c(G) = relab_{hc}(\overrightarrow{add}_{a,b}(c'_1(ADD_{RC}(G \oplus relab_k(ADD_S(t_1)))))),$$

which proves the result.

We finally note that it follows from the above proof that c' and a concrete p-graph H isomorphic to $t_{\mathbb{GP}}$ can also be defined directly from c as follows. Let u be the occurrence of x_1 in c. First, $c' = f_p(\cdots f_2(f_1(x_1))\cdots)$ where $f_1, f_2, \ldots, f_p$ are the unary operations that occur in this order on the path in $Syn(c)$ from u to the root. Second, $H° = cval(c[\varnothing])°$, cf. Section 2.5.2. Third, for each vertex x of H, we have $port_H(x) = (port_{cval(c[\varnothing])}(x), A(x), A'(x))$, where $A(x)$ (resp. $A'(x)$) is the set of all $b \in B$ such that there is an edge $x \to u$ (resp. $u \to x$) in the concrete p-graph $cval(c[\mathbf{b}])$, cf. Section 2.5.2. ∎

The result extends to (K, Λ)-labeled graphs and the corresponding terms with $|C| \leq |B| \cdot 2^{2|\Lambda| \cdot |B|}$ because we need sets similar to A and A' for each edge label.

We can now show that there is an algorithm to determine whether or not a p-graph G has clique-width at most k, for given G and k. The algorithm has to find out whether there exists a term over F_C^{VR} with value G, where C is chosen such that $|C| = k$ and $\pi(G) \subseteq C$ (see Definition 2.89). By Corollary 2.102 it needs only consider terms with $|V_G|$ occurrences of constant symbols. Moreover, by the $\oplus$-free case of Proposition 2.103, it may additionally restrict attention to terms t that satisfy the following property: if $c[s]$ is a subterm of t where $c \in Ctxt(F_C^{VR})$ is written without $\oplus$ and $s \in T(F_C^{VR})$, then $|c| \leq k^2 - k + 2$ (because such a subterm can be replaced by $relab_h(ADD_R(s))$ for some h and R). Since there are only finitely many terms satisfying both conditions, the algorithm can just check them all. The number of terms is exponential in the number of vertices of G (see Corollary 2.122 below). This problem is discussed in more detail in Section 6.2.

2.5.4 Properties of clique-width

We review some properties of clique-width and give some examples.

Proposition 2.104 For every labeled p-graph G, we have:

(1) $|\pi(G)| \leq cwd(G) \leq |V_G|$,

(2) if G is undirected, unlabeled and without ports (i.e., $\pi(G)$ is singleton), then $cwd(G) \leq |V_G| - k$ where k is the largest integer such that $2^k < |V_G| - k$.

Proof: (1) This fact is similar to Remark 2.55(c). As observed before, $|\pi(G)| \leq cwd(G)$ because $\pi(val(t)) \subseteq C$ for every $t \in F_C^{VR}$. For showing that $cwd(G) \leq |V_G|$ we let G be unlabeled to simplify the proof. Let C be a set of port labels that contains $\pi(G)$ such that there exists a bijection $k : V_G \to C$. For each x in V_G we let $c(x)$ be the constant $\mathbf{a}(x)$ (resp. $\mathbf{a}^\ell(x)$) if $a = k(x)$ and x has no loop (resp. if $a = k(x)$ and x has a loop), cf. Definition 2.92. We let $R := \{(a, b) \mid a, b \in C \text{ and } (k^{-1}(a), k^{-1}(b)) \in edg_G\}$. We let $h : C \to \pi(G)$ be such that $h(a) = port_G(k^{-1}(a))$. Then we have

$$G = relab_h(ADD_R(c(x_1) \oplus \cdots \oplus c(x_n))),$$

where $V_G = \{x_1, \ldots, x_n\}$ (and ADD_R is defined before Proposition 2.103).

The construction for (K, Λ)-labeled p-graphs is essentially the same. The constants $\mathbf{a}_M(x)$ specify the labels of a vertex x with port label a, the loops incident to it and their labels. Each edge addition operation adds a unique edge with its label. Hence, the same number of port labels is used as for an unlabeled graph.

(2) This result is proved in [Joha]. ∎

The following results are useful for giving upper- and lower-bounds to the clique-width of particular p-graphs. We recall from Definition 2.12 that $H \subseteq_i G$ means that $H = G[V_H]$, i.e., that H is an induced subgraph of G (with the same labels for its vertices and edges as in G). If G is a labeled p-graph, we let $core(G)$ be the unlabeled, loop-free and undirected p-graph $\langle core(G^\circ), port_G \rangle$; for the core of a graph, see Definition 2.9.

We generalize as follows the edge-complement $\overline{G}$ of a simple, loop-free, undirected graph G defined in Example 1.30. If G is a concrete (K, Λ)-labeled p-graph, then we let $\widetilde{G}$ be the p-graph with the same vertices, port labels and vertex labels as G; it has an edge $x \to y$ labeled by λ (possibly a loop, if $x = y$) if and only if G has no such edge. If G is undirected, loop-free and unlabeled, then, apart from loops and port labels, we have $\widetilde{G} = \overline{G}$ (we recall that an undirected edge is a pair of opposite directed edges).

The next proposition should be compared with Remark 2.55(d) and Corollary 2.60(1).

Proposition 2.105

(1) If $H \subseteq_i G$, then $cwd(H) \leq cwd(G)$.

(2) The addition and deletion of loops, the modification of vertex labels and of labels of loops do not modify the clique-width of a p-graph.

(3) For every p-graph G, we have $cwd(core(G)) \leq cwd(G)$.

(4) For every (K, Λ)-labeled p-graph G, we have $cwd(\widetilde{G}) \leq 2 \cdot cwd(G)$.

Proof: (1) Let H, G be concrete (K, Λ)-labeled p-graphs such that $H \subseteq_i G$. We let $G = cval(t_{Id})$ with $t \in T(F_{C,[K,\Lambda]}^{VR})$, cf. Definition 2.92. We have that $V_G = Occ_0(t)$ and Id is the identity on V_G. The p-graph H is obtained from G by removing some vertices and their incident edges. For every $x \in V_G - V_H$, we replace in t_{Id} the constant symbol $\mathbf{a}_M(x)$ which specifies vertex labels (in K), loops and their labels (in Λ) by the constant symbol $\varnothing$ (which denotes the empty graph). We obtain a term $t'_{Id'}$, where $t' \in T(F_{C,[K,\Lambda]}^{VR})$ and Id' is the identity on $V_H = Occ_0(t')$, the value of which is nothing but H (by Lemmas 2.95 and 2.96). It follows that $cwd(H) \leq cwd(G)$.

(2) This follows from Proposition 2.100(2), because the transformations of terms considered there do not modify the port labels. If G is transformed into H, then $cwd(H) \leq cwd(G)$. Since the inverse transformations are of the same nature, we have an equality.

(3) Let G be a p-graph defined by a term t in $T(F_C^{VR})$. Let t' be the term in $T(F_C^{VR})$ obtained from t by replacing every $\mathbf{a}^\ell$ by $\mathbf{a}$ and every $\overrightarrow{add}_{a,b}$ by $add_{a,b}$ (for $a, b \in C$), which corresponds to deleting loops and forgetting edge directions, respectively. It is clear that $val(t') = core(G)$, hence $cwd(core(G)) \leq cwd(G)$. The proof is the same for labeled graphs.

(4) The proof will be given at the end of Section 2.5.6. The case of loop-free undirected graphs is proved in Theorem 4.1 of [CouOla]. ∎

We review the clique-widths of some classical graphs, as we did in Example 2.56(3) for tree-width. The graph P_n is an undirected path with n vertices. We denote by $G_{n \times m}^u$ the undirected rectangular grid $und(G_{n \times m})$.

Proposition 2.106

(1) The following equalities hold:

$$cwd(K_n) = 2 \text{ if } n \geq 2, \ cwd(K_{n,m}) = 2 \text{ if } n, m \geq 1,$$
$$cwd(P_n) = 3 \text{ if } n \geq 4,$$
$$cwd(C_5) = cwd(C_6) = 3, \ cwd(C_n) = 4 \text{ for } n \geq 7,$$
$$cwd(G_{n \times n}^u) = n + 1 \text{ if } n \geq 2,$$
$$\text{and } m + 1 \leq cwd(G_{n \times m}^u) \leq m + 2 \text{ if } n > m \geq 3.$$

(2) A simple loop-free undirected graph is a cograph if and only if its clique-width is at most 2.

Proof: (1) The cases of K_n and $K_{n,m}$ follow from Examples 2.88(2) and 2.88(3) and Remark 2.107(1) below. For proving that $cwd(P_n) \leq 3$ consider the terms t_n such that $t_2 := add_{1,2}(\mathbf{1} \oplus \mathbf{2})$ and $t_{n+1} := relab_{3 \to 2}(relab_{2 \to 1}(add_{2,3}(t_n \oplus \mathbf{3})))$. It is clear that $val(t_n)^\circ = P_n$. Since $P_2 = K_{1,1}$ and $P_3 = K_{1,2}$, these two graphs have clique-width 2. It is easy to check that no term with two port labels can define P_4. Hence, $cwd(P_4) > 2$ and thus, we have $cwd(P_n) = 3$ if $n \geq 4$ by Proposition 2.105(1): $P_4 \subseteq_i P_n$. Since $C_3 = K_3$ and $C_4 = K_{2,2}$, these two graphs have clique-width 2. See [CouOla], Example 2.2, for C_n, $n \geq 5$. The results for grids are proved in [GolRot]. (The equality $cwd(G_{n \times m}^u) = m + 2$ if $n > m \geq 3$ is Conjecture 6.3 in this article.)

(2) We have seen in Example 2.88(3) that every cograph has clique-width at most 2. The other direction follows from Proposition 2.105(1) and the well-known fact (see [*BranLS]) that if G is not a cograph, then it has an induced subgraph isomorphic to P_4, so that $cwd(G) \geq 3$. ∎

Remark 2.107 (1) A p-graph G has clique-width 0 if and only if it is empty. It has clique-width 1 if and only if all its edges are loops and $|\pi(G)| = 1$.

(2) We have observed in Remark 2.55(d) that tree-width does not depend on edge directions. We have also proved in Proposition 2.47(3) that changing an edge direction or a label in a graph defined by a term in $T(F^{HR})$ or in $T(F_{[K,\Lambda]}^{HR})$ corresponds to changing a constant symbol in this term. For clique-width, the situation is similar for vertex labels (cf. Proposition 2.100(2)), but not for edge labels and edge directions because the operations $\overrightarrow{add}_{a,b}$ and $\overrightarrow{add}_{a,b,\lambda}$ define edge directions and/or edge labels simultaneously for a set of edges forming a directed complete bipartite graph. Hence the modification of a single edge direction or of a single edge label may necessitate a reorganization of the term defining the graph. We have already seen that in Example 2.91 for edge labels. The following example shows that clique-width depends on edge directions.

Example 2.108 Let G_0, G_1, G_2 be the following graphs:

$$G_0 = \bullet - \bullet - \bullet, \; G_1 = \bullet \to \bullet \leftarrow \bullet, \; G_2 = \bullet \to \bullet \to \bullet.$$

Clearly, $G_0 = P_3 = K_{1,2}$ and $cwd(G_0) = 2$. We have $G_1 = val((\mathbf{1} \oplus \mathbf{1}) \overrightarrow{\otimes} \mathbf{1})$ where the *complete directed join* of G and H is defined by $G \overrightarrow{\otimes} H = relab_{2 \to 1}(\overrightarrow{add}_{1,2}(G \oplus relab_{1 \to 2}(H)))$, thus $cwd(G_1) = 2$. Then $cwd(G_2) \leq |V_{G_2}| = 3$, and one can check that there is no term written with two port labels that defines G_2. Hence $cwd(G_2) = 3$. Since $cwd(G_2) > cwd(G_0) = cwd(G_1)$, this example shows that clique-width is preserved neither under reversal of edge directions[47] nor under forgetting edge directions. □

We have seen in Proposition 2.62 that the tree-width of a nonempty graph is the maximal tree-width of its connected components, of its 2-connected components and

---

[47] It is preserved under simultaneous reversal of all edge directions.

also of its 3-blocks. There exists a similar result for clique-width and the notion of a prime induced subgraph, relative to *modular decomposition*. Before we state it we need some definitions. They concern unlabeled graphs.

Definition 2.109 (Substitution to a vertex) Let G and H be disjoint concrete graphs[48] and u be a loop-free vertex of G. We let $G[u \leftarrow H]$ be the concrete graph K defined as follows:

$$V_K := V_{G-u} \cup V_H,$$
$$edg_K := edg_{G-u} \cup edg_H$$
$$\cup \{(v,w) \mid v \in V_{G-u},\ w \in V_H \text{ and } (v,u) \in edg_G\}$$
$$\cup \{(w,v) \mid v \in V_{G-u},\ w \in V_H \text{ and } (u,v) \in edg_G\}.$$

If G is a p-graph, we define the port mapping of K by

$$port_K(v) := \begin{cases} port_G(v) & \text{if } v \in V_{G-u}, \\ port_G(u) & \text{if } v \in V_H. \end{cases}$$

This definition is applicable if H is a p-graph but the port mapping of H does not play any role: we have $G[u \leftarrow H] = G[u \leftarrow H^\circ]$. We say that $G[u \leftarrow H]$ is obtained from G by the *substitution of H to the vertex u*. Informally, the vertex u is replaced by H and each vertex w of H is linked to each neighbor v of u in G in the same way as u is linked to v, and w is an a-port in $G[u \leftarrow H]$ if u is an a-port in G. We will define a term evaluating to $G[u \leftarrow H]$ from terms evaluating to G and to H.

We denote by $relab_{D \to a}$ the mapping $relab_h$ such that $h(d) = a$ for all $d \in D$ and $h(b) = b$ for all $b \in \mathcal{A} - D$.

Proposition 2.110

(1) Let G be the value of a term in $T(F_C^{cVR})$, let u be a loop-free vertex of G described by $\mathbf{a}(u)$ in t and let H be the value of a term t' in $T(F_D^{cVR})$ such that H and G are disjoint. Then $G[u \leftarrow H]$ is the value of the term $t[relab_{\pi(H) \to a}(t')/\mathbf{a}(u)]$ that belongs to $T(F_{C \cup D}^{cVR})$.[49]

(2) If $H \neq \varnothing$, then $cwd(G[u \leftarrow H]) = \max\{cwd(G), cwd(H^\circ)\}$.

Proof: (1) The equality $G[u \leftarrow H] = val(t'')$ where $t'' := t[relab_{\pi(H) \to a}(t')/\mathbf{a}(u)]$ can be proved by induction on the structure of t.

(2) By changing if necessary the port labels used in t', one can assume that $D \subseteq C$ if $|D| \leq |C|$ and that $C \subset D$ if $|C| < |D|$. It follows then that $|C \cup D| = \max\{|C|, |D|\}$ and that $cwd(G[u \leftarrow H]) \leq \max\{cwd(G), cwd(H^\circ)\}$ (because $cwd(H^\circ) = cwd(val(relab_{\pi(H) \to a}(t')))$).

---

[48] It actually suffices to assume that H and $G - u$ are disjoint, where $G - u$ denotes the induced subgraph $G[V_G - \{u\}]$.

[49] Here $t[t''/\mathbf{a}(u)]$ denotes the term obtained from t by replacing the occurrence of $\mathbf{a}(u)$ in t by the term t''.

For the opposite equality, we note that $H^\circ \subseteq_i G[u \leftarrow H]^\circ$ and that G is isomorphic to the subgraph of $G[u \leftarrow H]$ induced by $V_G \cup \{w\}$ where w is any vertex of H. Hence by Proposition 2.105(1) we have

$$cwd(H^\circ) \leq cwd(G[u \leftarrow H]^\circ) \leq cwd(G[u \leftarrow H]),$$

and $cwd(G) \leq cwd(G[u \leftarrow H])$, which gives the desired result. ∎

If $H = \varnothing$, then $G[u \leftarrow H] = G - u$ and we have $cwd(G - u) \leq cwd(G)$ by Proposition 2.105(1). We may have that $cwd(G[u \leftarrow H])$ is less than $cwd(G) = \max\{cwd(G), cwd(H^\circ)\}$.

Definition 2.111 (Modules and prime induced subgraphs) To simplify the presentation, we only consider graphs without loops. A *module* of a graph G is a set $M \subseteq V_G$ such that, for every $x, y \in M$ and every $z \in V_G - M$, we have

$$(x, z) \in edg_G \iff (y, z) \in edg_G, \text{ and}$$
$$(z, x) \in edg_G \iff (z, y) \in edg_G.$$

Thus, V_H is a module of $G[u \leftarrow H]$. A graph is *prime* if it has no other modules than its vertex set, each of its singletons and $\varnothing$. A *prime induced subgraph* is an induced subgraph that is prime. The graph P_3 is not prime but the graphs P_n are prime for all $n \geq 4$.

Proposition 2.112 For every graph G we have

$$cwd(G) = \max\{cwd(H) \mid H \text{ is a prime induced subgraph of } G\}.$$

Proof: The direction $\geq$ follows from Proposition 2.105(1). For the other direction, we use the fact that every directed graph G can be expressed by means of (nested) substitutions of graphs isomorphic to its prime induced subgraphs and of the particular graphs I_2 (two isolated vertices), $\overrightarrow{K_2} = \bullet \rightarrow \bullet$ and K_2.[50] (If G is undirected, the graph $\overrightarrow{K_2}$ is not used in this expression.) The other inequality follows from this result and Proposition 2.110(2). ∎

Example 2.113 Two graphs G and H are shown in Figure 2.12. We have $G = H[u \leftarrow I_2][v \leftarrow \overrightarrow{K_2}][w \leftarrow M]$ where $M = \bullet \rightarrow \bullet \rightarrow \bullet$. The graph H is defined by the term

$$t_H := relab_{\{2,3\} \rightarrow 1}(\overrightarrow{add}_{1,3}(relab_{3 \rightarrow 1}[\overrightarrow{add}_{2,1}(\overrightarrow{add}_{3,2}(\mathbf{1}(u) \oplus \mathbf{2}(v) \oplus \mathbf{3}(w)))] \oplus \mathbf{3}(x))).$$

---

[50] This is called the *modular decomposition* of G. The theory of modular decomposition is presented in [*EhrHR] and in [*MöhRad].

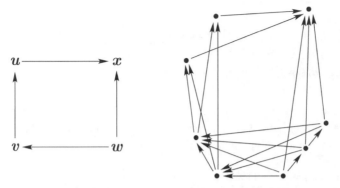

Figure 2.12 Graphs H (left) and G (right) of Example 2.113.

The p-graphs I_2, $\overrightarrow{K_2}$ and M of respective types $\{1\}$, $\{2\}$ and $\{3\}$ are the values of the terms $t_1 := 1 \oplus 1$, $t_2 := relab_{1\rightarrow 2}(\overrightarrow{add}_{1,2}(1 \oplus 2))$ and $t_3 := relab_{\{1,2\}\rightarrow 3}(\overrightarrow{add}_{1,2}(\overrightarrow{add}_{2,3}(1 \oplus 2 \oplus 3)))$. Then G is the value of the term $t_H[t_1/\mathbf{1}(u), t_2/\mathbf{2}(v), t_3/\mathbf{3}(w)]$ and, hence, of the term

$$t_G := relab_{\{2,3\}\rightarrow 1}(\overrightarrow{add}_{1,3}(relab_{3\rightarrow 1}[\overrightarrow{add}_{2,1}(\overrightarrow{add}_{3,2}(t_1 \oplus t_2 \oplus t_3))] \oplus 3)),$$

which belongs to $T(F_{[3]}^{\mathrm{VR}})$. It follows that $cwd(G) \leq 3$. Actually $cwd(G) = 3$ because H and M have clique-width 3. Note that H and M are induced subgraphs of G. $\qquad\square$

It follows from this proposition that an optimal clique-width term for a graph can be obtained from its modular decomposition and optimal clique-width terms for its prime induced subgraphs.

2.5.5 Comparisons between tree-width and clique-width

Tree-width, path-width and clique-width are functions on graphs that one can consider as graph complexity measures.[51] Certain graphs considered as basic have low values for these measures: paths (graphs of the form P_n) have path-width at most 1, trees have tree-width at most 1, cliques (the graphs K_n) have clique-width at most 2. These three complexity measures yield infinite hierarchies of graph families. In this section, we will compare tree-width and clique-width. Since clique-width is only defined for simple graphs, the comparisons will be stated for simple graphs (without ports or sources).

A *tournament* is an orientation of a clique, i.e., a directed graph obtained from a clique by choosing a direction for each of its edges.

---

[51] The term "complexity" refers to the number of basic operations from a given signature necessary to construct the considered graph. It follows that cliques are "difficult" to build with the operations of F^{HR} but "easy" to build with the operations of F^{VR}.

Proposition 2.114

(1) For every simple (K, Λ)-labeled graph G we have:

   (1.1) $cwd(G) \leq 2^{2|\Lambda| \cdot (twd(G)+1)} + twd(G) + 1$;

   (1.2) $cwd(G) \leq 3 \cdot 2^{twd(G)-1}$ if G is unlabeled and undirected;

   (1.3) $lcwd(G) \leq pwd(G) + 2$.

(2) For each k, there is a graph G of tree-width k and clique-width at least $2^{\lfloor k/2 \rfloor - 1}$.

(3) Cliques have clique-width at most 2 and unbounded tree-width. Tournaments have unbounded clique-width and tree-width.

Proof: (1) Inequality (1.1) will be proved in Section 4.3.4; the slightly stronger statement $cwd(G) \leq 2^{2twd(G)+2} + 1$ for G unlabeled and directed is proved in [CouOla].[52] Inequality (1.2) is proved in [CorRot]. We now prove inequality (1.3) for directed unlabeled graphs. Its extension to the other cases will be discussed afterwards.

Let G be a simple directed graph of pathwidth $k - 1$. By Proposition 2.85, it is the value of a slim term in $T(F^{HR}_{[k]})$. Let us say that a term t in $T(F^{HR})$ is simple if $val(t)$ is simple; note that if t is simple, then every subterm of t is simple. By induction on the structure of t, we transform every simple slim term t over $F^{HR}_{[k]}$ into a slim term t' over $F^{cVR}_{[0,k]}$ such that the p-graph $cval(t')$ is obtained from the s-graph $val(t)$ by turning each a-source into an a-port, and each internal vertex into a 0-port. Thus, if in particular t evaluates to G, then t' evaluates to the graph G of type $\{0\}$, that is, whose vertices are turned into 0-ports (G is without sources). Hence, $lcwd(G) \leq k + 1 = pwd(G) + 2$.

The inductive definition is as follows. It is straightforward to prove its correctness.

- If t is $\varnothing$, then $t' := \varnothing$. If t is $\mathbf{a}$ or $\mathbf{a}^\ell$, then t' is defined as $\mathbf{a}(x)$ or $\mathbf{a}^\ell(x)$ for some x. If t is $\overrightarrow{\mathbf{ab}}$, then $t' := \overrightarrow{add}_{a,b}(\mathbf{a}(x) \oplus \mathbf{b}(y))$ for some x and $y \neq x$.
- If t is $ren_h(t_1)$ where h is a permutation of $[k]$, then $t' := relab_h(t'_1)$.
- If t is $fg_a(t_1)$, then $t' := relab_{a \to 0}(t'_1)$. Similarly for fg_B.
- If t is $\mathbf{a} \parallel t_1$ and $a \notin \tau(t_1)$, then $t' := \mathbf{a}(x) \oplus t'_1$ for some x that is not a vertex of $cval(t'_1)$; otherwise $a \in \tau(t')$ and $t' := t'_1$.
- If t is $\mathbf{a}^\ell \parallel t_1$ and $a \notin \tau(t_1)$, then $t' := \mathbf{a}^\ell(x) \oplus t'_1$ for some x that is not a vertex of $cval(t'_1)$; otherwise $a \in \tau(t')$ and, by induction (and because t is simple), $cval(t'_1)$ has a unique a-port y, which is described by a unique occurrence in t'_1 of a constant symbol $\mathbf{b}(y)$ for some $b \in [k]$ (because of renamings, we may have $b \neq a$); then we obtain t' by replacing in t'_1 this symbol by $\mathbf{b}^\ell(y)$.
- If t is $\overrightarrow{\mathbf{ab}} \parallel t_1$, then we define t' as $\overrightarrow{add}_{a,b}(s')$ where $s := \mathbf{a} \parallel \mathbf{b} \parallel t_1$. (Note that by the fourth case, $t' = \overrightarrow{add}_{a,b}(t'_1)$ if $a, b \in \tau(t_1)$.)

The case of undirected graphs is similar. Edge labels do not require the use of new port labels (which the reader could expect from Example 2.91), because edges are added

---

52 Theorem 5.5 of this article states that $cwd(G) \leq 2^{2twd(G)+1} + 1$; however, the proof establishes the upper-bound $2^{2twd(G)+2} + 1$. See also Section 2.7 about this proof.

one by one. Hence, if in the last case t is $\overrightarrow{\mathbf{ab}}_\lambda \mathbin{/\!/} t_1$, then t' is defined as $\overrightarrow{add}_{a,b,\lambda}(t_1')$. (Clearly, the power of the operation $\overrightarrow{add}_{a,b}$ to create simultaneously several edges is not used in this construction.) Vertex labels and labeled loops are handled by obvious modifications of the one-before-last case. In fact, if t is $\mathbf{a}_\kappa \mathbin{/\!/} t_1$ and $a \notin \tau(t_1)$, then $t' := \mathbf{a}_{\{\kappa\}}(x) \oplus t_1'$; otherwise $a \in \tau(t_1)$ and t' is obtained from t_1' by replacing $\mathbf{b}_M(y)$ by $\mathbf{b}_{M\cup\{\kappa\}}(y)$, where y is the a-port of $cval(t_1')$. And similarly with $\mathbf{a}_\lambda^\ell$ instead of $\mathbf{a}_\kappa$. Hence, the same inequality holds for labeled graphs.

(2) is proved in [CorRot]. Hence, the exponential function in the comparison of tree-width and clique-width cannot be replaced by a linear or even a polynomial function.

(3) The case of cliques follows from Example 2.56(3) and Proposition 2.106. For tournaments, we use a counting argument: for each n there exist $2^{n(n-1)/2}$ (concrete) tournaments with vertex set $[n]$, but for each finite set C there are only $2^{O(n \cdot \log(n))}$ graphs with this vertex set that are the values of terms in $T(F_C^{cVR})$. This fact follows from Corollary 2.122 below (in Section 2.5.6). Hence, for each finite set C, some tournaments cannot be defined by terms in $T(F_C^{cVR})$. Another proof using monadic second-order transductions will be given in Section 7.2, Proposition 7.39(1). ∎

The proof of Inequality (1.2) given in [CorRot] extends easily to undirected edge-labeled graphs. If ℓ is the number of edge labels and $k = twd(G)$, we get

$$cwd(G) \leq (2^\ell - 1) \cdot 2^{\ell(k-1)} + 2^{\ell k} + 1 - (2^\ell - 1)^k,$$

which gives, more simply:

$$cwd(G) \leq 2^{\ell k + 1} - (2^\ell - 1)^k.$$

Directed graphs can be handled as undirected graphs with two edge labels. We obtain thus, with $\ell = 2$ in the first formula, and for $k \geq 2$:

$$cwd(G) \leq 7 \cdot 2^{2k-2} + 1 - 3^k.$$

These upper-bounds are different from (and better than) those arising from the proof of Section 4.3.4 because they are based on the characterization of graphs of tree-width at most k using the operations of H_k defined in Remark 2.84(4).

There is a striking difference between clique-width and tree-width regarding their linear versions: just compare (1.2) and (1.3). These results also show that every set of simple graphs of bounded tree-width has bounded clique-width, and not conversely. However, for particular types of graphs like planar graphs or graphs of degree at most d for any fixed d, clique-width and tree-width are *equivalent* in the sense that the same sets of simple planar graphs and of simple graphs of bounded degree have bounded tree-width and bounded clique-width. This is made precise in the following proposition.

If $p \geq 2$, we say that a graph G is *without* $K_{p,p}$ if $core(G)$ has no subgraph isomorphic to $K_{p,p}$. If G is labeled and λ is an edge label, we let G_λ be the subgraph of G consisting of its edges labeled by λ (and all its vertices). The following result is essentially from [GurWan].

Proposition 2.115 Let $p \geq 2$.

(1) For every simple graph G without $K_{p,p}$, we have

$$twd(G) \leq 3(p-1) \cdot cwd(G) - 1.$$

If G has degree at most p, then $twd(G) \leq 3p \cdot cwd(G) - 1$. If G is planar, then $twd(G) \leq 6 \cdot cwd(G) - 1$.

(2) If G is (K, Λ)-labeled, simple and G_λ is without $K_{p,p}$ for every $\lambda \in \Lambda$, then

$$twd(G) \leq 3(p-1) \cdot |\Lambda| \cdot cwd(G) - 1.$$

(3) The same results hold for path-width and linear clique-width, with 3 replaced by 6. $\qquad\qquad\square$

Before proving Proposition 2.115, we need some preliminary definitions and results (that will also be useful in Section 7.4). We first consider unlabeled graphs. We let G without $K_{p,p}$ be equal to $cval(t)$ with $t \in T(F_C^{\mathrm{VR}})$, so that $V_G = Occ_0(t)$. (We will use the notation and results of Section 2.5.2.) For $u \in N_t$, we let $G(u)$ denote the p-graph $cval(t)/u$ with vertex set $\{x \in V_G \mid x \leq_t u\}$. Up to port labels, $G(u)$ is a subgraph of G, i.e., $G(u)^\circ \subseteq G$. A vertex x of $G(u)$ belongs to the set of a-ports of $G(u)$, denoted by $Pt(u,a)$, if and only if $port_t(u,x) = a$ (cf. Definition 2.94). We say that a is *small at* u if $|Pt(u,a)| \leq p-1$. Otherwise it is *large at* u. We let $Ext(u,a)$ be the set of vertices in $V_G - V_{G(u)}$ that are adjacent (in G) to an a-port of $G(u)$. We let T be the syntactic tree of t and we will define a mapping $f : N_t = N_T \to \mathcal{P}(V_G)$ and prove that (T, f) is a tree-decomposition of G. We let:

$$f(u) := \{u\} \text{ if } u \in Occ_0(t); \text{ otherwise:}$$
$$f(u) := \bigcup\{Pt(u',a) \mid a \in C, u' \text{ is a son of } u \text{ and } |Pt(u',a)| \leq p-1\}$$
$$\cup\bigcup\{Ext(u,a) \mid a \in C, |Ext(u,a)| \leq p-1\}.$$

It is clear that $V_G \subseteq \bigcup\{f(u) \mid u \in N_T\}$. With these definitions and notation we have:

Lemma 2.116

(1) Every vertex of $Pt(u,a)$ is adjacent to every vertex of $Ext(u,a)$. Hence, one cannot have $|Pt(u,a)| \geq p$ and $|Ext(u,a)| \geq p$.

(2) Let $x \in V_G$, let Y be the set of vertices of G that are adjacent to x, and let u be the father of the leaf x of T. If $|Y| \leq p-1$, then $\{x\} \cup Y \subseteq f(u)$.

Proof: (1) All vertices of $Pt(u,a)$ are subject to the same relabelings and edge additions defined by operations at occurrences strictly above u in t. Hence every vertex of $Pt(u,a)$ is adjacent to every vertex of $Ext(u,a)$. (This follows easily from Lemmas 2.95 and 2.96.) We would get a subgraph of $core(G)$ isomorphic to $K_{p,p}$ if $Pt(u,a)$ and $Ext(u,a)$ would have both at least p elements.

(2) Since $Pt(x,a) = \{x\}$ where a is the port label of x in $G(x)$, so that $|Pt(x,a)| = 1 \leq p-1$, we have $Pt(x,a) \subseteq f(u)$ hence $x \in f(u)$. Let l be the port label of x in $G(u)$. Then $Ext(u,l) = Y \cap (V_G - V_{G(u)})$. Since $|Ext(u,l)| \leq |Y| \leq p-1$, we have $Ext(u,l) \subseteq f(u)$, and so $Y \cap (V_G - V_{G(u)}) \subseteq f(u)$. It remains to show that $Y \cap V_{G(u)} \subseteq f(u)$. Consider $y \in Y \cap V_{G(u)}$. Let u' be the son of u such that $y \leq_T u'$, and let h be the port label of y in $G(u')$. Then $y \in Pt(u',h)$ and $x \in Ext(u',h)$. Since every vertex of $Pt(u',h)$ is adjacent to every vertex of $Ext(u',h)$ by (1), we have $Pt(u',h) \subseteq Y$. Hence $|Pt(u',h)| \leq p-1$, and so $Pt(u',h) \subseteq f(u)$, and $y \in f(u)$. ∎

Proof of Proposition 2.115:
(1) The assertions about graphs of bounded degree and planar graphs follow from the main statement because if G has degree at most p then it is without $K_{p+1,p+1}$, and planar graphs are without $K_{3,3}$.

We now prove that (T, f) defined before the previous lemma is a tree-decomposition of G. This will give the first assertion since $|f(u)| \leq 3|C| \cdot (p-1)$ for every u. Some claims will be necessary.

Claim 2.115.1 Let $x \in Occ_0(t)$ and let $(u_n, u_{n-1}, \ldots, u_1)$ be the path in T from the root u_n to $u_1 = x$. Let $l_1, \ldots, l_n$ be the successive port labels of x in $G(u_1), G(u_2), \ldots, G(u_n) = G$. We have

$$\{x\} = Pt(u_1, l_1) \subseteq Pt(u_2, l_2) \subseteq \cdots \subseteq Pt(u_n, l_n), \tag{2.5}$$

and

$$Ext(u_1, l_1) \supseteq Ext(u_2, l_2) \supseteq \cdots \supseteq Ext(u_n, l_n) = \emptyset. \tag{2.6}$$

If m is the largest index such that l_m is small at u_m, we have $|Ext(u_j, l_j)| \leq p-1$ for every j, $m < j \leq n$.

Proof: The inclusions of (2.5) are clear from Lemma 2.95. We may have $Pt(u_i, l_i) \subset Pt(u_{i+1}, l_{i+1})$ but only if u_{i+1} is an occurrence of a disjoint union or of a relabeling. The inclusions $Ext(u_{i+1}, l_{i+1}) \subseteq Ext(u_i, l_i)$ of (2.6) follow from the fact that, by (2.5) and Lemma 2.116(1), a vertex y of G belongs to $Ext(u_i, l_i)$ if and only if $y \not\leq_T u_i$ and y is adjacent to x in G, and similarly for $i+1$. Note that we have an equality if u_{i+1} is an occurrence of an edge addition operation. If u_{i+1} is an occurrence of a relabeling that transforms some label l into l_{i+1} and if $Pt(u_i, l) \neq \emptyset$, then $Ext(u_{i+1}, l_{i+1}) = Ext(u_i, l_i) = Ext(u_i, l)$. By the definition of m, $|Pt(u_j, l_j)| \geq p$ for every j such that $m < j \leq n$. By Lemma 2.116(1), this implies $|Ext(u_j, l_j)| \leq p-1$ for all these indices. □

Let x, y be adjacent vertices of G (they are leaves of T). Let w be their least common ancestor in T, let $(u_q, u_{q-1}, \ldots, u_1)$ be the path in T from w to x where $u_q =$

$w, u_1 = x$ and let $l_1, \ldots, l_q$ be the corresponding port labels of x as in Claim 2.115.1; let $(v_{q'}, v_{q'-1}, \ldots, v_1)$ with $v_{q'} = w$ and $v_1 = y$, and $h_1, \ldots, h_{q'}$ be the analogous sequences relative to y. Note that w is an occurrence of $\oplus$ and that the edge between x and y is created by an edge addition at some occurrence w' between w and the root.

Claim 2.115.2 $Pt(u_{q-1}, l_{q-1}) \subseteq Ext(v_j, h_j)$ for every j, $1 \le j \le q' - 1$.

Proof: We have $x \in Pt(u_{q-1}, l_{q-1})$ and $y \in Ext(u_{q-1}, l_{q-1})$. By Lemma 2.116(1), the operation that creates the edge between x and y also creates edges between all vertices of $Pt(u_{q-1}, l_{q-1})$ and y. Hence $Pt(u_{q-1}, l_{q-1}) \subseteq Ext(v_j, h_j)$. $\qquad\square$

Claim 2.115.3 If x and y are adjacent in G, then $\{x, y\} \subseteq f(u)$ for some $u \in N_T$.

Proof: If x has degree at most $p - 1$, the claim follows from Lemma 2.116(2). For proving the general case, we use the notation of Claim 2.115.2. By this claim (for $j = q' - 1$) and by Lemma 2.116(1), it is not possible that $|Pt(u_{q-1}, l_{q-1})| \ge p$ and $|Pt(v_{q'-1}, h_{q'-1})| \ge p$. If $|Pt(u_{q-1}, l_{q-1})| \le p - 1$, then $Pt(u_{q-1}, l_{q-1}) \subseteq f(u_q) = f(w)$ and so $x \in f(w)$. If we have also $|Pt(v_{q'-1}, h_{q'-1})| \le p - 1$, then similarly, $y \in f(w)$. Next we consider the case where $|Pt(v_{q'-1}, h_{q'-1})| \ge p$. Let m' be the largest integer such that $|Pt(v_{m'}, h_{m'})| \le p - 1$. We have $m' < q' - 1$ and $x \in Ext(v_{m'+1}, h_{m'+1})$. By Claim 2.115.1, $|Ext(v_{m'+1}, h_{m'+1})| \le p - 1$ and so $Ext(v_{m'+1}, h_{m'+1}) \subseteq f(v_{m'+1})$, so that $x \in f(v_{m'+1})$. Since $|Pt(v_{m'}, h_{m'})| \le p - 1$, we have also $Pt(v_{m'}, h_{m'}) \subseteq f(v_{m'+1})$, hence $y \in f(v_{m'+1})$. The last possible case is when l_{q-1} is large at u_{q-1} and $h_{q'-1}$ is small at $v_{q'-1}$. The proof is the same by exchanging the roles of x and y. $\qquad\square$

Claim 2.115.4 The pair (T, f) satisfies the connectivity condition, hence is a tree-decomposition of G.

Proof: Let $x \in V_G$ and let $u_1, \ldots, u_n, l_1, \ldots, l_n$ and m be as in Claim 2.115.1. It follows from the definitions that $x \in f(u_i)$ for every i such that $1 \le i \le \min\{n, m + 1\}$ (we may have $m = n$), and $x \notin f(u_i)$ if $i > m + 1$. Assume now that $x \in f(v)$ for $v \notin \{u_1, \ldots, u_n\}$. This means that x is adjacent to y in $Pt(v, h)$, x belongs to $Ext(v, h)$ and $|Ext(v, h)| \le p - 1$. We let $q, v_1, \ldots, v_{q'}, h_1, \ldots, h_{q'}, m'$ be as in Claims 2.115.2 and 2.115.3. We have $v = v_j$ and $h = h_j$ for some j, $1 \le j \le q' - 1$. By Claim 2.115.2, $Pt(u_{q-1}, l_{q-1}) \subseteq Ext(v_j, h_j) = Ext(v, h)$. Hence l_{q-1} is small at u_{q-1}. We have $|Ext(v_j, h_j)| \le p - 1$ and by Claim 2.115.1, $Ext(v_{j'}, h_{j'}) \subseteq Ext(v_j, h_j)$ if $j' > j$, so that $Ext(v_{j'}, h_{j'}) \subseteq f(v_{j'})$ for every $j' = j, \ldots, q' - 1$. We also have $x \in Ext(v_{j'}, h_{j'})$ by Claim 2.115.2, and so $x \in f(v_{j'})$ for all these indices j'. Since $x \in f(u_q)$ (because l_{q-1} is small at u_{q-1} and u_q is the father of u_{q-1}), we have $x \in f(u)$ for every $u \in \{u_1, \ldots, u_{q-1}, u_q, v_{q'-1}, \ldots, v_j\}$ and this set induces the path between x and v in T. We have proved that every node u in $f^{-1}(x)$ belongs to a path linking x and u, all nodes of which are in $f^{-1}(x)$. Hence the connectivity condition holds. $\qquad\square$

We have defined a tree-decomposition (T, f) with $wd(T, f) \le 3|C| \cdot (p - 1) - 1$ from a term t over F_C^{VR} such that $G = val(t)$ is without $K_{p,p}$. Hence $twd(G) \le 3(p - 1) \cdot cwd(G) - 1$. This completes the proof of Proposition 2.115 for unlabeled graphs.

(2) If G has edge labels (vertex labels do not matter), it is the union of its subgraphs G_λ for all edge labels λ. The above proof yields for the graphs G_λ tree-decompositions (T, f_λ) that have the same underlying tree T. This tree is the syntactic tree of the given term t in $T(F_{C,[K,\Lambda]}^{VR})$ that evaluates to G. We get a tree-decomposition (T, f) of G by taking $f(u)$ equal to the union of the sets $f_\lambda(u)$ for each node u of T.

(3) Let us now assume that t is a slim term in $T(F_C^{VR})$, cf. Definition 2.89. Then the tree-decomposition (T, f) can be turned into a path-decomposition by contracting all edges (u,v) of T such that u is an occurrence of $\oplus$ in t, v is a leaf of t and the other son of u (i.e., the son $\neq v$) is not a leaf of t. If the two sons of u are leaves, one contracts only one of the two corresponding edges. This doubles (at most) the number of vertices in each box, by Lemma 2.64. This completes the proof of Proposition 2.115. ∎

We have seen in Example 2.108 that clique-width is sensitive to the reversal of edge directions while tree-width is not. This observation can be made more precise by the following result of [Cou95a]. Its proof which uses monadic second-order transductions will be sketched in Section 7.2, Proposition 7.40. We recall that $und^{-1}(G)$ is the set of orientations of G, i.e., the set of directed graphs G' such that $G = und(G')$, the graph obtained from G' by forgetting edge directions. Since G is assumed to be simple, the condition $und(G') = G$ implies that G' has no pairs of opposite edges.

Proposition 2.117 There exists a function $f : \mathcal{N} \to \mathcal{N}$ such that, for every simple undirected graph G, we have:

$$twd(G) \leq f(\max\{cwd(G') \mid G' \in und^{-1}(G)\}).$$

□

By this proposition, if a set of directed graphs has bounded clique-width and is closed under the reversal of directions of any set of edges, then it has bounded tree-width.

2.5.6 Variations on F^{VR}

The operations forming the signature F^{VR} have been chosen so as to be as simple and as few as possible. The signature F^{VR} has many variants that are equivalent in the sense that the same sets of graphs have bounded clique-width and bounded width with respect to these alternative signatures. We present here three of these variants. They are formulated for unlabeled graphs but their statements and proofs extend easily to labeled graphs. We will not detail these extensions.

Elementary relabelings

Clique-width is defined in Definition 2.89 with respect to a finite set $C \subseteq \mathcal{A}$ and the operations $\oplus, add_{a,b}, \overrightarrow{add}_{a,b}, relab_h$, where a, b belong to C and $h : C \to C$. The following proposition shows that this definition is equivalent to the original one of [CouOla] that is used in many subsequent works (see Section 2.7 for references). We define

$$F_C^{'\text{VRu}} := \{\oplus, add_{a,b}, relab_{a \to b}, \mathbf{a}, \mathbf{a}^\ell, \varnothing \mid a, b \in C, \ a \neq b\},$$

and

$$F_C^{'\text{VR}} := F_C^{'\text{VRu}} \cup \{\overrightarrow{add}_{a,b} \mid a, b \in C, \ a \neq b\}.$$

Proposition 2.118 The clique-width of a p-graph G is the minimal cardinality of a set C such that $G = val(t)$ for some term t in $T(F_C^{'\text{VR}})$ or in $T(F_C^{'\text{VRu}})$ if G is undirected.

Proof: We wish to replace an operation $relab_h$ for h in $[C \to C]_f$ by a composition of elementary relabelings $relab_{a \to b}$ for some a, b in C. This is not possible if h is a bijection that is not the identity. Consider $C = \{a, b\}$ with $h(a) = b$, $h(b) = a$. We need another port label c to express $relab_h$ as $relab_{c \to b} \circ relab_{b \to a} \circ relab_{a \to c}$. However, this *is* possible if h is not a bijection. For example, if $h(a) = b$, $h(b) = c$, $h(c) = a$ and $h(d) = a$, then $relab_h = relab_{d \to b} \circ relab_{b \to c} \circ relab_{c \to a} \circ relab_{a \to d} \circ relab_{d \to c}$. We omit the proof of this fact.

For completing the proof we observe that the operations $relab_h$ such that $h : C \to C$ is a bijection can be eliminated. We can use Equalities (4)–(8) and (12′) of Proposition 2.101 to eliminate them, as we have used some equalities of Proposition 2.48 to eliminate the operations ren_h in Proposition 2.49. ∎

We have just proved that clique-width is not modified by a certain restriction of the signature F^{VR}. We now examine the effect of extending it.

New edge addition operations

For a directed or undirected p-graph G and $a \in \mathcal{A}$, we let $add_{a,a}(G)$ be the p-graph G' defined by:

$$V_{G'} := V_G,$$
$$edg_{G'} := edg_G \cup \{(x, y) \mid x, y \in V_G, \ x \neq y, \ port_G(x) = port_G(y) = a\},$$
$$port_{G'} := port_G.$$

This new operation adds undirected edges, or equivalently pairs of opposite directed edges, but no loops. We let F_C^{iVR} be the signature F_C^{VR} augmented with the operations $add_{a,a}$ for all $a \in C$, and we let $cwd^i(G)$ be the corresponding width defined by

$$cwd^i(G) := \min\{|C| \mid G = val(t), \ t \in T(F_C^{\text{iVR}})\}.$$

Cliques are the values of the terms $add_{a,a}(\mathbf{a} \oplus \cdots \oplus \mathbf{a})$ written with a single port label. Hence $cwd^i(K_n) = 1$ whereas $cwd(K_n) = 2$ for $n \geq 2$. The graphs $K_n \oplus I_m$ are the only simple loop-free undirected graphs G such that $cwd^i(G) = 1$. (We recall that I_m has m vertices and no edge). We have $cwd^i(P_4) = 2 < cwd(P_4) = 3$ because $P_4 = relab_{2 \to 1}(add_{1,1}(add_{1,2}(\mathbf{1} \oplus \mathbf{2}) \oplus add_{1,2}(\mathbf{1} \oplus \mathbf{2})))$. We have $cwd^i(K_{n,m}) = cwd(K_{n,m}) = 2$ for $n, m \geq 2$.

Proposition 2.119 For every p-graph G, we have

$$cwd^i(G) \leq cwd(G) \leq 2 \cdot cwd^i(G).$$

Proof: Since cwd^i is defined in terms of more operations than cwd, the inequality $cwd^i(G) \leq cwd(G)$ is trivial. We now prove the other direction. Let G be such that $cwd^i(G) = k$. Without loss of generality we assume that the port labels used to define G are $1, \ldots, k$. We let F be the set of operations consisting of $\oplus$, $add_{i,j}$ for $i \in [k]$, $j \in [k] \cup \{k+i\}$, $relab_h$ where $h: [2k] \to [2k]$ and $\mathbf{i}$ and $\mathbf{i}^\ell$ for all $i \in [k]$. For $P \subseteq [k]$, we let ADD_P be the derived operation defined as $\bigcirc_{i \in P} add_{i,i}$, i.e., as the composition in any order of the operations $add_{i,i}$ for all i in P. (Hence $ADD_\emptyset$ is the identity.) We have the following equalities, for all i, j in $[k]$, $h: [k] \to [k]$, and all p-graphs G, H of type included in $[k]$:

(a) $ADD_P(G \oplus H) = relab_{g^{-1}}[(\bigcirc_{i \in P} add_{i,k+i})(ADD_P(G) \oplus relab_g(ADD_P(H)))]$,
   where $g: [k] \to [k+1, 2k]$ is defined by $g(j) := k + j$ for $j \in [k]$;
(b) $ADD_P(add_{i,i}(G)) = ADD_{P \cup \{i\}}(G)$;
(c) $ADD_P(\overrightarrow{add}_{i,j}(G)) = \overrightarrow{add}_{i,j}(ADD_P(G))$ if $i \neq j$ (and the same for $add_{i,j}$);
(d) $ADD_P(relab_h(G)) = relab_h(f(ADD_{P'}(G)))$,
   where $P' = \{i \mid h(i) \in P\}$ and f is the composition of all operations $add_{i,j}$ such that $i, j \in [k]$, $i \neq j$ and $h(i) = h(j) \in P$.

These equalities entail the following claim. Its proof is easy by induction on the structure of t, by using equalities (a)–(d).

Claim 2.119.1 For every term $t \in T(F^{iVR}_{[k]})$ and every $P \subseteq [k]$, there exists a term t' in $T(F)$, hence in $T(F^{VR}_{[2k]})$, such that $val(ADD_P(t)) = val(t')$. $\qquad\square$

By taking $P = \emptyset$, we obtain that $cwd(G) \leq 2 \cdot cwd^i(G)$. $\qquad\blacksquare$

Binary derived operations replacing unary operations

Let C be a finite set of port labels. Let us consider the derived operations defined as compositions of unary operations of F^{VR}_C. There are infinitely many such compositions but they only define finitely many different operations. In fact, we have shown in Proposition 2.103 that every context in $Ctxt(F^{VR}_C)$ that is written without $\oplus$, and hence with unary operations only, is equivalent to a context of the form $relab_h(ADD_R(x_1))$ with $h: C \to C$ and $R \subseteq C \times C$ (and ADD_R is the composition, in any order, of the operations $\overrightarrow{add}_{a,b}$ with $(a, b) \in R$, $a \neq b$). Obviously, we may also assume that $R \cap Id_C = \emptyset$, where $Id_C := \{(a,a) \mid a \in C\}$. There are $k^k \cdot 2^{k^2-k}$ contexts of that form, where $k = |C|$. We now show that they are all inequivalent.

Lemma 2.120 Let $c = relab_h(ADD_R(x_1))$ and $c' = relab_{h'}(ADD_{R'}(x_1))$ be contexts with $h, h' : C \to C$ and $R, R' \subseteq (C \times C) - Id_C$. Then $c_{\mathbb{GP}} = c'_{\mathbb{GP}}$ if and only if $h = h'$ and $R = R'$.

Proof: One direction is trivial. For the other direction, let us assume that $c_{\mathbb{GP}}(G) = relab_h(ADD_R(G)) = relab_{h'}(ADD_{R'}(G)) = c'_{\mathbb{GP}}(G)$ for every p-graph G. If $h(a) = b$ and $h'(a) = b' \neq b$, then $c_{\mathbb{GP}}(\mathbf{a}) = \mathbf{b}$ and $c'_{\mathbb{GP}}(\mathbf{a}) = \mathbf{b}'$ and this contradicts the hypothesis (because $\mathbf{b} \neq \mathbf{b}'$). Hence $h = h'$. If $(a,b) \in R - R'$, then $c_{\mathbb{GP}}(\mathbf{a} \oplus \mathbf{b})$ has one edge whereas $c'_{\mathbb{GP}}(\mathbf{a} \oplus \mathbf{b})$ has none. This is impossible, hence $R \subseteq R'$, and by symmetry we have an equality. ∎

For every $R \subseteq C \times C$ and every $h : C \to C$, we let $\otimes_{R,h}$ be the binary derived operation defined by $G \otimes_{R,h} H := relab_h(ADD_R(G \oplus H))$. We let $F_C^{\kappa\mathrm{VR}}$ be the "compact" signature consisting of the same constants as F_C^{VR} (including $\emptyset$) and of the binary operations $\otimes_{R,h}$ for all $R \subseteq C \times C$ and $h : C \to C$.

Proposition 2.121 Let $C \in \mathcal{P}_f(\mathcal{A})$. The same p-graphs are defined by $T(F_C^{\mathrm{VR}})$, by $T(F_C^{\kappa\mathrm{VR}})$ and by $T(F_C^{\kappa\mathrm{VR}} - \{\emptyset\}) \cup \{\emptyset\}$. The corresponding transformations of terms can be done in linear time.

Proof: Obviously, every term over $F_C^{\kappa\mathrm{VR}}$ can be turned into an equivalent term over F_C^{VR} by replacing every subterm $t_1 \otimes_{R,h} t_2$ by $relab_h(ADD_R(t_1 \oplus t_2))$. In the other direction, subterms $relab_h(t_1)$ can be replaced by $t_1 \otimes_{\emptyset,h} \emptyset$, subterms $\overrightarrow{add}_{a,b}(t_1)$ by $t_1 \otimes_{\{(a,b)\},Id} \emptyset$, where Id is the identity on C, and subterms $t_1 \oplus t_2$ by $t_1 \otimes_{\emptyset,Id} t_2$.

One can also transform a term t over $F_C^{\mathrm{VR}} - \{\emptyset\}$ (see Corollary 2.102) into an equivalent term t' over $F_C^{\kappa\mathrm{VR}} - \{\emptyset\}$. This can be shown by induction on the size of t. Let $t = c[t_1]$, where $c \in Ctxt(F_C^{\mathrm{VR}})$ is written without $\oplus$ and t_1 is either a constant symbol or of the form $t_2 \oplus t_3$. By Proposition 2.103, t is equivalent to a term $relab_h(ADD_R(t_1))$ with $h : C \to C$ and $R \subseteq C \times C$. If t_1 is a or a^ℓ and $h(a) = b$, then we define t' as $\mathbf{b}$ or $\mathbf{b}^\ell$. If $t_1 = t_2 \oplus t_3$, then, using the induction hypothesis, we define $t' := t'_2 \otimes_{R,h} t'_3$.

All these transformations can be done in linear time for fixed C. ∎

A term t over $F_C^{\kappa\mathrm{VR}} - \{\emptyset\}$ has length $2|V_{val(t)}| - 1$ because $F_C^{\kappa\mathrm{VR}}$ consists only of binary operations and constant symbols, and $V_{val(t)}$ is in bijection with the leaves of t. Such a term is in general smaller than those in $T(F_C^{\mathrm{VR}})$ that denote the same p-graph. However a data structure representing the operations of $F_C^{\kappa\mathrm{VR}}$ must, essentially, represent a p-graph with $|C|$ vertices for each occurrence of a binary function symbol $\otimes_{R,h}$, viz. the p-graph H with $H^\circ := \langle C, R \rangle$ and $port_H := h$. It is not clear that this would be really more efficient than storing a term from $T(F_C^{\mathrm{VR}})$ from which redundancies like $\overrightarrow{add}_{a,b}(\overrightarrow{add}_{a,b}(\cdots))$ are eliminated. This aspect is important for algorithmic applications to be considered in Chapter 6. The signature $F_C^{\kappa\mathrm{VR}}$ will be useful anyway for certain proofs in Chapter 7 (in particular for proving the Equationality Theorem, see the proof of Theorem 7.34). The following corollary is analogous to Corollary 2.74(2).

Corollary 2.122 For all $k, n > 0$, the number of concrete graphs of clique-width at most k with vertex set $[n]$ is less than $2^{n \cdot \log(n) + 2k^2 \cdot n}$.

Proof: For $k = 1$ the statement follows easily from Remark 2.107(1). Now let $k \geq 2$. We let $\sharp$ and $*$ be symbols of respective arities 0 and 2. The number C_n of terms in $T(\{*, \sharp\})$ having $n + 1$ occurrences of $\sharp$ is the n-th Catalan number. Its value is less than 2^{2n} (see, e.g., [*FlaSed]). The number $N_k(n)$ of terms in $T(F_{[k]}^{c\kappa \mathrm{VR}} - \{\varnothing\})$ that define concrete directed p-graphs (possibly with loops) with vertex set $[n]$ can be evaluated as $n! \cdot C_{n-1} \cdot c^n \cdot f^{n-1}$, where f is the number of inequivalent binary operations in $F_{[k]}^{\kappa \mathrm{VR}}$ and c is the number of constants in this set different from $\varnothing$. We have thus $c = 2k$, $f = k^k \cdot 2^{k^2 - k}$ and $C_{n-1} < 2^{2n-2}$. Since $n! < 2^{n \cdot \log(n)}$, we get $\log(N_k(n)) < n \cdot \log(n) + 2n - 2 + n \cdot \log(2k) + (n-1) \cdot (k \cdot \log(k) + k^2 - k) < n \cdot \log(n) + 2k^2 \cdot n$, which gives the result.[53] ∎

This corollary entails that the set of all graphs, and even that of undirected graphs has unbounded clique-width. Because otherwise, the number of concrete directed (or undirected) graphs with vertex set $[n]$ would be $2^{O(n \cdot \log(n))}$ whereas it is $2^{\Omega(n^2)}$. However, Proposition 2.106 is more precise for proving the unboundedness of clique-width over all graphs because it gives the exact clique-width of rectangular undirected grids.

We now use Proposition 2.121 to prove Statement (4) of Proposition 2.105.

Proof of Proposition 2.105(4): We will prove that the edge-complement $\widetilde{G}$ of a (directed) p-graph G has clique-width at most $2 \cdot cwd(G)$. The proof will extend easily to labeled p-graphs. Furthermore, we will give a linear-time algorithm that constructs a term $\widetilde{t}$ evaluating to $\widetilde{G}$ from a term t in $T(F_C^{\kappa \mathrm{VR}})$ that evaluates to G. For the purpose of readability we will denote $\widetilde{G}$ also by $ec(G)$.

Let be given $t \in T(F_C^{\kappa \mathrm{VR}})$. For every position u in t, we define $Edg(t \uparrow u) \subseteq C \times C$ by the following top-down induction:

- if u is the root, then $Edg(t \uparrow u) := \varnothing$;
- if u is a son of an occurrence w of $\otimes_{R,h}$,
  then $Edg(t \uparrow u) := R \cup h^{-1}(Edg(t \uparrow w))$,
  where $h^{-1}(S) := \{(a, b) \mid (h(a), h(b)) \in S\}$ for every $S \subseteq C \times C$.

The operations above u in t add to the p-graph $cval(t)/u$ the edges from x to y such that x is an a-port and y is a b-port of $cval(t)/u$, and $(a, b) \in Edg(t \uparrow u)$ with $a \neq b$.[54]

---

[53] The last inequality uses the fact that $2 + \log(2k) + k \cdot \log(k) + k^2 - k \leq 2k^2$ for every $k \geq 2$.

[54] Similar sets, satisfying the same property, can be defined for $t \in T(F_C^{\mathrm{VR}})$ by the following clauses:

if u is the root, then $Edg(t \uparrow u) := \varnothing$,

if u is a son of an occurrence w of $\oplus$, then $Edg(t \uparrow u) := Edg(t \uparrow w)$,

(For the notation $cval(t)/u$ see Section 2.5.2.) The sets $Edg(t \uparrow u)$ can be computed in linear time along a depth-first traversal of t starting from the root.

The construction of $\widetilde{t}$ will be based on the following obvious fact:

Claim 2.105.1 Let C' be disjoint with C, let G_1 and G_2 be p-graphs with $\pi(G_1) \subseteq C$ and $\pi(G_2) \subseteq C'$, and let $R \subseteq (C \times C') \cup (C' \times C)$. Then we have $ec(G_1 \otimes_{R,h} G_2) = ec(G_1) \otimes_{R^c,h} ec(G_2)$, where $R^c = ((C \times C') \cup (C' \times C)) - R$. □

In order to use it, we will change some port labels in the given term in such a way that, in every subterm $t_1 \otimes_{R,h} t_2$ of the modified term, we have $\pi(t_1) \subseteq C$ and $\pi(t_2) \subseteq C'$ where $C' := \{a' \mid a \in C\}$ is a disjoint copy of C. We define also:

$$p \text{ and } p^- : C \cup C' \to C \cup C' \text{ such that, for all } a \in C :$$

$$p(a) := p(a') := a' \text{ and } p^-(a') := p^-(a) := a.$$

For $R \subseteq C \times C$, we define:

$$R^p := \{(a,b'),(a',b) \mid (a,b) \in R, a \neq b\},$$
$$R^{pc} := ((C \times C') \cup (C' \times C)) - R^p$$
$$= \{(a,b'),(a',b) \mid a,b \in C, (a,b) \notin R, a \neq b\}$$
$$\cup \{(a,a'),(a',a) \mid a \in C\}.$$

We are ready to define $\widetilde{t} \in T(F_{C \cup C'}^{\kappa VR})$ from $t \in T(F_C^{\kappa VR})$. For each $u \in Pos(t)$, we replace the symbol f that occurs at u by a symbol f' of the same arity according to the following clauses:

if $f = \varnothing$,   then $f' := \varnothing$,

if $f = \mathbf{a}$,   then $f' := $ if u is a right son then $\mathbf{a'}^\ell$ else $\mathbf{a}^\ell$,

if $f = \mathbf{a}^\ell$,   then $f' := $ if u is a right son then $\mathbf{a'}$ else $\mathbf{a}$,

if $f = \otimes_{R,h}$,   then $f' := $ if u is a right son then $\otimes_{S,pohop^-}$ else $\otimes_{S,hop^-}$, where $S := (R \cup h^{-1}(Edg(t \uparrow u)))^{pc}$.

Note that $Pos(\widetilde{t}) = Pos(t)$.

The first statement of the next claim yields the result if we take $u = root_t$ because then $Edg(t \uparrow u) = \varnothing$.

---

if u is the son of an occurrence w of $\overrightarrow{add}_{a,b}$, then $Edg(t \uparrow u) := Edg(t \uparrow w) \cup \{(a,b)\}$,

if u is the son of an occurrence w of $add_{a,b}$, then $Edg(t \uparrow u) := Edg(t \uparrow w) \cup \{(a,b),(b,a)\}$,

if u is the son of an occurrence w of $relab_h$, then $Edg(t \uparrow u) := h^{-1}(Edg(t \uparrow w))$.

Claim 2.105.2

(1) For every $u \in Pos(t)$ that is the root or a left son:

$\pi(\tilde{t}/u) \subseteq C$ and $cval(\tilde{t})/u = ec(ADD_{Edg(t\uparrow u)}(cval(t)/u))$, and

(2) for every $u \in Pos(t)$ that is a right son:

$\pi(\tilde{t}/u) \subseteq C'$ and $cval(\tilde{t})/u = relab_p(ec(ADD_{Edg(t\uparrow u)}(cval(t)/u)))$.

Proof: For fixed t, by bottom-up induction on u. The cases where u is a leaf are clear from the definitions. Let u be an occurrence of $\otimes_{R,h}$ that is the root or a left son, and let u_1 and u_2 be the left and right son of u. Then, by the definition of $\tilde{t}$:

$$cval(\tilde{t})/u = relab_{hop^-}(ADD_S(cval(\tilde{t})/u_1 \oplus cval(\tilde{t})/u_2)),$$

where $S := (R \cup h^{-1}(Edg(t \uparrow u)))^{pc}$. On the other hand,

$ADD_{Edg(t\uparrow u)}(cval(t)/u)$

$= ADD_{Edg(t\uparrow u)}(relab_h(ADD_R(cval(t)/u_1 \oplus cval(t)/u_2)))$

$= relab_h(ADD_{h^{-1}(Edg(t\uparrow u))}(ADD_R(cval(t)/u_1 \oplus cval(t)/u_2)))$

$= relab_h(relab_{p^-}(ADD_{(h^{-1}(Edg(t\uparrow u))\cup R)^p}(G_1 \oplus G_2)))$

$= relab_{hop^-}(ADD_{(h^{-1}(Edg(t\uparrow u))\cup R)^p}(G_1 \oplus G_2)),$

where

$$G_1 := ADD_{h^{-1}(Edg(t\uparrow u)\cup R}(cval(t)/u_1)$$
$$= ADD_{Edg(t\uparrow u_1)}(cval(t)/u_1), \text{ and}$$
$$G_2 := relab_p(ADD_{h^{-1}(Edg(t\uparrow u)\cup R}(cval(t)/u_2))$$
$$= relab_p(ADD_{Edg(t\uparrow u_2)}(cval(t)/u_2)),$$

because $R \cup h^{-1}(Edg(t \uparrow u)) = Edg(t \uparrow u_1) = Edg(t \uparrow u_2)$. By the induction hypothesis, and the remark that $ec(relab_g(H)) = relab_g(ec(H))$ for all g and H, we have: $ec(G_1) = cval(\tilde{t})/u_1$ and $ec(G_2) = cval(\tilde{t})/u_2$. By Claim 2.105.1, and the same remark, we have:

$ec(ADD_{Edg(t\uparrow u)}(cval(t)/u))$

$= relab_{hop^-}(ADD_S(ec(G_1) \oplus ec(G_2)))$

$= relab_{hop^-}(ADD_S(cval(\tilde{t})/u_1 \oplus cval(\tilde{t})/u_2)),$

which is equal to $cval(\tilde{t})/u$, as was to be proved.

The proof is fully similar if u is an occurrence of $\otimes_{R,h}$ that is a right son. $\square$

Hence, for $t \in T(F_C^{\kappa VR})$ we have constructed a term $\widetilde{t}$ in $T(F_{C \cup C'}^{\kappa VR})$ that evaluates to the edge-complement of $cval(t)$. It follows that for every p-graph G, $cwd(\widetilde{G}) \leq 2 \cdot cwd(G)$. We have given a construction of $\widetilde{t}$ that can be done in linear time for each fixed set C.

For labeled graphs, we use one set $Edg(t \uparrow u)_\lambda$ for each edge label λ, and we use operations $\otimes_{R,h}$, where R is replaced by a family of binary relations on C, with one relation for each λ. We need no other auxiliary port labels than those in C'. Vertex labels are not modified by the construction. Hence, we still have $cwd(\widetilde{G}) \leq 2 \cdot cwd(G)$. ∎

2.6 Many-sorted graph algebras

In order to minimize the underlying formal framework, we have defined F^{HR} and F^{VR} as signatures with a single sort. We now define many-sorted variants of these signatures, and we will turn $\mathbb{JS}$ and $\mathbb{GP}$ into many-sorted algebras.

The basic idea is to formalize at an abstract level the notion of type of an s- or p-graph that has emerged naturally from the initial definitions: in the case of $\mathbb{JS}$, the type of an s-graph G is $\tau(G)$, the set of names of its sources and, in the case of $\mathbb{GP}$, the type of a p-graph G is $\pi(G)$, the set of labels of its ports. In each case, the class of all graphs with sources or ports is partitioned according to types, and the operations of the signatures are specialized accordingly. In the terminology of Section 1.2, the properties "G has type C" for $C \in \mathcal{P}_f(\mathcal{A})$ form inductive sets relative to each of the signatures F^{HR} and F^{VR}. The type of an s- or p-graph will be its sort in the many-sorted formal framework. This technicality will be useful for dealing with derived operations, for defining equation systems and for defining the notion of recognizability.

2.6.1 Many-sorted algebras

Algebras with a single sort have been defined in Section 2.1 and are subsumed by the following definitions.

Definition 2.123 (Many-sorted algebras) Let S be a set of *sorts*. A *functional S-signature* is a set F of *function symbols* where each $f \in F$ has an associated *input type* $\alpha(f)$ in $Seq(S)$ (the set of finite, possibly empty sequences of elements of S) and an *output type* $\sigma(f)$ in S. The length of $\alpha(f)$ is the *arity* of f and is denoted by $\rho(f)$. A *constant symbol* is a symbol of arity 0. We let $F_i := \{f \in F \mid \rho(f) = i\}$, $F_+ := F - F_0$ and $\rho(F) := \max\{\rho(f) \mid f \in F\}$. We say that F is *finite* if S and F are finite.

An *F-algebra* is an object $\mathbb{M} = \langle (M_s)_{s \in S}, (f_{\mathbb{M}})_{f \in F} \rangle$, where $(M_s)_{s \in S}$ is the family of *domains* of $\mathbb{M}$ and $(f_{\mathbb{M}})_{f \in F}$ is its family of *operations*. The (possibly empty) sets M_s are pairwise disjoint and for each f, $f_{\mathbb{M}}$ is a total function $M_{\alpha(f)} \to M_{\sigma(f)}$, where $M_{(s_1,\ldots,s_n)}$ denotes $M_{s_1} \times \cdots \times M_{s_n}$. We also say that f has *type* $s_1 \times \cdots \times s_n \to s$ if

$\alpha(f) = (s_1, \ldots, s_n)$ and $\sigma(f) = s$. If $\alpha(f) = ()$ (the empty sequence), then f is a constant symbol and $f_\mathbb{M}$ is an element of $M_{\sigma(f)}$ called a *constant* of $\mathbb{M}$.

If $\mathbb{M} = \langle (M_s)_{s \in \mathcal{S}}, (f_\mathbb{M})_{f \in F} \rangle$ is an F-algebra, we let M denote $\bigcup \{M_s \mid s \in \mathcal{S}\}$. The *sort* $\sigma(m)$ of an element m of M is the unique sort s such that $m \in M_s$. A subset of M is *homogenous* if its elements all have the same sort. Each operation $f_\mathbb{M}$ is a partial function $: M^{\rho(f)} \to M$. Dealing with partial functions is always a source of difficulties that is avoided by using sorts, whenever possible. But not every partial function $g : A^n \to A$ can be expressed as $f_\mathbb{A}$ for some many-sorted algebra $\mathbb{A}$. As a counter-example, consider $A = \{0, 1\}$ and $g : A^2 \to A$ such that $g(0, 0) = 0$, $g(1, 1) = 1$ and g is undefined for the other pairs of arguments.

Let F be an $\mathcal{S}$-signature (we will frequently omit "functional"). A *subsignature* H of F is an $\mathcal{S}'$-signature such that $\mathcal{S}' \subseteq \mathcal{S}$ and every symbol of H is a symbol of F with the same input and output types as with respect to F. We write this $H \sqsubseteq F$. If $\mathbb{M}$ is an F-algebra and $\mathbb{N}$ is an H-algebra, we say that $\mathbb{N}$ is a *subalgebra* of $\mathbb{M}$, which we denote by $\mathbb{N} \subseteq \mathbb{M}$, if H is a subsignature of F, $N_s \subseteq M_s$ for every sort s of H and $f_\mathbb{N} = f_\mathbb{M} \restriction N_{\alpha(f)}$ for every $f \in H$.

For an F-algebra $\mathbb{M}$ and a subsignature F' of F with set of sorts $\mathcal{S}'$, we define the subalgebra $\mathbb{M} \restriction F' := \langle (M_s)_{s \in \mathcal{S}'}, (f_\mathbb{M})_{f \in F'} \rangle$ of $\mathbb{M}$. For a set of sorts $\mathcal{S}'' \subseteq \mathcal{S}$, we define the subsignature $F \restriction \mathcal{S}'' := \{f \in F \mid \alpha(f) \in Seq(\mathcal{S}''), \sigma(f) \in \mathcal{S}''\}$ and the subalgebra $\mathbb{M} \restriction \mathcal{S}'' := \mathbb{M} \restriction (F \restriction \mathcal{S}'')$.

A *homomorphism* $h : \mathbb{M} \to \mathbb{N}$, where $\mathbb{M}$ and $\mathbb{N}$ are F-algebras, is a mapping $h : M \to N$ such that h maps M_s to N_s for each s and $h(f_\mathbb{M}(m_1, \ldots, m_{\rho(f)})) = f_\mathbb{N}(h(m_1), \ldots, h(m_{\rho(f)}))$ for every $(m_1, \ldots, m_{\rho(f)})$ in $M_{\alpha(f)}$.[55] If h is a bijection, it is an *isomorphism* and $\mathbb{M}, \mathbb{N}$ are *isomorphic* algebras.

The *Cartesian product* $\mathbb{M} \times \mathbb{N}$ of F-algebras $\mathbb{M}$ and $\mathbb{N}$ is defined as the F-algebra $\langle (M_s \times N_s)_{s \in \mathcal{S}}, (f_{\mathbb{M} \times \mathbb{N}})_{f \in F} \rangle$ with $f_{\mathbb{M} \times \mathbb{N}}((m_1, n_1), \ldots, (m_{\rho(f)}, n_{\rho(f)})) := (f_\mathbb{M}(m_1, \ldots, m_{\rho(f)}), f_\mathbb{N}(n_1, \ldots, n_{\rho(f)}))$ for every $(m_1, \ldots, m_{\rho(f)})$ in $M_{\alpha(f)}$ and $(n_1, \ldots, n_{\rho(f)})$ in $N_{\alpha(f)}$.

Definition 2.124 (Terms over many-sorted signatures) Let F be an $\mathcal{S}$-signature. For each s in $\mathcal{S}$, let L_s denote a subset of F^*. The following system of $|\mathcal{S}|$ equations:

$$\begin{cases} \vdots \\ L_s = \bigcup f \, L_{\alpha(f)[1]} L_{\alpha(f)[2]} \cdots L_{\alpha(f)[\rho(f)]}, \\ \vdots \end{cases}$$

where, in each equation, the union extends to all f in F such that $\sigma(f) = s$, has a unique solution because the unicity result of Proposition 3.15 is valid, even if the system has infinitely many equations. The component L_s of the solution is the set of

---

[55] The first requirement on h means that h is *sort-preserving*, i.e., that $\sigma(h(m)) = \sigma(m)$ for every $m \in M$.

terms over F of sort s, denoted by $T(F)_s$. If F is finite, then each language $T(F)_s$ is context-free. We let $T(F) := \bigcup \{T(F)_s \mid s \in \mathcal{S}\}$.

Even if F is infinite, these languages enjoy the same unambiguity of parsing as in the one-sorted case. The definition of the value of a term over F in an F-algebra $\mathbb{M}$ is the same as in the one-sorted case: the mapping $val_{\mathbb{M}}$ maps $T(F)_s$ to M_s for each s in $\mathcal{S}$, hence $val_{\mathbb{M}}$ is a sort-preserving mapping from $T(F)$ to M. We also denote $val_{\mathbb{M}}(t)$ by $t_{\mathbb{M}}$. Let $F' \subseteq F$. A subset of M is *generated by* F' if it is a subset of $val_{\mathbb{M}}(T(F'))$. The subalgebra of $\mathbb{M}$ generated by F' is the subalgebra $\mathbb{M}'$ with signature F' such that $M'_s = val_{\mathbb{M}}(T(F')_s)$ for every sort s of F'. (It is the smallest subalgebra of $\mathbb{M}$ with signature F'; it is empty if F' has no constant symbols.) In particular, we say that $\mathbb{M}$ is generated by F' if $M_s = val_{\mathbb{M}}(T(F')_s)$ for every $s \in \mathcal{S}$. We say that it is *finitely generated* if it is generated by a finite subsignature of F; this implies that $\mathcal{S}$ is finite.

We let $\mathbb{T}(F)$ denote the F-algebra $\langle (T(F)_s)_{s \in \mathcal{S}}, (f_{\mathbb{T}(F)})_{f \in F} \rangle$ of terms, where $f_{\mathbb{T}(F)}(t_1, \ldots, t_{\rho(f)}) := f t_1 \cdots t_{\rho(f)}$ for every f in F and $(t_1, \ldots, t_{\rho(f)})$ in $T(F)_{\alpha(f)}$. The evaluation mapping $val_{\mathbb{M}}$ is the unique homomorphism from $\mathbb{T}(F)$ to $\mathbb{M}$, and for this reason the algebra $\mathbb{T}(F)$ is called the *initial F-algebra*.

Definition 2.125 (Derived operations) A set X is $\mathcal{S}$-*sorted* if it is equipped with a mapping $\sigma : X \to \mathcal{S}$; for $x \in X$, we say that $\sigma(x)$ is the *sort* of x. If X is an $\mathcal{S}$-sorted set of variables and F is an $\mathcal{S}$-signature, we denote by $T(F, X)_s$ the set of terms $T(F \cup X)_s$, and $T(F, X) := T(F \cup X)$. Note that if t belongs to $T(F, X) - X$, then the sorts of the variables having occurrences in t are determined in a unique way. Hence, the mapping $\sigma : X \to \mathcal{S}$ need not be specified if $|X| \geq 2$ and all variables of X have occurrences in t.

Let the standard set of variables $X_n = \{x_1, \ldots, x_n\}$ be $\mathcal{S}$-sorted (but the following definitions extend in an obvious way to a set X of n variables, linearly ordered in some specified or implicit way). In every F-algebra $\mathbb{M}$, a term t in $T(F, X_n)_s$ defines a total function: $M_{\sigma(x_1)} \times \cdots \times M_{\sigma(x_n)} \to M_s$ denoted by $t_{\mathbb{M}}$ and called a *derived operation* of $\mathbb{M}$. Its definition is the same as for one-sorted algebras, and it will also be denoted by $\lambda m_1, \ldots, m_n \cdot t_{\mathbb{M}}(m_1, \ldots, m_n)$, or by $\lambda x_1, \ldots, x_n \cdot t$ if $\mathbb{M}$ is clear from the context. It is *linear* if t is linear (see Definition 2.5) and it is *strict* if every $x_i \in X_n$ has an occurrence in t. An *extended derived operation* is an n-ary function g defined from an $n + k$-ary derived operation $t_{\mathbb{M}}$ and k fixed elements $a_1, \ldots, a_k$ of M of appropriate sorts, by $g(m_1, \ldots, m_n) := t_{\mathbb{M}}(m_1, \ldots, m_n, a_1, \ldots, a_k)$. If $a_1, \ldots, a_k$ are the values of terms $t_1, \ldots, t_k \in T(F)$, then g is a derived operation defined by the term $t[t_i/x_{n+i}; i \in [k]]$ (cf. Definition 2.6 for the substitution of terms).

It will be useful to extend the signature of an algebra $\mathbb{M}$ by adding derived operations designated by new symbols. Here is a more precise definition. Let F be an $\mathcal{S}$-signature. A *derived operation declaration* is a 4-tuple $\langle g, w, s, t_g \rangle$ consisting of:

- a symbol g possibly in F, which is the *name* of the declared operation;
- a sequence w in $Seq(\mathcal{S})$, its *input type*, denoted by $\alpha(g)$;

- an element s of $\mathcal{S}$, its *output type*, denoted by $\sigma(g)$;
- and a term t_g in $T(F,X_n)_s$, called its *defining term*, written with the symbols of F and the standard set of variables $X_n = \{x_1,\ldots,x_n\}$ which is $\mathcal{S}$-sorted such that $(\sigma(x_1),\ldots,\sigma(x_n)) = \alpha(g)$.

If $g \in F$, we impose that $\alpha(g)$ and $\sigma(g)$ are as in F and that $t_g = g(x_1,\ldots,x_n)$. An operation $\langle g,\alpha(g),\sigma(g),t_g \rangle$ is *linear* if the term t_g is linear, i.e., if each variable x_i has at most one occurrence in t_g. This operation is *strict* if x_i has at least one occurrence in t_g for each $i = 1,\ldots,n$, where $n = |\alpha(g)|$.

A *derived signature* of F is a set H of derived operation declarations such that no two of them have the same name. (A variable x_i may have different sorts in different declarations.) It is an $\mathcal{S}$-signature. We say that H is *linear* and / or *strict* if all its operations are.

Let $\mathbb{M}$ be an F-algebra. For given H, and in order to simplify notation, we identify a derived operation declaration $\langle g,\alpha(g),\sigma(g),t_g \rangle$ with its name g, and we denote by $g_{\mathbb{M}}$ the derived operation $(t_g)_{\mathbb{M}} : M_{\alpha(g)} \to M_{\sigma(g)}$. From a derived signature H of F and an F-algebra $\mathbb{M}$, we obtain the H-algebra $\mathbb{M}_H = \langle (M_s)_{s \in \mathcal{S}}, (g_{\mathbb{M}})_{g \in H} \rangle$. Such an algebra is called a *derived algebra* of $\mathbb{M}$.

The next proposition shows that the derived operations of $\mathbb{M}_H$ are derived operations of $\mathbb{M}$. For every term t in $T(H,Y_m)_s$ (where Y_m is the $\mathcal{S}$-sorted standard set of variables $\{y_1,\ldots,y_m\}$), we let $\theta_H(t)$ in $T(F,Y_m)_s$ be the term obtained from t by substituting for each function symbol g in t its defining term t_g. Formally:

$$\theta_H(y_i) = y_i,$$

$$\theta_H(g(t_1,\ldots,t_n)) = t_g[\theta_H(t_1)/x_1,\ldots,\theta_H(t_n)/x_n],$$

where $i \in [m]$, $t_g \in T(F,X_n)_{\sigma(g)}$ and $t_j \in T(F,Y_m)_{\sigma(x_j)}$ for $j \in [n]$. We recall that $\cdots[\cdots/x_1,\ldots,\cdots/x_n]$ denotes the substitution of terms for (first-order) variables; its semantic meaning is explained in Definition 2.6. The mapping θ_H is a *second-order substitution*.[56] If H is linear, then the mapping θ_H can be computed in linear time. (By a bottom-up computation on t, one can construct the syntactic tree of $\theta_H(t)$ and then, by depth-first traversal of this tree, one can obtain $\theta_H(t)$ itself. Both steps can be done in linear time. For further processing, the syntactic tree of $\theta_H(t)$ is actually more useful than (the word) $\theta_H(t)$.)

Proposition 2.126 Let H be a derived signature of a signature F and let $\mathbb{M}$ be an F-algebra. For every term t in $T(H,Y_m)$, we have $t_{\mathbb{M}_H} = (\theta_H(t))_{\mathbb{M}}$.

Proof: Straightforward induction on the structure of t. ∎

---

[56] In Formal Language Theory, terms are usually called "trees," and the mapping θ_H is called a *tree homomorphism*, cf. [*Com+] Section 1.4. It reduces to a word homomorphism if $F = F_1$ and $H = H_1$ (cf. Definition 2.7). If $t_g = f(x_1,\ldots,x_{\rho(g)})$ for every g, then the mapping θ_H replaces function symbols by function symbols of the same arity. Such a mapping is called an *alphabetic relabeling*.

Definition 2.127 (Effectively given many-sorted algebras) We extend Definition 2.8 to many-sorted algebras. To be encoded and (semi-)effectively given, sets of sorts, signatures and algebras must be finite or countable.

An S-sorted signature F is *effectively given* if the sets S and F are effectively given, and the input and output type mappings α and σ are computable (and algorithms computing these mappings must be specified, cf. Definition 2.8). It implies that the arity mapping is computable.

An F-algebra $\mathbb{M}$ is *semi-effectively given* if the signature F is effectively given, the set M is semi-effectively given, the sort mapping $\sigma : M \to S$ is computable and the operations of $\mathbb{M}$ are computable in a uniform way by the mapping $\zeta_{\mathbb{M}}$ as in Definition 2.8 (and algorithms computing these mappings must be specified). It is *effectively given* if, additionally, M is effectively given (i.e., $|M|$ is specified) and the mapping $s \mapsto |M_s|$ from S to $\mathcal{N} \cup \{\omega\}$, which defines the cardinality of M_s for each s in S, is computable (and an algorithm computing it must be specified).

As in the one-sorted case, if F is effectively given, then the algebra $\mathbb{T}(F)$ is semi-effectively given and the mapping $val_{\mathbb{M}}$ is computable. If F generates $\mathbb{M}$, then a computable mapping can determine for each element of M a term in $T(F)$ that evaluates to it.

If $\mathbb{M}$ is (semi-)effectively given, then the domains M_s are decidable subsets of M because the sort mapping σ of $\mathbb{M}$ is computable, and hence they are (semi-)effectively given by their standard encodings (see Definition 2.8). From a specification of $\mathbb{M}$, an algorithm can be constructed that, on input $s \in S$, computes a specification of M_s.

It is easy to check that the Cartesian product of two (semi-)effectively given algebras $\mathbb{M}$ and $\mathbb{N}$ is (semi-)effectively given, using the standard encoding of $M \times N$. A specification of the product algebra can be constructed from those of $\mathbb{M}$ and $\mathbb{N}$.

A derived signature H of an S-signature F can be viewed as a pair $H = \langle H^\circ, \delta_H \rangle$, where H° is an ordinary S-signature and δ_H is a mapping : $H^\circ \to T(F, X \times S)$ such that $X \times S$ is the S-sorted set of variables with $X := \{x_1, x_2, \ldots\}$ and $\sigma(\langle x_i, s \rangle) = s$ for every $\langle x_i, s \rangle \in X \times S$, and such that for every $g \in H^\circ$ of type $s_1 \times \cdots \times s_n \to s$ we have $\delta_H(g) \in T(F, \{\langle x_1, s_1 \rangle, \ldots, \langle x_n, s_n \rangle\})_s$. If $\langle g, w, s, t_g \rangle$ is a derived operation declaration in H, then $g \in H^\circ$ with $\alpha(g) = w$ and $\sigma(g) = s$, and $\delta_H(g) = t_g[\langle x_1, w[1] \rangle / x_1, \ldots, \langle x_n, w[n] \rangle / x_n]$ with $n := |w|$; note that $(\delta_H(g))_{\mathbb{M}} = (t_g)_{\mathbb{M}}$ for every F-algebra $\mathbb{M}$. We say that the derived signature H is *effectively given* if F and H° are effectively given and δ_H is computable (and an algorithm computing it must be specified).

2.6.2 The many-sorted HR algebra

Although the operations of the algebra $\mathbb{JS}$ are total, we turn this algebra into a many-sorted one. We write definitions and statements for unlabeled graphs only. The extension to labeled graphs is straightforward since the only modifications for this extension concern the constant symbols and the associated basic graphs.

Definition 2.128 (Turning F^{HR} into a many-sorted signature) As in Section 2.3, we let $\mathcal{A}$ be the countable set of source names. We let $\mathcal{S} := \mathcal{P}_f(\mathcal{A})$ be the countable set of sorts. For each $C \in \mathcal{S}$, we let $\mathcal{JS}_C$ be the set[57] $\{G \in \mathcal{JS} \mid \tau(G) = C\}$. This set will be the domain of sort C of the many-sorted HR algebra. We now define the operations of that algebra:

(a) For $C, D \in \mathcal{S}$, we let $/\!/_{C,D}$ be the binary operation of type $C \times D \to C \cup D$ that is the restriction of $/\!/$ to $\mathcal{JS}_C \times \mathcal{JS}_D$.

(b) For $C \in \mathcal{S}$ and $B \subseteq C$, we let $fg_{B,C}$ of type $C \to C - B$ be the restriction of fg_B to $\mathcal{JS}_C$. (Note that for arbitrary B we have $fg_B \upharpoonright \mathcal{JS}_C = fg_{B \cap C, C}$.)

(c) If $C, D \in \mathcal{S}$ and h is a bijection : $C \to D$, then we let $ren_{h,C}$ of type $C \to D$ be the restriction to $\mathcal{JS}_C$ of $ren_{h'}$, where h' is the permutation of $C \cup D$ that extends h and is defined in Lemma 2.37.

(d) If a constant symbol in F^{HR} defines an s-graph of type C, then we let its sort be C.

Hence, we have defined an $\mathcal{S}$-signature that we will denote by F^{tHR}. The associated F^{tHR}-algebra $\mathbb{JS}^t$ has domains $\mathcal{JS}_C$ for each $C \in \mathcal{S}$ and its operations have been defined above as restrictions of operations of $\mathbb{JS}$. Clearly, both F^{tHR} and $\mathbb{JS}^t$ are effectively given. If E is a finite set of source labels, we let $\mathbb{JS}^t[E]$ be the subalgebra of $\mathbb{JS}^t$ with the finite set of sorts $\mathcal{P}(E)$, with domains $\mathcal{JS}_C$ for $C \subseteq E$, and the operations of F^{tHR} restricted to $C, D \subseteq E$ (cf. (a)–(d) above). To be precise, these operations are $/\!/_{C,D}, fg_{B,C}$ for $B \subseteq C$, $ren_{h,C}$ for h such that $h(C) \subseteq E$ and the constant symbols defining s-graphs of type included in E. This finite signature is denoted by F_E^{tHR}. It generates the subalgebra $\mathbb{JS}^{t,gen}[E]$ of $\mathbb{JS}^t[E]$ (cf. Remark 2.39(1)).

We will compare the derived operations of $\mathbb{JS}$ and of $\mathbb{JS}^t$, and, in particular, the terms of $T(F^{HR})$ and $T(F^{tHR})$ that all evaluate to s-graphs in $\mathcal{JS}$.

Definition 2.129 (Untyping) We recall that X_n is the standard set of variables $\{x_1, \ldots, x_n\}$, now assumed to be $\mathcal{P}_f(\mathcal{A})$-sorted. *Untyping* is the mapping $Unt :$ $T(F^{tHR}, X_n) \to T(F^{HR}, X_n)$ that omits subscripts relative to sorts. Formally:

$$Unt(t) := t \quad \text{if } t \text{ is a constant symbol or a variable (in } X_n),$$

$$Unt(t_1 /\!/_{C,D} t_2) := Unt(t_1) /\!/ Unt(t_2),$$

$$Unt(fg_{B,C}(t)) := fg_B(Unt(t)),$$

$$Unt(ren_{h,C}(t)) := ren_{h'}(Unt(t)),$$

where in the last case, h is a bijection : $C \to h(C)$ and h' is the permutation of $C \cup h(C)$ that extends h and is defined in Lemma 2.37.

---

[57] Do not confuse $\mathcal{JS}_C$ with $\mathcal{JS}[C] = \{G \in \mathcal{JS} \mid \tau(G) \subseteq C\}$, cf. Remark 2.39(1). Note that $\mathcal{JS}_\emptyset = \mathcal{JS}[\emptyset]$ is the set of all graphs.

Proposition 2.130 For every term t in $T(F^{tHR}, X_n)$ the mapping $Unt(t)_{\mathbb{JS}} : \mathcal{JS}^n \to \mathcal{JS}$ extends the mapping $t_{\mathbb{JS}^t} : \mathcal{JS}_{\sigma(x_1)} \times \cdots \times \mathcal{JS}_{\sigma(x_n)} \to \mathcal{JS}_{\sigma(t)}$. If $n = 0$, then $Unt(t)_{\mathbb{JS}}$ and $t_{\mathbb{JS}^t}$ are the same s-graph in $\mathcal{JS}_{\sigma(t)}$.

Proof: By induction on the structure of t; straightforward from the definitions. ∎

We now consider an inverse *typing* mapping.

Definition 2.131 (Typing) Let the variables $x_1, \ldots, x_n$ of X_n have respective sorts $\beta(x_1), \ldots, \beta(x_n)$ in $\mathcal{P}_f(\mathcal{A})$. We define two mappings on $T(F^{HR}, X_n)$, denoted by σ and Typ, whose respective values are a sort and a term in $T(F^{tHR}, X_n)$. They are defined as follows:

$$\sigma(x_i) := \beta(x_i),$$
$$\sigma(c) := \tau(c_{\mathbb{JS}}) \quad \text{if } c \text{ is a constant symbol,}$$
$$\sigma(t_1 /\!/ t_2) := \sigma(t_1) \cup \sigma(t_2),$$
$$\sigma(fg_B(t)) := \sigma(t) - B,$$
$$\sigma(ren_h(t)) := h(\sigma(t)),$$

$$Typ(t) := t \quad \text{if } t \text{ is a constant symbol or a variable,}$$
$$Typ(t_1 /\!/ t_2) := Typ(t_1) /\!/_{\sigma(t_1), \sigma(t_2)} Typ(t_2),$$
$$Typ(fg_B(t)) := fg_{B \cap \sigma(t), \sigma(t)}(Typ(t)),$$
$$Typ(ren_h(t)) := ren_{h', \sigma(t)}(Typ(t)),$$

where in the last case, $h' := h \restriction \sigma(t)$ (hence h' is a bijection : $\sigma(t) \to h(\sigma(t))$).

The term $Typ(t)$ may contain operations of the form $fg_{\varnothing,C}$ or $ren_{h,C}$ such that h is the identity on C. Such operations define identity mappings on $\mathcal{JS}_C$ and hence can be deleted from the term $Typ(t)$. Note that σ and Typ depend on the mapping β that assigns sorts to variables; we write them σ^β and Typ^β if we need to make this dependency explicit. Note also that $\sigma(t) = \tau(t)$ for $t \in T(F^{HR})$ (see Definition 2.35).

Proposition 2.132 Let β assign sorts to the variables of X_n. For every term t in $T(F^{HR}, X_n)$ and all s-graphs $G_1, \ldots, G_n$ of respective types $\beta(x_1), \ldots, \beta(x_n)$ we have:

(1) $\tau(t_{\mathbb{JS}}(G_1, \ldots, G_n)) = \sigma^\beta(t)$;
(2) $Typ^\beta(t) \in T(F^{tHR}, X_n)_{\sigma^\beta(t)}$;
(3) $Typ^\beta(t)_{\mathbb{JS}^t}(G_1, \ldots, G_n) = t_{\mathbb{JS}}(G_1, \ldots, G_n)$.

If $n = 0$, then the mappings σ^β and Typ^β do not depend on β, and we have:

(4) $\tau(t_{\mathbb{JS}}) = \sigma(t)$;
(5) $Typ(t) \in T(F^{tHR})_{\sigma(t)}$;
(6) $Typ(t)_{\mathbb{JS}^t} = t_{\mathbb{JS}}$.

Proof: As for the previous proof, by a straightforward induction on the structure of t. ∎

Note that $Unt(Typ(t)) = \bar{t}$ for every term t in $T(F^{HR})$, where $\bar{t}$ is the reduced term of Proposition 2.38. Note also that the mappings Unt and Typ can be computed in linear time by straightforward implementations of their inductive definitions.

We now consider the restriction of these definitions and results to s-graphs of type included in some finite subset E of $\mathcal{A}$. It is clear (by a quick inspection) that, for every finite set E of source names, all definitions of this section can be adapted (let us say *restricted*) to s-graphs of type included in E, to the finite signatures F_E^{HR} and F_E^{tHR}, and to the subalgebras $\mathbb{JS}[E]$ and $\mathbb{JS}^t[E]$ of $\mathbb{JS}$ and $\mathbb{JS}^t$ respectively. The associated typing and untyping mappings are easy to define and Propositions 2.130 and 2.132 extend to them. It follows that $val(T(F_C^{tHR})) = val(T(F_C^{HR}))$, hence, that the width of an s-graph is the same whether it is defined with respect to F^{tHR} or to F^{HR} (cf. Section 2.4.5).

Example 2.133 (Series-composition) We recall from Section 1.1.3 that, for directed s-graphs, series-composition maps $\mathcal{JS}_{[2]} \times \mathcal{JS}_{[2]}$ to $\mathcal{JS}_{[2]}$.[58] As observed in Section 1.4.2, it can be defined by the term $t = fg_3(ren_{2\leftrightarrow3}(x_1) /\!/ ren_{1\leftrightarrow3}(x_2))$ over F^{HR}. The corresponding binary derived operation $t_{\mathbb{JS}}$ applies to any two s-graphs G,H given as arguments. However, it corresponds to the graph-theoretical series-composition only if G and H both have type $\{1,2\}$. The term $t' = fg_4(ren_{2\leftrightarrow4}(x_1) /\!/ ren_{1\leftrightarrow4}(x_2))$ defines the same binary operation on $\mathcal{JS}_{[2]}$. But $t_{\mathbb{JS}} \neq t'_{\mathbb{JS}}$, as one checks by considering s-graphs of type $\{1,2,3\}$. With the above definitions, we can see that series-composition is $Typ^\beta(t)_{\mathbb{JS}^t}$ as well as $Typ^\beta(t')_{\mathbb{JS}^t}$, where β assigns sort $\{1,2\}$ to x_1 and to x_2. It is clear that $\sigma^\beta(t) = \sigma^\beta(t') = \{1,2\}$. The term $Typ^\beta(t)$ looks as follows:

$$fg_{3,\{1,2,3\}}(ren_{2\leftrightarrow3,\{1,2\}}(x_1) /\!/_{\{1,3\},\{2,3\}} ren_{1\leftrightarrow3,\{1,2\}}(x_2)).$$

Let us abbreviate $\{1,2\}$ by s. If H is the set of derived operation declarations consisting of $\langle \bullet, (s,s), s, Typ^\beta(t) \rangle$, $\langle /\!/, (s,s), s, x_1 /\!/_{s,s} x_2 \rangle$ and $\langle e, (), s, \overrightarrow{\mathbf{12}} \rangle$, then H is a derived signature of (the directed version of) F^{tHR}, and the (one-sorted) subalgebra $\mathbb{JS}_H^t \restriction \{s\}$ of the derived H-algebra $\mathbb{JS}_H^t$ is equal to the algebra $\mathbb{J}_2^d$ of Section 1.1.3. The domain of the subalgebra of $\mathbb{J}_2^d$ generated by its signature $H \restriction \{s\}$ is the set of series-parallel graphs. □

The following example anticipates Chapters 3 and 4, where we will study in detail the equation systems first introduced in Section 1.1.4.

Example 2.134 We consider the equation written with the series-composition operation $\bullet$ that is defined by the term t of Example 2.133 (*not* the one defined by $Typ^\beta(t)$):

$$X = (X /\!/ X) \cup (X \bullet X) \cup fg_1(X) \cup fg_2(X) \cup \{e\}, \tag{2.7}$$

---

[58] The set $\mathcal{JS}_{[2]} = \{G \in \mathcal{JS} \mid \tau(G) = \{1,2\}\}$ is equal to the set $\mathcal{J}_2^d$ of Section 1.1.3.

where e is the constant symbol $\overrightarrow{12}$ with value $1 \bullet \rightarrow \bullet 2$. The variable X ranges over sets of s-graphs. It is not hard to see that all graphs in the set L defined as the least solution in $\mathcal{P}(\mathcal{J}\mathcal{S})$ of Equation (2.7), have types $\emptyset$, $\{1\}$, $\{2\}$ or $\{1,2\}$. Since they have no other type like $\{1,2,3\}$, series-composition can be defined in this equation by t' as well as by t (these terms are defined in Example 2.133). If we impose a "type discipline," we cannot use this equation. Equation systems relative to many-sorted signatures are defined so that all terms in an equation have the same sort, and this is not the case for Equation (2.7). However, by using the method of Section 1.2.5 and the observation that types have an "inductive behavior," we can transform Equation (2.7) into a system of equations over F^{tHR}. In what follows we write 12 for $\{1,2\}$, 1 for $\{1\}$, 2 for $\{2\}$, and 0 for $\emptyset$. For $s_1, s_2 \in \{0,1,2,12\}$ we define $\bullet_{s_1,s_2} := Typ^{\beta}(t)$ with $\beta(x_i) := s_i$ for $i = 1,2$; it is the restriction of $\bullet$ to $\mathcal{J}\mathcal{S}_{s_1} \times \mathcal{J}\mathcal{S}_{s_2}$. For $s \in \{0,1,2,12\}$ we let $L_s := \{G \in L \mid \tau(G) = s\}$. These sets L_s satisfy the following equations:

$$L_{12} = L_{12} /\!/_{12,12} L_{12} \cup L_1 /\!/_{1,12} L_{12} \cup L_2 /\!/_{2,12} L_{12}$$
$$\cup L_0 /\!/_{0,12} L_{12} \cup L_1 /\!/_{1,2} L_2$$
$$\cup L_{12} \bullet_{12,12} L_{12} \cup L_{12} \bullet_{12,2} L_2 \cup L_1 \bullet_{1,12} L_{12} \cup L_1 \bullet_{1,2} L_2 \cup \{e\},$$

$$L_1 = L_1 /\!/_{1,1} L_1 \cup L_1 /\!/_{1,0} L_0$$
$$\cup L_{12} \bullet_{12,1} L_1 \cup L_{12} \bullet_{12,0} L_0 \cup L_1 \bullet_{1,1} L_1 \cup L_1 \bullet_{1,0} L_0 \cup fg_2(L_{12}).$$

In writing these equations we use the fact that $/\!/$ is commutative and we let set union have lower priority than the other binary operations. Similar equations can be written for L_2 and L_0. We obtain thus a "typed" system of four equations written over F^{tHR}, the least solution of which in $\mathcal{P}(\mathcal{J}\mathcal{S})$ is the 4-tuple (L_{12}, L_1, L_2, L_0). Then $L = L_{12} \cup L_1 \cup L_2 \cup L_0$. Equation (2.7) is clearly much shorter and more readable than this system. In general, for every untyped equation system, the types of the s-graphs it defines are not immediate from the syntax but can be computed by an algorithm. We will discuss this aspect in Chapter 4, Proposition 4.10. □

2.6.3 The many-sorted VR algebra

The definitions and results of the previous section can be adapted to the signature F^{VR}, to the algebra $\mathbb{GP}$ and to their restrictions F_E^{VR} and $\mathbb{GP}[E]$ for a finite set E of port labels in a straightforward way.

The relevant notion of type of a p-graph G is $\pi(G) = port_G(V_G)$ and will be the sort of G. Hence the set of sorts is $\mathcal{S} := \mathcal{P}_f(\mathcal{A})$, as in the case of $\mathbb{JS}^{\text{t}}$. For each $C \in \mathcal{S}$, we let $\mathcal{GP}_C$ be the set $\{G \in \mathcal{GP} \mid \pi(G) = C\}$. It is the domain of sort C of the many-sorted VR algebra. The operations of that algebra are obtained by restricting the operations of $\mathbb{GP}$ to the sets $\mathcal{GP}_C$, as in the case of $\mathbb{JS}$, cf. Definition 2.128.

Disjoint union $\oplus$ yields binary operations $\oplus_{C,D} : C \times D \to C \cup D$. The unary operations $add_{a,b}$ and $\overrightarrow{add}_{a,b}$ yield $add_{a,b,C}$ and $\overrightarrow{add}_{a,b,C}$ respectively, for all sets (sorts) C containing a and b. These operations have type $C \to C$. Since $\overrightarrow{add}_{a,b}(G) = G$ if $\{a,b\}$ is not included in $\pi(G)$, we need not consider $\overrightarrow{add}_{a,b,C}$ or $add_{a,b,C}$ for that case. The unary operation $relab_g$ for $g \in [\mathcal{A} \to \mathcal{A}]_f$ yields unary operations $relab_{h,C}$ with $h = g \upharpoonright C$, of type $C \to h(C)$; thus, there is an operation $relab_{h,C}$ for every mapping $h : C \to \mathcal{A}$. Finally the constants $\mathbf{a}$, $\mathbf{a}^\ell$ (used to define abstract p-graphs) and $\mathbf{a}(x)$, $\mathbf{a}^\ell(x)$ (for defining concrete p-graphs) are of sort $\{a\}$, and the constant $\varnothing$ has sort $\emptyset$.

We obtain in this way $\mathcal{S}$-signatures F^{tVR} and F^{tcVR} and a many-sorted F^{tVR}-algebra denoted by $\mathbb{GP}^t$. We also obtain subalgebras $\mathbb{GP}^t[E]$ and $\mathbb{GP}^{t,gen}[E]$ of $\mathbb{GP}^t$ by restriction to p-graphs with port labels in a set E, and in the latter case to p-graphs that are defined by terms over F_E^{tVR}. Definitions 2.129 and 2.131 and Propositions 2.130 and 2.132 have analogs that are easy to spell out. As in the case of the HR algebra, we have that $val(T(F_C^{tVR})) = val(T(F_C^{VR}))$, hence, that the clique-width of a p-graph is the same whether it is defined with respect to F^{tVR} or to F^{VR}. The extensions to labeled p-graphs are also straightforward.

2.7 References

The literature on tree-width, clique-width and related complexity measures expands every year with the addition of scores of articles and communications to conferences. Concerning tree-width, the articles by Bodlaender *et al.* [*Bod93, *Bod98, Bod96] and [BodGHK] cover the main results. The situation concerning clique-width is much less "stable". The article [*KamLM] surveys recent results about clique-width.

Whereas a graph class has bounded tree-width if and only if it excludes a fixed planar graph as a minor (Proposition 2.61), there is no such clear-cut result characterizing bounded clique-width. Undirected graphs excluding P_4 as an induced subgraph have clique-width at most 2. Many efforts have been devoted to finding generalizations of this result. The undirected graphs, every induced subgraph of which with q vertices has at most $q - 3$ induced paths isomorphic to P_4, have clique-width at most q if $q \geq 7$ and unbounded clique-width if $q \in \{4, 5, 6\}$ ([MakRot]).

A complete classification of graph classes defined by excluding one or more 4-vertex graphs as induced subgraphs is established in [BraELL]: in each case, either an upper-bound to the clique-width is given or a proof that the clique-width is unbounded. A similar classification is presented in [BraDLM] for graph classes defined by excluding one or more 1-vertex extensions of P_4. For having a class of bounded clique-width, one must exclude at least two such graphs, for instance $\mathbf{1} \oplus P_4$ and $\mathbf{1} \otimes P_4$, yielding clique-width at most 16. Excluding P_5 and $\mathbf{1} \otimes P_4$

yields clique-width at most 5 but excluding $1 \oplus P_4$ and P_5 yields no bound on the clique-width.

Unit interval graphs have unbounded clique-width [GolRot] but every hereditary proper class of unit interval graphs (i.e., closed under taking induced subgraphs and not containing all unit interval graphs) has bounded clique-width ([Loz]).

It is not yet completely clear what properties imply that a class of graphs has bounded clique-width. Lower-bounds to clique-width are difficult to establish: some tools are given in [CorRot, GolRot, MakRot]. By indirect arguments using monadic second-order transductions (Chapter 7), one can sometimes prove quickly that a class of graphs has unbounded clique-width without obtaining a specific lower-bound for the clique-width of particular graphs of this class.

The first article introducing the notion of a term evaluating to a graph and the algebraic view of context-free graph grammars is [BauCou], where such terms are called *graph expressions*. The grammars in [BauCou] are based on rewriting by hyperedge replacement; this notion will be presented in Section 4.1.5 (see also Section 4.6). The signature F^{HR}, first defined in [Cou93], simplifies the one defined in [BauCou] and yields the exact characterization of tree-width (Theorem 2.83). The one of [BauCou] yields no such characterization.

The notion of clique-width is implicit in [CouER] where graph operations are defined for giving an algebraic formulation of certain context-free graph grammars that are based on rewriting by hyperhandle replacement, where a hyperhandle is a hyperedge together with its incident vertices. It is shown in [CouER] that these grammars have the same power as context-free vertex replacement graph grammars (technically called C-edNCE graph grammars,[59] see [*EngRoz] for a detailed exposition). Clique-width is defined and studied in [CouOla]. The closely related notion of *NLC-width* (denoted by *nlcwd*) has been introduced in [Wan]. The graph operations defined in this article model the rewriting mechanism of NLC graph grammars.[60] NLC-width and clique-width are related by $cwd(G) \leq nlcwd(G) \leq 2 \cdot cwd(G)$, and these inequalities are proved by effective linear time transformations. It follows that the same sets of graphs have bounded clique-width and NLC-width.

A variant of clique-width can be defined by allowing vertices to have several port labels from a finite set C. This variant is used in [CouOla] as an auxiliary notion and, in a more important way, in [CouTwi] where a result corresponding to Proposition 2.103 is proved. The proof of Proposition 5.4 of [CouOla] is incorrect. It shows that, roughly speaking, if the "width" of a graph is k with respect to the operations of F^{VR} generalized so as to handle multiple labels, then its clique-width is at most 2^k. The idea is to encode a set of at most k labels attached to a vertex by a single label. An operation $add_{a,b}$ is then replaced by the composition of the

---

[59] The C stands for "context-free" (or "confluent") and the NCE stands for "neighborhood controlled embedding".

[60] The NLC graph grammars, where NLC stands for "node label controlled," are particular NCE graph grammars, see [*EngRoz].

operations $add_{A,B}$ such that $a \in A$ and $b \in B$, where A and B are (or encode) subsets of C. This is incorrect because one should have $A \neq B$ in $add_{A,B}$. Proposition 2.119 "fixes the bug" and yields the bound of 2^{k+1} instead of 2^k. This proposition is used for proving Theorem 5.5 of [CouOla] (cf. Proposition 2.114, Footnote 52) but, in the particular terms used in this proof, the operations $add_{A,A}$ do not occur, hence this proof is correct.

The notion of modular decomposition (also called *substitution decomposition*) is surveyed in [*MöhRad] and presented with another terminology in [*EhrHR]. A corresponding algebra of graphs is studied in [Cou96a], [CouWei] and [LodWei].

Many-sorted algebras are presented in detail in the book [*Wec]. Second-order substitutions are studied in the survey article [*Cou83].

3

Equational and recognizable sets in many-sorted algebras

This chapter deals with many-sorted algebras in general. We define and study their equational and recognizable sets. These algebras may have infinitely (but countably) many sorts and operations. The use of infinite signatures, even in the case of a single sort, is motivated by the intended applications to the graph algebras $\mathbb{JS}$ and $\mathbb{GP}$. In the framework of these two algebras, *equational* and *recognizable sets of graphs* will be studied in Chapter 4. We will pay special attention to effectivity questions: in view of the algorithmic applications, the equational and recognizable sets must be specified in finitary ways.

In Section 3.1 we define the equational sets of a many-sorted algebra; the definition is based on the well-known Least Fixed-Point Theorem; the context-free languages and the regular sets of terms form the motivating examples. An easy but fundamental result is that the image of an equational set under a homomorphism of algebras is equational. In Section 3.2 we present some tools to transform and simplify equation systems, and define derivation trees. We conclude with observations on the usefulness of equation systems and compare them with closely related specification devices. Section 3.3 is a review of definitions and basic results about automata on terms (usually called tree automata, and assumed to be known from, e.g., [*Com+] or [*GecSte]). Section 3.4 introduces the recognizable sets of a many-sorted algebra; several equivalent definitions are given in terms of homomorphisms, congruences and inductive sets of predicates. (The latter definition has a clear intuitive meaning for the verification of properties of the elements of equational sets.) The regular languages and the regular sets of terms form two motivating examples. Several types of effective specifications of recognizable sets are defined and compared. The main closure result states that the intersection of an equational and a recognizable set is equational, with an effective construction. It generalizes the classical result that the intersection of a context-free and a regular language is context-free, and is essential for applications to graphs. We conclude the section with a discussion of recognizability.

3.1 The equational sets of an algebra

Algebras and many-sorted algebras are defined in Sections 2.1 and 2.6.1.

3.1.1 Powerset algebras

In order to specify by equations certain subsets of algebras, we extend their operations to their powersets.

Definition 3.1 (Powerset algebras) Let F be an S-signature. We enlarge it into $F_{\cup}$ by adding, for every sort s in S, a new symbol $\cup_s$ of type $s \times s \to s$ and a new constant Ω_s of sort s. With an F-algebra $\mathbb{M}$, we associate its *powerset algebra*:

$$\mathcal{P}(\mathbb{M}) := \langle (\mathcal{P}(M_s))_{s \in S}, (f_{\mathcal{P}(\mathbb{M})})_{f \in F_{\cup}} \rangle,$$

where

$$\Omega_{s\mathcal{P}(\mathbb{M})} := \emptyset,$$
$$A_1 \cup_{s\mathcal{P}(\mathbb{M})} A_2 := A_1 \cup A_2 \text{ for } A_1, A_2 \subseteq M_s,$$
$$f_{\mathcal{P}(\mathbb{M})}(A_1, \ldots, A_k) := \{ f_{\mathbb{M}}(a_1, \ldots, a_k) \mid a_1 \in A_1, \ldots, a_k \in A_k \}$$
$$\text{for } A_i \subseteq M_{s_i} \text{ where } \alpha(f) = (s_1, \ldots, s_k).$$

Hence $\mathcal{P}(\mathbb{M})$ is an $F_{\cup}$-algebra (let us not worry that $\emptyset$ has several sorts[1]). An $F_{\cup}$-algebra of the form $\mathcal{P}(\mathbb{M})$ as above will be called a *powerset F-algebra*. Its operations are *monotone*, which means that if $f \in F_{\cup}$ has type $s_1 \times \cdots \times s_k \to s$ and $A_i \subseteq B_i \subseteq M_{s_i}$ for each i, then $f_{\mathcal{P}(\mathbb{M})}(A_1, \ldots, A_k) \subseteq f_{\mathcal{P}(\mathbb{M})}(B_1, \ldots, B_k)$. They are even *additive*: if $A_i = \bigcup_{j \in J} B_j$, then $f_{\mathcal{P}(\mathbb{M})}(A_1, \ldots, A_i, \ldots, A_k) = \bigcup_{j \in J} f_{\mathcal{P}(\mathbb{M})}(A_1, \ldots, A_{i-1}, B_j, A_{i+1}, \ldots, A_k)$.

We recall from Definition 2.5 that X_n, Y_k denote standard sets of variables, respectively $\{x_1, \ldots, x_n\}$ and $\{y_1, \ldots, y_k\}$. For every term $t \in T(F, X_n)_s$, where X_n is S-sorted and $s \in S$, we obtain a derived operation of $\mathbb{M}$, denoted by $t_{\mathbb{M}}$, and also a derived operation of $\mathcal{P}(\mathbb{M})$, denoted by $t_{\mathcal{P}(\mathbb{M})}$, that maps $\mathcal{P}(M_{\sigma(x_1)}) \times \cdots \times \mathcal{P}(M_{\sigma(x_n)})$ into $\mathcal{P}(M_s)$. The operations $t_{\mathcal{P}(\mathbb{M})}$ are monotone. This follows from the monotonicity of the operations of $\mathcal{P}(\mathbb{M})$ and an induction on the structure of t.

In the following lemma which relates these derived operations, we write $t = \widetilde{t}[x_{i_1}/y_1, \ldots, x_{i_k}/y_k]$ for some term $\widetilde{t}$ in $T(F, Y_k)$ such that $ListVar(\widetilde{t}) = y_1 \cdots y_k$, cf. Definition 2.6. There is actually a unique term $\widetilde{t}$ satisfying these conditions. The sort of y_j is necessarily $\sigma(x_{i_j})$.

Lemma 3.2 For all $A_1, \ldots, A_n$ with $A_i \subseteq M_{\sigma(x_i)}$ for $i = 1, \ldots, n$, we have:

(1) $t_{\mathcal{P}(\mathbb{M})}(A_1, \ldots, A_n) \supseteq \{ t_{\mathbb{M}}(a_1, \ldots, a_n) \mid a_1 \in A_1, \ldots, a_n \in A_n \}$;

---

[1] Strictly speaking, we have an empty set of each sort. Thus, the powersets $\mathcal{P}(M_s)$ are pairwise disjoint.

(2) equality holds in (1) if, for every $i = 1,\ldots,n$, $|A_i| \le 1$ whenever the variable x_i has more than one occurrence in t, and $A_i \ne \emptyset$ whenever x_i has no occurrence in t;

(3) $t_{\mathcal{P}(\mathbb{M})}(A_1,\ldots,A_n) = \emptyset$ if for some i, we have $A_i = \emptyset$ and x_i has at least one occurrence in t;

(4) $t_{\mathcal{P}(\mathbb{M})}(A_1,\ldots,A_n) = \{\widetilde{t}_{\mathbb{M}}(a_1,\ldots,a_k) \mid a_1 \in A_{i_1},\ldots,a_k \in A_{i_k}\}$.

Proof: By induction on the structure of t. For proving (4) we observe that $t_{\mathcal{P}(\mathbb{M})}(A_1,\ldots,A_n) = \widetilde{t}_{\mathcal{P}(\mathbb{M})}(A_{i_1},\ldots,A_{i_k})$ and we use (2). ∎

The inclusion in (1) may be strict: if $L \subseteq A^*$ is a language consisting of at least two words and $t = x_1 \cdot x_1$, then $\widetilde{t} = y_1 \cdot y_2$ (with $i_1 = i_2 = 1$) and

$$L \cdot L = t_{\mathcal{P}(\mathbb{W}(A))}(L) = \widetilde{t}_{\mathcal{P}(\mathbb{W}(A))}(L,L) = \{\widetilde{t}_{\mathbb{W}(A)}(w_1,w_2) \mid w_1,w_2 \in L\}$$

$$= \{w_1 \cdot w_2 \mid w_1,w_2 \in L\} \supset \{w^2 \mid w \in L\} = \{t_{\mathbb{W}(A)}(w) \mid w \in L\},$$

where the word algebra $\mathbb{W}(A)$ is defined in Definition 2.7. This also shows that, in general, $t_{\mathcal{P}(\mathbb{M})}$ is not additive.

Two terms t and t' in $T(F_{\cup},X_n)$ are *equivalent* in $\mathcal{P}(\mathbb{M})$, which is denoted by $t \equiv_{\mathcal{P}(\mathbb{M})} t'$, if $t_{\mathcal{P}(\mathbb{M})} = t'_{\mathcal{P}(\mathbb{M})}$. They are *$\mathcal{P}$-equivalent*, denoted by $t \equiv_{\mathcal{P}} t'$, if $t \equiv_{\mathcal{P}(\mathbb{M})} t'$ for every F-algebra $\mathbb{M}$.

Definition 3.3 (Polynomials) Let F be an $\mathcal{S}$-signature, let X be an $\mathcal{S}$-sorted set of variables and let $s \in \mathcal{S}$. A *monomial* is a term in $T(F,X)$. A *polynomial* of sort s is a term in $T(F_{\cup},X)_s$ of the form either Ω_s, or t_1, or $t_1 \cup_s \cdots \cup_s t_p$ for some pairwise distinct monomials $t_1,\ldots,t_p \in T(F,X)_s$ with $p \ge 2$. We denote by $Pol(F,X)_s$ the set of polynomials of sort s, built with $F, \cup_s, \Omega_s$ and X, and by $Pol(F,X)$ the union of the sets $Pol(F,X)_s$ for all sorts s. We use infix notation for the binary symbol $\cup_s$ and omit parentheses (as explained in Definition 2.4). We can do so since set union is associative. We will frequently omit the subscript s in $\cup_s$ and Ω_s.

If D is a finite subset of $T(F,X)_s$, $D = \{t_1,\ldots,t_q\}$, we will denote by $\bigcup D$ the polynomial $t_1 \cup t_2 \cup \cdots \cup t_q$. Since set union is commutative, we make the convention that $\bigcup D$ is the same polynomial for every enumeration of D. If $q = 0$, then $\bigcup D = \Omega$, and if $q = 1$, then $\bigcup D = t_1$. We will say in all cases that this polynomial is the *union* of the set of terms D, even if $|D| \le 1$. For every $t \in T(F_{\cup},X)$, we define the set of *monomials of* t, a subset of $T(F,X)$ called its *expansion* and denoted by $Mon(t)$:

$$Mon(x) := \{x\} \quad \text{for } x \in X,$$
$$Mon(\Omega) := \emptyset,$$
$$Mon(t_1 \cup t_2) := Mon(t_1) \cup Mon(t_2),$$
$$Mon(f(t_1,\ldots,t_k)) := \{f(m_1,\ldots,m_k) \mid m_i \in Mon(t_i), \; 1 \le i \le k\}.$$

Using the additivity of the operations of $\mathcal{P}(\mathbb{M})$, it is easy to show by induction on the structure of t that for all $A_1, \ldots, A_n$ with $A_i \subseteq M_{\sigma(x_i)}$ for $i = 1, \ldots, n$, we have

$$t_{\mathcal{P}(\mathbb{M})}(A_1, \ldots, A_n) = \bigcup \{ m_{\mathcal{P}(\mathbb{M})}(A_1, \ldots, A_n) \mid m \in Mon(t) \}.$$

Lemma 3.4 Every term t is $\mathcal{P}$-equivalent to the polynomial $\bigcup Mon(t)$. $\qquad\square$

Definition 3.5 (Homomorphisms of powerset algebras) Let F be an S-signature, $\mathbb{M}$ and $\mathbb{N}$ be two F-algebras and $h : \mathbb{M} \to \mathbb{N}$ be a homomorphism. We let $^{\mathcal{P}}h : \mathcal{P}(\mathbb{M}) \to \mathcal{P}(\mathbb{N})$ be defined by

$$^{\mathcal{P}}h(A) := h(A) = \{ h(a) \mid a \in A \} \subseteq N_s$$

for every $s \in S$ and $A \subseteq M_s$. It is a homomorphism of $F_\cup$-algebras. (We use the notation $^{\mathcal{P}}h$ to stress that this mapping is a homomorphism of $F_\cup$-algebras.)

3.1.2 Equation systems and equational sets

We now define equation systems and their solutions in powerset algebras.

Definition 3.6 (Equation systems) Let S be a set of sorts and F be an S-signature. Let X be a finite S-sorted set of variables. An *equation* is an ordered pair (x, t) such that $x \in X$ and $t \in T(F_\cup, X)_{\sigma(x)}$. It will be written $x = t$ and will be said to *define* x. An *equation system over F with *set of unknowns* X is a set of equations consisting of one equation that defines each unknown. The set of unknowns of an equation system S is denoted by $Unk(S)$. It will be useful to assume that the set $Unk(S)$ is linearly ordered. In most cases, it will be the standard set $X_n = \{x_1, \ldots, x_n\}$ ordered by increasing indices. The sorts of these unknowns are not fixed, they are specified by a mapping $\sigma : X_n \to S$, depending on the considered system. The generic equation system S over F is thus denoted by

$$S = \langle x_1 = t_1, \ldots, x_n = t_n \rangle \tag{3.1}$$

In examples, it will also be written as follows, in a more readable way:

$$\begin{cases} x_1 = t_1, \\ \quad \vdots \\ x_n = t_n. \end{cases}$$

Here is, for example, an equation system T (with $\rho(f) = 2$ and $\rho(a) = \rho(b) = 0$):

$$\begin{cases} x_1 = fx_1(x_2 \cup x_3) \cup a, \\ x_2 = x_1 \cup fx_1a, \\ x_3 = f(x_1 \cup x_2)x_3 \cup b. \end{cases}$$

We denote by $F(S)$ the finite subsignature of F consisting of the symbols having occurrences in S. Its set of sorts is the finite set $Sort(S)$ consisting of all sorts of S that occur as $\sigma(f)$ or in the sequence $\alpha(f)$ for some f in $F(S)$, together with the sorts of the unknowns of S. (We may have in S an equation $x = x \cup y$, where the common sort of x and y does not occur in the type of any symbol in $F(S)$. In this case, x defines the empty set because the corresponding component of the least solution of S is empty.) Hence, S is an equation system over $F(S)$; in fact, $F(S)$ is the smallest subsignature F' of F such that S is an equation system over F'. As an example, $F(T) = \{f, a, b\}$ for the above system T.

A *polynomial system* is an equation system whose equations have right-hand sides that are polynomials. A polynomial system is *uniform* if each monomial of the right-hand side of an equation has exactly one occurrence of a function symbol (hence is of the form $f(x_{i_1}, \ldots, x_{i_k})$ for $f \in F_k$). It is *quasi-uniform* if each such monomial has at most one occurrence of a function symbol (hence can also be an unknown).

Let $\mathbb{M}$ be an F-algebra. A *solution* of S in $\mathcal{P}(\mathbb{M})$ is an n-tuple $(L_1, \ldots, L_n)$ such that $L_i \subseteq M_{\sigma(x_i)}$ and $L_i = t_{i\mathcal{P}(\mathbb{M})}(L_1, \ldots, L_n)$ for each $i = 1, \ldots, n$. We say that $(L_1, \ldots, L_n)$ is an *oversolution* of S if $L_i \supseteq t_{i\mathcal{P}(\mathbb{M})}(L_1, \ldots, L_n)$ for each i. We will mainly be interested in the least solution (that turns out to be also the least oversolution), where n-tuples of sets are ordered component-wise: $(L_1, \ldots, L_n) \subseteq (L'_1, \ldots, L'_n)$ if and only if $L_j \subseteq L'_j$ for each j. A solution $(L_1, \ldots, L_n)$ is the *least solution* if and only if $(L_1, \ldots, L_n) \subseteq (L'_1, \ldots, L'_n)$ for every solution $(L'_1, \ldots, L'_n)$, and it is unique with this minimality property; similarly for the *least oversolution*.

If $S' = \langle x_1 = t'_1, \ldots, x_n = t'_n \rangle$ is another system over F such that $t_i \equiv_{\mathcal{P}} t'_i$ for each $i = 1, \ldots, n$, then S and S' have the same solutions and the same oversolutions in $\mathcal{P}(\mathbb{M})$, hence also the same least solution. In particular, Lemma 3.4 says that each term t in $T(F_\cup, X)$ is $\mathcal{P}$-equivalent to a polynomial. Hence, if we replace in an equation system S each right-hand side of an equation by an equivalent polynomial, we obtain a polynomial system that has the same solutions and oversolutions as S. For example, here is the (quasi-uniform) polynomial system T' associated with the above system T:

$$\begin{cases} x_1 = fx_1x_2 \cup fx_1x_3 \cup a, \\ x_2 = x_1 \cup fx_1a, \\ x_3 = fx_1x_3 \cup fx_2x_3 \cup b. \end{cases}$$

Hence there is no loss of generality in considering only polynomial systems, which we will do in the sequel.

It is convenient to order linearly the set of variables of an equation system, but this order is essentially irrelevant: if a system S' is obtained from a system S by permuting its equations, hence by changing the ordering of $Unk(S)$, then its solutions and its oversolutions are obtained from those of S by permuting their components. In particular, if we are interested in the set associated with a particular unknown, there is no loss of generality in assuming that this unknown is x_1.

We will be interested in the least solutions of equation systems, because we consider them as systems of mutually recursive definitions. An equation in languages, for instance $x = axx \cup b$ where a, b are letters and the unknown x denotes a language over $\{a, b\}$, can be seen as a recursive definition of a language. More generally, an equation system is a set of mutually recursive definitions. Such systems may have several solutions. For instance, the equation $y = yy \cup a$ has for solutions the languages $a^* (= \{a\}^*), a^+ (= aa^*), \{a, b\}^*$ and $\{a, bb\}^*$ among others. However, a^+ is the (unique) least solution. As in Semantics for defining the meaning of a recursive definition, we will take the least solution as *the (canonical) solution*.

With every system S as in (3.1), we associate a function $S_{\mathcal{P}(\mathbb{M})}$ that maps $\mathcal{P}(M_{\sigma(x_1)}) \times \cdots \times \mathcal{P}(M_{\sigma(x_n)})$ into itself and is defined by

$$S_{\mathcal{P}(\mathbb{M})}(\overline{L}) := (t_{1\mathcal{P}(\mathbb{M})}(\overline{L}), \ldots, t_{n\mathcal{P}(\mathbb{M})}(\overline{L})),$$

where $\overline{L}$ stands for $(L_1, \ldots, L_n)$. A solution (resp. an oversolution) of S in $\mathcal{P}(\mathbb{M})$ can also be defined as a solution of the equation $\overline{x} = S_{\mathcal{P}(\mathbb{M})}(\overline{x})$ (resp. of the inequation $\overline{x} \supseteq S_{\mathcal{P}(\mathbb{M})}(\overline{x})$) where $\overline{x}$ stands for $(x_1, \ldots, x_n)$. They are called respectively a *fixed-point* of the function $S_{\mathcal{P}(\mathbb{M})}$ and a *pre-fixed-point* of this function.

A sequence of n-tuples $(L_1^{(i)}, \ldots, L_n^{(i)})_{i \in \mathcal{N}}$ is *increasing*[2] if $(L_1^{(i)}, \ldots, L_n^{(i)}) \subseteq (L_1^{(i+1)}, \ldots, L_n^{(i+1)})$ for every $i \in \mathcal{N}$. It has a *least upper-bound* (for component-wise inclusion) denoted like a union and defined by

$$\bigcup_{i \in \mathcal{N}} (L_1^{(i)}, \ldots, L_n^{(i)}) := (\bigcup_{i \in \mathcal{N}} L_1^{(i)}, \ldots, \bigcup_{i \in \mathcal{N}} L_n^{(i)}).$$

The following result is well known (see the book [*ArnNiw] that studies in detail least and greatest fixed-points) but we will sketch its proof for completeness. It is an instance of the Least Fixed-Point Theorem.[3]

Theorem 3.7 Let S be an equation system over an S-signature F. Let $\mathbb{M}$ be an F-algebra. The sequence $(S_{\mathcal{P}(\mathbb{M})}^i(\emptyset, \ldots, \emptyset))_{i \in \mathcal{N}}$ is increasing. Its least upper-bound is the (unique) least solution, and also the (unique) least oversolution of S in $\mathcal{P}(\mathbb{M})$.

---

[2] Two consecutive elements may be equal; a sequence is *strictly increasing* if no two elements are equal.

[3] This theorem, usually attributed to Tarski and/or to Kleene, says that every monotone and ω-continuous function $: E \to E$ where E is an ω-complete partial order with a least element has a least fixed-point. See [LasNS] for a review of its uses in Semantics.

Proof: The partially ordered set $E := \mathcal{P}(M_{\sigma(x_1)}) \times \cdots \times \mathcal{P}(M_{\sigma(x_n)})$ has least element $\overline{\emptyset} := (\emptyset, \ldots, \emptyset)$ and is ω-complete, which means that every increasing sequence has a least upper-bound. We have noted above that the derived operations $t_{\mathcal{P}(M)}$ are monotone. It follows that the mapping $S_{\mathcal{P}(M)} : E \to E$ is monotone, which means that $S_{\mathcal{P}(M)}(\overline{L}) \subseteq S_{\mathcal{P}(M)}(\overline{L}')$ if $\overline{L}, \overline{L}' \in E$ and $\overline{L} \subseteq \overline{L}'$. The mapping $S_{\mathcal{P}(M)}$ is also ω-continuous, which means that $S_{\mathcal{P}(M)}(\bigcup_{i \in \mathcal{N}} \overline{L}^{(i)}) = \bigcup_{i \in \mathcal{N}} S_{\mathcal{P}(M)}(\overline{L}^{(i)})$ whenever $(\overline{L}^{(i)})_{i \in \mathcal{N}}$ is an increasing sequence. This fact can be proved for the operations $f_{\mathcal{P}(M)}$ (which are even additive), the derived operations $t_{\mathcal{P}(M)}$ and then the mapping $S_{\mathcal{P}(M)}$. The detailed proof is routine.

We have $\overline{\emptyset} \subseteq S_{\mathcal{P}(M)}(\overline{\emptyset})$, hence $S^i_{\mathcal{P}(M)}(\overline{\emptyset}) \subseteq S^{i+1}_{\mathcal{P}(M)}(\overline{\emptyset})$ for every i (by using i times the monotonicity of $S_{\mathcal{P}(M)}$). We let $\overline{A} := \bigcup_{i \in \mathcal{N}} S^i_{\mathcal{P}(M)}(\overline{\emptyset})$. By the ω-continuity of $S_{\mathcal{P}(M)}$, we have $\overline{A} = S_{\mathcal{P}(M)}(\overline{A})$, hence $\overline{A}$ is a solution, and also an oversolution. If $\overline{B}$ is any oversolution of S in $\mathcal{P}(M)$, then we have $S^i_{\mathcal{P}(M)}(\overline{\emptyset}) \subseteq \overline{B}$ for each i (easy proof by induction on i), so that $\overline{A} \subseteq \overline{B}$. Hence, $\overline{A}$ is the least oversolution of S in $\mathcal{P}(M)$ and also the least solution. ∎

We denote by $\mu\overline{x} \cdot S_{\mathcal{P}(M)}(\overline{x})$ the least solution (oversolution) of an equation system S in $\mathcal{P}(M)$, where $\overline{x}$ stands for some ordering of $Unk(S)$ (and not necessarily $Unk(S) = X_n$). Two systems S and S' with the same set of unknowns are *equivalent in* $\mathcal{P}(M)$ if $\mu\overline{x} \cdot S_{\mathcal{P}(M)}(\overline{x}) = \mu\overline{x} \cdot S'_{\mathcal{P}(M)}(\overline{x})$. This is the case in particular if $S_{\mathcal{P}(M)} = S'_{\mathcal{P}(M)}$, because then S and S' have the same solutions, and thus the same least solution in $\mathcal{P}(M)$.

Two systems S and S' as above are *equivalent* if they are equivalent in every powerset algebra $\mathcal{P}(M)$. We have already noted that every system is equivalent to a polynomial system. Transformations of systems into equivalent ones will be studied in Section 3.2.

Definition 3.8 (Equational sets of an algebra) We will denote by $\mu\overline{x} \cdot S_{\mathcal{P}(M)}(\overline{x}) \upharpoonright y$ the component of the tuple $\mu\overline{x} \cdot S_{\mathcal{P}(M)}(\overline{x})$ corresponding to the unknown y. (It is the i-th component if $y = x_i$ and $\overline{x} = (x_1, \ldots, x_n)$.) Such a set is said to be *defined by* (S, y) *in* $\mathcal{P}(M)$, or *in* M in order to simplify the notation. More generally, we say that a set $L \subseteq M$ is *defined by* (S, Y) *in* M if it is the union of the sets defined by (S, y) in M for all $y \in Y \subseteq Unk(S)$. We will also say that L is *defined by* S *in* M (without specifying Y). Note that if all unknowns in Y have the same sort, then L is defined by (S', z) in M where S' is obtained by adding to S the equation $z = \bigcup Y$.

An *equational set* of an F-algebra M is a set defined in M by an equation system over F. By the definitions, every equational set L can be defined by (S, Y) in M for some equational system S and some set Y consisting of unknowns of S with pairwise distinct sorts (and such that $Y = X_k$, $k \leq n$ and $Unk(S) = X_n$). If L is homogenous (i.e., all its elements have the same sort), then it can be defined by (S, z) in M for some equational system S and some unknown z of S (in particular $z = x_1$). We denote by **Equat**(M) the set of equational sets of M and by **Equat**(M)$_s$ for $s \in \mathcal{S}$, the set of those included in M_s. We will also say that a set in **Equat**($\mathbb{M}$) is $\mathbb{M}$-*equational* or *equational in* $\mathbb{M}$. Examples of equational sets have been given in Sections 1.1.1, 1.1.2 and 1.1.3.

If D is a finite subset of $T(F)_s$ for some sort s, then the equation $x = \bigcup D$ defines in $\mathcal{P}(\mathbb{M})$ the finite set of values in $\mathbb{M}$ of the terms of D. Every finite subset of M whose elements are defined by terms is thus equational in $\mathbb{M}$. (This fact will be restated in Proposition 3.45(2).)

An equation system S over F only uses the finite subsignature $F(S)$ of F (its set of sorts is $Sort(S) \subseteq S$). Hence, we need only study the equational sets of algebras over finite signatures, as shown in the next proposition. (By contrast, the infiniteness of the signature will be important for recognizable sets.)

Proposition 3.9 Let $\mathbb{M}$ be an F-algebra.

(1) Let S be an equation system over a subsignature F' of F and let $\mathbb{M}'$ be an F'-algebra such that $\mathbb{M}' \subseteq \mathbb{M}$. Then

$$\mu \overline{x} \cdot S_{\mathcal{P}(\mathbb{M}')}(\overline{x}) = \mu \overline{x} \cdot S_{\mathcal{P}(\mathbb{M})}(\overline{x}).$$

Moreover, every solution of S in $\mathcal{P}(\mathbb{M}')$ is a solution of S in $\mathcal{P}(\mathbb{M})$.

(2) If $\mathbb{M}' \subseteq \mathbb{M}$, then **Equat**($\mathbb{M}'$) $\subseteq$ **Equat**($\mathbb{M}$).

(3) Every equational set of $\mathbb{M}$ is equational in a finitely generated subalgebra of $\mathbb{M}$.

Proof: (1) Let $Unk(S) = \{x_1, \ldots, x_n\}$ and $E' := \mathcal{P}(M'_{\sigma(x_1)}) \times \cdots \times \mathcal{P}(M'_{\sigma(x_n)})$, cf. the proof of Theorem 3.7. If $\overline{L} \in E'$, then $S_{\mathcal{P}(\mathbb{M}')}(\overline{L}) = S_{\mathcal{P}(\mathbb{M})}(\overline{L})$ and this n-tuple belongs to E'. This implies the second statement. It also implies that, for every i, we have $S^i_{\mathcal{P}(\mathbb{M}')}(\overline{\emptyset}) = S^i_{\mathcal{P}(\mathbb{M})}(\overline{\emptyset})$ (by induction on i). Hence these two sequences have the same least upper-bound and $\mu \overline{x} \cdot S_{\mathcal{P}(\mathbb{M}')}(\overline{x}) = \mu \overline{x} \cdot S_{\mathcal{P}(\mathbb{M})}(\overline{x})$.

(2) Immediate from (1).

(3) Let L be defined by S in $\mathbb{M}$. We let $\mathbb{M}'$ be the subalgebra of $\mathbb{M}$ generated by the signature $F(S)$. It is a finitely generated subalgebra of $\mathbb{M}$. Since S is an equation system over $F(S)$, it follows from (1) that L is equational in $\mathbb{M}'$. ∎

Equational sets in semi-effectively given algebras

Let F be an effectively given $\mathcal{S}$-signature. The set of sorts $\mathcal{S}$ is thus also effectively given, cf. Definitions 2.8 and 2.127, and $\mathcal{S} \cup F$ is at most countable. An equation system S over F is specified in the obvious way as a set of equations $x = t$ with $x \in X$ and $t \in T(F_\cup, X)_{\sigma(x)}$, where we may assume that $X = X_n$ for some n. The set X_n is effectively given by its standard encoding (that encodes x_i as i) and an algorithm must be specified that computes the mapping $\sigma : X_n \rightarrow \mathcal{S}$ (usually by means of a table, of course). Assuming that each function symbol $\cup_s$ is encoded by the same integer as s, the terms t can be specified through the standard encoding of the semi-effectively given set $T(F_\cup, X_n)$, since $F_\cup$ is the disjoint union of F and $\{\cup_s \mid s \in \mathcal{S}\}$.

Let us consider the least solution of S in $\mathcal{P}(\mathbb{M})$, where $\mathbb{M}$ is a semi-effectively given F-algebra. By Lemma 3.2(4), each iterate $S^i_{\mathcal{P}(\mathbb{M})}(\overline{\emptyset})$ is a tuple of finite subsets of M. Since the derived operations of $\mathbb{M}$ are computable via an encoding ξ_M of the elements of M by integers, we can compute the tuple $S^i_{\mathcal{P}(\mathbb{M})}(\overline{\emptyset})$ (or rather the tuple of finite sets of integers that encode it via ξ_M; we will omit this precision in the sequel) by an algorithm that takes as input the system S, the integer i and the computable function $\zeta_\mathbb{M}$ of Definitions 2.8 and 2.127 that encodes the operations of $\mathbb{M}$.

We say that an n-tuple $(L_1, \ldots, L_n)$ in $\mathcal{P}(M)^n$ is *finite* if the set $L_1 \cup \cdots \cup L_n$ is finite.

Proposition 3.10 Let $\mathbb{M}$ be a semi-effectively given F-algebra and S be an equation system over F. If the least solution of S in $\mathcal{P}(\mathbb{M})$ is finite, then it can be computed.

Proof: The least solution of S in $\mathcal{P}(\mathbb{M})$ is the least upper-bound of the increasing sequence $S^i_{\mathcal{P}(\mathbb{M})}(\overline{\emptyset})$ for $i \geq 0$. Each tuple of this sequence is finite and, furthermore, computable. By the finiteness assumption, and since the sequence is increasing, there exists an integer i such that $S^{i+1}_{\mathcal{P}(\mathbb{M})}(\overline{\emptyset}) = S^i_{\mathcal{P}(\mathbb{M})}(\overline{\emptyset})$. The first one, let us denote it by i_0, can be computed. Then $S^j_{\mathcal{P}(\mathbb{M})}(\overline{\emptyset}) = S^{i_0}_{\mathcal{P}(\mathbb{M})}(\overline{\emptyset})$ for each $j \geq i_0$. Hence $\mu\overline{x} \cdot S_{\mathcal{P}(\mathbb{M})}(\overline{x}) = S^{i_0}_{\mathcal{P}(\mathbb{M})}(\overline{\emptyset})$ and this tuple can be computed. ∎

The finiteness of $\mu\overline{x} \cdot S_{\mathcal{P}(\mathbb{M})}(\overline{x})$ is guaranteed if each domain M_s such that s is the sort of an unknown of S is finite.

One might expect that the membership of an element of an effectively given algebra in an equational set given by an equation system is decidable. This is not the case. As a counter-example, take a computable total function $f : \mathcal{N} \to \mathcal{N}$ such that the membership of an integer in the set $f(\mathcal{N})$ is undecidable. Then consider the effectively given algebra $\mathbb{M} := \langle \mathcal{N}, f, Suc, 0 \rangle$, where Suc is the successor function. The set $f(\mathcal{N})$ is equational in $\mathbb{M}$ because $(f(\mathcal{N}), \mathcal{N})$ is the least solution in $\mathcal{P}(\mathbb{M})$ of the system $\langle x = f(y), y = 0 \cup Suc(y) \rangle$.

3.1.3 Context-free languages

We examine context-free grammars for several reasons. First because the notion of an equational set arises from the Least Fixed-Point Characterization of context-free languages and yields a generalization of context-free languages to arbitrary algebras, cf. Section 1.1.1. Second because the equational sets of terms are (particular) context-free languages and also "generate," as shown in Proposition 3.23(1) below, all equational sets. Finally, technical notions about derivation sequences will be useful for studying transformations of equation systems.

Definition 3.11 (Context-free grammars and languages) A *context-free grammar* over a *terminal* alphabet A is a triple $G = \langle A, X, R \rangle$ such that X is a finite set of variables (disjoint with A) called the *nonterminal* alphabet of G and R is a finite subset of $X \times (A \cup X)^*$ called the set of *derivation rules*. A rule (x, w) is also written $x \to w$ for readability. The *one-step derivation relation*, denoted by $\Rightarrow_G$, is defined as follows for u, u' in $(A \cup X)^*$:

$u \Rightarrow_G u'$ if and only if $u = vxv'$ and $u' = vwv'$ for some $v, v' \in (A \cup X)^*$ and some rule (x, w) in R.

The language *generated by G from* a word u in $(A \cup X)^*$ is $L(G, u) := \{w \in A^* \mid u \Rightarrow_G^* w\}$. In many cases an *initial nonterminal* $x \in X$ is specified and the language of interest $L(G, x)$ is called the *language generated* by the grammar.

A language $L \subseteq A^*$ is *context-free* if it is $L(G, x)$ for some context-free grammar G and some nonterminal x of G. Note that we allow the terminal alphabet A to be countably infinite. However, since the set of rules of a grammar is finite, we have $L \subseteq B^*$ for some finite subset B of A.

For a context-free grammar G and a set Y of nonterminals of G, we denote the language $\bigcup_{y \in Y} L(G, y)$ by $L(G, Y)$ and we say that it is generated by G from Y. Since the family of context-free languages is closed under union, every language $L(G, Y)$ is context-free.

Definition 3.12 (Equational sets in the free monoid) We recall from Definition 2.7 that, for every alphabet A, we let $F_A := \{\cdot, \epsilon\} \cup A$ be the (one-sorted) functional signature such that $\cdot$ is binary and the other symbols are nullary. We let $\mathbb{W}(A)$ be the F_A-algebra $\langle A^*, \cdot, \varepsilon, (a)_{a \in A} \rangle$ (it is actually the free monoid generated by A, augmented with constants). We use infix notation for $\cdot$ and we may omit parentheses (cf. Definition 2.4). The mapping $val_{\mathbb{W}(A)} : T(F_A) \to A^*$ that evaluates a term over F_A into a word over A (by interpreting ϵ as the empty word ε, a as the word a and $\cdot$ as concatenation) will be simply denoted by val. We extend it into a mapping $val : T(F_A, X) \to (A \cup X)^*$, where X is a set of variables, by evaluating x in X into the word x. For $t \in T(F_A, X)$, $val(t)$ is obtained from t by erasing all occurrences of the symbols $\cdot$ and ϵ.

With every equation system S over F_A, we associate the context-free grammar $G[S] := \langle A, X, R \rangle$ such that $X = Unk(S)$ and R is the set of rules $x \to w$ such that $w = val(m)$ for some monomial $m \in Mon(t)$ ($\subseteq T(F_A, X)$) and some equation $x = t$ of S. Conversely, for every context-free grammar G over A, there exists an equation system S over F_A such that $G[S] = G$. There are actually several such systems for a given grammar.

For an example of an equation system S over F_A see Example 1.2 (where ε should be changed into ϵ); the context-free grammar $G[S]$ is given in Example 1.1 (with nonterminals S and T instead of x_1 and x_2).

Theorem 3.13 (Least Fixed-Point Characterization) For every equation system $S = \langle x_1 = t_1, \ldots, x_n = t_n \rangle$ over F_A, we have

$$\mu \overline{x} \cdot S_{\mathcal{P}(\mathbb{W}(A))}(\overline{x}) = (L(G[S], x_1), \ldots, L(G[S], x_n)).$$

Proof sketch: We let $G := G[S] = \langle A, X, R \rangle$. We first prove that the sequence of languages $(L(G, x_1), \ldots, L(G, x_n))$ is an oversolution of S in $\mathcal{P}(\mathbb{W}(A))$. It suffices to observe that, if $x_i \to w$ belongs to R and $w = u_0 x_{i_1} u_1 \cdots u_{k-1} x_{i_k} u_k$ for $i_1, \ldots, i_k$ in $[n]$ and $u_0, u_1, \ldots, u_k$ in A^*, then

$$u_0 L(G, x_{i_1}) u_1 \cdots u_{k-1} L(G, x_{i_k}) u_k \subseteq L(G, w) \subseteq L(G, x_i).$$

This fact is a consequence of the if-direction of the following *Fundamental Property of Derivation Sequences* of context-free grammars.[4] For every $p \in \mathcal{N}$ and every word $v \in A^*$, we have

$$w \Rightarrow_G^p v \text{ if and only if there exist } p_1, \ldots, p_k \text{ in } \mathcal{N} \text{ and } v_1, \ldots, v_k \text{ in } A^* \text{ such that}$$
$$v = u_0 v_1 u_1 \cdots u_{k-1} v_k u_k, \, p = p_1 + \cdots + p_k \text{ and } x_{i_j} \Rightarrow_G^{p_j} v_j \text{ for each } j = 1, \ldots, k.$$

Hence, $t_{i\mathcal{P}(\mathbb{W}(A))}(L(G, x_1), \ldots, L(G, x_n)) \subseteq L(G, x_i)$ for each i.

For proving the opposite inclusion, we let $(L_1, \ldots, L_n)$ be any oversolution of S in $\mathcal{P}(\mathbb{W}(A))$. For every $j \in \mathcal{N}$, we have

$$\forall i \in [n], \forall v \in A^*, \forall p \in [0, j] \, (\text{if } x_i \Rightarrow_G^{p+1} v \text{ then } v \in L_i).$$

This can be proved by induction on j, using the only-if direction of the Fundamental Property of Derivation Sequences. It follows that $L(G, x_i) \subseteq L_i$ for every i, hence that $(L(G, x_1), \ldots, L(G, x_n))$ is the least oversolution of S in $\mathcal{P}(\mathbb{W}(A))$. This gives the result by Theorem 3.7. ∎

Corollary 3.14 A language over A is context-free if and only if it is equational in $\mathbb{W}(A)$.

Proof: Theorem 3.13 shows that every $\mathbb{W}(A)$-equational language is context-free. Let conversely $L = L(G, x)$ be a context-free language over A. There exists an equation system S such that $G[S] = G$. The component of its least solution that corresponds to x is $L(G, x)$ by Theorem 3.13, hence L is equational in $\mathbb{W}(A)$. ∎

Equation systems have in most cases several solutions, and we will only be interested in their least ones. However, some context-free grammars yield systems having unique solutions, and this observation shortens some proofs (in particular, that two grammars are equivalent when one is obtained from the other by a transformation

---

[4] The only-if direction of this important property is stated and proved in [*Har] as "technical lemma 3.3.1."

of the corresponding systems). It is also more satisfactory to know that an equation system defines completely a tuple of languages without having to take into account any minimization principle.

A context-free grammar is *strict* if the right-hand side of each derivation rule is the empty word or contains at least one terminal symbol.

Proposition 3.15 Let G be a strict context-free grammar over an alphabet A and nonterminal alphabet X_n, and S be an equation system such that $G[S] = G$. The n-tuple $(L(G,x_1),\dots,L(G,x_n))$ is the unique solution of S in $\mathcal{P}(\mathbb{W}(A))$.

Proof sketch: The n-tuple $(L(G,x_1),\dots,L(G,x_n))$ is the least solution of S in $\mathcal{P}(\mathbb{W}(A))$ by Theorem 3.13. If $(M_1,\dots,M_n)$ is any solution of S in $\mathcal{P}(\mathbb{W}(A))$, then $L(G,x_j) \subseteq M_j$ for each j. We prove by induction on $|v|$ that for all $i \in [n]$ and $v \in A^*$, if $v \in M_i$ then $v \in L(G,x_i)$. Let $v \in M_i$. Since $(M_1,\dots,M_n)$ is a solution of S, we have $v \in u_0 M_{i_1} u_1 \cdots u_{k-1} M_{i_k} u_k$ for some rule of G of the form $x_i \to u_0 x_{i_1} u_1 \cdots u_{k-1} x_{i_k} u_k$ with $u_0,\dots,u_k$ in A^*. Hence $v = u_0 v_1 u_1 \cdots u_{k-1} v_k u_k$ for some v_1 in $M_{i_1},\dots,v_k$ in M_{i_k}. Since G is strict, $|v_j| < |v|$ for each $j = 1,\dots,k$, and $v_j \in L(G,x_{i_j})$ by the induction hypothesis. Hence $v \in L(G,x_i)$. So, $(L(G,x_1),\dots,L(G,x_n))$ is the unique solution of S. ∎

We have defined two other algebras with domain A^* in Definition 2.7, viz. $\mathbb{W}_{left}(A)$ and $\mathbb{W}_{right}(A)$. The equational sets of these algebras are actually the regular languages over A, as we will prove in the next section.

3.1.4 Equational sets of terms

The next natural family of equational sets to consider is that of sets of terms. An immediate consequence of Proposition 3.9 is that a set of terms over a signature F is equational in $\mathbb{T}(F)$ if and only if it is equational in $\mathbb{T}(F')$ for some finite subsignature F' of F. This proposition is applicable because if $F' \subseteq F$, then $\mathbb{T}(F')$ is the subalgebra of $\mathbb{T}(F)$ generated by F'.

Proposition 3.16 Let F be a finite signature. For every sort s, the set of terms $T(F)_s$ is equational in $\mathbb{T}(F)$.

Proof: For every sort s we define an unknown x_s of sort s. We let its defining equation be $x_s = \bigcup D$, where D is the set of monomials of the form $f(x_{\alpha(f)[1]},\dots,x_{\alpha(f)[\rho(f)]})$ for all $f \in F$ such that $\sigma(f) = s$. We obtain an equation system S. It is clear that

$$\mu \bar{x} \cdot S_{\mathcal{P}(\mathbb{T}(F))}(\bar{x}) \upharpoonright x_s \subseteq T(F)_s$$

for every sort s. For the opposite inclusion, we have, for every term $t \in T(F)_s$ (we recall that $ht(t)$ is its height; a constant symbol has height 0):

$$t \in S_{\mathcal{P}(\mathbb{T}(F))}^{ht(t)+1}(\overline{\emptyset}) \upharpoonright x_s,$$

which is easily proved by induction on the structure of t. It follows that

$$T(F)_s \subseteq \mu\bar{x} \cdot S_{\mathcal{P}(\mathbb{T}(F))}(\bar{x}) \restriction x_s$$

for every sort s, which gives the result. ∎

We now consider particular context-free grammars, that will characterize equational sets of terms.

Definition 3.17 (Regular grammars over a signature) Let F be an $\mathcal{S}$-signature. A *regular grammar* over F is a context-free grammar over F (used here as an alphabet) of the form $G = \langle F, X, R \rangle$ such that:

(1) the nonterminal alphabet X is $\mathcal{S}$-sorted with sort mapping σ;
(2) each rule in R is a pair (x,t), where $t \in T(F,X)_{\sigma(x)}$, also written $x \to t$.

We say that G is *quasi-uniform* (resp. *uniform*) if each right-hand side of a rule has at most one (resp. exactly one) occurrence of a symbol in F.

We have $L(G,x) \subseteq T(F)_{\sigma(x)}$ for every nonterminal $x \in X$, because if $w \in T(F,X)_{\sigma(x)}$ and $w \Rightarrow_G w'$, then $w' \in T(F,X)_{\sigma(x)}$.

A set of terms is *regular over F* if it is $L(G,Y)$ for some regular grammar G over F and some set Y of nonterminals of G. Since the set of rules of a grammar is finite, a set of terms is regular over F if and only if it is regular over F' for some finite subsignature F' of F.

The *regular grammar $G(S)$ associated with an equation system S over F* is defined as $\langle F, Unk(S), R \rangle$, where R is the set of rules $x \to t$ such that $x \in Unk(S)$ and $t \in Mon(p)$, where $x = p$ is the equation of S that defines x. It is uniform or quasi-uniform if S is so. It is clear that, conversely, every regular grammar is $G(S)$ for some equation system S over the same signature. Hence, regular grammars are in bijective correspondence with polynomial equation systems.

The following theorem is similar to Theorem 3.13:

Theorem 3.18 For every equation system S over a signature F, we have

$$\mu\bar{x} \cdot S_{\mathcal{P}(\mathbb{T}(F))}(\bar{x}) = (L(G(S),x_1), \ldots, L(G(S),x_n)).$$

□

Easy modifications of the proof of Theorem 3.13 can yield a proof of this theorem. However, we will prove it as a corollary of Theorem 3.13 with the help of some notions to be introduced after a corollary and some remarks.

Corollary 3.19 Let F be a signature. A set of terms is regular over F if and only if it is equational in $\mathbb{T}(F)$. □

Although the equational sets of terms are the same as the regular ones, we will call them frequently "regular sets" to emphasize their language theoretical specifications by grammars and also by automata that we will define in Section 3.3.

Remark 3.20 (1) Every equational set of $\mathbb{T}(F)$ is a regular set of terms over F hence a context-free language over F. But a subset of $T(F)$ that is a context-free language over F need not be $\mathbb{T}(F)$-equational. For example, consider $F := \{f,g,a,b\}$ with $\rho(f) := 2$, $\rho(g) := 1$ and $\rho(a) := \rho(b) := 0$. The subset $\{fg^n ag^n b \mid n \geq 0\}$ of $T(F)$ is a context-free language but it is not $\mathbb{T}(F)$-equational because it is not a regular set of terms. This last fact will be proved in Example 3.54 by means of a "pumping argument."

A regular subset of $T(F)$ may be a context-free language over the alphabet F that is not a regular one: this is the case for the set $L \subseteq T(F)$ defined by the regular grammar over F (as above) with rules $x \to fafxa$ and $x \to a$ because $L = \{(faf)^n a^{n+1} \mid n \geq 0\}$.

A set of terms can be regular over a signature F and, at the same time, be *not* regular over a signature F' obtained by changing the arities of the symbols of F. For example, the set of terms fLb is regular (where L is as above), but it is not regular if f,a have arity 1 and b has arity 0 (although it is also a set of terms for the new arity mapping); this follows from (4) below.

(2) The context-free grammars $G(S)$ and $G[S]$ associated with an equation system S over F_A are not the same. For example, if S is the system of Example 1.2, then $G(S)$ is the regular grammar over F_A with the following rules (written for more readability in infix notation with parentheses):

$$x_1 \to a \cdot (x_1 \cdot x_2), \quad x_1 \to x_1 \cdot x_1, \quad x_1 \to a,$$

$$x_2 \to b \cdot ((x_2 \cdot x_1) \cdot x_2), \quad x_2 \to a, \quad x_2 \to c \quad \text{and} \quad x_2 \to \epsilon,$$

and $G[S]$ is the context-free grammar G over A of Example 1.1 with nonterminals x_1 and x_2 instead of S and T. Its rules are as follows:

$$x_1 \to ax_1x_2, \quad x_1 \to x_1x_1, \quad x_1 \to a,$$

$$x_2 \to bx_2x_1x_2, \quad x_2 \to a, \quad x_2 \to c, \quad \text{and} \quad x_2 \to \varepsilon.$$

(3) Let $G = \langle A,X,R \rangle$ be a context-free grammar and S be an equation system over F_A such that $G[S] = G$. Let $x \Rightarrow^*_{G(S)} t$ with $t \in T(F_A,X)$ be a derivation sequence relative to the regular grammar $G(S)$. Its image under the evaluation mapping *val*, i.e., the corresponding sequence of images by *val* (where *val* evaluates a term in $T(F_A,X)$ into a word over $A \cup X$) is a derivation sequence $x \Rightarrow^*_G val(t)$ of the word $val(t)$ relative to the context-free grammar G. Conversely every derivation sequence of G is the image under *val* of a derivation sequence of $G(S)$. Thus, $L(G[S],x) = val(L(G(S),x))$ for every nonterminal x.

To illustrate this observation, we use again the system S and the grammar $G = G[S]$ as in (2). The image under *val* of the derivation sequence of $G(S)$: $x_1 \Rightarrow a \cdot (x_1 \cdot x_2) \Rightarrow a \cdot ((x_1 \cdot x_1) \cdot x_2) \Rightarrow a \cdot ((x_1 \cdot x_1) \cdot \epsilon) \Rightarrow a \cdot ((a \cdot x_1) \cdot \epsilon) \Rightarrow a \cdot ((a \cdot a) \cdot \epsilon)$ is $x_1 \Rightarrow ax_1 x_2 \Rightarrow ax_1 x_1 x_2 \Rightarrow ax_1 x_1 \Rightarrow aax_1 \Rightarrow aaa$.

(4) The set of words over a finite alphabet A is in bijection (in two ways) with the set of terms over U_A, where U_A is the associated unary signature of Definition 2.7. Via these bijections, the regular languages over A correspond to the regular sets of terms over U_A, and the equational sets of the U_A-algebras $\mathbb{W}_{left}(A)$ and $\mathbb{W}_{right}(A)$ correspond to the equational sets of $\mathbb{T}(U_A)$. Hence, the equational sets of the algebras $\mathbb{W}_{left}(A)$ and $\mathbb{W}_{right}(A)$ are the regular languages over the alphabet A. $\quad\square$

In order to obtain Theorem 3.18 from Theorem 3.13 and Proposition 3.9(1), we make some technical observations about sorts. If F is an $\mathcal{S}$-signature, we let F^- be the one-sorted signature obtained by making all sorts identical. It follows that $T(F)$ defined as $\bigcup_{s \in \mathcal{S}} T(F)_s$ is a subset of $T(F^-)$. Every equation system over F is also one over F^-.

Lemma 3.21 For every equation system S over a signature F, we have that $\mu\bar{x} \cdot S_{\mathcal{P}(\mathbb{T}(F))}(\bar{x}) = \mu\bar{x} \cdot S_{\mathcal{P}(\mathbb{T}(F^-))}(\bar{x})$. Moreover, every solution of S in $\mathcal{P}(\mathbb{T}(F))$ is a solution of S in $\mathcal{P}(\mathbb{T}(F^-))$.

Proof: Let S be a system with $Unk(S) = X_n$ and $E := \mathcal{P}(T(F)_{\sigma(x_1)}) \times \cdots \times \mathcal{P}(T(F)_{\sigma(x_n)})$. It is clear from the definitions that if $\bar{L} = (L_1, \ldots, L_n)$ belongs to E, then $S_{\mathcal{P}(\mathbb{T}(F^-))}(\bar{L})$ also belongs to E and is equal to $S_{\mathcal{P}(\mathbb{T}(F))}(\bar{L})$. This implies the second statement. It also implies, by induction on i, that $S^i_{\mathcal{P}(\mathbb{T}(F))}(\bar{\emptyset}) = S^i_{\mathcal{P}(\mathbb{T}(F^-))}(\bar{\emptyset})$ for each $i \in \mathcal{N}$. This yields the result. $\quad\blacksquare$

This lemma proves that $\mathbf{Equat}(\mathbb{T}(F)) \subseteq \mathbf{Equat}(\mathbb{T}(F^-))$. Conversely, if L is equational in $\mathbb{T}(F^-)$ and is contained in $T(F)$, then it is equational in $\mathbb{T}(F)$. This fact will be proved below in Remark 3.38(4).

Let $S = \langle x_1 = p_1, \ldots, x_n = p_n \rangle$ be an equation system over an $\mathcal{S}$-signature F. We will use the regular grammar $G(S)$ over F. Since this grammar is context-free, there exists an equation system $\bar{S}$ over the signature $F_F = \{\cdot, \epsilon\} \cup F$ (cf. Definition 3.12) such that the associated context-free grammar $G[\bar{S}]$ is equal to $G(S)$. For example, if $S = \langle x = fxgxx \cup a \rangle$ (with f, g binary), then $G(S) = \langle \{f, g, a\}, \{x\}, \{x \rightarrow fxgxx, x \rightarrow a\} \rangle$ and $\bar{S}$ can be (it is not uniquely defined) the equation $x = (f \cdot x) \cdot ((g \cdot x) \cdot x) \cup a$ (we use infix notation for $\cdot$). With these definitions we have:

Lemma 3.22 For every system S over a signature F:

$$\mu\bar{x} \cdot S_{\mathcal{P}(\mathbb{T}(F))}(\bar{x}) = \mu\bar{x} \cdot \bar{S}_{\mathcal{P}(\mathbb{W}(F))}(\bar{x}).$$

Moreover, every solution of S in $\mathcal{P}(\mathbb{T}(F))$ is a solution of $\bar{S}$ in $\mathcal{P}(\mathbb{W}(F))$.

Proof: We have the following equalities:

$$\mu\bar{x} \cdot S_{\mathcal{P}(\mathbb{T}(F))}(\bar{x}) = \mu\bar{x} \cdot S_{\mathcal{P}(\mathbb{T}(F^-))}(\bar{x})$$
$$= \mu\bar{x} \cdot S_{\mathcal{P}(\mathbb{S}(F^-))}(\bar{x})$$
$$= \mu\bar{x} \cdot \overline{S}_{\mathcal{P}(\mathbb{W}(F))}(\bar{x}),$$

which follow respectively from Lemma 3.21, from Proposition 3.9(1) (because $\mathbb{T}(F^-) \subseteq \mathbb{S}(F^-)$) (cf. Definition 2.2) and from the fact that $S_{\mathcal{P}(\mathbb{S}(F^-))}$ and $\overline{S}_{\mathcal{P}(\mathbb{W}(F))}$ are the same function from $(\mathcal{P}(F^*))^n$ to itself ($n := |\bar{x}|$) by the definition of $\overline{S}$. ∎

Proof of Theorem 3.18: We have:

$$\mu\bar{x} \cdot S_{\mathcal{P}(\mathbb{T}(F))}(\bar{x}) = \mu\bar{x} \cdot \overline{S}_{\mathcal{P}(\mathbb{W}(F))}(\bar{x})$$
$$= (L(G[\overline{S}], x_1), \ldots, L(G[\overline{S}], x_n))$$
$$= (L(G(S), x_1), \ldots, L(G(S)), x_n)),$$

respectively by Lemma 3.22, by Theorem 3.13 and by the definition of $\overline{S}$. ∎

From Lemma 3.22 and Proposition 3.15 we obtain: if S is an equation system over F such that $G(S)$ is strict, then S has a unique solution in $\mathcal{P}(\mathbb{T}(F))$.

As another consequence of Lemma 3.22, consider the equation system S in the proof of Proposition 3.16. In Definition 2.124, $T(F)_s$ is defined as the x_s component of the unique solution of the equation system $\overline{S}$ in $\mathbb{W}(F)$ (and note that it has a unique solution by Proposition 3.15). Hence, by Lemma 3.22, $T(F)_s = \mu\bar{x} \cdot S_{\mathcal{P}(\mathbb{T}(F))}(\bar{x}) \restriction x_s$. This proves, again, Proposition 3.16.

3.1.5 Homomorphic images of equational sets

We now establish some algebraic properties of equational sets. We recall from Definition 2.124 that $val_{\mathbb{M}}$ denotes the mapping that associates with a term t in $T(F)$ its value $val_{\mathbb{M}}(t)$ in the F-algebra $\mathbb{M}$; it is the unique homomorphism from $\mathbb{T}(F)$ to $\mathbb{M}$.

Proposition 3.23 Let $\mathbb{M}, \mathbb{M}'$ be F-algebras and $h : \mathbb{M} \to \mathbb{M}'$ be a homomorphism.[5] We have:

(1) If S is an equation system over F, then

$$\mu\bar{x} \cdot S_{\mathcal{P}(\mathbb{M}')}(\bar{x}) = {}^{\mathcal{P}}h(\mu\bar{x} \cdot S_{\mathcal{P}(\mathbb{M})}(\bar{x})).$$

In particular:

$$\mu\bar{x} \cdot S_{\mathcal{P}(\mathbb{M})}(\bar{x}) = val_{\mathbb{M}}(\mu\bar{x} \cdot S_{\mathcal{P}(\mathbb{T}(F))}(\bar{x})).$$

---

[5] We recall from Definition 3.5 that ${}^{\mathcal{P}}h$ denotes the extension of h to subsets of M and that it is a homomorphism of $F_\cup$-algebras : $\mathcal{P}(\mathbb{M}) \to \mathcal{P}(\mathbb{M}')$. We let also ${}^{\mathcal{P}}h((L_1, \ldots, L_n)) := ({}^{\mathcal{P}}h(L_1), \ldots, {}^{\mathcal{P}}h(L_n))$.

(2) A set is equational in $\mathbb{M}'$ if and only if it is the image under h of an equational set in $\mathbb{M}$.

(3) A set L is equational in $\mathbb{M}$ if and only if $L = val_{\mathbb{M}}(K)$ for some regular set of terms $K \subseteq T(F)$. In particular, every equational set of $\mathbb{M}$ is generated by a finite subsignature of F.

(4) The set M_s is equational in $\mathbb{M}$ if and only if it is generated by a finite subsignature of F.

Proof: (1) Since $^P h$ is a homomorphism : $\mathcal{P}(\mathbb{M}) \to \mathcal{P}(\mathbb{M}')$, we have

$$^P h(f_{\mathcal{P}(\mathbb{M})}(\overline{L})) = f_{\mathcal{P}(\mathbb{M}')}(^P h(\overline{L})),$$

for every $f \in F$ and every tuple $\overline{L}$ of appropriate type. This equality extends to the derived operations $t_{\mathcal{P}(\mathbb{M})}$ and $t_{\mathcal{P}(\mathbb{M}')}$ (routine induction on the structure of t) and to the mappings $S_{\mathcal{P}(\mathbb{M})}$ and $S_{\mathcal{P}(\mathbb{M}')}$, that is, we have $^P h(S_{\mathcal{P}(\mathbb{M})}(\overline{L})) = S_{\mathcal{P}(\mathbb{M}')}(^P h(\overline{L}))$. By using an induction on i, we get that $^P h(S^i_{\mathcal{P}(\mathbb{M})}(\overline{\emptyset})) = S^i_{\mathcal{P}(\mathbb{M}')}(\overline{\emptyset})$ for each i. The result follows then since $^P h$ is additive.

(2) is immediate from (1).

(3) The first assertion follows from (2) and Corollary 3.19. The second assertion is a consequence of Proposition 3.9(3).

(4) Follows from (3) and Proposition 3.16. Note that if $M_s \subseteq val_{\mathbb{M}}(T(F'))$ then $M_s = val_{\mathbb{M}}(T(F')_s)$. ∎

As an example of the second statement of (1), consider an equation system S over F_A, for an alphabet A, and an unknown x of S. We know that $\mu\overline{x} \cdot S_{\mathcal{P}(\mathbb{W}(A))}(\overline{x}) \restriction x = L(G[S],x)$ from Theorem 3.13, and we know that $\mu\overline{x} \cdot S_{\mathcal{P}(\mathbb{T}(F_A))}(\overline{x}) \restriction x = L(G(S),x)$ from Theorem 3.18. In Remark 3.20(3) we have already seen that $L(G[S],x) = val_{\mathbb{W}(A)}(L(G(S),x))$.

As another example, consider the U_A-algebras $\mathbb{W}_{left}(A)$ and $\mathbb{W}_{right}(A)$. It is straightforward to show that a set of terms $L \subseteq T(U_A)$ is regular over U_A if and only if it is of the form $K\epsilon$ for some regular language K over the alphabet A. In that case $val_{\mathbb{W}_{left}(A)}(L) = K$, and $val_{\mathbb{W}_{right}(A)}(L)$ is the mirror image of K. Since the class of regular languages is closed under mirror image, it follows from Proposition 3.23(3) that the equational sets of the algebras $\mathbb{W}_{left}(A)$ and $\mathbb{W}_{right}(A)$ are the regular languages over the alphabet A, as we have already seen in Remark 3.20(4) (where the mentioned bijections are the mappings $val_{\mathbb{W}_{left}(A)}$ and $val_{\mathbb{W}_{right}(A)}$).

Corollary 3.24 For every equation system S over a signature F and every unknown y of S, the following are equivalent:

(1) $L(G(S),y) = \emptyset$;
(2) $\mu\overline{x} \cdot S_{\mathcal{P}(\mathbb{M})}(\overline{x}) \restriction y = \emptyset$ for some F-algebra $\mathbb{M}$;
(3) $\mu\overline{x} \cdot S_{\mathcal{P}(\mathbb{M})}(\overline{x}) \restriction y = \emptyset$ for every F-algebra $\mathbb{M}$.

These conditions are decidable if F is effectively given.

Proof: Theorem 3.18 and Proposition 3.23(1) entail that, for every F-algebra $\mathbb{M}$, we have $\mu\bar{x} \cdot S_{\mathcal{P}(\mathbb{M})}(\bar{x}) \upharpoonright y = val_{\mathbb{M}}(L(G(S),y))$. This implies the equivalence of (1), (2) and (3). Since $G(S)$ is a context-free grammar, the emptiness of $L(G(S),y)$ is decidable, which proves the last assertion.   ∎

3.1.6 Equational sets of commutative words

For each $k \geq 1$, we let $\mathbb{N}^k$ be the set of k-tuples of natural numbers equipped with componentwise addition. We denote by $\vec{u}$ the generic element of $\mathcal{N}^k$ and by $\vec{0}$ the *null* k-tuple $(0,\ldots,0)$ which is the unit of the monoid[6] $\mathbb{N}^k$. For each $i \in [k]$, we let a_i be a constant symbol denoting $(0,\ldots,0,1,0,\ldots,0)$ with a single 1 at the i-th position and $A := \{a_1,\ldots,a_k\}$. Hence, $\mathbb{N}^k$ is an F_A-algebra (with addition as operation $\cdot_{\mathbb{N}^k}$ and $\vec{0}$ as constant $\epsilon_{\mathbb{N}^k}$). It is isomorphic to the quotient algebra[7] of $\mathbb{W}(A)$ by the congruence generated by the equalities $a_i \cdot a_j = a_j \cdot a_i$ for all $i,j \in [k]$, hence is the *free commutative monoid* generated by A. With this identification, we can say that the unique homomorphism $h : \mathbb{W}(A) \to \mathbb{N}^k$ "makes the letters of A commute," in other words, only counts the numbers of occurrences of letters regardless of their positions: $h(w) = (n_1,\ldots,n_k)$ where $n_i = |w|_{a_i}$ for each i.

For $\lambda \in \mathcal{N}$ and $\vec{u} \in \mathcal{N}^k$, we let $\lambda \cdot \vec{u}$ (or just $\lambda\vec{u}$) denote $\vec{u} + \vec{u} + \cdots + \vec{u}$ with λ times $\vec{u}$. A subset L of $\mathcal{N}^k$ is *linear* (*affine* would be a better word, but "linear" is usual) if it is of the form $\{\vec{u} + \lambda_1\vec{v}_1 + \cdots + \lambda_p\vec{v}_p \mid \lambda_1,\ldots,\lambda_p \in \mathcal{N}\}$ for some $\vec{u},\vec{v}_1,\ldots,\vec{v}_p$ in $\mathcal{N}^k$. A set is *semi-linear* if it is a finite union of linear sets. Its *description* is the finite set of tuples $(\vec{u},\vec{v}_1,\ldots,\vec{v}_p)$ from which it is defined. From the description of a semi-linear set, one can check if it is finite, because a linear set described by $\vec{u},\vec{v}_1,\ldots,\vec{v}_p$ is finite if and only if all tuples $\vec{v}_i$ are null.

The following proposition reformulates a classical result of formal language theory.[8]

Proposition 3.25 For every $k \geq 1$, **Equat**$(\mathbb{N}^k)$ is the set of semi-linear subsets of $\mathcal{N}^k$. There exists an algorithm that transforms a description of a semi-linear set into an equation system over F_A defining this set in $\mathbb{N}^k$, and another one that converts an equation system over F_A into descriptions of the semi-linear sets that form its least solution in $\mathcal{P}(\mathbb{N}^k)$.

Proof: Let $L = L_1 \cup \cdots \cup L_n$ be a semi-linear set, where each L_i is linear. Each element $\vec{u}$ of $\mathcal{N}^k$ can be expressed as a sum of constant symbols. A linear set L_i of the form $\{\vec{u} + \lambda_1\vec{v}_1 + \cdots + \lambda_p\vec{v}_p \mid \lambda_1,\ldots,\lambda_p \in \mathcal{N}\} \subseteq \mathcal{N}^k$ is the least solution of the

---

[6] A monoid is a set equipped with an associative binary operation and a unit element.
[7] The formal definition of this standard notion is in Definition 3.63.
[8] Usually called Parikh's Theorem. The homomorphism $h : \mathbb{W}(A) \to \mathbb{N}^k$ is called the Parikh mapping.

equation $x_i = \vec{u} \cup (\vec{v}_1 + x_i) \cup \cdots \cup (\vec{v}_p + x_i)$. If S is the equation system consisting of these n equations, then L is defined by (S, X_n) in $\mathbb{N}^k$. This gives the first algorithm.

The opposite construction is done in a clean way in [AceEI] (on the basis of a previous proof in [Pil]) in terms of equation systems. It can also be done as follows in terms of context-free grammars. By Proposition 3.23(2) and Corollary 3.14, a set is equational in $\mathbb{N}^k$ if and only if it is the commutative image $h(L)$ of a context-free language $L \subseteq A^*$ where h is the above defined homomorphism $h : \mathbb{W}(A) \to \mathbb{N}^k$. A description of $h(L)$ can be constructed from a grammar defining L: an elementary proof is in Section 6.9 of [*Har]. Both constructions are effective.   ∎

Let g_F be a mapping : $F \to \mathcal{N}^k$ and let $g : T(F) \to \mathcal{N}^k$ be the mapping defined by

$$g(t) = \sum_{f \in F} |Occ(t, f)| \cdot g_F(f).$$

Then g is a homomorphism of F-algebras : $\mathbb{T}(F) \to \mathbb{P}$ where $\mathbb{P}$ is the F-algebra that is the derived algebra of $\mathbb{N}^k$ with operations defined by

$$f_{\mathbb{P}}(\vec{u}_1, \ldots, \vec{u}_{\rho(f)}) = g_F(f) + \vec{u}_1 + \cdots + \vec{u}_{\rho(f)}.$$

We call such a mapping g an *affine* mapping : $T(F) \to \mathcal{N}^k$.

Corollary 3.26 The image of a regular set of terms under an affine mapping is a semi-linear set, and a description of this set can be computed from the affine mapping and a regular grammar.

Proof: Let g be an affine mapping : $T(F) \to \mathcal{N}^k$ based on $g_F : F \to \mathcal{N}^k$, and consider $L := L(G, x)$, where $G = \langle F, X, R \rangle$ is a regular grammar and $x \in X$. For every $f \in F$, let w_f be a word in A^* such that $h(w_f) = g_F(f)$, and let $\pi : (F \cup X)^* \to (A \cup X)^*$ be the mapping such that, for $t \in (F \cup X)^*$, $\pi(t)$ is obtained from t by replacing every occurrence of every f by w_f (i.e., π is the unique monoid homomorphism such that $\pi(f) = w_f$ for $f \in F$). Obviously, $g(t) = h(\pi(t))$ for every $t \in T(F)$. Now define the context-free grammar $G' = \langle A, X, R' \rangle$ such that R' consists of all rules $x \to \pi(t)$ with $x \to t$ in R. Then $L(G', x) = \{\pi(t) \mid t \in L(G, x)\} = \pi(L)$. Hence $g(L) = h(\pi(L)) = h(L(G', x))$, i.e., $g(L)$ is the image under h of a context-free language over A. Thus, $g(L)$ is equational in $\mathbb{N}^k$ and hence semi-linear by the previous proposition. All constructions are effective.   ∎

An algebraic presentation of this proof will be given in Example 3.42(3).

3.2 Transformations of equation systems

We present some transformations of equation systems that normalize or simplify them while preserving the "main components" of their least solutions. We also establish some closure properties of the family of equational sets. All proofs are constructive (provided the considered signatures are effectively given), unless mentioned otherwise.

3.2.1 Unfolding

Unfolding is a very natural transformation of context-free grammars that we generalize to equation systems. Further transformations are based on it. In particular, we will obtain a generalization of the transformation that puts a context-free grammar in Chomsky normal form.

We say that a system S' is a *subsystem* of S, denoted by $S' \subseteq S$, if $Unk(S') \subseteq Unk(S)$ and each equation of S' is an equation of S.

Lemma 3.27 If $S' \subseteq S$, then $\mu\bar{y} \cdot S'_{\mathcal{P}(\mathbb{M})}(\bar{y}) \upharpoonright z = \mu\bar{x} \cdot S_{\mathcal{P}(\mathbb{M})}(\bar{x}) \upharpoonright z$ for every $z \in Unk(S')$.

Proof: Let S be an equation system of the form (3.1) with set of unknowns $Unk(S) = X_n = \{x_1,\dots,x_n\}$ and let $Y := Unk(S')$ be ordered as in $Unk(S)$, that is, $Y = \{x_{i_1},\dots,x_{i_p}\}$ with $i_1 < i_2 < \cdots < i_p$. For any n-tuple of sets $\bar{L} = (L_1,\dots,L_n)$, we let $\bar{L} \upharpoonright Y$ denote the p-tuple $(L_{i_1},\dots,L_{i_p})$. Then, for each $i \geq 0$, we have $S^i_{\mathcal{P}(\mathbb{M})}(\bar{\emptyset}) \upharpoonright Y = S'^i_{\mathcal{P}(\mathbb{M})}(\bar{\emptyset})$. This is easily proved by induction on i, and yields the result. ∎

Definition 3.28 (Equivalent systems) Let S and S' be two equation systems over F, and let $Z \subseteq Unk(S) \cap Unk(S')$. The systems S and S' are *Z-equivalent in* $\mathcal{P}(\mathbb{M})$ if $\mu\bar{x} \cdot S_{\mathcal{P}(\mathbb{M})}(\bar{x}) \upharpoonright z = \mu\bar{y} \cdot S'_{\mathcal{P}(\mathbb{M})}(\bar{y}) \upharpoonright z$ for every $z \in Z$. They are *Z-equivalent* if they are Z-equivalent in $\mathcal{P}(\mathbb{M})$ for every F-algebra $\mathbb{M}$. We delete the prefix Z when $Z = Unk(S) = Unk(S')$.

That two systems are equivalent in $\mathcal{P}(\mathbb{M})$ may depend on particular properties of $\mathbb{M}$. However, there exist syntactic criteria ensuring that two systems are equivalent. We first consider unfolding: this transformation consists in replacing (several times) in the right-hand sides of some equations of a system unknowns by their defining terms. It is a standard transformation of recursive definitions; it preserves the defined sets, i.e., the least solutions of the corresponding equation systems.

Definition 3.29 (Unfolding) Let $S = \langle x_1 = p_1,\dots,x_n = p_n \rangle$ be an equation system where $p_1,\dots,p_n$ are polynomials. We let $\Rightarrow_S$ be the rewriting relation on $T(F_\cup,X_n)$ defined by:[9]

$$t \Rightarrow_S t' \quad :\Longleftrightarrow \quad t = c[x_i] \text{ and } t' = c[p_i],$$

---

[9] It is the one-step derivation relation of the regular grammar $G = \langle F_\cup,X_n,R \rangle$ with $R = \{x_1 \to p_1,\dots,x_n \to p_n\}$. Note that this is *not* the regular grammar $G(S)$. The languages generated by G are of no interest.

for some $i \in [n]$ and some context c in $Ctxt(F_\cup, X_n)$ (cf. Definition 2.6). We say that $S' = \langle x_1 = q_1, \ldots, x_n = q_n \rangle$ is obtained from S by *unfolding* if $p_i \Rightarrow^*_S q_i$ for each $i = 1, \ldots, n$. We denote this by $S \nearrow S'$. The terms q_i are not necessarily polynomials. However, they can be expanded into polynomials by Lemma 3.4. If S'' is obtained from S' by this lemma, we also say that S'' is obtained from S by unfolding and we use the same notation $S \nearrow S''$. Note that S' and S'' are equivalent.

Example 3.30 We let S be the system

$$\begin{cases} x_1 = f(x_1, x_2) \cup a, \\ x_2 = g(x_1, x_1, x_2) \cup b. \end{cases}$$

The system S':

$$\begin{cases} x_1 = f(f(x_1, x_2) \cup a, g(x_1, x_1, x_2) \cup b) \cup a, \\ x_2 = g(f(x_1, x_2) \cup a, x_1, x_2) \cup b, \end{cases}$$

is obtained from S by unfolding. The right-hand sides of its equations can then be expanded into polynomials, which gives the following system S'' equivalent to S':

$$\begin{cases} x_1 = hf(f(x_1, x_2), g(x_1, x_1, x_2)) \cup f(a, g(x_1, x_1, x_2)) \\ \qquad \cup f(f(x_1, x_2), b) \cup f(a, b) \cup a, \\ x_2 = g(f(x_1, x_2), x_1, x_2) \cup g(a, x_1, x_2) \cup b. \end{cases}$$

Proposition 3.31 If a system S' is obtained from a system S by unfolding, then S and S' are equivalent. $\qquad\square$

We need for the proof two technical observations. Let $S = \langle x_1 = p_1, \ldots, x_n = p_n \rangle$ and let $t, t' \in T(F_\cup, X_n)$ be such that $t \Rightarrow^*_S t'$.

If $\overline{L}$ is a solution of S in $\mathcal{P}(\mathbb{M})$, then $t'_{\mathcal{P}(\mathbb{M})}(\overline{L}) = t_{\mathcal{P}(\mathbb{M})}(\overline{L})$: this is clear because, at any rewriting step, one replaces x_i having value L_i (where $\overline{L} = (L_1, \ldots, L_n)$) by p_i which has the same value in $\mathcal{P}(\mathbb{M})$ since $\overline{L}$ is a solution of S.

Our second observation is that if $\widetilde{t}$ is the unique term in $T(F_\cup, \{y_1, \ldots, y_k\})$ such that $ListVar(\widetilde{t}) = y_1 y_2 \cdots y_k$ and $t = \widetilde{t}[x_{i_r}/y_r; 1 \leq r \leq k]$ for some $i_1, \ldots, i_k$ in $[n]$, then $t' = \widetilde{t}[s_r/y_r; 1 \leq r \leq k]$, where for each r we have $x_{i_r} \Rightarrow^*_S s_r$.

This is a reformulation of the only-if direction of the Fundamental Property of Derivation Sequences (mentioned in the proof of Theorem 3.13) for the context-free grammar with rules $x_1 \to p_1, \ldots, x_n \to p_n$. We illustrate this fact with an example, where S is the system of Example 3.30.

Example 3.32 Let $t = f(x_1, h(x_1, x_2))$ and

$$t' = f(x_1, h(f(x_1, x_2) \cup a, g(x_1, f(x_1, x_2) \cup a, x_2) \cup b)).$$

Then we have $t \Rightarrow_S^* t'$ with $\widetilde{t}, y_1, y_2, y_3, s_1, s_2, s_3, i_1, i_2, i_3$ as follows:

$$\widetilde{t} = f(y_1, h(y_2, y_3)),$$
$$t = \widetilde{t}[x_1/y_1, x_1/y_2, x_2/y_3],$$
$$t' = \widetilde{t}[s_1/y_1, s_2/y_2, s_3/y_3],$$

with $s_1 = x_1$, $s_2 = f(x_1, x_2) \cup a$, $s_3 = g(x_1, f(x_1, x_2) \cup a, x_2) \cup b$, $i_1 = i_2 = 1$ and $i_3 = 2$.
It is clear that $x_1 \Rightarrow_S^* s_1$, $x_1 \Rightarrow_S^* s_2$ and $x_2 \Rightarrow_S^* s_3$.

Proof of Proposition 3.31: Let $S = \langle x_1 = p_1, \ldots, x_n = p_n \rangle$ and $S' = \langle x_1 = q_1, \ldots, x_n = q_n \rangle$ be such that $p_i \Rightarrow_S^* q_i$ for each i, so that $S \nearrow S'$. Let $\mathbb{M}$ be an F-algebra. Every solution $\overline{L}$ of S in $\mathcal{P}(\mathbb{M})$ is a solution of S' because if $p_i \Rightarrow_S^* q_i$, then $p_{i\mathcal{P}(\mathbb{M})}(\overline{L}) = q_{i\mathcal{P}(\mathbb{M})}(\overline{L})$. Hence

$$\mu\overline{x} \cdot S'_{\mathcal{P}(\mathbb{M})}(\overline{x}) \subseteq \mu\overline{x} \cdot S_{\mathcal{P}(\mathbb{M})}(\overline{x}).$$

We now prove the other direction. We let:

$$\overline{L}^{(j)} = (L_1^{(j)}, \ldots, L_n^{(j)}) := S_{\mathcal{P}(\mathbb{M})}^j(\overline{\emptyset}),$$

so that

$$\overline{L} = (L_1, \ldots, L_n) = \mu\overline{x} \cdot S_{P(\mathbb{M})}(\overline{x})$$

if $L_i := \bigcup_{j \in \mathcal{N}} L_i^{(j)}$ for each i. We let also:

$$\overline{L'} = (L'_1, \ldots, L'_n) := \mu\overline{x} \cdot S'_{\mathcal{P}(\mathbb{M})}(\overline{x}).$$

We will prove that $L_i \subseteq L'_i$ for all i by proving that, for all i and j, we have $L_i^{(j)} \subseteq L'_i$. We prove by induction on j the following stronger statement, from which the previous one is obtained by taking $t = x_i$.

Claim 3.31.1 For every $j \in \mathcal{N}$, every $i = 1, \ldots, n$ and every $t \in T(F_\cup, X_n)$ such that $x_i \Rightarrow_S^* t$, we have $L_i^{(j)} \subseteq t_{\mathcal{P}(\mathbb{M})}(\overline{L'})$.

Proof: This is clear for $j = 0$ because $L_i^{(0)} = \emptyset$. We consider the inductive step. We have $L_i^{(j+1)} = p_{i\mathcal{P}(\mathbb{M})}(\overline{L}^{(j)})$. The hypothesis $x_i \Rightarrow_S^* t$ yields two cases.

Case 1: $x_i \Rightarrow_S p_i \Rightarrow_S^* t$. If $\widetilde{p}_i \in T(F_\cup, \{y_1, \ldots, y_k\})$ with $ListVar(\widetilde{t}) = y_1 y_2 \cdots y_k$ and $p_i = \widetilde{p}_i[x_{i_r}/y_r; 1 \leq r \leq k]$, then we have $t = \widetilde{p}_i[s_r/y_r; 1 \leq r \leq k]$ where $x_{i_r} \Rightarrow_S^* s_r$ for each r. We must prove that

$$L_i^{(j+1)} = p_{i\mathcal{P}(\mathbb{M})}(\overline{L}^{(j)}) \subseteq t_{\mathcal{P}(\mathbb{M})}(\overline{L'}).$$

Since

$$p_{i\mathcal{P}(\mathbb{M})}(\overline{L}^{(j)}) = \widetilde{p}_{i\mathcal{P}(\mathbb{M})}(L_{i_1}^{(j)}, \ldots, L_{i_k}^{(j)})$$

and

$$t_{\mathcal{P}(\mathbb{M})}(\overline{L'}) = \widetilde{p}_{i\mathcal{P}(\mathbb{M})}(s_{1\mathcal{P}(\mathbb{M})}(\overline{L'}),\dots,s_{k\mathcal{P}(\mathbb{M})}(\overline{L'})),$$

and since the derived operations of $\mathcal{P}(\mathbb{M})$ are monotone (we have used this fact in Theorem 3.7), it suffices to prove that

$$L_{i_r}^{(j)} \subseteq s_{r\mathcal{P}(\mathbb{M})}(\overline{L'}) \quad \text{for every } r = 1,\dots,k.$$

But this follows from the induction hypothesis.

Case 2: $t = x_i$. We must prove that $L_i^{(j+1)} \subseteq t_{\mathcal{P}(\mathbb{M})}(\overline{L'}) = L_i'$. But $L_i' = q_{i\mathcal{P}(\mathbb{M})}(\overline{L'})$ (because $\overline{L'}$ is a solution of S') and $p_i \Rightarrow_S^* q_i$. So this follows from Case 1 with q_i instead of t. $\qquad\square$

This completes the proof of the claim and, thus, of the proposition. $\qquad\blacksquare$

Uniform and quasi-uniform systems are defined in Definition 3.6. We denote by $F(X_n)$ the set of terms in $T(F,X_n)$ having a single occurrence of a symbol in F (which is thus the leading symbol).

Proposition 3.33 For every equation system S, one can construct a uniform system that is $Unk(S)$-equivalent to S.

Proof: We first transform S into a quasi-uniform system S' by repeating the following transformation step as many times as necessary (i.e., as possible).

Transformation Q: Let $S = \langle x_1 = p_1,\dots,x_n = p_n \rangle$ be an equation system over F, where each p_i belongs to $Pol(F,X_n)$. Assume that for some i, some monomial m in $Mon(p_i)$ is not in $F(X_n) \cup X_n$.

Then $m = f(m_1,\dots,m_k)$ and at least one of the m_j's is not an unknown. For each such term m_j we introduce a new unknown y_j of the same sort as m_j, we introduce the new equation $y_j = m_j$ and we replace in m the term m_j by y_j. We let Y be the set of these new unknowns. We have transformed m into $m' \in F(X_n \cup Y)$, and the system S into a system $Q(S)$ with $Unk(Q(S)) = Unk(S) \cup Y$.

It is clear that $Q(S) \nearrow S'$, where S' is S augmented with the new equations $y_j = m_j$, so that $S \subseteq S'$. It follows then from Proposition 3.31 and Lemma 3.27 that S and $Q(S)$ are $Unk(S)$-equivalent.

Let S be the system that we want to transform into an equivalent quasi-uniform system. We define a sequence $S = S_0, S_1,\dots,S_r$ of equation systems such that $S_{i+1} = Q(S_i)$ for each $i < r$. This sequence terminates when it reaches a quasi-uniform system S_r. For each i we have $Unk(S) \subseteq Unk(S_i)$ and the systems S_{i+1} and S_i are $Unk(S_i)$-equivalent. Hence they are $Unk(S)$-equivalent. It remains to verify that a quasi-uniform system S_r is actually obtained.

For $m \in T(F,X_n)$, we define $\delta(m) :=$ if $m \in X_n$ then 0 else $|m|_F - 1$, where $|m|_F$ is the number of occurrences in m of symbols from F. For a polynomial $p = m_1 \cup m_2 \cup \cdots \cup m_s$ we let $\delta(p) := \delta(m_1) + \cdots + \delta(m_s)$ and $\delta(\Omega) := 0$. For an equation system $S = \langle x_1 = p_1,\dots,x_n = p_n \rangle$ we let $\delta(S) := \delta(p_1) + \cdots + \delta(p_n)$. It is clear that S

is quasi-uniform if and only if $\delta(S) = 0$ and that $\delta(Q(S)) < \delta(S)$. These facts ensure that a quasi-uniform system S_r is obtained in r steps with $r \leq \delta(S)$. We will call Q^* the transformation consisting in iterating Q as much as possible. It produces a quasi-uniform system.

Next we show how to transform a quasi-uniform system S into an equivalent uniform system without introducing new variables. For a quasi-uniform equation system $S = \langle x_1 = p_1, \ldots, x_n = p_n \rangle$ we denote by $C(S)$ the set of indices of unknowns that are monomials of S, formally $C(S) := \{i \in [n] \mid x_i \in Mon(p_j) \text{ for some } j \in [n]\}$. Note that S is uniform if and only if $C(S) = \emptyset$. The following transformation U constructs an equivalent quasi-uniform system $U(S)$ such that $|C(U(S))| < |C(S)|$. Consequently, as in the first part of the proof, the transformation U^* that consists in repeating U as much as possible leads to an equivalent uniform system, in at most n steps.

Transformation U: Let $S = \langle x_1 = p_1, \ldots, x_n = p_n \rangle$ be a quasi-uniform equation system, and assume that i belongs to $C(S)$. We first change p_i into p_i' by removing x_i from p_i in the case where x_i is a monomial of p_i; otherwise $p_i' = p_i$. It is clear that the resulting system $\widehat{S}$ has the same oversolutions as S in any powerset algebra $\mathcal{P}(\mathbb{M})$; thus S and $\widehat{S}$ are equivalent by Theorem 3.7. Now we construct $U(S) = \langle x_1 = p_1', \ldots, x_n = p_n' \rangle$ where, for every $j = 1, \ldots, n$ with $j \neq i$, polynomial p_j' is obtained by replacing in p_j the monomial x_i by p_i' in the case where x_i is a monomial of p_j; otherwise $p_j' = p_j$). Clearly, $\widehat{S} \nearrow U(S)$ and hence they are equivalent by Proposition 3.31. In the construction of $U(S)$, we can of course replace $m \cup m$ by m in a polynomial p_j'. It is clear that $C(U(S)) = C(S) - \{i\}$ and that $U(S)$ is quasi-uniform and equivalent to S. ∎

Example 3.34 We let S be the following system:

$$\begin{cases} x_1 = x_2 \cup f(x_1, g(x_1, x_2)) \cup a, \\ x_2 = x_1 \cup g(x_2, h(x_1, x_1)). \end{cases}$$

The first step of the construction produces the quasi-uniform system S':

$$\begin{cases} x_1 = x_2 \cup f(x_1, x_3) \cup a, \\ x_2 = x_1 \cup g(x_2, x_4), \\ x_3 = g(x_1, x_2), \\ x_4 = h(x_1, x_1). \end{cases}$$

The associated uniform system is S'':

$$\begin{cases} x_1 = f(x_1, x_3) \cup g(x_2, x_4) \cup a, \\ x_2 = f(x_1, x_3) \cup g(x_2, x_4) \cup a, \\ x_3 = g(x_1, x_2), \\ x_4 = h(x_1, x_1). \end{cases}$$

Note that since $x_1 \Rightarrow^*_{G(S)} x_2$ and $x_2 \Rightarrow^*_{G(S)} x_1$, the unknowns x_1 and x_2 are defined by the same equation in S''.

Remark 3.35 (1) The transformation U^* corresponds for context-free grammars to eliminating the rules of the form $x \to y$, where y is a nonterminal.

(2) A context-free grammar $G = \langle A, X, R \rangle$ is in *Chomsky normal form* (see, e.g., [*Har]) if $R \subseteq X \times (A \cup XX)$. For such a grammar, there is a uniform system S over $F_A - \{\epsilon\}$ such that $G[S] = S$. Conversely, if S is a uniform system over $F_A - \{\epsilon\}$, then $G[S]$ is in Chomsky normal form. The construction of the proof of Proposition 3.33 generalizes the one that puts a context-free grammar without rules $x \to \varepsilon$ into an equivalent one in Chomsky normal form. The rules $x \to \varepsilon$ can be eliminated by a specific step that is not covered by Proposition 3.33.

(3) In the algorithm of Proposition 3.33, the transformation U^* can be done before the transformation Q^*.

3.2.2 Simplifications of equation systems

We consider transformations of equation systems that delete unknowns hence that reduce the number of equations while preserving the main components of the least solution.

Definition 3.36 (Trim systems) An unknown x_i of a system $S = \langle x_1 = p_1, \ldots, x_n = p_n \rangle$ is *productive* if $L(G(S), x_i) \neq \emptyset$ or, by Corollary 3.24, the i-th component of the least solution of S in any powerset algebra $\mathcal{P}(\mathbb{M})$ is nonempty. We say that S is *trim* if all its unknowns are productive. For arbitrary S, we let $Y \subseteq Unk(S)$ be the set of its productive unknowns. We let $Trim(S)$ be the system obtained from S by deleting the equations $x_i = p_i$ for $x_i \in Unk(S) - Y$ and, in the remaining equations (we assume that $p_1, \ldots, p_n$ are polynomials), all monomials containing unknowns not in Y. Its construction is effective by Corollary 3.24.

Let $Z \subseteq Unk(S)$. An unknown u of S is *useful for* Z if there exists a derivation sequence $z \Rightarrow^*_{G(S)} t$ such that $z \in Z$ and u occurs in t. This implies that u is an unknown of every subsystem S' of S such that $Z \subseteq Unk(S')$. We let $Cut(S, Z)$ be the subsystem of S whose unknowns are those that are useful in S for Z. Thus, $Cut(S, Z)$ is the smallest subsystem S' of S such that $Z \subseteq Unk(S')$. Hence it can be constructed effectively. We let $Trim(S, Z) := Cut(Trim(S), Z \cap Unk(Trim(S)))$.

Proposition 3.37 The system $Trim(S)$ is trim and $Unk(Trim(S))$-equivalent to S. For every subset Z of $Unk(Trim(S))$, the system $Trim(S, Z)$ is trim, it is Z-equivalent to S, and all its unknowns are useful for Z. There exists an algorithm that constructs $Trim(S)$ and $Trim(S, Z)$. If S is (quasi-)uniform, then so are $Trim(S)$ and $Trim(S, Z)$.

Proof: Let $Y := Unk(Trim(S))$. It follows from the definitions that the regular grammar $G(Trim(S))$ is the one obtained from $G(S)$ by deleting the nonterminals in

$Unk(S) - Y$ (i.e., those that are not productive) and the rules where they occur. It follows that $L(G(Trim(S)),y) = L(G(S),y)$ for every $y \in Y$, whence we get that $Trim(S)$ is trim and, by Theorem 3.18 and Proposition 3.23(1), that S and $Trim(S)$ are Y-equivalent. The second assertion follows from Lemma 3.27 since $Trim(S,Z)$ is a subsystem of $Trim(S)$. ∎

Remark 3.38 (1) The transformation of S into $Trim(S,Z)$ corresponds for a context-free grammar $G = \langle A,X,R \rangle$ to reducing it, i.e., to removing the nonterminals that do not occur in any derivation sequence $z \Rightarrow_G^* w$ with $z \in Z$ and $w \in A^*$.

(2) Proposition 3.37 is useful for other purposes than reducing the size of an equation system. Here is an example. If u is an unknown of $Trim(S,Z)$, then $z \Rightarrow_{G(S)}^* t$ for some $z \in Z$ and some term t in $T(F,\{u\})$ that has a unique occurrence of u (hence $t \in Ctxt(F)$ if we use u as special variable for defining contexts) and for every w in $L(G(S),u)$, the term $t[w/u]$ belongs to $L(G(S),z)$. Let $\mathbb{M}$ be an F-algebra and d be any element of $\mu \bar{x} \cdot S_{\mathcal{P}(\mathbb{M})}(\bar{x}) \restriction u$. Then, by Theorem 3.18 and Proposition 3.23(1), $t_{\mathbb{M}}(d) \in \mu \bar{x} \cdot S_{\mathcal{P}(\mathbb{M})}(\bar{x}) \restriction z$.

(3) As an application of this fact, we prove a variant of Proposition 3.10. An operation f of an F-algebra $\mathbb{M}$ is *infinity-preserving* if for every $L_1,\ldots,L_{\rho(f)}$ such that $\emptyset \neq L_i \subseteq M_{\alpha(f)[i]}$ and at least one of the sets $L_1,\ldots,L_{\rho(f)}$ is infinite, the set $f_{\mathcal{P}(\mathbb{M})}(L_1,\ldots,L_{\rho(f)})$ is infinite. On integers, addition is infinity-preserving, but multiplication is not because $\{0\} \cdot L = \{0\}$ for every set L. We say that $\mathbb{M}$ is *infinity-preserving* if all its operations are.

If $\mathbb{M}$ is semi-effectively given and infinity-preserving, then every finite $\mathbb{M}$-equational set can be computed.

Here is the proof. Suppose that $L := \mu \bar{x} \cdot S_{\mathcal{P}(\mathbb{M})}(\bar{x}) \restriction x_1$ is finite. By Proposition 3.37 we may assume that S is trim and that all its unknowns are useful for $\{x_1\}$. We claim that each component $L_u := \mu \bar{x} \cdot S_{\mathcal{P}(\mathbb{M})}(\bar{x}) \restriction u$ of the least solution of S is finite. By the above remark (2), there is a term $t \in T(F,\{u\})$ with one occurrence of u, such that if d is in L_u then $t_{\mathbb{M}}(d) \in L$. Thus, if L_u would be infinite, then, since $\mathbb{M}$ is infinity-preserving, $t_{\mathcal{P}(\mathbb{M})}(L_u) = \{t_{\mathbb{M}}(d) \mid d \in L_u\} \subseteq L$ would be infinite. Hence the least solution of S in $\mathcal{P}(\mathbb{M})$ is finite, and it can be computed by Proposition 3.10.

(4) As another application, we prove the following converse of Lemma 3.21:

If $L \subseteq T(F)$ is equational in $\mathbb{T}(F^-)$, then it is equational in $\mathbb{T}(F)$.

By Theorem 3.18 and Propositions 3.33 and 3.37, we have that $L = L(G(S),Y)$ for a polynomial system $S = \langle x_1 = p_1,\ldots,x_n = p_n \rangle$ over F^- that is uniform and trim with all unknowns useful for $Y \subseteq X_n$. We will show that S is also an equation system over F, hence that L is equational in $\mathbb{T}(F)$. We first claim that there exist sorts $s_1,\ldots,s_n$ such that $L(G(S),x_i) \subseteq T(F)_{s_i}$ for $i = 1,\ldots,n$. The sort s_i will be defined as the sort of x_i. Obviously, $s_1 := s$. Now let $2 \leq i \leq n$, and let $w \in L(G(S),x_i)$. Since x_i is useful for $\{x_1\}$, there exists a derivation sequence $x_1 \Rightarrow_{G(S)}^* t$ such that the term $t \in T(F^-,\{x_i\})$

has a unique occurrence of x_i. Hence $t[w/x_i] \in L(G(S), x_1)$ and so $t[w/x_i] \in T(F)_s$. Then w is in $T(F)_r$ for a sort r that is determined from t: it is $r = \alpha(f)[j]$ where x_i occurs in t as j-th argument of some f in F. (Since S is uniform, we cannot have $t = x_i$.) Since r does not depend on w, the sort $s_i := r$ satisfies the claim, and is taken as sort of x_i. Now, each unknown of S has a sort. It remains to check that $Mon(p_i) \subseteq T(F, X_n)_{s_i}$ for $i = 1, \ldots, n$, i.e., that all monomials of S satisfy the sort constraints. Let $m \in Mon(p_i)$ and let $\widetilde{m}$ be the unique term in $T(F^-, Y_k)$ such that $ListVar(\widetilde{m}) = y_1 \cdots y_k$ and $m = \widetilde{m}[x_{i_r}/y_r; 1 \leq r \leq k]$ for $i_1, \ldots, i_k \in [n]$. Since S is trim, there exists a term $w_r \in L(G(S), x_{i_r})$ for every $r \in [k]$, and so $m \Rightarrow^*_{G(S)} \widetilde{m}[w_r/y_r; 1 \leq r \leq k] \in L(G(S), x_i)$. Then $w_r \in T(F)_{s_{i_r}}$ and $\widetilde{m}[w_r/y_r; 1 \leq r \leq k] \in T(F)_{s_i}$. This implies that $\widetilde{m}[x_{i_r}/y_r; 1 \leq r \leq k] \in T(F, X_n)_{s_i}$, i.e., that $m \in T(F, X_n)_{s_i}$. $\qquad\square$

One can eliminate an unknown x of an equation system that defines a finite set L: it suffices to replace it in the right-hand sides of all equations by finitely many terms that define L. The corresponding transformation (it uses unfolding) is opposite to the one done in Proposition 3.33: it eliminates certain equations but increases the sizes of the remaining ones.

Proposition 3.39 Every infinite homogenous equational set of an F-algebra $\mathbb{M}$ is defined by an equation system, each unknown of which defines an infinite subset of M.

Proof: It follows from Proposition 3.23(1) that if an equational set of $\mathbb{M}$ is finite, then it is the set of values of a finite set of terms.

Let S be an equation system of the form (3.1) with least solution $\overline{L} = (L_1, \ldots, L_n)$ in $\mathcal{P}(\mathbb{M})$ and such that the given infinite equational set equals L_i for some $i \in [n]$. Let $m \in [n]$ be such that L_m is finite. By the preliminary remark, $L_m = \{t_{\mathbb{M}} \mid t \in D\}$, where D is a finite subset of $T(F)$. Let S' be the system obtained from S by replacing the equation $x_m = t_m$ by $x_m = \bigcup D$. We claim that $\overline{L}$ is the least solution of S' in $\mathcal{P}(\mathbb{M})$. It is clear that it is a solution of S', hence that $\mu \overline{x} \cdot S'_{\mathcal{P}(\mathbb{M})}(\overline{x}) \subseteq \overline{L}$. For proving the converse, we observe that $S^i_{\mathcal{P}(\mathbb{M})}(\overline{\emptyset}) \subseteq S'^i_{\mathcal{P}(\mathbb{M})}(\overline{\emptyset})$ for every i (easy proof by induction on i). Hence $\overline{L} = \mu \overline{x} \cdot S_{\mathcal{P}(\mathbb{M})}(\overline{x}) \subseteq \mu \overline{x} \cdot S'_{\mathcal{P}(\mathbb{M})}(\overline{x})$, which gives the desired equality and proves the claim. It follows then from Proposition 3.31 that we can also replace each occurrence of x_m in the right-hand sides of the equations of S' by $\bigcup D$, and that we obtain a system S'' equivalent to S' and to S. But then we can delete the equation $x_m = \bigcup D$ from S'' and get a subsystem that is $(Unk(S) - \{x_m\})$-equivalent to S in $\mathcal{P}(\mathbb{M})$ by Lemma 3.27.

This step can be repeated for each unknown that defines a finite set. Since we assumed that some unknown defines an infinite set, we end up with a system having at least one equation. Its least solution is $(L_{i_1}, \ldots, L_{i_k})$, where $L_{i_1}, \ldots, L_{i_k}$ are the infinite components of $\overline{L}$ and $i_1 < \cdots < i_k$. $\qquad\blacksquare$

Remark 3.40 (1) If S is a system as in the previous proof such that the m-th component T_m of its least solution in $\mathcal{P}(\mathbb{T}(F))$ is finite, then the set L_m is finite by

Proposition 3.23(1), and we can take $D := T_m$ in order to eliminate x_m. However, the converse need not be true, as shown by the example of the equation system $\langle x_1 = x_1 \cdot x_2 \cup a, x_2 = x_2 \cdot x_2 \cup \epsilon \rangle$ associated with a context-free grammar with terminal alphabet $A = \{a\}$, to be solved in $\mathcal{P}(\mathbb{W}(A))$. In this case, $T_2 = T(\{\cdot, \epsilon\})$ is infinite but $L_2 \subseteq A^*$ consists only of the empty word.

(2) The construction described in the proof Proposition 3.39 is effective provided we know, for each $m \in [n]$, whether L_m is finite and, when it is, we also know a finite set of terms D that evaluate to its elements. The second condition is ensured when $\mathbb{M}$ is semi-effectively given and infinity-preserving, see Remark 3.38(3). Then we can compute the elements of L_m, and for each element d of L_m we can compute a term that evaluates to d by enumerating all terms of $T(F)$ and computing their value, cf. Definitions 2.8 and 2.127. The HR algebra $\mathbb{JS}$ and the VR algebra $\mathbb{GP}$ are effectively given and infinity-preserving, and it will be shown in Sections 4.1.4 and 4.3.3 that the finiteness problem is decidable for their equational sets. Thus, Proposition 3.39 is effective for these algebras. □

3.2.3 Using derived operations

We now examine the possibility of using in equation systems derived operations like series-composition in the equation that defines series-parallel graphs (cf. Sections 1.1.3 and 1.4.2). This is interesting for several reasons: equation systems are in this way shorter and more clear, inductive proofs like those considered in Section 1.2.3 use less auxiliary inductive hypotheses and we will obtain a general notion of derivation tree that generalizes the classical one for context-free grammars (cf. Section 1.1.5).

We recall from Definition 2.125 that if H is a set of derived operation declarations of an S-signature F, then for every F-algebra $\mathbb{M}$, we obtain an H-algebra $\mathbb{M}_H$. Its domains are those of $\mathbb{M}$ and its operations are the associated derived operations defined in $\mathbb{M}$ by H. A derived operation g is *linear* if its defining term t_g is linear (i.e., no variable occurs twice or more in it). It is *strict* if each variable y_i for $i \in [\rho(g)]$ has an occurrence in t_g. This condition implies that the corresponding function depends on all its $\rho(g)$ arguments. A derived signature H is *linear* and/or *strict* if all its operations are linear and/or strict.

Proposition 3.41 Let H be a linear derived signature of an S-signature F. For every F-algebra $\mathbb{M}$, we have **Equat**$(\mathbb{M}_H) \subseteq$ **Equat**$(\mathbb{M})$. If, furthermore, $F \subseteq H$, then **Equat**$(\mathbb{M}_H) =$ **Equat**$(\mathbb{M})$.

Proof: Let L in **Equat**$(\mathbb{M}_H)$ be nonempty. Let S be an equation system over H such that L is defined by (S, Y) in $\mathbb{M}_H$ for some $Y \subseteq Unk(S)$. By Proposition 3.33, we can assume that S is uniform. Let us write $S = \langle x_1 = p_1, \ldots, x_n = p_n \rangle$ where, for each i, $p_i = \cdots \cup g(x_{i_1}, \ldots, x_{i_k}) \cup \cdots$. We let $\widehat{p}_i := \cdots \cup t_g[x_{i_1}/y_1, \ldots, x_{i_k}/y_k] \cup \cdots$ where we assume that the linear term t_g defining g is in $T(F, Y_k)_{\sigma(x_i)}$ with $Y_k = \{y_1, \ldots, y_k\}$ and that $\sigma(y_j) = \sigma(x_{i_j})$ for each j. We let $\widehat{S}$ be the equation system $\langle x_1 = \widehat{p}_1, \ldots, x_n = \widehat{p}_n \rangle$ over F.

We first assume that H is strict. The proof is easier, and this particular case will be useful. By Lemma 3.2(1) we have

$$g_{\mathcal{P}(\mathbb{M}_H)}(L_{i_1}, \ldots, L_{i_k}) \subseteq t_g{}_{\mathcal{P}(\mathbb{M})}(L_{i_1}, \ldots, L_{i_k}), \tag{3.2}$$

for all sets $L_1, \ldots, L_n$ of appropriate sorts. However, since H is linear and strict each variable y_j has a unique occurrence in t_g so that by Lemma 3.2(2) we have an equality in (3.2) (even if some of the sets $L_1, \ldots, L_n$ are empty). It follows that $\widehat{S}_{\mathcal{P}(\mathbb{M})} = S_{\mathcal{P}(\mathbb{M}_H)}$ and that $\widehat{S}_{\mathcal{P}(\mathbb{M})}$ and $S_{\mathcal{P}(\mathbb{M}_H)}$ have the same least fixed-point. Hence, the set L is defined by $(\widehat{S}, Y)$ in $\mathbb{M}$ and belongs to **Equat**$(\mathbb{M})$.

We now consider the case where H is linear but not necessarily strict. By Proposition 3.37, we can assume that the system S is trim and still uniform. From inequality (3.2) we get that

$$S_{\mathcal{P}(\mathbb{M}_H)}(\overline{L}) \subseteq \widehat{S}_{\mathcal{P}(\mathbb{M})}(\overline{L}), \tag{3.3}$$

for every n-tuple $\overline{L} = (L_1, \ldots, L_n)$ of sets of appropriate sorts. It follows that

$$\mu \overline{x} \cdot S_{\mathcal{P}(\mathbb{M}_H)}(\overline{x}) \subseteq \mu \overline{x} \cdot \widehat{S}_{\mathcal{P}(\mathbb{M})}(\overline{x}).$$

However, it follows from Lemma 3.2(2) that $S_{\mathcal{P}(\mathbb{M}_H)}(\overline{L}) = \widehat{S}_{\mathcal{P}(\mathbb{M})}(\overline{L})$ if $\overline{L}$ is nonempty, i.e., if its components are all nonempty. Since S is trim, the tuple $\overline{K} := \mu \overline{x} \cdot S_{\mathcal{P}(\mathbb{M}_H)}(\overline{x})$ is nonempty. Hence we have $\overline{K} = S_{\mathcal{P}(\mathbb{M}_H)}(\overline{K}) = \widehat{S}_{\mathcal{P}(\mathbb{M})}(\overline{K})$ and so $\overline{K}$ is a solution of $\widehat{S}$ in $\mathcal{P}(\mathbb{M})$. Hence, by Theorem 3.7, we have $\mu \overline{x} \cdot \widehat{S}_{\mathcal{P}(\mathbb{M})}(\overline{x}) \subseteq \overline{K} = \mu \overline{x} \cdot S_{\mathcal{P}(\mathbb{M}_H)}(\overline{x})$, hence the equality which, as above, gives the desired conclusion.

The second assertion follows from Proposition 3.9(2) because $\mathbb{M}$ is a subalgebra of $\mathbb{M}_H$. ∎

The hypothesis that the derived operations are linear is essential as shown by the following example.

Example 3.42 (1) By Corollary 3.14, the context-free languages over $A := \{a, b\}$ are exactly the equational sets of $\mathbb{W}(A)$. Let us now enrich $\mathbb{W}(A)$ into $\mathbb{W}(A)_{sq}$ by adding the derived unary operation sq defined by $sq(u) := u \cdot u$ for all $u \in \{a, b\}^*$. The equation $x_1 = sq(x_1) \cup a$ defines the set $\{a^{2^n} \mid n \geq 0\}$ that is equational in $\mathbb{W}(A)_{sq}$. It is not a context-free language because $\{2^n \mid n \geq 0\}$ is not semi-linear (cf. Section 3.1.6), hence it is not equational in $\mathbb{W}(A)$.

(2) Next, we consider the system $S = \langle x = g(y, z), y = a, z = f(z) \rangle$ that is not trim, and we let g be defined by $t_g = f(y_1)$. The unknowns x and z of S define the empty set. If we replace g by its definition, we get the system $\langle x = f(y), y = a, z = f(z) \rangle$ for which x no longer defines the empty set. This shows that it is important to use a trim system if H is not strict.

(3) The proof of Corollary 3.26 can be presented in the following way. Let g be an affine mapping : $T(F) \to \mathcal{N}^k$, and let $\mathbb{P}$ be the associated F-algebra, as defined before Corollary 3.26. The operations of $\mathbb{P}$ are linear derived operations of $\mathcal{N}^k$ and $g = val_{\mathbb{P}}$. If $K \subseteq T(F)$ is regular, then $g(K)$ is equational in $\mathbb{P}$ by Proposition 3.23(3), hence equational in $\mathcal{N}^k$ by Proposition 3.41 and semi-linear by Proposition 3.25. $\square$

We now use derived signatures to generalize the algebraic definition of derivation trees that has been given informally in Example 1.3 (Section 1.1.5) for the context-free grammar G of Example 1.1.

Definition 3.43 (Derivation trees and parsing) Let $S = \langle x_1 = p_1, \ldots, x_n = p_n \rangle$ be an equation system over an S-signature F, where, for each i, $p_i = m_{i,1} \cup \cdots \cup m_{i,r_i}$ for some monomials $m_{i,j}$. For $1 \leq i \leq n$, $1 \leq j \leq r_i$, we let $\widetilde{m}_{i,j} \in T(F, \{y_1, \ldots, y_k\})$ be the (unique) term such that $ListVar(\widetilde{m}_{i,j}) = y_1 \cdots y_k$ and $m_{i,j} = \widetilde{m}_{i,j}[x_{i_1}/y_1, \ldots, x_{i_k}/y_k]$ for some $i_1, \ldots, i_k$. (For example, if $m_{i,j} = f(g(x_2,x_1), h(x_4,x_1))$, then $\widetilde{m}_{i,j} = f(g(y_1,y_2), h(y_3,y_4))$ and $(i_1,i_2,i_3,i_4) = (2,1,4,1)$.) This implies that $\sigma(y_l) = \sigma(x_{i_l})$ for each l.[10] We let $q_{i,j}$ be a function symbol of type $\sigma(y_1) \times \cdots \times \sigma(y_k) \to \sigma(x_i)$.

We obtain thus an S-signature $Q := \{q_{i,j} \mid 1 \leq i \leq n, 1 \leq j \leq r_i\}$, associated with the system S. We turn each $q_{i,j} \in Q$ into a linear and strict derived operation of F, defined as $\lambda y_1, \ldots, y_k \cdot \widetilde{m}_{i,j}$. And we let S' be the uniform system $\langle x_1 = p'_1, \ldots, x_n = p'_n \rangle$, where p'_i is obtained from p_i by replacing $m_{i,j}$ by $q_{i,j}(x_{i_1}, \ldots, x_{i_k})$.

For every F-algebra $\mathbb{M}$, we get a Q-algebra $\mathbb{M}_Q$ with the same sorts and the same domains (cf. Definition 2.125). It is a derived algebra of $\mathbb{M}$. By the definitions and Lemma 3.2(4), we have $S_{\mathcal{P}(\mathbb{M})} = S'_{\mathcal{P}(\mathbb{M}_Q)}$. For checking this fact we observe that

$$m_{i,j}\, \mathcal{P}(\mathbb{M})(A_1, \ldots, A_n) = \{\widetilde{m}_{i,j}(a_1, \ldots, a_k) \mid a_1 \in A_{i_1}, \ldots, a_k \in A_{i_k}\}$$
$$= q_{i,j}\, \mathcal{P}(\mathbb{M}_Q)(A_{i_1}, \ldots, A_{i_k}),$$

where the sets $A_1, \ldots, A_n$ may be empty, since Q is a strict and linear derived signature (cf. the proof of Proposition 3.41). Hence we have

$$\mu\bar{x} \cdot S'_{\mathcal{P}(\mathbb{M}_Q)}(\bar{x}) = \mu\bar{x} \cdot S_{\mathcal{P}(\mathbb{M})}(\bar{x}).$$

We now compare the solutions of S and S' in their respective term algebras. The least solution of S' in $\mathcal{P}(\mathbb{T}(Q))$ is an n-tuple of sets of terms $(D_1, \ldots, D_n)$ such that $D_i \subseteq T(Q)_{\sigma(x_i)}$ for each i. By Proposition 3.23(1) we have

$$\mu\bar{x} \cdot S'_{\mathcal{P}(\mathbb{M}_Q)}(\bar{x}) = (val_{\mathbb{M}_Q}(D_1), \ldots, val_{\mathbb{M}_Q}(D_n)).$$

If $d \in D_i$ and $val_{\mathbb{M}_Q}(d) = m$ (which implies that m belongs to $\mu\bar{x} \cdot S_{\mathcal{P}(\mathbb{M})}(\bar{x}) \restriction x_i$), then we say that d is a *derivation of tree of m relative* to (S, x_i). Each element of a set

---

[10] The sort of a variable y_l is not necessarily the same in different terms $\widetilde{m}_{i,j}$. We could waive this "difficulty" by using variables $y_{i,j,l}$ instead of y_l in $\widetilde{m}_{i,j}$, but we prefer lighter notation.

$\mu\bar{x} \cdot S_{\mathcal{P}(\mathbb{M})}(\bar{x}) \upharpoonright x_i$ has at least one derivation tree relative to (S, x_i). If it has a unique one, we will say that S is *unambigous* in $\mathbb{M}$. However, for dealing with graphs, we will almost never be able to use unambiguous systems.

The elements of D_i are called the *derivation trees of S relative to x_i*. A derivation tree $d \in D_i$ can be evaluated in the derived algebra $\mathbb{M}_Q$, but also in $\mathbb{M}$ as follows: by Proposition 2.126, we have $val_{\mathbb{M}_Q}(d) = val_{\mathbb{M}}(\theta_Q(d))$ where θ_Q is the second-order substitution : $T(Q) \to T(F)$ associated with the derived signature Q. It follows that

$$\mu\bar{x} \cdot S_{\mathcal{P}(\mathbb{M})}(\bar{x}) = (val_{\mathbb{M}}(\theta_Q(D_1)), \dots, val_{\mathbb{M}}(\theta_Q(D_n))).$$

In particular, the least solution of S in $\mathcal{P}(\mathbb{T}(F))$ is $(\theta_Q(D_1), \dots, \theta_Q(D_n))$, and $\theta_Q = val_{\mathbb{T}(F)_Q}$ by Proposition 2.126.

Let $\mathbb{M}$ be a semi-effectively given F-algebra and S be an equation system as above with least solution $(L_1, \dots, L_n)$ in $\mathcal{P}(\mathbb{M})$. A *parsing algorithm* for $(S, \mathbb{M})$ takes as input an unknown x_i of S and an element m of M and outputs either the answer that m is not in L_i or a derivation tree d of m relative to (S, x_i). In the latter case, one can compute in linear time in $|d|$ the term $\theta_Q(d) \in T(F)$ that evaluates to m (cf. the discussion before Proposition 2.126).

Example 3.44 We consider the equation

$$T = (T /\!/ (T /\!/ T)) \cup (T \bullet (T /\!/ T)) \cup e \tag{3.4}$$

over the algebra $\mathbb{J}_2^d$ in which we have defined series-parallel graphs in Section 1.1.3. It defines a particular set of series-parallel graphs. This equation can be written as

$$T = q_1(T, T, T) \cup q_2(T, T, T) \cup q_3, \tag{3.5}$$

where $q_1 := \lambda y_1, y_2, y_3 \cdot y_1 /\!/ (y_2 /\!/ y_3)$, $q_2 := \lambda y_1, y_2, y_3 \cdot y_1 \bullet (y_2 /\!/ y_3)$ and $q_3 := e$. Equation (3.4) defines a set of terms in $T(\{/\!/, \bullet, e\})$ containing, to take an example, the term $t = (e \bullet (e /\!/ e)) /\!/ (e /\!/ (e /\!/ (e /\!/ e)))$. The corresponding derivation tree is the term $d = q_1(q_2(q_3, q_3, q_3), q_3, q_1(q_3, q_3, q_3))$, and we have $t = \theta_Q(d)$, where θ_Q is the second-order substitution associated with the above definitions of q_1, q_2, q_3. $\qquad\square$

3.2.4 Closure properties of the class of equational sets

We prove some closure properties of the class of equational sets of general algebras. Further properties particular to graph algebras will be proved in Chapter 4.

Proposition 3.45 Let $\mathbb{M}$ be an F-algebra.

(1) If $L_1, L_2 \in \mathbf{Equat}(\mathbb{M})$, then $L_1 \cup L_2 \in \mathbf{Equat}(\mathbb{M})$.

(2) Every finite subset of $val_{\mathbb{M}}(T(F))$ is equational in $\mathbb{M}$.

(3) If $f \in F$ has type $s_1 \times s_2 \times \cdots \times s_k \to s$ and $L_i \in \mathbf{Equat}(\mathbb{M})_{s_i}$ for each $i = 1, \ldots, k$, then $f_{\mathcal{P}(\mathbb{M})}(L_1, \ldots, L_k) \in \mathbf{Equat}(\mathbb{M})_s$.

Proof: (1) Let L_1 be defined by (S_1, Y_1) in $\mathbb{M}$, and L_2 by (S_2, Y_2). By renaming the unknowns of one of the systems if needed, we may assume that $Unk(S_1)$ and $Unk(S_2)$ are disjoint. Let S be the union of S_1 and S_2. By Lemma 3.27, L_1 is defined by (S, Y_1) and L_2 is defined by (S, Y_2) in $\mathbb{M}$. Hence $L_1 \cup L_2$ is defined by $(S, Y_1 \cup Y_2)$.

(2) By (1), it suffices to prove this for a homogenous subset. If $t_1, \ldots, t_p \in T(F)_s$, then the set $\{t_{1\mathbb{M}}, \ldots, t_{p\mathbb{M}}\} = val_{\mathbb{M}}(\{t_1, \ldots, t_p\})$ is defined by the equation $x_1 = t_1 \cup_s \cdots \cup_s t_p$.

(3) For $k = 2$ the proof is similar to the one for (1), but we add to S the new equation $z = \bigcup\{f(y_1, y_2) \mid y_1 \in Y_1, y_2 \in Y_2\}$. Then $f_{\mathcal{P}(\mathbb{M})}(L_1, L_2)$ is defined by (S, z) in $\mathbb{M}$. The generalization to functions f of any positive arity is clear. The case $k = 0$ follows from (2) (we have in this case $f_{\mathcal{P}(\mathbb{M})} = \{f_{\mathbb{M}}\}$). ∎

3.2.5 Concluding remarks on equational sets

What are equation systems good for?

What are the benefits of defining a set L of words, graphs or other combinatorial objects by an equation system?

First, such a set L has a finite description that can be used by algorithms, e.g., for deciding its emptiness, for computing it in certain cases or for extracting numerical information; Proposition 3.10, Corollary 3.24 and Corollary 3.26 have given such algorithms. Second, every element of L has a denotation by at least one term over the signature of the relevant algebra (by Proposition 3.23(1)) and has a hierarchical structure expressed by its derivation trees and by the corresponding terms (cf. Definition 3.43). Third, certain universal properties of L can be proved inductively as we have seen in Proposition 1.6. This proposition will be generalized below in Proposition 3.91 (Section 3.4.7). Finally, the Filtering Theorem (Theorem 1.8 and Theorem 3.88 below) yields effective constructions of equation systems that define certain subsets of given equational sets.

Applications to graphical objects are developed by Drewes in [*Dre06]. He defines grammars that generate *pictures*, i.e., drawn graphs or images in an abstract setting. His "tree-based approach" is algebraic and the context-free sets he defines are equational.

Some difficulties

There are several problems raised by equation systems and equational sets. First, the parsing problem (cf. Definition 3.43): we have observed at the end of Section 3.1.2 that the membership problem may be undecidable for equational sets of effectively given algebras; the existence of efficient parsing algorithms depends on the considered algebras. For the equational sets of the HR and the VR algebras, the membership problem is always decidable but sometimes NP-complete (see Sections 4.1, 4.3 and 6.2). Every element of an equational set has at least one derivation tree. The nonambiguity of the corresponding system, that is, the case where each generated element has a unique derivation tree, is interesting, in particular for counting the elements of a given size (cf. the book [*FlaSed]). However, even for context-free languages the nonambiguity is a difficult notion: it is undecidable and certain context-free languages have no unambiguous grammar. Nothing can be said in general, and equational sets of graphs are very seldom defined by unambiguous equation systems. A last question concerns the closure of the class of equational sets under transformations that would generalize the rational transductions of words. Apart from the closure under homomorphisms, an easy but fundamental result proved in Proposition 3.23(1), not much can be said for general algebras. But a theory of graph transductions, based on monadic second-order logic, informally presented in Section 1.7, will be developed in Chapter 7. These transductions preserve the equationality of sets of graphs.

We now review some notions related to equation systems.

μ-calculus

The language of μ-*calculus*, studied in the book by Arnold and Niwinski [*ArnNiw], offers some notation for equational sets. For example, the least solution of the system $S = \langle x = f(x,y) \cup a, y = g(x,y) \cup b \rangle$ in every powerset $\{f,g,a,b\}$-algebra is denoted by the μ-term[11] $\mu(x,y) \cdot (f(x,y) \cup a, g(x,y) \cup b)$ of the *vectorial* μ-calculus (Section 2.7 of [*ArnNiw]). This term evaluates to $\mu(x,y) \cdot S_{\mathcal{P}(\mathbb{M})}(x,y)$ in each powerset $\{f,g,a,b\}$-algebra $\mathcal{P}(\mathbb{M})$.

Furthermore, each component of the least solution of an equation system is the value of a μ-term that only uses least fixed-points over variables, not over tuples of variables. For example, $\mu(x,y) \cdot S_{\mathcal{P}(\mathbb{M})}(x,y) \upharpoonright x$ is the value in $\mathcal{P}(\mathbb{M})$ of the μ-term $\mu x \cdot (f(x, \mu y \cdot (g(x,y) \cup b)) \cup a)$ (Lemma 1.4.2 of [*ArnNiw]).

However, the general μ-terms can be written with set intersection and a greatest fixed-point operator (that is dual in some sense to the least fixed-point one). Hence, they can also define sets that are not equational: for example, the intersection of two context-free languages (even without using greatest fixed-points). The emptiness of the corresponding sets is thus undecidable, whereas that of equational sets is decidable.

Rational expressions

The *rational* sets of a monoid $\mathbb{M}$ are the members of the least class of subsets of M that contains the finite sets and is closed under union, the multiplication of $\mathbb{M}$ extended to sets and denoted by $\cdot_{\mathcal{P}(\mathbb{M})}$, and the *star operation*, that is defined as follows (where $\cdot$

---

[11] The terms of the μ-calculus are called μ-*terms*.

denotes $\cdot_{\mathcal{P}(\mathbb{M})}$ and 1 denotes $1_{\mathbb{M}}$):

> if $L \subseteq M$, then L^* is the least subset X of M such that:
>
> $X = L \cdot X \cup \{1\}$
>
> (or equivalently $X = X \cdot L \cup \{1\}$ or $X = X \cdot X \cup L \cup \{1\}$).

Rational sets can thus be defined by μ-terms, but a regular expression like $a(ab^*)^* ac^*$ that defines a regular (also called *rational*) language is more readable than its translation into a μ-term.

The rational sets of a monoid $\mathbb{M}$ are equational in $\mathbb{M}$ and form in most cases a proper subclass of **Equat**$(\mathbb{M})$, denoted by **Rat**$(\mathbb{M})$. However, in a commutative monoid every equational set is rational (this is proved in [AceEI] and entails Proposition 3.25).

If $\mathbb{M}$ is the monoid of words over a finite alphabet, then **Rat**$(\mathbb{M})$ is the class of regular languages by a well-known result due to Kleene (cf. [*Eil, *Sak]). For this reason, rational sets are frequently presented in connection with finite automata.[12] However, **Rat**$(\mathbb{M})$ is more relevant to **Equat**$(\mathbb{M})$ than to the class **Rec**$(\mathbb{M})$ of recognizable sets that generalizes the class of regular languages. (We will define and study this class in Section 3.4.) More will be said about rational sets in Section 3.4.10.

3.3 Intermezzo on automata

Finite automata on terms correspond to regular grammars, and hence define the equational sets of terms. Deterministic automata have the same expressive power as the general ones, but, furthermore, correspond to finite congruences. Hence they define the recognizable sets of terms, which turn out to be the same as the equational sets. Recognizability has been introduced in Section 1.2 and will be the subject of Section 3.4. In this section, we review the definitions and properties of those automata on terms that are closely linked (in different ways) to the equational and recognizable sets in general algebras. Automata on terms are studied in detail in [*Com+] and [*GecSte].

The equational sets of the term algebra $\mathbb{T}(F)$ are important (by Proposition 3.23(1)) because, for every F-algebra $\mathbb{M}$, a subset L of M is equational if and only if it is $val_{\mathbb{M}}(K)$ for some equational set K of $\mathbb{T}(F)$. Hence, every term in K can be used as a linear notation for an element of L. Furthermore, if $t \in K$ and $val_{\mathbb{M}}(t) = m$, then t defines a kind of hierarchical decomposition of m because if $t = f(t_1, t_2)$, then m is the composition of $val_{\mathbb{M}}(t_1)$ and $val_{\mathbb{M}}(t_2)$ by $f_{\mathbb{M}}$ and the same holds for t_1, t_2 and all their subterms. Algorithms for checking properties of elements of equational sets can be based on such decompositions, as we will see for graphs in Chapter 6. Thus, we obtain linear and structured notations for the elements of equational sets.

---

[12] The book by Sakarovich [*Sak] interleaves finite automata, rationality and recognizability without emphasizing the fundamental differences between these three notions.

The relationships between recognizability and deterministic automata will be developed in Section 3.4.2. We review, and even slightly extend, the classical notion of an automaton to be run on terms.

3.3.1 Automata on terms

In this section and the next, sets of sorts and signatures are finite.

Definition 3.46 (Automata) Let F be a finite S-signature.[13] An F-*automaton* (or just *automaton* if F need not be specified)[14] is a 4-tuple $\mathscr{A} = \langle F, Q_{\mathscr{A}}, \delta_{\mathscr{A}}, Acc_{\mathscr{A}} \rangle$ such that $Q_{\mathscr{A}}$ is a finite or infinite S-sorted set called the set of *states* ($\sigma(q) \in S$ is the sort of $q \in Q_{\mathscr{A}}$), $Acc_{\mathscr{A}}$ is a subset of $Q_{\mathscr{A}}$ called the set of *accepting states* and $\delta_{\mathscr{A}}$ is a set of tuples called the set of *transition rules*, satisfying the following two conditions (we recall that F_k denotes the set of symbols of F of arity k):

(1) Each element of $\delta_{\mathscr{A}}$ is of one of the two possible forms:

    (1.1) $(q_1, \ldots, q_k, f, q)$ for $k \geq 0$, $q_1, \ldots, q_k, q \in Q_{\mathscr{A}}, f \in F_k$;
    (1.2) (q, q') for $q, q' \in Q_{\mathscr{A}}$; these pairs are called the ε-transitions.

(2) The sort mapping $\sigma : Q_{\mathscr{A}} \to S$ satisfies the following conditions:

    (2.1) f has type $\sigma(q_1) \times \cdots \times \sigma(q_k) \to \sigma(q)$ in any tuple of the form (1.1);
    (2.2) $\sigma(q) = \sigma(q')$ for every ε-transition (q, q').

For better readability, we will denote by $f[q_1, \ldots, q_k] \to_{\mathscr{A}} q$ (and by $f \to_{\mathscr{A}} q$ if $f \in F_0$) a transition rule of type (1.1) and by $q' \to_{\mathscr{A}} q$ an ε-transition. Each of these transition rules is said to *yield* q (the last component of the tuples).

A triple $\mathscr{A} = \langle F, Q_{\mathscr{A}}, \delta_{\mathscr{A}} \rangle$ where F, $Q_{\mathscr{A}}$ and $\delta_{\mathscr{A}}$ are as above is called an F-*semi-automaton*. Hence, it is just an F-automaton without accepting states. The definitions given below for semi-automata extend to automata in the obvious way.

A semi-automaton $\mathscr{A}$ is ε-*free* if it has no ε-transitions. It is *finite* if its set of states $Q_{\mathscr{A}}$ is finite (which implies that $\delta_{\mathscr{A}}$ is finite).

We denote by $\sharp\mathscr{A} := |Q_{\mathscr{A}}|$ the cardinality of $Q_{\mathscr{A}}$. We define the *size* of $\mathscr{A}$, denoted by $\|\mathscr{A}\|$, as the sum of weights of its transition rules where the weight of a rule relative to a function symbol f is $\rho(f) + 1$ and that of an ε-transition is 2. Hence we have

$$\|\mathscr{A}\| \leq 2 \cdot (\sharp\mathscr{A})^2 + \Sigma_{f \in F}(\rho(f) + 1) \cdot (\sharp\mathscr{A})^{\rho(f)+1}.$$

If the function symbols are at most binary, which will be frequently the case for the automata of Chapter 6, then $\|\mathscr{A}\| = O(|F| \cdot (\sharp\mathscr{A})^3)$.

The size of $\mathscr{A}$ is thus proportional to the space needed to store its transition rules as a list of tuples, where we assume that states use constant space. If the number

---

[13] We recall from Definition 2.123 that this implies that S is finite.
[14] These automata are frequently called bottom-up (or frontier-to-root) tree automata in the literature on tree language theory.

of states is very large (which will be the case in Chapter 6), then a state may need $\lceil \log(\sharp \mathscr{A}) \rceil$ bits to be stored.[15]

A *run* of an ε-free semi-automaton $\mathscr{A}$ on a term $t \in T(F)$ is a mapping $r : Pos(t) \to Q_{\mathscr{A}}$ such that:

(i) if u is an occurrence of a constant symbol f, then $f \to_{\mathscr{A}} r(u)$
(ii) if u is an occurrence of a function symbol $f \in F_k$, for $k > 0$, with sequence of sons[16] $u_1, \ldots, u_k$, then $f[r(u_1), \ldots, r(u_k)] \to_{\mathscr{A}} r(u)$.

If $\mathscr{A}$ is not ε-free, then $\to_{\varepsilon}^*$ is the reflexive and transitive closure of the set of ε-transitions. We let $\delta_{\mathscr{A}}^{\varepsilon}$ be the set of transition rules $f[q_1, \ldots, q_k] \to q$ such that $f[q_1, \ldots, q_k] \to_{\mathscr{A}} q'$ for some q' such that $q' \to_{\varepsilon}^* q$. We have $\delta_{\mathscr{A}} \subseteq \delta_{\mathscr{A}}^{\varepsilon}$. A run of $\mathscr{A}$ is defined as a run of the ε-free semi-automaton $\langle F, Q_{\mathscr{A}}, \delta_{\mathscr{A}}^{\varepsilon} \rangle$.

For $q \in Q_{\mathscr{A}}$, we let $L(\mathscr{A}, q)$ be the set of terms t in $T(F)$ on which there is a run r of $\mathscr{A}$ such that $r(root_t) = q$. It is clear that $L(\mathscr{A}, q) \subseteq T(F)_{\sigma(q)}$ for every q in $Q_{\mathscr{A}}$.

If $\mathscr{A}$ is an automaton, then a run r on t is *accepting* if $r(root_t)$ is an accepting state. We let $L(\mathscr{A}) := \bigcup_{q \in Acc_{\mathscr{A}}} L(\mathscr{A}, q) \subseteq T(F)$. It is the set of terms on which $\mathscr{A}$ has an accepting run. We say that $L(\mathscr{A})$ is the *language accepted* (or *recognized*) by $\mathscr{A}$. Two automata are *equivalent* if they accept the same language.

Regular grammars

We now relate automata with regular grammars and equation systems. With every finite F-semi-automaton $\mathscr{A} = \langle F, Q, \delta \rangle$, we associate a regular grammar $G(\mathscr{A})$ over F (without initial nonterminal) as follows: its set of nonterminals is Q and its rules are $q' \to q$ if $q \to_{\mathscr{A}} q'$ and $q \to f(q_1, \ldots, q_k)$ if $f[q_1, \ldots, q_k] \to_{\mathscr{A}} q$. It is clear that $L(G(\mathscr{A}), q) = L(\mathscr{A}, q)$ for every state q. If $\mathscr{A}$ is an automaton, then $L(\mathscr{A})$ is the union of the languages generated by $G(\mathscr{A})$ from its nonterminals that are the accepting states of $\mathscr{A}$, i.e., $L(\mathscr{A}) = L(G(\mathscr{A}), Acc_{\mathscr{A}})$.

By Definition 3.17, there is an equation system $S(\mathscr{A})$ such that $G(S(\mathscr{A})) = G(\mathscr{A})$. The grammar $G(\mathscr{A})$ and the equation system $S(\mathscr{A})$ are quasi-uniform. If $\mathscr{A}$ is ε-free, then they are uniform. Conversely, every quasi-uniform (resp. uniform) regular grammar is $G(\mathscr{A})$ for some finite semi-automaton (resp. some finite ε-free semi-automaton) $\mathscr{A}$, and similarly for $S(\mathscr{A})$. From Proposition 3.33 we get the following proposition[17] that subsumes Corollary 3.19 for a finite $\mathcal{S}$-signature F:

Proposition 3.47 Let F be finite and $L \subseteq T(F)$.

(1) The language L is accepted by a finite F-automaton if and only if it is regular over F, if and only if it is equational in $\mathbb{T}(F)$.

---

[15] Space efficient representations of automata are used in the software MONA: see [BasKla], [Hen+], [Kla].

[16] The term "son" refers to the syntactic tree of t, cf. Definition 2.14 in Section 2.2.

[17] We recall from Definition 3.8 that a set is defined by an equation system if it is a union of components of its least solution in the considered powerset algebra that are all of the same sort.

(2) It is accepted by an automaton with n states if and only if it is defined by a uniform equation system with n unknowns, if and only if it is generated by a uniform regular grammar with n nonterminals. $\square$

The term "regular" for a language will refer to its definition by finite automata as well as by regular grammars (in the case where F is finite).

Special types of automata

Definition 3.48 (Trim and reduced automata) A state q of a semi-automaton $\mathscr{A}$ is *accessible* if $L(\mathscr{A}, q) \neq \emptyset$, i.e., if it occurs in a run of $\mathscr{A}$ on some term, equivalently if it is a productive nonterminal of the grammar $G(\mathscr{A})$. We say that $\mathscr{A}$ is *trim* if its states are all accessible. An automaton $\mathscr{A}$ is *reduced* if each state belongs to the image of an accepting run, i.e., is *useful* for accepting some term.

Obviously, one can *trim* a semi-automaton $\mathscr{A}$ by trimming the corresponding equation system $S(\mathscr{A})$, i.e., by constructing $Trim(S(\mathscr{A}))$ and turning that system back into a semi-automaton. Similarly, one can *reduce* an automaton $\mathscr{A}$ by constructing $Trim(S(\mathscr{A}), Acc_{\mathscr{A}})$ and turning that back into an automaton. Thus, by Proposition 3.37, one can effectively transform each semi-automaton $\mathscr{A}$ into a trim semi-automaton $\mathscr{B}$ such that $Q_{\mathscr{B}} = \{q \in Q_{\mathscr{A}} \mid L(\mathscr{A}, q) \neq \emptyset\}$ and $L(\mathscr{B}, q) = L(\mathscr{A}, q)$ for every $q \in Q_{\mathscr{B}}$. Similarly, every automaton can be effectively transformed into an equivalent reduced automaton.

Definition 3.49 (Deterministic and complete automata) An F-semi-automaton $\mathscr{A}$ is *deterministic* if it is ε-free and, for every $q_1, \ldots, q_k$ and every $f \in F_k$, there is at most one state q such that $f[q_1, \ldots, q_k] \to_{\mathscr{A}} q$. It is *complete* if, for every $q_1, \ldots, q_k$ and every f in F_k such that $\alpha(f) = (\sigma(q_1), \ldots, \sigma(q_k))$, there is a state q such that $f[q_1, \ldots, q_k] \to_{\mathscr{A}} q$. If $\mathscr{A}$ is deterministic and complete, then its size $\|A\|$ (cf. Definition 3.46) is exactly $\Sigma_{f \in F}(\rho(f) + 1) \cdot (\sharp \mathscr{A})^{\rho(f)}$.

By the addition of at most one state of each sort (such states are usually called *sinks*), one can transform a deterministic semi-automaton $\mathscr{A}$ into a complete and deterministic one $\mathscr{B}$ such that $L(\mathscr{B}, q) = L(\mathscr{A}, q)$ for every state q of $\mathscr{A}$.

Let $\mathscr{A}$ be deterministic and complete: on each $t \in T(F)$ it has a unique run, which will be denoted by $run_{\mathscr{A}, t}$. If the transition relation is a *computable* function,[18] which means that for given function symbol f and states $q_1, \ldots, q_{\rho(f)}$ of appropriate sorts the *unique* state q such that $f[q_1, \ldots, q_{\rho(f)}] \to_{\mathscr{A}} q$ can be computed by an algorithm, then this run can be computed during a bottom-up traversal of t. This computation takes time $a \cdot |t|$ where a is an upper-bound to the time taken to perform a transition, that is, to find or compute from $f, q_1, \ldots, q_k$ the state q such that $f[q_1, \ldots, q_k] \to_{\mathscr{A}} q$. This value a is significant if the transition has to be computed, but also if the automaton

---

[18] This condition is not trivial since the set of states may be infinite (but should be effectively given).

is finite but is so large that some time (that is no longer considered as constant) is
required to find the appropriate transition in a table.[19]

We recall the classical *determinization* of automata (cf. Section 5 of [*GecSte] or
Theorem 1.1.9 of [*Com+]).

Proposition 3.50 For every finite automaton $\mathscr{A}$, one can construct a trim, complete
and deterministic finite automaton $\mathscr{B}$ that is equivalent to $\mathscr{A}$. □

We only recall that the states of $\mathscr{B}$ of sort s are sets of states of $\mathscr{A}$ of sort s and that
$|Q_{\mathscr{B},s}| \leq 2^{|Q_{\mathscr{A},s}|}$, where $Q_{\mathscr{A},s}$ is the set of states of sort s of $\mathscr{A}$ and similarly for $\mathscr{B}$.

Assuming fixed some determinization algorithm, we will denote $\mathscr{B}$ by $det(\mathscr{A})$.
In particular, we will assume that $det(\mathscr{A}) = \mathscr{A}$ if $\mathscr{A}$ is already trim, complete and
deterministic.

Remark 3.51 (1) For every many-sorted signature F, we let F^- be the one-sorted
signature obtained from F by forgetting sorts (cf. Lemma 3.21). Every F-automaton
$\mathscr{A}$ is an F^--automaton and, thus, every regular language $L \subseteq T(F)$ is regular over
F^-. Conversely, a regular language L over F^- such that $L \subseteq T(F)$, is regular over F
by Corollary 3.19 and the observations made in Remark 3.38(4).

(2) Automata on words are just particular automata on terms, because words over
an alphabet A correspond bijectively to terms over the unary functional signature U_A,
as explained in Definition 2.7. More precisely, a word $w := a_1 \cdots a_n$ with $a_1, \dots, a_n$
in A is the value of the term $t := a_n(\cdots(a_1(\epsilon))\cdots)$ in the U_A-algebra $\mathbb{W}_{right}(A)$.
The runs of a U_A-automaton $\mathscr{A}$ on the term t correspond bijectively to those of an
automaton $\mathscr{B}$ (with the same set of states) on the word w. We do not detail the obvious
correspondence between the automaton $\mathscr{A}$ on terms and the automaton $\mathscr{B}$ on words.
Thus, as already explained in Remark 3.20(4) and after Proposition 3.23, a language
over the alphabet A is regular if and only if the corresponding set of terms is regular
over U_A.[20] □

Allowing infinite sets of states in automata makes it possible to enrich them so that,
in addition to checking if a term t belongs to a regular language, an automaton can
compute some value attached to t. This feature will be used in Chapter 6, but we give
immediately an example.

Example 3.52 We let $F = \{f, a, b\}$, $\rho(f) = 2$, $\rho(a) = \rho(b) = 0$ and L be the set of
terms in $T(F)$ that have an even number of occurrences of a. The following automaton

---

[19] In most classical uses of automata, e.g., in compilation, the size of the input is much larger than the
number of states and the value a may be considered as a constant. But in Chapter 6, we will construct
automata that are much larger than their intended input terms.

[20] In the U_A-algebra $\mathbb{W}_{left}(A)$, the word w corresponds to the term $t' := a_1(\cdots(a_n(\epsilon))\cdots)$. In this case,
the runs of an automaton on w correspond to those of a "top-down (or root-to-frontier) automaton" on
t' (cf. Section 8.1.2).

$\mathscr{A}$ accepts L and computes $ht(t)$, the height of t. It is defined as follows:

$$
\begin{aligned}
Q_{\mathscr{A}} &:= \{q_0, q_1\} \times \mathcal{N}, \\
Acc_{\mathscr{A}} &:= \{q_0\} \times \mathcal{N},
\end{aligned}
$$

$\delta_{\mathscr{A}}$ is the set of rules:

$$
a \to (q_1, 0), \ b \to (q_0, 0), \ f[(q_i, n), (q_j, p)] \to (q_l, r)
$$

such that $r = \max\{n, p\} + 1$, $i, j, l \in \{0, 1\}$, $l = \mathrm{mod}_2(i + j)$.

It is clear that $\mathscr{A}$ is complete and deterministic, that $L(\mathscr{A}) = L$ and that, for $t \in L$, we have $ht(t) = n$ if and only if $t \in L(\mathscr{A}, (q_0, n))$. Hence, the unique run of $\mathscr{A}$ on a term t not only evaluates the relevant parity information (by means of the first components of the states) but it also evaluates the heights of the subterms of t. Whether t is accepted or not by $\mathscr{A}$ depends only on the parity information and not on the heights of subterms. Hence, the language $L(\mathscr{A})$ is regular. $\qquad\square$

3.3.2 Pumping arguments

The following proposition collects results proved by so-called *pumping arguments*. The *size* of a term is the number of its positions, or its length if it is considered as a word (without commas and parentheses, cf. Definition 2.3). We let F be a finite signature of maximal arity $\rho(F) = k \geq 1$, consisting of c constant symbols and d symbols of positive arity. The size of a term in $T(F)$ of height $m \geq 0$ (a constant has height 0) is bounded by $\flat_k(m)$, where

$$
\flat_k(m) := 1 + k + k^2 + \cdots + k^m.
$$

The number of terms of height at most m is bounded by $\natural_{k,c,d}(m)$ defined by

$$
\natural_{k,c,d}(m) := (c + d + 1)^{\flat_k(m)},
$$

which gives

$$
\natural_{k,c,d}(m) \leq (c + d + 1)^{k^{m+1}} \quad \text{for } k \geq 2,
$$

and

$$
\natural_{1,c,d}(m) \leq (c + d + 1)^{m+1} \quad \text{for } k = 1.
$$

However, to have better bounds, we can use the following instead:

$$
\natural_{k,c,d}(0) := c \quad \text{and} \quad \natural_{k,c,d}(m) := c^{k^m} \cdot (d+1)^{\flat_k(m-1)} \quad \text{for } m \geq 1.
$$

Proposition 3.53 Let F be a finite signature of maximal arity $k \geq 1$ having c constant symbols and d symbols of positive arity. Let $L \subseteq T(F)$ be nonempty and accepted by a finite F-automaton $\mathscr{A}$ with $m + 1$ states or defined by a uniform equation system with $m + 1$ unknowns ($m \geq 0$).

(1) The height of a term of minimal size in L is at most m and its size is at most $\flat_k(m)$.
(2) The finiteness of L is decidable. If L is finite, then the height of a term in L is at most m and its size is at most $\flat_k(m)$. Furthermore, L can be computed and its cardinality is at most $\natural_{k,c,d}(m)$.

Proof: By Proposition 3.47, we need only consider the case where $L = L(\mathscr{A})$. We use a classical argument (cf. Chapter 1 of [*Com+] or Proposition 5.2 of [*GecSte]). If $\mathscr{A}$ has an accepting run r on a term t such that $r(u) = r(v)$ for some positions u, v with $u <_t v$, then $L = L(\mathscr{A})$ contains a term smaller than t and infinitely many terms larger than t.

To prove this, we write $t = c[c'[t']]$ where $c, c' \in Ctxt(F)$, c' is not empty, $t' = t/u$ and $c'[t'] = t/v$. Then $c[t']$ belongs to $L(\mathscr{A})$ and is smaller than t, whereas all terms $c[c'[c'[\cdots[c'[t']]\cdots]]]$ with at least two copies of c' belong to $L(\mathscr{A})$, are pairwise different and larger than t.

The bounds on heights, sizes and numbers of terms in the statement follow immediately. It is straightforward to show that L is infinite if and only if it contains a term of height at least $m + 1$ and at most $2m + 2$ (by choosing c' such that the length of the path from the root to the context variable is at most $m + 1$). This implies the decidability of finiteness.[21]   ∎

By such arguments, one can prove that certain languages are not regular. Here is an example to be used later.

Example 3.54 We prove the claim made in Remark 3.20(1) that the language $L = \{fg^n ag^n b \mid n \geq 0\}$ (with $\rho(f) = 2$, $\rho(g) = 1$, $\rho(a) = \rho(b) = 0$) is not regular. If we assume it is, then it is accepted by a complete deterministic finite automaton (by Proposition 3.50). There exist two integers n and $m < n$ such that the unique runs of this automaton on the terms $g^m a$ and $g^n a$ yield the same state. It follows that the term $fg^m ag^n b$ is accepted by this automaton but does not belong to L. This contradicts the initial assumption. Hence L is not regular.

3.4 The recognizable sets of an algebra

In the previous three sections, we have defined and studied the equational sets of an algebra. These sets generalize the context-free languages, and also the regular languages (of words and terms). We now define the recognizable sets, and so generalize the characterization of regular languages formulated in terms of finite congruences.

3.4.1 Definitions and examples

An F-algebra $\mathbb{A}$ is *locally finite* if each domain A_s is finite. It is *finite* if, furthermore, its set of sorts S is finite (although F may be infinite).

---

[21] For another method to decide the finiteness of L, see Sections 4.1 and 4.3 of the next chapter.

Definition 3.55 (Recognizable sets of an algebra) Let $\mathbb{M}$ be an F-algebra. A *recognizable set* of $\mathbb{M}$ is a set $L \subseteq M$ of the form $L = h^{-1}(C)$ where $h : \mathbb{M} \to \mathbb{A}$ is a homomorphism of F-algebras, $\mathbb{A}$ is locally finite and $C \subseteq A$. We will say that h *witnesses* the recognizability of L. We may in addition require that h is surjective. If this is not the case, we replace $\mathbb{A}$ by its subalgebra that is the image of $\mathbb{M}$ under h, and the set C by its intersection with $h(M)$. The requirement that h is surjective does not change the notion of recognizability.

We denote by **Rec**($\mathbb{M}$) the set of recognizable sets of $\mathbb{M}$ and by **Rec**($\mathbb{M}$)$_s$ the set of those included in M_s for s in S. We will also say that a set in **Rec**($\mathbb{M}$) is $\mathbb{M}$-*recognizable* or *recognizable in* $\mathbb{M}$. Just as equational sets, recognizable sets need not be homogenous, i.e., they may have elements of different sorts; unlike equational sets, they may have elements of countably many different sorts.

Proposition 3.56

(1) Let $\mathbb{M}$ be an F-algebra and $\mathbb{M}'$ be an F'-algebra such that $\mathbb{M}' \subseteq \mathbb{M}$. If $L \in \textbf{Rec}(\mathbb{M})$, then $L \cap M' \in \textbf{Rec}(\mathbb{M}')$. In particular, $\textbf{Rec}(\mathbb{M}) \cap \mathcal{P}(M') \subseteq \textbf{Rec}(\mathbb{M}')$.

(2) Let $k : \mathbb{M}' \to \mathbb{M}$ be a homomorphism of F-algebras. If $L \in \textbf{Rec}(\mathbb{M})$, then $k^{-1}(L) \in \textbf{Rec}(\mathbb{M}')$.

(3) If H is a derived signature of F and $\mathbb{M}$ is an F-algebra, then $\textbf{Rec}(\mathbb{M}) \subseteq \textbf{Rec}(\mathbb{M}_H)$. If, furthermore, $F \subseteq H$, then $\textbf{Rec}(\mathbb{M}) = \textbf{Rec}(\mathbb{M}_H)$.

Proof: (1) Let the homomorphism $h : \mathbb{M} \to \mathbb{A}$ of F-algebras witness the recognizability of L in $\mathbb{M}$. Thus, $L = h^{-1}(C)$ with $C \subseteq A$. Let $\mathbb{A}' := \mathbb{A} \upharpoonright F'$, cf. Definition 2.123. Then the restriction h' of h to M' is a homomorphism : $\mathbb{M}' \to \mathbb{A}'$ that witnesses the recognizability of $L \cap M'$ in $\mathbb{M}'$, because $L \cap M' = (h')^{-1}(C \cap A')$.

(2) If h is as in (1), then the homomorphism $h \circ k : \mathbb{M} \to \mathbb{A}$ witnesses the recognizability of $k^{-1}(L)$, with the same C.

(3) Let $h : \mathbb{M} \to \mathbb{A}$ witness the recognizability in $\mathbb{M}$ of a set L. Since h is a homomorphism, we have $h(t_{\mathbb{M}}(m_1, \ldots, m_n)) = t_{\mathbb{A}}(h(m_1), \ldots, h(m_n))$ for every $t \in T(F, X_n)$ and $m_1, \ldots, m_n \in M$ of appropriate sorts. Hence, h is a homomorphism : $\mathbb{M}_H \to \mathbb{A}_H$. It witnesses the recognizability of L in $\mathbb{M}_H$, with the same C. The last assertion follows from (1) because then $\mathbb{M} \subseteq \mathbb{M}_H$. $\blacksquare$

Note that, by the first assertion of this proposition, if we enrich an algebra by adding new operations, then the class of recognizable sets decreases or remains the same. It remains the same if we add or delete constants. In fact, if $h : \mathbb{M} \to \mathbb{A}$ is a homomorphism and $c_{\mathbb{M}}$ is a new constant added to $\mathbb{M}$, then h remains a homomorphism if we define $c_{\mathbb{A}} := h(c_{\mathbb{M}})$.

We had a statement similar to the first assertion for equational sets in Proposition 3.9(2), but with an opposite inequality. The second assertion is similar to the one in Proposition 3.23(2), but in the opposite direction. Finally, a statement similar to the third assertion is in Proposition 3.41 but only for linear derived operations, and with an opposite inequality. Other results of this kind will be stated in Proposition 3.85.

Example 3.57 (1) Let A be the alphabet $\{a,b\}$. The language $L = a^*b^*$ is recognizable in the unary algebra $\mathbb{W}_{right}(A)$ (cf. Definition 2.7). To see this we consider the U_A-algebra $\mathbb{B}$ with domain $\{1,2,Error\}$, with constant $\epsilon_\mathbb{B} := 1$ and operations such that

$$a_\mathbb{B}(1) := 1, \qquad\qquad a_\mathbb{B}(2) := a_\mathbb{B}(Error) := Error,$$

$$b_\mathbb{B}(1) := b_\mathbb{B}(2) := 2, \quad b_\mathbb{B}(Error) := Error.$$

Then $L = h^{-1}(\{1,2\})$, where h is the unique homomorphism of $\mathbb{W}_{right}(A)$ (isomorphic to $\mathbb{T}(U_A)$) into $\mathbb{B}$. We have $h^{-1}(1) = a^*$, $h^{-1}(2) = a^*b^+$ and $h^{-1}(Error) = A^*baA^*$.

(2) We let A and L be as in (1). The language L is also recognizable in $\mathbb{W}(A)$ because $L = l^{-1}(\{0,1,2,3\})$, where l is the unique homomorphism of $\mathbb{W}(A)$ into the F_A-algebra $\mathbb{D}$ such that $D := \{0,1,2,3,Error\}$, $\epsilon_\mathbb{D} := 0$, $a_\mathbb{D} := 1$, $b_\mathbb{D} := 2$ and the operation $\cdot_\mathbb{D}$ is defined by

$0 \cdot_\mathbb{D} x := x \cdot_\mathbb{D} 0 := x$ for every x in D,
$1 \cdot_\mathbb{D} 1 := 1, \quad 1 \cdot_\mathbb{D} 2 := 3, \quad 1 \cdot_\mathbb{D} 3 := 3,$
$2 \cdot_\mathbb{D} 2 := 2, \quad 3 \cdot_\mathbb{D} 2 := 3,$ and
$x \cdot_\mathbb{D} y := Error$ in all other cases.

We have then $l^{-1}(0) = \{\varepsilon\}$, $l^{-1}(1) = a^+$, $l^{-1}(2) = b^+$, $l^{-1}(3) = a^+b^+$ and $l^{-1}(Error) = A^*baA^*$.

These two examples are particular cases of theorems that relate recognizability and finite automata. In particular, for every finite alphabet A, we have

$$\mathbf{Rec}(\mathbb{W}(A)) = \mathbf{Rec}(\mathbb{W}_{left}(A)) = \mathbf{Rec}(\mathbb{W}_{right}(A)),$$

and this is the class of regular languages over A (see [*Eil], or [*Sak], Chapter II, Theorem 2.3).

(3) We let $K := \{4 + 3 \cdot \lambda \mid \lambda \in \mathcal{N}\} \subseteq \mathcal{N}$. This set is recognizable in the monoid $\langle \mathcal{N}, +, 0 \rangle$ (which we also denote by $\mathbb{N}$) because $K = k^{-1}(\{4\})$ where k is the unique homomorphism of $\mathbb{N}$ into the monoid $\mathbb{E}$ such that $E := \{0,1,2,3,4\}$, $k(1) := 1$, the unit of $\mathbb{E}$ is 0 (so that $k(0) = 0$) and the addition in $\mathbb{E}$ is defined by

$$x +_\mathbb{E} y := \begin{cases} x + y & \text{if } x + y \leq 4, \\ x + y - 3 & \text{if } 4 < x + y \leq 7, \\ x + y - 6 & \text{if } x + y > 7. \end{cases}$$

We have then $k^{-1}(0) = \{0\}$, $k^{-1}(1) = \{1\}$, $k^{-1}(2) = \{2 + 3 \cdot \lambda \mid \lambda \in \mathcal{N}\}$, $k^{-1}(3) = \{3 + 3 \cdot \lambda \mid \lambda \in \mathcal{N}\}$ and $k^{-1}(4) = K$. This is a particular case of a characterization of the $\mathbb{N}$-recognizable sets (cf. Proposition 3.93 below in Section 3.4.8).

(4) Let S be a finite set of sorts and F be a (possibly infinite) S-signature. For each sort s in S the set $T(F)_s$ is recognizable in $\mathbb{T}(F^-)$ and so is $T(F)$. We recall, cf. Remark 3.51(1), that F^- is the one-sorted signature obtained from F by "forgetting

sorts". For proving this, we define an F^--algebra $\mathbb{S}_\perp$ with domain $\mathcal{S}_\perp := \mathcal{S} \cup \{\perp\}$ and operations $f_{\mathbb{S}_\perp}$ defined by

$$f_{\mathbb{S}_\perp}(s_1,\ldots,s_{\rho(f)}) := \begin{cases} \sigma(f) & \text{if } \alpha(f) = (s_1,\ldots,s_{\rho(f)}), \\ \perp & \text{otherwise.} \end{cases}$$

It is clear that the mapping $h : T(F^-) \to \mathcal{S}_\perp$ such that:

$$h(t) := \begin{cases} s & \text{if } t \in T(F)_s \text{ for } s \in \mathcal{S}, \\ \perp & \text{otherwise,} \end{cases}$$

is a homomorphism. Hence the sets $T(F)_s = h^{-1}(\{s\})$ and the set $T(F) = h^{-1}(\mathcal{S})$ are recognizable in $\mathbb{T}(F^-)$.

(5) Let $\mathbb{M}$ be an F-algebra for an $\mathcal{S}$-signature with $\mathcal{S}$ possibly infinite. Each domain M_s is recognizable in $\mathbb{M}$ and so is M. A similar statement for equationality is in Proposition 3.23(4). To prove this, we use the F-algebra $\mathbb{S}$ defined as $\mathbb{S}_\perp$ in the previous example but without $\perp$, and where $\{s\}$ is the domain of sort s for each s in $\mathcal{S}$. We omit the easy verification.

(6) If L is recognizable in $\mathbb{M}$, then $L \cap M_s$ is also recognizable in $\mathbb{M}$ for each sort s, because, if L is as in Definition 3.55, then we have $L \cap M_s = h^{-1}(C \cap A_s)$. The converse is false. Let $\mathcal{S} := \mathcal{N}$, $F := \{a, f, g_1, \ldots, g_n, \ldots\}$ with $\rho(a) := 0$, $\sigma(a) := 0$, f of type $0 \to 0$ and g_i of type $0 \to i$ for each $i \geq 1$. Let $L := \{g_i f^i a \mid i \geq 1\}$. For each $i \geq 0$ the set $L \cap T(F)_i$ is the singleton set $\{g_i f^i a\}$ or the empty set, hence is recognizable in $\mathbb{T}(F)$ (say by Theorem 3.62 below) but L is not. We will prove this last fact by means of congruences in Example 3.71 in Section 3.4.3.   □

3.4.2 Recognizable sets of terms

Before exposing the general properties of recognizable sets, we show that a set of terms over a finite signature is recognizable if and only if it is accepted by a finite, complete and deterministic automaton (cf. Section 1.2.4).

Definition 3.58 (Deterministic automata and algebras) Let F be a finite $\mathcal{S}$-signature, let $\mathbb{B} = \langle (B_s)_{s \in \mathcal{S}}, (f_\mathbb{B})_{f \in F} \rangle$ be an F-algebra and let $C \subseteq B$. We let $\mathcal{A} := \mathcal{A}(\mathbb{B}, C)$ be the complete and deterministic F-automaton such that $Q_\mathcal{A} := B$, $Acc_\mathcal{A} := C$ and

$$\delta_\mathcal{A} := \{f[b_1, \ldots, b_{\rho(f)}] \to f_\mathbb{B}(b_1, \ldots, b_{\rho(f)}) \mid f \in F, b_i \in B_{\alpha(f)[i]} \text{ for } i \in [\rho(f)]\}.$$

The set of states of $\mathcal{A}$ is $\mathcal{S}$-sorted with sort mapping σ such that $\sigma(b) = s$ if and only if $b \in B_s$.[22] We let $\mathcal{A}(\mathbb{B})$ be the corresponding semi-automaton.

---

[22] We recall that the sets B_s are pairwise disjoint.

Conversely, every complete and deterministic F-automaton $\mathscr{A}$ is $\mathscr{A}(\mathbb{B}, C)$ for some F-algebra $\mathbb{B}$ with $B_s := \{q \in Q_{\mathscr{A}} \mid \sigma(q) = s\}$ and $C := Acc_{\mathscr{A}}$. The following lemma is clear from the definitions.

Lemma 3.59 Let F be a finite signature and $\mathbb{B}$ be an F-algebra.

(1) For every term $t \in T(F)$, the unique run $run_{\mathscr{A}(\mathbb{B}),t} : Pos(t) \to B$ of the semi-automaton $\mathscr{A}(\mathbb{B})$ on t verifies $run_{\mathscr{A}(\mathbb{B}),t}(u) = val_{\mathbb{B}}(t/u)$ for all $u \in Pos(t)$. We have $L(\mathscr{A}(\mathbb{B}, C)) = val_{\mathbb{B}}^{-1}(C)$ for every subset C of B.

(2) The semi-automaton $\mathscr{A}(\mathbb{B})$ is trim if and only if the algebra $\mathbb{B}$ is generated by F. □

Note that $\mathbb{B}$ need not be finite for this lemma to be valid. If $\mathbb{B}$ is finite (equivalently, is locally finite since S is assumed finite) then $\mathscr{A}(\mathbb{B}, C)$ is a finite automaton, and vice versa. We obtain that a subset of $T(F)$ is recognizable in $\mathbb{T}(F)$ if and only if it is accepted by a finite F-automaton. Before stating this as a theorem that will extend Corollary 3.19 and Proposition 3.47(1), we consider sets of terms over infinite signatures. A recognizable set of $\mathbb{T}(F)$ such that F is infinite is not necessarily a subset of $T(F')$ for some finite subsignature F' of F because $T(F)$ itself is recognizable (by Example 3.57(5)). For the sake of comparison, we recall from Proposition 3.9 that an equational set of terms over an infinite signature F uses only a finite subsignature F' of F and is equational in $\mathbb{T}(F')$.

We first prove that the recognizability of a set of terms over a signature F' is the same when considered with respect to $\mathbb{T}(F')$ and to $\mathbb{T}(F)$, where F is any signature containing F' as a subsignature. Hence, the recognizability of a set of terms is an intrinsic property,[23] although this is not completely evident from the definition.

Proposition 3.60 Let F' be a subsignature of a signature F. A subset L of $T(F')$ is $\mathbb{T}(F)$-recognizable if and only if it is $\mathbb{T}(F')$-recognizable.

Proof: Let S' be the set of sorts of F'. The term algebra $\mathbb{T}(F')$ has no domain of sort s if $s \in S - S'$, and $T(F')_s \subseteq T(F)_s$ if $s \in S'$.

Since $\mathbb{T}(F') \subseteq \mathbb{T}(F)$, Proposition 3.56(1) implies that every $\mathbb{T}(F)$-recognizable subset of $T(F')$ is $\mathbb{T}(F')$-recognizable. Let us conversely assume that $L = val_{\mathbb{A}'}^{-1}(C)$ for a locally finite F'-algebra $\mathbb{A}'$ and $C \subseteq A'$. For every $s \in S$, let $\perp_s$ be a new element of sort s. Let $\mathbb{A}$ be the locally finite F-algebra with domains $A_s = A'_s \cup \{\perp_s\}$ for $s \in S'$, and $A_s = \{\perp_s\}$ for $s \in S - S'$, and with operations $f_{\mathbb{A}}$, for $f \in F$, defined as follows (where f has type $s_1 \times \cdots \times s_k \to s$): for all $a_1, \ldots, a_k \in A$ such that $a_i \in A_{s_i}$ for every $i \in [k]$, we let $f_{\mathbb{A}}(a_1, \ldots, a_k) := f_{\mathbb{A}'}(a_1, \ldots, a_k)$ if $a_1, \ldots, a_k \in A'$ and $f \in F'$, and $f_{\mathbb{A}}(a_1, \ldots, a_k) := \perp_s$ otherwise. Obviously, $val_{\mathbb{A}}(t) = val_{\mathbb{A}'}(t)$ for every $t \in T(F')$. Hence $L = val_{\mathbb{A}}^{-1}(C)$ is recognizable in $\mathbb{T}(F)$. ∎

---

[23] Here we assume that the arities of the function symbols that occur in the terms of the set are fixed, cf. Remark 3.20(1).

Example 3.61 (Recognizable sets of words over infinite alphabets) If A is a finite or infinite alphabet, then $\mathbb{W}(A)$ is the F_A-algebra $\langle A^*, \cdot, \varepsilon, (a)_{a \in A} \rangle$ where each letter a is a constant symbol denoting itself. (It is the free monoid generated by the set of constants A.) A recognizable subset of $\mathbb{W}(A)$ is a language $L \subseteq A^*$ of the form $h^{-1}(C)$, where h is a homomorphism $: \mathbb{W}(A) \to \mathbb{B} = \langle B, \cdot_{\mathbb{B}}, \epsilon_{\mathbb{B}}, (h(a))_{a \in A} \rangle$ such that $\langle B, \cdot_{\mathbb{B}}, \epsilon_{\mathbb{B}} \rangle$ is a finite monoid and $C \subseteq B$. If A is finite, then $L \subseteq A^*$ is recognizable in $\mathbb{W}(A)$ if and only if it is regular, see Example 3.57(2).

We now assume that A is (countably) infinite. For each $b \in B$, let $A_b := A \cap h^{-1}(b)$. We obtain a partition $\{A_b\}_{b \in B}$ of A. Consider each element b of B as a letter of a new finite alphabet $\overline{B}$. Let K be the set of words $w \in \overline{B}^*$ such that $k(w) \in C$ where k is the unique homomorphism $: \mathbb{W}(\overline{B}) \to \mathbb{B}$ extending the identity $: \overline{B} \to B$ (so that $k(b) = b$). It is clear that K is a recognizable language over a finite alphabet and that $L = \sigma(K)$ where σ is the substitution that replaces a letter b in a word in K by any letter in A_b (where different letters of A_b can be substituted at distinct occurrences of b).

This shows that recognizable sets of words over infinite alphabets can be described, via the substitution σ, in terms of regular languages hence of finite automata. A similar description could be done for recognizable subsets of $\mathbb{T}(F)$ where F is an infinite signature.

To complete the picture, we observe that if A is infinite and $L \in \mathbf{Rec}(\mathbb{W}(A))$, then $L \cap B^*$ is recognizable for every finite subset B of A (by Proposition 3.56(1) because $\mathbb{W}(B) \subseteq \mathbb{W}(A)$). However, the converse is false. Let A be the infinite alphabet $\{a, b_0, b_1, \ldots, b_n, \ldots\}$. The set $L := \{a^n b_n \mid n \in \mathcal{N}\}$ is not recognizable in $\mathbb{W}(A)$ (the proof will be given in Example 3.71) but $L \cap B^*$ is finite, hence recognizable for every finite set $B \subseteq A$. ☐

The following theorem extends Corollary 3.19 and Proposition 3.47(1):

Theorem 3.62 Let F be a signature and let $L \subseteq T(F)$. The following properties are equivalent:
(1) L is equational in $\mathbb{T}(F)$;
(2) L is regular over F;
(3) L is recognizable in $\mathbb{T}(F')$ for some finite subsignature F' of F;
(4) L is recognizable in $\mathbb{T}(F)$ and $L \subseteq T(F')$ for some finite subsignature F' of F.

If additionally F is finite, then the following two properties are equivalent to properties (1)–(4):

(5) L is recognizable in $\mathbb{T}(F)$;
(6) L is accepted by a finite F-automaton.

Proof: The equivalence of (1) and (2) follows from Corollary 3.19.

We now prove the implication (2) $\Longrightarrow$ (3). If $L \subseteq T(F)$ is regular, then $L \subseteq T(F')$ for some finite subsignature F' of F and (by Propositions 3.47 and 3.50) is defined by a complete and deterministic finite F'-automaton equal to $\mathscr{A}(\mathbb{B},C)$ for some $(\mathbb{B},C)$ (cf. Definition 3.58). It follows from Lemma 3.59 that $L = L(\mathscr{A}(\mathbb{B},C))$ is recognizable in $\mathbb{T}(F')$. The opposite implication follows from Definition 3.58 and Lemma 3.59: if $L = val_{\mathbb{B}}^{-1}(C)$, then $L = L(\mathscr{A}(\mathbb{B},C))$.

The equivalence of (3) and (4) is an immediate consequence of Proposition 3.60. If F is finite then (4) and (5) are obviously equivalent, and the equivalence of (2) and (6) follows from Proposition 3.47(1). ∎

Thus, for a finite signature, the families of equational, regular and recognizable sets of terms are all equal to the family of languages accepted by finite automata.

3.4.3 Recognizability and congruences

Definition 3.63 (Locally finite congruences) Let $\mathbb{M}$ be an F-algebra with set of sorts $\mathcal{S}$. An equivalence relation $\sim$ on M is *sort-preserving* if any two equivalent elements have the same sort. We write $m \sim_s m'$ if $m \sim m'$ and $\sigma(m) = \sigma(m') = s$. The *index* of a sort-preserving equivalence relation $\sim$ is the mapping $\gamma_\sim : \mathcal{S} \to \mathcal{N} \cup \{\omega\}$ such that $\gamma_\sim(s)$ is the number of equivalence classes of $\sim_s$. The equivalence $\sim$ is *locally finite* if $\gamma_\sim(s) \in \mathcal{N}$ for each s. If the set of sorts is finite, a sort-preserving equivalence relation is locally finite if and only if it is *finite* i.e., has finitely many classes. Finally, an equivalence relation *saturates* a subset of M if this set is a union of equivalence classes, not necessarily all of the same sort.

A *congruence* on $\mathbb{M}$ (or an *F-congruence* if we need to specify the relevant signature) is a sort-preserving equivalence relation on M such that, for every f in F of positive arity, the following holds (where $s_1 \times \cdots \times s_k \to s$ is the type of f): for every $m_1,\ldots,m_k, m'_1,\ldots,m'_k$ in M, if $m_1 \sim_{s_1} m'_1,\ldots,m_k \sim_{s_k} m'_k$ then[24]

$$f_{\mathbb{M}}(m_1,\ldots,m_k) \sim_s f_{\mathbb{M}}(m'_1,\ldots,m'_k).$$

If $\sim$ is a congruence on $\mathbb{M}$, the *quotient* F-algebra $\mathbb{M}/\sim$ is defined as $\langle (M_s/\sim_s)_{s\in\mathcal{S}}, (f_{\mathbb{M}/\sim})_{f\in F}\rangle$ where $f_{\mathbb{M}/\sim}([m_1],\ldots,[m_k]) := [f_{\mathbb{M}}(m_1,\ldots,m_k)]$.[25] The canonical surjective mapping $h_\sim$ such that $h_\sim(m) = [m]$ is an F-algebra homomorphism $: \mathbb{M} \to \mathbb{M}/\sim$. The quotient algebra $\mathbb{M}/\sim$ is locally finite if and only if the congruence $\sim$ is locally finite.

Proposition 3.64 Let $\mathbb{M}$ be an F-algebra. A subset L of M is recognizable in $\mathbb{M}$ if and only if it is saturated by a locally finite congruence on $\mathbb{M}$.

---

[24] If this preservation property holds for all operations, it also holds for all derived operations.
[25] For an equivalence relation $\sim$ on a set A, we denote by $[a]_\sim$ (or just $[a]$) the equivalence class of $a \in A$, and $C/\sim := \{[a]_\sim \mid a \in C\}$ for every $C \subseteq A$. We also denote $C/\sim$ by $[C]_\sim$ (or just $[C]$).

Proof: Let $L = h^{-1}(C)$, where $h : \mathbb{M} \to \mathbb{A}$ is a homomorphism and $\mathbb{A}$ is locally finite. The equivalence relation on M defined by $m \sim^h m'$ if and only if $h(m) = h(m')$ is a locally finite congruence and L is the union of the equivalence classes $h^{-1}(a)$ for $a \in C$.

Conversely, if $L \subseteq M$ is a union of classes of a locally finite congruence $\sim$, then we let $\mathbb{A} := \mathbb{M}/\sim$ be the quotient algebra with domains $A_s = M_s/\sim_s$. It is locally finite and $L = h_\sim^{-1}([L]_\sim)$ where $[L]_\sim \subseteq A$ and $h_\sim$ is the canonical surjective homomorphism $: \mathbb{M} \to \mathbb{A}$. ∎

We will say that a locally finite congruence $\sim$ that saturates L, *witnesses* the recognizability of L.

We now associate with every set a canonical congruence that saturates it.

Definition 3.65 (Syntactic congruence) Let $\mathbb{M}$ be an F-algebra and $L \subseteq M$. We define on M an equivalence relation as follows: $m \approx^L m'$ if and only if m and m' have the same sort, say s, and for every linear[26] term $c \in T(F, Y_p)$, where $p \geq 1$, $Y_p = \{y_1, \ldots, y_p\}$ and $\sigma(y_1) = s$, we have, for all $m_i \in M_{\sigma(y_i)}$, $i = 2, \ldots, p$:

$$c_{\mathbb{M}}(m, m_2, \ldots, m_p) \in L \iff c_{\mathbb{M}}(m', m_2, \ldots, m_p) \in L.$$

It is clear that $\approx^L$ is a sort-preserving equivalence relation. It saturates L: by taking $c = y_1$, we get that $m \in L$ and $m \approx^L m'$ imply $m' \in L$. If F generates $\mathbb{M}$ one can take $p = 1$ to define $\approx^L$ (cf. the remarks on extended derived operations in Definition 2.125); in other words, it suffices to consider contexts $c \in Ctxt(F)$.

The equivalence $\approx^L$ (denoted by $\approx$ if L is clear from the context) is called the *syntactic congruence* of L. The index of $\approx^L$, denoted by γ_L, is called the *recognizability index* of L. The following proposition generalizes a well-known result for regular (word) languages.

Proposition 3.66 Let $\mathbb{M}$ be an F-algebra and $L \subseteq M$.
(1) The equivalence relation $\approx^L$ is a congruence on $\mathbb{M}$. The set L is recognizable in $\mathbb{M}$ if and only if this congruence is locally finite.
(2) The syntactic congruence $\approx^L$ of a recognizable set L is the unique congruence $\approx$ that saturates L and is such that $\gamma_\approx(s) \leq \gamma_\sim(s)$ for every sort s and every congruence $\sim$ that witnesses the recognizability of L.

---

[26] We recall that a term in $T(F, Y_p)$ is linear if each variable of Y_p has most one occurrence. Actually, by the proof of the next proposition, the equivalence relation $\approx^L$ is the same if we omit the restriction to linear terms.

Proof: (1) We write $\approx$ instead of $\approx^L$. If $f \in F_k$, $m_1 \approx m'_1, \ldots, m_k \approx m'_k$ and $\alpha(f) = (\sigma(m_1), \ldots, \sigma(m_k))$ then

$$f_{\mathbb{M}}(m_1, m_2, \ldots, m_k) \approx f_{\mathbb{M}}(m'_1, m_2, \ldots, m_k),$$
$$f_{\mathbb{M}}(m'_1, m_2, m_3, \ldots, m_k) \approx f_{\mathbb{M}}(m'_1, m'_2, m_3, \ldots, m_k),$$
$$\vdots$$
$$f_{\mathbb{M}}(m'_1, m'_2, \ldots, m'_{k-1}, m_k) \approx f_{\mathbb{M}}(m'_1, \ldots, m'_{k-1}, m'_k),$$

which gives $f_{\mathbb{M}}(m_1, \ldots, m_k) \approx f_{\mathbb{M}}(m'_1, \ldots, m'_k)$. Hence, $\approx$ is a congruence. It saturates L and (by Proposition 3.64) L is recognizable if $\approx$ is locally finite.

Conversely, let L be a union of classes of a locally finite congruence $\sim$ on $\mathbb{M}$. We claim that, for all all $m, m' \in M$, we have

$$m \sim m' \implies m \approx m'.$$

Let m and m' be such that $m \sim m'$. For all $c \in T(F, Y_p)$ and all $m_2, \ldots, m_p$, we have:

$$c_{\mathbb{M}}(m, m_2, \ldots, m_p) \in L \implies c_{\mathbb{M}}(m', m_2, \ldots, m_p) \in L. \tag{3.6}$$

This is so because $\sim$ is a congruence, so that for $c, m_2, \ldots, m_p$ as above we have $c_{\mathbb{M}}(m, m_2, \ldots, m_p) \sim c_{\mathbb{M}}(m', m_2, \ldots, m_p)$, and hence (3.6) holds, since $\sim$ saturates L. It follows from the claim that each equivalence class of $\approx$ is a union of equivalence classes of $\sim$ and so $\gamma_{\approx}(s) \leq \gamma_{\sim}(s)$ for every sort s. Since $\sim$ is locally finite, the same holds for $\approx$.

(2) The minimality property of $\approx^L$ was shown in the proof of (1). For proving the unicity, we consider any congruence $\sim$ with the same minimality property. Each equivalence class of $\approx_s$ is a union of equivalence classes of $\sim_s$. If $\sim \neq \approx$, then $\gamma_{\approx}(s) < \gamma_{\sim}(s)$ for some s, hence $\sim$ is not minimal. Hence $\sim = \approx$, as was to be proved. ∎

Example 3.67 (Recognizable languages of words) Let $L \subseteq A^*$, where A is a finite alphabet. We will examine the syntactic congruences $\approx^L$ relative to the three algebras of words that we have defined in Definition 2.7. We first consider the F_A-algebra $\mathbb{W}(A)$. Since $\mathbb{W}(A)$ is generated by F_A, it suffices to consider contexts. For every context c in $Ctxt(F_A)$ there exist words x and y in A^* such that $c_{\mathbb{W}(A)}(w) = xwy$ for every w in A^*. For every x and y in A^*, the function $\lambda w \in A^* \cdot xwy$ is $c_{\mathbb{W}(A)}$ for some c in $Ctxt(F_A)$. It follows that the equivalence relation $\approx^L$ is such that, for $w, w' \in A^*$:

$$w \approx^L w' \text{ if and only if } xwy \in L \Longleftrightarrow xw'y \in L \text{ for all } x, y \in A^*,$$

which is the standard definition of the syntactic congruence of a language. This congruence is finite if and only if L is defined by a finite automaton (see Theorem 2.3 in Chapter II of [*Sak] or see [*Eil] for the constructions that establish this equivalence).

We now consider the algebra $\mathbb{W}_{right}(A)$ that is generated by its signature U_A. The contexts in $Ctxt(U_A)$ are the terms of the form $c = xu$ with $x \in A^*$ (and u is the context variable). Such a context defines the function $\lambda w \in A^* \cdot wy$, where y is the mirror image of x, and every such function is defined by a context. It follows that for every language $L \subseteq A^*$ the syntactic congruence $\approx^L_{right}$ relative to this algebra satisfies for all $w, w' \in A^*$:

$$w \approx^L_{right} w' \text{ if and only if } wy \in L \Longleftrightarrow w'y \in L \text{ for all } y \in A^*.$$

Its finiteness, i.e., the recognizability of L in $\mathbb{W}_{right}(A)$ is equivalent to its definability by a finite automaton.

The function defined by $c = xu$ in $\mathbb{W}_{left}(A)$ is $\lambda w \in A^* \cdot xw$. The syntactic congruence $\approx^L_{left}$ relative to this algebra satisfies for all $w, w' \in A^*$:

$$w \approx^L_{left} w' \text{ if and only if } xw \in L \Longleftrightarrow xw' \in L \text{ for all } x \in A^*,$$

and its finiteness is equivalent to regularity. The last two characterizations of regular languages are due to Myhill and Nerode. (See [*Sak], Theorem 2.3, Chapter II for their proofs in a language theoretic framework.) □

We now consider the recognizable sets of an algebra that is generated by its signature.

Lemma 3.68 Let $h : \mathbb{M} \to \mathbb{N}$ be a homomorphism of F-algebras, where $\mathbb{M}$ and $\mathbb{N}$ are generated by F. Let $L \subseteq N$ and $K = h^{-1}(L) \subseteq M$. Then, for all m and m' in M, $m \approx^K m'$ if and only if $h(m) \approx^L h(m')$. The homomorphism $g : \mathbb{M}/\approx^K \to \mathbb{N}/\approx^L$ induced by h, is an isomorphism, and $g([K]_{\approx^K}) = [L]_{\approx^L}$. The sets K and L have the same recognizability index, i.e., $\gamma_K = \gamma_L$.

Proof: Since $\mathbb{N}$ is generated by F, h is surjective. Let m and m' be such that $m \approx^K m'$. For proving that $h(m) \approx^L h(m')$, we consider an arbitrary context c in $Ctxt(F)$. We have $c_\mathbb{N}(h(m)) = h(c_\mathbb{M}(m))$. Hence, if $c_\mathbb{N}(h(m)) \in L$, we have $c_\mathbb{M}(m) \in K$ and thus $c_\mathbb{M}(m') \in K$ and $c_\mathbb{N}(h(m')) \in L$. Since F generates $\mathbb{N}$, the syntactic congruence of L can be defined in terms of contexts, hence we have $h(m) \approx^L h(m')$. That $h(m) \approx^L h(m')$ implies $m \approx^K m'$ can be proved similarly. We get from h a homomorphism: $\mathbb{M} \to \mathbb{N}/\approx^L$ that factorizes through $\mathbb{M}/\approx^K$ and induces an isomorphism g by the first assertion. Since g is defined by $g([m]) := [h(m)]$, we have $g([K]_{\approx^K}) = [L]_{\approx^L}$. Since g is sort-preserving, there is a bijection from $M_s/\approx^K_s$ to $N_s/\approx^L_s$ for each sort s. Thus, the recognizability indices of K and L are the same. ■

Proposition 3.69 Let $\mathbb{M}$ be an F-algebra generated by F and let $L \subseteq M$. The set L and the language $val_\mathbb{M}^{-1}(L)$ have the same recognizability index. Thus, L is recognizable in $\mathbb{M}$ if and only if $val_\mathbb{M}^{-1}(L)$ is recognizable in $\mathbb{T}(F)$.

Proof: Since F generates both $\mathbb{M}$ and $\mathbb{T}(F)$, we can apply Lemma 3.68 to the homomorphism $val_{\mathbb{M}} : \mathbb{T}(F) \to \mathbb{M}$. Hence $K := val_{\mathbb{M}}^{-1}(L)$ and L have the same recognizability index. The second assertion is an immediate consequence of the first assertion and Proposition 3.66(1). ∎

Remark 3.70 (1) Lemma 3.68 still holds if, instead of assuming that $\mathbb{M}$ and $\mathbb{N}$ are generated by F, we only assume that the homomorphism h is surjective. The proof is essentially the same. It uses linear terms in $T(F, Y)$ instead of contexts in $Ctxt(F)$ (cf. Definition 3.65).

(2) Proposition 3.69 is somehow "dual" to Proposition 3.23(3), which characterizes the $\mathbb{M}$-equational sets as the images under $val_{\mathbb{M}}$ of the recognizable sets of terms over the relevant finite subsignature (see Theorem 3.62(2–4)). The present characterization of the recognizable sets of $\mathbb{M}$ is also based on the recognizable sets of terms (over a possibly infinite signature), but uses their inverse images under the mapping $val_{\mathbb{M}}$. Similarly, Lemma 3.68 is "dual" to Proposition 3.23(2).

(3) In the statement of Proposition 3.69, it is not enough to assume that L (instead of $\mathbb{M}$) is generated by F in order to have $L \in \mathbf{Rec}(\mathbb{M})$ if $val_{\mathbb{M}}^{-1}(L) \in \mathbf{Rec}(\mathbb{T}(F))$. Here is a counter-example. We take $\mathbb{Z}$ consisting of the integers with $F_2 = \{-\}$ and $F_0 = \{0\}$ (we have $x + y = x - (0 - y)$). Then $\mathbb{Z}$ is not generated by F, but $\{0\}$ is. The set $val_{\mathbb{Z}}^{-1}(0)$ is equal to $T(F)$, hence is recognizable in $\mathbb{T}(F)$ but $\{0\}$ is not recognizable in $\mathbb{Z}$, which we prove as follows. Assume it is, with witnessing congruence $\sim$. At least one of the equivalence classes of $\sim$ is infinite. Since $\{0\}$ is saturated, $[0]_{\sim} = \{0\}$. Hence, there exist n, p in $\mathbb{Z}$ such that $n < p$ with $p \sim n \not\sim 0$. Since $\sim$ is a congruence, this implies that $p - n \sim n - n = 0$. Hence, $[0]_{\sim}$ would contain $p - n \neq 0$, which contradicts the fact that $[0]_{\sim} = \{0\}$. □

Example 3.71 We have claimed in Example 3.57(6) that the language $L = \{g_i f^i a \mid i \geq 1\}$ is not recognizable. This follows from Proposition 3.66(1) and the observation that the congruence $\approx^L$ has infinitely many equivalence classes of sort 0. This is so because, for all i, j with $0 < i < j$, we have $g_i f^i a \in L$ and $g_i f^j a \notin L$ so that $f^i a \not\approx^L f^j a$.

The proof is similar for the language $L = \{a^n b_n \mid n \geq 0\}$ of Example 3.61, cf. Example 3.67. For every $i > 0$ and $j > i$ we have $a^i \not\approx^L a^j$ because $a^i \cdot b_i \in L$ and $a^j \cdot b_i \notin L$. □

Our aim is now to review the classical relation between syntactic congruences and minimal automata.

Definition 3.72 (Minimal automata) Let $\mathbb{M}$ be a finitely generated F-algebra. For every (locally) finite congruence $\sim$ on $\mathbb{M}$, the F-semi-automaton $\mathscr{A}(\mathbb{M}/\sim)$ is finite, trim, complete and deterministic. Conversely, every F-semi-automaton $\mathscr{A}$ with these properties is of this form, and more precisely is (up to renaming of states) $\mathscr{A}(\mathbb{T}(F)/\sim)$, where $\sim$ is defined by $t \sim t'$ if and only if t and t' belong to $L(\mathscr{A}, q)$ for some $q \in Q_{\mathscr{A}}$.

If $L \subseteq M$ is recognizable, then the F-automaton $\mathscr{A}(\mathbb{M}/{\approx}^L, [L])$ is called the *minimal automaton* of L and is denoted by $\mathscr{M}(L)$. By Lemma 3.68 (applied to $val_{\mathbb{M}}$), it is[27] the automaton $\mathscr{A}(\mathbb{T}(F)/{\approx}^K, [K])$ where $K := val_{\mathbb{M}}^{-1}(L)$. Thus, $\mathscr{M}(L)$ is also the minimal automaton of $val_{\mathbb{M}}^{-1}(L)$, and it accepts the language $val_{\mathbb{M}}^{-1}(L)$.

Proposition 3.73 Let F be a finite signature and K be a recognizable set of $\mathbb{T}(F)$.

(1) If $\mathscr{A}$ is any (possibly infinite) trim, complete and deterministic automaton recognizing K, then it is of the form $\mathscr{A}(\mathbb{B}, C)$ for some F-algebra $\mathbb{B}$ and some set $C \subseteq B$, and furthermore $\mathscr{A}(\mathbb{B}/{\approx}^C, [C])$ is the minimal automaton of K.

(2) The minimal automaton of K is the unique complete and deterministic F-automaton with a minimal number of states that recognizes K. Its number of states is $\sum_{s \in S} \gamma_K(s)$, where S is the set of sorts of F.

Proof: (1) By Lemma 3.59(2), the algebra $\mathbb{B}$ is generated by F. Hence, Lemma 3.68 is applicable to $h := val_{\mathbb{B}}$ and $L := C$, and yields the result (because $K = val_{\mathbb{B}}^{-1}(C)$).

(2) Let $\mathscr{A}$ be a complete and deterministic F-automaton with a minimal number of states that recognizes K. It is trim because otherwise, it can be replaced by a smaller complete and deterministic one that recognizes K. Hence, $\mathscr{A} = \mathscr{A}(\mathbb{B}, C)$ with $\mathbb{B}$ and C as in (1). We have a canonical surjective homomorphism h from $\mathbb{B}$ to $\mathbb{B}/{\approx}^C$. The automaton $\mathscr{A}(\mathbb{B}/{\approx}^C, [C])$ is, by (1), the same as $\mathscr{M}(K)$. If h is not an isomorphism, then the automaton $\mathscr{A}(\mathbb{B}, C)$ does not have a minimal number of states. Hence, $\mathscr{A}$ is the same as $\mathscr{M}(K)$, and $\mathscr{M}(K)$ is the unique (up to renaming of states) complete and deterministic F-automaton with a minimal number of states that recognizes K. The last assertion follows from its definition. ∎

Assertion (1) of this proposition shows how the minimal automaton can be obtained from any finite, trim, complete and deterministic automaton recognizing K, hence from any finite automaton by Proposition 3.50. A minimization algorithm taking time $O(\|\mathscr{A}\|^2)$ (for $\mathscr{A}$ deterministic) is described in [CarDF].

3.4.4 Effective recognizability

Effectively given one-sorted and many-sorted algebras have been defined in Definitions 2.8 and 2.127, and an application to equational sets has been given in Section 3.1.2. We now extend these definitions to recognizability.

Throughout this section we assume that every signature F is effectively given, with a fixed encoding ξ_F (and a fixed encoding ξ_S of its set of sorts S).[28]

Definition 3.74 (Effectively recognizable sets) Let $\mathbb{M}$ be an F-algebra. A set L in **Rec**$(\mathbb{M})$ is *effectively recognizable* (resp. *semi-effectively recognizable*) if $L =$

---

[27] Here and in Proposition 3.73 we do not distinguish two automata that are the same up to a renaming of states. Formally, we would have to define isomorphism of automata.

[28] Recall from Definition 2.8 that, for an encoded set A, ξ_A is the bijection : $A \to enc_A \subseteq \mathcal{N}$ that encodes the elements of A as natural numbers.

$h^{-1}(C)$, where $h : \mathbb{M} \to \mathbb{A}$ is a homomorphism to a locally finite and effectively given (resp. semi-effectively given) F-algebra $\mathbb{A}$ and $C \subseteq A$, and furthermore:

(1) C is a decidable subset of A (cf. Definition 2.8), and

(2) $\mathbb{M}$ is semi-effectively given and the homomorphism h is computable (cf. Definition 2.8).

These conditions imply that L is a decidable subset of M: for $m \in M$, one computes $h(m)$ and checks whether it is in C (or more precisely, if m is given by its code number $m' = \xi_M(m)$, one computes the integer $\widetilde{h}(m') = \xi_A(h(\xi_M^{-1}(m'))) = \xi_A(h(m))$ and one checks whether it encodes an element of C).

If Condition (2) is dropped in the above definition of an effectively recognizable set L, then we will say that L is *effectively term-recognizable*. In this case it is decidable for any term t in $T(F)$ whether $val_\mathbb{M}(t)$ is in L: since $h \circ val_\mathbb{M} = val_\mathbb{A}$, one computes $val_\mathbb{A}(t)$, cf. Definition 2.8, and checks whether it is in C. Moreover, as shown below in Proposition 3.76(4), if F is finite then a finite F-automaton can be constructed that implements this algorithm.

Effectively recognizable sets (and their variants) will be used as input and output of algorithms. Such a set L is specified by the specification of $\mathbb{A}$, an algorithm that decides membership in C and, if Condition (2) holds, a specification of $\mathbb{M}$ and an algorithm that computes h.

In the definition of an effectively recognizable set (and its two variants), we may assume that every domain A_s of $\mathbb{A}$ is nonempty. In fact, if $\mathbb{A}$ is semi-effectively given, then we define the F-algebra $\mathbb{A}'$ such that $A'_s := A_s \cup \{\perp_s\}$ and, for every $f \in F_k$ and $a_1, \ldots, a_k \in A'$ of appropriate sorts, $f_{\mathbb{A}'}(a_1, \ldots, a_k) := f_\mathbb{A}(a_1, \ldots, a_k)$ if $a_1, \ldots, a_k \in A$ and $f_{\mathbb{A}'}(a_1, \ldots, a_k) := \perp_{\sigma(f)}$ otherwise. If A is encoded by ξ_A and the set $\mathcal{S}$ of sorts of F is encoded by $\xi_\mathcal{S}$, then we encode A' as follows: $\xi_{A'}(a) := 2 \cdot \xi_A(a)$ for $a \in A$ and $\xi_{A'}(\perp_s) := 2 \cdot \xi_\mathcal{S}(s) + 1$ for $s \in \mathcal{S}$. It is easy to verify that $\mathbb{A}'$ is semi-effectively given with this encoding $\xi_{A'}$. If $\mathbb{A}$ is effectively given, then so is $\mathbb{A}'$, because $|A'| = \omega$ if $|\mathcal{S}| = \omega$ and $|A'| = \sum_{s \in \mathcal{S}}(|A_s| + 1)$ if $\mathcal{S}$ is finite. If Conditions (1) and (2) hold for $\mathbb{A}$, then they also hold for $\mathbb{A}'$ (with the same h and C).

The following lemma shows that every finitely generated subalgebra of a semi-effectively given algebra satisfies Condition (2) above.

Lemma 3.75

(1) Let $\mathbb{M}$ be an F-algebra generated by F and let $h : \mathbb{M} \to \mathbb{A}$ be a homomorphism. If $\mathbb{M}$ and $\mathbb{A}$ are semi-effectively given, then h is computable.

(2) Let $\mathbb{M}$ be a semi-effectively given F-algebra and let $\mathbb{M}'$ be the subalgebra of $\mathbb{M}$ generated by a finite subsignature F' of F.[29] Then $\mathbb{M}'$ is semi-effectively given and the mapping $in_{M', M}$ is computable.[30]

---

[29] Here we assume that F' is effectively given by its standard encoding as a decidable subset of F (and similarly for its set of sorts), see Definition 2.8.

[30] For sets B and A with $B \subseteq A$, the inclusion mapping $in_{B,A} : B \to A$ is defined by $in_{B,A}(b) = b$ for all $b \in B$.

Proof: (1) The image $h(m)$ of any element m of $\mathbb{M}$ is equal to $val_\mathbb{A}(t)$ for any t in $val_\mathbb{M}^{-1}(m)$. Since $\mathbb{M}$ is semi-effectively given and generated by F, some term t in $val_\mathbb{M}^{-1}(m)$ can be computed as observed in Definition 2.8. Then $\xi_A(val_\mathbb{A}(t)) = \xi_A(h(m))$ can also be computed since $\mathbb{A}$ is semi-effectively given, which completes the proof.

(2) Since F is effectively given, the set $T(F)$ is semi-effectively given and $T(F')$ is a decidable subset of $T(F)$. Let $t_0, t_1, \ldots$ be an enumeration of the elements of $T(F')$ such that the mapping $i \mapsto t_i$ is computable. We define the encoding $\xi_{M'}(m)$ of an element m of M' to be the smallest integer i such that $val_\mathbb{M}(t_i) = m$. Since $\mathbb{M}$ is semi-effectively given, the mapping $val_\mathbb{M}$ and hence the mapping $in_{M',M}$ and its inverse are computable. From this it is easy to see that $\mathbb{M}'$ is semi-effectively given with encoding $\xi_{M'}$. ∎

This lemma is meant to be constructive. For instance, for the first assertion this means that from the specifications of $\mathbb{M}$ and $\mathbb{A}$ an algorithm can be constructed that computes h. In what follows, all such results are meant to be constructive.

If F is finite and $\mathbb{A}$ is locally finite, then it is finite (because S is finite, cf. Definition 2.123). Hence, if $\mathbb{A}$ is given as a finite set A with its operations in tabular form, then it is effectively given (and vice versa). This shows that every recognizable set L that is specified by a given finite algebra $\mathbb{A}$ and a subset C of A, is effectively term-recognizable (and vice versa). By Lemma 3.75(1), if $\mathbb{M}$ is semi-effectively given and generated by F, then L is even effectively recognizable and hence a decidable subset of M. It will be shown in Section 4.2.3 that there exist homogenous recognizable sets of the HR algebra $\mathbb{JS}^t$ that are *not* decidable subsets of $\mathcal{JS}$ and hence cannot be specified as (semi-)effectively (term-)recognizable sets.

We now formulate an effective version of Proposition 3.56. For the notion of an effectively given derived signature see Definition 2.127.

Proposition 3.76

(1) Let $\mathbb{M}$ be an F-algebra and $\mathbb{M}'$ be an F'-algebra such that $\mathbb{M}' \subseteq \mathbb{M}$. Let S' be a decidable subset of S, and let the mappings $in_{S',S}$ and $in_{F',F}$ be computable.[31] If L is effectively term-recognizable in $\mathbb{M}$, then $L \cap M'$ is effectively term-recognizable in $\mathbb{M}'$.

Moreover, let $\mathbb{M}$ and $\mathbb{M}'$ be semi-effectively given and the mapping $in_{M',M}$ be computable. If L is (semi-) effectively recognizable in $\mathbb{M}$, then $L \cap M'$ is (semi-) effectively recognizable in $\mathbb{M}'$.

(2) Let $k : \mathbb{M}' \to \mathbb{M}$ be a homomorphism of F-algebras. If L is effectively term-recognizable in $\mathbb{M}$, then $k^{-1}(L)$ is effectively term-recognizable in $\mathbb{M}'$.

---

[31] Here S' and S are the sets of sorts of F' and F, respectively. Note that we do *not* assume that S' is encoded by its standard encoding as a subset of S (see Definition 2.8), and similarly for F' and M'.

Moreover, let $\mathbb{M}$ and $\mathbb{M}'$ be semi-effectively given and k be computable. If L is (semi-)effectively recognizable in $\mathbb{M}$, then $k^{-1}(L)$ is (semi-) effectively recognizable in $\mathbb{M}'$.

(3) Let H be an effectively given derived signature of F and $\mathbb{M}$ be an F-algebra. If L is effectively term-recognizable in $\mathbb{M}$, then L is effectively term-recognizable in $\mathbb{M}_H$. If L is (semi-) effectively recognizable in $\mathbb{M}$, then L is (semi-) effectively recognizable in $\mathbb{M}_H$. Furthermore, if $F \subseteq H$ and $in_{F,H}$ is computable, then the converse implications also hold.

Additionally we state the following special case of (2):

(4) Let $\mathbb{M}$ be an F-algebra with finite F. If L is effectively term-recognizable in $\mathbb{M}$, then a finite F-automaton recognizing the language $val_{\mathbb{M}}^{-1}(L)$ can be computed from a specification of L.

Proof: The proof is based on that of Proposition 3.56.

(1) Let $L = h^{-1}(C)$ with $h : \mathbb{M} \to \mathbb{A}$ and $C \subseteq A$, as in Definition 3.74. We may assume that the domains A_s of $\mathbb{A}$ are nonempty. Let $\mathbb{A}' := \mathbb{A} \restriction F'$. Since membership in S' is decidable, so is membership in A' (because $A' = \{a \in A \mid \sigma(a) \in S'\}$). Hence, to make $\mathbb{A}'$ semi-effectively given, we can use its standard encoding as a decidable subset of A. Since $in_{S',S}^{-1}$ and $in_{F',F}$ are computable,[32] it is easy to verify that $\mathbb{A}'$ is semi-effectively given. If $\mathbb{A}$ is effectively given, then so is $\mathbb{A}'$. In fact, the mapping $s' \mapsto |A_{s'}'|$ is computable because it is the composition of $in_{S',S}$ and of the mapping $s \mapsto |A_s|$. By the above assumption, $|A'|$ is infinite if S' is infinite. If S' is finite, then $|A'| = \sum_{s' \in S'} |A_{s'}'|$. Hence $|A'|$ can be computed.

The restriction h' of h to M' is a homomorphism : $\mathbb{M}' \to \mathbb{A}'$ and $L \cap M' = (h')^{-1}(C \cap A')$. Since C is a decidable subset of A, the set $C \cap A'$ is obviously one of A'. Since $h' = h \circ in_{M',M}$, it is computable if $in_{M',M}$ and h are.

(2) This is obvious from the proof of Proposition 3.56(2).

(3) In general, if H is effectively given and $\mathbb{M}$ is (semi-) effectively given, then $\mathbb{M}_H$ is (semi-) effectively given (with respect to the same encoding of M): to compute $g_{\mathbb{M}_H}(m_1, \ldots, m_k)$, first compute $\delta_H(g)$ and then use $\zeta_{\mathbb{M}}$ to compute $\delta_H(g)_{\mathbb{M}}(m_1, \ldots, m_k)$. The result is now immediate from the proof of Proposition 3.56(3).

(4) Let $L = h^{-1}(C)$ for an effectively given finite algebra $\mathbb{A}$, a homomorphism $h : \mathbb{M} \to \mathbb{A}$ and a decidable subset C of A. Since $\mathbb{A}$ is finite and effectively given, it can be computed (or more precisely, an algebra $\mathbb{A}'$ isomorphic to $\mathbb{A}$, with $A' = enc_A$, can be computed). Since membership in C is decidable, C can also be computed. Hence the finite automaton $\mathscr{A}(\mathbb{A}, C)$ can be computed that recognizes $val_{\mathbb{M}}^{-1}(L)$ by Lemma 3.59 (because $val_{\mathbb{A}} = h \circ val_{\mathbb{M}}$). ∎

---

[32] In general, if A, B are encoded sets with $B \subseteq A$ and $in_{B,A}$ is computable (i.e., $\xi_A \circ \xi_B^{-1}$ is computable), then so is $in_{B,A}^{-1}$ (i.e., $\xi_B \circ \xi_A^{-1}$ is): for $b \in B$, $\xi_B(b)$ is the smallest integer i such that $\xi_A(\xi_B^{-1}(i)) = \xi_A(b)$.

Definition 3.77 (Decidable and effectively given congruences) Let $\mathbb{M}$ be a semi-effectively given F-algebra. An equivalence relation $\sim$ on M is *decidable* if it is a decidable subset of $M \times M$.[33] A congruence $\sim$ on $\mathbb{M}$ is *effectively given* if it is decidable, $|M/\sim|$ is given, and the index $\gamma_\sim$ is computable.[34]

We now state an effective version of Proposition 3.64. Also, we examine conditions that ensure that semi-effective recognizability implies effective recognizability (the converse implication being trivial).

Proposition 3.78 Let $\mathbb{M}$ be a semi-effectively given F-algebra and let $L \subseteq M$.
(1) L is semi-effectively recognizable if and only if membership in L is decidable[35] and L is saturated by a locally finite and decidable congruence.
(2) If membership in L is decidable and L is saturated by a locally finite and effectively given congruence, then L is effectively recognizable.
(3) Let $\mathbb{M}$ be finitely generated by F. Then:

   (3.1) if $\sim$ is a decidable and finite congruence, then $\sim$ is effectively given and the finite quotient algebra $\mathbb{M}/\sim$ is effectively given;

   (3.2) if L is semi-effectively recognizable, then it is effectively recognizable.

Proof: (1) The "only if" direction is clear from the definitions and the proof of Proposition 3.64: since h is computable, the congruence $\sim^h$ is decidable. For proving the "if" direction, we let $\sim$ be a locally finite and decidable congruence that saturates L. From the proof of Proposition 3.64 we know that $L = h_\sim^{-1}(C)$ where $h_\sim$ is the canonical surjective homomorphism : $\mathbb{M} \to \mathbb{A} := \mathbb{M}/\sim$ and $C := \{[m]_\sim \mid m \in L\}$.

In order to make the locally finite algebra $\mathbb{A}$ semi-effectively given, we need an encoding ξ_A of the elements of $A = M/\sim$ by integers. For each i in enc_M, we let $h(i)$ be the smallest integer $j \in enc_M$ such that $\xi_M^{-1}(j) \sim \xi_M^{-1}(i)$. Since $\sim$ is decidable (and enc_M is a decidable subset of $\mathcal{N}$), the mapping $h : enc_M \to enc_M$ is computable. We now define the encoding ξ_A by $\xi_A([m]_\sim) := h(\xi_M(m))$ for every $m \in M$. Since h is computable, $enc_A = \{i \in enc_M \mid h(i) = i\}$ is a decidable subset of $\mathcal{N}$. Let us show that, with this encoding, the algebra $\mathbb{A}$ is semi-effectively given (cf. Definition 2.8). An integer $i \in \mathcal{N}$ encodes an element of A_s (through ξ_A) if and only if i encodes an element of M of sort s (through ξ_M), and this can be determined by an algorithm because $\mathbb{M}$ is semi-effectively given. (Sorts are handled through the encoding ξ_S). If i_0 encodes an element f of F (through ξ_F) and $i_1, \ldots, i_p$ encode elements of A, with $p := \rho(f)$, then $\widetilde{\zeta}_\mathbb{A}(i_0, i_1, \ldots, i_p) = h(\widetilde{\zeta}_\mathbb{M}(i_0, i_1, \ldots, i_p))$. Hence $\zeta_\mathbb{A}$ is computable and $\mathbb{A}$ is semi-effectively given.

---

[33] In other words, the mapping $f : M \times M \to \{True, False\}$ such that $m \sim m'$ if and only if $f(m, m') = True$ is computable.

[34] Recall from Definition 3.63 that $\gamma_\sim$ is the mapping $s \mapsto |M_s/\sim_s|$. The phrase "$|M/\sim|$ is given" means that $|M/\sim|$ is part of the specification of an effectively given congruence $\sim$ (which also consists of algorithms to decide $\sim$ and to compute $\gamma_\sim$). A decidable equivalence relation $\sim$ is specified by an algorithm that decides $\sim$.

[35] Since this equivalence is meant to be constructive, a membership algorithm for L must be specified.

It remains to show Conditions (1) and (2) of Definition 3.74. Membership in C is decidable, because membership in L is: if i encodes an element $[m]$ of A, then $[m] \in C$ if and only if i encodes an element of L. The homomorphism $h_\sim$ is computable: if i encodes an element m of M, then $h(i)$ encodes $h_\sim(m) = [m]$.

(2) The proof is the same as in (1). If $\sim$ is effectively given, then $\mathbb{A}$ is effectively given, because $|A| = |M/\sim|$ and $|A_s| = |M_s/\sim_s| = \gamma_\sim(s)$ for every sort s.

(3.1) We now assume that $\mathbb{M}$ is generated by its finite signature F, hence has finitely many sorts, and that $\sim$ is a decidable and finite congruence. As before, let $\mathbb{A} := \mathbb{M}/\sim$. Obviously, the finite algebra $\mathbb{A}$ is also generated by F and, as shown in the proof of (1), it is semi-effectively given. It suffices to compute $|A_s|$ for every sort s, because then $|A|$ is the sum of those cardinalities and hence $\sim$ and $\mathbb{A}$ are effectively given. Let S be the equation system constructed in the proof of Proposition 3.16 such that $\mu \bar{x} \cdot S_{\mathcal{P}(\mathbb{T}(F))}(\bar{x}) \restriction x_s = T(F)_s$ for every sort s. Since $\mathbb{A}$ is generated by F, we have that $A_s = val_{\mathbb{A}}(T(F)_s)$ which equals $\mu \bar{x} \cdot S_{\mathcal{P}(\mathbb{A})}(\bar{x}) \restriction x_s$ by Proposition 3.23(1). Since $\mathbb{A}$ is semi-effectively given, the least solution of S in $\mathcal{P}(\mathbb{A})$ can be computed by Proposition 3.10, and so A_s and $|A_s|$ can be computed for every s.

(3.2) This is immediate from (1), (2) and (3.1). ∎

Remark 3.79 (1) Without its first two hypotheses (F is finite and generates $\mathbb{M}$), Assertion (3.1) of Proposition 3.78 is false. Consider for example the set $\mathcal{N}$ equipped with one constant symbol with value i for each $i \in \mathcal{N}$. Let $h : \mathcal{N} \to \mathcal{N}$ be a computable mapping such that $h(\mathcal{N})$ is finite. The algebra $\mathbb{M} := \langle \mathcal{N}, (i)_{i \in \mathcal{N}} \rangle$ is generated (trivially) by its signature F that consists only of constant symbols. The equivalence relation defined by $n \sim m$ if and only if $h(n) = h(m)$ is decidable and has finitely many classes, but one cannot determine the number of classes, because the knowledge of an algorithm computing h is not enough to determine the cardinality of the set $h(\mathcal{N})$. The same holds if F is replaced by a finite subsignature of it, hence that does not generate the corresponding algebra.

(2) We now consider a similar case. We assume that h is a computable mapping : $\mathcal{N} \to \{0, 1\}$. Hence h is a computable homomorphism : $\mathbb{M} \to \mathbb{A}$ where $\mathbb{A} := \langle \{0, 1\}, (h(i))_{i \in \mathcal{N}} \rangle$. Then $\mathbb{A}$ is effectively given but we cannot decide if h is surjective. Even if we know that a decidable congruence has at most two classes, we cannot compute its number of classes. This shows that we cannot restrict $\mathbb{A}$ in an effective manner so as to make h surjective, as in the general case (cf. Definition 3.55). It is the reason that the converse direction of implication (2) of Proposition 3.78 does not hold; one easily shows that it *does* hold when the witnessing homomorphism is surjective.

(3) We use Proposition 3.78(1) to show that for semi-effectively recognizable sets Proposition 3.76(1) holds without its requirements on S', S, F' and F. Let L be a decidable subset of M and let $\sim$ be a decidable congruence on $\mathbb{M}$ witnessing the recognizability of L. Then $\sim' := \sim \cap (M' \times M')$ is a congruence on $\mathbb{M}'$ that witnesses the recognizability of $L \cap M'$ (it is locally finite because $|M'_s/\sim'_s| \leq |M_s/\sim_s|$ for every

sort s of F'). Since $in_{M',M}$ is computable, $\sim'$ is decidable and $L \cap M'$ is a decidable subset of M'. □

Proposition 3.80 Let $\mathbb{M}$ be a finitely generated and semi-effectively given F-algebra and let $L \subseteq M$ be recognizable. From an integer $p \geq \sharp \mathscr{M}(L)$ and a membership algorithm for L, one can compute a finite F-automaton that recognizes the language $val_{\mathbb{M}}^{-1}(L)$.

Proof: Let $\mathbb{M}$ and L be as in the statement. By Proposition 3.73(2), there exists a complete and deterministic F-automaton $\mathscr{A}$ with at most p states such that $L(\mathscr{A}) = val_{\mathbb{M}}^{-1}(L)$. We will use the set $Ctxt(F)$ of contexts over F, defined as terms in $T(F, \{x\})$.

Claim 3.80.1 For every $m \in M$, there exists a complete and deterministic $(F \cup \{x\})$-automaton $\mathscr{B}$ with $\sharp \mathscr{A}$ states such that, for every $c \in Ctxt(F)$, we have $c \in L(\mathscr{B})$ if and only if $c_{\mathbb{M}}(m) \in L$.

Proof: Let t be a term in $T(F)$ that evaluates to m. Let q be the unique state of $\mathscr{A}$ such that $t \in L(\mathscr{A}, q)$. Let $\mathscr{B}$ be the $(F \cup \{x\})$-automaton obtained from $\mathscr{A}$ by the addition of the transition rule $x \to q$. It is clear that $L(\mathscr{B}) \cap Ctxt(F)$ is the set of contexts c such that $c[t] \in L(\mathscr{A})$, i.e., such that $c_{\mathbb{M}}(m) \in L$. □

Claim 3.80.2 For every $m, m' \in M$, we have $m \approx^L m'$ if and only if

$$c_{\mathbb{M}}(m) \in L \Longleftrightarrow c_{\mathbb{M}}(m') \in L \text{ for every } c \in Ctxt(F) \text{ of height at most } 3 \cdot p^2.$$

Proof: Let $\mathscr{B}$ and $\mathscr{B}'$ be associated with m and m' by the previous claim. Then $K := Ctxt(F) \cap (L(\mathscr{B}) - L(\mathscr{B}'))$ is the set of contexts in $Ctxt(F)$ such that $c_{\mathbb{M}}(m) \in L$ but $c_{\mathbb{M}}(m') \notin L$. Since $Ctxt(F)$ is recognized by a complete and deterministic $(F \cup \{x\})$-automaton with 3 states of each sort, K can be recognized by such an automaton with at most $3 \cdot p^2$ states (the detailed construction is easy to work out with the product construction of automata recalled below in the proof of Proposition 3.85). If K is nonempty, it contains a context of height at most $3 \cdot p^2$ by Proposition 3.53. By considering in a similar way the set $Ctxt(F) \cap (L(\mathscr{B}') - L(\mathscr{B}))$, we get the result. □

We complete the proof of the proposition as follows. There are finitely many contexts in $Ctxt(F)$ of height at most $3 \cdot p^2$. Since $\mathbb{M}$ is semi-effectively given, $c_{\mathbb{M}}(m)$ can be computed for each context c, and since we have a membership algorithm for L, we can decide if $c_{\mathbb{M}}(m) \in L$. It follows from Claim 3.80.2 that the syntactic congruence $\approx^L$ of L is decidable. Hence, by Propositions 3.66(1), 3.78(1) and 3.78(3.2), L is effectively recognizable and so the required automaton can be computed by Proposition 3.76(4). In fact, following the line of this proof, the finite F-automaton $\mathscr{A}(\mathbb{M}/\approx^L, [L])$ is computed, which is the minimal automaton $\mathscr{M}(L)$ of L by definition. ■

Here is a concrete application of this proposition (given as Theorem 6.28 in [*DowFel]). Let $L \subseteq A^*$ be a regular language that is recognized by an *unknown*, possibly nondeterministic automaton with at most p states. Hence, some deterministic and complete automaton with at most 2^p states recognizes L. From p and any membership algorithm for L, one can compute by the previous proposition a finite automaton recognizing L. Hence, provided we know an upper-bound p, we get from an *arbitrary membership algorithm* a linear-time one. (This follows from Proposition 3.80 applied to the algebra $\mathbb{M} = \mathbb{W}_{right}(A)$, cf. Remark 3.51(2).)

In Section 3.4.10 we will review the concrete and effective ways to specify recognizable sets in semi-effectively given algebras.

3.4.5 Inductive predicates

Recognizability can be formulated in a more intuitive way in terms of inductive sets of properties. This notion has been presented in Section 1.2, but in place of the informal term "property," we will use the term "predicate," from logic. We will characterize the homogenous recognizable sets in terms of inductive sets of predicates.

Definition 3.81 (Inductive sets of predicates) A *predicate* on a set E is a mapping $: E \to \{True, False\}$. Let $\mathbb{M}$ be an F-algebra with set of sorts S. A *family of predicates* on $\mathbb{M}$ is an indexed set $(\widehat{p})_{p \in P}$, given with a mapping $\alpha : P \to S$ such that each $\widehat{p}$ is a predicate on $M_{\alpha(p)}$. We call $\alpha(p)$ the *(input) type* of p, and for every $s \in S$ we denote by P_s the set $\{p \in P \mid \alpha(p) = s\}$. Such a family will also be denoted by P. The notation P for a family of predicates refers to the set of symbols P, to the type mapping α and to the predicates $\widehat{p}$ for p in P. The family P is *locally finite* if for each $s \in S$ the set P_s is finite. We say that P is f-*inductive* where $f \in F_n$, if for every $p \in P$ of type $s = \sigma(f)$ there exist $k_1, \ldots, k_n$ in $\mathcal{N}$, a Boolean function B with $(k_1 + \cdots + k_n)$ arguments and a sequence $(p_{1,1}, \ldots, p_{1,k_1}, p_{2,1}, \ldots, p_{2,k_2}, \ldots, p_{n,k_n})$ of $k_1 + \cdots + k_n$ predicates in P, such that, if the type of f is $s_1 \times s_2 \times \cdots \times s_n \to s$, then:

(1) $\alpha(p_{i,j}) = s_i$ for all $i = 1, \ldots, n$ and $j = 1, \ldots, k_i$; and
(2) for all $m_1 \in M_{s_1}, \ldots, m_n \in M_{s_n}$ we have:

$$\widehat{p}(f_{\mathbb{M}}(m_1, \ldots, m_n)) = B(\widehat{p}_{1,1}(m_1), \ldots, \widehat{p}_{1,k_1}(m_1), \widehat{p}_{2,1}(m_2), \ldots, \widehat{p}_{n,k_n}(m_n)).$$

Formally, we let $\overline{k}$ denote the sequence $(k_1, \ldots, k_n)$ and we let $Y_{\overline{k}}$ be the set of Boolean variables $y_{i,j}$ for $1 \le i \le n$ and $1 \le j \le k_i$. Then B is a term in $T(\{\vee, \wedge, \neg, True, False\}, Y_{\overline{k}})$ and, in the above expression of $\widehat{p}(f_{\mathbb{M}}(m_1, \ldots, m_n))$, the variable $y_{i,j}$ takes value $\widehat{p}_{i,j}(m_i)$. The tuple $(B, p_{1,1}, \ldots, p_{1,k_1}, p_{2,1}, \ldots, p_{n,k_n})$ is called a *splitting of p relative to f*.[36]

Intuitively, this means that the validity of p for any element of M of the form $f_{\mathbb{M}}(m_1, \ldots, m_n)$ where $m_1, \ldots, m_n \in M$, can be computed from the truth values of finitely many predicates of P for $m_1, \ldots, m_n$. This computation can be done by a Boolean expression that depends only on p and f. For $F' \subseteq F$, we say that P is

---

[36] The integers $k_1, \ldots, k_n$ are known from the indices of the variables $y_{i,j}$ occurring in B.

F'-*inductive* if it is f-inductive for every f in F'. For $p \in P$, we let $sat(\mathbb{M}, p) :=$ $\{m \in M_{\alpha(p)} \mid \widehat{p}(m) = True\}$.

Example 3.82 Let us consider the 2-colorability of series-parallel graphs, already detailed in Chapter 1, Example 1.11. The equalities (1.10) yield the following splitting of σ relative to $\bullet$:

$$\sigma(G \bullet H) = (\sigma(G) \wedge \sigma(H)) \vee (\delta(G) \wedge \delta(H)),$$

formally expressed by the 5-tuple $((y_{1,1} \wedge y_{2,1}) \vee (y_{1,2} \wedge y_{2,2}), \sigma, \delta, \sigma, \delta)$. Thus, $k_1 = k_2 = 2$. Equalities (1.10) show that $\{\sigma, \delta\}$ is an $\{/\!/, \bullet\}$-inductive family of predicates on the algebra $\mathbb{J}_2^d$. $\square$

Proposition 3.83 Let $\mathbb{M}$ be an F-algebra and $s \in \mathcal{S}$. A subset L of M_s is recognizable in $\mathbb{M}$ if and only if $L = sat(\mathbb{M}, p_0)$ for some predicate p_0 of type s belonging to a locally finite F-inductive family of predicates on $\mathbb{M}$.

Proof: "Only if." Let $L = h^{-1}(C) \subseteq M_s$, for some homomorphism $h : \mathbb{M} \to \mathbb{A}$ where $\mathbb{A}$ is a locally finite F-algebra and $C \subseteq A_s$. We let $P = A \cup \{p_0\}$. Each element a of A_u has type u (considered as a member of P) and p_0 has type s. For $m \in M_u$ and $a \in A_u$, we let $\widehat{a}(m) := True$ if and only if $h(m) = a$, and, for $m \in M_s$, we let $\widehat{p}_0(m) := True$ if and only if $h(m) \in C$.

We check that the family P is F-inductive. For all n-ary $f \in F$, all elements $m_1, \ldots, m_n$ of M of appropriate sorts and all $a \in A_{\sigma(f)}$, we have

$$\widehat{a}(f_{\mathbb{M}}(m_1, \ldots, m_n)) = True$$
$$\iff \quad h(f_{\mathbb{M}}(m_1, \ldots, m_n)) = a$$
$$\iff \quad f_{\mathbb{A}}(h(m_1), \ldots, h(m_n)) = a.$$

Hence

$$\widehat{a}(f_{\mathbb{M}}(m_1, \ldots, m_n)) = \bigvee \widehat{a_1}(m_1) \wedge \cdots \wedge \widehat{a_n}(m_n),$$

where the disjunction ranges over all n-tuples $(a_1, \ldots, a_n)$ in A^n such that $f_{\mathbb{A}}(a_1, \ldots, a_n) = a$. (For $c \in F_0$, we have $\widehat{a}(c_{\mathbb{M}}) = True$ if and only if $c_{\mathbb{A}} = a$). We have $\widehat{p}_0(m) = \bigvee_{a \in C} \widehat{a}(m)$ for all $m \in M$. The splitting of $\widehat{a}$ relative to f is thus the tuple $(B, p_{1,1}, \ldots, p_{n,k})$ where $(a_1^{(j)}, \ldots, a_n^{(j)})_{1 \leq j \leq k}$ is an enumeration of the set $\{(a_1, \ldots, a_n) \mid f_{\mathbb{A}}(a_1, \ldots, a_n) = a\}$, B is the Boolean expression $\bigvee_{1 \leq j \leq k}(y_{1,j} \wedge \cdots \wedge y_{n,j})$ and $p_{i,j} = a_i^{(j)}$. Hence P is F-inductive. It is clear that it is locally finite. We have $L = sat(\mathbb{M}, p_0)$. Hence L satisfies the required property.

"If." Let P be a locally finite F-inductive family of predicates. Let $L = sat(\mathbb{M}, p_0)$ for some $p_0 \in P$. For $u \in \mathcal{S}$, we let $\sim_u$ be the equivalence relation on M_u defined by $m \sim_u m'$ if and only if $\widehat{p}(m) = \widehat{p}(m')$ for every $p \in P_u$. It follows from Condition 2 of Definition 3.81 that $\sim$ is a congruence. It has at most $2^{|P_u|}$ equivalence classes of each sort u. The set $L = sat(\mathbb{M}, p_0)$ is a union of equivalence classes, hence is recognizable by Proposition 3.64. ∎

Let P be a locally finite F-inductive family of predicates on a semi-effectively given F-algebra $\mathbb{M}$. It is *computable* if the set P is effectively given and the mapping α, the mapping $(p, m) \mapsto \widehat{p}(m)$ (for $p \in P$ and $m \in M_{\alpha(p)}$) and the mapping $s \mapsto |P_s|$ are computable. It is *effectively F-inductive* if, in addition, there is a computable mapping that associates with every $p \in P$ and $f \in F$ such that $\alpha(p) = \sigma(f)$, a splitting of p relative to f.

Corollary 3.84 Let $\mathbb{M}$ be a semi-effectively given F-algebra and let p_0 belong to a locally finite F-inductive family of predicates P on $\mathbb{M}$. If P is computable, then $sat(\mathbb{M}, p_0)$ is semi-effectively recognizable. If P is effectively F-inductive, then $sat(\mathbb{M}, p_0)$ is effectively recognizable.

Proof: To prove the first assertion we use Proposition 3.78(1). Since membership in $sat(\mathbb{M}, p_0)$ is obviously decidable, it suffices to show that the locally finite congruence $\sim$ from the "if" part of the proof of Proposition 3.83 is decidable. This follows from the fact that for every sort u the finite set P_u can be computed: enumerate the elements p of P, test whether $\alpha(p) = u$, and halt when $|P_u|$ elements of P_u have been found.

For proving the second assertion, we let P be effectively F-inductive. We construct a locally finite F-algebra $\mathbb{A}$ and a homomorphism h from $\mathbb{M}$ to $\mathbb{A}$. For each s, we define A_s as the set of all mappings : $P_s \to \{True, False\}$. For a constant symbol $c \in F$, we define $c_{\mathbb{A}}$ as the mapping $p \mapsto \widehat{p}(c_{\mathbb{M}})$ for p in $P_{\sigma(c)}$. For $f \in F$ of arity $n > 0$ and $g_1, \dots, g_n$ respectively in $A_{s_1}, \dots, A_{s_n}$, where $\alpha(f) = (s_1, \dots, s_n)$, we define $f_{\mathbb{A}}(g_1, \dots, g_n)$ as the mapping that maps each p in $P_{\sigma(f)}$ to

$$B(g_1(p_{1,1}), \dots, g_1(p_{1,k_1}), g_2(p_{2,1}), \dots, g_n(p_{n,k_n})),$$

where $(B, p_{1,1}, \dots, p_{1,k_1}, p_{2,1}, \dots, p_{n,k_n})$ is the computed splitting of p relative to f. It is clear that $\mathbb{A}$ is effectively given[37] and locally finite. We let $h : M_s \to A_s$ be defined as follows: $h(m)$ is the function that maps p to $\widehat{p}(m)$. It is clear that h is computable and that, for each p, $sat(\mathbb{M}, p) = h^{-1}(C_p)$ where C_p is the set of mappings g such that $g(p) = True$. Obviously, membership in C_p is decidable. It remains to prove that h is a homomorphism. We have for all objects $f, p, m_1, \dots, m_n$ of relevant types:

$$\begin{aligned}
&h(f_{\mathbb{M}}(m_1, \dots, m_n))(p) \\
&= \widehat{p}(f_{\mathbb{M}}(m_1, \dots, m_n)) \\
&= B(\widehat{p}_{1,1}(m_1), \dots, \widehat{p}_{1,k_1}(m_1), \widehat{p}_{2,1}(m_2), \dots, \widehat{p}_{n,k_n}(m_n)) \\
&= B(h(m_1)(p_{1,1}), \dots, h(m_1)(p_{1,k_1}), h(m_2)(p_{2,1}), \dots, h(m_n)(p_{n,k_n})) \\
&= f_{\mathbb{A}}(h(m_1), \dots, h(m_n))(p),
\end{aligned}$$

---

[37] The set A can be viewed as the decidable subset of $Seq(P)$ consisting of all sequences $(p_1, \dots, p_r)$ such that $\alpha(p_i) = \alpha(p_j)$ and $\xi_P(p_i) < \xi_P(p_j)$ for all $i < j$.

which gives the result.   ∎

In fact, it can easily be shown by inspection of the "only if" part of the proof of Proposition 3.83, that a subset L of M_s is effectively recognizable in $\mathbb{M}$ if and only if $L = sat(\mathbb{M}, p_0)$ for some predicate p_0 of type s belonging to a locally finite effectively F-inductive family of predicates on $\mathbb{M}$. The corresponding statement for semi-effectively recognizable sets and computable F-inductive families only holds in one direction, due to the requirement that the mapping $s \mapsto |P_s|$ must be computable.

If the signature F of $\mathbb{M}$ is finite and if P is, as in Corollary 3.84, locally finite and effectively F-inductive, then a finite F-automaton as in Proposition 3.76(4) can be constructed for $val_{\mathbb{M}}^{-1}(L)$ where $L := sat(\mathbb{M}, p_0)$. If $\mathbb{M}$ is moreover generated by F, then it suffices that P is computable (by Proposition 3.78(3.2)).

3.4.6 Closure properties

We now consider some closure properties of the family of recognizable sets. We already know from Propositions 3.56(2) and 3.76(2) that it is closed under inverse homomorphisms. Extended derived operations are defined in Definition 2.125.

Proposition 3.85

(1) Let F be an S-signature and $\mathbb{M}$ be an F-algebra. The family of sets **Rec**$(\mathbb{M})$ is closed under union, intersection and difference; it contains the set $\bigcup_{s \in S'} M_s$ for every set of sorts $S' \subseteq S$ (in particular, the empty set and the set M). If $g : M_u \to M_s$ is a unary extended derived operation and $L \in$ **Rec**$(\mathbb{M})$, then $g^{-1}(L) \in$ **Rec**$(\mathbb{M})_u$.

(2) Let $\mathbb{M}, \mathbb{M}'$ be F-algebras. If $L \in$ **Rec**$(\mathbb{M})$ and $L' \in$ **Rec**$(\mathbb{M}')$, then $L \times L' \in$ **Rec**$(\mathbb{M} \times \mathbb{M}')$.

Proof: (1) Let $L, L' \in$ **Rec**$(\mathbb{M})$ be defined as $h^{-1}(C)$ and $h'^{-1}(C')$ respectively for homomorphisms $h : \mathbb{M} \to \mathbb{A}$ and $h' : \mathbb{M} \to \mathbb{A}'$ to locally finite F-algebras $\mathbb{A}$ and $\mathbb{A}'$, and for subsets C and C' of A and A' respectively. Then $L \cup L' = h''^{-1}((C \times A') \cup (A \times C'))$ where h'' is the homomorphism $: \mathbb{M} \to \mathbb{A} \times \mathbb{A}'$ such that $h''(m) = (h(m), h'(m))$. We have also $L \cap L' = h''^{-1}(C \times C')$ and $L - L' = h''^{-1}((C \times A') - (A \times C'))$. These three sets are $\mathbb{M}$-recognizable because $\mathbb{A} \times \mathbb{A}'$ is locally finite.[38] Let $\mathbb{S}$ be the unique F-algebra with domain $\{s\}$ for $s \in S$ and let h be the unique homomorphism $: \mathbb{M} \to \mathbb{S}$. Then $\bigcup_{s \in S'} M_s = h^{-1}(S')$ and hence is $\mathbb{M}$-recognizable, cf. Example 3.57(5).

Consider now an extended derived operation defined, for $m \in M$, by $g(m) := t_{\mathbb{M}}(m, m_2, \ldots, m_p)$ for some term t in $T(F, X_p)_s$ with $\sigma(x_1) = u$ and $m_2, \ldots, m_p$ in M of appropriate sorts. Let $L, h, \mathbb{A}, C$ be as above. Let $g' : A_u \to A_s$ be defined by $g'(a) := t_{\mathbb{A}}(a, h(m_2), \ldots, h(m_p))$. We have $g' \circ h = h \circ g$ (both are mappings $: M_u \to A_s$).

---

[38] We have observed in Section 3.4.2 that a complete deterministic F-semi-automaton is nothing but an F-algebra. The use of $\mathbb{A} \times \mathbb{A}'$ corresponds to the classical *product* of automata (cf. Section 6.3.1 below, Proposition 7.1 of [*GecSte] or Section 1.3 of [*Com+]).

Since $L = h^{-1}(C)$, we obtain that $g^{-1}(L) = g^{-1}(h^{-1}(C)) = h^{-1}(g'^{-1}(C)) = h^{-1}(C')$, where $C' = g'^{-1}(C) \subseteq A_u$. Hence $g^{-1}(L)$ is recognizable in $\mathbb{M}$.

(2) Let L and L' be defined as in (1), except that $h' : \mathbb{M}' \to \mathbb{A}'$. Then $L \times L' = h''^{-1}(C \times C')$, where h'' is now the homomorphism : $\mathbb{M} \times \mathbb{M}' \to \mathbb{A} \times \mathbb{A}'$ such that $h''(m, m') = (h(m), h'(m'))$. ∎

With the hypotheses and notations of the previous proposition, we have the following corollary:

Corollary 3.86 If L and L' are effectively recognizable, then $L \cup L'$, $L \cap L'$, $L - L'$, $g^{-1}(L)$ and $L \times L'$ are effectively recognizable. The analogous statements hold for effectively term-recognizable sets and for semi-effectively recognizable sets.

Proof: The algebra $\mathbb{A} \times \mathbb{A}'$ is (semi-)effectively given if $\mathbb{A}$ and $\mathbb{A}'$ are so (see Definition 2.8), and membership in $C \times A'$, $A \times C'$ and $g'^{-1}(C)$ is decidable if that holds for C and C'. Moreover, h'' is computable if h and h' are so. We omit the details. ∎

These propositions generalize some well-known closure properties of the family of regular (recognizable) word languages. However, the closure under concatenation of regular languages does not extend to the closure of recognizable sets under the operations of the considered algebra: it is not always true that if $f \in F_k$ and $L_1 \in \mathbf{Rec}(\mathbb{M})_{s_1}, \ldots, L_k \in \mathbf{Rec}(\mathbb{M})_{s_k}$ where $\alpha(f) = (s_1, \ldots, s_k)$, then the set $f_{\mathcal{P}(\mathbb{M})}(L_1, \ldots, L_k)$ is in $\mathbf{Rec}(\mathbb{M})$. Singleton sets are not always recognizable. (These facts can be contrasted with Assertions (2) and (3) of Proposition 3.45 about equational sets.) We give here some counterexamples.

Example 3.87 An example of a singleton that is not recognizable is given in Remark 3.70(3). For an example of nonclosure under an operation, we consider again the algebra $\mathbb{W}(A)_{sq}$ which is the free monoid $\mathbb{W}(A)$ augmented with the squaring function: $sq(u) = u \cdot u$, cf. Example 3.42(1). It follows from Proposition 3.56(3) that $\mathbf{Rec}(\mathbb{W}(A)_{sq}) = \mathbf{Rec}(\mathbb{W}(A))$. However the language $sq(L) = \{ab^n ab^n \mid n \geq 0\}$, that is the image of the recognizable (i.e., regular) language ab^* under sq is not recognizable. Hence the operation sq does not preserve recognizability in $\mathbb{W}(A)_{sq}$. Algebraic conditions on an algebra $\mathbb{M}$ ensuring that $\mathbf{Rec}(\mathbb{M})$ is closed under its operations are considered in [Cou94a].

3.4.7 The Filtering Theorem

The following theorem (stated as Theorem 1.8 in Section 1.2.3) extends the classical result that the intersection of a context-free language and a regular one is context-free. Associated with the Recognizability Theorem, even in its weak form stated as Theorem 1.21 in Section 1.4.3, its yields the decidability of the monadic second-order satisfiability problems for the VR- and HR-equational sets of graphs.

Theorem 3.88 (Filtering Theorem, algebraic version) Let $\mathbb{M}$ be an F-algebra, let $L \in \mathbf{Equat}(\mathbb{M})$ and $K \in \mathbf{Rec}(\mathbb{M})$. Then $L \cap K \in \mathbf{Equat}(\mathbb{M})$. If K is semi-effectively recognizable or effectively term-recognizable, then an equation system defining $L \cap K$ can be constructed from one defining L.

Proof: First, we prove that we can assume that F is finite and that K is effectively term-recognizable. By (the proof of) Proposition 3.9(3), there is a finite subsignature F' of F such that L is equational in the subalgebra $\mathbb{M}'$ of $\mathbb{M}$ that is generated by F' (which is an F'-algebra): this signature F' is $F(S)$ where S is any equation system that defines L. The set $K' := K \cap M'$ is recognizable in $\mathbb{M}'$ by Proposition 3.56(1). If K is effectively term-recognizable in $\mathbb{M}$, then so is K' in $\mathbb{M}'$ by Proposition 3.76(1). If K is semi-effectively recognizable in $\mathbb{M}$, then so is K' in $\mathbb{M}'$ by Lemma 3.75(2) and Proposition 3.76(1) (see also Remark 3.79(3)) and even effectively recognizable in $\mathbb{M}'$ by Proposition 3.78(3.2). Since $L \subseteq M'$, we have that $L \cap K = L \cap K'$. Thus, it suffices to prove that $L \cap K'$ is $\mathbb{M}'$-equational, because then it is $\mathbb{M}$-equational, with the same equation system, by Proposition 3.9(1).

In the remainder of the proof we assume that F is finite, and we only consider the assumption that K is effectively term-recognizable. Let $\mathbb{M}$ be an F-algebra, let L be $\mathbb{M}$-equational and defined by an equation system S over F, and let K be $\mathbb{M}$-recognizable. Then $L = val_{\mathbb{M}}(R)$ for the equational set of terms R defined by S in $\mathcal{P}(\mathbb{T}(F))$, by Proposition 3.23(1). The set $R' := val_{\mathbb{M}}^{-1}(K)$ is recognizable in $\mathbb{T}(F)$ by Proposition 3.56(2). Then $L \cap K = val_{\mathbb{M}}(R) \cap K = val_{\mathbb{M}}(R \cap val_{\mathbb{M}}^{-1}(K)) = val_{\mathbb{M}}(R \cap R')$. The set $R \cap R'$ is regular by Theorem 3.62 and Proposition 3.85(1). Hence $L \cap K$ is $\mathbb{M}$-equational by Proposition 3.23(3).

We now assume that K is effectively term-recognizable in $\mathbb{M}$. Then a finite F-automaton $\mathscr{B} = \mathscr{A}(\mathbb{B}, Acc_{\mathscr{B}})$ recognizing R' can be constructed by (the proof of) Proposition 3.76(4). It remains to prove that from S and $\mathscr{B}$ an equation system S' can be constructed that defines $R \cap R'$ in $\mathbb{T}(F)$, and hence defines $L \cap K$ in $\mathbb{M}$ by Proposition 3.23(1). We will give two, closely related, proofs. The first one is very short but its drawback is that it gives no direct construction of the equation system S'. Our second proof details such a construction.

First construction: Since Theorem 3.62 is constructive, a finite F-automaton $\mathscr{A}(\mathbb{A}, C)$ recognizing R can be computed from S. Hence, the set $R \cap R'$ is recognized by the F-automaton $\mathscr{A}(\mathbb{A} \times \mathbb{B}, C \times Acc_{\mathscr{B}})$, cf. the proof of Proposition 3.85(1). An equation system S' defining $R \cap R'$ can be computed from this automaton, again because Theorem 3.62 is constructive.

Second construction: We let S be of the general form $\langle x_1 = p_1, \ldots, x_n = p_n \rangle$ where the p_i's are polynomials and L is defined by (S, Y) in $\mathbb{M}$ for some $Y \subseteq X_n$. We recall from Theorem 3.18 that the least solution of S in $\mathcal{P}(\mathbb{T}(F))$ is the n-tuple

$(L(G(S), x_1), \dots, L(G(S), x_n))$, where $G(S)$ is the regular grammar associated with S (see Definition 3.17, Section 3.1.4). Thus, $R = L(G(S), Y)$.

For every $i \in [n]$ and every state $q \in Q_{\mathscr{B}}$ such that $\sigma(q) = \sigma(x_i)$ (where $Q_{\mathscr{B}} = B$), we let $T_{i,q} := L(G(S), x_i) \cap L(\mathscr{B}, q)$. Our objective is to construct the system S' with set of unknowns $Unk(S') := \{x_{i,q} \mid i \in [n], q \in Q_{\mathscr{B}}, \sigma(q) = \sigma(x_i)\}$ such that $L(G(S'), x_{i,q}) = T_{i,q}$ for each $x_{i,q}$ in $Unk(S')$. We let $\sigma(x_{i,q}) := \sigma(x_i)$. The construction of S' is essentially the product of two finite automata, as in the first construction.

For each $x_{i,q}$ in $Unk(S')$, we will define a polynomial $p_{i,q}$ in $Pol(F, Unk(S'))$ intended to be the right-hand side of the equation of S' that defines $x_{i,q}$. First, for every monomial $m \in T(F, X_n)$ of a right-hand side of S and every $q \in Q_{\mathscr{B}}$ such that $\sigma(q) = \sigma(m)$, we define as follows a polynomial $p(m, q)$: if $m = t[x_{i_1}/y_1, \dots, x_{i_k}/y_k]$, where $t \in T(F, Y_k)$ and $ListVar(t) = y_1 \cdots y_k$, then we let $p(m, q) := \bigcup D$, where D is the set of all terms $t[x_{i_1, q_1}/y_1, \dots, x_{i_k, q_k}/y_k]$ such that $q_1, \dots, q_k \in Q_{\mathscr{B}}$ and $t_{\mathbb{B}}(q_1, \dots, q_k) = q$.[39]

Then, if $p_i = m_1 \cup \cdots \cup m_r$ (for $i \in [n]$) we define $p_{i,q}$ as $p(m_1, q) \cup \cdots \cup p(m_r, q)$ whenever $q \in Q_{\mathscr{B}}$ and $\sigma(q) = \sigma(p_i) = \sigma(x_i)$. It is clear that S' can be constructed from S and $\mathscr{B}$.

Claim 3.88.1 For every $x_{i,q}$ in $Unk(S')$ we have $T_{i,q} = L(G(S'), x_{i,q})$.

Proof sketch: One can verify that the sets $T_{i,q}$ form an oversolution of S' in $\mathcal{P}(\mathbb{T}(F))$.[40] It follows from the Least Fixed-Point Theorem (Theorem 3.7) that $L(G(S'), x_{i,q}) \subseteq T_{i,q}$ for each $x_{i,q}$ in $Unk(S')$.

We complete the proof by proving (by induction on j) that $T_i^{(j)} \cap L(\mathscr{B}, q) \subseteq L(G(S'), x_{i,q})$ for all $x_{i,q} \in Unk(S')$ and $j \in \mathcal{N}$, where $(T_1^{(j)}, \dots, T_n^{(j)}) := (S_{\mathcal{P}(\mathbb{T}(F))})^j(\emptyset)$. All proofs are routine from the definitions. $\square$

Hence the system S' defines in $\mathbb{T}(F)$ the sets of terms $T_{i,q}$. It follows that

$$R \cap R' = L(G(S), Y) \cap R'$$
$$= \bigcup \{L(G(S), x_\ell) \cap L(\mathscr{B}, q) \mid x_\ell \in Y, q \in Acc_{\mathscr{B}}\}$$
$$= \bigcup \{T_{\ell, q} \mid x_\ell \in Y, q \in Acc_{\mathscr{B}}, \sigma(q) = \sigma(x_\ell)\}.$$

Thus, $R \cap R'$ is defined by (S', X) in $\mathbb{T}(F)$ where

$$X := \{x_{\ell, q} \mid x_\ell \in Y, q \in Acc_{\mathscr{B}}, \sigma(q) = \sigma(x_\ell)\}.$$

Then $L \cap K$ is defined by (S', X) in $\mathbb{M}$. $\blacksquare$

---

[39] Equivalently, $run_{\mathscr{B}', t}(root_t) = q$, where the automaton $\mathscr{B}'$ is obtained from $\mathscr{B}$ by adding the transition rules $y_1 \to q_1, \dots, y_k \to q_k$.

[40] And even a solution. If the system S has no monomial equal to an unknown, then the regular grammar $G(S')$ is strict and the system S' has a unique solution in $\mathcal{P}(\mathbb{T}(F))$ by Proposition 3.15 (cf. the observation after Lemma 3.22). Hence the sets $T_{i,q}$ form this unique solution and we have the desired equalities. But the proof of this fact is essentially the same as the one for the general case.

Corollary 3.89 Let $\mathbb{M}$ be an F-algebra where F is finite and let $(L_1, \ldots, L_n)$ be the least solution in $\mathcal{P}(\mathbb{M})$ of an equation system S over F with $Unk(S) = X_n$. Let $h : \mathbb{M} \to \mathbb{B}$ be a homomorphism into a finite F-algebra $\mathbb{B}$. There exists an equation system that defines the sets $L_i \cap h^{-1}(q)$ for $i \in [n]$ and $q \in B$ with $\sigma(q) = \sigma(x_i)$. It can be constructed from S and $\mathbb{B}$, if $\mathbb{B}$ is effectively given.

Proof: One can perform the second construction in the proof of the Filtering Theorem, i.e., construct the system S' from S and the automaton $\mathcal{B} := \mathcal{A}(\mathbb{B}, Acc_{\mathcal{B}})$, for any $Acc_{\mathcal{B}}$ (because the construction does not depend on $Acc_{\mathcal{B}}$). Taking $L = L_i$ and so $Y = \{i\}$, and taking $K = h^{-1}(q)$ and so $R' = val_{\mathbb{M}}^{-1}(h^{-1}(q)) = val_{\mathbb{B}}^{-1}(q) = L(\mathcal{B})$ with $Acc_{\mathcal{B}} = \{q\}$, the proof shows that $L \cap K$ is defined by $(S', x_{i,q})$ in $\mathbb{M}$. ∎

Remark 3.90 (1) The system S' constructed in the second proof of the Filtering Theorem has at most $n \cdot \sharp \mathcal{B}$ equations but it is not necessarily trim, even if S is assumed to be. Trimming is interesting in several respects. First, this operation results in a system defining $L \cap K$ of smaller (or at least not larger) size than S'. Second, in the case of a system constructed by Corollary 3.89, it produces the set of unknowns $x_{i,q}$ such that $L_i \cap h^{-1}(q) \neq \emptyset$ (by Corollary 3.24). This will help to construct proofs of facts of the form $L_i \subseteq K$ for sets K that are recognizable in $\mathbb{M}$, as we did in Proposition 1.6 (see Proposition 3.91 below). We have proved in a similar way in Section 1.2.5 that no series-parallel graph can satisfy the two 2-coloring properties σ and δ of Example 1.11.

(2) From the above observation on the number of equations, and the second construction in the proof of the Filtering Theorem, we get (with the notation of that proof) that $L \cap K$ is nonempty if and only if it contains an element defined by a term of height at most $n \cdot \sharp \mathcal{B}$ where n is the number of equations of a uniform system that defines L. This fact follows from Proposition 3.53(1).

(3) We may have a computable homomorphism $h : \mathbb{M} \to \mathbb{B}$ without knowing if $h^{-1}(q) = \emptyset$ for given q in B (cf. Remark 3.79(2)). The second construction in the proof of the Filtering Theorem works nevertheless but we have $L_i \cap h^{-1}(q) = \emptyset$ if $q \notin h(M)$. As discussed in (1), we can determine the pairs (i, q) such that $L_i \cap h^{-1}(q) = \emptyset$, hence for each i, the finite set $h(L_i)$.

(4) The proof of the Filtering Theorem defines L as well as K (to simplify the discussion we assume that F is finite) by finite, possibly nondeterministic, automata on terms, but it does not extend to the case where K is equational. Let L and S be as in the second construction in the proof, with $L = val_{\mathbb{M}}(L(G(S), x_1))$. Assume that $K = val_{\mathbb{M}}(L(G(\overline{S}), x_1))$ for another equation system $\overline{S}$ with set of unknowns X_m. One can build a system S' that defines in $\mathbb{T}(F)$ the sets $L(G(S), x_i) \cap L(G(\overline{S}), x_j)$ for $i \in [n]$ and $j \in [m]$. Then $val_{\mathbb{M}}(L(G(S), x_i) \cap L(G(\overline{S}), x_j)) \subseteq val_{\mathbb{M}}(L(G(S), x_i)) \cap val_{\mathbb{M}}(L(G(\overline{S}), x_j))$ but the inclusion may be strict, hence the system S' does not define in general the set $L \cap K$. This is not a surprise because we know that the intersection of two context-free languages is not always a context-free language. □

The Filtering Theorem is stated in Chapter 1 (Theorem 1.8) in terms of finite inductive sets of properties instead of recognizable sets. By Proposition 3.83, this is equivalent to the formulation of Theorem 3.88 (cf. Theorem 1.12). In Chapter 5, we will state a logical formulation of the Filtering Theorem where the recognizable sets K are defined by monadic second-order sentences.

We now generalize Proposition 1.6 and Corollary 1.9 and we show how inclusions of the form $L \subseteq K$ where L is equational and K is recognizable can be proved (by fixed-point induction, see Section 1.2).

Proposition 3.91

(1) Let L be equational in an algebra $\mathbb{M}$ and be defined by (S, X_μ) for some equation system S with $Unk(S) = X_n$ and some $\mu \in [n]$. Let K be recognizable in $\mathbb{M}$. Then $L \subseteq K$ if and only if there exists an n-tuple $(K_1, \ldots, K_n)$ of $\mathbb{M}$-recognizable sets that is an oversolution of S and is such that $K_\ell \subseteq K$ for every $\ell \in [\mu]$.

(2) If, furthermore, K is effectively recognizable and S is given, one can decide whether $L \subseteq K$, and, if this is the case, one can construct the effectively recognizable sets $K_1, \ldots, K_n$. The analogous statements hold for effective term-recognizability and for semi-effective recognizability.

Proof: (1) Let $(L_1, \ldots, L_n)$ be the least solution of S in $\mathcal{P}(\mathbb{M})$, with $L = \bigcup_{\ell \in [\mu]} L_\ell$. Let K be defined as $h^{-1}(C)$, where h is a homomorphism of $\mathbb{M}$ into a locally finite algebra $\mathbb{B}$ and $C \subseteq B$.

Let us assume that $L \subseteq K$. For each $i = 1, \ldots, n$, we define the recognizable set $K_i := h^{-1}(h(L_i))$. In other words, K_i is the union of the sets $h^{-1}(q)$ such that $L_i \cap h^{-1}(q) \neq \emptyset$, for $q \in B$. Since K is a union of pairwise disjoint sets $h^{-1}(q)$, we have $L_\ell \subseteq K_\ell \subseteq K$ for every $\ell \in [\mu]$ (and $K - \bigcup_{\ell \in [\mu]} K_\ell$ is the union of the sets $h^{-1}(q)$ such that $q \in C$ and $L \cap h^{-1}(q) = \emptyset$). It remains to prove that for each i, we have

$$p_i \mathcal{P}(\mathbb{M})(K_1, \ldots, K_n) \subseteq K_i,$$

where p_i is the polynomial that is the right-hand side of the equation of S that defines x_i. It suffices actually to prove

$$t_{\mathcal{P}(\mathbb{M})}(K_1, \ldots, K_n) \subseteq K_i,$$

for each monomial t of p_i. Let us write $t = \widetilde{t}[x_{i_1}/y_1, \ldots, x_{i_k}/y_k]$ where $\widetilde{t} \in T(F, Y_k)$ and $ListVar(\widetilde{t}) = y_1 \cdots y_k$. By Lemma 3.2(4), an element m of $t_{\mathcal{P}(\mathbb{M})}(K_1, \ldots, K_n)$ is of the form $\widetilde{t}_{\mathbb{M}}(m_1, \ldots, m_k)$ with $m_j \in K_{i_j}$ for $j = 1, \ldots, k$. For each such j, there is in L_{i_j} an element m'_j such that $h(m'_j) = h(m_j)$. Then, the element of M defined as $m' := \widetilde{t}_{\mathbb{M}}(m'_1, \ldots, m'_k)$ belongs to $t_{\mathcal{P}(\mathbb{M})}(L_1, \ldots, L_n)$ and hence to L_i since t is a monomial of the equation that defines x_i. Since h is a homomorphism, we have $h(m') = h(m)$. Hence m belongs to K_i. This shows that $(K_1, \ldots, K_n)$ is an oversolution of S.

The opposite implication follows from Theorem 3.7: if $(K_1, \ldots, K_n)$ is an over-solution of S, then $L_i \subseteq K_i$ for each i, and since $K_\ell \subseteq K$ for every $\ell \in [\mu]$, we get $L = \bigcup_{\ell \in [\mu]} L_\ell \subseteq K$ as desired.

(2) Assume now that K is effectively recognizable and is defined as above. By Corollary 3.86, $M - K$ is effectively recognizable (M is effectively recognizable by the proof of Proposition 3.85(1)). Hence, an equation system for the set $L \cap (M - K)$ can be constructed by Theorem 3.88 and one can decide if that set is empty, i.e., if $L \subseteq K$ (by Corollary 3.24). For each $i \in [n]$ and each element q of B, the set L_i is defined by S and the set $h^{-1}(q)$ is effectively recognizable. Hence, by the same procedure, it can be decided whether $L_i \cap h^{-1}(q) \neq \emptyset$, i.e., whether $q \in h(L_i)$. Thus, K_i is effectively recognizable. The same proof holds in the other two cases. $\blacksquare$

Hence, as soon as some correct n-tuple $(K_1, \ldots, K_n)$ has been guessed, it remains to prove properties of the forms $t_{\mathcal{P}(\mathbb{M})}(K_1, \ldots, K_n) \subseteq K'$ and $K' \subseteq K$ for recognizable sets $K, K', K_1, \ldots, K_n$ to complete the proof.

We conclude this section by comparing the equational and the recognizable sets of an algebra.

Corollary 3.92 A domain M_s of an F-algebra $\mathbb{M}$ is generated by a finite subsignature of F if and only if $\mathbf{Rec}(\mathbb{M})_s \subseteq \mathbf{Equat}(\mathbb{M})_s$.

Proof: Let M_s be generated by a finite subsignature of F. Then $M_s \in \mathbf{Equat}(\mathbb{M})_s$ by Proposition 3.23(4). For every set $K \in \mathbf{Rec}(\mathbb{M})_s$ we have $K = M_s \cap K \in \mathbf{Equat}(\mathbb{M})_s$ by Theorem 3.88. Assume conversely that $\mathbf{Rec}(\mathbb{M})_s \subseteq \mathbf{Equat}(\mathbb{M})_s$. Since M_s is recognizable by Proposition 3.85(1), it is equational, hence it is generated by a finite subsignature of F, again by Proposition 3.23(4). $\blacksquare$

The inclusion $\mathbf{Rec}(\mathbb{M})_s \subseteq \mathbf{Equat}(\mathbb{M})_s$ may be false without the hypothesis that M_s is finitely generated. To prove this it suffices to consider an algebra $\mathbb{M}$ with a countably infinite domain M and a signature F consisting of one constant symbol to denote each element. Hence, $\mathbb{M}$ is generated by F but not by any finite subset of F. The set M is recognizable but it is not equational.

Since some context-free languages are not regular, we do not always have $\mathbf{Equat}(\mathbb{M})_s \subseteq \mathbf{Rec}(\mathbb{M})_s$. The families $\mathbf{Rec}(\mathbb{M})_s$ and $\mathbf{Equat}(\mathbb{M})_s$ are thus incomparable in general.

3.4.8 Recognizable sets of commutative words

We have described in Section 3.1.6 the equational sets of the additive monoid $\mathbb{N}^k = \langle \mathcal{N}^k, +, \vec{0} \rangle$, where $\vec{0} = (0, \ldots, 0)$. We now describe its recognizable sets. A subset of $\mathcal{N}$ is *ultimately periodic* if it is of the form $P \cup \{u + \lambda \cdot v \mid u \in Q, \lambda \in \mathcal{N}\}$ where $P, Q \in \mathcal{P}_f(\mathcal{N})$ and $v \in \mathcal{N}$.

Proposition 3.93

(1) A subset of $\mathcal{N}$ is recognizable in $\mathbb{N}$ if and only if it is ultimately periodic if and only if it is equational in $\mathbb{N}$.

(2) If $k \geq 2$, a set is recognizable in $\mathbb{N}^k$ if and only if it is a finite union of Cartesian products $R_1 \times \cdots \times R_k$ such that $R_1, \ldots, R_k \in \mathbf{Rec}(\mathbb{N})$.

Proof sketch: (1) The length is an isomorphism from $\mathbb{W}(\{a\})$ to the additive monoid $\mathbb{N} = \langle \mathcal{N}, +, 0 \rangle$. Using this fact, the first equivalence is proved in [*Eil], Proposition 1.1 of Chapter V, by means of deterministic finite automata over the alphabet $\{a\}$. The second equivalence is a special case of Proposition 3.25.

(2) The second assertion follows from the fact that for every two monoids $\mathbb{M}_1$ and $\mathbb{M}_2$, $\mathbf{Rec}(\mathbb{M}_1 \times \mathbb{M}_2)$ is the set of finite unions of sets $R_1 \times R_2$ where $R_1 \in \mathbf{Rec}(\mathbb{M}_1)$ and $R_2 \in \mathbf{Rec}(\mathbb{M}_2)$ (Proposition 12.2 of Chapter III of [*Eil]). One direction of this equality is immediate from Proposition 3.85(2). The proof of the other direction can be sketched as follows: let $L \in \mathbf{Rec}(\mathbb{M}_1 \times \mathbb{M}_2)$ with finite congruence $\sim$; then $L = \{(m, m') \mid (m, \epsilon_{\mathbb{M}_2}) \sim (p, \epsilon_{\mathbb{M}_2})$ and $(\epsilon_{\mathbb{M}_1}, m') \sim (\epsilon_{\mathbb{M}_1}, p')$ for some $(p, p') \in L\}$, hence L is a union of sets $L_p \times L_{p'}$ where $L_p = \{m \in M_1 \mid (m, \epsilon_{\mathbb{M}_2}) \sim (p, \epsilon_{\mathbb{M}_2})\}$ and $L_{p'} = \{m \in M_2 \mid (\epsilon_{\mathbb{M}_1}, m) \sim (\epsilon_{\mathbb{M}_1}, p')\}$; there are finitely many such sets $L_p, L_{p'}$ and they are recognizable. ∎

It follows that the linear set $\{(n, n) \mid n \in \mathcal{N}\}$ is equational in $\mathbb{N} \times \mathbb{N}$ but is not recognizable in this algebra.

Remark 3.94 That $\mathbf{Rec}(\mathbb{M}_1 \times \mathbb{M}_2)$ is the set of finite unions of products of recognizable sets of $\mathbb{M}_1$ and $\mathbb{M}_2$ is true for monoids but not for arbitrary algebras. Let $\mathbb{N}' := \langle \mathcal{N}, Suc, 0 \rangle$ where $Suc(n) = n + 1$ for all $n \in \mathcal{N}$. Then, the set $\{(n, n) \mid n \in \mathcal{N}\}$ is recognizable in $\mathbb{N}' \times \mathbb{N}'$ with congruence defined by $(n, n') \sim (p, p')$ if and only if $n = n'$ and $p = p'$ or $n \neq n'$ and $p \neq p'$. Hence $\mathbf{Rec}(\mathbb{N}' \times \mathbb{N}') \supset \mathbf{Rec}(\mathbb{N} \times \mathbb{N})$. However, $\mathbf{Rec}(\mathbb{N}') = \mathbf{Rec}(\mathbb{N})$ because $\mathbb{N}$ and $\mathbb{N}'$ are respectively isomorphic to $\mathbb{W}(\{a\})$ and to $\mathbb{W}_{right}(\{a\})$, which have the same recognizable set as recalled in Example 3.57(2).

3.4.9 Decidability questions

It is well known that the emptiness problem for finite automata is decidable, that is, there exists an algorithm that decides whether the set of terms accepted by a given finite automaton is empty or not: see Chapter 1 of [*Com+] or Proposition 5.3 of [*GecSte]. This is actually a consequence of the equivalence of regular grammars and automata, established in Proposition 3.47. However, this result does not extend to general recognizable sets, not even to those that are effectively recognizable. This should be contrasted with Proposition 3.24 relative to equational sets.

Proposition 3.95 The emptiness of an effectively recognizable set is not decidable in general, not even if the algebra has a single sort. It is decidable if the signature is

finite and generates the considered algebra (even for semi-effectively recognizable
and effectively term-recognizable sets).

Proof: In order to prove the first assertion, we give two counterexamples. We let
F be the infinite one-sort signature consisting of a constant symbol a and of unary
function symbols f_n, for all $n \in \mathcal{N}$. Let g be a computable mapping $: \mathcal{N} \to \{0, 1\}$. Let
$\mathbb{A}$ be the finite F-algebra associated with g by

$$A = \{0, 1\}, \quad a_{\mathbb{A}} = 0, \quad f_{n\mathbb{A}}(1) = 1, \quad f_{n\mathbb{A}}(0) = g(n).$$

The set $L := val_{\mathbb{A}}^{-1}(\{1\}) \subseteq \mathbb{T}(F)$ is effectively recognizable in $\mathbb{T}(F)$. It is clear that
$L \neq \emptyset$ if and only if $g(n) = 1$ for some $n \in \mathcal{N}$, and this not decidable.

Here is the second example. We let F' be reduced to the constant symbol a and
$\mathbb{M} := \langle \mathcal{N}, a_{\mathbb{M}} \rangle$ with $a_{\mathbb{M}} := 0$. This algebra is not generated by F'. Let $\mathbb{A}$ and g be as
above. The mapping h such that $h(0) = 0$ and $h(i) = g(i)$ if $i \geq 1$, is a homomorphism
$: \mathbb{M} \to \mathbb{A}$. Hence $h^{-1}(\{1\})$ is effectively recognizable in $\mathbb{M}$. It is nonempty if and only
if $g(i) = 1$ for some $i \geq 1$. And this is not decidable.

The decidability assertion is an immediate consequence of Theorem 3.88 and
Propositions 3.23(4) and 3.24 (because if $K \in \mathbf{Rec}(\mathbb{M})$, then $K = \emptyset$ if and only if
$M_s \cap K = \emptyset$ for every sort s). The two counter-examples show that none of the two
hypotheses can be omitted. ∎

3.4.10 Concluding remarks on recognizability

The two results that motivate the introduction and the study of recognizability are the
linear-time membership algorithm (where linearity is meant with respect to the size
of a term denoting the considered object) and the Filtering Theorem. Both of them
are based on the local finiteness properties that characterize recognizability.

We now review the concrete and effective ways to specify recognizable sets and
the uses of such specifications. Then we discuss when the notion of recognizability
becomes void, and finally we come back to rational sets and regular expressions
already discussed in Section 3.2.5.

Specification of recognizable sets

Let $\mathbb{M}$ be an algebra that is generated by its signature F, and let F be effectively
given. We distinguish two cases.

The first case is when F is finite, hence $\mathbb{M}$ is finitely generated. Then a subset L of
$\mathbb{M}$ is recognizable if and only if the language $val_{\mathbb{M}}^{-1}(L)$ is regular (by Proposition 3.69
and Theorem 3.62). The most convenient way to specify L is by a finite complete
and deterministic F-automaton $\mathscr{A}$ recognizing $val_{\mathbb{M}}^{-1}(L)$. It is decidable for a term
$t \in T(F)$ if $val_{\mathbb{M}}(t)$ is in L: one just runs $\mathscr{A}$ on t. If $\mathbb{M}$ is semi-effectively given, then
membership in L is decidable by an algorithm using two steps: first one looks for a

term $t \in T(F)$ that evaluates[41] to the given element m of M. We know that a (possibly inefficient) algorithm exists for finding such t (see Definitions 2.8 and 2.127). Then, by running $\mathscr{A}$ on t, one obtains the answer that m belongs to L or not. The emptiness problem for L is decidable since it reduces to the emptiness problem for the language $L(\mathscr{A})$. The effectivity statement of the Filtering Theorem and its Corollary 3.89 are applicable.

Such an automaton can be constructed from the following data:

- a specification of L as a semi-effectively recognizable or effectively term-recognizable set[42] by a finite algebra $\mathbb{A}$ and a subset C of A (by Propositions 3.76(4) and 3.78(3.2)); or
- a membership algorithm for L and a decidable finite congruence witnessing the recognizability of L (by Proposition 3.78); or
- a membership algorithm for L and an upper-bound to $\gamma_L(s)$ for each sort s (by Proposition 3.80);[43] or
- a computable finite F-inductive family of predicates (by Corollary 3.84).

The second case is when F is infinite. Let L be semi-effectively recognizable or effectively term-recognizable in $\mathbb{M}$. Then, for every finite subsignature F' of F, the set $L \cap M'$ is effectively recognizable in the subalgebra $\mathbb{M}'$ of $\mathbb{M}$ generated by F' (by the first paragraph of the proof of Theorem 3.88). Hence, by the first case, one can construct a finite F'-automaton $\mathscr{A}_{F'}$ that recognizes the language $val_{\mathbb{M}'}^{-1}(L \cap M')$, which equals $val_{\mathbb{M}}^{-1}(L) \cap T(F')$. It is decidable for a term $t \in T(F)$ if $val_{\mathbb{M}}(t)$ is in L: letting F' be the (finite) minimal subsignature of F such that $t \in T(F')$, one gets the desired answer by running $\mathscr{A}_{F'}$ on t. If $\mathbb{M}$ is semi-effectively given, then membership in L is decidable: given m in M, an algorithm can find a term t in $T(F)$ that evaluates to m (cf. Definitions 2.8 and 2.127); then one decides whether $val_{\mathbb{M}}(t)$ is in L. The emptiness problem is undecidable (Proposition 3.95); however, the two constructions in the proof of the Filtering Theorem can be done by algorithms.

The automaton $\mathscr{A}_{F'}$ can be constructed from the same data as in the finite case, with "finite" replaced by "locally finite". For the third kind of data, an algorithm must be specified that computes an upper-bound to $\gamma_L(s)$ for each sort s of F. Then we can use Proposition 3.80 (applied to $\mathbb{M}'$ and $L \cap M'$) because, by Remark 3.79(3), $\gamma_{L \cap M'}(s) \leq \gamma_L(s)$ for every $s \in \mathcal{S}'$ (where $\mathcal{S}'$ is the set of sorts of F') and hence $\sharp \mathcal{M}(L \cap M') \leq \Sigma_{s \in \mathcal{S}'} \gamma_L(s)$.

Let us finally observe that if $\mathbb{M}$ is semi-effectively given (hence countable) but not generated by its signature, then one can enrich the signature by adding one constant

---

[41] *Any such term will give the correct answer. The corresponding parsing problem is less constrained than for equational sets.*

[42] In the first case, $\mathbb{M}$ and $\mathbb{A}$ must be semi-effectively given; the homomorphism : $\mathbb{M} \to \mathbb{A}$ need not be specified, by Lemma 3.75(1). In the second case, $\mathbb{A}$ must be effectively given; $\mathbb{M}$ need not be semi-effectively given. In both cases a membership algorithm for C must be specified.

[43] Recall that γ_L is the recognizability index of L, cf. Definition 3.65.

symbol for denoting each element: the recognizable sets are the same (cf. the remark after Proposition 3.56), and we can obtain effective constructions under the conditions of the second case.

Recognizability is void in certain algebras

The notion of recognizability depends on the relevant algebra. It may become void in two opposite cases. If the algebraic structure is "poor," for example in the absence of functions of positive arity, then every set L is recognizable (the corresponding congruence has two classes, L and its complement). The notion of recognizability becomes trivial.

The opposite case is when the algebraic structure is so rich that there are very few recognizable sets. If one enriches an algebra by adding new operations, one gets fewer (or the same) recognizable sets. For example, consider the set $\mathcal{N}$ of nonnegative integers equipped with the successor and the predecessor functions (where the predecessor is defined by $pred(0) = 0, pred(n+1) = n$). The only recognizable sets are $\mathcal{N}$ and the empty set. Indeed, if $\sim$ is a congruence and if $n \sim n+p$ for some $n \geq 0, p > 0$, then by using $n + p - 1$ times the function $pred$, we find that $0 \sim 1$. It follows (using the successor function repeatedly) that any two integers n and $n+1$ are equivalent, hence any two integers are equivalent.

As another example, consider the monoid $\mathbb{W}(\{a,b\})$ of words over two letters. Let us add a unary operation, the *circular shift*, defined by: $sh(\varepsilon) = \varepsilon$ and $sh(aw) = wa$, $sh(bw) = wb$, for every word w. The language ba^* is no longer recognizable in this algebra, denoted by $\mathbb{W}(\{a,b\})_{sh}$. Here is the proof. Assume it is with finite congruence $\sim$. We have $a^m b \sim a^{m+n} b$ for some m and $n > 0$. Then $ba^m = sh^m(a^m b) \sim sh^m(a^{m+n}b) = a^n ba^m$ but this word is not in ba^*. Hence, ba^* is not saturated by any finite congruence on $\mathbb{W}(\{a,b\})_{sh}$. However, recognizability does not degenerate completely since every commutative language L that is recognizable in $\mathbb{W}(\{a,b\})$ remains recognizable in $\mathbb{W}(\{a,b\})_{sh}$. To see this, let $\approx^L$ be the syntactic congruence of L. We recall from Example 3.67 that for $w, w' \in \{a,b\}^*$:

$$w \approx^L w' \text{ if and only if } xwy \in L \Longleftrightarrow xw'y \in L \text{ for all } x,y \in \{a,b\}^*.$$

It is finite since L is recognizable. Since L is commutative,[44] it is also a congruence for $sh(w)$: if $w \approx^L w'$ and $xsh(w)y \in L$ then $xwy \in L$, hence $xw'y$ and $xsh(w')y$ belong to L; this proves that $sh(w) \approx^L sh(w')$.

Rational sets

We have defined the rational sets of monoids in Section 3.2.5. More generally, star operations, and thus regular expressions and sets that can be called rational, can be

---

[44] This means that if $v \in L$ and $h(v) = h(w)$ then $w \in L$, where h is the homomorphism : $\mathbb{W}(\{a,b\}) \to \mathbb{N} \times \mathbb{N}$ defined in Section 3.1.6.

defined whenever we have binary associative operations like $\oplus, \otimes, /\!/, \bullet$ (cf. Chapter 1). Such operations have been defined and investigated for planar directed acyclic graphs in [BosDW] and for *pictures* in [*GiaRes] (a *picture* in this work is a directed vertex-labeled rectangular grid, cf. Example 2.56(3) and Section 5.2.3; they are different from those considered in [*Dre06]). In most cases, the corresponding class of rational sets contains properly the one of recognizable sets.

3.5 References

The equational and the recognizable sets of a one-sorted algebra have been defined by Mezei and Wright in [MezWri]. Many results presented in this chapter, in particular Proposition 3.23 and Theorem 3.62 (restricted to finite signatures), are due to them. This chapter is otherwise based on [*Cou86], [*Cou96b], [*Cou97], but the notions of effectivity are new. Recognizability is extended so as to allow elements of different sorts in the same recognizable set.

Theorem 3.7 is Proposition 1.4.5 of the book [*ArnNiw]. The preservation of sets of solutions and of least solutions under unfoldings and related transformations is studied in detail in [*Cou86]. More general equation systems, called *regular systems of equations*, are studied in [*Cou83], [*Cou86] and [*Cou90a] with motivations from the semantics of recursive programs. Regular grammars and automata (cf. Section 3.3) are studied in [*GecSte] and [*Com+].

The recognizable and the equational sets of $\mathbb{R}$, the algebra of rooted trees are characterized in terms of appropriate finite automata and of equivalent logical sentences in [*Lib06], [BonTal], [Cou89a] and [*Cou96b]. Chapter 8 of [*Com+] also considers rooted *ordered unranked trees* in which each node has a sequence of sons of unbounded length.

4

Equational and recognizable sets of graphs

The general algebraic notions of both equational and recognizable sets have been defined in Chapter 3. In this chapter, we apply them to the graph algebras of Chapter 2 and give examples that are interesting from the graph theoretic point of view. We also review the consequences of the results of Chapter 3 and establish some properties that are particular to graph algebras.

In Section 4.1, we study the equational sets of the algebra $\mathbb{JS}$; these are called the HR-equational sets. We establish decidability results for membership, emptiness and finiteness, and show some relationships between tree-width and HR-equationality. We compute the types of the elements of an HR-equational set and prove that the equational sets of the one-sorted algebra $\mathbb{JS}$ are the same as those of the many-sorted algebra $\mathbb{JS}^t$. The HR-equational sets are also defined by the hyperedge replacement graph grammars: this grammatical characterization was actually their original definition.

In Section 4.2 we study the recognizable sets of the many-sorted algebra $\mathbb{JS}^t$, called the HR-recognizable sets. We generalize to them the characterization of regular languages due to Myhill and Nerode, and also prove that there are uncountably many HR-recognizable sets of graphs.

In Section 4.3, we study similarly the equational sets of the one-sorted algebra $\mathbb{GP}$, which are called the VR-equational sets. We establish for them decidability results similar to those for HR-equational sets. The VR-equational sets are related to clique-width in the same way that the HR-equational sets are to tree-width. Every HR-equational set of simple graphs is VR-equational: this is not surprising since the clique-width of a simple graph is bounded in terms of its tree-width but this observation does not replace a proof. The converse implication holds for sparse graphs.

In Section 4.4, we study the recognizable sets of the many-sorted algebra $\mathbb{GP}^t$, which are called the VR-recognizable sets. We generalize to them the characterization of regular languages by Myhill and Nerode. Every VR-recognizable set of graphs is HR-recognizable and the converse implication holds for the sets of simple graphs that are sparse. These implications are similar but opposite to those that relate the different types of equational sets.

In Chapter 5, we will prove that the sets of graphs defined by monadic second-order sentences are recognizable: this will give an easy way to specify recognizable sets and to use this notion.

Before starting the formal exposition, let us state as a rule of thumb that most of the definitions and results apply to labeled graphs (possibly with sources or ports) with essentially the same proofs as for unlabeled graphs (or s-graphs or p-graphs). By "labeled" graphs, s-graphs or p-graphs, we mean (K, Λ)-labeled graphs, s-graphs or p-graphs, for some finite sets K and Λ. In order to facilitate reading, we will not always specify these technical details. The phrase "(labeled) graphs" in statements is intended to remind the reader of this convention.

4.1 HR-equational sets of simple graphs

We recall from Section 2.3 that the acronym HR refers to the (one-sorted) algebra of (labeled) graphs with (and without) sources (called s-graphs), denoted by $\mathbb{JS}$. Its signature F^{HR} is countable and consists of the following operations:

- the parallel-composition denoted by $/\!/$;
- the operation fg_B that "forgets" the sources with names in a finite set B, (i.e., turns these sources into internal vertices);
- the operation ren_h that renames sources according to a finite permutation h of the set of source names;
- constant symbols denoting edges, loops and isolated vertices; the eventual vertex and edge labels are specified by the constant symbols.

We denote by $F^{\mathrm{HR}}_{[K,\Lambda]}$ the variant of F^{HR} with which (K, Λ)-labeled graphs can be defined as values of terms, and the corresponding algebra by $\mathbb{JS}_{[K,\Lambda]}$. The many-sorted variants of these algebras are denoted by $\mathbb{JS}^{\mathrm{t}}$ and $\mathbb{JS}^{\mathrm{t}}_{[K,\Lambda]}$, see Section 2.6.2. We will recall more details about them when appropriate.

4.1.1 HR equation systems

The operations of F^{HR} can be used to write equation systems according to the definitions of Section 3.1, from which we get the notion of an HR-equational set of s-graphs. By "a set of s-graphs," we mean a set of abstract s-graphs. If a set of concrete s-graphs L is specified in one way or another, we say that L is HR-equational if the set $\{[G]_{iso} \mid G \in L\}$ is HR-equational.

Definition 4.1 (HR-equational sets of s-graphs) A set of s-graphs is *HR-equational* if it is equational in the algebra $\mathbb{JS}$. We now review the relevant specializations of some definitions of Chapter 3. An *HR equation system* is an equation system over the signature F^{HR}. Its general form is (without loss of generality by Lemma 3.4, see

Definition 3.6) $S = \langle x_1 = p_1, \ldots, x_n = p_n \rangle$, where each p_i is a polynomial over the signature F^{HR}, i.e., $p_i \in Pol(F^{HR}, X_n)$. According to Section 3.1.2, we will denote its least solution in the powerset algebra $\mathcal{P}(\mathbb{JS})$ by $\mu \bar{x} \cdot S_{\mathcal{P}(\mathbb{JS})}(\bar{x})$ (with $\bar{x}$ denoting $(x_1, \ldots, x_n)$). In order to get closer to the usual notation for context-free grammars, we will denote by $L(S, x_i)$ the component $\mu \bar{x} \cdot S_{\mathcal{P}(\mathbb{JS})}(\bar{x}) \restriction x_i$ of $\mu \bar{x} \cdot S_{\mathcal{P}(\mathbb{JS})}(\bar{x})$. We will call it the *set defined by* (S, x_i), or *generated by* S *from* x_i. We will denote by $L_{Term}(S, x_i)$ the corresponding set when S is solved in the powerset term algebra $\mathcal{P}(\mathbb{T}(F^{HR}))$. By Theorem 3.18, $L_{Term}(S, x_i) = L(G(S), x_i)$, where $G(S)$ is the regular grammar associated with S. For $Y \subseteq Unk(S)$, we denote the set $\bigcup_{y \in Y} L(S, y)$ by $L(S, Y)$, and similarly for $L_{Term}(S, Y)$. Since $\mathbb{JS}$ is a one-sorted algebra, every HR-equational set is of the form $L(S, x)$ for some HR equation system S and some unknown x of S.

For defining HR-equational sets of (K, Λ)-labeled s-graphs, we use the signature $F^{HR}_{[K, \Lambda]}$, the algebra $\mathbb{JS}_{[K, \Lambda]}$ and the term algebra $\mathbb{T}(F^{HR}_{[K, \Lambda]})$ instead of F^{HR}, $\mathbb{JS}$ and $\mathbb{T}(F^{HR})$ respectively.

Every HR equation system S is an equation system over the finite signature $F(S)$ and hence over F^{HR}_C (or $F^{HR}_{C, [K, \Lambda]}$) for some finite set of source names $C \subseteq \mathcal{A}$.[1] We denote the smallest such C by $\mathcal{A}(S)$ (so that $Sort(S) \subseteq \mathcal{P}(\mathcal{A}(S))$). It is the set of source names that occur in S and it can be computed as the union of the sets $\mu(t)$ such that t is a monomial in an equation of S, where Definition 2.35 is generalized to terms with variables in the obvious way (adding $\mu(x) = \emptyset$ to the rules that compute $\mu(t)$, for every variable x). Thus, $F(S) \subseteq F^{HR}_{\mathcal{A}(S)}$ (or $F(S) \subseteq F^{HR}_{\mathcal{A}(S), [K, \Lambda]}$).

The system S has the same least solution in the powerset algebras of $\mathbb{T}(F(S))$ and $\mathbb{T}(F^{HR})$ (or $\mathbb{T}(F^{HR}_{[K, \Lambda]})$), by Proposition 3.9(1). We obtain the following result from Proposition 3.23(1), where we denote by *val* the mapping $val_{\mathbb{JS}}$ (or $val_{\mathbb{JS}_{[K, \Lambda]}}$) that evaluates a term into the corresponding s-graph.

Proposition 4.2 For every HR equation system S and every unknown x of S, we have $L(S, x) = val(L_{Term}(S, x))$ and $L_{Term}(S, x) \subseteq T(F(S))$. □

It follows that every s-graph G belonging to an HR-equational set $L(S, x)$ is the value of some term in $T(F(S))$. Such a term yields a linear notation for G. Furthermore, it can be used as input to an algorithm checking certain properties of G or computing certain values attached to it, in polynomial time, as we will see in Chapter 6.

It follows also from Proposition 3.9(1) that an HR-equational set of s-graphs L is equational in the subalgebras $\mathbb{JS}[\mathcal{A}(S)]$ and $\mathbb{JS}^{gen}[\mathcal{A}(S)]$ (cf. Remark 2.39(1)), where S is an equation system that defines L, and similarly in the labeled case. We will use this fact in the proofs of Propositions 4.10 and 4.13(3) below.

All closure results for equational sets established in Sections 3.1.5 and 3.2.4 hold for the HR-equational sets. For example, the mapping *und* that forgets edge directions

---

[1] This uses the obvious fact that every finite subsignature of F^{HR} is a subsignature of F^{HR}_C for some finite set C (and similarly for the labeled case). The definition of the signature $F(S)$ is in Definition 3.6.

(cf. Definition 2.9) is a homomorphism: $\mathbb{JS}^d \to \mathbb{JS}^u$. It follows from Proposition 3.23(2) that $und(L)$ is HR-equational if L is an HR-equational set of directed s-graphs.

We now present or recall a few examples.

Example 4.3 (1) **Series-parallel graphs:** The set of (directed) series-parallel graphs is defined by the equation

$$x = (x /\!/ x) \cup fg_3(ren_{2\leftrightarrow3}(x) /\!/ ren_{1\leftrightarrow3}(x)) \cup \overrightarrow{12}.$$

The second term in the right-hand side of this equation is the expanded expression of the monomial $x \bullet x$, written with $\bullet$ (series-composition), a derived operation of $\mathbb{JS}$ (cf. Example 2.133, Section 2.6.2). The undirected series-parallel graphs are obtained similarly with **12** instead of $\overrightarrow{12}$.

(2) **Rooted trees:** The root of a tree is defined as its 1-source. (We recall from Definition 2.13 that edge directions follow from the choice of a root.) The defining equation is

$$y = (y /\!/ y) \cup fg_2(\overrightarrow{12} /\!/ ren_{1\leftrightarrow2}(y)) \cup \mathbf{1}.$$

(3) **Trees:** Trees, as defined in Definition 2.13, form the set $L(S,u)$, where S is the following system:

$$\begin{cases} u = fg_1(z), \\ z = (z /\!/ z) \cup fg_2(\mathbf{12} /\!/ ren_{1\leftrightarrow2}(z)) \cup \mathbf{1}. \end{cases}$$

(4) **Syntactic trees of terms:** We consider the syntactic trees of the terms in $T(F)$ where $F := \{f,g,a,b\}$, $\rho(f) := 2$, $\rho(g) := 1$, $\rho(a) := \rho(b) := 0$. Each node of such a tree T has a label in F; each edge has also a label, $\ell := 1$ or $r := 2$, used to indicate whether it leads to the first (left) or to the second (right) son. The defining equation is

$$t = (\mathbf{1}_f /\!/ fg_2(\overrightarrow{12}_\ell /\!/ ren_{1\leftrightarrow2}(t) /\!/ fg_2(\overrightarrow{12}_r /\!/ ren_{1\leftrightarrow2}(t))))$$
$$\cup (\mathbf{1}_g /\!/ fg_2(\overrightarrow{12}_\ell /\!/ ren_{1\leftrightarrow2}(t))) \cup \mathbf{1}_a \cup \mathbf{1}_b.$$

The constant symbols $\mathbf{1}_\kappa$ and $\overrightarrow{12}_\lambda$ denote respectively an isolated vertex that is a 1-source labeled by κ and an edge from a 1-source to a 2-source labeled by λ. (We have defined labeled s-graphs in Chapter 2, Definitions 2.11 and 2.32.)

(5) **Ladders:** By a ladder, we mean an s-graph of the form shown in Figure 4.1. The equation that defines ladders is $y = \mathbf{12} \cup f(y)$, where f is the operation that adds one step to a ladder. It can be defined by the following term:[2]

$$f(u) = ren_{2\leftrightarrow4}(ren_{1\leftrightarrow3}(fg_{12}(u /\!/ \mathbf{13} /\!/ \mathbf{24} /\!/ \mathbf{34}))),$$

---

[2] In examples, we write fg_{12} or fg_{abc} for $fg_{\{1,2\}}$ or $fg_{\{a,b,c\}}$ respectively.

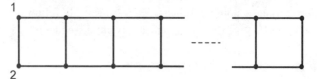

Figure 4.1 A ladder.

where u denotes any s-graph. However, we will only use $f(u)$ for graphs u of type $\{1,2\}$. One can do better, with fewer source names:

$$f(u) = ren_{2\leftrightarrow3}(fg_2[ren_{1\leftrightarrow3}(fg_1(u \mathbin{/\!/} \mathbf{13})) \mathbin{/\!/} \mathbf{23}]) \mathbin{/\!/} \mathbf{12}.$$

As we will see in Section 4.1.2, the second expression of the function f yields a better estimate of the maximal tree-width of a ladder.

(6) **Cycles:** The set of cycles C_n, $n \geq 2$, is defined as $L(T,w)$, where T is the following system:

$$\begin{cases} w = fg_{12}(v \mathbin{/\!/} \mathbf{12}), \\ v = \mathbf{12} \cup (v \bullet \mathbf{12}), \end{cases}$$

and $\bullet$ is series-composition, see (1). The last equation can also be written as follows by expanding the definition of $\bullet$ and replacing $ren_{1\leftrightarrow3}(\mathbf{12})$ by $\mathbf{23}$:

$$v = \mathbf{12} \cup fg_3(ren_{2\leftrightarrow3}(v) \mathbin{/\!/} \mathbf{23}).$$

Note that if we replace in T the monomial $v \bullet \mathbf{12}$ by $v \bullet v$, then the same sets are generated.

(7) **Outerplanar graphs:** A *biconnected outerplanar* graph consists of one cycle C_n for some $n \geq 2$, called the *external cycle*, and of edges that can be drawn without crossings inside a circle representing the external cycle. It is thus planar. The set of biconnected outerplanar graphs is $L(S,u)$, where S is the equation system:

$$\begin{cases} u = fg_{12}(v \mathbin{/\!/} \mathbf{12}_a), \\ v = \mathbf{12}_a \cup (\mathbf{12} \mathbin{/\!/} v) \cup (v \bullet v). \end{cases}$$

The edges labeled by a (which are defined by the constant symbol $\mathbf{12}_a$) are those of the external cycle. Note that $L(S,v)$ is a set of undirected series-parallel graphs (disregarding the edge label a).

(8) **Graphs of bounded tree-width:** Theorem 2.83 has characterized the s-graphs of tree-width at most $k - 1$ and of type included in $[k]$ as those defined by terms in $T(F_{[k]}^{\mathrm{HR}})$, where $F_{[k]}^{\mathrm{HR}}$ is the finite subsignature of F^{HR} consisting of the operations

using source names in $[k]$. It follows then from Propositions 3.16, 3.9(1) and 3.23(2) that the set $TWD(\leq k-1,[k])$ is HR-equational.[3]

For directed s-graphs, the set $TWD(\leq k-1,[k])$ is defined, for $k \geq 2$, by the equation

$$x = (x \,/\!/\, x) \cup \left(\bigcup_{B \subseteq [k]} fg_B(x) \right) \cup \left(\bigcup_{h \in Perm_f([k])} ren_h(x) \right) \cup \overrightarrow{12} \cup 1 \cup 1^\ell \cup \varnothing.$$

Since this equation contains the terms $ren_h(x)$, we need not use the constant symbols $\overrightarrow{ab}$ for $(a,b) \neq (1,2)$ because they can be defined by terms of the form $ren_h(\overrightarrow{12})$. For example, $\overrightarrow{32} = ren_h(\overrightarrow{12})$, where $h(1) = 3$, $h(2) = 2$, $h(3) = 1$. For generating undirected s-graphs, we replace $\overrightarrow{12}$ by 12.

We have proved in Proposition 2.49 that the operations ren_h are dispensable. This fact shows that the above equation is equivalent to the system of two equations (we could write it as a single equation but it is clearer in this way):

$$\begin{cases} x = y \cup (x \,/\!/\, x) \cup (\bigcup_{B \subseteq [k]} fg_B(x)), \\ y = \left(\bigcup_{a,b \in [k], a \neq b} \overrightarrow{ab} \right) \cup \left(\bigcup_{a \in [k]} a \right) \cup (\bigcup_{a \in [k]} a^\ell) \cup \varnothing. \end{cases}$$

If we wish to define the set $TWD(\leq k-1)$ of graphs of tree-width at most $k-1$, we use the additional equation $z = fg_{[k]}(x)$. A bit more generally, for each finite set of source names C, the set $TWD(\leq k-1,C)$ of s-graphs of tree-width at most $k-1$ and of type included in C is HR-equational. For $|C| = k$ this follows from Theorem 2.83 as for $C = [k]$. For $|C| \neq k$ we use Proposition 3.45. If $|C| > k$, then $TWD(\leq k-1,C)$ is the union of all $TWD(\leq k-1,D)$ with $D \subseteq C$ and $|D| = k$, and if $|C| < k$, then $TWD(\leq k-1,C) = fg_{D-C}(TWD(\leq k-1,D))$ for any set D with $C \subseteq D$ and $|D| = k$.

(9) **Graphs of bounded path-width:** The characterization of graphs of path-width at most $k-1$ given in Proposition 2.85 can be translated into an HR equation system. From this proposition we get immediately the equation

$$x = y \cup (x \,/\!/\, y) \cup \left(\bigcup_{B \subseteq [k]} fg_B(x) \right) \cup \left(\bigcup_{h \in Perm_f([k])} ren_h(x) \right),$$

to which we add the equations of (8) that define z and y. As in (8), we can omit in the equation defining x the terms $ren_h(x)$.

(10) **Graphs of bounded bandwidth:** A simple, loop-free undirected graph G has bandwidth at most k where $k \geq 1$ if its vertex set can be enumerated as $\{v_1, \dots, v_n\}$ in such a way that for every edge $v_i - v_j$ with $i < j$ we have $j - i \leq k$. The *bandwidth* of G is defined as the smallest such k. The set of graphs of bandwidth $\leq k$ that have at

---

[3] See Definition 2.53 for $TWD(\leq k,C)$ and $TWD(\leq k)$.

least $k+1$ vertices is $L(S,x)$ where S is the HR equation system:

$$
\begin{cases}
x = fg_{[k]}(y), \\
y = b \cup f(z), \\
z = y \parallel e, \\
b = g_1 \cup \cdots \cup g_m, \\
e = e_1 \cup \cdots \cup e_p.
\end{cases}
$$

In this system, $\{g_1, \ldots, g_m\}$ is a set of terms denoting the simple loop-free undirected s-graphs with $k+1$ vertices and of type $[k]$ (cf. Proposition 2.33), f is the derived operation defined by the term $ren_h(fg_1(x_1))$, where $h(1) := k+1$, $h(i) := i-1$ for $i = 2, \ldots, k+1$, and $e_1, \ldots, e_p$ are the terms $\mathbf{k}'$ (we let $k' := k+1$) and $\parallel_{i \in I} i\mathbf{k}'$, for all nonempty subsets I of $[k]$. Hence $p = 2^k$. The unknown y defines s-graphs of type $[k]$. The unknown e defines s-graphs of several types: each subset of $[k+1]$ that contains $k+1$ is the type of some s-graph in $L(S,e)$.

Since the system S defines slim terms, the graphs of bandwidth at most k have path-width at most k (by Propositions 2.85 and 4.2).

(11) **Graphs of bounded cyclic bandwidth:** An undirected loop-free graph G has *cyclic bandwidth* at most k, where $k \geq 1$, if its set of vertices can be enumerated as $\{v_0, \ldots, v_n\}$ in such a way that, for every edge $v_i - v_j$ with $0 \leq i < j \leq n$, we have $j - i \leq k$ or $i + n + 1 - j \leq k$. See Figure 4.2 for an example with $n = 7$ and $k = 3$.

It is not hard to construct for each k an HR equation system that defines the set $CB(\leq k)$ of simple graphs of cyclic bandwidth at most k. The problem of recognizing if a graph belongs to $CB(\leq k)$ is NP-complete for each $k \geq 2$ [LeuVW]. This shows that certain HR-equational sets have an NP-complete membership problem.

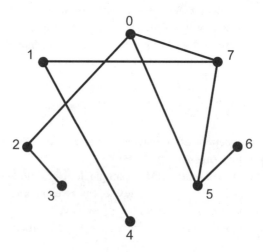

Figure 4.2 A graph of cyclic bandwidth ≤ 3.

(12) **Context-free languages:** Every word over a finite alphabet A can be considered as a directed vertex-labeled graph: this graph is a directed path, each vertex of which has a label from A. The empty word is an isolated vertex with no label (we can also label this vertex by ϵ). Hence a language can be considered as a set of labeled graphs. With this convention, every context-free language is HR-equational, but some HR-equational languages like $\{a^n b^n c^n \mid n \geq 0\}$ or $\{ww \mid w \in A^*\}$ are not context-free. The HR-equational languages will be studied in Chapter 8. $\square$

Definition 4.4 (Derivation trees and parsing) The notion of derivation tree, defined in Definition 3.43 (Section 3.2.3) for general equation systems, is applicable to HR equation systems.

The *parsing problem* for (S, x), where S is an HR equation system and x is one of its unknowns, consists, for a given (labeled) s-graph G, of finding a derivation tree of it or reporting that no such tree exists, i.e., that G does not belong to $L(S, x)$.

Example 4.3(11) has shown that this problem can be NP-complete for certain HR equation systems. This fact should be contrasted with the case of context-free languages, since every context-free language has a cubic membership algorithm. We will come back to the parsing problem in Chapter 6.

Let S be an HR equation system. If G has a derivation tree d relative to S, then this tree can be transformed by a second-order substitution into a term t in $T(F^{\mathrm{HR}})$ or in $T(F^{\mathrm{HR}}_{[K,\Lambda]})$ (or more precisely in $T(F(S))$) that evaluates to G. This transformation can be done in time proportional to the size of d (cf. Definition 3.43). The size of G, defined as the number of vertices and edges, is $O(|t|)$. More precisely (cf. Section 2.3.2), $|V_G| \leq 2 \cdot |Occ_0(t)|$ and $|E_G| \leq |Occ_0(t)|$, where $Occ_0(t)$ is the set of occurrences of constant symbols in t different from $\varnothing$. Furthermore, $|t| = \Theta(|d|)$ if S is uniform. These facts follow from the various definitions.

Proposition 4.5

(1) The membership problem for an HR-equational set is decidable.
(2) The emptiness of an HR-equational set (given by an equation system) is decidable.
(3) If a nonempty HR-equational set is defined by a uniform system with n unknowns, then it contains an s-graph of size at most $3 \cdot 2^{n-1}$.

Proof: (1) The proof is similar to the one of Proposition 3.10. Let S be a uniform HR equation system with set of unknowns X_n and let $L = L(S, x_1)$. For a given s-graph H, we define M as the finite set of s-graphs G of type included in $\mathcal{A}(S)$ such that $G^\circ \subseteq H^\circ$. If $\overline{L} = (L_1, \ldots, L_n)$ is an n-tuple of sets of s-graphs, we let $\overline{L} \cap M$ denote the n-tuple $(L_1 \cap M, \ldots, L_n \cap M)$. Obviously, H belongs to L if and only if it belongs to the first component of $\mu \overline{x} \cdot S_{\mathcal{P}(\mathbb{JS})}(\overline{x}) \cap M$. Since the sequence $S^i_{\mathcal{P}(\mathbb{JS})}(\overline{\varnothing})$ is increasing (cf. the Least Fixed-Point Theorem, Theorem 3.7), so is the sequence $S^i_{\mathcal{P}(\mathbb{JS})}(\overline{\varnothing}) \cap M$.

Claim 4.5.1 For every $i \in \mathcal{N}$, if

$$S^{i+1}_{\mathcal{P}(\mathbb{JS})}(\bar{\emptyset}) \cap M = S^i_{\mathcal{P}(\mathbb{JS})}(\bar{\emptyset}) \cap M,$$

then

$$S^{i+2}_{\mathcal{P}(\mathbb{JS})}(\bar{\emptyset}) \cap M = S^{i+1}_{\mathcal{P}(\mathbb{JS})}(\bar{\emptyset}) \cap M.$$

Proof: It is clear from the definitions that, for all s-graphs G and K of type included in $\mathcal{A}(S)$, if $G \parallel K \in M$ then $G, K \in M$, and if $f(G) \in M$, where f is a unary operation of F^{HR}, then $G \in M$. Hence, for every tuple $\bar{L}$ with $L_i \subseteq \mathcal{JS}[\mathcal{A}(S)]$, we have $S_{\mathcal{P}(\mathbb{JS})}(\bar{L}) \cap M \subseteq S_{\mathcal{P}(\mathbb{JS})}(\bar{L} \cap M)$.

It suffices to prove that $S^{i+2}_{\mathcal{P}(\mathbb{JS})}(\bar{\emptyset}) \cap M \subseteq S^{i+1}_{\mathcal{P}(\mathbb{JS})}(\bar{\emptyset})$. Let $\bar{L} := S^i_{\mathcal{P}(\mathbb{JS})}(\bar{\emptyset})$ and assume that $S_{\mathcal{P}(\mathbb{JS})}(\bar{L}) \cap M = \bar{L} \cap M$. Then we have, by the initial observation and the assumption, that

$$S_{\mathcal{P}(\mathbb{JS})}(S_{\mathcal{P}(\mathbb{JS})}(\bar{L})) \cap M \subseteq S_{\mathcal{P}(\mathbb{JS})}(S_{\mathcal{P}(\mathbb{JS})}(\bar{L}) \cap M) = S_{\mathcal{P}(\mathbb{JS})}(\bar{L} \cap M),$$

which is included in $S_{\mathcal{P}(\mathbb{JS})}(\bar{L})$ by the monotonicity of $S_{\mathcal{P}(\mathbb{JS})}$, as was to be proved. $\qquad\square$

In order to decide if H belongs to L, we compute the sequence of n-tuples of finite sets $S^i_{\mathcal{P}(\mathbb{JS})}(\bar{\emptyset})$ for $i := 1, 2, \ldots$ until one of the following two cases holds:

(a) either H belongs to the first component of $S^i_{\mathcal{P}(\mathbb{JS})}(\bar{\emptyset})$; or
(b) $S^{i+1}_{\mathcal{P}(\mathbb{JS})}(\bar{\emptyset}) \cap M = S^i_{\mathcal{P}(\mathbb{JS})}(\bar{\emptyset}) \cap M$ and Case (a) does not hold.

In Case (a) we return a positive answer and in Case (b) a negative one. This algorithm terminates because M is finite and the sequence $S^i_{\mathcal{P}(\mathbb{JS})}(\bar{\emptyset})$ is increasing. It is correct because if it terminates in Case (b), then, by the Claim, $S^j_{\mathcal{P}(\mathbb{JS})}(\bar{\emptyset}) \cap M = S^i_{\mathcal{P}(\mathbb{JS})}(\bar{\emptyset}) \cap M$ for every $j > i$, hence $\mu \bar{x} \cdot S_{\mathcal{P}(\mathbb{JS})}(\bar{x}) \cap M = S^i_{\mathcal{P}(\mathbb{JS})}(\bar{\emptyset}) \cap M$, so that H does not belong to L.

We do not claim that this is an efficient algorithm. The existence of polynomial algorithms for the membership problem will be discussed in Section 6.2.

(2) This is a special case of Corollary 3.24.

(3) We use Proposition 4.2. Let $L = L(S, x_1)$ be nonempty and defined by a uniform system S. Let t be a term in $L_{Term}(S, x_1)$ of minimal size. Its height is at most $n - 1$ by Proposition 3.53 (applied to $F(S)$), and it has at most 2^{n-1} occurrences of constant symbols (or leaves). Since, as observed above, each such occurrence describes at most one edge and at most two vertices, the size of the s-graph $val(t)$ is at most $3 \cdot 2^{n-1}$. $\qquad\blacksquare$

We conclude this section with a small technical observation.

Remark 4.6 Let L be an HR-equational set defined as $L(S,x_1)$ for some equation system S and let C be the union of the types of the s-graphs in L. We have $C \subseteq \mathcal{A}(S)$.

The source names in $\mathcal{A}(S) - C$ are auxiliary and can be changed by others. More precisely, if h is a finite permutation of $\mathcal{A}$ (the set of all source names) that is the identity on C, then the equation system S' obtained from S by replacing, in each operation symbol that occurs in S, each source name a by $h(a)$ is such that $\mathcal{A}(S') = h(\mathcal{A}(S))$. It is clear that, for each unknown x of S, we have $L(S',x) = ren_h(L(S,x))$. Hence, $L(S',x_1) = L(S,x_1) = L$ since h is the identity on C. In particular, if L is a set of graphs, one can define it by a system S such that $\mathcal{A}(S) = [k]$ for some k. This remark extends Remark 2.39(2). $\qquad\square$

4.1.2 HR-equational sets and tree-width

We have proved in Proposition 2.76 that if $G = val(t)$, where $t \in T(F_C^{HR})$ or $t \in T(F_{C,[K,\Lambda]}^{HR})$, which means that the source names used in the operations occurring in t belong to C, then the tree-width of G is at most $|C| - 1$. This result has consequences for HR-equational sets. We recall that if S is an HR equation system, then $F(S) \subseteq F_{\mathcal{A}(S)}^{HR}$ (or $F(S) \subseteq F_{\mathcal{A}(S),[K,\Lambda]}^{HR}$). Hence, by Proposition 4.2, we obtain the following result:

Proposition 4.7 Every HR-equational set of (labeled) s-graphs has bounded tree-width. More precisely, for every HR equation system S and $x \in Unk(S)$, we have $twd(L(S,x)) \leq |\mathcal{A}(S)| - 1$. $\qquad\square$

This proposition gives an upper-bound on the tree-width of any s-graph in $L(S,x)$. The upper-bound depends on the system S and is not necessarily the exact one, defined as $twd(L(S,x)) := \max\{twd(G) \mid G \in L(S,x)\}$. In the example of ladders (Example 4.3(5)), the equation based on the first expression of f yields the upper-bound $3 (= 4 - 1)$, whereas the second expression of f yields the upper-bound $2 (= 3 - 1)$, which is the exact one. Two natural questions can be raised:

(1) Can one compute $twd(L(S,x))$ from an HR equation system S and x in $Unk(S)$?
(2) Can every HR-equational set of s-graphs L be defined as $L(S,x)$ for some HR equation system S such that $|\mathcal{A}(S)| - 1 = twd(L)$?

The answers to these questions are respectively yes and no. The first answer is based on the fact that $L(S,x)$ is equational in the many-sorted algebra $\mathbb{JS}^t$ (as will be shown in Proposition 4.13) and the fact that each set[4] $TWD(\leq k, \mathcal{A}(S))$ is effectively recognizable in $\mathbb{JS}^t$, i.e., is effectively HR-recognizable (as will be shown in Proposition 4.31(7) using the Recognizability Theorem to be proved in Section 5.3).

---

[4] We recall from Definition 2.53 that $TWD(\leq k, C)$ is the set of s-graphs of tree-width at most k and of type included in C.

From a system S, we have an upper-bound ℓ on the tree-width of the s-graphs in $L(S,x)$. By Proposition 3.91(2) one can test, for each $k < \ell$, if all s-graphs in $L(S,x)$ have tree-width at most k, which gives the exact value $twd(L(S,x))$. Actually, this algorithm is intractable for reasons discussed after Proposition 4.31.

We now give a counter-example showing the negative answer to the second question.

Counter-example 4.8 We consider the set L of odd paths with a 1-source as the middle vertex. Examples of such paths are $\bullet 1$, $\bullet - \bullet 1 - \bullet$, $\bullet - \bullet - \bullet 1 - \bullet - \bullet$. These s-graphs have tree-width 1 and can be expressed as $P /\!/ P$, where P is any path one end of which is the 1-source. However, a system of equations based on this description, say $S := \langle x = y /\!/ y, \ y = ren_{1 \leftrightarrow 2}(fg_1(y /\!/ 12)) \cup 1 \rangle$, defines too many elements (L is a proper subset of $L(S,x)$), and no finite set of equations similar to this one can impose that two paths to be connected by their ends have equal length. However, $L = L(T,x)$, where T is the system:

$$\begin{cases} x = ren_{1 \leftrightarrow 3}(fg_{12}(\mathbf{23} /\!/ \mathbf{13} /\!/ y)) \cup \mathbf{1}, \\ y = (\mathbf{1} /\!/ \mathbf{2}) \cup f'(y), \end{cases}$$

and $L(T,y)$ is the set of ladders without "steps" (cf. Example 4.3(5)). In this system, f' is defined by $f'(y) := ren_{2 \leftrightarrow 3}(fg_2[ren_{1 \leftrightarrow 3}(fg_1(y /\!/ \mathbf{13})) /\!/ \mathbf{23}])$. But no system of equations using only two source names can exactly define L. (We do not give the technical proof of this fact.) $\qquad\square$

Remark 4.9 (1) We know from Example 4.3(8) that the set of graphs of tree-width at most k is HR-equational, but what about the set $TWD(k)$ of those of tree-width *exactly* k? This set is also HR-equational. In fact, it can be expressed as follows:

$$TWD(k) = \{G \in TWD(\leq k) \mid G \notin TWD(\leq k - 1)\},$$

hence as the intersection of a $\mathbb{JS}^t$-equational and a $\mathbb{JS}^t$-recognizable set (see the discussion on Question (1) above; the complement of $TWD(\leq k - 1)$ is $\mathbb{JS}^t$-recognizable by Proposition 3.85). Thus, it is $\mathbb{JS}^t$-equational by the Filtering Theorem (Theorem 3.88) and hence HR-equational by Proposition 4.13 below. Although an HR equation system defining $TWD(k)$ can be constructed, the corresponding algorithm is intractable (see the discussion after Proposition 4.31). Similar observations hold for the set $PWD(k)$ of graphs of path-width exactly k.

(2) Proposition 4.7 and Example 4.3(8) show that a set L of (labeled) graphs may not be HR-equational for two reasons: either it has unbounded tree-width or it has bounded tree-width, but has a too complex internal structure for being described by a finite set of equations. In the first case, illustrated by the set of planar graphs (Example 2.56(4)), L is not a subset of any HR-equational set. In the second case, illustrated by the set of paths of length 2^n for some n (see Proposition 4.16 below),

L is a subset of an HR-equational set, the set of graphs of tree-width at most k (and labeled with the same labels as the graphs in L). In the case of context-free languages, which are the equational sets of free monoids, only the second case can occur because the set A^* is context-free for each finite alphabet A.

4.1.3 Type analysis of HR equation systems

In Examples 2.134, 4.3(8) and 4.3(10), we have given examples of HR equation systems that define s-graphs, not all of the same type.[5] We now show how to compute the (finite) set $\tau(L)$ of types of the s-graphs of an HR-equational set L by using the preservation of least solutions of equation systems under homomorphisms (Proposition 3.23(1)).

Proposition 4.10 For every HR-equational set L given by an HR equation system S, the finite set $\tau(L) \subseteq \mathcal{P}(\mathcal{A}(S))$ can be computed.

As a preparation for the proof, we turn the type mapping $\tau : \mathcal{JS} \to \mathcal{P}_f(\mathcal{A})$ into a homomorphism : $\mathbb{JS} \to \mathbb{T}$ where $\mathbb{T}$ is an F^{HR}-algebra called the *type algebra*, that we now define. Its domain is $\mathcal{P}_f(\mathcal{A})$. Its operations are defined as follows, for $X, Y \subseteq \mathcal{A}$:

$$X \parallel_{\mathbb{T}} Y := X \cup Y,$$
$$fg_{B\mathbb{T}}(X) := X - B, \quad \text{where } B \in \mathcal{P}_f(\mathcal{A}),$$
$$ren_{h\mathbb{T}}(X) := h(X), \quad \text{where } h \text{ is a finite permutation of } \mathcal{A}, \text{ and}$$
$$c_{\mathbb{T}} := \tau(c_{\mathbb{JS}}), \quad \text{where } c \in F_0^{\mathrm{HR}}.$$

If C is a subset of $\mathcal{A}$, we denote by $\mathbb{T}[C]$ the subalgebra of $\mathbb{T}$ with signature F_C^{HR} obtained by replacing $\mathcal{A}$ by C in the above definitions.

Lemma 4.11 The mapping τ is a homomorphism from the F^{HR}-algebra $\mathbb{JS}$ to $\mathbb{T}$. Its restriction to $\mathbb{JS}[C]$ is a homomorphism to $\mathbb{T}[C]$.

Proof: From Section 2.3.1, we know that $\tau(G \parallel H) = \tau(G) \cup \tau(H)$, $\tau(fg_B(G)) = \tau(G) - B$ and $\tau(ren_h(G)) = h(\tau(G))$. This gives the result. ∎

Proof of Proposition 4.10: We let L be defined by S, and $\mathcal{A}(S)$ be the set of source names occurring in S (cf. Definition 4.1). By the definitions we have

$$(\tau(L(S,x_1)),\ldots,\tau(L(S,x_n))) = \tau(\mu\bar{x} \cdot S_{\mathcal{P}(\mathbb{JS})}(\bar{x})).$$

By Proposition 3.9(1), S has the same least solution in $\mathcal{P}(\mathbb{JS})$ and in $\mathcal{P}(\mathbb{JS}[\mathcal{A}(S)])$. Hence:

$$\tau(\mu\bar{x} \cdot S_{\mathcal{P}(\mathbb{JS})}(\bar{x})) = \tau(\mu\bar{x} \cdot S_{\mathcal{P}(\mathbb{JS}[\mathcal{A}(S)])}(\bar{x})).$$

---

[5] The *type* $\tau(G)$ of an s-graph G is the set of names of its sources; we let $\tau(L) := \{\tau(G) \mid G \in L\}$.

Since τ is a homomorphism from $\mathbb{JS}[\mathcal{A}(S)]$ to $\mathbb{T}[\mathcal{A}(S)]$ by Lemma 4.11, Proposition 3.23(1) implies that the above tuple is equal to $\mu\bar{x} \cdot S_{\mathcal{P}(\mathbb{T}[\mathcal{A}(S)])}(\bar{x})$, which is computable by Proposition 3.10 since its components are all finite (as they are subsets of the finite set $\mathcal{P}(\mathcal{A}(S))$).[6] ∎

Note that $L(S,x_i) = \emptyset$ if and only if $\tau(L(S,x_i))$ is empty.[7] This algorithm yields also an emptiness test as a by-product. Another consequence of Lemma 4.11 is that the set $\mathcal{JS}_C$ of s-graphs of type equal to some finite subset C of $\mathcal{A}(S)$ is recognizable in the algebra $\mathbb{JS}[\mathcal{A}(S)]$. Hence, by the Filtering Theorem (Theorem 3.88) applied to $\mathbb{M} := \mathbb{JS}[\mathcal{A}(S)]$, if L is HR-equational then $L \cap \mathcal{JS}_C$ is HR-equational for every $C \in \tau(L)$.

Definition 4.12 (Typed HR equation systems) A set of (labeled) s-graphs is *homogenous* if all its elements have the same type. An HR-equational set is thus a finite union of homogenous HR-equational sets. A *typed HR equation system* is an equation system S over the many-sorted signature F^{tHR} (or $F^{\text{tHR}}_{[K,\Lambda]}$). Such systems define the $\mathbb{JS}^t$-equational sets, which are also finite unions of homogenous sets because the sort of an s-graph G in the many-sorted algebra $\mathbb{JS}^t$ is its type $\tau(G)$. We define $L(S,x_i)$, $L_{Term}(S,x_i)$, $L(S,Y)$ and $L_{Term}(S,Y)$ as in Definition 4.1, for $\mathbb{JS}^t$ and $\mathbb{T}(F^{\text{tHR}})$ instead of $\mathbb{JS}$ and $\mathbb{T}(F^{\text{HR}})$.

Although they are not solved in the same algebra, an HR equation system and a typed HR equation system will be said to be *equivalent* if they have the same least solution, respectively in $\mathcal{P}(\mathbb{JS})$ and in $\mathcal{P}(\mathbb{JS}^t)$.

We will prove that a set of s-graphs is HR-equational (i.e., $\mathbb{JS}$-equational) if and only if it is $\mathbb{JS}^t$-equational. That means that the HR equation systems and the typed HR equation systems define the same sets of s-graphs.

An HR equation system S is *homogenous* if each set $L(S,x)$ for $x \in Unk(S)$ is homogenous. We will prove that a trim and homogenous equation system can be transformed into an equivalent typed HR equation system. We extend to HR equation systems the typing and untyping transformations of terms described in Section 2.6.2. If S is a typed HR equation system of the general form $\langle x_1 = p_1,\ldots,x_n = p_n \rangle$, we let $Unt(S)$ be the system obtained from S by replacing, in each polynomial p_i, a monomial t by $Unt(t)$ (cf. Definition 2.129). We will prove that S and $Unt(S)$ are equivalent. Conversely, let S be a homogenous and trim HR equation system (cf. Definition 3.36). We define the type $\tau(x)$ of an unknown x of S as the common type of all s-graphs in $L(S,x)$. Then, we transform S into a typed system $Typ(S)$ by replacing in each right-hand side of an equation of S a monomial t by $Typ^\tau(t)$, where the typing of Definition 2.131 is done with τ as type assignment (the mapping β in that definition).

---

[6] Proposition 3.10 is applicable because the finite algebra $\mathbb{T}[\mathcal{A}(S)]$ is effectively given. In fact, $F^{\text{HR}}_{\mathcal{A}(S)}$ and $\mathbb{T}[\mathcal{A}(S)]$ can be computed from S (and F^{HR}).

[7] Not to be confused with $\tau(L(S,x_i)) = \{\emptyset\}$, which implies that $L(S,x_i)$ is not empty and contains only graphs (without sources).

An HR equation system S is *typable* if $S = Unt(T)$ for some typed HR equation system T. If S is trim, then $T = Typ(S)$. If an HR equation system S is trim and homogenous, then it can be typed (i.e., $Typ(S)$ is defined), but it need not be typable because $Unt(Typ(S))$ need not be equal to S. For a typable HR equation system S, the sets $L_{Term}(S, x_i)$ consist of reduced terms (cf. Proposition 2.38), i.e., the operations fg_B and ren_h are only applied to terms t such that $B \subseteq \tau(t)$ and h is a permutation of $\tau(t) \cup h(\tau(t))$ respectively. It follows from the next proposition that every HR-equational set can be defined by a typable HR equation system.

Proposition 4.13

(1) A typed HR equation system S is equivalent to its associated typable system $Unt(S)$. For every unknown x, $L_{Term}(Unt(S), x)$ consists of reduced terms.
(2) If S is a trim and homogenous HR equation system, then $Typ(S)$ is a typed system, and it is equivalent to S.
(3) Every HR-equational set is defined by a typed HR equation system that can be effectively constructed from any HR equation system defining the considered set.
(4) A set of s-graphs is HR-equational if and only if it is $\mathbb{JS}^t$-equational.

Proof: (1) Since Unt is the restriction to $T(F^{tHR})$ of a word homomorphism $: \mathbb{W}(F^{tHR}) \to \mathbb{W}(F^{HR})$, cf. Definition 2.129, it is immediate from a comparison of the context-free grammars $G(S)$ and $G(Unt(S))$ that $L_{Term}(Unt(S), x) = Unt(L_{Term}(S, x))$ for every unknown x. This implies the first statement by Proposition 3.23(1) (cf. Proposition 4.2) and Proposition 2.130. It also implies the second statement, because $Unt(t)$ is reduced for every term $t \in F^{tHR}$, as one easily shows by induction on the structure of t.

(2) We first prove that $Typ(S)$ is typed: we need only verify that if t is a monomial of the right-hand side of an equation $x_i = p_i$ of S, then the sort of $Typ^\tau(t)$ is $\tau(x_i)$. Let G_j be an (arbitrary) element of $L(S, x_j)$ for each $j = 1, \ldots, n$; its type is $\tau(x_j)$. By Proposition 2.132(1,2) the sort of $Typ^\tau(t)$ equals the type of the s-graph $t_{\mathbb{JS}}(G_1, \ldots, G_n)$. Since $t_{\mathbb{JS}}(G_1, \ldots, G_n)$ belongs to $L(S, x_i)$, it has type $\tau(x_i)$.

For proving that S and $Typ(S)$ have the same least solution, we observe that, for every $i \in \mathcal{N}$, we have

$$S^i_{\mathcal{P}(\mathbb{JS})}(\overline{\emptyset}) = Typ(S)^i_{\mathcal{P}(\mathbb{JS}^t)}(\overline{\emptyset}).$$

This is easily proved by induction on i by using Proposition 2.132(3) (and Lemma 3.2(4)). By taking the least upper-bounds over all i of both sides of the equation, we get by Theorem 3.7 the equality of the least solutions of S and $Typ(S)$.

(3) Let L be HR-equational and nonempty, defined as $L(S, x)$ for an HR equation system S. Since S has the same least solution in $\mathcal{P}(\mathbb{JS})$ and in $\mathcal{P}(\mathbb{JS}[\mathcal{A}(S)])$, and since the type mapping τ is a homomorphism from $\mathbb{JS}[\mathcal{A}(S)]$ to $\mathbb{T}[\mathcal{A}(S)]$ (as in the proof of Proposition 4.10), we can apply Corollary 3.89 and trim the resulting system by using Proposition 3.37. We obtain a trim equation system S' with an unknown y_C for every

$y \in Unk(S)$ and $C \subseteq \mathcal{A}(S)$ such that $C \in \tau(L(S,y))$; moreover, $L(S',y_C)$ is the set of s-graphs in $L(S,y)$ of type C. It is homogenous, hence we get by (2) an equivalent typed system $Typ(S')$. Then L, initially defined as $L(S,x)$, is $L(Typ(S'),Y)$, where Y consists of all unknowns $x_C \in Unk(S')$.

(4) The proof follows directly from (1) and (3).   ∎

The construction in the proof of (3) has already been sketched in Example 2.134. The typed system resulting from it can be much larger than the original one.

Example 4.14 (A large typed system) Let E_n for $n \geq 2$ be the equation $x = fg_1(x) \cup$ $\cdots \cup fg_n(x) \cup s_n$, where s_n is the term $\overrightarrow{\mathbf{12}} \; / \!/ \; \overrightarrow{\mathbf{23}} \; / \!/ \; \cdots \; / \!/ \; \overrightarrow{\mathbf{mn}}$ with $m = n - 1$. The size of E_n is $O(n)$ but the size of the corresponding typed system S_n is $O(2^n)$. In fact, we have $L(E_n,x) = \{val(fg_B(s_n)) \mid B \subseteq [n]\}$, hence it consists of 2^n s-graphs of pairwise different types. The system S_n has thus one equation for each element of $L(E_n,x)$. □

All these definitions and results extend to labeled graphs.

4.1.4 Sizes of graphs and the finiteness problem

Definition 4.15 (Size functions) We recall from Definition 2.9 that the size of a graph G is the integer $\|G\| := |V_G| + |E_G|$. We define the *size* of an s-graph G as $\|G^\circ\|$. We will prove that if L is an HR-equational set, then the set $\|L\| := \{\|G\| \mid G \in L\}$ is semi-linear and hence its finiteness can be tested. As shown in Section 3.1.6, the semi-linear sets are the equational sets of $\mathbb{N}$, the free commutative monoid $\langle \mathcal{N}, +, 0, 1 \rangle$ generated by the constant 1. We will use the auxiliary size function defined by $\|G\|_i := \|G\| - |\tau(G)|$. Hence, $\|G\|_i$ is the number of internal vertices of G plus its number of edges. This function satisfies the following properties:

$$\|G \;/ \!/\; H\|_i = \|G\|_i + \|H\|_i, \tag{4.1}$$

$$\|fg_B(G)\|_i = \|G\|_i + |B \cap \tau(G)|, \tag{4.2}$$

$$\|ren_h(G)\|_i = \|G\|_i. \tag{4.3}$$

Proposition 4.16 For every HR-equational set L, the set $\|L\|$ is semi-linear. A description of $\|L\|$ can be computed from a system that defines L.

Proof: If L is defined by a system S, then $fg_{\mathcal{A}(S)}(L)$ is an HR-equational set of graphs by Proposition 3.45(3), and $\|L\| = \|fg_{\mathcal{A}(S)}(L)\| = \|fg_{\mathcal{A}(S)}(L)\|_i$. Thus, in what follows, it suffices to show that $\|L\|_i$ is semi-linear. Moreover, by Proposition 4.13 (and the fact that the class of semi-linear sets is closed under union), we may assume that L is defined by (S,x) where S is a typable HR equation system and $x \in Unk(S)$.

By Proposition 4.2 and Corollary 3.19, $L = val(K)$, where K is the regular set of terms $L_{Term}(S,x)$, which, by Proposition 4.13(1), consists of reduced terms. We will use Corollary 3.26 of Section 3.1.6. Taking $F := F^{HR}$, let $g : T(F) \to \mathcal{N}$ be the affine mapping defined by $g_F(/\!/) := 0$, $g_F(fg_B) := |B|$, $g_F(ren_h) := 0$ and $g_F(c) := \|c\|_i$ for every constant symbol c. Then $g(t) = \|val(t)\|_i$ for every reduced term $t \in T(F^{HR})$ by Equalities (4.1), (4.2) and (4.3) above. Hence $g(K) = \|L\|_i$ and $\|L\|_i$ is semi-linear by Corollary 3.26. ∎

Example 4.17 We let S be the typable system of Example 4.3(3) that defines trees. In order to have a more clear construction, we turn it into the following typable and uniform system S':

$$\begin{cases} u = fg_1(z), \\ z = (z \,/\!/\, z) \cup fg_2(z') \cup \mathbf{1}, \\ z' = \mathbf{12} \,/\!/\, z'', \\ z'' = ren_{1 \leftrightarrow 2}(z), \end{cases}$$

where u has type $\emptyset$, z has type $\{1\}$, z' has type $\{1,2\}$ and z'' has type $\{2\}$. We follow the proof of Corollary 3.26 as presented in Example 3.42(3). By the proof of Proposition 3.41 we obtain from S' the following system $\widehat{S'}$, to be solved in $\mathcal{P}(\mathbb{N})$:

$$\begin{cases} u = z + 1, \\ z = (z + z) \cup (z' + 1) \cup 0, \\ z' = 1 + z'', \\ z'' = z. \end{cases}$$

Every unknown that defines a set of s-graphs L in S', defines the set of integers $\|L\|_i$ in $\widehat{S'}$. By unfolding and by removing unknowns that are not useful for $\{u\}$ (Sections 3.2.1 and 3.2.2, and in particular Propositions 3.31 and 3.37), one can simplify $\widehat{S'}$ into the following system:

$$\begin{cases} u = z + 1, \\ z = (z + z) \cup (1 + z + 1) \cup 0. \end{cases}$$

This equation system defines in $\mathbb{N}$ the semi-linear set $\|L(S, u)\| = \{2n + 1 \mid n \in \mathcal{N}\}$. ☐

Remarks 4.18 (1) It is important that systems are typable so that terms are reduced and hence Equality (4.2) can be used to define an affine mapping.

(2) We will generalize this result[8] by replacing the sizes of s-graphs of an HR-equational set by tuples of integers $(|X_1|, \ldots, |X_k|)$, where $X_1, \ldots, X_k$ are sets of vertices and/or edges defined by a monadic second-order formula. In particular, the set of cardinalities of the vertex sets of the graphs of an HR-equational set is semi-linear.

---

[8] See Theorem 7.42 and the end of Section 7.4.

This fact can be proved by modifying the size functions of Definition 4.15: it suffices not to count edges. The proof of Proposition 4.16 is then the same. $\square$

Proposition 4.19 There is an algorithm that decides the finiteness of an HR-equational set L given by an equation system S; if L is finite, it can be computed from S. Furthermore, if L is finite and S is uniform and typable, then the size of an s-graph in L is at most $3 \cdot 2^{n-1}$ and the cardinality of L is at most $(|F(S)|+1)^{2^n}$, where n is the number of unknowns of S.

Proof: We consider unlabeled s-graphs; the proof for (K, Λ)-labeled s-graphs is entirely similar. Let L be an HR-equational set. Since $\tau(L)$ is finite, L is finite if and only if $\|L\|$ is finite, and this can be decided by Proposition 4.16 (because one can check from the description of $\|L\|$ as a semi-linear set if that set is finite).

Let us now assume that L is finite. To prove that L can be computed, it suffices to show, by Remark 3.38(3) and Proposition 4.13(4), that the effectively given algebra $\mathbb{JS}^t$ is infinity-preserving. A homogenous set of s-graphs L' is finite if and only if $\|L'\|_i$ is finite. Thus, it follows from Equalities (4.1), (4.2) and (4.3) that all the operations of $\mathbb{JS}^t$ are infinity-preserving.[9]

To prove the last assertion, we use a pumping argument similar to the one in the proof of Proposition 3.53. Let $X \subseteq Unk(S)$ be such that $L = L(S, X)$ and let $K := L_{Term}(S, X)$. Then $L = val(K)$ and $K \subseteq T(F(S))$ by Proposition 4.2. Note that K need not be finite. Let $\mathscr{A}$ be a finite $F(S)$-automaton with n states that accepts K, see Proposition 3.47(1). Since S is typable, we may assume that all terms in $L(\mathscr{A}, q)$ have the same type, for each $q \in Q_{\mathscr{A}}$.

Claim 4.19.1 For every s-graph G in L there exists a term $t \in K$ of height at most $n - 1$ such that $val(t) = G$.

Proof: Let $t \in K$ be such that $val(t) = G$. If $ht(t) \geq n$, then $\mathscr{A}$ has an accepting run r on a term t such that $r(u) = r(v)$ for some positions u, v with $u <_t v$. Hence $t = c[c'[t']]$, where $c, c' \in Ctxt(F(S))$, c' is not empty, $t' = t/u$, $c'[t'] = t/v$ and t/u and t/v have the same type. By Proposition 2.51, there exist an s-graph H and operations ren_h and fg_B such that $c'_{\mathbb{JS}}(G') = ren_h(fg_B(G' /\!/ H))$ for every s-graph G' with $\tau(G') \subseteq \mathcal{A}(S)$. Then $\tau(H) \subseteq \tau(G')$ and H has no edges and no internal vertices, because otherwise all terms $c[c'[c'[\cdots [c'[t']]\cdots]]]$ belong to K and the sizes of their values in $\mathcal{JS}$ are unbounded. Hence $c[t']$ belongs to K, is smaller than t and has value G (because $c'[t']$ has the same value as t'). Repeating this procedure one obtains the required term. $\square$

By this claim, the size of G is at most $3 \cdot 2^{n-1}$ by the same argument as in the proof of Proposition 4.5. Moreover, the cardinality of L is at most equal to the number of terms in $T(F(S))$ of height $\leq n - 1$, which is at most $\natural_{k,c,d}(n-1) \leq (c+d+1)^{k^n}$ by

---

[9] We note that the algebra $\mathbb{JS}$ is also infinity-preserving, by a slightly more complicated argument (if $\tau(L)$ is infinite, where L is one of the arguments of an operation, then $\tau(M)$ is infinite where M is the result).

the remarks preceding Proposition 3.53. In the present case, $k = 2$ and $c + d = |F(S)|$ from which the required upper-bound is obtained. ∎

4.1.5 Hyperedge replacement

The purpose of this section is to describe the link between graph grammars defined in terms of rewritings and HR equation systems. We do not intend to develop the theory of context-free graph grammars based on graph rewritings. This theory is presented in detail in the handbook [*Roz].

We will show that the monomials forming the polynomials p_i of a typed HR equation system $S = \langle x_1 = p_1, \ldots, x_n = p_n \rangle$ can be seen as certain hypergraphs with sources, and that the s-graphs defined by S can be obtained by finite sequences of rewritings based on replacements of hyperedges by such hypergraphs, like the words generated by a context-free grammar are defined by sequences of replacements of letters by words.

We first explain this construction from the algebraic point of view of Chapter 3, cf. Remark 3.20(3). Let L be an equational set of an F-algebra $\mathbb{M}$ defined by an equation system S. Every element m of L is defined by a derivation sequence relative to the regular grammar $G(S): x \Rightarrow t_1 \Rightarrow \cdots \Rightarrow t_n$ such that x is a nonterminal symbol of $G(S)$ (hence an unknown of S), $t_1, \ldots, t_n$ are terms in $T(F, Unk(S))$, $t_n \in T(F)$ and $m = val_{\mathbb{M}}(t_n)$. In the case of words, i.e., of the F_A-algebra $\mathbb{M} = \mathbb{W}(A)$, the mapping $val_{\mathbb{M}}$ extends in a natural way into a homomorphism : $\mathbb{T}(F_A \cup Unk(S)) \to \mathbb{W}(A \cup Unk(S))$, so that the image of the derivation sequence $x \Rightarrow t_1 \Rightarrow \cdots \Rightarrow t_n$ under this mapping is a derivation sequence of the context-free grammar $G[S]$. Therefore, the words of an equational set of $\mathbb{W}(A)$ are produced by sequences of rewritings of words, and not only of terms.

In a general F-algebra $\mathbb{M}$, the main task in order to get a similar notion of generation by rewriting is to define, for every equation system S, a superalgebra $\mathbb{M}'$ of $\mathbb{M}$ and a value mapping $T(F \cup Unk(S)) \to M'$ that is a homomorphism. The appropriate superalgebra of $\mathbb{JS}^t$ is an F^{tHR}-algebra of hypergraphs with sources.

Definition 4.20 (Hypergraphs with sources) We do not define general hypergraphs but only the very particular ones needed for defining *hyperedge replacement*. We let $\mathcal{S} := \mathcal{P}_f(\mathcal{A})$ be our usual set of sorts and we let U be a finite $\mathcal{S}$-sorted set of variables,[10] i.e., each $x \in U$ has a sort $\tau(x) \in \mathcal{S}$. A U-*hypergraph* is a 5-tuple $H = \langle V_H, E_H, vert_H, lab_H, slab_H \rangle$ where:

V_H is the set of vertices;
E_H is the set of edges and hyperedges;
lab_H is a partial function : $E_H \to U$, the domain of which is the set of hyperedges;

---

[10] The symbols in U are nullary. We call them *variables* because they will be the unknowns of equation systems associated with regular grammars having U as set of nonterminals.

vert$_H$ is a mapping with domain E_H satisfying the following conditions:

- if e is an edge (in this case $lab_H(e)$ is undefined[11]) then $vert_H(e)$ is, as in Definition 2.9, a set of 1 or 2 vertices (if H is defined as *undirected*) or an ordered pair of (possibly equal) vertices (if H is defined as *directed*);
- if $lab_H(e) = x \in U$ then $vert_H(e)$ is an injective mapping from $\tau(x)$ to V_H, and we call $\tau(e) := \tau(x)$ the *type* of e;

$slab_H : V_H \to \mathcal{A}$ is as in Definition 2.24.

The elements of E_H that have a label can be seen as hyperedges: their sets of vertices have cardinalities not restricted to 1 or 2. Their vertices are referred to by elements of $\mathcal{A}$. A hyperedge of type $\emptyset$ has no vertex. The notations $Src(H)$, $\tau(H)$, src_H are defined from $slab_H$ as in Definition 2.24. Hence H has sources unless its type $\tau(H)$ is empty.

Definition 4.21 (Hyperedge replacement) We describe an operation that generalizes the replacement of a nonterminal symbol by a word in a derivation sequence of a context-free grammar.

Let G and H be U-hypergraphs (both directed or both undirected). If $e \in E_G$ is a hyperedge of type $\tau(H)$, then we define as follows the U-hypergraph $G[e \leftarrow H]$ resulting from the *substitution of H for e in G*. We assume G and H disjoint (if they are not we take disjoint copies in the usual way). We let

$$G[e \leftarrow H] := [(G - e) \oplus fg_{\tau(H)}(H)]/\approx,$$

where $G - e$ is G minus the hyperedge e, and $\approx$ is the equivalence relation on $V_G \cup V_H$ generated by the set of pairs $\{(vert_G(e)(a), src_H(a)) \mid a \in \tau(H)\}$. We denote by $\oplus$ the disjoint union; it is well defined since the two arguments are disjoint and of disjoint types. If $\tau(H) = \tau(e) = \emptyset$, we have $G[e \leftarrow H] = (G - e) \oplus H$. The type of $G[e \leftarrow H]$ is $\tau(G)$.

The transformation of G into $G[e \leftarrow H]$ is also called a *hyperedge replacement*. Figure 4.3 shows an example where $\tau(e) = \tau(H) = [3]$ and $vert_G(e)$ maps 1, 2 and 3 respectively to u, v and w. The hyperedge e is shown as directed for readability.

Definition 4.22 (Hyperedge replacement graph grammars) A *hyperedge replacement graph grammar*, an *HR grammar* for short, is a pair $\Gamma = \langle U, R \rangle$ consisting of a finite S-sorted set U of variables called *nonterminal symbols* and a finite set R of rules where a *rule* is a pair (x, H) such that $x \in U$ and H is a U-hypergraph of type $\tau(x)$. The one step derivation relation on U-hypergraphs is defined by

$$G \Rightarrow G' \text{ if and only if } G' = G[e \leftarrow H],$$

---

[11] Or belongs to Λ if we consider (K, Λ)-labeled graphs; the reader will make easily the necessary modifications for the case of labeled graphs.

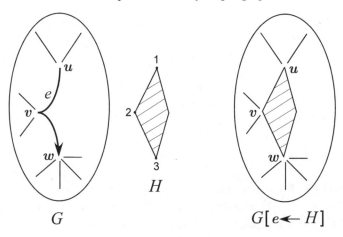

Figure 4.3 Hyperedge replacement.

where e is a hyperedge of G labeled by some x and $(x, H) \in R$. For each $x \in U$, we let **x** denote the concrete hypergraph of type $\tau(x)$ with a unique hyperedge e labeled by x such that $V_{\mathbf{x}} = \tau(x)$ and $vert_{\mathbf{x}}(e)(a) = src_{\mathbf{x}}(a) = a$ for every a in $\tau(x)$. We view **x** also as a constant symbol and denote by **U** the set of these symbols for all x in U. The set generated from $x \in U$ by Γ is the set $L(\Gamma, x)$ of s-graphs (i.e., of U-hypergraphs without hyperedges) G such that $\mathbf{x} \Rightarrow^* G$. For $Y \subseteq U$, we denote the set $\bigcup_{y \in Y} L(\Gamma, y)$ by $L(\Gamma, Y)$. We call such a set an *HR set of s-graphs*.

Our objective is to prove that the HR-equational and the HR sets of s-graphs are the same. We need a few more definitions and results.

Definition 4.23 (Terms denoting hypergraphs) The operations of F^{tHR} (defined in Definitions 2.25, 2.27, 2.30, 2.31 and 2.128) extend in an obvious way to hypergraphs. Hence every term t in $T(F^{\text{tHR}} \cup \mathbf{U})$ evaluates to a U-hypergraph $val(t)$. The detailed analysis of the mapping val done in Section 2.3.2 extends to these terms. In particular, when a concrete hypergraph G is isomorphic to $val(t)$ for some term t in $T(F^{\text{tHR}} \cup \mathbf{U})$, we have a bijection w from E_G to the set of occurrences in t of constant symbols of the form **ab**, $\overrightarrow{\mathbf{ab}}$, $\mathbf{a}^\ell$ or **x** (for $a, b \in \mathcal{A}$ and $x \in U$), cf. condition (1) of Definition 2.43. The next proposition establishes that the mapping val commutes with substitutions in terms and in hypergraphs.

Proposition 4.24 Let $t, s \in T(F^{\text{tHR}} \cup \mathbf{U})$, let $x \in U$ and let c be a context in $Ctxt(F^{\text{tHR}} \cup \mathbf{U})$ such that $t = c[\mathbf{x}]$ and $\tau(x) = \tau(s)$.[12] Let e be the corresponding hyperedge labeled by x, in a concrete hypergraph G isomorphic to $val(t)$. We have $val(c[s]) = G[e \leftarrow val(s)]$.

---

[12] The notion of context is defined in Definition 2.6. Writing $t = c[\mathbf{x}]$ corresponds to distinguishing an occurrence of **x** in t.

Proof: Straightforward induction on the structure of t based on the following facts (where, for readability, the types of the operations are not shown):

(1) $\mathbf{x}[e \leftarrow H] = H$,
(2) $fg_B(G)[e \leftarrow H] = fg_B(G[e \leftarrow H])$,
(3) $ren_h(G)[e \leftarrow H] = ren_h(G[e \leftarrow H])$,
(4) $(G \parallel G')[e \leftarrow H] = G[e \leftarrow H] \parallel G'$ if $e \in E_G$,

which hold for all G, G', H, e, fg_B, ren_h of appropriate types. ∎

Definition 4.25 (From HR equation systems to HR grammars and vice versa)
Let $S = \langle x_1 = p_1, \ldots, x_n = p_n \rangle$ be a typed HR equation system. Let $G(S)$ be the corresponding regular grammar (cf. Section 3.1.4, Definition 3.17). Its rules are the pairs (x_i, m) such that m is one of the monomials of p_i; they form the set $R(S)$. We let $\Gamma[S] := \langle U, R \rangle$ be the HR grammar such that $U := \{x_1, \ldots, x_n\}$ and

$$R := \{(x, val(m[\mathbf{x_1}/x_1, \ldots, \mathbf{x_n}/x_n])) \mid (x, m) \in R(S)\}.$$

Conversely, for every HR grammar $\Gamma = \langle U, R \rangle$, there exists a typed HR equation system S such that $\Gamma[S] = \Gamma$. To define it, it is sufficient to construct a term $m_H \in T(F^{tHR} \cup U)$ denoting H for each right-hand side of a rule (x_i, H) in R, cf. Propositions 2.33 and 2.132. Such a system S is not defined in a unique way. The grammar $\Gamma[S]$ is analogous to the context-free grammar $G[S]$ of Definition 3.12.

Lemma 4.26 Let S be a typed HR equation system and $\Gamma[S]$ be the associated HR grammar.
(1) If $x_i \Rightarrow t_1 \Rightarrow \cdots \Rightarrow t_p$ is a derivation sequence of the regular grammar $G(S)$, then $\mathbf{x_i} \Rightarrow val(t_1) \Rightarrow \cdots \Rightarrow val(t_p)$ is a derivation sequence of the HR grammar $\Gamma[S]$.
(2) Conversely, for every derivation sequence $\mathbf{x_i} \Rightarrow H_1 \Rightarrow \cdots \Rightarrow H_p$ of $\Gamma[S]$, there exists a derivation sequence $x_i \Rightarrow t_1 \Rightarrow \cdots \Rightarrow t_p$ of $G(S)$ such that $H_j = val(t_j)$ for each $j = 1, \ldots, p$.

Proof: Straightforward from the definitions and Proposition 4.24. ∎

Proposition 4.27
(1) Let S and $\Gamma[S]$ be as in Definition 4.25. The least solution of S in $\mathcal{P}(\mathbb{JS}^t)$ is $(L(\Gamma[S], x_1), \ldots, L(\Gamma[S], x_n))$.
(2) A set of s-graphs is HR if and only if it is HR-equational.

Proof: (1) Let $(L_1, \ldots, L_n)$ be the least solution of S in $\mathcal{P}(\mathbb{JS}^t)$. We know from Propositions 3.18 and 3.23(1) that each component L_i is $\{val(t) \mid t \in L(G(S), x_i)\}$. It

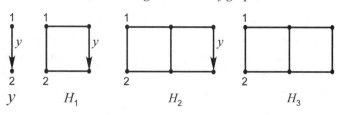

Figure 4.4 Ladders.

follows from Lemma 4.26(1) that $L_i \subseteq L(\Gamma[S], x_i)$. The opposite inclusion follows similarly from Lemma 4.26(2).

(2) Immediate consequence of (1) and the two constructions of Definition 4.25, by Proposition 4.13(4).   ∎

Example 4.28 We use the example of ladders defined in Example 4.3(5), and the typed equation system $E := \langle y = \mathbf{12} \cup f(y) \rangle$ with $\tau(y) = \{1, 2\}$. The corresponding HR grammar $\Gamma[E]$ consists of the two rules $(y, \mathbf{12})$ and (y, H) where H is the $\{y\}$-hypergraph defined by $f(\mathbf{y})$ and shown as H_1 in Figure 4.4. An example of a derivation sequence of $G(E)$ is $y \Rightarrow f(y) \Rightarrow f(f(y)) \Rightarrow f(f(\mathbf{12}))$. The corresponding derivation sequence of $\Gamma[E]$ is $\mathbf{y} \Rightarrow H_1 \Rightarrow H_2 \Rightarrow H_3$, where the $\{y\}$-hypergraphs $\mathbf{y}, H_1, H_2, H_3$ are shown in Figure 4.4.   □

We leave to the interested reader the easy task of extending the definitions and results of this section to labeled graphs.

4.2 HR-recognizable sets of graphs

The notion of HR-recognizability is a particular case of the general algebraic notion defined and studied in Chapter 3, applied to the *many-sorted* algebra $\mathbb{JS}^t$ (where the sort of an s-graph G is its type $\tau(G)$) and not to the algebra $\mathbb{JS}$.

4.2.1 Definitions and first examples

In this section, we detail the definitions and we give examples of sets of s-graphs that are HR-recognizable and of sets that are not.

Definition 4.29 (HR-recognizable sets of s-graphs) An equivalence relation $\sim$ on the one-sorted F^{HR}-algebra $\mathbb{JS}$ is *type-preserving* if every two equivalent (labeled) s-graphs G, G' have the same sort, i.e., $\tau(G) = \tau(G')$. We call $\tau(G)$ the *type* of the equivalence class $[G]_\sim$. A type-preserving equivalence relation is *locally finite* if it has finitely many classes of each type. Hence, a type-preserving congruence $\sim$ on $\mathbb{JS}$

is a union of equivalence relations $\sim_C$ on the domains $\mathcal{JS}_C$ and (by Definition 3.63) a congruence on the many-sorted F^{tHR}-algebra $\mathbb{JS}^{\mathrm{t}}$ defined in Section 2.6.2.

A set of s-graphs is *HR-recognizable* if it is recognizable in the many-sorted algebra $\mathbb{JS}^{\mathrm{t}}$. We denote by $\mathbf{Rec}(\mathbb{JS}^{\mathrm{t}})$ the family of HR-recognizable sets.[13] Since $\mathbb{JS}^{\mathrm{t}}$ is effectively given, the notions of an effectively HR-recognizable, a semi-effectively HR-recognizable and an effectively HR-term-recognizable set make sense and are instances of the general definitions. By Proposition 2.33 and Lemma 3.75(1), a set is effectively HR-recognizable if and only if it is effectively HR-term-recognizable. The notion of HR-recognizability can also be characterized in terms of the one-sorted algebra $\mathbb{JS}$: it follows from Proposition 3.64 that a set of s-graphs is HR-recognizable if and only if it is saturated by a type-preserving and locally finite congruence on $\mathbb{JS}$. The HR-recognizable sets of (K, Λ)-labeled s-graphs are defined similarly with respect to the algebra $\mathbb{JS}^{\mathrm{t}}_{[K,\Lambda]}$.

If L is a subset of $\mathcal{JS}[E]$ for some finite set E of source names, that is, if all s-graphs in L have a type included in E, then its recognizability can be considered with respect to $\mathbb{JS}^{\mathrm{t}}$ or to $\mathbb{JS}^{\mathrm{t}}[E]$. If L is HR-recognizable, then it is also recognizable in $\mathbb{JS}^{\mathrm{t}}[E]$ by Proposition 3.56(1). If, furthermore, $L \subseteq \mathcal{JS}^{\mathrm{gen}}[E]$, then it is also recognizable in $\mathbb{JS}^{\mathrm{t,gen}}[E]$ by the same proposition. We will come back to such comparisons in Proposition 4.39 below.

The many-sorted algebra $\mathbb{JS}^{\mathrm{t}}[E]$ has finitely many sorts. Hence, if L is recognizable in $\mathbb{JS}^{\mathrm{t}}[E]$, it is also recognizable in $\mathbb{JS}[E]$ because every locally finite congruence on $\mathbb{JS}^{\mathrm{t}}[E]$ is a finite congruence on $\mathbb{JS}[E]$. Let us conversely assume that $L \subseteq \mathcal{JS}[E]$ is recognizable in $\mathbb{JS}[E]$ with finite congruence $\sim$. This congruence may not be type-preserving. We define from it the equivalence relation $\approx$ by: $G \approx H$ if and only if $\tau(G) = \tau(H)$ and $G \sim H$. Since the s-graphs in $\mathcal{JS}[E]$ have finitely many possible types (the subsets of E), the type-preserving equivalence relation $\approx$ is finite. It is a congruence by Lemma 4.11. It saturates L and hence witnesses its recognizability in the algebra $\mathbb{JS}^{\mathrm{t}}[E]$. Thus, $\mathbb{JS}^{\mathrm{t}}[E]$ and $\mathbb{JS}[E]$ have the same recognizable sets. By Proposition 3.78(1), they also have the same semi-effectively recognizable sets. The same proof gives that $\mathbb{JS}^{\mathrm{t,gen}}[E]$ and $\mathbb{JS}^{\mathrm{gen}}[E]$ have the same recognizable sets and the same semi-effectively recognizable sets, and, hence, by Proposition 3.78(3), also the same effectively recognizable sets.[14]

All closure results for recognizable sets established in Proposition 3.56 and Section 3.4.6 hold for the HR-recognizable sets. As noticed when considering the HR-equational sets, the mapping *und* that forgets edge directions is a homomorphism : $\mathbb{JS}^{\mathrm{t,d}} \to \mathbb{JS}^{\mathrm{t,u}}$. It follows from Proposition 3.56(2) that $und^{-1}(L)$ is HR-recognizable if L is an HR-recognizable set of directed s-graphs.

---

[13] As for HR-equational sets, HR-recognizability concerns sets of abstract (labeled) s-graphs. A set L of concrete (labeled) s-graphs is HR-recognizable if and only if the set $\{[G]_{iso} \mid G \in L\}$ is HR-recognizable. If we specify a congruence on concrete (labeled) s-graphs, we impose that two isomorphic (labeled) s-graphs are equivalent.

[14] This can also be shown for $\mathbb{JS}^{\mathrm{t}}[E]$ and $\mathbb{JS}[E]$, by a proof based on Definition 3.74.

Congruences witnessing recognizability are usually not easy to construct, even for "simple" sets like the set of paths. The reason is that one has to describe congruence classes of countably many types and, usually, many classes of each type. Our main tool for proving that a set of (labeled) graphs is recognizable is to characterize it as the set of (finite) models of a sentence of monadic second-order logic. For some sets given below we can define a congruence and for the others we rest on their monadic second-order definability and on the Recognizability Theorem (Section 5.3) to claim that they are recognizable.

Example 4.30 (1) **Graphs with an even number of edges or of vertices:** For the set L_e of s-graphs with an even number of edges, the congruence is easy to define. We let

$$G \sim_e G' \text{ if and only if } \tau(G) = \tau(G') \text{ and } |E_G| - |E_{G'}| \text{ is even.}$$

That $\sim_e$ is indeed a congruence follows from the fact that $|E_{G /\!/ G'}| = |E_G| + |E_{G'}|$ and $E_{f(G)} = E_G$ for every unary operation f. There are two equivalence classes of each type. Since $\sim$ is effectively given (as one easily checks), L_e is effectively recognizable by Proposition 3.78(2).

For the set L_v of s-graphs with an even number of vertices, the definition is similar:

$$G \sim_v G' \text{ if and only if } \tau(G) = \tau(G') \text{ and } |V_G| - |V_{G'}| \text{ is even.}$$

This equivalence relation is a congruence because $|V_{G /\!/ G'}| = |V_G| + |V_{G'}| - |\tau(G) \cap \tau(G')|$ and $V_{f(G)} = V_G$ for every unary operation f. Again we have two equivalence classes of each type, and L_v is effectively recognizable.

If, in the definition of $\sim_e$, we omit the condition $\tau(G) = \tau(G')$, then we get a congruence for the algebra $\mathbb{JS}$ that has two classes (but is not type-preserving). The set of s-graphs with an even number of edges is thus recognizable in the one-sorted algebra $\mathbb{JS}$. This is not the same for $\sim_v$ as one checks easily. The set of s-graphs with an even number of vertices is not recognizable in $\mathbb{JS}$. This explains why we define recognizability with respect to the many-sorted algebra $\mathbb{JS}^t$. If we would define it with respect to $\mathbb{JS}$, then many "simple" sets of s-graphs would not be recognizable.

(2) **Connected graphs:** In Section 4.1.3 we have characterized the type function τ as a homomorphism of $\mathbb{JS}$ into an F^{HR}-algebra with domain $\mathcal{P}_f(\mathcal{A})$. We now refine this function. We let the *connectedness type* of an s-graph G be defined by $\theta(G) := \tau(CC(G))$, i.e., $\theta(G)$ is the multiset of types of the s-graphs in $CC(G)$, the set of connected components of G. In $\theta(G)$, we limit to 2 the number of occurrences of $\emptyset$. Hence $\emptyset \in \theta(G)$ if and only if G has at least one connected component without sources, and $\emptyset$ has 2 occurrences in $\theta(G)$ if and only if G has at least two such connected components. The elements of $\theta(G)$ form a partition of $\tau(G)$. The multiset character of $\theta(G)$ is limited to at most two occurrences of $\emptyset$. The following statement is easy to prove:

There exist mappings $/\!/^\theta, fg_B^\theta, ren_h^\theta$ such that for all s-graphs G, H we have

$$\theta(G /\!/ H) = \theta(G) /\!/^\theta \theta(H),$$

$$\theta(fg_B(G)) = fg_B^\theta(\theta(G)),$$

$$\theta(ren_h(G)) = ren_h^\theta(\theta(G)),$$

where B and h range over $\mathcal{P}_f(\mathcal{A})$ and $Perm_f(\mathcal{A})$ respectively.

The desired congruence is defined by $G \sim G'$ if and only if $\theta(G) = \theta(G')$. Since $\tau(G)$ can be defined from $\theta(G)$, this congruence is type-preserving. It is locally finite since the number of occurrences of $\emptyset$ is bounded. More precisely, it has $3 \cdot B(|C|)$ classes of type C where $B(k)$ is the k-th Bell number[15] that counts the number of partitions (with nonempty sets) of $[k]$ if $k > 0$ and $B(0) = 1$. The set of connected s-graphs of type $C \in \mathcal{P}_f(\mathcal{A})$ is $\theta^{-1}(\{C\})$. The set of disconnected graphs is $\{G \in \mathcal{JS} \mid \tau(G) = \emptyset \text{ and } \theta(G) = \{\emptyset, \emptyset\}\}$. These sets are effectively recognizable by Proposition 3.78(2).

We can replace in the definition of the mapping θ any value $\theta(G)$ containing $\emptyset$ and another set, either empty or not, by the unique value "NC" meaning "not connected." The values of this alternative function, call it $\theta'(G)$, can be the empty set (for the empty graph), $\{\emptyset\}$, NC or a partition of $\tau(G)$ with nonempty sets. A congruence $\sim'$ can be defined by $G \sim' G'$ if and only if $\tau(G) = \tau(G')$ and $\theta'(G) = \theta'(G')$. We obtain a congruence with less classes of each type that witnesses also the recognizability of the set of connected s-graphs. It has 3 classes of type $\emptyset$ and $B(|C|) + 1$ classes of type C if C is not empty.

(3) **One graph:** Let H be an s-graph. To show that the singleton $\{H\}$ is effectively HR-recognizable, we define $\chi(G)$ for an s-graph G by

$$\chi(G) = \begin{cases} G & \text{if } G^\circ \subseteq H^\circ, \\ \perp & \text{otherwise.} \end{cases}$$

We define $\sim$ by $G \sim G'$ if and only if $\tau(G) = \tau(G')$ and $\chi(G) = \chi(G')$. It is a congruence because G° is a subgraph of $(G /\!/ G')^\circ, fg_B(G)^\circ$ and $ren_h(G)^\circ$. Clearly, $\{H\} = \{G \in \mathcal{JS} \mid \tau(G) = \tau(H), \chi(G) = H\}$, hence is saturated by $\sim$. It is effectively recognizable by Proposition 3.78(2). Thus, by Corollary 3.86, all finite sets are effectively HR-recognizable. □

The last example implies Proposition 4.5(1), stating that the membership problem is decidable for every HR-equational set L: since $\{H\}$ is effectively recognizable, so is the set $\mathcal{JS} - \{H\}$ (by Corollary 3.86) and one can decide whether $L \subseteq \mathcal{JS} - \{H\}$ by Propositions 3.91(2) and 4.13(4), i.e., whether H does not belong to L. This algorithm

---

[15] We have $k^{k/2} \leq B(k) \leq k!$ for every $k \geq 8$.

uses the Filtering Theorem (Theorem 3.88). Its application uses essentially the same computations as the algorithm in the proof of Proposition 4.5(1).

Proposition 4.31 The following sets are effectively HR-recognizable (for all k):
(1) the set of graphs of degree at most k;
(2) the sets of trees and of paths;
(3) the set of k-colorable graphs;
(4) the set of biconnected graphs;
(5) the set of graphs having at most k connected components;
(6) the set of planar graphs;
(7) the sets $TWD(\leq k, C)$ for all finite sets C;[16]
(8) any minor-closed set of simple, loop-free, undirected graphs such that the excluded minors are known;
(9) the set of directed graphs having a Hamiltonian circuit (a directed cycle going through all vertices).

Proof: Each of these sets is characterized by a monadic second-order sentence. The Recognizability Theorem (Section 5.3) entails that they are all HR-recognizable. They are effectively so, since the sentences can be constructed.

Graphs of degree at most k are actually characterized by first-order sentences, one for each k. For the monadic second-order characterization of Case (6), see Corollary 1.15. For Case (7), see Section 5.2.2 (Proposition 5.11). A set as in Case (8) is characterized among the simple, loop-free and undirected graphs by finitely many excluded minors (see the observations after Proposition 2.21), so that it is characterized by a monadic second-order sentence using edge quantifications (by Corollary 1.14; the edge set quantifications are needed to express that the considered graph is simple, and only for this condition). If the excluded minors are known, the sentence can be constructed. The set of Hamiltonian graphs is characterized by a monadic second-order sentence using edge set quantifications. ∎

As the reader may guess, the corresponding classes of labeled graphs are also effectively HR-recognizable.

For proving Proposition 4.31, the direct construction of congruences is feasible for Cases (1)–(5) and (9) along the lines of the previous examples. It would be extremely complicated (and uninteresting to work out) for Cases (6) and (7).

The proof of Case (7) is based on the characterizations of the sets $TWD(\leq k)$ by finite sets of excluded minors (cf. Corollary 2.60(2) in Section 2.4.2), hence by a monadic second-order sentence (cf. Corollary 1.14). From this sentence one can easily build a monadic second-order sentence characterizing the set $TWD(\leq k, C)$, as shown in the proof of (P7) of Proposition 5.11. Finally, the Recognizability Theorem is used to construct a specification of $TWD(\leq k, C)$ as an effectively HR-recognizable

---

[16] We recall from Definition 2.53 that $TWD(\leq k, C)$ denotes the set of s-graphs of tree-width at most k and of type included in C.

set. Although such a specification can be constructed, the corresponding algorithm is intractable for two reasons: first, because the excluded minors characterizing the set $TWD(\leq k)$ are numerous and presumably very large for $k > 3$ (see [Lag] and [Din]) and, second, because the constructions of the Recognizability Theorem are intractable, even for relatively "small" monadic second-order formulas.

We now present examples of sets that are not HR-recognizable, together with some tools that allow us to establish these facts.

Proposition 4.32 The following sets are not HR-recognizable:

L_1: the set of odd paths with the middle vertex as 1-source;[17]
L_2: the set of graphs of the form $G \oplus H$ with $|V_G| = |V_H|$;
L_3: the set of graphs of the form $G \oplus G$.

Proof: We first recall that the set of terms

$$N = \{fg^n ag^n b \mid n \geq 0\} \subseteq T(\{f,g,a,b\}) \tag{4.4}$$

(where $\rho(f) = 2, \rho(g) = 1, \rho(a) = \rho(b) = 0$) is not regular, hence is not recognizable in the algebra of terms, as proved in Example 3.54. We will also use the fact that if $H := \{f,g,a,b\}$ is a derived signature of a signature F, $\mathbb{M}$ is an F-algebra and $L \subseteq M$ is recognizable in $\mathbb{M}$, then $\{t \in T(H) \mid val_{\mathbb{M}_H}(t) \in L\}$ is a recognizable, hence regular, set of terms by Theorem 3.62 and Assertions (2) and (3) of Proposition 3.56. We will use this fact for $\mathbb{M} := \mathbb{JS}^t$.

For proving the three statements, we need only replace in the set N defined by (4.4) the operations f,g,a,b by appropriate derived operations.

Case of L_1: We let a and b denote $\mathbf{1}$, we let g denote the derived operation such that $g(u) := ren_{1 \leftrightarrow 2}(fg_1(u /\!/ \mathbf{12}))$ and we let f denote $/\!/$. We let $val : T(\{f,g,a,b\}) \to \mathcal{JS}$ be the associated value mapping. Let K be the regular set of terms $\{fg^n ag^m b \mid n,m \geq 0\}$. If L_1 was HR-recognizable, then $val^{-1}(L_1) \subseteq T(\{f,g,a,b\})$ would be recognizable, hence regular, and so would $K \cap val^{-1}(L_1)$ by Proposition 3.85(1). But this set is N, which is not regular. Hence L_1 is not HR-recognizable.

Cases of L_2, L_3: The proofs are similar. Instead of $/\!/$, we take for f the binary derived operation $f(u,v) := fg_1(u) /\!/ fg_1(v)$. ∎

These examples prove that the families of HR-recognizable and HR-equational sets are incomparable: the set L_1 is HR-equational (cf. Counter-example 4.8) but not HR-recognizable, whereas the set of planar graphs is HR-recognizable (cf. Proposition 4.31(6)) but not HR-equational (cf. Remark 4.9(2)). This fact should be contrasted with the case of words: every recognizable, hence regular, language is context-free, hence is equational in $\mathbb{W}(A)$. However, we find again the familiar situation with

---

[17] This set is also used in Counter-example 4.8.

the following fact: every HR-recognizable set of graphs of bounded tree-width is HR-equational, as will be shown in Corollary 4.38.

The recognizability index can also be used to prove that certain sets are or are not HR-recognizable.

Definition 4.33 (The recognizability index of a set of s-graphs) The index of a type-preserving equivalence relation $\sim$ has been defined in Definition 3.63, Section 3.4.3, for general many-sorted algebras. It is the function $\gamma_\sim$ that defines for each sort s the number (possibly ω) of equivalence classes of $\sim$ of sort s. If $\sim$ is a congruence of the many-sorted HR algebra $\mathbb{JS}^t$, then for each $C \in \mathcal{P}_f(\mathcal{A})$, the number $\gamma_\sim(C)$ depends only on $|C|$.[18] This is so because if $C' \in \mathcal{P}_f(\mathcal{A})$ has cardinality $|C|$ and h is a finite permutation of $\mathcal{A}$ such that $h(C) = C'$, then for all G, G' of type C we have $ren_h(G) \sim ren_h(G')$ if $G \sim G'$ (because $\sim$ is a congruence). Since ren_h has for inverse $ren_{h^{-1}}$, we also have the opposite implication. Hence, ren_h is a bijection from the equivalence classes of type C to those of type C' and thus, $\gamma_\sim(C) = \gamma_\sim(C')$. We define $\gamma_\sim(k) := \gamma_\sim(C)$, where C is any set of cardinality k. We also call $\gamma_\sim$ the *index* of the congruence $\sim$.

The *HR-recognizability index of* a set L of (labeled) s-graphs is the mapping $\gamma_{\approx^L}$, denoted simply by γ_L, where $\approx^L$ is the syntactic congruence of L with respect to $\mathbb{JS}^t$ (cf. Definition 3.65). Hence, L is HR-recognizable if and only if $\gamma_L(k) \in \mathcal{N}$ for every k (by Proposition 3.66(1)). If this is the case, then $\gamma_L(k) \leq \gamma_\sim(k)$ for every k, if $\sim$ is any congruence witnessing the recognizability of L (by Proposition 3.66(2)).

The HR-recognizability index of a set L is thus a measure of its complexity as an HR-recognizable set. (Another recognizability index will be defined in a similar way with respect to the VR algebra, cf. Section 4.4.) It is relevant to the recognition of HR-recognizable sets of s-graphs by finite automata on terms since it defines the number of states of their associated minimal automata (cf. Section 3.4.3, Propositions 3.69 and 3.73(2)). Going back to Example 4.30, we can see that $\gamma_L(k) \leq 2$ for the sets of Example 4.30(1). For the set L of connected graphs (Example 4.30(2)) we get $\gamma_L(0) = 3$ and $\gamma_L(k) \leq B(k) + 1$ for $k > 0$. For the set L_1 of Proposition 4.32 we get $\gamma_{L_1}(1) = \omega$, and thus another proof that this set is not HR-recognizable.

4.2.2 A simpler HR-recognizability criterium

We give a characterization of HR-recognizability that is similar to the characterization by Myhill and Nerode of the regular languages (cf. [*Sak], Theorem 2.3 of Chapter II). It makes some proofs of recognizability or of nonrecognizability easier because it involves "simple" graph operations, related to the well-understood notion of vertex separator.

---

[18] We have noted a similar fact in Remark 4.6 for the HR-equational sets.

We consider (labeled) s-graphs of type $[k]$ for $k \geq 0$. We denote by $\square_k$ the following derived operation of $\mathbb{JS}^t$:

$$G \square_k H := fg_{[k]}(G \mathbin{/\!/} H),$$

for G, H both of type $[k]$. Thus, $G \square_k H$ has type $\emptyset$. We let $\mathcal{JN} := \bigcup_{k \geq 0} \mathcal{JN}_k$, where $\mathcal{JN}_k$ is the set of (labeled) s-graphs of type $[k]$. We let $\mathbb{JN}^t$ be the corresponding many-sorted algebra. Its set of sorts is $\mathcal{N}$ and its signature $\{\square_k \mid k \in \mathcal{N}\}$ is denoted by $F_\square$.

For every subset L of $\mathcal{JN}_0$, we let $\equiv_L$ be the equivalence relation on $\mathcal{JN}$ defined by

$$G \equiv_L G' \text{ if and only if there exists } k \in \mathcal{N} \text{ such that}$$

$$\tau(G) = \tau(G') = [k] \text{ and } G \square_k H \in L \iff G' \square_k H \in L \text{ for all } H \in \mathcal{JN}_k.$$

We will compare three equivalences associated with L that will turn out to be essentially the same: its syntactic congruence relative to $\mathbb{JS}^t$, denoted by $\approx^L$; its syntactic congruence relative to $\mathbb{JN}^t$, denoted by $\sim_L$; and the equivalence $\equiv_L$.

Theorem 4.34 For every set L of (labeled) graphs:

(1) the equivalence $\equiv_L$ is the syntactic congruence $\sim_L$ of L with respect to $\mathbb{JN}^t$;
(2) L is HR-recognizable if and only if it is recognizable in $\mathbb{JN}^t$ if and only if the equivalence $\equiv_L$ is locally finite; the HR-recognizability index of L is equal to the index of $\equiv_L$.

Proof: (1) It is clear that $G \sim_L G'$ implies $G \equiv_L G'$ because the latter congruence is defined like the syntactic congruence of L with respect to $\mathbb{JN}^t$, but with terms c of the particular form $y_1 \square_k y_2$. (We use the notation of Definition 3.65 where syntactic congruences are defined.)

For the opposite implication, we observe that for every linear term c in $T(F_\square, Y_p)$ that has an occurrence of y_1, if $G \in \mathcal{JN}_k$, $H_2, \ldots, H_p \in \mathcal{JN}$ and the type of $K := c_{\mathbb{JN}^t}(G, H_2, \ldots, H_p)$ is $\emptyset$, then $K = G \square_k H$ for some s-graph H that depends only on $H_2, \ldots, H_p$. We show this by induction on the structure of c. If $c = y_1$, we can take $H := \emptyset$. If $c = y_1 \square_m e$ (and hence $m = k$), we can take $H := e_{\mathbb{JN}^t}(G, H_2, \ldots, H_p)$, which depends only on $H_2, \ldots, H_p$ because e has no occurrence of y_1. If $c = d \square_m e$ and $d \neq y_1$ has an occurrence of y_1, then $K = d_{\mathbb{JN}^t}(G, H_2, \ldots, H_p) \square_m H_e$, where $H_e := e_{\mathbb{JN}^t}(G, H_2, \ldots, H_p)$ depends only on $H_2, \ldots, H_p$. Since the type of $d_{\mathbb{JN}^t}(G, H_2, \ldots, H_p)$ is $\emptyset$ (and hence $m = 0$), we obtain by induction that $K = (G \square_k H_d) \square_0 H_e$, where H_d depends only on $H_2, \ldots, H_p$. But $(G \square_k H_d) \square_0 H_e = G \square_k (H_d \mathbin{/\!/} H_e)$ (by the definition of the operations of $F_\square$ and Equality (18) after Proposition 2.48). Hence we can take $H := H_d \mathbin{/\!/} H_e$. Since $\square_m$ is commutative, it suffices to consider these cases.

Since the elements of L have type $\emptyset$, this observation implies that if $G \equiv_L G'$ and $c_{\mathbb{JN}^t}(G, H_2, \ldots, H_p) \in L$, then $c_{\mathbb{JN}^t}(G', H_2, \ldots, H_p) \in L$. Hence, $G \equiv_L G'$ implies $G \sim_L G'$.

(2) By (1) and Proposition 3.66(1), it suffices to show the equality of the two indices. Since the set $\mathcal{A}$ of source names is linearly ordered, for every set $C \subset \mathcal{A}$ of cardinality k there exists an order preserving bijection $h_C : C \to [k]$.

Let $\sim_{ren}$ be the type-preserving equivalence relation on $\mathcal{JS}$ defined by $G \sim_{ren} G'$ if and only if $\tau(G) = \tau(G')$ and $ren_{h_{\tau(G)}}(G) \sim_L ren_{h_{\tau(G)}}(G')$. The equivalence classes of $\sim_{ren}$ of type C are in bijection by ren_{h_C} with those of $\sim_L$ of type $[k]$ where $k := |C|$. Hence, $\sim_{ren}$ has the same index as $\sim_L$, and so the same index as $\equiv_L$ by (1).

We now prove that $\sim_{ren}$ is the syntactic congruence $\approx^L$ of L with respect to $\mathbb{JS}^t$. Let G, G' be such that $G \sim_{ren} G'$, let C be their common type, let $k := |C|$ and let c be a context in $Ctxt(F^{tHR})$ such that $c(G) \in L$. By Propositions 2.130 and 2.51 (for c of sort $\emptyset$ and $A = C$: one can take $A' = A$ and ren_h can be omitted) we have $c_{\mathbb{JS}^t}(G) = fg_C(G /\!/ H)$ for some s-graph H of type $\tau(G) = C$. Then we have $c_{\mathbb{JS}^t}(G) = ren_{h_C}(G) \square_k ren_{h_C}(H)$, and also $c_{\mathbb{JS}^t}(G') = ren_{h_C}(G') \square_k ren_{h_C}(H)$. By these equalities, since $G \sim_{ren} G'$, $c_{\mathbb{JS}^t}(G) \in L$ implies $c_{\mathbb{JS}^t}(G') \in L$. This proves that $G \approx^L G'$. Now assume that $G \approx^L G'$ with C and k as before. By (1), it suffices to show that $ren_{h_C}(G) \equiv_L ren_{h_C}(G')$. This is obvious because for every $H \in \mathcal{JN}_k$ we have $ren_{h_C}(G) \square_k H = c_{\mathbb{JS}^t}(G, H)$ for the term $c(y_1, y_2) = fg_{[k]}(ren_{h_C}(y_1) /\!/ y_2)$, and similarly for G'. Hence, $\sim_{ren}$ is equal to $\approx^L$. ∎

As an example, we get that the singleton $\{H\}$ is HR-recognizable if H is a graph, with a similar proof as in Example 4.30(3): if G and G' are s-graphs of type $[k]$ such that $G \sim G'$, then $G \equiv_{\{H\}} G'$.

We get also that every finite set of graphs is HR-recognizable. Another less immediate consequence is that the set of graphs of bandwidth at most k is not HR-recognizable for $k \geq 2$: see Theorem 6.79 of [*DowFel].

4.2.3 Uncountably many HR-recognizable sets

Since we allow an HR-recognizable set to have elements of countably many different sorts, it is not so surprising that there are uncountably many HR-recognizable sets. For example: let I consist of the s-graphs of type $[n]$ with n vertices and no edges, for $n > 0$. Every subset of I is HR-recognizable as one checks easily by considering the equivalence $\sim$ such that $G \sim G'$ if and only if $\tau(G) = \tau(G')$ and, either $G = G' \in I$ or G and G' do not belong to I. However, there are also uncountably many (homogenous) HR-recognizable sets of graphs. This prevents any attempt of characterizing these sets by finite graph automata or logical formulas (as this is possible for words, terms and trees).[19]

---

[19] The reason is of course that there are only countably many finite graph automata and logical formulas. For the same reason, "most" HR-recognizable sets are not effectively HR-recognizable (nor semi-effectively HR-recognizable).

For proving this fact we give a condition ensuring that a set of (labeled) graphs is HR-recognizable. It is formulated with the operations $\square_k$ of Theorem 4.34 (which is used to prove the next lemma).

Condition *REC*: Let $L \subseteq \mathcal{JN}_0$ be a set of (labeled) graphs. For each $k \geq 0$ there exists a finite set $B_k \subseteq \mathcal{JN}_k$ such that:

$$\text{if } G \in L \text{ and } G = H \square_k K, \text{ then at least one of } H \text{ and } K \text{ is in } B_k.$$

Lemma 4.35 Every set L of connected (labeled) graphs that satisfies Condition *REC* is HR-recognizable.

Proof: Since all graphs in L are connected and $\square_0$ is the disjoint union of graphs, we may assume that $B_0 = \{\varnothing\}$. (The case $k = 0$ of *REC* is satisfied whenever L is a set of connected graphs.)

We let $\sim$ be the binary relation on graphs in $\mathcal{JN}$ defined as follows: $H \sim H'$ if and only if there exists $k \in \mathcal{N}$ such that $\tau(H) = \tau(H') = [k]$ and

either $H, H' \in B_k$ and $H = H'$
or $H, H' \notin B_k$ and for every $K \in B_k$, $H \square_k K \in L \Longleftrightarrow H' \square_k K \in L$.

The relation $\sim$ is a type-preserving equivalence relation. It has at most $|B_k| + 2^{|B_k|}$ classes of type $[k]$, hence is locally finite. The three classes of type $\varnothing$ are $B_0 = \{\varnothing\}$, $L - B_0$ and $\mathcal{JN}_0 - (L \cup B_0)$. Hence, the set L is saturated by $\sim$. It remains to check that $\sim$ is a congruence on the algebra $\mathbb{JN}^t$.

Let us prove that $H \square_k K \sim H' \square_k K'$, where H, H', K, K' are of type $[k]$, $H \sim H'$ and $K \sim K'$. If one of them, say $H \square_k K$, equals $\varnothing$, then both H and K are the empty graph and hence (by the definition of $\sim$) so are H', K' and $H' \square_k K'$. If $H \square_k K$ and $H' \square_k K'$ are not empty, then we need only prove that $H \square_k K \in L$ implies $H' \square_k K' \in L$. Assume $H \square_k K \in L$ and $H, K \in B_k$: then $H' = H$ and $K' = K$ by the definition of $\sim$ and $H' \square_k K' = H \square_k K \in L$. Otherwise, by Condition *REC*, exactly one of H and K, say K, is in B_k. Hence $K' = K$. Since $H \sim H'$ we have $H, H' \notin B_k$. Hence, $H \square_k K \in L$ implies $H' \square_k K \in L$ by the definition of $\sim$. Then $H' \square_k K'$ belongs to L because $K' = K$.   ∎

If Condition *REC* holds for a set L, then it holds for all its subsets. Hence, if it holds for an infinite set, there are, by Lemma 4.35, uncountably many recognizable sets of graphs.

Proposition 4.36 Every set of cliques and every set of square grids is HR-recognizable. So is every set of complete bipartite graphs $K_{n,n}$.

Proof: We first consider cliques. If $K_n = G \square_k H$, where $\tau(G) = \tau(H) = [k]$, then at least one of G and H has no internal vertices. We let U_k be the finite set of simple loop-free undirected s-graphs in $\mathcal{JN}_k$ that have no internal vertices. We obtain the result by Lemma 4.35, by taking $B_k := U_k$.

Next we consider the complete bipartite graphs $K_{n,n}$. If $K_{n,n} = G \square_k H$, where $\tau(G) = \tau(H) = [k]$ and A, B are such that $K_{n,n} = A \otimes B$ (cf. Section 2.4.2 for this notation), then, either at least one of G and H has no internal vertices, or at least one of A and B is a set of sources of G (and of H). The latter implies $n \leq k$, and there are only finitely many possible s-graphs G and H in such a decomposition, forming a set D_k. We obtain the result by Lemma 4.35 by taking $B_k := U_k \cup D_k$ where U_k is as above.

Finally we consider the case of square grids. Lemma 2.17 of [Cou90b] establishes that if $G_{n \times n} = G \square_k H$, where $\tau(G) = \tau(H) = [k]$ and $n \geq 2k+3$, then at least one of G and H has at most $k + k^2$ vertices. We let B_k be the set of simple loop-free s-graphs in $\mathcal{JN}_k$ of degree at most 4 with at most $(2k+2)^2$ vertices. Hence, if $G_{n \times n} = G \square_k H$ and $\tau(G) = \tau(H) = [k]$, then either $n \leq 2k+2$ and $|V_G|, |V_H| \leq (2k+2)^2$, or $n \geq 2k+3$ and at least one of G and H has at most $k + k^2 < (2k+2)^2$ vertices. It is clear that B_k is finite, which gives the result, again with Lemma 4.35. ∎

If A is a set of positive integers, the set $L_A := \{K_{1,n} \mid n \in A\}$ is HR-recognizable if and only if A is recognizable in $\mathbb{N}$. (If A is recognizable, then L_A is CMS-definable by Proposition 5.25, hence HR-recognizable by the Recognizability Theorem (Theorem 5.68). Conversely, let L_A be HR-recognizable with syntactic congruence $\sim$. For every positive integer n, let $g(n)$ be the s-graph $K_{1,n}$ with an r-source of degree n; the equivalence relation on $\mathcal{N}$ defined by $n \equiv m$ if and only if $g(n) \sim g(m)$ witnesses the recognizability of A.) Hence, in the previous proposition, one cannot replace the set of graphs $K_{n,n}$ by the set of all complete bipartite undirected graphs.

4.2.4 HR-recognizability and bounded tree-width

For further reference, we make explicit the following consequence of the algebraic version of the Filtering Theorem (Theorem 3.88, Section 3.4.7):

Theorem 4.37 The intersection $L \cap K$ of an HR-equational set L and an HR-recognizable set K is HR-equational. If K is semi-effectively HR-recognizable and L is given by an HR equation system, then an equation system defining $L \cap K$ can be constructed.

Proof: Immediate from Proposition 4.13 and the Filtering Theorem, which says that the intersection of a $\mathbb{JS}^t$-equational set with a $\mathbb{JS}^t$-recognizable set is $\mathbb{JS}^t$-equational. ∎

Corollary 4.38 Let L be a set of (labeled) s-graphs of type included in a finite set C. If L is HR-recognizable and has tree-width at most k, then it is HR-equational.

Proof: Let L be as in the statement. The set $TWD(\leq k, C)$ is HR-equational (Example 4.3(8)). Since $L = TWD(\leq k, C) \cap L$, L is also HR-equational by Theorem 4.37. ∎

We now compare HR-recognizability to recognizability in a finitely generated algebra.

Proposition 4.39 Let $k \in \mathcal{N}$ and let C be a set of source labels of cardinality $k + 1$. Let L be a set of (labeled) s-graphs of type included in C and of tree-width at most k. Then L is HR-recognizable if and only if it is recognizable in the F_C^{tHR}-algebra $\mathbb{JS}^{\text{t,gen}}[C]$, if and only if $\gamma_L(n)$ is finite for every $n \leq k + 1$.

Proof: First observe that $L \subseteq \mathcal{J}S^{\text{t,gen}}[C]$. This follows from the algebraic characterization of tree-width given in Theorem 2.83 and from the discussion after Proposition 2.132. (The domains of $\mathbb{JS}^{\text{t,gen}}[C]$ consist of the graphs of tree-width at most k and of type included in C.)

Let L be HR-recognizable. It is, by definition, recognizable in $\mathbb{JS}^{\text{t}}$, hence by Proposition 3.56(1), it is recognizable in the subalgebra $\mathbb{JS}^{\text{t,gen}}[C]$ of $\mathbb{JS}^{\text{t}}$ generated by the subsignature F_C^{tHR} of F^{tHR} (see Section 2.3.1, Remark 2.39(1)). Furthermore, the value of its HR-recognizability index $\gamma_L(n)$ is finite for every n, hence in particular for every $n \leq k + 1$.

We now discuss the converse implications. If $\gamma_L(n)$ is finite for each $n \leq k + 1$, then the restriction of the syntactic congruence $\approx^L$ to $\mathcal{J}S^{\text{t,gen}}[C]$ is finite and hence witnesses the recognizability of L in the algebra $\mathbb{JS}^{\text{t,gen}}[C]$. It remains to prove that the recognizability of L in $\mathbb{JS}^{\text{t,gen}}[C]$ implies its HR-recognizability. The proof, done in [CouLag], is quite complicated. ∎

We will say a few words on HR-recognizable sets of unbounded tree-width. If L is an HR-recognizable set of graphs, then each set $L \cap TWD(\leq k)$ is HR-recognizable by Proposition 3.85(1), because $TWD(\leq k)$ is HR-recognizable by Proposition 4.31. The converse implication is actually not true, as shown below.

Proposition 4.40 There exists a set of graphs L that is not HR-recognizable but is such that $L \cap TWD(\leq k)$ is HR-recognizable for each k.

Proof: We take $L = \{G_{n \times n} \oplus G_{n \times n} \mid n \geq 1\}$. Since $twd(G_{n \times n}) = twd(G_{n \times n} \oplus G_{n \times n}) = n$ (by Example 2.56(3) and Proposition 2.62), each set $L \cap TWD(\leq k)$ is finite, hence HR-recognizable (by Example 4.30(3)). However, the syntactic congruence $\approx^L$ of L has infinitely many classes of type $\emptyset$ because for every n, $G_{n \times n} \approx^L H$ if and only if $H = G_{n \times n}$ as one checks easily. Hence, L is not HR-recognizable. ∎

4.3 VR-equational sets of simple graphs

We now consider the equational sets of the algebra $\mathbb{GP}$, defined in Chapter 2, Section 2.5.1. Its elements are simple graphs or, more generally, simple (K, Λ)-labeled graphs, where K is a finite set of vertex labels and Λ is a finite set of edge labels. Their vertices have other labels, called *port labels*, belonging to a (countably infinite) set $\mathcal{A}$. Port labels are used by unary operations reviewed below to add edges to a graph. Graphs with ports are called p-graphs to recall this particular type of labeling. As for HR-equational sets, most of the results can be stated for $((K, \Lambda)$-)labeled graphs and their proofs are not more difficult than for unlabeled graphs. We will not detail the extensions of the proofs to labeled graphs.

4.3.1 VR equation systems

A set of (labeled) p-graphs is *VR-equational* if it is equational in the algebra $\mathbb{GP}$. We recall that the operations of $\mathbb{GP}$ are:

- the disjoint union, denoted by $\oplus$;
- the operations $\overrightarrow{add}_{a,b}$ and $add_{a,b}$ that add, respectively, directed and undirected, unlabeled edges[20] between vertices with port labels a and b; for adding edges labeled by λ, we use $\overrightarrow{add}_{a,b,\lambda}$ and $add_{a,b,\lambda}$;
- the operations $relab_h$ that modify port labels by using a mapping $h : \mathcal{A} \to \mathcal{A}$ which is the identity outside of a finite subset of $\mathcal{A}$;
- constant symbols denoting isolated vertices, possibly with incident loops; the possible labels of vertices and loops are specified by these symbols.

The notion of a VR equation system S and the notations $L(S,x)$, $L_{Term}(S,x)$, $F(S)$, $\mathcal{A}(S)$ are similar to those relative to HR equation systems. They cover the case of labeled graphs as well. We will let *val* denote the evaluation mapping $val_{\mathbb{GP}}$ (or $val_{\mathbb{GP}_{[K,\Lambda]}}$) from terms in $T(F^{VR})$ (or in $T(F^{VR}_{[K,\Lambda]})$) to p-graphs (or to labeled p-graphs). The following proposition is similar to Proposition 4.2:

Proposition 4.41 For every VR equation system S and every unknown x of S, we have $L(S,x) = val(L_{Term}(S,x))$ and $L_{Term}(S,x) \subseteq T(F(S))$. $\qquad\square$

The remarks about parsing and its algorithmic difficulty, derivation trees and the corresponding terms made in Definition 4.4 extend in an obvious way, and we do not repeat them. We only observe that if G is the value of a term t in $T(F^{VR}_{[K,\Lambda]})$, then $|V_G| = |Occ_0(t)|$, $|E_G| \leq |Occ_0(t)|^2$ if G is unlabeled and $|E_G| = O(|Occ_0(t)|^2)$ if G is labeled. ($Occ_0(t)$ is the set of occurrences of constant symbols in t different from $\varnothing$.) The following proposition is similar to Proposition 4.5 and is proved in the same way.

Proposition 4.42
(1) The membership problem for a VR-equational set is decidable.
(2) The emptiness of a VR-equational set (given by an equation system) is decidable.
(3) If a nonempty VR-equational set is defined by a uniform system with n unknowns, then it contains a p-graph with at most 2^{n-1} vertices. $\qquad\square$

---

[20] In this setting, an undirected edge is a pair of opposite directed edges.

The technical observation made in Remark 4.6 about changing source labels extend to port labels in an obvious way.

We now give new examples and review some already given ones.

Example 4.43 (1) **Cographs:** The set of cographs is defined by the equation (already considered in Sections 1.1.2 and 1.4.1)

$$x = (x \oplus x) \cup (x \otimes x) \cup \mathbf{1},$$

where the complete join $\otimes$ is the derived operation of $\mathbb{GP}^u$, the subalgebra of $\mathbb{GP}$ consisting of undirected p-graphs, defined by $x \otimes y := relab_{2\to1}(add_{1,2}(x \oplus relab_{1\to2}(y)))$. The constant symbol $\mathbf{1}$ denotes a graph with one vertex that is the 1-port (all vertices of a graph have port label 1). Without the monomial $x \oplus x$, this equation defines the set of cliques K_n, $n \geq 1$. These sets are not HR-equational, by Proposition 4.7 (recall from Example 2.56(3) that K_n has tree-width $n-1$).

(2) **Directed cographs:** The set of directed cographs is defined similarly by the equation

$$y = (y \oplus y) \cup (y \overrightarrow{\otimes} y) \cup (y \otimes y) \cup \mathbf{1},$$

where $\overrightarrow{\otimes}$ is the derived operation of $\mathbb{GP}$ defined by $x \overrightarrow{\otimes} y := relab_{2\to1}(\overrightarrow{add}_{1,2}(x \oplus relab_{1\to2}(y)))$.

(3) **Rooted trees:** The sets of rooted trees (already considered in Example 4.3(2)) and of nonempty rooted forests are defined by the equation system S:

$$\begin{cases} r = \mathbf{1} \cup relab_{3\to2}(\overrightarrow{add}_{1,3}(\mathbf{1} \oplus relab_{1\to3}(f))), \\ f = r \cup (f \oplus f). \end{cases}$$

An alternative system using the empty graph $\varnothing$ is S' as follows:

$$\begin{cases} r = relab_{3\to2}(\overrightarrow{add}_{1,3}(\mathbf{1} \oplus relab_{1\to3}(f'))), \\ f' = \varnothing \cup (r \oplus f'). \end{cases}$$

We have $L(S,r) = L(S',r)$ and $L(S',f') = L(S,f) \cup \{\varnothing\}$.

(4) **Trees:** The set of trees (cf. Example 4.3(3)) is $L(S'',t)$, where S'' is the equation system:

$$\begin{cases} t = relab_{2\to1}(r), \\ r = \mathbf{1} \cup relab_{3\to2}(\overrightarrow{add}_{1,3}(\mathbf{1} \oplus relab_{1\to3}(f))), \\ f = r \cup (f \oplus f). \end{cases}$$

The other sets considered in Example 4.3 are also VR-equational sets because every HR-equational set of simple graphs is VR-equational as we will prove later (Theorem 4.49).

(5) **Graphs of bounded clique-width:** For each k, the set $CWD(\leq k)$ of graphs of clique-width at most k is VR-equational: the proof is the same as for the sets $TWD(\leq k - 1)$ in Example 4.3(8). To be precise, Definition 2.89 yields, for $k \geq 1$, the following equation that defines the set of undirected p-graphs G of clique-width at most k such that $\pi(G) \subseteq [k]$:

$$x = (x \oplus x) \cup \left(\bigcup add_{a,b}(x) \right) \cup \left(\bigcup relab_h(x) \right) \cup \mathbf{1} \cup \mathbf{1}^\ell \cup \varnothing,$$

where the unions extend to all $a, b \in [k]$ with $a < b$ and all mappings h in $[[k] \to [k]]_f$. A 1-port with an incident loop is denoted by the constant symbol $\mathbf{1}^\ell$. By Proposition 2.118, one can replace the operations $relab_h$ by $relab_{a \to b}$ for all $a, b \in [k]$, $a \neq b$. For generating directed p-graphs, one replaces $add_{a,b}$ by $\overrightarrow{add}_{a,b}$. Finally, for generating graphs, we add the equation $y = relab_h(x)$, where h is such that $h(i) := 1$ for all $i \in [k]$. However, we cannot claim, as in Remark 4.9(1), that the sets of graphs of clique-width exactly k are VR-equational because these classes are not known to be monadic second-order logic definable, except for $k \leq 2$.

Slightly more generally, the set $CWD(\leq k, C)$ of p-graphs of clique-width at most k and of type included in C is VR-equational for every $k \in \mathcal{N}$ and every finite set C of port labels. So are the corresponding sets of labeled graphs. For each $k \geq 2$, the set $CWD(\leq k)$ is VR-equational but not HR-equational since it contains all cliques.

Linear clique-width can be handled in the same way as path-width in Example 4.3(9). □

In Section 2.5.6, we have defined variants of the signature F^{VR}, obtained either by removing certain operations, by adding new operations or by replacing certain operations by (linear) derived ones. These new operations act on the same objects, the (labeled) p-graphs. Equation systems can be written over these alternative signatures, and they might define equational sets that are not exactly the VR-equational. However, they still define the VR-equational sets in all cases.

We first consider the case of $F^{\kappa\mathrm{VR}} := \bigcup_{C \in \mathcal{P}_f(\mathcal{A})} F_C^{\kappa\mathrm{VR}}$. This signature is linear derived from F^{VR}, but F^{VR} is also linear derived from it: the operation $\oplus$ is in both signatures and for every unary operation f of F^{VR}, we have $f(G) = g(G, \varnothing)$, where g is the binary operation of $F^{\kappa\mathrm{VR}}$ defined by the term $f(x_1 \oplus x_2)$, cf. the proof of Proposition 2.121. It follows from Proposition 3.41 that the equational sets are the same for the two algebras.

A similar argument holds for the signature F'^{VR}, because every relabeling can be expressed as a composition of elementary relabelings (by using at most one auxiliary new port label, cf. Section 2.5.6, Proposition 2.118). For the signature $F^{i\mathrm{VR}}$ it will be proved as a corollary of the Equationality Theorem for the VR algebra (Theorem 7.36) which gives a logical characterization of the class of VR-equational sets, although it would not be difficult to give a direct proof by a transformation of equation systems.

4.3.2 VR-equational sets and clique-width

For every VR equation system S, we define $\mathcal{A}(S)$ as the smallest subset C of $\mathcal{A}$ such that the signature $F(S)$ is included in F_C^{VR} (or in $F_{C,[K,\Lambda]}^{VR}$ if labeled graphs are to be defined). (If S is written with derived operations like $\otimes$ and $\overrightarrow{\otimes}$, we replace these operations by their definitions in order to determine the set $\mathcal{A}(S)$.) For the very same reasons as in Proposition 4.7 we have:

Proposition 4.44 Every VR-equational set of (labeled) p-graphs has bounded clique-width. More precisely, for every VR equation system S and $x \in Unk(S)$, we have $cwd(L(S,x)) \le |\mathcal{A}(S)|$.

A set of (labeled) graphs is not VR-equational if, either it has unbounded clique-width or has bounded clique-width but has an internal structure too complicated to be described exactly by an equation system. We made a similar observation about HR-equational sets in Remark 4.9(2), and the same examples illustrate the two cases: the set of planar graphs for the first case (because square grids have unbounded clique-width by Proposition 2.106(1)) and, for the second one, any set of paths whose lengths do not form a semi-linear set of integers (by Proposition 4.47(1) below).

4.3.3 Type analysis of VR equation systems

We recall that the *type* of a (labeled) p-graph G is the set $\pi(G)$ of its port labels. If L is a set of (labeled) p-graphs, then $\pi(L) := \{\pi(G) \mid G \in L\}$. The proof of Proposition 4.10 can be adapted to the present case in a straightforward manner and we get:

Proposition 4.45 For every VR-equational set L given by a VR equation system S, the finite set $\pi(L) \subseteq \mathcal{P}(\mathcal{A}(S))$ can be computed. $\qquad \square$

To compare the VR-equational sets with the equational sets of the many-sorted VR algebra $\mathbb{GP}^t$ of Section 2.6.3, typed VR equation systems can be defined as in Definition 4.12, together with all the related terminology in that definition.

Proposition 4.46 Every VR equation system can be effectively transformed into an equivalent typed VR equation system, and vice versa. A set of p-graphs is VR-equational if and only if it is $\mathbb{GP}^t$-equational. $\qquad \square$

The proofs are easily adapted from those of Proposition 4.13 (Section 4.1.3). A typed VR equation system can be exponentially larger than an untyped one defining the same set. Example 4.14 can be adapted to prove this fact.

For determining if a VR-equational set is finite, we can use the same method as in Section 4.1.4 for HR systems with the following adaptation: we need only count vertices, and we let $|G| := |V_G|$. (A different size measure on graphs is used in Definitions 2.9 (Section 2.2) and 4.15 (Section 4.1.4).) We have for all G, H, a, b, λ and h of the relevant types $|G \oplus H| = |G| + |H|$, $|add_{a,b}(G)| = |add_{a,b,\lambda}(G)| = |\overrightarrow{add}_{a,b}(G)| = |\overrightarrow{add}_{a,b,\lambda}(G)| = |relab_h(G)| = |G|$, $|c| = 1$ for every constant c that is not $\varnothing$ and $|\varnothing| = 0$.

A set of simple (abstract) p-graphs L with vertex and edge labels in finite sets is finite if and only if the set $\{|G| \mid G \in L\}$ is finite. Hence the methods used for Propositions 4.16 and 4.19 give the following results:

Proposition 4.47

(1) For every VR-equational set L, the set $\{|G| \mid G \in L\}$ is semi-linear. A description of it can be computed from a system that defines L.
(2) There is an algorithm that decides the finiteness of a VR-equational set L given by an equation system S; if L is finite, it can be computed from S.
(3) If L is finite and defined by a uniform and typable equation system S, then $|G| \leq 2^{n-1}$ for every G in L, and the cardinality of L is at most $(|F(S)| \cdot 2^{a^2 \cdot \ell} + 1)^{2^n}$, where $a := |\mathcal{A}(S)|$, ℓ is the number of edge labels and n is the number of unknowns of S.

Proof: (1) The proof is analogous to the one of Proposition 4.16.

(2) The proof is analogous to the one of the first statement of Proposition 4.19.

(3) The proof is similar to the one of the last statement of Proposition 4.19, but in the Claim it uses Proposition 2.103 instead of Proposition 2.51. Moreover, the Claim is now the following:

Claim 4.47.1 For every p-graph G in L there exists a term $t \in K$ of height at most $n - 1$ such that $|val(t)| = |G|$.

Proof: Using the same terminology as in the proof of Claim 4.19.1, we obtain that $c'_{\mathbb{GP}}(G') = relab_h(ADD_R(G' \oplus H))$ and that H must be empty, and hence $c'_{\mathbb{GP}}(val(t')) = ADD_R(val(t'))$ because $c'[t']$ and t' have the same type. Hence $|val(c[t'])| = |G|$ by the equations for $|\cdot|$ ($|G \oplus H| = |G| + |H|$ etc.) stated above. □

By this claim, the size of G is at most 2^{n-1} by the same argument as in the proof of Proposition 4.42. The upper-bound for $|L|$ is obtained in the same way as in the proof of Proposition 4.19, taking into account that, according to the proof of the claim, G can be obtained from t by adding an operation of the form ADD_R at each node of t and then applying val. Note that $R \subseteq \mathcal{A}(S) \times \mathcal{A}(S) \times \Lambda$ where Λ is the set of edge labels. ∎

For a VR-equational set L, the set $\|L\|$ (cf. Definition 4.15) need not be semi-linear: just consider for L the set of cliques.

4.3.4 Comparison with the HR-equational sets

VR-equational sets only contain simple (labeled) graphs (and p-graphs), whereas HR-equational sets may contain (labeled) graphs (and s-graphs) with multiple edges. We will compare the families of VR-equational sets and of HR-equational sets of simple (labeled) graphs. The following proposition shows that one can distinguish in an effective way the HR equation systems that define simple graphs.

Proposition 4.48 There exists an algorithm that, for every HR equation system S and every unknown x of S, decides if $L(S,x)$ consists only of simple s-graphs, and, if this is not the case, that transforms S into an HR equation system S' such that $L(S',x)$ is the set of simple s-graphs in $L(S,x)$.

Proof: It is not difficult to prove that the sets of simple, directed (or undirected), unlabeled (or (K,Λ)-labeled) s-graphs are effectively HR-recognizable, by constructing congruences as in Examples 4.30 (for the details, see the proof of Theorem 4.58). The existence of the desired algorithms follows then from Proposition 3.91(2) and the Filtering Theorem (Theorem 4.37). ∎

Here is the main theorem of this section.

Theorem 4.49 Every HR-equational set L of simple (labeled) graphs is VR-equational. A VR equation system defining L can be constructed from an HR equation system defining it. ☐

For doing the proof, we introduce some technical notions that will also be useful for comparing HR- and VR-recognizable sets of simple (labeled) graphs.

Definition 4.50 (The border and the interior of an s-graph) We first give definitions for directed unlabeled s-graphs. The set of simple directed s-graphs is denoted by $\mathcal{JS}^s$. For an s-graph G and $X \subseteq V_G$, we denote by $G[X]$ the s-graph $\langle G^\circ[X], slab_G \upharpoonright X \rangle$. In words, $G[X]$ is the induced subgraph of G with vertex set X.

The *border* of an s-graph $G \in \mathcal{JS}^s$ is the s-graph $\beta(G) := G[Src(G)]$ (it is also in $\mathcal{JS}^s$). Clearly we have $\tau(\beta(G)) = \tau(G)$. Note that $\beta(G)$ is also a p-graph, with $port_G = slab_G$.

We define also the *interior* of G as the p-graph $Int(G) \in \mathcal{GP}$ that consists of $G^\circ[Int_G]$ (we recall that Int_G is $V_G - Src(G)$, the set of internal vertices of G) equipped with the following mapping $port_{Int(G)}$:

$$port_{Int(G)}(u) := \langle A, A' \rangle, \text{ where}$$

$$A := \{a \in \tau(G) \mid \text{ there is in } G \text{ an edge } u \to src_G(a)\},$$

$$A' := \{a \in \tau(G) \mid \text{ there is in } G \text{ an edge } src_G(a) \to u\}.$$

Hence $\pi(Int(G))$ is a subset of $\mathcal{P}(\tau(G)) \times \mathcal{P}(\tau(G))$.[21] If $|\tau(G)| \le N$, then $|\tau(\beta(G)) \cup \pi(Int(G))| \le N + 2^{2N}$. Note that $\beta(G) = G$ and $Int(G) = \varnothing$ if $V_G = Src(G)$. Moreover, if G is a graph, i.e., $\tau(G) = \emptyset$, then $\beta(G) = \varnothing$ and $Int(G) = G$.[22]

Note that an s-graph G can be reconstructed from $\beta(G)$ and $Int(G)$, because we have

$$G = \text{RELAB}(\text{ADD}(\beta(G) \oplus Int(G))),$$

where ADD and RELAB are derived operations (respectively compositions of edge additions and of relabelings) that are easy to define.

The validity of the following facts, intended to show that β and Int are "almost" homomorphisms, is clear from the definitions.

Let $G, H \in \mathcal{JS}^s$. If $G \,/\!/\, H$ is simple, then

$$\beta(G \,/\!/\, H) = \beta(G) \,/\!/\, \beta(H) \text{ and } Int(G \,/\!/\, H) = Int(G) \oplus Int(H). \qquad (4.5)$$

If h is a finite permutation of $\mathcal{A}$, then

$$\beta(ren_h(G)) = ren_h(\beta(G)) \text{ and } Int(ren_h(G)) = \text{RELAB}_h(Int(G)), \qquad (4.6)$$

where $\text{RELAB}_h := relab_g$ such that $g(\langle A, A' \rangle) = \langle h(A), h(A') \rangle$ for all $A, A' \subseteq \tau(\beta(G))$. Next we consider the source forgetting operations:

$$\beta(fg_B(G)) = \beta(G)[Src(G) - src_G(B)] \text{ and}$$

$$Int(fg_B(G)) = \text{RELAB}_B(\text{ADD}_{R(B)}(Int(G) \oplus \beta(G)[src_G(B)])), \qquad (4.7)$$

where $src_G(B) := \{src_G(b) \mid b \in B \cap \tau(G)\}$, $\text{ADD}_{R(B)}$ is as defined before Proposition 2.103 with

$$R(B) := \bigcup_{A, A' \subseteq \tau(\beta(G)), b \in B} \{(\langle A, A' \rangle, b) \mid b \in A\} \cup \{(b, \langle A, A' \rangle) \mid b \in A'\},$$

and $\text{RELAB}_B := relab_k$ such that, for $A, A' \subseteq \tau(\beta(G))$ and $b \in B$,

$$k(\langle A, A' \rangle) := \langle A - B, A' - B \rangle \text{ and } k(b) := \langle A_b, A'_b \rangle \text{ with}$$

$$A_b := \{a \in \tau(G) - B \mid src_G(b) \to src_G(a) \text{ is an edge in } \beta(G)\},$$

$$A'_b := \{a \in \tau(G) - B \mid src_G(a) \to src_G(b) \text{ is an edge in } \beta(G)\}.$$

---

[21] As in the proof of Proposition 2.103, we assume that the elements of this set are appropriately encoded as port labels in $\mathcal{A}$ that are not in $\tau(G)$. Formally this can be achieved by an injective computable mapping $\alpha : \mathcal{A} \cup (\mathcal{P}_f(\mathcal{A}) \times \mathcal{P}_f(\mathcal{A})) \to \mathcal{A}$. Then $port_{Int(G)}(u)$ is formally defined as $\alpha(\langle A, A' \rangle)$ with A, A' as above. When $\beta(G)$ is viewed as a p-graph, we can assume formally that its source names are encoded by α (and hence can be distinguished from the encoded port labels of $Int(G)$). Finally, we assume that $\alpha(\langle \emptyset, \emptyset \rangle) = 1$, the default port label.

[22] We recall that $\varnothing$ is a constant symbol denoting the empty graph.

It should be noted that $RELAB_h$, $ADD_{R(B)}$ and $RELAB_B$ appear to depend on G, but actually, they only depend on $\beta(G)$.

Finally, if G is defined by a constant symbol then

$$\beta(G) = G \text{ and } Int(G) = \varnothing. \tag{4.8}$$

For dealing with undirected graphs, just one set A is needed instead of two sets A and A' (since every undirected edge is a pair of opposite directed edges, the corresponding sets A and A' are equal). The derived operation $ADD_{R(B)}$ can use operations $add_{A,b}$ instead of $\overrightarrow{add}_{\langle A,A'\rangle,b}$ and $\overrightarrow{add}_{b,\langle A,A'\rangle}$. If $|\tau(G)| \leq N$, then $|\tau(\beta(G)) \cup \pi(Int(G))| \leq N + 2^N$. We will explain later the modifications required to handle labeled graphs.

We illustrate these definitions and facts with an example concerning undirected graphs. Figure 4.5 shows undirected s-graphs G, H. One can check on this example that $\beta(G \parallel H) = \beta(G) \parallel \beta(H)$ and $Int(G \parallel H) = Int(G) \oplus Int(H)$ and that

$$Int(fg_3(G)) = relab_{3 \rightarrow \{2\}}(relab_{\{2,3\} \rightarrow \{2\}}(add_{\{2,3\},3}(Int(G) \oplus 3))).$$

Proof of Theorem 4.49: Let L be an HR-equational set of simple directed graphs, given as $L(S,X)$ where S is a uniform HR equation system (cf. Proposition 3.33) and $X \subseteq Unk(S)$. By Proposition 3.37 we may also assume that S is trim with all unknowns useful for X. This implies that all s-graphs defined by S are simple: consider G in $L(S,y)$ and a derivation sequence of $G(S)$ from some x in X to a term t in $T(F^{HR}, \{y\})$, where y has a unique occurrence; then $t_{\mathbb{JS}}(G) \in L(S,x) \subseteq L$. Since the operations of F^{HR} preserve multiple edges and L consists only of simple graphs, G must be simple.

The mapping β behaves "homomorphically" (by (4.5), (4.6) and (4.7) of Definition 4.50) and there are only finitely many possible graphs $\beta(G)$ (up to isomorphism) such that $\tau(G) = \tau(\beta(G)) \subseteq \mathcal{A}(S)$. Hence, we can assume that $\beta(G) = \beta(G')$ for every $G, G' \in L(S,y)$ and every $y \in Unk(S)$ because, by using Corollary 3.89 of the Filtering Theorem, we can transform S into an equivalent system satisfying this property if this is not already the case.[23] We denote by $\beta(y)$ the common value $\beta(G)$ of the s-graphs G in $L(S,y)$.

The next step is a transformation of S into a VR equation system S^{Int} such that $Unk(S^{Int}) := Unk(S)$ and $L(S^{Int},y) = \{Int(G) \mid G \in L(S,y)\}$ for every $y \in Unk(S)$. Since L is a set of graphs, $Int(G) = G$ for every $G \in L$ and so L equals $L(S^{Int},X)$ and hence is VR-equational.

It remains to explain how S^{Int} is obtained from S. The general equation of S is of the form $x = \cdots \cup m \cup \cdots$, where each m is a monomial with only one function

---

[23] The mapping h defined by $h(G) = \beta(G)$ if G is simple and $h(G) = \perp$ otherwise, is a homomorphism from $\mathbb{JS}[\mathcal{A}(S)]$ to a finite algebra with domain $\{G \in \mathcal{JS}^s \mid \tau(G) \subseteq \mathcal{A}(S), Int_G = \emptyset\} \cup \{\perp\}$, cf. the proof of Proposition 4.13(3).

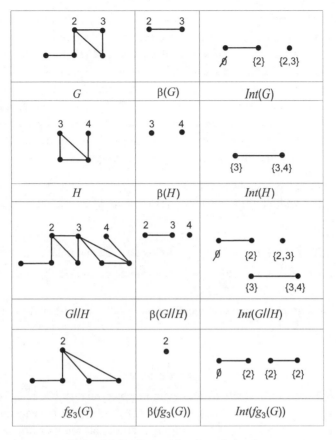

Figure 4.5 Borders and interiors of s-graphs.

symbol because S is uniform. We define the corresponding equation of S^{Int} as $x^{Int} = \cdots \cup m^{Int} \cup \cdots$, where m^{Int} is as follows:

if $m = y \parallel z$, then $m^{Int} := y \oplus z$,

if $m = ren_h(y)$, then $m^{Int} := \mathrm{RELAB}_h(y)$,

if $m = fg_B(y)$, then $m^{Int} := \mathrm{RELAB}_B(\mathrm{ADD}_{R(B)}(y \oplus t_B))$,

if m is defined by a constant symbol, then $m^{Int} := \varnothing$.

In this definition, the operations RELAB_h, RELAB_B and $\mathrm{ADD}_{R(B)}$ are as in Definition 4.50 (with $\beta(y)$ instead of $\beta(G)$), and t_B is a term in $T^{\mathrm{VR}}_{\tau(\beta(y)) \cup B}$ denoting the p-graph $\beta(y)[src_{\beta(y)}(B)]$. This construction concludes the proof. Its correctness is based on Equalities (4.5)–(4.8).

If $N = |\mathcal{A}(S)|$, then this construction defines a system S^{Int} with $|\mathcal{A}(S^{Int})| \leq N + 2^{2N}$. The same construction works for undirected graphs but uses at most $N + 2^N$ port labels as noted above. We now consider the case of (K, Λ)-labeled graphs. The labels of vertices and loops are specified by the constant symbols that define graphs with a unique vertex. These richer constants are handled as in the unlabeled case, see (4.8). For each edge label, we need one pair of sets $\langle A, A' \rangle$, where A and A' are as in the case of directed unlabeled graphs. Hence, for a set Λ of edge labels, we need at most $N + 2^{2|\Lambda| \cdot N}$ port labels for directed graphs and at most $N + 2^{|\Lambda| \cdot N}$ for undirected ones. The above construction is easy to adapt.   ∎

As a consequence, we obtain a bound for the clique-width of a simple graph in terms of its tree-width (cf. Proposition 2.114). By taking $N = k + 1$, we get for G of tree-width at most k:

$$cwd(G) \leq k + 1 + 4^{k+1} \qquad \text{if } G \text{ is directed and unlabeled, and}$$
$$cwd(G) \leq k + 1 + 4^{|\Lambda| \cdot (k+1)} \qquad \text{if } G \text{ is directed and } (K, \Lambda)\text{-labeled.}$$

For undirected graphs, we get respectively the smaller upper-bounds $k + 1 + 2^{k+1}$ and $k + 1 + 2^{|\Lambda| \cdot (k+1)}$.

Sparse graphs

Although VR equation systems are strictly more powerful than HR equation systems (this is proved by Examples 4.43(1), 4.43(2) and 4.43(5)), they are equally powerful for generating languages, sets of trees (through appropriate representations as labeled graphs) and sets of planar graphs. We now present a necessary and sufficient condition that subsumes these cases. Let $p \in \mathcal{N}$. A graph G is *p-sparse* if $|E_G| \leq p \cdot |V_G|$. A set of graphs is *p-sparse* if all its graphs are p-sparse. It is *uniformly p-sparse* if all subgraphs of its graphs are p-sparse.

Recall from Proposition 2.115 that an unlabeled graph G is without $K_{n,n}$ if $core(G)$ has no subgraph isomorphic to $K_{n,n}$; if G is labeled, we say that it is without $K_{n,n}$ if G_λ (the subgraph of G consisting of its edges labeled by λ) is without $K_{n,n}$ for every edge label λ.

Theorem 4.51 Let L be a VR-equational set of (labeled) graphs. The following conditions are equivalent:

(1) L is HR-equational;
(2) L has bounded tree-width;
(3) L is uniformly p-sparse for some $p \in \mathcal{N}$;
(4) L is p-sparse for some $p \in \mathcal{N}$;
(5) there exists $n \in \mathcal{N}$ such that every graph in L is without $K_{n,n}$.

Proof: We have the following implications:

(1) $\Longrightarrow$ (2) by Proposition 4.7.

(2) $\Longrightarrow$ (3): The set of subgraphs of graphs in L has bounded tree-width by Corollary 2.60(1). Every simple loop-free undirected graph G of tree-width at most k is k-sparse by Corollary 2.74(1) (for $|V_G| \geq k+1$; if $|V_G| \leq k$, then $|E_G| \leq |V_G| \cdot (|V_G|-1)/2 < k \cdot |V_G|$). Hence, a simple directed graph with edge labels from a set of at most ℓ labels (and possibly labeled loops), is $(2k+1) \cdot \ell$-sparse.

(3) $\Longrightarrow$ (1) follows from the Equationality Theorems for the VR and the HR algebra proved in Sections 7.2 and 7.4, and the Sparseness Theorem for graphs in Section 9.4.1. See Section 9.4.3 for the proof.

(3) $\Longrightarrow$ (4) is immediate.

(5) $\Longrightarrow$ (2) follows from Proposition 4.44, because for G with edge labels from a set of at most ℓ labels and without $K_{n,n}$, we have $twd(G) \leq 3(n-1) \cdot \ell \cdot cwd(G) - 1$ by Proposition 2.115.

(4) $\Longrightarrow$ (5): The proof is based on the following claim.

Claim 4.51.1 For every VR-equational set L of (labeled) graphs there exists $d \in \mathcal{N}$ such that the following holds:

if H is an induced subgraph of some graph G in L, then there exists a graph $\widetilde{G} \in L$ such that $|V_{\widetilde{G}}| \leq d \cdot |V_H|$ and H is isomorphic to an induced subgraph of $\widetilde{G}$.

Proof: Let L be a VR-equational set of graphs. We write the proof for unlabeled directed graphs, but it is essentially the same for labeled and/or undirected graphs. By Proposition 4.41, $L = val(K)$ for some regular language $K \subseteq T(F(S))$, where S is a VR equation system and $F(S) \subseteq F_{\mathcal{A}(S)}^{\mathrm{VR}}$. We let $\mathscr{A} = \mathscr{A}(\mathbb{B}, C)$ be a complete deterministic finite $F(S)$-automaton recognizing K (cf. Definition 3.58 and Theorem 3.62).

For a context $c \in Ctxt(F(S))$ we define $c_\varnothing$ to be the context obtained from c by changing all constant symbols in c into $\varnothing$. By Equalities (3), (6) and (11) of Proposition 2.101 and by (the last sentence of) Proposition 2.103, every derived operation $(c_\varnothing)_{\mathbb{GP}}$ is of the form $G \mapsto relab_h(ADD_R(G))$, and hence there are $k^k \cdot 2^{k^2-k}$ such operations, where $k = |\mathcal{A}(S)|$ (cf. Lemma 2.120). We define two contexts $c, c' \in Ctxt(F(S))$ to be *equivalent* if $c_{\mathbb{B}} = c'_{\mathbb{B}}$ and $(c_\varnothing)_{\mathbb{GP}} = (c'_\varnothing)_{\mathbb{GP}}$. This equivalence relation has at most $\sharp\mathscr{A}^{\sharp\mathscr{A}} \cdot k^k \cdot 2^{k^2-k}$ classes. For every context c, we let $\overline{c}$ be a context equivalent to c with a minimal number of leaves. Moreover, we let b be the maximal number of leaves of all contexts $\overline{c}$. Finally, we define the required constant d as $2b+1$. Note that d depends on $\mathcal{A}(S)$ and $\mathscr{A}$ only.

Let $G \in L$. Without loss of generality, we assume that $G = cval(t)$ for some $t \in K$. Its vertex set is thus $Occ_0(t)$ (see Section 2.5.2 for the definitions of $cval(t)$ and $Occ_0(t)$). Let $H = G[X]$, where $X \subseteq Occ_0(t)$, and let $n := |V_H| = |X|$. Our aim is to transform t into a term $\widetilde{t} \in K$ such that H is isomorphic to an induced subgraph of $val(\widetilde{t})$ and $\widetilde{t}$ has at most $d \cdot n$ leaves.

Consider nodes u and v of t such that u is strictly below v, and such that for every $x \in X$, either x is below u (we may have $x = u$) or the least common ancestor of u and x is strictly above v. We will say that (u, v) is an *X-free pair* of nodes. Let c be the context such that $t = (t \uparrow v)[c[t/u]]$. Informally, c is the part of the term t "between" u and v, and none of its leaves is in X. We denote c by $c_{[u,v]}$ if we need to indicate u and v.

Suppose that c' is a context equivalent to c, and that t' is obtained from t by changing c into c', i.e., $t' := (t \uparrow v)[c'[t/u]]$. Since $c'_{\mathbb{B}} = c_{\mathbb{B}}$, the term t' is accepted by $\mathscr{A}$.[24] Hence $t' \in K$ and $val(t') \in L$. By the proof of Proposition 2.105(1), $H = cval(t)[X] = cval(t_X)$ where the term t_X is obtained from t by changing the label of every leaf of t that is not in X into $\varnothing$. Since, by assumption, the leaves of c are not in X and $(c_\varnothing)_{\mathrm{GP}} = (c'_\varnothing)_{\mathrm{GP}}$, this implies that H is isomorphic to an induced subgraph of $val(t')$. (To be precise, we have that $t_X = (t_X \uparrow v)[c_\varnothing[t_X/u]]$ because the leaves of c are not in X. Let $s := (t_X \uparrow v)[c'_\varnothing[t_X/u]]$. Then $val(t_X) = val(s)$ because $(c_\varnothing)_{\mathrm{GP}} = (c'_\varnothing)_{\mathrm{GP}}$. Let $X' := Occ_0(s)$. Then $s = t'_{X'}$ (which is defined in the same way as t_X). Again by the proof of Proposition 2.105(1), $cval(t')[X'] = cval(t'_{X'})$ and so $cval(s)$ is an induced subgraph of $cval(t')$. Hence, $H = val(t_X) = val(s)$ is isomorphic to an induced subgraph of $val(t')$.)

In particular, if the number of leaves of c is larger than b, then we can change c into $\bar{c}$, which has at most b leaves. Since the resulting term has fewer leaves than t, repetition of this transformation (as long as possible) leads to a term $\tilde{t} \in K$ such that H is isomorphic to a subgraph of $val(\tilde{t}) \in L$ and the number of leaves of $c_{[u,v]}$ is at most b for every X-free pair (u, v) of nodes of $\tilde{t}$, where X is any subset of $Occ_0(\tilde{t})$ such that H is isomorphic to $cval(\tilde{t})[X]$.

We let Y be the set of nodes of $\tilde{t}$ consisting of X and the least common ancestors of any two leaves in X. Then, we consider the pairs (u, v) such that $u \in Y$ and v is a maximal proper ancestor of u not in Y (thus, no node of Y is on the path from v to the father of u). There are at most $|Y| = 2n - 1$ such pairs. Each of them is X-free and so the number of leaves of $c_{[u,v]}$ is at most b. Hence $\tilde{t}$ has at most $n + (2n - 1)b < (2b + 1)n = dn$ leaves.

Thus, if $\tilde{G} := cval(\tilde{t})$, then $\tilde{G} \in L$, H is isomorphic to an induced subgraph of $\tilde{G}$, and $|V_{\tilde{G}}| = |Occ_0(\tilde{t})| \leq d \cdot |V_H|$. (We do not claim that $\tilde{G}$ is isomorphic to an induced subgraph of G.) $\qquad\square$

This claim implies that if a graph G of L has a subgraph isomorphic to $K_{n,n}$, then there exists $\tilde{G} \in L$ such that $|V_{\tilde{G}}| \leq 2dn$ and $|E_{\tilde{G}}| \geq n^2$. Hence, if there is such a graph for every n, then L is not p-sparse for any p, and (4) implies (5). $\qquad\blacksquare$

The Semi-Linearity Theorem of Chapter 7 (Theorem 7.42) implies that Condition (5) is decidable. So one can decide if a VR-equational set of graphs is HR-equational.

---

[24] By Lemma 3.59, because $val_{\mathbb{B}}(t) = (t \uparrow v)_{\mathbb{B}}(c_{\mathbb{B}}(val_{\mathbb{B}}(t/u)))$ and similarly for t' and c'.

The class of graphs of degree at most $2p$ is p-sparse. The class of simple graphs embeddable in the plane or in a fixed surface is p-sparse for some p (with $p = 3$ for loop-free planar graphs, see [*Die], Corollary 4.2.10). The same holds more generally for the class of graphs that do not contain a fixed graph as a minor. We will prove in Section 9.4.3 that if a set of graphs of bounded clique-width is uniformly p-sparse for some p, then it has bounded tree-width. In Theorem 4.51 we need only assume that L is p-sparse for some p in order to obtain, not only that it has bounded tree-width but, furthermore, that it is HR-equational.

4.4 VR-recognizable sets of simple graphs

The VR-recognizable sets of graphs are defined as the recognizable sets of the many-sorted algebra $\mathbb{GP}^t$. The following definition is similar to Definition 4.29 for HR-recognizable sets.

4.4.1 Definitions and examples

Definition 4.52 (VR-recognizable sets of p-graphs) A set of p-graphs is *VR-recognizable* if it is recognizable in the many-sorted algebra $\mathbb{GP}^t$, i.e., if it is a union of classes, possibly of different sorts, of a type-preserving and locally finite congruence on the one-sorted F^{VR}-algebra $\mathbb{GP}$. The sort of a p-graph G relative to $\mathbb{GP}^t$ is its type $\pi(G)$, defined as the set of its port labels.

For (K, Λ)-labeled p-graphs, the definitions are the same with $F^{VR}_{[K,\Lambda]}$ instead of F^{VR} and $\mathbb{GP}^t_{[K,\Lambda]}$ instead of $\mathbb{GP}^t$.

If $L \subseteq \mathcal{GP}[E]$ for some finite set E of port labels, that is, such that all p-graphs in L have a type included in E, then its recognizability can be considered with respect to $\mathbb{GP}^t$ or to $\mathbb{GP}^t[E]$. If L is VR-recognizable, then it is also recognizable in $\mathbb{GP}^t[E]$ by Proposition 3.56(1). If, furthermore, $L \subseteq \mathcal{GP}^{gen}[E]$, then it is also recognizable in $\mathbb{GP}^{t,gen}[E]$ by the same proposition.

It is straightforward to show, as in Definition 4.29, that the algebras $\mathbb{GP}^{t,gen}[E]$ and $\mathbb{GP}^{gen}[E]$ have the same (effectively) recognizable sets.

The *VR-recognizability index* of a set of p-graphs L is the mapping γ_L such that $\gamma_L(k)$, for $k \in \mathcal{N}$, is the number of equivalence classes of type C of the syntactic congruence $\approx^L$ of L with respect to $\mathbb{GP}^t$, where C is any set of port labels of cardinality k. By using bijective relabelings (like in Definition 4.33), we can prove that this number is well defined.

Many results are similar to the corresponding ones for HR-recognizable sets. In particular, the instantiation to the VR algebra of the algebraic version of the Filtering Theorem:

Theorem 4.53 The intersection $L \cap K$ of a VR-equational set L and a VR-recognizable set K is VR-equational. If K is semi-effectively VR-recognizable and L is given by a VR equation system, then an equation system defining $L \cap K$ can be constructed. $\square$

Its proof is fully similar to that of Theorem 4.37 relative to the HR algebra (using Proposition 4.46 instead of Proposition 4.13).

We now give some examples of VR-recognizable sets and define congruences that witness their recognizability.

Example 4.54 (1) **Graphs with an even number of vertices:** The set L_v of simple graphs having an even number of vertices is VR-recognizable. For proving this we define the type-preserving equivalence relation $\sim$ by: $G \sim H$ if and only if G and H have the same type and $|V_G|$ and $|V_H|$ have the same parity, i.e., $|V_G| - |V_H|$ is even. The verification that $\sim$ is a congruence on $\mathbb{GP}$ witnessing the VR-recognizability of L_v is as for Example 4.30(1).

(2) **Graphs with an even number of edges:** We now prove that the set L_e of simple graphs having an even number of directed edges is VR-recognizable. The construction of a type-preserving congruence that saturates L_e is slightly more complicated than in the previous example. For distinct port labels a and b and a p-graph G, we let $edg_G(a,b)$ be the set of edges from an a-port to a b-port. (We recall that $port_G$ is the port mapping of G, from vertices to port labels). We define a type-preserving equivalence relation $\equiv$ by:

$G \equiv H$ if and only if G and H have the same type and:

(1) $|edg_G|$ and $|edg_H|$ have the same parity;
(2) $|edg_G(a,b)|$ and $|edg_H(a,b)|$ have the same parity for every two distinct port labels a and b;
(3) $|port_G^{-1}(a)|$ and $|port_H^{-1}(a)|$ have the same parity for every port label a.

This equivalence is locally finite and it saturates L_e. The verification that it is a congruence uses observations similar to those used in Example 4.30(1). Here are some others, involving $|edg_G(a,b)|$ and $|port_H^{-1}(a)|$.

If $H = \overrightarrow{add}_{a,b}(G)$ then $|edg_H| = |edg_G| + |port_G^{-1}(a)| \cdot |port_G^{-1}(b)| - |edg_G(a,b)|$. It follows that the parity of $|edg_H|$ is a function of the parities of the cardinalities of the right-hand side of this equality. (This equality has motivated the introduction of the sets $edg_G(a,b)$.) We also have: $|edg_H(a,b)| = |port_G^{-1}(a)| \cdot |port_G^{-1}(b)|$ and $|edg_H(c,d)| = |edg_G(c,d)|$ for $(c,d) \neq (a,b)$.

If $H = relab_h(G)$, then $|edg_H(a,b)|$ is the sum of all $|edg_H(c,d)|$ such that $h(c) = a$ and $h(d) = b$.

The set L_e is thus VR-recognizable. This fact follows also from the Recognizability Theorem (Section 5.3.9, Theorem 5.68) because a graph has an even number of directed edges if and only if the number of its vertices having an odd indegree is even.

This property is thus C_2MS-expressible. (For a generalization of this observation, see [Cou10].)

(3) **Planar graphs and other MS-definable sets:** The sets of simple planar graphs, of cographs, of trees, of rooted trees are VR-recognizable: this follows from the Recognizability Theorem to be proved in the next chapter. In fact, for simple graphs, the sets (1)–(8) of Proposition 4.31 are all (effectively) VR-recognizable.

(4) **Connected graphs:** It is not difficult to define a congruence that witnesses the VR-recognizability of the set of connected simple graphs and that has a bit less than 2^{2^k} classes of type C, where C is a set of k port labels. We will prove that no such congruence can have less than 2^{2^k} classes of type C if C has $2k + 1$ elements. This gives a lower-bound to the VR-recognizability index of the set of connected graphs. Our proof only uses undirected graphs, but the lower-bound applies *a fortiori* to all (simple) graphs.

We let $C := [0, 2k]$ for $k > 0$. For every nonempty subset A of C we define

$$t_A := (\bigcirc_{a,b \in A, a \neq b} add_{a,b})(\bigoplus_{a \in A} \mathbf{a}).$$

The p-graph $val(t_A)$ is a clique of type A and each a in A is the port label of a unique vertex.

Let Z be a nonempty subset of $\mathcal{P}_k([2k])$, defined as the set of subsets of $[2k]$ of cardinality k. We let $\bigcup Z$ be the union of the sets in Z and we define

$$s_Z := t_{C - \bigcup Z} \oplus \bigoplus_{A \in Z} t_A.$$

The p-graph $H_Z := val(s_Z)$ has type C and is not connected (0 belongs to no set in Z and Z is not empty).

We now define some contexts intended to show that distinct p-graphs H_Z and $H_{Z'}$ are not equivalent for any congruence $\sim$ that witnesses the VR-recognizability of the set of connected p-graphs. For every nonempty subset B of $[2k]$, we define

$$c_B := (\bigcirc_{b \in B} add_{0,b})(w),$$

where w is the special variable used to define contexts. It is clear that $val(c_B[s_Z])$ is connected if and only if $B \cap A \neq \emptyset$ for every $A \in Z$.

We let $\mathcal{Z}$ be the set of all nonempty subsets of $\mathcal{P}_k([2k])$. It is thus the set of all sets Z as defined above. We have $|\mathcal{P}_k([2k])| > 2^k$ hence $|\mathcal{Z}| > 2^{2^k}$, by routine computations.

Let us now consider two distinct elements Z and Z' of $\mathcal{Z}$. We first assume that some set A belongs to $Z - Z'$ and we let $B := [2k] - A$. Then $val(c_B[s_Z])$ is not connected because $B \cap A = \emptyset$. On the other hand, $val(c_B[s_{Z'}])$ is connected because $B \cap A' \neq \emptyset$ for every $A' \in Z'$. In fact, since all sets in Z and Z' have cardinality k, each set A' in Z' must contain an element not in A (otherwise $A' \subseteq A$ so that $A' = A$, but we have assumed that $A \notin Z'$). For a set $A \in Z' - Z$ we get a similar fact by exchanging the roles of Z and Z'.

It follows that the p-graphs H_Z for $Z \in \mathcal{Z}$ are all of type C and are pairwise inequivalent with respect to any congruence that witnesses the VR-recognizability of the set of connected p-graphs.

What about the set of connected graphs without ports, equivalently of type $\{0\}$? By replacing c_B by $(\bigcirc_{b \in [2k]} relab_{b \to 0})(c_B)$, we get the same result for the VR-recognizability of connected graphs. Hence, in both cases we have $\gamma_L(2k+1) > 2^{2^k}$ for every $k > 0$.

Hence, we have a double-exponential lower-bound for the VR-recognizability index of the set of connected graphs. In Example 4.30(2), we have given the much lower upper-bound $k! + 1$ for the HR-recognizability index of the same set (allowing multiple edges, but they do not matter).

(5) **Sets that are not VR-recognizable:** The set of graphs $K_{n,n}$ for all $n > 0$ is not VR-recognizable. The proof uses the same tools as those of Proposition 4.32: we let the constant symbols a and b denote $\mathbf{1}$, the unary operation g denote $\lambda x \cdot (x \oplus \mathbf{1})$ and f denote the complete join $\otimes$. Note that this set is HR-recognizable (Proposition 4.36). Note also that this set is VR-equational, with the following equation system:

$$\begin{cases} x = relab_{2 \to 1}(add_{1,2}(y)), \\ y = (\mathbf{1} \oplus \mathbf{2}) \cup (y \oplus y), \end{cases}$$

but that it is not HR-equational (Proposition 4.7). $\qquad\square$

By a proof similar to that of Example 4.30(3), one can prove that each singleton set consisting of a p-graph (and hence each finite set of p-graphs) is effectively VR-recognizable.

Remark 4.55 (On variants of the VR operations) We observed at the end of Section 4.3.1 that the variants of the signature F^{VR} defined in Section 2.5.6 yield the same classes of equational sets. They also yield the same classes of recognizable sets. For the signatures $F^{\kappa VR}$ and F'^{VR} it is proved in the same way as at the end of Section 4.3.1, using Theorem 3.56(3). For the signature F^{iVR}, this follows from a result of [CouWei] (Theorem 4.5) saying that if we add to the signature F^{tVR} unary operations on p-graphs defined by quantifier-free formulas (we will define these operations in Section 5.3.2), then the class of recognizable sets of the obtained algebra is still the class of VR-recognizable sets. This result applies to F^{iVR}, and to its many-sorted version. $\qquad\square$

4.4.2 A simpler VR-recognizability criterium

We will give a characterization of VR-recognizability analogous to that of HR-recognizability of Theorem 4.34 that extends the characterization of the regular languages due to Myhill and Nerode. We define an operation on p-graphs similar to the operation $\square_k$ on s-graphs, but not commutative. This operation is derived from Proposition 2.103 that characterizes the operations on p-graphs defined by contexts.

Definition 4.56 (Some binary derived operations) Let $(A_1, A_1'), (A_2, A_2'), \ldots$ be an enumeration of the set $\mathcal{P}_f(\mathcal{N}) \times \mathcal{P}_f(\mathcal{N})$ such that for every $k \in \mathcal{N}$ the sequence $(A_1, A_1'), \ldots, (A_{2^{2k}}, A_{2^{2k}}')$ is an enumeration of $\mathcal{P}([k]) \times \mathcal{P}([k])$. If G and H are p-graphs of types respectively included in $[k]$ and in $[k + 1, k + 2^{2k}]$, we define

$$G \ltimes_k H := relab_1(ADD_{R_k}(G \oplus H)),$$

where $relab_1$ is the relabeling that transforms every port label in $[k + 2^{2k}]$ into the default port label 1,

$$R_k := \{(k+j, i) \mid j \in [2^{2k}], i \in A_j\} \cup \{(i, k+j) \mid j \in [2^{2k}], i \in A_j'\},$$

and ADD_R is defined before Proposition 2.103. The asymmetric symbol $\ltimes_k$ represents the fact that the right argument can have more port labels than the left one. Here is the reason. If a context takes a p-graph G such that $\pi(G) \subseteq [k]$ as argument, then it may add new vertices to G, and edges between such a new vertex x and those of G. For each such x, these edges depend only on the port labels of the vertices of G. Hence, the edge additions defined by the considered context can be specified by a pair (A, A') of subsets of $[k]$ attached to each such vertex x (as shown in the proof of Proposition 2.103). Instead of using pairs of subsets of $[k]$ as port labels, we use integers in $[k + 1, k + 2^{2k}]$. Informally, in an expression $G \ltimes_k H$, the argument H represents a context taking G as argument. This explains the asymmetry of the operation $\ltimes_k$. (The symmetry of $\square_k$ is thus a particular property of the HR algebra).

Our aim is to get a criterium for VR-recognizability similar to the one of Theorem 4.34 for HR-recognizability. We define, for a set L of simple graphs and $k \in \mathcal{N}$, an equivalence relation on p-graphs of type $[k]$ as follows:

$G \sim_{L,k} G'$ if and only if for every p-graph H of type included in $[k + 1, k + 2^{2k}]$, we have

$$G \ltimes_k H \in L \iff G' \ltimes_k H \in L.$$

Theorem 4.57 A set L of simple graphs is VR-recognizable if and only if the equivalences $\sim_{L,k}$ are finite for all k.

Proof: For convenience we write $\sim_k$ instead of $\sim_{L,k}$. The equivalences $\sim_k$ are defined as the syntactic congruence $\approx^L$, but in terms of particular contexts. Hence, for every two p-graphs G and G' of type $[k]$, $G \approx^L G'$ implies $G \sim_k G'$. Hence, if L is VR-recognizable, the equivalences $\sim_k$ are all finite.

For proving the converse, we first define $\sim'_k$ as the equivalence relation on graphs of type $[k]$ such that

$$G \sim'_k G' \text{ if and only if } ADD_{R_1}(G) \sim_k ADD_{R_1}(G') \text{ for every } R_1 \subseteq [k] \times [k].$$

Then $\sim'_k$ is finite if $\sim_k$ is finite. It has at most $p \cdot 2^{k^2}$ classes if $\sim_k$ has p classes.

Let L be a set of simple graphs such that the equivalences $\sim_k$, whence also the equivalences $\sim'_k$ are finite. Let B be a finite set of k port labels and c be a context in $Ctxt(F^{VR})$ such that $c(G)$ has type $\{1\}$ for every p-graph G of type B. If follows from Proposition 2.103 that there exists a finite set C of port labels disjoint with B, a binary relation $S \subseteq (B \times B) \cup (C \times B) \cup (B \times C)$ and a p-graph H of type included in C such that

$$c(G) := c_{\mathbb{GP}}(G) = relab_1(ADD_S(G \oplus H)),$$

for every p-graph G of type B. (See Definition 4.56 for $relab_1$.)

We let h_B be a bijection of B to $[k]$ and we define $h_C : C \to [k+1, k+2^{2k}]$ as follows. For $a \in C$, $h_C(a) := k+j$ such that $A_j = \{h_B(b) \mid b \in B \text{ and } (a,b) \in S\}$ and $A'_j = \{h_B(b) \mid b \in B \text{ and } (b,a) \in S\}$. We let R be the image of S under the mapping $h := h_B \cup h_C$ (i.e., $R := \{(h(a_1), h(a_2)) \mid (a_1, a_2) \in S\}$), and $R_1 := R \cap ([k] \times [k])$. Then we have for every p-graph G of type B:

$$c(G) = (ADD_{R_1}(relab_{h_B}(G))) \ltimes_k relab_{h_C}(H).$$

Let G and G' be p-graphs of type B, with B of cardinality k, such that $relab_{h_B}(G) \sim'_k relab_{h_B}(G')$, and c be a context expressed as above in terms of R, R_1, h_B and h_C. Then $ADD_{R_1}(relab_{h_B}(G)) \sim_k ADD_{R_1}(relab_{h_B}(G'))$. It follows that $c(G) \in L$ implies $c(G') \in L$. This is true for all contexts, so we get $G \approx^L G'$.

Hence $G \approx^L G'$ if $\pi(G) = \pi(G')$ and $relab_{h_{\pi(G)}}(G) \sim'_k relab_{h_{\pi(G)}}(G')$, where $k = |\pi(G)|$. This implies that the number of equivalence classes of $\approx^L$ of sort B is not larger than the number of equivalence classes of $\sim'_k$, hence is finite. Hence, L is VR-recognizable. ∎

The extension to labeled graphs is easy. It uses the expression of context of Proposition 2.103 with one relation R for each edge label.

4.4.3 Comparison with the HR-recognizable sets

We now compare the HR-recognizable sets of simple graphs and the VR-recognizable sets, as we did in Section 4.3.4 for the corresponding notions of equational sets. The proof of this theorem uses notions and notation introduced in Definition 4.50.

Theorem 4.58 Every VR-recognizable set L of (labeled) graphs is HR-recognizable. Moreover, if L is effectively VR-recognizable then it is effectively HR-recognizable, and similarly for semi-effective recognizability.

Proof: We first prove the first statement, using congruences. We let L be a VR-recognizable set of (unlabeled) graphs and $\sim$ be a type-preserving and locally finite congruence on the algebra $\mathbb{GP}$ that saturates L. We define as follows a binary relation $\approx$ on s-graphs:

> $G \approx H$ if and only if $\tau(G) = \tau(H)$ and, either G and H have both multiple edges or they are both simple, $\beta(G) = \beta(H)$ and $Int(G) \sim Int(H)$.

In this definition, β and Int are the border and interior functions of Definition 4.50 respectively. It is clear that $\approx$ is a type-preserving and locally finite equivalence relation on $\mathbb{JS}$. We first prove that it saturates L. Let G and H be such that $G \approx H$ and $G \in L$. Since G is a graph, we get $G = Int(G) \sim Int(H) = H$. Hence $H \in L$, because $\sim$ saturates L.

It remains to prove that $\approx$ is a congruence for the operations of F^{HR}. Let $G, G', H, H' \in \mathcal{JS}$ with $G \approx G'$ and $H \approx H'$.

(1) If G is not simple, the same holds for G' and for $G \,\|\, H$ and $G' \,\|\, H'$. Hence $G \,\|\, H \approx G' \,\|\, H'$ because $\tau(G \,\|\, H) = \tau(G' \,\|\, H')$. The proof is similar if any of G', H or H' is not simple.

(2) If G, G', H, H' are all simple and $G \,\|\, H$ is not simple, then $\beta(G) \,\|\, \beta(H)$ is not simple. But the same holds for $\beta(G') \,\|\, \beta(H')$, hence $G' \,\|\, H'$ is not simple and $G \,\|\, H \approx G' \,\|\, H'$.

(3) We now assume that $G, G', H, H', G \,\|\, H, G' \,\|\, H'$ are all simple. We have, by using (4.5) of Definition 4.50,

$$\beta(G \,\|\, H) = \beta(G) \,\|\, \beta(H) = \beta(G') \,\|\, \beta(H') = \beta(G' \,\|\, H'),$$

since $\beta(G) = \beta(G')$ and $\beta(H) = \beta(H')$. We also have

$$Int(G \,\|\, H) = Int(G) \oplus Int(H) \sim Int(G') \oplus Int(H') = Int(G' \,\|\, H'),$$

since $Int(G) \sim Int(G')$, $Int(H) \sim Int(H')$ and $\sim$ is a congruence. Hence $G \,\|\, H \approx G' \,\|\, H'$.

The proofs that $G \approx G'$ implies $ren_h(G) \approx ren_h(G')$ and $fg_B(G) \approx fg_B(G')$ are fully similar, by using (4.6) and (4.7) of Definition 4.50 respectively. Hence L is HR-recognizable.

To prove the second statement (including another proof of the first statement), let $L = h_{VR}^{-1}(C)$ where h_{VR} is a homomorphism from $\mathbb{GP}^t$ to a locally finite F^{tVR}-algebra $\mathbb{A}$ and $C \subseteq A$. We have to define a locally finite F^{tHR}-algebra $\mathbb{B}$, a homomorphism h_{HR} from $\mathbb{JS}^t$ to $\mathbb{B}$, and $C' \subseteq B$, such that $L = h_{HR}^{-1}(C')$. We define $B := B_1 \cup B_2$, where $B_1 := \{\perp_D \mid D \in \mathcal{P}_f(\mathcal{A})\}$ and B_2 is the set of all pairs (β, a) such that β is a simple s-graph with $Int_\beta = \emptyset$ and $a \in A$ has a sort included in $\mathcal{P}(\tau(\beta)) \times \mathcal{P}(\tau(\beta))$. The sort of $\perp_D$ is D and the sort of (β, a) is $\tau(\beta)$. Furthermore, we define $h_{\mathrm{HR}}(G) = (\beta(G), h_{\mathrm{VR}}(Int(G)))$ if G is simple, and $h_{\mathrm{HR}}(G) = \perp_{\tau(G)}$ otherwise. Finally, $C' := \{(\emptyset, a) \mid a \in C\}$. Since L consists of simple graphs, $L = h_{\mathrm{HR}}^{-1}(C')$. It is straightforward to show, using (4.5), (4.6) and (4.7) of Definition 4.50 (together with the fact that h_{VR} is a homomorphism), that B can be turned into a F^{tHR}-algebra $\mathbb{B}$ such that h_{HR} is a homomorphism from $\mathbb{JS}^{\mathrm{t}}$ to $\mathbb{B}$. We leave the details to the reader.[25] By definition of B, the algebra $\mathbb{B}$ is locally finite. Assume now that $\mathbb{A}$ is (semi-)effectively given. Since B_2 is a decidable subset of $\mathcal{JS} \times A$, the set B can be encoded by a standard encoding such that $\mathbb{B}$ is (semi-)effectively given; again, we omit the details. If C is a decidable subset of A, then C' is a decidable subset of B, and if h_{VR} is computable, then so is h_{HR}.

The extension of this proof to labeled graphs is straightforward (cf. the end of the proof of Theorem 4.49). ∎

The following result, proved in [CouWei] (Theorem 6.2 and Proposition 6.4), is strikingly similar to Theorem 4.51.

Theorem 4.59 Let L be an HR-recognizable set of simple graphs. Each of the following conditions implies that L is VR-recognizable:

(1) L has bounded tree-width;
(2) L is uniformly p-sparse for some $p \in \mathcal{N}$;
(3) there exists $n \in \mathcal{N}$ such that every graph in L is without $K_{n,n}$.

4.5 HR- and VR-equational and recognizable sets

For sets of simple graphs, we have the implications of Table 4.1. Their converses do not hold, with the set $\{K_{n,n} \mid n > 0\}$ as a counter-example (see Example 4.54(5)). For sets of simple graphs without $K_{n,n}$ for some n, the two implications in Table 4.1 are equivalences. In both these cases, the family of HR-equational sets of graphs is incomparable with that of HR-recognizable sets, and the same holds for VR-equational and VR-recognizable sets. To be precise, the set L_1 of Proposition 4.32 is HR-equational but not HR-recognizable, and the set of planar graphs is VR-recognizable but not VR-equational. However, for sets of simple graphs of bounded tree-width (these sets are without some $K_{n,n}$), we have the equivalences and the implication (without converse) shown in Table 4.2.[26] It is similar to the proper inclusion of the family of regular languages (of words) in that of context-free languages.

Monadic second-order definable sets of graphs and monadic second-order transductions, to be studied in Chapters 5 and 7, will enrich the landscape.

---

[25] As one example, for a bijection $h : D \to E$, we define $(ren_{h,D})_{\mathbb{B}}(\perp_D) := \perp_E$ and $(ren_{h,D})_{\mathbb{B}}(\beta, a) := (ren_h(\beta), \mathrm{RELAB}_h(a))$, where $\mathrm{RELAB}_h := (relab_g)_{\mathbb{A}}$ such that $g(\langle A, A' \rangle) = \langle h(A), h(A') \rangle$ for all $A, A' \subseteq D$, cf. (4.6) of Definition 4.50.

[26] See Corollary 4.38.

Table 4.1 *Sets of simple graphs.*

HR-equational	$\Longrightarrow$	VR-equational
HR-recognizable	$\Longleftarrow$	VR-recognizable

Table 4.2 *Sets of simple graphs of bounded tree-width.*

HR-recognizable	$\Longleftrightarrow$	VR-recognizable
	$\Downarrow$	
HR-equational	$\Longleftrightarrow$	VR-equational

4.6 References

The family of HR-equational sets of graphs was first defined in terms of graph rewritings based on *hyperedge replacement* in [BauCou] and in [*Hab]. The analogous definition of VR-equational sets is more complicated because the *context-freeness* of a given *vertex replacement* rewriting system is not immediately visible from the syntax and has to be checked by an algorithm. We refer the reader to [*EngRoz]. The article [Cou87] defines an abstract (axiomatic) notion of context-free graph grammar applied to certain grammars defined by vertex replacement.

The finiteness problem for HR-equational sets can be solved (see [*DreKH] or [*Hab]) by means of a *pumping lemma* that generalizes the pumping lemma for context-free languages (cf. Proposition 3.53). We use instead the Semi-Linearity Theorem (known as Parikh's Theorem) for context-free languages (Proposition 3.25). We will further generalize it in Chapter 7 (Theorem 7.42).

Theorem 4.49 is proved in essentially the same way in [*Eng94] (the proof is formulated in terms of transductions).

The implication (4) $\Longrightarrow$ (1) of Theorem 4.51 was first proved in [Cou95b] for directed edge-labeled graphs (in a more complicated way).

The notion of an HR-recognizable set of graphs is based on a particular algebra, the algebra $\mathbb{JS}$ and its many-sorted version. Similar or related algebraic structures on finite graphs can be defined. In many cases, they yield the same notion of a recognizable set of graphs (cf. Theorem 4.34 and Remark 4.55). Other results of this kind are proved in [Cou94a], [CouLag], [CouMak] and [CouWei]. The article [BluCou06] defines as *equivalent* two signatures of graph operations that act on the same graphs and yield the same classes of equational and recognizable sets. It studies variants of F^{HR} and F^{VR} that are equivalent to them.

Certain HR equation systems have a unique solution in the algebra $\mathbb{JS}$. Such systems are used in [CouSén] in order to establish that if an HR-equational set of graphs is minor-closed, then one can compute its finite set of excluded minors from an HR equation system defining this set. The opposite construction will be presented in Chapter 5 (Application 5.72). Both use monadic second-order logic.

Directed cographs are investigated in [CrePau]. This article characterizes them by eight excluded induced subgraphs.

5

Monadic second-order logic

This chapter defines monadic second-order logic, and shows how it can be used to formalize the expression of graph properties. Monadic second-order formulas can also be used to express the properties of sets of vertices and/or edges in graphs. The Recognizability Theorem (Section 5.3.8) implies that a class of graphs characterized as the class of (finite) models of a monadic second-order sentence is VR-recognizable, and that it is HR-recognizable (but not necessarily VR-recognizable) if the sentence is written with edge set quantifications.

It follows that the monadic second-order satisfiability problem for the class of graphs of tree-width or of clique-width at most k is decidable for each k. Applications to the construction of fixed-parameter tractable algorithms and other algorithmic consequences will be developed in Chapter 6.

Although our first applications concern graphs, we will give the definitions and prove the main results for the general case of relational structures because proofs are not more difficult, and the corresponding results apply in a uniform manner to labeled graphs represented, in several ways, by binary relational structures. However, relational structures with n-ary relations for $n \geq 2$ are actually interesting by themselves. As we have shown in Section 1.9, they are useful to formalize betweenness and cyclic ordering (and for relational databases).

In Sections 5.1 and 5.2, we review basic notions of logic, define monadic second-order logic, and give some results that delineate its expressive power. In Section 5.3, we establish the Recognizability Theorem and in Section 5.4 develop its consequences regarding the decidability of monadic second-order satisfiability problems. In Section 5.5, we present logical characterizations of (certain) recognizable sets of trees. In Section 5.6, we give upper-bounds to the number of inequivalent formulas of bounded quantifier-height.

5.1 Relational structures and logical languages

We will not discuss in detail syntactic questions such as the renaming of bound variables. Rather, we will give examples and rest on the intended meanings of formulas. Our review of definitions will fix notation.

5.1.1 Relational structures

Definition 5.1 (Relational signatures and structures) We will use logical structures with relations of positive arity and constants, specified by relation symbols and nullary function symbols respectively. Constants will be useful for representing the sources of s-graphs. We will not use functions of positive arity.

A *relational signature* is a finite set $\mathcal{R}$ of symbols such that each symbol $R \in \mathcal{R}$ has an associated natural number called its *arity*, denoted by $\rho(R)$. We let $\mathcal{R}_0$ be the set of those of arity 0 and call them *constant symbols*. We let $\mathcal{R}_i$ be the set of symbols of arity $i \geq 1$ and call them *relation symbols*. We let $\mathcal{R}_+ := \bigcup \{\mathcal{R}_i \mid i \geq 1\}$ and $\rho(\mathcal{R})$ be the maximal arity of a symbol in $\mathcal{R}$. A relational signature $\mathcal{R}$ is *binary* if $\rho(\mathcal{R}) \leq 2$.

A *concrete $\mathcal{R}$-structure* is a tuple $S = \langle D_S, (R_S)_{R \in \mathcal{R}_+}, (c_S)_{c \in \mathcal{R}_0} \rangle$ such that D_S is a finite, possibly empty, set called the *domain*[1] of S, each R_S is a $\rho(R)$-ary relation on D_S, i.e., a subset of $D_S^{\rho(R)}$ called the *interpretation* of R, and each c_S is an element of D_S called the *interpretation of c*. A *(relational) structure* is a concrete $\mathcal{R}$-structure for some relational signature $\mathcal{R}$. An element c_S is also called a *constant* of S. Several constant symbols may have the same interpretation. We say that S is *constant-separated* if $c_S \neq d_S$ for all $c, d \in \mathcal{R}_0, c \neq d$. An *empty structure*, i.e., a structure S with $D_S = \emptyset$, is denoted $\varnothing$. The domain D_S may be empty only if $\mathcal{R}_0 = \emptyset$.

A *substructure* of a concrete $\mathcal{R}$-structure $S = \langle D_S, (R_S)_{R \in \mathcal{R}_+}, (c_S)_{c \in \mathcal{R}_0} \rangle$ is a concrete $\mathcal{R}$-structure of the form $S[X]$ for some $X \subseteq D_S$ containing all constants of S and defined as follows:

$$D_{S[X]} = X,$$
$$c_{S[X]} = c_S \text{ for all } c \in \mathcal{R}_0,$$
$$R_{S[X]} = R_S \cap X^{\rho(R)} \text{ for all } R \in \mathcal{R}_+.$$

We say that $S[X]$ is the substructure of S *induced by X*.

Let S, S' be concrete $\mathcal{R}$-structures. An *isomorphism* from S to S' is a bijection $h : D_S \to D_{S'}$ such that $h(c_S) = c_{S'}$ for every $c \in \mathcal{R}_0$ and, for every R in $\mathcal{R}_+$ and every $d_1, \ldots, d_{\rho(R)}$ in D_S, we have $(h(d_1), \ldots, h(d_{\rho(R)})) \in R_{S'}$ if and only if $(d_1, \ldots, d_{\rho(R)}) \in R_S$. We denote by $S \simeq S'$ the existence of an isomorphism from S to S' and say that S and S' are *isomorphic*. As for graphs (cf. Chapter 2) we define an *abstract $\mathcal{R}$-structure* as the isomorphism class $[S]_{iso}$ of a concrete $\mathcal{R}$-structure S. We denote by $STR^c(\mathcal{R})$ the set of concrete $\mathcal{R}$-structures and by $STR(\mathcal{R})$ the set of abstract ones. By a *class of structures*, we mean a set of concrete structures closed under isomorphism. It defines a set of abstract structures.

---

[1] The domain of an $\mathcal{R}$-structure is also frequently called its *universe* [*Lib04]. Since we only consider finite structures the term "domain" is more appropriate. It is also customary to forbid empty domains. Allowing or forbidding empty domains makes a difference for proof systems: $\exists x. (x = x)$ is true in nonempty domains only. Since we will not use any proof system (we only use logical formulas to describe classes of graphs and more generally of relational structures), allowing empty structures is harmless. Relational signatures and structures will always be finite as opposed to functional signatures and algebras.

We will use relational structures to represent finite combinatorial objects such as words, terms and graphs of various types. By a *representation*, we mean a mapping from a set of objects to a set of concrete or abstract structures. Isomorphic objects (in particular graphs) must be mapped to isomorphic structures. A representation is *faithful* if the converse also holds: two objects are isomorphic (hence, equal in the cases of terms and words) if and only if their representing structures are isomorphic.

We now define several representations to be used in all further chapters.

Example 5.2 (Words, graphs and terms as relational structures) (1) We first consider words over a finite alphabet A. We let $\mathcal{W}_A := \{suc\} \cup \{lab_a \mid a \in A\}$ be the relational signature such that $\rho(suc) = 2$ and $\rho(lab_a) = 1$ for every a in A. For every word w over A, we let $\lfloor w \rfloor$ be the concrete $\mathcal{W}_A$-structure[2] such that

$$D_{\lfloor w \rfloor} := [n] \text{ if } w \text{ has length } n,$$
$$suc_{\lfloor w \rfloor} := \{(1,2),(2,3),\ldots,(n-1,n)\},$$
$$i \in lab_{a\lfloor w \rfloor} \text{ if and only if } w[i] = a.$$

Hence, the elements of $D_{\lfloor w \rfloor}$ are the positions of the word w and $suc_{\lfloor w \rfloor}(x,y)$ holds if and only if y is the successor of x in w. If u is the factor (also called subword) of a word w starting at position i and ending at position j, then $\lfloor u \rfloor$ is isomorphic to the substructure of $\lfloor w \rfloor$ induced by the interval $[i,j]$, which is a subset of $D_{\lfloor w \rfloor}$.

The empty word ε is thus represented by the empty structure. This may be inconvenient in certain situations. We can also let $\lfloor \varepsilon \rfloor$ be the structure with a singleton domain and empty relations suc and lab_a. We will specify in each case if a variant of the initially given definition is used. The same properties of words can be expressed by first-order or by monadic second-order sentences via both representations.

More generally, we say that two representations are *equivalent* if the same properties of the represented objects can be expressed by sentences of the logical language of interest,[3] monadic second-order logic in most cases.

Yet another representation of words is possible. Let $\mathcal{W}_A^\ell$ be obtained from $\mathcal{W}_A$ by replacing suc by $\leq$, intended to denote the linear order of positions in a word w. We let $\lfloor w \rfloor^\ell$ be the concrete $\mathcal{W}_A^\ell$-structure with domain $[n]$, where $n = |w|$, and such that $lab_{a\lfloor w \rfloor^\ell} := lab_{a\lfloor w \rfloor}$ for each $a \in A$ and $\leq_{\lfloor w \rfloor^\ell}$ is the natural order on $[n]$. The two representations of words $\lfloor w \rfloor^\ell$ and $\lfloor w \rfloor$ are equivalent for monadic second-order logic, but not for first-order logic (a classical fact, see [*Tho97a]). Fewer properties of words can be expressed by first-order sentences through $\lfloor w \rfloor$ than through $\lfloor w \rfloor^\ell$.

---

[2] Instead of $[n]$, we could take as domain of $\lfloor w \rfloor$ the interval $[0, n-1]$ and we get an isomorphic structure. Since isomorphic structures satisfy the same logical formulas, by Lemma 5.4 below, this alternative representation yields the same logical expression of properties of words.

[3] The notion of monadic second-order transduction, to be studied in Chapter 7, can formalize the transformation of a representation into another equivalent one and the corresponding translations of formulas. But it is not necessary to use it at this stage, because the translations of formulas are easy to write directly.

All representations we have defined above are faithful: different words are represented by nonisomorphic structures.

For $L \subseteq A^*$, we denote by $\lfloor L \rfloor$ the class of concrete structures that are isomorphic to $\lfloor w \rfloor$ for some $w \in L$ (and similarly for $\lfloor L \rfloor^\ell$).

(2) We now explain how graphs can be represented by relational structures, in such a way that simple graphs are faithfully represented. We let $\mathcal{R}_s := \{edg\}$ with *edg* binary. With a concrete directed graph G, we associate the concrete $\mathcal{R}_s$-structure $\lfloor G \rfloor := \langle V_G, edg_G \rangle$, where the domain[4] of $\lfloor G \rfloor$ is its vertex set V_G, and $(x,y) \in edg_G$ if and only if there is an edge from x to y. We do the same if G is undirected and let $(x,y) \in edg_G$ if and only if there is an edge linking x and y: it follows that edg_G is in this case symmetric. Every concrete $\mathcal{R}_s$-structure represents a graph and every concrete $\mathcal{R}_s$-structure S with symmetric relation edg_S represents an undirected graph. The structures representing isomorphic graphs are isomorphic. If a graph G is defined as an abstract graph, then we let $\lfloor G \rfloor$ also denote the corresponding abstract structure. In the sequel, we will distinguish between concrete and abstract structures representing graphs only when necessary.

The structure $\lfloor G \rfloor$ does not contain information concerning the multiplicity of edges. It is faithful for simple graphs,[5] but is not for all graphs. We will define in Section 5.2.5 a faithful representation of all graphs. However, for expressing logically graph properties such as connectivity that do not depend on the multiplicity of edges, the structure $\lfloor G \rfloor$ is sufficient (cf. Section 1.3.1).

For directed graphs, we will sometimes use *suc* instead of *edg*. We will call y a *successor* of x, and x a *predecessor* of y if $(x,y) \in suc_G$. For rooted forests we will use *son* instead of *edg*, cf. Definition 2.13.

For representing a p-graph G of type[6] $\pi(G) \subseteq C$, where C is a finite subset of $\mathcal{A}$, we will use the signature $\mathcal{R}_{s,C} := \{edg\} \cup \{lab_a \mid a \in C\}$, with lab_a unary for every a, and the concrete $\mathcal{R}_{s,C}$-structure $\lfloor G \rfloor_C := \langle V_G, edg_G, (lab_{aG})_{a \in C} \rangle$, where $lab_{aG}(x)$ holds if and only if x is an a-port. Every p-graph is by definition simple, hence this representation is faithful for the p-graphs in $\mathcal{GP}[C]$. If $C \subseteq D \subseteq \mathcal{A}$ (with D finite), then $\lfloor G \rfloor_C$ and $\lfloor G \rfloor_D$ are equivalent representations for the p-graphs in $\mathcal{GP}[C]$, for first-order and monadic second-order logic ($\lfloor G \rfloor_D$ is obtained from $\lfloor G \rfloor_C$ by adding the empty sets lab_{aG} for $a \in D - C$). If G has type C, we denote $\lfloor G \rfloor_C$ also simply by $\lfloor G \rfloor$. The mapping $G \mapsto \lfloor G \rfloor$ is a faithful representation of all p-graphs. A structure S in $STR(\mathcal{R}_{s,C})$ is of the form $\lfloor G \rfloor_C$ for some p-graph G of type included in C (of type C) if and only if the sets lab_{aS} form a partition[7] of D_S (a partition no part of which is empty, respectively).

---

[4] We denote $D_{\lfloor G \rfloor}$ by V_G and $edg_{\lfloor G \rfloor}$ by edg_G.

[5] To be precise, for simple graphs G and H, a mapping $: V_G \to V_H$ is an isomorphism from G to H if and only if it is one from $\lfloor G \rfloor$ to $\lfloor H \rfloor$.

[6] The type of a p-graph G is the set $\pi(G)$ of its port labels. The type of an s-graph G is the set $\tau(G)$ of its source labels.

[7] We allow the sets forming a partition of a set to be empty.

For representing an s-graph G of type $\tau(G) \subseteq C$ (in a faithful way if it is simple), we use the above signature and let $lab_{aG}(x)$ hold if and only if x is an a-source of G. A structure S in $STR(\mathcal{R}_{s,C})$ is of the form $\lfloor G \rfloor_C$ for some s-graph G of type included in C (of type C) if and only if the sets lab_{aS} are empty or singletons (or are all singletons, respectively). As for p-graphs, $\lfloor G \rfloor_C$ is also denoted by $\lfloor G \rfloor$ if G has type C. For s-graphs of type C we will also need a variant of $\lfloor G \rfloor$ in which the sources are viewed as constants. Thus, we use the relational signature $\mathcal{R}_{ss,C} := \{edg\} \cup C$, where each element of C is a constant symbol. If $G = \langle V_G, edg_G, src_G \rangle$, we let $\lfloor G \rfloor$ be $\langle V_G, edg_G, (a_G)_{a \in C} \rangle$ with $a_G := src_G(a)$ for each a in C. (These representations are equivalent for both first-order logic and monadic second-order logic.) For this variant, a structure S in $STR(\mathcal{R}_{ss,C})$ is of the form $\lfloor G \rfloor$ for some s-graph G of type C if and only if it is constant-separated. A substructure of the $\mathcal{R}_{ss,C}$-structure $\lfloor G \rfloor$ representing an s-graph G is thus $\lfloor G[X] \rfloor$ for a set of vertices X containing all sources; hence it represents an induced subgraph of G (cf. Definition 2.12) having the same sources as G. Note that $\mathcal{R}_{ss,C}$ cannot be used for s-graphs of type properly included in C, because every constant symbol in C must have an interpretation.

These representations can be adapted so as to represent (K, Λ)-labeled graphs, p-graphs and s-graphs (faithfully if they are simple). We recall from Definition 2.11 that K is the set of vertex labels, Λ is the set of edge labels and that $(K \cup \Lambda) \cap \mathcal{A} = \emptyset$. We use unary relations $lab_{\kappa G}$ for $\kappa \in K$ in the obvious way to represent the vertex labels of a graph G. For each $\lambda \in \Lambda$, we use a binary relation $edg_{\lambda G}$ consisting of all pairs (x, y) such that there is in G an edge labeled by λ from x to y (or between x and y if G is undirected). Hence, a labeled p-graph G of type C is faithfully represented by $\lfloor G \rfloor := \langle V_G, (edg_{\lambda G})_{\lambda \in \Lambda}, (lab_{aG})_{a \in K \cup C} \rangle$. The corresponding relational signatures are denoted by $\mathcal{R}_{s,[K,\Lambda]}$, $\mathcal{R}_{s,C,[K,\Lambda]}$ and $\mathcal{R}_{ss,C,[K,\Lambda]}$. (This notation emphasizes the role of C in contrast to that of K.)

We recall from Definition 2.11 that binary $\mathcal{R}$-structures without constants correspond bijectively to simple directed $(\mathcal{R}_1, \mathcal{R}_2)$-labeled graphs.[8] We will frequently identify such a graph G with the relational structure $\lfloor G \rfloor$ that represents it faithfully.

The representation $\lfloor w \rfloor$ of a word $w \in A^*$, initially defined in (1), can alternatively be described as $\lfloor G(w) \rfloor$, where $G(w)$ is the directed $(A, \emptyset)$-labeled path with $V_{G(w)} = [|w|]$, $edg_{G(w)} = \{(i, i+1) \mid i \in [|w|-1]\}$ and vertex i has label $w[i]$.

If L is a set of (labeled) concrete or abstract graphs, we denote by $\lfloor L \rfloor$ the set of concrete or abstract structures $\lfloor G \rfloor$ for $G \in L$. If L is a set of (labeled) p-graphs or s-graphs of type included in C, for some finite $C \subseteq \mathcal{A}$, we denote by $\lfloor L \rfloor_C$ the set of structures $\lfloor G \rfloor_C$ for $G \in L$. Moreover, if C is the smallest such set, i.e., if C is the union of all (finitely many) types of the elements of L, then we also denote $\lfloor L \rfloor_C$ by $\lfloor L \rfloor$. Note that $\lfloor L \rfloor$ is not equal to $\{\lfloor G \rfloor \mid G \in L\}$, except when L is homogenous. Note also that $\lfloor L \rfloor$ is defined only when L has *bounded type*, i.e., $L \subseteq \mathcal{GP}[C]$ or $L \subseteq \mathcal{JS}[C]$

---

[8] To be precise this only holds if $\mathcal{R}_2 \neq \emptyset$, due to the convention in Definition 2.11 that $(K, \emptyset)$-labeled graphs are graphs with unlabeled edges.

for some finite $C \subseteq \mathcal{A}$; this suffices for our purposes, because all VR-equational and HR-equational sets have bounded type (with $C = \mathcal{A}(S)$ for some VR or HR equation system S).

(3) We now define faithful representations of terms by relational structures. For every finite functional signature F, we let $\mathcal{R}_F$ be the relational signature $\mathcal{R}_{s,[F,[\rho(F)]]}$ (with *edg* changed into *son*), i.e.,

$$\mathcal{R}_F = \{son_i \mid 1 \le i \le k\} \cup \{lab_f \mid f \in F\},$$

where $k = \rho(F)$, son_i is binary and lab_f is unary. Recall from Definition 2.14 that the syntactic tree $Syn(t)$ of a term t over F is an $(F, [\rho(F)])$-labeled rooted tree. With $t \in T(F)$ we associate the concrete $\mathcal{R}_F$-structure $\lfloor t \rfloor := \lfloor Syn(t) \rfloor$. So,

$$\lfloor t \rfloor = \langle N_t, (son_{it})_{1 \le i \le k}, (lab_{ft})_{f \in F} \rangle,$$

where the domain of $\lfloor t \rfloor$ is the set N_t of nodes of $Syn(t)$ (which is the set $Pos(t)$ of positions of t, cf. Definitions 2.3 and 2.14), $(u, v) \in son_{it}$ if and only if v is the i-th son of u (we use the terminology of trees), and $lab_{ft} = Occ(t, f)$. This is obviously a faithful representation: as discussed after Definition 2.14 the mapping $t \mapsto [Syn(t)]_{iso}$ is injective.

For some proofs it is important (or convenient) to add the constant symbol rt to $\mathcal{R}_F$, which, in every structure $\lfloor t \rfloor$, is interpreted as the root of $Syn(t)$, i.e., $\lfloor t \rfloor = \langle N_t, (son_{it})_{1 \le i \le k}, (lab_{ft})_{f \in F}, rt_t \rangle$ with $rt_t := root_t$ (the first position of t). In this way we obtain an equivalent representation, because the root of t is the unique element v of N_t that does not satisfy $son_{it}(u, v)$ for any $i \in [k]$ and $u \in N_t$. This representation corresponds to viewing the syntactic tree of t as a labeled s-graph with one source (its root). As for s-graphs, rt can also be taken as a unary symbol such that $rt_t(u)$ holds if and only if u is the root of t.

A variant, equivalent to the above defined representation (with or without rt), is as follows: instead of k binary "son" relations, we use one binary relation *son* and k unary "brother" relations br_i defined, for each $i = 1, \ldots, k$, by $br_{it}(v) :\Longleftrightarrow \exists u . son_{it}(u, v)$.[9] This obviously corresponds to a variant of $Syn(t)$ with unlabeled edges and $(F \cup [\rho(F)])$-labeled nodes, such that each node has one label from F and one label from $[\rho(F)]$.

As for words (where $\lfloor \varepsilon \rfloor$ can be chosen in different ways) we will use a single notation, here $\mathcal{R}_F$ and $\lfloor t \rfloor$, for these slightly different relational signatures and structures. We will specify which is used if necessary.

In Definition 2.7, we have defined two bijections μ from A^* (for a finite alphabet A) to $T(U_A)$, where U_A is a functional signature containing one constant symbol (corresponding to the empty word) and a unary function symbol for each a in A. We deduce

---

[9] For representing terms defining cographs, the relations br_i are not necessary as we have seen in Example 1.38 of Section 1.7.1 because the operations $\oplus$ and $\otimes$ are commutative.

from it two other faithful representations of words: $w \in A^*$ is represented by $\lfloor \mu(w) \rfloor$. The empty word is represented by a nonempty structure. These representations are equivalent to the representation $\lfloor w \rfloor$ defined in (1) for both first-order and monadic second-order logic.

As for words, if L is a set of terms over F, we denote by $\lfloor L \rfloor$ the class of concrete structures that are isomorphic to $\lfloor t \rfloor$ for some $t \in L$.

We now relate the representations of the subterms and the contexts of t with $\lfloor t \rfloor$, cf. Definition 2.6. If u is a position of a term $t \in T(F)$, we define $\lfloor t \rfloor / u$ to be the $\mathcal{R}_F$-structure $\lfloor Syn(t)/u \rfloor$. As discussed after Definition 2.14, $Syn(t/u) \simeq Syn(t)/u$ and hence the structure $\lfloor t/u \rfloor$ is isomorphic to $\lfloor t \rfloor / u$ (which is a substructure of $\lfloor t \rfloor$ if we use the signature $\mathcal{R}_F$ with binary relations son_i). For defining contexts as terms, we use w as special variable. We represent a context $t \uparrow u$ by the $\mathcal{R}_{F \cup \{w\}}$-structure $\lfloor t \rfloor \uparrow u := \lfloor Syn(t) \uparrow u \rfloor$. Since $Syn(t \uparrow u) \simeq Syn(t) \uparrow u$, the structure $\lfloor t \uparrow u \rfloor$ is isomorphic to $\lfloor t \rfloor \uparrow u$ (which is *not* a substructure of $\lfloor t \rfloor$ because u has a different label). In the case where $\mathcal{R}_F$ and $\mathcal{R}_{F \cup \{w\}}$ contain the constant symbol rt, we define $rt_{\lfloor t \rfloor / u} := u$ (which implies that $\lfloor t \rfloor / u$ is not any more a substructure of $\lfloor t \rfloor$) and $rt_{\lfloor t \rfloor \uparrow u} := rt_t$.

(4) We now explain how to represent a graph G and one of its tree-decompositions by a single relational structure. This can be done in several ways.

If (T, f) is a tree-decomposition of a graph G such that, without loss of generality, $N_T \cap V_G = \emptyset$, we let $S_0(G, T, f)$ be the relational structure $\langle V_G \cup N_T, edg_G, son_T, box_f \rangle$ that expands $\lfloor G \rfloor$ by adding new domain elements (the nodes of T) and the binary relations son_T and box_f; the latter one is the set of pairs (u, x) such that $u \in N_T$ and $x \in f(u)$. By using this relational structure, one can express by logical formulas properties of tree-decompositions, for example that every box is a clique (cf. Definition 2.71 about chordal graphs and their tree-decompositions).

It may be interesting to expand $\lfloor G \rfloor$ in a more economical way, by adding only new relations. It easily follows from Proposition 2.67(2) that every nonempty graph has an optimal 1-downwards increasing tree-decomposition (T, f) such that $|f(root_T)| = 1$. In this case, the mapping $x \mapsto \max(f^{-1}(x))$ is a bijection from V_G to N_T, cf. Remark 2.55(a). Hence, we can identify V_G and N_T and simplify $S_0(G, T, f)$ into $S_1(G, T, f) := \langle V_G, edg_G, son_T, box_f \rangle$.

Every normal tree-decomposition (cf. Example 2.56(6)) satisfies the above properties (it can still be optimal by Corollary 2.73(3)). If (T, f) is normal, then one can even omit box_f and use $S_2(G, T, f) := \langle V_G, edg_G, son_T \rangle$ because the mapping f can be determined from edg_G and son_T as follows:

$$f(u) :=$$

$$\{u\} \cup \{w \mid (w, u) \in son_T^* \text{ for some } (v, w) \in edg_G \cup edg_G^{-1} \text{ such that } (u, v) \in son_T^*\},$$

for every $u \in N_T = V_G$, hence box_f can be expressed by a monadic second-order formula.

If G is k-chordal, one can even define son_T for some normal tree T by a monadic second-order formula depending on k (there is one such formula for each k). This means that an optimal tree-decomposition (T, f) of G represented by $S_1(G, T, f)$ can be defined "inside" the structure $\lfloor G \rfloor$. We will prove this fact in Proposition 9.56(2). We will explain in Section 7.6 the consequences of results of this type. $\square$

Definition 5.3 (Sizes of structures) We will only consider finite relational structures over finite signatures. Hence, just as (finite) graphs, structures can be inputs of algorithms. The time and space complexities of such algorithms will be analyzed in terms of the sizes of the input structures. We define first the *size* of a relational signature $\mathcal{R}$ as

$$\|\mathcal{R}\| := 1 + \Sigma_{R \in \mathcal{R}}(1 + \rho(R)).$$

Since the empty signature must be somehow described, its size is not 0. We now define the *size* of an $\mathcal{R}$-structure S as

$$\|S\| := \|\mathcal{R}\| + |D_S| + \Sigma_{R \in \mathcal{R}_+} \rho(R) \cdot |R_S|.$$

The size of the empty structure over the empty signature is also positive. These definitions are essentially the ones of [*FluGro] and [*Lib04], where they are justified by explicit encodings of structures by words of length $O(\|S\|)$. Let us see what these definitions yield when applied to the representations of words, graphs and terms. For a word w over A represented by the $\mathcal{W}_A$-structure $\lfloor w \rfloor$ we have $\|\lfloor w \rfloor\| = \Theta(1 + |w| + |A|)$, but for the representation by a $\mathcal{W}_A^\ell$-structure, we have $\|\lfloor w \rfloor^\ell\| = \Theta(1 + |w|^2 + |A|)$ which justifies to prefer the first representation. For a p-graph G of type $\pi(G) = C$ represented by the $\mathcal{R}_{s,C}$-structure $\lfloor G \rfloor$, we have $\|\lfloor G \rfloor\| = \Theta(1 + |V_G| + |edg_G| + |C|) = O(|G|^2)$. For a simple s-graph G of type $\tau(G) = C$ represented by the $\mathcal{R}_{ss,C}$-structure $\lfloor G \rfloor$, we have $\|\lfloor G \rfloor\| = \Theta(1 + |V_G| + |edg_G| + |C|) = \Theta(\|G\|)$. For a term t in $T(F)$ (where F is finite and $\rho(F)$ is the maximal arity of its symbols) represented by the $\mathcal{R}_F$-structure $\lfloor t \rfloor$, we have $\|\lfloor t \rfloor\| = \Theta(|t| + |F| + \rho(F))$.

5.1.2 First-order logic

We let $\mathcal{V}_0$ be a countable alphabet of lowercase letters called *first-order variables*. Let $\mathcal{R}$ be a relational signature.[10] A *term* is either a variable (from $\mathcal{V}_0$) or a constant symbol (from $\mathcal{R}_0$). The atomic formulas are $s = t$, $R(t_1, \ldots, t_{\rho(R)})$ for $R \in \mathcal{R}_+$ and terms $s, t, t_1, \ldots, t_{\rho(R)}$, and the Boolean constants *True* and *False*. The *first-order formulas over* $\mathcal{R}$ are formed from atomic formulas with the propositional connectives $\wedge, \vee, \neg, \Rightarrow, \Leftrightarrow$, and quantifications $\exists x$ and $\forall x$ for $x \in \mathcal{V}_0$.

---

[10] For discussing syntax, a relational signature is frequently called a *vocabulary*. We prefer the term "relational signature," which refers to the types of objects described by formulas. Furthermore, it is similar to the term "functional signature" used for algebras.

Let $\mathcal{X} \subseteq \mathcal{V}_0$. We will denote by $\mathrm{FO}(\mathcal{R}, \mathcal{X})$ the set of first-order formulas over $\mathcal{R}$ with free variables in $\mathcal{X}$. In order to specify the free variables that may occur in a formula φ, we will write it $\varphi(x_1, \ldots, x_n)$ if $\varphi \in \mathrm{FO}(\mathcal{R}, \{x_1, \ldots, x_n\})$. (Some variables in $\{x_1, \ldots, x_n\}$ may have no free occurrence in φ.) If $\varphi(x_1, \ldots, x_n)$ is a first-order formula over $\mathcal{R}$, if $S \in STR^c(\mathcal{R})$, and if $d_1, \ldots, d_n \in D_S$, then we write $S \models \varphi(d_1, \ldots, d_n)$ or $(S, d_1, \ldots, d_n) \models \varphi$ to mean that φ is true in S if x_i is given the value d_i for $i = 1, \ldots, n$. If φ has no free variables, it is said to be *closed*, equivalently to be a *sentence*. It describes a property of S and not of tuples of elements of S. We will denote by $sat(S, \varphi, (x_1, \ldots, x_n))$ the set $\{(d_1, \ldots, d_n) \in D_S^n \mid S \models \varphi(d_1, \ldots, d_n)\}$. It is important to specify $x_1, \ldots, x_n$ as arguments of sat for two reasons: first to specify the order of the variables, and second because the list of variables is not fixed in a unique way by φ. If φ is $edg(x, y)$, we may need to consider $sat(\lfloor G \rfloor, edg(x, y), (u, x, y))$ or $sat(\lfloor G \rfloor, edg(x, y), (x, y, z, u))$ or $sat(\lfloor G \rfloor, edg(x, y), (y, x))$. If $n = 0$, then $sat(S, \varphi, ()) = \emptyset$ if $S \models \neg \varphi$, and $sat(S, \varphi, ()) = \{()\}$ if $S \models \varphi$, where $()$ denotes the empty tuple. If $S \models \varphi$, then S *satisfies* φ or *is a model* of φ.[11] We denote by $\mathrm{MOD}(\varphi)$ the class of models of φ. (It is a class by Lemma 5.4 below.) But in many cases, we will also denote by $\mathrm{MOD}(\varphi)$ the corresponding set of abstract structures.

Sentences express properties of a structure S whereas formulas express properties of elements of S specified as values of the free variables. For example, the formula $\varphi(x)$ over $\mathcal{R}_s$ defined as:

$$\forall y_1, y_2, y_3 [edg(x, y_1) \wedge edg(x, y_2) \wedge edg(x, y_3) \Rightarrow (y_1 = y_2 \vee y_1 = y_3 \vee y_2 = y_3)]$$

expresses[12] that the vertex x of the considered directed graph G has outdegree at most 2. The sentence $\forall x . \varphi(x)$ expresses that G has outdegree at most 2. We now give an example concerning words. We let $A = \{a, b, c\}$. A formula θ can be constructed in such a way that, for every word $w \in A^*$,

$$\lfloor w \rfloor \models \theta \text{ if and only if } w \in ab^*c.$$

Here is θ:

$$\exists x [lab_a(x) \wedge \forall y (\neg suc(y, x))]$$
$$\wedge \forall x [(lab_a(x) \vee lab_b(x)) \Rightarrow \exists y (suc(x, y) \wedge (lab_b(y) \vee lab_c(y)))]$$
$$\wedge \forall x [lab_c(x) \Rightarrow \forall y (\neg suc(x, y))].$$

Note that θ has models that are not representations of words.[13] So, θ characterizes $\lfloor ab^*c \rfloor$ as a subset of $\lfloor A^* \rfloor$; it does not characterize the abstract structures representing the words of ab^*c among those of $STR(\mathcal{W}_A)$.

---

[11] We define $S \models \varphi$ only if φ is a sentence.
[12] As usual, $\forall y_1, y_2, y_3$ abbreviates $\forall y_1 \forall y_2 \forall y_3$.
[13] For example, S with $D_S = \{1, 2, 3\}$, $suc_S = \{(1, 2), (2, 3), (3, 2)\}$, $lab_{aS} = \{1\}$, $lab_{bS} = \{2, 3\}$, $lab_{cS} = \emptyset$.

5.1.3 Second-order logic

Although we will only use its fragment called *monadic second-order logic*, we describe second-order logic for completeness. We will use a countable set of variables V_ω consisting of first-order variables as in Section 5.1.2 and of *relation variables*, denoted by uppercase letters $X, Y, X_1, \ldots, X_m$. Each relation variable has an arity which is a positive integer ($\rho(X)$ is the arity of X), and there are countably many variables of each arity.

Let $\mathcal{R}$ be a relational signature. We now define the *second-order formulas over* $\mathcal{R}$. The atomic formulas are: *True, False*, $s = t$, $R(t_1, \ldots, t_n)$ and $X(t_1, \ldots, t_n)$ where $s, t, t_1, \ldots, t_n$ are terms, $R \in \mathcal{R}$, $X \in V_\omega$ and $n = \rho(R) = \rho(X) \geq 1$. The formulas are constructed from the atomic formulas with the propositional connectives and the quantifications $\exists x, \forall x, \exists X, \forall X$ for first-order and relation variables. (Instead of a formal syntax we give examples below.) For $\mathcal{X} \subseteq V_\omega$, we will denote by $SO(\mathcal{R}, \mathcal{X})$ the set of second-order formulas over $\mathcal{R}$ with free variables in $\mathcal{X}$, and by $FO(\mathcal{R}, \mathcal{X})$ the subset of those that have only first-order quantifications. The notation $\varphi(X_1, \ldots, X_m, x_1, \ldots, x_n)$ indicates that the free variables of φ belong to $\{X_1, \ldots, X_m, x_1, \ldots, x_n\}$.

The *size* $|\varphi|$ and the *quantifier-height* $qh(\varphi)$ of a formula φ are defined inductively as follows:

$$|\varphi| := 1, qh(\varphi) := 0 \text{ if } \varphi \text{ is atomic},$$
$$|\neg\varphi| := |\varphi| + 1, qh(\neg\varphi) := qh(\varphi),$$
$$|\varphi \wedge \psi| := |\varphi \vee \psi| := |\varphi \Rightarrow \psi| := |\varphi \Leftrightarrow \psi| := |\varphi| + |\psi| + 1,$$
$$qh(\varphi \wedge \psi) := qh(\varphi \vee \psi) := qh(\varphi \Rightarrow \psi) := qh(\varphi \Leftrightarrow \psi) := \max\{qh(\varphi), qh(\psi)\},$$
$$|\exists X.\varphi| := |\forall X.\varphi| := |\exists x.\varphi| := |\forall x.\varphi| := |\varphi| + 1,$$
$$qh(\exists X.\varphi) := qh(\forall X.\varphi) := qh(\exists x.\varphi) := qh(\forall x.\varphi) := qh(\varphi) + 1.$$

It is clear that $qh(\varphi) = 0$ if φ is quantifier-free and that $|\varphi|$ is the number of subformulas of φ. (A subformula is counted once for each of its occurrences.)

We now consider the semantics of a formula $\varphi(X_1, \ldots, X_m, x_1, \ldots, x_n)$. If $S \in STR^c(\mathcal{R})$, if $E_1, \ldots, E_m$ are relations on D_S of respective arities $\rho(X_1), \ldots, \rho(X_m)$ and if $d_1, \ldots, d_n \in D_S$, then the notation $S \models \varphi(E_1, \ldots, E_m, d_1, \ldots, d_n)$ means that φ is true in S for the values $E_1, \ldots, E_m$ of $X_1, \ldots, X_m$ and $d_1, \ldots, d_n$ of $x_1, \ldots, x_n$ respectively. We let $sat(S, \varphi, (X_1, \ldots, X_m), (x_1, \ldots, x_n))$ be the set of $(m+n)$-tuples $(E_1, \ldots, E_m, d_1, \ldots, d_n)$ of relations on D_S and of elements of this set such that $S \models \varphi(E_1, \ldots, E_m, d_1, \ldots, d_n)$. Let $\mathcal{X} := \{X_1, \ldots, X_m, x_1, \ldots, x_n\}$. An $\mathcal{X}$-*assignment in* S is a mapping γ with domain $\mathcal{X}$ such that $\gamma(X_i)$ is a $\rho(X_i)$-ary relation on D_S for each $i = 1, \ldots, m$ and $\gamma(x_j) \in D_S$ for $j = 1, \ldots, n$. It is just another way of formalizing a tuple $(E_1, \ldots, E_m, d_1, \ldots, d_n)$. We will also use the notation $(S, \gamma) \models \varphi$ instead of $S \models \varphi(E_1, \ldots, E_m, d_1, \ldots, d_n)$ where $\gamma(X_i) = E_i$ and $\gamma(x_j) = d_j$. Thus, $sat(S, \varphi, (X_1, \ldots, X_m), (x_1, \ldots, x_n))$ can be viewed as the set of all $\mathcal{X}$-assignments γ such that $(S, \gamma) \models \varphi$. The notions of sentence, of satisfaction and of model defined in Section 5.1.2 extend in the obvious way.

Here are some examples. We let $\beta(X)$ be the formula in $SO(\emptyset, \{X\})$:

$$\forall x, y, z([X(x,y) \wedge X(x,z) \Rightarrow y = z] \wedge [X(x,z) \wedge X(y,z) \Rightarrow x = y]),$$

where X is binary. It expresses that X is a functional relation, and that the corresponding function is injective. The following formula $\alpha(X)$ expresses that X is an automorphism[14] of a simple graph G represented by the structure $\lfloor G \rfloor$ in $STR(\mathcal{R}_s)$. Here is $\alpha(X)$:

$$\beta(X) \wedge \forall x \exists y . X(x,y) \wedge \forall x \exists y . X(y,x) \wedge \forall x, y, x', y' [edg(x,y)$$
$$\wedge ((X(x,x') \wedge X(y,y')) \vee (X(x',x) \wedge X(y',y))) \Rightarrow edg(x',y')].$$

Hence, G has a nontrivial automorphism if and only if $\lfloor G \rfloor$ satisfies the sentence[15]

$$\exists X[\alpha(X) \wedge \exists x, y(x \neq y \wedge X(x,y))].$$

Consider now the formula $\gamma(Y_1, Y_2)$ in $SO(\emptyset, \{Y_1, Y_2\})$:

$$\exists X[\beta(X) \wedge \forall x\{(Y_1(x) \Leftrightarrow \exists y . X(x,y)) \wedge (Y_2(x) \Leftrightarrow \exists y . X(y,x))\}].$$

It expresses that there exists a bijection (the binary relation defined by X) between the sets Y_1, Y_2, handled as unary relations. Hence, the following second-order sentence characterizes the nonregular language $\{a^n b^n \mid n \geq 1\}$ as a subset of $\{a, b\}^*$:

$$\exists x[lab_a(x) \wedge \forall y(\neg suc(y,x))]$$
$$\wedge \forall x[lab_a(x) \Rightarrow \exists y(suc(x,y) \wedge (lab_a(y) \vee lab_b(y)))]$$
$$\wedge \forall x, y[lab_b(x) \wedge suc(x,y) \Rightarrow lab_b(y)]$$
$$\wedge \exists Y_1, Y_2[\gamma(Y_1, Y_2) \wedge \forall x[lab_a(x) \Leftrightarrow Y_1(x)] \wedge \forall x[lab_b(x) \Leftrightarrow Y_2(x)]].$$

This is an example of a language that cannot be characterized by a monadic second-order sentence (a particular type of second-order formula to be defined in the next section). Note that the sentences of the first three lines characterize the regular language $a^+ b^+$.

5.1.4 Monadic second-order logic

The main logical language used in this book is *monadic second-order logic*, which lies between first-order and second-order logic.

All formulas will be written with variables from a set $\mathcal{V}_1$ consisting of the countable set $\mathcal{V}_0$ of first-order variables and the countably many relation variables of arity one

---

[14] An *automorphism* of G is an isomorphism of G to itself.
[15] We write $x \neq y$ instead of $\neg x = y$ in order to get shorter and more readable formulas.

from $\mathcal{V}_\omega$. Since a relation with one argument is nothing but a set, these relation variables will be called *set variables*.

Let $\mathcal{R}$ be a relational signature. A *monadic second-order formula* (MS formula in short) *over* $\mathcal{R}$ is a second-order formula written with $\mathcal{R}$ and the variables of $\mathcal{V}_1$: the quantified and free variables are first-order or set variables. In order to get more readable formulas, we will write $x \in X$ and $x \notin X$ instead of $X(x)$ and $\neg X(x)$, where X is a set variable. For $\mathcal{X} \subseteq \mathcal{V}_1$, we will denote by $MS(\mathcal{R}, \mathcal{X})$ the set of monadic second-order formulas over $\mathcal{R}$, with free variables in $\mathcal{X}$. Occasionally, monadic second-order formulas may have free relation variables of any arity, but their quantifications apply to first-order and set variables only.

In Section 1.3.1 we have constructed monadic second-order sentences expressing that a graph is connected and that it is 3-vertex colorable. We now give other examples. The sentence $\delta \in MS(\{suc\}, \emptyset)$ given below expresses that a word in A^* has odd length. Since this property does not depend on letters, δ does not use the unary relations lab_a for $a \in A$. Here is δ:

$$\exists X [\delta'(X) \wedge \exists x(x \in X \wedge \forall y(\neg suc(x,y)))],$$

where $\delta'(X)$ is

$$\exists x [x \in X \wedge \forall y(\neg suc(y,x))] \wedge \forall x,y[suc(x,y) \Rightarrow (x \in X \Leftrightarrow y \notin X)].$$

For every nonempty word $w \in A^*$, there is a unique set $X \subseteq D_{\lfloor w \rfloor}$ satisfying δ': the elements of X are the odd positions in w. The sentence δ expresses that the last position is odd, i.e., that the considered word has odd length.

The next lemma is a basic fact of model theory. If h is a mapping $: D \to D'$ and $E \subseteq D^k$, we define $h(E) := \{(h(d_1), \ldots, h(d_k)) \mid (d_1, \ldots, d_k) \in E\}$.

Lemma 5.4 Let $h : S \to T$ be an isomorphism of concrete $\mathcal{R}$-structures. If $\varphi \in SO(\mathcal{R}, \{X_1, \ldots, X_m, x_1, \ldots, x_n\})$, then

$$sat(T, \varphi, (X_1, \ldots, X_m), (x_1, \ldots, x_n)) = h(sat(S, \varphi, (X_1, \ldots, X_m), (x_1, \ldots, x_n))),$$

i.e., is the set of tuples $(h(E_1), \ldots, h(E_m), h(d_1), \ldots, h(d_n))$ such that the tuple $(E_1, \ldots, E_m, d_1, \ldots, d_n)$ belongs to $sat(S, \varphi, (X_1, \ldots, X_m), (x_1, \ldots, x_n))$.

If φ is a sentence, then $S \models \varphi$ if and only if $T \models \varphi$.

Proof: By induction on the structure of φ. ∎

From the last assertion, we get that if φ is a sentence ($n = m = 0$; () is the empty sequence of variables), then $sat(S, \varphi, (), ())$ is well defined if S is an abstract structure; it is then equal to $\{()\}$ or to $\emptyset$.

5.1.5 Logical definitions of properties of relational structures

Let $\mathcal{L}$ be a logical language, typically first-order logic second-order logic or monadic second-order logic, designated respectively by FO, SO or MS, or one of the extensions of monadic second-order logic defined below. Let $\mathcal{R}$ be a relational signature. We denote by $\mathcal{L}(\mathcal{R},\emptyset)$ the corresponding set of sentences over $\mathcal{R}$. If $\varphi \in \mathcal{L}(\mathcal{R},\emptyset)$, then the set of concrete $\mathcal{R}$-structures S such that $S \models \varphi$, denoted by $\mathrm{MOD}(\varphi)$, is closed under isomorphism by Lemma 5.4, hence is a class of structures. We say that such a class is $\mathcal{L}$-*definable*. If $C \subseteq STR^c(\mathcal{R})$ is a class of $\mathcal{R}$-structures, then the class $\{S \in C \mid S \models \varphi\} = \mathrm{MOD}(\varphi) \cap C$ is said to be $\mathcal{L}$-*definable with respect to C*.[16]

We say that a property P of structures S in $STR^c(\mathcal{R})$ is $\mathcal{L}$-*expressible* if, for some sentence φ of $\mathcal{L}(\mathcal{R},\emptyset)$, it is equivalent to $S \models \varphi$ for all S in $STR^c(\mathcal{R})$. Thus, a class of structures $\mathcal{D} \subseteq STR^c(\mathcal{R})$ is $\mathcal{L}$-definable if and only if the membership in $\mathcal{D}$ of a concrete $\mathcal{R}$-structure is $\mathcal{L}$-expressible.

A property P of tuples $(E_1,\ldots,E_m,d_1,\ldots,d_n)$ in structures S of $STR^c(\mathcal{R})$ is $\mathcal{L}$-*expressible* if there exists a formula φ in $\mathcal{L}(\mathcal{R},\{X_1,\ldots,X_m,x_1,\ldots,x_n\})$ such that, for all $S \in STR^c(\mathcal{R})$ and all $(E_1,\ldots,E_m,d_1,\ldots,d_n)$ of appropriate types, $P(S,E_1,\ldots,E_m,d_1,\ldots,d_n)$ holds if and only if $S \models \varphi(E_1,\ldots,E_m,d_1,\ldots,d_n)$.

Similar definitions hold for $\mathcal{L}$-expressibility *with respect to a class $C \subseteq STR^c(\mathcal{R})$*: in the above, just replace $STR^c(\mathcal{R})$ by C.

We adapt these definitions to finite combinatorial objects such as words, terms, partial orders, graphs, cyclic orders (cf. Section 1.9) represented by relational structures. Let C be a class of such objects and *rep* be a mapping : $C \to STR^c(\mathcal{R})$ with the natural requirement that isomorphic objects are mapped to isomorphic structures. We say that *rep* is *faithful* if, conversely, for all $H, K \in C$, H is equal or isomorphic to K if $rep(H)$ is isomorphic to $rep(K)$.

We say that a property P of the elements of C is $\mathcal{L}$-*expressible via rep* if there exists a sentence φ in $\mathcal{L}(\mathcal{R},\emptyset)$ such that H has property P if and only if $rep(H) \models \varphi$, for every $H \in C$ (and similarly for properties of tuples, when relevant). In the case where *rep* is faithful, and *only in this case*, we say that a set $L \subseteq C$ is $\mathcal{L}$-*definable via rep* if membership in L is $\mathcal{L}$-expressible via *rep*.

To take an example, although the representation of a graph G by $\lfloor G \rfloor$ is not faithful (it does not distinguish multiple edges), connectedness, a property that does not depend on the multiplicity of edges, is MS-expressible via this representation. (This example has been considered in Section 1.3.1.) But we will not say that the set of connected graphs is MS-definable. However, we *can* say that the set of simple connected graphs is MS-definable: the representation $\lfloor \cdot \rfloor$ is faithful on the class C of simple graphs.

---

[16] If C is $\mathcal{L}$-definable, then a class of structures $\mathcal{D} \subseteq C$ is $\mathcal{L}$-definable with respect to C if and only if it is $\mathcal{L}$-definable, and the same holds for $\mathcal{L}$-expressibility defined next. See Corollary 5.12 for a use of this observation.

The expressibility of P and the definability of L are always *with respect to* some reference class C that we will sometimes leave implicit. We will usually omit the mention of the representation *rep* for words, terms and graphs, whenever it equals $\lfloor \cdot \rfloor$ as defined in Example 5.2. For a set L of p-graphs (or simple s-graphs) of bounded type, the default representation is $\lfloor \cdot \rfloor_C$ where C is the union of all types of the elements of L.[17] In each case, several variants of these default representations are considered in Example 5.2, which are all $\mathcal{L}$-equivalent for every $\mathcal{L} \in \{\text{FO}, \text{MS}, \text{SO}\}$ in the following precise sense. For another mapping $rep' : C \rightarrow STR^c(\mathcal{R}')$, we say that *rep* and *rep'* are $\mathcal{L}$-*equivalent* if the same properties of the elements of C are $\mathcal{L}$-expressible via *rep* and via *rep'*, effectively (i.e., the corresponding transformations of sentences in $\mathcal{L}(\mathcal{R}, \emptyset)$ and $\mathcal{L}(\mathcal{R}', \emptyset)$ are computable).

For a set $L \subseteq C$, let $rep(L)$ be the class of structures in $STR^c(\mathcal{R})$ that are isomorphic to some structure $rep(H)$ with $H \in L$. It can easily be shown that L is $\mathcal{L}$-definable via *rep* (with respect to C) if and only if $rep(L)$ is $\mathcal{L}$-definable with respect to $rep(C)$.[18] If the class $rep(C)$ of all structures that represent objects in C (modulo isomorphism) is $\mathcal{L}$-definable, then L is $\mathcal{L}$-definable via *rep* if and only if $rep(L)$ is $\mathcal{L}$-definable (see Footnote 16). For example, if C is the class $\mathcal{GP}[C]$ of p-graphs of type included in C, then the class of structures $\lfloor C \rfloor_C$ is MS-definable (cf. Example 5.2(2)). Hence, a set L of p-graphs is MS-definable if and only if $\lfloor L \rfloor$ is MS-definable (see Example 5.2(2) for the definition of $\lfloor L \rfloor$).

The hierarchy of logical languages $\text{FO} \subset \text{MS} \subset \text{SO}$ yields a corresponding hierarchy of graph properties: every FO-expressible property is MS-expressible (implicitely, via $\lfloor \cdot \rfloor$) but not conversely. In Sections 5.2.5 and 5.2.6, we will introduce intermediate languages between MS and SO and compare their expressive power with that of MS.

Why considering relational structures with constants?

The root of the syntactic tree of a term and the sources of an s-graph have been defined in Examples 5.2(2) and 5.2(3) as constants of their representing structures, but we have observed that these constants can be replaced by unary relations denoting singleton sets. This observation can be extended to all relational structures with constants. For every relational signature $\mathcal{R}$ with $\mathcal{R}_0 \neq \emptyset$, we let $\mathcal{R}_* := \mathcal{R}_+ \cup Lab_{\mathcal{R}}$ where $Lab_{\mathcal{R}} := \{lab_a \mid a \in \mathcal{R}_0\}$ is a new[19] set of unary relation symbols. For every $S \in STR^c(\mathcal{R})$, we let S_* be the structure in $STR^c(\mathcal{R}_*)$ such that $lab_{aS_*} := \{a_S\}$ for every $a \in \mathcal{R}_0$ and $R_{S_*} = R_S$ for every $R \in \mathcal{R}_+$. We denote by $STR^c_*(\mathcal{R}_*)$ the class of structures in $STR^c(\mathcal{R}_*)$ such that the sets denoted by the relation symbols lab_a in $Lab_{\mathcal{R}}$ are singletons. The mapping $S \mapsto S_*$ is a bijection between (concrete or abstract) $\mathcal{R}$-structures and $\mathcal{R}_*$-structures satisfying this condition. Every first-order,

---

[17] Note that $\mathcal{L}$-definability is only defined for sets of p-graphs of bounded type, and similarly for simple s-graphs.

[18] In both directions, the same sentence can be used. Faithfulness of *rep* is needed in the if direction.

[19] i.e., disjoint from $\mathcal{R}$.

second-order or monadic second-order formula over $\mathcal{R}$ can be rewritten into one over $\mathcal{R}_*$ expressing in S_* the same property as the given formula[20] in S. However, this transformation may increase the quantifier-height: the atomic formula $R(a,x)$ where a is a constant symbol and x is a variable is transformed into $\exists y(lab_a(y) \wedge R(y,x))$. In the proof of the Recognizability Theorem, it will be important to express the operations on p-graphs and s-graphs in terms of the disjoint union of structures and of transformations specified by quantifier-free formulas. In order to express in this way the parallel-composition, a basic operation of the HR algebra, we cannot handle sources of graphs with unary relations, hence, we must use constants.

5.1.6 Decidability questions

Let $\mathcal{R}$ be a relational signature, let $\mathcal{C} \subseteq STR^c(\mathcal{R})$ and let $\mathcal{L}$ be a logical language as in the previous section. The $\mathcal{L}$-theory of $\mathcal{C}$ is the following set of sentences:

$$Th_{\mathcal{L}}(\mathcal{C}) := \{\varphi \in \mathcal{L}(\mathcal{R},\emptyset) \mid S \models \varphi \text{ for every } S \in \mathcal{C}\}.$$

We say that the $\mathcal{L}$-theory of $\mathcal{C}$ is *decidable* if some algorithm can test the membership in this set of any given sentence belonging to $\mathcal{L}(\mathcal{R},\emptyset)$. The $\mathcal{L}$-*satisfiability problem for* $\mathcal{C}$ is the problem of deciding whether a given sentence of $\mathcal{L}(\mathcal{R},\emptyset)$ belongs to the following set:

$$Sat_{\mathcal{L}}(\mathcal{C}) := \{\varphi \in \mathcal{L}(\mathcal{R},\emptyset) \mid S \models \varphi \text{ for some } S \in \mathcal{C}\}.$$

We will always consider logical languages $\mathcal{L}$ such that $\mathcal{L}(\mathcal{R},\emptyset)$ is closed under negation. Hence, the $\mathcal{L}$-theory of $\mathcal{C}$ is decidable if and only if its $\mathcal{L}$-satisfiability problem is decidable, because $Sat_{\mathcal{L}}(\mathcal{C}) = \{\varphi \in \mathcal{L}(\mathcal{R},\emptyset) \mid \neg\varphi \notin Th_{\mathcal{L}}(\mathcal{C})\}$.

Similar definitions hold for a class $\mathcal{C}$ of objects via a representation $rep : \mathcal{C} \to STR^c(\mathcal{R})$. We use the same default representations as in Section 5.1.5. As an example, for a finite functional signature F and a set of terms $L \subseteq T(F)$: $Th_{\mathcal{L}}(L) := \{\varphi \in \mathcal{L}(\mathcal{R}_F,\emptyset) \mid \lfloor t \rfloor \models \varphi \text{ for every } t \in L\} = Th_{\mathcal{L}}(\lfloor L \rfloor)$.

Theorem 5.5 The first-order theory of the class of graphs, of the class of graphs of degree at most 3, and of the class of planar graphs of degree at most 3 are undecidable. Thus, so are their monadic second-order and their second-order theories.[21]   □

The first assertion of this theorem is due to Trakhtenbrot. See Section 5.7 for references. We will exhibit in Section 5.4 some conditions on a set of graphs $\mathcal{C}$ ensuring that its MS-theory and its MS-satisfiability problem are decidable. It is thus

---

[20] Conversely, every first-order, second-order or monadic second-order formula expressing a property of structures in $STR^c_*(\mathcal{R}_*)$ can be rewritten into an equivalent formula over $\mathcal{R}$ that is respectively first-order, second-order or monadic second-order. See Lemma 7.5 for the details.

[21] These results concern only finite graphs. However, the first-order theory of the class of all finite and infinite graphs is also undecidable. See any textbook on first-order logic.

desirable to separate the decidable cases from the undecidable ones. The following result about grids is a basic tool for obtaining undecidability results for monadic second-order theories.

We define a *rectangular grid* as a directed graph isomorphic, for some positive integers n and m, to the graph $G_{n \times m}$ defined as follows (cf. Example 2.56(3)). We let its vertex set $V_{G_{n \times m}}$ be $[0, n-1] \times [0, m-1]$ and its edges be the pairs $((i,j),(i',j'))$ such that $0 \le i \le i' \le n-1$, $0 \le j \le j' \le m-1$ and, either $i' = i+1$ and $j' = j$, or $i' = i$ and $j' = j+1$. A *square grid* is a graph isomorphic to $G_{n \times n}$. We denote $und(G_{n \times m})$ by $G^{u}_{n \times m}$, cf. Proposition 2.106.

Theorem 5.6 The monadic second-order satisfiability problem for any set of graphs containing graphs isomorphic to $G_{n \times n}$ or to $G^{u}_{n \times n}$ for infinitely many integers n is undecidable.

Proof: We first consider the case of directed graphs. One can construct a monadic second-order sentence γ that defines the square grids with respect to the class of simple directed graphs (see Proposition 5.14(1) in Section 5.2.3). Consider a Turing machine M with total alphabet $A_M = \{a_1, \ldots, a_m\}$ (input letters, states, end markers, blank symbol) and an initial configuration $w_M \in A_M^*$. If $(U_1, \ldots, U_m)$ is a partition of the set of vertices of a square grid $G_{n \times n}$, then one can view it as defining a sequence of n words of length n in A_M^*. Each line of the grid encodes a word, where a line of $G_{n \times n}$ is an induced subgraph with set of vertices $[0, n-1] \times \{j\}$ for some j and the $(i+1)$-th letter of this word is a_k if and only if $(i,j) \in U_k$. The property of being a line can be expressed by an MS formula by Proposition 5.14(2).

A partition $(U_1, \ldots, U_m)$ of $V_{G_{n \times n}}$ encodes a computation sequence of M if the first line encodes the initial configuration w_M (with blank symbols to the right), and every two consecutive lines encode configurations that form a correct transition of M. If some line encodes a configuration with an accepting state, then we get from $G_{n \times n}$ and the sequence of configurations encoded by $(U_1, \ldots, U_m)$ a halting computation of M.

There is an algorithm that constructs, for every Turing machine M, an MS formula $\psi_M(X_1, \ldots, X_m)$ such that for every n, $\lfloor G_{n \times n} \rfloor \models \psi_M(U_1, \ldots, U_m)$ if and only if $(U_1, \ldots, U_m)$ encodes on $G_{n \times n}$ a halting computation of M. It follows that if the considered set of graphs $\mathcal{C}$ contains infinitely many square grids, then M halts if and only if some graph in $\mathcal{C}$ satisfies the sentence φ_M:

$$\gamma \wedge \exists X_1, \ldots, X_m . \psi_M,$$

if and only if φ_M belongs to $Sat_{MS}(\mathcal{C})$. Hence, for each set $\mathcal{C}$ of directed graphs containing infinitely many square grids, the halting problem of Turing machines reduces to the MS-satisfiability problem for $\mathcal{C}$. The latter problem is thus undecidable.

The proof is the same for a set of undirected graphs with the help of Proposition 5.14(3). ∎

Remark 5.7 Let us assume that in this construction, the machine M is deterministic and that ψ_M is defined in such a way that, when M halts, the square grid on which its unique halting computation is encoded is of minimal size. It follows that there exists at most one relational structure (up to isomorphism, it is then $\lfloor G_{n \times n} \rfloor$ for some n) satisfying φ_M. Hence, since the halting problem for deterministic Turing machines is undecidable, even if we know that a monadic second-order sentence φ has only finitely many models up to isomorphism, we cannot construct this set by an algorithm that takes all sentences φ as input. □

We have just discussed the problem of verifying if a given sentence is true in some structure of a (fixed) set. We now introduce briefly the problem of verifying if a given sentence is true in a single given structure. This problem, called the *model-checking problem*, is trivially decidable since structures are by definition finite and the validity of a sentence is an effective notion. Its algorithmic complexity depends on the considered structures and on the logical language of the given sentence. This issue is the subject of Descriptive Complexity (see Section 5.7 for references). We only quote two results relevant to the algorithmic applications to be studied in Chapter 6:

(1) For each first-order sentence φ, one can check if $S \models \varphi$ in time $O(|\varphi| \cdot \|S\|^w)$, where w is the *width* of φ defined as the maximum number of free variables of a subformula of φ (Proposition 6.6 in [*Lib04] or Proposition 4.24 in [*FluGro]).
(2) Each level of the polynomial hierarchy contains a model-checking problem for some monadic second-order sentence that is complete for polynomial reductions (Proposition 11 in [MakPnu]).

These results distinguish sharply first-order from monadic second-order logic with respect to the complexity of model-checking. Furthermore, the second result shows that monadic second-order logic is, by itself, of no help for obtaining efficient model-checking algorithms over all graphs. However, by restricting the tree-width or the clique-width of input graphs or relational structures, we will obtain polynomial algorithms, as we will see in Chapters 6 and 9.

5.2 Graph properties expressible in monadic second-order logic

In the previous section and in Chapter 1, we have given several examples of graph properties expressed by logical formulas (via $\lfloor \cdot \rfloor$). We now review some tools that help to construct formulas. Since they work for second-order logic, we present them for this language. We will also review some negative results that show that certain properties are not expressible in a particular language.

In order to shorten notation, we will use $\bar{x}$ to denote tuples of pairwise distinct first-order variables (typically $\bar{x} = (x_1, \ldots, x_n)$) and $\overline{X}$ to denote tuples of pairwise distinct relation variables (typically $\overline{X} = (X_1, \ldots, X_m)$). In such cases, we will denote by $\overline{d}$ an n-tuple $(d_1, \ldots, d_n)$ of values in the domain D_S of a concrete structure S, where d_i is taken as value of x_i for each i. These values need not be pairwise distinct. We will denote similarly by $\overline{E}$ an m-tuple of relations on D_S of respective arities $\rho(X_1), \ldots, \rho(X_m)$, which need not be pairwise distinct. We will also regard $\overline{x}$ and $\overline{X}$ as the corresponding sets of variables by using notation like $y \in \overline{x}$ or $z \in \overline{X} \cup \overline{x}$. With this convention, $SO(\mathcal{R}, \overline{X} \cup \overline{x})$ denotes $SO(\mathcal{R}, \{X_1, \ldots, X_m, x_1, \ldots, x_n\})$. If $\varphi \in SO(\mathcal{R}, \overline{X} \cup \overline{x})$, then the use of the notation $S \models \varphi(\overline{E}, \overline{d})$ implies that $\overline{E}$ is an m-tuple of relations on D_S of appropriate arities and that $\overline{d} \in D_S^n$.

Two formulas φ and φ' in $SO(\mathcal{R}, \overline{X} \cup \overline{x})$ are *equivalent*, denoted $\varphi \equiv \varphi'$, if for all concrete $\mathcal{R}$-structures S, for all relations $E_1, \ldots, E_m$ on D_S of respective arities $\rho(X_1), \ldots, \rho(X_m)$ and all tuples $\overline{d} \in D_S^n$, we have

$$S \models \varphi(\overline{E}, \overline{d}) \text{ if and only if } S \models \varphi'(\overline{E}, \overline{d}).$$

Two equivalent formulas express the same property of the relevant tuples $(\overline{E}, \overline{d})$ in every structure $S \in STR^c(\mathcal{R})$. Equivalence can be relativized to a subset $\mathcal{C}$ of $STR^c(\mathcal{R})$: equivalent formulas express the same property of the relevant tuples in the structures of $\mathcal{C}$.

5.2.1 Substitutions and relativization

If φ is a formula in $SO(\mathcal{R}, \mathcal{X})$, $x_1, \ldots, x_n$ are pairwise distinct first-order variables in $\mathcal{X}$ and $t_1, \ldots, t_n$ are terms,[22] then we denote by $\varphi[t_1/x_1, \ldots, t_n/x_n]$ the result of the substitution of $t_1, \ldots, t_n$ for $x_1, \ldots, x_n$. If some of the terms $t_1, \ldots, t_n$ are variables then, before substitution is done, some bound variables of φ may have to be *renamed*. For example, if $\varphi(x_1, x_2, x_3)$ is

$$\forall y[edg(x_1, y) \vee edg(y, x_2)] \wedge \forall u[edg(x_1, u) \Rightarrow edg(u, x_3)],$$

then $\varphi[y/x_1, u/x_2, y/x_3]$ is (or can be, because bound variables can be renamed in several ways):

$$\forall y'[edg(y, y') \vee edg(y', u)] \wedge \forall u[edg(y, u) \Rightarrow edg(u, y)].$$

In many cases, we will simplify the notation $\varphi[t_1/x_1, \ldots, t_n/x_n]$ into $\varphi(t_1, \ldots, t_n)$. Clearly, if φ is monadic second-order or first-order, then so is $\varphi[t_1/x_1, \ldots, t_n/x_n]$. This transformation satisfies the following property, which shows its semantic meaning:

Lemma 5.8 Let φ belong to $SO(\mathcal{R}, \{X_1, \ldots, X_m, x_1, \ldots, x_n, u_1, \ldots, u_q\})$, let $t_1, \ldots, t_n$ be terms and let $y_1, \ldots, y_p$ be the variables occurring in these terms. Let

---

[22] We recall that a term is either a first-order variable or a constant symbol (we do not use function symbols of positive arity for writing logical formulas).

$\psi(X_1,\ldots,X_m,y_1,\ldots,y_p,u_1,\ldots,u_q)$ be the formula $\varphi[t_1/x_1,\ldots,t_n/x_n]$. For every concrete $\mathcal{R}$-structure S, for all relations $E_1,\ldots,E_m$ on D_S of respective arities $\rho(X_1),\ldots,\rho(X_m)$ and for all $\overline{d} \in D_S^p$ and $\overline{f} \in D_S^q$, we have

$$S \models \psi(\overline{E},\overline{d},\overline{f}) \text{ if and only if } S \models \varphi(\overline{E},d_1',\ldots,d_n',\overline{f}),$$

where for all $i = 1,\ldots,n$, we let $d_i' := a_S$ if $t_i = a$ and a belongs to $\mathcal{R}_0$, and $d_i' := d_j$ if $t_i = y_j$. $\qquad\qquad\square$

We now define substitutions to relation symbols of formulas intended to define the corresponding relations. Let $\varphi \in SO(\mathcal{R},\overline{X} \cup \overline{U} \cup \overline{x})$, where $\overline{X} = (X_1,\ldots,X_p)$, $\overline{U} = (U_1,\ldots,U_q)$ and $\overline{x} = (x_1,\ldots,x_n)$. For each $i = 1,\ldots,p$, we let $\psi_i \in SO(\mathcal{R},\overline{U} \cup \overline{x} \cup \overline{w}_i)$, where $\overline{w}_i$ is a sequence of $\rho(X_i)$ pairwise distinct first-order variables with $\overline{x} \cap \overline{w}_i = \emptyset$. We let $\varphi[\lambda\overline{w}_1 \cdot \psi_1/X_1,\ldots,\lambda\overline{w}_p \cdot \psi_p/X_p]$ be the formula in $SO(\mathcal{R},\overline{U} \cup \overline{x})$ constructed as follows:

- by renaming some bound variables if necessary, one defines a formula φ' equivalent to φ where no variable of $\overline{X} \cup \overline{U} \cup \overline{x}$ is bound;
- then, one replaces every atomic formula $X_i(u_1,\ldots,u_{\rho(X_i)})$ of φ' (where the terms $u_1,\ldots,u_{\rho(X_i)}$ need not be pairwise distinct) by the formula $\psi_i[u_1/y_1,\ldots,u_{\rho(X_i)}/y_{\rho(X_i)}]$ (which belongs to $SO(\mathcal{R},\overline{U} \cup \overline{x})$), where $\overline{w}_i = (y_1,\ldots,y_{\rho(X_i)})$.

It is clear that $\varphi[\lambda\overline{w}_1 \cdot \psi_1/X_1,\ldots,\lambda\overline{w}_p \cdot \psi_p/X_p]$ is first-order (resp. monadic second-order) if φ and the formulas ψ_i are first-order (resp. monadic second-order). With this notation, we have the following lemma:

Lemma 5.9 Let $S \in STR^c(\mathcal{R})$, let $\overline{E}$ be a q-tuple of relations on D_S of appropriate arities and let $\overline{d} \in D_S^n$. For each $i = 1,\ldots,p$, let T_i be the $\rho(X_i)$-ary relation on D_S defined by

$$(a_1,\ldots,a_{\rho(X_i)}) \in T_i \text{ if and only if } S \models \psi_i(\overline{E},\overline{d},a_1,\ldots,a_{\rho(X_i)}).$$

Then
$$S \models \varphi[\lambda\overline{w}_1 \cdot \psi_1/X_1,\ldots,\lambda\overline{w}_p \cdot \psi_p/X_p](\overline{E},\overline{d}) \text{ if and only if}$$
$$S \models \varphi(T_1,\ldots,T_p,\overline{E},\overline{d}).$$

Proof: Straightforward verification from the definition, by using an induction on the structure of φ. $\qquad\blacksquare$

We use the λ-notation $\lambda\overline{w}_i \cdot \psi_i$ because ψ_i defines (for fixed values of $\overline{U} \cup \overline{x}$) a mapping $: D_S^{\rho(X_i)} \to \{True,False\}$. We will also simplify the notation $\varphi[\lambda\overline{w}_1 \cdot \psi_1/X_1,\ldots,\lambda\overline{w}_p \cdot \psi_p/X_p]$ into $\varphi[Z_1/X_1,\ldots,Z_p/X_p]$ (or just $\varphi(Z_1,\ldots,Z_p)$) in the case where, for each i, Z_i is a relation variable such that $\rho(Z_i) = \rho(X_i)$, ψ_i is $Z_i(y_1,\ldots,y_{\rho(X_i)})$ and $\overline{w}_i = (y_1,\ldots,y_{\rho(X_i)})$.

We now present the relativization of a formula to a substructure.

Lemma 5.10 Let $\varphi \in SO(\mathcal{R}, \{Y_1, \ldots, Y_m, x_1, \ldots, x_n\})$ and let X be a set variable not in $\{Y_1, \ldots, Y_m\}$. A formula φ' in $SO(\mathcal{R}, \{Y_1, \ldots, Y_m, X, x_1, \ldots, x_n\})$ can be constructed such that for every $S \in STR^c(\mathcal{R})$, every m-tuple of relations $\overline{E} = (E_1, \ldots, E_m)$ on D_S with $E_i \subseteq D_S^{\rho(Y_i)}$, every subset A of D_S that contains a_S for each $a \in \mathcal{R}_0$, and every $\overline{d} \in A^n$:

$$S \models \varphi'(\overline{E}, A, \overline{d}) \text{ if and only if } S[A] \models \varphi(\overline{E'}, \overline{d}),$$

where $E_i' := E_i \cap A^{\rho(Y_i)}$ for each i. If φ is first-order or monadic second-order, then φ' is first-order or monadic second-order respectively.

Proof: We can assume that X does not occur in φ. (If X is bound in φ, then we rename its bound occurrences.) We let φ' be associated with φ by the following inductive definition, where φ may have free variables of all types (relation variables as well as first-order variables):

$\psi' = \psi$ for every atomic formula ψ,
$(\psi_1 \ op \ \psi_2)' = \psi_1' \ op \ \psi_2'$ where $op \in \{\wedge, \vee, \Rightarrow, \Leftrightarrow\}$,
$(\neg \psi)' = \neg \psi'$,
$(\forall x. \psi)' = \forall x [x \in X \Rightarrow \psi']$ and $(\exists x. \psi)' = \exists x [x \in X \wedge \psi']$,
$(\forall Y. \psi)' = \forall Y. \psi'$ and $(\exists Y. \psi)' = \exists Y. \psi'.$[23]

In each case, and by induction, we get that $qh(\psi') = qh(\psi)$.   ∎

The formula φ', called the *relativization* of φ to X, will be denoted by $\varphi \upharpoonright X$. Semantically, it expresses the relativization to the substructure induced by X of the property expressed by φ. Its quantifier-height is the same as that of φ.

5.2.2 Transitive closure and path properties

We now consider the logical expression of the reflexive and transitive closure T^* of a binary relation T. In Section 1.3.1, we have already defined the monadic second-order formula $TC[R; x, y]$:

$$\forall X [x \in X \wedge \forall u, v(u \in X \wedge R(u, v) \Rightarrow v \in X) \Rightarrow y \in X].$$

In every structure S, this formula defines the binary relation R_S^*. One can also use it for defining the relation T^* if T is not a relation of the considered structure but is defined by some formula, for example by $\alpha(u, v)$ equal to $edg(u, v) \vee edg(v, u)$. The relation T may be expressed by a formula with free variables besides u and v.

---

[23] Since set variables, such as Y, are always used in formulas of the form $x \in Y$, we need not specify that $Y \subseteq X$ (and similarly for arbitrary relation variables). This construction must be modified if we use set predicates as we will do in Section 5.2.6, cf. Remark 5.26.

Such a formula can be $\beta(u,v,Y)$ defined as $edg(u,v) \wedge u \in Y \wedge v \in Y$ to take another example about graphs. Then, the formula $(TC[R;x,y])[\lambda u,v \cdot \alpha/R]$, more simply denoted by $TC[(\lambda u,v \cdot \alpha);x,y]$ and defined as the result of the substitution of the formula α for the relation symbol R in $TC[R;x,y]$, expresses that the vertices x and y are the ends of a possibly empty undirected path (cf. Lemma 5.9). The formula $x \in Y \wedge y \in Y \wedge TC[(\lambda u,v \cdot \beta);x,y]$ expresses that there is a directed Y-*path* from x to y, i.e., a directed path from x to y in the subgraph induced by Y.

Note that $TC[(\lambda u,v \cdot \gamma);x,y]$ is a monadic second-order formula if γ is a monadic second-order formula. With these tools, we can express by monadic second-order formulas many properties of paths in graphs. In the next lemma, we write formulas over the signature $\mathcal{R}_s$, intended to express some basic graph properties.[24]

Proposition 5.11 There exist monadic second-order formulas expressing in $\lfloor G \rfloor$ the following properties of vertices x, y and of a set of vertices X of a directed graph G:

P1: G is strongly connected;
P2: $x,y \in X$ and there is a directed X-path from x to y;
P3: X is a connected component of G;
P4: G, assumed to be simple, is a rooted tree;[25]
P5: for G assumed without circuit, X is the set of vertices of a directed path from x to y;
P6: G is planar.

For every finite set C of source labels and every positive integer k, the following property of an s-graph G of type included in C is monadic second-order expressible in $\lfloor G \rfloor_C$:

P7: G has tree-width at most k.[26]

Note that, except P4, these properties do not depend on the multiplicity of edges. They can be expressed logically for all graphs via the representation $\lfloor \cdot \rfloor$.

Proof: We will denote by φ_i, $i = 1,\ldots,6$, the formula expressing property Pi, in a concrete $\mathcal{R}_s$-structure $S = \lfloor G \rfloor$.

(1) φ_1 is the formula $\forall x,y. TC[edg;x,y]$. It is trivially true in the empty structure that represents the empty graph. An empty directed graph is strongly connected. (For the same reason, we defined an empty graph as connected, cf. Definition 2.9.)

(2) φ_2 is $x \in X \wedge TC[(\lambda u,v \cdot \beta);x,y]$, where β is $edg(u,v) \wedge u \in X \wedge v \in X$.

---

[24] See Chapter 2 for graph theoretical terminology.
[25] More precisely, this means that the property of being a rooted tree is MS-expressible with respect to the class $\mathcal{C}$ of simple directed graphs (see Section 5.1.5). Similarly, in the next property P5, $\mathcal{C}$ is the class of directed graphs without circuit.
[26] In other words, the set of s-graphs $TWD(\leq k, C)$ is MS-definable. For the definition of $\lfloor G \rfloor_C$ see Example 5.2(2).

(3) φ_3 is $\exists y[y \in X \wedge \forall x(x \in X \Leftrightarrow TC[(\lambda u, v \cdot \alpha); x, y])]$ where α is $edg(u,v) \vee edg(v,u)$.

(4) Let G be simple. That it has no circuits (and hence, no loops) can be expressed by the sentence γ:

$$\forall x, y(TC[edg; x, y] \Rightarrow \neg edg(y,x)).$$

Then φ_4 is the sentence

$$\gamma \wedge \exists x \forall y. TC[edg; x, y] \wedge \forall x, y, z(edg(x,z) \wedge edg(y,z) \Rightarrow x = y),$$

expressing that G has no circuits, that every vertex is reachable from some vertex (the root) by a directed path and that every vertex is of indegree at most 1. (We recall that rooted trees have edges directed from the root towards the leaves.) Note that if we add to φ_4 the condition that every vertex has outdegree at most one, we obtain a sentence φ_4' expressing that G is a directed path.

(5) The formula $\varphi_2(x,y,X)$ expresses that x and y belong to X and there is a directed X-path from x to y. Hence the formula $\varphi_2'(x,y,X)$ defined as[27]

$$\varphi_2(x,y,X) \wedge \forall Y[Y \subseteq X \wedge \varphi_2(x,y,Y) \Rightarrow X = Y],$$

expresses that x and y belong to X and that X is the set of vertices $\{z_0, z_1, \ldots, z_k\}$ of a directed path $x = z_0 \rightarrow z_1 \rightarrow z_2 \rightarrow \cdots \rightarrow z_{k-1} \rightarrow z_k = y$ that is minimal in the sense that there is no edge $z_i \rightarrow z_j$ for $i+1 < j$. Hence, this path cannot be shortened by a "shortcut"; there may exist several minimal paths of different lengths from x to y. If $x = y$, then $X = \{x\}$. This construction is valid even if G has circuits.

We now want to construct a formula $\varphi_5(x,y,X)$ describing all directed paths from x to y, not only the minimal ones. We assume that G has no circuits (otherwise the construction is not possible, see Proposition 5.13 below). In order to express that X is linearly ordered by the relation $edg^*_{G[X]}$, we let $\varphi_5'(X)$ be

$$\forall x, y[x \in X \wedge y \in X \Rightarrow \varphi_2(x,y,X) \vee \varphi_2(y,x,X)].$$

Let $\varphi_5(x,y,X)$ be the formula defined as

$$x \in X \wedge y \in X \wedge \varphi_5'(X) \wedge \forall z \in X[\varphi_2(x,z,X) \wedge \varphi_2(z,y,X)].$$

It says that X is linearly ordered by $edg^*_{G[X]}$ with first element x and last element y. If $x = y$, then $X = \{x\}$. This is equivalent to the fact that X is the set of vertices of a directed path from x to y because the graph G is finite and without circuit.

(6) This has been proved in Corollary 1.15 by means of the characterization of planar graphs in terms of excluded minors and the formalization of minor inclusion by MS-formulas.

---

[27] For readability, $Y \subseteq X$ replaces $\forall y[y \in Y \Rightarrow y \in X]$, $X = Y$ replaces $\forall x[x \in X \Leftrightarrow x \in Y]$ and $\varphi_2(x,y,Y)$ stands for $\varphi_2[Y/X]$, cf. Section 5.2.1 for substitutions of variables.

(7) For Property P7, we first consider graphs (without sources, i.e., $C = \emptyset$ and $\lfloor G \rfloor_C = \lfloor G \rfloor$). For each integer k, we know by Corollary 2.60(2) (Section 2.4.2) that there is a finite set of excluded minors characterizing the graphs of tree-width at most k. Corollary 1.14 gives the desired sentence $\varphi_{7,k}$.

We now consider an s-graph G of type C' included in C, represented by the $\mathcal{R}_{s,C}$-structure $\lfloor G \rfloor_C$. We have noted in Section 2.4.2 that it has tree-width at most k if and only if the graph $fg_{C'}(und(G) \parallel K_{C'})$ has tree-width at most k, hence, if and only if $\lfloor fg_{C'}(und(G) \parallel K_{C'}) \rfloor \models \varphi_{7,k}$. The latter condition is equivalent to $\lfloor G \rfloor_C \models \varphi'_{7,k,C}$ where $\varphi'_{7,k,C}$ is the sentence $\varphi_{7,k}[\lambda u, v \cdot \theta / edg]$ and θ is

$$edg(u,v) \vee \bigvee_{a,b \in C, a \neq b} (lab_a(u) \wedge lab_b(v)).$$

Thus, $\varphi'_{7,k,C}$ expresses property P7.   ∎

Note that sentence $\varphi_{7,k}$ does not specify any tree-decomposition of width k. However, it follows from Example 5.2(4) that, for each k, one can construct a second-order sentence (not a monadic one) ψ_k of the form $\exists X. \theta(X)$ where X is a binary relation variable and $\theta(X)$ is a monadic second-order formula, such that ψ_k is true in $\lfloor G \rfloor$ if and only if G has a normal tree-decomposition (T,f) of width at most k. The binary relation that is the intended value of X is son_T to be used in the structure $S_2(G,T,f) = \langle V_G, edg_G, son_T \rangle$ (cf. Example 5.2(4)). The formula $\theta(X)$ is of the form $\theta_1(X) \wedge \theta_2(X)$, where $\theta_1(X)$ expresses that son_T determines a rooted tree T that is normal for G, and $\theta_2(X)$ expresses that $|f(u)| \leq k+1$ for every $u \in N_T$. The formula $\theta_2(X)$ uses a monadic second-order formula $\mu(X,u,w)$ that defines in $\lfloor G \rfloor$, for every binary relation son_T satisfying $\theta_1(X)$, the binary relation $box_f = \{(u,w) \mid w \in f(u)\}$, cf. Example 5.2(4). We will discuss in Section 7.6 the possibility of replacing ψ_k by a *monadic* second-order sentence that can specify at least one tree-decomposition of width k of any graph of tree-width at most k.[28]

At the end of Section 5.1.2, we have shown that the regular language ab^*c is MS-definable, i.e., that the class $\lfloor ab^*c \rfloor$ of $\mathcal{W}_A$-structures that represent the words of ab^*c is MS-definable with respect to the class of structures representing words. We now consider the MS-definability of the latter class.

Corollary 5.12 For every alphabet A, the class of $\mathcal{W}_A$-structures $\lfloor A^* \rfloor$ is MS-definable. Similarly, for every finite functional (possibly many-sorted) signature F, the class of $\mathcal{R}_F$-structures $\lfloor T(F) \rfloor$ is MS-definable.

---

[28] It is not known yet whether the class $CWD(\leq k)$ of graphs of clique-width at most k is monadic second-order definable. Since it is not closed under taking minors, the proof for tree-width cannot extend. However, this class is closed under taking induced subgraphs, hence is characterized by a set of minimal excluded induced subgraphs. This set is $\{P_4\}$ for $k = 2$ and is infinite (and not yet characterized) for $k \geq 3$. If this set would be monadic second-order definable, then so would be $CWD(\leq k)$. The monadic second-order definability of the class of comparability graphs is provable in this way (see [Cou06a], Lemma 5.1).

Proof: Easy constructions using the sentences φ_4' and φ_4 of the proof of Proposition 5.11, respectively for words and terms. Such constructions can be done for the variants of $\lfloor w \rfloor$ and $\lfloor t \rfloor$ defined in Example 5.2. We omit the details.   ∎

It follows that for every language $L \subseteq A^*$, the class of $\mathcal{W}_A$-structures $\lfloor L \rfloor$ is MS-definable if and only if it is MS-definable with respect to the class $\lfloor A^* \rfloor$. Hence, L is MS-definable if and only if $\lfloor L \rfloor$ is MS-definable. A similar statement holds for terms.

We now review some graph properties that are not MS-expressible.[29]

A directed (resp. undirected) graph is Hamiltonian if it has at least two vertices and a circuit (resp. a cycle) going through all vertices. A perfect matching is a set of undirected edges that link each vertex of X to a single vertex of Y and vice versa, where $\{X, Y\}$ is a bipartition of the vertex set of the graph.

Proposition 5.13 The following properties are not MS-expressible:
(1) Two sets X and Y have equal cardinality.
(2) A simple graph is Hamiltonian, or has a perfect matching.
(3) In a simple directed graph, a set of vertices X is that of a directed path from x to y.
(4) A simple graph has a nontrivial automorphism.

Proof: We will use the result for words corresponding to Theorem 1.16 (see Section 1.10): if $L \subseteq \{a, b\}^*$ is the set of words w such that $\lfloor w \rfloor \models \varphi$ where φ is an MS sentence, then L is a regular language.[30]

(1) Assume that we have a formula $\psi \in \mathrm{MS}(\emptyset, \{X, Y\})$ such that for every set S and every $V, W \subseteq S$:

$$S \models \psi(V, W) \text{ if and only if } Card(V) = Card(W).$$

Then the MS sentence $\psi[\lambda x \cdot lab_a(x)/X, \lambda x \cdot lab_b(x)/Y]$ belonging to $\mathrm{MS}(\mathcal{W}_{\{a,b\}}, \emptyset)$ characterizes (as a subset of $\{a, b\}^*$) the language E of words having as many occurrences of a and b. This language is not regular, so we get a contradiction with the above result.

(2) With every word $w \in \{a, b\}^+$ represented by the structure $\lfloor w \rfloor = \langle [n], suc_{\lfloor w \rfloor}, lab_{a \lfloor w \rfloor}, lab_{b \lfloor w \rfloor} \rangle$, we associate the graph K_w with set of vertices $[n]$ and an edge from i to j if and only if $i \in lab_{a \lfloor w \rfloor}$ and $j \in lab_{b \lfloor w \rfloor}$ or vice versa. Hence K_w is a complete bipartite directed graph. It is Hamiltonian if and only if w belongs to the language E already used in (1).

Let us now assume the existence of a sentence η in $\mathrm{MS}(\mathcal{R}_s, \emptyset)$ that defines the Hamiltonian graphs among the simple directed ones. The FO formula $\mu(x_1, x_2)$ equal to $(lab_a(x_1) \wedge lab_b(x_2)) \vee (lab_b(x_1) \wedge lab_a(x_2))$ defines in $\lfloor w \rfloor$ the edges of K_w (note

---

[29] Recall from Section 5.1.5 that the default representation of a graph G is $\lfloor G \rfloor$.
[30] We will actually reprove this result in a more general setting in Section 5.3.9, cf. Corollary 5.66.

that $D_{\lfloor w \rfloor} = V_{K_w}$). It follows that for every word $w \in \{a,b\}^*$ of length at least 2, we have

$$w \in E \text{ if and only if } K_w \text{ is Hamiltonian if and only if } \lfloor w \rfloor \models \eta[\lambda x_1, x_2 \cdot \mu/edg].$$

This would imply that the words in E of length at least 2 form a regular language, which is not the case. This contradiction completes the proof.

The proof is the same for undirected graphs and for perfect matchings because a complete bipartite graph $K_{n,m}$ has a Hamiltonian cycle or a perfect matching if and only if $n = m$.

(3) Assume that we have a formula $\varphi \in MS(\mathcal{R}_s, \{x,y,X\})$ expressing that, in a simple directed graph, X is the set of vertices of a directed path from x to y. Then the MS sentence

$$\exists X [\forall u (u \in X) \wedge \exists x, y (\varphi(x,y,X) \wedge x \neq y \wedge edg(y,x))]$$

expresses that the considered graph assumed to have at least three vertices is Hamiltonian. This is not possible by (2), hence no such formula φ exists.

(4) The proof is similar to that of (2). With every word of the form $a^n bcd^m$ with $n \geq 2$, $m \geq 2$, we associate the graph $H_{n,m}$ with vertex set $[-n, m+1]$ and undirected edges between i and $i+1$ for every $i \in [-n, m-1]$ and between 0 and $m+1$. (The vertices in $[-n, -1]$ correspond to the occurrences of a, those in $[1,m]$ to the occurrences of d, 0 is the occurrence of b and $m+1$ is that of c.) This graph has a nontrivial automorphism if and only if $n = m$, if and only if the given word belongs to the nonregular language $\{a^n bcd^n \mid n \geq 2\}$. Hence, no monadic second-order sentence can express that a graph has a nontrivial automorphism. We omit a detailed writing of the formula analogous to $\mu(x_1, x_2)$ in (2).   ∎

In Remark 5.21 we will prove in a similar way that the property for a graph to have a spanning tree of degree at most 3 is not MS-expressible.

5.2.3 A worked example: the definition of square grids

The *rectangular grid* $G_{n \times m}$ is defined in Section 5.1.6 as the graph with set of vertices $V := [0, n-1] \times [0, m-1]$ and set of directed edges consisting of the pairs $((i,j), (i',j'))$ such that $(i,j), (i',j') \in V$ and either $i' = i$ and $j' = j+1$ or $i' = i+1$ and $j' = j$. If $x \to y$ in $G_{n \times m}$, we say that y is a *successor* of x and that x is a *predecessor* of y.

The *north-*, *west-*, *south-* and *east-borders* of $G_{n \times m}$ are the following sets of vertices:

$$X_n = [0, n-1] \times \{m-1\}, \ X_w = \{0\} \times [0, m-1],$$
$$X_s = [0, n-1] \times \{0\}, \text{ and } X_e = \{n-1\} \times [0, m-1].$$

A set of the form $[0, n-1] \times \{j\}$ is called a *line* of $G_{n \times m}$.

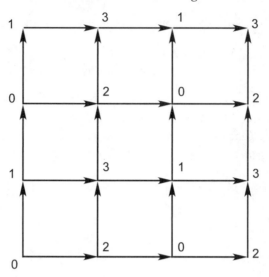

Figure 5.1　The well-coloring of the grid $G_{4\times4}$.

The *well-coloring* of $G_{n\times m}$ is the 4-tuple of sets of vertices Y_0, Y_1, Y_2, Y_3 such that

$$(i,j) \in Y_0 \text{ if and only if } i \equiv 0 \text{ and } j \equiv 0 \quad (\text{mod } 2),$$
$$(i,j) \in Y_1 \text{ if and only if } i \equiv 0 \text{ and } j \equiv 1 \quad (\text{mod } 2),$$
$$(i,j) \in Y_2 \text{ if and only if } i \equiv 1 \text{ and } j \equiv 0 \quad (\text{mod } 2),$$
$$(i,j) \in Y_3 \text{ if and only if } i \equiv 1 \text{ and } j \equiv 1 \quad (\text{mod } 2).$$

Figure 5.1 shows the well-coloring of $G_{4\times4}$, where a vertex is colored by k if it belongs to Y_k.

All these notions will be used in the following proposition in order to prove that the class of *square grids*, i.e., of graphs isomorphic to $G_{n\times n}$ for some $n \geq 2$, is MS-definable, and to identify their lines. These constructions are used in the proof of Theorem 5.6.

We let $\mathcal{W}$ be the set of set variables $\{X_n, X_w, X_s, X_e, Y_0, Y_1, Y_2, Y_3\}$, and $\mathcal{W}'$ be $\mathcal{W} \cup \{U_a, U_b, U_c\}$.

Proposition 5.14

(1)　There exists a formula θ in $\mathrm{MS}(\mathcal{R}_s, \mathcal{W})$ such that, for every simple directed graph G and every mapping $\nu : \mathcal{W} \to \mathcal{P}(V_G)$, we have $(\lfloor G \rfloor, \nu) \models \theta$ if and only if G is isomorphic to the square grid $G_{n\times n}$ for some $n \geq 2$ by a bijection $h : V_G \to V_{G_{n\times n}}$ such that $h(\nu(X_n))$, $h(\nu(X_w))$, $h(\nu(X_s))$ and $h(\nu(X_e))$ are the four borders of $G_{n\times n}$ and $h(\nu(Y_0))$, $h(\nu(Y_1))$, $h(\nu(Y_2))$ and $h(\nu(Y_3))$ form the well-coloring of $G_{n\times n}$.

There is a sentence γ in $MS(\mathcal{R}_s, \emptyset)$ that defines the square grids with respect to the class of simple directed graphs.

(2) There exists a formula χ in $MS(\mathcal{R}_s, \mathcal{W} \cup \{Z\})$ such that for every $(\mathcal{W} \cup \{Z\})$-assignment ν in $\lfloor G \rfloor$ that satisfies θ, we have

$(\lfloor G \rfloor, \nu) \models \chi$ if and only if $h(\nu(Z))$ is a line of $G_{n \times n}$ (where h is as in (1)).

(3) These results extend to simple undirected graphs and the undirected square grids $G^u_{n \times n} := und(G_{n \times n})$. One constructs formulas θ' in $MS(\mathcal{R}_s, \mathcal{W}')$, γ' in $MS(\mathcal{R}_s, \emptyset)$ and χ' in $MS(\mathcal{R}_s, \mathcal{W}' \cup \{Z\})$. A $\mathcal{W}'$-assignment ν' in $\lfloor G \rfloor$ that satisfies θ' defines an isomorphism from G to $G^u_{n \times n}$, the sentence γ' defines the undirected square grids with respect to the class of simple undirected graphs and the formula χ' satisfies the same property as in (2), for θ' and $G^u_{n \times n}$.

Proof: (1) We let θ_0 express the following conditions concerning a simple directed graph G given with sets of vertices $X_n, X_w, X_s, X_e, Y_0, Y_1, Y_2, Y_3$. These conditions are intended to characterize the grids $G_{2n \times 2n}$ for $n \geq 1$:

(1.1) G is connected and has no circuits;

(1.2) Y_0, Y_1, Y_2, Y_3 form a partition of V_G and G is bipartite with bipartition $\{Y_0 \cup Y_3, Y_1 \cup Y_2\}$; assuming this we will call a vertex, a successor of y or a predecessor of y that belongs to Y_i, an *i-vertex*, an *i-successor* of y or an *i-predecessor* of y respectively;

(1.3) every vertex has at most one *i*-successor and at most one *i*-predecessor for each *i*;

(1.4) $G[X_n]$ is a directed path of even length consisting alternatively of 1- and 3-vertices; its first vertex is a 1-vertex that is the unique element of $X_n \cap X_w$; its last vertex is a 3-vertex that is the unique element of $X_n \cap X_e$; the condition that $G[X_n]$ is a directed path is MS-expressible by Proposition 5.11 (see P4) and Lemma 5.10; the other conditions are easy to formalize;

(1.4') $G[X_e]$ is a directed path of even length consisting alternatively of 2- and 3-vertices; its first vertex is a 2-vertex which is the unique element of $X_s \cap X_e$; its last vertex is a 3-vertex which is the unique element of $X_n \cap X_e$;

(1.4'') and (1.4''') state similar conditions on the sets X_s and X_w, see Figure 5.1;

(1.5) a 1-vertex in X_n has a 3-successor and no 0-successor; a 3-vertex in $X_n - X_e$ has a 1-successor and no 2-successor; the 3-vertex in $X_n \cap X_e$ has no successors;

(1.5'), (1.5'') and (1.5''') state similar properties of X_e, X_s, X_w, see Figure 5.1; in particular a 1-vertex in X_w has a 0-predecessor and no 3-predecessor; a 0-vertex in $X_w - X_s$ has a 1-predecessor and no 2-predecessor; the 0-vertex in $X_w \cap X_s$ has no predecessors;

(1.6) for each vertex x in $V_G - (X_n \cup X_e)$ there exist y, z, u such that $y \neq z$, $x \to y$, $x \to z$, $y \to u$ and $z \to u$;

(1.7) the unique directed path from the vertex in $X_s \cap X_w$ with vertices having colors $0, 2, 3, 1, 0, 2, 3, 1, \ldots$ in this order reaches the unique vertex in $X_n \cap X_e$; this

condition can be expressed with the help of the transitive closure construction
of Section 5.2.2.

Conditions (1.1)–(1.6) characterize the well-colored rectangular grids of the form
$G_{2n \times 2m}$ for $n \geq 1, m \geq 1$. In particular, Conditions (1.4) and (1.5) express how the
borders must be. Conditions (1.3) and (1.6) express how the neighbourhood of a vertex
not on any border must be. The alternation of colors in Condition (1.7) specifies an
"ascending staircase" from left to right. This condition holds if and only if $m = n$. If
$n = m = 1$ the path defined by this condition has only three vertices, with respective
colors $0, 1, 3$.

It is not hard to modify this construction in order to characterize the grids of the
form $G_{(2n+1) \times (2n+1)}$ for $n \geq 1$ by a formula θ_1. Hence, we take $\theta_0 \vee \theta_1$ as formula θ.
As sentence γ, we take $\exists \overline{W}.\theta$.

(2) A line of G given with $X_n, X_w, \ldots, Y_3$ as above is a set $L \subseteq V_G$ such that:

(2.1) $G[L]$ is a directed path from a vertex in X_w to a vertex in X_e;
(2.2) either $L \subseteq Y_0 \cup Y_2$ or $L \subseteq Y_1 \cup Y_3$.

This gives the construction of χ.

(3) We now extend this construction to simple undirected graphs. Let G be so and let
$(U_a, U_b, U_c) \in \mathcal{P}(V_G)^3$ define a proper vertex 3-coloring of G.[31] We let $G(U_a, U_b, U_c)$
be the simple directed graph G' such that:

$$V_{G'} := V_G,$$
$$E_{G'} := \{(u, v) \mid u - v \text{ is an edge of } G \text{ and}$$
$$(u \in U_a \wedge v \in U_b) \vee (u \in U_b \wedge v \in U_c) \vee (u \in U_c \wedge v \in U_a)\}.$$

It is clear that $G = und(G')$, and $G_{n \times m} = und(G_{n \times m})(U_a, U_b, U_c)$ for some proper
vertex 3-coloring (U_a, U_b, U_c). Such a coloring of $und(G_{4 \times 4})$ is shown in Figure 5.2.
Thus, G is isomorphic to $G_{n \times n}^{u}$ for some $n \geq 2$ if and only if there exists a proper vertex
3-coloring (U_a, U_b, U_c) such that $G(U_a, U_b, U_c)$ is isomorphic to $G_{n \times n}$ for some $n \geq 2$.

Hence in order to extend the constructions of (1) and (2) to simple undirected
graphs, we let θ' be the formula

$$\kappa(U_a, U_b, U_c) \wedge \theta[(\lambda u, v \cdot \alpha) / edg],$$

where κ expresses that (U_a, U_b, U_c) defines a proper vertex 3-coloring of the given
graph G (cf. Chapter 1, Section 1.3.1) and α is the following formula, which directs
the edges of G according to the coloring defined by (U_a, U_b, U_c):

$$edg(u, v) \wedge [(u \in U_a \wedge v \in U_b) \vee (u \in U_b \wedge v \in U_c) \wedge (u \in U_c \wedge v \in U_a)].$$

---

[31] *Proper* means that two adjacent vertices have different colors.

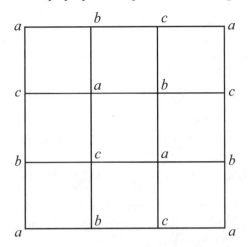

Figure 5.2 A proper 3-coloring of $G^{\mathrm{u}}_{4\times 4}$.

Hence, $\theta[(\lambda u, v \cdot \alpha)/edg]$ expresses that $G(U_a, U_b, U_c)$ is isomorphic to $G_{n\times n}$ for some n. The desired sentence γ' is then $\exists \overline{\mathcal{W}'}.\theta$, and thus the formula χ' is $\chi[(\lambda u, v \cdot \alpha)/edg]$.   ∎

Directed, vertex-labeled rectangular grids, called *pictures*, can be considered as 2-dimensional words and studied with tools of logic and automata theory. See [*GiaRes] for a survey of such a study.

5.2.4 Monadic second-order definability of regular languages

We prove that every regular language (cf. Theorem 3.62) is monadic second-order definable. The converse of this statement will be proved in the Recognizability Theorem (Theorem 5.64 and Corollary 5.67). Together, these results establish Theorem 1.16.

Theorem 5.15 Let F be a finite functional signature. For every finite F-automaton $\mathscr{A}$, one can construct a monadic second-order sentence φ in $MS(\mathcal{R}_F, \emptyset)$ such that $L(\mathscr{A}) = \{t \in T(F) \mid \lfloor t \rfloor \models \varphi\}$.

Proof: Let $\mathscr{A} = \langle F, Q_{\mathscr{A}}, \delta_{\mathscr{A}}, Acc_{\mathscr{A}} \rangle$ be a finite F-automaton as in Definition 3.46. Without loss of generality, we can assume that $\mathscr{A}$ is ε-free and that $Q_{\mathscr{A}} = [n]$. A term t in $T(F)$ is represented by the $\mathcal{R}_F$-structure

$$\lfloor t \rfloor := \langle N_t, (son_{it})_{i\in[k]}, (lab_{ft})_{f\in F}, rt_t \rangle,$$

where $N_t = Pos(t)$ and $k = \rho(F)$ (cf. Example 5.2(3); the variants of this representation can be used as well).

We must express the existence on t of an accepting run $r : Pos(t) \to Q_{\mathscr{A}}$ of $\mathscr{A}$. Such a run will be represented by the n-tuple $(U_1, \dots, U_n)$ of subsets of $N_t = Pos(t)$ such that $U_i = r^{-1}(i)$. An arbitrary n-tuple $(U_1, \dots, U_n) \in \mathcal{P}(N_t)^n$ represents an accepting run if and only if the following conditions hold:

(C1) $\{U_1, \dots, U_n\}$ is a partition[32] of N_t;
(C2) rt_t belongs to U_i for some $i \in Acc_{\mathscr{A}}$;
(C3) for each occurrence u of a constant symbol f, if u belongs to U_i, then $f \to_{\mathscr{A}} i$;
(C4) for each occurrence u of $f \in F_p$ with $p \geq 1$, if u belongs to U_i and its sons $u_1, \dots, u_p$ belong respectively to $U_{j_1}, \dots, U_{j_p}$, then $f[j_1, \dots, j_p] \to_{\mathscr{A}} i$.

The conjunction of conditions (C1)–(C4) is expressible by a formula ψ in $FO(\mathcal{R}_F, \{X_1, \dots, X_n\})$ (where a set U_i as above will be the value of the variable X_i). The desired sentence φ is thus $\exists X_1, \dots, X_n. \psi$.   ∎

Remark 5.16 (1) The size of φ is bounded by $O((\sharp\mathscr{A})^2 + \|\mathscr{A}\|)$, where $\|\mathscr{A}\|$ is the size of $\mathscr{A}$, defined as its weighted number of transition rules (cf. Definition 3.46).

(2) Theorem 5.15 states that $L(\mathscr{A})$ is MS-definable with respect to $T(F)$ (implicitly via the faithful representation $\lfloor \cdot \rfloor$). If α_F is the monadic second-order sentence that defines $\lfloor T(F) \rfloor$ (cf. Corollary 5.12), then $MOD(\varphi \wedge \alpha_F) = \lfloor L(\mathscr{A}) \rfloor$ where φ is constructed from $\mathscr{A}$ by Theorem 5.15. Hence, the class of structures $\lfloor L(\mathscr{A}) \rfloor$ is also MS-definable.

(3) Words are in bijection with terms over a unary functional signature, and automata on words can be seen as particular automata on terms as observed in Section 3.3.1, Remark 3.51(2). Moreover, as observed in Example 5.2(3), the representations $\lfloor w \rfloor$ and $\lfloor \mu(w) \rfloor$ are equivalent (where w is a word and $\mu(w)$ the corresponding term). Hence, the construction of Theorem 5.15 can be adapted to finite automata on words in a straightforward manner.[33]

5.2.5 Edge set quantifications

As already noted, the representation of a graph by a relational structure as we have done up to now, is faithful only for simple graphs. It does not allow the expression of graph properties that depend on the multiplicity of edges nor of certain properties of simple graphs of the form "there exists a set of edges such that..." such as the property of being Hamiltonian or of having a perfect matching, see Proposition 5.13. We now define another representation, where the edges are elements of the domain, that is faithful for all graphs and is appropriate for expressing logically certain properties of graphs that depend on the multiplicity of

---

[32] Some sets U_i may be empty.
[33] Thomas has proved in [Tho82] that a sentence φ as in this theorem can be constructed with a single existential set quantification, however large n may be.

edges. Furthermore, it allows the expression of more properties of simple graphs by using monadic second-order sentences. For a graph G, the corresponding structure is denoted $\lceil G \rceil$.

Definition 5.17 (Representation by incidence graphs) Incidence graphs have been considered in Chapter 2, Example 2.56(5). We will distinguish between the cases of directed and undirected graphs.[34]

If G is a concrete undirected graph, then its incidence graph $Inc(G)$ is the simple directed bipartite graph with vertex set $V_G \cup E_G$ and edges $e \to v$, for every $e \in E_G$ and $v \in V_G$ such that v is an end vertex of e. We let $\lceil G \rceil := \lfloor Inc(G) \rfloor$. It will be convenient to denote the adjacency relation of $Inc(G)$ by the binary relation symbol *in* instead of *edg*, in order to distinguish it from the adjacency relation of G. Hence $\lceil G \rceil := \langle V_G \cup E_G, in_G \rangle$ with $in_G := \{(e,v) \mid v$ is an end vertex of $e\}$, and this structure belongs to $STR^c(\mathcal{R}_m^u)$, where $\mathcal{R}_m^u := \{in\}$.

If G is directed, its incidence graph $Inc(G)$ is also simple, directed and bipartite, but with labeled edges. Its vertex set is $V_G \cup E_G$ and its edges are $e \to v$ labeled by 1 if $e \in E_G$ has tail v and $e \to v$ labeled by 2 if $e \in E_G$ has head v. If e is a loop, we have two edges $e \to v$, with labels 1 and 2. We let then $\lceil G \rceil := \lfloor Inc(G) \rfloor :=$ $\langle V_G \cup E_G, in_{1G}, in_{2G} \rangle$, where in_{1G} and in_{2G} define the edges labeled respectively by 1 and 2.[35] This is an $\mathcal{R}_m^d$-structure, where $\mathcal{R}_m^d := \{in_1, in_2\}$.[36]

Two graphs G and G' are isomorphic if and only if the structures $\lceil G \rceil$ and $\lceil G' \rceil$ are isomorphic, hence this representation is faithful for all graphs. In a structure S representing a graph G, the edges are the elements x of D_S that satisfy the formula $\exists y. in_1(x,y)$ or the formula $\exists y. in(x,y)$ depending on whether S belongs to $STR^c(\mathcal{R}_m^d)$ or to $STR^c(\mathcal{R}_m^u)$. In the former case, $S = \lceil G \rceil$ for some directed graph G if and only if in_{1S} and in_{2S} are two functional relations with the same domain, call it E, and $in_{1S} \cup in_{2S} \subseteq E \times (D_S - E)$. In the latter case, $S = \lceil G \rceil$ for some undirected graph G if and only if $|\{y \in D_S \mid (x,y) \in in_S\}| \leq 2$ for every $x \in D_S$ and $in_S \subseteq E \times (D_S - E)$ where $E = \{x \in D_S \mid (x,y) \in in_S$ for some $y\}$. In both cases, the conditions are first-order expressible. If they hold, then $E_G = E$ and $V_G = D_S - E$; in the directed case, $vert_G(x) = (y_1, y_2)$ if and only if $(x, y_1) \in in_{1S}$ and $(x, y_2) \in in_{2S}$, and in the undirected case, $vert_G(x) = \{y \in D_S \mid (x,y) \in in_S\}$. Thus, the class of $\mathcal{R}_m^d$-structures $\{\lceil G \rceil \mid G$ is a directed graph$\}$ is FO-definable, and similarly for the undirected case (cf. Corollary 5.12).

---

[34] We do not consider here an undirected edge as a pair of opposite directed edges. This convention is natural and convenient if graphs are handled as binary relations on sets of vertices, but it is not if graphs are handled as sets of edges that share vertices, and this is what we are doing.

[35] $Inc(G)$ can be formalized as a $(\{\kappa\}, \{1,2\})$-labeled graph, where κ is a default vertex label. Its representation $\lfloor Inc(G) \rfloor$ is as in Example 5.2(2), without the useless unary relation $lab_{\kappa G}$ (and with in_i instead of edg_i).

[36] The subscript "m" recalls that the considered representation is faithful for graphs with multiple edges. The superscripts "d" and "u" stand for "directed" and "undirected" respectively.

A variant of $\lceil G \rceil$ will be useful. We let $\mathcal{R}_m := \{in, lab_{Edge}\}$. For a directed graph G, we define $\lceil G \rceil \in STR^c(\mathcal{R}_m)$ such that[37]

$$D_{\lceil G \rceil} := V_G \cup E_G,$$
$$lab_{Edge\lceil G \rceil} := E_G,$$
$$in_{\lceil G \rceil} := \{(v,e) \mid e \in E_G, v \text{ is the tail of } e\}$$
$$\cup \{(e,v) \mid e \in E_G, v \text{ is the head of } e\}.$$

If G is undirected, we let $D_{\lceil G \rceil}$ and $lab_{Edge\lceil G \rceil}$ be as above and $in_{\lceil G \rceil}$ be $\{(e,v) \mid e \in E_G$ and v is an end vertex of $e\}$. In this case the relation lab_{Edge} is redundant but keeping it allows us to use the same signature for directed and undirected graphs.

With this variant, the relational structure representing a directed or undirected graph G is, up to the change of edg into in, of the form $\lfloor H \rfloor$ for some vertex-labeled (rather than edge-labeled) directed graph H. For algorithmic applications (cf. Chapter 6) it is better to have few relation symbols of arity more than 1. In fact, H is an obvious variant of $Inc(G)$.

We now extend these definitions to s-graphs. In order to represent s-graphs of type $\tau(G) \subseteq C$ (the type $\tau(G)$ of G is the set of names of its sources), where C is a finite subset of $\mathcal{A}$, we expand the three above defined signatures by the set $\{lab_a \mid a \in C\}$ of unary symbols. Additionally, to represent s-graphs of type C we expand them by the set C of constant symbols, cf. Example 5.2(2). We let $\mathcal{R}^d_{m,C} := \mathcal{R}^d_m \cup \{lab_a \mid a \in C\}$ and $\mathcal{R}^d_{ms,C} := \mathcal{R}^d_m \cup C$, and similarly[38] for $\mathcal{R}^u_m$ and $\mathcal{R}_m$. Let G be an s-graph with $\tau(G) \subseteq C$. In the $\mathcal{R}^d_{m,C}$-structure $\lceil G \rceil_C$, we let $lab_a(x)$ hold if and only if x is an a-source of G. This representation is faithful for the s-graphs in $\mathcal{JS}^d[C]$. Similar to the case of p-graphs, if $C \subseteq D \subseteq \mathcal{A}$, then $\lceil G \rceil_C$ and $\lceil G \rceil_D$ are equivalent representations for the s-graphs in $\mathcal{JS}^d[C]$, for first-order and monadic second-order logic. Now let $\tau(G) = C$. Then we denote $\lceil G \rceil_C$ also by $\lceil G \rceil$. The mapping $G \mapsto \lceil G \rceil$ is a faithful representation of all s-graphs. Additionally, in the $\mathcal{R}^d_{ms,C}$-structure $\lceil G \rceil$, the interpretation of a is the a-source of G, for every $a \in C$; since no two source names designate the same vertex, this variant of $\lceil G \rceil$ is constant-separated (cf. Section 5.1.1). The two representations $\lceil G \rceil$ are equivalent for both first-order and monadic second-order logic. Similar definitions hold for $\mathcal{R}^u_m$ and $\mathcal{R}_m$.

For an s-graph G, we define $Inc(G)$ as the simple s-graph such that $Inc(G)^\circ := Inc(G^\circ)$ and $slab_{Inc(G)} := slab_G$. Thus, $Inc(G)$ has the same sources as G, which are all in V_G. From the above definitions and those in Example 5.2(2) for simple s-graphs, it is clear that $\lceil G \rceil_C = \lfloor Inc(G) \rfloor_C$ and $\lceil G \rceil = \lfloor Inc(G) \rfloor$ (provided we take "matching" variants).

---

[37] We use the notation $\lceil G \rceil$ as for the first representation, because the two representations are very similar and are equivalent for both first-order and monadic second-order logic.

[38] In the subscripts of these signatures "m", "C" and "s" recall that multiple edges are represented, that C is a set of source labels and that the sources are constants rather than unary relations, respectively.

Vertex and edge labels are represented in a natural way by additional unary relations. That is, if λ is an edge label, then $e \in lab_{\lambda \lceil G \rceil}$ if and only if e is an edge with label λ. The corresponding signatures are denoted by adding the subscript $[K, \Lambda]$. Technical details will be given when necessary.

If L is a set of (labeled) graphs, then $\lceil L \rceil$ denotes the corresponding set of structures $\lceil G \rceil$ for $G \in L$. If L is a set of (labeled) s-graphs of type included in C, for some finite $C \subseteq \mathcal{A}$, then $\lceil L \rceil_C$ denotes the set of structures $\lceil G \rceil_C$ for $G \in L$. In particular, if C is the union of all (finitely many) types of the elements of L, then we also denote $\lceil L \rceil_C$ by $\lceil L \rceil$. Note that $\lceil L \rceil$ is defined only when L has bounded type, i.e., $L \subseteq \mathcal{JS}[C]$ for some finite $C \subseteq \mathcal{A}$. Note also that $\lceil L \rceil = \lfloor Inc(L) \rfloor$, where $Inc(L) := \{Inc(G) \mid G \in L\}$.

The properties of a graph G can be expressed logically, either via its representation by $\lceil G \rceil$ with domain $V_G \cup E_G$, or via the initially defined representation $\lfloor G \rfloor$ with domain V_G. The representation $\lfloor G \rfloor$ allows only quantification over vertices, sets of vertices, and relations on vertices (according to the logical language we consider), whereas the representation $\lceil G \rceil$ also allows quantification over edges, sets of edges and relations on edges.

We define the MS_2-*expressible* graph properties as those that are MS-expressible via the representation of G by $\lceil G \rceil$. They are expressed by formulas over the signatures $\mathcal{R}_m^d$, $\mathcal{R}_m^u$ or $\mathcal{R}_m$, depending on the chosen variant of the representation $\lceil G \rceil$. The subscript 2 recalls that these formulas use quantifications over two types of objects. (This point is formalized below, cf. the discussion before Proposition 5.20.) An MS_1-expressible graph property is just an MS-expressible one, i.e., based on the standard representation $\lfloor \cdot \rfloor$ that we have considered up to now. The subscript 1 will be used to insist that formulas use only quantifications over vertices and sets of vertices. The notion of MS_2-*definability* of a set of graphs is defined accordingly (which is possible because the representation $\lceil \cdot \rceil$ is faithful on the class of all graphs). For the MS_2-definability of a set L of s-graphs of bounded type, the representation $\lceil \cdot \rceil_C$ is used, where C is the union of all types of the elements of L. Since the class of structures $\lceil \mathcal{JS}[C] \rceil_C$ is MS-definable, L is MS_2-definable if and only if $\lceil L \rceil$ is MS-definable (analogously to the MS-definability of a set of p-graphs of bounded type).

Similarly, we have FO_2-, FO_1-, SO_2- or SO_1-expressible graph properties. We will also speak of FO_2- or SO_2-definable sets of graphs, and of FO_1-, MS_1- or SO_1-definable sets of simple graphs. These definitions extend to s-graphs and p-graphs in obvious ways.

For every graph G, the relation edg_G is definable in $\lceil G \rceil$ by a first-order formula. It follows that for each $\mathcal{L} \in \{FO, MS, SO\}$, a graph property is $\mathcal{L}_2$-expressible if it is $\mathcal{L}_1$-expressible. Trivially, this does not hold in the other direction, see Proposition 5.19(1) below. Proposition 5.20 compares these different types of expressibility for properties of simple graphs.

The notations MS_1 and MS_2 refer both to monadic second-order formulas, but written over different relational signatures. The corresponding relational structures represent graphs in different (and nonequivalent) ways. For an arbitrary $\mathcal{R}$-structure S, the MS formulas over $\mathcal{R}$ generalize the MS_1 formulas for graphs (a simple graph is considered as a binary relation on its vertex set), and they form the "natural" monadic second-order language for expressing properties of S (cf. Section 5.1). In Section 9.2, we will define the incidence graph $Inc(S)$ of an $\mathcal{R}$-structure S. The MS formulas over the relational signature of $\lfloor Inc(S) \rfloor$, which is a faithful representation of S, will be called the MS_2 formulas. They generalize to relational structures the MS_2 formulas initially defined for graphs.

Example 5.18 (Chordal graphs) It follows from Proposition 2.72 that the chordal graphs are the graphs in the class $\mathcal{C}$ of nonempty, connected, simple, loop-free and undirected graphs that have no induced cycle with at least four vertices. They are also the graphs in $\mathcal{C}$ having a perfect spanning tree. By the first characterization, the class of chordal graphs is MS_1-definable (with respect to the class of simple graphs). However, this characterization uses forbidden configurations and it says nothing of the perfect spanning trees of the graphs recognized as chordal. All efficient algorithms on chordal graphs (see Section 2.4.4) are based on perfect elimination orders, which can be obtained easily from perfect spanning trees.

A perfect spanning tree of a simple loop-free undirected graph G can be defined as a pair (E, r) such that:

(1) E is a set of edges of G forming a spanning tree U;
(2) the pair $T := (U, r)$ is a normal spanning tree of G (cf. Definition 2.13);
(3) T is a perfect spanning tree (cf. Definition 2.71).

These conditions can easily be expressed by an MS_2 formula $\varphi(X, z)$, using some of the formulas of Proposition 5.11. To express Conditions (1) and (2), we need an MS_2-expression of the edge relation son_T of T (where $son_T(x, y)$ holds if and only if y is a son of x). It can be obtained in terms of the root r of T and the set of edges E of the tree U (which is $und(T)$). In fact, y is a son of x in T if and only if there exists an edge $e \in E$ such that (i) x and y are the ends of e and (ii) x and r belong to the same connected component of $U - e$. It follows that the relation $\leq_T$ can be expressed by an MS_2 formula in terms of E and r, and so can Conditions (1) and (2).

Hence a simple, loop-free and undirected graph G is chordal if and only if $\lceil G \rceil \models \exists X, z . \varphi(X, z)$. From every pair (E, r) that satisfies $\varphi(X, z)$ in $\lceil G \rceil$, we can specify by auxiliary monadic second-order formulas the perfect spanning tree T and the associated (optimal) normal tree-decomposition, cf. Corollary 2.73(3) and Example 5.2(4). Hence, there exists a monadic second-order transduction (cf. Sections 1.7 and 1.8) that associates with $\lceil G \rceil$, for any chordal graph G, an optimal tree-decomposition of this graph (Proposition 9.56(1)). As already discussed in Example 5.2(4), a similar result holds for $\lfloor G \rfloor$ when G is k-chordal, for each k (Proposition 9.56(2)). $\square$

Proposition 5.19 The following graph properties are MS_2-expressible but are not MS_1-expressible:

(1) a graph is simple,
(2) in a directed graph, X is a set of edges (resp. W is a set of vertices) forming a directed path from x to y, where $x \neq y$,
(3) a graph is Hamiltonian.

Proof: (1) is clear from the definitions.

(2) We consider a directed graph G. For every $U \subseteq E_G$, we let $V(U)$ denote the set of vertices incident with at least one edge in U. We let $\theta(x,y,Y)$ be the first-order formula expressing the following conditions:

(2.1) x,y are vertices and $x \neq y$;
(2.2) Y is a set of edges;
(2.3) x is incident with exactly one edge in Y and is its tail; similarly, y is incident with exactly one edge in Y and is its head;
(2.4) each vertex in $V(Y) - \{x,y\}$ is incident with exactly two edges in Y, it is the tail of one and the head of the other.

Since G is finite, this formula says that Y is the union of the set X of edges of a path from x to y and of the sets of edges of some circuits disconnected from this path. Hence, X is characterized as the least subset Z of Y such that $\theta(x,y,Z)$ holds. It follows that the desired MS_2 formula[39] is the following ψ:

$$\theta(x,y,X) \vee \forall Z[Z \subset X \Rightarrow \neg\theta(x,y,Z)].$$

There is, of course, no MS_1 formula equivalent to ψ because such a formula has no free variables denoting sets of edges.

To express that W is the set of vertices of a path from x to y, we use the formula $\varphi(x,y,W)$ defined as $\exists X(\psi(x,y,X) \wedge \text{``}W = V(X)\text{''})$. The quantification over sets of edges is thus crucial in φ since we have proved in Proposition 5.13(3) that no formula in $MS(\mathcal{R}_s, \{x,y,W\})$ equivalent to φ does exist.

(3) For directed graphs, this follows from (2) and the proof of Proposition 5.13(3), and from Proposition 5.13(2). The proof for undirected graphs is similar. ∎

For writing and using MS_2 formulas it is often convenient to assume that the variables are *typed* in the following way: a first-order variable written $x{:}e$ (instead of x) will always take values in the set of edges and a first-order variable $x{:}v$ will always take values in the set of vertices. Similarly, a set variable of the form $X{:}e$ will denote sets of edges and a set variable $X{:}v$ will denote sets of vertices. The set of variables is thus $\mathcal{V}_1 \times \{e,v\}$). The atomic formulas must be "correctly typed," e.g., $in(x{:}e,y{:}v)$

---

[39] $Z \subset X$ replaces $\forall z[z \in Z \Rightarrow z \in X] \wedge \exists x[x \in X \wedge x \notin Z]$.

and $x{:}e \in X{:}e$. Obviously, every typed MS_2 sentence is equivalent (with respect to all structures of the form $\lceil G \rceil$) to an ordinary (untyped) MS_2 sentence, that one can construct by an easy induction. For example, a subformula $\exists x{:}e.\, \psi$ is replaced by $\exists x(\exists y.\, in(x,y) \wedge \psi[x/x{:}e])$, assuming without loss of generality that the variable $x{:}v$ does not occur in ψ.

Conversely, every MS_2 sentence φ can be transformed into an equivalent typed one. The transformation is as follows (assuming without loss of generality that no variable occurs more than once in a quantifier of φ). First, every subformula $\exists x.\, \psi$ of φ is replaced by $\exists x{:}e.\, \psi[x{:}e/x] \vee \exists x{:}v.\, \psi[x{:}v/x]$, and similarly for $\forall x$ and $\wedge$. Thus, both types are considered for each first-order variable. Second, incorrectly typed atomic subformulas such as $in(x{:}e, y{:}e)$ are replaced by *False*. Finally, every subformula $\exists X.\, \psi$ is replaced by $\exists X{:}e, X{:}v.\, \psi'$ (and similarly for $\forall X$), where ψ' is obtained from ψ by replacing all atomic subformulas of the form $x{:}e \in X$ and $x{:}v \in X$ respectively by $x{:}e \in X{:}e$ and $x{:}v \in X{:}v$. Thus, the variable $X{:}e$ is interpreted as the set of edges in X and $X{:}v$ as the set of vertices in X. The correctness of this transformation is clear. Similar transformations can be made for FO_2 and SO_2 sentences. In particular, the type of an n-ary relation variable is an element of $\{e, v\}^n$; thus, in the transformation of an untyped SO_2 sentence into a typed one, each n-ary relation variable is replaced by 2^n typed relation variables of arity n.

Proposition 5.20 Let P be a property of simple (labeled) graphs.[40]

(1) P is FO_2-expressible if and only if it is FO_1-expressible.
(2) P is SO_2-expressible if and only if it is SO_1-expressible.
(3) P is MS_2-expressible if it is MS_1-expressible; the converse does not hold.

Proof: As observed above, every $\mathcal{L}_1$-expressible graph property is $\mathcal{L}_2$-expressible, for each $\mathcal{L} \in \{FO, SO, MS\}$. Then, obviously, this also holds for properties of simple graphs. That the converse does not hold for monadic second-order logic follows from Propositions 5.13(2) and 5.19(3). The converse holds for FO because a quantification of the form "there exists an edge e such that $\ldots$" can be replaced by a quantification of the form "there exist vertices x and y that form an edge such that $\ldots$." The proof is similar for SO: quantification over n-ary relations on edges is replaced by quantification over $2n$-ary relations on vertices. We omit the technical details. ∎

Remark 5.21 Let P be the property that a simple undirected graph G has a spanning tree of degree at most 3. This property is MS_2-expressible but not MS_1-expressible. We prove this last fact by the following construction similar to those of Proposition 5.13. We let A and B be two disjoint sets and $G(A, B)$ be the

---

[40] Thus, in statements (1)–(3), expressibility is with respect to the class $\mathcal{C}$ of simple (labeled) graphs.

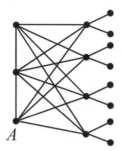

Figure 5.3 The graph $G(A,B)$ for $|A| = 3, |B| = 4$.

graph G such that

> $G[A]$ is a path,
>
> $V_G = A \cup (B \times \{0,1,2\})$ and
>
> $edg_G = edg_{G[A]} \cup E \cup E^{-1}$, where
>
> $E = (A \times (B \times \{0\})) \cup \{((x,0),(x,i)) \mid x \in B, i \in \{1,2\}\}.$

Figure 5.3 shows $G(A,B)$ for $|A| = 3$ and $|B| = 4$. In general, the graph $G(A,B)$ has a spanning tree of degree at most 3 if and only if $|A| + 2 \geq |B|$. If the considered property would be MS_1-expressible, the language $L = \{a^n b^m \mid n,m \geq 1, n+2 \geq m\}$ would be regular – which is not the case. The detailed proof can be done with MS-transductions, which are presented in Section 1.7 and will be studied in Chapter 7: one defines a monadic second-order transduction that transforms $\lfloor a^n b^m \rfloor$ into $\lfloor G(A,B) \rfloor$ for all $n,m \geq 1$ and all sets A,B such that $|A| = n$ and $|B| = m$, and then the Backwards Translation Theorem (already presented in Section 1.7) yields a contradiction with the assertion that property P is MS_1-expressible. $\qquad\square$

The next theorem describes classes of simple graphs on which MS_1 and MS_2 formulas are equally expressive. For these graphs, the representations $\lfloor G \rfloor$ and $\lceil G \rceil$ are thus equivalent with respect to the expression of their properties in MS logic. A graph G is *uniformly p-sparse* if $|E_{G[X]}| \leq p \cdot |X|$ for every $X \subseteq V_G$ (cf. Theorem 4.51). Every simple loop-free undirected planar graph is uniformly 3-sparse (Corollary 4.2.10 in [*Die]). Every undirected graph of degree at most k and every simple loop-free undirected graph of tree-width at most k is uniformly k-sparse (clear for the former case; by Corollary 2.74(1) for the latter case, cf. the proof of Theorem 4.51).

Theorem 5.22 (Sparseness Theorem) Let C be a class of simple and uniformly p-sparse (labeled) graphs, for some $p > 0$. A property is MS_2-expressible with respect to C if and only if it is MS_1-expressible with respect to C. There are effective translations between the defining sentences of both types. $\qquad\square$

The translation of an MS_1 formula into an MS_2 formula that defines the same simple graphs or the same property of vertex sets of simple graphs is easy and does not depend on p. The proof of the opposite translation is quite long and will be given in Section 9.4.

To apply the above result to words and terms, we identify a word w over a finite alphabet A with the vertex-labeled graph $G(w)$ defined in Example 5.2(2), and a term t over a finite signature F with the labeled graph $Syn(t)$. Thus, we define $\lceil w \rceil := \lceil G(w) \rceil$ and $\lceil t \rceil := \lceil Syn(t) \rceil$ and, just as for graphs, define MS_1 and MS_2 formulas for words w to be MS formulas interpreted in $\lfloor w \rfloor$ and in $\lceil w \rceil$ respectively, and similarly for terms. Since $G(w)$ and $Syn(t)$ are graphs of bounded degree, the following result is immediate from Theorem 5.22. However, as an example of the proof of Theorem 5.22, we also give a direct proof.

Corollary 5.23 The same properties of words, of terms and of labeled (rooted) trees can be expressed by MS_1 and by MS_2 formulas.

Proof: Let us consider words w over a finite alphabet A and typed MS_2 sentences φ over $\mathcal{R}^d_{m,[K,\Lambda]}$ with $K = A$ and $\Lambda = \emptyset$. The idea is to replace each edge of $G(w)$ by its head; thus, every edge is encoded by a vertex and every set of edges by a set of vertices. To transform φ into an equivalent MS_1 sentence, we change every atomic subformula $in_1(x{:}e, y{:}v)$ into $suc(y,x)$, and every atomic subformula $in_2(x{:}e, y{:}v)$ into $y = x$, then we drop the types of variables (all variables will denote vertices).

For terms t over a finite signature F we use the same encoding of edges by their heads. Let φ be a typed MS_2 sentence over $\mathcal{R}^d_{m,[K,\Lambda]}$ with $K = F$ and $\Lambda = [\rho(F)]$. To describe the transformation of φ it is convenient to use the representation $\lfloor t \rfloor = \langle N_t, son_t, (lab_{ft})_{f \in F}, (br_{it})_{1 \leq i \leq \rho(F)} \rangle$, see Example 5.2(3). The atomic subformulas $in_j(x{:}e, y{:}v)$ are changed as above, with *son* instead of *suc*, and every atomic subformula $lab_i(x{:}e)$ is changed into $br_i(x)$.

For rooted trees a similar transformation can be used as for terms, and for trees the transformed formula should start by choosing a root (thus turning the tree into a rooted one). We omit the details.   ∎

A table comparing the expressive power of MS_1 and MS_2 formulas, and their extensions with cardinality predicates will be given in Section 5.2.7.

5.2.6 Cardinality predicates

We now introduce an extension of monadic second-order logic called *counting monadic second-order logic*.

Definition 5.24 (Counting monadic second-order logic) For integers p, q with $q > p \geq 0$ and $q \geq 2$, and for a set variable X, we let $Card_{p,q}(X)$ be a new atomic formula expressing that $|X| \equiv p \pmod{q}$. We denote by $CMS(\mathcal{R}, \mathcal{X})$ the extension of MS($\mathcal{R}, \mathcal{X}$) that uses the atomic formulas $Card_{p,q}(X)$ in addition to those previously defined. The notation C_rMS will be used (instead of CMS) to specify the sets of formulas written with the set predicates $Card_{p,q}$ such that $q \leq r$. Hence, C_0MS, C_1MS and MS designate the same formulas.

"Monadic second-order logic" (written in full) will refer in the sequel to counting monadic second-order logic as well as to the language defined in Section 5.1.4. The notations MS, C_rMS and CMS will specify the absence or the possible use of set predicates $Card_{p,q}$. For expressing graph properties, the notations C_rMS$_1$ and C_rMS$_2$ will be used in the obvious way. For every r, the same properties of words, terms and labeled trees can be expressed by C_rMS$_1$ and by C_rMS$_2$-formulas, which generalizes[41] Corollary 5.23.

It is easy to write, for each nonnegative integer p, a formula in FO($\emptyset, \{X\}$) expressing that a set X has cardinality p. It follows that each atomic formula $Card_{p,q}(X)$, for $p \neq 0$, is equivalent to

$$\exists Y \ [Y \subseteq X \wedge \text{“}X - Y \text{ has cardinality } p\text{”} \wedge Card_{0,q}(Y)].$$

Hence, without loss of expressive power, we could define CMS formulas by using only the atomic formulas $Card_{0,q}(X)$ expressing that $|X|$ is a multiple of q. However, having all formulas $Card_{p,q}(X)$ will be useful in some proofs such as the one of Lemma 5.38.

We now show an example of the use of CMS formulas.[42] Let L be a subset of $\mathcal{N}^k$ and let $Card_L(U_1, \ldots, U_k)$ be the property of subsets $U_1, \ldots, U_k$ of a set D that $(|U_1|, \ldots, |U_k|) \in L$. The following proposition will help to obtain a logical characterization of recognizable sets of rooted trees (Section 5.5):

Proposition 5.25 If a set L is recognizable in $\mathbb{N}^k$, then the property $Card_L$ of k-tuples of sets is CMS-expressible.

Proof: We have seen in Section 3.4.8 that **Rec**($\mathbb{N}^k$) is the set of finite unions of Cartesian products $L_1 \times \cdots \times L_k$, where $L_i = \{a_i + \lambda \cdot b_i \mid \lambda \in \mathcal{N}\}$ for some $a_i, b_i \in \mathcal{N}$.[43] That $|U_i|$ belongs to L_i means that $|U_i| = a_i$ if $b_i = 0$ and that $|U_i| \geq a_i$ if $b_i = 1$. These properties are first-order expressible. If $b_i \geq 2$, then $|U_i|$ belongs to L_i if and only if U_i contains a set Y such that $|Y|$ is a multiple of b_i and $|U_i - Y| = a_i$. This property is expressible by a C_{b_i}MS-formula. The result follows. ∎

Remark 5.26 In the proof of Lemma 5.10, which constructs the relativization of a second-order formula φ to a set X, if the formula φ contains set predicates $Card_{p,q}$,

---

[41] Theorem 5.22 actually holds, for every r, for C_rMS$_2$- and C_rMS$_1$-expressibility.
[42] The expression of *vertex-minor inclusion* (a variant of minor inclusion relevant to clique-width) is another, much more complicated one. See [CouOum].
[43] This is clear for $k = 1$, and hence also for $k \geq 2$ because $A \times (B \cup C) = (A \times B) \cup (A \times C)$ for arbitrary sets A, B, C.

we must replace the transformation rules for set quantification with the following:

$$(\forall Y.\psi)' = \forall Y(\forall u(u \in Y \Rightarrow u \in X) \Rightarrow \psi'),$$

$$(\exists Y.\psi)' = \exists Y(\forall u(u \in Y \Rightarrow u \in X) \wedge \psi').$$

Hence, we have $qh(\varphi') \leq qh(\varphi) + 1$ (and not $qh(\varphi') = qh(\varphi)$ as in Lemma 5.10). $\square$

CMS formulas versus MS formulas

Every formula $\varphi \in MS(\emptyset, \{X_1, \ldots, X_m\})$ is equivalent to a finite disjunction of conjunctions of conditions of the form $|Y_1 \cap Y_2 \cap \cdots \cap Y_m| = n$ or $|Y_1 \cap Y_2 \cap \cdots \cap Y_m| > n$ with $n \in \mathcal{N}$ and, for each $i = 1, \ldots, m$, the set Y_i is either X_i or $D - X_i$ where D is the domain of the considered structure (this is proved in [Cou90b] by quantifier elimination). In particular, the property that a set has an even cardinality is not MS-expressible. (Another proof is given in Proposition 7.12 of [*Lib04].) Hence, the CMS formulas have strictly more expressive power than the MS formulas. However, they are equally expressive on linearly ordered (and linearly orderable) structures, as we will see next.

Lemma 5.27 For every $q \geq 2$, one can construct a formula $\gamma_q(X)$ belonging to $MS(\{\leq\}, \{X\})$ to express, in every[44] linearly ordered set $(D, \leq_D)$, that the cardinality of a subset of D denoted by X is a multiple of q.

Proof: One writes a formula γ_q expressing that, if X is not empty, then there exist sets $Y_0, Y_1, \ldots, Y_{q-1}$ forming a partition of X and such that the first element of X is in Y_1, the last element of X is in Y_0 and for every two elements x, y of X, if $x \in Y_p$ and y is the successor of x for the restriction to X of the order $\leq_D$, then $y \in Y_{p'}$, where $p' \equiv p+1 \pmod{q}$. Clearly, if $X = \{x_1, \ldots, x_n\}$ with $x_k <_D x_{k+1}$ for all $k \in [n-1]$, then $Y_p = \{x_i \mid i \in [n], i \equiv p \pmod{q}\}$ for every $p \in [0, q-1]$. ∎

Lemma 5.27 extends to any class $\mathcal{C}$ of structures on which a linear order is definable by MS formulas. We now define precisely this notion.

Definition 5.28 (MS-orderable classes of structures) Let $\mathcal{R}$ be a relational signature, let $\theta_0 \in MS(\mathcal{R}, \{X_1, \ldots, X_m\})$ and $\theta_1 \in MS(\mathcal{R}, \{X_1, \ldots, X_m, x_1, x_2\})$. The pair (θ_0, θ_1) *orders* (implicitly *linearly*) a structure $S \in STR^c(\mathcal{R})$ if the following conditions hold:

$$S \models (\exists X_1, \ldots, X_m.\theta_0) \wedge \forall X_1, \ldots, X_m(\theta_0 \Rightarrow \widehat{\theta_1}),$$

---

[44] The formula γ_q uses a linear order to express a property of X that does not depend on any linear order. We say that such a formula is *order-invariant*.

where $\widehat{\theta}_1$ is the formula in $MS(\mathcal{R}, \{X_1, \ldots, X_m\})$ defined as

$$\forall x.\theta_1(x,x) \wedge \forall x,y(\theta_1(x,y) \wedge \theta_1(y,x) \Rightarrow x = y) \wedge$$

$$\forall x,y,z(\theta_1(x,y) \wedge \theta_1(y,z) \Rightarrow \theta_1(x,z)) \wedge \forall x,y(\theta_1(x,y) \vee \theta_1(y,x)).$$

This means that, for every m-tuple $(E_1, \ldots, E_m)$ satisfying θ_0, the binary relation defined by θ_1, where $X_1, \ldots, X_m$ take respectively the values $E_1, \ldots, E_m$, is a linear order on D_S, and that there is at least one such m-tuple in S.

A set $\mathcal{C} \subseteq STR^c(\mathcal{R})$ is MS-*orderable* if there exists a pair (θ_0, θ_1) that orders all its structures. A set of graphs $\mathcal{C}$ is MS_1-orderable (resp. MS_2-orderable) if the set $\lfloor \mathcal{C} \rfloor$ (resp. the set $\lceil \mathcal{C} \rceil$) is MS-orderable.

It is clear that if (θ_0, θ_1) orders S, then the pair $(\widehat{\theta}_1, \theta_1)$ orders the class $MOD(\exists X_1, \ldots, X_m.\widehat{\theta}_1)$ that contains S. Hence, θ_1 is enough for ordering a set of structures $\mathcal{C}$ because $\widehat{\theta}_1$ can replace θ_0. However, it is in general more clear to use the conditions on $X_1, \ldots, X_m$ defined by a formula θ_0 than by $\widehat{\theta}_1$.

Proposition 5.29 Let $\mathcal{C}$ be an MS-orderable class of concrete $\mathcal{R}$-structures. The CMS-expressible properties of the structures in $\mathcal{C}$ are MS-expressible.

Proof: Let $(\theta_0(X_1, \ldots, X_m), \theta_1(X_1, \ldots, X_m, x_1, x_2))$ be a pair of formulas that orders the structures of $\mathcal{C}$. Let $\varphi \in CMS(\mathcal{R}, \emptyset)$ express some property of the structures of $\mathcal{C}$. We can assume that the variables $X_1, \ldots, X_m$ have no occurrence in φ and that, for every atomic formula $Card_{p,q}(Y)$, we have $p = 0$. For each q, we let γ_q' be $\gamma_q[\lambda x_1, x_2 \cdot \theta_1 / \leq]$, where γ_q is constructed in Lemma 5.27. This formula belongs to $MS(\mathcal{R}, \{X, X_1, \ldots, X_m\})$. For every $S \in \mathcal{C}$, for every $E_1, \ldots, E_m \subseteq D_S$ satisfying θ_0 and for every $E \subseteq D_S$, we have $S \models \gamma_q'(E, E_1, \ldots, E_m)$ if and only if $|E|$ is a multiple of q.

We obtain φ' from φ by replacing every atomic formula $Card_{0,q}(Y)$ by $\gamma_q'(Y, X_1, \ldots, X_m)$ (i.e., by $\gamma_q'[Y/X]$, which is the result of the substitution of Y to X in γ_q'). We let then φ'' be the formula of $MS(\mathcal{R}, \emptyset)$:

$$\exists X_1, \ldots, X_m[\theta_0(X_1, \ldots, X_m) \wedge \varphi'(X_1, \ldots, X_m)].$$

If $S \models \varphi''$, then $S \models \theta_0(E_1, \ldots, E_m)$ and $S \models \varphi'(E_1, \ldots, E_m)$ for some $E_1, \ldots, E_m$, hence $S \models \varphi$ by the construction of φ'. Conversely, if $S \models \varphi$, then there exist $E_1, \ldots, E_m$ satisfying θ_0, hence $S \models \varphi'(E_1, \ldots, E_m)$. Thus we have $S \models \varphi''$. Hence, φ is equivalent to φ'' in every structure S in $\mathcal{C}$. ∎

This result extends to formulas with free variables. Understanding "exactly" which sets of structures are MS-orderable is an open problem. Here are some partial answers.

Proposition 5.30 Let $d \in \mathcal{N}_+$. The class of rooted trees of outdegree at most d is MS_1-orderable. More generally, the class of rooted forests consisting of at most d rooted trees of outdegree at most d is MS_1-orderable.

Proof: Let T be a rooted tree given by the structure $\langle N_T, son_T \rangle$, where $son_T(x,y)$ holds if and only if y is a son of x. We denote by $\geq_T$ the reflexive and transitive closure of son_T, cf. Definition 2.13. A partition $(N_1, N_2, \ldots, N_d)$ of N_T is *good* if no two sons of any node belong to the same set N_i; intuitively it defines a linear order on the sons of each node of T (through the natural order on $[d]$). If T has outdegree at most d (i.e., if each node has at most d sons) then N_T has a good partition in d (possibly empty) sets. From a good partition $(N_1, \ldots, N_d)$ of N_T, we can define the following linear order:

$x \leq y$ if and only if either $x \geq_T y$ or there exist z, x', y' such that $son_T(z, x')$, $x' \geq_T x$, $son_T(z, y')$, $y' \geq_T y$, $x' \in N_i$, $y' \in N_j$ and $i < j$.

It is not hard to see that $\leq$ is a linear order on N_T; in fact, it corresponds to the usual pre-order tree traversal (for rooted trees for which a linear order is given on the sons of each node). An FO_1 formula $\theta_0(X_1, \ldots, X_d)$ can express that a given d-tuple $(N_1, \ldots, N_d)$ of subsets of N_T is a good partition and an MS_1 formula $\theta_1(X_1, \ldots, X_d, x_1, x_2)$ can express that $x_1 \leq x_2$ holds where X_i takes the value N_i (for $i = 1, \ldots, d$) and $\leq$ is defined from $(N_1, \ldots, N_d)$. (One can write θ_1 with the help of the formula φ_2 of Proposition 5.11.)

Hence, (θ_0, θ_1) defines a linear order of N_T (and even a topological order) for every rooted tree T of outdegree at most d. The proof for forests is similar except that in the definition of a good partition, we require that the roots of two trees of the forest do not belong to the same set N_i. Then, we let $x_1 < x_2$ if x_1 and x_2 belong to trees T and T' with roots respectively in N_i and N_j such that $i < j$. We omit the details. ∎

Corollary 5.31 For each $d \in \mathcal{N}$, the class of simple graphs that have a spanning tree of degree at most d is MS_2-orderable.

Proof: Let G be a simple graph that has a spanning tree of degree at most d. We will define a linear order on V_G by means of sets $E, R, N_1, \ldots, N_d$ such that:

(1) E is a set of edges forming a spanning tree U of G;
(2) R consists of a unique vertex r;
(3) $(N_1, \ldots, N_d)$ is a good partition of V_G, where "good" is relative to the rooted tree $T := (U, r)$.

These conditions can be expressed in $\lceil G \rceil$ by an MS formula, and from $(N_1, \ldots, N_d)$ one obtains a linear order of V_G by using Proposition 5.30. For applying it, we need an MS formula for the relation son_T in terms of E and r, which can be obtained as described in Example 5.18. From a linear order $\leq$ on V_G, one defines a lexicographic order on $V_G \times V_G$, whence on E_G, since G is assumed to be simple. If G is undirected, we consider an edge $\{x, y\}$ as a pair (x, y) with $x \leq y$ and use the lexicographic order on $V_G \times V_G$. ∎

Example 5.32 (1) The class of simple 3-connected planar graphs is MS_1-orderable: these graphs have spanning trees of degree at most 3 ([Bar]), hence they are MS_2-orderable. But for planar graphs, MS_2 formulas without free variables denoting edges and sets of edges (cf. the notion of a typed formula defined before Proposition 5.20) can be replaced by MS_1 formulas (cf. Remark 5.21 and Theorem 5.22).

(2) The set of cliques $\{K_n \mid n \geq 1\}$ is MS_2-orderable. It is not MS_1-orderable because otherwise the class of finite sets (edgeless graphs) would be MS_1-orderable and the set predicates $Card_{0,q}(X)$ would be MS-expressible. We know that this is not the case. □

5.2.7 Expressive power of monadic second-order languages

The expressive power of the monadic second-order languages MS_1, MS_2, CMS_1 and CMS_2 is compared in Figure 5.4: the inclusions represented from bottom-up, are proper. The expressive powers of CMS_1 and MS_2 are incomparable. These facts are proved by some examples listed below. They also hold for the expressibility of properties of simple graphs.

For expressing properties of words over a fixed finite alphabet, of terms over a fixed finite signature and of trees of bounded degree, all these four languages are equivalent by Corollaries 5.23 and 5.31. There are some intermediate cases, in particular that of simple uniformly k-sparse graphs (for each fixed k), shown in Figure 5.5. The equivalences follow from Theorem 5.22. Figures 5.6 and 5.7 show other situations. Here are the examples establishing proper inclusions and incomparabilities in these tables.

Example 5.33 The set of *odd stars*, i.e., of graphs $K_{1,2n+1}$ for $n \geq 0$, realizes the proper inclusions of Figure 5.5. It is uniformly 1-sparse and CMS_1-definable, but it is not MS_1-definable, otherwise the even cardinality set predicate would be monadic second-order expressible.

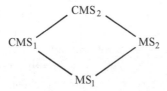

Figure 5.4 The four monadic second-order languages.

$$CMS_1 \ \equiv \ CMS_2$$
$$|$$
$$MS_1 \ \equiv \ MS_2$$

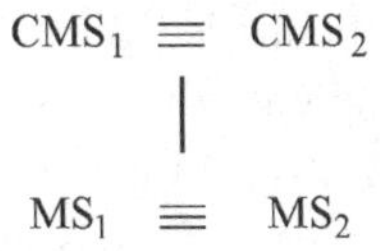

Figure 5.5 The case of simple uniformly k-sparse graphs.

$$CMS_2 \ \equiv \ MS_2$$
$$|$$
$$CMS_1 \ \equiv \ MS_1$$

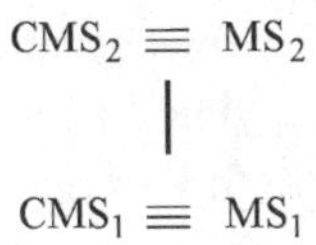

Figure 5.6 The case of joined paths.

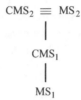

$$\text{CMS}_2 \equiv \text{MS}_2$$

$$\text{CMS}_1$$

$$\text{MS}_1$$

Figure 5.7 The case of labeled cliques.

The set of *joined paths*, i.e., of graphs $P_n \otimes P_m$ for $n, m \geq 1$ is MS_1-orderable, hence, $\text{CMS}_1 \equiv \text{MS}_1$ and $\text{CMS}_2 \equiv \text{MS}_2$ for it (i.e., for expressing the properties of its graphs). Its subset $\{P_n \otimes P_n \mid n \geq 1\}$ is MS_2-definable but not MS_1-definable (similar proof as for Proposition 5.13(2)).

We let LC be the set of $(\{a\}, \emptyset)$-labeled cliques (a vertex has label a or no label, an edge has no label). It is MS_2-orderable, hence $\text{CMS}_2 \equiv \text{MS}_2$ for it. But its subset K consisting of the labeled cliques that have the same number of unlabeled vertices and of vertices labeled by a is MS_2-definable but not CMS_1-definable (by usual arguments).

These examples show that all inclusions of Figure 5.4 are proper. The set of odd stars and the set K show that MS_2 and CMS_1 are incomparable (even for expressing properties of simple graphs only). □

5.3 Monadic second-order logic and recognizability

This section is devoted to the proof of the Recognizability Theorem, a corner-stone of the theory developed in this book. Informally, *every* CMS-*definable set of graphs or of relational structures is recognizable* with respect to an appropriate algebra.

The most technical part of the proof (in the next section) is that of the Splitting Theorem for $\oplus$, the union of disjoint graphs or structures. This theorem says that, for every MS formula φ, the set of assignments that satisfy φ in the union of two disjoint structures S and T can be determined from the sets of those that satisfy auxiliary formulas of no larger quantifier-height in S and in T. Up to equivalence, there are finitely many formulas of bounded quantifier-height with a fixed set of free variables. If φ is a sentence, we obtain a finite set of auxiliary sentences, hence a finite set of properties that is inductive with respect to $\oplus$ (cf. Section 1.2.3). This theorem generalizes a result by Feferman and Vaught for first-order logic and is a (very) special case of the Composition Theorem by Shelah. (See [*Mak] and [Tho97b] for detailed discussions and consequences of these results.) An easy similar result for unary operations defined by quantifier-free formulas without constant symbols, called the Backwards Translation Theorem, yields the Recognizability Theorem for the VR algebra. The proof of that theorem for the HR algebra needs structures with constants (for representing the sources of s-graphs) and presents small technical difficulties.

We will actually prove the Recognizability Theorem for a many-sorted algebra of relational structures with constants denoted by STR.

This long section is organized as follows. We first prove the Splitting Theorem for unions of disjoint relational structures, possibly with constants. Next, we define (unary) quantifier-free operations on structures. Initially we mainly consider restricted quantifier-free operations that can only delete constants from their input structures (they cannot delete "ordinary" elements). For them, we then prove the Backwards Translation Theorem. As a motivation for these definitions, we show that the operations of the VR and of the HR algebra are, or can be expressed in terms of, disjoint union and restricted quantifier-free operations. Then, we extend the Splitting Theorem to derived operations (on structures) built with disjoint union and restricted quantifier-free operations. We apply it to the computation of "bounded theories" (see Section 5.3.6 for definitions). Next we define the algebra STR of relational structures, with the operation of disjoint union and the quantifier-free operations, and we prove the Recognizability Theorem for the subalgebra with restricted quantifier-free operations. This restricted form of the theorem yields the cases of the algebras of words and terms, and of the VR and HR algebras of graphs. Finally, we extend the Recognizability Theorem to STR.

5.3.1 The Splitting Theorem for unions of disjoint concrete structures

Throughout this subsection, we assume that $\mathcal{R}$, $\mathcal{R}'$, $\mathcal{R}''$ are relational signatures such that $\mathcal{R}_0 = \mathcal{R}'_0 \cup \mathcal{R}''_0$, $\mathcal{R}'_0 \cap \mathcal{R}''_0 = \emptyset$ and $\mathcal{R}_+ = \mathcal{R}'_+ = \mathcal{R}''_+$.

Definition 5.34 (Disjoint union of structures, a special case) Let S and T be concrete $\mathcal{R}'$- and $\mathcal{R}''$-structures respectively, that are *disjoint*, i.e., such that $D_S \cap D_T = \emptyset$. Their *union*[45] is the $\mathcal{R}$-structure denoted by $S \oplus T$ (or by $S \oplus_{\mathcal{R}',\mathcal{R}''} T$ if we need to specify $\mathcal{R}'$ and $\mathcal{R}''$) such that

$$D_{S \oplus T} := D_S \cup D_T,$$
$$R_{S \oplus T} := R_S \cup R_T \quad \text{for each } R \in \mathcal{R}_+,$$
$$a_{S \oplus T} := \begin{cases} a_S & \text{if } a \in \mathcal{R}'_0, \\ a_T & \text{if } a \in \mathcal{R}''_0. \end{cases}$$

As for graphs (cf. Definition 2.23), we extend this definition to abstract structures: the *disjoint union* $S \oplus T$ of $S \in STR(\mathcal{R}')$ and $T \in STR(\mathcal{R}'')$ is the isomorphism class of $S' \oplus T'$, where S' and T' are any two disjoint concrete structures respectively isomorphic to S and T.

---

[45] A more general union, for which we will waive the restriction that $\mathcal{R}'_+ = \mathcal{R}''_+$, will be defined in Example 5.44. The Splitting Theorem will extend to it.

We will consider C_rMS formulas of bounded quantifier-height. If $r, h \in \mathcal{N}$, S is a relational signature and $\mathcal{X}$ is a finite set of first-order and set variables, then we denote by C_rMS$^h(S, \mathcal{X})$ the set of formulas of counting monadic second-order logic over S, with free variables in $\mathcal{X}$, of quantifier-height at most h and written with the set predicates $Card_{p,q}(X)$ for $2 \leq q \leq r$, $0 \leq p < q$. If $r \leq 1$, there are no such predicates and we write MS^h instead of C_0MS^h or C_1MS^h. Similarly, we write FO^h for first-order logic. If $h = 0$, we obtain quantifier-free formulas: they may have free set variables.

The sets C_rMS$^h(S, \mathcal{X})$ are infinite, but they are finite up to logical equivalence (this equivalence is defined at the beginning of Section 5.2), hence we can view them as finite sets. This important point will be discussed in detail in Section 5.6. Furthermore, we will write formulas without universal quantifications. A formula of the form $\forall Y_1, \ldots, Y_p. \varphi$ will be replaced by $\neg \exists Y_1, \ldots, Y_p. \neg \varphi$. This transformation modifies neither the variables, nor their number of occurrences nor the quantifier-height. Similarly, we will not use $\Rightarrow$, $\Leftrightarrow$, *False* but only $\wedge, \vee, \neg$, and *True*.

Let φ be a formula in C_rMS$^h(\mathcal{R}, \mathcal{X})$ with $\mathcal{X} = \{X_1, \ldots, X_m, z_1, \ldots, z_n\}$. For any two disjoint concrete structures S and T as above, we will express the set

$$sat(S \oplus T, \varphi, (X_1, \ldots, X_m), (z_1, \ldots, z_n))$$

of $\mathcal{X}$-assignments that satisfy φ in $S \oplus T$ (defined in Section 5.1.3; recall that this set is undefined for abstract structures unless $m = n = 0$) in terms of sets of the forms

$$sat(S, \psi, (X_1, \ldots, X_m), (x_1, \ldots, x_k)) \text{ and } sat(T, \theta, (X_1, \ldots, X_m), (y_1, \ldots, y_\ell)),$$

for finitely many C_rMS formulas ψ and θ of quantifier-height at most that of φ, where $\{x_1, \ldots, x_k\}$ and $\{y_1, \ldots, y_\ell\}$ are subsets of $\{z_1, \ldots, z_n\}$. We will call such an expression a *splitting* of the set $sat(S \oplus T, \varphi, (X_1, \ldots, X_m), (z_1, \ldots, z_n))$. We will also refer to it as a *splitting of* φ because it is actually a transformation of formulas that does not depend on S and T. To make the proof easier to read, we first consider the case of first-order formulas, which is actually of independent interest.

We need more notation. Sequences of pairwise distinct set and first-order variables are denoted respectively by $\overline{X}$ and $\overline{z}$ (see the beginning of Section 5.2). These sequences may be empty and are then denoted by ε instead of $()$. The use of the notation $\overline{z} = \overline{x}\overline{y}$ will assume that $\overline{x}$ and $\overline{y}$ have no variable in common (shortly denoted by $\overline{x} \cap \overline{y} = \emptyset$, since we also view $\overline{z}$ as denoting the set of variables occurring in the sequence $\overline{z}$).

Let $\overline{X}$ denote $(X_1, \ldots, X_m)$ and $\overline{z}$ denote $(z_1, \ldots, z_n)$. If $m = 0$ (i.e., φ has no free set variables), then we will write $sat(S, \varphi, \overline{z})$ instead of $sat(S, \varphi, \varepsilon, \overline{z})$. If $n = 0$, we will write $sat(S, \varphi, \overline{X})$ instead of $sat(S, \varphi, \overline{X}, \varepsilon)$. If $m = n = 0$ (i.e., φ is a sentence), then we will write $sat(S, \varphi)$ instead of $sat(S, \varphi, \varepsilon, \varepsilon)$.

If $\bar{z} = (z_1, z_2, \ldots, z_n)$ and π is a permutation of $[n]$, we let $\pi(\bar{z})$ denote the sequence $(z_{\pi(1)}, z_{\pi(2)}, \ldots, z_{\pi(n)})$, and similarly for $\bar{a} \in D^n$ (where D is any set). If $A \subseteq D^n$, then we define

$$perm_\pi(A) := \{(a_{\pi(1)}, \ldots, a_{\pi(n)}) \mid (a_1, \ldots, a_n) \in A\} = \{\pi(\bar{a}) \mid \bar{a} \in A\}.$$

It follows that if $\bar{z} = (z_1, \ldots, z_n)$, then

$$sat(S, \varphi, \pi(\bar{z})) = perm_\pi(sat(S, \varphi, \bar{z})).$$

If $S \oplus T$ is a (well-defined) concrete structure and φ has its free variables in $\bar{x}\bar{y}$ (i.e., in $\bar{x} \cup \bar{y}$ with $\bar{x} \cap \bar{y} = \emptyset$), we define

$$sat(S \oplus T, \varphi, \bar{x}; \bar{y}) \subseteq D_S^{|\bar{x}|} \times D_T^{|\bar{y}|}$$

as the subset of $sat(S \oplus T, \varphi, \bar{x}\bar{y})$ consisting of its tuples with the first $|\bar{x}|$ components in D_S and the last $|\bar{y}|$ components in D_T.[46]

If π is a permutation of $[k + \ell]$ that applies to a sequence of first-order variables $\bar{x}\bar{y} = (x_1, \ldots, x_k)(y_1, \ldots, y_\ell) = (x_1, \ldots, x_k, y_1, \ldots, y_\ell)$, then we say that it *preserves the relative orders of variables in $\bar{x}$ and in $\bar{y}$* if $\pi(i) < \pi(j)$ whenever, either $1 \leq i < j \leq k$ or $k + 1 \leq i < j \leq k + \ell$. In other words, π interleaves the sequences $\bar{x}$ and $\bar{y}$.

Lemma 5.35 We have $sat(S \oplus T, \varphi, \bar{z}) = \biguplus perm_\pi(sat(S \oplus T, \varphi, \bar{x}; \bar{y}))$, where the union[47] extends to all triples $(\pi, \bar{x}, \bar{y})$ such that $\bar{z} = \pi(\bar{x}\bar{y})$ and π preserves the relative orders of variables in $\bar{x}$ and in $\bar{y}$.[48]

Proof: We first observe that if $\bar{z} = \pi(\bar{x}\bar{y})$, then

$$perm_\pi(sat(S \oplus T, \varphi, \bar{x}; \bar{y})) \subseteq perm_\pi(sat(S \oplus T, \varphi, \bar{x}\bar{y}))$$

$$= sat(S \oplus T, \varphi, \pi(\bar{x}\bar{y}))$$

$$= sat(S \oplus T, \varphi, \bar{z}).$$

This gives the inclusion from right to left. We now prove the other one.

We let $\bar{z} = (z_1, \ldots, z_n)$. For[49] $\bar{a} = (a_1, \ldots, a_n) \in sat(S \oplus T, \varphi, \bar{z})$, we let the *profile* of $\bar{a}$ be $\{i \in [n] \mid a_i \in D_S\}$. For each subset I of $[n]$ of cardinality $k \leq n$ there exists a unique permutation π_I of $[n]$ such that $\pi_I(\bar{z}) = (z_{i_1}, \ldots, z_{i_k}, z_{j_1}, \ldots, z_{j_\ell})$, where $i_1, \ldots, i_k$ is an increasing enumeration of I, and $j_1, \ldots, j_\ell$ is an increasing enumeration of $[n] - I$

---

[46] This notation is ambiguous because a structure can be expressed as $S \oplus T$ in several ways. It stands for the precise notation $sat(S, T, \varphi, \bar{x}; \bar{y})$, which is less readable because it does not refer to $\oplus$.

[47] The notation $X = Y_1 \uplus \cdots \uplus Y_k$ means that $X = Y_1 \cup \cdots \cup Y_k$ and that $Y_1, \ldots, Y_k$ are pairwise disjoint.

[48] In this lemma and the next, $\bar{x}$ and/or $\bar{y}$ may be ε. See also Footnote 46 above.

[49] The notation $\bar{a} = (a_1, \ldots, a_n)$ for $a_i \in D_{S \oplus T}$ does not assume that $a_1, \ldots, a_n$ are pairwise distinct.

(with $\ell = n - k$). Hence, $\pi_I^{-1}(\bar{x}\bar{y}) = \bar{z}$ and the permutation π_I^{-1} preserves the relative orders of variables in $\bar{x}$ and in $\bar{y}$, where $\bar{x} := (z_{i_1},\ldots,z_{i_k})$ and $\bar{y} := (z_{j_1},\ldots,z_{j_\ell})$.[50]

Let $\bar{a} = (a_1,\ldots,a_n)$ belong to $sat(S \oplus T,\varphi,\bar{z})$, and let I be its profile. Then $\pi_I(\bar{a}) = \bar{b}\bar{c}$ where $\bar{b} = (a_{i_1},\ldots,a_{i_k}) \in D_S^k$ and $\bar{c} = (a_{j_1},\ldots,a_{j_\ell}) \in D_T^\ell$ (because $D_S \cap D_T = \emptyset$). It follows that $\bar{b}\bar{c} \in sat(S \oplus T,\varphi,\bar{x};\bar{y})$ where $\pi_I(\bar{z}) = (z_{i_1},\ldots,z_{i_k},z_{j_1},\ldots,z_{j_\ell})$ and $\bar{x} := (z_{i_1},\ldots,z_{i_k})$ and $\bar{y} := (z_{j_1},\ldots,z_{j_\ell})$. Hence $\bar{a} \in perm_{\pi_I^{-1}}(sat(S \oplus T,\varphi,\bar{x};\bar{y}))$. This gives the inclusion from left to right. It remains to check that the union in the right-hand side is disjoint.

Since $D_S \cap D_T = \emptyset$, the profile of a tuple $\bar{a}$ in $perm_\pi(sat(S \oplus T,\varphi,\bar{x};\bar{y}))$ is $\{i \in [n] \mid z_i \in \bar{x}\}$. Observe that the triple $(\pi,\bar{x},\bar{y})$ is completely determined by the set $\bar{x}$, hence different such triples have different sets $\bar{x}$ and thus correspond to different profiles of tuples in $D_{S \oplus T}^n$. The sets $perm_\pi(sat(S \oplus T,\varphi,\bar{x};\bar{y}))$ are therefore pairwise disjoint. ∎

This lemma entails that we need only split the sets $sat(S \oplus T,\varphi,\bar{x};\bar{y})$.

Lemma 5.36 For every $\varphi \in FO^h(R,\bar{x}\bar{y})$, there exist $p \in \mathcal{N}$ and p pairs of formulas (θ_i,ψ_i) with $\theta_i \in FO^h(R',\bar{x})$ and $\psi_i \in FO^h(R'',\bar{y})$ such that, for all disjoint concrete R'- and R''-structures, respectively S and T, we have[51]

$$sat(S \oplus T,\varphi,\bar{x};\bar{y}) = \bigcup_{1 \leq i \leq p} sat(S,\theta_i,\bar{x}) \times sat(T,\psi_i,\bar{y}).$$

Proof: By induction on the structure of φ. We let $k := |\bar{x}|$ and $\ell := |\bar{y}|$, and we let sat denote $sat(S \oplus T,\varphi,\bar{x};\bar{y})$. We have the following cases:

Case 1: φ is *True*. We take $p = 1$ and θ_1,ψ_1 both equal to *True*.

Case 2: φ is $s = t$ (s,t are terms, i.e., variables or constant symbols). We have three subcases:

$$sat = \begin{cases} sat(S,\varphi,\bar{x}) \times D_T^\ell & \text{if } s,t \in \bar{x} \cup R_0',\ ^{52} \\ D_S^k \times sat(T,\varphi,\bar{y}) & \text{if } s,t \in \bar{y} \cup R_0'', \\ \emptyset & \text{otherwise,} \end{cases}$$

which give respectively $p = 1$ and $(\theta_1,\psi_1) = (\varphi,True)$ in the first case, $p = 1$ and $(\theta_1,\psi_1) = (True,\varphi)$ in the second case and $p = 0$ in the last case.

Case 3: φ is $R(t_1,\ldots,t_n)$. We have again three subcases:

$$sat = \begin{cases} sat(S,\varphi,\bar{x}) \times D_T^\ell & \text{if } t_1,\ldots,t_n \in \bar{x} \cup R_0', \\ D_S^k \times sat(T,\varphi,\bar{y}) & \text{if } t_1,\ldots,t_n \in \bar{y} \cup R_0'', \\ \emptyset & \text{otherwise,} \end{cases}$$

---

[50] We may have $I = \emptyset$ or $I = [n]$. In these cases, π_I is the identity, $\bar{x} = \varepsilon$ in the first case and $\bar{y} = \varepsilon$ in the second case.

[51] If $\bar{x} = \varepsilon$, then $sat(S,\theta_i) = sat(S,\theta_i,\varepsilon)$ is either $\emptyset$ if $S \models \neg\theta_i$ or $\{\varepsilon\}$ if $S \models \theta_i$. If S is the empty structure, then $sat(S,\theta_i,\bar{x}) = \emptyset$ if $|\bar{x}| \geq 1$.

[52] If $T = \emptyset$, then $D_T^\ell = \emptyset$ if $\ell \geq 1$ and $D_T^0 = \{\varepsilon\}$.

which give the same pairs (θ_i, ψ_i) as in the previous case.

Case 4: φ is $\varphi_1 \vee \varphi_2$. We have

$$sat = sat(S \oplus T, \varphi_1, \bar{x}; \bar{y}) \cup sat(S \oplus T, \varphi_2, \bar{x}; \bar{y}).$$

The induction hypothesis yields the result.

Case 5: φ is $\varphi_1 \wedge \varphi_2$. In this case:[53]

$$sat = sat(S \oplus T, \varphi_1, \bar{x}; \bar{y}) \cap sat(S \oplus T, \varphi_2, \bar{x}; \bar{y}).$$

We note that if $A_i, A'_j \subseteq D_S^k$ and $B_i, B'_j \subseteq D_T^\ell$, then

$$\left(\bigcup_{1 \le i \le p} A_i \times B_i \right) \cap \left(\bigcup_{1 \le j \le p'} A'_j \times B'_j \right) = \bigcup_{\substack{1 \le i \le p \\ 1 \le j \le p'}} \left(A_i \cap A'_j \right) \times \left(B_i \cap B'_j \right).$$

Hence, if $(\theta_i, \psi_i)_{i \in [p]}$ is associated by the induction hypothesis with φ_1 and $(\theta'_j, \psi'_j)_{j \in [p']}$ with φ_2, we get for φ the family

$$(\theta_i \wedge \theta'_j, \psi_i \wedge \psi'_j)_{(i,j) \in [p] \times [p']}.$$

Case 6: φ is $\neg \varphi_1$. We have

$$sat = (D_S^k \times D_T^\ell) - sat(S \oplus T, \varphi_1, \bar{x}; \bar{y})$$

and, by using induction,

$$sat(S \oplus T, \varphi_1, \bar{x}; \bar{y}) = \bigcup_{1 \le i \le p} sat(S, \theta_i, \bar{x}) \times sat(T, \psi_i, \bar{y}).$$

For subsets A_i of D_S^k and B_i of D_T^ℓ, where $1 \le i \le p$, we have

$$(D_S^k \times D_T^\ell) - \bigcup_{1 \le i \le p} A_i \times B_i = \bigcup_{I \subseteq [p]} \left(\bigcap_{i \in I} (D_S^k - A_i) \right) \times \left(\bigcap_{j \in [p] - I} (D_T^\ell - B_j) \right).$$

It follows that

$$sat = \bigcup_{I \subseteq [p]} sat(S, \bigwedge_{i \in I} \neg \theta_i, \bar{x}) \times sat(T, \bigwedge_{j \in [p] - I} \neg \psi_j, \bar{y}),$$

where $\bigwedge_{i \in J} \alpha_i$ is replaced by *True* if $J = \emptyset$.

---

[53] Since conjunction can be expressed in terms of disjunction (Case 4) and negation (Case 6), we could omit this case. However, the direct construction given in this case is more clear than the one that uses a combination of the constructions of Cases 4 and 6.

In the particular case where φ is $\neg True$ (which is equivalent to *False*) this computation yields $p = 2$, but we can also take $p = 0$.

Case 7: φ is $\exists u. \varphi_1$. We have

$$sat = pr_1(sat(S \oplus T, \varphi_1, \bar{x}u; \bar{y})) \cup pr_2(sat(S \oplus T, \varphi_1, \bar{x}; \bar{y}u)),$$

where $pr_1 : (D_S^{k+1} \times D_T^{\ell}) \to (D_S^k \times D_T^{\ell})$ is the projection that "eliminates" the $(k + 1)$-th component and $pr_2 : (D_S^k \times D_T^{\ell+1}) \to (D_S^k \times D_T^{\ell})$ similarly "eliminates" the last one.

We have by induction

$$sat(S \oplus T, \varphi_1, \bar{x}u; \bar{y}) = \bigcup_{1 \leq i \leq p} sat(S, \theta_i, \bar{x}u) \times sat(T, \psi_i, \bar{y}).$$

Note that

$$pr_1(sat(S, \theta_i, \bar{x}u) \times sat(T, \psi_i, \bar{y})) = sat(S, \exists u. \theta_i, \bar{x}) \times sat(T, \psi_i, \bar{y}).$$

We have also by induction

$$sat(S \oplus T, \varphi_1, \bar{x}; \bar{y}u) = \bigcup_{1 \leq j \leq p'} sat(S, \theta'_j, \bar{x}) \times sat(T, \psi'_j, \bar{y}u),$$

and we have

$$pr_2(sat(S, \theta'_j, \bar{x}) \times sat(T, \psi'_j, \bar{y}u)) = sat(S, \theta'_j, \bar{x}) \times sat(T, \exists u. \psi'_j, \bar{y}).$$

It follows that

$$sat(S \oplus T, \exists u. \varphi_1, \bar{x}; \bar{y})$$
$$= \bigcup_{1 \leq i \leq p} sat(S, \exists u. \theta_i, \bar{x}) \times sat(T, \psi_i, \bar{y}) \cup \bigcup_{1 \leq j \leq p'} sat(S, \theta'_j, \bar{x}) \times sat(T, \exists u. \psi'_j, \bar{y}).$$

Using induction, we see that $qh(\theta_i), qh(\psi_i) \leq qh(\varphi_1)$ and $qh(\theta'_j), qh(\psi'_j) \leq qh(\varphi_1)$. Hence, we have $qh(\psi) \leq 1 + qh(\varphi_1) = qh(\varphi)$, where ψ is any of $\exists u. \theta_i, \psi_i, \theta'_j$, or $\exists u. \psi'_j$.

This completes the proof. ∎

We can now state the Splitting Theorem for first-order formulas (with $\mathcal{R}, \mathcal{R}'$ and $\mathcal{R}''$ as at the beginning of this subsection).

Proposition 5.37 For every $\varphi \in FO^h(\mathcal{R}, \bar{z})$, there exists a finite family of 5-tuples $(\pi_i, \theta_i, \psi_i, \bar{x}_i, \bar{y}_i)_{1 \leq i \leq p}$ such that, for all disjoint concrete $\mathcal{R}'$- and $\mathcal{R}''$-structures, respectively S and T, we have

$$sat(S \oplus T, \varphi, \bar{z}) = \biguplus_{1 \leq i \leq p} perm_{\pi_i}(sat(S, \theta_i, \bar{x}_i) \times sat(T, \psi_i, \bar{y}_i)),$$

where, for each i, $\theta_i \in \mathrm{FO}^h(\mathcal{R}', \bar{x}_i)$, $\psi_i \in \mathrm{FO}^h(\mathcal{R}'', \bar{y}_i)$, $\bar{z} = \pi_i(\bar{x}_i\bar{y}_i)$ and π_i is a permutation that preserves the relative orders of variables in $\bar{x}_i$ and in $\bar{y}_i$.

Proof: We use Lemma 5.35, the observation that $perm_\pi(A \cup B) = perm_\pi(A) \cup perm_\pi(B)$, and a stronger version of Lemma 5.36 where the union is disjoint, that we establish as follows.

Assume we have an expression of $sat(S \oplus T, \varphi, \bar{x}; \bar{y})$ as in Lemma 5.36 in terms of θ_i and ψ_i for $p \geq 2$. For every nonempty subset I of $[p]$ we let θ_I be $\bigwedge_{i \in I} \theta_i \wedge \bigwedge_{i \in [p]-I} \neg\theta_i$ and similarly ψ_I (with the ψ_i's). It is then clear that

$$sat(S \oplus T, \varphi, \bar{x}; \bar{y}) = \biguplus_{\substack{I \cap J \neq \emptyset, \\ I \cup J \subseteq [p]}} sat(S, \theta_I, \bar{x}) \times sat(T, \psi_J, \bar{y}).$$

Lemma 5.35 yields $sat(S \oplus T, \varphi, \bar{z}) = \biguplus perm_\pi(sat(S \oplus T, \varphi, \bar{x}; \bar{y}))$, where the disjoint union extends to all triples $(\pi, \bar{x}, \bar{y})$ such that $\bar{z} = \pi(\bar{x}\bar{y})$ and π preserves the relative orders of variables in $\bar{x}$ and in $\bar{y}$. From the above equality we get

$$perm_\pi(sat(S \oplus T, \varphi, \bar{x}; \bar{y})) = \biguplus_{\substack{I \cap J \neq \emptyset, \\ I \cup J \subseteq [p]}} perm_\pi(sat(S, \theta_I, \bar{x}) \times sat(T, \psi_I, \bar{y})),$$

and the union is actually disjoint because each $perm_\pi$ is injective on the set $D_S^{|\bar{x}|} \times D_T^{|\bar{y}|}$. This gives the desired result. ∎

Our next objective is to extend this proposition to monadic second-order logic. Let φ belong to $\mathrm{C}_r\mathrm{MS}^h(\mathcal{R}, \overline{X} \cup \bar{x}\bar{y})$ where $\overline{X} = (X_1, \ldots, X_m)$, $\bar{x} = (x_1, \ldots, x_k)$ and $\bar{y} = (y_1, \ldots, y_\ell)$. If S, T are disjoint concrete $\mathcal{R}'$- and $\mathcal{R}''$-structures respectively, we let $sat(S \oplus T, \varphi, \overline{X}, \bar{x}; \bar{y})$ be the set of tuples $(\overline{A}, \bar{a}, \bar{b})$ such that $A_i \subseteq D_S \cup D_T$ for $i \in [m]$, $a_i \in D_S$ for $i \in [k]$, $b_i \in D_T$ for $i \in [\ell]$ and $S \oplus T \models \varphi(\overline{A}, \bar{a}, \bar{b})$, where $\overline{A} = (A_1, \ldots, A_m)$, $\bar{a} = (a_1, \ldots, a_k)$ and $\bar{b} = (b_1, \ldots, b_\ell)$. In other words, $sat(S \oplus T, \varphi, \overline{X}, \bar{x}; \bar{y})$ is the intersection of $sat(S \oplus T, \varphi, \overline{X}, \bar{x}\bar{y})$ with $\mathcal{P}(D_{S \oplus T})^m \times D_S^k \times D_T^\ell$.

For combining sets $sat(S, \theta, \overline{X}, \bar{x})$ and $sat(T, \psi, \overline{X}, \bar{y})$, we need a more complex operation than the Cartesian product used for first-order formulas. If $\widetilde{A} = (A_1, \ldots, A_m, a_1, \ldots, a_k) \in \mathcal{P}(D)^m \times D^k$ and $\widetilde{B} = (B_1, \ldots, B_m, b_1, \ldots, b_\ell) \in \mathcal{P}(D')^m \times D'^\ell$, we define the $(m + k + \ell)$-tuple

$$\widetilde{A} \boxtimes \widetilde{B} := (A_1 \cup B_1, \ldots, A_m \cup B_m, a_1, \ldots, a_k, b_1, \ldots, b_\ell).$$

If $\widehat{A} \subseteq \mathcal{P}(D)^m \times D^k$ and $\widehat{B} \subseteq \mathcal{P}(D')^m \times D'^\ell$, we let

$$\widehat{A} \boxtimes \widehat{B} := \{\widetilde{A} \boxtimes \widetilde{B} \mid \widetilde{A} \in \widehat{A}, \widetilde{B} \in \widehat{B}\} \subseteq \mathcal{P}(D \cup D')^m \times D^k \times D'^\ell.$$

We will also need to permute first-order components of assignments. If $\widehat{A} \subseteq \mathcal{P}(D)^m \times D^n$ and π is a permutation of $[n]$, then we let

$$perm_\pi(\widehat{A}) := \{(A_1,\ldots,A_m,a_{\pi(1)},\ldots,a_{\pi(n)}) \mid (A_1,\ldots,A_m,a_1,\ldots,a_n) \in \widehat{A}\}.$$

Hence, if $\overline{X} = (X_1,\ldots,X_m)$ and $\overline{z} = (z_1,\ldots,z_n)$, we have

$$sat(S,\varphi,\overline{X},\pi(\overline{z})) = perm_\pi(sat(S,\varphi,\overline{X},\overline{z})).$$

The following lemma extends Lemma 5.36.

Lemma 5.38 For every formula φ in $C_rMS^h(\mathcal{R},\overline{X} \cup \overline{x}\overline{y})$, there exists a family $(\theta_i,\psi_i)_{1 \le i \le p}$ of pairs of formulas such that $\theta_i \in C_rMS^h(\mathcal{R}',\overline{X} \cup \overline{x})$, $\psi_i \in C_rMS^h(\mathcal{R}'',\overline{X} \cup \overline{y})$ and for all disjoint concrete $\mathcal{R}'$- and $\mathcal{R}''$-structures, respectively S and T, we have

$$sat(S \oplus T,\varphi,\overline{X},\overline{x};\overline{y}) = \biguplus_{1 \le i \le p} sat(S,\theta_i,\overline{X},\overline{x}) \boxtimes sat(T,\psi_i,\overline{X},\overline{y}).$$

Proof: We let as above $m := |\overline{X}|$, $k := |\overline{x}|$ and $\ell := |\overline{y}|$. We first prove a weak version where $\biguplus$ is replaced by $\cup$, by induction on the structure of φ. Many steps will be as in the proof of Lemma 5.36. We let sat denote $sat(S \oplus T,\varphi,\overline{X},\overline{x};\overline{y})$.

Case 1: φ is *True* or $s = t$ or $R(t_1,\ldots,t_n)$. If φ is *True*, we have $p = 1$ and θ_1,ψ_1 both equal to *True* as in the first case of Lemma 5.36. If φ is $s = t$, we have

$$sat = \begin{cases} sat(S,\varphi,\overline{X},\overline{x}) \boxtimes (\mathcal{P}(D_T)^m \times D_T^\ell) & \text{if } s,t \in \overline{x} \cup \mathcal{R}_0', \\ (\mathcal{P}(D_S)^m \times D_S^k) \boxtimes sat(T,\varphi,\overline{X},\overline{y}) & \text{if } s,t \in \overline{y} \cup \mathcal{R}_0'', \\ \varnothing & \text{otherwise,} \end{cases}$$

and similarly if φ is $R(t_1,\ldots,t_n)$, with $t_1,\ldots,t_n$ replacing s,t. This gives the same splittings of sat as in Cases 2 and 3 of Lemma 5.36.

Case 2: φ is $t \in Y$ (and Y belongs to $\overline{X}$). Then

$$sat = \begin{cases} sat(S,\varphi,\overline{X},\overline{x}) \boxtimes (\mathcal{P}(D_T)^m \times D_T^k) & \text{if } t \in \overline{x} \cup \mathcal{R}_0', \\ (\mathcal{P}(D_S)^m \times D_S^\ell) \boxtimes sat(T,\varphi,\overline{X},\overline{y}) & \text{if } t \in \overline{y} \cup \mathcal{R}_0'', \end{cases}$$

which gives the same first two forms of splitting of the case where φ is $s = t$ (see Case 1).

Case 3: φ is $Card_{n,q}(Y)$ (with Y in $\overline{X}$). We recall that $Card_{n,q}(C)$ means $|C| \equiv n$ (mod q). For every set $C = A \uplus B$ we have $Card_{n,q}(C) \Longleftrightarrow \bigvee Card_{i,q}(A) \wedge Card_{j,q}(B)$, where the disjunction extends to all pairs (i,j) such that $0 \leq i,j < l$ and $i + j \equiv n$ (mod q). It follows that

$$sat = \bigcup sat(S, Card_{i,q}(Y), \overline{X}, \overline{x}) \boxtimes sat(T, Card_{j,q}(Y), \overline{X}, \overline{y}),$$

where the union extends to all pairs (i,j) as above.

Case 4: φ is $\varphi_1 \vee \varphi_2$. As in Case 4 of Lemma 5.36.

Case 5: φ is $\varphi_1 \wedge \varphi_2$. We note that for $\widehat{A}_i, \widehat{A}'_j \subseteq \mathcal{P}(D_S)^m \times D_S^k$ and $\widehat{B}_i, \widehat{B}'_j \subseteq \mathcal{P}(D_T)^m \times D_T^\ell$ we have

$$\left(\bigcup_{1 \leq i \leq p} \widehat{A}_i \boxtimes \widehat{B}_i \right) \cap \left(\bigcup_{1 \leq j \leq p'} \widehat{A}'_j \boxtimes \widehat{B}'_j \right) = \bigcup_{\substack{1 \leq i \leq p \\ 1 \leq j \leq p'}} (\widehat{A}_i \cap \widehat{A}'_j) \boxtimes (\widehat{B}_i \cap \widehat{B}'_j),$$

because for all i,j as above:

$$(\widehat{A}_i \boxtimes \widehat{B}_i) \cap (\widehat{A}'_j \boxtimes \widehat{B}'_j) = (\widehat{A}_i \cap \widehat{A}'_j) \boxtimes (\widehat{B}_i \cap \widehat{B}'_j).$$

Let us prove this (dropping the subscripts). If $(\overline{E}, \overline{d}) \in (\widehat{A} \boxtimes \widehat{B}) \cap (\widehat{A}' \boxtimes \widehat{B}')$, then $(\overline{E}, \overline{d}) = \widetilde{A} \boxtimes \widetilde{B} = \widetilde{A}' \boxtimes \widetilde{B}'$ for some $\widetilde{A} \in \widehat{A}$, $\widetilde{A}' \in \widehat{A}'$, $\widetilde{B} \in \widehat{B}$ and $\widetilde{B}' \in \widehat{B}'$. Since $D_S \cap D_T = \emptyset$, this implies that, if $\overline{E} = (E_1, \ldots, E_m)$ and $\overline{d} = \overline{a}\overline{b}$ with $\overline{a} \in D_S^k$ and $\overline{b} \in D_T^\ell$, then $\widetilde{A} = \widetilde{A}' = (E_1 \cap D_S, \cdots, E_m \cap D_S, \overline{a})$ and similarly $\widetilde{B} = \widetilde{B}'$. Hence $(\overline{E}, \overline{d}) \in (\widehat{A} \cap \widehat{A}') \boxtimes (\widehat{B} \cap \widehat{B}')$, and we have $(\widehat{A} \boxtimes \widehat{B}) \cap (\widehat{A}' \boxtimes \widehat{B}') \subseteq (\widehat{A} \cap \widehat{A}') \boxtimes (\widehat{B} \cap \widehat{B}')$. The other inclusion is obvious.

Hence, we have the same splitting as in Case 5 of Lemma 5.36.

Case 6: φ is $\neg\varphi_1$. As in Case 6 of Lemma 5.36 with the following fact:

$$(\mathcal{P}(D_S)^m \times D_S^k) \boxtimes (\mathcal{P}(D_T)^m \times D_T^\ell) - \bigcup_{1 \leq i \leq p} \widehat{A}_i \boxtimes \widehat{B}_i$$

$$= \bigcup_{I \subseteq [p]} \left(\bigcap_{i \in I} ((\mathcal{P}(D_S)^m \times D_S^k) - \widehat{A}_i) \right) \boxtimes \left(\bigcap_{j \in [p]-I} ((\mathcal{P}(D_T)^m \times D_T^\ell) - \widehat{B}_j) \right).$$

Case 7: φ is $\exists u. \varphi_1$. As in Case 7 of Lemma 5.36.

Case 8: φ is $\exists U. \varphi_1$. Let $(\theta_i, \psi_i)_{1 \leq i \leq p}$ be associated with φ_1, by induction. We have

$$sat = pr(sat(S \oplus T, \varphi_1, \overline{X}U, \overline{x}; \overline{y})),$$

where the projection $pr : \mathcal{P}(D)^{m+1} \times D^n \to \mathcal{P}(D)^m \times D^n$ "eliminates" the $(m+1)$-th component. It is clear that for $\widehat{A} \subseteq \mathcal{P}(D)^{m+1} \times D^k$ and

$\widehat{B} \subseteq \mathcal{P}(D')^{m+1} \times D'^{\ell}$, we have $pr(\widehat{A} \boxtimes \widehat{B}) = pr(\widehat{A}) \boxtimes pr(\widehat{B})$. Then:

$$sat = pr(\bigcup_i sat(S, \theta_i, \overline{X}U, \overline{x}) \boxtimes sat(T, \psi_i, \overline{X}U, \overline{y}))$$
$$= \bigcup_i pr(sat(S, \theta_i, \overline{X}U, \overline{x})) \boxtimes pr(sat(T, \psi_i, \overline{X}U, \overline{y}))$$
$$= \bigcup_i sat(S, \exists U. \theta_i, \overline{X}, \overline{x}) \boxtimes sat(T, \exists U. \psi_i, \overline{X}, \overline{y}).$$

This gives the desired result with the family $(\exists U. \theta_i, \exists U. \psi_i)_{1 \le i \le p}$.

We have constructed splittings in terms of unions. The stronger statement with disjoint union is obtained as in the proof of Proposition 5.37. ∎

The final word is given by the following theorem, which extends Proposition 5.37 to monadic second-order formulas:

Theorem 5.39 (Splitting Theorem, the basic case) Let $\mathcal{R}$, $\mathcal{R}'$ and $\mathcal{R}''$ be relational signatures such that $\mathcal{R} = \mathcal{R}' \cup \mathcal{R}''$, $\mathcal{R}'_0 \cap \mathcal{R}''_0 = \emptyset$ and $\mathcal{R}_+ = \mathcal{R}'_+ = \mathcal{R}''_+$. For every r and h in $\mathcal{N}$ and every formula φ in $C_r\mathrm{MS}^h(\mathcal{R}, \overline{X} \cup \overline{z})$, there exists a finite family of 5-tuples $(\pi_i, \theta_i, \psi_i, \overline{x}_i, \overline{y}_i)_{1 \le i \le p}$ such that, for all disjoint concrete $\mathcal{R}'$- and $\mathcal{R}''$-structures, respectively S and T, we have

$$sat(S \oplus T, \varphi, \overline{X}, \overline{z}) = \biguplus_{1 \le i \le p} perm_{\pi_i}(sat(S, \theta_i, \overline{X}, \overline{x}_i) \boxtimes sat(T, \psi_i, \overline{X}, \overline{y}_i)),$$

where, for each i, $\theta_i \in C_r\mathrm{MS}^h(\mathcal{R}', \overline{X} \cup \overline{x}_i)$, $\psi_i \in C_r\mathrm{MS}^h(\mathcal{R}'', \overline{X} \cup \overline{y}_i)$, $\overline{z} = \pi_i(\overline{x}_i \overline{y}_i)$ and π_i is a permutation that preserves the relative orders of variables in $\overline{x}_i$ and in $\overline{y}_i$.

Proof: As in Lemma 5.35, and with the same proof, we have

$$sat(S \oplus T, \varphi, \overline{X}, \overline{z}) = \biguplus perm_\pi(sat(S \oplus T, \varphi, \overline{X}, \overline{x}; \overline{y})),$$

where the union extends to all triples $(\pi, \overline{x}, \overline{y})$ such that $\overline{z} = \pi(\overline{x}\overline{y})$ and π preserves the ordering of variables in $\overline{x}$ and in $\overline{y}$.

Lemma 5.38 expresses each set $sat(S \oplus T, \varphi, \overline{X}, \overline{x}; \overline{y})$ as a disjoint union of sets $sat(S, \theta_i, \overline{X}, \overline{x}) \boxtimes sat(T, \psi_i, \overline{X}, \overline{y})$. Since $perm_\pi(\widehat{A} \uplus \widehat{B}) = perm_\pi(\widehat{A}) \uplus perm_\pi(\widehat{B})$ for $\widehat{A}, \widehat{B} \subseteq \mathcal{P}(D)^m \times D^n$, we get the stated result. ∎

The family $(\pi_i, \theta_i, \psi_i, \overline{x}_i, \overline{y}_i)_{1 \le i \le p}$ will be called a *splitting* of φ relative to the operation of disjoint union.

Remark 5.40 (1) The proof of Theorem 5.39 is effective. It gives an algorithm that constructs from φ, $\overline{X}$ and $\overline{z}$ a splitting of φ relative to $\oplus$. This construction depends on $\mathcal{R}$, $\mathcal{R}'_0$ and $\mathcal{R}''_0$. The integers r and h can be taken as the minimal ones such that $\varphi \in C_r\mathrm{MS}^h(\mathcal{R}, \overline{X} \cup \overline{z})$.

(2) If $\bar{z}$ is empty, then we have a simpler formulation of Theorem 5.39 that avoids permutations of first-order variables:

$$sat(S \oplus T, \varphi, \overline{X}) = \biguplus_{1 \leq i \leq p} sat(S, \theta_i, \overline{X}) \boxtimes sat(T, \psi_i, \overline{X}).$$

If φ is a sentence, we have an even simpler formulation which is also meaningful for abstract structures S and T:[54] $S \oplus T \models \varphi$ if and only if there exists i such that $S \models \theta_i$ and $T \models \psi_i$, and, furthermore, if it exists, this index is unique (which corresponds to the disjoint union used in the theorem). □

We now develop an example relative to cographs. These graphs are constructed from isolated vertices by disjoint union and complete join $\otimes$ (see Section 1.1.2).

Example 5.41 We examine how Lemmas 5.35 and 5.36 extend to the operation $\otimes$ by generalizing what has been done for $\oplus$. Lemma 5.35 obviously holds for $\otimes$. Case 3 of Lemma 5.36 reduces to the following one where G and H are graphs (we identify G and H with $\lfloor G \rfloor$ and $\lfloor H \rfloor$ respectively):

Case 3 (for $\oplus$): φ is $edg(s,t)$; if $sat = sat(G \oplus H, \varphi, \bar{x}; \bar{y})$ then

$$sat = \begin{cases} sat(G, \varphi, \bar{x}) \times V_H^{|\bar{y}|} & \text{if } s, t \in \bar{x}, \\ V_G^{|\bar{x}|} \times sat(H, \varphi, \bar{y}) & \text{if } s, t \in \bar{y}, \\ \varnothing & \text{otherwise (i.e., if } s \in \bar{x} \text{ and } t \in \bar{y} \text{ or if } s \in \bar{y} \text{ and } t \in \bar{x}). \end{cases}$$

For the operation $\otimes$, we need only modify Case 3 as follows:

Case 3 (for $\otimes$): φ is $edg(s,t)$; if $sat' = sat(G \otimes H, \varphi, \bar{x}; \bar{y})$ then

$$sat' = \begin{cases} sat(G, \varphi, \bar{x}) \times V_H^{|\bar{y}|} & \text{if } s, t \in \bar{x}, \\ V_G^{|\bar{x}|} \times sat(H, \varphi, \bar{y}) & \text{if } s, t \in \bar{y}, \\ V_G^{|\bar{x}|} \times V_H^{|\bar{y}|} & \text{if } s \in \bar{x} \text{ and } t \in \bar{y}, \text{ or if } t \in \bar{x} \text{ and } s \in \bar{y}. \end{cases}$$

This gives the following splitting (taking into account permutations of variables):

$$\begin{aligned} sat(G \otimes H, edg(x,y), xy) = \; & sat(G, edg(x,y), xy) \uplus sat(H, edg(x,y), xy) \\ & \uplus sat(G, True, x) \times sat(H, True, y) \\ & \uplus sat(H, True, x) \times sat(G, True, y).^{55} \end{aligned} \tag{5.1}$$

All other cases are as in Lemma 5.36.

---

[54] The set $sat(S, \varphi, \overline{X}, \bar{z})$ is undefined if S is an abstract structure, unless the sequences $\overline{X}$ and $\bar{z}$ are both empty, which implies that φ is a sentence (cf. Lemma 5.4).
[55] Note that $sat(G, True, x)$ is empty if $G = \varnothing$.

We now give a complete construction for the first-order formula $\varphi(x,y)$ defined as $edg(x,y) \vee \varphi_1(x,y)$, where $\varphi_1(x,y)$ is $\exists z(edg(x,z) \wedge edg(y,z))$. We have

$$sat(G \oplus H, \varphi, xy) = sat(G \oplus H, edg(x,y), xy) \cup sat(G \oplus H, \varphi_1, xy).$$

In order to compute $sat(G \oplus H, \varphi_1, xy)$, we let $\varphi_2(x,y,z)$ denote $edg(x,z) \wedge edg(y,z)$. For all disjoint sequences of variables $\overline{u}$ and $\overline{v}$ such that $\overline{u}\,\overline{v}$ enumerates the free variables of φ_2, we have

$$sat(G \oplus H, \varphi_2, \overline{u}; \overline{v}) = sat(G \oplus H, edg(x,z), \overline{u}; \overline{v}) \cap sat(G \oplus H, edg(y,z), \overline{u}; \overline{v}).$$

There are $8 \,(= 2^3)$ types of splitting for φ_2 since this formula has three free variables. We get in particular:

$$sat(G \oplus H, \varphi_2, xyz; \varepsilon)$$
$$= sat(G \oplus H, edg(x,z), xyz; \varepsilon) \cap sat(G \oplus H, edg(y,z), xyz; \varepsilon)$$
$$= sat(G, edg(x,z), xyz) \cap sat(G, edg(y,z), xyz).$$

This set can be written as a disjoint union of sets like $sat(G, edg(x,z) \wedge \neg edg(y,z), xyz)$ but it is simpler to write it as

$$sat(G \oplus H, \varphi_2, xyz; \varepsilon) = sat(G, \varphi_2, xyz).$$

By a similar computation, we get

$$sat(G \oplus H, \varphi_2, \varepsilon; xyz) = sat(H, \varphi_2, xyz).$$

The six other cases are as follows:

$$sat(G \oplus H, \varphi_2, xy; z), \quad sat(G \oplus H, \varphi_2, xz; y), \quad sat(G \oplus H, \varphi_2, x; yz),$$
$$sat(G \oplus H, \varphi_2, yz; x), \quad sat(G \oplus H, \varphi_2, y; xz), \quad sat(G \oplus H, \varphi_2, z; xy),$$

and each of them yields the empty set. It follows that

$$sat(G \oplus H, \varphi_1, xy) = sat(G, \varphi_1, xy) \uplus sat(H, \varphi_1, xy).$$

Since we have

$$sat(G \oplus H, edg(x,y), xy) = sat(G, edg(x,y), xy) \uplus sat(H, edg(x,y), xy),$$

we get

$$sat(G \oplus H, \varphi, xy) = sat(G, edg(x,y), xy) \cup sat(G, \varphi_1, xy)$$
$$\cup sat(H, edg(x,y), xy) \cup sat(H, \varphi_1, xy)$$
$$= sat(G, \varphi, xy) \uplus sat(H, \varphi, xy).$$

The unions in the first equality are not disjoint (are not $\uplus$) because $edg(x,y)$ and $\varphi_1(x,y)$ may hold simultaneously.

For the sake of comparison, we now consider the splitting of the same formula relative to $\otimes$. We have

$$sat(G \otimes H, \varphi, xy) = sat(G \otimes H, edg(x,y), xy) \cup sat(G \otimes H, \varphi_1, xy). \qquad (5.2)$$

The splitting of $sat(G \otimes H, edg(x,y), xy)$ has been given above in Equality (5.1). The one for φ_2 yields again eight cases. Similar to $\oplus$ we have

$$sat(G \otimes H, \varphi_2, \overline{u}; \overline{v}) = sat(G \otimes H, edg(x,z), \overline{u}; \overline{v}) \cap sat(G \otimes H, edg(y,z), \overline{u}; \overline{v}).$$

We obtain (by omitting some computations) the following equalities:

$$sat(G \otimes H, \varphi_2, xyz; \varepsilon) = sat(G, \varphi_2, xyz),$$
$$sat(G \otimes H, \varphi_2, \varepsilon; xyz) = sat(H, \varphi_2, xyz),$$
$$sat(G \otimes H, \varphi_2, xy; z) = V_G \times V_G \times V_H,$$
$$sat(G \otimes H, \varphi_2, xz; y) = sat(G, edg(x,z), xz) \times V_H,$$
$$sat(G \otimes H, \varphi_2, x; yz) = V_G \times sat(H, edg(y,z), yz),$$
$$sat(G \otimes H, \varphi_2, yz; x) = sat(G, edg(y,z), yz) \times V_H,$$
$$sat(G \otimes H, \varphi_2, y; xz) = V_G \times sat(H, edg(x,z), xz),$$
$$sat(G \otimes H, \varphi_2, z; xy) = V_G \times V_H \times V_H.$$

We get the splitting for φ_1 from the general scheme

$$sat(G \otimes H, \exists z. \varphi_2, \overline{u}; \overline{v}) = pr(sat(G \otimes H, \varphi_2, \overline{u}z; \overline{v}) \cup pr'(sat(G \otimes H, \varphi_2, \overline{u}; \overline{v}z)),$$

where pr and pr' are the appropriate projections.[56] We obtain

---

[56] Since we allow G, H to be empty, we cannot replace $sat(H, \exists z. True, \varepsilon)$ by $sat(H, True, \varepsilon)$, because $sat(H, \exists z. True, \varepsilon) = \emptyset$ if $H = \emptyset$ and is $\{\varepsilon\}$ otherwise. The projection mapping $pr : A \times B \to A$ does not satisfy the rule $pr(X \times B) = X$ for all $X \subseteq A$ because this equality is false if $B = \emptyset$ and $X \neq \emptyset$.

$$sat(G \otimes H, \varphi_1, xy) = sat(G, \varphi_1, xy) \cup sat(H, \varphi_1, xy)$$

$$\cup\, sat(G, True, xy) \times sat(H, \exists z.True, \varepsilon)$$

$$\cup\, sat(G, \exists z.edg(x,z), x) \times sat(H, True, y)$$

$$\cup\, sat(G, True, x) \times sat(H, \exists z.edg(y,z), y)$$

$$\cup\, sat(H, True, x) \times sat(G, \exists z.edg(y,z), y)$$

$$\cup\, sat(H, \exists z.edg(x,z), x) \times sat(G, True, y)$$

$$\cup\, sat(H, True, xy) \times sat(G, \exists z.True, \varepsilon). \tag{5.3}$$

Let us continue our computations. According to Equality (5.2), if we now take the union of the right-hand sides of (5.1) and (5.3), we get a splitting of $sat(G \otimes H, \varphi, xy)$. However, this expression can be simplified because, e.g.,

$$sat(G, \exists z.edg(x,z), x) \times sat(H, True, y) \subseteq sat(G, True, x) \times sat(H, True, y).$$

We obtain finally the following expression:

$$sat(G \otimes H, \varphi, xy) = sat(G, \varphi, xy) \cup sat(H, \varphi, xy)$$

$$\cup\, sat(G, True, x) \times sat(H, True, y)$$

$$\cup\, sat(H, True, x) \times sat(G, True, y)$$

$$\cup\, sat(G, True, xy) \times sat(H, \exists z.True, \varepsilon)$$

$$\cup\, sat(H, True, xy) \times sat(G, \exists z.True, \varepsilon).$$

We will see in Section 5.3.5 (Example 5.54) another proof of the Splitting Theorem for $\otimes$. $\qquad\qquad\qquad\qquad\qquad\qquad\qquad\qquad\qquad\qquad\qquad\qquad\qquad\qquad\square$

5.3.2 Quantifier-free transformations of structures

The transformation of a graph into its edge-complement defined in Chapter 1, Section 1.7.1, is an example of a quantifier-free transformation of relational structures. We now give the formal definition, for which we need some notation.

Let $\mathcal{R}$ and $\mathcal{R}'$ be relational signatures and $\mathcal{X}$ be a set of first-order variables. Then $QF(\mathcal{R}, \mathcal{X})$ denotes the set of quantifier-free formulas written with $\mathcal{R}$ and free variables in $\mathcal{X}$. (It is equal to $C_0MS^0(\mathcal{R}, \mathcal{X}) = FO^0(\mathcal{R}, \mathcal{X})$.)

Our objective is to specify functions : $STR(\mathcal{R}) \rightarrow STR(\mathcal{R}')$ by formulas in $QF(\mathcal{R}, \mathcal{X})$. A *transduction* is a subset of $STR(\mathcal{R}) \times STR(\mathcal{R}')$ for some relational signatures $\mathcal{R}$ and $\mathcal{R}'$, that we consider also as a mapping from $STR(\mathcal{R})$ to $\mathcal{P}(STR(\mathcal{R}'))$.

Monadic second-order transductions (defined in Section 1.7) will be studied in Chapter 7. In this chapter, we will consider particular monadic second-order transductions that are total functions : $STR(\mathcal{R}) \to STR(\mathcal{R}')$ and that are specified by quantifier-free formulas. We will call them *operations* because they will mainly be used as unary operations of algebras of graphs and relational structures.

Definition 5.42 (Quantifier-free operations) We first consider the particular but already useful case where $\mathcal{R}'_0 = \emptyset$. A *quantifier-free operation definition scheme*[57] (a *QFO definition scheme* in short) of type $\mathcal{R} \to \mathcal{R}'$ is a tuple of quantifier-free formulas $\mathcal{D} = \langle \delta, (\theta_R)_{R \in \mathcal{R}'} \rangle$ such that $\delta \in \mathrm{QF}(\mathcal{R}, \{x\})$ and $\theta_R \in \mathrm{QF}(\mathcal{R}, \{x_1, \ldots, x_{\rho(R)}\})$ for every $R \in \mathcal{R}'$. We call δ the *domain formula* of $\mathcal{D}$ and the formulas θ_R its *relation formulas*.

The associated mapping $f : STR^c(\mathcal{R}) \to STR^c(\mathcal{R}')$, denoted by $\widehat{\mathcal{D}}$, is defined as follows. If S is a concrete $\mathcal{R}$-structure, then $f(S)$ is the concrete $\mathcal{R}'$-structure such that

$$D_{f(S)} := sat(S, \delta, x),$$

$$R_{f(S)} := sat(S, \varphi_R, (x_1, \ldots, x_{\rho(R)})),$$

where φ_R is the formula $\theta_R(x_1, \ldots, x_{\rho(R)}) \wedge \delta(x_1) \wedge \cdots \wedge \delta(x_{\rho(R)})$.

If S' is isomorphic to S, then $\widehat{\mathcal{D}}(S')$ and $\widehat{\mathcal{D}}(S)$ are isomorphic. Hence the mapping $\widehat{\mathcal{D}}$ is well defined : $STR(\mathcal{R}) \to STR(\mathcal{R}')$, i.e., on abstract structures. We call it a *quantifier-free operation*, but this is only a special case, the general definition will be given shortly after some remarks and examples.

We have $\widehat{\mathcal{D}}(\varnothing) = \varnothing$ (where $\varnothing$ denotes the empty structure of any signature without constants) and we may have $\widehat{\mathcal{D}}(S) = \varnothing$ for S nonempty.

The mapping $\lfloor G \rfloor \mapsto \lfloor \overline{G} \rfloor$, where $\overline{G}$ is the edge-complement of a simple undirected loop-free graph G (see Example 1.30), is clearly a quantifier-free operation. The same holds for the slightly different edge-complement considered in Proposition 2.105 and for the mapping $\lfloor w \rfloor \mapsto \lfloor \widetilde{w} \rfloor$, where $\widetilde{w}$ is the mirror image of a word w: $\delta(x)$ is *True*, $\theta_{suc}(x_1, x_2)$ is $suc(x_2, x_1)$, and $\theta_{lab_a}(x_1)$ is $lab_a(x_1)$.

We now extend the notion of QFO definition scheme to the case where $\mathcal{R}'$ has constants. A *QFO definition scheme* of type $\mathcal{R} \to \mathcal{R}'$ is a tuple of formulas $\mathcal{D} = \langle \delta, (\theta_R)_{R \in \mathcal{R}'_+}, (\kappa_{c,d})_{c \in \mathcal{R}_0, d \in \mathcal{R}'_0} \rangle$, where the formulas δ, θ_R are as in the above special case and the formulas $\kappa_{c,d}$ are quantifier-free sentences over $\mathcal{R}$ called *constant defining sentences*, intended to express that $d_{f(S)} = c_S$ whenever $S \models \kappa_{c,d}$. Hence, $\mathcal{R}'_0$ must be empty if $\mathcal{R}_0$ is empty.

In order to ensure that $f(S) = \widehat{\mathcal{D}}(S)$ is a well-defined structure in $STR^c(\mathcal{R}')$ for every $S \in STR^c(\mathcal{R})$, we require that the sentences $\bigvee_{e \in \mathcal{R}_0} \kappa_{e,d}$, $\kappa_{c,d} \Rightarrow \delta(c)$ and $\kappa_{c,d} \Rightarrow \neg \kappa_{c',d}$ are true in every structure S in $STR(\mathcal{R})$, for all $d \in \mathcal{R}'_0$ and all c, c' in $\mathcal{R}_0$ with $c \neq c'$. The first of these sentences (or rather its validity in S) ensures that every constant $d_{f(S)}$

---

[57] More powerful *quantifier-free definition schemes, QF-definition schemes* in short, will be defined in Chapter 7, Definition 7.2.

for d in $\mathcal{R}'$ is defined as the value in S of some constant symbol e belonging to $\mathcal{R}_0$; the sentences $\kappa_{c,d} \Rightarrow \delta(c)$ ensure that the interpretation c_S of $c \in \mathcal{R}_0$ intended to be the interpretation of d in $f(S)$ is actually in the domain of $f(S)$; the last sentences ensure that $d_{f(S)}$ is defined as c_S for at most one c. These sentences ensure collectively that each d in $\mathcal{R}'_0$ has a unique well-defined interpretation in $\widehat{\mathcal{D}}(S)$ for every $S \in STR(\mathcal{R})$.

By a *quantifier-free operation* (a *QF operation* in short), we mean a function $\widehat{\mathcal{D}} : STR^c(\mathcal{R}) \to STR^c(\mathcal{R}')$ (hence also $: STR(\mathcal{R}) \to STR(\mathcal{R}')$) defined by a QFO definition scheme $\mathcal{D}$. The set of these operations is denoted by F_1^{QF}. It is the set of unary operations of a functional signature F^{QF} to be defined below in Definition 5.62.

We say that a QFO definition scheme $\mathcal{D}$ and the corresponding operation $\widehat{\mathcal{D}}$ are *domain preserving* if the domain formula $\delta(x)$ is equivalent to *True*. This implies that $\widehat{\mathcal{D}}(S)$ has the same domain as S for every structure S. The following notion is slightly more general (this technicality will be useful to deal with operations like the parallel-composition of s-graphs represented by relational structures with constants). We say that $\mathcal{D}$ and $\widehat{\mathcal{D}}$ are *quasi-domain preserving* if the domain formula is equivalent to a formula of the form $\bigwedge_{c \in \mathcal{R}_0}(\alpha_c \Rightarrow x \neq c)$, where α_c is a quantifier-free sentence. This implies that $D_S - D_{\widehat{\mathcal{D}}(S)} = \{c_S \mid S \models \alpha_c\}$. Thus, only constants can be deleted from D_S. If α_c is *False* for all c, then $\mathcal{D}$ is domain preserving. If S is constant-separated, then $c_S \in D_S - D_{\widehat{\mathcal{D}}(S)}$ if and only if $S \models \alpha_c$. If it is not, we may have c_S in $D_S - D_{\widehat{\mathcal{D}}(S)}$ with $S \models \neg\alpha_c$ but c_S is nevertheless not in $D_{\widehat{\mathcal{D}}(S)}$ because $c'_S = c_S$ and $S \models \alpha_{c'}$ for some $c' \neq c$.

For qualifying operations, we will shorten "domain preserving quantifier-free" into DP-QF and "quasi-domain preserving quantifier-free" into QDP-QF. The corresponding definition schemes will be called the DP-QFO and the QDP-QFO definition schemes.

Remark 5.43 (1) The conditions to be verified by the constant defining sentences of a QFO definition scheme are decidable because they are expressed by quantifier-free sentences. Such a sentence is false in some $\mathcal{R}$-structure if and only if it is false in some finite $\mathcal{R}$-structure, if and only if it is false in an $\mathcal{R}$-structure whose domain has cardinality at most $|\mathcal{R}_0|$ (as the reader can check easily, and much quicker than by looking into [*BörGG] or Chapter 9 of [*BradMan]; this is true even if $\mathcal{R}_0 = \emptyset$). It follows that the notion of a QFO definition scheme is effective, that is, one can decide if it satisfies the required semantical conditions. For a similar reason, one can decide if two quantifier-free formulas are equivalent. Hence, one can decide if two QFO definition schemes define the same function on all concrete structures (which necessarily implies that they define the same function on all abstract structures).

Furthermore, for each pair $(\mathcal{R}, \mathcal{R}')$ there are finitely many QF operations of type $\mathcal{R} \to \mathcal{R}'$. That is because every quantifier-free formula is a Boolean combination of atomic formulas. Since there are finitely many atomic formulas (with variables in a fixed finite set $\mathcal{X}$), and since every quantifier-free formula is equivalent to one in disjunctive normal form, there are finitely many pairwise inequivalent quantifier-free formulas (with variables in $\mathcal{X}$). This will be discussed in detail below in Section 5.6, Corollary 5.95.

A quantifier-free formula with one free variable x is false in some $\mathcal{R}$-structure if and only if it is false in an $\mathcal{R}$-structure whose domain has cardinality at most $|\mathcal{R}_0| + 1$ (immediate from the above similar remark for sentences). Hence, the notion of a DP-QFO definition scheme is also effective. Since there are only finitely many pairwise inequivalent sentences α_c, there are only finitely many pairwise inequivalent formulas of the form $\bigwedge_{c \in \mathcal{R}_0} (\alpha_c \Rightarrow x \neq c)$, and for each of these formulas it can be decided whether it is equivalent to the domain formula $\delta(x)$. This shows the effectivity of the notion of a QDP-QFO definition scheme.

(2) In Definition 5.42, we do not require that the input structure S is constant-separated, i.e., that $c_S \neq c'_S$ for every two different constant symbols c and c'. But even if S is constant-separated, a definition scheme may construct $f(S)$ with $d_{f(S)} = d'_{f(S)}$ where $d \neq d'$. In order to ensure that $\widehat{\mathcal{D}}$ transforms constant-separated structures into constant-separated ones (which will be useful for dealing with s-graphs), we need only require additionally that the sentences $\kappa_{c,d} \Rightarrow \neg\kappa_{c,d'}$ are true in every $S \in STR(\mathcal{R})$, for all $c \in \mathcal{R}_0$ and all $d, d' \in \mathcal{R}'_0$ with $d' \neq d$. This can be decided by the observations made in (1).

(3) Let us comment about the role of constant symbols. Consider a constant symbol a and a binary relation R. Let f be the QF operation that transforms an $\{R, a\}$-structure S by adding to R_S all pairs (d, d') such that (d, a_S) and (a_S, d') are in R_S. It is not a QF operation over $\{R, a\}_*$-structures, i.e., over $\{R, a\}$-structures where a is replaced by a unary relation lab_a (cf. Section 5.1.5) to be interpreted by a singleton set. We use constants in relational structures in order to handle such cases.

Example 5.44 The *natural inclusion* $\iota_{\mathcal{R},\mathcal{R}'} : STR^c(\mathcal{R}) \to STR^c(\mathcal{R}')$, where[58] $\mathcal{R} \subseteq \mathcal{R}'$ and $\mathcal{R}_0 = \mathcal{R}'_0$, is defined by $\mathcal{D} := \langle True, (\theta_R)_{R \in \mathcal{R}'_+}, (\kappa_{c,d})_{c,d \in \mathcal{R}_0} \rangle$ such that

$$\theta_R \text{ is } \begin{cases} R(x_1, \ldots, x_{\rho(R)}) & \text{if } R \in \mathcal{R}_+, \\ False & \text{if } R \in \mathcal{R}'_+ - \mathcal{R}_+, \end{cases}$$

and, for all $c, d \in \mathcal{R}_0$:

$$\kappa_{c,d} \text{ is } \begin{cases} True & \text{if } d = c, \\ False & \text{if } d \neq c. \end{cases}$$

Thus, $R_{\widehat{\mathcal{D}}(S)} = R_S$ for $R \in \mathcal{R}_+$ and $R_{\widehat{\mathcal{D}}(S)} = \emptyset$ for $R \in \mathcal{R}'_+ - \mathcal{R}_+$, and $\widehat{\mathcal{D}}(S)$ has the same constants as S.

We recall that the union $S \oplus T$ of disjoint structures S in $STR^c(\mathcal{R}')$ and T in $STR^c(\mathcal{R}'')$ has been defined under the condition that $\mathcal{R}'$ and $\mathcal{R}''$ have the same relation

---

[58] When we discuss several signatures $\mathcal{R}$ and $\mathcal{R}'$, we assume, unless otherwise specified, that a symbol in $\mathcal{R} \cap \mathcal{R}'$ has the same arity with respect to both of them.

symbols (of positive arity) and no constant symbol in common. We now waive the former condition (we still assume that $\mathcal{R}'_0 \cap \mathcal{R}''_0 = \emptyset$) and define

$$S \oplus_{\mathcal{R}',\mathcal{R}''} T := \iota_{\mathcal{R}',\mathcal{R}'_1}(S) \oplus \iota_{\mathcal{R}'',\mathcal{R}''_1}(T),$$

where $\mathcal{R}'_1 := \mathcal{R}'_0 \cup \mathcal{R}'_+ \cup \mathcal{R}''_+$ and $\mathcal{R}''_1$ is defined similarly as $\mathcal{R}''_0 \cup \mathcal{R}'_+ \cup \mathcal{R}''_+$. We thus obtain a concrete $(\mathcal{R}' \cup \mathcal{R}'')$-structure $S \oplus T$. We will see that this extended union still satisfies the Splitting Theorem.

This extension also applies to the disjoint union of abstract structures, that is then a derived operation expressed in terms of the disjoint union of Definition 5.34 and the QF operations of Definition 5.42.

We will frequently write $\oplus$ for $\oplus_{\mathcal{R}',\mathcal{R}''}$.

The natural inclusion is actually an inverse of the following *reduction operation*.[59] If $\mathcal{R}' \subseteq \mathcal{R}$, the operation $fg_{\mathcal{R}-\mathcal{R}'} : STR^c(\mathcal{R}) \to STR^c(\mathcal{R}')$ deletes the relations and constants in $\mathcal{R} - \mathcal{R}'$. Its definition scheme is easy to write. As for the natural inclusion, it is domain preserving. We will write fg_R for $fg_{\{R\}}$. □

Here is an important example of a QDP-QF operation.

Example 5.45 (Fusion of constants) Let a and b be distinct constant symbols of a relational signature $\mathcal{R}$. We define $fuse_{a,b} : STR(\mathcal{R}) \to STR(\mathcal{R})$ as the transformation of abstract structures that fuses a_S and b_S and reorganizes the tuples accordingly. If $a_S = b_S$, then $fuse_{a,b}(S) = S$. Concretely, b_S is deleted if $a_S \neq b_S$. This operation is our main example of a QDP-QF operation.

Formally, for $S \in STR^c(\mathcal{R})$, we define $fuse_{a,b}(S)$ as the concrete structure S' such that

$$D_{S'} := \text{if } a_S = b_S \text{ then } D_S \text{ else } D_S - \{b_S\}.$$

For every $d \in \mathcal{R}_0$

$$d_{S'} := \text{if } d_S = b_S \text{ then } a_S \text{ else } d_S,$$

or equivalently

$$d_{S'} := \text{if } a_S = b_S \vee d_S \neq b_S \text{ then } d_S \text{ else } a_S.$$

For every $R \in \mathcal{R}_+$:

$$R_{S'} := \{(\widehat{e}_1, \ldots, \widehat{e}_{\rho(R)}) \mid (e_1, \ldots, e_{\rho(R)}) \in R_S\},$$

$$\text{where } \widehat{e} := \text{if } e = b_S \text{ then } a_S \text{ else } e.$$

---

[59] In Model Theory, $fg_{\mathcal{R}-\mathcal{R}'}(S)$ is called a *reduct* of S. This operation generalizes the source forgetting operation on s-graphs (Definition 2.30, Section 2.3.1).

This mapping can be defined by the QFO definition scheme $\langle \delta, (\theta_R)_{R \in \mathcal{R}_+}, (\kappa_{c,d})_{c,d \in \mathcal{R}_0} \rangle$ such that $\delta(x)$ is $a = b \vee (a \neq b \wedge x \neq b)$ and:

$\kappa_{a,d}$ is *True*   if $d \in \{a,b\}$,

$\kappa_{c,d}$ is *False*   if $d \in \{a,b\}$ and $c \in \mathcal{R}_0 - \{a\}$,

$\kappa_{a,d}$ is $d = b$ if $d \in \mathcal{R}_0 - \{a,b\}$,

$\kappa_{d,d}$ is $d \neq b$ if $d \in \mathcal{R}_0 - \{a,b\}$,

$\kappa_{c,d}$ is *False*   if $d \in \mathcal{R}_0 - \{a,b\}$ and $c \in \mathcal{R}_0 - \{a,d\}$.

Note that $\kappa_{b,b}$ is *False*. It remains to define the formulas $\theta_R(x_1,\ldots,x_{\rho(R)})$ for $R \in \mathcal{R}_+$. We first define them with existential quantifications for the purpose of clarity. Then we will eliminate these quantifications. We let $n = \rho(R)$. The relation $R_{fuse_{a,b}(S)}$ is defined in S by $\delta(x_1) \wedge \cdots \wedge \delta(x_n) \wedge \psi_R(x_1,\ldots,x_n)$, where $\psi_R(x_1,\ldots,x_n)$ is

$$\exists y_1,\ldots,y_n[R(y_1,\ldots,y_n) \wedge \bigwedge_{1 \leq i \leq n} (y_i = b \wedge x_i = a) \vee (y_i \neq b \wedge x_i = y_i)].$$

This formula is equivalent to

$$\bigvee_{I \subseteq [n]} \exists y_1,\ldots,y_n[R(y_1,\ldots,y_n) \wedge (\bigwedge_{i \in I} y_i = b \wedge x_i = a) \wedge (\bigwedge_{i \notin I} y_i \neq b \wedge x_i = y_i)],$$

hence to

$$\bigvee_{I \subseteq [n]} [R(x_1^I,\ldots,x_n^I) \wedge (\bigwedge_{i \in I} x_i = a) \wedge (\bigwedge_{i \notin I} x_i \neq b)],$$

where x_i^I stands for b if $i \in I$ and for x_i if $i \notin I$. Hence, we obtain a quantifier-free formula $\theta_R(x_1,\ldots,x_n)$ equivalent to $\psi_R(x_1,\ldots,x_n)$.

The operation $fuse_{a,b}$ is QDP-QF because an element is deleted only if it is the value of b and the domain formula of its definition scheme is equivalent to $\bigwedge_{c \in \mathcal{R}_0} (\alpha_c \Rightarrow x \neq c)$, where α_c is $a \neq b$ if $c = b$ and is *False* otherwise.

We will also use its variant $ffuse_{a,b}$ defined as $fg_b \circ fuse_{a,b}$ that fuses a and b and then forgets b. It maps $STR(\mathcal{R})$ to $STR(\mathcal{R} - \{b\})$. It is also QDP-QF and preserves constant-separation, which is not the case for $fuse_{a,b}$. Its definition scheme is obtained from that of $fuse_{a,b}$ by omitting the formulas $\kappa_{c,b}$ for all c.

Note that the operations $fuse_{a,b}$ and $fuse_{b,a}$ are the same on abstract structures but not on concrete ones, because if $a_S \neq b_S$ in the concrete input structure S, then b_S is deleted by the former whereas a_S is deleted by the latter.   □

5.3.3 Backwards translation with respect to QF operations

The main result of this section is the Backwards Translation Theorem (Theorem 5.47). We first prove a weak form of it that entails the Recognizability Theorem for the VR algebra.

Proposition 5.46 Let $\mathcal{D}$ be a DP-QFO definition scheme of type $\mathcal{R} \to \mathcal{R}'$ where $\mathcal{R}'_0 = \emptyset$. For every formula φ in $C_r \mathrm{MS}^h(\mathcal{R}', \overline{X} \cup \overline{y})$, there exists a formula μ in $C_r \mathrm{MS}^h(\mathcal{R}, \overline{X} \cup \overline{y})$ such that, for every concrete $\mathcal{R}$-structure S:[60]

$$sat(\widehat{\mathcal{D}}(S), \varphi, \overline{X}, \overline{y}) = sat(S, \mu, \overline{X}, \overline{y}).$$

Proof: We have $D_S = D_{\widehat{\mathcal{D}}(S)}$. Hence, when constructing μ intended to define in S the assignments that satisfy φ in $\widehat{\mathcal{D}}(S)$, we need not put in μ the condition that the considered values assigned to variables are in the domain of $\widehat{\mathcal{D}}(S)$. This will be necessary in other cases to be considered later. The proof is by induction on the structure of φ.

Case 1: φ is *True* or $u = v$ or $u \in Y$ or $Card_{p,q}(Y)$, with u, v in $\overline{y}$ and Y in $\overline{X}$. We take μ equal to φ.

Case 2: φ is $R(u_1, \ldots, u_n)$ with $R \in \mathcal{R}'_+$ and $u_1, \ldots, u_n \in \overline{y}$. The variables $u_1, \ldots, u_n$ need not be pairwise distinct. We have $R_{\widehat{\mathcal{D}}(S)} = sat(S, \theta_R, \varepsilon, (x_1, \ldots, x_n))$ hence, by Lemma 5.8:

$$sat(\widehat{\mathcal{D}}(S), \varphi, \overline{X}, \overline{y}) = sat(S, \theta_R[u_1/x_1, \ldots, u_n/x_n], \overline{X}, \overline{y}),$$

and we take μ equal to $\theta_R[u_1/x_1, \ldots, u_n/x_n]$. This formula is quantifier-free.

Case 3: φ is $\varphi_1 \vee \varphi_2$ or $\varphi_1 \wedge \varphi_2$ or $\neg \varphi_1$. Assuming μ_1, μ_2 associated with φ_1, φ_2 by induction, we take μ to be $\mu_1 \vee \mu_2$ or $\mu_1 \wedge \mu_2$ or $\neg \mu_1$ respectively.

Case 4: φ is $\exists u.\varphi_1$ or $\exists U.\varphi_1$. Assuming μ_1 associated with φ_1 we take μ to be respectively $\exists u.\mu_1$ or $\exists U.\mu_1$. The correctness of this step follows from the fact that $\mathcal{D}$ is domain preserving. ∎

We now consider a QDP-QFO definition scheme $\mathcal{D}$. For every set variable X, we denote by $\overline{\delta}(X)$ the quantifier-free formula $\bigwedge_{c \in \mathcal{R}_0} (\alpha_c \Rightarrow c \notin X)$. It is clear that for every subset A of D_S, we have, $S \models \overline{\delta}(A)$ if and only if $A \subseteq D_{\widehat{\mathcal{D}}(S)}$.

Theorem 5.47 (Backwards Translation Theorem)

(1) Let $\mathcal{D}$ be a QDP-QFO definition scheme of type $\mathcal{R} \to \mathcal{R}'$. For every $r, h \in \mathcal{N}$ and every formula φ in $C_r \mathrm{MS}^h(\mathcal{R}', \overline{X} \cup \overline{y})$, there exists ψ in $C_r \mathrm{MS}^h(\mathcal{R}, \overline{X} \cup \overline{y})$ such

---

[60] Note that S and $\widehat{\mathcal{D}}(S)$ are concrete structures. A set $sat(S, \varphi, \overline{U}, \overline{u})$ is well defined only if S is a concrete structure, or if S is abstract and $\overline{U} = \overline{u} = \varepsilon$, i.e., if φ is a sentence.

that, for every concrete $\mathcal{R}$-structure S, we have:

$$sat(\widehat{\mathcal{D}}(S),\varphi,\overline{X},\overline{y}) = sat(S,\psi,\overline{X},\overline{y}).$$

(2) If $\mathcal{D}$ is a QFO definition scheme, then this statement holds for φ in $\mathrm{FO}^h(\mathcal{R}',\overline{y})$ with ψ in $\mathrm{FO}^h(\mathcal{R},\overline{y})$.

Proof: (1) The definition of ψ will use an auxiliary formula μ associated with φ by the same induction as in Proposition 5.46. It satisfies the following property: $sat(\widehat{\mathcal{D}}(S),\varphi,\overline{X},\overline{y})$ is the set of all elements of $sat(S,\mu,\overline{X},\overline{y})$ that are $(\overline{X} \cup \overline{y})$-assignments in $\widehat{\mathcal{D}}(S)$, or in other words $sat(\widehat{\mathcal{D}}(S),\varphi,\overline{X},\overline{y}) = sat(S,\mu,\overline{X},\overline{y}) \cap sat(\widehat{\mathcal{D}}(S),True,\overline{X},\overline{y})$.

Case 1: φ is *True* or $Card_{p,q}(Y)$ or $s \in Y$ or $s = t$, with $s,t \in \overline{y} \cup \mathcal{R}'_0$ and $Y \in \overline{X}$.

If φ is *True* or $Card_{p,q}(Y)$, then μ is φ.

If φ is $s \in X$ and s belongs $\overline{y}$, then μ is φ; if φ is $s \in X$ and $s = d \in \mathcal{R}'_0$, then μ is $\bigvee_{c \in \mathcal{R}_0}(\kappa_{c,d} \wedge c \in X)$.

If φ is $s = t$, we have several cases. If s and t belong to $\overline{y}$, then μ is φ; if s belongs to $\overline{y}$ and $t = d \in \mathcal{R}'_0$, then μ is $\bigvee_{c \in \mathcal{R}_0}(\kappa_{c,d} \wedge s = c)$; the case where t belongs to $\overline{y}$ and $s = d \in \mathcal{R}'_0$ is similar; if $s = d \in \mathcal{R}'_0$ and $t = d' \in \mathcal{R}'_0$, then we let μ be $\bigvee_{c,c' \in \mathcal{R}_0}(\kappa_{c,d} \wedge \kappa_{c',d'} \wedge c = c')$.

In all cases, μ is quantifier-free.

Case 2: φ is $R(t_1,\ldots,t_n)$ with $R \in \mathcal{R}'_+$ and $t_1,\ldots,t_n \in \overline{y} \cup \mathcal{R}'_0$. For readability, we only consider the case where $n = 2$, and let φ be $R(s,t)$ with s and t in $\overline{y} \cup \mathcal{R}'_0$.

If $s,t \in \overline{y}$, then μ is $\theta_R[s/x_1,t/x_2]$. If $s \in \overline{y}$ and $t = d \in \mathcal{R}'_0$, then μ is $\bigvee_{c \in \mathcal{R}_0}(\kappa_{c,d} \wedge \theta_R[s/x_1,c/x_2])$. The case where $t \in \overline{y}$ and $s = d \in \mathcal{R}'_0$ is similar. If $s = d \in \mathcal{R}'_0$ and $t = d' \in \mathcal{R}'_0$, then we define μ as $\bigvee_{c,c' \in \mathcal{R}_0}(\kappa_{c,d} \wedge \kappa_{c',d'} \wedge \theta_R[c/x_1,c'/x_2])$.

The construction of μ in the general case is similar, but lengthy to write down. In all cases the formula μ is quantifier-free, because, in particular, the formulas θ_R are quantifier-free.

Case 3: φ is $\varphi_1 \wedge \varphi_2$ or $\varphi_1 \vee \varphi_2$ or $\neg\varphi_1$. As in Proposition 5.46.

Case 4: φ is $\exists u.\varphi_1$ or $\exists U.\varphi_1$. Assuming that μ_1 has been constructed from φ_1 by induction, we take μ to be respectively $\exists u(\delta(u) \wedge \mu_1)$ or $\exists U(\overline{\delta}(U) \wedge \mu_1)$. This definition is correct (using induction) and μ has quantifier-height $1 + qh(\mu_1)$ which is at most $1 + qh(\varphi_1) = qh(\varphi)$, because the formulas $\delta(u)$ and $\overline{\delta}(U)$ are quantifier-free.

The desired formula ψ is the conjunction of μ, of the formulas $\delta(y_i)$ for every y_i in $\overline{y}$, and of the formulas $\overline{\delta}(X_i)$ for every X_i in $\overline{X}$. This completes the proof of the first assertion.

(2) If $\varphi \in \mathrm{FO}^h(\mathcal{R}',\overline{y})$, then the formulas $\overline{\delta}(X)$ are not needed in this construction. Hence it applies even if $\mathcal{D}$ is not quasi-domain preserving. ∎

We will denote by $\varphi^{\mathcal{D}}$ the formula ψ constructed by backwards translation applied to φ and $\mathcal{D}$.

Remark 5.48 (1) As for Theorem 5.39, the proof of Theorem 5.47 is effective. It gives an algorithm that constructs ψ from φ and $\mathcal{D}$.

(2) If φ is a sentence, then $\varphi^{\mathcal{D}}$ is also a sentence and we have

$$\widehat{\mathcal{D}}(S) \models \varphi \text{ if and only if } S \models \varphi^{\mathcal{D}},$$

and this is meaningful for an abstract structure S.

(3) It is important for proving the Recognizability Theorem that Backwards Translation does not increase the quantifier-height. This concern motivates the technicality about QDP-QF operations and the particular treatment of constants (cf. Remark 5.43(3)).

(4) We consider the QDP-QF operation $\mathit{fuse}_{a,b}$ applied to a concrete structure S such that $a_S \neq b_S$. The structure $T := \mathit{fuse}_{a,b}(S)$ is constructed from S by deleting b_S, whereas it can also be seen as a quotient of S. That is, we have a surjective mapping $h : D_S \to D_T$ that identifies a_S and b_S and is the identity otherwise. The relation R_T consists of all tuples $(h(a_1), \ldots, h(a_n))$ such that the tuple $(a_1, \ldots, a_n)$ is in R_S, and for every constant symbol c, we have $h(c_S) = c_T$. To simplify the discussion, let us consider in Theorem 5.47 a formula φ with a single free variable y. Letting $\mathcal{D}$ define $\mathit{fuse}_{a,b}$, we do not have

$$\mathit{sat}(S, \varphi^{\mathcal{D}}, \varepsilon, y) = h^{-1}(\mathit{sat}(\mathit{fuse}_{a,b}(S), \varphi, \varepsilon, y))$$

as one might think, but we have

$$\mathit{sat}(S, \varphi^{\mathcal{D}}, \varepsilon, y) = \{d \in D_S - \{b_S\} \mid \mathit{fuse}_{a,b}(S) \models \varphi(d)\}.$$

$\square$

The following proposition deals with compositions of QF operations.

Proposition 5.49 If $f : STR^c(\mathcal{R}) \to STR^c(\mathcal{R}')$ and $g : STR^c(\mathcal{R}') \to STR^c(\mathcal{R}'')$ are QF operations, then so is $g \circ f : STR^c(\mathcal{R}) \to STR^c(\mathcal{R}'')$. If f and g are domain preserving or are quasi-domain preserving, then $g \circ f$ has the same property.

Proof: Let f be defined by $\mathcal{D} = \langle \delta, (\theta_R)_{R \in \mathcal{R}'_+}, (\kappa_{c,d})_{c \in \mathcal{R}_0, d \in \mathcal{R}'_0} \rangle$ and g be defined by $\mathcal{D}' = \langle \delta', (\theta'_R)_{R \in \mathcal{R}''_+}, (\kappa'_{d,e})_{d \in \mathcal{R}'_0, e \in \mathcal{R}''_0} \rangle$. We define

$$\mathcal{D}'' := \langle \delta'', (\theta''_R)_{R \in \mathcal{R}''_+}, (\kappa''_{c,e})_{c \in \mathcal{R}_0, e \in \mathcal{R}''_0} \rangle.$$

We need a formula δ'' such that $\mathit{sat}(\widehat{\mathcal{D}}(S), \delta', x) = \mathit{sat}(S, \delta'', x)$. It is given by the second assertion of Theorem 5.47 with $h = 0$, so we can take δ'' equal to $\delta'^{\mathcal{D}}$. This theorem gives us also, for each $R \in \mathcal{R}''_n$, $n \geq 1$, a formula θ''_R defined as $\theta'^{\mathcal{D}}_R$ such that $sat(\widehat{\mathcal{D}}(S), \theta'_R, \bar{x}) = sat(S, \theta''_R, \bar{x})$ where $\bar{x} = (x_1, \ldots, x_n)$. Finally, we let $\kappa''_{c,e}$ be $\bigvee_{d \in \mathcal{R}'_0} \kappa_{c,d} \wedge \kappa'^{\mathcal{D}}_{d,e}$, where, by Remark 5.48(2), $S \models \kappa'^{\mathcal{D}}_{d,e}$ if and only if $\widehat{\mathcal{D}}(S) \models \kappa'_{d,e}$. It follows from Theorem 5.47(2) that $\mathcal{D}''$ defines $g \circ f$.

We now examine the case where f and g are quasi-domain preserving. It is clear that for every structure S in $STR^c(\mathcal{R})$, the elements in $D_{g \circ f(S)} - D_S$ are constants of S, because if such an element is deleted by g, then it is a constant of $f(S)$, hence a constant of S, and if it is deleted by f it is also a constant of S. However, the notion of a QDP-QF operation is defined by a particular syntactic form of the domain formula. We now check that δ'' has the required form. Let $\delta(x)$ be $\bigwedge_{c \in \mathcal{R}_0} (\alpha_c \Rightarrow x \neq c)$ and $\delta'(x)$ be $\bigwedge_{d \in \mathcal{R}'_0} (\beta_d \Rightarrow x \neq d)$. We can take $\delta''(x)$ to be $\bigwedge_{c \in \mathcal{R}_0} (\gamma_c \Rightarrow x \neq c)$, where γ_c is $\alpha_c \vee \bigvee_{d \in \mathcal{R}'_0} (\kappa_{c,d} \wedge \beta_d^{\mathcal{D}})$.

The case of DP-QF operations is clear. ∎

5.3.4 The VR and HR graph operations

We apply the above defined notions to describe the operations of the VR and the HR graph algebras. In particular, the unary operations are all QF operations.

Application 5.50 (The operations of the VR algebra $\mathbb{GP}^t$) The many-sorted algebra $\mathbb{GP}^t$ has been defined in Section 2.6.3. Our aim is to express its operations in terms of operations on relational structures. A p-graph G of type C (with $C \in \mathcal{P}_f(\mathcal{A})$) is faithfully represented by the $\mathcal{R}_{s,C}$-structure $\lfloor G \rfloor = \langle V_G, edg_G, (lab_{cG})_{c \in C} \rangle$ (see Example 5.2(2)) and we have

$$\lfloor G \oplus_{C,D} H \rfloor = \lfloor G \rfloor \oplus_{\mathcal{R}_{s,C}, \mathcal{R}_{s,D}} \lfloor H \rfloor,$$

where $C = \pi(G)$ and $D = \pi(H)$. Here we use the extended disjoint union of structures defined in Example 5.44; note that $\mathcal{R}_{s,C} \cup \mathcal{R}_{s,D} = \mathcal{R}_{s,C \cup D}$. We now formalize the unary VR operations as quantifier-free operations.

If C is a finite set of port labels, $\pi(G) = C$ and $a, b \in C$, then

$$\lfloor \overrightarrow{add}_{a,b,C}(G) \rfloor = \widehat{\mathcal{D}}_{a,b,C}(\lfloor G \rfloor),$$

where $\mathcal{D}_{a,b,C}$ is the DP-QFO definition scheme $\langle True, \theta_{edg}, (\theta_{lab_c})_{c \in C} \rangle$ of type $\mathcal{R}_{s,C} \to \mathcal{R}_{s,C}$ such that

$$\begin{aligned} \theta_{edg}(x_1, x_2) \quad &\text{is} \quad edg(x_1, x_2) \vee (lab_a(x_1) \wedge lab_b(x_2)), \\ \theta_{lab_c}(x_1) \quad &\text{is} \quad lab_c(x_1), \text{ for each } c \in C. \end{aligned}$$

If $h : C \to \mathcal{A}$ and $D = h(C)$ we have

$$\lfloor relab_{h,C}(G) \rfloor = \widehat{\mathcal{D}}_{h,C}(\lfloor G \rfloor)$$

for every p-graph G of type $\pi(G) = C$, where $\mathcal{D}_{h,C}$ is the DP-QFO definition scheme $\langle True, \theta_{edg}, (\theta_{lab_d})_{d \in D} \rangle$ of type $\mathcal{R}_{s,C} \to \mathcal{R}_{s,D}$ such that $\theta_{edg}(x_1, x_2)$ is $edg(x_1, x_2)$ and $\theta_{lab_d}(x_1)$ is $\bigvee_{c \in h^{-1}(d)} lab_c(x_1)$ for each $d \in D$.

The structures corresponding to the constant graphs of $\mathbb{GP}^t$ have domains with at most one element. The extension to operations on labeled graphs (cf. Definition 2.90) is straightforward.

Application 5.51 (The unary operations of the HR algebra $\mathbb{JS}^t$) We now consider in a similar way the operations of the many-sorted algebra $\mathbb{JS}^t$ defined in Section 2.6.2. An s-graph G of type C is faithfully represented here by the constant-separated $\mathcal{R}_{ms,C}$-structure $\lceil G \rceil = \langle V_G \cup E_G, in_G, lab_{EdgeG}, (a_G)_{a \in C} \rangle$ (see Section 5.2.5), where a_G is the a-source of G.

If $B \subseteq C$, the operation $fg_{B,C}$ takes as argument an s-graph of type C and produces an s-graph of type $C - B$. We have

$$\lceil fg_{B,C}(G) \rceil = \widehat{\mathcal{D}}_{B,C}(\lceil G \rceil),$$

where $\mathcal{D}_{B,C} := \langle True, \theta_{in}, \theta_{lab_{Edge}}, (\kappa_{a,b})_{a \in C, b \in C-B} \rangle$ of type $\mathcal{R}_{ms,C} \to \mathcal{R}_{ms,C-B}$ and

$$\theta_{in}(x_1, x_2) \quad \text{is} \quad in(x_1, x_2),$$
$$\theta_{lab_{Edge}}(x_1) \quad \text{is} \quad lab_{Edge}(x_1),$$
$$\kappa_{a,b} \quad \text{is} \quad \begin{cases} True & \text{if } a = b, \\ False & \text{if } a \neq b. \end{cases}$$

Let h be a bijection $: C \to D$ where $C, D \in \mathcal{P}_f(\mathcal{A})$. The operation $ren_{h,C}$ of type $C \to D$ is defined by the scheme $\mathcal{D}_{h,C} := \langle True, \theta_{in}, \theta_{lab_{Edge}}, (\kappa_{a,b})_{a \in C, b \in D} \rangle$ of type $\mathcal{R}_{ms,C} \to \mathcal{R}_{ms,D}$ where θ_{in} and $\theta_{lab_{Edge}}$ are as above and

$$\kappa_{a,b} \quad \text{is} \quad \begin{cases} True & \text{if } h(a) = b, \\ False & \text{if } h(a) \neq b. \end{cases}$$

The QFO definition schemes $\mathcal{D}_{B,C}$ and $\mathcal{D}_{h,C}$ are domain preserving and they preserve constant-separation. Parallel-composition will be defined below in Example 5.55, in terms of disjoint union ($\oplus$) and the QDP-QF operations $ffuse_{a,b}$ of Example 5.45.

The reader may recall that we have defined in Definition 4.50 a kind of simulation of the operations of the HR algebra by derived operations of the VR algebra. Why not resting on it instead of defining particular operations for the HR algebra? There are several reasons. First, this simulation is quite complicated and it describes graphs of tree-width at most k by VR operations using a number of port labels that is exponential in k (and we know by Proposition 2.114(2) that this blow-up is unavoidable). Second, it is defined for simple graphs and would have to be adapted to graphs with multiple edges. Third, graph classes of bounded tree-width are frequently considered for algorithmic purposes and it is thus useful to consider them more directly than via cumbersome encodings. And, finally, the encoding of Definition 4.50 works for graphs but not for relational structures of bounded tree-width, a case that we will consider in Chapter 9.

5.3.5 The Splitting Theorem for derived operations

We will show that the Splitting Theorem (Theorem 5.39) extends to all linear and strict derived operations, built with disjoint union and QDP-QF operations (as defined in Example 5.44 and Definition 5.42 respectively). For sentences, it extends to *all* derived operations.

Definition 5.52 (n-ary QDP-QF operations) We first define binary operations. Let $\mathcal{R}^{(1)},\ldots,\mathcal{R}^{(4)},\mathcal{R}$ be relational signatures such that $\mathcal{R}_0^{(3)} \cap \mathcal{R}_0^{(4)} = \emptyset$ (and according to the convention of Footnote 58 in Example 5.44, a symbol belonging to two of them has the same arity with respect to both). Let $f : STR(\mathcal{R}^{(1)}) \times STR(\mathcal{R}^{(2)}) \to STR(\mathcal{R})$ be defined by

$$f(S,T) := h(g(S) \oplus_{\mathcal{R}^{(3)},\mathcal{R}^{(4)}} g'(T)),$$

where g, g' and h are QDP-QF operations such that $g : STR(\mathcal{R}^{(1)}) \to STR(\mathcal{R}^{(3)})$, $g' : STR(\mathcal{R}^{(2)}) \to STR(\mathcal{R}^{(4)})$ and $h : STR(\mathcal{R}^{(3)} \cup \mathcal{R}^{(4)}) \to STR(\mathcal{R})$.

We call such a mapping f a *binary QDP-QF operation*[61] and a *binary DP-QF operation* if g, g' and h are DP-QF operations. Its definition uses only the finitely many symbols of $\mathcal{R}^{(1)} \cup \cdots \cup \mathcal{R}^{(4)} \cup \mathcal{R}$. Note that every binary linear and strict derived operation f that is built with disjoint union and QDP-QF operations, can be written as above because the family of QDP-QF operations is closed under composition (Proposition 5.49) and contains the identity on every $STR(\mathcal{R})$ (it is $\iota_{\mathcal{R},\mathcal{R}}$, see Example 5.44).

In general, for $n \geq 1$, an n-ary mapping $f : STR(\mathcal{R}^{(1)}) \times \cdots \times STR(\mathcal{R}^{(n)}) \to STR(\mathcal{R})$ is called an *n-ary QDP-QF operation* if it is a linear and strict derived operation that is built with disjoint unions $\oplus_{\mathcal{R}',\mathcal{R}''}$ and QDP-QF operations. Clearly, by Proposition 5.49, the unary QDP-QF operations are the same as the QDP-QF operations. Every n-ary QDP-QF operation f with $n > 2$ can be expressed as a composition of binary QDP-QF operations. To be precise, it can be defined (recursively) as $f(S_1,\ldots,S_n) := h(g_1(S_1,\ldots,S_m),g_2(S_{m+1},\ldots,S_n))$, where h is a binary QDP-QF operation, g_1 is an m-ary QDP-QF operation, with $1 \leq m \leq n - 1$, and g_2 is an $(n-m)$-ary QDP-QF operation. Similar definitions hold for n-ary DP-QF operations.

These operations act as well on concrete structures, provided the argument structures are pairwise disjoint (the Splitting Theorem to be stated concerns concrete structures).

---

[61] We will not use the full term *binary quasi-domain preserving quantifier-free operation*!

Before proving that the Splitting Theorem extends to these operations, we give some examples.

Example 5.53 (Series-composition) Series-composition has been defined as a binary mapping $\bullet$ on directed s-graphs of type $\{1,2\}$. Via the faithful representation of an s-graph G by $\lceil G \rceil$ in $STR(\mathcal{R}_{ms,\{1,2\}}) = STR(\{in, lab_{Edge}, 1, 2\})$, (1 and 2 are constant symbols), we have

$$\lceil G \bullet H \rceil = \lceil G \rceil \bullet_s \lceil H \rceil, \text{ where}$$

$$\bullet_s : STR(\mathcal{R}_{ms,\{1,2\}}) \times STR(\mathcal{R}_{ms,\{1,2\}}) \to STR(\mathcal{R}_{ms,\{1,2\}})$$

is the binary operation such that

$$S \bullet_s T := fg_{\{3,4\}}(fuse_{3,4}(ren_{2\leftrightarrow3}(S) \oplus ren_{1\leftrightarrow4}(T))).$$

It is a binary QDP-QF operation since by Example 5.45, Application 5.51 and Proposition 5.49 the unary operations $ren_{2\leftrightarrow3}$, $ren_{1\leftrightarrow4}$ and $fg_{\{3,4\}} \circ fuse_{3,4}$ ($= fg_3 \circ ffuse_{3,4}$) are QDP-QF operations that preserve constant-separation. (This is necessary for them to represent operations on s-graphs). We now review the types of the operations used in this definition:

$$\oplus : \mathcal{R}_{ms,\{1,3\}} \times \mathcal{R}_{ms,\{2,4\}} \to \mathcal{R}_{ms,\{1,2,3,4\}},$$

$$ren_{2\leftrightarrow3} : \mathcal{R}_{ms,\{1,2\}} \to \mathcal{R}_{ms,\{1,3\}},$$

$$ren_{1\leftrightarrow4} : \mathcal{R}_{ms,\{1,2\}} \to \mathcal{R}_{ms,\{2,4\}},$$

$$fuse_{3,4} : \mathcal{R}_{ms,\{1,2,3,4\}} \to \mathcal{R}_{ms,\{1,2,3,4\}},$$

$$fg_{\{3,4\}} : \mathcal{R}_{ms,\{1,2,3,4\}} \to \mathcal{R}_{ms,\{1,2\}}.$$

The definition would be the same with $\{in_1, in_2\}$ instead of $\{in, lab_{Edge}\}$, that is, if $\lceil G \rceil$ is defined alternatively as an $\{in_1, in_2, 1, 2\}$-structure.   □

Example 5.54 (The complete join) The complete join of simple undirected graphs can be expressed by

$$\lfloor G \otimes H \rfloor = create(nat(\lfloor G \rfloor) \oplus mark(\lfloor H \rfloor)).$$

In this writing, $\lfloor G \rfloor, \lfloor H \rfloor \in STR(\{edg\})$, the operation *nat* is $\iota_{\{edg\},\{edg,m\}}$ that expands the structure $\lfloor G \rfloor$ with the new empty unary relation m (cf. Example 5.44), the operation *mark* expands $\lfloor H \rfloor$ with the new unary relation m equal to the domain of the argument structure, and the operation *create* adds undirected edges between any two vertices x and y such that $\neg m(x) \wedge m(y)$ holds, and then deletes m. The types of these

DP-QF operations are:

$$mark, nat : STR(\{edg\}) \rightarrow STR(\{edg, m\}),$$

$$create : STR(\{edg, m\}) \rightarrow STR(\{edg\}).$$

Hence the complete join is a binary DP-QF operation. □

Example 5.55 (Parallel-composition of structures) Let $\mathcal{R}, \mathcal{R}'$ be two relational signatures. We denote by $\parallel_{\mathcal{R}, \mathcal{R}'}$, or simply by $\parallel$, the operation : $STR(\mathcal{R}) \times STR(\mathcal{R}') \rightarrow STR(\mathcal{R} \cup \mathcal{R}')$ described informally as follows: $S \parallel T$ is obtained from the union of disjoint concrete structures S and T by fusing the constants a_S and a_T for every a in $\mathcal{R}_0 \cap \mathcal{R}'_0$. We call $S \parallel T$ the *parallel-composition* of S and T.

Let $A := \mathcal{R}_0 \cap \mathcal{R}'_0$ and let $A' := \{a' \mid a \in A\}$ be a set of auxiliary constant symbols disjoint from $\mathcal{R}_0 \cup \mathcal{R}'_0$ (we assume of course that $a' \neq b'$ if $a \neq b$). For structures $S \in STR(\mathcal{R})$ and $T \in STR(\mathcal{R}')$ the parallel-composition operation can be expressed by

$$S \parallel_{\mathcal{R}, \mathcal{R}'} T = ffuse(S \oplus_{\mathcal{R}, \mathcal{R}''} ren(T)),$$

where the DP-QF operation *ren* changes each $a \in A$ into a' (its definition scheme is similar to the one of ren_h in Application 5.51), $\mathcal{R}''$ is the relational signature $(\mathcal{R}' - A) \cup A'$ and *ffuse* is the operation $fg_{A'} \circ fuse$, where *fuse* is the composition in any order of the QDP-QF operations $fuse_{a,a'}$ for all $a \in A$.[62] Hence, this operation is a binary QDP-QF operation. This operation is also defined on disjoint concrete structures. It is not commutative on concrete structures whereas it is on abstract ones.

For s-graphs G and H of respective types C and D, faithfully represented by the constant-separated structures $\lceil G \rceil \in STR(\mathcal{R}_{ms,C})$ and $\lceil H \rceil \in STR(\mathcal{R}_{ms,D})$, we have $\lceil G \parallel_{C,D} H \rceil = \lceil G \rceil \parallel_{\mathcal{R}_{ms,C}, \mathcal{R}_{ms,D}} \lceil H \rceil$.[63] Hence the parallel-composition operations of the many-sorted algebra $\mathbb{JS}^t$ are binary QDP-QF operations. □

Example 5.56 (Concatenation of linearly ordered structures) Let us consider a relational signature $\mathcal{R}$ containing the binary symbol $\leq$. For $S, T \in STR(\mathcal{R})$ we denote by $S \overrightarrow{\oplus} T$ the structure $S \oplus T$ where, furthermore, we let $x \leq y$ hold for every $x \in D_S$ and $y \in D_T$. By an easy modification of the construction of Example 5.54, we obtain that $\overrightarrow{\oplus} : STR(\mathcal{R}) \times STR(\mathcal{R}) \rightarrow STR(\mathcal{R})$ is a binary DP-QF operation.

We recall from Example 5.2(1) that $\lfloor w \rfloor^\ell$ is the representation of a word $w \in A^*$ that uses the order of occurrences of letters. We have $\lfloor v \cdot w \rfloor^\ell = \lfloor v \rfloor^\ell \overrightarrow{\oplus} \lfloor w \rfloor^\ell$. We will use this observation in the proof of Corollary 5.66.

If we construct $\mathcal{R}$-structures by using the *ordered disjoint union* $\overrightarrow{\oplus}$, basic structures with a single element $*$ satisfying $* \leq *$, and quantifier-free operations that do not

---

[62] Note that *ffuse* is also the composition in any order of the operations $ffuse_{a,a'}$ for all $a \in A$.
[63] For handling directed graphs, replacing $\{in, lab_{Edge}\}$ by $\{in_1, in_2\}$ would not change this equality, and its consequences remain valid.

modify the relation defined by $\leq$, then all structures constructed by these operations are linearly ordered by $\leq$. $\qquad\qquad\square$

We now state the extension of Theorem 5.39 to binary QDP-QF operations. Its statement is exactly the same as the one of Theorem 5.39, with $\oplus$ replaced by an arbitrary binary QDP-QF operation f (and with $\mathcal{R},\mathcal{R}'$ replaced by $\mathcal{R}^{(1)},\mathcal{R}^{(2)}$, which is of course irrelevant).

Theorem 5.57 (Splitting Theorem, for binary QDP-QF operations) Let f be a binary QDP-QF operation : $STR(\mathcal{R}^{(1)}) \times STR(\mathcal{R}^{(2)}) \to STR(\mathcal{R})$. For every r and h in $\mathcal{N}$ and every formula φ in $C_r\mathrm{MS}^h(\mathcal{R},\overline{X}\cup\overline{z})$, there exists a finite family of 5-tuples $(\pi_i,\theta_i,\psi_i,\overline{x}_i,\overline{y}_i)_{1\leq i\leq p}$ such that, for all disjoint concrete $\mathcal{R}^{(1)}$- and $\mathcal{R}^{(2)}$-structures, respectively S and T, we have

$$sat(f(S,T),\varphi,\overline{X},\overline{z}) = \biguplus_{1\leq i\leq p} perm_{\pi_i}(sat(S,\theta_i,\overline{X},\overline{x}_i) \boxtimes sat(T,\psi_i,\overline{X},\overline{y}_i)),$$

where, for each i, $\theta_i \in C_r\mathrm{MS}^h(\mathcal{R}^{(1)},\overline{X}\cup\overline{x}_i)$, $\psi_i \in C_r\mathrm{MS}^h(\mathcal{R}^{(2)},\overline{X}\cup\overline{y}_i)$, $\overline{z} = \pi_i(\overline{x}_i\overline{y}_i)$ and π_i is a permutation that preserves the relative orders of variables in $\overline{x}_i$ and in $\overline{y}_i$.

Proof: We let f be defined by $f(S,T) := h(g(S) \oplus_{\mathcal{R}^{(3)},\mathcal{R}^{(4)}} g'(T))$. By Example 5.44 and Proposition 5.49 we may assume that $\mathcal{R}_+^{(3)} = \mathcal{R}_+^{(4)}$. By Theorems 5.39 and 5.47, we have

$$sat(f(S,T),\varphi,\overline{X},\overline{z}) = sat(g(S) \oplus g'(T),\varphi',\overline{X},\overline{z})$$

$$= \biguplus_{1\leq i\leq p} perm_{\pi_i}(sat(g(S),\theta_i',\overline{X},\overline{x}_i) \boxtimes sat(g'(T),\psi_i',\overline{X},\overline{y}_i))$$

$$= \biguplus_{1\leq i\leq p} perm_{\pi_i}(sat(S,\theta_i,\overline{X},\overline{x}_i) \boxtimes sat(T,\psi_i,\overline{X},\overline{y}_i)),$$

where φ' is constructed from φ by backwards translation relative to the unary operation h (Theorem 5.47). The sequences of variables, the formulas and the permutations in the second equality are constructed by the Splitting Theorem (Theorem 5.39) applied to φ'. The formulas θ_i and ψ_i in the last equality are constructed by backwards translation applied to the formulas θ_i' and the unary operation g, and to the formulas ψ_i' and the unary operation g', respectively. $\qquad\blacksquare$

This theorem extends to n-ary QDP-QF operations of all arities in a straightforward way. We will however not need that extension. Instead, we will use Corollary 5.60 below.

Remark 5.58 (1) By Remarks 5.40(1) and 5.48(1), the proof of Theorem 5.57 is effective. It gives an algorithm that constructs from φ, $\overline{X}$, $\overline{z}$ and f a splitting of φ

relative to f. If f is defined by $f(S,T):=h(g(S)\oplus_{\mathcal{R}^{(3)},\mathcal{R}^{(4)}}g'(T))$, then the definition schemes of g, g' and h are given as input to the algorithm.

(2) If $\bar{z}=\varepsilon$, then the statement of Theorem 5.57 is simpler. No permutations are needed and we have

$$sat(f(S,T),\varphi,\overline{X}) = \biguplus_{1\le i\le p} sat(S,\theta_i,\overline{X}) \boxtimes sat(T,\psi_i,\overline{X}).$$

(3) If φ is a sentence, then we obtain the following statement:

$$f(S,T)\models\varphi \text{ if and only if } S\models\theta_i \text{ and } T\models\psi_i \text{ for some } i\in[p],$$

and, furthermore, the index i is unique when it exists. This statement holds for abstract structures S and T.

If, furthermore, $\mathcal{R}^{(1)}=\mathcal{R}^{(2)}$, we can define $g(S):=f(S,S)$ for an abstract $\mathcal{R}^{(1)}$-structure S. Then we have

$$g(S)\models\varphi \text{ if and only if } S\models\bigvee_{i\in[p]}\theta_i\wedge\psi_i.$$

More generally, the Splitting Theorem extends to sentences and to derived operations on abstract structures built with disjoint union and QDP-QF operations that are not linear. However, we will not need this extension for proving recognizability results because, by Proposition 3.56(3), if a set is recognizable in an F-algebra, then it is recognizable in any derived algebra of it, defined with derived operations that need not be linear.

(4) If φ is a sentence and T is a fixed structure, then the unary mapping defined by $k(S):=f(S,T)$ satisfies the following:

$$k(S)\models\varphi \text{ if and only if } S\models\theta,$$

where θ is the disjunction of the sentences θ_i such that $T\models\psi_i$. $\qquad\square$

5.3.6 Computing bounded theories

According to the definition given in Section 5.1.6, the $\mathcal{L}$-theory $Th_{\mathcal{L}}(\{S\})$ of a structure S (where $\mathcal{L}$ denotes a logical language) is the set of sentences of this language that are valid in S. This is usually an infinite set. Our objective is to compute theories of structures in certain cases where they are finite because of syntactic restrictions, typically on the quantifier-height.

Definition 5.59 (Bounded theories) In Section 5.6, we will give an algorithm that transforms an arbitrary formula φ in $C_r\mathrm{MS}^h(\mathcal{R},\mathcal{X})$ into an equivalent formula $\widehat{\varphi}$ that is a kind of normal form of φ. (We have $\widehat{\psi} = \psi$ if $\psi = \widehat{\varphi}$.) This formula also belongs to $C_r MS^h(\mathcal{R}, \mathcal{X})$. If $\mathcal{X}$ is finite, then the set

$$\widehat{C_r MS^h}(\mathcal{R}, \mathcal{X}) \ := \ \{\widehat{\varphi} \mid \varphi \in C_r MS^h(\mathcal{R}, \mathcal{X})\}$$

is finite and can be computed from r, h, $\mathcal{R}$ and $\mathcal{X}$ (Proposition 5.94). We define the (h,r)-*bounded (CMS) theory* of a concrete or abstract $\mathcal{R}$-structure S as the set $\widehat{Th(S, \mathcal{R}, h, r)}$ of sentences in $C_r MS^h(\mathcal{R}, \emptyset)$ that are valid in S, i.e.,

$$Th(S, \mathcal{R}, h, r) := \{\varphi \in \widehat{C_r MS^h}(\mathcal{R}, \emptyset) \mid S \models \varphi\}$$
$$= \{\widehat{\varphi} \mid \varphi \in C_r MS^h(\mathcal{R}, \emptyset), S \models \varphi\}.$$

The set $Th(S, \mathcal{R}, h, r)$ is finite and computable from S, $\mathcal{R}$, h and r, although not in a tractable way. This is a specialization of the notion of theory introduced in Section 5.1.6.

More generally, if $X_k = \{x_1, \ldots, x_k\}$ is the standard set of first-order variables, with the standard order $(x_1, \ldots, x_k)$, and $\bar{a} = (a_1, \ldots, a_k)$ is a k-tuple of elements of the domain of a concrete structure S, we let

$$Th(S, \mathcal{R}, h, r, \bar{a}) \ := \ \{\varphi \in \widehat{C_r MS^h}(\mathcal{R}, X_k) \mid S \models \varphi(\bar{a})\}$$
$$= \ \{\widehat{\varphi} \mid \varphi \in C_r MS^h(\mathcal{R}, X_k), \bar{a} \in sat(S, \varphi, (x_1, \ldots, x_k))\}.$$

An easy consequence of Theorems 5.47 and 5.57 is the following:

Corollary 5.60 Let $f : STR(\mathcal{R}^{(1)}) \times \cdots \times STR(\mathcal{R}^{(n)}) \to STR(\mathcal{R})$ be an n-ary QDP-QF operation. For every $h, r, k_1, \ldots, k_n, k \in \mathcal{N}$ with $k = k_1 + \cdots + k_n$, there exists a computable mapping Z_f:

$$\mathcal{P}(\widehat{C_r MS^h}(\mathcal{R}^{(1)}, X_{k_1})) \times \cdots \times \mathcal{P}(\widehat{C_r MS^h}(\mathcal{R}^{(n)}, X_{k_n})) \to \mathcal{P}(\widehat{C_r MS^h}(\mathcal{R}, X_k)),$$

such that, for all pairwise disjoint concrete $\mathcal{R}^{(1)}$-, ..., $\mathcal{R}^{(n)}$-structures, respectively $S_1, \ldots, S_n$, and every k-tuple $\bar{a}_1 \cdots \bar{a}_n$ of elements of $D_{f(S_1, \ldots, S_n)}$ such that $\bar{a}_1 \in (D_{S_1})^{k_1}, \ldots, \bar{a}_n \in (D_{S_n})^{k_n}$, we have

$$Th(f(S_1, \ldots, S_n), \mathcal{R}, h, r, \bar{a}_1 \cdots \bar{a}_n) = Z_f(Th(S_1, \mathcal{R}^{(1)}, h, r, \bar{a}_1), \ldots, Th(S_n, \mathcal{R}^{(n)}, h, r, \bar{a}_n)).$$

When necessary, we will use the more precise notation $Z_{f, h, r, (k_1, \ldots, k_n)}$ for Z_f.

A similar statement holds for abstract structures $S_1, \ldots, S_n$ and empty tuples $\bar{a}_1, \ldots, \bar{a}_n$.

Proof: We first consider the case $n = 1$. Let $\mathcal{D}$ be the definition scheme of the QDP-QF operation f. Let S be a concrete $\mathcal{R}^{(1)}$-structure and let $\bar{a}$ be a k-tuple of elements of $D_{f(S)}$. It has to be shown that $Th(f(S),\mathcal{R},h,r,\overline{a})$ can be computed from $Th(S,\mathcal{R}^{(1)},h,r,\overline{a})$. We have

$$Th(f(S),\mathcal{R},h,r,\overline{a}) = \{\varphi \in \widehat{C_r MS^h}(\mathcal{R},X_k) \mid \widehat{\varphi^{\mathcal{D}}} \in Th(S,\mathcal{R}^{(1)},h,r,\overline{a})\},$$

where $\varphi^{\mathcal{D}}$ is the backwards translation of φ relative to $\mathcal{D}$ (cf. Theorem 5.47). In order to prove the inclusion $\supseteq$, we consider φ in the set to the right. If $\overline{a} \in sat(S,\varphi^{\mathcal{D}},(x_1,\ldots,x_k)) = sat(S,\varphi^{\mathcal{D}},(x_1,\ldots,x_k))$, then Theorem 5.47 implies that $\overline{a} \in sat(f(S),\varphi,(x_1,\ldots,x_k))$ and $\varphi \in Th(f(S),\mathcal{R},h,r,\overline{a})$. The proof is similar for the other inclusion. Hence, we can define

$$Z_f(\Theta) := \{\varphi \in \widehat{C_r MS^h}(\mathcal{R},X_k) \mid \widehat{\varphi^{\mathcal{D}}} \in \Theta\},$$

for every $\Theta \subseteq \widehat{C_r MS^h}(\mathcal{R}^{(1)},X_k)$, and Z_f is the desired function. Note that $Z_f(\Theta)$ is defined for every subset Θ of $\widehat{C_r MS^h}(\mathcal{R}^{(1)},X_k)$, even if Θ is not a theory: Θ may be empty or contain contradictory formulas.

The proof for $n = 2$ is similar to the one for $n = 1$. Let f be a binary QDP-QF operation, let S_1,S_2 be concrete structures, and let $\overline{a}_1 \in (D_{S_1})^{k_1}$ and $\overline{a}_2 \in (D_{S_2})^{k_2}$. We consider a formula $\varphi \in \widehat{C_r MS^h}(\mathcal{R},X_{k_1+k_2})$. By Theorem 5.57 (considering only identity permutations), a finite family $(\theta_i,\psi_i)_{\leq i \leq p}$ can be computed, with $\theta_i \in C_r MS^h(\mathcal{R}^{(1)},X_{k_1})$ and $\psi_i \in C_r MS^h(\mathcal{R}^{(2)},\{x_{k_1+1},\ldots,x_{k_1+k_2}\})$, such that $\overline{a}_1\overline{a}_2 \in sat(f(S_1,S_2),\varphi,(x_1,\ldots,x_{k_1+k_2}))$ if and only if there exists $i \in [p]$ with $\overline{a}_1 \in sat(S_1,\theta_i,(x_1,\ldots,x_{k_1}))$ and $\overline{a}_2 \in sat(S_2,\psi_i,(x_{k_1+1},\ldots,x_{k_1+k_2}))$. Hence, for $\Theta \subseteq \widehat{C_r MS^h}(\mathcal{R}^{(1)},X_{k_1})$ and $\Psi \subseteq \widehat{C_r MS^h}(\mathcal{R}^{(2)},X_{k_2})$, we can define $Z_f(\Theta,\Psi)$ such that φ is in $Z_f(\Theta,\Psi)$ if and only if $\widehat{\theta_i} \in \Theta$ and $\widehat{\psi_i'} \in \Psi$ for some $i \in [p]$, where ψ_i' is defined as $\psi_i[x_1/x_{k_1+1},\ldots,x_{k_2}/x_{k_1+k_2}]$.

Finally, we consider an n-ary QDP-QF operation with $n > 2$. Let f be defined by $f(S_1,\ldots,S_n) := h(g_1(S_1,\ldots,S_m),g_2(S_{m+1},\ldots,S_n))$. Then we can define $Z_f(\Theta_1,\ldots,\Theta_n) := Z_h(Z_{g_1}(\Theta_1,\ldots,\Theta_m),Z_{g_2}(\Theta_{m+1},\ldots,\Theta_n))$. We omit the details.

The computability of Z_f follows from Remarks 5.48(1) and 5.58(1). The result for abstract structures and empty sequences is an immediate consequence of the result for concrete structures. ∎

It should be noted that, in fact, the mappings Z_f are *uniformly* computable: there is an algorithm that on input $(f,\Theta_1,\ldots,\Theta_n)$ computes $Z_f(\Theta_1,\ldots,\Theta_n)$ as output (cf. the definition of an effectively given algebra in Definition 2.8). The n-ary QDP-QF operation f is specified by a definition scheme (in the unary case), by three definition schemes (in the binary case), or by the way it is built from binary QDP-QF operations (together with their specifications).

The following application will be useful in Section 7.2 for proving the Equationality Theorem, and in Section 8.5 to prove that monadic second-order transductions of terms can be simulated by tree-walking transducers.

Application 5.61 (Operations on terms) Let F be a fixed finite functional signature. The concrete $\mathcal{R}_F$-structure $\lfloor t \rfloor = \langle N_t, (son_{it})_{1 \leq i \leq k}, (lab_{ft})_{f \in F}, rt_t \rangle$ that represents faithfully a term $t \in T(F)$ is defined in Example 5.2(3). Without rt_t, it equals $\lfloor Syn(t) \rfloor$. Thus, its domain is the set of positions of t, hence the set of nodes of the syntactic tree $Syn(t)$ of t (see Definition 2.14), denoted by N_t. The constant rt_t is the root of t (denoted by $root_t$); it will be needed in what follows, but can as well be represented by a unary relation symbol rt such that $rt_t(u)$ holds if and only if $u = root_t$.

Our objective is to express the theory $Th(\lfloor f(t,t') \rfloor, \mathcal{R}_F, h, r, \overline{a}\overline{b})$ in terms of $Th(\lfloor t \rfloor, \mathcal{R}_F, h, r, \overline{a})$ and $Th(\lfloor t' \rfloor, \mathcal{R}_F, h, r, \overline{b})$ where $\overline{a}$ and $\overline{b}$ are sequences of positions in t and t' respectively. For this purpose we will express $\lfloor f(t,t') \rfloor$ as $f^*(\lfloor t \rfloor, \lfloor t' \rfloor)$ for some appropriate binary operation f^*. A technical problem comes from the fact that the concrete structures $\lfloor t \rfloor$ and $\lfloor t' \rfloor$ are not disjoint, whereas Corollary 5.60 is stated for disjoint structures. To overcome it, we can use for $t \in T(F)$ and $u \in N_t$, the structure $\lfloor t \rfloor / u$ defined in Example 5.2(3). Without rt, it equals $\lfloor Syn(t)/u \rfloor$. It is the concrete $\mathcal{R}_F$-structure isomorphic to $\lfloor t/u \rfloor$ whose domain is $N_t/u = \{v \in N_t \mid v \leq_t u\}$; the relations of $\lfloor t \rfloor / u$ are the restrictions to its domain of those of $\lfloor t \rfloor$, and $rt_{\lfloor t \rfloor / u} = u$.

For $t \in T(F)$ and $u \in N_t$, if $\overline{a}$ is a k-tuple of elements of $D_{\lfloor t \rfloor / u}$, we will use the shorter (hopefully more clear) notation $Th(t, \downarrow u, h, r, \overline{a})$ instead of $Th(\lfloor t \rfloor / u, \mathcal{R}_F, h, r, \overline{a})$. In particular, we will use $Th(t, \downarrow root_t, h, r, \overline{a})$ to denote $Th(\lfloor t \rfloor, \mathcal{R}_F, h, r, \overline{a})$.

Let $f \in F_2$. We consider the mapping that associates the term $t = f(t_1, t_2)$ with $t_1, t_2 \in T(F)$. It can be expressed by

$$\lfloor t \rfloor = f^*(\lfloor t \rfloor / u_1, \lfloor t \rfloor / u_2),$$

where u_1 and u_2 are the left and right sons of the root of t and f^* is the binary operation on disjoint concrete relational structures S, T defined by

$$f^*(S, T) := f_1(\mathbf{rt}, p(S, T)),$$

where f_1 and p are the following binary DP-QF operations:

$$f_1(U, W) := f_1'(U \oplus W)$$

$$p(S, T) := ren_{rt \to s}(S) \oplus ren_{rt \to s'}(T). \tag{5.4}$$

These operations are defined with the help of the auxiliary constant symbols s and s', and the concrete $\{rt\}$-structure $\mathbf{rt}$ having the unique domain element 1: in the use of f^* to define $\lfloor t \rfloor$, this element is taken as the root of t. The unary DP-QF operation $ren_{rt \to s}$ renames rt into s, with $\kappa_{rt,s}$ taken equal to $True$ in its definition scheme. The unary DP-QF operation f_1' (depending on f) is defined by the following formulas:

$$\theta_{son_1}(x_1, x_2) \quad \text{is} \quad son_1(x_1, x_2) \vee (x_1 = rt \wedge x_2 = s),$$

$$\theta_{son_2}(x_1, x_2) \quad \text{is} \quad son_2(x_1, x_2) \vee (x_1 = rt \wedge x_2 = s'),$$

$$\theta_{son_i}(x_1, x_2) \quad \text{is} \quad son_i(x_1, x_2) \text{ if } i \geq 3,$$

$$\theta_{lab_f}(x_1) \quad \text{is} \quad lab_f(x_1) \vee x_1 = rt,$$

$$\theta_{lab_g}(x_1) \quad \text{is} \quad lab_g(x_1) \text{ if } g \in F - \{f\},$$

$$\kappa_{rt,rt} \quad \text{is} \quad True,$$

$$\kappa_{s,rt}, \kappa_{s',rt} \quad \text{are} \quad False.$$

Since the structure $f^*(S, T)$ is intended to belong to $STR^c(\mathcal{R}_F)$ but $s, s' \notin \mathcal{R}_F$, there are no formulas $\kappa_{c,s}$ and $\kappa_{c,s'}$ in the definition scheme of f_1'. Note also that if rt is a unary relation symbol, then $x_1 = rt$ should be replaced by $rt(x_1)$, and the constant defining sentences should be replaced by the definition of $\theta_{rt}(x_1)$ as $rt(x_1)$.

For $h, r, \ell, k, k' \in \mathcal{N}$, we will construct a mapping Z_{f^*} such that, for every term t in $T(F)$ with leading symbol f, every $\overline{d} \in (D_{\mathbf{rt}})^\ell$, every $\overline{a} \in (D_{\lfloor t \rfloor / u_1})^k$ and every $\overline{b} \in (D_{\lfloor t \rfloor / u_2})^{k'}$, we have[64]

$$Th(t, \downarrow root_t, h, r, \overline{d}\,\overline{a}\,\overline{b}) = Z_{f^*}(Th(t, \downarrow u_1, h, r, \overline{a}), Th(t, \downarrow u_2, h, r, \overline{b})).$$

If $t = f(t_1, t_2)$ and $\ell = 0$ (so that $\overline{d} = \varepsilon$), this equality implies that

$$Th(\lfloor t \rfloor, \mathcal{R}_F, h, r, \overline{a}\,\overline{b}) = Z_{f^*}(Th(\lfloor t_1 \rfloor, \mathcal{R}_F, h, r, \overline{a'}), Th(\lfloor t_2 \rfloor, \mathcal{R}_F, h, r, \overline{b'})),$$

where $\overline{a}$ corresponds[65] to a sequence of positions $\overline{a'}$ of t_1, and $\overline{b}$ to a sequence of positions $\overline{b'}$ of t_2.

For defining Z_{f^*} we use the fact that, by the Equalities (5.4), the mapping f_2 defined by $f_2(U, S, T) := f_1(U, p(S, T))$ is a ternary DP-QF operation, such that $f^*(S, T) = f_2(\mathbf{rt}, S, T)$. Thus, by Corollary 5.60, we have

$$Th(t, \downarrow root_t, h, r, \overline{d}\,\overline{a}\,\overline{b})$$

$$= Z_{f_2, h, r, (\ell, k, k')}(Th(\mathbf{rt}, \{rt\}, h, r, \overline{d}), Th(t, \downarrow u_1, h, r, \overline{a}), Th(t, \downarrow u_2, h, r, \overline{b})).$$

Hence, we obtain Z_{f^*} from $Z_{f_2, h, r, (\ell, k, k')}$ by using the fixed set $Th(\mathbf{rt}, \{rt\}, h, r, \overline{d})$. To be precise, we define

$$Z_{f^*}(\Theta, \Theta') := Z_{f_2, h, r, (\ell, k, k')}(Th(\mathbf{rt}, \{rt\}, h, r, \overline{d}), \Theta, \Theta'),$$

for all $\Theta \subseteq \widehat{C_r MS^h}(\mathcal{R}_F, X_k)$ and $\Theta' \subseteq \widehat{C_r MS^h}(\mathcal{R}_F, X_{k'})$. If we need to be more precise, we denote this function by $Z_{f^*, h, r, (\ell, k, k')}$.

This result extends to function symbols f of arbitrary arities. Here we have used a ternary DP-QF operation f_2 with one fixed argument. For a symbol f of arity n, we need an $(n + 1)$-ary operation f_2 with one fixed argument. That operation can be

---

[64] There is a unique tuple $\overline{d}$ in $D_{\mathbf{rt}}^\ell$, hence $\overline{d}$ need not occur in the right-hand side.

[65] Via the unique isomorphism of $\lfloor Syn(t) \rfloor / u_1$ and $\lfloor Syn(t/u_1) \rfloor$ described after Definition 2.14, which is also the unique isomorphism of $\lfloor t \rfloor / u_1$ and $\lfloor t / u_1 \rfloor = \lfloor t_1 \rfloor$, and similarly for $\overline{b}$.

expressed as a composition of binary DP-QF operations, and hence is an $(n+1)$-ary DP-QF operation, to which Corollary 5.60 can be applied.

We now consider in a similar way the insertion of a term $f(t_1,t_2)$ in a context. We recall that a context $c \in Ctxt(F)$ is defined as a term in $T(F \cup \{w\})$, where w is a (special) variable that has a unique occurrence in c. A context is represented by an $\mathcal{R}_{F \cup \{w\}}$-structure. For $t \in T(F)$ and $u \in N_t$, the structure $\lfloor t \rfloor \uparrow u$ is defined as $\lfloor Syn(t) \rfloor \uparrow u$ with, additionally, $rt_{\lfloor t \rfloor \uparrow u} = rt_t$. It is the concrete $\mathcal{R}_{F \cup \{w\}}$-structure isomorphic to $\lfloor t \uparrow u \rfloor$ whose domain is $N_t \uparrow u = N_t - \{v \in N_t \mid v <_t u\}$; the relations of $\lfloor t \rfloor \uparrow u$ are the restrictions to its domain of those of $\lfloor t \rfloor$, except that the label of u is changed into w. See Definition 2.6, Definition 2.14 and Example 5.2(3) for the details. If $\overline{d}$ is an ℓ-tuple of elements of $D_{\lfloor t \rfloor \uparrow u}$, we will abbreviate $Th(\lfloor t \rfloor \uparrow u, \mathcal{R}_{F \cup \{w\}}, h, r, \overline{d})$ by $Th(t, \uparrow u, h, r, \overline{d})$.

Let $f \in F_2$, $c \in Ctxt(F)$ and $t, t_1, t_2 \in T(F)$ be such that $t = c[f(t_1,t_2)]$. Let u be the occurrence of f in t such that $c = t \uparrow u$, and let u_1 and u_2 be its left and right sons. Hence, $t/u_1 = t_1$ and $t/u_2 = t_2$. We have $D_{\lfloor t \rfloor \uparrow u} = N_t \uparrow u$, $D_{\lfloor t \rfloor / u_1} = N_t/u_1$ and $D_{\lfloor t \rfloor / u_2} = N_t/u_2$. It follows that $D_{\lfloor t \rfloor} = N_t$ is the disjoint union of the sets $D_{\lfloor t \rfloor \uparrow u}$, $D_{\lfloor t \rfloor / u_1}$ and $D_{\lfloor t \rfloor / u_2}$. Furthermore, we have

$$\lfloor t \rfloor = f^{\dagger}(\lfloor t \rfloor \uparrow u, \lfloor t \rfloor / u_1, \lfloor t \rfloor / u_2),$$

where $f^{\dagger}$ is the ternary DP-QF operation defined by

$$f^{\dagger}(U, S, T) := f_1^{\dagger}(U \oplus ren_{rt \to s}(S) \oplus ren_{rt \to s'}(T)),$$

s, s' are new constant symbols and $f_1^{\dagger}$ is the unary DP-QF operation of type $\mathcal{R}_{F \cup \{w\}} \cup \{s, s'\} \to \mathcal{R}_F$ that has exactly the same definition scheme as the unary DP-QF operation f_1' discussed before, except that the atomic formula $x_1 = rt$ is replaced by $lab_w(x_1)$.[66] We use here a ternary DP-QF operation, as in the case of $t = f(t_1,t_2)$, but not with a fixed argument.

Application of Corollary 5.60 to $f^{\dagger}$ shows that, for $h, r, \ell, k, k' \in \mathcal{N}$, we can construct a ternary function $Z_{f^{\dagger}}$ on sets of formulas such that, for all $\overline{d} \in (D_{\lfloor t \rfloor \uparrow u})^{\ell}$, $\overline{a} \in (D_{\lfloor t \rfloor / u_1})^k$ and $\overline{b} \in (D_{\lfloor t \rfloor / u_2})^{k'}$, we have

$$Th(t, \downarrow root_t, h, r, \overline{d}\,\overline{a}\,\overline{b})$$

$$= Z_{f^{\dagger}}(Th(t, \uparrow u, h, r, \overline{d}), Th(t, \downarrow u_1, h, r, \overline{a}), Th(t, \downarrow u_2, h, r, \overline{b})). \tag{5.5}$$

---

[66] In the case where rt is a unary relation symbol, the atomic formula $rt(x_1)$ is replaced by $lab_w(x_1)$, except that the definition of $\theta_{rt}(x_1)$ as $rt(x_1)$ remains unchanged.

It implies that

$$Th(\lfloor t \rfloor, \mathcal{R}_F, h, r, \overline{d}\,\overline{a}\,\overline{b})$$

$$= Z_{f^\dagger}(Th(\lfloor c \rfloor, \mathcal{R}_{F \cup \{w\}}, h, r, \overline{d}'), Th(\lfloor t_1 \rfloor, \mathcal{R}_F, h, r, \overline{a}'), Th(\lfloor t_2 \rfloor, \mathcal{R}_F, h, r, \overline{b}')),$$

where $\overline{d}$ corresponds to a sequence $\overline{d}'$ of positions of the context c and $\overline{a}$ to a sequence $\overline{a}'$ of positions of t_1, and similarly for $\overline{b}$, cf. Footnote 65. If necessary, we use the more precise notation $Z_{f^\dagger, h, r, (\ell, k, k')}$ for $Z_{f^\dagger}$.

In the same situation, we also have

$$\lfloor t \rfloor / u = f^*(\lfloor t \rfloor / u_1, \lfloor t \rfloor / u_2),$$

with f^* defined as before, except that $D_{\mathbf{rt}} = \{u\}$. Since $D_{\mathbf{rt}}$ is a singleton, the fixed theory $Th(\mathbf{rt}, \{rt\}, h, r, \overline{d})$ does not depend on $D_{\mathbf{rt}}$, and so the same mapping Z_{f^*} is obtained. So, we have

$$Th(t, \downarrow u, h, r, \overline{d}\,\overline{a}\,\overline{b}) = Z_{f^*}(Th(t, \downarrow u_1, h, r, \overline{a}), Th(t, \downarrow u_2, h, r, \overline{b})), \tag{5.6}$$

where Z_{f^*} is $Z_{f^*, h, r, (\ell, k, k')}$. There are also mappings $Z_{f^{(1)}}$ and $Z_{f^{(2)}}$ such that

$$Th(t, \uparrow u_1, h, r, \overline{d}\,\overline{a}\,\overline{b}) = Z_{f^{(1)}}(Th(t, \uparrow u, h, r, \overline{d}), Th(t, \downarrow u_2, h, r, \overline{b})),$$

$$Th(t, \uparrow u_2, h, r, \overline{d}\,\overline{a}\,\overline{b}) = Z_{f^{(2)}}(Th(t, \uparrow u, h, r, \overline{d}), Th(t, \downarrow u_1, h, r, \overline{a})). \tag{5.7}$$

To obtain $Z_{f^{(1)}}$ we define the binary operation $f^{(1)}$ such that

$$\lfloor t \rfloor \uparrow u_1 = f^{(1)}(\lfloor t \rfloor \uparrow u, \lfloor t \rfloor \downarrow u_2) = f_1^{(1)}(\lfloor t \rfloor \uparrow u \oplus \mathbf{s} \oplus ren_{rt \to s'}(\lfloor t \rfloor \downarrow u_2)),$$

where $\mathbf{s}$ is the concrete $\{s\}$-structure with domain $\{u_1\}$ and $f_1^{(1)}$ is the unary DP-QF operation of type $\mathcal{R}_{F \cup \{w\}} \cup \{s, s'\} \to \mathcal{R}_{F \cup \{w\}}$ that has the same definition scheme as $f_1^\dagger$ with additionally $\theta_{lab_w}(x_1)$ taken equal to $x_1 = s$. The mapping $Z_{f^{(2)}}$ is obtained in a similar way.

As before, these results easily extend to function symbols of arbitrary arity.

5.3.7 The many-sorted algebra of relational structures

The Recognizability Theorem to be stated in the next section (Theorem 5.75) says that every CMS-definable class of relational structures is recognizable. We first define the relevant algebra.

Definition 5.62 (The algebra $\mathbb{STR}$ of relational structures) We let S be the set of relational signatures. It is countable because we assume that each relational signature is a finite subset of a countable (effectively given) set of symbols containing countably many symbols of each arity. We define $\mathbb{STR}$ as the $\mathcal{S}$-sorted algebra with domain $STR(\mathcal{R})$ for each $\mathcal{R} \in \mathcal{S}$. The (functional) signature F^{QF} of this algebra consists of the following symbols; we also define the operations they denote:

(a) for every $\mathcal{R}, \mathcal{R}' \in \mathcal{S}$ such that $\mathcal{R}_0 \cap \mathcal{R}'_0 = \emptyset$, the binary symbol $\oplus_{\mathcal{R},\mathcal{R}'}$ of type $\mathcal{R} \times \mathcal{R}' \rightarrow \mathcal{R} \cup \mathcal{R}'$; it denotes the disjoint union of an $\mathcal{R}$-structure and an $\mathcal{R}'$-structure (extending the operation of Definition 5.34 as explained in Example 5.44);

(b) for every $\mathcal{R}, \mathcal{R}' \in \mathcal{S}$ and every QFO definition scheme $\mathcal{D}$ of type $\mathcal{R} \rightarrow \mathcal{R}'$ (this implies that $\mathcal{R}'_0 = \emptyset$ if $\mathcal{R}_0 = \emptyset$), the unary symbol $qf_{\mathcal{D},\mathcal{R},\mathcal{R}'}$ of the same type; it denotes the unary QF operation $\widehat{\mathcal{D}} : STR(\mathcal{R}) \rightarrow STR(\mathcal{R}')$ associated with $\mathcal{D}$;

(c) the constant symbols $\varnothing_{\mathcal{R}}$ (only if $\mathcal{R}_0 = \emptyset$), to denote the empty $\mathcal{R}$-structure, and $\Diamond_{\mathcal{B},\mathcal{R}}$ where $\mathcal{R}_0 \subseteq \mathcal{B} \subseteq \mathcal{R}$, to denote the $\mathcal{R}$-structure S with a single element $*$ such that $a_S := *$ if $a \in \mathcal{R}_0$, $R_S := \{(*, \ldots, *)\}$ if $R \in \mathcal{B} - \mathcal{R}_0$ and $R_S := \emptyset$ if $R \in \mathcal{R} - \mathcal{B}$.

We will frequently omit the subscripts $\mathcal{R}$ and $\mathcal{R}'$ in the symbols $\oplus_{\mathcal{R},\mathcal{R}'}$, $\varnothing_{\mathcal{R}}$ and $\Diamond_{\mathcal{B},\mathcal{R}}$. Instead of the symbol $qf_{\mathcal{D},\mathcal{R},\mathcal{R}'}$ we will use the operation $\widehat{\mathcal{D}}$, in accordance with our custom not to distinguish an operation from the symbol denoting it.

The notation F^{QF} emphasizes the important role of quantifier-free operations. The signature F^{QF} contains unary operations that are not quasi-domain preserving.

We now define several subalgebras of $\mathbb{STR}$. If we restrict the unary operations to those that are quasi-domain preserving, we obtain a subsignature F^{QF}_{pres} of F^{QF} and a subalgebra $\mathbb{STR}_{pres}$ of $\mathbb{STR}$. Both algebras have the same sorts and the same domains. Every n-ary QDP-QF operation is a (linear and strict) derived operation of the F^{QF}-algebra $\mathbb{STR}_{pres}$.

We let $\mathbb{STR}_{sep}$ be the subalgebra of $\mathbb{STR}$ obtained by taking as domain of sort $\mathcal{R}$ the set $STR_{sep}(\mathcal{R})$ of constant-separated structures in $STR(\mathcal{R})$ and by restricting the unary operations to those that preserve constant-separation. The corresponding signature is denoted by F^{QF}_{sep}. One can combine these two restrictions to obtain the $F^{QF}_{sep,pres}$-algebra $\mathbb{STR}_{sep,pres}$.

We let $\mathcal{S}_{nc}$ be the subset of $\mathcal{S}$ consisting of relational signatures without constant symbols. The associated structures are constant-separated in a trivial way. We obtain thus a subalgebra $\mathbb{STR}_{nc}$ of $\mathbb{STR}_{sep}$ with set of sorts $\mathcal{S}_{nc}$ and functional signature F^{QF}_{nc} that is a subsignature of F^{QF}_{sep}. Its domain of sort $\mathcal{R}$ is $STR(\mathcal{R})$. If we now restrict F^{QF}_{nc} to $F^{QF}_{nc,pres}$ by keeping only the unary DP-QF operations, we obtain the $\mathcal{S}_{nc}$-sorted algebra $\mathbb{STR}_{nc,pres}$.

All these algebras are effectively given (see Definitions 2.8 and 2.127). The discussion in Definition 2.32 of the fact that the HR algebra $\mathbb{JS}$ is effectively given extends to $\mathbb{STR}$ and its subalgebras: it is clear that its set of sorts is effectively given; so are the definition schemes of its operations and of those of its subalgebras by Remarks 5.43(1) and 5.43(2). That two definition schemes can define the same operation is no problem, cf. Remark 5.58(1) and the paragraph after Corollary 5.60.

Lemma 5.63 The algebra $\mathbb{STR}$ is generated by $F_{\text{pres}}^{\text{QF}}$, the algebra $\mathbb{STR}_{\text{sep}}$ is generated by $F_{\text{pres,sep}}^{\text{QF}}$ and the algebra $\mathbb{STR}_{\text{nc}}$ is generated by $F_{\text{nc,pres}}^{\text{QF}}$. $\qquad\square$

Hence, by Lemma 3.75(1), a set is effectively recognizable in these algebras if and only if it is effectively term-recognizable (just as for the HR and VR algebras).

The proof of Lemma 5.63 is similar to those of Proposition 2.33 showing that the HR algebra is generated by its signature, and of Proposition 2.104(1) doing the same for the VR algebra. Every term t in $T(F^{\text{QF}})$ evaluates into a concrete structure denoted by $cval(t)$ whose domain is a subset of $Occ_0(t)$ defined as the set of occurrences in t of constant symbols other than $\varnothing_{\mathcal{R}}$ (they are leaves of the syntactic tree of t). The corresponding abstract structure is denoted by $val(t)$. The formal definition of $cval(t)$ is a straightforward generalization of the corresponding one given in Section 2.5.2 for terms in $T(F^{\text{VR}})$.

In Section 9.3, we will define complexity measures on relational structures that can be considered as generalizations of clique-width. They are related with the generation of relational structures by particular subsignatures of $F_{\text{pres}}^{\text{QF}}$.

If we identify a p-graph G of type $\pi(G) = C$ with the $\mathcal{R}_{\text{s},C}$-structure $\lfloor G \rfloor = \langle V_G, edg_G, (lab_{aG})_{a \in C} \rangle$ (and C with $\mathcal{R}_{\text{s},C}$), then, since we have observed in Application 5.50 that the unary operations of F^{tVR} are DP-QF operations, we obtain that the many-sorted algebra $\mathbb{GP}^{\text{t}}$ is a subalgebra of $\mathbb{STR}_{\text{nc,pres}}$.[67] Recall from Example 5.2(2) that a structure S in $STR(\mathcal{R}_{\text{s},C})$ is $\lfloor G \rfloor$ for some graph G such that $\pi(G) = C$ if and only if the sets lab_{aS} for $a \in C$ form a partition of D_S that consists of nonempty sets. This condition is first-order expressible.

5.3.8 The Recognizability Theorem for $\mathbb{STR}_{\text{pres}}$

We first prove the Recognizability Theorem for the algebra $\mathbb{STR}_{\text{pres}}$, and then extend it to $\mathbb{STR}$.

Theorem 5.64 Let $\mathcal{R}$ be a relational signature. For every sentence φ in $\text{CMS}(\mathcal{R}, \emptyset)$, the set $\text{MOD}(\varphi) \subseteq STR(\mathcal{R})$ of models of φ is effectively recognizable in the algebra $\mathbb{STR}_{\text{pres}}$.

Proof: We let r and h be such that $\varphi \in C_r\text{MS}^h(\mathcal{R}, \emptyset)$. For each relational signature $\mathcal{R}' \in S$, we let $\sim_{\mathcal{R}'}$ be the equivalence relation on $STR(\mathcal{R}')$ defined by $S \sim_{\mathcal{R}'} T$ if and only if $Th(S, \mathcal{R}', h, r) = Th(T, \mathcal{R}', h, r)$. This condition means that $S \models \psi$ if and only if $T \models \psi$, for every sentence ψ in $C_r\text{MS}^h(\mathcal{R}', \emptyset)$.

Since all the operations of $\mathbb{STR}_{\text{pres}}$ are binary or unary QDP-QF operations, it follows from Corollary 5.60 that the equivalence relations $\sim_{\mathcal{R}'}$ (which are well defined on abstract structures) form a congruence $\sim$ on $\mathbb{STR}_{\text{pres}}$. This congruence is locally

---

[67] Note that the constants of $\mathbb{GP}^{\text{t}}$ are also constants of $\mathbb{STR}_{\text{nc,pres}}$: $\varnothing$ is $\varnothing_{\mathcal{R}_{\text{s},\emptyset}}$, $\mathbf{a}$ is $\Diamond_{\mathcal{B}, \mathcal{R}_{\text{s},\{a\}}}$ with $\mathcal{B} = \{lab_a\}$, and $\mathbf{a}^\ell$ is $\Diamond_{\mathcal{B}', \mathcal{R}_{\text{s},\{a\}}}$ with $\mathcal{B}' = \{edg, lab_a\}$.

finite because the set $\mathcal{P}(\widehat{C_r\mathrm{MS}^h}(\mathcal{R}',\emptyset))$ is finite for every $\mathcal{R}' \in \mathcal{S}$ (by Proposition 5.94 in Section 5.6 below), it is decidable because the mapping $S \mapsto Th(S,\mathcal{R}',h,r)$ is computable, and it saturates $\mathrm{MOD}(\varphi)$. The membership in $\mathrm{MOD}(\varphi)$ of any given $\mathcal{R}$-structure S is decidable. This proves the semi-effective recognizability of $\mathrm{MOD}(\varphi)$ by Proposition 3.78(1) in Section 3.4.4.

We now establish the effective recognizability by a different proof, because we cannot use Proposition 3.78 to get the effectivity from the semi-effectivity. Following Definition 3.74, we construct a locally finite $F_{\mathrm{pres}}^{\mathrm{QF}}$-algebra $\mathbb{A}$ and a homomorphism κ from $\mathrm{STR}_{\mathrm{pres}}$ to $\mathbb{A}$. The set of sorts is $\mathcal{S}$. For each $\mathcal{R}' \in \mathcal{S}$, we let $A_{\mathcal{R}'}$ be the powerset of $\widehat{C_r\mathrm{MS}^h}(\mathcal{R}',\emptyset)$ and κ be the mapping that associates with each $S \in STR(\mathcal{R}')$ its theory $\kappa(S) := Th(S,\mathcal{R}',h,r)$.[68] We use Corollary 5.60 to define the operations of $\mathbb{A}$ in such a way that κ is a homomorphism. If f is a QDP-QF operation of type $\mathcal{R}' \to \mathcal{R}''$, then we let $f_{\mathbb{A}}$ be the mapping $Z_{f,h,r,0}$. It follows from Corollary 5.60 that, for every $\mathcal{R}'$-structure S, we have $Th(f(S),\mathcal{R}'',h,r) = f_{\mathbb{A}}(Th(S,\mathcal{R}',h,r))$, hence $\kappa(f(S)) = f_{\mathbb{A}}(\kappa(S))$. The proof is similar for disjoint union, which is a binary DP-QF operation: if $f = \oplus_{\mathcal{R}',\mathcal{R}''}$, then we let $f_{\mathbb{A}}$ be the mapping $Z_{f,h,r,(0,0)}$.

Hence, we have a locally finite algebra $\mathbb{A}$, and κ is a homomorphism into it. If we define $C \subseteq A$ by $C := \{\Theta \in A_{\mathcal{R}} \mid \widehat{\varphi} \in \Theta\}$, then it is easy to see that $\mathrm{MOD}(\varphi) = \kappa^{-1}(C)$. Since κ is computable and C is a decidable subset of A, it remains to show that $\mathbb{A}$ is effectively given. Using the encoding of the effectively given countable set of relation symbols (see Definition 5.62) it is straighforward to encode the set of all CMS sentences, and hence the set of its finite subsets, of which A is a decidable subset: as observed in Definition 5.59, the set $\widehat{C_r\mathrm{MS}^h}(\mathcal{R}',\emptyset)$ can be computed from r, h and $\mathcal{R}'$. From this we obtain an encoding of A, with a computable sort mapping. The previous observation also shows that the mapping $\mathcal{R}' \mapsto |A_{\mathcal{R}'}|$ is computable (and obviously, $|A| = \omega$). The operations of $\mathbb{A}$ are uniformly computable (cf. Definition 2.8) because the mappings Z_f are uniformly computable (as observed after Corollary 5.60). This shows that $\mathbb{A}$ is effectively given.   ∎

In Theorem 5.96 of Section 5.6 we will compute an upper-bound to the recognizability index of $\mathrm{MOD}(\varphi)$.

Example 5.65 We give an example showing that the Recognizability Theorem fails to hold if we add to the signature of $\mathrm{STR}_{\mathrm{pres}}$ a single unary operation that is defined by a definition scheme involving a single formula with a single first-order quantifier.

We let $\mathcal{R}$ be the relational signature $\mathcal{R}_{s,[2]} = \{edg, lab_1, lab_2\}$. Every p-graph G of type $[2]$ is represented by the $\mathcal{R}$-structure $\lfloor G \rfloor$. Since the many-sorted VR algebra $\mathbb{GP}^{\mathrm{t}}$ is a subalgebra of $\mathrm{STR}_{\mathrm{pres}}$ (as discussed at the end of Section 5.3.7), we will use the operations of $\mathbb{GP}^{\mathrm{t}}$, without showing their types.

---

[68] To ensure that the domains of $\mathbb{A}$ are pairwise disjoint, we assume that the sets $\mathrm{CMS}(\mathcal{R}',\emptyset)$ are pairwise disjoint, i.e., that a sentence is always given together with a relational signature.

We define $f : STR(\mathcal{R}) \rightarrow STR(\mathcal{R})$ as the operation that transforms a structure S into $f(S)$ by modifying only the relation edg_S; the relation $edg_{f(S)}$ is defined in S by the formula

$$edg(x_1, x_2) \vee (x_1 = x_2 \wedge \exists x_3 [edg(x_1, x_3) \wedge edg(x_3, x_3)]).$$

It transforms a p-graph G by adding loops to some vertices: in $f(G)$, a vertex has a loop if and only if it has a loop in G or it is the tail of a directed edge whose head has a loop.

We let STR_{pres}^{f} be the algebra STR_{pres} with the additional unary operation f, and we let φ be the first-order sentence $\forall x. edg(x, x)$. We claim that the set $MOD(\varphi) \subseteq STR(\mathcal{R})$ is not recognizable in STR_{pres}^{f}. To prove this claim, we define, for $i = 1, 2$, the unary derived VR operation g_i of type $[i] \rightarrow [2]$ such that for every p-graph G of type $[i]$:

$$g_i(G) := relab_{3 \rightarrow 1}(relab_{1 \rightarrow 2}(\overrightarrow{add}_{3,1}(3 \oplus G))).$$

We let $STR_{\text{pres}}^{f,g}$ be the algebra STR_{pres}^{f} to which the operations g_1 and g_2 are added; it is a derived algebra of STR_{pres}^{f}. Then, we define $t_{n,m}$ as the term $f^n g_2^m g_1 1^\ell$. It is clear that the term $g_2^m g_1 1^\ell$ defines the p-graph $1 \rightarrow 2 \rightarrow 2 \rightarrow \cdots 2 \rightarrow 2 \circlearrowright$ of type $[2]$ with $m + 2$ vertices and a loop on the last vertex. Then, $\lfloor val(t_{n,m}) \rfloor \models \varphi$ if and only if $n \geq m + 1$. If $MOD(\varphi) \subseteq STR(\mathcal{R})$ would be recognizable in STR_{pres}^{f}, then it would be recognizable in $STR_{\text{pres}}^{f,g}$ by Proposition 3.56(3), and, by Propositions 3.56(2) and 3.60, the set of terms $t_{n,m}$ such that $\lfloor val(t_{n,m}) \rfloor \models \varphi$ would be recognizable in $\mathbb{T}(\{f, g_1, g_2, 1^\ell\})$, hence regular by Theorem 3.62. But this is not the case. $\square$

Before generalizing Theorem 5.64 to the algebra STR, we present some of its consequences.

5.3.9 Recognizable languages and recognizable sets of graphs

We first consider sets of words. A word $w \in A^*$ is faithfully represented by the structure $\lfloor w \rfloor^\ell$ defined in Example 5.2(1), which uses the order of occurrences of letters.[69]

Corollary 5.66 Let A be a finite alphabet. If a language $L \subseteq A^*$ is MS-definable, then it is effectively recognizable in $\mathbb{W}(A)$, hence regular.

Proof: Let α_A be the MS sentence over $\mathcal{W}_A^\ell := \{\leq\} \cup \{lab_a \mid a \in A\}$ (where each relation lab_a is unary) that defines the class $\lfloor A^* \rfloor^\ell$. (Corollary 5.12 holds for $\lfloor \cdot \rfloor^\ell$.) Let $L \subseteq A^*$ be MS-definable. It is the set of words w such that $\lfloor w \rfloor$ satisfies an MS

---

[69] The structure $\lfloor w \rfloor$, also defined in Example 5.2(1), uses a successor relation suc. The order relation of $\lfloor w \rfloor^\ell$ is the reflexive and transitive closure of suc. It follows that the same properties of words are MS-expressible via these two faithful representations, i.e., they are MS-equivalent.

sentence φ over $\mathcal{W}_A$, hence also the set of those such that $\lfloor w \rfloor^\ell$ satisfies an MS sentence φ' over $\mathcal{W}_A^\ell$. It follows that $\lfloor L \rfloor^\ell = \mathrm{MOD}(\varphi' \wedge \alpha_A)$.

For words $v, w \in A^*$, we have $\lfloor v \cdot w \rfloor^\ell = \lfloor v \rfloor^\ell \overrightarrow{\oplus} \lfloor w \rfloor^\ell$, where $\overrightarrow{\oplus}$ is the derived operation of $\mathrm{STR}_{\mathrm{pres}}$ defined in Example 5.56. Hence, if we identify every word w in A^* with the $\mathcal{W}_A^\ell$-structure $\lfloor w \rfloor^\ell$ (or more precisely, with the corresponding abstract structure), then the monoid of words $\mathbb{W}(A)$ is a subalgebra $\mathbb{H}$ of a derived algebra of $\mathrm{STR}_{\mathrm{pres}}$. Since $\lfloor L \rfloor^\ell$ is effectively recognizable in $\mathrm{STR}_{\mathrm{pres}}$ by Theorem 5.64, we get by Assertions (1) and (3) of Proposition 3.76 that it is effectively recognizable in $\mathbb{H}$. Hence L is effectively recognizable in $\mathbb{W}(A)$ and thus regular (cf. Example 3.57(2)).   ∎

We now consider terms and give a proof in our setting of the well-known result stated as Theorem 1.16 (cf. Theorem 5.15).

Corollary 5.67 Let F be a finite functional signature. If a language $L \subseteq T(F)$ is MS-definable, then it is effectively recognizable in $\mathbb{T}(F)$, hence regular.

Proof: We first assume that F is one-sorted. We have shown in Section 5.3.6 (Equalities (5.4) in Application 5.61) that the operations on terms, $(t_1, \ldots, t_k) \mapsto f(t_1, \ldots, t_k)$, for $f \in F_k$, are derived operations of $\mathrm{STR}_{\mathrm{pres}}$, via the representation of a term t by the $\mathcal{R}_F$-structure $\lfloor t \rfloor$. Hence, identifying t and $\lfloor t \rfloor$, we have that $\mathbb{T}(F)$ is a subalgebra of a derived algebra of $\mathrm{STR}_{\mathrm{pres}}$. If $L \subseteq T(F)$ is MS-definable, then $\lfloor L \rfloor$ is MS-definable by Corollary 5.12, hence effectively recognizable in $\mathrm{STR}_{\mathrm{pres}}$ by Theorem 5.64. By a proof similar to that of the previous corollary, we obtain that L is effectively recognizable in $\mathbb{T}(F)$ and thus regular by Theorem 3.62.

Now let F be many-sorted, and let F^- be the one-sorted signature obtained from F by making all sorts identical. From the above, we get that L is effectively recognizable in $\mathbb{T}(F^-)$. Then it is effectively recognizable in $\mathbb{T}(F)$ by Theorem 3.62 and Remark 3.38(4) (which are both effective), hence regular over F.   ∎

The classical proof of this result, based on the construction of a finite automaton recognizing $L \subseteq T(F)$ from a monadic second-order formula that defines it, will be given in Section 6.3.3.

The case of graph algebras

We now apply the Recognizability Theorem to the typed algebras $\mathbb{GP}^{\mathrm{t}}$ and $\mathbb{JS}^{\mathrm{t}}$ which are also subalgebras of derived algebras of $\mathrm{STR}_{\mathrm{pres}}$.

We recall that a p-graph of type $\pi(G) \subseteq C$ is faithfully represented by the $\mathcal{R}_{\mathrm{s},C}$-structure $\lfloor G \rfloor_C$, and that an s-graph G of type $\tau(G) \subseteq C$ (it can have multiple edges) is faithfully represented by the $\mathcal{R}_{\mathrm{m},C}$-structure $\lceil G \rceil_C$ (we also recall that $\mathcal{R}_{\mathrm{m},C} = \{in, lab_{Edge}\} \cup \{lab_a \mid a \in C\}$, where in is binary and the other symbols are unary, cf. Section 5.2.5). In order to represent vertex and edge labels, we use additional unary relation symbols in $\lceil G \rceil_C$ and additional unary and binary relation symbols in $\lfloor G \rfloor_C$. For each finite set C, the class of structures of the form

$\lfloor G \rfloor_C$ (resp. $\lceil G \rceil_C$) for some p-graph (resp. s-graph) G of type included in C is FO-definable. It follows (cf. the proof of Corollary 5.66 for a similar fact) that a set L of (labeled) p-graphs of bounded type is CMS-definable if and only if the set of structures $\lfloor L \rfloor = \lfloor L \rfloor_C = \{\lfloor G \rfloor_C \mid G \in L\}$ is CMS-definable, where C is the union of all types of the elements of L. A similar statement holds for the CMS$_2$-definability of a set of (labeled) s-graphs of bounded type.

Theorem 5.68 (Recognizability Theorem for graph algebras)

(1) Every CMS-definable set of (labeled) p-graphs is effectively VR-recognizable.
(2) Every CMS$_2$-definable set of (labeled) s-graphs is effectively HR-recognizable.

Proof: The proofs are essentially the same for labeled and unlabeled graphs.

(1) Let L be a CMS-definable set of p-graphs of bounded type. By the above remarks, we have $\lfloor L \rfloor = \lfloor L \rfloor_C = \mathrm{MOD}(\varphi)$ for some sentence φ in CMS$(\mathcal{R}_{\mathrm{s},C}, \emptyset)$.

We first show that for every $D \subseteq C$, the set $\{G \in L \mid \pi(G) = D\}$ is CMS-definable. Let φ_D in CMS$(\mathcal{R}_{\mathrm{s},D}, \emptyset)$ be obtained from φ by changing every atomic subformula $lab_a(x)$, $a \in C - D$, into *False*. Then

$$\mathrm{MOD}(\varphi_D) = \{\lfloor G \rfloor_D \mid G \in L, \pi(G) \subseteq D\},$$

and so

$$\mathrm{MOD}\Big(\varphi_D \wedge \bigwedge_{a \in D} \exists x.lab_a(x)\Big) = \{\lfloor G \rfloor_D \mid G \in L, \pi(G) = D\}.$$

Hence $\{G \in L \mid \pi(G) = D\}$ is CMS-definable. Since L is a finite union of such sets, this implies, by Corollary 3.86, that we may assume from now on that L is homogenous, and so $\lfloor L \rfloor = \{\lfloor G \rfloor \mid G \in L\} = \mathrm{MOD}(\varphi)$.

As observed in Section 5.3.7, if G is identified with $\lfloor G \rfloor$, then the many-sorted VR algebra $\mathbb{GP}^{\mathrm{t}}$ is a subalgebra of $\mathbb{STR}_{\mathrm{nc,pres}}$, hence of $\mathbb{STR}_{\mathrm{pres}}$. So, $\lfloor L \rfloor$ is effectively recognizable in $\mathbb{STR}_{\mathrm{pres}}$ by Theorem 5.64, hence, by Proposition 3.76(1), it is effectively recognizable in that subalgebra, and thus L is effectively recognizable in $\mathbb{GP}^{\mathrm{t}}$. Proposition 3.76(1) is applied to $\mathbb{M}' = \mathbb{GP}^{\mathrm{t}}$ and $\mathbb{M} = \mathbb{STR}_{\mathrm{pres}}$, $in_{\mathcal{S}',\mathcal{S}}$ is the mapping $C \mapsto \mathcal{R}_{\mathrm{s},C}$, $in_{\mathcal{F}',\mathcal{F}}$ is the mapping that maps each VR operation to the corresponding operation of $\mathbb{STR}_{\mathrm{pres}}$ as defined in Application 5.50, and $in_{\mathcal{M}',\mathcal{M}}$ is the mapping $G \mapsto \lfloor G \rfloor$. These mappings are computable, and $\{\mathcal{R}_{\mathrm{s},C} \mid C \in \mathcal{P}_f(\mathcal{A})\}$ is a decidable subset of the set of sorts of $\mathbb{STR}_{\mathrm{pres}}$. Note that, although G is identified with $\lfloor G \rfloor$, they are encoded by different elements of $\mathcal{N}$ in $\mathbb{GP}^{\mathrm{t}}$ and $\mathbb{STR}_{\mathrm{pres}}$.

(2) By the same proof as above (replacing $\lfloor \cdot \rfloor$ by $\lceil \cdot \rceil$), we may assume that L is a homogenous set of s-graphs, and so $\lceil L \rceil = \{\lceil G \rceil \mid G \in L\}$. And then, we may also assume that $\lceil G \rceil$ is an $\mathcal{R}_{\mathrm{ms},C}$-structure rather than an $\mathcal{R}_{\mathrm{m},C}$-structure, i.e., the sources of G are represented by constants rather than by unary relations (we recall that $\mathcal{R}_{\mathrm{ms},C} = \{in, lab_{Edge}\} \cup C$, where each element of C is a constant symbol, cf. Section 5.2.5).

Identifying G with $\lceil G \rceil$, the many-sorted HR algebra $\mathbb{JS}^t$ is a subalgebra of a derived algebra of $\mathbb{STR}_{pres}$. This follows from Application 5.51 and Example 5.55, where it is shown that, with this identification, the operations of $\mathbb{JS}^t$ are binary QDP-QF and unary DP-QF operations, respectively, and hence (linear) derived operations of $\mathbb{STR}_{pres}$. Moreover, the definition schemes of the binary and unary QDP-QF operations can be computed from the HR operations, and hence the involved derived signature is effectively given. Theorem 5.64 and Assertions (1) and (3) of Proposition 3.76 can thus be used. ∎

We now consider CMS-definable sets of p-graphs of bounded clique-width, and similarly, CMS_2-definable sets of s-graphs of bounded tree-width. A set L of p-graphs is of bounded clique-width and of bounded type if and only if $L \subseteq val(T_C^{VR})$ for some $C \in \mathcal{P}_f(\mathcal{A})$, and a similar statement holds for s-graphs (see Definition 2.89 and Theorem 2.83(1), respectively).

Corollary 5.69 (Weak Recognizability Theorem for graph algebras)
(1) Every CMS-definable set L of (labeled) p-graphs of bounded clique-width is effectively recognizable in the algebra $\mathbb{GP}^{gen}[C]$, where C is any finite subset of $\mathcal{A}$ such that $L \subseteq \mathcal{GP}^{gen}[C] = val(T_C^{VR})$ (i.e., such that $cwd(L) \leq |C|$ and $\pi(G) \subseteq C$ for every $G \in L$).
(2) Every CMS_2-definable set L of (labeled) s-graphs of bounded tree-width is effectively recognizable in the algebra $\mathbb{JS}^{gen}[C]$, where C is any finite subset of $\mathcal{A}$ such that $L \subseteq \mathcal{JS}^{gen}[C] = val(T_C^{HR})$ (i.e., such that $twd(L) \leq |C| - 1$ and $\tau(G) \subseteq C$ for every $G \in L$).

Proof: We first prove (2). As observed after Proposition 2.132, $val(T_C^{HR}) = val(T_C^{tHR})$, and hence $L \subseteq \mathcal{JS}^{t,gen}[C]$. Since L is effectively recognizable in $\mathbb{JS}^t$ by Theorem 5.68, it is also effectively recognizable in the subalgebra $\mathbb{JS}^{t,gen}[C]$ by Proposition 3.76(1). Since $\mathbb{JS}^{t,gen}[C]$ has finitely many sorts, the set L is also effectively recognizable in $\mathbb{JS}^{gen}[C]$ as noted in Chapter 4, Definition 4.29.

The proof of (1) is the same. The remark about effective recognizability in the one-sorted algebra $\mathbb{GP}^{gen}[C]$ is in Definition 4.52. ∎

Filtering theorems: logical versions

An "algebraic" Filtering Theorem has been proved in Section 3.4.7. Here, we state "logical" Filtering Theorems for graph algebras, together with the related proof technique of fixed-point induction (as formulated at the end of Section 1.2.2).

Corollary 5.70 (Filtering Theorem for the VR algebra) For every VR-equational set L and every CMS-definable set K of (labeled) p-graphs, we have:
(1) The set $L \cap K$ is VR-equational, and an equation system for it can be constructed from an equation system defining L and a sentence defining K.

(2) The emptiness and the finiteness of $L \cap K$ can be decided. If $L \cap K$ is finite, it can be computed.[70]

(3) Let L be defined by (S, X_μ) for some typed VR equation system S with $Unk(S) = X_n$ and some $\mu \in [n]$. Then $L \subseteq K$ if and only if there exists an n-tuple $(K_1, \ldots, K_n)$ of CMS-definable sets of (labeled) p-graphs that is an oversolution of S and is such that $K_\ell \subseteq K$ for every $\ell \in [\mu]$. If $L \subseteq K$ (which can be decided by (1) and (2)), then one can construct CMS sentences that define $K_1, \ldots, K_n$.

Proof: (1) is immediate from Theorems 5.68 and 4.53, and (2) from Propositions 4.42(2) and 4.47(2).

The "if" direction of (3) follows immediately from Theorem 5.68 and Proposition 3.91(1), for $\mathbb{M} := \mathbb{GP}^t$. Let $\lfloor K \rfloor = \mathrm{MOD}(\varphi)$ for a sentence φ in $C_r \mathrm{MS}^h(\mathcal{R}_{s,D}, \emptyset)$ with $D \in \mathcal{P}_f(\mathcal{A})$. For deciding $L \subseteq K$, we first use Proposition 4.45 to decide if $L \subseteq \mathcal{GP}[D]$, and then we use (1) and (2) to decide if $L \cap (\mathcal{GP}[D] - K)$ is empty (note that $\mathcal{GP}[D] - K$ is defined by $\neg \varphi$).

To prove the "only if" direction of (3), let $(L_1, \ldots, L_n)$ be the least solution of S in $\mathcal{P}(\mathbb{GP}^t)$ and suppose that $L = L_1 \cup \cdots \cup L_\mu \subseteq K$. Let D_j be the type of the elements of L_j, for every $j \in [n]$. We first prove that we may assume that $L = L_i$ and that all elements of K have type D_i, for some $i \in [n]$. Obviously, $L_\ell \subseteq \{G \in K \mid \pi(G) = D_\ell\}$ for every $\ell \in [\mu]$. As shown in the proof of Theorem 5.68, $\{G \in K \mid \pi(G) = D_\ell\}$ is CMS-definable. For every $\ell \in [\mu]$, let $(K_1^\ell, \ldots, K_n^\ell)$ be an n-tuple of CMS-definable sets that is an oversolution of S such that $K_\ell^\ell \subseteq \{G \in K \mid \pi(G) = D_\ell\}$. Then it is straightforward to verify that the n-tuple $(K_1, \ldots, K_n) := (\bigcap_{\ell \in [\mu]} K_1^\ell, \ldots, \bigcap_{\ell \in [\mu]} K_n^\ell)$ satisfies the requirements.

So, let $L = L_1$ and let all elements of K be of type $D_1 = D$. It follows from Proposition 3.9 that every $\lfloor L_i \rfloor$ is equational in $\mathbb{STR}_{\mathrm{pres}}$. By Theorem 5.64, $\lfloor K \rfloor$ is effectively recognizable in $\mathbb{STR}_{\mathrm{pres}}$. In what follows we identify $\lfloor L_i \rfloor$ with L_i and $\lfloor K \rfloor$ with K; note that since K is homogenous, $\lfloor K \rfloor = \{\lfloor G \rfloor \mid G \in K\}$, and similarly for L_i. It is shown in the proof of Theorem 5.64 that $K = \kappa^{-1}(C)$ for a homomorphism $\kappa : \mathbb{STR}_{\mathrm{pres}} \to \mathbb{A}$ and $C \subseteq A$. Moreover, $C = \{\Theta \in A_{\mathcal{R}_{s,D}} \mid \widehat{\varphi} \in \Theta\}$ and, for every $\mathcal{R}'$-structure S, $A_{\mathcal{R}'}$ is the powerset of $\widehat{C_r \mathrm{MS}^h(\mathcal{R}', \emptyset)}$ and $\kappa(S) = Th(S, \mathcal{R}', h, r)$. We now apply Proposition 3.91(1) for $\mathbb{M} := \mathbb{STR}_{\mathrm{pres}}$. According to its proof, the n-tuple $(K_1, \ldots, K_n)$ can be taken such that K_i is the union of all sets $\kappa^{-1}(q)$ such that $L_i \cap \kappa^{-1}(q) \neq \emptyset$ for $q \in A$. Obviously, it suffices to take $q \in A_{\mathcal{R}_{s,D_i}}$ and hence the union is finite. For every $q \subseteq \widehat{C_r \mathrm{MS}^h(\mathcal{R}', \emptyset)}$, the set $\kappa^{-1}(q)$ consists of all $\mathcal{R}'$-structures S such that $Th(S, \mathcal{R}', h, r) = q$, and hence it is CMS-definable by the conjunction of all sentences in q. So, K_i is CMS-definable by a finite disjunction φ_i of such sentences, and φ_i can be constructed because the nonemptiness of $L_i \cap \kappa^{-1}(q)$ can be decided by (2): L_i is VR-equational and $\kappa^{-1}(q)$ is CMS-definable. Note that K_i consists

---

[70] The set $L \cap K$ is a set of abstract p-graphs. Computing this set means producing one concrete graph from each isomorphism class of an abstract graph in $L \cap K$.

of p-graphs: if $G \in L_i \cap \kappa^{-1}(q)$ then $q = Th(\lfloor G \rfloor, \mathcal{R}_{s,D_i}, h, r)$ contains a sentence expressing that $\lfloor G \rfloor$ is the representation of a p-graph.   ∎

The results and their proofs are fully similar for the HR algebra and CMS_2-definability. We state them for easy reference.

Corollary 5.71 (Filtering Theorem for the HR algebra) For every HR-equational set L and every CMS_2-definable set K of (labeled) s-graphs, we have:
(1) The set $L \cap K$ is HR-equational, and an equation system for it can be constructed from an equation system defining L and a sentence defining K.
(2) The emptiness and the finiteness of $L \cap K$ can be decided. If $L \cap K$ is finite, then it can be computed.
(3) Let L be defined by (S, X_μ) for some typed HR equation system S with $Unk(S) = X_n$ and some $\mu \in [n]$. Then $L \subseteq K$ if and only if there exists an n-tuple $(K_1, \ldots, K_n)$ of CMS_2-definable sets of (labeled) s-graphs that is an oversolution of S and is such that $K_\ell \subseteq K$ for every $\ell \in [\mu]$. If $L \subseteq K$ (which can be decided by (1) and (2)), then one can construct CMS_2 sentences that define $K_1, \ldots, K_n$.

Proof: (1) follows from Theorems 5.68 and 4.37, and (2) from Propositions 4.5(2) and 4.19. In the proof of the "if" direction of (3), Proposition 4.10 is used instead of Proposition 4.45. For the "only if" direction, we observe that $\lceil L_i \rceil$ is equational in $\mathbb{STR}_{pres}$ by Proposition 3.9 and Proposition 3.41: the operations of $\mathbb{JS}^t$ are linear derived operations of $\mathbb{STR}_{pres}$, as mentioned in the proof of Theorem 5.68(2).   ∎

We now present an application to the construction of an HR equation system that defines a set of graphs characterized by excluded minors.

Application 5.72 (From excluded minors to HR equation systems) Let be given a finite set Ω of simple, loop-free undirected graphs, and let $L := Forb(\Omega)$ be the set of graphs that do not contain any of them as a minor. We know from the proof of Proposition 2.61 that L has bounded tree-width if and only if some planar graph belongs to Ω. Since we want to find an HR equation system for defining L, we assume that Ω contains some planar graph P (otherwise the construction is impossible by Proposition 4.7).

We also know from the proof of Proposition 2.61 that $twd(Forb(\{P\})) \leq f(P)$, where f is a computable function. It follows that

$$L = TWD(\leq f(P)) \cap Forb(\Omega).$$

The set $Forb(\Omega)$ is MS-definable by Corollary 1.14: it is defined by the conjunction of the sentences $\neg MINOR_H$ for all $H \in \Omega$ (cf. Proposition 4.31(8)). Hence, by Corollary 5.71(1) and Example 4.3(8) (that constructs an HR equation system for $TWD(\leq k)$, where k is a given integer), the set L is HR-equational and a system defining it can be constructed.   □

Noncomputability results for congruences

By Theorems 5.64 and 5.68, several locally finite congruences on the VR and the HR algebras are associated with monadic second-order sentences:

(1) the congruence $\sim^h$ that defines two p-graphs or two s-graphs as equivalent if they have the same type and satisfy the same MS sentences of quantifier-height at most h (we take $r = 0$ in the proof of Theorem 5.64);
(2) the syntactic congruence $\approx^{\mathrm{MOD}(\varphi)}$ of the set of p-graphs or s-graphs of the same type that satisfy an MS sentence φ.

Clearly, $\sim^h$ refines $\approx^{\mathrm{MOD}(\varphi)}$ if the quantifier-height of φ is at most h. In the following proposition, we only consider congruences relative to the VR algebra (and for unlabeled graphs), but the same negative results hold for the HR algebra, and the proofs are essentially the same.

Proposition 5.73
(1) For each integer h the equivalence relation $\sim^h$ on p-graphs is decidable, but there is no algorithm that computes, for any given $h \in \mathcal{N}$ and any given finite set C of port labels, the number of equivalence classes of $\sim^h$ on p-graphs of type C.
(2) It is not decidable whether $G \approx^{\mathrm{MOD}(\varphi)} H$ for any given p-graphs G and H and any given sentence φ.

Proof: (1) The equivalence relation $\sim^h$ is decidable (and even uniformly decidable, for varying h) by the observations in Definition 5.59.

We now prove that if, for each h, we can compute the number of equivalence classes of $\sim^h$ of type $\{1\}$ (the type of nonempty graphs without ports, we let 1 be a default port label), then we can decide the monadic second-order satisfiability problem. Let φ be an MS sentence, h be its quantifier-height and let N_h be the number of equivalence classes of $\sim^h$ of type $\{1\}$. Let $G_1, G_2, \ldots, G_i, \ldots$ be an effective enumeration of all simple nonempty graphs. Since $\sim^h$ is decidable, one can compute the increasing sequence $n_1 < n_2 < \cdots < n_i < \cdots$ such that $n_1 = 1$ and, for each $i > 1$, n_i is the smallest integer $j > n_{i-1}$ such that G_j is not $\sim^h$-equivalent to any graph G_k for $k < j$. This sequence has exactly N_h elements. Since we know this value, we can determine when its computation is terminated. It follows that φ is satisfied in some simple graph if and only if it is in some of the graphs $G_{n_1}, G_{n_2}, \ldots, G_m$, where $m := n_{N_h}$, or in the empty graph. This can be decided and we get a contradiction with Theorem 5.5.

(2) We will use the proof of Theorem 5.6 and Remark 5.7. Let $L := \mathrm{MOD}(\varphi_M)$, where M is a deterministic Turing machine and L is empty if M does not halt or consists of one directed square grid (up to isomorphism) if M halts. Let H be any nonempty simple directed graph. If $L = \emptyset$, then for every simple directed graph G, we have $G \oplus H \in L$ if and only if $G \oplus \emptyset \in L$. If $L = \{K\}$, then this equivalence does not hold (take $G = K$). More generally, we get that $H \approx^L \emptyset$ if and only if M does not halt, which is not decidable. This reduction establishes the result. ∎

Remark 5.74 The situation is different if we consider congruences relative to finitely generated graph algebras, like those of graphs of tree-width or clique-width at most some fixed k. In these cases, the decidable congruence $\sim^h$ is effectively given by Proposition 3.78(3.1) of Section 3.4.4, which means (cf. Definition 3.77) that an algorithm can compute the cardinality of each congruence class. The index of $\sim^h$ is thus computable. It is an upper-bound to that of $\approx^{\mathrm{MOD}(\varphi)}$ (where h is the quantifier-height of φ). By Claim 3.80.2, this congruence is decidable and its index is computable, as for $\sim^h$, by Proposition 3.78(3.1).

5.3.10 Handling general quantifier-free operations

We will prove the Recognizability Theorem for the algebra $\mathbb{STR}$. Its unary operations are all QF operations, even those that are not quasi-domain preserving.

Theorem 5.75 (Recognizability Theorem for $\mathbb{STR}$) For every relational signature $\mathcal{R}$ and every sentence φ in $\mathrm{CMS}(\mathcal{R},\emptyset)$, the set $\mathrm{MOD}(\varphi) \subseteq STR(\mathcal{R})$ is effectively recognizable in $\mathbb{STR}$.   □

This theorem is a consequence of Theorem 5.64 and the following one:

Theorem 5.76 Every set of relational structures that is recognizable, semi-effectively recognizable or effectively recognizable in $\mathbb{STR}_{\mathrm{pres}}$ has the same property in $\mathbb{STR}$.

To prove this, we need some definitions and lemmas. We recall that, if all constants of a structure S are in a subset E of its domain, then $S[E]$ denotes the substructure of S with domain E. (Its signature is the same as that of S.)

Definition 5.77 (Domain restricting operations) A QF operation $f : STR^c(\mathcal{R}) \to STR^c(\mathcal{R})$ is *domain restricting* if, for some formula δ in $\mathrm{QF}(\mathcal{R},\{x\})$, we have $f(S) = S[sat(S,\delta,x)]$ for every $S \in STR^c(\mathcal{R})$. This implies that $S \models \delta(c)$ for every $c \in \mathcal{R}_0$ and every $S \in STR^c(\mathcal{R})$. These operations are those defined by definition schemes whose domain formula δ is equivalent to a formula of the form $\delta'(x) \vee \bigvee_{c \in \mathcal{R}_0} x = c$ for some $\delta' \in QF(\mathcal{R},\{x\})$. The definition scheme $\mathcal{D} = \langle \delta, (\theta_R)_{R \in \mathcal{R}_+}, (\kappa_{c,d})_{c,d \in \mathcal{R}_0} \rangle$ is thus completely specified by δ. Its other formulas are $\theta_R(x_1,\ldots,x_n)$ equal to $R(x_1,\ldots,x_n)$, $\kappa_{c,c}$ equal to *True* for each $c \in \mathcal{R}_0$ and $\kappa_{c,d}$ equal to *False* for distinct $c,d \in \mathcal{R}_0$. By Remark 5.43(1) there are, for each $\mathcal{R}$, finitely many domain restricting QF operations $: STR^c(\mathcal{R}) \to STR^c(\mathcal{R})$. Their definition schemes are obtained by considering only formulas $\delta' \in \mathrm{QF}(\mathcal{R},\{x\})$ that are in disjunctive normal form.

Lemma 5.78 Let $\mathcal{R}$ and $\mathcal{R}'$ be relational signatures. Every QF operation $f : STR^c(\mathcal{R}) \to STR^c(\mathcal{R}')$ can be expressed as $g \circ h$, where h is a domain restricting QF operation $: STR^c(\mathcal{R}) \to STR^c(\mathcal{R})$ and g is a QDP-QF operation $: STR^c(\mathcal{R}) \to STR^c(\mathcal{R}')$.

Proof: Let f have definition scheme $\langle \delta, (\theta_R)_{R \in \mathcal{R}'_+}, (\kappa_{c,d})_{c \in \mathcal{R}_0, d \in \mathcal{R}'_0} \rangle$. Let $S \in STR^c(\mathcal{R})$ and $S' := f(S) \in STR^c(\mathcal{R}')$. Let E be such that

$$\{c_S \mid c \in \mathcal{R}_0\} \cup sat(S, \delta, x) \subseteq E \subseteq D_S.$$

We have $D_{f(S)} \subseteq E$ because, by definition, $D_{f(S)} = sat(S, \delta, x)$. Since $c_S \in E$ for each $c \in \mathcal{R}_0$, the structure $S[E]$ belongs to $STR^c(\mathcal{R})$. We have

$$f(S[E]) = f(S). \tag{5.8}$$

We next check this equality. For every $d \in E$ we have $S \models \delta(d)$ if and only if $S[E] \models \delta(d)$ by the definition of $S[E]$ and because δ is quantifier-free. Hence the domains of $f(S)$ and $f(S[E])$ are the same. For the same reasons, we have, for all $k \geq 1$, all $R \in \mathcal{R}'_k$, all $d_1, \ldots, d_k$ in E and all $c \in \mathcal{R}_0$ and $d \in \mathcal{R}'_0$:

$$
\begin{aligned}
S \models \theta_R(d_1, \ldots, d_k) \quad &\text{if and only if} \quad S[E] \models \theta_R(d_1, \ldots, d_k), \text{ and} \\
S \models \kappa_{c,d} \quad &\text{if and only if} \quad S[E] \models \kappa_{c,d}.
\end{aligned}
$$

This proves Equality (5.8). It follows that $f = g(h(S))$ for every $S \in STR^c(\mathcal{R})$, where h is the domain restricting QF operation with domain formula $\delta(x) \vee \bigvee_{c \in \mathcal{R}_0} x = c$ and g is the QDP-QF operation defined by the definition scheme $\langle \delta', (\theta_R)_{R \in \mathcal{R}'_+}, (\kappa_{c,d})_{c \in \mathcal{R}_0, d \in \mathcal{R}'_0} \rangle$, where δ' is $\bigwedge_{c \in \mathcal{R}_0} (\neg \delta(c) \Rightarrow x \neq c)$. The structure $h(S)$ is $S[E]$, where $E := D_{f(S)} \cup \{c_S \mid c \in \mathcal{R}_0\}$ so that by Equality (5.8), $f(h(S)) = f(S)$. But f and g have the same effect on $h(S)$ because the same elements of E satisfy δ and δ'. ∎

Lemma 5.79 Let $\mathcal{R}'$ and $\mathcal{R}''$ be relational signatures such that $\mathcal{R}'_0 \cap \mathcal{R}''_0 = \emptyset$. Let $f : STR^c(\mathcal{R}' \cup \mathcal{R}'') \to STR^c(\mathcal{R}' \cup \mathcal{R}'')$ be a domain restricting QF operation. There exist two families of pairs $(\chi''_i, g'_i)_{1 \leq i \leq p'}$ and $(\chi'_j, g''_j)_{1 \leq j \leq p''}$ such that, for every i and j, $\chi'_j \in QF(\mathcal{R}', \emptyset)$, $\chi''_i \in QF(\mathcal{R}'', \emptyset)$, g'_i and g''_j are domain restricting operations of respective types $\mathcal{R}' \to \mathcal{R}'$ and $\mathcal{R}'' \to \mathcal{R}''$, and for every two disjoint concrete $\mathcal{R}$- and $\mathcal{R}''$-structures, respectively S and T:

(1) there exist a unique $i \in [p']$ such that $T \models \chi''_i$ and a unique $j \in [p'']$ such that $S \models \chi'_j$;

(2) we have $f(S \oplus T) = g'_i(S) \oplus g''_j(T)$ where i, j are as in (1).

Proof: One might think that, for every f, S and T as in the statement we have $f(S \oplus T) = f(S) \oplus f(T)$. This is actually true if $\mathcal{R}'_0 = \mathcal{R}''_0 = \emptyset$, but not in general. Consider for example the operation f with domain formula $\delta(x)$ equal to

$$\big(R(x) \wedge R'(a) \wedge R'(b)\big) \vee \big(R(x) \wedge \neg R'(a) \wedge \neg R'(b)\big) \vee \big(\neg R(x) \wedge R'(a) \wedge \neg R'(b)\big).$$

There is no pair of domain restricting QF operations (g, h) such that, for every concrete disjoint $\{R, R', a\}$-structure S and $\{R, R', b\}$-structure T, we have $f(S \oplus T) = g(S) \oplus h(T)$, because the set $D_{f(S \oplus T)} \cap D_S$ depends on whether $R'(b)$ holds or not in T, and is not a function of S alone.

The formulas χ'_j, χ''_i and the domain formulas of the operations g'_i and g''_j will be constructed by Proposition 5.37 (the Splitting Theorem for first-order formulas) applied to the quantifier-free formula $\delta(x)$ of the definition scheme of f. (Since we want g'_i and g''_j to be domain restricting, the other formulas of their definition schemes are fixed, cf. Definition 5.77.) Proposition 5.37 also holds for the general disjoint union defined in Example 5.44, by a proof fully similar to the one of Theorem 5.57. By Proposition 5.37, one can find $n, m \in \mathcal{N}$ and construct quantifier-free formulas $\theta_1(x), \ldots, \theta_n(x), \theta_{n+1}, \ldots, \theta_{n+m}$ and $\psi_1, \ldots, \psi_n, \psi_{n+1}(x), \ldots, \psi_{n+m}(x)$ of the following forms:

$\theta_i \in \mathrm{QF}(\mathcal{R}', \{x\})$ and $\psi_i \in \mathrm{QF}(\mathcal{R}'', \emptyset)$ for each $i \in [n]$,

$\theta_{n+j} \in \mathrm{QF}(\mathcal{R}', \emptyset)$ and $\psi_{n+j} \in \mathrm{QF}(\mathcal{R}'', \{x\})$ for each $j \in [m]$,

such that for every two disjoint concrete structures S and T of the appropriate types:

$$\{d \in D_S \mid S \oplus T \models \delta(d)\} = \bigcup_{i \in [n]} \{d \in D_S \mid S \models \theta_i(d) \text{ and } T \models \psi_i\}, \text{ and}$$

$$\{d \in D_T \mid S \oplus T \models \delta(d)\} = \bigcup_{j \in [m]} \{d \in D_T \mid S \models \theta_{n+j} \text{ and } T \models \psi_{n+j}(d)\}.$$

For every $I \subseteq [n]$, we define the sentence χ''_I to be $\bigwedge_{i \in I} \psi_i \wedge \bigwedge_{i \in [n]-I} \neg \psi_i$ and the domain restricting QF operation g'_I of type $\mathcal{R}' \to \mathcal{R}'$ to have domain formula $\bigvee_{i \in I} \theta_i(x)$. Similarly, for every $J \subseteq [m]$, we let χ'_J be the sentence $\bigwedge_{j \in J} \theta_{n+j} \wedge \bigwedge_{j \in [m]-J} \neg \theta_{n+j}$, and the domain restricting operation g''_J of type $\mathcal{R}'' \to \mathcal{R}''$ to have domain formula $\bigvee_{j \in J} \psi_{n+j}(x)$. Clearly, for all S and T as in the statement:

(1) there exist a unique $I \subseteq [n]$ such that $T \models \chi''_I$ and a unique $J \subseteq [m]$ such that $S \models \chi'_J$;

(2) we have $f(S \oplus T) = g'_I(S) \oplus g''_J(T)$, where I and J are as in (1).

We have the desired families $(\chi''_i, g'_i)_{1 \leq i \leq p'}$ and $(\chi'_j, g''_j)_{1 \leq j \leq p''}$ with $p' = 2^n$ and $p'' = 2^m$. $\blacksquare$

Proof of Theorem 5.76: Let L be $\mathrm{STR}_{\mathrm{pres}}$-recognizable, with congruence $\equiv$ witnessing its recognizability.

First, we define from it another equivalence relation. For $S, T \in STR(\mathcal{R})$ we let $S \sim_\mathcal{R} T$ if and only if $S \equiv_\mathcal{R} T$ and $S \sim^0 T$, where $S \sim^0 T$ means that $Th(S, \mathcal{R}, 0, 0) = Th(T, \mathcal{R}, 0, 0)$ (i.e., S and T satisfy the same quantifier-free sentences). Since, as shown in the proof of Theorem 5.64, $\sim^0$ is a (decidable) locally finite congruence, the equivalence $\sim$ is also a locally finite congruence witnessing the $\mathrm{STR}_{\mathrm{pres}}$-recognizability of L.

We now define from $\sim$ another equivalence relation $\approx$ by letting $S \approx_{\mathcal{R}} T$ if and only if $f(S) \sim_{\mathcal{R}} f(T)$ for every domain restricting QF operation f of type $\mathcal{R} \to \mathcal{R}$. It is clear that $S \approx_{\mathcal{R}} T$ implies $S \sim_{\mathcal{R}} T$ (because the identity on $STR(\mathcal{R})$ is domain restricting) and hence L is a union of classes of $\approx$. Each equivalence relation $\approx_{\mathcal{R}}$ has at most $p \cdot q$ classes, where p is the number of classes of $\sim_{\mathcal{R}}$ and q is the number of domain restricting QF operations of type $\mathcal{R} \to \mathcal{R}$, cf. Definition 5.77. Hence the equivalence relation $\approx$ is locally finite. It remains to prove that it is a congruence on STR.

We first consider disjoint union. We let $S, S_1 \in STR(\mathcal{R}')$ and $T, T_1 \in STR(\mathcal{R}'')$ with $S \approx_{\mathcal{R}'} S_1$ and $T \approx_{\mathcal{R}''} T_1$. We claim that $S \oplus T \approx_{\mathcal{R}' \cup \mathcal{R}''} S_1 \oplus T_1$. The goal is to prove that $f(S \oplus T) \sim f(S_1 \oplus T_1)$ for every domain restricting QF operation f of type $\mathcal{R}' \cup \mathcal{R}'' \to \mathcal{R}' \cup \mathcal{R}''$. Consider one of them, f, to which we apply Lemma 5.79. Since $S \approx_{\mathcal{R}'} S_1$, we have $S \sim_{\mathcal{R}'} S_1$ and hence S and S_1 satisfy the same quantifier-free sentences. The same holds for T and T_1. Hence we have indices i and j (cf. Lemma 5.79) for which $T \models \chi_i''$ and $S \models \chi_j'$, so that we also have $T_1 \models \chi_i''$ and $S_1 \models \chi_j'$. Hence we have $f(S \oplus T) = g_i'(S) \oplus g_j''(T)$ and also $f(S_1 \oplus T_1) = g_i'(S_1) \oplus g_j''(T_1)$. We have $g_i'(S) \sim g_i'(S_1)$ and $g_j''(T) \sim g_j''(T_1)$ since $S \approx_{\mathcal{R}'} S_1$ and $T \approx_{\mathcal{R}''} T_1$. Hence $g_i'(S) \oplus g_j''(T) \sim g_i'(S_1) \oplus g_j''(T_1)$, i.e., $f(S \oplus T) \sim_{\mathcal{R}' \cup \mathcal{R}''} f(S_1 \oplus T_1)$. Hence $S \oplus T \approx S_1 \oplus T_1$ as was to be proved.

We now consider $S, S_1 \in STR(\mathcal{R})$ such that $S \approx S_1$, and a QF operation k of type $\mathcal{R} \to \mathcal{R}'$. We must prove that $k(S) \approx k(S_1)$. Let f be domain restricting of type $\mathcal{R}' \to \mathcal{R}'$. The mapping $f \circ k$ is a QF operation of type $\mathcal{R} \to \mathcal{R}'$ by Proposition 5.49. Let us apply Lemma 5.78. We have $f(k(S)) = g(h(S))$ and $f(k(S_1)) = g(h(S_1))$. Since $S \approx S_1$, we have $h(S) \sim h(S_1)$, hence $g(h(S)) \sim g(h(S_1))$, since g is quasi-domain preserving and $\sim$ is a congruence on STR_{pres}. Hence the equivalence $\approx$ is a congruence on STR. This completes the proof of recognizability.

If $\equiv$ is a decidable congruence, then so are $\sim$ and $\approx$ (cf. Corollary 5.95). Hence the result for semi-effective recognizability follows from Proposition 3.78(1).

The proof for effective recognizability is essentially the same as the above proof, using locally finite algebras rather than congruences. It also proves the previous results, but since it is less transparent, we have kept the above proof with congruences.

Assume that L is effectively STR_{pres}-recognizable, and let $L = \beta^{-1}(C)$, where β is a homomorphism from STR_{pres} to a locally finite, effectively given $F_{\mathrm{pres}}^{\mathrm{QF}}$-algebra $\mathbb{B}$ and $C \subseteq B_{\mathcal{R}}$. Let κ and $\mathbb{A}$ be as in the proof of Theorem 5.64 (with $h = r = 0$).

We define the locally finite F^{QF}-algebra $\mathbb{D}$ such that $D_{\mathcal{R}}$ is the set of mappings from $\mathrm{DRQF}_{\mathcal{R}}$ to $B_{\mathcal{R}} \times A_{\mathcal{R}}$, where $\mathrm{DRQF}_{\mathcal{R}}$ is the finite set of domain restricting QF operations of type $\mathcal{R} \to \mathcal{R}$ (more precisely, it is the set of their definition schemes). For $d \in D$ and $h \in \mathrm{DRQF}_{\mathcal{R}}$, we define $d_1(h) \in B$ and $d_2(h) \in A$ such that $d(h) = \langle d_1(h), d_2(h) \rangle$. The operations of $\mathbb{D}$ are defined as follows. Let k be a QF operation of type $\mathcal{R} \to \mathcal{R}'$ and let $f \in \mathrm{DRQF}_{\mathcal{R}'}$. We apply Proposition 5.49 and Lemma 5.78 to obtain $h \in \mathrm{DRQF}_{\mathcal{R}}$ and a QDP-QF operation g such that $f \circ k = g \circ h$. Then we define the operation $k_{\mathbb{D}}$ such that $k_{\mathbb{D}}(d)(f) := g_{\mathbb{B} \times \mathbb{A}}(d(h))$ for every $d \in D_{\mathcal{R}}$. Now consider disjoint union $\oplus$ of type $\mathcal{R}' \times \mathcal{R}'' \to \mathcal{R}' \cup \mathcal{R}''$ and let $f \in DRQF_{\mathcal{R}' \cup \mathcal{R}''}$. Then, using Lemma 5.79, we define the operation $\oplus_{\mathbb{D}}$ such that $(d' \oplus_{\mathbb{D}} d'')(f) := d'(g_i') \oplus_{\mathbb{B} \times \mathbb{A}} d''(g_j'')$ for all $d' \in D_{\mathcal{R}'}$ and $d'' \in D_{\mathcal{R}''}$, provided $\chi_i'' \in d_2''(id_{\mathcal{R}''})$ and $\chi_j' \in d_2'(id_{\mathcal{R}'})$, where $id_{\mathcal{R}}$ is the identity on $STR(\mathcal{R})$. If i and j are not unique, then $(d' \oplus_{\mathbb{D}} d'')(f)$ can be defined arbitrarily. Since both $\mathbb{B}$ and $\mathbb{A}$ are effectively given and $DRQF_{\mathcal{R}}$ can be computed, the algebra $\mathbb{D}$ is effectively given.

We define the mapping δ from $\mathbb{STR}$ to $\mathbb{D}$ such that

$$\delta(S)(h) := \langle \beta(h(S)), \kappa(h(S)) \rangle,$$

for every $\mathcal{R}$-structure S and every $h \in DRQF_{\mathcal{R}}$. Then $L = \delta^{-1}(E)$ for $E := \{d \in D_{\mathcal{R}} \mid d_1(id_{\mathcal{R}}) \in C\}$. It is immediate from the definitions that δ is a homomorphism. ∎

5.4 Decidable monadic second-order theories

We have presented undecidability results in Section 5.1.6. Here we present decidability results for sets of graphs of bounded tree-width or clique-width. They extend some "classical" results: the monadic second-order theory of a regular language (of words or of terms) is decidable, by reduction to emptiness problems for languages defined by finite automata. The links between monadic second-order logic and finite automata were discovered in the 1960s with the motivation of finding classes of finite and infinite structures having decidable monadic second-order theories.

Theorem 5.80
(1) Every VR-equational set of (labeled) p-graphs has a decidable CMS-theory, equivalently a decidable CMS-satisfiability problem.
(2) Every HR-equational set of (labeled) s-graphs has a decidable CMS_2-theory, equivalently a decidable CMS_2-satisfiability problem.

Proof: (1) Let L be a VR-equational set given by an equation system, with $\lfloor L \rfloor = \lfloor L \rfloor_C$, and let φ be a CMS sentence in $CMS(\mathcal{R}_{s,C})$. The set of p-graphs $K := \{G \in \mathcal{GP}[C] \mid \lfloor G \rfloor_C \models \varphi\}$ is CMS-definable. By Corollary 5.70(2), one can test whether $L \cap K$ is empty, which solves the CMS-satisfiability problem. By Corollary 5.70(3), one can test whether $L \subseteq K$, i.e., whether every p-graph in L satisfies φ, hence whether φ belongs to the CMS-theory of L. Hence the CMS-theory of L is decidable.

(2) The proofs are similar for HR-equational sets and CMS_2 sentences by using Corollary 5.71. ∎

Corollary 5.81
(1) For each k, the CMS-theory of the set of simple (labeled) graphs of clique-width at most k and its CMS-satisfiability problem are decidable.

(2) For each k, the CMS_2-theory of the set of (labeled) graphs of tree-width at most k and its CMS_2-satisfiability problem are decidable.

Proof: Immediate consequence of the previous theorem and the results of Chapter 4 (Examples 4.3(8) and 4.43(5)) showing that the sets $CWD(\leq k)$ and $TWD(\leq k)$ and the corresponding sets of (K, Λ)-labeled graphs (for fixed pairs (K, Λ) of sets of labels) are respectively VR- and HR-equational. ∎

In Section 7.5 we will state converse results establishing that only sets of graphs of bounded tree-width and clique-width can have decidable MS_2- and C_2MS-theories respectively.

5.5 Logical characterization of recognizability

We have proved in Theorem 5.15 and in Corollaries 5.66 and 5.67 the result stated in Theorem 1.16 that a language (of terms, and of words as a special case) is regular if and only if it is MS-definable. It can be restated and generalized in terms of CMS-definability and recognizability in STR.

We will identify a term $t \in T(F)$ and the structure $\lfloor t \rfloor$ that represents it (faithfully). Similarly, we will identify L and $\lfloor L \rfloor$ in the following statement and its proof.

Theorem 5.82 Let F be a finite functional signature. The following properties of a language $L \subseteq T(F)$ are equivalent:
(1) L is recognizable in STR;
(2) L is recognizable in $T(F)$;
(3) L is regular over F;
(4) L is MS-definable;
(5) L is CMS-definable.

The same equivalences hold if $L \subseteq A^*$ for some finite alphabet A, with $T(F)$ replaced by $W(A)$, $W_{left}(A)$ or $W_{right}(A)$.

The translations between sentences and automata are effective.

Proof: Since the representation $\lfloor \cdot \rfloor$ is faithful and by Corollary 5.12, a set $L \subseteq T(F)$ is MS-definable (or CMS-definable) if and only if the class $\lfloor L \rfloor$ is.
(1) $\Longrightarrow$ (2) follows (for one-sorted F) from Proposition 3.56 because $T(F)$ is a subalgebra of a derived algebra of STR (by Application 5.61), cf. the proof of Corollary 5.67.
(2) $\Longrightarrow$ (3): the regularity of $L \subseteq T(F)$ is equivalent to its recognizability in $T(F)$ (cf. Theorem 3.62).
(3) $\Longrightarrow$ (4) is proved by a translation of a finite automaton $\mathscr{A}$ into an MS sentence that defines $L(\mathscr{A})$ (cf. Theorem 5.15).

(4) $\Longrightarrow$ (5) is trivial; (5) $\Longrightarrow$ (4) is proved in Proposition 5.29, using the fact that the structures representing terms are MS-orderable.

(5) $\Longrightarrow$ (1) by the Recognizability Theorem (Theorem 5.75). Since A^* is isomorphic to $T(U_A)$, where U_A is the unary signature of Definition 2.7, these results specialize to subsets of A^*. The algebras $\mathbb{W}(A)$, $\mathbb{W}_{left}(A)$ and $\mathbb{W}_{right}(A)$ have the same recognizable sets.   ∎

A natural question is whether there exists a similar characterization for sets of graphs. One could hope that a set of graphs is HR-recognizable if and only if it CMS_2-definable, but we have proved in Proposition 4.36 that there are uncountably many HR-recognizable sets of graphs. This result forbids any such characterization, as well as any characterization in terms of "finite graph automata" (because different HR-recognizable sets would necessarily correspond to different formulas or automata, and logical languages as well as families of finite automata are countable). Logical characterizations can only exist if recognizability is relativized to particular graph classes. (However, this cardinality argument does not exclude a logical characterization of the effectively HR-recognizable sets of graphs.) The following theorem concerns trees and is one motivation (among others) for introducing counting monadic second-order logic. The more general case of sets of graphs of bounded tree-width will be discussed in Section 7.6.

We consider rooted trees. We view a rooted tree t as an s-graph with one source: the root of t. It can be identified with the $\{son, rt\}$-structure $\lfloor t \rfloor$. Thus, $t = \langle N_t, son_t, rt_t \rangle$, where $\rho(son) = 2$, $\rho(rt) = 0$ and rt_t is the root of t. The set of sons of a node is not ordered. We turn the set of rooted trees into an $\{/\!/, ext, *\}$-algebra $\mathbb{R}$ where $/\!/$ (the parallel-composition) glues two trees at their roots, and ext is the unary operation that extends a tree by attaching an edge at the root and making the resulting new vertex into the new root. The constant $*$ denotes the tree reduced to a root. We let F_{rt} be the functional signature $\{/\!/, ext, *\}$ (the subscript "rt" means "rooted trees").

Clearly, $\mathbb{R}$ is a subalgebra of a derived algebra of $\mathbb{JS}^t$, cf. Example 4.3(3): the derived operation ext is defined by the term $fg_2(\mathbf{12} /\!/ ren_{1 \leftrightarrow 2}(x_1))$, where $\mathbf{1}$ stands for rt. It is also a subalgebra of a derived algebra of $\mathbb{STR}$, because for simple s-graphs G, Application 5.51 and Example 5.55 also work for $\mathcal{R}_{ss,C}$-structures $\lfloor G \rfloor$ instead of $\mathcal{R}_{ms,C}$-structures $\lceil G \rceil$.

If t is a rooted tree, we denote by $Unr(t)$ the unrooted and undirected tree obtained from t by forgetting the root and the directions of edges. Thus, $Unr(t) = und(fg_{rt}(t))$.

Theorem 5.83 For every set L of rooted trees, the following properties are equivalent:

(1) L is recognizable in $\mathbb{STR}$;
(2) L is recognizable in $\mathbb{JS}^t$;

(3) L is recognizable in $\mathbb{R}$;

(4) L is CMS-definable.

If L is a set of (undirected) trees, then we have the equivalences of (1), (2), (4) and the following property:

(3') The set $Unr^{-1}(L)$ is recognizable in $\mathbb{R}$.

Proof: The implications (1) $\Longrightarrow$ (3) and (2) $\Longrightarrow$ (3) follow from Proposition 3.56 and previous remarks. Implications (4) $\Longrightarrow$ (1) and (4) $\Longrightarrow$ (2) follow from the Recognizability Theorem (Theorems 5.75 and 5.68(2), respectively).

We prove implication (3) $\Longrightarrow$ (4). Let L be recognizable in $\mathbb{R}$. There exists a surjective homomorphism $h : \mathbb{R} \to \mathbb{Q}$, where $\mathbb{Q}$ is a finite F_{rt}-algebra such that $L = h^{-1}(Q_0)$ for some subset Q_0 of Q (the domain of the algebra $\mathbb{Q}$). The surjectivity of h implies that $/\!/_{\mathbb{Q}}$ is associative and commutative with unit element $*_{\mathbb{Q}}$ because the same properties hold in $\mathbb{R}$.

Let Q be enumerated as $\{q_1, \ldots, q_k\}$. Let $t \in \mathbb{R}$ and $u_1, \ldots, u_n$ be the sons of the root of t. Let $m(t) \in \mathcal{N}^k$ be the k-tuple $(m_1, \ldots, m_k)$ such that, for each $j \in [k]$, m_j is the cardinality of $\{i \in [n] \mid h(t/u_i) = q_j\}$.

Then h can be factorized as $\ell \circ m$ for some function $\ell : \mathcal{N}^k \to Q$ because

$$h(t) = h(ext(t/u_1) /\!/ \cdots /\!/ ext(t/u_n))$$
$$= ext_{\mathbb{Q}}(h(t/u_1)) /\!/_{\mathbb{Q}} \cdots /\!/_{\mathbb{Q}} ext_{\mathbb{Q}}(h(t/u_n))$$
$$= (m_1 \,\square\, ext_{\mathbb{Q}}(q_1)) /\!/_{\mathbb{Q}} \cdots /\!/_{\mathbb{Q}} (m_k \,\square\, ext_{\mathbb{Q}}(q_k)),$$

where, for $r \in \mathcal{N}$ and $q \in Q$, we let $r \,\square\, q := q /\!/_{\mathbb{Q}} \cdots /\!/_{\mathbb{Q}} q$ with r times q if $r \geq 1$ and $0 \,\square\, q = *_{\mathbb{Q}}$. Hence $h(t)$ can be expressed as a fixed function of $m(t)$. Furthermore, $\ell(\vec{0}) = *_{\mathbb{Q}}$ and $\ell(m + m') = \ell(m) /\!/_{\mathbb{Q}} \ell(m')$ for all $m, m' \in \mathcal{N}^k$ as one checks easily. Hence ℓ is a homomorphism : $\langle \mathcal{N}^k, +, \vec{0} \rangle \to \mathbb{Q}$ and $\ell^{-1}(q)$ is recognizable in $\mathbb{N}^k$ for each $q \in \mathbb{Q}$. From these remarks and Proposition 5.25, we get the existence of a CMS sentence φ that defines L. This sentence can be written as follows:

$$\exists X_1, \ldots, X_k \; [\text{``}(X_1, \ldots, X_k) \text{ is a partition of the set of nodes''}$$

$$\wedge (\bigvee_{q_i \in Q_0} rt \in X_i)$$

$\wedge$ "for every $i \in [k]$ and every $u \in X_i$, if S is the set of sons of u, then

the k-tuple $(|S \cap X_1|, \ldots, |S \cap X_k|)$ belongs to $\ell^{-1}(q_i)$"].

We now consider the case where L is a set of trees. Then, we have (4) $\Longrightarrow$ (1) and (1) $\Longrightarrow$ (2). For proving (2) $\Longrightarrow$ (3'), assume that L is recognizable in $\mathbb{JS}^t$. Let R be the set of all rooted trees; it is recognizable in $\mathbb{JS}^t$. Since $Unr^{-1}(L) = R \cap fg_{rt}^{-1}(und^{-1}(L))$, it is recognizable in $\mathbb{JS}^t$ by Proposition 3.85(1) (for the intersection, and for fg_{rt}^{-1} because fg_{rt} is an operation of the algebra $\mathbb{JS}^t$) and by Proposition 3.56(2) (for und^{-1} because und is a homomorphism from the directed version of $\mathbb{JS}^t$ to its undirected version). Hence, $Unr^{-1}(L)$ is recognizable in $\mathbb{R}$ by the implication (2) $\Longrightarrow$ (3) for rooted trees. For proving (3$'$) $\Longrightarrow$ (4), we assume that $Unr^{-1}(L)$ is recognizable in $\mathbb{R}$. It is thus CMS-definable, from which we get easily that L is CMS-definable.　■

Remark 5.84 (1) Finite automata, called $\mathbb{R}$-automata, that run bottom-up on rooted trees are studied in [BonTal], [Cou89a], [*Cou96b] and [*Lib06]. Rooted trees are called in these works *unranked, unordered trees*, because terms are (inadequately) called trees. These automata have infinitely many transition rules, because rooted trees have unbounded degree, but their rules are described in finitary ways. In the proof of the previous theorem, the sets $\ell^{-1}(q)$ describe the transition rules of a finite $\mathbb{R}$-automaton with set of states Q. Since the sets $\ell^{-1}(q)$ are recognizable in $\mathbb{N}^k$, they can be described by CMS formulas by Proposition 5.25. The corresponding automaton with states $q_1,\ldots,q_k$ works as follows: if the sons $u_1,\ldots,u_n$ of a node u are in states $p_1,\ldots,p_n$ respectively, if m_j for $j=1,\ldots,k$ is the number of indices i such that $p_i = q_j$ and if $(m_1,\ldots,m_k) \in \ell^{-1}(q)$, then q is a possible state for u. The accepting states are those of Q_0, cf. the proof that (3) implies (4). This proof shows that if a set of rooted trees is recognizable in $\mathbb{R}$, then it is accepted by an $\mathbb{R}$-automaton. The $\mathbb{R}$-automata constructed in this proof are actually deterministic and complete because ℓ is a mapping from $\mathcal{N}^k$ to Q. More details on these automata can be found in the above quoted references.

(2) The set of odd stars $\{K_{1,2n+1} \mid n \geq 0\}$ (i.e., of rooted trees where the root is the node of degree $2n+1$) is recognizable in $\mathbb{R}$ (the construction of a congruence is easy), and C_2MS-definable. It is not MS-definable (cf. Examples 5.33).　　□

We now generalize this result. For convenience we only consider one-sorted functional signatures.

Definition 5.85 (Terms over associative and commutative operations) An *AC-signature* is a pair (F,H), where F is a finite one-sorted functional signature and $H \subseteq F$ is a set of symbols of arity 2. An F-algebra $\mathbb{M}$ is an (F,H)-*AC-algebra* if the operations $f_{\mathbb{M}}$ are associative and commutative for all $f \in H$. For example, the algebra $\mathbb{R}$ of rooted trees defined above is an $(F_{\mathrm{rt}}, \{/\!/\})$-AC-algebra.

We let $\mathbb{T}_{AC}(F,H)$ denote the quotient F-algebra $\mathbb{T}(F)/\equiv_{AC(H)}$, where $\equiv_{AC(H)}$ is the congruence on $\mathbb{T}(F)$ generated by the associativity and commutativity laws for the operations of H. For every (F,H)-AC-algebra $\mathbb{M}$, there exists a unique homomorphism $h : \mathbb{T}_{AC}(F,H) \to \mathbb{M}$ (see [*BaaNip], [*DersJou] or [*Wec] for these classical facts). It follows that if $\mathbb{M}$ is an (F,H)-AC-algebra generated by its signature F, then a set $L \subseteq M$ is recognizable in $\mathbb{M}$ if and only if $h^{-1}(L) \subseteq T(F)/\equiv_{AC(H)}$ is recognizable in $\mathbb{T}_{AC}(F,H)$. This follows from Lemma 3.68, in the same way as Proposition 3.69.

Our objective is to obtain a logical characterization of the recognizable sets of the algebra $\mathbb{T}_{AC}(F,H)$ similar to those of Theorems 5.82 and 5.83. We first define a faithful representation of the elements of $T(F)/\equiv_{AC(H)}$. For each term $t \in T(F)$, we let $\widehat{t}$ be the term obtained by *flattening* t as follows. For every symbol $f \in H$ we introduce new symbols f_m of arity m, for every $m \geq 2$. For flattening t, we first replace every occurrence of $f \in H$ in t by f_2, and then we iterate as many times as possible the replacement of a subterm of the form $f_n(s_1, \ldots, s_{i-1}, f_k(t_1, \ldots, t_k), s_{i+1}, \ldots, s_n)$ by $f_{n-1+k}(s_1, \ldots, s_{i-1}, t_1, \ldots, t_k, s_{i+1}, \ldots, s_n)$.

For example if $F = \{g, \oplus, \otimes, a, b, c\}$, $H = \{\oplus, \otimes\}$ and

$$t_0 = \oplus(a, \otimes(\oplus(g(a,b), \oplus(b,c)), \otimes(g(b,a), \otimes(a,c)))),$$

then

$$\widehat{t_0} = \oplus_2(a, \otimes_4(\oplus_3(g(a,b), b, c), g(b,a), a, c)).$$

The term $\widehat{t}$ is uniquely defined and any two terms t and t' such that $\widehat{t} = \widehat{t'}$ are $\equiv_{AC(H)}$-equivalent, but the converse does not hold because this construction does not take into account the commutativity of the operations of H. We have $t \equiv_{AC(H)} t'$ if and only if $\widehat{t}$ and $\widehat{t'}$ are equivalent by the congruence that permutes the arguments of the operations of H. In the above example, t_0 is $\equiv_{AC(H)}$-equivalent to any term t_1 such that

$$\widehat{t_1} = \oplus_2(\otimes_4(a, c, g(b,a), \oplus_3(b, c, g(a,b))), a).$$

For $t \in T(F)$, we will represent the equivalence class $[t]_{\equiv_{AC(H)}}$ by a relational structure $S(t)$ obtained as follows from $\lfloor \widehat{t} \rfloor$, where $\widehat{t}$ is the flattened term corresponding to t. We let $k := \rho(F)$, we let F' be the set of all symbols that have an occurrence in $\widehat{t}$, and $k' := \rho(F')$. (Both F' and k' depend on t.) The term $\widehat{t}$ can be represented, according to Example 5.2(3), by the relational structure $\lfloor \widehat{t} \rfloor = \langle N_{\widehat{t}}, son_{\widehat{t}}, (lab_{f\widehat{t}})_{f \in F'}, (br_{i\widehat{t}})_{1 \leq i \leq k'}, rt_{\widehat{t}} \rangle$, that we transform into

$$S(t) := \langle N_{\widehat{t}}, son_{\widehat{t}}, (lab^*_{f\widehat{t}})_{f \in F}, (br^*_{i\widehat{t}})_{1 \leq i \leq k}, rt_{\widehat{t}} \rangle,$$

by letting

$$lab^*_{f\widehat{t}} := \begin{cases} lab_{f\widehat{t}} & \text{if } f \in F - H, \\ \bigcup_{f_m \in F'} lab_{f_m\widehat{t}} & \text{if } f \in H, \end{cases}$$

$$br^*_{i\widehat{t}} := \{u \in br_{i\widehat{t}} \mid \text{ the label of the father of } u \text{ in } \widehat{t} \text{ is in } F - H\}.$$

It is clear that $S(t) \in STR(\mathcal{R}_F)$ because the unary relations br_i for $i > k$ have been eliminated and the subscripts of the symbols in H have been dropped.

Example 5.86 We let $F = \{g, \oplus, \otimes, a, b, c\}$ and $H = \{\oplus, \otimes\}$, as before. The two flattened terms

$$\widehat{t} = g(\oplus_3(a, g(b,c), a), \otimes_3(\oplus_2(c,a), a, b))$$

and

$$\widehat{t'} = g(\oplus_3(a, a, g(b,c)), \otimes_3(a, b, \oplus_2(a,c)))$$

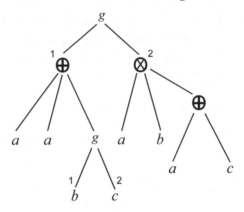

Figure 5.8 A structure $S(t)$.

represent the same element s of $T(F)/\equiv_{AC(H)}$. The structure $\lfloor s \rfloor = S(t) = S(t')$ is shown in Figure 5.8. $\qquad\square$

Lemma 5.87 For any two terms $t, t' \in T(F)$, we have $t \equiv_{AC(H)} t'$ if and only if $S(t)$ and $S(t')$ are isomorphic.

Proof sketch: The effect of the transformation of $\lfloor \hat{t} \rfloor$ into $S(t)$ is to forget the order of arguments of the function symbols that belong to H. The result follows by routine arguments. $\qquad\blacksquare$

For $s = [t]_{\equiv_{AC(H)}}$ belonging to $T(F)/\equiv_{AC(H)}$, we define $\lfloor s \rfloor$ as $S(t)$. This structure is well defined by the previous lemma. It makes now sense to say that a subset of $T(F)/\equiv_{AC(H)}$ is CMS-definable, or recognizable with respect to an algebra of relational structures, because we can identify an element s of $T(F)/\equiv_{AC(H)}$ and its representing structure $\lfloor s \rfloor$.

Theorem 5.88 Let (F, H) be an AC-signature. The following properties of a subset of $T(F)/\equiv_{AC(H)}$ are equivalent:
(1) it is recognizable in $\mathbb{STR}$;
(2) it is recognizable in $\mathbb{T}_{AC}(F, H)$;
(3) it is CMS-definable.

Proof: (1) $\Longrightarrow$ (2). Each operation of $\mathbb{T} := \mathbb{T}_{AC}(F, H)$ is, via the identification of $s \in T \ (= T(F)/\equiv_{AC(H)})$ and $\lfloor s \rfloor \in STR(\mathcal{R}_F)$, a derived operation of $\mathbb{STR}$. For $f \in F - H$, we have that $f_{\mathbb{T}}$ is f^*, as defined in Application 5.61 (with some obvious changes, because we use a variant representation of terms by structures). For $f \in H$ and $s, s' \in T$, we have that $\lfloor f_{\mathbb{T}}(s, s') \rfloor = ext_f(\lfloor s \rfloor) \mathbin{/\!/} ext_f(\lfloor s' \rfloor)$, where $\mathbin{/\!/}$ is parallel-composition of structures (cf. Example 5.55) and ext_f is the QF operation such that $ext_f(S) = S$ if $S \models lab_f(rt)$ and $ext_f(S) = ren_{f_1 \to f}(f_1^*(S))$ otherwise. This gives the result, as in the proof of Corollaries 5.66 and 5.67 concerning words and terms.

The implication (3) $\Longrightarrow$ (1) follows from the Recognizability Theorem.

It remains to prove (2) $\Longrightarrow$ (3). We extend the proof of (3) $\Longrightarrow$ (4) of Theorem 5.83. Let $L \subseteq T$ be $\mathbb{T}$-recognizable and $h : \mathbb{T} \to \mathbb{Q}$ be a surjective homomorphism to a finite F-algebra $\mathbb{Q}$ such that $L = h^{-1}(Q_0)$ for some $Q_0 \subseteq Q$. Since $f_\mathbb{T}$ is associative and commutative for every $f \in H$, the surjectivity of h ensures that $f_\mathbb{Q}$ is so. Hence $\mathbb{Q}$ is an (F,H)-AC-algebra. We enumerate Q as $\{q_1,\ldots,q_k\}$ and, for every $f \in H$ and $q_{i_1}, q_{i_2}, \ldots, q_{i_n} \in Q$, $n \geq 3$, we extend $f_\mathbb{Q}$ by defining recursively

$$f_\mathbb{Q}(q_{i_1}, q_{i_2}, \ldots, q_{i_n}) := f_\mathbb{Q}(q_{i_1}, f_\mathbb{Q}(q_{i_2}, \ldots, q_{i_n})).$$

We denote by $\mathcal{N}^k_{\geq 2}$ the set of k-tuples $(m_1,\ldots,m_k)$ in $\mathcal{N}^k$ such that $m_1 + \cdots + m_k \geq 2$, and for each $f \in H$, we define the mapping $\ell_f : \mathcal{N}^k_{\geq 2} \to Q$ by

$$\ell_f(m_1,\ldots,m_k) := f_\mathbb{Q}(q_1,\ldots,q_1,q_2,\ldots,q_2,\ldots,q_k,\ldots,q_k),$$

where, for each $i \in [k]$, the sequence of arguments of $f_\mathbb{Q}$ has m_i occurrences of q_i.

For $t \in T(F)$, we let $[t]$ denote $[t]_{\equiv_{AC(H)}}$. If $s = [t] \in T$ and $\widehat{t} = f_n(\widehat{t_1}, \ldots, \widehat{t_n})$ with $f \in H$ (which implies that f is the leading symbol of t but not of any of $t_1,\ldots,t_n$), we define $m(s) := (m_1,\ldots,m_k)$, where $m_j := |\{i \in [n] \mid h([t_i]) = q_j\}|$ for each $j = 1,\ldots,k$. Then we have $h(s) = \ell_f(m(s))$, by the definitions and by the associativity and commutativity of $f_\mathbb{Q}$.

Our objective is to prove that each set $\ell_f^{-1}(q_i)$ is recognizable in $\mathbb{N}^k$. We let $\sim$ be the equivalence relation on $\mathcal{N}^k$ defined as follows: $m \sim m'$ if and only if either $m, m' \in \mathcal{N}^k_{\geq 2}$ and $\ell_f(m) = \ell_f(m')$, or m and m' are both not in $\mathcal{N}^k_{\geq 2}$ and they are equal. This equivalence relation has at most $2k+1$ classes and saturates each set $\ell_f^{-1}(q_i)$. We now prove that it is a congruence on $\mathbb{N}^k$. If $m, m', r, r' \in \mathcal{N}^k$ and $m \sim m'$ and $r \sim r'$, we must prove that $m + r \sim m' + r'$. Assume first that m, m', r, r' belong to $\mathcal{N}^k_{\geq 2}$. Then so do $m + r$ and $m' + r'$. Since $f_\mathbb{Q}$ is associative and commutative, we have $\ell_f(m + r) = f_\mathbb{Q}(\ell_f(m), \ell_f(r))$ and similarly for m' and r'. But $\ell_f(m) = \ell_f(m')$ and $\ell_f(r) = \ell_f(r')$, hence we get $f_\mathbb{Q}(\ell_f(m), \ell_f(r)) = f_\mathbb{Q}(\ell_f(m'), \ell_f(r'))$ and $m + r \sim m' + r'$. For the other cases, the verifications are similar. Each set $L_i := \ell_f^{-1}(q_i)$ is thus recognizable in $\mathbb{N}^k$, hence each property $Card_{L_i}$ is CMS-expressible by Proposition 5.25.

As in the proof of Theorem 5.83, we construct a CMS sentence φ that defines $L = \{s \in T \mid h(s) \in Q_0\}$. The construction combines those of Theorems 5.15 and 5.83. Let $s = [t]$ be represented by the $\mathcal{R}_F$-structure $S(t)$ whose domain is the set $N_{\widehat{t}}$ of nodes of the flattened term $\widehat{t}$. The sentence φ is constructed as $\exists X_1,\ldots,X_k.\theta$, where θ expresses that $(X_1,\ldots,X_k)$ is a partition of $N_{\widehat{t}}$, that the root belongs to some X_i such that $q_i \in Q_0$ and that for every $u \in N_{\widehat{t}}, u \in X_i$ if and only if $h([t/u']) = q_i$, where $u' \in N_t$ is such that $\widehat{t/u'} = \widehat{t}/u$.

The latter condition is expressed by the following equivalent one:

for every u, if u is an occurrence of f with sons $u_1, \ldots, u_n$ and if $u_j \in X_{i_j}$ for each j, then $u \in X_i$ where i is such that $q_i = f_{\mathbb{Q}}(q_{i_1}, \ldots, q_{i_n})$.

For the expression of the last equality, there are two cases:

(a) either $f \in F - H, n = \rho(f)$, and the condition is written in a straightforward way, as in Theorem 5.15;

(b) or $f \in H$, and in this case, letting $(m_1, \ldots, m_k)$ count the numbers of states of the set $\{q_{i_1}, \ldots, q_{i_n}\}$ equal to $q_1, \ldots, q_k$ respectively, we need only check that $(m_1, \ldots, m_k) \in \ell_f^{-1}(q_i)$, and we do this with the help of Proposition 5.25 as in the proof that (3) implies (4) in Theorem 5.83.

Note that in Case (a), u has an ordered sequence of sons, and a son is the i-th element of this sequence if it belongs to br_{it}^*. In Case (b), it has a set (not a sequence) of at least two sons, of unbounded size.   ∎

Remark 5.89   (1) The proof of (2) $\Longrightarrow$ (3) can be seen as the construction of a finite deterministic automaton that runs bottom-up on the syntactic trees of flattened terms. Some nodes of these trees (the occurrences of the symbols in $F - H$) have a sequence of sons of bounded length and the others have a set of sons of unbounded (finite) cardinality (cf. Remark 5.84(1)).

(2) As in Theorem 5.83, we have the additional equivalent condition of recognizability in $\mathbb{JS}^t$: It is implied by (3). However, it is not straightforward to prove that for a subset of $T(F)/\equiv_{AC(H)}$ (viewed as a set of labeled s-graphs such as the one in Figure 5.8), its recognizability in $\mathbb{JS}^t$ implies its recognizability in $\mathbb{T}$, because $\mathbb{T}$ is not a subalgebra of a derived algebra of $\mathbb{JS}^t$: if $f \in H$, the operation $(s, s) \mapsto f_{\mathbb{T}}(s, s')$ has four cases, depending on whether the roots of s and s' are labeled f, so that it is not a derived operation. However, more complex derived operations involving conditional statements are defined in [BluCou06] (they are called *finite state derived operations*) that can handle these types of situations. They "preserve recognizability" as do our derived operations (cf. Proposition 3.56(3)) and one can prove by using them that if $L \subseteq T(F)/\equiv_{AC(H)}$, then $L \in \mathbf{Rec}(\mathbb{JS}^t)$ implies $L \in \mathbf{Rec}(\mathbb{T})$.

5.6 Equivalences of logical formulas

In this section, we present some equivalences and transformations of logical formulas from which we can make precise in which sense the sets of first-order and of monadic second-order formulas over a finite signature, of quantifier-height at most some given integer and having free variables in a finite set, can be considered as finite. From these notions, we will bound the indices of the congruences constructed for proving the Recognizability Theorem. However, we will obtain better upper-bounds in the next chapter.

5.6.1 Boolean formulas

We let $p_1,\ldots,p_n$ be (standard) Boolean variables and B_n be the set of terms $T(\{\wedge,\vee,\neg,\Rightarrow,\Leftrightarrow,True,False\},\{p_1,\ldots,p_n\})$, called *Boolean terms*. We write $b \equiv b'$ if b and b' in B_n are *equivalent*, i.e., if they define the same Boolean function.

It is well known that b and b' are equivalent if and only if b can be transformed into b' by the laws of Boolean calculus like $\neg\neg p \equiv p$, $p \vee p \equiv p$, $p \Rightarrow q \equiv \neg p \vee q$, $p \wedge (q \vee r) \equiv (p \wedge q) \vee (p \wedge r)$ and $p \vee \neg p \equiv True$, to take a few examples. We assume that a strict linear order $<$ on $\bigcup_{n \geq 0} B_n$ is fixed. (It can be defined as a lexicographic ordering on formulas considered as words; we need not specify it precisely.)

For each n and each $b \in B_n$, there is a unique formula $\widetilde{b}$ in B_n that is equivalent to b and is in disjunctive normal form in such a way that:

- disjuncts are ordered by increasing order with respect to $<$;
- in each disjunct, each variable from $\{p_1,\ldots,p_n\}$ occurs exactly once and variables occur in increasing order of indices.

Let us take some examples. If $b = \neg\neg p_1$ and $n = 1$, then $\widetilde{b} = p_1$, but if $n = 2$, we have $\widetilde{b} = (p_1 \wedge p_2) \vee (p_1 \wedge \neg p_2)$. If $n = 3$ and $b = p_1 \wedge (p_2 \vee \neg p_3)$, then $\widetilde{b} = (p_1 \wedge p_2 \wedge p_3) \vee (p_1 \wedge p_2 \wedge \neg p_3) \vee (p_1 \wedge \neg p_2 \wedge \neg p_3)$.

Let us now assume that $L \subseteq \mathrm{CMS}(\mathcal{R},\mathcal{X})$ is a set of formulas whose outermost symbol is not in $\{\wedge,\vee,\neg,\Rightarrow,\Leftrightarrow,True,False\}$. We denote by $B_n[L]$ the set of formulas of the form $b(\varphi_1,\ldots,\varphi_n)$ (which abbreviates $b[\varphi_1/p_1,\ldots,\varphi_n/p_n]$), where $b \in B_n$, $\varphi_1,\ldots,\varphi_n \in L$ and for each i, φ_i is substituted for p_i in b. It is clear that $B_n[L] \subseteq C_r\mathrm{MS}^h(\mathcal{R},\mathcal{X})$ if $L \subseteq C_r\mathrm{MS}^h(\mathcal{R},\mathcal{X})$. We also assume that $\mathrm{CMS}(\mathcal{R},\mathcal{X})$ is linearly ordered by some strict ordering $<$.

Definition 5.90 (Boolean normalization) If $\varphi = b(\varphi_1,\ldots,\varphi_n) \in B_n[L]$, then we denote by $\widetilde{\varphi}$ the formula in $B_k[L]$ constructed as follows, for some appropriate $k \leq n$:
- we enumerate $\{\varphi_1,\ldots,\varphi_n\}$ as $\{\varphi_{i_1},\ldots,\varphi_{i_k}\}$ with $\varphi_{i_1} < \cdots < \varphi_{i_k}$ and $k \leq n$ (we have $k < n$ if $\varphi_i = \varphi_j$ for some i and $j \neq i$);
- we write accordingly φ as $b'(\varphi_{i_1},\ldots,\varphi_{i_k})$ for some $b' \in B_k$;
- we let $\widetilde{\varphi}$ be $\widetilde{b'}(\varphi_{i_1},\ldots,\varphi_{i_k})$.

Note that $\widetilde{\varphi} = \varphi$ if $\varphi = \widetilde{\psi}$ for some ψ. We call $\widetilde{\varphi}$ a *normalized Boolean combination of formulas*.

We let $B[L] := \bigcup_{n \geq 0} B_n[L]$ and $\widetilde{B}[L] := \{\widetilde{\varphi} \mid \varphi \in B[L]\}$. We let $\beta(m)$ be the number $\Sigma_{0 \leq p \leq m} 2^{2^p} \cdot \binom{m}{p}$. It will be enough to note that $2^{2^m} < \beta(m) < 2^{2^{m+1}}$.

With these hypotheses and notation we have:

Proposition 5.91
(1) If $\varphi \in B[L]$, then $\widetilde{\varphi} \equiv \varphi$ and $\widetilde{\varphi}$ is obtained from φ by applying the laws of Boolean calculus.
(2) If $|L| = m \in \mathcal{N}$, then $|\widetilde{B}[L]| \leq \beta(m)$ and $\widetilde{B}[L]$ can be computed from L.

Proof: (1) Clear from the definition and the classical results of Boolean calculus. Note that the same formulas of L appear as subformulas of φ and of $\widetilde{\varphi}$.

(2) For each finite set $\{\varphi_1, \ldots, \varphi_p\}$ of pairwise distinct formulas, one can build 2^{2^p} formulas in $\widetilde{B}[\{\varphi_1, \ldots, \varphi_p\}]$ containing each formula φ_i as a subformula; they are the formulas $\widetilde{b}(\varphi_1, \ldots, \varphi_p)$ for all b in B_p. Hence:

$$|\widetilde{B}[L]| = \Sigma_{K \subseteq L} 2^{2^{|K|}} = \beta(|L|),$$

by the definition of β. It is clear that one can (in principle, not practically) compute the finite set $\widetilde{B}[L]$ if the finite set L is given. ∎

5.6.2 Monadic second-order formulas

In this section, $\mathcal{X}$ will always denote a finite subset of the countable set $\mathcal{V}_1$ of first-order and set variables (cf. Section 5.1.4). Furthermore, r and h denote elements of $\mathcal{N}$, and $\mathcal{R}$ is a fixed relational signature.

The semantic equivalence $\equiv$ on formulas of $\mathrm{CMS}(\mathcal{R}, \mathcal{X})$ (defined at the beginning of Section 5.2) is undecidable.[71] Our objective is to define a syntactic equivalence $\approx$ that refines $\equiv$ (i.e., such that $\varphi \approx \psi$ implies $\varphi \equiv \psi$), is decidable and is finite over each set $C_r \mathrm{MS}^h(\mathcal{R}, \mathcal{X})$.

We let $\sim$ be the equivalence relation on $\mathrm{CMS}(\mathcal{R}, \mathcal{X})$ generated by the following elementary transformation rules and their inverses (where u denotes a first-order or a set variable):

(1) application of a law of Boolean calculus like the replacement of $\varphi \vee \varphi$ by φ or of $\varphi \wedge (\psi \vee \psi')$ by $(\varphi \wedge \psi) \vee (\varphi \wedge \psi')$;

(2) replacement of $\forall u. \varphi$ by $\neg \exists u. \neg \varphi$;

(3) renaming of bound variables, e.g., replacement of $\exists u. \varphi$ by $\exists w. \varphi[w/u]$ if w has no free occurrence in φ.

If ψ is obtained from φ by rule (3) only, we write $\varphi \equiv_\alpha \psi$ and say that φ and ψ are *α-equivalent*. It is clear that $\varphi \sim \psi$ implies $\varphi \equiv \psi$. Moreover, if $\varphi \sim \psi$ and $\varphi \in C_r \mathrm{MS}^h(\mathcal{R}, \mathcal{X})$ then $\psi \in C_r \mathrm{MS}^h(\mathcal{R}, \mathcal{X})$. For example, let φ be the sentence

$$\forall X \neg \exists y (y \in X \wedge (\exists z. R(y, z) \vee \exists z'. R(y, z'))),$$

and ψ be the sentence

$$\neg \exists X, y (y \in X \wedge \exists z. R(y, z)).$$

---

[71] A consequence of Theorem 5.5 and the other classical undecidability results. The equivalences of formulas with respect to finite and infinite, and to finite and countable structures instead of with respect to finite ones (which is our definition of equivalence), are also undecidable.

We have $\varphi \sim \psi$. On the other hand, $\exists y(y \neq y \vee R(x,y))$ is not $\sim$-equivalent to $\exists y. R(x,y)$ because, although $y \neq y \vee R(x,y)$ is $\equiv$-equivalent to $R(x,y)$, these formulas are not equivalent with respect to $\sim$.

We now define a transformation of formulas, such that if φ is transformed into $\widehat{\varphi}$, then $\widehat{\varphi} \sim \varphi$ and thus $\widehat{\varphi} \equiv \varphi$. We assume that a linear order is fixed on the set $\mathcal{V}_1$ of first-order and set variables.

Definition 5.92 (Normalization of formulas) With each $\varphi \in \mathrm{CMS}(\mathcal{R},\mathcal{X})$, we associate a formula $\widehat{\varphi} \in \mathrm{CMS}(\mathcal{R},\mathcal{X})$ by the following algorithm:

Step 1: We transform φ into φ' written without universal quantifiers by using rule (2) of the definition of $\sim$. Hence, we obtain φ' in $\mathrm{CMS}(\mathcal{R},\mathcal{X})$ that is $\sim$-equivalent to φ. In what follows we still denote φ' by φ for more readability.

Step 2: For every finite subset $\mathcal{Y}$ of $\mathcal{V}_1$ that is disjoint with $\mathcal{X}$ and contains at least $qh(\varphi)$ first-order variables and at least $qh(\varphi)$ set variables, we transform φ into $\widehat{\varphi}\mathcal{Y}$, by induction on $|\varphi|$ (the size of φ, see Section 5.1.3):

Case 1: If φ is atomic, we let $\widehat{\varphi}\mathcal{Y}$ be φ.

Case 2: If φ is $b(\varphi_1,\ldots,\varphi_n)$, where $b \in B_n$, $b \notin \{p_1,\ldots,p_n\}$ and each formula φ_i is either atomic or existential (i.e., of the form $\exists u. \psi$), then, abbreviating $b(\widehat{\varphi}_1\mathcal{Y},\ldots,\widehat{\varphi}_n\mathcal{Y})$ by β, we let $\widehat{\varphi}\mathcal{Y}$ be $\widetilde{\beta}$ by using the Boolean normalization of Definition 5.90.

Case 3: If φ is $\exists u. \psi$, where u is a first-order variable and $\psi \in C_r\mathrm{MS}^{h-1}(\mathcal{R},\mathcal{X}\cup\{u\})$, we define $\widehat{\varphi}\mathcal{Y}$ as $\exists y_1. \widehat{\psi[y_1/u]}_{\mathcal{Y}-\{y_1\}}$, where y_1 is the first first-order variable in $\mathcal{Y}$ (with respect to the order on $\mathcal{V}_1$).

Case 4: If φ is $\exists U. \psi$, where U is a set variable and $\psi \in C_r\mathrm{MS}^{h-1}(\mathcal{R},\mathcal{X}\cup\{U\})$, we define similarly $\widehat{\varphi}\mathcal{Y}$ as $\exists Y_1. \widehat{\psi[Y_1/U]}_{\mathcal{Y}-\{Y_1\}}$, where Y_1 is the first set variable in $\mathcal{Y}$.

It can easily be shown, simultaneously with the above definition, that if $\varphi \equiv_\alpha \varphi'$, then $\widehat{\varphi}\mathcal{Y} = \widehat{\varphi'}\mathcal{Y}$.[72] That implies that $\widehat{\varphi}\mathcal{Y}$ is well defined in Cases 3 and 4 (note that the substitutions $\psi[y_1/u]$ and $\psi[Y_1/U]$ are only defined modulo α-equivalence).

Step 3: We define $\widehat{\varphi}$ as $\widehat{\varphi}\mathcal{Y}$, where $\mathcal{Y}$ consists of the first $qh(\varphi)$ first-order variables that are not in $\mathcal{X}$ and the first $qh(\varphi)$ set variables that are not in $\mathcal{X}$.

Let us take an example. Let φ be the sentence of quantifier-height 3 expressing that an undirected graph is not connected:

$$\exists X\big(\exists x. x \in X \ \wedge \ \exists y. y \notin X \ \wedge \ \forall u,v(u \in X \wedge edg(u,v) \Rightarrow v \in X)\big).$$

After eliminating $\forall$ and $\Rightarrow$ and using Boolean laws, we obtain the equivalent sentence

$$\exists X\big(\exists x. x \in X \ \wedge \ \exists y. y \notin X \ \wedge \ \neg\exists u,v(u \in X \wedge edg(u,v) \wedge v \notin X)\big),$$

---

[72] For instance, in Case 3, if $\varphi \equiv_\alpha \varphi'$ then φ' is $\exists u'. \psi'$ with $\psi \equiv_\alpha \psi'[u/u']$. Hence $\psi[y_1/u] \equiv_\alpha \psi'[y_1/u']$ and hence $\widehat{\varphi}\mathcal{Y} = \widehat{\varphi'}\mathcal{Y}$ by induction.

and after renaming its bound variables, we get the sentence $\widehat{\varphi}$ as follows:

$$\exists Y_1 \big(\exists y_1 . y_1 \in Y_1 \ \wedge \ \exists y_1 . y_1 \notin Y_1 \ \wedge \ \neg \exists y_1 , y_2 (y_1 \in Y_1 \wedge edg(y_1, y_2) \wedge y_2 \notin Y_1) \big),$$

where y_1, y_2 are the first two first-order variables of $\mathcal{V}_1$ and Y_1 is its first set variable.

Proposition 5.93 For every $\varphi \in C_r \mathrm{MS}^h(\mathcal{R}, \mathcal{X})$, we have:

(1) $\widehat{\varphi} \in C_r \mathrm{MS}^h(\mathcal{R}, \mathcal{X})$ and the set of bound variables of $\widehat{\varphi}$ is $\{y_i \mid 1 \leq i \leq h_1\} \cup \{Y_i \mid 1 \leq i \leq h_2\}$ for some integers $h_1, h_2 \in [0, qh(\varphi)]$, where y_i is the i-th first-order variable of $\mathcal{V}_1 - \mathcal{X}$ and Y_i its i-th set variable;

(2) $\widehat{\varphi} \sim \varphi$ and thus, $\widehat{\varphi} \equiv \varphi$;

(3) the mapping $\varphi \mapsto \widehat{\varphi}$ is idempotent, i.e., if $\varphi = \widehat{\psi}$ for some ψ, then $\widehat{\varphi} = \varphi$.

Proof: We prove the same assertions for each $\widehat{\varphi}_{\mathcal{Y}}$ (where, in Assertion (1), $\mathcal{V}_1 - \mathcal{X}$ is replaced by $\mathcal{Y}$), by induction on $|\varphi|$. We can assume that φ is written without universal quantifiers by the first step of the algorithm of Definition 5.92. In Case 1, φ is quantifier-free and the result is obvious. In Case 2, the result follows by induction and Proposition 5.91(1). In Case 3, $\widehat{\psi[y_1/u]}_{\mathcal{Y} - \{y_1\}} \sim \psi[y_1/u]$ by induction. The variable y_1 does not occur free in φ, because $\mathcal{Y} \cap \mathcal{X} = \emptyset$. Hence $\widehat{\varphi}_{\mathcal{Y}}$ is $\sim$-equivalent to $\exists y_1 . \widehat{\psi[y_1/u]}$ (by the definition of $\sim$) and to φ by renaming bound variables. If $\varphi = \widehat{\beta}_{\mathcal{Y}}$ for some formula β, then $u = y_1$ and $\psi = \widehat{\theta}_{\mathcal{Y} - \{y_1\}}$ for some θ, and so $\widehat{\varphi}_{\mathcal{Y}}$ is $\exists y_1 . \widehat{\psi}_{\mathcal{Y} - \{y_1\}}$ by definition, which equals φ by induction. The proof is similar for Case 4. ∎

For each $r, h, \mathcal{R}, \mathcal{X}$, we define

$$\widehat{C_r \mathrm{MS}}^h(\mathcal{R}, \mathcal{X}) := \{\widehat{\varphi} \mid \varphi \in C_r \mathrm{MS}^h(\mathcal{R}, \mathcal{X})\}.$$

We let $f(r, \mathcal{R}, n)$ be a computable upper-bound to the number of atomic formulas in $C_r \mathrm{MS}(\mathcal{R}, \mathcal{X})$ for any $\mathcal{X}$ with at most n variables. Without loss of generality, we can assume that f is monotone in n, i.e., that $f(r, \mathcal{R}, n) \leq f(r, \mathcal{R}, n+1)$. We define $g(h, r, \mathcal{R}, n) \in \mathcal{N}$ by the following induction:

$$g(0, r, \mathcal{R}, n) = \beta(f(r, \mathcal{R}, n)),$$

$$g(h+1, r, \mathcal{R}, n) = \beta(3 \cdot g(h, r, \mathcal{R}, n+1)),$$

where β is the function of Definition 5.90; the function g is thus monotone in n. We recall that $2^{2^m} < \beta(m) < 2^{2^{m+1}}$. An easy calculation (by using an induction on h) yields:[73]

$$\exp(2h+2, f(r, \mathcal{R}, n+h)) < g(h, r, \mathcal{R}, n) < \exp(2h+2, f(r, \mathcal{R}, n+h) + h + 1).$$

---

[73] The function $\exp : \mathcal{N}^2 \to \mathcal{N}$ is defined by $\exp(0, n) = n$ and $\exp(d+1, n) = 2^{\exp(d, n)}$.

Proposition 5.94 For each $r, h, \mathcal{R}, \mathcal{X}$, we have

$$|\widehat{C_r MS^h}(\mathcal{R}, \mathcal{X})| \leq g(h, r, \mathcal{R}, |\mathcal{X}|),$$

and the finite set $\widehat{C_r MS^h}(\mathcal{R}, \mathcal{X})$ can be computed from $r, h, \mathcal{R}, \mathcal{X}$.

Proof: We prove this by induction on h for every set $\widehat{C_r MS^h}y(\mathcal{R}, \mathcal{X}) := \{\widehat{\varphi}y \mid \varphi \in C_r MS^h(\mathcal{R}, \mathcal{X})\}$.

If $h = 0$, then the formulas of $C_r MS^0(\mathcal{R}, \mathcal{X})$ are Boolean combinations of atomic formulas. Hence $\widehat{C_r MS^0}y(\mathcal{R}, \mathcal{X}) = \widetilde{B}[L]$, where L is the set of atomic formulas in $C_r MS(\mathcal{R}, \mathcal{X})$. By Proposition 5.91(2), this set can be computed and has cardinality at most $\beta(f(r, \mathcal{R}, |\mathcal{X}|)) = g(0, r, \mathcal{R}, |\mathcal{X}|)$ by the definitions of f and g.

If $h \geq 0$, then a formula φ of $C_r MS^{h+1}(\mathcal{R}, \mathcal{X})$ is a Boolean combination of (atomic or existential) formulas in $C_r MS^h(\mathcal{R}, \mathcal{X})$ and of formulas of the form $\exists u . \psi$, where $\psi \in C_r MS^h(\mathcal{R}, \mathcal{X} \cup \{u\})$ and u is a first-order or a set variable. Hence, $\widehat{\varphi}y$ is a normalized Boolean combination θ of (atomic or existential) formulas in $\widehat{C_r MS^h}y(\mathcal{R}, \mathcal{X})$ and of formulas of the form $\exists y_1 . \psi'$ or $\exists Y_1 . \psi''$, where ψ' (resp. ψ'') is in $\widehat{C_r MS^h}y_{-\{y_1\}}(\mathcal{R}, \mathcal{X} \cup \{y_1\})$ (resp. $\widehat{C_r MS^h}y_{-\{Y_1\}}(\mathcal{R}, \mathcal{X} \cup \{Y_1\})$). Vice versa, every such formula θ equals $\widehat{\varphi}y$ for some φ in $C_r MS^{h+1}(\mathcal{R}, \mathcal{X})$ (if θ is $\exists y_1 . \psi'$ as above, then $\widehat{\theta}y = \exists y_1 . \widehat{\psi'}y_{-\{y_1\}} = \exists y_1 . \psi'$ by idempotency, see Proposition 5.93(3), and so $\theta = \widehat{\theta}y$). By Proposition 5.91(2), this set of formulas θ can be computed and has cardinality at most $g(h, r, \mathcal{R}, |\mathcal{X}|) + 2 \cdot g(h, r, \mathcal{R}, |\mathcal{X}| + 1) \leq 3 \cdot g(h, r, \mathcal{R}, |\mathcal{X}| + 1)$ (because g is monotone in its last argument), which gives $|\widehat{C_r MS^{h+1}}y(\mathcal{R}, \mathcal{X})| \leq g(h + 1, r, \mathcal{R}, |\mathcal{X}|)$. ∎

Corollary 5.95 For every $\mathcal{R}$ and $\mathcal{R}'$, there are finitely many QF operations of type $\mathcal{R} \to \mathcal{R}'$. A finite set $DS_{\mathcal{R}, \mathcal{R}'}$ of QFO definition schemes can be computed such that $\{\widehat{\mathcal{D}} \mid \mathcal{D} \in DS_{\mathcal{R}, \mathcal{R}'}\}$ is the set of all QF operations of type $\mathcal{R} \to \mathcal{R}'$.

Proof: A definition scheme for such an operation consists of $p := 1 + |\mathcal{R}'_+| + |\mathcal{R}_0| \times |\mathcal{R}'_0|$ formulas with at most $\rho(\mathcal{R}')$ free variables, all quantifier-free ones, i.e., formulas in $C_0 MS^0(\mathcal{R}, X_{\rho(\mathcal{R}')})$. There are thus at most $g(0, 0, \mathcal{R}, \rho(\mathcal{R}'))^p = \beta(f(0, \mathcal{R}, \rho(\mathcal{R}')))^p$ inequivalent definition schemes. The definition schemes that consist of formulas in $\widehat{C_0 MS^0}(\mathcal{R}, X_{\rho(\mathcal{R}')})$ form the required set $DS_{\mathcal{R}, \mathcal{R}'}$. ∎

We now evaluate $f(r, \mathcal{R}, n)$ in some concrete cases. The atomic formulas written with a finite set $\mathcal{X}$ of first-order and set variables, a relational signature $\mathcal{R}$ and cardinality predicates $Card_{p,q}(X)$ for $p < q \leq r$ are of the following forms:

> *True, False,* $s = t$, $t \in X$, $Card_{p,q}(X)$ and $R(t_1, \ldots, t_{\rho(R)})$
>
> for $s, t, t_1, \ldots, t_{\rho(R)} \in \mathcal{X} \cup \mathcal{R}_0$, $R \in \mathcal{R}_+$ and $X \in \mathcal{X}$.

Hence we can take as bounding function:

$$f(r, \mathcal{R}, n) = 2 + (n+c)^2 + n(n+c) + nr^2 + \sum_{R \in \mathcal{R}_+} (n+c)^{\rho(R)},$$

where $n = |\mathcal{X}|$ and $c = |\mathcal{R}_0|$. (A more precise estimate could be obtained by counting separately first-order and set variables, by considering $s = t$ as identical to $t = s$, etc., but would not be very useful because of the predominant influence of the quantifier-height in Proposition 5.94.)

For handling directed s-graphs G of type C and represented by $\lceil G \rceil$, we can use the relational signature $\mathcal{R} = \mathcal{R}^d_{\text{ms},C}$ consisting of the two binary relation symbols in_1 and in_2 and $k := |C|$ constant symbols. We get $f(r, \mathcal{R}, n) = 3(n+k)^2 + n(n+k) + nr^2 + 2$. If we use the alternative signature $\{in, lab_{Edge}\}$, where we replace one binary relation by a unary one, we obtain the slightly smaller function $f(r, \mathcal{R}, n) = 2(n+k)^2 + n(n+k) + nr^2 + n + k + 2$.

For handling p-graphs with k port labels we use a signature $\mathcal{R}$ consisting of one binary relation symbol and k unary ones. Thus, $f(r, \mathcal{R}, n) = 3n^2 + nr^2 + nk + 2$. From this observation and the computation in the proof of Corollary 5.95, we get that there are at most $\beta(2k + 14)^{m+1}$ QF operations transforming p-graphs of type C of cardinality k into p-graphs of type D of cardinality m.

5.6.3 Numbers of formulas and recognizability indices

Theorem 5.64 has established that for every sentence φ in $\text{CMS}(\mathcal{R}, \emptyset)$ the class of structures $\text{MOD}(\varphi) \subseteq STR(\mathcal{R})$ is recognizable in $\mathbb{STR}_{\text{pres}}$. The following theorem bounds its recognizability index (Section 3.4.3). We will use the function f that bounds the number of atomic formulas (cf. Proposition 5.94).

Theorem 5.96 Let $\mathcal{R}$ be a relational signature and $\varphi \in C_r\text{MS}(\mathcal{R}, \emptyset)$ for some $r \in \mathcal{N}$. The recognizability index of $\text{MOD}(\varphi)$ with respect to the algebra $\mathbb{STR}_{\text{pres}}$ is bounded by the function that maps each sort $\mathcal{R}'$ to

$$\exp(2h + 3, f(r, \mathcal{R}', h) + h + 1),$$

where $h := qh(\varphi)$.

Proof: By Proposition 5.93 and the proof of Theorem 5.64, the sort-preserving equivalence relation $\approx$ on the algebra $\mathbb{STR}_{\text{pres}}$ defined by $S \approx_{\mathcal{R}'} T$ if and only if for every sentence ψ in $\widetilde{C_r\text{MS}^h}(\mathcal{R}', \emptyset)$, we have

$$S \models \psi \text{ if and only if } T \models \psi,$$

is a congruence on $\mathbb{STR}_{pres}$ that saturates $MOD(\varphi)$. It has thus at most 2^N classes of sort $\mathcal{R}'$, where $N := |\widehat{C_r MS^h}(\mathcal{R}', \emptyset)|$. The result follows then from Proposition 5.94. ∎

Let us make this more concrete by bounding the recognizability indices relative to the HR and the VR algebras that have been defined in Chapter 4 (cf. Definitions 4.33 and 4.52). Since the many-sorted VR and HR algebras are subalgebras of derived algebras of $\mathbb{STR}$, the recognizability index of a homogenous set of p-graphs or s-graphs is at most that of the corresponding set of structures, for all sorts of those subalgebras (cf. the proofs of Propositions 3.56 and 3.64). The upper-bound on the recognizability index of a class $MOD(\varphi)$ we have given depends on the quantifier-height of φ by an iterated exponentiation. Similar lower-bounds are known, see [StoMey] and [Wey]. However, the recognizability index γ of a homogenous CMS- or CMS$_2$-definable set of p-graphs or s-graphs, is a *finitely iterated exponential function*,[74] because the nesting of exponentiations depends only on the quantifier-height of the defining sentence, and the number of atomic formulas is bounded by a polynomial in the number of port or source labels.

In the following corollary, the size of a p-graph is its number of vertices, and the size of an s-graph is its number of vertices and edges.

Corollary 5.97 For every CMS-definable set of p-graphs L there exists a finitely iterated exponential function $f_L : \mathcal{N} \to \mathcal{N}$ such that, for every $k \in \mathcal{N}$, if L_k is the set of p-graphs in L of clique-width at most k then:

(1) if $L_k \neq \emptyset$, then it contains a p-graph of size at most $f_L(k)$;
(2) if L_k is finite, then every p-graph in L_k has size at most $f_L(k)$.

Similar results holds for CMS$_2$-definable sets of s-graphs and tree-width.

Proof: We first prove this for HR. Since every CMS$_2$-definable set of s-graphs is a finite union of homogenous CMS$_2$-definable sets of s-graphs, as shown in the proof of Theorem 5.68, we may assume that L is homogenous. Let E be the type of its elements, and let $TWD(\leq k - 1, = E)$ be the set of all s-graphs of type E and tree-width at most $k - 1$. By Example 4.3(8) and the proof of Proposition 4.13, there is a uniform typed HR equation system S that generates the set $TWD(\leq k - 1, = E)$, and it has one unknown of type D for each $D \subseteq C$, where C is such that $E \subseteq C$ and $|C| = k$. Thus, $TWD(\leq k - 1, = E)$ is equational in the finitely generated algebra $\mathbb{JS}^{t,gen}[C]$. By Proposition 3.56(1), $L_k = L \cap TWD(\leq k - 1, C)$ is recognizable in that algebra. By the Filtering Theorem (Theorem 3.88), $L_k = TWD(\leq k - 1, = E) \cap L$

---

[74] That is, $\gamma(k) \leq \exp(d, k)$ for every k, for some fixed d. Such functions are said to be *elementary*, which is a rather counter-intuitive terminology!

is also equational in that algebra. By the proof of Theorem 3.88, L_k is generated by a uniform typed HR equation system with $\sharp B_k$ unknowns, where B_k is a complete and deterministic F_C^{tHR}-automaton recognizing $\{t \in T(F_C^{\mathrm{tHR}}) \mid val(t) \in L_k\}$. Let $\lambda x \cdot \exp(d, x)$ be an upper-bound to the recognizability index of L in $\mathbb{JS}^{\mathrm{t}}$, which is also an upper-bound to the recognizability index of L_k in $\mathbb{JS}^{\mathrm{t,gen}}[C]$. By Propositions 3.69 and 3.73(2), $\sharp B_k$ is at most $\sum_{D \subseteq [k]} \exp(d, |D|) \le 2^k \cdot \exp(d, k) \le \exp(d+1, k)$. Hence, by Proposition 4.13(1), L_k is generated by a uniform typable HR equation system with at most $\exp(d+1, k)$ unknowns. Assertions (1) and (2) now follow from Propositions 4.5(3) and 4.19 respectively.

The proof for CMS-definable sets of p-graphs is similar, using VR-equational sets, Example 4.43(5) and Proposition 4.46, and Propositions 4.42(3) and 4.47(3).   ■

The numbers of formulas are more than "cosmological" ([StoMey]), and practical implementations are not directly possible. Better upper-bounds will be given in Chapter 6. They are based on an alternative proof of the Weak Recognizability Theorem.

The sentence expressing connectivity has quantifier-height 3. This gives, for graphs of tree-width $\le k$ an HR-recognizability index of the form $\exp(9, p(k))$ for some polynomial p, whereas, by a direct analysis, we found in Example 4.30(2) an upper-bound of $O(k!)$.

Remark 5.98 The reader may ask why we do not use *prenex* formulas, that is, formulas of the form $\forall \overline{w}_1 \, \exists \overline{w}_2 \, \forall \overline{w}_3 \cdots \theta$, where $\overline{w}_1, \overline{w}_2, \ldots$ are possibly empty sequences of first-order and set variables and θ is quantifier-free. There would be no loss of generality in doing so because every formula φ can be transformed into an equivalent prenex formula $\widehat{\varphi}$. For doing this, one uses iteratively rules like the one transforming $\exists x.\theta \wedge \exists y.\theta'$ into $\exists x, y(\theta \wedge \theta')$, where x and y are distinct variables respectively not free in θ' and in θ, together with renamings of bound variables.[75]

The size of $\widehat{\varphi}$ is at most one more than that of φ but its quantifier-height may increase in a nonlinear way. There exist formulas φ_n in $\mathrm{FO}(\{edg\}, \{x, y\})$ of size $O(n^2)$ and quantifier-height n such that $\widehat{\varphi}_n$ has quantifier-height $O(n^2)$. One takes

---

[75] Since we allow empty structures, some usual transformations of formulas fail to preserve equivalence. For example the sentence $\exists x(x = x) \vee \forall y(y = y)$ is not equivalent to the sentence $\exists x(x = x \vee \forall y(y = y))$ because it is true in the empty structure whereas the latter sentence is not. Nevertheless, every monadic second-order formula φ has an equivalent monadic second-order formula $\widehat{\varphi}$ in prenex form that we construct as follows. We let $\widetilde{\varphi}$ be the "usual" prenex form of φ, equivalent to φ in all nonempty structures. The formulas φ and $\widetilde{\varphi}$ have the same free variables. If φ has free first-order variables, we let $\widehat{\varphi}$ be $\widetilde{\varphi}$. Otherwise, if $\varnothing \models \varphi(\varnothing, \ldots, \varnothing)$, we let $\widehat{\varphi}$ be $\forall x.\widetilde{\varphi}$, where x is a first-order variable; if $\varnothing \models \neg\varphi(\varnothing, \ldots, \varnothing)$, we let $\widehat{\varphi}$ be $\exists x.\widetilde{\varphi}$. In the first case, φ and $\widehat{\varphi}$ cannot be satisfied in the empty structure because their free first-order variables cannot have any value. Hence φ is equivalent to $\widetilde{\varphi}$ also in the empty structure. In the second case, it is not hard to check that φ and $\widehat{\varphi}$ are equivalent in all structures including the empty one.

$\varphi_n(x,y)$ expressing that for every m, $1 \le m \le n$, there is a directed walk of length $m + 1$ in the considered graph from vertex x to vertex y. □

5.7 References

Descriptive complexity is covered in the books by Libkin [*Lib04], Ebbinghaus and Flum [*EbbFlu] and Immermann [*Imm].

Theorem 5.5 stating the undecidability of the first-order satisfiability problem for finite graphs is due to Trakhtenbrot. See [*Lib04] or [*EbbFlu] for the proof. Stronger versions result from restrictions to particular classes of graphs. The case of graphs of degree at most 3 is proved in [Wil], that of planar graphs of degree at most 4 is proved in [Her] and that of planar graphs of degree at most 3 is stated with a handwaving proof in [GarSha]. Theorem 5.6 is due to Seese ([See91]).

That MS_2-formulas are no more expressive than MS_1-formulas for expressing the properties of graphs in certain classes of graphs is proved in [Cou94b] and [Cou03]. *Guarded second-order* formulas defined in [GräHO] are equivalent to our MS_2-formulas. (The equivalence is valid for all structures; see Section 7 of [GräHO].)

The definability by monadic second-order formulas of linear orderings of the vertex sets of graphs from particular classes is studied in [Cou96a]. (Such a definition for all graphs by a unique monadic second-order transduction is impossible.) Some of these results yield logical characterizations of recognizability for subsets of certain graph classes and for *traces*. Traces are partially ordered sets with labeled elements that represent equivalence classes of words with respect to congruences generated by commutativity rules of the form $ab \equiv ba$, where a, b are letters; the theory of traces is developed in the book [*DiekRoz]. Lemma 5.27 entails that every CMS formula can be translated into an order-invariant MS formula written with an additional binary relation $\le$ intended to be interpreted by linear orderings. But some order-invariant MS sentences are not equivalent to CMS sentences ([GanRub]).

The Splitting Theorem for disjoint unions of structures follows the paradigm of a result by Feferman and Vaught. This theorem, its history and its numerous variants and consequences are presented by Makowsky in [*Mak]. The proof of the Splitting Theorem given in this chapter is both more general and (hopefully) more clear than the original versions published in [Cou87], [Cou90b], [CouMos]. The name "Splitting Theorem" refers to a polynomial-time algorithm that computes the Tutte polynomial of the graphs of tree-width bounded by a given integer. The relevance to the present situation is explained in [*Mak].

Other proofs of the Weak Recognizability Theorem and its algorithmic applications will be presented in Chapter 6.

Apart from finite automata on terms, finite automata on labeled rooted trees (called *unordered, unranked* trees) and on labeled rooted *unranked, ordered* trees have been defined and compared with monadic second-order logic and related languages. We refer the reader to [*Cou96b], [*Lib06], [BonTal], [Cou89a] and to Chapter 8 of [*Com+].

The monadic second-order orderability of graphs (cf. Proposition 5.30 and Corollary 5.31) is studied in [BluCou11].

The results of Section 5.6 improve some results from [CouWei].

6

Algorithmic applications

This chapter is devoted to the algorithmic applications of the Recognizability Theorem (Theorem 5.75, Section 5.3.10), one of the main results of this book. We will prove the existence of fixed-parameter tractable algorithms with tree-width and clique-width as parameters that solve the model-checking problem for monadic second-order sentences, or that count or list the answers to queries expressed by monadic second-order formulas with first-order free variables. These algorithms are based on graph decompositions of different types, that are formalized as terms over the signatures F^{HR} and F^{VR}. They use two main constructions. The first one is the parsing of the input graph, that is, the computation of a term in $T(F_{[k]}^{HR})$ or in $T(F_{[k]}^{VR})$, for given values of k, that evaluates to this graph. The second one is a kind of "compilation" of the given monadic second-order formula into a finite automaton over the signature $F_{[k]}^{HR}$ or $F_{[k]}^{VR}$.[1]

Both constructions raise difficult algorithmic problems. Concerning the second one, the existence of a finite automaton follows from the Weak Recognizability Theorem (Corollary 5.69), but the proof gives no usable algorithm for constructing automata. We give an alternative construction in Section 6.3 that is essentially the one done by Büchi in the early 1960s and implemented by Klarlund and his team ([BasKla], [Hen+], [Kla]) in the software MONA. We will present some improvements tending to lower the sizes of the automata constructed from monadic second-order formulas. Even if automata cannot be constructed in practice from logical descriptions and by general algorithms, it is useful to know that they exist. This existence motivates the research of alternative constructions based on combinatorial properties rather than on logical formulas.

Section 6.1 reviews terminology and states the results that follow from the Recognizability Theorem. Section 6.2 reviews the main parsing algorithms. Sections 6.3 and 6.4 are devoted to the constructions of automata and their uses for model-checking

---

[1] This construction is similar to the transformation of a regular expression into a finite automaton, see [*AhoLSU] or [*Cre].

and other algorithmic problems. We will also compare several proofs of the Recognizability Theorem.

6.1 Fixed-parameter tractable algorithms for model-checking

We first recall from Chapter 1 the definition of a fixed-parameter tractable algorithm. We generalize it slightly by allowing the parameters to have other values than integers.

Definition 6.1 (Fixed-parameter tractable algorithms) A *decision problem* is defined as a pair $(\mathcal{C}, P)$ consisting of an effectively given set $\mathcal{C}$ and a computable mapping $P : \mathcal{C} \to \{True, False\}$, i.e., P is a decidable property of the elements of $\mathcal{C}$. We assume that each element d of $\mathcal{C}$ has a computable size $|d|$ in $\mathcal{N}$. We also assume that $\mathcal{C}$ is *parametrized*, i.e., that it is equipped with a computable function $p : \mathcal{C} \to \mathcal{N}$ such that $p(d) \leq |d|$ for every $d \in \mathcal{C}$.

An algorithm that decides whether $P(d) = True$ for every element d of $\mathcal{C}$ is *fixed-parameter tractable (FPT) with respect to p* if its computation time is bounded, for every input d, by an expression of the form $f(p(d)) \cdot |d|^c$ for some fixed computable function $f : \mathcal{N} \to \mathcal{N}$ and some fixed positive integer c.[2] A decision problem is *fixed-parameter tractable with respect to* some parameter p if it has an algorithm that is fixed-parameter tractable with respect to p. If $c = 1$, 2 or 3, we say that the algorithm (hence also the decision problem it solves) is *fixed-parameter linear*, *fixed-parameter quadratic* or *fixed-parameter cubic*, respectively, with respect to p.

If $\mathcal{C}$ is a set of graphs, the size of an input graph G is usually its number of vertices and edges, denoted by $\|G\|$, or, in many cases, its number of vertices, denoted by $|G|$. The parameter can be the degree of G, its tree-width or its clique-width.[3] Other examples can be found in the books [*DowFel] and [*FluGro], which present in detail the theory of fixed-parameter tractability.

In model-checking problems (see the next definition), the input is, typically, a pair consisting of a sentence φ and a relational structure. In most applications, such a structure represents a graph G, but we have seen other cases in Section 1.9. The parameter is frequently, for a model-checking problem on (labeled) graphs, the integer $twd(G) + |\varphi|$.

However, we will extend the initial definition so as to allow the parameter function p to take its values in any effectively given set $\mathcal{D}$, and not only in $\mathcal{N}$. We still require that $p : \mathcal{C} \to \mathcal{D}$ and $f : \mathcal{D} \to \mathcal{N}$ are computable functions (which is meaningful on effectively given sets). In the above case we will use the parameter $(twd(G), \varphi)$ that is more natural than $twd(G) + |\varphi|$. Since, for (labeled) graphs, there are only finitely many sentences of size bounded by a given integer (up to renamings of

---

[2] We use the convention that the computation time 0, for example if $|d| = 0$ or if $f(k) := 100 \cdot k^3$ and $p(d) = 0$, indicates a computation taking constant time.

[3] In the case of tree-width, one should take the parameter $twd(G) + 1$ to meet the requirement that $p(d) \in \mathcal{N}$.

bound variables, and of vertex and edge labels), it is equivalent to say that the time taken by an algorithm is $g(twd(G), \varphi) \cdot \|G\|^c$ for some computable function g or $f(twd(G) + |\varphi|) \cdot \|G\|^c$ for some computable function f, because if we know g, we can obtain f as follows:[4]

$$f(p) := \max\{g(k, \varphi) \mid k \leq p, |\varphi| \leq p - k\},$$

and the parameters $(twd(G), \varphi)$ and $twd(G) + |\varphi|$ are both computable from the input (φ, G). So, we get equivalent definitions for fixed-parameter tractability because there is no condition on the time necessary to compute the function f. A similar argument holds for all parameters considered in this chapter.

It is important to specify the parameter when saying that a problem is FPT. For example, the problem of deciding whether a graph G satisfies an MS sentence φ is FPT with respect to $(\varphi, twd(G))$, but neither with respect to φ nor to $(\varphi, Deg(G))$, where $Deg(G)$ is the degree of G (unless $P = NP$, since 3-colorability is MS-expressible and NP-complete on graphs of degree at most 4: see [*GarJoh], problem GT4).

Our definitions differ from those of [*DowFel] and [*FluGro] on some minor points: Downey and Fellows put the value of the parameter as an explicit part of the input. In other words, the input of an algorithm parametrized by p is the pair $(d, p(d))$ instead of d. Our definition corresponds in this case to their *strongly uniform fixed-parameter tractability* (cf. Definition 2.4 in [*DowFel]). Flum and Grohe require that the parameter function p be computable in polynomial time. This is not the case of tree-width (as a mapping from graphs to integers) and they waive the polynomial-time constraint in Section 11.4 of [*FluGro]. Our objective is not to develop, or even to discuss, the theory of fixed-parameter tractability, but to have simple definitions that cover the cases we will present (see Theorems 6.3 and 6.4). We refer the reader to the books cited above for detailed discussions of the definitions.

Definition 6.2 (Model-checking) Let $\mathcal{L}$ be a logical language, typically first-order logic or monadic second-order logic, and let $\mathcal{C}$ be a set of relational structures. The *model-checking problem for* $(\mathcal{C}, \mathcal{L})$, denoted by $MC(\mathcal{C}, \mathcal{L})$, is defined as follows:

Input: An $\mathcal{R}$-structure S in $\mathcal{C}$ and a sentence φ in $\mathcal{L}(\mathcal{R}, \emptyset)$.
Question: Is it true that $S \models \varphi$?

We will analyze model-checking algorithms in terms of the sizes of the input structures: the size of S, denoted by $\|S\|$, is defined in Definition 5.3 and compared there to the size of the word, the term or the graph it may represent. The parameters will consist of the sentence and of the tree-width or the clique-width of S. The tree-width and the clique-width of a binary structure can be taken as those of the corresponding

---

[4] Under the natural assumption that $g(k, \varphi) = g(k, \varphi')$ if φ' is obtained from φ by a renaming of bound variables, and/or of vertex and edge labels.

labeled graph. General definitions will be given in Chapter 9. In the present chapter, we will only consider binary relational structures that represent words, terms and graphs (via the representation $\lfloor \cdot \rfloor$ for the model-checking of CMS sentences, and via $\lceil \cdot \rceil$ for CMS$_2$ ones).

Theorem 6.3

(1) For every finite functional signature F, the problem MC($T(F)$, CMS) is fixed-parameter linear with respect to (F, φ), where φ is the input sentence.

(2) For every triple of finite, pairwise disjoint sets of labels (C, K, Λ), the problem of checking whether $\lceil val(t) \rceil_C \models \varphi$ for $t \in T(F^{\mathrm{HR}}_{C,[K,\Lambda]})$ and $\varphi \in \mathrm{CMS}_2(\mathcal{R}_{\mathrm{m},C,[K,\Lambda]}, \emptyset)$ is fixed-parameter linear with respect to (C, K, Λ, φ).

(3) For every triple of finite, pairwise disjoint sets of labels (C, K, Λ), the problem of checking whether $\lfloor val(t) \rfloor_C \models \varphi$ for $t \in T(F^{\mathrm{VR}}_{C,[K,\Lambda]})$ and $\varphi \in \mathrm{CMS}(\mathcal{R}_{\mathrm{s},C,[K,\Lambda]}, \emptyset)$ is fixed-parameter linear with respect to (C, K, Λ, φ).

In these statements, we use parameters that are tuples of finite objects rather than integers. We do not consider F, C, K, Λ and φ as fixed, but as parts of the input and of the parameter.[5]

Proof: (1) Let F be a finite functional signature and φ belong to $\mathrm{CMS}(\mathcal{R}_F, \emptyset)$. By Corollary 5.67, one can construct a finite deterministic automaton $\mathscr{A}$ that recognizes the set of terms t in $T(F)$ such that $\lfloor t \rfloor \models \varphi$. This being done, say in time $f_1(F, \varphi)$, then for any given $t \in T(F)$, one need only run $\mathscr{A}$ on t to get the answer. This answer is obtained in time $a \cdot |t|$, where a is an upper-bound to the time necessary to perform a transition. This time depends on $\mathscr{A}$, hence on F and φ. The global time for getting the result is thus of the form $f_1(F, \varphi) + f_2(F, \varphi) \cdot |t|$.

(2) For every C, K, Λ and φ as in the statement, one can construct by the Weak Recognizability Theorem for the HR algebra (Corollary 5.69), and by Proposition 3.76(2) and Theorem 3.62, a finite deterministic automaton $\mathscr{A}$ that recognizes the set of terms $val^{-1}(L) \subseteq T(F^{\mathrm{HR}}_{C,[K,\Lambda]})$, where L is the set of s-graphs $\{G \in \mathcal{JS}^{\mathrm{gen}}_{[K,\Lambda]}[C] \mid \lceil G \rceil_C \models \varphi\}$. Then, for any given term $t \in T(F^{\mathrm{HR}}_{C,[K,\Lambda]})$, one need only run $\mathscr{A}$ on t to find out whether $\lceil val(t) \rceil_C \models \varphi$. As above, the computation time is of the form $f_1(C, K, \Lambda, \varphi) + f_2(C, K, \Lambda, \varphi) \cdot |t|$.

(3) Similar to Case (2) with the VR algebra. ∎

The algorithms that prove Assertions (2) and (3) of this theorem take terms as inputs and not graphs. If we want to check if $\lceil G \rceil \models \varphi$ for a given graph G, we need to parse it first, that is, to find a term $t \in T(F^{\mathrm{HR}}_{C,[K,\Lambda]})$ such that $G = val(t)$ (and similarly for Case (3)). Parsing algorithms will be reviewed in Section 6.2. More concrete constructions of automata than those based on the proof of the Recognizability Theorem (Theorem 5.64) will be given in Section 6.3.

---

[5] So, more precisely, Statement (1) says that the problem MC($\bigcup_F T(F)$, CMS) is fixed-parameter linear with respect to (F, φ), where the union extends over all finite functional signatures F.

From Theorem 6.3 and by using the algorithms reviewed in Section 6.2 below, we get the following result on the model-checking of graphs.

Theorem 6.4

(1) For all finite sets K and Λ of vertex and edge labels, the model-checking problem $MC(\mathcal{J}\mathcal{S}_{[K,\Lambda]}[\emptyset], CMS_2)$ is fixed-parameter linear with respect to $(twd(G), \varphi)$, where G is the input graph and φ is the input sentence.

(2) For all finite sets K and Λ of vertex and edge labels, the model-checking problem $MC(\mathcal{G}\mathcal{P}_{[K,\Lambda]}[\emptyset], CMS)$ is fixed-parameter cubic with respect to $(cwd(G), \varphi)$, where G is the input graph and φ is the input sentence.

In these statements, we consider K and Λ as parts of the input,[6] but not as parts of the parameter.

Proof: (1) Let G be a (K, Λ)-labeled graph with n vertices and edges (i.e., $n = |V_G| + |E_G|$). We first show that it suffices to prove the result for parameter $(twd(G), K, \Lambda, \varphi)$ instead of $(twd(G), \varphi)$. The validity of φ in $\lceil G \rceil$ depends only on the labels from $K \cup \Lambda$ that actually occur in φ. More precisely, if K_φ and Λ_φ denote the sets of those labels, then $\lceil G \rceil \models \varphi$ if and only if $\lceil G' \rceil \models \varphi$, where G' is obtained from G by removing all labels that are not in $K_\varphi \cup \Lambda_\varphi$. Note that G' has the same tree-width as G, cf. Remark 2.55(d). If it takes time at most $f'(twd(G'), K_\varphi, \Lambda_\varphi, \varphi) \cdot n$ to determine whether $\lceil G' \rceil \models \varphi$, then it takes time at most $f(twd(G), \varphi) \cdot n$ to determine whether $\lceil G \rceil \models \varphi$, where $f(k, \varphi) := f'(k, K_\varphi, \Lambda_\varphi, \varphi)$. The function f is computable because the sets K_φ and Λ_φ of useful labels are computable from φ.

We now prove the result for parameter $(twd(G), K, \Lambda, \varphi)$. We know from the proof of Theorem 1.25 that an algorithm can compute an optimal tree-decomposition of the unlabeled undirected graph $core(G)$ in time at most $f(twd(G)) \cdot n$ for some computable function f.[7] The linear time algorithm of Theorem 2.83 transforms this tree-decomposition into a term s in $T(F^{HR}_{[twd(G)+1]})$ that evaluates to $core(G)$; this algorithm computes also a corresponding witness. The algorithms of Proposition 2.47 (Assertions (1) and (2)) transform s into a term t in $T(F^{HR}_{[twd(G)+1],[K,\Lambda]})$ that evaluates to G; they take time $O(|K \cup \Lambda| \cdot |s|)$, which is $O(|K \cup \Lambda| \cdot f(twd(G)) \cdot n)$. The result follows then from Theorem 6.3(2).

(2) The proof for parameter $(cwd(G), K, \Lambda, \varphi)$ is similar to the one in (1), but the corresponding parsing problem is more difficult. In Section 6.2.3 we will prove the existence of an algorithm that, for a simple (K, Λ)-labeled graph G with n vertices, computes in time $g(cwd(G), K, \Lambda) \cdot n^3$ a term in $F^{VR}_{[h_\Lambda(G)],[K,\Lambda]}$ that evaluates to G, where $h_\Lambda(G)$ is a positive integer that can be computed from G (this integer is bounded

---

[6] So, more precisely, Statement (1) says that the problem $MC(\bigcup_{K,\Lambda} \mathcal{J}\mathcal{S}_{[K,\Lambda]}[\emptyset], CMS_2)$ is fixed-parameter linear with respect to $(twd(G), \varphi)$, where the union extends over all finite sets K and Λ of vertex and edge labels. Note that $\bigcup_{K,\Lambda} \mathcal{J}\mathcal{S}_{[K,\Lambda]}[\emptyset]$ is the class of all labeled graphs.

[7] Recall from Remark 2.55(d) that $twd(core(G)) = twd(G)$.

in terms of $cwd(G)$ and Λ, but is independent of K; see the proof of Proposition 6.8).
This will complete the proof with the help of Theorem 6.3(3).

The proof for parameter $(cwd(G), \varphi)$ is also similar to the one in (1). In this case,
G' is obtained from G by removing all vertex labels in $K - K_\varphi$ and all edges that have
labels in $\Lambda - \Lambda_\varphi$. Since $cwd(G') \leq cwd(G)$, the function f can now be defined from
f' by $f(k, \varphi) := \max\{f'(m, K_\varphi, \Lambda_\varphi, \varphi) \mid 0 \leq m \leq k\}$. ∎

It is straightforward to generalize Theorem 6.4 to s-graphs in $\mathcal{JS}_{[K,\Lambda]}[C]$ and
p-graphs in $\mathcal{GP}_{[K,\Lambda]}[C]$, via the representations $\lceil \cdot \rceil_C$ and $\lfloor \cdot \rfloor_C$ respectively (where C
belongs to the input but not to the parameter). In fact, the source and port labels of
C can just be viewed as additional vertex labels, and Theorem 6.4 can be applied to
the resulting graphs in $\mathcal{JS}_{[C\cup K,\Lambda]}[\emptyset]$ and $\mathcal{GP}_{[C\cup K,\Lambda]}[\emptyset]$.

Definition 6.5 (Property-checking; counting and optimizing) Theorems 6.3 and
6.4 provide algorithms for model-checking, that is, for verifying that a given relational
structure (in particular, one representing a graph) satisfies a property expressed by a
given sentence in some fixed logical language.

Property-checking is the problem of verifying that a given assignment in a given
relational structure satisfies a given formula, hence that the elements specified by the
assignment satisfy a certain property. We will see in Section 6.4 that the algorithms
of Theorems 6.3 and 6.4 extend to the property-checking problem.

We now define monadic second-order counting functions. For convenience we
only consider formulas of which all free variables are set variables. Let be given
a relational signature $\mathcal{R}$ and a monadic second-order formula φ with set of free
variables $\mathcal{X} := \{X_1, \dots, X_n, Y_1, \dots, Y_p\}$. Let S be a concrete $\mathcal{R}$-structure and γ be
an $\{Y_1, \dots, Y_p\}$-assignment in it. We let $\sharp sat(S, \varphi, (X_1, \dots, X_n), \gamma)$ be the number of
n-tuples $(U_1, \dots, U_n)$ in $\mathcal{P}(D_S)^n$ such that $(S, \gamma') \models \varphi$, where γ' is the $\mathcal{X}$-assignment
in S that extends γ and is such that $\gamma'(X_j) := U_j$ for each $j \in [n]$. If $p = 0$, we
drop γ from the notation; in this case $\sharp sat(S, \varphi, (X_1, \dots, X_n))$ is the cardinality of
$sat(S, \varphi, (X_1, \dots, X_n))$. We call the function that maps γ to $\sharp sat(S, \varphi, (X_1, \dots, X_n), \gamma)$
for fixed φ and S a *monadic second-order counting function*; its computation is a
monadic second-order counting problem.

Finally, we define optimizing functions in a similar way. We let $\mathcal{R}$, φ, $\mathcal{X}$, S and
γ be as above, with $n = 1$, so $\mathcal{X} := \{X, Y_1, \dots, Y_p\}$. We define $Maxsat(S, \varphi, X, \gamma)$ as
the maximum cardinality of a subset U of D_S such that $(S, \gamma') \models \varphi$, where γ' is the
$\mathcal{X}$-assignment in S that extends γ and is such that $\gamma'(X) := U$. As before, we drop
γ from the notation if $p = 0$; in that case $Maxsat(S, \varphi, X)$ is the maximal cardinality
of the sets in $sat(S, \varphi, X)$. We call the function that maps γ to $Maxsat(S, \varphi, X, \gamma)$
for fixed φ and S, or the one defined similarly with minimum instead of maximum,
a *monadic second-order optimizing function*; its computation is a *monadic second-order optimizing problem*.

In Section 6.4.3 we will extend Theorems 6.3 and 6.4 to the computation of monadic
second-order counting and optimizing functions by FPT algorithms.

6.2 Decomposition and parsing algorithms

The algorithms of Theorem 6.4 need to parse the input graphs, i.e., to compute appropriate terms evaluating to them. We will review three types of algorithms: those that construct tree-decompositions, easily convertible into terms over F^{HR} or $F^{HR}_{[K,\Lambda]}$, those that construct such terms from derivation trees for graphs defined by HR equation systems and those that construct terms over F^{VR} or $F^{VR}_{[K,\Lambda]}$.

6.2.1 Constructing tree-decompositions

Since the tree-width of a graph depends neither on the directions and multiplicities of its edges, nor on the existence of loops nor on its possible vertex and edge labels, we will only consider simple, loop-free, undirected graphs. As discussed in the proof of Theorem 6.4(1), if G is a (K, Λ)-labeled graph, every tree-decomposition of $core(G)$ (cf. Remark 2.55(d)) is a tree-decomposition of G and can be converted in linear time into a term in $T(F^{HR}_{[K,\Lambda]})$ evaluating to G (by Theorem 2.83 and Proposition 2.47).

The problem of deciding if $twd(G) \leq k$ for a given pair (G,k) is NP-complete and can be solved in time $O(n^{k+2})$, where $n = |V_G|$ ([ArnCP]), and the corresponding algorithm constructs a tree-decomposition of width at most k if $twd(G) \leq k$. In the sequel we will only quote *decomposition algorithms* for this problem, which construct tree-decompositions of width at most k, as opposed to *recognition algorithms*, which only decide whether $twd(G) \leq k$, say, by checking the absence of finitely many minors (Corollary 2.60) or by other means like performing sequences of reductions as in [ArnCPS].

Bodlaender has given in [Bod96] a decomposition algorithm that takes time $2^{32k^3} \cdot n$. It is actually not implementable but it entails Theorems 1.25(2) and 6.4(1). Efficient linear algorithms exist for $k \leq 3$. The article [BodFKKT] contains two decomposition algorithms that are more likely to be usable, even if they are not linear. The first one takes time $f(n,k) \cdot 2^n$ and space $g(n,k) \cdot 2^n$ for some polynomials f and g. It has been tested with success for nonrandom graphs with at most 50 vertices and tree-width at most 35 arising from concrete problems. Another algorithm is given that takes time $f'(n,k) \cdot 3^n$ and polynomial space for some polynomial f' (but no implementation is reported). Some other algorithms that have been implemented and tested are presented in [BacBod06], [BacBod07], and [VEBK]. A survey can be found in [*Bod06].

Approximation algorithms have also been developed. Flum and Grohe give in [*FluGro] (Proposition 11.14) an algorithm that constructs for every graph G a tree-decomposition of width at most $4 \cdot twd(G) + 1$ in time $O(3^{3 \cdot twd(G)} \cdot twd(G) \cdot n^2)$. This algorithm is based on work by Reed who has given in [Ree] a similar algorithm taking time $O(n \cdot \log(n))$ for producing a tree-decomposition of width at most $3 \cdot twd(G) + 2$. Even if the constructed tree-decomposition is not optimal (i.e., has a width that is not minimal), the algorithm is usable for solving the model-checking problem of Theorem 6.4(1) for graphs of tree-width bounded by a fixed value. The article [BouKMT] gives a polynomial-time algorithm that constructs for every graph G a tree-decomposition of width at most $twd(G)(560 + 80 \cdot \log(twd(G)))$. However, due to the large constants, this algorithm is not practically usable. Another polynomial-time approximation algorithm that is not usable is given in [FeiHL]: it produces a tree-decomposition of width $O(twd(G) \cdot \sqrt{\log(twd(G))})$.

For each k, there is an algorithm that decides in time $O(n)$ if a simple graph G with n vertices has path-width at most k, and if this is the case, it outputs a path-decomposition of width at most k. The article [BodKlo] (Theorem 6.1) constructs, for any two integers k and m, an algorithm doing that from a tree-decomposition of G of width at most m. Hence, if G only is given, one first computes by using the algorithm of [Bod96] a tree-decomposition of this graph of width at most k, to which the algorithm of [BodKlo] is applied in order to obtain if possible a path-decomposition of width at most k. If no such tree-decomposition exists, the path-width of G is more than k. The problem of deciding if a graph has path-width at most k is NP-complete if k is part of the input ([ArnCP], [BodGHK]).

6.2.2 Parsing with respect to HR equation systems

The second method for defining a term in $T(F^{HR})$ that evaluates to a given graph G consists in using an HR equation system S that defines a set L to which G belongs. If we know how G is generated by S, that is, if we know a derivation tree of G relative to S, then (cf. Definition 4.4) we can transform this tree in linear time into a term in $T(F^{HR}_{\mathcal{A}(S)})$ that evaluates to G, where $\mathcal{A}(S)$ is a set of source labels that depends only on S (with $|\mathcal{A}(S)|$ possibly larger than $twd(G) + 1$, see Proposition 4.7). Using Theorem 6.3, this gives what we want for model-checking.[8] Otherwise, we have to rest on a parsing algorithm for the system S. The situation is here not very encouraging. First, by the NP-completeness result of [ArnCP], we know that there is no general parsing algorithm running in time $f(S) \cdot n^c$ for fixed c (unless P = NP). But even worse, certain HR-equational sets of graphs have an NP-complete membership problem. An example is the set of graphs of cyclic bandwidth at most 2 (by [LeuVW], see Example 4.3(11)). This is also the case of *string grammars with disconnection* ([LanWel]), and these grammars can be seen as defining HR-equational sets of disconnected graphs of degree at most 2. However, imposing connectedness is not enough to modify the situation ([*DreKH], Theorems 2.7.1 and 2.7.2). Lautemann has proved in [Lau] (also [*DreKH], Theorem 2.7.7) that an HR-equational[9] set of graphs L has a polynomial-time parsing algorithm if it satisfies the following condition:

---

[8] Flow-graphs of structured C programs have tree-width at most 6 [Thop]. The corresponding tree-decompositions can be obtained easily from the derivation trees of the programs, relative to the grammar of the C language.

[9] These results are formulated in terms of HR grammars (hyperedge replacement graph grammars), which correspond closely to HR equation systems, as we have seen in Section 4.1.5.

L is defined by an HR equation system S such that, if $k = |\mathcal{A}(S)|$ (hence k is the number of source labels used to write S), then there exists a constant a such that for every $G \in L$:

$$Sep(G,k) \leq a \cdot \log(|V_G|),$$

where, for $k \in \mathcal{N}$, $Sep(G,k)$ is the maximum number of connected components of $G[V_G - X]$ for any $X \subseteq V_G$ such that $|X| \leq k$.

The exponent of the polynomial in the time bound of this algorithm depends on S. Let us give an illustration. If G is a connected graph of degree at most d, we have $Sep(G,k) \leq k \cdot d$ for every k. Since the set of connected graphs of degree at most d and of tree-width at most $k - 1$ is defined by an HR equation system S with $|\mathcal{A}(S)| = k$, by Example 4.3(8) and (the proof of) Corollary 5.71(1), this result and Proposition 2.62 give a polynomial-time parsing algorithm for graphs of tree-width at most $k - 1$ and degree at most d.

Other polynomial-time parsing algorithms for HR grammars are presented in [*DreKH]. For each k, the set of graphs of path-width at most k is HR-equational, and the algorithm of [BodKlo] applies to these sets.

6.2.3 Graphs of bounded clique-width

The parsing problem relative to the signature F^{VR} is even more difficult than that relative to the signature F^{HR}. First, the problem of deciding whether $cwd(G) \leq k$ for given (G,k) (with G simple and undirected) is NP-complete ([FelRRS]). For $k = 2$, this verification can be done in time $O(|V_G| + |E_G|)$ because a simple, loop-free undirected graph has clique-width at most 2 if and only if it is a cograph (cf. Proposition 2.106(2)) so that one can use the algorithm of [CorPS]. It can be done in cubic time for $k = 3$ ([CorHLRR]). For other fixed values of k, it is not known if the problem is polynomial or NP-complete. Furthermore, there is presumably no polynomial-time approximation algorithm since the article [FelRRS] establishes the following:

If for some real number ε such that $0 \leq \varepsilon < 1$ there exists a polynomial-time algorithm that computes, for every simple undirected graph G of minimum degree 3, an integer $f(G)$ such that $|f(G) - cwd(G)| \leq |V_G|^\varepsilon$, then $P = NP$.

In order to overcome this difficulty, Oum and Seymour introduced the notion of rank-width in [OumSey].

Definition 6.6 (Rank-width) A *cubic tree* is a tree T whose nodes are all of degree 3 or 1. The latter ones are called the *leaves*; the set of leaves is denoted by *Leaves*(T). Let G be a simple loop-free undirected graph, described by a symmetric $V_G \times V_G$ *adjacency matrix* A_G with coefficients in $\{0,1\}$ such that $A_G[v,w] = 1$

if and only if v and w are the ends of an edge of G. A *layout*[10] of G is a pair (T,f), where T is a cubic tree and f is a bijection : $V_G \rightarrow Leaves(T)$. For each edge e of T, the two connected components of $T - e$ (the forest obtained from T by removing e) induce a bipartition (X_e, Y_e) of $Leaves(T)$, hence a bipartition $(X'_e, Y'_e) = (f^{-1}(X_e), f^{-1}(Y_e))$ of V_G. We let $rk(e,T,f)$ be the rank over $GF(2)$ (the field with the two elements 0 and 1) of $A_G[X'_e, Y'_e]$, the rectangular $X'_e \times Y'_e$ submatrix of A_G (it describes the edges of G that link the vertices of X'_e and Y'_e). Then we define the *rank* of (T,f) as $rk(T,f) := \max\{rk(e,T,f) \mid e \in E_T\}$ and the *rank-width* of G as $rwd(G) := \min\{rk(T,f) \mid (T,f) \text{ is a layout of } G\}$.

For every simple loop-free undirected graph G, we have, by [OumSey]:

$$rwd(G) \leq cwd(G) \leq 2^{rwd(G)+1} - 1.$$

More precisely, this article establishes that from every term t in $F^{VR}_{[k]}$ defining a graph G, one can construct in linear time a layout (T,f) of G such that $rk(T,f) \leq k$. One defines T from the syntactic tree of t by contracting some edges so as to make it cubic. Conversely, from every such layout one can construct in quadratic time a term in $T(F^{VR}_{[2^{k+1}-1]})$ that evaluates to G, together with a corresponding witness. Hence clique-width and rank-width are equivalent complexity measures on simple loop-free undirected graphs: the same sets of such graphs have bounded clique-width and bounded rank-width. The problem of deciding whether $rwd(G) \leq k$ for given (G,k) is NP-hard, as for clique-width (see the survey [*GotHOS]). However, Hliněný and Oum constructed in [HliOum] a parsing algorithm[11] running in time $g_0(k) \cdot n^3$ (for some fixed function g_0) that decides if a graph G with n vertices has rank-width at most k. The algorithm produces a layout of rank at most k if there exists one. Hence, one can obtain from it a term in $T(F^{VR}_{[2^{k+1}-1]})$ that evaluates to G (with a witness); since every layout of G has size $O(n)$, this takes time $O(n^2)$. Thus, by the proof of Theorem 1.25 and using Theorem 6.3(3), this algorithm gives the proof of Theorem 6.4(2) for simple loop-free undirected unlabeled graphs.

We now extend this result to larger classes of graphs. First, we recall that clique-width does not depend on vertex labels and loops: by Proposition 2.100, if one adds or modifies vertex labels, or if one adds or deletes loops in a graph G that is the value of a term t, then one can transform t in linear time into a term t' that evaluates to the modified graph G', just by modifying some constant symbols of t. It is important to note that one need not parse G'. The situation is different for edge labels and edge directions.

---

[10] Such a pair is usually called a "rank-decomposition." However, its definition is not based on the rank of any matrix. The notion of rank is used to associate a value with a layout, and this value measures its complexity. Other values can be associated with layouts, see for example [Cou04].

[11] As the linear algorithm of Bodlaender [Bod96] for deciding tree-width, this algorithm is not implementable.

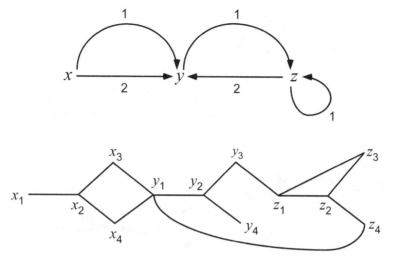

Figure 6.1 An edge-labeled graph G and its encoding $B_{[2]}(G)$.

Definition 6.7 (Encodings of directed and labeled graphs) We will encode simple directed graphs, and more generally, simple edge-labeled directed and undirected graphs by simple vertex-labeled undirected graphs.

We first consider the case of simple directed graphs. For such a graph G, we define $B(G)$ as the simple vertex-labeled undirected graph[12] such that

$$V_{B(G)} := V_G \times [3],$$
$$E_{B(G)} := \{(v,i) - (v,i+1) \mid v \in V_G, i \in [2]\}$$
$$\cup \{(v,3) - (w,1) \mid v \to w \text{ in } G\},$$
$$\gamma_{B(G)}((v,i)) := i \text{ for every } (v,i) \in V_{B(G)}.$$

Here is the decoding transformation. If H is a vertex-labeled undirected graph of the form $B(G)$ for some G, with labeling function $\gamma_H : V_H \to \{1,2,3\}$, then G can be recovered from H as follows:

(a) one directs every edge linking a vertex v with label 3 and a vertex w with label 1 from v to w;

(b) one contracts the other edges (they link a vertex with label 2 and one with label 1 or 3).

The mappings B and B^{-1} are FO-transductions. It follows from Corollary 7.38(2) (also stated after Corollary 1.43) that there exist two computable integer functions p

---

[12] If we use $V_G \times [0,3]$ instead of $V_G \times [3]$, and if we replace in $E_{B(G)}$ the edge $(v,1) - (v,2)$ by the two edges $(v,1) - (v,0)$ and $(v,0) - (v,2)$, then we get a bipartite graph $B(G)$. This variant is useful for proving the result of [CouOum] discussed in Section 7.5, Theorem 7.55(2). The same observation holds for the encoding $B_\wedge$ defined below.

and $\bar{p}$ such that, for every simple directed graph G:

$$cwd(B(G)) \leq p(cwd(G)) \quad \text{and} \quad cwd(G) \leq \bar{p}(cwd(B(G))).$$

We now generalize this construction into an encoding of simple, directed edge-labeled graphs. We let the set Λ of edge labels be $\{\lambda_1,\ldots,\lambda_m\}$ and not contain 1 and 2. We define a mapping B_Λ (it is an FO-transduction) that associates with every directed $(\emptyset,\Lambda)$-labeled graph G an $([m+2],\emptyset)$-labeled undirected graph $B_\Lambda(G)$ as follows:

$$V_{B_\Lambda(G)} := V_G \times [m+2],$$
$$E_{B_\Lambda(G)} := \{(v,1)-(v,2) \mid v \in V_G\}$$
$$\cup \{(v,2)-(v,i)) \mid v \in V_G, i \in \{3,\ldots,m+2\}\}$$
$$\cup \{(v,2+i)-(w,1) \mid v \to w \text{ is an edge of } G \text{ with label } \lambda_i\},$$
$$\gamma_{B_\Lambda(G)}((v,i)) := i \text{ for every vertex } (v,i) \text{ of } B_\Lambda(G).$$

Figure 6.1 shows a directed $(\emptyset,[2])$-labeled graph G and its encoding $B_{[2]}(G)$ by a $([4],\emptyset)$-labeled graph. A vertex (v,i) of $B_{[2]}(G)$ is represented by v_i.

As in the case of the mapping B (which is B_Λ for $m=1$), we have

$$cwd(B_\Lambda(G)) \leq p_\Lambda(cwd(G)) \quad \text{and} \quad cwd(G) \leq \bar{p}_\Lambda(cwd(B_\Lambda(G))),$$

for some computable functions p_Λ and $\bar{p}_\Lambda$. Furthermore, focusing on B_Λ^{-1} and the second inequality, it follows from Corollary 7.38(3) that there is an algorithm that transforms a term t in $T(F^{\mathrm{VR}}_{[k],[[m+2],\emptyset]})$ that evaluates to $B_\Lambda(G)$ into one in $T(F^{\mathrm{VR}}_{[\bar{p}_\Lambda(k)],[\emptyset,\Lambda]})$ that evaluates to G, in time $g_1(k,\Lambda) \cdot |t|$ for some computable function g_1. Thus, the algorithm is fixed-parameter linear with respect to (k,Λ).

The following proposition will yield the parsing step of the proof of Theorem 6.4(2):

Proposition 6.8 There exists an algorithm that, for every simple (K,Λ)-labeled graph G with n vertices, computes in time at most $g(cwd(G),K,\Lambda) \cdot n^3$ a term over $F^{\mathrm{VR}}_{[h_\Lambda(G)],[K,\Lambda]}$ evaluating to G, where g is a computable function and $h_\Lambda(G)$ is a positive integer that can be computed from G.

Proof: We first consider directed unlabeled graphs. Let be given a simple directed graph G with n vertices, and an integer k. We construct the vertex-labeled undirected graph $B(G)$. We run $\mathcal{A}$, the algorithm of [HliOum], on input $(B(G),k)$. It takes time at most $g_0(k) \cdot n^3$. There are two possible outputs:

   (i) The answer that $rwd(B(G)) > k$.
   (ii) A layout of $B(G)$ of rank at most k.

In Case (ii), another algorithm can construct in time $O(n^2)$ a term t in $T(F^{\mathrm{VR}}_{[2^{k+1}-1],[[3],\emptyset]})$ that evaluates to $B(G)$, and in time $g_1(\emptyset,k) \cdot |t|$ a term in

$T(F^{VR}_{[\bar{p}(2^{k+1}-1)]})$ that evaluates to G. Altogether this takes time at most $g_2(k) \cdot n^3$ for some computable function g_2. Case (i) cannot occur if $p(cwd(G)) \leq k$ because $rwd(B(G)) \leq cwd(B(G)) \leq p(cwd(G))$.

Hence, if we run $\mathcal{A}$ with input $(B(G),k)$ for $k = 1,2,3,\ldots$, we must find a value k for which the output is a layout of $B(G)$ of rank at most k. Let $m(G)$ be the first such k. Then $m(G) \leq p(cwd(G))$ by the above observation. Hence, we obtain a term in $T(F^{VR}_{[h(G)]})$ that evaluates to G, where $h(G) := \bar{p}(2^{m(G)+1} - 1)$. This term is obtained in time at most $(\sum_{k=1}^{p(cwd(G))} g_2(k)) \cdot n^3$. Note that the integer $h(G)$ is bounded by $f(cwd(G))$ for a computable function f.

For $(\emptyset, \Lambda)$-labeled graphs, the proof is the same with the encoding B_Λ in place of B. In time bounded by $(\sum_{k=1}^{p_\Lambda(cwd(G))} g_2(\Lambda,k)) \cdot n^3$, we obtain a term in $T(F^{VR}_{[h_\Lambda(G)],[\emptyset,\Lambda]})$ that evaluates to G, where $h_\Lambda(G) := \bar{p}_\Lambda(2^{m(G)+1} - 1)$. Finally, this proof extends easily to simple (K, Λ)-labeled graphs because we know that clique-width does not depend on vertex labels, and (by Proposition 2.100), that a term in $T(F^{VR}_{[\emptyset,\Lambda]})$ that defines an edge-labeled directed graph can be easily transformed into one in $T(F^{VR}_{[K,\Lambda]})$ that also specifies the vertex labels. Note that an upper-bound to $h_\Lambda(G)$ can be computed from $cwd(G)$ and Λ. ∎

The reader may ask about the parsing problem for VR equation systems. By Theorem 4.49, which establishes that every HR-equational set is VR-equational, all difficulties met with HR equation systems also occur with VR ones. However, polynomial-time recognition algorithms exist for certain restricted classes of graphs of bounded clique-width. All these algorithms construct terms in $T(F^{VR}_{[k]})$ for the recognized graphs (see [BraDLM], [BraELL], [EspGW] and [MakRot]) but they only work for undirected graphs without edge labels.

6.3 Monadic second-order formulas compiled into finite automata

Theorems 6.3 and 6.4 entail the existence of FPT algorithms for monadic second-order model-checking problems on terms and on graphs of bounded tree-width or clique-width. They are consequences of the Weak Recognizability Theorem (Corollary 5.69). However, although its proof is effective, the states of the finite automata resulting from it are (or encode) huge sets of formulas. Hence, they do not give tractable implementations.

The purpose of this section is to give another translation of monadic second-order sentences into finite automata that is hopefully usable for not too complex sentences and for "small" bounds on tree-width or clique-width, or for "small" signatures in the case of terms.[13] However, the difficulty is unavoidable as soon as one wishes a model-checking algorithm that works for every CMS sentence, as proved by Frick and Grohe in [FriGro04]. They established the following theorem.[14]

---

[13] For terms, this translation (presented in Section 6.3.3) is the classical one used to prove Theorem 1.16.

[14] We recall that $\exp : \mathcal{N} \times \mathcal{N} \to \mathcal{N}$ is defined by $\exp(0,n) = n$, $\exp(d+1,n) = 2^{\exp(d,n)}$.

Theorem 6.9 If there exist $c,d \in \mathcal{N}$ such that, for every MS sentence φ and every word $w \in \{0,1\}^*$, one can decide if $\lfloor w \rfloor \models \varphi$ in time $\exp(d,|\varphi|) \cdot |w|^c$, then $P = NP$. $\square$

We will nevertheless try to get applicable cases of Theorem 6.3. We first review (from Section 3.3) some definitions concerning finite automata and we also present some new constructions.

In this section, we will only consider finite, one-sorted functional signatures and finite, ε-free automata over such signatures. We will call them simply *functional signatures* and *automata*.

6.3.1 Automata

We recall (from Section 3.3) that an F-automaton is defined as a four-tuple $\mathcal{A} = \langle F, Q_{\mathcal{A}}, \delta_{\mathcal{A}}, Acc_{\mathcal{A}} \rangle$ (signature, states, transition rules, accepting states). It accepts (or recognizes) the language $L(\mathcal{A}) \subseteq T(F)$. Without $Acc_{\mathcal{A}}$, we have an F-semi-automaton. The language $L(\mathcal{A},q)$ is the set of terms recognized by $\mathcal{A}$ with q as unique accepting state. The number of states of $\mathcal{A}$ is denoted by $\sharp\mathcal{A}$. A transition rule in $\delta_{\mathcal{A}}$ is of the form $f[q_1,\ldots,q_{\rho(f)}] \to q$ with $f \in F$ and $q,q_1,\ldots,q_{\rho(f)} \in Q_{\mathcal{A}}$.

We will denote by $det(\mathcal{A})$ the trim, complete and deterministic automaton constructed from an automaton $\mathcal{A}$ (cf. Proposition 3.50). Its set of states is (or is in bijection with) a subset of $\mathcal{P}(Q_{\mathcal{A}})$.

For every complete and deterministic automaton $\mathcal{A}$, the automaton $\overline{\mathcal{A}}$ defined from $\mathcal{A}$ by taking $Q_{\mathcal{A}} - Acc_{\mathcal{A}}$ as set of accepting states recognizes the language $T(F) - L(\mathcal{A})$.

Products of automata and Boolean operations

If $\mathcal{A}$ and $\mathcal{B}$ are two F-semi-automata, their *product* is the F-semi-automaton $\mathcal{A} \times \mathcal{B} := \langle F, Q_{\mathcal{A}} \times Q_{\mathcal{B}}, \delta_{\mathcal{A} \times \mathcal{B}} \rangle$, where $\delta_{\mathcal{A} \times \mathcal{B}}$ is defined as the set of rules $f[(p_1,q_1),\ldots,(p_k,q_k)] \to (p,q)$ with $f[p_1,\ldots,p_k] \to_{\mathcal{A}} p$ and $f[q_1,\ldots,q_k] \to_{\mathcal{B}} q$.

If $\mathcal{A}$ and $\mathcal{B}$ are automata, then we define the F-automaton $\mathcal{A} \cap \mathcal{B}$ as $\mathcal{A} \times \mathcal{B}$ equipped with set of accepting states $Acc_{\mathcal{A}} \times Acc_{\mathcal{B}}$. Clearly, $L(\mathcal{A} \cap \mathcal{B}) = L(\mathcal{A}) \cap L(\mathcal{B})$. We define similarly $\mathcal{A} \cup \mathcal{B}$ as $\mathcal{A} \times \mathcal{B}$ with set of accepting states $(Acc_{\mathcal{A}} \times Q_{\mathcal{B}}) \cup (Q_{\mathcal{A}} \times Acc_{\mathcal{B}})$. If $\mathcal{A}$ and $\mathcal{B}$ are complete, then $L(\mathcal{A} \cup \mathcal{B}) = L(\mathcal{A}) \cup L(\mathcal{B})$.

We now generalize this construction so as to build an automaton that accepts a Boolean combination of the languages accepted by given automata $\mathcal{A}_1,\ldots,\mathcal{A}_m$.

Let $L_1,\ldots,L_m \subseteq T(F)$ be accepted by complete and deterministic F-automata, respectively $\mathcal{A}_1,\ldots,\mathcal{A}_m$. Let $L := b(L_1,\ldots,L_m) \subseteq T(F)$ be defined from $L_1,\ldots,L_m$ by a term $b \in T(\{\cap,\cup,\neg\},\{x_1,\ldots,x_m\})$ (where $\overline{M}$ denotes $T(F) - M$). We define $\mathcal{C}$ as the F-semi-automaton $\mathcal{A}_1 \times (\cdots \times (\mathcal{A}_{m-1} \times \mathcal{A}_m) \cdots)$. For convenience, we identify its set of states with $Q_{\mathcal{A}_1} \times \cdots \times Q_{\mathcal{A}_{m-1}} \times Q_{\mathcal{A}_m}$ and we denote $\mathcal{C}$ by $\mathcal{A}_1 \times \cdots \times \mathcal{A}_{m-1} \times \mathcal{A}_m$. It is complete and deterministic. We now define from b a set of accepting states for $\mathcal{C}$.

We let Acc be the set of m-tuples $(q_1, \ldots, q_m) \in Q_{\mathscr{C}}$ such that $b(v_1, \ldots, v_m) = True$, where v_j is *True* if $q_j \in Acc_{\mathscr{A}_j}$ and *False* otherwise. Here, b is evaluated on truth values with $\cup, \cap$ and $^-$ interpreted respectively as $\vee, \wedge$ and $\neg$. We let $\mathscr{B}$ be the F-automaton consisting of the semi-automaton $\mathscr{C}$ and Acc as set of accepting states. It is clear that $L(\mathscr{B}) = L$. Note that $\mathscr{B} = \overline{\mathscr{A}_1}$ if $m = 1$ and $b(L_1) = \overline{L_1}$.

If b is *positive*, i.e., written without the complementation operation $^-$, then this construction is valid provided the automata $\mathscr{A}_i$ are complete; they need not be deterministic. The automaton $\mathscr{B}$ is then complete.

In all cases, $\sharp \mathscr{B} = \sharp \mathscr{A}_1 \times \cdots \times \sharp \mathscr{A}_m$, but $\mathscr{B}$ may be replaced by a smaller automaton by reduction or by minimization in the case where it is deterministic. Hence, we have proved the following proposition:

Proposition 6.10 Let F be a functional signature. If $L_1, \ldots, L_m$ are regular subsets of $T(F)$, recognized by complete and deterministic automata with, respectively, $n_1, \ldots, n_m$ states, then every Boolean combination of $L_1, \ldots, L_m$ is recognized by a complete and deterministic automaton with at most $n_1 \times \cdots \times n_m$ states. The same holds for a positive Boolean combination of languages $L_1, \ldots, L_m$ and automata that are complete but not necessarily deterministic. $\qquad\square$

Direct and inverse images of automata

We will need certain images and inverse images of languages recognized by automata. Let H be a functional signature (possibly $H = F$), and $h : H \to F$ be an arity preserving mapping, i.e., such that $\rho(h(f)) = \rho(f)$ for every $f \in H$. For every $t \in T(H)$, we let $h(t) \in T(F)$ be the term obtained from t by replacing f by $h(f)$ at each of its occurrences. The mapping h on terms is an alphabetic relabeling (cf. the paragraph before Proposition 2.126).

If $L \subseteq T(H)$, then $h(L) := \{h(t) \mid t \in L\}$. If $\mathscr{A}$ is an H-semi-automaton, we let $h(\mathscr{A})$ be the F-semi-automaton obtained from $\mathscr{A}$ by replacing each transition rule $f[q_1, \ldots, q_{\rho(f)}] \to q$ by $h(f)[q_1, \ldots, q_{\rho(f)}] \to q$. Clearly, $h(L(\mathscr{A}, q)) = L(h(\mathscr{A}), q)$ for every state q. In general, $h(\mathscr{A})$ is not deterministic, even if $\mathscr{A}$ is. If $\mathscr{A}$ is an automaton, then $L(h(\mathscr{A})) = h(L(\mathscr{A}))$, where $h(\mathscr{A})$ has the same accepting states as $\mathscr{A}$. We call this transformation the *direct image construction*.

We now consider inverse images. If $K \subseteq T(F)$, then $h^{-1}(K) := \{t \in T(H) \mid h(t) \in K\}$. If $\mathscr{A}$ is an F-semi-automaton or an F-automaton, then we let the H-semi-automaton or H-automaton $h^{-1}(\mathscr{A})$ be obtained by putting in $h^{-1}(\mathscr{A})$ a transition rule $f[q_1, \ldots, q_{\rho(f)}] \to q$ if and only if $h(f)[q_1, \ldots, q_{\rho(f)}] \to_{\mathscr{A}} q$. We have $L(h^{-1}(\mathscr{A}), q) = h^{-1}(L(\mathscr{A}, q))$ for every state q, and $L(h^{-1}(\mathscr{A})) = h^{-1}(L(\mathscr{A}))$ if $\mathscr{A}$ is an automaton. Note that $h^{-1}(\mathscr{A})$ is deterministic (resp. complete) if $\mathscr{A}$ is so. We call this transformation the *inverse image construction*.

Proposition 6.11 Let $h : H \to F$ be an arity preserving mapping between functional signatures.

(1) If $L \subseteq T(H)$ is recognized by an H-automaton with n states, then $h(L)$ is recognized by an F-automaton with n states.

(2) If $K \subseteq T(F)$ is recognized by an F-automaton $\mathscr{A}$ with n states, then $h^{-1}(K)$ is recognized by an H-automaton $\mathscr{B}$ with n states. If $\mathscr{A}$ is complete and deterministic, then so is $\mathscr{B}$. ☐

Counting and optimization based on automata

The following constructions will be used in Section 6.4 for computing monadic second-order counting and optimizing functions (see Definition 6.5).

Let H, F be functional signatures and $h : H \rightarrow F$ be arity preserving. Let $\mathscr{A}$ be a complete and deterministic H-semi-automaton. For $t \in T(F)$ and $q \in Q_{\mathscr{A}}$ we define

$$\sharp_{\mathscr{A}, h^{-1}}(t, q) := |h^{-1}(t) \cap L(\mathscr{A}, q)|$$

and $\sharp_{\mathscr{A}, h^{-1}}(t)$ as the function $\lambda q \in Q_{\mathscr{A}} \cdot \sharp_{\mathscr{A}, h^{-1}}(t, q)$. If, furthermore, $\mathscr{A}$ is an automaton, then, for every $t \in T(F)$, we have

$$|h^{-1}(t) \cap L(\mathscr{A})| = \sum_{q \in Acc_{\mathscr{A}}} \sharp_{\mathscr{A}, h^{-1}}(t, q).$$

For $J \subseteq H$ and $s \in T(H)$, we denote by $|s|_J$ the number of occurrences in s of symbols belonging to J. For $t \in T(F)$ and $q \in Q_{\mathscr{A}}$, we define

$$Max_{\mathscr{A}, h^{-1}, J}(t, q) := \max\{|s|_J \mid s \in h^{-1}(t) \cap L(\mathscr{A}, q)\}.$$

and $Max_{\mathscr{A}, h^{-1}, J}(t)$ as the function $\lambda q \in Q_{\mathscr{A}} \cdot Max_{\mathscr{A}, h^{-1}, J}(t, q)$. (The maximum of an empty set of integers is $-\infty$.) It follows that for every t, we have

$$\max\{|s|_J \mid s \in h^{-1}(t) \cap L(\mathscr{A})\} = \max\{Max_{\mathscr{A}, h^{-1}, J}(t, q) \mid q \in Acc_{\mathscr{A}}\}.$$

With these hypotheses and notations, we have the following result:

Proposition 6.12 Let $h : H \rightarrow F$ be an arity preserving mapping between functional signatures, let $J \subseteq H$ and $\mathscr{A}$ be a complete and deterministic H-automaton. The functions $\sharp_{\mathscr{A}, h^{-1}}$ and $Max_{\mathscr{A}, h^{-1}, J}$ are computable in linear time by an induction on the structure of the input term.

Proof: We describe the inductive computation, assuming that all function symbols f have arity 0 or 2. The extension to other arities is straightforward. We fix h and $\mathscr{A}$, and we replace $\sharp_{\mathscr{A}, h^{-1}}(t, q)$ by $\sharp(t, q)$ to simplify the notation.

For $f \in F_0$, $\sharp(f, q)$ is the number of symbols f' in $h^{-1}(f)$ such that $f' \rightarrow_{\mathscr{A}} q$. We can rewrite this into

$$\sharp(f) = D_f,$$

where D_f is a function (with finite domain $Q_{\mathscr{A}}$) that can be determined from h and $\mathscr{A}$ in time $O(\sharp\mathscr{A})$, hence in time that does not depend on the input term.

For $f \in F_2$ and $t_1, t_2 \in T(F)$ we have

$$\sharp(f(t_1,t_2),q) = \sum_{(q_1,q_2,f') \in R} \sharp(t_1,q_1) \cdot \sharp(t_2,q_2),$$

where R is the set of triples (q_1,q_2,f') in $Q_{\mathscr{A}} \times Q_{\mathscr{A}} \times H_2$ such that $f'[q_1,q_2] \to_{\mathscr{A}} q$ and $f' \in h^{-1}(f)$. This equality is true because

$$h^{-1}(f(t_1,t_2)) \cap L(\mathscr{A},q) = \biguplus_{(q_1,q_2,f') \in R} f'\big(h^{-1}(t_1) \cap L(\mathscr{A},q_1), h^{-1}(t_2) \cap L(\mathscr{A},q_2)\big),$$

and the union is disjoint because $\mathscr{A}$ is deterministic.[15] Hence we have

$$\sharp(f(t_1,t_2)) = D_f(\sharp(t_1),\sharp(t_2)),$$

for all $f \in F_2$ and $t_1, t_2 \in T(F)$, where D_f is a fixed function of type:

$$[Q_{\mathscr{A}} \to \mathcal{N}] \times [Q_{\mathscr{A}} \to \mathcal{N}] \to [Q_{\mathscr{A}} \to \mathcal{N}].$$

This function can be computed from h and $\mathscr{A}$ in time $O((\sharp\mathscr{A})^2)$ if each arithmetic operation is considered as evaluable in constant time. This gives the computation time $O(|t|)$ for fixed automaton $\mathscr{A}$ and fixed $h : H \to F$.

A similar construction can be done for computing $Max_{\mathscr{A},h^{-1},J}$. We fix $\mathscr{A}$, h and J, and we write $Max(t,q)$ instead of $Max_{\mathscr{A},h^{-1},J}(t,q)$.

If $f \in F_0$, and $H_0(f,q)$ is the set of symbols f' in $h^{-1}(f)$ such that $f' \to_{\mathscr{A}} q$, then

$$Max(f,q) = \begin{cases} 1 & \text{if } H_0(f,q) \cap J \neq \emptyset, \\ 0 & \text{if } H_0(f,q) \cap J = \emptyset \text{ and } H_0(f,q) \neq \emptyset, \\ -\infty & \text{if } H_0(f,q) = \emptyset. \end{cases}$$

If $f \in F_2$ and $t_1, t_2 \in T(F)$, then

$$Max(f(t_1,t_2),q) = \max\{\varepsilon_{J,f'} + Max(t_1,q_1) + Max(t_2,q_2) \mid (q_1, q_2, f') \in R\},$$

where $\varepsilon_{J,f'} := \texttt{if } f' \in J \texttt{ then } 1 \texttt{ else } 0$, and R is as above for the computation of $\sharp(t,q)$. The union need not be disjoint here because we take a maximum and not a sum, hence the definition is the same if $\mathscr{A}$ is not deterministic. ∎

Remark 6.13 (1) In the previous proposition, we have assumed that the arithmetic operations are done in constant time. For handling large integers, it is more accurate to assume that they take time proportional to the logarithms of their arguments. In

---

[15] For sets of terms T_1, T_2, we let $f(T_1, T_2)$ denote $\{f(t_1,t_2) \mid t_1 \in T_1, t_2 \in T_2\}$.

the computation of $\sharp_{\mathscr{A},h^{-1}}(t)$, we have $\sharp_{\mathscr{A},h^{-1}}(s,q) \leq d^n$ if s is any subterm of t, n is the size of t and d is the maximal cardinality of $h^{-1}(f)$ for $f \in F$. Hence, the computation time is multiplied by $n \cdot \log(d)$ and becomes quadratic instead of linear, if the logarithmic cost of arithmetical operations is used. For optimizing functions, the results and the intermediate values are bounded by n. Hence, the computation time is multiplied by $\log(n)$.

(2) The proof of Proposition 6.12 can be seen as a construction of complete and deterministic automata with infinite sets of states (the sets of functions from $Q_{\mathscr{A}}$ to $\mathcal{N}$ or to $\mathcal{N} \cup \{-\infty\}$). Each state is a function having a finite domain, hence is a finitely encodable object. This automaton is thus effectively given, i.e., all its components are effectively given. The transition functions are computable in constant time if arithmetic operations have unit cost. They are computable in linear time (for $\sharp_{\mathscr{A},h^{-1}}$) or in logarithmic time (for $Max_{\mathscr{A},h^{-1},J}$) if arithmetic operations have logarithmic cost. $\qquad\square$

Runs of nondeterministic automata

Let $\mathscr{A}$ be an F-semi-automaton. For $t \in T(F)$, we denote by $run^*_{\mathscr{A}}(t)$ the set of states of the form $r(root_t)$ for some run r of $\mathscr{A}$ on t. This set is nonempty if $\mathscr{A}$ is complete and it has a single element if $\mathscr{A}$ is complete and deterministic (cf. Definition 3.49). If d is a positive integer we say that $\mathscr{A}$ is d-*nondeterministic* if, for all $f \in F$ and $q_1,\ldots,q_{\rho(f)} \in Q_{\mathscr{A}}$, there are at most d states q such that $f[q_1,\ldots,q_{\rho(f)}] \to_{\mathscr{A}} q$. The integer d measures the *degree of nondeterminism* of $\mathscr{A}$. A 1-nondeterministic semi-automaton is deterministic.

In the next proposition, $\mathscr{A}$ is d-nondeterministic and a is an upper-bound to the time necessary to determine the i-th state q such that $f[q_1,\ldots,q_{\rho(f)}] \to_{\mathscr{A}} q$ for $i \in [d]$ or to find that the number of such states q is less than i, so that there is no i-th state (we can assume that $Q_{\mathscr{A}}$ is linearly ordered). We will say that $\mathscr{A}$ has *transition time* bounded by a.

Proposition 6.14 Let $\mathscr{A}$ be a d-nondeterministic F-semi-automaton and let a bound its transition time. For every term $t \in T(F)$, one can determine the set $run^*_{\mathscr{A}}(t)$ in time bounded by $a \cdot d \cdot (\sharp\mathscr{A})^{\rho(F)} \cdot |t|$.

Proof: We use a bottom-up computation to compute the set $run^*_{\mathscr{A}}(t/u)$ (of cardinality at most $\sharp\mathscr{A}$) for each node u of the syntactic tree of t. At a node u that is an occurrence of a symbol of arity $k \geq 0$, there are at most $d \cdot (\sharp\mathscr{A})^k$ transitions, and for each of them, it takes time a to find the resulting state q. This gives the result. $\qquad\blacksquare$

Remark 6.15 In the particular case where $\rho(F) = 2$ and the nondeterminism of transitions is limited to constant symbols, we get in the previous proposition the better bound $a \cdot (d + (\sharp\mathscr{A})^2) \cdot |t|$ on the computation time.

6.3.2 Normalizing monadic second-order formulas

We will need some normalization lemmas for monadic second-order formulas. As in Section 5.3 the formal constructions will use formulas without universal quantifications. The expression $\forall X . \varphi$ will be used as a shorthand for $\neg \exists X . \neg \varphi$. Similarly for $\Rightarrow$, $\Leftrightarrow$ and *False*. In order to facilitate formal constructions, variables will be of the forms X_i and x_j for $i, j \in \mathcal{N}_+$. Relational signatures will be without constant symbols.

Definition 6.16 (Formulas without first-order variables) Let $\mathcal{R}$ be a relational signature without constant symbols. For each $n \in \mathcal{N}$, we let $\mathcal{X}_n := \{X_1, \ldots, X_n\}$ and $\mathcal{X}_\omega := \{X_i \mid i \in \mathcal{N}_+\}$. In order to write formulas without first-order variables, we will use the following atomic formulas, whose meanings are specified unless they are clear:

True,

$Card_{p,q}(X_i)$ for $0 \le p < q$ and $q \ge 2$,

$X_i \subseteq X_j$,

$X_i = \emptyset$,

$Sgl(X_i)$, to mean that X_i denotes a singleton set,

$R(X_{i_1}, \ldots, X_{i_{\rho(R)}})$ for $R \in \mathcal{R}$, to mean that $X_{i_1} = \{d_1\}, \ldots, X_{i_{\rho(R)}} = \{d_{\rho(R)}\}$

for some $(d_1, \ldots, d_{\rho(R)}) \in R_S$ (where S is the considered $\mathcal{R}$-structure).

We could express $X_i = \emptyset$ and $Sgl(X_i)$, respectively, by $\forall Y (Y \subseteq X_i \Rightarrow X_i \subseteq Y)$ and $\forall Z (Z \subseteq X_i \Rightarrow [X_i \subseteq Z \vee \forall Y (Y \subseteq Z \Rightarrow Z \subseteq Y)])$, at the cost of introducing more quantifiers. Since these formulas are used frequently, it is convenient to take them as atomic formulas. The equality $X_i = X_j$ is expressed by $X_i \subseteq X_j \wedge X_j \subseteq X_i$ but it could also be added as another atomic formula.

We will denote by $C_r MS'(\mathcal{R}, \mathcal{X})$ (for $r \in \mathcal{N}$ and $\mathcal{X} \subseteq \mathcal{X}_\omega$) the set of monadic second-order formulas that have their free variables in $\mathcal{X}$ and are written with these atomic formulas for $i, j, i_1, \ldots, i_{\rho(\mathcal{R})} \in \mathcal{N}_+$ and $q \le r$, the Boolean connectives and existential quantifications on set variables. We will write $CMS'(\mathcal{R}, \mathcal{X})$ if we do not wish to bound q in the atomic formulas $Card_{p,q}(X_i)$.

We will denote by $mfv(\varphi)$ the maximal index $i \in \mathcal{N}_+$ of a free variable in a formula φ in $CMS'(\mathcal{R}, \mathcal{X}_\omega)$.

The following lemma says that the languages $C_r MS$ and $C_r MS'$ have the same expressive power.

Lemma 6.17

(1) Every formula φ in $C_r MS(\mathcal{R}, \{X_1, \ldots, X_p, x_{p+1}, \ldots, x_n\})$ can be translated into a formula ψ in $C_r MS'(\mathcal{R}, \{X_1, \ldots, X_p, X_{p+1}, \ldots, X_n\})$ of the same quantifier-height

and of size $O(|\varphi|)$, such that for every concrete $\mathcal{R}$-structure S and every n-tuple $(A_1, \ldots, A_n)$ of subsets of D_S:

$S \models \psi(A_1, \ldots, A_n)$ if and only if

there exist $d_{p+1}, \ldots, d_n \in D_S$ such that

$A_{p+1} = \{d_{p+1}\}, \ldots, A_n = \{d_n\}$ and $S \models \varphi(A_1, \ldots, A_p, d_{p+1}, \ldots, d_n)$.

(2) Conversely, every formula ψ in $C_r MS'(\mathcal{R}, \mathcal{X}_n)$ is equivalent to a formula in $C_r MS(\mathcal{R}, \mathcal{X}_n)$ of quantifier-height at most $qh(\psi) + \max\{2, \rho(\mathcal{R})\}$.

Proof: (1) Without loss of generality, we can assume that there is no index $i > 1$ such that the variables X_i and x_i both occur in φ. (Otherwise, we rename some bound variables to ensure this condition.) Then we take ψ equal to

$$Sgl(X_{p+1}) \wedge \cdots \wedge Sgl(X_n) \wedge \overline{\varphi},$$

where $\overline{\varphi}$ is obtained as follows from φ by induction on its structure:

(i) we replace the atomic formulas $x_i \in X_j$ and $x_i = x_j$ by $X_i \subseteq X_j$, and those of the form $R(x_{i_1}, \ldots, x_{i_k})$ by $R(X_{i_1}, \ldots, X_{i_k})$;
(ii) we replace $\exists x_i . \theta$ by $\exists X_i (Sgl(X_i) \wedge \overline{\theta})$.

The quantifier-height of ψ is that of φ. We have $|\psi| \leq 3 \cdot |\varphi| + 2(n-p)$. The correctness of the construction is clear. By the initial condition, no conflict of variable can occur from the replacement of x_i by X_i.

(2) The converse construction is straightforward. The quantifier-height may increase because the atomic formula $X_i \subseteq X_j$ is replaced by $\neg \exists x(x \in X_i \wedge x \notin X_j)$, the atomic formula $Sgl(X_i)$ is replaced by $\sigma(X_i)$ defined as $\exists x(x \in X_i \wedge \forall y(y \in X_i \Rightarrow x = y))$, and $R(X_{i_1}, \ldots, X_{i_k})$ is replaced by $\sigma(X_{i_1}) \wedge \cdots \wedge \sigma(X_{i_k}) \wedge \exists x_1, \ldots, x_k(x_1 \in X_{i_1} \wedge \cdots \wedge x_k \in X_{i_k} \wedge R(x_1, \ldots, x_k))$. ∎

An example will be given below (Example 6.23).

Remark 6.18 In the last clause of Definition 6.16, we can extend relations to have sets as arguments in other ways. Instead of $R(X_{i_1}, \ldots, X_{i_k})$, we can use:

$R^{\forall}(X_{i_1}, \ldots, X_{i_k})$ to mean:

$(d_1, \ldots, d_k) \in R_S$ for every $d_1 \in X_{i_1}, \ldots, d_k \in X_{i_k}$, and

$R^{\exists}(X_{i_1}, \ldots, X_{i_k})$ to mean:

$(d_1, \ldots, d_k) \in R_S$ for some $d_1 \in X_{i_1}, \ldots, d_k \in X_{i_k}$.

We can also combine both types of extensions and, for R binary, define $R^{\forall \exists}(X, Y)$ to mean: for every d in X there exists d' in Y such that $(d, d') \in R_S$. Lemma 6.17 remains valid with these variants. For example, the formula $R^{\forall}(X_i)$ is replaced by

$\neg\exists x(x \in X_i \wedge \neg R(x))$. These variants may simplify the constructions of automata to be developed below. □

The following notions will help to bound the sizes of automata constructed from formulas.

Definition 6.19 (Block quantifier-height and Boolean arity) A Boolean term b has *arity* $m \geq 1$ if it is in $T(\{\wedge, \vee, \neg\}, \{p_1, \dots, p_m\})$ and each variable p_i has at least one occurrence in b. It is *positive* if b is written without $\neg$. We denote by $Bool_m$ the set of Boolean terms (positive or not) of arity m. We will use the notation $b(\varphi_1, \dots, \varphi_m)$ for $b[\varphi_1/p_1, \dots, \varphi_m/p_m]$, where $\varphi_1, \dots, \varphi_m$ are logical formulas.

Let $\mathcal{R}$ be fixed (it will not appear in the notation). We let $\mathcal{L}$ denote the set of all CMS' (or CMS) formulas over $\mathcal{R}$, and we define subsets of $\mathcal{L}$ as follows:

$\mathcal{A}$ is the set of atomic formulas;

$\mathcal{B}_0$ is the set of formulas $b(\alpha_1, \dots, \alpha_m)$ such that $m \in \mathcal{N}_+$, $b \in Bool_m$ and $\alpha_1, \dots, \alpha_m$ are formulas in $\mathcal{A}$;

$\mathcal{B}_i$, for $i \geq 1$, is the set of formulas $b(\varphi_1, \dots, \varphi_m)$ such that $m \in \mathcal{N}_+$, $b \in Bool_m$ and $\varphi_1, \dots, \varphi_m$ are formulas in $\mathcal{E}_i \cup \dots \cup \mathcal{E}_1 \cup \mathcal{A}$ such that at least one of them is in $\mathcal{E}_i$;

$\mathcal{E}_i$, for $i \geq 1$, is the set of formulas $\exists \overline{X}.\varphi$, where $\overline{X}$ is a nonempty sequence of set variables, φ belongs to $\mathcal{B}_{i-1}$ and φ does not begin with a quantification.[16]

It is clear that each formula of $\mathcal{L}$ belongs to one and only one of the sets $\mathcal{B}_i$. Since the term p_1 is in $Bool_1$, we have $\mathcal{A} \subseteq \mathcal{B}_0$ and $\mathcal{E}_i \subseteq \mathcal{B}_i$ for $i \geq 1$.

The *block quantifier-height* of a formula φ is the integer i such that $\varphi \in \mathcal{B}_i$. We denote it by $bqh(\varphi)$. Its *Boolean arity* is the maximal arity m of a Boolean term b such that $b(\varphi_1, \dots, \varphi_m)$ is a subformula of φ for some pairwise distinct formulas $\varphi_1, \dots, \varphi_m$. We denote it by $ba(\varphi)$. For example, $bqh(\varphi) = 3$ and $ba(\varphi) = 2$ for φ defined as

$$\forall X_1 \exists X_3, X_4(edg(X_1, X_2) \wedge \exists X_5(edg(X_3, X_5) \wedge \neg edg(X_4, X_1))).$$

We will need another normalization lemma. It is a simple variant of the normalization in Proposition 5.93 (Section 5.6), involving renaming of bound variables only.

Definition 6.20 (n-normal formulas) For each $n \in \mathcal{N}$, we define certain formulas of CMS'$(\mathcal{R}, \mathcal{X}_\omega)$ as *n-normal* by structural induction:
(1) an atomic formula φ is *n-normal* if its variables are all in $\mathcal{X}_n$;
(2) a Boolean combination of formulas $\theta_1, \dots, \theta_k$ is *n-normal* if each formula θ_i is *n-normal*;
(3) a formula of the form $\exists X_i.\theta$ is *n-normal* if $i = n+1$ and θ is $(n+1)$-normal.

---

[16] We recall that $\forall X_1 \cdots$ is a shorthand for $\neg \exists X_1 \neg \cdots$, and that, e.g., $\exists X_1, X_2.\varphi$ abbreviates $\exists X_1 \exists X_2.\varphi$. Similarly, $\forall X_1, X_2.\varphi$ abbreviates $\forall X_1 \forall X_2.\varphi$, which, in turn, we will view as an abbreviation of $\neg \exists X_1 \exists X_2.\neg \varphi$ (not of $\neg \exists X_1 \neg \neg \exists X_2.\neg \varphi$).

If a formula has at least one quantifier, it is n-normal for at most one integer n determined by the leftmost quantified variable that must be X_{n+1}.

Lemma 6.21 If a formula φ is n-normal, then its set of quantified variables is $\{X_{n+1},\ldots,X_{n+qh(\varphi)}\}$ and its free variables are all in $\mathcal{X}_n$.

Proof: Straightforward by induction on the structure of φ. ∎

Lemma 6.22 Every formula φ in $C_r\mathrm{MS}'(\mathcal{R},\mathcal{X}_n)$ can be transformed into an equivalent n-normal formula $\widehat{\varphi}$ in $C_r\mathrm{MS}'(\mathcal{R},\mathcal{X}_n)$ by a renaming of bound variables.

Proof: It is clear that $\widehat{\varphi}$ can be obtained from φ by an appropriate renaming of its bound variables into the variables $X_{n+1},\ldots,X_{n+qh(\varphi)}$. Formally, the construction of $\widehat{\varphi}$ is by induction on the size of φ.

(1) If φ is atomic, we let $\widehat{\varphi}$ be φ.

(2) If φ is $\neg\varphi_1$ or $\varphi_1 \vee \varphi_2$ or $\varphi_1 \wedge \varphi_2$, then we let $\widehat{\varphi}$ be respectively $\neg\widehat{\varphi_1}$ or $\widehat{\varphi_1} \vee \widehat{\varphi_2}$ or $\widehat{\varphi_1} \wedge \widehat{\varphi_2}$.

(3) If φ is $\exists X_{n+1}.\theta$, then we let $\widehat{\varphi}$ be $\exists X_{n+1}.\widehat{\theta}$, where $\widehat{\theta}$ is the $(n+1)$-normal formula obtained from θ by the induction hypothesis.

(4) If φ is $\exists X_k.\theta$ with $k \neq n+1$, then we let θ' be $\theta[X_{n+1}/X_k]$. Note that X_{n+1} is not free in θ. The substitution of X_{n+1} to X_k may necessitate some renaming of bound variables in θ, but θ' has the same size as θ and its free variables are in $\mathcal{X}_{n+1}$. We let then $\widehat{\varphi}$ be $\exists X_{n+1}.\widehat{\theta'}$, where $\widehat{\theta'}$ is $(n+1)$-normal and obtained by the induction hypothesis. ∎

Hence, $\widehat{\varphi}$ has the same quantifier-height, block quantifier-height and Boolean arity as φ, and the same free variables.

Example 6.23 For expressing that X is the vertex set of a connected component of an undirected graph, we let $\varphi(X)$ be the following CMS formula written with first-order variables, universal quantifications and implications (and with some atomic CMS' formulas):

$$X \neq \emptyset \wedge \forall u,v(edg(u,v) \wedge u \in X \Rightarrow v \in X)$$
$$\wedge \neg\exists Y[Y \subseteq X \wedge \neg(X \subseteq Y) \wedge Y \neq \emptyset \wedge \forall u,v(edg(u,v) \wedge u \in Y \Rightarrow v \in Y)].$$

Here is an equivalent formula θ in $\mathrm{CMS}'(\mathcal{R}_s,\{X\})$:

$$X \neq \emptyset \wedge \neg\exists U,V(Sgl(U) \wedge Sgl(V) \wedge edg(U,V) \wedge U \subseteq X \wedge \neg(V \subseteq X))$$
$$\wedge \neg\exists Y[Y \subseteq X \wedge \neg(X \subseteq Y) \wedge Y \neq \emptyset$$
$$\wedge \neg\exists U,V(Sgl(U) \wedge Sgl(V) \wedge edg(U,V) \wedge U \subseteq Y \wedge \neg(V \subseteq Y))].$$

By replacing X by X_1 and by renaming the bound variables, we obtain the following 1-normal formula $\widehat{\theta}$:

$$X_1 \neq \emptyset$$
$$\wedge \neg \exists X_2, X_3 (Sgl(X_2) \wedge Sgl(X_3) \wedge edg(X_2, X_3) \wedge X_2 \subseteq X_1 \wedge \neg(X_3 \subseteq X_1))$$
$$\wedge \neg \exists X_2 [X_2 \subseteq X_1 \wedge \neg(X_1 \subseteq X_2) \wedge X_2 \neq \emptyset$$
$$\wedge \neg \exists X_3, X_4 (Sgl(X_3) \wedge Sgl(X_4) \wedge edg(X_3, X_4) \wedge X_3 \subseteq X_1 \wedge \neg(X_4 \subseteq X_1))].$$

These formulas have quantifier-height 3 and block quantifier-height 2 (they belong to $\mathcal{B}_2$). The first one has Boolean arity 4 and the last two have Boolean arity 5. The last one can be simplified by deleting $Sgl(X_2) \wedge Sgl(X_3)$ and $Sgl(X_3) \wedge Sgl(X_4)$, which are implied respectively by $edg(X_2, X_3)$ and $edg(X_3, X_4)$. The resulting formula has Boolean arity only 4.

6.3.3 Monadic second-order formulas on terms

For every (finite) functional signature F, we will use the relational signature $\mathcal{R}_F := \{son_i \mid 1 \leq i \leq k\} \cup \{lab_f \mid f \in F\}$ where k is $\rho(F)$, defined as $\max\{\rho(f) \mid f \in F\}$. A term t in $T(F)$ is faithfully represented by the $\mathcal{R}_F$-structure $\lfloor t \rfloor :=$ $\langle N_t, (son_{it})_{1 \leq i \leq k}, (lab_{ft})_{f \in F}\rangle$. Its domain N_t is the set of nodes of the syntactic tree of t. The set lab_{ft} is the set of occurrences of f in t and son_{it} is the set of pairs (u, v) such that v is the i-th son of u. Slightly different but equivalent relational structures have been introduced in Example 5.2(3) for representing terms.[17]

If φ is a sentence in $CMS(\mathcal{R}_F, \emptyset)$ or in $CMS'(\mathcal{R}_F, \emptyset)$, then we define

$$L_\varphi := \{t \in T(F) \mid \lfloor t \rfloor \models \varphi\}.$$

We will prove that L_φ is regular (this is one implication of Theorem 1.16 also proved in Corollary 5.67, Section 5.3) by constructing an F-automaton $\mathscr{A}_\varphi$ that recognizes L_φ. This construction will use an induction on the structure of φ, and thus we will need to define sets L_φ associated with formulas φ and not only with sentences.

We recall from Definition 6.16 that $\mathcal{X}_n$ denotes $\{X_1, \ldots, X_n\}$ and that $\mathcal{X}_\omega$ denotes the union of the sets $\mathcal{X}_n$.

Definition 6.24 (a) Let D be a set and γ be a mapping : $\mathcal{X}_n \to \mathcal{P}(D)$. The *characteristic vector relative to* γ of an element u of D is the n-tuple $\overline{w}_\gamma(u) = (w_1, \ldots, w_n) \in \{0, 1\}^n$ such that, for each i, we have $w_i = 1$ if and only if $u \in \gamma(X_i)$.

(b) If σ is a mapping $\mathcal{X}_n \to \mathcal{X}_m$, called a *substitution*, and γ is an assignment : $\mathcal{X}_m \to \mathcal{P}(D)$, then $\overline{w}_{\gamma \circ \sigma}(u) = h_\sigma(\overline{w}_\gamma(u))$, where h_σ is the mapping : $\{0, 1\}^m \to \{0, 1\}^n$ such

---

[17] In Example 5.2(3), we have defined a relational signature with a constant symbol rt for denoting the root. We will not use this symbol in this chapter. We have also defined a relational signature consisting of a single binary *son* relation and of unary relations br_i for $1 \leq i \leq k$ meaning that the considered node is the i-th son of its father. The corresponding representations of terms are equivalent for expressing properties of terms by first-order or monadic second-order formulas.

that $h_\sigma(w_1,\ldots,w_m) = (w_{\sigma'(1)},\ldots,w_{\sigma'(n)})$ and $\sigma(X_i) = X_{\sigma'(i)}$ for each $i = 1,\ldots,n$. The verification is straightforward from the definitions. For example, if $n = 4$, $m = 6$ and σ maps X_1 to X_2, X_2 to X_6 and, finally, X_3 and X_4 to X_3, then $h_\sigma(w_1,\ldots,w_6) = (w_2, w_6, w_3, w_3)$. Hence, if γ is an assignment : $\mathcal{X}_6 \to \mathcal{P}(D)$, if $U_i = \gamma(X_i)$ for $i = 1,\ldots,6$ and $W_i = \gamma(\sigma(X_i))$ for $i = 1,\ldots,4$, then we have $u \in W_1 \cap \overline{W_2} \cap W_3 \cap W_4$ if and only if $\overline{w}_{\gamma \circ \sigma}(u) = (1,0,1,1)$ if and only if $\overline{w}_\gamma(u) = (*,1,1,*,*,0)$, where $*$ means 0 or 1, if and only if $u \in U_2 \cap U_3 \cap \overline{U_6}$.

(c) For every $n \in \mathcal{N}$, we let $F^{(n)}$ be the signature $F \times \{0,1\}^n$, where, for every $w \in \{0,1\}^n$, the symbol (f,w) has arity $\rho(f)$. Let $t \in T(F)$. For every assignment $\gamma : \mathcal{X}_n \to \mathcal{P}(N_t)$, we let $t * \gamma$ be the term in $T(F^{(n)})$ such that $N_{t*\gamma} = N_t$, its relations son_i for $i = 1,\ldots,\rho(F)$ are the same as those of t, and $u \in Occ(t * \gamma, (f,w))$ if and only if $u \in Occ(t,f)$ and $w = \overline{w}_\gamma(u)$. The function symbol that occurs at an occurrence u of a symbol f in t is thus $(f,\overline{w}_\gamma(u))$ in $t * \gamma$. Obviously, every term in $T(F^{(n)})$ is of the form $t * \gamma$ for some t and γ.

(d) If $P(X_1,\ldots,X_n)$ is a property of sets of positions of terms in $T(F)$, we define $L_{P(X_1,\ldots,X_n)}$ as the language

$$\{t * \gamma \mid t \in T(F), \gamma : \mathcal{X}_n \to \mathcal{P}(N_t) \text{ and } P(\gamma(X_1),\ldots,\gamma(X_n)) \text{ is true for } t\}.$$

(e) If $\varphi \in \mathrm{CMS}'(\mathcal{R}_F, \mathcal{X}_n)$, we define

$$L_{\varphi,n} := \{t * \gamma \mid t \in T(F), \gamma : \mathcal{X}_n \to \mathcal{P}(N_t) \text{ and } (\lfloor t \rfloor, \gamma) \models \varphi\}.$$

Hence, $L_{\varphi,n}$ is defined if and only if $n \geq mfv(\varphi)$, where $mfv(\varphi)$ denotes the maximal index of a free variable in φ. We let L_φ denote $L_{\varphi,mfv(\varphi)}$. If φ and φ' are formulas with free variables in $\mathcal{X}_n$, then they are equivalent with respect to $\lfloor T(F) \rfloor$ if and only if $L_{\varphi,n} = L_{\varphi',n}$. $\qquad\square$

We will prove that the language $L_{P(X_1,\ldots,X_n)}$ is regular if and only if P is monadic second-order expressible. The "only if" direction follows from Theorem 5.15.[18] In the other direction, we will construct for every formula φ in $\mathrm{CMS}'(\mathcal{R}_F, \mathcal{X}_\omega)$ an automaton $\mathcal{A}_\varphi$ that recognizes the language L_φ. We will use an induction on the structure of φ, and for some subformulas ψ of φ we will have to construct automata $\mathcal{A}_{\psi,m}$ that recognize $L_{\psi,m}$ for $m > mfv(\psi)$: consider for example $\varphi = \psi \wedge \theta$ with ψ and θ such that $mfv(\psi) = 3$ and $mfv(\theta) = 5$. Then $L_\psi \cap L_\theta$ is always empty because $F^{(3)} \cap F^{(5)} = \emptyset$, hence it is not equal to L_φ in general. Clearly, $L_\varphi = L_{\psi,5} \cap L_\theta$. The same happens when φ expresses P and $mfv(\varphi) < n$ (which means that P does

---

[18] If $L_{P(X_1,\ldots,X_n)}$ is defined by an MS sentence φ over $\mathcal{R}_{F^{(n)}}$ (in which the variables $X_1,\ldots,X_n$ are not quantified), then P is monadic second-order expressible by the MS formula $\psi(X_1,\ldots,X_n)$ over $\mathcal{R}_F$ that is obtained from φ by changing every atomic subformula $lab_{(f,w)}(x)$ into the formula $lab_f(x) \wedge (\bigwedge_{w_i=1} x \in X_i) \wedge (\bigwedge_{w_i=0} x \notin X_i)$.

not depend on $X_{mfv(\varphi)+1},\ldots,X_n$): then an automaton $\mathscr{A}_{\varphi,n}$ must be constructed that recognizes $L_{\varphi,n}$.

This fact motivates the next lemma for which we define some notation. For every substitution $\sigma : \mathcal{X}_n \to \mathcal{X}_m$ and every formula φ in CMS$'(\mathcal{R}_F,\mathcal{X}_n)$, we let φ^σ be the formula $\varphi[\sigma(X_1)/X_1,\ldots,\sigma(X_n)/X_n]$ in CMS$'(\mathcal{R}_F,\mathcal{X}_m)$. We extend h_σ into an arity preserving mapping : $F^{(m)} \to F^{(n)}$ by letting $h_\sigma(f,w) = (f,h_\sigma(w))$.

Lemma 6.25 For every formula φ in CMS$'(\mathcal{R}_F,\mathcal{X}_n)$ and every substitution $\sigma : \mathcal{X}_n \to \mathcal{X}_m$, we have $L_{\varphi^\sigma,m} = h_\sigma^{-1}(L_{\varphi,n})$.

Proof: Let t be a term in $T(F)$ and γ be an assignment : $\mathcal{X}_m \to \mathcal{P}(N_t)$. Since φ^σ is $\varphi[\sigma(X_1)/X_1,\ldots,\sigma(X_n)/X_n]$, we obtain from Lemma 5.9 that $(\lfloor t \rfloor,\gamma) \models \varphi^\sigma$ if and only if $(\lfloor t \rfloor,\gamma \circ \sigma) \models \varphi$. Thus, $t*\gamma \in L_{\varphi^\sigma,m}$ if and only if $t*(\gamma \circ \sigma) \in L_{\varphi,n}$. From the observation made in Definition 6.24(b) that $h_\sigma(\overline{w}_\gamma(u)) = \overline{w}_{\gamma \circ \sigma}(u)$ for every u in the relevant set, here N_t, we get that $h_\sigma(t*\gamma) = t*(\gamma \circ \sigma)$. Hence, $t*\gamma \in L_{\varphi^\sigma,m}$ if and only if $t*\gamma \in h_\sigma^{-1}(L_{\varphi,n})$. Since every term in $T(F^{(m)})$ is of the form $t*\gamma$, this shows that $L_{\varphi^\sigma,m} = h_\sigma^{-1}(L_{\varphi,n})$. ∎

We continue with the example in Definition 6.24(b) of the substitution σ such that $h_\sigma(w_1,\ldots,w_6) = (w_2,w_6,w_3,w_3)$: from the language L_φ where φ is $X_1 \subseteq X_2 \wedge edg(X_3,X_4)$, we get $L_\psi = h_\sigma^{-1}(L_\varphi)$, where ψ is $X_2 \subseteq X_6 \wedge edg(X_3,X_3)$.

As another example, if $\varphi \in$ CMS$'(\mathcal{R}_F,\mathcal{X}_n)$ and $m > n$, then $L_{\varphi,m} = h_\sigma^{-1}(L_{\varphi,n})$, where h_σ is the mapping that removes the last $m-n$ Booleans of each symbol of $F^{(m)}$ (i.e., $\sigma(X_i) = X_i$ for $i = 1,\ldots,n$ and $h_\sigma(w_1,\ldots,w_m) = (w_1,\ldots,w_n)$). Hence, from an automaton recognizing $L_{\varphi,n}$ one can construct an automaton recognizing $L_{\varphi,m}$ by the inverse image construction (Proposition 6.11(2)).

The inductive construction of automata

Theorem 6.26 For every functional signature F and every formula φ in the set CMS$'(\mathcal{R}_F,\mathcal{X}_\omega)$, one can construct a complete and deterministic automaton $\mathscr{A}_\varphi$ that recognizes the language $L_\varphi \subseteq T(F^{(mfv(\varphi))})$. □

We will later bound the size of $\mathscr{A}_\varphi$ in terms of φ and F. We now give the proof of the theorem.

Let φ be as in the statement of the theorem. We have $L_\varphi = L_{\widehat{\varphi},mfv(\varphi)}$, where $\widehat{\varphi}$ is the $mfv(\varphi)$-normal formula constructed from φ by Lemma 6.22. Hence, it suffices to restrict attention to normal formulas and to construct, for every $n \in \mathcal{N}$ and every n-normal formula ψ in CMS$'(\mathcal{R}_F,\mathcal{X}_n)$, an automaton $\mathscr{A}_{\psi,n}$ that recognizes the language $L_{\psi,n}$. Then $\mathscr{A}_\varphi$ is the automaton $det(\mathscr{A}_{\widehat{\varphi},mfv(\varphi)})$. To construct $\mathscr{A}_{\psi,n}$ we will use an induction on the structure of ψ.

Table 6.1 *Transition rules of $\mathscr{A}_{Sgl(X_1)}$.*

Transition rules	Conditions
$(c, 0) \to 0$ $(c, 1) \to 1$	$\rho(c) = 0$
$(f, \alpha)[q_1, \ldots, q_{\rho(f)}] \to 0$	$\alpha = q_1 = \cdots = q_{\rho(f)} = 0$
$(f, \alpha)[q_1, \ldots, q_{\rho(f)}] \to 1$	exactly one of $\alpha, q_1, \ldots, q_{\rho(f)}$ is 1, the others being 0
$(f, \alpha)[q_1, \ldots, q_{\rho(f)}] \to Error$	all other cases

Construction 6.27 (Automata for atomic formulas) The atomic formulas are of the following forms:

$\qquad$ (0) $\quad$ *True*,

$\qquad$ (1) $\quad$ $Card_{p,q}(X_i)$, $X_i = \emptyset$, $Sgl(X_i)$, $lab_f(X_i)$,

$\qquad$ (2) $\quad$ $X_i \subseteq X_j$, $son_s(X_i, X_j)$,

with $f \in F$ and $1 \le s \le \rho(F)$.

For the constant *True* of type (0) (without free variables), we have $L_{True,n} = T(F^{(n)})$, hence the automaton is trivial to build (and actually not needed because *True* can easily be eliminated from formulas).

We need only construct automata recognizing the sets $L_{\psi,1}$, where ψ is of type (1) with $i = 1$, and the sets $L_{\psi,2}$ where ψ is of type (2) with $i = 1, j = 2$. For the other cases, we can use an appropriate substitution σ by Lemma 6.25 and the inverse image construction for automata (Proposition 6.11(2)).

All automata will be complete and deterministic. We will only detail two cases, the other ones being very similar.

Case 1: The automaton $\mathscr{A}_{Sgl(X_1)}$.

We let $Q := \{0, 1, Error\}$ be the set of states. The automaton must check, for every term over $F^{(1)}$, that one and only one occurrence of a symbol in that term belongs to $F \times \{1\}$. This is of course straightforward, but we take advantage of this very simple case to explain how we will specify automata. The transition rules are in Table 6.1. The automaton is complete and deterministic, and 1 is its unique accepting state. By the last line of Table 6.1, we have $(f, \alpha)[\ldots, Error, \ldots] \to Error$.

Convention: A state named *Error* is never accepting (but some other states may not be accepting). For every transition $(f, \alpha)[\ldots, Error, \ldots] \to q$ we have $q = Error$, hence

Table 6.2 *Meanings of the states of* $\mathscr{A}_{Sgl(X_1)}$.

State q	Property P_q
0	$X_1 = \emptyset$
1	$Sgl(X_1)$
Error	$Card(X_1) \geq 2$

a recognized "error" propagates bottom-up. All transitions not listed yield *Error*. By using such states, we can make automata complete. Reducing these automata would eliminate the states *Error* and make them incomplete, but we have seen in Proposition 6.10 that complete automata are necessary for defining Boolean combinations of languages. Of course, when we apply the complementation operation to a complete deterministic automaton having a state *Error*, since this state becomes accepting in the new automaton, we rename it in order to satisfy this convention. □

In order to make the construction of $\mathscr{A}_{Sgl(X_1)}$ clear and its correctness proof easy (but we will not do that proof), we specify in Table 6.2 the meanings of the different states of the constructed automaton $\mathscr{A} := \mathscr{A}_{Sgl(X_1)}$. That is, for every term $t * \gamma \in T(F^{(1)})$:

$$t * \gamma \in L(\mathscr{A}, q) \text{ if and only if } (\lfloor t \rfloor, \gamma(X_1)) \text{ satisfies } P_q. \tag{6.1}$$

In Table 6.2, we write X_1 instead of $\gamma(X_1)$ for the purpose of clarity. We will do the same below in all similar tables. Here is an equivalent formulation of Equivalence (6.1). If r is the unique run of $\mathscr{A}$ on a term $t * \gamma$ and $u \in N_t$, then $r(u) = q$ if and only if $(\lfloor t \rfloor/u, \gamma(X_1) \cap (N_t/u))$ satisfies P_q. (We recall from Example 5.2(3) that $\lfloor t \rfloor/u$ is the $\mathcal{R}_F$-structure $\lfloor Syn(t)/u \rfloor$, with domain N_t/u: the set of descendants of u in t, including u.) Property P_q is thus the *meaning* of the state q. It describes its role with respect to the property of t to be checked. The three properties listed in this table are mutually exclusive and cover all cases: this means that they specify the behavior of a complete and deterministic automaton. All automata specified below by similar tables will be complete and deterministic.

It is now quite easy, by using an induction on the structure of t, to prove Equivalence (6.1). In fact, technically it is more convenient to prove the equivalent formulation of Equivalence (6.1) by bottom-up induction on the node u.

Case 2: The automaton $\mathscr{A}_{son_2(X_1,X_2)}$.

It has four states: $0, 2, Ok, Error$, whose meanings are shown in Table 6.3. In this table, r denotes the root of the syntactic tree of the considered term t (or of $Syn(t)/u$ in the equivalent formulation of Equivalence (6.1)). The accepting state is Ok. The

Table 6.3 *Meanings of the states of $\mathscr{A}_{son_2(X_1,X_2)}$.*

State q	Property P_q
0	$X_1 = X_2 = \emptyset$
2	$X_1 = \emptyset, X_2 = \{r\}$
Ok	$son_2(X_1, X_2)$
Error	all other cases

Table 6.4 *Transition rules of $\mathscr{A}_{son_2(X_1,X_2)}$.*

Rules	Conditions
$(f,00)[q_1,\ldots,q_{\rho(f)}] \to 0$ $(f,01)[q_1,\ldots,q_{\rho(f)}] \to 2$	$\rho(f) \geq 0$, $q_1 = q_2 = \cdots = q_{\rho(f)} = 0$
$(f,10)[q_1,\ldots,q_{\rho(f)}] \to Ok$	$\rho(f) \geq 2$, $q_2 = 2$, $q_1 = q_3 = \cdots = q_{\rho(f)} = 0$
$(f,00)[q_1,\ldots,q_{\rho(f)}] \to Ok$	$\rho(f) \geq 1$ and for some i, $q_i = Ok$ and $q_j = 0$ for all $j \neq i$

transition rules are shown in Table 6.4. In this table (and in the similar ones below), we shorten $(0,0)$ into 00, $(0,1)$ into 01, etc. The automaton is complete and deterministic.

Construction 6.28 (Automata for Boolean compositions) We will consider formulas in the sets $\mathcal{B}_i$. Recall that we restrict attention to normal formulas. We let ψ be an n-normal formula of the form $b(\theta_1,\ldots,\theta_m)$ for some Boolean term b. The formulas θ_i are thus also n-normal. We assume that, for each i, we have constructed the automaton $\mathscr{A}_{\theta_i,n}$ recognizing $L_{\theta_i,n}$. We have $L_{\psi,n} = b(L_{\theta_1,n},\ldots,L_{\theta_m,n})$, where the Boolean operations $\vee$, $\wedge$ and $\neg$ are interpreted as union, intersection and complementation with respect to $T(F^{(n)})$. The language $L_{\psi,n}$ is thus recognized by an automaton $\mathscr{A}_{\psi,n}$ obtained from $det(\mathscr{A}_{\theta_1,n}),\ldots,det(\mathscr{A}_{\theta_m,n})$ by Proposition 6.10. If b is positive, i.e., without negation, we need not determinize the automata $\mathscr{A}_{\theta_1,n},\ldots,\mathscr{A}_{\theta_m,n}$. We need only assume they are complete.

Construction 6.29 (Automata for quantified formulas) We now consider formulas in the sets $\mathcal{E}_i$. We let ψ be n-normal and of the form $\exists X_{n+1},X_{n+2},\ldots,X_m.\theta$ where θ is m-normal and does not begin with a quantification. (Hence, θ is atomic or is a Boolean combination of formulas). We assume that an automaton $\mathcal{A}_{\theta,m}$ has already been constructed.

We let $h : F^{(m)} \to F^{(n)}$ be the arity preserving mapping that associates $(f,(w_1,\ldots,w_n))$ in $F^{(n)}$ with $(f,(w_1,\ldots,w_m))$ in $F^{(m)}$ by erasing the last $m - n$

Booleans. It is clear from the definitions that $L_{\psi,n} = h(L_{\theta,m})$. Hence, $L_{\psi,n} = L(h(\mathcal{A}_{\theta,m}))$ by the direct image construction for automata (Proposition 6.11(1)). We can take $\mathcal{A}_{\psi,n}$ equal to $h(\mathcal{A}_{\theta,m})$ or to $det(h(\mathcal{A}_{\theta,m}))$ in order to have a complete and deterministic automaton.

Proof of Theorem 6.26: The theorem follows immediately from Constructions 6.27 to 6.29. ∎

Bounding the sizes of the constructed automata

Our aim is to derive from these constructions an upper-bound to $\sharp\mathcal{A}_\varphi$, the number of states of the complete and deterministic automaton $\mathcal{A}_\varphi$. This will give a bound on its size $\|\mathcal{A}_\varphi\|$ (cf. Definition 3.46 in Section 3.3) since it is at most

$$2^{mfv(\varphi)} \cdot |F| \cdot (\rho(F) + 1) \cdot (\sharp\mathcal{A}_\varphi)^{\rho(F)+1},$$

where $\rho(F)$ is the maximal arity of F (and of $F^{(mfv(\varphi))}$).

In the following evaluation, we will assume that complete and deterministic automata are constructed at each step. Minimizing them reduces their sets of states, but we cannot evaluate how much. Constructions 6.27 to 6.29 distinguish three types of formulas: the atomic formulas in $\mathcal{A}$, those in the sets $\mathcal{B}_i$ and those in the sets $\mathcal{E}_i$. We will bound $\sharp\mathcal{A}_\varphi$ in terms of (i) the maximum of $\sharp\mathcal{A}_\psi$ for the atomic subformulas ψ of φ, (ii) the Boolean arity of φ and (iii) its block quantifier-height, by considering in turn each of these constructions.

Atomic formulas, cf. Construction 6.27: By Lemma 6.25 and Proposition 6.11(2) (about substitutions and the inverse image construction), we need only construct complete and deterministic automata for atomic formulas with variables either X_1 or X_1 and X_2, because by these results we can transform these automata into complete and deterministic ones with the same sets of states, over the larger signatures $F^{(n)}$. The automata for *True* have one state, those for $X_i = \emptyset$ and $X_i \subseteq X_j$ have two states, those for $Sgl(X_i)$ and $lab_f(X_i)$ (for $f \in F$) have three states, those for $son_s(X_i, X_j)$ (for $1 \leq s \leq \rho(F)$) have four states, and finally, the automata $\mathcal{A}_{Card_{p,q}(X_1)}$ have q states. Thus, for a formula φ in $C_r MS'(\mathcal{R}_F, \mathcal{X}_\omega)$, we have $\sharp\mathcal{A}_\psi \leq \max\{r, 4\}$ for every atomic subformula ψ of φ.

Boolean combinations, cf. Construction 6.28: We use the product of complete and deterministic automata $\mathcal{A}_{\theta_1,n}, \ldots, \mathcal{A}_{\theta_m,n}$ for constructing an automaton $\mathcal{A}_{\psi,n}$ for ψ of the form $b(\theta_1, \ldots, \theta_m)$. Since we may assume that $\theta_1, \ldots, \theta_m$ are pairwise distinct, we obtain by Proposition 6.10 a complete and deterministic automaton such that

$$\sharp\mathcal{A}_{\psi,n} \leq \sharp\mathcal{A}_{\theta_1,n} \times \cdots \times \sharp\mathcal{A}_{\theta_m,n} \leq N^{ba(\psi)},$$

if $\sharp\mathcal{A}_{\theta_i,n} \leq N$ for each i.

Existential quantifications, cf. Construction 6.29: If φ is $\exists \overline{X}.\theta$, then $\mathscr{A}_{\psi,n}$ is obtained from $\mathscr{A}_{\theta,m}$ by the direct image construction (with $m := n + |\overline{X}|$). By Proposition 6.11(1), this construction produces a nondeterministic automaton with the same number of states. It has degree of nondeterminism 2^{m-n} (cf. Section 6.3.1). Since we determinize automata at each step, we get

$$\sharp \mathscr{A}_{\psi,n} \leq 2^{\sharp \mathscr{A}_{\theta,m}}.$$

Corollary 6.30 Let φ be a formula of block quantifier-height h and Boolean arity m. Let a be the maximal number of states of a complete and deterministic automaton constructed for an atomic subformula of φ. The number of states of the automaton $\mathscr{A}_\varphi$, and of the automata constructed during the construction of $\mathscr{A}_\varphi$, is bounded by $\exp(h, m(a^m + h))$.

Proof: By Lemma 6.22 and the proof of Theorem 6.26, it suffices to prove this for every n-normal formula ψ and the automaton $\mathscr{A}_{\psi,n}$.

We let $E : \mathcal{N} \times \mathcal{N}_+ \times \mathcal{N}_+ \to \mathcal{N}_+$ be the total function such that $E(0,x,m) = x$ and $E(h+1,x,m) = 2^{m \cdot E(h,x,m)}$ for all $h \in \mathcal{N}$ and $x,m \in \mathcal{N}_+$. It is easy to see that E is monotonic in each of its arguments. By elementary calculations,[19] we can see that $E(h,x,m) \leq \exp(h, m(x + h))$ for all h,x,m. Hence it suffices to prove the following claim.

Claim 6.30.1 If ψ belongs to $\mathcal{B}_h$, if $m = ba(\psi)$ and a is as in the statement of the corollary (for ψ), then we have $\sharp \mathscr{A}_{\psi,n} \leq E(h,a^m,m)$.

Proof: If $h = 0$, then $\psi \in \mathcal{B}_0$ and $\sharp \mathscr{A}_{\psi,n} \leq a^m = E(0,a^m,m)$.

If ψ belongs to $\mathcal{E}_{h+1}$ and is $\exists \overline{X}.\theta$ with $\theta \in \mathcal{B}_h$, then $\sharp \mathscr{A}_{\psi,n} \leq 2^{E(h,a^m,m)}$, since $\sharp \mathscr{A}_{\theta,n+|\overline{X}|} \leq E(h,a^m,m)$.

If $\psi \in \mathcal{B}_{h+1}$, then $\psi = b(\theta_1,\ldots,\theta_{m'})$, $m' \leq m$, with $\theta_1,\ldots,\theta_{m'} \in \mathcal{E}_{h+1} \cup \cdots \cup \mathcal{E}_1 \cup \mathcal{A}$, hence

$$\sharp \mathscr{A}_{\psi,n} \leq (\max\{\sharp \mathscr{A}_{\theta_1,n},\ldots,\sharp \mathscr{A}_{\theta_{m'},n}\})^m$$
$$\leq (2^{E(h,a^m,m)})^m = E(h+1,a^m,m).$$

We have $\sharp \mathscr{A}_{\theta_i,n} \leq 2^{E(h,a^m,m)}$ because E is monotonic in each of its arguments.

This completes the proof of the claim and, thus, of the corollary. ∎

Remark 6.31 (1) The bound $\exp(h, m(a^m + h))$ is also an upper-bound to the number of states of the minimal automaton of the language L_φ, hence of its recognizability index (cf. Proposition 3.73(2)). This upper-bound is better than the one obtained from

---

[19] We first observe that $m \cdot \exp(h,x) \leq \exp(h,x+m)$ if $h > 0$, and then we prove the inequality by induction on h.

Theorem 5.96 in Section 5.6 (for sentences φ), because it has less levels of exponentiation. However, the implementation of the algorithm of Theorem 6.26 remains problematic due to the lack of memory that occurs during the computation.

(2) For $n \geq mfv(\varphi)$, the language $L_{\varphi,n}$ is equal to $L_{P(X_1,\ldots,X_n)}$, where $P(X_1,\ldots,X_n)$ is the property expressed by φ (cf. Definition 6.24). By Proposition 3.73(1), minimization of the automaton $\mathscr{A}_{\varphi,n}$ leads to the (unique) minimal automaton $\mathscr{M}(P) :=$ $\mathscr{M}(L_{P(X_1,\ldots,X_n)})$ that recognizes the language $L_{P(X_1,\ldots,X_n)}$. However, $\mathscr{A}_{\varphi,n}$ may be of much larger size than $\mathscr{M}(P)$. Hence, even if $\mathscr{M}(P)$ is of reasonable size, its computation may fail due to the too large size of $\mathscr{A}_{\varphi,n}$. $\qquad\square$

Corollary 6.30 gives upper-bounds to the sizes of the automata constructed by the algorithm used to prove Theorem 6.26. We will present two notions intended to facilitate effective constructions of automata: the direct computation of automata for frequently used properties and the use of Boolean set terms in the writing of formulas.

Precomputed automata

Remark 6.31(2) motivates the following idea. For expressing properties of terms (handled as syntactic trees) certain MS-expressible notions are frequently used and can be "precompiled" into "small" automata. This will reduce the nesting level of determinizations in the construction of the automaton for a formula using these properties. We only give one example.

Least common ancestor: For $t \in T(F)$, we let $Lca(X_1,X_2)$ be the property of sets $X_1, X_2 \subseteq N_t$ defined by:

$X_1 \neq \emptyset$ and $X_2 = \{u\}$, where u is the least common ancestor of the nodes in X_1,

which we denote by $u = lca(X_1)$. It can be expressed by a monadic second-order formula $\varphi(X_1,X_2)$ (we will give it below) but this formula is not needed for the following construction. We define directly (without using φ) a complete and deterministic automaton $\mathscr{A}$ that recognizes the language $L_{Lca(X_1,X_2)}$.

Its states are $0, 1, Ok$ and $Error$ with meanings defined by Table 6.5. The accepting state is Ok. The corresponding transition rules are given in Table 6.6, with $k = \rho(f)$.

The property $Lca(X_1,X_2)$ is expressed by $\varphi(X_1,X_2)$ defined as

$$X_1 \neq \emptyset \wedge Ub(X_2,X_1) \wedge \forall X_3(Ub(X_3,X_1) \Rightarrow Ub(X_3,X_2)),$$

where $Ub(X,Y)$ means that $X = \{u\}$ and u is an ancestor of each element of Y (possibly $u \in Y$). The latter property can be expressed by the formula

$$Sgl(X) \wedge \forall Z(X \subseteq Z \wedge \forall V, W[V \subseteq Z \wedge son(V,W) \Rightarrow W \subseteq Z] \Rightarrow Y \subseteq Z),$$

Table 6.5 *Meanings of the states of the automaton $\mathcal{A}$.*

State q	Property P_q
0	$X_1 = X_2 = \emptyset$
1	$X_1 \neq \emptyset \wedge X_2 = \emptyset$
Ok	$X_1 \neq \emptyset \wedge X_2 = \{lca(X_1)\}$
Error	all other cases

Table 6.6 *Transition rules of the automaton $\mathcal{A}$.*

Transition rules	Conditions
$(f,00)[q_1,\ldots,q_k] \to 0$	$k \geq 0$, $q_1 = q_2 = \cdots = q_k = 0$
$(f,00)[q_1,\ldots,q_k] \to Ok$	$k \geq 1$ and for some i, $q_i = Ok$ and $q_j = 0$ for $j \neq i$
$(f,00)[q_1,\ldots,q_k] \to 1$	$k \geq 1$ and for some i, $q_i = 1$ and $q_j \in \{0,1\}$ for $j \neq i$
$(f,01)[q_1,\ldots,q_k] \to Ok$	$k \geq 2$, at least two of $q_1,\ldots,q_k$ are 1, the others being 0
$(f,11)[q_1,\ldots,q_k] \to Ok$ $(f,10)[q_1,\ldots,q_k] \to 1$	$k \geq 0$, $q_i \in \{0,1\}$ for all i

where $son(V,W)$ stands for the disjunction of the atomic formulas $son_i(V,W)$ for $i = 1,\ldots,\rho(F)$.

The constructed automaton is actually minimal (because for any two states q and q', there exist terms $t \in L(\mathcal{A},q)$, $t' \in L(\mathcal{A},q')$ and a context $c \in Ctxt(F^{(2)})$ such that $c[t] \in L_{Lca(X_1,X_2)}$ and $c[t'] \notin L_{Lca(X_1,X_2)}$ or vice versa; cf. Section 3.4.3, Proposition 3.73(1)). However, the application to φ of the general constructions of Theorem 6.26 involves lengthy and space-consuming computations. Hence, for applying them, as done for example in the software MONA ([Hen+, Kla]), it seems preferable to use monadic second-order formulas constructed with more atomic formulas than those listed in Construction 6.27. In this way, we extend the syntax of monadic second-order formulas without extending their expressive power.

Boolean set terms

Another method for lowering the block quantifier-heights of formulas consists in writing them with Boolean set terms. We first give examples. Assume that we need to express the set property[20] $X_1 - X_2 \subseteq X_3$. It is equivalent to $X_1 \subseteq X_2 \cup X_3$, hence can be expressed by the formula $\varphi(X_1, X_2, X_3)$ defined as

$$\forall U(X_2 \subseteq U \wedge X_3 \subseteq U \Rightarrow X_1 \subseteq U),$$

but it cannot be expressed by a quantifier-free formula in CMS'$(\emptyset, \{X_1, X_2, X_3\})$. However, $L_\varphi = h^{-1}(L_{X_1 \subseteq X_2})$, where h is the mapping : $\{0,1\}^3 \to \{0,1\}^2$ such that, for every $i, j, k \in \{0,1\}$, we have $h((i,j,k)) := (i, \max\{j,k\})$.

Similarly, if φ' is intended to express that the set $X_1 \cap X_2$ has even cardinality, then we have $L_{\varphi'} = h^{-1}(L_{Card_{0,2}(X_1)})$, where $h : \{0,1\}^2 \to \{0,1\}$ is such that $h((i,j)) := \min\{i,j\}$.

The following definition applies to other relational signatures than $\mathcal{R}_F$.

Definition 6.32 (Boolean set terms) Let $\mathcal{R}$ be a relational signature without constant symbols.

A *Boolean set term* is a term in $T(B, \mathcal{X}_n)$, where $B := \{\cup, \cap, ^-\}$ and $^-$ is the unary operation of complementation. We extend the language CMS'$(\mathcal{R}, \mathcal{X}_n)$ by allowing atomic formulas of the forms $Card_{p,q}(t_1)$, $t_1 \subseteq t_2$, $t_1 = \emptyset$, $Sgl(t_1)$ and $R(t_1, \ldots, t_{\rho(R)})$, where $t_1, t_2, \ldots, t_{\rho(R)}$ are Boolean set terms. Examples of Boolean set terms are X_2, $X_3 \cap \overline{X_1}$ and $(X_3 \cap \overline{X_1}) \cup (\overline{X_3} \cap X_4)$. Their interpretations and those of the atomic formulas built with them is clear from the syntax. We denote by CMS$^{Bool}(\mathcal{R}, \mathcal{X}_n)$ the corresponding set of formulas. It contains CMS'$(\mathcal{R}, \mathcal{X}_n)$.

A *Boolean substitution* is a mapping $\sigma : \mathcal{X}_n \to T(B, \mathcal{X}_m)$. For every formula $\varphi \in \text{CMS}^{Bool}(\mathcal{R}, \mathcal{X}_n)$, we let φ^σ be the formula $\varphi[\sigma(X_1)/X_1, \ldots, \sigma(X_n)/X_n]$ in CMS$^{Bool}(\mathcal{R}, \mathcal{X}_m)$.

We now apply this definition to the formulas that express properties of terms. Let $t \in T(F)$ and $\gamma : \mathcal{X}_m \to \mathcal{P}(N_t)$. It is clear (from similar results in Section 5.2.1) that $(\lfloor t \rfloor, \gamma) \models \varphi^\sigma$ if and only if $(\lfloor t \rfloor, \delta) \models \varphi$, where δ is the assignment : $\mathcal{X}_n \to \mathcal{P}(N_t)$ defined by $\delta(X_i) := \sigma(X_i)_{\mathbb{P}(N_t)}(\gamma(X_1), \ldots, \gamma(X_m))$ and $\mathbb{P}(N_t)$ is the Boolean algebra of subsets of N_t. It is easy to check that $\overline{w}_\delta(u) = h_\sigma(\overline{w}_\gamma(u))$ for every $u \in N_t$, where h_σ is the mapping : $\{0,1\}^m \to \{0,1\}^n$ such that, for every $w \in \{0,1\}^m$, we have $h_\sigma(w) := (\sigma(X_1)_{\mathbb{B}}(w), \ldots, \sigma(X_n)_{\mathbb{B}}(w))$ and $\mathbb{B}$ is the Boolean algebra $\{0,1\}$. As before, we extend h_σ into an arity preserving mapping : $F^{(m)} \to F^{(n)}$ by letting

---

[20] For a more complicated example, consider $X \subseteq Y \cup (Z - V)$. It is expressed by $\forall U[Y \subseteq U \wedge Z - V \subseteq U \Rightarrow X \subseteq U]$. Since $Z - V \subseteq U$ is expressed by $\forall W(U \subseteq W \wedge V \subseteq W \Rightarrow Z \subseteq W)$, the property $X \subseteq Y \cup (Z - V)$ is expressed by the formula $\forall U[Y \subseteq U \wedge \forall W(U \subseteq W \wedge V \subseteq W \Rightarrow Z \subseteq W) \Rightarrow X \subseteq U]$ of (block) quantifier-height 2.

$h_\sigma(f, w) = (f, h_\sigma(w))$. The following lemma generalizes Lemma 6.25 and its proof is the same.

Lemma 6.33 For every formula φ in $\mathrm{CMS}^{Bool}(\mathcal{R}_F, \mathcal{X}_n)$ and every Boolean substitution $\sigma : \mathcal{X}_n \to T(B, \mathcal{X}_m)$, we have $L_{\varphi^\sigma, m} = h_\sigma^{-1}(L_{\varphi, n})$.

By applying this lemma to the atomic formulas φ of Construction 6.27, we can extend Theorem 6.26 to formulas in $\mathrm{CMS}^{Bool}(\mathcal{R}_F, \mathcal{X}_n)$. Formulas in this set are of lower block quantifier-height than the equivalent formulas in $\mathrm{CMS}'(\mathcal{R}_F, \mathcal{X}_n)$. Since the inverse image construction (cf. Proposition 6.11(2)) does not increase the number of states, using them will necessitate less determinizations of intermediate automata than if we start from the equivalent formulas that do not use Boolean set terms. The expression of Corollary 6.30 that bounds the number of states of the constructed automata is the same but it will be applied to formulas of smaller block quantifier-height (the value h in Corollary 6.30).

6.3.4 Monadic second-order properties of graphs of bounded clique-width

We will construct automata for checking the monadic second-order properties of simple (unlabeled) p-graphs of bounded clique-width. Extending the constructions to simple (K, Λ)-labeled p-graphs of bounded clique-width will be straightforward.

In the next subsection, we will do the same for graphs of bounded tree-width defined by terms over the signature F_C^{HR}. We first present the construction for p-graphs of bounded clique-width defined by terms over F_C^{VR} because it is simpler. (However, the parsing problem for the VR algebra is more difficult, as we have seen in Section 6.2.)

Let $C := \{a, b, \dots\}$ be a finite set of port labels. A simple p-graph G of type $\pi(G) \subseteq C$ is represented faithfully by the $\mathcal{R}_{s,C}$-structure $\lfloor G \rfloor_C = \langle V_G, edg_G, (lab_{aG})_{a \in C} \rangle$, where some of the sets lab_{aG} may be empty.

We will denote these p-graphs by terms over the finite functional signature F_C^{VR}. We recall that it consists of the function symbols $\oplus$, $\overrightarrow{add}_{a,b}$, $add_{a,b}$ and $relab_h$ such that $a, b \in C$, $a \neq b$ and $h \in [C \to C]$, and of the constant symbols $\varnothing$, $\mathbf{a}$ and $\mathbf{a}^\ell$ for $a \in C$. We let $\mathbf{C} := \{\mathbf{a}, \mathbf{a}^\ell \mid a \in C\}$. Although the constant symbol $\varnothing$ that denotes the empty graph can be eliminated from the terms that denote nonempty graphs, it will be useful in certain constructions.

Every term t in $T(F_C^{\mathrm{VR}})$ evaluates to a concrete p-graph $G = cval(t) = cval(t_{Id})$ of clique-width at most $|C|$, with vertex set V_G equal to $Occ_0(t)$, the set of occurrences in t of the constant symbols different from $\varnothing$ (cf. Section 2.5.2). Hence, in order to encode an assignment $\gamma : \mathcal{X}_n \to \mathcal{P}(V_G)$ in a term t evaluating to G, we need only attach Booleans to the constant symbols in $\mathbf{C}$. We will denote by $F_C^{\mathrm{VR}(n)}$ the signature F_C^{VR}, where $\mathbf{C}$ is replaced by the set of constant symbols $\mathbf{C}^{(n)}$ defined as $\{(\mathbf{c}, w) \mid \mathbf{c} \in \mathbf{C}, w \in \{0, 1\}^n\}$. Hence, we use a slight modification of the notation of Definition 6.24(c). For $t \in T(F_C^{\mathrm{VR}})$ and $\gamma : \mathcal{X}_n \to \mathcal{P}(Occ_0(t))$ we define $t * \gamma$ to be the

Table 6.7 *Transition rules of* $\mathscr{B}_{Sgl(X_1)}$.

Transition rules	Conditions
$\varnothing \to 0$	
$(\mathbf{c},0) \to 0$	$\mathbf{c} \in \mathbf{C}$
$(\mathbf{c},1) \to 1$	
$f[0] \to 0, f[1] \to 1$	f is $\overrightarrow{add}_{a,b}$ or $add_{a,b}$ or $relab_h$
$\oplus[0,0] \to 0$	
$\oplus[0,1] \to 1$	
$\oplus[1,0] \to 1$	

term in $T(F_C^{VR(n)})$ obtained from t by replacing $\mathbf{c}$ by $(\mathbf{c}, \overline{w}_\gamma(u))$ at each occurrence u of $\mathbf{c} \in \mathbf{C}$. Every term in $T(F_C^{VR(n)})$ is of the form $t * \gamma$.

For $\varphi \in CMS'(\mathcal{R}_{s,C}, \mathcal{X}_n)$, we define $L_{C,\varphi,n}^{VR}$ as the set of terms $t * \gamma$ in $T(F_C^{VR(n)})$ such that $(\lfloor cval(t)\rfloor_C, \gamma) \models \varphi$ and we denote the set $L_{C,\varphi,mfv(\varphi)}^{VR}$ by $L_{C,\varphi}^{VR}$. We will also use the notation $L_{C,P(X_1,...,X_n)}^{VR}$ for $L_{C,\varphi,n}^{VR}$ if the n-ary property P is expressed by φ and $mfv(\varphi) \leq n$. In most cases, the set C will be fixed by the context and we will omit the subscript C in the notation $L_{C,\varphi,n}^{VR}$ and the related ones. The exponent VR is intended to distinguish these languages from the languages $L_{\varphi,n}$ of the previous section.

There is an obvious variant of Lemma 6.25: $L_{C,\varphi^\sigma,m}^{VR} = h_\sigma^{-1}(L_{C,\varphi,n}^{VR})$ for every formula φ in $CMS'(\mathcal{R}_{s,C}, \mathcal{X}_n)$ and every substitution $\sigma : \mathcal{X}_n \to \mathcal{X}_m$, where h_σ is extended into an arity preserving mapping : $F_C^{VR(m)} \to F_C^{VR(n)}$ by letting $h_\sigma(\mathbf{c}, w) = (\mathbf{c}, h_\sigma(w))$ for $\mathbf{c} \in \mathbf{C}$ and $h_\sigma(f) = f$ for $f \notin \mathbf{C}^{(m)}$. The proof is similar to the one of Lemma 6.25.

By an inductive construction similar to that of Section 6.3.3, we can construct, for each formula φ (assumed without loss of generality to be $mfv(\varphi)$-normal), an automaton $\mathscr{B}_{C,\varphi}$ (denoted by $\mathscr{B}_\varphi$ if C is specified by the context) that recognizes the language $L_{C,\varphi}^{VR}$. For formulas that are not atomic, obvious variants of Constructions 6.28 and 6.29 can be used.[21] We will only construct complete and deterministic automata for atomic formulas. Furthermore, we will only do that for formulas with variables X_1 and X_2 because substitutions and inverse images give the general case by (the variant of) Lemma 6.25 and Proposition 6.11(2).

Construction 6.34 (Automata for atomic formulas over $\mathcal{R}_{s,C}$**)** For constructing $\mathscr{B}_{Sgl(X_1)}$, we only modify Table 6.1 appropriately, which gives Table 6.7. The accepting state is 1. By our convention on missing transitions, we also have the transition $\oplus[1,1] \to Error$, which is not listed in the table.

---

[21] In Construction 6.29, the arity preserving mapping $h : F_C^{VR(m)} \to F_C^{VR(n)}$ erases the last $m - n$ Booleans associated to every $\mathbf{c} \in \mathbf{C}$ and is the identity on $F_C^{VR(m)} - \mathbf{C}^{(m)}$.

Table 6.8 *Meanings of the states of* $\mathscr{B}_{edg(X_1,X_2)}$.

State q	Property P_q
0	$X_1 = X_2 = \emptyset$
Ok	$X_1 = \{v_1\}, X_2 = \{v_2\}, edg_{cval(t)}(v_1,v_2)$ for some v_1, v_2 in $V_{cval(t)}$
$a(1)$	$X_2 = \emptyset, X_1 = \{v_1\}, port_{cval(t)}(v_1) = a$
$a(2)$	$X_1 = \emptyset, X_2 = \{v_2\}, port_{cval(t)}(v_2) = a$
ab	$X_1 = \{v_1\}, X_2 = \{v_2\}, port_{cval(t)}(v_1) = a,$ $port_{cval(t)}(v_2) = b, v_1 \neq v_2, \neg edg_{cval(t)}(v_1,v_2)$
$Error$	all other cases

Similarly, the automata for $Card_{p,q}(X_1)$ (with q states), $X_1 \subseteq X_2$ and $X_1 = \emptyset$ (both with two states) are straightforward to build. We can also take as basic the automaton for $X_1 = X_2$ (also with two states).

We now construct the automaton $\mathscr{B} := \mathscr{B}_{edg(X_1,X_2)}$. Its set of states is

$$Q := \{0, Ok, Error\} \cup \{a(1), a(2), ab \mid a, b \in C, a \neq b\},$$

with accepting state Ok. The meanings of these states are described in Table 6.8, where, similarly to Equivalence (6.1), we have, for every term $t * \gamma \in T(F_C^{\mathrm{VR}(2)})$:

$$t * \gamma \in L(\mathscr{B}, q) \text{ if and only if } (\lfloor cval(t) \rfloor_C, \gamma(X_1), \gamma(X_2)) \text{ satisfies } P_q, \qquad (6.2)$$

and we recall that $\gamma(X_1)$ and $\gamma(X_2)$ are written X_1 and X_2 in the table for better readability.

The number of states is $k^2 + k + 3$, where $k = |C|$. The transition rules are in Table 6.9. Among the missing transitions, we mention $\oplus[Ok, Ok] \to Error$, $(\mathbf{a}, 11) \to Error$ if $a \in C$ and $relab_h[ab] \to Error$ if $h(a) = h(b)$. The table specifies $O(k^4)$ transitions, with (or without) counting the transitions to $Error$.

Equivalence (6.2) can be proved by induction on the structure of t, using Lemmas 2.95 and 2.96. Technically it is more convenient to prove its equivalent formulation by bottom-up induction on u: if r is the unique run of $\mathscr{B}$ on a term $t * \gamma$ and $u \in N_t$, then $r(u) = q$ if and only if $(\lfloor cval(t)/u \rfloor_C, \gamma(X_1) \cap (N_t/u), \gamma(X_2) \cap (N_t/u))$ satisfies P_q.[22] Then $cval(t)$ must be replaced by $cval(t)/u$ in Table 6.8 (and $port_{cval(t)/u}(v_i)$ can be replaced by $port_t(u, v_i)$, see Definition 2.94).

---

[22] Recall from Section 2.5.2 that $cval(t)/u$ denotes the p-graph $cval(t_{Id}/u)$, and note that $\gamma(X_i) \cap (N_t/u) = \gamma(X_i) \cap V_{cval(t)/u}$.

Table 6.9 *The transition rules of $\mathscr{B}_{edg(X_1,X_2)}$.*

Transition rules	Conditions
$\varnothing \to 0$	
$(\mathbf{c},00) \to 0$	
$(\mathbf{c},10) \to a(1)$	$\mathbf{c}$ is $\mathbf{a}$ or $\mathbf{a}^\ell$
$(\mathbf{c},01) \to a(2)$	
$(\mathbf{a}^\ell,11) \to Ok$	
$relab_h[q] \to q$	$q \in \{0,Ok\}$
$relab_h[a(1)] \to b(1)$	$b = h(a)$
$relab_h[a(2)] \to b(2)$	
$relab_h[ab] \to cd$	$c = h(a), d = h(b), c \neq d$
$\overrightarrow{add}_{a,b}[q] \to q$	$q \in Q - \{ab\}$
$\overrightarrow{add}_{a,b}[ab] \to Ok$	
$add_{a,b}[q] \to q$	$q \in Q - \{ab,ba\}$
$add_{a,b}[ab] \to Ok$	
$add_{b,a}[ab] \to Ok$	
$\oplus[a(1),b(2)] \to ab$	$a \neq b$
$\oplus[b(2),a(1)] \to ab$	
$\oplus[a(2),b(1)] \to ba$	
$\oplus[b(1),a(2)] \to ba$	
$\oplus[q,0] \to q$	$q \in Q$
$\oplus[0,q] \to q$	

Table 6.9 contains also transitions for the operations $add_{a,b}$ that add pairs of opposite directed edges, which we also consider as undirected edges. If we restrict this automaton to the signature F_C^{VRu} that generates only undirected graphs (by deleting the transitions relative to the operations $\overrightarrow{add}_{a,b}$), then any two states ab and ba can be identified as the reader will check easily. Hence, the number of states is (slightly) reduced to $k(k-1)/2 + 2k + 3$.

The automaton for $lab_a(X_1)$ can be constructed in a similar, but easier way, with $k+2$ states. We can also take as basic the automaton for $lab_a^\vee(X_1)$ (cf. Remark 6.18), with 2^k states. $\qquad\square$

The constructions of automata are done for generic sets C. That is, if we replace C by another set in bijection with it by h, then the corresponding automata are obtained by replacing everywhere in the states, in the transitions and the accepting states of the original ones, each $a \in C$ by $h(a)$. In particular, the numbers of states and transitions depend only on the cardinality of the considered set C.

We obtain the following theorem:

Theorem 6.35 For every finite set C of port names and every formula φ in $\text{CMS}'(\mathcal{R}_{s,C}, \mathcal{X}_\omega)$, one can construct a complete and deterministic automaton $\mathcal{B}_{C,\varphi}$ that recognizes the language $L_{C,\varphi}^{\text{VR}} \subseteq T(F_C^{\text{VR}(mfv(\varphi))})$. This automaton has at most $\exp(h, m(a^m + h))$ states, where $h = bqh(\varphi)$, $m = ba(\varphi)$, and a is the maximal number of states of a complete and deterministic automaton (over $F_C^{\text{VR}(1)}$ and $F_C^{\text{VR}(2)}$) constructed for an atomic subformula of φ. □

For sentences φ, this theorem provides an alternative proof of Theorem 6.3(3). By Proposition 3.69, it entails the Weak Recognizability Theorem for the VR algebra (Corollary 5.69(1)) that also follows from the (full) Recognizability Theorem for the VR algebra (Theorem 5.68(1)). However, the (full) Recognizability Theorem for the VR algebra does not follow from it (see Section 6.4.6 below).

Remark 6.36 We can also prove the first statement of Theorem 6.35 without constructing the automaton $\mathcal{B}_{edg(X_1,X_2)}$ and by using instead Theorem 6.26. It suffices to show that the edge relation of $G = cval(t)$ for $t \in T(F_C^{\text{VR}(n)})$ is definable in the relational structure $\lfloor t \rfloor$ by an MS formula $\eta(x,y)$. It follows that every formula φ in $\text{CMS}'(\mathcal{R}_{s,C}, \mathcal{X}_n)$, with $n := mfv(\varphi)$, can be translated into a formula ψ in $\text{CMS}'(\mathcal{R}_{F_C^{\text{VR}}}, \mathcal{X}_n)$ with $mfv(\psi) = n$ such that, for every term $t \in T(F_C^{\text{VR}(n)})$ and every mapping $\gamma : \mathcal{X}_n \to \mathcal{P}(N_t)$, we have

$$(\lfloor t \rfloor, \gamma) \models \psi(X_1, \ldots, X_n) \text{ if and only if}$$

$$\gamma(X_1), \ldots, \gamma(X_n) \text{ are subsets of } Occ_0(t)$$

$$\text{and } (\lfloor cval(t) \rfloor_C, \gamma) \models \varphi(X_1, \ldots, X_n).$$

Hence $L_{C,\varphi}^{\text{VR}} = L_\psi$ and the automata $\mathcal{B}_{C,\varphi}$ and $\mathcal{A}_\psi$ are equivalent. This (short) proof is not satisfactory because the formula $\eta(x,y)$ is fairly complicated. It must express that x, y are occurrences of constant symbols, for example $\mathbf{a}$ and $\mathbf{b}$, and that they have a common ancestor u that is an occurrence of an edge creating operation $\overrightarrow{add}_{c,d}$, and that the port labels a and b are transformed into c and d by the port relabeling operations that occur on the paths in t between x and u, and between y and u (cf. the function $port_t$ used in Definition 2.94 and Lemma 2.95; these relabelings are actually handled by the rules in Table 6.9 relative to the operations $relab_h$.) The construction of η will be done in the proof that the evaluation mapping $\lfloor t \rfloor \mapsto \lfloor cval(t) \rfloor_C$ from terms to p-graphs is a monadic second-order transduction (Proposition 7.30, Section 7.2), so that the transformation of φ into ψ is a special case of the Backwards Translation Theorem (Theorem 7.10). It follows that ψ has a larger quantifier-height than φ. Using this proof would yield a longer computation, although the minimal automata of $\mathcal{B}_{C,\varphi}$ and

$\mathscr{A}_\psi$ are the same (up to renaming of states). The construction of $\mathscr{B}_{C,edg(X_1,X_2)}$ can be seen as that of a precomputed automaton (cf. Section 6.3.3) for η.

More tools for constructing automata

At the end of Section 6.3.3, we have presented two notions intended to facilitate the construction of automata: the direct construction of automata for frequently used properties and the use of Boolean set terms in the writing of formulas. These two notions are applicable to the present case. Our next objective is to enrich the tool box with a construction of automata for relativized formulas. We will also detail the direct construction of an automaton for paths, analogous to what we did above for the least common ancestor function.

Construction 6.37 (Automata for relativized formulas) We recall from Section 5.2.1 that if φ is a sentence, then $\varphi \restriction X_1$ is a formula expressing that the substructure induced by the set denoted by X_1 satisfies the property expressed by φ, and we fix a set C. We will prove that the inverse image construction applied to $\mathscr{B}_\varphi$ yields an automaton equivalent to $\mathscr{B}_{\varphi \restriction X_1}$. Slightly more generally, we consider a property of vertex sets $P(X_1,\dots,X_n)$. We let $Q(X_1,\dots,X_{n+1})$ be the property such that, for all sets of vertices $X_1,\dots,X_{n+1}$ of a graph G, the property $Q(X_1,\dots,X_{n+1})$ is true if and only if $P(X_1 \cap X_{n+1},\dots,X_n \cap X_{n+1})$ is true in the induced subgraph $G[X_{n+1}]$. Then we will construct an automaton recognizing $L^{VR}_{Q(X_1,\dots,X_{n+1})}$ from one recognizing $L^{VR}_{P(X_1,\dots,X_n)}$. This construction will be used in the case where P is a monadic second-order property. We present it in terms of properties rather than in terms of formulas to stress that it does not depend on the structure of formulas.

We define h as the arity preserving mapping : $F^{VR(n+1)}_C \to F^{VR(n)}_C$ such that, for every $\mathbf{c} \in C$ and $w \in \{0,1\}^n$, we have $h((\mathbf{c},w0)) := \varnothing$ and $h((\mathbf{c},w1)) := (\mathbf{c},w)$, and $h(f) := f$ for $f \notin \mathbf{C}^{(n)}$. With these hypotheses and notation we obtain the following lemma:

Lemma 6.38 We have $L^{VR}_{Q(X_1,\dots,X_{n+1})} = h^{-1}(L^{VR}_{P(X_1,\dots,X_n)})$. Thus, if an automaton $\mathscr{B}_{P(X_1,\dots,X_n)}$ recognizes the language $L^{VR}_{P(X_1,\dots,X_n)}$, then the automaton $h^{-1}(\mathscr{B}_{P(X_1,\dots,X_n)})$ recognizes the language $L^{VR}_{Q(X_1,\dots,X_{n+1})}$.

Proof: Let $t * \gamma$ belong to $T(F^{VR(n+1)}_C)$ and $G = cval(t)$. Then $h(t * \gamma) = t' * \gamma'$, where, by the definitions, t' evaluates to $G' := G[\gamma(X_{n+1})]$ (cf. the proof of Proposition 2.105(1)) and γ' is the assignment : $\mathcal{X}_n \to \mathcal{P}(V_{G'})$ such that $\gamma'(X_i) = \gamma(X_i) \cap \gamma(X_{n+1})$ for each i. It follows that $t * \gamma \in L^{VR}_{Q(X_1,\dots,X_{n+1})}$ if and only if $t' * \gamma' \in L^{VR}_{P(X_1,\dots,X_n)}$. ∎

This result shows the usefulness of the symbol $\varnothing$ to denote the empty graph, for which we have defined transitions in Tables 6.7 and 6.9.

Construction 6.39 (Precomputed automata for path properties) The following construction concerns undirected graphs (it can easily be extended to directed paths in directed graphs). For an undirected graph G, we let $Path(X_1, X_2)$ mean that $X_1 \subseteq X_2$, $|X_1| = 2$ and there is a path in $G[X_2]$ that links the two vertices of X_1. It is monadic second-order expressible.

We will construct an automaton $\mathscr{B}_{C, Path(X_1, X_2)}$ that recognizes the language $L^{VR}_{C, Path(X_1, X_2)}$, without using the logical expression of $Path(X_1, X_2)$. We take C of cardinality at least 2 because otherwise, the graphs generated by F^{VR}_C have no edges apart from loops (cf. Remark 2.107(1)) and $Path(X_1, X_2)$ is always false. We need some auxiliary notions.

Let G be an undirected p-graph of type $\pi(G) = \{\, port_G(x) \mid x \in V_G \} \subseteq C$. For $x \in V_G$, we let[23]

$$\alpha(G, x) := \{\, port_G(y) \mid y \in V_G \text{ and } x -^*_G y \} \subseteq C,$$

and

$$\beta(G) := \{(\, port_G(x), port_G(y)) \mid x, y \in V_G \text{ and } x -^*_G y \} \subseteq C \times C.$$

Hence $\beta(G)$ is a symmetric and reflexive relation on the set $\pi(G)$. In particular, $\beta(G)$ determines $\pi(G)$. The relation $\beta(G)$ is not necessarily transitive.

We will prove that the functions α and β can be computed inductively on the structure of a term over F^{VR}_C that evaluates to G. If $h : C \to C$ and $B \subseteq C \times C$, then $h(B) := \{(h(a), h(b)) \mid (a, b) \in B\}$. We extend the composition of binary relations (denoted by $\cdot$) to $A \subseteq C$ and $B \subseteq C \times C$ by letting

$$A \odot B := \{b \in C \mid (a, b) \in B \text{ for some } a \in A\}.$$

This set is the image of A by the relation B, usually denoted $B(A)$. If $B' \subseteq C \times C$ then we write $A \odot B \cdot B'$ for $A \odot (B \cdot B')$, which equals $(A \odot B) \cdot B'$. For $a, b \in C$ with $a \neq b$, we let

$$a \circledast b := \{(a, a), (a, b), (b, a), (b, b)\}.$$

Claim 6.39.1

(1) $\alpha(G \oplus H, x) = \begin{cases} \alpha(G, x) & \text{if } x \in V_G, \\ \alpha(H, x) & \text{if } x \in V_H. \end{cases}$

(2) $\beta(G \oplus H) = \beta(G) \cup \beta(H)$.

(3) $\alpha(add_{a,b}(G), x)$
    $=$ `if` $a, b \in \pi(G)$ `then` $\alpha(G, x) \cup \alpha(G, x) \odot (a \circledast b) \cdot \beta(G)$ `else` $\alpha(G, x)$.

(4) $\beta(add_{a,b}(G))$
    $=$ `if` $a, b \in \pi(G)$ `then` $\beta(G) \cup \beta(G) \cdot (a \circledast b) \cdot \beta(G)$ `else` $\beta(G)$.

(5) $\alpha(relab_h(G), x) = h(\alpha(G, x))$.

---

[23] Since $x -_G y$ means that x and y are adjacent, $x -^*_G y$ means that x and y are equal or linked by a path.

(6) $\beta(relab_h(G)) = h(\beta(G))$.
(7) $\alpha(\mathbf{a},x) = \alpha(\mathbf{a}^\ell,x) = \{a\}$.
(8) $\beta(\mathbf{a}) = \beta(\mathbf{a}^\ell) = \{(a,a)\}$.
(9) $\beta(\varnothing) = \emptyset$.

Proof: The verifications are easy from the definitions. We only sketch the proof of the inclusion $\subseteq$ in (3).

If c is in $\alpha(add_{a,b}(G),x)$, then either it is in $\alpha(G,x)$ or there exists a path from x to a c-port z that uses one or more edges added to G by $add_{a,b}$. If this path contains only one such edge and goes through an a-port u and immediately after through a b-port w, we have $a \in \alpha(G,x)$ and $(b,c) \in \beta(G)$ and thus $c \in \alpha(G,x) \odot (a \circledast b) \cdot \beta(G)$ because $(a,b) \in a \circledast b$. If this path contains several such edges, the first one being $u - w$ and the last one being $u' - w'$, where u and u' are a-ports and w and w' are b-ports, then there is also an edge between u and w' and the previous case gives the result. If $u - w$ and $u' - w'$ are as above except that w' is an a-port and u' is a b-port, then there is also an edge between w and w'. Then $a \in \alpha(G,x)$ and $(a,c) \in \beta(G)$ and thus $c \in \alpha(G,x) \odot (a \circledast b) \cdot \beta(G)$ because $(a,a) \in a \circledast b$. $\qquad\square$

We now construct the automaton $\mathcal{B}_{C,Path(X_1,X_2)}$ with its set of states Q defined as

$$\{Ok, Error\}$$
$$\cup \{(0,B) \mid B \subseteq C \times C\}$$
$$\cup \{(1,A,B) \mid \emptyset \neq A \subseteq C, B \subseteq C \times C\}$$
$$\cup \{(2,\{A,A'\},B) \mid A,A' \subseteq C, A \neq \emptyset, A' \neq \emptyset, B \subseteq C \times C\},$$

with accepting state Ok. The meanings of these states are described in Table 6.10. In this table, we denote by $G(t)$ the graph $cval(t)$, and by $G(t,X_2)$ the graph $cval(t)[\gamma(X_2)]$ where γ is the assignment encoded by the considered term (cf. Equivalences (6.1) and (6.2)). Equivalently, we could write $G(u)$ to denote $cval(t)/u$, and $G(u,X_2)$ to denote $cval(t)/u[\gamma(X_2) \cap (N_t/u)]$.

The transition rules are shown in Table 6.11, where we use the following auxiliary functions:

$$f(B,a,b) := \text{ if } \{(a,a),(b,b)\} \subseteq B \text{ then } B \cup (B \cdot (a \circledast b) \cdot B) \text{ else } B,$$

$$g(A,B,a,b) := \text{ if } \{(a,a),(b,b)\} \subseteq B \text{ then } A \cup (A \odot (a \circledast b) \cdot B) \text{ else } A.$$

These definitions reflect respectively Properties (4) and (3) of Claim 6.39.1. This completes the construction of $\mathcal{B}_{C,Path(X_1,X_2)}$.

As for the previously constructed automata, we omit the correctness proof. Hence, we admit that all computations of the semi-automaton $\mathcal{B}_{C,Path(X_1,X_2)}$ (as defined by Table 6.11) satisfy the properties of Table 6.10. This table shows that some states

Table 6.10 *Meanings of the states of $\mathscr{B}_{C,Path(X_1,X_2)}$.*

State q	Property P_q
$(0,B)$	$X_1 = \emptyset$, $B = \beta(G(t,X_2))$
$(1,A,B)$	$X_1 = \{v\} \subseteq X_2$, $A = \alpha(G(t,X_2),v)$, $B = \beta(G(t,X_2))$
$(2,\{A,A'\},B)$	$X_1 = \{v,v'\} \subseteq X_2$, $v \neq v'$, $A = \alpha(G(t,X_2),v)$, $A' = \alpha(G(t,X_2),v')$, $B = \beta(G(t,X_2))$, there is no path between v and v' in $G(t,X_2)$
Ok	$Path(X_1,X_2)$ holds in $G(t)$
Error	all other cases

Table 6.11 *The transition rules of $\mathscr{B}_{C,Path(X_1,X_2)}$.*

Transition rules	Conditions
$\emptyset \to (0,\emptyset)$ $(\mathbf{c},00) \to (0,\emptyset)$ $(\mathbf{c},01) \to (0,\{(a,a)\})$ $(\mathbf{c},11) \to (1,\{a\},\{(a,a)\})$	$\mathbf{c} \in \{\mathbf{a},\mathbf{a}^\ell\}$
$relab_h[Ok] \to Ok$ $relab_h[(0,B)] \to (0,h(B))$ $relab_h[(1,A,B)] \to (1,h(A),h(B))$ $relab_h[(2,\{A,A'\},B)] \to (2,\{h(A),h(A')\},h(B))$	
$add_{a,b}[Ok] \to Ok$ $add_{a,b}[(0,B)] \to (0,B')$ $add_{a,b}[(1,A,B)] \to (1,D,B')$ $add_{a,b}[(2,\{A,A'\},B)] \to (2,\{D,D'\},B')$	$B' = f(B,a,b)$ $D = g(A,B,a,b)$ $D' = g(A',B,a,b)$ $(A \odot (a \circledast b) \cdot B) \cap A' = \emptyset$
$add_{a,b}[(2,\{A,A'\},B)] \to Ok$	$(A \odot (a \circledast b) \cdot B) \cap A' \neq \emptyset$
$\oplus[Ok,(0,B)] \to Ok$ $\oplus[(0,B),Ok] \to Ok$ $\oplus[(0,B),(0,B')] \to (0,B'')$ $\oplus[(0,B),(1,A,B')] \to (1,A,B'')$ $\oplus[(1,A,B'),(0,B)] \to (1,A,B'')$ $\oplus[(1,A,B),(1,A',B')] \to (2,\{A,A'\},B'')$ $\oplus[(0,B),(2,\{A,A'\},B')] \to (2,\{A,A'\},B'')$ $\oplus[(2,\{A,A'\},B'),(0,B)] \to (2,\{A,A'\},B'')$	$B'' = B \cup B'$

cannot occur in any computation. For example, by the definition of the function β, the component B of each accessible state must be a symmetric and reflexive relation. All states for which B is not so are inaccessible and, consequently, are eliminated if we trim the automaton (cf. Section 3.3.1, Definition 3.48). The state $(1, \{a\}, \emptyset)$ is also inaccessible. The set Q has cardinality $2 + 2^{k^2} + (2^k - 1)2^{k^2} + (2^k - 1)^2 2^{k^2} = 2 + (2^{2k} - 2^k + 1)2^{k^2}$ (where $k = |C| \geq 2$). The cardinality of the set of accessible states is somewhat less than that, but it lies between $2^{k(k-1)/2}$ and 2^{k^2+2} as one can easily check. Determining its exact value is of no interest because the corresponding minimal automaton matters more than this one, even after trimming.

Let us now see what would give the expression of the property $Path(X_1, X_2)$ by the MS formula:

$$\forall x[x \in X_1 \Rightarrow x \in X_2]$$
$$\wedge \exists x, y(x \in X_1 \wedge y \in X_1 \wedge x \neq y \wedge \forall z(z \in X_1 \Rightarrow x = z \vee y = z)$$
$$\wedge \forall X_3[x \in X_3 \wedge \forall u, v(u \in X_3 \wedge u \in X_2 \wedge v \in X_2 \wedge edg(u, v) \Rightarrow v \in X_3) \Rightarrow y \in X_3])$$

of quantifier-height 5 (its construction uses Property P2 of Proposition 5.11). Its translation into a formula without first-order variables has the same quantifier-height. The given construction of $\mathcal{B}_{C,Path(X_1,X_2)}$ avoids thus lengthy computations. As already noted in Remark 6.31(2), the minimal automaton equivalent to $\mathcal{B}_{C,Path(X_1,X_2)}$ depends only on the property $Path(X_1, X_2)$. It is thus the same as the one that could be obtained from any monadic second-order expression of this property, provided the computations do not abort due to lack of memory.

Remark 6.40 (Some computing experiments) The automaton of Table 6.11 has been implemented and minimized successfully[24] for $|C| = 2, 3, 4$. (All automata discussed in this remark are complete and deterministic). For $|C| = 3$, the initial trim automaton has 214 states. The corresponding minimal automaton has 125 states and more than 16 500 transition rules that do not yield the state *Error*. For $|C| = 4$, the minimal automaton has 2197 states.

For $C := \{a, b\}$, the initial trim automaton $\mathcal{A} := \mathcal{B}_{C,Path(X_1,X_2)}$ has 25 states and 372 transition rules; its minimal automaton has 12 states and 137 rules. The minimization algorithm has determined the following 12 equivalence classes of states of $\mathcal{A}$. We first list the nine singleton classes:

> *Ok, Error,*
>
> $(0, \emptyset)$, $(0, \{(a, a)\})$, $(0, \{(b, b)\})$,
>
> $(1, \{a\}, \{(a, a)\})$, $(1, \{b\}, \{(b, b)\})$,
>
> $(2, \{\{a\}, \{a\}\}, \{(a, a)\})$, $(2, \{\{b\}, \{b\}\}, \{(b, b)\})$.

---

[24] All computations have been done by Durand with her software AUTOWRITE [Dur] that trims, reduces, determinizes and minimizes automata on terms.

There is one class with two elements:

$$(0,\{(a,a),(b,b)\}), \ (0,a \circledast b),$$

one class with five elements:

$$(1,\{a\},\{(a,a),(b,b)\}), \ (1,\{b\},\{(a,a),(b,b)\}),$$
$$(1,\{a\},a \circledast b), \ (1,\{b\},a \circledast b),$$
$$(1,\{a,b\},a \circledast b),$$

and one class with nine elements:

$$(2,\{\{a\},\{a\}\},\{(a,a),(b,b)\}), \ (2,\{\{b\},\{b\}\},\{(a,a),(b,b)\}),$$
$$(2,\{\{a\},\{a\}\},a \circledast b), \ (2,\{\{b\},\{b\}\},a \circledast b),$$
$$(2,\{\{a\},\{b\}\},\{(a,a),(b,b)\}),$$
$$(2,\{\{a\},\{b\}\},a \circledast b),$$
$$(2,\{\{a\},\{a,b\}\},a \circledast b), \ (2,\{\{b\},\{a,b\}\},a \circledast b), \ \text{and}$$
$$(2,\{\{a,b\},\{a,b\}\},a \circledast b).$$

By Proposition 2.106(2), the simple loop-free undirected graphs of clique-width at most 2 are the cographs that we have already discussed in Sections 1.1.2 and 1.4.1, and in Example 4.43(1). These graphs are the p-graphs of type $\{a\}$ generated by **a**, $\oplus$ and the complete join $\otimes$, a derived operation of $\mathbb{GP}^u$. Only six of the states of the minimal automaton of $\mathscr{A}$ are needed for the minimal $\{\oplus, \otimes, \mathbf{a}\}$-automaton. These states are *Ok*, *Error*, $(0,\emptyset)$, $(0,\{(a,a)\})$, $(1,\{a\},\{(a,a)\})$ and $(2,\{\{a\},\{a\}\},\{(a,a)\})$, which are also states of $\mathscr{A}$. The corresponding transition rules are those of $\mathscr{A}$ for **a** and $\oplus$ together with, for $\otimes$, the following ones:

$$\otimes [(1,\{a\},\{(a,a)\}), \ (1,\{a\},\{(a,a)\})] \rightarrow Ok,$$
$$\otimes [(2,\{\{a\},\{a\}\},\{(a,a)\}), \ (0,\{(a,a)\})] \rightarrow Ok, \text{ and}$$
$$\otimes [(0,\{(a,a)\}), \ (2,\{\{a\},\{a\}\},\{(a,a)\})] \rightarrow Ok,$$

and all other rules for $\otimes$ are as for $\oplus$, for example:

$$\otimes[(0,\{(a,a)\}), \ (1,\{a\},\{(a,a)\})] \rightarrow (1,\{a\},\{(a,a)\}),$$

which is like:

$$\oplus[(0,\{(a,a)\}), \ (1,\{a\},\{(a,a)\})] \rightarrow (1,\{a\},\{(a,a)\}).$$

It would be interesting to simplify in a similar way the functional signature that generates graphs of clique-width 3 and more. This might yield workable automata for graphs of larger clique-width than 2. □

Hopefully tractable monadic second-order properties

Our aim is here to identify classes of monadic second-order graph properties for which the associated automata have manageable sizes. These properties are expressible without quantifier alternation in terms of some "basic" monadic second-order properties.

Definition 6.41 Let $\mathcal{P}$ be a set of pairwise inequivalent formulas in $CMS'(\mathcal{R}_s, \mathcal{X}_\omega)$ such that the set of free variables of any $\varphi \in \mathcal{P}$ is $\mathcal{X}_{mfv(\varphi)}$. The integer $mfv(\varphi)$ is called the *arity* of the property P defined by φ and is denoted by $\rho(P)$.

We let $\mathcal{P}^{Bool}$ be the set of formulas in $CMS^{Bool}(\mathcal{R}_s, \mathcal{X}_\omega)$ obtained from those of $\mathcal{P}$ by substitutions of Boolean set terms (cf. Definition 6.32). We let $\mathcal{EB}(\mathcal{P}^{Bool})$ be the set of sentences of the form $\exists \overline{X}.\theta$ such that θ is a Boolean combination of formulas in $\mathcal{P}^{Bool}$ and $\overline{X}$ is a sequence of set variables.

We assume that, for each φ in $\mathcal{P}$ and k in $\mathcal{N}_+$, we have constructed (or some algorithm can construct) a complete and deterministic $F_{[k]}^{VR(mfv(\varphi))}$-automaton $\mathcal{B}_{[k],\varphi}$ recognizing $L_{[k],\varphi}^{VR}$, with $N(k,\varphi)$ states. This number is also denoted by $N(k,P)$ if P is the property corresponding to φ. If a formula α is obtained from $\varphi \in \mathcal{P}$ by a substitution $\mathcal{X}_{mfv(\varphi)} \to T(B, \mathcal{X}_n)$ of Boolean set terms, then an automaton recognizing $L_{[k],\alpha,n}^{VR}$, with $N(k,\alpha) = N(k,\varphi)$ states, can be constructed for it by (the obvious variant of) Lemma 6.33 and Proposition 6.11(2).

Since the formulas of $\mathcal{P}$ are over $\mathcal{R}_s$, the properties that they express do not depend on port labels (they are properties of graphs, not of p-graphs). Hence, the graphs to be checked can be constructed with sets of port labels $[k]$ instead of sets C of cardinality k without loss of generality or efficiency. The $F_C^{VR(mfv(\varphi))}$-automata associated with such sets C are all "isomorphic" to $\mathcal{B}_{[k],\varphi}$. We use here the remark before Theorem 6.35.

Proposition 6.42 Let $\mathcal{P} \subseteq CMS'(\mathcal{R}_s, \mathcal{X})$ be as in Definition 6.41. Let φ in $\mathcal{EB}(\mathcal{P}^{Bool})$ be a sentence of the form $\exists X_1, \ldots, X_n.\theta$, where θ is a Boolean combination of formulas $\alpha_1, \ldots, \alpha_m$ in $\mathcal{P}^{Bool}$. For each k, there exists a nondeterministic $F_{[k]}^{VR}$-automaton $\mathcal{B}_{[k],\varphi}$ with at most $N := N(k,\alpha_1) \times \cdots \times N(k,\alpha_m)$ states that recognizes the language $L_{[k],\varphi}^{VR}$. Membership in $L(\mathcal{B}_{[k],\varphi})$ of any term $t \in T(F_{[k]}^{VR})$ can be checked in time $m \cdot a \cdot (2^n + N^2) \cdot |t|$, where a is an upper-bound to the transition time of the automata for $\alpha_1, \ldots, \alpha_m$.

Proof: For each i, the formula α_i is obtained from some $\varphi_i \in \mathcal{P}$ by a substitution of Boolean set terms. From the complete deterministic automaton $\mathcal{B}_{[k],\varphi_i}$ for φ_i with $N(k,\varphi_i)$ states, we obtain, by the inverse image construction, a complete deterministic automaton for α_i with the same number of states, recognizing $L^{VR}_{[k],\alpha_i,n}$. From the automata for $\alpha_1,\ldots,\alpha_m$ a complete and deterministic automaton for θ can be obtained by the product construction of Proposition 6.10. It recognizes $L^{VR}_{[k],\theta,n}$ and has N states (but after trimming, it can have less states). A nondeterministic automaton $\mathscr{B}_{[k],\varphi}$ for φ with the same number of states is then obtained from it by a direct image construction (as in Construction 6.29, by Proposition 6.11(1)). That automaton has degree of nondeterminism 2^n and is only nondeterministic on constant symbols. The evaluation of the computation time uses Proposition 6.14 and Remark 6.15. ∎

Example 6.43 (Automata for some basic graph properties) In the following, we define a set of basic monadic second-order graph properties that we denote by $\mathcal{P}$ (like the corresponding set of formulas).

In Definition 6.41, we have given no syntactic constraints on the formulas of $\mathcal{P}$. In order to get applicable instances of Proposition 6.42, we will only (or mainly) consider formulas φ such that $N(k,\varphi) = 2^{O(k^2)}$. Imposing on the numbers $N(k,\varphi)$ a polynomial bound, or even the bound $2^{O(k)}$, would eliminate too many basic and useful graph properties. Furthermore, the automata for the important property of connectivity have $2^{2^{\Theta(k)}}$ states. A similar situation holds for the property of being a forest.

Example 6.43(1) (A set $\mathcal{P}$ of basic graph properties) We let $\mathcal{P}$ consist of the properties defined by the atomic formulas, the property $Path(X_1,X_2)$ that has been defined in Construction 6.39, together with the following properties:

$Disj(X_1,\ldots,X_p)$ expresses that the vertex sets $X_1,\ldots,X_p$ are pairwise disjoint;

$Part(X_1,\ldots,X_p)$ expresses that $X_1,\ldots,X_p$ form a partition of the vertex set of the considered graph (where some of the sets $X_1,\ldots,X_p$ may be empty);

$Card_{\leq p}(X_1)$ expresses that the set X_1 has cardinality at most p;

St expresses that the considered graph is *stable*, i.e. that its only edges are loops;

$Clique$ expresses that any two distinct vertices of the considered graph are linked by an undirected edge (equivalently, by two opposite directed edges);

$Link(X_1,X_2)$ expresses that there exists an edge linking a vertex of X_1 to one of X_2 (hence, $Link$ is $edg^{\exists\exists}$, cf. Remark 6.18);

$Dom(X_1,X_2)$ expresses that X_2 *dominates* X_1, i.e., that X_1 and X_2 are disjoint and every vertex of X_1 is the head of an edge with tail in X_2;

$InDeg_p(X_1,X_2)$ expresses that every vertex in X_1 is the head of exactly p edges with tail in X_2 (X_1 and X_2 need not be disjoint and a loop incident with $x \in X_1$ counts if $x \in X_2$);

$Conn$ expresses that the considered graph is connected; and

$NoCycle$ expresses that the considered undirected graph is a forest (has no cycle, in particular, no loop).

We let $ConnIfDeg_d$ express connectedness for a graph assumed to have degree at most d, and similarly, $NoCycleDeg_d$ express that a graph of degree at most d is a forest.

From St, which is a property of the considered graph, one gets by relativization (Construction 6.37) the corresponding property $St(X_1)$ of the subgraph induced by a set of vertices X_1. Similarly, we obtain the relativized properties $Clique(X_1)$, $Conn(X_1)$, $NoCycle(X_1)$, $ConnIfDeg_d(X_1)$ and $NoCycleDeg_d(X_1)$. We also add these properties to $\mathcal{P}$. By Lemma 6.38 and Proposition 6.11(2), an $F_{[k]}^{VR}$-automaton for property P can easily be transformed into an $F_{[k]}^{VR(1)}$-automaton for the property $P(X_1)$, with the same number of states.

Table 6.12 shows upper-bounds to the numbers of states $N(k,P)$ for these properties.[25] The values come from constructions of complete and deterministic automata that are not necessarily minimal. We will not give detailed descriptions of these automata. We will only define their sets of states and describe their meanings, as we did previously (cf. Table 6.10). The transitions will be easy to construct from these descriptions. We will use the notation of Table 6.10: $G(t)$ denotes $cval(t)$ and $G(t,X_i)$ denotes $cval(t)[\gamma(X_i)]$. We order the automata by increasing number of states.

The properties $Disj(X_1,\ldots,X_p)$ and $Part(X_1,\ldots,X_p)$: The $F_{[k]}^{VR(p)}$-automata for these properties are straightforward to construct; they have two states but their sizes depend of course on p and k.

The property $Card_{\leq p}(X_1)$: The states are $0,1,\ldots,p$ and *Error*. The accepting states are $0,1,\ldots,p$. The meanings of the states are described by the following conditions:

$$P_i \quad : \quad G(t) \text{ has } i \text{ vertices.}$$
$$P_{Error} \quad : \quad G(t) \text{ has more than } p \text{ vertices.}$$

Stability: The states are *Error* and A for each $A \subseteq [k]$. All states are accepting except *Error*. Their meanings are described by the following conditions:

$$P_A \quad : \quad A = \pi(G(t)) \text{ and } G(t) \text{ is stable.}$$
$$P_{Error} \quad : \quad G(t) \text{ is not stable.}$$

Since we use the constant symbol $\varnothing$ (it is useful for the relativization construction), we may have $G(t) = \varnothing$ and $A = \emptyset$.

---

[25] The use of Θ indicates that we also have a lower-bound for the minimal automaton.

Table 6.12 *Some graph properties.*

Property P	$N(k,P)$
$X_1 = \emptyset$	2
$X_1 \subseteq X_2$	2
$Disj(X_1,\ldots,X_p)$	2
$Part(X_1,\ldots,X_p)$	2
$Sgl(X_1)$	3
$Card_{p,q}(X_1)$	q
$Card_{\leq p}(X_1)$	$p+2$
$edg(X_1,X_2)$	k^2+k+3
St	2^k+1
$Link(X_1,X_2)$	$2^{2k}+1$
$Dom(X_1,X_2)$	$2^{2k}+1$
$InDeg_p(X_1,X_2)$	$(p+2)^{2k}$
$Path(X_1,X_2)$	2^{k^2+2}
$Clique$	$2^{\Theta(k^2)}$
$ConnIfDeg_d$	$2^{d \cdot k^2}$
$NoCycleIfDeg_d$	$2^{d^2 \cdot k^2}$
$Conn$	$2^{2^{\Theta(k)}}$
$NoCycle$	$2^{2^{O(k)}}$

The property $Link(X_1,X_2)$: The states are Ok and the pairs (A,B) for all $A,B \subseteq [k]$. The accepting state is Ok. Their meanings are as follows:

$$P_{(A,B)} \quad : \quad A = \pi(G(t,X_1)), B = \pi(G(t,X_2)) \text{ and}$$
$$Link(X_1,X_2) \text{ does not hold in } G(t).$$
$$P_{Ok} \quad : \quad Link(X_1,X_2) \text{ holds in } G(t).$$

Domination, i.e., the property $Dom(X_1,X_2)$: The states are $Error$ and the pairs (A,B) for all $A,B \subseteq [k]$. The accepting states are the pairs $(\emptyset,B)$ (note that $Error$ is not the only nonaccepting state). Their meanings are as follows:

$$P_{(A,B)} \quad : \quad X_1 \cap X_2 = \emptyset, B = \pi(G(t,X_2)) \text{ and}$$
$$A \text{ is the set of port labels in } G(t) \text{ of the vertices } x \text{ in } X_1$$
$$\text{that are not the head of an edge with tail in } X_2.$$
$$P_{Error} \quad : \quad X_1 \cap X_2 \neq \emptyset.$$

Bounded indegree: An automaton for $InDeg_p(X_1, X_2)$, where $p \geq 0$, with less than $(p+2)^{2k}$ states is constructed and proved to be correct in [CouDur11]. It is intended to run on irredundant terms.

A term t in $T(F_{[k]}^{\mathrm{VR}(n)})$ is *irredundant* if, for every occurrence w of an operation $\overrightarrow{add}_{a,b}$ (or $add_{a,b}$) with son w_1, the p-graph $cval(t)/w_1$ has no edge from an a-port to a b-port (or between an a-port and a b-port). The edge addition operations strictly below w that create these edges can be removed: the term obtained in this way is equivalent to t. Hence, every term $t \in T(F_{[k]}^{\mathrm{VR}(n)})$ can be transformed into an equivalent irredundant term s in $T(F_{[k]}^{\mathrm{VR}(n)})$ as follows: if $\overrightarrow{add}_{a,b}$ or $add_{a,b}$ has an occurrence u in t such that $(a,b) \in Edg(t \uparrow u)$ (this set is defined in the proof of Proposition 2.105(4), at the end of Section 2.5.6), then this edge addition operation can be removed at u (or replaced by the identity operation $relab_{Id}$). Since the sets $Edg(t \uparrow u)$ can be computed in linear time, the term s can also be constructed in linear time.

It follows that it suffices to construct automata that work as desired on irredundant terms. For $InDeg_p(X_1, X_2)$ and for the property *NoCycle* considered below, such automata are simpler to construct (and smaller) than those intended to accept arbitrary terms.

The property Clique: The states are the pairs (A, R) for all $A \subseteq [k]$ and binary relations R on A. The accepting states are the pairs $(A, \emptyset)$ (we consider $\emptyset$ as a clique). Their meanings are as follows:

$$P_{(A,R)} \quad : \quad A = \pi(G(t)) \text{ and } R \text{ is the set of pairs } (a,b) \in A \times A$$
$$\text{such that there exist in } G(t) \text{ a vertex } x \text{ with port label } a$$
$$\text{and a vertex } y \text{ with port label } b \text{ such that } (x,y) \notin edg_G.$$

The number of states is less than 2^{k^2+k}. The minimal automaton has $2^{\Theta(k^2)}$ states as one can check easily.

Connectivity: An automaton for this property must have many states: we know from Example 4.54(4) that there exists no automaton with 2^{2^m} states or less that checks the connectivity of graphs of clique-width at most $k = 2m + 1$. Our construction will be similar to the proof of HR-recognizability in Example 4.30(2). We will use the following notions and notation relative to a p-graph G of type included in $[k]$:

$CC(G)$ is the set of its connected components,
$\pi(CC(G))$ is the multiset of the types $\pi(H)$ for all $H \in CC(G)$.

We denote by $|M|$ the cardinality of a multiset M and by $Set(M)$ the corresponding set (so that $Set(M) \subseteq M$ and $|Set(M)| \leq |M|$). For a multiset M, we define $Set^\dagger(M)$

as the multiset consisting of two occurrences of d (and nothing else) if M consists of at least two occurrences of d, and as $Set(M)$ otherwise.

We are ready to define the set of states Q of the automaton for connectivity. We let E be the set of nonempty subsets of $[k]$ and Q be the set of multisets of the form $Set^\dagger(M)$, where M is a multiset of elements of E. The meanings of these multisets are described by the following equality:

$$P_M \quad : \quad M = Set^\dagger(\pi(CC(G(t)))).$$

The accepting states are the singletons. Examples of states are $\{\{1,2\},\{1,2\}\}$ and $\{\{1\},\{1,2\},\{2,3,4,6\}\}$. They may correspond to a p-graph $G(t)$ having three connected components, all of type $\{1,2\}$ in the first case, and four connected components of types $\{1\}, \{1\}, \{1,2\}$ and $\{2,3,4,6\}$ in the second one. The state $\{\{1,2\}\}$ is accepting and corresponds to a connected p-graph of type $\{1,2\}$.

The number of states is thus $2^{|D|} + |D| = 2^{2^k-1} + 2^k - 1$. However, if we want only to verify the connectivity of graphs of degree at most d, we can use a smaller automaton: we replace Q by the set Q_d of states $M \in Q$ such that each element of $[k]$ belongs to at most d sets of M, to which we add a state *Error*. Hence, each state M in Q_d is a set of at most dk elements from E. It follows that its cardinality is less than $2^{d \cdot k^2}$. The corresponding automaton may reject a term that defines a connected graph of degree larger than d.

Detailed constructions and correctness proofs can be found in [CouDur11].

Forests: The construction of an automaton with $2^{2^{O(k)}}$ states for the property *NoCycle* saying that an undirected graph has no cycles, i.e., that it is a forest, is also quite complicated. As for connectivity, if we want to verify the absence of cycles in graphs of degree at most d, we can use a smaller automaton, with $2^{d^2 \cdot k^2}$ states. The reader will find these constructions in [CouDur11]. They work for irredundant terms.

Letting $\mathcal{P}$ be the above set of basic properties, we now give a few examples of graph properties in $\mathcal{EB}(\mathcal{P}^{Bool})$.

Example 6.43(2) (Monadic second-order vertex partitioning problems) We first consider colorability problems for undirected graphs; vertex p-colorability can be expressed as follows:

$$\exists X_1, \ldots, X_p \, (Part(X_1, \ldots, X_p) \wedge St(X_1) \wedge \cdots \wedge St(X_p)).$$

A vertex p-coloring defined by $X_1, \ldots, X_p$ is *achromatic* ([Bod89]) if there is an edge between any two distinct sets X_i and X_j. The existence of such a coloring is thus expressed, for an undirected graph, by

$$\exists X_1, \ldots, X_p \, (Part(X_1, \ldots, X_p) \wedge \cdots St(X_i) \cdots \wedge \cdots Link(X_i, X_j) \cdots),$$

where the conjunctions extend to all $1 \leq i < j \leq p$.

A vertex p-coloring of an undirected graph G, defined by $X_1, \ldots, X_p$, is *acyclic* (cf. [*JenTof]) if each induced graph $G[X_i \cup X_j]$ is acyclic, i.e., is a forest. The existence of an acyclic p-coloring is thus expressed by

$$\exists X_1, \ldots, X_p \, (Part(X_1, \ldots, X_p) \wedge \cdots St(X_i) \cdots \wedge \cdots NoCycle(X_i \cup X_j) \cdots),$$

where the conjunctions extend to all $1 \leq i < j \leq p$.

These three properties are in $\mathcal{EB}(\mathcal{P}^{Bool})$, where $\mathcal{P}$ is as in Example 6.43(1). For p-coloring, achromatic p-coloring and acyclic p-coloring, we obtain nondeterministic automata with respectively $2^{O(p \cdot k)}$, $2^{O(p^2 \cdot k)}$ and $2^{p^2 \cdot 2^{O(k)}}$ states, by Proposition 6.42.

We now consider domination problems constrained by particular monadic second-order properties. The sentence

$$\exists X (\theta(X) \wedge Dom(\overline{X}, X)),$$

where $\overline{X}$ denotes the complement of X, expresses that there exists a set X that dominates all other vertices (cf. Example 6.43(1)) while satisfying a property specified by θ. This sentence is thus in $\mathcal{EB}((\mathcal{P} \cup \{\theta\})^{Bool})$.

A graph has *domatic number* at most p if its vertex set can be partitioned into p sets such that each of these sets dominates all other vertices. That is,

$$\exists X_1, \ldots, X_p \, (Part(X_1, \ldots, X_p) \wedge Dom(\overline{X}_1, X_1) \wedge \cdots \wedge Dom(\overline{X}_p, X_p)).$$

Minor inclusion can also be considered as a vertex partitioning problem. We recall from Lemma 1.13 and Corollary 1.14 that an undirected graph G contains a fixed simple loop-free undirected graph H as a minor if and only if it satisfies the following sentence MINOR_H:

$$\exists X_1, \ldots, X_p \, (Disj(X_1, \ldots, X_p) \wedge Conn(X_1) \wedge \cdots \wedge Conn(X_p)$$
$$\wedge X_1 \neq \emptyset \wedge \cdots \wedge X_p \neq \emptyset \wedge \cdots Link(X_i, X_j) \wedge \cdots),$$

where $\{v_1, \ldots, v_p\} = V_H$ and there is in MINOR_H one formula $Link(X_i, X_j)$ for each edge $\{v_i, v_j\}$ of H. We obtain a nondeterministic automaton with $2^{p \cdot 2^{O(k)} + p^2 \cdot O(k)}$ states for checking that H is a minor of the given graph; we can use a nondeterministic automaton with $2^{O(d \cdot p \cdot k^2 + p^2 \cdot k)}$ states if the input graph has degree at most d.

Example 6.43(3) (Perfect graphs) A (simple, loop-free and undirected) graph G is *perfect* if, for every induced subgraph H of G, the chromatic number is equal to the clique number, i.e., the smallest number p such that H is vertex p-colorable equals the maximum size of a clique of H.

We first consider chordal graphs: they are all perfect ([*Gol]). A nonempty connected (simple, loop-free and undirected) graph is chordal (Section 2.4.4, Proposition 2.72) if and only if it does not contain an induced cycle C_n for any $n \geq 4$. The existence of such a cycle is expressed as follows:

$$\exists X, Y (St(X \cap Y) \wedge Path(X \cap Y, X) \wedge Path(X \cap Y, Y)).$$

This example is not intended to provide an efficient algorithm for testing chordality, but to illustrate the use of set terms to get a compact formula. The validity of $Path(X \cap Y, X)$ implies that $X \cap Y$ has cardinality 2. Hence the condition $St(X \cap Y)$ can be replaced by $St_2(X \cap Y)$, where $St_2(Z)$ means that Z is stable and has two elements. This is interesting because a complete and deterministic automaton for this property with only $3 + k(k+1)/2$ states (instead of $2^k + 1$ for $St(Z)$) can be easily constructed; it is similar to the one for $edge(X_1, X_2)$. Hence, by this observation, we get (before minimization), a smaller automaton for the nonchordality of nonempty, connected, simple, loop-free and undirected graphs.

The definition of perfect graphs is not monadic second-order expressible (because the fact that two sets have the same cardinality is not) but their characterization established in [ChuRST] in terms of excluded holes and antiholes is. A *hole* is an induced cycle of odd length at least 5 and an *antihole* is the edge-complement of a hole. A graph has a hole if and only if it satisfies the following sentence:

$$\exists X, Y, Z, U, V (Disj(X, Y, Z, U, V) \wedge St(X) \wedge St(Y) \wedge X \neq \emptyset \wedge Y \neq \emptyset$$

$$\wedge edge(Z, U) \wedge edge(U, V) \wedge \neg edge(Z, V)$$

$$\wedge Deg_2(X, V \cup Y) \wedge Deg_2(Y, X \cup Z) \wedge Deg_2(V, U \cup X) \wedge Deg_2(Z, Y \cup U)$$

$$\wedge \neg Link(X, Z \cup U) \wedge \neg Link(Y, U \cup V)),$$

where $Deg_2(X_1, X_2)$ means that each vertex in X_1 is adjacent to exactly two vertices in X_2. However, since we consider loop-free graphs, $Deg_2(X_1, X_2)$ is equivalent to $InDeg_2(X_1, X_2)$ (an undirected edge is a pair of opposite directed edges). Thus, the property of having a hole (for a loop-free undirected graph) is in $\mathcal{EB}(\mathcal{P}^{Bool})$.

Every term $t \in T(F_{[k]}^{VR})$ can be transformed in linear time (for fixed k) into an equivalent irredundant one t' in $T(F_{[k]}^{VR})$ and into an irredundant term $\bar{t} \in T(F_{[2k]}^{VR})$ that defines the edge-complement of the graph $val(t)$, see the proof of Proposition 2.105(4). We obtain that a graph $val(t)$ is perfect if and only if the $F_{[2k]}^{VR}$-automaton for holes rejects both t' and $\bar{t}$.

The algorithm of [ChuCLSV] can test if G is perfect in time $O(n^9)$ (where n is the number of vertices). From this logical expression of holes, we get a fixed-parameter *cubic* algorithm for testing perfectness, with clique-width as parameter.

Tests conducted by Durand are reported in [CouDur11].

6.3.5 Monadic second-order properties of graphs of bounded tree-width

In the previous section, we have constructed automata from monadic second-order formulas for graphs of bounded clique-width. We have given in this way an alternative proof of the Weak Recognizability Theorem for the VR algebra and a fixed-parameter tractable algorithm for the model-checking problem of CMS sentences, the parameter being the clique-width of the given graph (if the sentence to be checked is fixed). Our objective is here to do the same for the HR algebra and the model-checking problem for CMS_2 sentences with respect to tree-width. Before embarking on constructions of automata, let us look at what can be obtained from the results and constructions of Section 6.3.4.

First, we sketch a proof that the model-checking problem $MC(\mathcal{JS}_{[K,\Lambda]}[\emptyset], CMS)$ is fixed-parameter linear with respect to $(twd(G), \varphi)$, where G is the input graph and φ is the input sentence (this is a weakening of Theorem 6.4(1) where the richer language CMS_2 is used). We can derive this result from Theorem 6.35 (or Theorem 6.3(3)) relative to the VR algebra as follows. Let G be a labeled graph given as $val(t)$ for a term t over $F^{HR}_{[K,\Lambda]}$. By the proof of Theorem 4.49, one can transform this term in linear time into a term t' over $F^{VR}_{[K,\Lambda]}$ that evaluates to the same graph. Theorem 6.35 gives the result, with the help of the parsing step described in the proof of Theorem 6.4(1). This proof has two drawbacks: first, it does not apply to sentences with edge set quantifications and second, it transforms a term in $T(F^{HR}_{[k],[K,\Lambda]})$ into a term in $T(F^{VR}_{[m],[K,\Lambda]})$ where m is exponential in k (cf. the discussion in Section 5.3.4).

Another possibility consists in observing that if G is a (K, Λ)-labeled graph of tree-width k, then its incidence graph $Inc(G)$ is a $(K \cup \Lambda, \emptyset)$-labeled simple directed graph of tree-width at most $\max\{2, k\}$ (cf. Example 2.56(5)). Hence, the previous reduction is applicable, which answers the first objection but not the second one. However, the incidence graph of such a graph of path-width k has path-width at most $k + 1$ (easy) and clique-width at most $k + 3$ (by Proposition 2.114(1.3)). Hence, the two drawbacks we are discussing disappear for graphs G of bounded path-width given by slim terms over $F^{HR}_{[k+1],[K,\Lambda]}$. We have observed at the end of Section 6.2.1 that path-decompositions of width at most any fixed k can be computed in linear time (for graphs of path-width at most k). From such a decomposition, we can compute one for $Inc(G)$. Proposition 2.85 shows that it can be converted in linear time into a slim term over $F^{HR}_{[k+2],[K \cup \Lambda, \emptyset]}$, and Proposition 2.114(1.3) shows that this term can be turned into one over $F^{VR}_{[k+3],[K \cup \Lambda, \emptyset]}$ that defines $Inc(G)$. The necessary terms can thus be constructed in linear time. The constructions of automata of Section 6.3.4 can be adapted (without the exponential increase of their sizes due to Proposition 2.114(2)) and applied to the model-checking of CMS_2 sentences. This observation is developed in [Cou10].

Since graphs of bounded tree-width are frequently considered, and since the above reductions use very large sets of port labels, it is useful to have a direct inductive construction of automata, along the lines of Section 6.3.4. As for graphs of bounded clique-width, the main task will be to specify the relevant signature (analogous to $F^{\text{VR}(n)}$) and to construct automata for the atomic formulas. The other steps will use products, direct and inverse images of automata (cf. Section 6.3.1 and Constructions 6.28 and 6.29).

We will only construct automata that check properties of directed unlabeled graphs (without sources). The constructions will extend easily to undirected and/or labeled graphs with sources. In particular, the source names of an s-graph can be specified as vertex labels. (See the remark after Theorem 6.4, the remarks in Section 5.1.5 about replacing constant symbols by unary relations, and Remark 6.51 below.)

Definition 6.44 (A variant of the signature F^{HR}) Let C be a finite set of source labels. We let F'^{HRd}_C be the derived signature of F^{HRd}_C consisting of the following constant and operation symbols:

- the constant symbols $\varnothing$, $\mathbf{a}^\ell$ and $\overrightarrow{\mathbf{ab}}$ for $a, b \in C$ such that $a \neq b$;
- the parallel composition $/\!/$;
- the source renaming operations ren_h such that h is a permutation of C;
- the unary (derived) operations miv_a, for $a \in C$, defined by:

$$miv_a(G) := fg_a(\mathbf{a} /\!/ G).$$

The operation miv_a *makes an internal vertex* by using the source label a: if $a \in \tau(G)$, then $miv_a(G) = fg_a(G)$; otherwise $miv_a(G)$ is $G \oplus fg_a(\mathbf{a})$, i.e., is G augmented with a new isolated internal vertex. In both cases, the number of internal vertices is increased by one. Note that $\tau(miv_a(G)) = \tau(G) - \{a\}$. The other symbols have the same interpretation as in $\mathbb{JS}^d$.

For a term t in $T(F'^{\text{HRd}}_C)$, the s-graph $val(t)$ is well defined since F'^{HRd}_C is a derived signature of F^{HRd}_C. Note that the isolated vertices of $val(t)$ cannot be sources. Apart from that, F^{HRd}_C and F'^{HRd}_C generate the same s-graphs. The type $\tau(t)$ of t is defined to be $\tau(val(t))$ and $\tau(u) := \tau(t/u)$ for $u \in N_t$. We denote by $T(F'^{\text{HRd}}_C)_\varnothing$ the set of terms t in $T(F'^{\text{HRd}}_C)$ such that $\tau(t) = \varnothing$, i.e., such that $val(t)$ has no sources.

If G is a concrete s-graph such that $G \simeq val(t)$ for a term t in $T(F'^{\text{HRd}}_C)$, then there are bijections between E_G and the set $Occ_0(t)$ of occurrences in t of constant symbols different from $\varnothing$, and between Int_G, the set of internal vertices of G, and the set $Occ_1(t)$ of occurrences in t of the unary symbols miv_a. This fact is made more precise in the following proposition that uses definitions from Section 2.3.2 and some new definitions.

We define for every term t in $T(F'^{\text{HRd}}_C)$ a canonical (concrete) s-graph $c'val(t)$, to be distinguished from the canonical s-graph $cval(t)$ associated with every t in $T(F^{\text{HRd}})$. Let t be a term in $T(F'^{\text{HRd}}_C)$. Without loss of generality, we assume that $C \cap N_t = \varnothing$. For every $u \in N_t$, we denote $Occ_i(t) \cap N_t/u$ by $Occ_i(t)/u$ and we define

a concrete s-graph $G(u)$ with edge set $Occ_0(t)/u$, vertex set $(Occ_1(t)/u) \cup \tau(u)$ and
a-source a for every $a \in \tau(u)$, by the following bottom-up induction:

If u is an occurrence of $\varnothing$, then $G(u) := \varnothing$.

If u is an occurrence of $\mathbf{a}^\ell$, then $G(u)$ has vertex a (that belongs to $\tau(u)$) with the
   incident loop u.

If u is an occurrence of $\overrightarrow{\mathbf{ab}}$, then $G(u)$ has vertices a and b and u is an edge from
   a to b.

If u is an occurrence of $/\!/$ with sons u_1 and u_2, then $G(u)$ is the concrete graph
   $G(u_1) /\!/ G(u_2)$ (it is easy to check that $G(u)$ is well defined).

If u is an occurrence of ren_h with son u_1, then $G(u)$ is the concrete graph
   $ren_h(G(u_1))$ with every vertex $a \in \tau(u_1)$ renamed into $h(a)$.[26]

If u is an occurrence of miv_a with son u_1, there are two cases. If $a \in \tau(u_1)$, then
   $G(u)$ is the concrete graph $fg_a(G(u_1))$ with vertex a renamed into u. If $a \notin \tau(u_1)$,
   then $G(u)$ is $G(u_1)$ augmented with u as a new internal and isolated vertex.

We will denote $G(root_t)$ by $c'val(t)$. It is a concrete s-graph isomorphic to $val(t)$.
It has edge set $Occ_0(t)$ and vertex set $Occ_1(t) \cup \tau(t)$. If $t \in T(F_C'^{HRd})_\varnothing$ (which means
that $val(t)$ is a graph), then $V_{c'val(t)} = Occ_1(t)$. If G is a concrete graph isomorphic
to $c'val(t)$ by an isomorphism $w : G \rightarrow c'val(t)$, then w can be viewed as a bijection
$: V_G \cup E_G \rightarrow Occ_1(t) \cup Occ_0(t)$. We call such a bijection a *simple witness* of the
isomorphism of G and $val(t)$. ("Simple" because it is simpler than the witnesses
defined in Section 2.3.2, and to distinguish it from them.)

Proposition 6.45

(1) The same graphs are defined by the terms in $T(F_C'^{HRd})_\varnothing$ and in $T(F_C^{HRd})_\varnothing$.

(2) Let s be a term in $T(F_C^{HRd})_\varnothing$ that defines a concrete graph G, and w be a canonical
    witness of $G \simeq val(s)$. One can compute in linear time (for fixed C) an equivalent
    term t in $T(F_C'^{HRd})_\varnothing$ and a simple witness w' of $G \simeq val(t)$.

(3) Let L be a set of graphs and $\mathcal{A}$ be an automaton with N states such that, for every
    term t in $T(F_C'^{HRd})_\varnothing$, we have $t \in L(\mathcal{A})$ if and only if $val(t) \in L$. From $\mathcal{A}$, an
    automaton $\mathcal{B}$ with $2^{|C|} \cdot N$ states can be constructed that recognizes the set of
    terms t in $T(F_C^{HRd})_\varnothing$ such that $val(t) \in L$. If $\mathcal{A}$ is complete and deterministic,
    then so is $\mathcal{B}$.

Proof: (1) Since $F_C'^{HRd}$ is a derived signature of F_C^{HRd}, the s-graphs it defines are also
defined by terms over F_C^{HRd}. The opposite implication, for graphs (without sources),
follows from (2) that we now prove.

(2) Let $s \in T(F_C^{HRd})_\varnothing$ and w be a canonical witness of $G \simeq val(s)$. In two steps, we
will transform s into t, and w into w'.

---

[26] If G is a concrete graph and g is a bijection between V_G and some set V, then the graph obtained
by *renaming* every $v \in V_G$ into $g(v)$ is the unique graph H with $V_H = V$ such that (g, Id) is an
isomorphism $: G \rightarrow H$, where Id is the identity on E_G.

Step 1: At each occurrence u in s of an operation fg_B, we replace this operation by the composition of the operations fg_a for all a in $B \cap \tau(u_1)$, where u_1 is the son of u (in particular, fg_B is removed if $B \cap \tau(u_1) = \emptyset$). We obtain in this way an equivalent term s_1 and, easily, a corresponding canonical witness w_1.

Step 2: We obtain the desired term t by replacing in s_1 every constant symbol **a** by $\emptyset$ and every operation symbol fg_a by miv_a.[27] Since t is obtained from s_1 by replacements of unary operations, $N_t = N_{s_1}$. It is straightforward to show, by bottom-up induction on $u \in N_{s_1}$, that $val(t/u)$ is obtained from $val(s_1/u)$ by removing all isolated sources. In particular, if $s_1/u = fg_a(s_1/u_1)$ and the a-source of $val(s_1/u_1)$ is isolated, then $a \notin \tau(t/u_1)$ and so the operation miv_a adds an isolated vertex to $val(t/u_1)$. Hence $val(t) = val(s_1)$ and $G \simeq val(t)$. It remains to define a simple witness w' of this fact. Recall that $N_t = N_{s_1}$. We define $w'(e) := w_1(e)$ if $e \in E_G$.

By the definitions of Section 2.3.2, if $x \in V_G$, then $w_1(x) = (u, a)$, where $a \in \tau(s_1/u)$ and $(u, a) \in X_{\max}$, i.e., u is the unique maximal node of s_1 in the set $\{u' \in N_{s_1} \mid (u', a') \approx_{s_1} (u, a)$ for some $a' \in C\}$. The node u cannot be the root of s_1 because $\tau(s_1) = \tau(s) = \emptyset$. The father v of u is necessarily an occurrence of fg_a in s_1, hence of miv_a in t. We take it as $w'(x)$.

A formal proof of the correctness of w' can be based on a proof that $cval(s_1)$ is isomorphic to $c'val(t)$. By bottom-up induction on $u \in N_t$ it can be shown that the concrete graph that is obtained from $((Exp(s_1) \restriction u)/\approx_{s_1,u})X_{\max}$ (see the proof of Proposition 2.41) by removing its isolated sources, is isomorphic to $G(u)$. For edges, the isomorphism is the identity on $Occ_0(t)/u$. For vertices, it maps $(u, a) \in X_{\max}$ to a if u is the root, and to the father of u otherwise.

To optimize the construction one can, by using Equalities (3), (5) and (9) of Proposition 2.48, eliminate from s all occurrences of $\emptyset$ (cf. Proposition 2.49). Using Equalities (3) and (9), the occurrences of $\emptyset$ in t can be removed except those that are directly below some miv_a.

All these computations can be done in time $O(|s|)$.

(3) Let $\mathscr{A}$ be an $F_C'^{\mathrm{HRd}}$-automaton with the stated property. We construct an F_C^{HRd}-automaton $\mathscr{B}$ that, for each input term s of type $\emptyset$, simulates the automaton $\mathscr{A}$ on input t, where t is defined from s as in the proof of (2). The states of $\mathscr{B}$ are of the form (A, q) where $A \subseteq C$ and q is a state of $\mathscr{A}$. The final states are all $(\emptyset, q)$ with $q \in Acc_{\mathscr{A}}$. In the first component of its state, $\mathscr{B}$ computes the type of s. For an input symbol $\emptyset$, $\mathbf{a}^\ell$, $\overrightarrow{\mathbf{ab}}$, $/\!/$ or ren_h, the automaton $\mathscr{B}$ just simulates $\mathscr{A}$ (and updates the type). For an input symbol $\mathbf{a}$, the automaton $\mathscr{B}$ simulates $\mathscr{A}$ for the input symbol $\emptyset$ (and computes the type $\{a\}$). If $\mathscr{B}$ arrives in state (A, q) at an input symbol fg_B, then it simulates $\mathscr{A}$ (in state q) on the consecutive input symbols $miv_{a_1}, \ldots, miv_{a_m}$, where $\{a_1, \ldots, a_m\} = B \cap A$ (and it updates the type to $A - B$). In the particular case where $m = 0$, the state q is unchanged. The detailed construction of $\mathscr{B}$ is omitted. ∎

---

[27] For the replacement of **a** by $\emptyset$ see the observation after Proposition 2.49.

Example 6.46 We let $C := \{a,b,c,d\}$ and

$$s := fg_{a,b,d\,1}(\overrightarrow{\mathbf{ab}}_3 \,\|_2\, fg_{c\,4}(\mathbf{c}_6 \,\|_5\, (\overrightarrow{\mathbf{cb}}_8 \,\|_7\, fg_{a,d\,9}(\mathbf{a}_{11} \,\|_{10}\, \mathbf{d}_{12})))).$$

The indices 1 to 12 indicate the positions of the term s (in Polish prefix notation). The concrete graph $G \simeq val(s)$, with vertices u,v,x,y,z, is

$$\bullet_u \longrightarrow \bullet_v \longleftarrow \bullet_x \quad \bullet_y \quad \bullet_z .$$

Its canonical witness w maps vertices as follows:

$$w(u) = (5,c), \; w(v) = (2,b), \; w(x) = (2,a),$$
$$w(y) = (10,a), \; w(z) = (10,d),$$

and maps the edges (u,v) (defined by the symbol $\overrightarrow{\mathbf{cb}}$) and (x,v) (defined by the symbol $\overrightarrow{\mathbf{ab}}$) to 8 and 3 respectively.

We apply to s the transformation of Step 1 in the proof of Proposition 6.45(2) and obtain the term

$$s_1 := fg_{a\,1}(fg_{b\,2}(\overrightarrow{\mathbf{ab}}_4 \,\|_3\, fg_{c\,5}(\mathbf{c}_7 \,\|_6\, (\overrightarrow{\mathbf{cb}}_9 \,\|_8\, fg_{a\,10}(fg_{d\,11}(\mathbf{a}_{13} \,\|_{12}\, \mathbf{d}_{14})))))$$

and the canonical witness w_1 that maps vertices as follows:

$$w_1(u) = (6,c), \; w_1(v) = (3,b), \; w_1(x) = (2,a),$$
$$w_1(y) = (11,a), \; w_1(z) = (12,d),$$

and maps the edges (u,v) and (x,v) to 9 and 4 respectively. This term is in the left part of Figure 6.2. Note for example that $w_1(u) = (6,c)$: vertex u is defined by $\mathbf{c}$ at position 7 and by $\overrightarrow{\mathbf{cb}}$ at position 9; it is made internal by fg_c at position 5 and is thus represented by $w_1(u) = (6,c)$, where 6 is the son of 5 and, therefore, the highest position where u is still a source.

We now transform s_1 according to Step 2 in the proof of Proposition 6.45(2). Then, we get

$$t := miv_{a\,1}(miv_{b\,2}(\overrightarrow{\mathbf{ab}}_4 \,\|_3\, miv_{c\,5}(\varnothing_7 \,\|_6\, (\overrightarrow{\mathbf{cb}}_9 \,\|_8\, miv_{a\,10}(miv_{d\,11}(\varnothing_{13} \,\|_{12}\, \varnothing_{14}))))),$$

with the simple witness w' that maps vertices as follows:

$$w'(u) = 5, \; w'(v) = 2, \; w'(x) = 1,$$
$$w'(y) = 10, \; w'(z) = 11$$

and, as w_1, maps the edges (u,v) and (x,v) to 9 and 4 respectively. This term is in the right-hand side of Figure 6.2. Vertex u is now represented by $w'(u) = 5$. Note that u is not a source at position 5. Since $\mathbf{c}$ is replaced at position 7 by $\varnothing$, vertex u is only defined by $\overrightarrow{\mathbf{cb}}$ at position 9. The two internal vertices y and z defined in s and in s_1 by $\mathbf{a}$ and $\mathbf{d}$ are now defined by miv_a and miv_d at positions 10 and 11 in t.

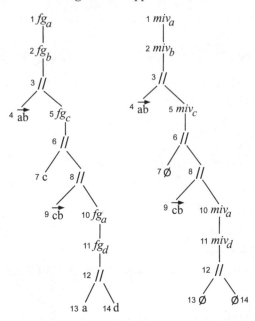

Figure 6.2 The transformation of Proposition 6.45(2).

We can remove two occurrences of $\varnothing$ from t, and obtain the term

$$t' := miv_{a1}(miv_{b2}(\overrightarrow{\mathbf{ab}}_4 /\!\!/_3 \; miv_{c5}(\overrightarrow{\mathbf{cb}}_9 /\!\!/_8 \; miv_{a10}(miv_{d11}(\varnothing_{12})))))),$$

with the same witness w' (but, to obtain the correct positions, one must subtract 2 from the numbers 8 to 12). Note that the positions marked 1, 2, 5, 10, 11 in t' correspond (by the simple witness) to vertices, and the positions marked 4 and 9 to edges. The positions marked 3, 8 and 12, which are occurrences of $/\!\!/$ and $\varnothing$, describe no vertex or edge of G.

One more remark: if in term s we replace the constant symbol $\overrightarrow{\mathbf{ab}}$ by $\varnothing$, then, clearly, the corresponding edge disappears, but also its end vertex x, which may be undesired. This is so because x *and its incident edge* are specified by the same symbol $\overrightarrow{\mathbf{ab}}$. If we do this replacement in the term t (or in t'), then vertex x remains because it is specified by the first occurrence of miv_a.

More generally, the terms t in $T(F_C'^{\mathrm{HRd}})$ specify the vertices of the graphs they define by unary operations and the edges by constant symbols. If, in such a term, we replace by $\varnothing$ some constant symbols, we delete the corresponding edges, but delete no vertices. This remark will be useful for Proposition 6.50. $\qquad\square$

The CMS properties of directed graphs G are expressed by sentences of $\mathrm{CMS}'(\{edg\},\emptyset)$ interpreted in the structures $\lfloor G \rfloor$. Their CMS_2 properties are expressed by sentences of $\mathrm{CMS}'(\{in_1, in_2\},\emptyset)$ interpreted in the structures $\lceil G \rceil =$

$\lfloor Inc(G) \rfloor$, cf. Section 5.2.5. We recall that $edg_{\lfloor G \rfloor}(x_1, x_2)$ holds if and only if $in_{1\lceil G \rceil}(x_3, x_1)$ and $in_{2\lceil G \rceil}(x_3, x_2)$ hold for some element x_3 (that is necessarily an edge of G).

Let t be a term in $T(F_C'^{HR})$. For building automata associated with the CMS_2 formulas expressing properties of the concrete graph $G := c'val(t)$ (cf. Definition 6.44), we need to encode in t the assignments $\gamma : \mathcal{X}_n \to \mathcal{P}(V_G \cup E_G)$. We let $F_C'^{HRd(n)}$ be the signature obtained from F_C^{HRd} by replacing the symbols miv_a, $\overrightarrow{ab}$ and a^ℓ respectively by (miv_a, w), $(\overrightarrow{ab}, w)$ and (a^ℓ, w), for all $w \in \{0,1\}^n$, of the same arity. For t in $T(F_C'^{HRd})$ and $\gamma : \mathcal{X}_n \to \mathcal{P}(Occ_0(t) \cup Occ_1(t))$ we define $t * \gamma$ to be the term in $T(F_C'^{HRd(n)})$ obtained from t by replacing f by $(f, \overline{w}_\gamma(u))$ at each occurrence u of $f = miv_a, f = \overrightarrow{ab}$ or $f = a^\ell$. Every term in $T(F_C'^{HRd(n)})$ is of the form $t * \gamma$. Note that, in general, γ is an assignment : $\mathcal{X}_n \to \mathcal{P}(Int_G \cup E_G)$, where G is the concrete s-graph $c'val(t)$.

If $\varphi \in CMS'(\{in_1, in_2\}, \mathcal{X}_n)$, we let $L_{C,\varphi,n}^{HR}$ be the set of terms $t * \gamma$ in $T(F_C'^{HRd(n)})_\emptyset$ such that $(\lceil c'val(t) \rceil, \gamma) \models \varphi$. Our objective is to construct an automaton $\mathcal{C}_{C,\varphi,n}$ recognizing this set in the following sense:

$$L_{C,\varphi,n}^{HR} = L(\mathcal{C}_{C,\varphi,n}) \cap T(F_C'^{HRd(n)})_\emptyset.$$

Note that $\mathcal{C}_{C,\varphi,n}$ is not determined in a unique way by this equality. This condition is sufficient because we can assume that only terms in $T(F_C'^{HRd(n)})_\emptyset$ will be given as input to $\mathcal{C}_{C,\varphi,n}$. However, the language $T(F_C'^{HRd(n)})_\emptyset$ is recognized by an automaton with 2^k states, where $k = |C|$. Hence, any automaton $\mathcal{C}_{C,\varphi,n}$ as above with N states can be replaced by an automaton with $2^k \cdot N$ states that recognizes $L_{C,\varphi,n}^{HR}$. Moreover, in the special case where φ is a sentence, it follows from Proposition 6.45(3) that $\mathcal{C}_{C,\varphi,0}$ can be transformed into an automaton with $2^k \cdot N$ states that recognizes the set of terms t in $T(F_C^{HRd})_\emptyset$ such that $\lceil val(t) \rceil \models \varphi$.

We now present the inductive construction of the automata $\mathcal{C}_{C,\varphi,n}$. It is not hard to check that if $L_{C,\varphi,n}^{HR} = L(\mathcal{C}_{C,\varphi,n}) \cap T(F_C'^{HRd(n)})_\emptyset$ and $L_{C,\psi,n}^{HR} = L(\mathcal{C}_{C,\psi,n}) \cap T(F_C'^{HRd(n)})_\emptyset$, then

$$L_{C,\varphi\vee\psi,n}^{HR} = (L(\mathcal{C}_{C,\varphi,n}) \cup L(\mathcal{C}_{C,\psi,n})) \cap T(F_C'^{HRd(n)})_\emptyset,$$

$$L_{C,\varphi\wedge\psi,n}^{HR} = L(\mathcal{C}_{C,\varphi,n}) \cap L(\mathcal{C}_{C,\psi,n}) \cap T(F_C'^{HRd(n)})_\emptyset, \text{ and}$$

$$L_{C,\neg\varphi,n}^{HR} = (T(F_C'^{HRd(n)}) - L(\mathcal{C}_{C,\varphi,n})) \cap T(F_C'^{HRd(n)})_\emptyset.$$

If θ is φ^σ, where σ is a substitution of variables or of Boolean set terms (cf. Lemmas 6.25 and 6.33), which transforms $\varphi \in CMS'(\{in_1, in_2\}, \mathcal{X}_n)$ into $\varphi^\sigma \in CMS'(\{in_1, in_2\}, \mathcal{X}_m)$, then

$$L_{C,\theta,m}^{HR} = h_\sigma^{-1}(L(\mathcal{C}_{C,\varphi,n})) \cap T(F_C'^{HRd(m)})_\emptyset,$$

where h_σ is defined in these lemmas, extended to an arity preserving mapping :
$F_C^{\prime HRd(m)} \to F_C^{\prime HRd(n)}$ in the obvious way (cf. Section 6.3.4). If θ is $\exists X_{n+1},\ldots,X_m.\varphi$,
then

$$L_{C,\theta,n}^{HR} = h(L(\mathscr{C}_{C,\varphi,m})) \cap T(F_C^{\prime HRd(n)})_\emptyset,$$

where $h : F_C^{\prime HRd(m)} \to F_C^{\prime HRd(n)}$ deletes the last $m-n$ Booleans from each symbol
that encodes an assignment. The last two equalities hold because $\tau(h_\sigma(t)) = \tau(t)$ for
every term $t \in T(F_C^{\prime HRd(m)})$ and similarly for h.

Hence, we need only construct automata for the atomic formulas. The other constructions, which use products and direct or inverse images of automata, are similar or identical to those of Sections 6.3.3 and 6.3.4.

Construction 6.47 (Incidence relations) Our aim is to construct, for each finite set
C of source labels, a complete and deterministic $F_C^{\prime HRd(2)}$-automaton $\mathscr{C}_{C,in_1(X_1,X_2)}$ that
recognizes (in the above particular sense) the language $L_{C,in_1(X_1,X_2)}^{HR}$.[28]

As set of states, we take $Q := C \cup \{0, Ok, Error\}$ of cardinality $k+3$ and Ok as
accepting state. The meanings of the states are described in Table 6.13. Although the
automaton to be constructed will have to operate on terms $t \in T(F_C^{\prime HRd(2)})$ of type
$\tau(t) = \emptyset$, the meanings of its states must be specified with respect to terms of all types.
We will use the same convention as before regarding X_1 and X_2. The vertices and
edges of the concrete graph $c'val(t)$ (for $\tau(t) = \emptyset$) are nodes of t (cf. Definition 6.44).
An edge e of $c'val(t)$ is a leaf and its ends x and y are both on the path between
the root and the leaf e. Hence, the ends of an edge of $c'val(t)$ are *above* this edge
in t; for $\tau(t) \neq \emptyset$, x and y can be in $\tau(t)$: the set of sources of $c'val(t)$. When the
properties P_q are formulated (equivalently) for a node u of t, $c'val(t)$ must be replaced
in Table 6.13 by $G(u)$, as defined in Definition 6.44, i.e., P_q should be satisfied in
$(\lceil G(u) \rceil_C, \gamma(X_1) \cap (N_t/u), \gamma(X_2) \cap (N_t/u))$.

Table 6.14 shows its transition rules. The construction of $\mathscr{C}_{C,in_2(X_1,X_2)}$ is of course
similar.

The automata for the other atomic formulas are similar to this one or to those of
Section 6.3.4. As for terms and graphs in Sections 6.3.3 and 6.3.4, it may be useful
to precompile automata for frequently used properties. We now give some examples.

Construction 6.48 (Precompiled automata) (1) Let *Vertices*(X) and *Edges*(X)
mean respectively that X is a set of vertices and that X is a set of edges. These
first-order set predicates can be useful for shortening the writing of CMS_2 formulas. Complete and deterministic automata for them, that have two states, are
straightforward to construct.

---

[28] We write $L_{C,\varphi}^{HR}$ instead of $L_{C,\varphi,mfv(\varphi)}^{HR}$ and similarily for $\mathscr{C}_{C,\varphi}$.

Table 6.13 *Meanings of the states of $\mathcal{C}_{C,\,in_1(X_1,X_2)}$.*

State q	Property P_q
0	$X_1 = X_2 = \emptyset$
Ok	$X_1 = \{e\}, X_2 = \{v\}$ for some $e \in E_{c'val(t)}$, where $v \in Int_{c'val(t)}$ and v is the tail of e
a	$X_2 = \emptyset, X_1 = \{e\}$ for some $e \in E_{c'val(t)}$ such that the tail of e is the a-source of $c'val(t)$
Error	all other cases

Table 6.14 *Transition rules of $\mathcal{C}_{C,\,in_1(X_1,X_2)}$.*

Transition rules	Conditions
$\emptyset \rightarrow 0$	
$(\overrightarrow{\mathbf{ab}}, 00) \rightarrow 0$	
$(\overrightarrow{\mathbf{ab}}, 10) \rightarrow a$	
$(\mathbf{a}^\ell, 00) \rightarrow 0$	
$(\mathbf{a}^\ell, 10) \rightarrow a$	
$(miv_a, 00)[q] \rightarrow q$	$q \in Q$
$(miv_a, 01)[a] \rightarrow Ok$	$a \in C$
$ren_h[0] \rightarrow 0$	
$ren_h[Ok] \rightarrow Ok$	
$ren_h[a] \rightarrow h(a)$	$a \in C$
$/\!/[0, q] \rightarrow q$	
$/\!/[q, 0] \rightarrow q$	$q \in Q$

(2) Next consider the construction of $\mathcal{C}_{C,edg(X_1,X_2)}$. We have:

$$edg(X_1, X_2) :\Longleftrightarrow \exists X_3 (in_1(X_3, X_1) \wedge in_2(X_3, X_2)).$$

It follows that by using inverse images, product and direct image, we can get a nondeterministic automaton with $(k+3)^2$ states, hence a deterministic one having $2^{(k+3)^2}$ states (before trimming). We could construct a deterministic automaton with less states. Rather than doing it, we show that every deterministic automaton must have at least $2^{k(k-1)}$ states. Hence, a specific construction would not be very useful.

For each set R of pairs (a,b) of distinct elements of C, we denote by $/\!/R$ the term $/\!/_{(a,b)\in R}(\overrightarrow{\mathbf{ab}},00)$ belonging to $T(F'^{\mathrm{HRd}(2)}_C)$. It is the parallel-composition of the constant symbols $(\overrightarrow{\mathbf{ab}},00)$ for all $(a,b)\in R$ (with $/\!/\varnothing := \varnothing$). For $a,b\in C$ with $a\neq b$, we let $c_{a,b}$ be the context in $Ctxt(F'^{\mathrm{HRd}(2)}_C)$ defined as the composition (in any order) of the unary operations $(miv_a,10)$, $(miv_b,01)$ and $(miv_d,00)$ for every $d\in C-\{a,b\}$. We have $c_{a,b}[/\!/R]\in L^{\mathrm{HR}}_{C,edg(X_1,X_2)}$ if and only if $(a,b)\in R$. This shows that every complete and deterministic automaton recognizing $L^{\mathrm{HR}}_{C,edg(X_1,X_2)}$ must "know" R exactly after having traversed from bottom-up a term $/\!/R$. Hence, it must have at least $2^{k(k-1)}$ different states.

(3) Next we consider subgraphs and induced subgraphs. Let G be an s-graph and X be a subset of $V_G\cup E_G$. The property $Sub(X)$ is defined by

every edge in X has its (possibly equal) ends in X,

i.e., it means that G has a subgraph H such that $V_H = V_G\cap X$ and $E_H = E_G\cap X$ (cf. Definitions 2.12 and 2.25; its sources are those of G that are in X). Furthermore, we denote by $Src_G(X)$ the set of sources of G that are end vertices of edges in X. We define an $F'^{\mathrm{HRd}(1)}_C$-automaton $\mathscr{C}_{C,Sub(X_1)}$ having the states $Error$ and A for all subsets A of C. Their meanings for a term t are as follows (with the same notation as in Table 6.13):

$$P_A \quad : \quad Sub(X_1\cup Src_{c'val(t)}(X_1)) \text{ is true and } A = Src_{c'val(t)}(X_1),$$

$$P_{Error} \quad : \quad Sub(X_1\cup Src_{c'val(t)}(X_1)) \text{ is false, i.e., some edge in } X_1$$
$$\text{has an end that is an internal vertex of } c'val(t) \text{ not in } X_1.$$

The condition on A is meaningful because each element a of $\tau(t)$ is the a-source of the concrete s-graph $c'val(t)$ (cf. Definition 6.44). The accepting state is $\varnothing$. Table 6.15 shows the transition rules.

We now construct the more complicated automaton for the property $Ind(X)$ defined as follows, for an s-graph G and a subset X of $V_G\cup E_G$:

X is the set of edges and vertices of an induced subgraph of G (cf. Definition 2.12), i.e., $Sub(X)$ holds and every edge of G having its ends in X is in X.

The latter requirement is equivalent to $NoEdge((V_G\cap X)\cup(E_G-X))$, where, for an s-graph G and a subset X of $V_G\cup E_G$, the property $NoEdge(X)$ means:

no edge in X has its ends in X.

We first construct the automaton $\mathscr{C}_{C,NoEdge(X_1)}$. Its states are $Error$ and the triples (A,B,R) such that $A,B\subseteq C$ and R is a set consisting of subsets of C of cardinality 2. This automaton has $2^{2k+k(k-1)/2}+1$ states whose meanings are as follows:

Table 6.15 *Transition rules of* $\mathscr{C}_{C,Sub(X_1)}$.

Transition rules	Conditions
$\varnothing \to \varnothing$	
$(\overrightarrow{\mathbf{ab}},0) \to \varnothing$	
$(\overrightarrow{\mathbf{ab}},1) \to \{a,b\}$	
$(\mathbf{a}^{\ell},0) \to \varnothing$	
$(\mathbf{a}^{\ell},1) \to \{a\}$	
$(miv_a,0)[A] \to A$	$a \notin A$
$(miv_a,1)[A] \to A - \{a\}$	
$ren_h[A] \to h(A)$	
$/\!/[A,B] \to A \cup B$	

P_{Error} : some edge in X_1 has its two (possibly equal) ends in X_1.
$P_{(A,B,R)}$ is the conjunction of four conditions:

(1) P_{Error} does not hold;
(2) $A = \{a \in \tau(t) \mid a$ is the vertex incident with a loop of $c'val(t)$ that belongs to $X_1\}$;
(3) $B = \{b \in \tau(t) \mid b$ is an end of an edge of $c'val(t)$ that belongs to X_1 and has its other end in $X_1\}$;
(4) R is the set of sets $\{a,b\}$ such that $a,b \in \tau(t)$, $a \neq b$, and a and b are the ends of an edge of $c'val(t)$ that belongs to X_1.

The accepting state is $(\varnothing,\varnothing,\varnothing)$. Table 6.16 shows the transitions.

For obtaining $\mathscr{C}_{C,Ind(X_1)}$, we let h be the mapping $F_C^{\prime HRd(1)} \to F_C^{\prime HRd(1)}$ such that, for every $a,b \in C$ and $i \in \{0,1\}$: $h((\overrightarrow{\mathbf{ab}},i)) := (\overrightarrow{\mathbf{ab}},1-i)$ and $h((\mathbf{a}^{\ell},i)) := (\mathbf{a}^{\ell},1-i)$, and that does not modify the other symbols. It is clear (cf. Lemma 6.33 on substitutions of Boolean set terms) that

$$L_{C,Ind(X_1)}^{HR} = L_{C,Sub(X_1)}^{HR} \cap h^{-1}(L_{C,NoEdge(X_1)}^{HR}),$$

so that we obtain easily $\mathscr{C}_{C,Ind(X_1)}$ as a product of $\mathscr{C}_{C,Sub(X_1)}$ and an inverse image of $\mathscr{C}_{C,NoEdge(X_1)}$. It has $2^{3k+k(k-1)/2} + 1$ states.

Construction 6.49 (Relativizations) We now show that relativizations can be obtained along the lines of Construction 6.37, with some more technical details. We consider a property $P(X_1,\ldots,X_n)$ of sets of edges and vertices. We let $Q(X_1,\ldots,X_{n+1})$

Table 6.16 *Transition rules of* $\mathscr{C}_{C,NoEdge(X_1)}$.

Transition rules	Conditions
$\varnothing \to (\varnothing,\varnothing,\varnothing)$	
$(\overrightarrow{\mathbf{ab}},0) \to (\varnothing,\varnothing,\varnothing)$	
$(\overrightarrow{\mathbf{ab}},1) \to (\varnothing,\varnothing,\{\{a,b\}\})$	
$(\mathbf{a}^\ell,0) \to (\varnothing,\varnothing,\varnothing)$	
$(\mathbf{a}^\ell,1) \to (\{a\},\varnothing,\varnothing)$	
$(miv_a,0)[(A,B,R)] \to (A',B',R')$	$A' = A - \{a\}$, $B' = B - \{a\}$, R' is R without the pairs containing a
$(miv_a,1)[(A,B,R)] \to (A,B',R')$	$a \notin A \cup B$, $B' = B \cup \{b \mid \{a,b\} \in R\}$, R' is R without the pairs containing a
$ren_h[(A,B,R)] \to (h(A),h(B),h(R))$	
$/\!/[(A,B,R),(A',B',R')] \to (A'',B'',R'')$	$A'' = A \cup A'$, $B'' = B \cup B'$, $R'' = R \cup R'$

be the property such that, for all sets of edges and vertices $X_1,\ldots,X_{n+1}$ of a graph G (without sources), the set X_{n+1} satisfies $Sub(X_{n+1})$, hence defines a subgraph H of G, and $P(X_1 \cap X_{n+1},\ldots,X_n \cap X_{n+1})$ is true in H. We will construct an automaton $\mathscr{C}_{C,Q(X_1,\ldots,X_{n+1})}$ from $\mathscr{C}_{C,P(X_1,\ldots,X_n)}$ with the help of the automaton $\mathscr{C}_{C,Sub(X_1)}$ of Construction 6.48(3).

We let h be the mapping : $F_C'^{HRd(n+1)} \to F_C'^{HR(n)}$ such that, for every $a,b \in C$ with $a \neq b$ and every $w \in \{0,1\}^n$:

$$h((\overrightarrow{\mathbf{ab}},w0)) := \varnothing,$$

$$h((\overrightarrow{\mathbf{ab}},w1)) := (\overrightarrow{\mathbf{ab}},w),$$

$$h((\mathbf{a}^\ell,w0)) := \varnothing,$$

$$h((\mathbf{a}^\ell,w1)) := (\mathbf{a}^\ell,w),$$

$$h((miv_a,w0)) := ren_{Id} \text{ (where } Id \text{ is the identity} : C \to C), \text{ and}$$

$$h((miv_a,w1)) := (miv_a,w).$$

The other symbols are not modified by h. We use the "neutral" constant symbol $\varnothing$ in order to replace the constant symbols that specify the deleted edges, and the "neutral" unary symbol ren_{Id} (denoting the identity operation) in order to replace the unary symbols $(miv_a, w0)$ that specify the deleted vertices. We let also $h' : F_C'^{HRd(n+1)} \rightarrow F_C'^{HR(1)}$ delete the first n Booleans of each sequence.

With these definitions and notation we have:

Proposition 6.50 $L^{HR}_{C,Q(X_1,...,X_{n+1})} = L^{HR}_{C,Sub(X_{n+1})} \cap h^{-1}(L^{HR}_{C,P(X_1,...,X_n)})$ and thus $\mathscr{C}_{C,Q(X_1,...,X_{n+1})}$ can be defined as $h'^{-1}(\mathscr{C}_{C,Sub(X_1)}) \cap h^{-1}(\mathscr{C}_{C,P(X_1,...,X_n)})$. Similar facts hold for induced subgraphs, with $Ind(X_{n+1})$ instead of $Sub(X_{n+1})$.

Proof: The proof is similar to that of Lemma 6.38 except that we must take into account that X_{n+1} defines a subgraph.

Claim 6.50.1 Let $t \in T(F_C'^{HRd})_\varnothing$, $G := c'val(t)$ and H be a subgraph of G. Let $U := V_H \cup E_H$, so that $Sub(U)$ holds in G. Then $H = c'val(h(t * \gamma))$, where γ is the $\mathcal{X}_1$-assignment such that $\gamma(X_1) = U$.

Proof: Let $t' := h(t * \gamma)$ and $H' := c'val(t')$. Note that G, H and H' are concrete graphs with vertices and edges belonging to $N_t = N_{t'}$ by the definitions. It is clear that H and H' have the same edges (i.e., $E_H = E_{H'}$, but we will verify that the incidences are the same) and the same internal vertices.

Let us check that the incidences in H and H' are the same. Let e be an edge of H' brought in by an occurrence u (in U) of a constant symbol $\overrightarrow{\mathbf{ab}}$ or $\mathbf{a}^\ell$. Its tail x in H' (introduced as the a-source in the corresponding basic graphs) is made internal by some operation miv_c, precisely at occurrence x. Some renamings on the path in t' between u and x have transformed a into c. These renamings are not modified by h. Let y be the tail of e in H. If $y \neq x$, then y is an occurrence (below x) of some operation miv_d that has been changed into ren_{Id} by h. But this implies that y is not in U whereas it is the tail of the edge $e \in U$. This contradicts the fact that H is a subgraph of G. Hence $y = x$. The same argument works for the heads of the edges of H'.

Since the incidences in H and H' are the same, all edges of H' are incident with internal vertices of H'. Thus, H' has no sources (because sources of any $c'val(t')$ cannot be isolated). Hence H' has the same vertices as H, and so $H' = H$. This proves the claim. $\qquad\square$

We now prove the proposition. Let $t * \gamma$ belong to $T(F_C'^{HRd(n+1)})_\varnothing$ and $G := c'val(t)$. Moreover, let $\gamma(X_{n+1})$ be the set of vertices and edges of a subgraph H of G, i.e., let $t * \gamma$ belong to $L^{HR}_{C,Sub(X_{n+1})}$. Then $h(t * \gamma) = t' * \gamma'$, where $c'val(t') = H$ (by the claim) and γ' is the assignment : $\mathcal{X}_n \rightarrow \mathcal{P}(E_H \cup V_H)$ such that $\gamma'(X_i) = \gamma(X_i) \cap \gamma(X_{n+1})$ for each i. It follows that $t * \gamma \in L^{HR}_{C,Q(X_1,...,X_{n+1})}$ if and only if $t' * \gamma' \in L^{VR}_{C,P(X_1,...,X_n)}$. Hence $L^{HR}_{C,Q(X_1,...,X_{n+1})} = L^{HR}_{C,Sub(X_{n+1})} \cap h^{-1}(L^{HR}_{C,P(X_1,...,X_n)})$. The assertion about automata follows immediately.

The proof is the same for induced subgraphs. $\qquad\blacksquare$

Remark 6.51 (Labeled graphs with sources) We adapt the above definitions to (K, Λ)-labeled graphs. We will use the constant symbols $\overrightarrow{\mathbf{ab}}_\lambda$ and $\mathbf{a}_\lambda^\ell$ for defining directed edges and loops labeled by $\lambda \in \Lambda$, and, for $a \in C$ and $\alpha = \{\kappa_1, \ldots, \kappa_p\} \subseteq K$, the derived operations $miv_{a,\alpha}$ defined by

$$miv_{a,\alpha}(G) := fg_a(\mathbf{a}_{\kappa_1} /\!\!/ \cdots /\!\!/ \mathbf{a}_{\kappa_p} /\!\!/ G).$$

We get a signature $F_{C,[K,\Lambda]}^{\prime HRd}$. Note that for a term t in $T(F_{C,[K,\Lambda]}^{\prime HRd})$ the sources of $val(t)$ do not have labels in K.

Proposition 6.45 can be extended to labeled graphs. In the proof of Statement (2), the desired term t is obtained by replacing in s_1 every occurrence of a constant symbol **a** by $\varnothing$, and every occurrence u of an operation symbol fg_a by $miv_{a,\alpha}$, where α is the set of labels of the a-source of $val(s_1/u_1)$ and u_1 is the son of u. In the proof of Statement (3), the automaton $\mathscr{B}$ has states (A, α, q), where α is a mapping $: A \to \mathcal{P}(K)$. For input term s, it computes the set of labels $\alpha(a) \subseteq K$ of each a-source of $val(s)$. Thus, $\mathscr{B}$ has $(2^{|K|} + 1)^{|C|} \cdot N$ states.

Constructions 6.47, 6.48 and 6.49 extend in a straightforward manner to the construction of automata for checking the CMS_2 properties of (K, Λ)-labeled graphs of bounded tree-width. They yield fixed-parameter linear model-checking algorithms with respect to tree-width because the parsing problem is not more difficult for labeled graphs than for unlabeled ones, as we have observed in the proof of Theorem 6.4(1), cf. Proposition 2.47(2).

For checking the properties of (K, Λ)-labeled s-graphs, we convert source names into vertex labels. If C is a finite set of source names, we let $\overline{C} := \{\overline{a} \mid a \in C\}$ be a set of vertex labels disjoint from $K \cup \Lambda$ and from the set $\mathcal{A}$ of source names. Every (K, Λ)-labeled s-graph G of type $A = \{a_1, \ldots, a_s\} \subseteq C$, defined by a term $t \in T(F_{C,[K,\Lambda]}^{\prime HRd})$, is converted into the $(\overline{C} \cup K, \Lambda)$-labeled graph $\overline{G}$ defined by the term

$$miv_{a_1, \{\overline{a}_1\} \cup \alpha_1}(\cdots (miv_{a_s, \{\overline{a}_s\} \cup \alpha_s}(t) \cdots)),$$

where α_i is the set of vertex labels of the a_i-source of G. It is then routine work to transform a CMS_2 sentence expressing a property of G, to be interpreted in $\lceil G \rceil_C$, into one that expresses the same property of $\overline{G}$, to be interpreted in $\lceil \overline{G} \rceil$. Hence, all constructions done in this section and the associated algorithms extend to (K, Λ)-labeled s-graphs of bounded tree-width. $\square$

We conclude this section by observing that the constructions of automata for checking properties of graphs of bounded tree-width are more complicated than those for graphs of bounded clique-width (developed in Section 6.3.4). For graphs of bounded path-width (and also for larger classes as shown in [Cou10]), the constructions of Section 6.3.4 can be adapted to the model-checking of CMS_2 properties as explained in the introduction of this section.

6.4 Other monadic second-order problems solved with automata

In the previous section, we have constructed automata for solving model-checking problems. We will show how these constructions can be used or adapted for solving related algorithmic problems. Whereas model-checking problems concern properties specified by monadic second-order sentences, the problems considered in this section will be specified by monadic second-order formulas with free variables.

6.4.1 Property-checking problems

Let $\mathcal{R}$ be a relational signature without constant symbols and φ be a formula in $CMS(\mathcal{R}, \overline{X}, \overline{z})$, where $\overline{X} = (X_1, \ldots, X_n)$ and $\overline{z} = (z_1, \ldots, z_m)$. We define new unary relations $P_1, \ldots, P_n$ and $Q_1, \ldots, Q_m$ intended to represent $\gamma(X_1), \ldots, \gamma(X_n)$ and $\{\gamma(z_1)\}, \ldots, \{\gamma(z_m)\}$ respectively for some $(\overline{X} \cup \overline{z})$-assignment γ in an $\mathcal{R}$-structure. We let $\mathcal{P} := \{P_1, \ldots, P_n, Q_1, \ldots, Q_m\}$ and we transform the formula φ into a sentence $\widehat{\varphi} \in CMS(\mathcal{R} \cup \mathcal{P}, \emptyset)$ as follows. Without loss of generality, we assume that $X_1, \ldots, X_n, z_1, \ldots, z_m$ have no bound occurrence in φ.

Step 1, concerning $X_1, \ldots, X_n$. We replace every atomic formula of the form $x \in X_i$ by $P_i(x)$ (where x is some z_j or some bound first-order variable) and every atomic formula $Card_{p,q}(X_i)$ by $\exists X(Card_{p,q}(X) \wedge \forall x(P_i(x) \Leftrightarrow x \in X))$.

Step 2, concerning $z_1, \ldots, z_m$. We replace every atomic formula θ having among its free variables $z_{i_1}, \ldots, z_{i_s}$, listed in such a way that $1 \le i_1 < i_2 < \cdots < i_s \le m$, by the formula $\exists z_{i_1}, \ldots, z_{i_s}(Q_{i_1}(z_{i_1}) \wedge \cdots \wedge Q_{i_s}(z_{i_s}) \wedge \theta)$. The first-order variables not in $\overline{z}$ are not modified by this transformation.

We obtain a sentence φ' (the variables $X_1, \ldots, X_n, z_1, \ldots, z_m$ are no longer free). We then let $\widehat{\varphi}$ be

$$\varphi' \wedge \bigwedge_{1 \le i \le m} \exists x(Q_i(x) \wedge \forall y(Q_i(y) \Rightarrow y = x)).$$

The quantifier-height of $\widehat{\varphi}$ is larger than that of φ by a constant at most $\max\{2, \rho(\mathcal{R})\}$. It has first-order variables that can be eliminated by Lemma 6.17 (Section 6.3.2). With these definitions and notation, we have:

Lemma 6.52 For every concrete $\mathcal{R}$-structure S and every $(\overline{X} \cup \overline{z})$-assignment γ in S we have

$$(S, \gamma) \models \varphi \quad \text{if and only if} \quad S + \gamma \models \widehat{\varphi},$$

where $S + \gamma$ is the $(\mathcal{R} \cup \mathcal{P})$-structure that expands S by $\gamma(X_i)$ as interpretation of P_i and $\{\gamma(z_j)\}$ as interpretation of Q_j for $i = 1, \ldots, n$ and $j = 1, \ldots, m$. $\square$

The validity of this lemma is clear from the construction. We now apply the lemma to terms and graphs.

If $S = \lfloor t \rfloor$ for a term t, then γ specifies nodes and sets of nodes of its syntactic tree and $S + \gamma$ is equivalent to $\lfloor t * \gamma \rfloor$ (this technical point will be discussed in Chapter 7,

Definition 7.20). Sentences over $S + \gamma$ are translatable into equivalent sentences over $\lfloor t * \gamma \rfloor$ (and vice versa).

For the application to graphs, there are two cases. If $S = \lfloor G \rfloor$, then γ specifies vertices and sets of vertices. Then, $S + \gamma = \lfloor G' \rfloor$, where G' is obtained from G by attaching new labels to its vertices. The corresponding transformation of terms in $T(F^{\mathrm{VR}})$ is easy (cf. Proposition 2.100): we need only modify certain constant symbols. If $S = \lceil G \rceil$, then γ specifies vertices, edges and sets thereof. Then, $S + \gamma = \lceil G' \rceil$, where G' is obtained from G by attaching new labels to its vertices and to its edges.[29] The corresponding transformation of terms in $T(F^{\mathrm{HR}})$ is also easy (cf. Proposition 2.47(2)). In both cases, we need not perform another parsing of the given graph G and the corresponding transformations of terms can be done in linear time.

Hence, Theorems 6.3 and 6.4 extend to the corresponding property-checking problems. The sizes of the data include those of the input assignment γ, defined by $\|\gamma\| := m + n + \Sigma_{j \in [n]} |\gamma(X_j)|$, and the verification that γ satisfies a given monadic second-order formula can be done within the same time bounds. The size of γ is at most $m + n \cdot |D_S|$, hence it is at most linear in the number of elements of the considered structure S.

The above proof of the extension of Theorems 6.3 and 6.4 to property-checking is based on Lemma 6.52 and Theorems 6.3 and 6.4. Alternatively, we can use the automata constructed in the previous sections. Let φ be a formula in $\mathrm{CMS}'(\mathcal{R}_F, \mathcal{X}_n)$. Then $(\lfloor t \rfloor, \gamma) \models \varphi$ if and only if $t * \gamma \in L_{\varphi,n}$. Now let φ be a formula in $\mathrm{CMS}'(\mathcal{R}_{s,C}, \mathcal{X}_n)$, let t be a term in $T(F_C^{\mathrm{VR}})$ such that $G \simeq val(t)$ and let w be a witness of that. Then $(\lfloor G \rfloor_C, \gamma) \models \varphi$ if and only if $t * \gamma' \in L_{C,\varphi,n}^{\mathrm{VR}}$, where $\gamma'(X_i) := w(\gamma(X_i))$ for $i \in [n]$. Similarly, if φ is a formula in $\mathrm{CMS}'(\mathcal{R}_m, \mathcal{X}_n)$ and t is a term in $T(F_C'^{\mathrm{HRd}})_{\emptyset}$ such that $G \simeq val(t)$ with simple witness w, then $(\lceil G \rceil, \gamma) \models \varphi$ if and only if $t * \gamma' \in L_{C,\varphi,n}^{\mathrm{HR}}$, with the same definition of γ'.

6.4.2 Listing and selection problems for monadic second-order queries

Definition 6.53 (Listing and selection problems) The *listing problem* consists in computing the set $sat(S, \varphi, (X_1, \ldots, X_n)) \subseteq \mathcal{P}(D_S)^n$ for a given concrete $\mathcal{R}$-structure S and a given formula φ in $\mathrm{CMS}'(\mathcal{R}, \mathcal{X}_n)$. In this context, the formula φ is also called a *query*.

The size of the output $A = sat(S, \varphi, (X_1, \ldots, X_n))$ is defined as

$$\|A\| := \sum_{(U_1, \ldots, U_n) \in A} (n + |U_1| + \cdots + |U_n|),$$

---

[29] There are two small technical problems here, that can easily be solved. Since several labels may be needed for one edge, these must be coded as one label. Also, a label may be on both a vertex and an edge, so these labels must be distinguished from each other (because K and Λ must be disjoint).

and must be distinguished from its cardinality $|A|$. It can be exponential in $|D_S|$. However, we have $\|A\| \leq n \cdot |A| \cdot (1 + |D_S|)$. An algorithm for a listing problem will be said to operate in *linear time* if its computation time is linear in the size of the input plus the size of the output.

The *selection problem* consists in computing one n-tuple belonging to the set $sat(S, \varphi, (X_1, \ldots, X_n))$ for a given concrete $\mathcal{R}$-structure S and a given formula φ in $\mathrm{CMS}'(\mathcal{R}, \mathcal{X}_n)$, or in reporting that this set is empty. The size of the output is at most $n \cdot (1 + |D_S|)$, hence, the evaluation of the time taken by an algorithm for this problem will be done in terms of the size of the input, as usual.

We first present algorithms for these problems in the case of monadic second-order queries on terms. These algorithms operate in linear time. We let F be a finite functional signature, φ be a formula in $\mathrm{CMS}'(\mathcal{R}_F, \mathcal{X}_n)$ and $\mathscr{A}$ be the associated complete and deterministic $F^{(n)}$-automaton $\mathscr{A}_{\varphi,n}$ that recognizes $L_{\varphi,n}$. We let $h : F^{(n)} \to F$ be the arity preserving mapping that deletes the Booleans of the symbols in $F^{(n)}$. Note that $N_{h(s)} = N_s$ for every $s \in T(F^{(n)})$. Let $t \in T(F)$. By the definitions of Section 6.3.3, an n-tuple $(U_1, \ldots, U_n)$ of subsets of N_t belongs to $sat(\lfloor t \rfloor, \varphi, (X_1, \ldots, X_n))$ if and only if there is a term s in $L(\mathscr{A}, q) \subseteq T(F^{(n)})$ such that $h(s) = t$, q is an accepting state and, for each $i \in [n]$, U_i is the set of occurrences in s of a symbol (f, w) such that $w[i] = 1$.

The following definitions generalize this fact. Instead of $F^{(n)}$, we take an arbitrary finite functional signature H and an arity preserving mapping $h : H \to F$. We let $\mathscr{A}$ be a complete and deterministic H-automaton. We let $\overline{J}$ be an n-tuple $(J_1, \ldots, J_n)$ of subsets of H. For every $s \in T(H)$, we let $N_{\overline{J}}(s)$ denote the n-tuple $(U_1, \ldots, U_n)$ of subsets of N_s such that $U_i := \{v \in N_s \mid lab_s(v) \in J_i\}$ (in other words, U_i is the set of occurrences v of symbols in J_i). More generally, if $u \in N_s$, we let $N_{\overline{J}}(s)/u$ denote the n-tuple $(U_1, \ldots, U_n)$ of subsets of N_s/u such that $U_i := \{v \in N_s \mid v \leq_s u \text{ and } lab_s(v) \in J_i\}$ (it is the set of occurrences v of symbols in J_i that are below u).

For $t \in T(F)$, $u \in N_t$ and $q \in Q_{\mathscr{A}}$, we let

$$sat_{\overline{J}}(u, q) := \{N_{\overline{J}}(s)/u \mid s \in h^{-1}(t) \cap L(\mathscr{A}, q)\},$$

and for $P \subseteq Q_{\mathscr{A}}$, we let

$$sat_{\overline{J}}(u, P) := \lambda q \in P \cdot sat_{\overline{J}}(u, q).$$

Finally, we let

$$sat_{\overline{J}}(t) := \bigcup \{sat_{\overline{J}}(root_t, q) \mid q \in Acc_{\mathscr{A}}\},$$

and we have

$$sat_{\overline{J}}(t) = \{N_{\overline{J}}(s) \mid s \in h^{-1}(t) \cap L(\mathscr{A})\}.$$

For $q \neq q'$, the sets $sat_{\overline{J}}(u, q)$ and $sat_{\overline{J}}(u, q')$ are disjoint, because the automaton $\mathscr{A}$ is deterministic, hence the union in the definition of $sat_{\overline{J}}(t)$ is a disjoint union. Since

$N_{h(s)} = N_s$ for every $s \in T(H)$, the set $sat_{\mathcal{J}}(u,q)$ consists of tuples $(U_1, \ldots, U_n)$ with $U_i \subseteq N_t/u$, and so $sat_{\mathcal{J}}(t) \subseteq \mathcal{P}(N_t)^n$.

With these definitions and notations, we have the following proposition from [FluFG].

Proposition 6.54
(1) There is an algorithm that computes, for every term t in $T(F)$, the set $sat_{\mathcal{J}}(t)$ in time $O(|t| + \|sat_{\mathcal{J}}(t)\|)$.
(2) There is an algorithm that computes, for every term t in $T(F)$, some element of $sat_{\mathcal{J}}(t)$, or reports that this set is empty, in time $O(|t|)$.

Proof: For simplicity we assume that F and H only contain symbols of arity 0 and 2. The proofs extend easily to the general case.

(1) We first show that the mappings $sat_{\mathcal{J}}(u, Q_{\mathcal{A}})$ can be computed by bottom-up induction on the node u. To do that, we recall the definition of the operation $\boxtimes$ from Section 5.3.1 (after Proposition 5.37). Let D and D' be disjoint sets. If $\overline{U} = (U_1, \ldots, U_n) \in \mathcal{P}(D)^n$ and $\overline{V} = (V_1, \ldots, V_n) \in \mathcal{P}(D')^n$, then $\overline{U} \boxtimes \overline{V}$ is the n-tuple $(U_1 \cup V_1, \ldots, U_n \cup V_n)$. (In this case, $\boxtimes$ is nothing but componentwise disjoint union.) For $A \subseteq \mathcal{P}(D)^n$ and $B \subseteq \mathcal{P}(D')^n$, we have defined $A \boxtimes B$ as $\{\overline{U} \boxtimes \overline{V} \mid \overline{U} \in A, \overline{V} \in B\}$. Note that $A \boxtimes B = \emptyset$ if and only if $A = \emptyset$ or $B = \emptyset$.

For a node u of t that is an occurrence of $f \in F_2$, with sons u_1 and u_2, we have

$$sat_{\mathcal{J}}(u,q) = \biguplus \{U_{\mathcal{J},f'} \boxtimes sat_{\mathcal{J}}(u_1,q_1) \boxtimes sat_{\mathcal{J}}(u_2,q_2) \mid h(f') = f, f'[q_1,q_2] \to_{\mathcal{A}} q\},$$
(6.3)

where $U_{\mathcal{J},f'}$ is the singleton $\{(U_1, \ldots, U_n)\}$ such that $U_i = \{u\}$ if $f' \in J_i$ and $U_i = \emptyset$ if $f' \notin J_i$.

For a node u that is an occurrence of $f \in F_0$ (hence is a leaf), we have

$$sat_{\mathcal{J}}(u,q) = \biguplus \{U_{\mathcal{J},f'} \mid h(f') = f, f' \to_{\mathcal{A}} q\}.$$
(6.4)

These equalities show that the mappings $sat_{\mathcal{J}}(u, Q_{\mathcal{A}})$ (for all $u \in N_t$) can be computed bottom-up. It remains to prove that the set $sat_{\mathcal{J}}(t)$ can be computed in linear time. Since $sat_{\mathcal{J}}(t)$ is the union of the sets $sat_{\mathcal{J}}(root_t, q)$ for $q \in Acc_{\mathcal{A}}$, it suffices to prove that the mapping $sat_{\mathcal{J}}(root_t, Acc_{\mathcal{A}})$ can be computed in linear time. We will use two ideas:

(a) to compute only those (nonempty) sets $sat_{\mathcal{J}}(u,q)$ that are needed for the computation of some (nonempty) set $sat_{\mathcal{J}}(root_t, p)$ for some $p \in Acc_{\mathcal{A}}$;
(b) to compute $A \boxtimes B$ in an efficient way.

First we discuss (a). For $t \in T(F)$ and $u \in N_t$, we define $call(u) \subseteq Q_{\mathcal{A}}$ by the following top-down induction on the syntactic tree of t:

(i) for the root of t, we define $call(root_t)$ as the set of all accepting states q of $\mathcal{A}$ such that $sat_{\mathcal{J}}(root_t, q) \neq \emptyset$;

(ii) for a node u of t that is an occurrence of some $f \in F_2$, with sons u_1 and u_2, we define $call(u_1)$ as the set of states p of $\mathscr{A}$ for which there exist states q_1 and q_2, a symbol f' in $h^{-1}(f)$ and a state $q \in call(u)$ such that $p = q_1, f'[q_1, q_2] \to_{\mathscr{A}} q$ and $sat_{\mathcal{J}}(u_i, q_i) \neq \emptyset$ for $i = 1$ and $i = 2$; we define similarly $call(u_2)$ with the condition $p = q_2$ instead of $p = q_1$.

Equalities (6.3) and (6.4) form a recursive definition of the sets $sat_{\mathcal{J}}(u, q)$. We can see that $call(u)$ consists of all states q such that the nonempty set $sat_{\mathcal{J}}(u, q)$ is called during the recursive computation of some nonempty set $sat_{\mathcal{J}}(root_t, p)$ with $p \in Acc_{\mathscr{A}}$. Thus, in order to compute the set $sat_{\mathcal{J}}(t)$, it suffices to compute the mappings $sat_{\mathcal{J}}(u, call(u))$ for all nodes u. Let us abbreviate $sat_{\mathcal{J}}(u, call(u))$ into $sat(u)$. Thus,

$$sat(u) = \lambda q \in call(u) \cdot sat_{\mathcal{J}}(u, q).$$

Note that, for an occurrence u of $f \in F_2$ with sons u_1 and u_2, the mapping $sat(u)$ can be computed from the mappings $sat(u_1)$ and $sat(u_2)$ by using Equality (6.3); moreover, each set $sat_{\mathcal{J}}(u_i, q_i)$ with $q_i \in call(u_i)$ is needed in the computation of some set $sat_{\mathcal{J}}(u, q)$ with $q \in call(u)$.

Since $sat_{\mathcal{J}}(u, q) = \emptyset$ if and only if $\sharp_{\mathscr{A}, h^{-1}}(t/u, q) = 0$, we can use Proposition 6.12 to determine, for all u and q, which of the sets $sat_{\mathcal{J}}(u, q)$ are nonempty. We use for that a bottom-up computation[30] on t that takes linear time. From this preliminary computation and the definition of $call(u)$, we can compute all sets $call(u)$ in linear time by a top-down computation on t. Finally, the mappings $sat(u) = sat_{\mathcal{J}}(u, call(u))$ can be computed by a bottom-up computation on t. It remains to prove that this computation can be done in linear time too.

Second we discuss (b).[31] We first note that the union of two disjoint sets can be computed in constant time: just add the sets[32]. Now we consider the computation of $A \boxtimes B$, for nonempty sets $A \subseteq \mathcal{P}(D)^n$ and $B \subseteq \mathcal{P}(D')^n$ such that $D \cap D' = \emptyset$. When computing $\overline{U} \boxtimes \overline{V}$, for $\overline{U} \in A$ and $\overline{V} \in B$, we make a copy of both $\overline{U}$ and $\overline{V}$ (in time $\|\overline{U}\| + \|\overline{V}\| = \|\overline{U} \boxtimes \overline{V}\| + n$) and compute their componentwise union (in constant time). However, when $\overline{U}$ is used for the last time we do not copy $\overline{U}$ but we use $\overline{U}$ itself, and similarly for $\overline{V}$. In this way $A \boxtimes B$ is computed in time

$$O(1 + 2n \cdot |A \boxtimes B| + \|A \boxtimes B\|_0 - \|A\|_0 - \|B\|_0),$$

where $\|A\|_0 := \Sigma_{(U_1, \dots, U_n) \in A}(|U_1| + \dots + |U_n|)$, i.e., $\|A\| = \|A\|_0 + n \cdot |A|$. It is straightforward to show that this is $O(1 + \|A \boxtimes B\|_0 - \|A\|_0 - \|B\|_0)$, provided $\|A\|_0 \neq 0$ and $\|B\|_0 \neq 0$.[33] If $\|A\|_0 = 0$, i.e., A is the singleton $\{(\emptyset, \dots, \emptyset)\}$, then $A \boxtimes B = B$ (and

---

[30] Note that in this computation we can actually work with Booleans and replace multiplication and addition by $\vee$ and $\wedge$ respectively (because $x \cdot y = 0$ if and only if $x = 0$ or $y = 0$, and $x + y = 0$ if and only if $x = 0$ and $y = 0$).

[31] This presentation corrects some imprecisions of [FluFG].

[32] A set can be represented by a linked list with an additional pointer to the end of the list. To compute the union of two disjoint sets, the end of the first set is linked to the beginning of the second set. Note that this destroys the given sets.

[33] To be precise, $|A \boxtimes B| = |A| \cdot |B| \leq 2 + \|A \boxtimes B\|_0 - \|A\|_0 - \|B\|_0$.

similarly, if $\|B\|_0 = 0$ then $A \boxtimes B = A$). Hence, $A \boxtimes B$ can be computed in time $O(1 + \|A \boxtimes B\|_0 - \|A\|_0 - \|B\|_0)$, for arbitrary nonempty sets A and B. Note that A and B are no longer available after the computation of $A \boxtimes B$.

A similar argument holds for arbitrary expressions that are written with the operations $\uplus$ and $\boxtimes$. As an example, we can compute $E = (A \boxtimes B) \uplus (A \boxtimes C)$ in time $O(1 + \|E\|_0 - \|A\|_0 - \|B\|_0 - \|C\|_0)$: first we make a copy A' of A (in time $\|A\|$), and then we compute $A' \boxtimes B$ and $A \boxtimes C$ as indicated above. From this fact it is clear that, for every node u with sons u_1 and u_2, we can compute $sat(u)$ from $sat(u_1)$ and $sat(u_2)$ in time $O(1 + \|sat(u)\|_0 - \|sat(u_1)\|_0 - \|sat(u_2)\|_0)$, where $\|sat(u)\|_0$ is defined as $\sum_{q \in call(u)} \|sat_{\bar{J}}(u,q)\|_0$. Since $sat(u)$ can be computed in constant time for a leaf u, this shows that the set $sat(root_t)$ can be computed in time $O(|t| + \|sat(root_t)\|_0)$, and hence that the set $sat_{\bar{J}}(t)$ can be computed in time $O(|t| + \|sat_{\bar{J}}(t)\|_0)$, which is $O(|t| + \|sat_{\bar{J}}(t)\|)$.

(2) We use an easy variant of the proof of (1). Instead of computing all tuples of $sat_{\bar{J}}(u,q)$, $q \in call(u)$, we compute only one of them. For each u this computation takes constant time, because for tuples $\overline{U}$ and $\overline{V}$ the tuple $\overline{U} \boxtimes \overline{V}$ can be computed in constant time. ∎

The method of [FluFG] used in the above proof is based on the direct translation of CMS formulas into automata described in Sections 6.3.3–6.3.5. An important ingredient is the efficient use of the operation $\boxtimes$. This operation arises in the statement of the Splitting Theorem and also when, as above, we compute the sets $sat(\lfloor t \rfloor, \varphi, (X_1, \ldots, X_n))$ with automata.

In Definition 6.1, fixed-parameter tractable algorithms are for decision problems, hence, they output only *True* or *False*. For the next theorem, we extend the definition by requiring that the computation time is bounded, for every input d with corresponding output e, by an expression of the form $f(p(d)) \cdot (|d|^c + |e|)$, where f is a fixed computable function, c is a fixed positive integer and $|e|$ is an appropriate notion of size for e. Note that this size may be exponentially larger than $|d|$ for the listing problem. This is not the case for the selection problem, and the computation time can be bounded by $f(p(d)) \cdot |d|^c$ as in Definition 6.1. If c is 1 (or 3), we say that the algorithm is fixed-parameter linear (or cubic respectively).

Theorem 6.55

(1) There exist fixed-parameter linear algorithms for solving the monadic second-order listing and selection problems on terms. The parameter is (F, φ), where the given term is in $T(F)$ and the given formula is φ.

(2) There exist fixed-parameter cubic algorithms for solving the listing and selection problems for simple (K, Λ)-labeled graphs and CMS' queries. The parameter is $(cwd(G), \varphi)$, where the given (K, Λ)-labeled graph is G and the given formula is φ. The size of the input graph is its number of vertices.

(3) There exist fixed-parameter linear algorithms for solving the listing and selection problems for (K, Λ)-labeled graphs and CMS'_2 queries. The parameter is

$(twd(G), \varphi)$ where the given (K, Λ)-labeled graph is G and the given formula is φ. The size of the input graph G is $\|G\|$.

Proof: (1) Let φ be a formula in $\text{CMS}'(\mathcal{R}_F, \mathcal{X}_n)$. We apply Proposition 6.54 by letting $H := F^{(n)}$, h be the arity preserving mapping $: F^{(n)} \to F$ that deletes the Booleans, $\overline{J} := (J_1, \ldots, J_n)$ with $J_i := \{(f, (w_1, \ldots, w_n)) \in H \mid w_i = 1\}$, and $\mathscr{A}$ be the complete and deterministic $F^{(n)}$-automaton $\mathscr{A}_{\varphi,n}$ that recognizes $L_{\varphi,n}$ (as constructed in Section 6.3.3). Then $sat(\lfloor t \rfloor, \varphi, (X_1, \ldots, X_n)) = \{N_{\overline{J}}(s) \mid s \in h^{-1}(t) \cap L(\mathscr{A})\} = sat_{\overline{J}}(t)$.

(2) and (3): the proofs are similar, with the constructions in Sections 6.3.4 and 6.3.5 of automata recognizing $L_{C,\varphi,n}^{\text{VR}}$ and $L_{C,\varphi,n}^{\text{HR}}$ (cf. the last paragraph of Section 6.4.1). As shown in the proof of Theorem 6.4, $C = [h_\Lambda(G)]$ and $C = [twd(G) + 1]$ respectively.   ∎

6.4.3 Monadic second-order counting and optimizing functions

Monadic second-order counting and optimizing functions have been defined in Definition 6.5. By using the counting and optimizing automata defined in Section 6.3.1, we can prove the following theorem. Detailed formulations are as for Theorem 6.55.

Theorem 6.56 Monadic second-order counting and optimization functions defined by CMS' formulas can be computed by fixed-parameter linear algorithms on terms and by fixed-parameter cubic algorithms on graphs for clique-width as the parameter. Such functions defined by CMS'_2 formulas can be computed by fixed-parameter linear algorithms on graphs for tree-width as the parameter.

Proof: We first consider the evaluation of counting functions on syntactic trees of terms in $T(F)$. Let φ be a formula in $\text{CMS}'(\mathcal{R}_F, \mathcal{X}_n \cup \{Y_1, \ldots, Y_p\})$. We first assume that $p = 0$. We let H, h, $\overline{J}$ and $\mathscr{A}$ be as in the proof of Theorem 6.55(1). Then the integer $\sharp sat(\lfloor t \rfloor, \varphi, (X_1, \ldots, X_n))$ to be computed is the cardinality of the set $\{N_{\overline{J}}(s) \mid s \in h^{-1}(t) \cap L(\mathscr{A})\}$. Since $N_{\overline{J}}(s) \neq N_{\overline{J}}(s')$ for distinct elements s and s' of $h^{-1}(t)$, we obtain that:

$$\sharp sat(\lfloor t \rfloor, \varphi, (X_1, \ldots, X_n)) = |h^{-1}(t) \cap L(\mathscr{A})| = \sum_{q \in Acc_{\mathscr{A}}} \sharp_{\mathscr{A}, h^{-1}}(t, q).$$

We can then apply Proposition 6.12. For the case of $\sharp sat(\lfloor t \rfloor, \varphi, (X_1, \ldots, X_n), \gamma)$, where $p > 0$ and γ is an $\{Y_1, \ldots, Y_p\}$-assignment in $S := \lfloor t \rfloor$, we apply this algorithm to $S + \gamma$, as we have done in Section 6.4.1.

For computing optimizing functions for terms, we also use Proposition 6.12. We use the same notation as above, with $n = 1$, and we write X for X_1 and J for $J_1 = $

$\{(f,1) \mid f \in F\}$. For $p = 0$ we obtain that

$$
\begin{aligned}
Maxsat(\lfloor t \rfloor, \varphi, X) &= \max\{|s|_J \mid s \in h^{-1}(t) \cap L(\mathscr{A})\} \\
&= \max\{Max_{\mathscr{A}, h^{-1}, J}(t, q) \mid q \in Acc_{\mathscr{A}}\}.
\end{aligned}
$$

These proofs extend to graphs, as in the proof of Theorem 6.55 (Assertions (2) and (3)).   ∎

The results of Theorems 6.55 and 6.56 for terms and graphs of bounded tree-width are proved in [CouMos] as consequences of the Splitting Theorem for n-ary QDP-QF operations (Theorem 5.57). This version of the theorem is needed for the graphs defined by the operations of the HR algebra because the parallel-composition fuses constants of its argument structures. Since the corresponding splitting is expressed in terms of disjoint unions, the cardinalities of the considered sets can be obtained by additions. More general evaluations are described in [CouMos]. They can also be implemented on top of deterministic automata, as in the proofs of Theorems 6.55 and 6.56.

6.4.4 Other algorithmic applications

(1) Enumeration problems

An *enumeration algorithm* of a set A of the form $sat(S, \varphi, (X_1, \ldots, X_n))$ outputs one by one and without repetition the elements of A. The listing algorithms of Theorem 6.55 are actually enumeration algorithms, but here we are interested in the *delay*, that is, in the computation time between two outputs. An enumeration algorithm has *linear delay* if this time is linear in the size of the next output. Linear delay enumeration algorithms of sets $sat(\lfloor G \rfloor, \varphi, (X_1, \ldots, X_n))$ for CMS formulas φ and graphs G of bounded clique-width are constructed in [Bag] and [Cou09]. Both algorithms are based on the automata constructed in Sections 6.3.3–6.3.5, but they need an additional step: the terms t that define the considered graphs G must be *a-balanced*, i.e., have height at most $a \cdot \log(|V_G|)$ for fixed a. We refer to [CouVan], [CouTwi] and [CouKan] for the transformations of terms in $T(F_C^{HR})$ or in $T(F_C^{VR})$ into a-balanced equivalent ones, for fixed values of a, but at the cost of using more source or port labels. These results give an alternative proof of Theorem 6.55.

(2) Reduction

Every CMS-definable set of graphs of tree-width at most k (where k is known) can be *recognized by reduction*. This means that there exists a finite set of Noetherian graph rewriting rules and a finite set of *accepting* irreducible graphs. The input graph G is rewritten until an irreducible graph H is obtained. The graph G is recognized if and only if H is accepting. The graph G need not be of tree-width at most k, and need not be parsed. If its tree-width is more than k, some irreducible graph that is not accepting will be reached. A positive answer does not yield a tree-decomposition of the input graph.

This result gives different fixed-parameter linear graph algorithms for the problem considered in Theorem 6.4(1). The general theorem is proved in the article [ArnCPS] and the algorithm is improved in [BodvAnt].

(3) Labeling schemes

A *labeling scheme* for checking a property $P(X_1,\ldots,X_n)$ of sets of vertices $X_1,\ldots,X_n$ of the graphs of a fixed class consists of two algorithms $\mathcal{A}$ and $\mathcal{B}$. Algorithm $\mathcal{A}$ attaches labels to the vertices of a given graph G of the class in such a way that the validity of $P(X_1,\ldots,X_n)$ can be checked by algorithm $\mathcal{B}$ that takes as input the labels of the vertices of the elements of $X_1,\ldots,X_n$ (and nothing else from G). The objective is to use labels of size $O(\log(|V_G|))$. The article [CouVan] gives a meta-theorem for constructing labeling schemes for checking monadic second-order properties $P(X_1,\ldots,X_n)$ for graphs of bounded clique-width. It is based on the translation of monadic second-order formulas into automata described in Sections 6.3.3 and 6.3.4. The general construction involves huge constants. A direct and more efficient construction for a particular problem is done in [CouTwi]. The result of [CouVan] is also useful as an intermediate step of the construction of labeling schemes for graphs of unbounded clique-width: in particular, for the MS property $Sep(u,v,X)$ that two vertices u and v of a planar graph are separated by a set of vertices X [CouGKT] and for first-order properties of certain graphs, that are locally of bounded clique-width [CouGK].

6.4.5 Optimality results

Let us look again at Theorem 6.4, which we restate in a simpler form as follows:[34]

Result 1: One can check in time $f(\varphi,k) \cdot n$ the validity of a CMS_2 sentence φ on a simple graph G with n vertices that belongs to any class $\mathcal{C}$ of tree-width at most k (for some fixed function f).

Result 2: One can check in time $g(\varphi,k) \cdot n^3$ the validity of a CMS sentence φ on a simple graph G with n vertices that belongs to any class $\mathcal{C}$ of clique-width at most k (for some fixed function g).

A natural question is whether and how one can improve these results. We will review three types of improvement.

(1) Alternative algorithms

A significant improvement of Result 2 could come from the discovery of a better parsing algorithm for graphs of bounded clique-width (or rank-width), see Section 6.2.3.

---

[34] For Result 1, recall from the proof of Theorem 4.51 that every simple graph of bounded tree-width is p-sparse for some p, i.e., has at most $p \cdot |V_G|$ edges.

For fixed φ and k, Result 1 can be achieved in logarithmic space instead of linear time ([ElbJT]; the proof does not extend to Result 2).

(2) More model-checking problems?

That is, can one treat more problems, by writing them in more powerful logical languages? The answer is yes: Results 1 and 2 can be extended to sentences of the form $\exists \overline{X}(\alpha \wedge \varphi)$, where φ is a monadic second-order formula and α is a Boolean combination of arithmetical conditions like $|Y| < a \cdot |Z| + b \cdot |Z'|$, where a and b are real numbers and Y, Z and Z' are variables from $\overline{X}$. This extension follows from Theorem 6.55 (or rather from a modification of its proof) and has already been considered in [ArnLS], [CouMos] and [ElbJT] for graphs of bounded tree-width. The methods of [CouMos] and [ElbJT] work for graphs of bounded clique-width (see [Rao]).

Some MS_2 problems that are not CMS-expressible can nevertheless be solved in time $O(n^{f(k)})$ for graphs of clique-width at most k ([Wan]); however, there is little hope of solving all MS_2 problems in polynomial time for graphs of clique-width 2: if $EXPTIME \neq NEXPTIME$, then there exists an MS_2 problem that is not solvable in polynomial time on cliques ([CouMakR]).

(3) Covering larger graph classes

Several results limit the possibilities of expanding the classes covered by Result 1. The following is proved in [MakMar], under the assumption that $P \neq NP$:

> if every problem expressible by an existential MS sentence (i.e., a sentence written as a sequence of existential quantifications followed by a first-order formula) is solvable in polynomial time on a class of graphs $\mathcal{C}$ closed under taking minors, or even, under taking topological minors, then $\mathcal{C}$ has bounded tree-width.

For each positive number a, we denote by $\mathcal{TW}_a$ the class of graphs G of tree-width at most $\log^a(|V_G|)$. Then, the following is proved in [KreTaz] under the Exponential Time Hypothesis:

> if $a \geq 29$, then some MS_2 problems have no polynomial time algorithm on $\mathcal{TW}_a$.

The Exponential Time Hypothesis states that problem SAT cannot be solved in subexponential time, i.e. in time $2^{o(n)}$.

Since there are also (unnatural) classes of graphs of unbounded clique-width (hence also of unbounded tree-width, that are even closed under taking subgraphs) for which every MS problem is solvable in polynomial time ([MakMar]), results like these two ones cannot be proved for arbitrary graph classes. This fact can be contrasted with the result of [See91], which says that every class of graphs having a decidable MS_2 theory has bounded tree-width (see Theorem 7.55 for its proof).

6.4.6 Comparing some proofs of the Recognizability Theorem

Motivated by the practical construction of automata, we have given in Theorem 6.35 an alternative proof of the Weak Recognizability Theorem (Corollary 5.69) for the VR algebra (and similarly for the HR algebra in Section 6.3.5). Let us raise a natural question:

Does Theorem 6.35 give another proof of the Recognizability Theorem for the VR algebra (Theorem 5.68)?

The answer is no. Theorem 6.35 establishes that, for every CMS sentence φ, for each k the set of graphs $\{G \mid cwd(G) \leq k$ and $\lfloor G \rfloor \models \varphi\}$ is recognizable in the subalgebra of $\mathbb{GP}$ consisting of the p-graphs that are the values of terms over $F_{[k]}^{VR}$. This does not imply that $L_\varphi := \{G \mid \lfloor G \rfloor \models \varphi\}$ is VR-recognizable, as observed in Proposition 4.40 for the HR case (of which the proof is also valid for the VR case).

However, in [Eng91] a description is given of MS- and MS_2-definable sets of labeled graphs (of unbounded clique-width and tree-width) by certain *regular expressions*. These expressions use basic sets of graphs (corresponding to the atomic formulas of the considered sentences) that somehow generalize the *local regular languages*. They also use the Boolean operations on sets of graphs, corresponding to the Boolean connectives in formulas, relabelings (corresponding to existential quantifications, cf. Construction 6.29) and inverse relabelings (for handling substitutions of variables, cf. Lemma 6.25). The basic sets of graphs are recognizable (appropriate congruences are easy to define for proving that) and the operations on sets used in these regular expressions preserve recognizability. The case of relabelings needs some more work but can be proved with the technique used in [BluCou06] (Proposition 58). One obtains in this way an alternative proof of the Recognizability Theorem. After working out the details, the reader can decide if it is simpler or not than the one presented in Chapter 5, based on the Splitting Theorem and on several technical lemmas. Furthermore, it is not clear how to obtain from this proof the extensions to the counting and optimization functions considered in Section 6.4.3. It seems also to require more work for handling the QDP-QF operations used in the algebra $\mathbb{STR}$.

6.5 References

There are numerous articles that give polynomial-time recognition algorithms for classes of structured graphs, that are, in most cases, classes of bounded tree-width or clique-width. Hedetniemi lists 238 titles in his bibliography [*Hed] published in 1994.

The theory of fixed-parameter tractability is exposed in the books by Downey and Fellows [*DowFel] and by Flum and Grohe [*FluGro], but these books do not present algorithms for classes of graphs of bounded clique-width. Fixed-parameter tractable algorithms for MS_1 properties and clique-width as parameter are studied in [CouMakR] and [*CouMR]. The surveys by Grohe [*Gro] and Kreutzer [*Kre] give algorithmic meta-theorems for first-order logic on graph classes of unbounded tree-width and clique-width that satisfy some structural constraints (relative to bounded tree-width).

Makowsky surveys in [*Mak] the meta-algorithmic uses of the Splitting Theorem for graph classes of bounded tree-width or clique-width, with applications to the computation of graph polynomials.

Bodlaender has published several articles about *implemented and tested* algorithms for computing tree-width: [BacBod06], [BacBod07], [*Bod06], [BodKos10], [BodKos11] and [VEBK].

For computing or approximating rank-width the known algorithms are given in [OumSey], [Oum08] and [HliOum] (an exact exponential algorithm is also given in [Oum09]). No implementations have been reported. There exists an algorithm that is fixed-parameter linear with respect to tree-width for computing the clique-width of a graph. However, it is not useful for model-checking, because if we have a tree-decomposition of width at most k, we can use it directly, without going through a term over F^{VR} to define the given graph.

Since we have presented rank-width in this chapter, we mention here that, for each k, the graphs of rank-width at most k are characterized by finitely many excluded *vertex-minors*, hence by a C_2MS sentence ([CouOum] and [Oum05]). These results are extended to directed graphs in [KanRao].

The recognition problems for HR and VR equation systems are discussed respectively in the book chapters [*DreKH] and [*EngRoz] where numerous references can be found.

Constructions of linear-time algorithms from logical descriptions and for graphs of bounded tree-width can be found in [Bor] and [BorPT]. Monadic second-order logic on finite and infinite words and terms is implemented in the software MONA, developed by Klarlund and others and described in many articles, in particular [BasKla], [Hen+] and [Kla]. Using it for checking MS graph properties is not successful because the automata of Section 6.3.4 are too large to be computable in practice as soon as the parameter k that bounds clique-width is more than 3 (see [Sog]). Other practically oriented approaches aiming at handling large fragments of monadic second-order logic are proposed in [Cou10], [CouDur10] and [CouDur11] and, with a different approach using games, in [KneLan]. Another research direction consists in developing algorithms for particular monadic second-order problems: see, e.g., [BuiTV], [GotPWa], [GotPWb] and [GanHli].

The ideas and tools of this section seem to be useful in applied computer science. They are quoted in works on computational linguistics [Kep], database querying [BeeEKM], logistics [MooSpie], compilation [Thop], telecommunications [McDRee], quantum computing [VNMDB], and perhaps yet in other fields.

7

Monadic second-order transductions

Monadic second-order transductions are transformations of relational structures specified by monadic second-order formulas. They can be used to represent transformations of graphs and related combinatorial structures via appropriate representations of these objects by relational structures, as shown in the examples discussed in Section 1.7.1.

These transductions are important for several reasons. First, because they are useful tools for constructing monadic second-order formulas with the help of the Backwards Translation Theorem (Theorem 7.10). Second, because by means of the Equationality Theorems (Theorems 7.36 and 7.51) they yield logical characterizations of the HR- and VR-equational sets of graphs that are independent of the signatures F^{HR} and F^{VR}. From these characterizations, we get short proofs that certain sets of graphs have bounded, or unbounded, tree-width or clique-width.

They also play the role of transducers in formal language theory: the image of a VR-equational set of graphs under a monadic second-order transduction is VR-equational, and a similar result holds for HR-equational sets and monadic second-order transductions that transform incidence graphs. Finally, the decidability of the monadic second-order satisfiability problem for a set of structures $\mathcal{C}$ implies the decidability of the same problem for its image $\tau(\mathcal{C})$ under a monadic second-order transduction τ. Hence, monadic second-order transductions make it possible to relate decidability and undecidability results concerning monadic second-order satisfiability problems for graphs and relational structures.

Section 7.1 presents the definitions and the fundamental properties. Section 7.2 is devoted to the Equationality Theorem for the VR algebra, one of the main results of this book. Section 7.3 presents the monadic second-order transductions of graphs represented by their incidence graphs. Section 7.4 establishes the Equationality Theorem for the HR algebra. Sections 7.5 and 7.6 present some applications, respectively to the decidability of monadic second-order satisfiability problems and to the logical characterization of recognizability.

In this chapter (and the next) all functional signatures are assumed to be one-sorted.

7.1 Definitions and basic properties

7.1.1 Transductions of relational structures

Definition 7.1 (Transductions) Let $\mathcal{R}$ and $\mathcal{R}'$ be relational signatures. A *concrete transduction* τ of type $\mathcal{R} \to \mathcal{R}'$ is a binary relation on concrete structures. More precisely, it is a set of pairs (S, S'), where S and S' are concrete $\mathcal{R}$- and $\mathcal{R}'$-structures respectively, hence it is a subset of $STR^c(\mathcal{R}) \times STR^c(\mathcal{R}')$. The associated *transduction* of abstract structures is the set of pairs $([S]_{iso}, [S']_{iso})$ for (S, S') in τ, hence a subset of $STR(\mathcal{R}) \times STR(\mathcal{R}')$, also denoted by τ. If we wish to be precise, we will denote it by $[\tau]_{iso}$. We say that two concrete transductions τ and τ' are *isomorphic* if $[\tau]_{iso} = [\tau']_{iso}$. In proofs, transductions will be defined between concrete structures, but the equality of two transductions and their properties will be understood in terms of the associated relations on abstract structures, except otherwise specified.

The *image* of a (concrete) $\mathcal{R}$-structure S under τ is the set of (concrete) $\mathcal{R}'$-structures $\tau(S) := \{T \mid (S, T) \in \tau\}$. In this way, τ can also be viewed as a mapping : $STR^c(\mathcal{R}) \to \mathcal{P}(STR^c(\mathcal{R}'))$ or : $STR(\mathcal{R}) \to \mathcal{P}(STR(\mathcal{R}'))$. The image $\tau(\mathcal{C})$ of a set $\mathcal{C}$ of (concrete) $\mathcal{R}$-structures is the union of the sets $\tau(S)$ for S in $\mathcal{C}$. The *domain* $Dom(\tau)$ of a (concrete) transduction τ is the set of (concrete) $\mathcal{R}$-structures S such that $\tau(S)$ is not empty.[1]

If τ and τ' are transductions (or concrete transductions) of respective types $\mathcal{R} \to \mathcal{R}'$ and $\mathcal{R}' \to \mathcal{R}''$, then their *composition* is the following transduction (or concrete transduction):

$$\tau \cdot \tau' := \{(S, T) \mid (S, U) \in \tau \text{ and } (U, T) \in \tau' \text{ for some } U\}.$$

It is of type $\mathcal{R} \to \mathcal{R}''$. Note that $(\tau \cdot \tau')(S) = \tau'(\tau(S))$ for every $\mathcal{R}$-structure S. The *inverse* of τ is the transduction (or the concrete transduction) of type $\mathcal{R}' \to \mathcal{R}$ defined by

$$\tau^{-1} := \{(T, S) \mid (S, T) \in \tau\}.$$

We will view a concrete transduction τ of type $\mathcal{R} \to \mathcal{R}'$ that is functional also as a partial function : $STR^c(\mathcal{R}) \to STR^c(\mathcal{R}')$, and similarly for transductions. For this reason, if τ and τ' are functional (and only in this case), we will sometimes denote by $\tau' \circ \tau$ the composition $\tau \cdot \tau'$. Note that the transduction associated with a concrete transduction that is not functional may be functional (see Example 7.3(1) below). If a transduction (or concrete transduction) τ is functional, we will often write $\tau(S) = T$ instead of $\tau(S) = \{T\}$.

For describing a concrete transduction τ of type $\mathcal{R} \to \mathcal{R}'$, we will frequently declare its type as $STR^c(\mathcal{R}) \to STR^c(\mathcal{R}')$, or even as $\mathcal{C} \to \mathcal{C}'$ in the case where $\tau \subseteq \mathcal{C} \times \mathcal{C}' \subseteq STR^c(\mathcal{R}) \times STR^c(\mathcal{R}')$. A similar notation will be used for transductions.

---

[1] The domain of τ should not be confused with the domain of a particular structure S in the domain of τ.

Definition 7.2 (Monadic second-order definition schemes and transductions)
Let $\mathcal{R}$ and $\mathcal{R}'$ be two relational signatures such that $\mathcal{R}'_0 = \emptyset$. (This restriction will be lifted later.) Let also $\mathcal{W}$ be a finite set of set variables called *parameters*. A *definition scheme of type* $\mathcal{R} \to \mathcal{R}'$ with set of parameters $\mathcal{W}$ is a tuple of formulas of the form $\mathcal{D} = \langle \chi, (\delta_i)_{i \in [k]}, (\theta_w)_{w \in \mathcal{R}' \circledast [k]} \rangle$ for some $k > 0$ such that:

> $\chi \in \mathrm{CMS}(\mathcal{R}, \mathcal{W})$ – this formula is called the *precondition*;
>
> $\delta_i \in \mathrm{CMS}(\mathcal{R}, \mathcal{W} \cup \{x\})$ for each $i \in [k]$ – these formulas are called the *domain formulas*;
>
> $\theta_w \in \mathrm{CMS}(\mathcal{R}, \mathcal{W} \cup \{x_1, \dots, x_{\rho(R)}\})$ for each $w \in \mathcal{R}' \circledast [k]$,
>> where $\mathcal{R}' \circledast [k]$ is defined as $\{(R, \mathbf{i}) \mid R \in \mathcal{R}', \mathbf{i} \in [k]^{\rho(R)}\}$ – these formulas are called the *relation formulas*.[2]

We write these formulas $\chi_\mathcal{D}, \delta_{i\mathcal{D}}, \theta_{w\mathcal{D}}$ if we need to specify $\mathcal{D}$. The *quantifier-height* $qh(\mathcal{D})$ of $\mathcal{D}$ is defined as the maximal quantifier-height of its formulas.

We now define the transduction of type $\mathcal{R} \to \mathcal{R}'$ associated with $\mathcal{D}$. Let S be a concrete $\mathcal{R}$-structure and let γ be a $\mathcal{W}$-assignment in S. We say that $\mathcal{D}$ defines the concrete $\mathcal{R}'$-structure T from (S, γ) if:

(i) $(S, \gamma) \models \chi$;

(ii) $D_T := \{(a, i) \in D_S \times [k] \mid (S, \gamma) \models \delta_i(a)\}$;

(iii) for each $n > 0$ and $R \in \mathcal{R}'_n$:
$$R_T := \{((a_1, i_1), \dots, (a_n, i_n)) \in D_T^n \mid (S, \gamma) \models \theta_{R, i_1, \dots, i_n}(a_1, \dots, a_n)\}.$$

By $(S, \gamma) \models \theta_w(a_1, \dots, a_n)$, we mean $(S, \gamma') \models \theta_w$, where γ' is the assignment extending γ such that $\gamma(x_i) = a_i$ for each $i = 1, \dots, n$ (and similarly for $(S, \gamma) \models \delta_i(a)$). The structure T is uniquely determined by $\mathcal{D}, S$ and γ whenever it is defined, i.e., whenever $(S, \gamma) \models \chi$. Therefore, we can use a functional notation and we will denote T by $\widehat{\mathcal{D}}(S, \gamma)$. The associated relation between concrete structures, defined by

$$\widehat{\mathcal{D}} := \{(S, \widehat{\mathcal{D}}(S, \gamma)) \mid \gamma \text{ is a } \mathcal{W}\text{-assignment in } S \text{ such that } (S, \gamma) \models \chi\},$$

is the *concrete transduction defined by* $\mathcal{D}$. It is called a *concrete monadic second-order transduction*. Thus $\widehat{\mathcal{D}}(S)$ denotes $\{\widehat{\mathcal{D}}(S, \gamma) \mid (S, \gamma) \models \chi \text{ for some } \gamma\}$. A *monadic second-order transduction* is the transduction $[\widehat{\mathcal{D}}]_{iso}$ of abstract structures associated with a concrete monadic second-order transduction $\widehat{\mathcal{D}}$. It will also be denoted by $\widehat{\mathcal{D}}$. Note that if τ is a concrete transduction, then we can say that it is a monadic second-order transduction (as opposed to being a concrete one), meaning that the associated transduction $[\tau]_{iso}$ of abstract structures is a monadic second-order transduction. This is in accordance with the fact that we also use τ to denote $[\tau]_{iso}$.

If $\tau \subseteq STR(\mathcal{R}) \times STR(\mathcal{R}')$ and $\mathcal{C} \subseteq Dom(\tau)$, we will say that τ is a *monadic second-order transduction on* $\mathcal{C}$ if there exists a monadic second-order transduction τ' of type

---

[2] We will usually write $w = (R, i_1, \dots, i_{\rho(R)})$ for $w = (R, (i_1, \dots, i_{\rho(R)}))$ and $\theta_{R, i_1, \dots, i_{\rho(R)}}$ for $\theta_{(R, i_1, \dots, i_{\rho(R)})}$.

$\mathcal{R} \to \mathcal{R}'$ such that $\tau'(S) = \tau(S)$ for every S in $\mathcal{C}$. Note that we do not require that the domain of τ' is $\mathcal{C}$. By Theorem 7.16 established below, we can require this if $\mathcal{C}$ is CMS-definable.

We recall that QF, FO, MS, CMS, $C_r MS^h$, MS_2, etc., refer to fragments, extensions or variants of first-order and (counting) monadic second-order logic. If $\mathcal{L}$ is one of them, we call $\mathcal{D}$ an *$\mathcal{L}$-definition scheme* if all its formulas are in the corresponding fragment, and then we call $\widehat{\mathcal{D}}$ an *$\mathcal{L}$-transduction*.[3] The term "monadic second-order transduction" does not distinguish MS, CMS, MS_2, etc. An FO- or a QF-definition scheme may use set variables as parameters: a parameter Y can be used in the atomic formulas $x \in Y$ and $Card_{p,q}(Y)$, but no set variable is quantified in an FO- or a QF-definition scheme.

If S is $\varnothing_\mathcal{R}$, the empty structure in $STR^c(\mathcal{R})$, then the only possible assignment is $\gamma_\emptyset$ such that $\gamma_\emptyset(Y) = \emptyset$ for every $Y \in \mathcal{W}$, and, either $\widehat{\mathcal{D}}(\varnothing_\mathcal{R}, \gamma_\emptyset)$ is undefined or it is $\varnothing_{\mathcal{R}'}$.

By Lemma 5.4, if $S, S' \in STR^c(\mathcal{R})$ and h is an isomorphism : $S \to S'$, then the structures in $\widehat{\mathcal{D}}(S)$ are isomorphic to those in $\widehat{\mathcal{D}}(S')$. More precisely, if γ is a $\mathcal{W}$-assignment in S and γ' is defined by $\gamma'(Y) = h(\gamma(Y))$ for every $Y \in \mathcal{W}$, then $\widehat{\mathcal{D}}(S', \gamma')$ is defined if and only if $\widehat{\mathcal{D}}(S, \gamma)$ is defined, and these structures are isomorphic by the mapping h' such that $h'((a,i)) = (h(a),i)$ for every (a,i) in the domain of $\widehat{\mathcal{D}}(S, \gamma)$. It follows that $\widehat{\mathcal{D}}([S]_{iso})$ is $\{[T]_{iso} \mid T \in \widehat{\mathcal{D}}(S')\}$, where S' is any concrete structure isomorphic to S. (Definition 7.1 defines $\widehat{\mathcal{D}}([S]_{iso})$ as $\{[T]_{iso} \mid T \in \widehat{\mathcal{D}}(S')$ for some $S' \in [S]_{iso}\}$.)

If a definition scheme $\mathcal{D}$ has parameters (i.e., $\mathcal{W} \neq \emptyset$), but if for every S in $STR^c(\mathcal{R})$, any two structures in $\widehat{\mathcal{D}}(S)$ are isomorphic, then we say that $\mathcal{D}$ and $\widehat{\mathcal{D}}$ are *parameter-invariant*. In this case $\widehat{\mathcal{D}}$ defines a partial function on abstract structures. Parameter-invariance is undecidable: it can be guaranteed by a particular construction of $\mathcal{D}$ but cannot be checked on arbitrary definition schemes.

If $\mathcal{W} = \emptyset$, we say that $\mathcal{D}$ and $\widehat{\mathcal{D}}$ are *parameterless*. Parameterless transductions are functional, already on concrete structures.

We will refer to the integer k by saying that $\mathcal{D}$ and $\widehat{\mathcal{D}}$ are *k-copying*. If $k = 1$, we say that $\mathcal{D}$ and $\widehat{\mathcal{D}}$ are *noncopying*. A noncopying definition scheme has the simpler form $\langle \chi, \delta, (\theta_R)_{R \in \mathcal{R}'} \rangle$, where we let δ denote δ_1 and θ_R denote $\theta_{R,1,...,1}$. The domain of $T := \widehat{\mathcal{D}}(S, \gamma)$ is defined as a subset of D_S and not of $D_S \times [1]$. If $\mathcal{D}$ is noncopying and the domain formula δ is equivalent to *True*, then we say that $\mathcal{D}$ and $\widehat{\mathcal{D}}$ are *domain-preserving*. It implies that $D_T = D_S$. We say that $\mathcal{D}$ and $\widehat{\mathcal{D}}$ are *domain-extending* if $k > 1$ and some domain formula δ_i is equivalent to *True*. Then the domain of any $T := \widehat{\mathcal{D}}(S, \gamma)$ contains $D_S \times \{i\}$, hence, a copy of D_S. In most cases, we will have this for $i = 1$.

---

[3] In Section 5.3.2 (Definition 5.42) we have defined the notion of a quantifier-free operation definition scheme, a QFO definition scheme in short, that defines a QF operation. Every such operation is a QF-transduction. See Example 7.3(2) and (the paragraph after) Definition 7.6 below.

In some statements of Section 9.4, we will consider concrete functional transductions $f : STR^c(\mathcal{R}) \to STR^c(\mathcal{R}')$ such that, for every S in some set $\mathcal{C} \subseteq Dom(f) \subseteq STR^c(\mathcal{R})$, we have $D_S \subseteq D_{f(S)}$; moreover, if h is an isomorphism from S_1 to S_2, then there is an isomorphism from $f(S_1)$ to $f(S_2)$ that extends h. We will say that such f is a *domain-extending monadic second-order transduction on* $\mathcal{C}$ if there exists a concrete domain-extending monadic second-order transduction τ of type $\mathcal{R} \to \mathcal{R}'$ such that $\mathcal{C} \subseteq Dom(\tau)$ and, for some i, the domain formula δ_i is equivalent to *True* and, for every $S \in \mathcal{C}$ and $T \in \tau(S)$, there is an isomorphism from T to $f(S)$ that maps (a, i) to a for every a in D_S. For an example see Example 7.8 below.

If $\mathcal{D}$ is k-copying, the domain of $\widehat{\mathcal{D}}(S, \gamma)$ is defined above as a subset of $D_S \times [k]$. In some constructions, it will be convenient to define it as a subset of $D_S \times I$ for a fixed index set I of cardinality k. Then $[k]$ can be replaced by I in the above definitions.

Every monadic second-order transduction τ is of *linear size increase*. We mean by this that there exists an integer k such that $T \in \tau(S)$ implies $|D_T| \leq k \cdot |D_S|$. This observation has been stated in Chapter 1 as Fact 1.37. It entails that certain graph transductions[4] are not monadic second-order transductions. For example, the line graph transduction[5] is not a monadic second-order transduction because the number of vertices of $Line(K_n)$ is $n(n-1)/2$, whereas that of K_n is n (cf. Example 1.45).

If τ is a monadic second-order transduction defined as $\widehat{\mathcal{D}}$, its domain is characterized by the formula $\exists Y_1, \ldots, Y_q . \chi_{\mathcal{D}}$, where $Y_1, \ldots, Y_q$ are the parameters, hence is $C_r MS$-definable if τ is a $C_r MS$-transduction. (A more general fact will be proved in Corollary 7.12.) The *image* of τ defined as the set $\tau(STR(\mathcal{R}))$ of abstract structures, is not always CMS-definable. For a counter-example, just consider the transduction that transforms a simple graph G into $G \oplus G$. It is a 2-copying monadic second-order transduction. We know that the set of graphs $G \oplus G$ is not VR-recognizable (by Proposition 4.32, since VR-recognizability implies HR-recognizability by Theorem 4.49), hence it is not monadic second-order definable by the Recognizability Theorem for the VR algebra (Theorem 5.68). Other counterexamples will be given below (Remark 7.23). As a consequence of Theorem 7.10 (the Backwards Translation Theorem, cf. Remark 7.13(3)) this fact proves that the inverse of a CMS-transduction is not always a CMS-transduction. By contrast, the class of rational transductions ([*Sak, Niv]) is closed under inverse.

For $\mathcal{L} \in \{CMS, MS, FO, \ldots\}$, an $\mathcal{L}$-transduction τ is *invertible* if its inverse τ^{-1} is an $\mathcal{L}$-transduction. This condition does not imply that τ is injective. (We will also use in Definition 7.9 injective and invertible CMS-transductions.) If a CMS-transduction

---

[4] Unless otherwise stated, a graph transduction is a relation between simple graphs such that a graph G is faithfully represented by the structure $\lfloor G \rfloor$.

[5] The line graph of G, denoted by $Line(G)$, is the undirected graph having E_G as vertex set and an edge between u and v if and only if these edges of G have at least one common end vertex.

is invertible, there are two positive rational numbers k and k' such that $k' \cdot |D_S| \leq |D_T| \leq k \cdot |D_S|$ whenever $T \in \tau(S)$.

Example 7.3 (1) **Quotient structures:** Let us assume that a CMS formula $\varphi(x,y)$ defines on each structure S of a class $C \subseteq STR^c(\mathcal{R})$ an equivalence relation $\approx$. Such a relation may be the equivalence relation generated by a relation $E \subseteq D_S \times D_S$ defined in each $S \in C$ by a CMS formula $\psi(x,y)$ (in particular, because the reflexive and transitive closure of a CMS-definable relation is CMS-definable, cf. Section 5.2.2).

We define the quotient structure $S/\approx$ as the $\mathcal{R}$-structure $\langle D_S/\approx, (R_{S/\approx})_{R\in\mathcal{R}} \rangle$ where $R_{S/\approx} := \{([a_1], \ldots, [a_{\rho(R)}]) \mid (a_1, \ldots, a_{\rho(R)}) \in R_S\}$ and $[a]$ denotes the equivalence class of a (cf. Example 1.35 and Definition 2.15).

We must actually define $S/\approx$ as $[T]_{iso}$ for a concrete structure T. As domain D_T of T, we take a cross-section of $\approx$ as value of a parameter Y and we specify the relations R_T accordingly. This can be done by a noncopying definition scheme $\mathcal{D}$ with one parameter Y; its precondition expresses that Y denotes a cross-section of $\approx$ (for a structure S in C). For each S in C, all structures in $\widehat{\mathcal{D}}(S)$ are isomorphic to one another. Hence, the total function $S \mapsto S/\approx$ from C to $STR(\mathcal{R})$ is a CMS-transduction on C, and the definition scheme $\mathcal{D}$ is parameter-invariant. The concrete transduction $\widehat{\mathcal{D}}$ is not functional.

(2) **Quantifier-free operations:** Quantifier-free operations : $STR^c(\mathcal{R}) \rightarrow STR^c(\mathcal{R}')$ where $\mathcal{R}'_0 = \emptyset$ are defined in Section 5.3.2 (Definition 5.42) by means of QFO definition schemes which are parameterless noncopying QF-definition schemes with *True* as precondition. Note that if $\mathcal{R}_0 = \emptyset$ and $\mathcal{D}$ is quantifier-free and parameterless, then its precondition can only be *True* or *False*.

(3) **Transductions that order linearly:** In Definition 5.28 we have defined the notion of a pair of MS formulas $(\alpha(X_1, \ldots, X_m), \beta(X_1, \ldots, X_m, x_1, x_2))$ that orders linearly the structures of a set C of concrete $\mathcal{R}$-structures. Such a pair can be made into a noncopying definition scheme $\mathcal{D}$ of type $\mathcal{R} \rightarrow \mathcal{R} \cup \{\leq\}$ with parameters $X_1, \ldots, X_m$: its precondition is α, its domain formula is *True*, θ_R is $R(x_1, \ldots, x_{\rho(R)})$ for each $R \in \mathcal{R}_+$ and its relation formula $\theta_{\leq}$ is β. The corresponding MS-transduction $\widehat{\mathcal{D}}$ expands each structure S in C by a linear order (in one or several ways). If $S \in STR^c(\mathcal{R}) - C$, then $\widehat{\mathcal{D}}(S)$ may be empty or contain expansions of S by binary relations that are not linear orders of D_S.

(4) **Second-order substitutions:** Let F and H be finite functional signatures and, for each g in H, let t_g be a term in $T(F, \{x_1, \ldots, x_{\rho(g)}\})$. The associated second-order substitution $\theta_H : T(H) \rightarrow T(F)$ replaces each g in H by the term t_g that is its definition over F; a formal definition is in Section 2.6.1, Definition 2.125. We say that θ_H is *linear* if each term t_g is linear, that is, if no variable has more than one occurrence in t_g. We say that θ_H is *strict* if each term t_g is strict, that is, if each variable x_i has at least one occurrence in t_g for $i = 1, \ldots, \rho(g)$, and we say that it is *nonerasing* if no term t_g is a variable.

If θ_H is linear, then it is an MS-transduction. More precisely, the set of pairs $(\lfloor t \rfloor, \lfloor \theta_H(t) \rfloor)$ such that $t \in T(H)$ is an MS-transduction.[6] Terms are represented here by relational structures without constants. It is a straightforward exercise to construct a definition scheme for θ_H that is parameterless and k-copying, where k is the maximal number of occurrences of symbols from F in a term t_g. The hypothesis that the terms t_g are linear cannot be removed. If g is unary with $t_g = f(x_1, x_1)$ and a is a constant symbol with $t_a = a$, then the size of $\theta_H(g^n(a))$ is $2^{n+1} - 1$, hence the corresponding transduction is not of linear size increase, so that it cannot be a monadic second-order transduction.

If θ_H is linear, strict and nonerasing, then it is an invertible FO-transduction. Again, it is a straightforward exercise to construct a noncopying definition scheme for θ_H^{-1}, with $|H|$ parameters (see also Examples 8.3 and 8.5). □

Remark 7.4 (Effectivity of definitions) Certain conditions on a definition scheme $\mathcal{D}$ of type $\mathcal{R} \rightarrow \mathcal{R}'$ are not decidable. Typical examples are whether it is domain-preserving, parameter-invariant or produces only structures of the form S_* for $S \in STR(\mathcal{R}')$ in the case where $\mathcal{R}'_0 \neq \emptyset$ (cf. Section 5.1.5 and below Section 7.1.2). Another undecidable property is whether $Dom(\widehat{\mathcal{D}}) = \emptyset$. These facts are due to the undecidability of the satisfiability problem for FO and MS sentences (Theorem 5.5). To the opposite, all particular properties of QF operations defined in Definition 5.42 are decidable from their definition schemes, because the absence of quantifiers makes the satisfiability problem decidable. □

7.1.2 Monadic second-order transductions producing structures with constants

In Definition 7.2, we have defined monadic second-order transductions of type $\mathcal{R} \rightarrow \mathcal{R}'$, where $\mathcal{R}'$ has no constant symbols. We now waive this restriction. The definition scheme of a transduction producing structures T with constants must have some way to specify them as pairs (u, i), where i is an integer and u is in the domain of the input structure. Rather than rewriting all definitions, we will rest on the particular case of Definition 7.2 by replacing each constant symbol by a new unary relation. This method has been discussed in Section 5.1.5.

Formally, for every relational signature $\mathcal{R}$ we let $\mathcal{R}_* := \mathcal{R}_+ \cup Lab_\mathcal{R}$, where $Lab_\mathcal{R} := \{lab_a \mid a \in \mathcal{R}_0\}$ is a new set of unary relations. (Unless otherwise specified, when we discuss several signatures $\mathcal{R}, \mathcal{R}', \ldots$ we will always assume that $(\mathcal{R} \cup \mathcal{R}' \cup \cdots) \cap (Lab_\mathcal{R} \cup Lab_{\mathcal{R}'} \cup \cdots) = \emptyset$.)

For every $S \in STR^c(\mathcal{R})$, we let S_* be the structure in $STR^c(\mathcal{R}_*)$ such that $lab_{aS_*} = \{a_S\}$ for every $a \in \mathcal{R}_0$ and $R_{S_*} = R_S$ for every $R \in \mathcal{R}_+$. We denote by $STR^c_*(\mathcal{R}_*)$ the class of concrete $\mathcal{R}_*$-structures such that the relations lab_a are interpreted as

---

[6] By Definitions 7.1 and 7.2, this means that the transduction of abstract structures $[\{(\lfloor t \rfloor, \lfloor \theta_H(t) \rfloor) \mid t \in T(H)\}]_{iso} = \{([\lfloor t \rfloor]_{iso}, [\lfloor \theta_H(t) \rfloor]_{iso}) \mid t \in T(H)\}$ is an MS-transduction.

singletons, and by $STR_*(\mathcal{R}_*)$ the corresponding set of abstract structures. The first-order sentence α defined as the conjunction of the sentences

$$\exists x.\, lab_a(x) \wedge \forall x,y(lab_a(x) \wedge lab_a(y) \Rightarrow x=y),$$

for all a in $\mathcal{R}_0$, characterizes $STR_*^c(\mathcal{R}_*)$. The mapping $S \mapsto S_*$ is a bijection $: STR^c(\mathcal{R}) \rightarrow STR_*^c(\mathcal{R}_*)$ and also a parameterless and domain-preserving QF-transduction. (It is even a QF operation since it is total, but we will not use it as an operation of a functional signature.) Its inverse associates with $T \in STR_*^c(\mathcal{R}_*)$ a structure in $STR^c(\mathcal{R})$ denoted by $T_\dagger$. If $\mathcal{R}_0 = \emptyset$, we have $\mathcal{R}_* = \mathcal{R}$ and $S = S_*$ for every $S \in STR^c(\mathcal{R})$ and thus, $STR_*^c(\mathcal{R}_*) = STR^c(\mathcal{R})$. Similar statements hold for abstract structures.

The following lemma says that the same first-order and monadic second-order properties of an $\mathcal{R}$-structure S can be expressed in S_* or, directly, in S. We let r be a nonnegative integer and $\mathcal{X}$ be a finite set of first-order and set variables.

Lemma 7.5 For every formula φ in $C_r MS(\mathcal{R}, \mathcal{X})$ there exists a formula ψ in $C_r MS(\mathcal{R}_*, \mathcal{X})$ such that, for every $S \in STR^c(\mathcal{R})$ and every $\mathcal{X}$-assignment γ in S, we have $(S, \gamma) \models \varphi$ if and only if $(S_*, \gamma) \models \psi$. Conversely, for every formula ψ in $C_r MS(\mathcal{R}_*, \mathcal{X})$ there exists a formula φ in $C_r MS(\mathcal{R}, \mathcal{X})$ satisfying the above equivalence. The same properties hold for FO instead of $C_r MS$.

Proof: For proving the first assertion, we construct ψ from φ by the following steps.

Step 1: We replace every atomic formula $R(t_1, \ldots, t_n)$, where each t_i is a constant symbol or a variable, by the formula

$$\exists x_1, \ldots, x_n(R(x_1, \ldots, x_n) \wedge \bigwedge_{i \in [n]} x_i = t_i).$$

Step 2: For every $a, b \in \mathcal{R}_0$, every first-order variable x and every set variable X, we replace certain atomic formulas as follows:

$$x = a \text{ and } a = x \text{ by } lab_a(x),$$
$$a = b \text{ by } \exists x(lab_a(x) \wedge lab_b(x)),$$
$$a \in X \text{ by } \exists x(lab_a(x) \wedge x \in X).$$

The verification that ψ satisfies the requested property is straightforward by induction on the structure of φ.

For the opposite direction, we construct φ from ψ by replacing $lab_a(x)$ by $x = a$ for every first-order variable x and every $a \in \mathcal{R}_0$.

The first transformation may increase the quantifier-height, the second preserves it. Both transform first-order formulas into first-order formulas. ∎

We will denote by φ_* the formula ψ constructed from φ, and by $\psi_\dagger$ the formula φ constructed from ψ. The second transformation of this proof is actually an application of the Backwards Translation Theorem for QF operations (Proposition 5.46).

Definition 7.6 (Transductions relating structures with constants) Let $\mathcal{R}$ and $\mathcal{R}'$ be two relational signatures, possibly with constant symbols. A *definition scheme* $\mathcal{D}$ *of type* $\mathcal{R} \to \mathcal{R}'$ is a definition scheme $\mathcal{D}_*$ of type $\mathcal{R} \to \mathcal{R}'_*$ such that $\widehat{\mathcal{D}_*} \subseteq STR^c(\mathcal{R}) \times STR^c_*(\mathcal{R}'_*)$. If $S \in STR^c(\mathcal{R})$ and γ is a W-assignment in S where W is the set of parameters of $\mathcal{D}_*$, then we define $\widehat{\mathcal{D}}(S, \gamma)$ as $\widehat{\mathcal{D}_*}(S, \gamma)_\dagger$. The concrete transduction of type $\mathcal{R} \to \mathcal{R}'$ defined by $\mathcal{D}$ is defined as $\widehat{\mathcal{D}} := \{(S, T_\dagger) \mid (S, T) \in \widehat{\mathcal{D}_*}\}$, and it is called a *concrete monadic second-order transduction*. The corresponding transduction of abstract structures is $[\widehat{\mathcal{D}}]_{iso}$, and it is called a *monadic second-order transduction*. Properties of a definition scheme $\mathcal{D}$ (and of the associated transduction), such as to be parameterless, k-copying, parameter-invariant, etc., are defined as the corresponding properties of $\mathcal{D}_*$.

Note that the property that $\widehat{\mathcal{D}_*} \subseteq STR^c(\mathcal{R}) \times STR^c_*(\mathcal{R}'_*)$ is not decidable for an arbitrary definition scheme $\mathcal{D}_*$. However, it can be guaranteed syntactically as we will see in Remark 7.13(2) after proving the fundamental Backwards Translation Theorem.

We now extend the remark of Example 7.3(2). Every QF operation of type $\mathcal{R} \to \mathcal{R}'$ is a parameterless noncopying QF-transduction with precondition *True*, although a QFO definition scheme (cf. Definition 5.42) is not necessarily a QF-definition scheme since it may contain constant defining sentences $\kappa_{c,d}$ for $c \in \mathcal{R}_0$ and $d \in \mathcal{R}'_0$. However, a QFO definition scheme can be translated into a QF-definition scheme $\mathcal{D}_*$ by taking as formula θ_{lab_d} (for $d \in \mathcal{R}'_0$) the formula $\bigvee_{c \in \mathcal{R}_0}(\kappa_{c,d} \wedge x_1 = c)$. The conditions on the constant defining sentences in Definition 5.42 ensure that lab_d is interpreted as a singleton.

If τ is a transduction of type $\mathcal{R} \to \mathcal{R}'$, then we let $\tau_* := \{(S, T_*) \mid (S, T) \in \tau\}$ and, for having a symmetric notion, we define also $\tau_{**} := \{(S_*, T_*) \mid (S, T) \in \tau\}$. These three transductions are equivalent in the following sense:

Lemma 7.7 The transduction τ is a C_rMS-transduction if and only if τ_* is a C_rMS-transduction if and only if τ_{**} is a C_rMS-transduction. The same statement holds for FO-transductions.

Proof: The first equivalence holds by definition. We now prove the second equivalence. Let us assume that τ_* is a C_rMS-transduction : $STR^c(\mathcal{R}) \to STR^c_*(\mathcal{R}'_*)$ with definition scheme $\mathcal{D}_*$. Then so is $\tau_{**} : STR^c_*(\mathcal{R}_*) \to STR^c_*(\mathcal{R}'_*)$ by the following construction. We let $\mathcal{D}'$ be the definition scheme of type $\mathcal{R}_* \to \mathcal{R}'_*$ obtained from $\mathcal{D}_*$ by replacing each of its formulas φ by the formula φ_* constructed in Lemma 7.5. Then, we change the precondition of $\mathcal{D}'$ into $\chi_{\mathcal{D}'} \wedge \alpha$, where α is the first-order formula expressing that a structure in $STR^c(\mathcal{R}_*)$ belongs to $STR^c_*(\mathcal{R}_*)$. This gives a definition scheme $\mathcal{D}_{**}$ of type $\mathcal{R}_* \to \mathcal{R}'_*$ such that $\widehat{\mathcal{D}_{**}} = \tau_{**}$.

A definition scheme for τ_* is obtained from one for τ_{**} by replacing each of its formulas ψ by the formula $\psi_\dagger$ constructed in Lemma 7.5. ∎

The first equivalence also holds for QF-transductions, by definition, but the second equivalence does not. If τ_* is a QF-transduction, then the corresponding transduction

τ_{**} is not necessarily one. If the domain formula of τ_* is $R(x) \vee R(c)$, where $c \in \mathcal{R}_0$, then the domain formula of τ_{**} must be equivalent to $R(x) \vee \exists y (lab_c(y) \wedge R(y))$, hence is not equivalent to a quantifier-free formula (cf. Section 5.1.5).

As can be verified from the proof of Lemma 7.7, all the usual properties of definition schemes (except being a QF-definition scheme) carry over. Thus, an alternative to Definition 7.6 is to define a definition scheme $\mathcal{D}$ of type $\mathcal{R} \to \mathcal{R}'$ to be a definition scheme $\mathcal{D}_{**}$ of type $\mathcal{R}_* \to \mathcal{R}'_*$ such that $\widehat{\mathcal{D}_{**}} \subseteq STR_*^c(\mathcal{R}_*) \times STR_*^c(\mathcal{R}'_*)$. Then $\widehat{\mathcal{D}}(S, \gamma)$ should be defined as $\widehat{\mathcal{D}_{**}}(S_*, \gamma)_\dagger$, and $\widehat{\mathcal{D}}$ as $\{(S_\dagger, T_\dagger) \mid (S, T) \in \widehat{\mathcal{D}_{**}}\}$.

7.1.3 Transductions of words, terms and graphs

We have defined transductions that transform relational structures. Via relational structures that represent faithfully words, terms and labeled p-graphs or s-graphs (cf. Section 5.1.1, *faithfully* means that isomorphic structures represent identical or isomorphic objects), we can specify transformations of such objects by monadic second-order transductions that transform their representing structures. These transformations of objects will also be called *monadic second-order transductions*.

Specifically, we have defined in Section 5.1.1 structures $\lfloor w \rfloor$, $\lfloor t \rfloor$, $\lfloor G \rfloor$ and $\lfloor G \rfloor_C$ that represent faithfully, respectively, a word w in A^*, a term t in $T(F)$, a simple labeled graph G and a p-graph or a simple s-graph G of type included in C. To take a representative example, a relation $\tau \subseteq T(F) \times A^*$ is said to be an MS-transduction if the relation $\tau' := \{(\lfloor t \rfloor, \lfloor w \rfloor) \mid (t, w) \in \tau\}$ is an MS-transduction of type $\mathcal{R}_F \to \mathcal{W}_A$.[7] Properties of τ', such as being k-copying, parameterless, parameter-invariant, etc., are transferred from τ' to τ in the obvious way. We omit a formal definition of an $\mathcal{L}$-transduction of objects, which would be similar to the definition of $\mathcal{L}$-definability of a set of objects in Section 5.1.5.

A *graph transduction* is a set of pairs (G, H) such that G and H are, possibly labeled, graphs. Similarly we define p-graph and s-graph transductions. If τ is, e.g., a monadic second-order p-graph transduction, then both the domain $Dom(\tau)$ and the image $Dom(\tau^{-1})$ of τ are sets of p-graphs of bounded type, because the p-graphs G in $Dom(\tau)$ must be represented as $\lfloor G \rfloor_C$ for a fixed C, and similarly for $Dom(\tau^{-1})$. If necessary, we will make precise the relational signatures used for representing the considered graphs, p-graphs, or s-graphs. Such a precision may be necessary since we have defined variants of the basic representations. The quantifier-free operations on graphs, p-graphs or s-graphs, in particular the edge-complement, the edge addition operation $\overrightarrow{add}_{a,b,C}$ and the source fusion $fuse_{a,b}$ (cf. Sections 5.3.2 and 5.3.4), are examples of monadic second-order graph (p-graph or s-graph) transductions. We now consider an example concerning forests.

---

[7] This means that $[\tau']_{iso}$ is an MS-transduction, cf. Footnote 6 in Example 7.3(4).

Example 7.8 (The incidence graph of a forest) The incidence graph of a simple undirected graph G is the directed graph $Inc(G)$ with vertex set $V_G \cup E_G$ and edge relation $in_G := \{(e,v) \mid e \in E_G$ and v is an end vertex of $e\}$, cf. Definition 5.17. We will show that the graph transduction Inc is an MS-transduction on the class $\mathcal{C}$ of forests, more precisely, that the mapping $\lfloor G \rfloor \mapsto \lfloor Inc(G) \rfloor$ is a 2-copying domain-extending MS-transduction on the class $\lfloor \mathcal{C} \rfloor$ (cf. the proof of Corollary 5.23).

Its definition scheme uses a parameter Y; its precondition χ expresses that G is a forest and that Y denotes a set of vertices A that has one and only one element in each connected component of G: this means that A can be chosen as a set of roots. We let (G,A) denote the corresponding rooted, and thus directed, forest.

The domain formulas δ_1 and δ_2 are respectively *True* and $x \notin Y$. A vertex u of G in $D_{\lfloor Inc(G) \rfloor}$ will correspond to the pair $(u,1)$, and an edge e will correspond to the pair $(u,2)$, where u is the head of e in (G,A). Every vertex not in A is thus the head of a unique edge, which justifies the definition of δ_2. We now complete the construction by defining the relation formulas: $\theta_{in,2,1}$ is $x_2 = x_1 \vee$ "$x_2 \to_{(G,A)} x_1$" and $\theta_{in,1,1}$, $\theta_{in,1,2}$, $\theta_{in,2,2}$ are *False*. This transduction is parameter-invariant. □

Definition 7.9 (Strongly equivalent representations) In Section 5.1.1, we have also defined variants of the above representations that are all faithful. The notion of a monadic second-order transduction of objects like words, terms and graphs may depend on their chosen representations. We say that two representations $\lfloor w \rfloor$ and $\lfloor w \rfloor'$ of a word w in A^* are *strongly MS-equivalent* if the relation $\alpha := \{(\lfloor w \rfloor, \lfloor w \rfloor') \mid w \in A^*\}$ and its inverse $\beta := \alpha^{-1}$ are injective MS-transductions (hence, α and β are invertible). More precisely, and in general, let $\mathcal{C}$ be a class of objects and let $rep : \mathcal{C} \to STR^c(\mathcal{R})$ and $rep' : \mathcal{C} \to STR^c(\mathcal{R}')$ be two representations, i.e., mappings that map isomorphic objects to isomorphic structures. Then rep and rep' are *strongly $\mathcal{L}$-equivalent* (for $\mathcal{L} \in \{MS, FO, \dots\}$) if the relation $\alpha := \{(rep(H), rep'(H)) \mid H \in \mathcal{C}\}$ is an injective $\mathcal{L}$-transduction on $rep(\mathcal{C})$ and $\beta := \alpha^{-1}$ is an (injective) $\mathcal{L}$-transduction on $rep'(\mathcal{C})$.[8]

Let $\tau \subseteq T(F) \times A^*$ and $\tau' := \{(\lfloor t \rfloor, \lfloor w \rfloor) \mid (t,w) \in \tau\}$ be as above, and let $\tau'' := \{(\lfloor t \rfloor, \lfloor w \rfloor') \mid (t,w) \in \tau\}$. Then we have $\tau'' = \tau' \cdot \alpha$ and $\tau' = \tau'' \cdot \beta$. Since the composition of two monadic second-order transductions is a monadic second-order transduction, as we will prove below (Theorem 7.14), we get that τ' is a monadic second-order transduction if and only if τ'' is a monadic second-order transduction. Hence, the replacement of $\lfloor w \rfloor$ by $\lfloor w \rfloor'$ to represent a word w does not modify the notion of monadic second-order transduction between words and terms, or between words and words, words and graphs, etc. The same argument holds in the general case.

---

[8] This syntactic definition is more restricted than the semantical definition of $\mathcal{L}$-equivalence given in Example 5.2 and Section 5.1.5. If *rep* and *rep'* are strongly MS-equivalent, then they are MS-equivalent, by the Backwards Translation Theorem to be proved in Section 7.1.4. The converse does not hold, as shown below.

It can be verified that all variants of representations discussed in Example 5.2 are strongly MS-equivalent, with one exception: representing the empty word by the empty structure is MS-equivalent but not strongly MS-equivalent to representing it by a nonempty structure (we considered these two possibilities in Example 5.2(1)) because the image of the empty structure $\varnothing_\mathcal{R}$ under an MS-transduction of type $\mathcal{R} \rightarrow \mathcal{R}'$ is necessarily the empty structure $\varnothing_{\mathcal{R}'}$. If this point is important, we will specify in statements concerning transductions of words which type of representation is used. This problem does not occur for terms and graphs, because no term is represented by the empty structure and the empty graph is always represented by the empty structure.

In addition to the standard representation of a graph G by the relational structure $\lfloor G \rfloor$ with domain V_G (cf. Example 5.2(2)), we have also defined a faithful representation by the structure $\lceil G \rceil = \lfloor Inc(G) \rfloor$ with domain $V_G \cup E_G$ (cf. Section 5.2.5). The representations $\lfloor G \rfloor$ and $\lceil G \rceil$ are not strongly equivalent for all graphs (but they are for forests by Example 7.8) hence, we get several types of graph transductions, depending on the choice of $\lfloor G \rfloor$ or $\lceil G \rceil$. This point has been discussed in detail in Section 1.8, and we will go back to it in Section 7.3 and in Chapter 9. In this section and the next one (i.e., Sections 7.1 and 7.2), we will only consider simple graphs, and a graph G will always be represented by (and even identified with) $\lfloor G \rfloor$.

7.1.4 The fundamental property of monadic second-order transductions

A definition scheme $\mathcal{D}$ defines a transduction between structures but it also yields a translation of formulas. The following theorem says that if $T = \widehat{\mathcal{D}}(S, \gamma)$, then each monadic second-order property of T can be expressed by a monadic second-order formula over S. (The particular case of QF operations, more precisely of QDP-QF operations, has been proved in Theorem 5.47.) The usefulness of monadic second-order transductions is largely based on this fact.

Let $\mathcal{D}$ be a k-copying definition scheme of type $\mathcal{R} \rightarrow \mathcal{R}'$ with set of parameters $\mathcal{W}$. Given a set $\mathcal{X}$ of set variables disjoint from $\mathcal{W}$, we introduce new set variables $X^{(i)}$ for $X \in \mathcal{X}$ and $i \in [k]$, and we let $\mathcal{X}^{(k)} := \{X^{(i)} \mid X \in \mathcal{X}, i \in [k]\}$.

Let $S \in STR^c(\mathcal{R})$. For every mapping $\eta : \mathcal{X}^{(k)} \rightarrow \mathcal{P}(D_S)$, we define $\eta^{[k]} : \mathcal{X} \rightarrow \mathcal{P}(D_S \times [k])$ by

$$\eta^{[k]}(X) := (\eta(X^{(1)}) \times \{1\}) \cup \cdots \cup (\eta(X^{(k)}) \times \{k\}).$$

Let $\mathcal{Y} = \{y_1, \ldots, y_n\}$ be a set of first-order variables, linearly ordered as indicated. For a mapping $\mu : \mathcal{Y} \rightarrow D_S$ and an n-tuple $\mathbf{i} = (i_1, \ldots, i_n) \in [k]^n$, we denote by $\mu_\mathbf{i} : \mathcal{Y} \rightarrow D_S \times [k]$ the mapping such that $\mu_\mathbf{i}(y_j) := (\mu(y_j), i_j)$. If $k = 1$, then we identify $D_S \times [1]$ with D_S and $\mu_{(1,\ldots,1)}$ with μ.

Note that, even if $T = \widehat{\mathcal{D}}(S, \gamma)$ is well defined (for a $\mathcal{W}$-assignment γ), the mapping $\eta^{[k]}$ is not necessarily an $\mathcal{X}$-assignment in T because $\eta^{[k]}(X)$ may not be a subset of the domain of T. A similar observation holds for $\mu_{\mathbf{i}}$.

Theorem 7.10 (Backwards Translation Theorem) Let $\mathcal{D}$ be a k-copying C_rMS-definition scheme of type $\mathcal{R} \to \mathcal{R}'$ with set of parameters $\mathcal{W}$. Let $\mathcal{X}$ be a finite set of set variables and $\mathcal{Y} = \{y_1, \ldots, y_n\}$ be a set of first-order variables. For every $\beta \in C_r$MS$(\mathcal{R}', \mathcal{X} \cup \mathcal{Y})$ and $\mathbf{i} \in [k]^n$, one can construct a formula $\beta_{\mathbf{i}}^{\mathcal{D}} \in C_rMS(\mathcal{R}, \mathcal{W} \cup \mathcal{X}^{(k)} \cup \mathcal{Y})$ such that for every $S \in STR^c(\mathcal{R})$, every $\mathcal{W}$-assignment γ, every $\mathcal{X}^{(k)}$-assignment η, and every $\mathcal{Y}$-assignment μ, all of them in S, we have:

$(S, \gamma \cup \eta \cup \mu) \models \beta_{\mathbf{i}}^{\mathcal{D}}$ if and only if

$\qquad \widehat{\mathcal{D}}(S, \gamma)$ is defined,

$\qquad \eta^{[k]} \cup \mu_{\mathbf{i}}$ is an $(\mathcal{X} \cup \mathcal{Y})$-assignment in $\widehat{\mathcal{D}}(S, \gamma)$, and

$\qquad (\widehat{\mathcal{D}}(S, \gamma), \eta^{[k]} \cup \mu_{\mathbf{i}}) \models \beta$.

The quantifier-height of $\beta_{\mathbf{i}}^{\mathcal{D}}$ is at most $k \cdot qh(\beta) + qh(\mathcal{D}) + 1$. $\qquad\qquad \square$

Before giving the proof, we present the construction of $\beta_{\mathbf{i}}^{\mathcal{D}}$ and a lemma, for the case where $\mathcal{R}'$ has no constant symbols. The extension to the general case will be straightforward.

We first define a formula $\beta_{\mathbf{i}}^{\star\mathcal{D}}$, by an inductive construction on the structure of β. As in the constructions of Chapters 5 and 6 and without loss of generality, we will assume that β contains neither universal quantifications, nor the Boolean connectives $\Rightarrow$ and $\Leftrightarrow$, nor the atomic formula *False*. Let $\mathcal{R}'_0 = \emptyset$. We let $\mathcal{D}$ be $\langle \chi, (\delta_i)_{i \in [k]}, (\theta_w)_{w \in \mathcal{R}' \otimes [k]} \rangle$ with set of parameters $\mathcal{W}$. For every $i \in [k]$ and every set variable X, we denote by $\bar{\delta}_i(X)$ the formula $\forall x (x \in X \Rightarrow \delta_i)$.

For every m and n, every β in C_rMS$(\mathcal{R}', \{X_1, \ldots, X_m, y_1, \ldots, y_n\})$, and every $\mathbf{i} \in [k]^n$, we define as follows a formula $\beta_{\mathbf{i}}^{\star\mathcal{D}}$ in C_rMS$(\mathcal{R}, \{X_1, \ldots, X_m\}^{(k)} \cup \{y_1, \ldots, y_n\})$:

- *True*$_{\mathbf{i}}^{\star\mathcal{D}}$ is *True*;
- $(y_j = y_{j'})_{\mathbf{i}}^{\star\mathcal{D}}$ is $y_j = y_{j'}$ if $\mathbf{i}[j] = \mathbf{i}[j']$ and is *False* otherwise (where $\mathbf{i}[j]$ is the j-th component of the tuple $\mathbf{i}$);
- $(y_j \in X_\ell)_{\mathbf{i}}^{\star\mathcal{D}}$ is $y_j \in X_\ell^{(\mathbf{i}[j])}$;
- $(R(y_{j_1}, \ldots, y_{j_{\rho(R)}}))_{\mathbf{i}}^{\star\mathcal{D}}$ is $\theta_w[y_{j_1}/x_1, \ldots, y_{j_{\rho(R)}}/x_{\rho(R)}]$, with $w := (R, \mathbf{i}[j_1], \ldots, \mathbf{i}[j_{\rho(R)}])$;
- $(Card_{p,q}(X_\ell))_{\mathbf{i}}^{\star\mathcal{D}}$ is $\bigvee \bigwedge_{i \in [k]} Card_{p_i, q}(X_\ell^{(i)})$, where the disjunction extends to all k-tuples $(p_1, \ldots, p_k) \in [0, q-1]^k$ such that $p_1 + \cdots + p_k \equiv p \pmod{q}$;
- $(\neg\beta)_{\mathbf{i}}^{\star\mathcal{D}}$ is $\neg\beta_{\mathbf{i}}^{\star\mathcal{D}}$, $(\beta \vee \gamma)_{\mathbf{i}}^{\star\mathcal{D}}$ is $\beta_{\mathbf{i}}^{\star\mathcal{D}} \vee \gamma_{\mathbf{i}}^{\star\mathcal{D}}$, $(\beta \wedge \gamma)_{\mathbf{i}}^{\star\mathcal{D}}$ is $\beta_{\mathbf{i}}^{\star\mathcal{D}} \wedge \gamma_{\mathbf{i}}^{\star\mathcal{D}}$;
- $(\exists y. \beta)_{\mathbf{i}}^{\star\mathcal{D}}$ is $\bigvee_{i \in [k]} \exists y (\delta_i[y/x] \wedge \beta_{\mathbf{i}i}^{\star\mathcal{D}})$ where $\mathbf{i}i$ denotes $(i_1, \ldots, i_n, i)$ if $\mathbf{i} = (i_1, \ldots, i_n)$ (note that $\beta_{\mathbf{i}}^{\star\mathcal{D}}$ has its free variables in the set $\{X_1, \ldots, X_m\}^{(k)} \cup \{y_1, \ldots, y_n, y\}$);

- $(\exists X.\beta)_{\mathbf{i}}^{\star\mathcal{D}}$ is $\exists X^{(1)},\ldots,X^{(k)}(\bigwedge_{i\in[k]}\overline{\delta}_i(X^{(i)})\wedge\beta_{\mathbf{i}}^{\star\mathcal{D}})$ (note that $\beta_{\mathbf{i}}^{\star\mathcal{D}}$ has its free variables in the set $\{X_1,\ldots,X_m,X\}^{(k)}\cup\{y_1,\ldots,y_n\}$).

The quantifier-height of $\beta_{\mathbf{i}}^{\star\mathcal{D}}$ is at most $k\cdot qh(\beta)+qh(\mathcal{D})+1$.

We let then $\beta_{\mathbf{i}}^{\mathcal{D}}$ be

$$\beta_{\mathbf{i}}^{\star\mathcal{D}}\wedge\chi\wedge\bigwedge_{i\in[k],\ell\in[m]}\overline{\delta}_i(X_\ell^{(i)})\wedge\bigwedge_{j\in[n]}\delta_{\mathbf{i}[j]}[y_j/x].$$

We will drop the subscript $\mathbf{i}$, writing $\beta^{\star\mathcal{D}}$ and $\beta^{\mathcal{D}}$, whenever $[k]^n$ is a singleton, i.e., either $k=1$ and $n\geq1$, and so $\mathbf{i}=(1,\ldots,1)$, or $n=0$, and so $\mathbf{i}=()$. If β is a sentence, then $\beta^{\mathcal{D}}$ is $\beta^{\star\mathcal{D}}\wedge\chi$.

We call the family $\beta^{\mathcal{D}}:=(\beta_{\mathbf{i}}^{\mathcal{D}})_{\mathbf{i}\in[k]^n}$ the *backwards translation* of β relative to the definition scheme $\mathcal{D}$. We will also use the notation $\tau^{\#}(\beta)$ for $\beta^{\mathcal{D}}$, where $\tau=\widehat{\mathcal{D}}$.[9]

Lemma 7.11 For every concrete $\mathcal{R}$-structure S, every $\mathcal{W}$-assignment γ in S such that $\widehat{\mathcal{D}}(S,\gamma)$ is defined, every $\{X_1,\ldots,X_m\}^{(k)}$-assignment η in S such that $\eta^{[k]}$ is an $\{X_1,\ldots,X_m\}$-assignment in $\widehat{\mathcal{D}}(S,\gamma)$, every $\{y_1,\ldots,y_n\}$-assignment μ in S and every $\mathbf{i}\in[k]^n$ such that $\mu_{\mathbf{i}}$ is a $\{y_1,\ldots,y_n\}$-assignment in $\widehat{\mathcal{D}}(S,\gamma)$, we have

$$(S,\gamma\cup\eta\cup\mu)\models\beta_{\mathbf{i}}^{\star\mathcal{D}}\text{ if and only if }(\widehat{\mathcal{D}}(S,\gamma),\eta^{[k]}\cup\mu_{\mathbf{i}})\models\beta.$$

Proof: The proof is a straightforward verification by induction on the structure of β. We only check a few cases. We let $T:=\widehat{\mathcal{D}}(S,\gamma)$.

If β is $y_j=y_{j'}$, then $\mu_{\mathbf{i}}(y_j)=(\mu(y_j),\mathbf{i}[j])$ and similarly for j'; hence $\mu_{\mathbf{i}}(y_j)=\mu_{\mathbf{i}}(y_{j'})$ in T if and only if $\mu(y_j)=\mu(y_{j'})$ and $\mathbf{i}[j]=\mathbf{i}[j']$, hence if and only if $(y_j=y_{j'})_{\mathbf{i}}^{\star\mathcal{D}}$ holds in (S,μ).

If β is $R(y_{j_1},\ldots,y_{j_p})$, then the p-tuple $(\mu_{\mathbf{i}}(y_{j_1}),\ldots,\mu_{\mathbf{i}}(y_{j_p}))$, which equals $((\mu(y_{j_1}),\mathbf{i}[j_1]),\ldots,(\mu(y_{j_p}),\mathbf{i}[j_p]))$, belongs to R_T if and only if $\theta_w(\mu(y_{j_1}),\ldots,\mu(y_{j_p}))$ holds in (S,γ), where $w=(R,\mathbf{i}[j_1],\ldots,\mathbf{i}[j_p])$, hence if and only if $(R(y_{j_1},\ldots,y_{j_p}))_{\mathbf{i}}^{\star\mathcal{D}}$ holds in $(S,\gamma\cup\mu)$.

If β is $\exists X.\beta'$, then this formula holds in $(T,\eta^{[k]}\cup\mu_{\mathbf{i}})$ if and only if there exists $A\subseteq D_T$ such that $(T,\eta^{[k]}\cup\mu_{\mathbf{i}})\models\beta'(A)$ and A can be written $(A^{(1)}\times\{1\})\cup\cdots\cup(A^{(k)}\times\{k\})$, for some $A^{(i)}\subseteq D_S$. The conditions $A^{(i)}\times\{i\}\subseteq D_T$ can be expressed by the validity of $\overline{\delta}_i(X^{(i)})$ with $A^{(i)}$ as value of $X^{(i)}$ for each i. By using the induction hypothesis for β', we have

$$(T,\eta^{[k]}\cup\mu_{\mathbf{i}})\models\beta'(A)\text{ if and only if }(S,\gamma\cup\eta\cup\mu)\models\beta_{\mathbf{i}}'^{\star\mathcal{D}}(A^{(1)},\ldots,A^{(k)}),$$

so that

$$(T,\eta^{[k]}\cup\mu_{\mathbf{i}})\models\beta\text{ if and only if }(S,\gamma\cup\eta\cup\mu)\models\beta_{\mathbf{i}}^{\star\mathcal{D}}.$$

The other cases are easy to establish. ∎

---

[9] The notation $\tau^{\#}(\beta)$ is slightly imprecise because $\beta^{\mathcal{D}}$ depends on $\mathcal{D}$, not only on τ. A transduction may have different definition schemes.

Proof of Theorem 7.10: We first assume that $\mathcal{R}'$ has no constant symbols and we let $\mathcal{D}$ be as in the previous lemma. As stated above, the formula $\beta_{\mathbf{i}}^{\mathcal{D}}$ is the following:

$$\chi \wedge \bigwedge_{i\in[k],\ell\in[m]} \overline{\delta}_i(X_\ell^{(i)}) \wedge \bigwedge_{j\in[n]} \delta_{\mathbf{i}[j]}[y_j/x] \wedge \beta_{\mathbf{i}}^{\star\mathcal{D}}.$$

Its conjuncts express respectively that $T := \widehat{\mathcal{D}}(S,\gamma)$ is defined, that $\eta^{[k]}$ is an $\{X_1,\ldots,X_m\}$-assignment in T, that $\mu_{\mathbf{i}}$ is a $\{y_1,\ldots,y_n\}$-assignment in T and, by Lemma 7.11, that $(T,\eta^{[k]} \cup \mu_{\mathbf{i}}) \models \beta$. Hence, $\beta_{\mathbf{i}}^{\mathcal{D}}$ satisfies the required property.

We now consider the case where $\mathcal{R}'$ may have constant symbols. We apply the above construction to the definition scheme $\mathcal{D}_*$ and the formula β_* constructed from β in Lemma 7.5. We obtain that:

$(S,\gamma \cup \eta \cup \mu) \models (\beta_*)_{\mathbf{i}}^{\mathcal{D}_*}$ if and only if

$\quad \widehat{\mathcal{D}}_*(S,\gamma)$ is defined,

$\quad \eta^{[k]} \cup \mu_{\mathbf{i}}$ is an $(\mathcal{X} \cup \mathcal{Y})$-assignment in $\widehat{\mathcal{D}}_*(S,\gamma)$ and

$\quad (\widehat{\mathcal{D}}_*(S,\gamma), \eta^{[k]} \cup \mu_{\mathbf{i}}) \models \beta_*.$

With Definition 7.6 and Lemma 7.5 we get the desired result by taking $\beta_{\mathbf{i}}^{\mathcal{D}} := (\beta_*)_{\mathbf{i}}^{\mathcal{D}_*}$ since we have $\widehat{\mathcal{D}}_*(S,\gamma) = \widehat{\mathcal{D}}(S,\gamma)_*$. Note that if $\mathcal{R}'_0 = \emptyset$, we get the same formulas as in the first case because $\mathcal{D}_* = \mathcal{D}$ and $\beta_* = \beta$. $\blacksquare$

We recall that C_0MS designates the same formulas as MS, hence the following corollary also concerns MS-transductions.

Corollary 7.12 Let τ be a C_rMS-transduction of type $\mathcal{R} \to \mathcal{R}'$. For every C_rMS-definable subclass $\mathcal{C}$ of $STR^c(\mathcal{R}')$, the subclass $\tau^{-1}(\mathcal{C})$ of $STR^c(\mathcal{R})$ is C_rMS-definable.

Proof: Let τ be specified by a C_rMS-definition scheme $\mathcal{D}$ with set of parameters $\mathcal{W} = \{Y_1,\ldots,Y_p\}$. (We need not distinguish the case where $\mathcal{R}'$ has constant symbols.) Let $\mathcal{C} := MOD(\beta) = \{T \in STR^c(\mathcal{R}') \mid T \models \beta\}$ for some sentence β in $C_rMS(\mathcal{R}',\emptyset)$. Then, $S \in STR^c(\mathcal{R})$ belongs to $\tau^{-1}(\mathcal{C})$ if and only if it has an image in $MOD(\beta)$ under $\widehat{\mathcal{D}}$, hence, by Theorem 7.10, if and only if it belongs to the class $MOD(\exists Y_1,\ldots,Y_p.\beta^{\mathcal{D}})$. This shows that $\tau^{-1}(\mathcal{C})$ is C_rMS-definable. $\blacksquare$

Remark 7.13 (1) If $\mathcal{D}$ is parameterless and β is a sentence, then we have

$$\widehat{\mathcal{D}}(S) \models \beta \text{ if and only if } S \models \beta^{\mathcal{D}}.$$

A similar property holds for certain transductions that are not monadic second-order transductions, in particular, for the *unfolding* of a graph into a possibly infinite tree (see Section 7.7). However, for monadic second-order transductions, we have more than the above equivalence. To simplify the discussion, we assume that $\mathcal{D}$ is

k-copying, parameterless and that $n = 0$ (the considered formula β has no first-order variables). By Theorem 7.10, we have a bijection between the set $sat(\widehat{\mathcal{D}}(S), \beta, \overline{\mathcal{X}})$ and the set $sat(S, \beta^{\mathcal{D}}, \overline{\mathcal{X}^{(k)}})$, where $\overline{\mathcal{X}}$ is a tuple that enumerates $\mathcal{X}$ and similarly for $\mathcal{X}^{(k)}$. Such a bijection is impossible for the unfolding, by a simple cardinality argument.

(2) Let $\mathcal{D}$ be a definition scheme of type $\mathcal{R} \to \mathcal{R}'_*$. Let α be the first-order sentence that characterizes the class $STR^c_*(\mathcal{R}'_*)$ and $\alpha^{\mathcal{D}}$ be its backwards translation. Then, if we replace the precondition of $\mathcal{D}$ by $\alpha^{\mathcal{D}}$, we obtain a definition scheme $\mathcal{D}'$ such that $\widehat{\mathcal{D}}' = \widehat{\mathcal{D}} \cap (STR^c(\mathcal{R}) \times STR^c_*(\mathcal{R}'_*))$, hence a definition scheme of type $\mathcal{R} \to \mathcal{R}'$. It is equivalent to $\mathcal{D}$ for producing structures that encode $\mathcal{R}'$-structures correctly.

(3) By Corollary 7.12, the image of a C_rMS-definable set of structures under an invertible C_rMS-transduction is C_rMS-definable. This is not true in general, even for an injective transduction (cf. Remark 7.23). $\qquad\qquad\square$

7.1.5 Constructions of monadic second-order transductions

The following theorem is, with Corollary 7.12, the second main consequence of the Backwards Translation Theorem.[10]

Theorem 7.14 (Composition of transductions)
(1) The composition $\tau \cdot \tau'$ of two C_rMS-transductions $\tau : \mathcal{R} \to \mathcal{R}'$ and $\tau' : \mathcal{R}' \to \mathcal{R}''$ is a C_rMS-transduction. If τ and τ' are both parameterless and/or both noncopying, then so is $\tau \cdot \tau'$. If τ is k-copying and is defined with q parameters, and if τ' is k'-copying and is defined with q' parameters, then $\tau \cdot \tau'$ is (kk')-copying and is defined with $q + kq'$ parameters.
(2) If τ and τ' as above are concrete transductions and at least one of them is noncopying, then $\tau \cdot \tau'$ is also a concrete C_rMS-transduction.
(3) If τ and τ' as above are both domain-preserving or both domain-extending, then so is $\tau \cdot \tau'$. If one of them is domain-preserving and the other is domain-extending, then $\tau \cdot \tau'$ is domain-extending.

Proof: We first consider the special case where $\mathcal{R}'$ and $\mathcal{R}''$ have no constant symbols. Since the proof is a bit technical, we consider four cases.

Case 1: τ and τ' are both noncopying.

We let τ and τ' be the concrete transductions defined respectively by the definition schemes $\mathcal{D} = \langle \chi, \delta, (\theta_R)_{R \in \mathcal{R}'} \rangle$ with set of parameters $\mathcal{W} := \{Y_1, \ldots, Y_q\}$ and $\mathcal{D}' = \langle \chi', \delta', (\theta'_R)_{R \in \mathcal{R}''} \rangle$ with set of parameters $\mathcal{W}' := \{Z_1, \ldots, Z_{q'}\}$ disjoint with $\mathcal{W}$. We build as follows a definition scheme $\mathcal{D}'' = \langle \chi'', \delta'', (\theta''_R)_{R \in \mathcal{R}''} \rangle$:

---

[10] We do not call it the Composition Theorem because this terminology is widely used for another result. See the introduction of Section 5.3.

(i) Its set of parameters is $\mathcal{W} \cup \mathcal{W}'$.

(ii) The precondition χ'' is $(\chi')^{\mathcal{D}}$, with free variables in $\mathcal{W} \cup \mathcal{W}'$. By Theorem 7.10, χ'' expresses, for a $(\mathcal{W} \cup \mathcal{W}')$-assignment γ in $S \in STR^c(\mathcal{R})$, that $T := \widehat{\mathcal{D}}(S, \gamma \restriction \mathcal{W})$ is defined, $\gamma \restriction \mathcal{W}'$ is a $\mathcal{W}'$-assignment in T, and $(T, \gamma \restriction \mathcal{W}') \models \chi'$, which means that $U := \widehat{\mathcal{D}}'(T, \gamma \restriction \mathcal{W}')$ is defined.

(iii) We have $D_U \subseteq D_T \subseteq D_S$ since τ and τ' are noncopying. We define the domain formula δ'' as $(\delta')^{\mathcal{D}}$. Its free variables are in $\{x\} \cup \mathcal{W} \cup \mathcal{W}'$. Again by Theorem 7.10, for every $(\mathcal{W} \cup \mathcal{W}')$-assignment γ satisfying χ'' and every $a \in D_S$, we have $(S, \gamma) \models \delta''(a)$ if and only if $a \in D_T$ and $(T, \gamma \restriction \mathcal{W}') \models \delta'(a)$, i.e., if and only if $a \in D_U$. Hence δ'' characterizes D_U as a subset of D_S (that depends on γ assumed to satisfy χ'').

(iv) It remains to define the relation formulas θ_R'' for $R \in \mathcal{R}''$. Each relation R_U for $R \in \mathcal{R}''$ is defined by θ_R' in T, hence by $(\theta_R')^{\mathcal{D}}$ in S (using an argument similar to the one in (iii)). Hence we need only take θ_R'' to be $(\theta_R')^{\mathcal{D}}$ for $R \in \mathcal{R}'$. We have θ_R' in $C_r MS(\mathcal{R}', \mathcal{W}' \cup \{x_1, \ldots, x_{\rho(R)}\})$ and θ_R'' in $C_r MS(\mathcal{R}, \mathcal{W} \cup \mathcal{W}' \cup \{x_1, \ldots, x_{\rho(R)}\})$.

It is clear from the construction of $\mathcal{D}''$ that it defines the concrete transduction $\tau \cdot \tau'$. This proves Assertions (1) and (2) for this case. To prove Assertion (3), we observe that instead of defining δ'' as $(\delta')^{\mathcal{D}}$, we can as well define it as the weaker formula $\delta \wedge (\delta')^{\star \mathcal{D}}$ and use Lemma 7.11 instead of Theorem 7.10. Thus, if both δ and δ' are *True*, then δ'' is (equivalent to) *True*. We note that, similarly, θ_R'' can as well be defined as $(\theta_R')^{\star \mathcal{D}}$.

Case 2: τ is noncopying and τ' is k'-copying ($k' > 1$). The construction is similar, again for concrete transductions: for $\mathcal{D}' = \langle \chi', (\delta_i')_{i \in [k']}, (\theta_w)_{w \in \mathcal{R}'' \circledast [k']} \rangle$, we let

$$\mathcal{D}'' := \langle \chi'', (\delta_i'')_{i \in [k']}, (\theta_w'')_{w \in \mathcal{R}'' \circledast [k']} \rangle,$$

where χ'' is $(\chi')^{\mathcal{D}}$, δ_i'' is $(\delta_i')^{\mathcal{D}}$ for each i (or it is $\delta_i \wedge (\delta_i')^{\star \mathcal{D}}$), and θ_w'' is $(\theta_w')^{\mathcal{D}}$ for each $w \in \mathcal{R}'' \circledast [k']$. As in Case 1, $\mathcal{D}''$ defines the concrete transduction $\tau \cdot \tau'$.

Case 3: τ is k-copying and τ' is k'-copying with $k, k' > 1$. We only prove Assertion (1). If $T := \widehat{\mathcal{D}}(S, \gamma)$ and $U := \widehat{\mathcal{D}}'(T, \gamma')$, we have $D_T \subseteq D_S \times [k]$ and $D_U \subseteq D_T \times [k'] \subseteq (D_S \times [k]) \times [k']$. We will prove that $\tau \cdot \tau'$ is a monadic second-order transduction. It is not a concrete one because D_U is not a subset of any $D_S \times [m]$. However, since we only want to construct structures isomorphic to those of $(\tau \cdot \tau')(S)$, we will consider $(D_S \times [k]) \times [k']$ as equal to $D_S \times ([k] \times [k'])$.

The set of parameters for $\mathcal{D}''$ will be $\mathcal{W}'' := \mathcal{W} \cup \mathcal{W}'^{(k)} = \{Y_1, \ldots, Y_q\} \cup \{Z_j^{(i)} \mid 1 \leq j \leq q', 1 \leq i \leq k\}$. For each $\mathcal{W}$-assignment γ in S, a $\mathcal{W}'$-assignment γ' in $T := \widehat{\mathcal{D}}(S, \gamma)$ is described in terms of a $\mathcal{W}'^{(k)}$-assignment η in S with $\gamma' = \eta^{[k]}$. We let the precondition χ'' be $(\chi')^{\mathcal{D}}$. Its free variables are in $\mathcal{W}''$. Then, by Theorem 7.10, $(S, \gamma \cup \eta) \models \chi''$ if and only if $U := \widehat{\mathcal{D}}'(T, \gamma')$ is defined.

The definition scheme $\mathcal{D}''$ will construct from S a structure with domain included in $D_S \times ([k] \times [k'])$, easily replaceable by $D_S \times [kk']$. Hence, we must define the domain formulas $\delta''_{i,j}$ for $(i,j) \in [k] \times [k']$ such that for γ and η satisfying χ'', we have, for each $a \in D_S$,

$$(a,(i,j)) \in D_U \text{ if and only if } (S, \gamma \cup \eta) \models \delta''_{i,j}(a).$$

We define $\delta''_{i,j}$ as $(\delta'_j)^{\mathcal{D}}_{(i)}$ (or as $\delta_i \wedge (\delta'_j)^{\star\mathcal{D}}_{(i)}$ for the domain-extending case). Assuming that $\gamma \cup \eta$ satisfies χ'', we have (again by Theorem 7.10, or Lemma 7.11), for $a \in D_S$,

$$(S, \gamma \cup \eta) \models (\delta'_j)^{\mathcal{D}}_{(i)}(a) \text{ if and only if } (a,i) \in D_T \text{ and } (T, \gamma') \models \delta'_j(a,i),$$

i.e., if and only if $((a,i),j) = (a,(i,j)) \in D_U$.

Finally for each $w = (R, (i_1,j_1), \ldots, (i_p,j_p))$ where $R \in \mathcal{R}''$ and $p = \rho(R)$, we define the relation formula θ''_w as $(\theta'_{R,j_1,\ldots,j_p})^{\mathcal{D}}_{(i_1,\ldots,i_p)}$. For $a_1, \ldots, a_p \in D_S$:

$$(S, \gamma \cup \eta) \models (\theta'_{R,j_1,\ldots,j_p})^{\mathcal{D}}_{(i_1,\ldots,i_p)}(a_1, \ldots, a_p) \text{ if and only if}$$

$$(a_q,i_q) \in D_T \text{ for each } q \in [p] \text{ and } (T, \gamma') \models \theta'_{R,j_1,\ldots,j_p}((a_1,i_1), \ldots, (a_p,i_p)),$$

i.e., if and only if $((a_1,(i_1,j_1)), \ldots, (a_p,(i_p,j_p))) \in R_U$.

It is clear from the construction of $\mathcal{D}''$ that it defines the abstract transduction $\tau \cdot \tau'$. We observe, however, that it "almost" defines the concrete transduction $\tau \cdot \tau'$, in the following sense. For fixed integers $k,k' > 1$ let θ be the binary relation $\{(((d,i),j),(d,k(i-1)+j)) \mid i \in [k], j \in [k']\}$, where d is any element of a set D. This relation defines a bijection from $(D \times [k]) \times [k']$ to $D \times [kk']$. For a concrete structure U with $D_U \subseteq (D \times [k]) \times [k']$ for some set D, we define $\overline{\theta}(U)$ to be the unique concrete structure V with $D_V \subseteq D \times [kk']$ such that $\theta \restriction D_U$ is an isomorphism from U to V. The definition scheme $\mathcal{D}''$ constructed in the proof of Case 3 can easily be modified into one that defines the concrete transduction $\tau \cdot \tau' \cdot \overline{\theta}$ (where τ and τ' are concrete k-copying and, respectively, k'-copying monadic second-order transductions).

Case 4: τ is k-copying with $k > 1$ and τ' is noncopying. If $T := \widehat{\mathcal{D}}(S, \gamma)$ and $U := \widehat{\mathcal{D}'}(T, \gamma')$, we have $D_U \subseteq D_T \subseteq D_S \times [k]$. The construction given for Case 3 is easily adapted. As in Cases 1 and 2, $\mathcal{D}''$ defines the concrete transduction $\tau \cdot \tau'$.

We finally consider the general case where $\mathcal{R}'$ and $\mathcal{R}''$ may have constant symbols. It is actually an immediate consequence of the special case and Lemma 7.7 because $(\tau \cdot \tau')_* = \tau_* \cdot \tau'_{**}$. (We also have, with more symmetry, $(\tau \cdot \tau')_{**} = \tau_{**} \cdot \tau'_{**}$ which gives the same proof.) ∎

Remark 7.15 (QF operations and QF-transductions) (1) We have observed in Example 7.3(2) and after Definition 7.6 that the quantifier-free operations (cf. Section 5.3.2) are total functions and, actually, QF-transductions of a particular form: they are parameterless, noncopying and have a precondition equal (or equivalent) to *True*. Proposition 5.49 (Section 5.3.2) has shown that the composition of two quantifier-free operations is a quantifier-free operation. It can be considered as a special case of Theorem 7.14.

(2) More generally than QF operations, we have defined QF- and FO-transductions as MS-transductions with definition schemes that use only QF and FO formulas respectively. Results like Theorems 7.10 and 7.14 and their corollaries have particular special cases for FO-transductions and FO-definable sets of structures. However, we will not detail the corresponding statements because, for the questions studied in this book, knowing that a class of structures is FO-definable rather than MS-definable brings no additional results. The fact that a transduction is an FO-transduction rather than a CMS-transduction indicates that it is "not too complicated," and we will make such observations only by passing. However, QF-transductions (or rather QF operations) have particularly important properties that we have used to prove the Recognizability Theorem in Chapter 5, and that MS-transductions do not share. Example 5.65 has shown that this theorem can fail to hold if only one unary operation that is an FO-transduction is added to the signature F^{VR}. $\qquad\square$

We continue by presenting other techniques to construct monadic second-order transductions.

Theorem 7.16 (Restriction Theorem) Let τ be a (concrete) C_rMS-transduction of type $\mathcal{R} \to \mathcal{R}'$, let $\alpha \in C_r\text{MS}(\mathcal{R},\emptyset)$ and $\beta \in C_r\text{MS}(\mathcal{R}',\emptyset)$. Then the relation $\tau \cap (\text{MOD}(\alpha) \times \text{MOD}(\beta))$ is a (concrete) C_rMS-transduction. If τ is k-copying and is defined with q parameters, then the same holds for $\tau \cap (\text{MOD}(\alpha) \times \text{MOD}(\beta))$.

Proof: We first consider the case where $\mathcal{R}'$ has no constant symbols. We let $\mathcal{D} = \langle \chi, (\delta_i)_{i \in [k]}, (\theta_w)_{w \in \mathcal{R}' \circledast [k]} \rangle$ be a definition scheme defining a concrete transduction τ. We construct $\mathcal{D}'$ from $\mathcal{D}$ by replacing χ by $\alpha \wedge \beta^{\mathcal{D}}$. (We recall that $\beta^{\mathcal{D}}$ is $\chi \wedge \beta^{\star\mathcal{D}}$.) It is clear from Theorem 7.10 that $\mathcal{D}'$ defines the concrete transduction $\tau \cap (\text{MOD}(\alpha) \times \text{MOD}(\beta))$.

In the case where $\mathcal{R}'$ has constant symbols, we observe that

$$(\tau \cap (\text{MOD}(\alpha) \times \text{MOD}(\beta)))_* = \tau_* \cap (\text{MOD}(\alpha) \times \text{MOD}(\beta_*)),$$

hence the result follows from the first case. $\qquad\blacksquare$

We denote by $\Diamond$ the structure in $STR^c(\emptyset)$ consisting of the single element $*$.

Proposition 7.17 Let S be an $\mathcal{R}$-structure having k elements. There exists a parameterless k-copying QF-transduction τ of type $\emptyset \to \mathcal{R}$ such that $\tau(\Diamond)$ consists of a single structure isomorphic to S.

Proof: The construction of a definition scheme for τ is easy. The domain of $\tau(\Diamond)$ will be $\{*\} \times [k]$ and all formulas of the definition scheme will be *True* or *False*. $\qquad\blacksquare$

We now consider the union of two transductions of the same type.

Proposition 7.18 If τ and τ' are (concrete) C_rMS-transductions of type $\mathcal{R} \rightarrow \mathcal{R}'$, then $\tau \cup \tau'$ is a (concrete) C_rMS-transduction.

Proof: Since $(\tau \cup \tau')_* = \tau_* \cup \tau'_*$, it suffices to consider the case where $\mathcal{R}'$ has no constant symbols.

Let $\mathcal{D}$ and $\mathcal{D}'$ be definition schemes for the concrete transductions τ and τ' that are respectively k- and k'-copying. We first assume that $k < k'$. We define from $\mathcal{D} = \langle \chi, \delta_1, \ldots, \delta_k, (\theta_w)_{w \in \mathcal{R} \circledast [k]} \rangle$ the k'-copying definition scheme

$$\mathcal{E} := \langle \chi, \delta_1, \ldots, \delta_k, \mathit{False}, \ldots, \mathit{False}, (\theta_{\mathcal{E}w})_{w \in \mathcal{R} \circledast [k']} \rangle.$$

Its domain formula δ_j is *False* if $k < j \leq k'$. The relation formula $\theta_{\mathcal{E}w}$ is θ_w if $w \in \mathcal{R} \circledast [k]$ and *False* if $w \in \mathcal{R} \circledast [k'] - \mathcal{R}' \circledast [k]$ (but it can be any formula by Definition 7.2, since the formulas δ_j are equal to *False* for $j > k$). It is clear that the concrete transductions defined by $\mathcal{E}$ and $\mathcal{D}$ are equal.

Hence we assume that $\mathcal{D}$ and $\mathcal{D}'$ are both k-copying, with sets of parameters $\mathcal{W}$ and $\mathcal{W}'$ respectively. (The sets $\mathcal{W}$ and $\mathcal{W}'$ need not be disjoint.) We construct a k-copying definition scheme $\mathcal{D}''$ by taking $\mathcal{W} \cup \mathcal{W}' \cup \{Z\}$, where $Z \notin \mathcal{W} \cup \mathcal{W}'$, as set of parameters and by defining $\chi_{\mathcal{D}''}$ as $(Z = \emptyset \wedge \chi_{\mathcal{D}}) \vee (Z \neq \emptyset \wedge \chi_{\mathcal{D}'})$ and similarly for the domain and relation formulas of $\mathcal{D}''$, so that for every $\mathcal{W}''$-assignment γ in a nonempty structure $S \in STR^c(\mathcal{R})$, we have:

$$\widehat{\mathcal{D}''}(S, \gamma) = \begin{cases} \widehat{\mathcal{D}}(S, \gamma \restriction \mathcal{W}) & \text{if } \gamma(Z) = \emptyset, \\ \widehat{\mathcal{D}'}(S, \gamma \restriction \mathcal{W}') & \text{if } \gamma(Z) \neq \emptyset. \end{cases}$$

There is a difficulty if $S = \emptyset_{\mathcal{R}}$, however, because the condition $\gamma(Z) \neq \emptyset$ cannot be satisfied. It follows that the above definition is incorrect in the only case where $\widehat{\mathcal{D}}(\emptyset_{\mathcal{R}}) = \emptyset$ and $\widehat{\mathcal{D}'}(\emptyset_{\mathcal{R}}) = \{\emptyset_{\mathcal{R}'}\}$. In this case, we just exchange the conditions $Z = \emptyset$ and $Z \neq \emptyset$ in the definition of the formulas of $\widehat{\mathcal{D}''}$. With this modification, $\mathcal{D}''$ defines the concrete transduction $\tau \cup \tau'$ as wanted.

The statement for transductions follows immediately. ∎

These two propositions entail that every finite set of $\mathcal{R}$-structures is the image of $\diamond$ under a monadic second-order transduction.

7.1.6 Some particular monadic second-order transductions

We will consider "labeling" transductions and "copying" transductions.

Labeling transductions

We know from Example 5.2(2) that vertex labels of a graph (including source and port labels) can be handled as unary relations in the structure that represents the graph. In general, we can view unary relations of a structure as labels attached to the elements of the structure. We will consider monadic second-order transductions that manipulate such labels. Since unary relations are sets, there is also a close relationship between labels and set variables, in particular parameters.

Definition 7.19 (Expansions by unary relations) Let $\mathcal{R}$ be a relational signature and $\mathcal{U}$ be a set of unary relation symbols in $\mathcal{R}$. For every $S \in STR^c(\mathcal{R})$, we let $fg_{\mathcal{U}}(S)$ be the corresponding reduct (cf. Example 5.44) in $STR^c(\mathcal{R} - \mathcal{U})$: it is the structure obtained by deleting (or "forgetting") from S the relations that interpret the symbols in $\mathcal{U}$. The mapping $fg_{\mathcal{U}}$ is a parameterless, noncopying and domain-preserving QF-transduction.

We let $exp_{\mathcal{U}}$ be its inverse, the transduction of type $(\mathcal{R} - \mathcal{U}) \rightarrow \mathcal{R}$ that associates with S the set of structures S' such that $fg_{\mathcal{U}}(S') = S$. These structures S' are called the *expansions* of S by the set $\mathcal{U}$ of unary relations. The transduction $exp_{\mathcal{U}}$ is a noncopying and domain-preserving QF-transduction with $\mathcal{U}$ as set of parameters: each $U \in \mathcal{U}$ is made into a parameter and the interpretation U_T of U in the output structure T is defined as $\gamma(U)$, where γ is the $\mathcal{U}$-assignment in the input structure from which T is defined, i.e., the relation formula θ_U is $x_1 \in U$. We will denote by $\mathcal{U}_{\text{par}}$ the set $\mathcal{U}$ considered as a set of parameters to distinguish it from the initial set of unary relation symbols included in $\mathcal{R}$.

Definition 7.20 (Monadic second-order relabelings) Let $\mathcal{R}$ and $\mathcal{R}'$ be relational signatures such that $\mathcal{R}_i = \mathcal{R}'_i$ for $i \neq 1$, i.e., that may differ only by their unary relation symbols. A concrete monadic second-order transduction τ of type $\mathcal{R} \rightarrow \mathcal{R}'$ is called a *monadic second-order relabeling* if it is domain-preserving and, for every T in $\tau(S)$, we have $a_T = a_S$ if $a \in \mathcal{R}_0 = \mathcal{R}'_0$ and $R_T = R_S$ if R has arity at least 2 (which implies $R \in \mathcal{R} \cap \mathcal{R}'$). Such a transduction can only modify, add and/or delete unary relations.

A transduction τ of type $\mathcal{R} \rightarrow \mathcal{R}'$ is a monadic second-order relabeling if and only if τ_{**} has a domain-preserving definition scheme whose relation formula θ_R is $R(x_1, \ldots, x_{\rho(R)})$ if $R \in \mathcal{R}'_* - \mathcal{R}'_1$. In such a definition scheme, there is no constraint on the precondition and on the relation formulas θ_R if R is unary and not of the form lab_a for any $a \in \mathcal{R}'_0$.

It is clear from the definitions that the composition of two monadic second-order relabelings is a monadic second-order relabeling. It is also clear that the composition of definition schemes used in the proof of Theorem 7.14 (Case 1) preserves the above syntactic property (provided θ''_R is defined as $(\theta'_R)^{\star D}$ rather than $(\theta'_R)^{D}$).

The transductions $fg_{\mathcal{U}}$ and $exp_{\mathcal{U}}$ of Definition 7.19 are monadic second-order relabelings.

We now consider two particular monadic second-order relabelings of terms. Let F and F' be two finite functional signatures. In the following definition, the associated relational signatures $\mathcal{R}_F$ and $\mathcal{R}_{F'}$ use the binary relations son_i (as opposed to the binary relation son and the unary relations br_i). Their only unary relations are thus the relations lab_f for f in F or in F'. A relation $\tau \subseteq T(F) \times T(F')$ is a *monadic second-order relabeling* if the corresponding relation $\{(\lfloor t \rfloor, \lfloor t' \rfloor) \mid (t, t') \in \tau\}$ is a monadic second-order relabeling of type $\mathcal{R}_F \to \mathcal{R}_{F'}$. Hence, if $(t,t') \in \tau$, the term t' is obtained from t by the replacement of function symbols from F by function symbols from F' of the same arity, in a way that is not uniform as in an alphabetic relabeling (cf. the paragraph before Proposition 2.126, and cf. Section 6.3.1), but is specified by monadic second-order formulas.

Here is our second particular monadic second-order relabeling of terms. Again, F is a finite functional signature. Let $\mathcal{U} := \{U_1, \ldots, U_n\}$ be a set of unary relation symbols disjoint from $\mathcal{R}_F$, ordered as indicated. If $t \in T(F)$ and $(A_1, \ldots, A_n)$ is an n-tuple of subsets of N_t, we denote by $\lfloor t \rfloor + (A_1, \ldots, A_n)$ the expansion of $\lfloor t \rfloor$ into the $(\mathcal{R}_F \cup \mathcal{U})$-structure such that $A_1, \ldots, A_n$ are the interpretations of $U_1, \ldots, U_n$ respectively (cf. Lemma 6.52 in Section 6.4.1 for the notation $S + \gamma$). We recall from Definition 6.24(c) in Section 6.3.3 that the pair of a term t and an n-tuple $(A_1, \ldots, A_n)$ as above is encoded by a term $t * (A_1, \ldots, A_n)$ in $T(F^{(n)})$, where u is an occurrence of (f,w) belonging to $F^{(n)}$ (w is a sequence of n Booleans) if and only if it is an occurrence of f in t and, for each $i = 1, \ldots, n$, $w[i] = 1$ if and only if $u \in A_i$. The mapping : $T(F^{(n)}) \to STR^c(\mathcal{R}_F \cup \mathcal{U})$ that maps $\lfloor t * (A_1, \ldots, A_n) \rfloor$ to $\lfloor t \rfloor + (A_1, \ldots, A_n)$ is a monadic second-order relabeling. It defines a bijection between $\lfloor T(F^{(n)}) \rfloor$ and $exp_{\mathcal{U}}(\lfloor T(F) \rfloor)$ that we denote by $\mu_{F,n}$. Its inverse is also a monadic second-order relabeling, which holds for all monadic second-order relabelings as will be shown in Proposition 7.22(1).

We will also use the monadic second-order relabeling $\pi_{F,n} := \mu_{F,n} \cdot fg_{\mathcal{U}}$. It represents, in terms of relational structures, the alphabetic relabeling of terms : $T(F^{(n)}) \to T(F)$ that deletes the Booleans from the symbols (f,w) of $F^{(n)}$.

Definition 7.21 (Labelings by bounded theories) A bounded theory (Definition 5.59 in Section 5.3.6) is a subset Φ of a fixed finite set $\mathcal{L}$ of formulas. We will make each such set into a unique symbol, to be used as a label.

Let $r, h \in \mathcal{N}$ and let $\mathcal{R}$ be a relational signature. We recall from Proposition 5.93 (Section 5.6) that every formula in $C_r MS^h(\mathcal{R}, \{x\})$ is equivalent to a formula belonging to the finite set $\mathcal{L} := \widehat{C_r MS^h}(\mathcal{R}, \{x\})$ of formulas that are in a certain "normal form." For $S \in STR^c(\mathcal{R})$ and $a \in D_S$, we define

$$\Phi(a) := \{\varphi \in \mathcal{L} \mid S \models \varphi(a)\},$$

i.e., $\Phi(a) = Th(S, \mathcal{R}, h, r, a)$ as defined in Definition 5.59. The set $\mathcal{P}(\mathcal{L})$ in which $\Phi(a)$ takes its values is finite and depends only on r, h and $\mathcal{R}$. For each subset Φ of $\mathcal{L}$, we let U_Φ be a new unary relation symbol. These symbols form a finite set $\mathcal{U}$. For each $S \in STR^c(\mathcal{R})$, we let $\widehat{S}$ be the unique expansion of S into an $(\mathcal{R} \cup \mathcal{U})$-structure satisfying the following sentence Ψ:

$$\bigwedge_{\Phi \subseteq \mathcal{L}} \forall x \Big(U_\Phi(x) \Leftrightarrow \bigwedge_{\varphi \in \Phi} \varphi \wedge \bigwedge_{\varphi \in \mathcal{L} - \Phi} \neg\varphi \Big).$$

It expresses that $U_{\Phi_{\widehat{S}}}$ consists of all elements a of D_S such that $\Phi(a) = \Phi$. The mapping $S \mapsto \widehat{S}$ is a $C_r MS^h$-transduction : $STR(\mathcal{R}) \to STR(\mathcal{R} \cup \mathcal{U})$ where each unary relation

U_Φ of $\mathcal{U}$ is defined by a formula of quantifier-height at most h. It is a parameterless (hence functional) monadic second-order relabeling.

We now prove that the class of monadic second-order relabelings is closed under inverse, which is not true of all monadic second-order transductions. In the following statement, F and F' denote as usual finite functional signatures.

Proposition 7.22

(1) Every monadic second-order relabeling is invertible; its inverse is a monadic second-order relabeling.
(2) The image of a CMS-definable class of structures under a monadic second-order relabeling is CMS-definable.
(3) The image of a regular language over F under a monadic second-order relabeling from $T(F)$ to $T(F')$ is regular over F'.

Proof: (1) We first give the idea. If a transduction deletes or modifies unary relations, its inverse can be expressed as a monadic second-order transduction using parameters that describe the (possible) initial values of these relations.

Let τ be a monadic second-order relabeling of type $\mathcal{R} \to \mathcal{R}'$ and let τ_{**} be defined by $\mathcal{D} := \langle \chi, \mathit{True}, (\theta_R)_{R\in\mathcal{R}'_*} \rangle$, where θ_R is $R(x_1,\ldots,x_{\rho(R)})$ for every R in $\mathcal{R}'_* - \mathcal{R}'_1$. Let W be the set of parameters of $\mathcal{D}$. For defining the inverse of τ_{**} (this will suffice because $(\tau_{**})^{-1} = (\tau^{-1})_{**}$), we take $W' := W \cup (\mathcal{R}_1)_{\mathrm{par}}$ as set of parameters, where $(\mathcal{R}_1)_{\mathrm{par}}$ is the set $\mathcal{R}_1$ considered as a set of parameters (cf. Definition 7.19).

The definition scheme for $(\tau_{**})^{-1}$ is then $\mathcal{D}' := \langle \chi', \mathit{True}, (\theta'_R)_{R\in\mathcal{R}_*} \rangle$, where θ'_R is $R(x_1,\ldots,x_{\rho(R)})$ for each R in $\mathcal{R}_*$ and the precondition χ' is defined as

$$\chi \wedge \bigwedge_{R\in\mathcal{R}'_1} \forall x_1 (R(x_1) \Leftrightarrow \theta_R).$$

Note that in the formulas of $\mathcal{D}'$, every atomic formula $U(x)$ with $U \in \mathcal{R}_1$ can be replaced by $x \in U$ by our syntactical convention for set variables (which does not hold for unary relation symbols). In particular, θ'_U then becomes $x_1 \in U$ for each $U \in \mathcal{R}_1$, which means that the unary relation symbol U is interpreted as the value of the parameter U. The precondition χ' ensures that, in a given $\mathcal{R}_*$-structure T, if the values of the parameters are taken as the relations U_S for $U \in \mathcal{R}_1$ in an $\mathcal{R}_*$-structure S where the other relations are as in T, then $(S,T) \in \widehat{\mathcal{D}}$. The other formulas of $\mathcal{D}'$ ensure that S and T have the same domain and the same relations defined by the symbols not in $\mathcal{R}_1$ or $\mathcal{R}'_1$.

(2) Follows from (1) and the Backwards Translation Theorem (Corollary 7.12).

(3) Follows from (2) and the equivalence between regularity and monadic second-order definability for languages over a finite functional signature (Theorem 5.82). ∎

Remark 7.23 We have already noted in Definition 7.2 and Remark 7.13(3) that the image of a CMS-definable class under a monadic second-order transduction is not always CMS-definable, hence that the inverse of a monadic second-order transduction is not always a monadic second-order transduction. We give three examples of such cases for monadic second-order transductions that only modify a single relation of arity 2.

First, we consider the mapping on p-graphs of type $\{a, b\}$ that deletes the edge relation. It transforms the structure $\lfloor (ab)^n \rfloor$ for $n \geq 1$ that represents the word $(ab)^n$ into the structure representing the edgeless p-graph $\mathbf{a} \oplus \cdots \oplus \mathbf{a} \oplus \mathbf{b} \oplus \cdots \oplus \mathbf{b}$ with an equal number of occurrences of $\mathbf{a}$ and $\mathbf{b}$. These structures do not form an MS-definable set by Proposition 5.13(1). The same proof shows that they do not form a CMS-definable set. However, the language $(ab)^+$ is regular, hence MS-definable by Theorem 5.15.

As another example, we consider the mapping *und* on graphs that forgets edge directions. Hence it redefines $edg(x_1, x_2)$ by $edg(x_1, x_2) \vee edg(x_2, x_1)$. We let $L :=$ $\{G_n \mid n \geq 1\}$, where G_n is the graph with vertices $u_1, \ldots, u_n, v_1, \ldots, v_n$ and edges $u_i \to v_i$ and $v_j \to u_i$ for all $i, j \in [n]$ with $j \neq i$. The set of graphs L is FO-definable, but $und(L)$ is the set $\{K_{n,n} \mid n \geq 1\}$, which is not MS-definable by the proof of Proposition 5.13(2) and also not CMS-definable by an easy extension of that proof.

As last example, we take the monadic second-order transduction from words to words that transforms a word $a_1 a_2 a_3 \cdots a_{2n}$ (where $a_1, a_2, \ldots, a_{2n}$ are letters) into the word $a_1 a_3 \cdots a_{2n-1} a_2 a_4 \cdots a_{2n}$. It is a parameterless domain-preserving transduction that transforms the regular, hence MS-definable, language $(ab)^+$ into the nonregular language $\{a^n b^n \mid n \geq 1\}$ which is not CMS-definable. This example shows that adding conditions of connectivity and bounded degree does not change the situation. $\qquad\square$

The next two propositions show that monadic second-order transductions can be factorized as compositions of transductions of particular types. We first show that every CMS-transduction can be decomposed into a (very simple) monadic second-order relabeling and a parameterless CMS-transduction.

If $\mathcal{W}$ is a set of set variables, we will denote by $\mathcal{W}_{\mathrm{rel}}$ the same set viewed as a set of unary relation symbols. The monadic second-order relabeling $\pi_{F,n} : T(F^{(n)}) \to T(F)$ associated with a finite functional signature F has been defined at the end of Definition 7.20.

Proposition 7.24

(1) Every concrete C_rMS-transduction τ of type $\mathcal{R} \to \mathcal{R}'$ with set of parameters $\mathcal{W}$ can be expressed as $exp_{\mathcal{W}_{\mathrm{rel}}} \cdot \tau'$, where τ' is a concrete parameterless C_rMS-transduction of type $\mathcal{R} \cup \mathcal{W}_{\mathrm{rel}} \to \mathcal{R}'$.

(2) Every concrete C_rMS-transduction $\tau : T(F) \to STR^c(\mathcal{R}')$ with n parameters can be expressed as $(\pi_{F,n})^{-1} \cdot \tau''$, where τ'' is the concrete parameterless

C_rMS-transduction : $T(F^{(n)}) \to STR^c(\mathcal{R}')$ defined by

$$\tau''(\lfloor t * (A_1,\ldots,A_n)\rfloor) := \tau(\lfloor t \rfloor,(A_1,\ldots,A_n)).$$

Proof: (1) We let τ' be defined by $\tau'(S+\gamma) := \tau(S,\gamma)$ for every concrete $\mathcal{R}$-structure S and every W-assignment γ in S, where $S+\gamma$ is the expansion of S into the concrete $(\mathcal{R} \cup W_{rel})$-structure such that $\gamma(Y)$ is the interpretation of $Y \in W_{rel}$ (see Lemma 6.52). The formulas of a parameterless definition scheme for τ' are obtained from those of one for τ by replacing, for every $Y \in W$, each atomic formula $Card_{p,q}(Y)$ by $\exists Z(\forall x(x \in Z \Leftrightarrow Y_i(x)) \wedge Card_{p,q}(Z))$ (and replacing each atomic formula $s \in Y$ by $Y(s)$, where s is a first-order variable or a constant symbol).

(2) We let τ'' be defined by $\tau''(\lfloor t * \gamma \rfloor) := \tau(\lfloor t \rfloor,\gamma)$ for every $t \in T(F)$ and $\gamma :$ $\mathcal{X}_n \to \mathcal{P}(N_t)$, where $\mathcal{X}_n = \{X_1,\ldots,X_n\}$ is the set of parameters (cf. Definition 6.24(c)). Let τ' be defined as in (1), with $\mathcal{R} := \mathcal{R}_F$ and $W := \mathcal{X}_n$. Then $\tau'' = \mu_{F,n} \cdot \tau'$, where $\mu_{F,n}$ is defined at the end of Definition 7.20, and this equality holds for the concrete transductions. By Theorem 7.14(2) (about the compositions of concrete monadic second-order transductions), we get the desired result. Theorem 7.14(2) is applicable because $\mu_{F,n}$ is noncopying, so that we have an equality of concrete transductions. ∎

Copying transductions

We now show that every parameterless CMS-transduction can be decomposed into a very simple copying transduction and a parameterless noncopying CMS-transduction. To this aim we define, for each $k > 1$, a particular parameterless k-copying monadic second-order transduction that transforms a structure $S \in STR^c(\mathcal{R})$ by adding to it $k-1$ copies of each element of its domain. The i-th copy of an element is related to it by a new binary relation B_i. Finally, a new unary relation P_i collects all i-th copies of the elements of D_S.

Definition 7.25 (The $copy_k$-transduction) For $k > 1$, we let $\mathcal{C}_k := \{P_i \mid 1 \le i \le k\} \cup \{B_i \mid 1 < i \le k\}$ be a set of relation symbols such that each P_i is unary and each B_i is binary. We assume that this set is disjoint with the other relational signatures $\mathcal{R}, \mathcal{R}',\ldots$ that we will consider. For each relational signature $\mathcal{R}$, we define as follows the mapping $copy_k : STR^c(\mathcal{R}) \to STR^c(\mathcal{R} \cup \mathcal{C}_k)$. It transforms a structure $S = \langle D_S, (R_S)_{R \in \mathcal{R}} \rangle$ into the structure $T := copy_k(S)$ with domain $D_T := D_S \times [k]$ and relations and constants defined by:

$$
\begin{aligned}
R_T &:= \{((a_1,1),\ldots,(a_{\rho(R)},1)) \mid (a_1,\ldots,a_{\rho(R)}) \in R_S\} \ \text{ for } R \in \mathcal{R}_+, \\
c_T &:= (c_S,1) \ \text{ for } c \in \mathcal{R}_0, \\
P_{iT} &:= D_S \times \{i\} \ \text{ for } 1 \le i \le k, \\
B_{iT} &:= \{((a,1),(a,i)) \mid a \in D_S\} \ \text{ for } 1 < i \le k.
\end{aligned}
$$

It is clear that $copy_k$ is a parameterless k-copying QF-transduction. The sets P_{iT} are FO-definable from the relations B_{iT}. Hence, we could omit them, and the next proposition would remain valid, but having them will make formulas more clear.

Proposition 7.26 For every concrete parameterless k-copying C_rMS-transduction τ of type $\mathcal{R} \to \mathcal{R}'$, there exists a concrete parameterless noncopying C_rMS-transduction μ of type $\mathcal{R} \cup \mathcal{C}_k \to \mathcal{R}'$ such that $\tau = copy_k \cdot \mu$ and every structure in $Dom(\mu)$ is isomorphic to one in $copy_k(STR^c(\mathcal{R}))$.

Proof: A structure U in $STR(\mathcal{R} \cup \mathcal{C}_k)$ of the form $copy_k(S)$ satisfies the following conditions:

(1) the sets $P_{1U}, \ldots, P_{kU}$ form a partition of the domain;
(2) every constant belongs to P_{1U} and for every $R \in \mathcal{R}_+$ and every tuple $\bar{a}$ in R_U, we have $\bar{a} \subseteq P_{1U}$, i.e., all components of $\bar{a}$ are in P_{1U};
(3) each relation B_{iU} defines a bijection between P_{1U} and P_{iU}.

We let β be the first-order formula expressing the conjunction of Conditions (1)–(3). Each structure $U \in STR(\mathcal{R} \cup \mathcal{C}_k)$ satisfying β is isomorphic to $copy_k(S)$, where S is $fg_{\mathcal{C}_k}(U[P_{1U}])$.

We denote by $\varphi \upharpoonright P_1$ the relativization of a formula φ to the set that is the interpretation of P_1, i.e., $\varphi \upharpoonright P_1$ is the formula $\varphi \upharpoonright X$ in which every atomic formula $x \in X$ is replaced by $P_1(x)$ (and X is a set variable that is not free in φ, cf. Lemma 5.10 in Section 5.2.1).

Suppose that τ is defined by $\mathcal{D} = \langle \chi, (\delta_i)_{i \in [k]}, (\theta_w)_{w \in \mathcal{R}'_* \otimes [k]} \rangle$. We construct a definition scheme $\mathcal{E} = \langle \chi', \delta', (\theta'_R)_{R \in \mathcal{R}'_*} \rangle$ for the transduction μ. Its precondition χ' has to express in the considered $(\mathcal{R} \cup \mathcal{C}_k)$-structure U that there is some structure S such that U is isomorphic to $copy_k(S)$ and $S \models \chi$. We define χ' to be $\beta \wedge (\chi \upharpoonright P_1)$.

The domain formula δ' must define the set of all elements (a, i) in D_U such that $S \models \delta_i(a)$. This can be done by defining $\delta'(x)$ as

$$\left(P_1(x) \Rightarrow (\delta_1 \upharpoonright P_1)(x) \right) \wedge \bigwedge_{2 \leq i \leq k} \left(P_i(x) \Rightarrow \exists y (B_i(y, x) \wedge (\delta_i \upharpoonright P_1)(y)) \right).$$

Finally, we must construct relation formulas θ'_R for $R \in \mathcal{R}'_*$. We use the relations B_i to obtain a copy of a given tuple that lies in $D_S \times \{1\}$. Letting $n := \rho(R)$, we have

$$((a_1, i_1), \ldots, (a_n, i_n)) \in R_{\widehat{\mathcal{D}}(S)} \text{ if and only if } S \models \theta_{R, i_1, \ldots, i_n}(a_1, \ldots, a_n).$$

For fixed $i_1, \ldots, i_n$, we can express this by the formula $\gamma_{i_1, \ldots, i_n}(x_1, \ldots, x_n)$ defined as

$$\exists y_1, \ldots, y_n \left(\bigwedge_{1 \leq j \leq n} B_{i_j}(y_j, x_j) \wedge (\theta_{R, i_1, \ldots, i_n} \upharpoonright P_1)(y_1, \ldots, y_n) \right),$$

where $B_{i_j}(y_j, x_j)$ is the formula $y_j = x_j \wedge P_1(x_j)$ if $i_j = 1$. Therefore, we can define $\theta'_R(x_1, \ldots, x_n)$ as

$$\bigvee_{1 \leq i_1, \ldots, i_n \leq k} \left(\bigwedge_{1 \leq j \leq n} P_{i_j}(x_j) \wedge \gamma_{i_1, \ldots, i_n}(x_1, \ldots, x_n) \right).$$

∎

Together with Proposition 7.24, we have a factorization of an arbitrary monadic second-order k-copying transduction τ as $\tau = exp_{\mathcal{U}} \cdot copy_k \cdot \mu$, where μ is parameterless and noncopying. If $k = 1$, the factor $copy_k$ disappears. If τ is parameterless, the factor $exp_{\mathcal{U}}$ disappears. This factorization is useful to prove that inverse monadic second-order transductions preserve recognizability (cf. [BluCou06], Theorem 51).

In Section 7.2 we will consider monadic second-order transductions from terms and trees to p-graphs, and we will obtain a logical characterization of the VR-equational sets of p-graphs. Chapter 8 will be devoted to the study of monadic second-order transductions that transform terms and words. Automata-based transducers will be defined, and proved to be equivalent to parameterless monadic second-order transductions. These equivalences generalize the equivalence of finite automata and monadic second-order sentences for the definition of languages (sets of terms or words) stated in Theorem 1.16 and proved in Chapter 5.

We continue with some technical (but useful) notions.

7.1.7 Comparing sets of structures via monadic second-order transductions

Let $\mathcal{C} \subseteq STR(\mathcal{R})$ and $\mathcal{C}' \subseteq STR(\mathcal{R}')$ be two sets of structures. We say that $\mathcal{C}'$ is *generated* by $\mathcal{C}$ if $\mathcal{C}' = \tau(\mathcal{C})$ for some MS-transduction of type $\mathcal{R} \to \mathcal{R}'$ and we denote this by $\mathcal{C}' \sqsubseteq \mathcal{C}$. By Theorem 7.14 (about the composition of CMS-transductions), this relation is transitive, hence it is a quasi-order. We say that $\mathcal{C}$ and $\mathcal{C}'$ are *equivalent* if $\mathcal{C} \sqsubseteq \mathcal{C}'$ and $\mathcal{C}' \sqsubseteq \mathcal{C}$. They are equivalent if and only if they generate the same sets of structures.

If $\mathcal{C}$ and $\mathcal{C}'$ are equivalent, then $\mathcal{C}' = \tau(\mathcal{C})$ and $\mathcal{C} = \tau'(\mathcal{C}')$ for MS-transductions τ and τ'. This is implied by the stronger condition that τ and τ' are MS-transductions such that τ is an injection and τ' is its inverse. We have met such a condition when defining strongly equivalent representations of combinatorial objects by relational structures in Definition 7.9.

A finite functional signature F is *weak* if $T(F)$ is finite. For it not to be weak, F must contain at least one constant symbol and at least one symbol of arity at least 1. It is *fat* if it contains at least one constant symbol and at least one symbol of arity at least 2. A term t is *slim* if, for any of its subterms of the form $f(t_1, \ldots, t_k)$, at most one of $t_1, \ldots, t_k$ is not a constant symbol (or is not a variable if we consider terms with variables). We denote by $Slim(F)$ the set of slim terms in $T(F)$, cf. Proposition 2.85.

Proposition 7.27

(1) The following sets of structures[11] are pairwise equivalent: paths, directed paths, A^* for every nonempty finite alphabet A and $Slim(F)$ for every finite signature F that is not weak.

(2) The same holds for the following sets: trees, rooted trees, binary rooted trees and $T(F)$ for every finite fat signature F.

Proof: (1) For transforming a path into a directed path, if suffices to select one of its ends (the MS-definition scheme uses a parameter for doing that). A monadic second-order formula can express the corresponding directions of edges. The inverse transformation is straightforward.

For transforming a directed path into a nonempty word over A, one uses a monadic second-order relabeling that chooses one letter for each position. The empty word can easily be obtained from, e.g., P_1 by an MS-transduction. Then Proposition 7.18 can be used. Again, the inverse transformation is straightforward.

For a finite signature F, let H be the unary finite signature such that $H_0 = F_0$ and H_1 consists of (or is in bijection with) the tuples of the form $\langle f, i, a_1, \dots, a_{k-1} \rangle$ for $k \geq 1$, $f \in F_k$, $i \in [k]$ and $a_1, \dots, a_{k-1} \in F_0$. Note that H is weak if and only if F is weak. There is a surjective mapping α from $T(H)$ to $Slim(F)$, defined as follows: $\alpha(c) := c$ for $c \in H_0$, and $\alpha(\langle f, i, a_1, \dots, a_{k-1} \rangle(t)) := f(a_1, \dots, a_{i-1}, \alpha(t), a_i, \dots, a_{k-1})$. Since α is a linear, strict and nonerasing second-order substitution, it is an invertible FO-transduction by Example 7.3(4). Hence, $Slim(F)$ and $T(H) = Slim(H)$ are equivalent. Note that α is not injective, e.g., $f(a_1, a_2) = \alpha(\langle f, 1, a_2 \rangle(a_1)) = \alpha(\langle f, 2, a_1 \rangle(a_2))$.

For a unary finite signature F, there is an obvious bijection between $T(F)$ and the words in the set $(F_+)^* F_0 \subseteq F^*$, where F is viewed as a finite alphabet. This bijection and its inverse are MS-transductions. Thus, there is an MS-transduction that transforms F^* into $T(F)$, and one that transforms $T(F)$ into $(F_+)^*$.

(2) For transforming a tree into a rooted tree, it suffices to choose a root and, from this choice, to determine the corresponding direction of the edges. Every rooted tree T can be obtained from a large enough binary rooted tree B by choosing a set U of nodes containing the root (to be the nodes of T) and by contracting all edges of B with their head not in U (cf. Definition 2.15 in Section 2.2 about edge contractions). This can be formalized by means of an MS-transduction with one parameter intended to specify U (cf. Example 7.3(1)). Additional parameters can specify node labels that determine the symbol from F attached to each node, and, except for the root, its rank in the sequence of brothers. Using Theorem 7.16, it can be checked that the resulting labeled tree is in $T(F)$. Hence, one can obtain $T(F)$ for each finite signature F by contracting edges in some way and by choosing new node labels. Conversely, if F is

---

[11] A path is a graph of the form P_n for some $n \geq 1$. We identify P_n with the structure $\lfloor P_n \rfloor$ and similarly for words and terms.

fat, then the rooted trees can be obtained from $T(F)$ in the same way as they were obtained from the binary rooted trees.

The detailed constructions of definition schemes are omitted because they are straightforward. ∎

Note that the transductions that relate paths and directed paths define inverse bijections on the corresponding abstract structures, but not on the concrete ones because a concrete path can be directed in two ways.

The set of paths is not equivalent to the set of trees because, by Corollary 7.49 below, the set of trees would have bounded path-width, which is not the case (see Definition 2.57).

The following proposition[12] implies that sets are not equivalent to paths. By sets we mean ∅-structures, i.e., graphs without edges. The nodes of a (rooted) tree that are not leaves are its *nonleaf* nodes (not to be confused with the internal nodes of a rooted tree whose root is defined as a source).

Proposition 7.28 Let τ be a CMS-transduction of type $\emptyset \to \{son\}$ that produces rooted (and thus directed) trees. The output trees have a bounded number of nonleaf nodes.

Proof: Let τ be k-copying with p parameters. Let $m := k \cdot 2^p$. Let us assume by contradiction that there is a rooted tree in $\tau(D)$ for some set $D \in STR(\emptyset)$, which is defined from values $D_1, \ldots, D_p$ of the parameters (with $D_i \subseteq D$) and that has more than m nonleaf nodes. There exist two such nodes equal to (a, i) and to (b, i) for some $i \in [k]$ and some elements a and $b \neq a$ of D that belong to the same sets in $\{D_1, \ldots, D_p\}$. The node (a, i) has a son, say (c, j). But the tuples $(D_1, \ldots, D_p, a, c)$ and $(D_1, \ldots, D_p, b, c)$ satisfy the same monadic second-order formulas by the choice of (a, b). This implies that (c, j) is also a son of (b, i), hence that $a = b$. This contradiction completes the proof. ∎

This proposition concerns for example the FO-transduction that makes a set of n elements (with $n > 2$) into the rooted tree $(K_{1,n-1}, r)$, where the root r is the unique vertex of degree $n - 1$. A parameter selects a vertex r to be the root and the relation son is defined by $son(x, y) :\Longleftrightarrow x = r \wedge y \neq x$.

Since directed paths are particular directed trees, Proposition 7.28 shows that a fixed monadic second-order transduction can define from all sets only paths of bounded length.

7.1.8 Evaluation of monadic second-order transductions

The following result is an application of the fixed-parameter algorithms for the problems considered in Sections 6.4.1 and 6.4.2. As usual, F is a finite functional signature and $\mathcal{R}$ a relational signature.

---

[12] It will be used in Section 7.6. It is generalized in [BluCou10].

Theorem 7.29 Let τ be a CMS-transduction : $T(F) \to STR^c(\mathcal{R})$ defined with a set of parameters $\mathcal{W}$. Given a term t in $T(F)$ and a $\mathcal{W}$-assignment γ in $\lfloor t \rfloor$, one can decide in time $O(|t|)$ whether the structure $S := \tau(t, \gamma)$ is defined; if it is, one can compute it in time $O(|t| + \|S\|)$. If only t is given, one can decide in time $O(|t|)$ whether t belongs to $Dom(\tau)$, and if this is the case, one can compute in time $O(|t| + \|S\|)$ a structure S in $\tau(t)$.

Proof: Let be given a definition scheme $\mathcal{D}$ for τ, with set of parameters $\mathcal{W}$. Given a term t in $T(F)$ and a $\mathcal{W}$-assignment γ in $\lfloor t \rfloor$, one can decide in time $O(|t|)$ whether $(\lfloor t \rfloor, \gamma) \models \chi_{\mathcal{D}}$ by the results of Section 6.4.1 showing that Theorem 6.3(1) extends to property-checking problems. The structure $S := \tau(t, \gamma)$ is defined if and only if the answer is positive.

If this is the case, in order to compute S, we must compute the sets defined by the other formulas of $\mathcal{D}$. Together with Lemma 6.17, Theorem 6.55(1) shows that, for every formula φ with free variables in $\mathcal{W} \cup \{x_1, \ldots, x_n\}$, one can compute the set $A := sat(\lfloor t \rfloor + \gamma, \varphi, (x_1, \ldots, x_n))$ in time $O(|t| + \|A\|)$. It follows that S can be computed in time $O(|t| + \|S\|)$. (See Definition 5.3 for $\|S\|$ and Definition 6.53 for $\|A\|$.)

If γ is not given, then one can decide in time $O(|t|)$ if $(\lfloor t \rfloor, \gamma) \models \chi_{\mathcal{D}}$ for some $\mathcal{W}$-assignment γ, and, by Theorem 6.55(1), one can compute such γ in time $O(|t|)$ if t belongs to the domain of τ. Then we apply to this γ the algorithm of the first case. ∎

Note that the time to compute S is polynomial in $|t|$: if $k := \rho(\mathcal{R})$, then $\|S\| = O(|t|^k)$.

The evaluation of (parameterless) monadic second-order graph transductions is discussed in Remarks 7.31(2) and 8.19(3).

7.2 The Equationality Theorem for the VR algebra

The main result of this section is a logical characterization of the VR-equational sets of (simple) graphs as the images of trees under monadic second-order graph transductions.[13] This theorem links structural aspects (algebraic descriptions of certain sets of graphs) and logical ones. It shows how tightly monadic second-order logic is linked with a certain type of graph structuring. There will be a companion theorem for HR-equational sets. These Equationality Theorems link our two representations of graphs by relational structures with our two graph algebras. They show the robustness of all these notions.

Let us discuss this in some more detail. Instead of trees we consider terms, cf. Proposition 7.27(2). Graphs can be denoted by terms in two ways, either by terms over the signature F^{VR} of the algebra $\mathbb{GP}$ or over the signature F^{HR} of the algebra

---

[13] A simple graph G is identified with its representing structure $\lfloor G \rfloor$.

JS. Hence, we have two mappings from terms to graphs, and we will prove that they are both monadic second-order transductions. It follows that every equational set of graphs is the image of a regular set of terms under a monadic second-order transduction. For an HR-equational set L, this is true in a strong sense: the set of *incidence graphs* of the graphs in L is the image of a regular set of terms under a monadic second-order transduction. Since the regular set is MS-definable, it can be incorporated into the monadic second-order transduction.

Two converse results will be proved: if the image of a set of terms $T(F)$ under a monadic second-order transduction is a set of graphs, then that set is VR-equational, and similarly, if it is the set of incidence graphs of a set of graphs L, then L is HR-equational. This provides us with characterizations of the VR- and HR-equational sets in terms of monadic second-order transductions, that are independent of the signatures F^{VR} and F^{HR}. These characterizations show the robustness of the whole theory. In particular, they yield easy proofs that for some variants of the signature F^{VR} the equational sets of the corresponding algebras are the same (and similarly for F^{HR}). They also yield, easily, that the classes of VR- and of HR-equational sets of graphs are closed under monadic second-order graph transductions of the relevant types: direct constructions from given equation systems and definition schemes would be impossible to write down in a readable way; all difficulties are concentrated in the Equationality Theorems.

Let us now start the proof of the Equationality Theorem for the VR algebra (the theorem relative to the HR algebra will be proved in Section 7.4). We first show that, for each finite set C of port labels, the mapping *val* (defined in Section 2.5, Definition 2.87) that associates a p-graph with every term in $T(F_C^{VR})$, is an MS-transduction.[14] Since the domain of a monadic second-order transduction must be a class of structures over a finite relational signature, and since F^{VR} is infinite, we cannot hope to have such a transduction with domain $T(F^{VR})$. Hence, we consider the finite subsignatures F_C^{VR} of F^{VR} for finite sets C.

For a p-graph G of type $\pi(G) \subseteq C$, we have defined in Example 5.2(2) the faithful representation $\lfloor G \rfloor_C$ as the $\mathcal{R}_{s,C}$-structure $\langle V_G, edg_G, (lab_{aG})_{a\in C} \rangle$, where lab_{aG} is the set of a-ports of G. If $\pi(G) = C$, then $\lfloor G \rfloor_C$ is also denoted $\lfloor G \rfloor$. Moreover, if L is a set of p-graphs of bounded type, then $\lfloor L \rfloor$ denotes the set $\{\lfloor G \rfloor_D \mid G \in L\} \subseteq STR^c(\mathcal{R}_{s,D})$, where D is the smallest set of port labels such that $\pi(G) \subseteq D$ for every $G \in L$. Note that if $D \subseteq C$, then the representations $\lfloor G \rfloor_D$ and $\lfloor G \rfloor_C$ of a p-graph such that $\pi(G) \subseteq D$ are strongly FO-equivalent (Definition 7.9). In fact, $\lfloor G \rfloor_C = \iota_{\mathcal{R}_{s,D},\mathcal{R}_{s,C}}(\lfloor G \rfloor_D)$ and $\lfloor G \rfloor_D = fg_{\mathcal{U}}(\lfloor G \rfloor_C)$, where $\mathcal{U} := \mathcal{R}_{s,C} - \mathcal{R}_{s,D} = \{lab_a \mid a \in C - D\}$, and both these transductions are QF operations, see Example 5.44.

Proposition 7.30 For each finite set C of port labels, the mapping $\lfloor t \rfloor \mapsto \lfloor val(t) \rfloor_C$ with domain $\lfloor T(F_C^{VR}) \rfloor$ is a parameterless noncopying MS-transduction of type

---

[14] The special case of cographs has been presented in Section 1.7.1 (Example 1.38).

$\mathcal{R}_{F_C^{VR}} \to \mathcal{R}_{s,C}$. If $D \subseteq C$, then the mapping $\lfloor t \rfloor \mapsto \lfloor val(t) \rfloor_D$ for $\pi(val(t)) \subseteq D$ (or for $\pi(val(t)) = D$) is also a parameterless noncopying MS-transduction, of type $\mathcal{R}_{F_C^{VR}} \to \mathcal{R}_{s,D}$. These results extend to labeled graphs.

Proof: We will show that the mapping $\lfloor t \rfloor \mapsto \lfloor cval(t) \rfloor_C$ is a concrete parameterless noncopying MS-transduction. This result follows from Lemmas 2.95 and 2.96 (Section 2.5.2) and we will use the notation of these lemmas. Let be given t in $T(F_C^{VR})$ and $G := cval(t)$ be the associated concrete graph. The vertex set of G is the set $Occ_0(t)$ of occurrences in t of constant symbols different from $\varnothing$. That can be expressed by a quantifier-free domain formula. The edges and ports of G are defined by relation formulas as follows:

(a) If $x \in V_G$, there is a loop incident with x if and only if x is an occurrence of $\mathbf{c}^\ell$ for some $c \in C$. This condition is expressible by a quantifier-free formula interpreted in $\lfloor t \rfloor$.

(b) If $x, y \in V_G$ with $x \neq y$, then $(x,y) \in edg_G$ if and only if there exist $a, b \in C$ and an occurrence u above x and y of one of the symbols $\overrightarrow{add}_{a,b}$, $add_{a,b}$ or $add_{b,a}$ such that $port_t(u,x) = a$ and $port_t(u,y) = b$. This condition is expressible by an MS formula with the help of the following auxiliary formulas: for every port label a there exists an MS formula $\varphi_a(x_1,x_2)$ such that, for all x, u in N_t, if $x \in Occ_0(t)$ and $x \leq_t u$, then

$$\lfloor t \rfloor \models \varphi_a(u,x) \text{ if and only if } port_t(u,x) = a.$$

We will describe later its construction.

(c) The port mapping of G satisfies $port_G(x) = port_t(root_t, x)$ for every $x \in V_G$. Thus, $port_G(x) = a$ if and only if $\lfloor t \rfloor \models \exists y(\varphi_a(y,x) \wedge \text{``}y \text{ is the root''})$.

Hence, to complete the proof that the edges and port labels of G can be determined by MS formulas interpreted in $\lfloor t \rfloor$, we need only describe the formulas $\varphi_a(x_1,x_2)$. That is similar to the proof of Theorem 5.15. By Lemma 2.95, the condition $port_t(u,x) = a$ (for $x \in Occ_0(t)$ and $x \leq_t u$) is equivalent to the existence of a family of sets of nodes $(X_c)_{c \in C}$ such that

(1) it is a partition of the set $[x,u] := \{w \mid x \leq_t w \leq_t u\}$;
(2) if x is an occurrence of $\mathbf{c}$ or of $\mathbf{c}^\ell$, then it belongs to X_c;
(3) for every $w \in [x,u]$ and every $c \in C$, if $w \in X_c$ is an occurrence of $relab_h$, then its son belongs to X_d for some d such that $h(d) = c$;
(4) for every $w \in [x,u]$ and every $c \in C$, if $w \in X_c$ is an occurrence of $\oplus$ or of an edge addition operation, then its son that belongs to $[x,u]$ is in X_c;
(5) $u \in X_a$.

Since these conditions are expressible by a monadic second-order formula, the proof of the first statement is complete.

Let τ be the mapping $\lfloor t \rfloor \mapsto \lfloor val(t) \rfloor_C$ with domain $T(F_C^{VR})$. Let β be a formula over $\mathcal{R}_{s,C}$ expressing that a p-graph has type included in D (or has type D). Then the required mapping is $(\tau \cap (\mathrm{MOD}(\mathit{True}) \times \mathrm{MOD}(\beta))) \cdot \mathit{fg}_{\mathcal{U}}$, where $\mathcal{U} := \{\mathit{lab}_a \mid a \in C - D\}$. The second statement now follows from Theorems 7.16 and 7.14.

The extension to (K, Λ)-labeled graphs is straightforward. ∎

Remark 7.31 (1) From this proposition, we obtain a short proof of the Weak Recognizability Theorem (Corollary 5.69(1)), which, by Proposition 3.69, states that the set of terms in $T(F_C^{\mathrm{VR}})$ that evaluate to a p-graph satisfying a CMS sentence φ is regular: by Proposition 7.30 and Backwards Translation (Corollary 7.12), this set of terms is CMS-definable. Hence it is regular by Theorem 6.26. For the algorithmic applications described in Chapter 6 this proof is not satisfactory because it produces a sentence of quantifier-height $qh(\varphi) + O(|C|)$. For this reason, we have given in Chapter 6 direct constructions of automata over F_C^{VR} rather than constructions based on Backwards Translation, cf. Remark 6.36.

(2) We also obtain a generalization of the model-checking result of Theorem 6.4(2) to the evaluation of parameter-invariant monadic second-order graph transductions:

There exists a fixed-parameter cubic algorithm that computes $\tau(G)$, for a given parameter-invariant monadic second-order graph transduction τ and a given labeled graph G.[15] The parameter of the algorithm is $(cwd(G), \mathcal{D})$, where $\mathcal{D}$ is the definition scheme for τ.

The proof is as follows. As shown in Proposition 6.8, a term t over $F_{[h_\Lambda(G)],[K,\Lambda]}^{\mathrm{VR}}$ can be computed in cubic time such that $val(t) = G$. By Theorem 7.14 and Proposition 7.30 (for $D = \emptyset$), $val \cdot \tau$ is a parameter-invariant monadic second-order transduction from terms to graphs. Thus, by Theorem 7.29, the graph $\tau(G) = (val \cdot \tau)(t)$ can be computed in time $O(|t| + \|\tau(G)\|)$.

As discussed after Theorem 6.4, this result also holds for p-graphs. □

Corollary 7.32 For every VR-equational set of labeled p-graphs L, there exist a fat finite functional signature F, consisting of binary and nullary symbols, and a parameterless noncopying MS-transduction τ such that $L = \tau(T(F))$.[16]

Proof: Let L be a VR-equational set of (K, Λ)-labeled p-graphs and let D be the smallest set that includes the types of all p-graphs in L. We know from Proposition 4.41 and Corollary 3.19 (see also Proposition 3.23(3)) that $L = val(R)$, where R is a regular subset of $T(F_{C,[K,\Lambda]}^{\mathrm{VR}})$ for some finite set C that includes D. By Proposition 7.30, the mapping $\lfloor t \rfloor \mapsto \lfloor val(t) \rfloor_D$ for $\pi(val(t)) \subseteq D$ is an MS-transduction. So is its restriction to $\lfloor R \rfloor$, by the Restriction Theorem (Theorem 7.16), because R is MS-definable (by Theorem 5.15). Let τ be the resulting MS-transduction. Then $\lfloor L \rfloor = \lfloor L \rfloor_D = \tau(\lfloor T(F_{C,[K,\Lambda]}^{\mathrm{VR}}) \rfloor)$. The fat signature $F_{C,[K,\Lambda]}^{\mathrm{VR}}$ has unary symbols, but

---

[15] For the notion of an FPT algorithm with output, see the paragraph before Theorem 6.55. The size of the input graph G can be taken as $|G| = |V_G|$; the size of the output graph is $\|\tau(G)\|$.

[16] More precisely, this means that $\lfloor L \rfloor = \tau(\lfloor T(F) \rfloor)$.

it can be replaced by a fat finite signature F consisting only of binary and nullary symbols by Proposition 7.27(2).   ∎

Corollary 7.33 If a set of (K, Λ)-labeled p-graphs of bounded type has bounded clique-width or bounded linear clique-width, then it is included in the image of the set of trees or, respectively, of the set of paths under a monadic second-order p-graph transduction.

Proof: If a set L of (K, Λ)-labeled p-graphs of bounded type has bounded clique-width, then it is included in the set $M := val(T(F^{VR}_{C,[K,\Lambda]}))$ for some finite set C. By Proposition 7.30, $\lfloor M \rfloor = \lfloor M \rfloor_C$ is the image of the set of terms $T := T(F^{VR}_{C,[K,\Lambda]})$ under an MS-transduction. By Proposition 7.27(2) the set T is equivalent to the set of trees. The result follows.

If L as above has bounded linear clique-width, then by the definition of linear clique-width (Definition 2.89), it is included in the set $val(Slim(F^{VR}_{C,[K,\Lambda]}))$ for some finite set C. By Proposition 7.27(1), the set $Slim(F^{VR}_{C,[K,\Lambda]})$ is equivalent to the set of paths. With Proposition 7.30, the result follows.   ∎

Our objective is now to establish the converses of these two corollaries. For proving the next theorem we will need a technical lemma.

Theorem 7.34 Let F be a finite functional signature, let K and Λ be finite sets of vertex and edge labels and let D be a finite set of port labels. Let $\tau : T(F) \to \mathcal{GP}_{[K,\Lambda]}[D]$ be a monadic second-order transduction, given by a definition scheme of type $\mathcal{R}_F \to \mathcal{R}_{s,D,[K,\Lambda]}$.

(1) One can construct a finite set C of port labels that includes D and an invertible MS-transduction $\mu : T(F) \to T(F^{VR}_{C,[K,\Lambda]})$ such that $\tau = \mu \cdot val$. If τ is parameterless, then so is μ.

(2) One can construct a regular language $M \subseteq T(F^{VR}_{C,[K,\Lambda]}) - \{\varnothing\}) \cup \{\varnothing\}$ such that $\tau(T(F)) = val(M)$. The set $\tau(T(F))$ is VR-equational.

Furthermore, if $Dom(\tau) \subseteq Slim(F)$, then $\mu(T(F)) \subseteq Slim(F^{VR}_{C,[K,\Lambda]})$ and M is a set of slim terms.   □

The elements of $\tau(T(F))$ are (K, Λ)-labeled p-graphs G of type included in D, represented by the structures $\lfloor G \rfloor_D$ in $STR(\mathcal{R}_{s,D,[K,\Lambda]})$. The (auxiliary) port labels in $C - D$ will be useful to build the output graphs, but they do not occur in the definition scheme of τ.

We denote by $Leaves(t)$ the set of leaves of the syntactic tree of a term t, i.e., the set of occurrences of constant symbols in t.

Lemma 7.35 Let F be a finite functional signature and $\tau : T(F) \to STR(\mathcal{R})$ be a monadic second-order transduction. One can construct a finite functional signature

F', a monadic second-order transduction $\tau' : T(F') \to STR(\mathcal{R})$ and an invertible MS-transduction $\pi : T(F) \to T(F')$ such that $\tau = \pi \cdot \tau'$ and $\tau'(T(F')) = \tau(T(F))$, and the following conditions hold:

(1) F' contains only nullary and binary symbols;

(2) τ' is parameterless and noncopying; and

(3) for every $t \in Dom(\tau')$, the domain of the concrete $\mathcal{R}$-structure $\tau'(t)$ is a subset of $Leaves(t)$.

Moreover, if τ is parameterless, then so is π. Furthermore, if $Dom(\tau) \subseteq Slim(F)$, then F' and τ' can be constructed in such a way that $Dom(\tau') \subseteq Slim(F')$.

Proof: We may assume that $\mathcal{R}_0 = \emptyset$. In three steps we transform (F, τ) into (F', τ'), which satisfies the required properties (except for the assertion about slim terms).

Step 1: Eliminating parameters If τ is parameterless, there is nothing to do. Otherwise, we let τ be specified by a definition scheme with n parameters forming the set $\mathcal{X}_n$. By Proposition 7.24(2), we can write τ as $(\pi_{F,n})^{-1} \cdot \tau_1$ for the parameterless monadic second-order transduction $\tau_1 : T(F^{(n)}) \to STR(\mathcal{R})$ such that $\tau_1(\lfloor t * \gamma \rfloor) := \tau(\lfloor t \rfloor, \gamma)$ for every $t \in T(F)$ and $\gamma : \mathcal{X}_n \to \mathcal{P}(N_t)$. The definition of τ_1 implies that $\tau_1(T(F^{(n)})) = \tau(T(F))$. The MS-transduction $\pi_{F,n}$ is invertible since it is a monadic second-order relabeling.

Step 2: Making the transduction noncopying and satisfy Condition (3) Let $(F^{(n)}, \tau_1)$ be obtained by the first step. Since the composition of two invertible MS-transductions is invertible (by Theorem 7.14), it suffices to continue the proof with $(F^{(n)}, \tau_1)$ instead of (F, τ). If τ_1 is noncopying and Condition (3) holds, there is nothing to do. Otherwise, to simplify the notation, we denote $(F^{(n)}, \tau_1)$ by (F, τ). Let τ be k-copying. (We may have $k = 1$ if Condition (3) is not satisfied.) We define $\overline{F}$ as the functional signature $F \cup \{*\}$ with arity mapping $\overline{\rho}$ such that $\overline{\rho}(*) := 0$ and $\overline{\rho}(f) := k + \rho(f)$ for each $f \in F$. We let $\lambda : T(F) \to T(\overline{F})$ be defined inductively as follows:

$$\lambda(f(t_1,\ldots,t_n)) = f(*,\ldots,*,\lambda(t_1),\ldots,\lambda(t_n)),$$

with k occurrences of $*$. We let also $\overline{\lambda}$ be the mapping $: \mathcal{N} \to \mathcal{N}$ defined by $\overline{\lambda}(u) := u + k(u - 1)$. It maps each position u of t to its *corresponding* position in $\lambda(t)$. (Positions are defined as positive integers.) In particular, if u is an occurrence in t of some symbol, then $\overline{\lambda}(u)$ is an occurrence of the same symbol in $\lambda(t)$. We give an example. Let $t = f(a, g(b), a)$ and $k = 2$. Then $\lambda(t) = f(*, *, a(*, *), g(*, *, b(*, *)), a(*, *))$. The positions of t are 1, 2, 3, 4, 5 and their corresponding positions in $\lambda(t)$ are respectively 1, 4, 7, 10, 13.

The mapping λ is injective. Since it is a linear, strict and nonerasing second-order substitution, it is an invertible parameterless $(k + 1)$-copying MS-transduction, cf. Example 7.3(4). Its inverse λ^{-1} has a definition scheme with particular properties described in the following claim; the construction is easy.

Claim 7.35.1 The partial function λ^{-1} has a parameterless noncopying MS-definition scheme $\mathcal{E}$ of type $\mathcal{R}_{\overline{F}} \to \mathcal{R}_F$ such that for every $t \in T(F)$, if S is the concrete structure $\lfloor \lambda(t) \rfloor$, then the domain of the structure $\widehat{\mathcal{E}}(S)$ is the subset $\overline{\lambda}(N_t)$ of D_S and $\overline{\lambda}$ is the unique isomorphism between $\lfloor t \rfloor$ and $\widehat{\mathcal{E}}(S)$.   □

We let $\overline{\tau} := \lambda^{-1} \cdot \tau$. This implies that $\overline{\tau}(T(\overline{F})) = \tau(T(F))$ and, since λ is injective, that $\tau = \lambda \cdot \overline{\tau}$. We now construct a definition scheme for $\overline{\tau}$ that satisfies Conditions (2) and (3). Let τ be given by a definition scheme $\mathcal{D} := \langle \chi, (\delta_i)_{i \in [k]}, (\theta_w)_{w \in \mathcal{R} \circledast [k]} \rangle$.

Claim 7.35.2 The partial function $\overline{\tau}$ has a parameterless noncopying definition scheme of type $\mathcal{R}_{\overline{F}} \to \mathcal{R}$ that satisfies Condition (3).

Proof: By Theorem 7.14 the concrete monadic second-order transduction $\widehat{\mathcal{E}} \cdot \widehat{\mathcal{D}}$ is parameterless. It transforms a concrete $\mathcal{R}_{\overline{F}}$-structure $S = \lfloor \lambda(t) \rfloor$ into a concrete $\mathcal{R}$-structure U as follows:

(i) First, by Claim 7.35.1, $\widehat{\mathcal{E}}$ transforms S into T isomorphic to $\lfloor t \rfloor$ such that D_T is the subset $\overline{\lambda}(N_t)$ of D_S.

(ii) Then τ transforms T into U such that $D_U \subseteq D_T \times [k] = \overline{\lambda}(N_t) \times [k]$.

We will construct a parameterless noncopying definition scheme $\mathcal{H}$ for $\overline{\tau}$ such that the concrete transduction $\widehat{\mathcal{H}}$ transforms S into U' isomorphic to U, in such a way that an element (u, i) of D_U is replaced in $D_{U'}$ by u_i, the i-th son of u: this is possible because every node u of $\lambda(t)$ of the form $\overline{\lambda}(w)$ has at least k sons. Furthermore, the first k ones are occurrences of the constant symbol $*$, hence, Condition (3) will be satisfied.

We construct $\mathcal{H} = \langle \chi', \delta', (\theta'_R)_{R \in \mathcal{R}} \rangle$. Its precondition χ' is defined as $\chi^{\mathcal{E}}$ (cf. Theorem 7.10), which guarantees that $\widehat{\mathcal{H}}$ and $\widehat{\mathcal{E}} \cdot \widehat{\mathcal{D}}$ have the same domain.

For $i \in [k]$ and u, v in $N_{\lambda(t)}$, the condition $\lfloor \lambda(t) \rfloor \models son_i(u, v)$ means that v will replace in $D_{U'}$ the pair $(\overline{\lambda}^{-1}(u), i)$ of D_U. The domain formula δ' is thus defined as

$$\bigvee_{i \in [k]} \exists y (son_i(y, x) \wedge (\delta_i)^{\mathcal{E}}[y/x]).$$

(In every structure $\lfloor \lambda(t) \rfloor$, every x satisfying this formula is an occurrence of $*$, hence corresponds to a leaf of $\lambda(t)$.)

For each $R \in \mathcal{R}_n$, the relation formula θ'_R is

$$\bigvee_{i \in [k]^n} \exists y_1, \dots, y_n \left(\left(\bigwedge_{j \in [n]} son_{i[j]}(y_j, x_j) \right) \wedge (\theta_{R,\mathbf{i}})^{\mathcal{E}}[y_1/x_1, \dots, y_n/x_n] \right).$$

It is clear that $\widehat{\mathcal{H}}(\lfloor \lambda(t) \rfloor) \simeq \tau(t)$ for every $t \in T(F)$. The precondition of $\mathcal{H}$ ensures that $\widehat{\mathcal{H}}(S)$ is defined only if $S \simeq \lfloor \lambda(t) \rfloor$, and we have noted that $\mathcal{H}$ satisfies Condition (3). This completes the proof of the claim.   □

Hence, the pair $(\overline{\tau}, \overline{F})$ satisfies all requirements of the lemma, except possibly Condition (1).

Step 3: Making function symbols nullary or binary If $(\overline{\tau}, \overline{F})$ constructed by the previous step satisfies Condition (1), i.e., if $\overline{F}$ consists of nullary and binary symbols, there is nothing to do. Otherwise, to simplify the notation, we denote $(\overline{\tau}, \overline{F})$ by (τ, F). We define F' as the signature $F \cup \{@, \bot\}$ with arity mapping ρ' such that $\rho'(f) := 0$ if $\rho(f) = 0$ or $f = \bot$, and $\rho'(f) := 2$ if $\rho(f) \neq 0$ or $f = @$. We let $\beta : T(F) \to T(F')$ be defined inductively as follows:

$$\begin{aligned}
\beta(f) &= f \quad \text{if } \rho(f) = 0, \\
\beta(f(t_1)) &= f(\beta(t_1), \bot) \quad \text{if } \rho(f) = 1, \\
\beta(f(t_1, \ldots, t_n)) &= f(\beta(t_1), @(\beta(t_2), \ldots @(\beta(t_{n-1}), @(\beta(t_n), \bot)) \cdots)) \\
&\qquad \text{with } n - 1 \text{ occurrences of } @, \quad \text{if } \rho(f) = n \geq 2.
\end{aligned}$$

The mapping β is injective. Just as λ in Step 2, it is a linear, strict and nonerasing second-order substitution and hence, an invertible parameterless MS-transduction. For its inverse β^{-1}, we can construct a definition scheme with particular properties.

Claim 7.35.3 The partial function β^{-1} has a parameterless noncopying MS-definition scheme $\mathcal{B}$ of type $\mathcal{R}_{F'} \to \mathcal{R}_F$ such that, for every $t \in T(F)$, if S is the concrete structure $\widehat{\mathcal{B}}(\lfloor \beta(t) \rfloor)$ and h is the (unique) isomorphism $: S \to \lfloor t \rfloor$, then $h^{-1}(Leaves(t)) \subseteq Leaves(\beta(t))$.

Proof: We sketch the construction of the required definition scheme $\mathcal{B}$. It has a precondition expressing that the input structure is isomorphic to $\lfloor \beta(t) \rfloor$ for some $t \in T(F)$. For an input structure $\lfloor \beta(t) \rfloor$, we let S be the intended output structure. Its domain D_S is defined as the set of occurrences of the symbols of F (which excludes the occurrences of $@$ and $\bot$). The set $lab_{f\,S}$ for $f \in F$ is the set of occurrences of f in $\beta(t)$. Finally, if $u, v \in D_S$, we define (u, v) to belong to $son_{i\,S}$ if and only if:

> $v \leq_{\beta(t)} u$ and
> the directed path in $Syn(\beta(t))$ from u to v consists of $i + 1$ nodes,
> $i - 1$ of which are occurrences of $@$.

It is clear from this construction of $\mathcal{B}$ that, by the isomorphism h^{-1}, the leaves of t correspond to leaves of $\beta(t)$ (to be precise, to the occurrences of the symbols of F_0 in $\beta(t)$). $\qquad\square$

To complete the construction of Step 3, we let τ' be the concrete transduction $\beta^{-1} \cdot \tau$, where β^{-1} is defined by the scheme $\mathcal{B}$ of the previous claim. As in Step 2, this implies that $\tau'(T(F')) = \tau(T(F))$ and, since β is injective, that $\tau = \beta \cdot \tau'$. For the transduction τ', the conditions achieved by the previous two steps are preserved. Condition (2) follows from Theorem 7.14 because both β^{-1} and τ are parameterless and noncopying. For Condition (3), we observe that $\widehat{\mathcal{B}}$ transforms the structure $\lfloor \beta(t) \rfloor$

for every $t \in T(F)$ into a structure S isomorphic to $\lfloor t \rfloor$, by an isomorphism $h : S \to \lfloor t \rfloor$. By the observations made in Definition 7.2, $\tau(S)$ is isomorphic to $\tau(\lfloor t \rfloor)$ by $h' := h \restriction D_{\tau(S)}$. We have

$$D_{\tau(S)} = h'^{-1}(D_{\tau(\lfloor t \rfloor)}) \subseteq h^{-1}(Leaves(t)) \subseteq Leaves(\beta(t)).$$

Hence, τ' transforms a term t' of the form $\beta(t)$ into a concrete structure with domain (the set $D_{\tau(S)}$ in the above proof) included in $Leaves(t')$.

Since F' satisfies Condition (1), this completes the proof of Step 3 and of the first assertion of the lemma. It remains to consider the assertion about slim terms.

If the signature F is unary, then the signature constructed in Step 1 is also unary. The function symbols constructed in Step 2 have arity at most $k + 1$, and Step 3 constructs a binary signature. It is clear that if $t \in T(F)$, then the term $\beta(\lambda(t))$ is slim. Hence, the domain of the transduction τ' consists of slim terms.

In the more general case where $Dom(\tau) \subseteq Slim(F)$ we precede Step 1 by a preliminary Step 0 in which we replace (F, τ) by (H, τ_0), where H is unary, as follows. Let H and α be defined as in the third paragraph of the proof of Proposition 7.27. Hence, the surjective mapping $\alpha : T(H) \to Slim(F)$ is an invertible MS-transduction of type $\mathcal{R}_H \to \mathcal{R}_F$. Now let $\tau_0 := \alpha \cdot \tau$. Then $\tau_0(T(H)) = \tau(T(F))$, and $\tau = \alpha^{-1} \cdot \tau_0$ because α is a mapping (and so $\alpha^{-1} \cdot \alpha$ is the identity on $Slim(F)$). Note that α is not injective. If τ is parameterless, then we can make α injective by restricting it to the MS-definable set of terms $t \in T(H)$ such that if $\langle f, i, a_1, \ldots, a_{k-1} \rangle(a)$ is a subterm of t with $a \in H_0$, then $i = 1$. Then the resulting α^{-1} is a parameterless MS-transduction with the same properties as above.   ∎

Proof of Theorem 7.34:   We will first prove Assertion (1) and then use it to prove Assertion (2). Assertion (1) actually entails that the set $\tau(T(F))$ is VR-equational. Let $N := \mu(T(F)) \subseteq T(F_{C,[K,\Lambda]}^{\mathrm{VR}})$. Then $\tau(T(F)) = val(N)$, because $\tau = \mu \cdot val$. Since μ^{-1} is an MS-transduction, $N = Dom(\mu^{-1})$ is MS-definable (see Definition 7.2), hence it is regular by Corollary 5.67. It follows that $\tau(T(F)) = val(N)$, where N is a regular set of terms over $F_{C,[K,\Lambda]}^{\mathrm{VR}}$, hence $\tau(T(F))$ is VR-equational by Proposition 3.23(3).

We now prove Assertion (1). We will assume that F and τ satisfy the three conditions of Lemma 7.35 (without the primes). This is not a loss of generality because this lemma shows how to replace, if necessary, a pair (F, τ) by one that satisfies these conditions. So, we assume that F has only symbols of arity 0 and 2, that τ is parameterless and noncopying, and that for every $t \in Dom(\tau)$, the vertex set of the concrete p-graph $\tau(t)$ is a subset of $Leaves(t)$.

We will first give the proof for unlabeled p-graphs, and later extend it to the labeled case. Moreover, we will first prove Assertion (1) for $F_C^{\kappa\mathrm{VR}}$ instead of F_C^{VR}. Later on, we will replace it by F_C^{VR}. The derived signature $F_C^{\kappa\mathrm{VR}}$ of F_C^{VR} was defined in Section 2.5.6. It has only constant symbols and binary operations defined as compositions of $\oplus$ and of unary operations of F_C^{VR}. The operations of $F_C^{\kappa\mathrm{VR}}$ are $\otimes_{R,h}$ for

$R \subseteq C \times C$ and $h : C \to C$; they are defined by $G \otimes_{R,h} H := relab_h(ADD_R(G \oplus H))$, where ADD_R is the composition (in any order) of the operations $\overrightarrow{add}_{a,b}$ for all pairs (a,b) in R such that $a \neq b$. Its constant symbols are $\varnothing$, $\mathbf{c}$ and $\mathbf{c}^\ell$ for $c \in C$. We will use the definitions of Section 2.5.2, which can be adapted to $F^{\kappa VR}$ in a straightforward way.

Let us first give an overview of the construction. We will define a set C of port labels including D and a parameterless monadic second-order relabeling $\mu : T(F) \to T(F_C^{\kappa VR})$ such that $Dom(\mu) = Dom(\tau) \subseteq T(F)$ and, for every $t \in Dom(\tau)$, the concrete p-graph $\tau(t)$ is equal to $cval(\mu(t))$.[17] The port labels in $C - D$ will be bounded theories (cf. Definitions 5.59 and 7.21). The pairs of port labels $(a,b) \in R$ such that $\otimes_{R,h}$ occurs at u in $\mu(t)$ will trigger the creation of the edges between the vertices of $\tau(t)$ below u_1 and those below u_2, where u_1 and u_2 are the two sons of u. The precise definitions of R and h that form the operation $\otimes_{R,h}$ occurring at a node u will be based on Application 5.61 where bounded theories are computed. These R and h will be denoted by R_u and h_u. The constant symbol occurring at a leaf x in $\mu(t)$ will be denoted by c_x.

The transductions τ and μ take terms as inputs. We know from Section 5.2.6 (Propositions 5.29 and 5.30) that cardinality predicates bring no expressive power to MS formulas on terms. Hence, without loss of generality, we will only consider MS formulas, i.e., monadic second-order formulas without cardinality predicates.

We will denote the concrete p-graph $\tau(t)$, of type included in D, by $G(t)$. Thus, we will define μ such that $G(t) = cval(\mu(t))$. Letting D be fixed, we will identify $G(t)$ with the concrete $\mathcal{R}_{s,D}$-structure $\widehat{\mathcal{D}}(\lfloor t \rfloor) = \lfloor G(t) \rfloor_D$, where $\mathcal{D} = \langle \chi, \delta, \theta_{edg}, (\theta_{lab_d})_{d \in D} \rangle$ is the given definition scheme $\mathcal{D}$ of τ. The precondition χ of $\mathcal{D}$ will also be the precondition of the definition scheme of μ. We will define the label c_x of a leaf x of t in $\mu(t)$ to be $\varnothing$ if and only if $\lfloor t \rfloor \models \delta(x)$, thus guaranteeing that $V_{G(t)} = Occ_0(\mu(t))$.

In order to simplify the exposition of the main part of the proof, we will additionally assume that every p-graph $G(t)$ is without loops. This restriction will be lifted at the end of the proof.

The (technical) proof will consist of five steps and some final steps. We need some notation in addition to that of Definition 5.59 and Application 5.61 (Section 5.3.6) which the reader is assumed to have in mind. As in Application 5.61, the relational signature $\mathcal{R}_F$ has a constant symbol for the root of the represented terms. We let q be the maximum quantifier-height of the relation formulas of $\mathcal{D}$. We define the following finite sets of formulas:

$$L_2 := \widehat{MS}^q(\mathcal{R}_F, \{x_1, x_2\}),$$
$$L_1 := \widehat{MS}^q(\mathcal{R}_F, \{x_1\}),$$
$$L_0 := \widehat{MS}^q(\mathcal{R}_F, \emptyset),$$
$$L_0' := \widehat{MS}^q(\mathcal{R}_{F \cup \{w\}}, \emptyset),$$

---

[17] See Definition 7.20 for the notion of a monadic second-order relabeling of terms and note that μ is invertible by Proposition 7.22(1). Note also that $N_{\mu(t)} = N_t$.

where w is the special variable used to define contexts. For a term t in $T(F)$ and a node $u \in N_t$, we let $First(u) := 1$ if u is the left son of its father or u is the root, and $First(u) := 2$ otherwise.

We now describe the main construction for an input term t in $T(F)$ such that $G(t)$ is defined (i.e., $t \in Dom(\tau)$).

Step 1: With every nonleaf node u in N_t with sons u_1 and u_2, we associate

$$\Theta(u) := (\Theta_0(u), \Theta_1(u), \Theta_2(u)),$$

where $\Theta_0(u) := Th(t, \uparrow u, q, 0, \varepsilon) \subseteq L'_0$ and $\Theta_i(u) := Th(t, \downarrow u_i, q, 0, \varepsilon) \subseteq L_0$ for $i = 1, 2$. (See Application 5.61 for notation.) This information is MS-expressible: for each Θ in $\mathcal{P}(L'_0) \times \mathcal{P}(L_0) \times \mathcal{P}(L_0)$, there is an MS formula $\Psi_\Theta(x)$ such that, for every nonleaf node u of t, we have $\lfloor t \rfloor \models \Psi_\Theta(u)$ if and only if $\Theta(u) = \Theta$.

We only give the construction of the formula $\Psi_\Phi(x)$ such that, if u is a nonleaf node, then $t \models \Psi_\Phi(u)$ if and only if $\Theta_0(u) = \Phi$. For every subset Φ of L'_0, we define $\Psi_\Phi(x)$ as

$$\exists X \left(\forall y (y \in X \Leftrightarrow \neg(y < x)) \wedge \bigwedge_{\varphi \in \Phi} \widetilde{\varphi}(X, x) \wedge \bigwedge_{\varphi \in L'_0 - \Phi} \neg\widetilde{\varphi}(X, x) \right),$$

where $y < x$ is a formula expressing that y is strictly below x, and where, for each φ, the formula $\widetilde{\varphi}(X, x)$ is obtained from $\varphi \restriction X$, the relativization of φ to X (cf. Section 5.2.1), by the replacement of each atomic formula $lab_w(z)$ by $z = x$ and each atomic formula $lab_f(z)$ (where $f \in F$) by $z \neq x \wedge lab_f(z)$. It is clear that $\lfloor t \rfloor \models \Psi_\Phi(u)$ if and only if $\lfloor t \rfloor \uparrow u \models \varphi$ for every φ in Φ and $\lfloor t \rfloor \uparrow u \models \neg\varphi$ for every φ in $L'_0 - \Phi$.

A similar construction can be done for the MS-expression of $\Theta_1(u)$ and $\Theta_2(u)$.

Step 2: We define the set of port labels C as $C' \cup D$, where $C' := \mathcal{P}(L_1) \times \{1, 2\}$ and we assume that $C' \cap D = \emptyset$. We define a partial function $p : N_t \times N_t \to C$ as follows. For $u \in N_t$ and $x \in V_{G(t)}$ such that $x \leq_t u$:

$$p(root_t, x) := port_{G(t)}(x) \in D,$$

$$p(u, x) \quad := (Th(t, \downarrow u, q, 0, x), First(u)) \in C' \quad \text{if } u \neq root_t.$$

Note that $p(u, x) \in D$ if and only if u is the root (and then $p(u, x)$ is defined for every $x \in V_{G(t)}$). We have defined an element of C' as a pair (Θ, i), where Θ is any subset of L_1, but only those such that $\Theta = Th(t, \downarrow u, q, 0, x)$ for some t, u and x will occur in the operation symbols of the terms in $\mu(T(F))$. For instance, the subsets of L_1 that contain a formula and its negation cannot be of the form $Th(t, \downarrow u, q, 0, x)$.

Step 3: The objective is now to define $\mu(t)$ in such a way that the partial function $port_{\mu(t)}$ (defined in Definition 2.94) is the function p of Step 2, i.e., such that $port_{cval(\mu(t))/u}(x) = p(u, x)$ for every $u \in N_t$ and $x \in V_u := \{x \in V_{G(t)} \mid x \leq_t u\}$. For this purpose, we will define a family of mappings $h_u : C \to C$, associated with the nonleaf nodes u of t such that, for each such node and each leaf x in V_u, we have

$$p(u,x) = (h_u \circ h_{v_1} \circ \cdots \circ h_{v_n})(p(x,x)),$$

where $(u,v_1,\ldots,v_n,x)$ is the path in $Syn(t)$ from u to x. (We have $n = 0$ if u is the father of x). The values $h_u(d)$ for $d \in D$ will not be needed, so we define (arbitrarily) $h_u(d) := d$. The values $h_u(c')$ for $c' \in C'$ are defined with the use of Application 5.61. We let f be the binary function symbol at u in N_t, and u_1 and u_2 be the sons of u.

Case 1: u is not the root. By Equation (5.6) in Application 5.61, if $x \leq_t u_1$, we have

$$Th(t, \downarrow u, q, 0, x) = Z_{f^*, q, 0, (0,1,0)}(Th(t, \downarrow u_1, q, 0, x), Th(t, \downarrow u_2, q, 0, \varepsilon)),$$

and, if $x \leq_t u_2$,

$$Th(t, \downarrow u, q, 0, x) = Z_{f^*, q, 0, (0,0,1)}(Th(t, \downarrow u_1, q, 0, \varepsilon), Th(t, \downarrow u_2, q, 0, x)).$$

We have $Th(t, \downarrow u_i, q, 0, \varepsilon) = \Theta_i(u)$ for $i = 1,2$ by Step 1. We define h_u such that $h_u(C') \subseteq C'$ as follows:

$$h_u((\Theta, 1)) := (Z_{f^*, q, 0, (0,1,0)}(\Theta, \Theta_2(u)), First(u)) \text{ and}$$
$$h_u((\Theta, 2)) := (Z_{f^*, q, 0, (0,0,1)}(\Theta_1(u), \Theta)), First(u)).$$

The function h_u is defined for all pairs (Θ, i) in C', even for those such that Θ is empty or contains contradictory formulas.

Case 2: $u = root_t$. We let $h'_u : C \to C'$ be the function h_u defined as in Case 1. If $(u, v_1, \ldots, v_n, x)$ is the directed path in $Syn(t)$ from u to x, then

$$(Th(t, \downarrow u, q, 0, x), 1) = (h'_u \circ h_{v_1} \circ \cdots \circ h_{v_n})(p(x,x)).$$

We can obtain $port_{G(t)}(x)$ from $Th(t, \downarrow u, q, 0, x)$ (that is $Th(t, \downarrow root_t, q, 0, x)$) because $port_{G(t)}(x) = d$ if and only if $\lfloor t \rfloor \models \theta_{lab_d}(x)$ if and only if $\widehat{\theta}_{lab_d} \in Th(t, \downarrow root_t, q, 0, x)$. Note that by the assumption that $\tau(t)$ is a p-graph, there is a unique $d \in D$ with these properties for each vertex x of $G(t)$. Hence we can define h_u such that $h_u(C') \subseteq D$ as follows:

$$h_u((\Theta, i)) := d \in D \text{ such that } \widehat{\theta}_{lab_d} \in \Theta' \text{ where } (\Theta', 1) = h'_u((\Theta, i)).$$

If there are several such d, or none, we take for $h_u((\Theta, i))$ an arbitrary, but fixed, element of D. For the pairs (Θ, i) arising from a term t, there is a unique such d. The definition of h_u handles the other cases to ensure that h_u is a total function. It follows that

$$port_{G(t)}(x) = (h_u \circ h_{v_1} \circ \cdots \circ h_{v_n})(p(x,x)).$$

This ends Case 2 and the definition of h_u.

Since the sets $\Theta_i(u)$ (defined in Step 1) are MS-expressible, the same holds for the functions h_u attached to the nonleaf nodes u of t. We mean by this that for each mapping $h : C \to C$, there is an MS formula $\Psi_h(x)$ such that $\lfloor t \rfloor \models \Psi_h(u)$ if and only if $h_u = h$.

We now define the labels c_x in $\mu(t)$ for every leaf x of t. If $x \in V_{G(t)}$ is not the root, we let c_x be the constant symbol $\mathbf{c}$, where $c = (Th(t, \downarrow x, q, 0, x), First(x)) = p(x, x)$. If $x \in V_{G(t)}$ is the unique node of t, we let c_x be $\mathbf{d}$, such that $port_{G(t)}(x) = d$. If x is a leaf of t that is not in $V_{G(t)}$, we let c_x be $\varnothing$.

These constant symbols are MS-expressible: the set $Th(t, \downarrow x, q, 0, x)$ only depends on the label of x in t, x is in $V_{G(t)}$ if and only if $\lfloor t \rfloor \models \delta(x)$, and $port_{G(t)}(x) = d$ if and only if $\lfloor t \rfloor \models \theta_{lab_d}(x)$.

Hence, we have defined the constant symbol c_x to be attached by μ at each leaf x of t and the mapping h_u of the new label to be attached at each nonleaf node u. This achieves the goal of representing p as the partial function $port_{\mu(t)}$. We have thus $port_{G(t)}(x) = p(root_t, x) = port_{\mu(t)}(root_t, x) = port_{cval(\mu(t))}(x)$ for each vertex x of $G(t)$. It remains to complete the definition in such a way that $\mu(t)$ evaluates to $G(t)$, i.e., that $G(t)^\circ = cval(\mu(t))^\circ$.

Step 4: We will define for each nonleaf node u of t a relation $R_u \subseteq C' \times C'$ such that the operation $\otimes_{R_u, h_u}$ attached to this node in $\mu(t)$ has the effect of creating the edges between the vertices of $G(t)$ below u_1 and those below u_2 and only these edges. This implies that $(cval(\mu(t))/u)^\circ = (G(t)[V_u])^\circ$, where $V_u := \{x \in V_{G(t)} \mid x \leq_t u\}$ as above; note that this equation also holds if u is a leaf x, by the definition of c_x.

If $x, y \in V_{u_1} \cup V_{u_2} \subseteq V_{G(t)}$, then $(x, y) \in edg_{G(t)}$ if and only if $\lfloor t \rfloor \models \theta_{edg}(x, y)$. Let $x \in V_{u_1}$ and $y \in V_{u_2}$. We know from Equation (5.5) in Application 5.61 that $Th(t, \downarrow root_t, q, 0, xy) \subseteq L_2$ is equal to

$$Z_{f^\dagger, q, 0, (0,1,1)}(Th(t, \uparrow u, q, 0, \varepsilon), Th(t, \downarrow u_1, q, 0, x), Th(t, \downarrow u_2, q, 0, y)).$$

We have $Th(t, \uparrow u, q, 0, \varepsilon) = \Theta_0(u)$ by the definition in Step 1. Hence we can define $R_u := R_{1u} \cup R_{2u}$, where R_{1u} is the set of pairs $((\Theta, 1), (\Theta', 2))$ such that $\widehat{\theta}_{edg} \in Z_{f^\dagger, q, 0, (0,1,1)}(\Theta_0(u), \Theta, \Theta')$ and R_{2u} is the set of pairs $((\Theta', 2), (\Theta, 1))$ such that $\widehat{\theta} \in Z_{f^\dagger, q, 0, (0,1,1)}(\Theta_0(u), \Theta, \Theta')$, where θ is $\theta_{edg}[x_2/x_1, x_1/x_2]$.

Again, as above for the h_u's and the c_x's, the relations R_u can be expressed by MS formulas.

Step 5: Here is the final definition of $\mu(t)$: at a nonleaf node u, we attach the operation $\otimes_{R_u, h_u}$ (where R_u and h_u are defined respectively in Steps 4 and 3); at a leaf x, we attach the constant symbol c_x (cf. Step 3).

From previous observations made in Steps 3 and 4, we get that $\mu(t)$ is a monadic second-order relabeling. Its domain formula is the one of $\mathcal{D}$.

It is clear from the construction that the port labels and the edges of $G(t)$ are defined correctly by $\mu(t)$. Hence $G(t) = cval(\mu(t))$ for every $t \in Dom(\tau) = Dom(\mu)$.

Final steps: A few things remain to be done to complete the proof of Assertion (1).

(a) First, we waive a restriction made initially to simplify the proof. Let us consider the case where $G(t)$ may have loops: a vertex x has a loop if and only if $\lfloor t \rfloor \models \theta_{edg}(x,x)$. In this case, we let c_x be $\mathbf{c}^\ell$ or $\mathbf{d}^\ell$ where c and d are defined as in Step 3.

(b) Then, we consider the case of (K, Λ)-labeled graphs. We handle vertex labels as we did for loops: we modify in the appropriate way the constant symbols c_x. Since vertex labels are MS-expressible in $\lfloor t \rfloor$, the new c_x's are still determined by MS formulas. For defining Λ-labeled edges, we replace in the operations $\otimes_{R,h}$ the binary relation R on port labels by a ternary relation $R \subseteq C \times C \times \Lambda$ (cf. the paragraph before Proposition 2.103). The extension is then straightforward. Labeled loops at a vertex x are defined as its labels, in the constant symbols c_x.

(c) We can replace $F^{\kappa VR}_{C,[K,\Lambda]}$ by $F^{VR}_{C,[K,\Lambda]}$. By the definition of the operations $\otimes_{R,h}$, there is a linear, strict and nonerasing second-order substitution $\theta : T(F^{\kappa VR}_{C,[K,\Lambda]}) \rightarrow T(F^{VR}_{C,[K,\Lambda]})$ such that t and $\theta(t)$ are equivalent terms for every t in $T(F^{\kappa VR}_C)$ (cf. Proposition 2.121). Then θ is a parameterless invertible MS-transduction by Example 7.3(4). Clearly, θ preserves the slimness of terms. Hence, if $\mu' : T(F) \rightarrow T(F^{\kappa VR}_{C,[K,\Lambda]})$ is the monadic second-order relabeling constructed above, then the MS-transduction $\mu := \mu' \cdot \theta : T(F) \rightarrow T(F^{VR}_{C,[K,\Lambda]})$ is the required one, i.e., we have $\tau = \mu \cdot val$.

This completes the proof of Assertion (1).

Let us now prove Assertion (2). In the first paragraph of this proof, we have shown that $\tau(T(F)) = val(N)$ for the regular language $N = \mu(T(F)) \subseteq T(F^{VR}_{C,[K,\Lambda]})$. Since we have defined $c_x := \varnothing$ if $x \in Leaves(t) - V_{G(t)}$ in Step 3 above, the terms $\mu(t)$ in N may contain occurrences of $\varnothing$. If $\tau(t)$ is not the empty graph, one can replace $\mu(t)$ by an equivalent term without $\varnothing$ (Corollary 2.102). By the next claim, one can construct a regular language M included in $T(F^{VR}_{C,[K,\Lambda]} - \{\varnothing\}) \cup \{\varnothing\}$ such that $val(M) = val(N)$. The proof given below for $T(F^{VR}_{C,[K,\Lambda]})$ can be adapted so as to work for $T(F^{\kappa VR}_{C,[K,\Lambda]})$.

Claim 7.34.1 Let N be a regular subset of $T(F^{VR}_{C,[K,\Lambda]})$. One can construct a regular subset M of $T(F^{VR}_{C,[K,\Lambda]} - \{\varnothing\}) \cup \{\varnothing\}$ such that $val(M) = val(N)$. If N consists of slim terms, then so does M.

Proof: For convenience, we abbreviate $F^{VR}_{C,[K,\Lambda]}$ by F. We will construct a finite functional signature H, a parameterless monadic second-order relabeling $\alpha : T(F) \rightarrow T(H)$ and a linear second-order substitution $\beta : T(H) \rightarrow T(F - \{\varnothing\}) \cup \{\varnothing\}$ such that t and $\beta(\alpha(t))$ are equivalent terms for every t in $T(F)$. Then we define $M := \beta(\alpha(N))$. Clearly, $val(M) = val(N)$. Moreover, M is regular by Proposition 7.22(3) (for α) and by Corollary 3.19 and Proposition 3.41 (for β).

We define H such that $H_0 := F_0$, $H_1 := F_1 \times \{0,1\}$ and $H_2 := \{\oplus\} \times \{0,1\}^2$. Let t be a term in $T(F)$. For every node u in N_t we define $em(u) := 0$ if $val(t/u) = \varnothing$ and $em(u) := 1$ otherwise. Since $val(t/u) = \varnothing$ if and only if all leaves of t below

u are occurrences of $\varnothing$, the property $em(u) = 0$ is MS-expressible. Thus, α can be constructed such that if u is an occurrence of $f \in F$, then α relabels u with $\langle f, em(u) \rangle$ if $f \in F_1$ and with $\langle f, em(u_1), em(u_2) \rangle$ if $f = \oplus$, where u_1 and u_2 are the sons of u (the labels of the leaves stay the same).

We define $\beta : T(H) \to T(F)$ by interpreting the symbols of H as derived operations of F, defined by the following terms with variables: each constant is defined by itself; for $f \in F_1$, the symbol $\langle f, 1 \rangle$ is defined by $f(x_1)$ and the symbol $\langle f, 0 \rangle$ by $\varnothing$; finally, $\langle \oplus, 1, 1 \rangle$ is defined by $\oplus(x_1, x_2)$, $\langle \oplus, 1, 0 \rangle$ by x_1, $\langle \oplus, 0, 1 \rangle$ by x_2, and $\langle \oplus, 0, 0 \rangle$ by $\varnothing$. Note that β is linear, but neither strict nor nonerasing.

Using the relevant equations in Proposition 2.101, it can easily be shown by induction on the structure of $t \in T(F)$ that t and $\beta(\alpha(t))$ are equivalent terms, and that $\beta(\alpha(t))$ is in $T(F - \{\varnothing\}) \cup \{\varnothing\}$. □

Finally, we have $\tau(T(F)) = val(M)$ where M is a regular language included in $T(F_{C,[K,\Lambda]}^{\mathrm{VR}}) - \{\varnothing\}) \cup \{\varnothing\}$. This completes the proof of Theorem 7.34. ∎

Let us stress the following consequence of Step 4: the operations of positive arity of a term $\mu(t)$ in $\mu(T(F)) \subseteq T(F_{C,[K,\Lambda]}^{\kappa \mathrm{VR}})$ are binary, and they create edges between their two disjoint argument p-graphs, but not inside these p-graphs. This implies that an edge linking two vertices is necessarily created by such an operation occurring at the least common ancestor in t of the two leaves corresponding to these vertices.

We denote by TREES the set of (undirected, unrooted) trees.

Theorem 7.36 (Equationality Theorem for the VR algebra) Let L be a set of possibly labeled p-graphs of bounded type. The following conditions are equivalent:
(1) L is VR-equational;
(2) L is the image of TREES under an MS-transduction;[18]
(3) L is the image of TREES under a CMS-transduction.

Proof: (1) $\Longrightarrow$ (2) is immediate from Corollary 7.32 and Proposition 7.27(2).

(3) $\Longrightarrow$ (1). Let $\lfloor L \rfloor = \tau(\lfloor \mathrm{TREES} \rfloor)$, where τ is a CMS-transduction. Hence $\lfloor L \rfloor = \tau'(\lfloor T(F) \rfloor)$, where F is some finite fat signature (because then $T(F)$ is equivalent to TREES by Proposition 7.27(2)) and τ' is a CMS-transduction. (For any such F, some transduction τ' can be constructed.) By Theorem 7.34, the set L is VR-equational. ∎

In statements (2) and (3) one can replace TREES by $T(F)$ for any finite fat signature F or by any other equivalent set like that of binary rooted trees, by Proposition 7.27(2). Theorem 7.36 yields a characterization of the VR-equational sets of p-graphs that is based on logic, in contrast with their algebraic (or, grammatical) definition based on least solutions of equation systems. This characterization makes no a priori use of the signature F^{VR}, but establishes that it (or a variant of it) is the right one.

---

[18] More precisely, this means (in the case where the p-graphs in L are (K, Λ)-labeled) that there is an MS-transduction τ of type $\mathcal{R}_s \to \mathcal{R}_{s,D,[K,\Lambda]}$ such that $\lfloor L \rfloor = \tau(\lfloor \mathrm{TREES} \rfloor)$, where D is the smallest set of port labels such that $\pi(G) \subseteq D$ for every $G \in L$.

As a first corollary of Theorem 7.36 we obtain that the alternative signature F^{iVR}, introduced before Proposition 2.119, leads to the same equational sets as F^{VR}. The signature F^{iVR} is obtained from F^{VR} by allowing $a = b$ for the edge addition operations $add_{a,b}$. Thus, every VR-equational set is also iVR-equational (i.e., equational for the F^{iVR}-algebra). It is clear that Section 2.5.2 also holds for F^{iVR}, and hence so do Proposition 7.30 and Corollary 7.32. From the resulting Corollary 7.32 we obtain that for every set iVR-equational set L, the set of structures $\lfloor L \rfloor$ is the image of $T(F)$ under an MS-transduction, for some fat signature F. It now follows that L is VR-equational from Proposition 7.27.

As a second corollary of Theorem 7.36 we obtain that the class of VR-equational sets of p-graphs is (effectively) closed under monadic second-order p-graph transductions. This generalizes the logical version of the Filtering Theorem for the VR algebra (Corollary 5.70(1)): the identity Id_K on a CMS-definable set of p-graphs K is clearly a CMS-transduction (with a precondition defining K) and $Id_K(L) = L \cap K$ for every set of p-graphs L.

Corollary 7.37 The image of a VR-equational set of labeled p-graphs under a monadic second-order p-graph transduction is VR-equational.

Proof: Let L be VR-equational and let $\lfloor L \rfloor = \lfloor L \rfloor_D$. Then $\lfloor L \rfloor_D = \tau(\text{TREES})$ for some MS-transduction τ, by Theorem 7.36 (for convenience, we write TREES instead of $\lfloor \text{TREES} \rfloor$). Let θ be a monadic second-order p-graph transduction, represented by a CMS-transduction θ' of type $\mathcal{R}_{s,D} \to \mathcal{R}_{s,E}$ (as usual we first consider unlabeled graphs). Then:

$$\lfloor \theta(L) \rfloor_E = \theta'(\lfloor L \rfloor_D) = \theta'(\tau(\text{TREES})) = (\tau \cdot \theta')(\text{TREES}).$$

Since, for every $E' \subseteq E$, the representations $\lfloor \cdot \rfloor_{E'}$ and $\lfloor \cdot \rfloor_E$ are strongly MS-equivalent (Definition 7.9), there is an MS-transduction α such that $\lfloor \theta(L) \rfloor = (\tau \cdot \theta' \cdot \alpha)(\text{TREES})$. Hence $\theta(L)$ is VR-equational by Theorems 7.14 and 7.36. The proof is the same for labeled graphs. ∎

As a third corollary we obtain a logical characterization of the sets of p-graphs of bounded clique-width.

Corollary 7.38

(1) A set of labeled p-graphs of bounded type has bounded clique-width if and only if it is included in the image of TREES under a monadic second-order p-graph transduction (MS or CMS).
(2) For every monadic second-order p-graph transduction τ there exists a computable function $f_\tau : \mathcal{N} \to \mathcal{N}$ such that, for every (possibly labeled) p-graph G and every (possibly labeled) p-graph H in $\tau(G)$, we have $cwd(H) \leq f_\tau(cwd(G))$.

(3) For every parameterless monadic second-order graph transduction θ there exists a computable function $f_\theta : \mathcal{N} \to \mathcal{N}$ and an algorithm[19] that transforms a term t in $T(F^{VR}_{[k],[K,\Lambda]})$ (k is any positive integer) into a term t' in $T(F^{VR}_{[f_\theta(k)],[K',\Lambda']})$ that evaluates to $\theta(val(t))$, provided $\theta(val(t))$ is defined. This algorithm operates in time $g(\theta,k) \cdot |t|$ for some computable function g.

Proof: (1) The only-if direction was proved in Corollary 7.33 but is also an immediate consequence of Theorem 7.36 because the set L of (K,Λ)-labeled p-graphs of clique-width at most k and of type included in a finite set D is VR-equational (Example 4.43(5)). The if direction is immediate from Theorem 7.36 because every VR-equational set has bounded clique-width (Proposition 4.44).

(2) We apply Corollary 7.37 to the VR-equational set L of (1). One can construct (at least in principle) a VR equation system defining its image L' under τ, and from it, one can determine some integer k' that bounds the clique-width of the p-graphs in L'. Hence we can take $f_\tau(k) := k'$. This proves the existence of the desired computable function f_τ. Note that $f_\tau(k)$ is computable by an algorithm that takes as input k and a definition scheme of τ.

(3) Let be given a parameterless monadic second-order graph transduction θ that maps (K,Λ)-labeled graphs to (K',Λ')-labeled ones. We will apply Theorem 7.34(1), with $D := \emptyset$, to the parameterless MS-transduction $\tau := val \cdot \theta$ that transforms a term t in $T(F^{VR}_{[k],[K,\Lambda]})$ into a (K',Λ')-labeled graph. The set C defined by this theorem can be taken equal to $[k']$ and we let $f_\theta(k) := k'$. The mapping μ resulting from this theorem is a parameterless MS-transduction from terms to terms and so, $t' := \mu(t)$ can be computed in linear time in $|t|$ (Theorems 7.29 and 8.14). The construction of μ takes time $g_1(\theta,k)$ and the computation of $\mu(t)$ takes time $g_2(\theta,k) \cdot |t|$ for fixed computable functions g_1 and g_2. Since $\tau = \mu \cdot val$, we have $val(t') = val(\mu(t)) = \tau(t) = \theta(val(t))$. ∎

This corollary establishes the fact used in the proof of Proposition 6.8 (in Section 6.2.3) that relates the clique-width of an edge-labeled directed graph G and that of its encoding by a vertex-labeled undirected graph $B_\Lambda(G)$, where Λ is the set of edge labels:

$$cwd(B_\Lambda(G)) \leq p_\Lambda(cwd(G)) \quad \text{and} \quad cwd(G) \leq \overline{p}_\Lambda(cwd(B_\Lambda(G))).$$

The functions p_Λ and $\overline{p}_\Lambda$ exist and are computable by Assertion (2) of Corollary 7.38 and the remark that B_Λ is an invertible parameterless FO-transduction. Assertion (3) yields the linear time algorithms for transforming a term witnessing that $cwd(G) \leq k$ into one witnessing that $cwd(B_\Lambda(G)) \leq p_\Lambda(k)$, and similarly for the other inequality.

Assertion (2) of Corollary 7.38 can also give quick proofs that certain sets of graphs have bounded or unbounded clique-width. Here are some examples. Let C be the class

---

[19] Since the proof is effective, there is a single algorithm that takes as input θ, k and t and produces the integer $f_\theta(k)$ and the term t'.

of nonempty, connected, simple and loop-free undirected graphs. A graph G in $\mathcal{C}$ is a *split graph* if $V_G = A \cup B$, $A \cap B = \emptyset$, A is stable ($G[A]$ has no edges) and B is a clique in G. (If A and B are both nonempty, there are edges between A and B.) Split graphs are chordal (cf. Definition 2.71). For every graph G and $k > 1$, we let $und(G)^k$ be the simple undirected graph with vertex set V_G and edges between any two distinct vertices at distance exactly k in $und(G)$.

Proposition 7.39

(1) The classes of split graphs, of chordal graphs and of tournaments have unbounded clique-width.
(2) There exists a computable function $f : \mathcal{N} \times \mathcal{N} \to \mathcal{N}$ such that, for every graph G and $k > 1$, we have $cwd(und(G)^k) \leq f(k, cwd(G))$.

Proof: (1) Let $\mathcal{S}$ be the class of split graphs. Let τ be the MS-transduction that takes a simple undirected graph G, finds sets A and B witnessing that it is a split graph and removes A and all edges $u - v$ of $G[B]$ such that $u - x - v$ for some vertex x in A of degree 2. If G is not a split graph, then there is no output. Then $\tau(\mathcal{S})$ is the class of all simple loop-free undirected graphs, and this class has unbounded clique-width. (This follows from Corollary 2.122 and also from Proposition 2.106(1).) So does $\mathcal{S}$ by Corollary 7.38(2). The same also holds for chordal graphs. Hence, although they can be constructed as trees of cliques (cf. Section 2.4.4), and cliques have clique-width at most 2, chordal graphs have unbounded clique-width.

We now describe an FO-transduction τ that transforms the class of tournaments into the class of simple loop-free undirected graphs. Let G be a tournament (i.e., an orientation of a clique) and X be a set of vertices such that, for every $x \in X$, there are exactly two edges with tail x and head in $V_G - X$. Then, we define $\tau(G, X)$ as the graph H such that $V_H = V_G - X$ and $y -_H z$ if and only if $x \to_G y$ and $x \to_G z$ for some $x \in X$. Every simple loop-free undirected graph H is $\tau(G, X)$ for some tournament G such that $V_G \supseteq V_H$ and $|X| = |E_H|$. Hence, we obtain that tournaments have unbounded clique-width. We have already proved this fact in a different way by a counting argument (Proposition 2.114(3)).

(2) It suffices to note that, for each $k > 1$, the mapping $G \mapsto und(G)^k$ is an FO-transduction and to apply Corollary 7.38(2). ∎

The following proposition shows that the unbounded clique-width of tournaments can be viewed as a consequence of the unbounded tree-width of cliques. It was already stated in Proposition 2.117.

Proposition 7.40 There exists a computable function $g : \mathcal{N} \to \mathcal{N}$ such that, for every simple undirected graph G, we have $twd(G) \leq g(\max\{cwd(H) \mid H \in und^{-1}(G)\})$. If a class of simple directed graphs is closed under arbitrary changes of edge directions and has bounded clique-width, then it has also bounded tree-width.

Proof: We will use the following claim whose proof is easy to extract from that of Lemma 2.3 of [Cou95a].

Claim 7.40.1 One can construct a monadic second-order graph transduction τ such that, for every undirected graph G and every odd integer $n \geq 3$, if the undirected rectangular grid $G^{\mathrm{u}}_{(2n-1)\times(n+1)}$ is a minor of G, then G has an orientation H such that $G^{\mathrm{u}}_{n\times(n+1)} \in \tau(H)$. □

Let k be an integer and G be a simple undirected graph, every orientation of which has clique-width at most k. Let m be the smallest odd integer larger than 3 and the integers $f_\tau(i)$ for all $i \in [k]$, where τ is constructed by Claim 7.40.1 and f_τ is the computable function of Corollary 7.38(2).

Then G does not contain $G^{\mathrm{u}}_{(2m-1)\times(m+1)}$ as a minor because otherwise, by the claim, it would have an orientation H such that $G^{\mathrm{u}}_{m\times(m+1)} \in \tau(H)$; then $cwd(H) \leq k$, so that, by Corollary 7.38(2), we have $cwd(G^{\mathrm{u}}_{m\times(m+1)}) \leq m$, but we know from Proposition 2.106(1) that $cwd(G^{\mathrm{u}}_{m\times(m+1)}) \geq m+1$. Hence, G does not contain $G^{\mathrm{u}}_{(2m-1)\times(m+1)}$ as a minor and thus, by Proposition 2.61, its tree-width is at most $g(k) := f(G^{\mathrm{u}}_{(2m-1)\times(m+1)})$, where f is a fixed computable function. Hence g is the required computable function.

The second assertion follows immediately. ∎

Linear equation systems and linear clique-width

A context-free grammar is linear if the right-hand side of each rule contains at most one occurrence of a nonterminal ([*Har, *Ber]). Similarly, we say that a (polynomial) equation system S is *linear* if each monomial in the right-hand side of each equation of S contains at most one occurrence of an unknown. A *linear equational* set of an algebra is a set defined in that algebra by a linear equational system. A set of p-graphs is *linear VR-equational* if it is linear equational in the algebra $\mathbb{GP}$, and a *linear HR-equational* set of s-graphs is defined similarly.

For $k \geq 1$, we say that a term t is *k-slim* if, for each of its subterms of the form $f(t_1,\ldots,t_k)$, at most one of $t_1,\ldots,t_k$ has size $|t_i| > k$. We denote by $Slim_k(F)$ the set of all k-slim terms over a signature F; note that $Slim_1(F) = Slim(F)$. It is straightforward to prove that for every linear equation system S over F and every $y \in Unk(S)$, we have $\mu\bar{x} \cdot S_{\mathcal{P}(\mathbb{T}(F))}(\bar{x}) \upharpoonright y \subseteq Slim_k(F)$, where k is the maximal size of a monomial in the right-hand side of an equation of S; the proof is by fixed-point induction (Proposition 3.91 with $K = K_i = Slim_k(F)$).

The linear clique-width $lcwd(G)$ of a graph G has been defined in Definition 2.89 with respect to slim terms, whereas clique-width is defined with respect to arbitrary terms over F^{VR}. It is to clique-width what path-width is to tree-width. We let PATHS denote the set of paths.

Theorem 7.41

(1) A set of labeled p-graphs of bounded type is linear VR-equational if and only if it is the image of PATHS under an MS-transduction (or, a CMS-transduction).

(2) A set of labeled p-graphs of bounded type has bounded linear clique-width if and only if it is included in the image of PATHS under a monadic second-order p-graph transduction (MS or CMS).

(3) For every monadic second-order p-graph transduction τ, there exists a computable function $f_\tau^{lin} : \mathcal{N} \to \mathcal{N}$ such that for every (possibly labeled) p-graph G and every (possibly labeled) p-graph H in $\tau(G)$, we have $lcwd(H) \leq f_\tau^{lin}(lcwd(G))$.

Proof: (1) Let L be a linear VR-equational set of (K, Λ)-labeled p-graphs. As in the proof of Corollary 7.32, we obtain that $L = val(R)$, where R is a regular set of terms over $F := F_{C,[K,\Lambda]}^{VR}$ and, by the above observation, $R \subseteq Slim_k(F)$ for some $k \geq 1$. Then $\lfloor L \rfloor = \tau(\lfloor Slim_k(F) \rfloor)$ for an MS-transduction τ. Generalizing the proof of Proposition 7.27(1) (taking $a_i \in T(F)$ such that $|a_i| \leq k$), it is clear that $Slim_k(F)$ is equivalent to PATHS.

Now let $\lfloor L \rfloor = \mu(\lfloor PATHS \rfloor)$ for a CMS-transduction μ. Then, by Proposition 7.27(1), $L = \tau(Slim(F))$ for some finite functional signature F and some monadic second-order transduction $\tau : T(F) \to \mathcal{GP}_{[K,\Lambda]}[D]$. Since it is clear that $Slim(F)$ is MS-definable, we may assume that $Dom(\tau) \subseteq Slim(F)$ by Theorem 7.16, and hence $L = \tau(T(F))$. By Theorem 7.34, $L = val(M)$ for a regular set $M \subseteq Slim(F')$ with $F' := F_{C,[K,\Lambda]}^{VR}$. Let S be an equation system over F' and let $Y \subseteq Unk(S)$ such that (S, Y) defines M in $\mathbb{T}(F')$ and L in $\mathbb{GP}$. By Propositions 3.39 and 3.37, we may assume that each unknown of S defines an infinite set of terms and is useful for Y. Then we can show that S is a linear equation system, and hence L is linear VR-equational. In fact, assume that S is not linear. Then the regular grammar $G(S)$, defined in Definition 3.17, has a rule $x \to t$ such that t has at least two occurrences of an unknown. Since x is useful for Y and S is trim, there exist $y \in Y$, $c \in Ctxt(F')$ and $s \in T(F', Unk(S))$ such that $y \Rightarrow_{G(S)}^* c[s]$, s has exactly two occurrences of an unknown: u_1 of x_1 and u_2 of x_2 (possibly $x_1 = x_2$), and $root_s$ is the least common ancestor of u_1 and u_2 in s, cf. Remarks 3.38(1) and 3.38(2). Since x_i defines an infinite set of terms, there exist terms t_1 and t_2 (possibly equal) in $T(F')$ such that $x_i \Rightarrow_{G(S)}^* t_i$ and $|t_i| > 1$. Then $y \Rightarrow_{G(S)}^* s' := c[s[t_1/x_1, t_2/x_2]]$ and so $s' \in M$. However, s' is not slim because the root of its subterm $s[t_1/x_1, t_2/x_2]$ has two sons that are not constant symbols. This contradicts the fact that M consists of slim terms.

(2) The only-if direction was proved in Corollary 7.33. In order to prove the converse, we let L be a subset of $\mu(PATHS)$ for some monadic second-order p-graph transduction μ. This implies, as shown in the second part of (1), that $L \subseteq val(M)$ for a regular set $M \subseteq Slim(F_{C,[K,\Lambda]}^{VR})$. It follows that L has linear clique-width at most $|C|$.

(3) Let τ be a monadic second-order p-graph transduction that transforms (K, Λ)-labeled graphs into (K', Λ')-labeled ones. Let L be the set of (K, Λ)-labeled graphs of linear clique-width at most k and of type included in a finite set D. It is the image of $Slim(F_{C'}^{VR})$ under the MS-transduction *val* (by Proposition 7.30) for some C' of cardinality k that includes D. Hence, $\tau(L)$ is the image of $Slim(F_{C',[K,\Lambda]}^{VR})$ under the MS-transduction $val \cdot \tau$. Hence, as in (2), its linear clique-width is bounded by a value computable from k and a definition scheme of τ. ∎

Note that the first statement of this theorem implies that Corollary 7.37 also holds for linear VR-equational sets.

The Semi-Linearity Theorem for VR-equational sets

We fix $n \geq 1$ and we let $\overline{X}$ denote the n-tuple of set variables $(X_1, \ldots, X_n)$. If γ is an $\overline{X}$-assignment in a structure S, we let

$$\overrightarrow{\#}\gamma := (|\gamma(X_1)|, \ldots, |\gamma(X_n)|) \in \mathcal{N}^n.$$

If $S \in STR^c(\mathcal{R})$ and $\varphi \in CMS(\mathcal{R}, \overline{X})$, we let

$$\overrightarrow{\#} sat(S, \varphi, \overline{X}) := \{\overrightarrow{\#}\gamma \mid (\gamma(X_1), \ldots, \gamma(X_n)) \in sat(S, \varphi, \overline{X})\},$$

and if $L \subseteq STR^c(\mathcal{R})$, we let

$$\overrightarrow{\#} sat(L, \varphi, \overline{X}) := \bigcup \{\overrightarrow{\#} sat(S, \varphi, \overline{X}) \mid S \in L\} \subseteq \mathcal{N}^n.$$

For the notion of a semi-linear subset of $\mathcal{N}^n$ see Section 3.1.6.

Theorem 7.42 If L is a regular set of terms, a VR-equational set of graphs or, more generally, the image of TREES under a CMS-transduction from trees to $\mathcal{R}$-structures, then the set $\overrightarrow{\#} sat(L, \varphi, \overline{X})$ associated with a CMS formula φ in $CMS(\mathcal{R}, \overline{X})$ is semi-linear. A description of this set can be constructed.

Proof: We first consider the case of a regular set of terms $L \subseteq T(F)$. It suffices to prove the result for $L = T(F)$: since L is MS-definable, there is a sentence ψ such that $L = \{t \in T(F) \mid \lfloor t \rfloor \models \psi\}$, and hence $\overrightarrow{\#} sat(L, \varphi, \overline{X}) = \overrightarrow{\#} sat(T(F), \psi \wedge \varphi, \overline{X})$.

We know from Section 6.3.3 that the set $L_{\varphi,n}$ defined as $\{t * \gamma \mid t \in T(F), (\gamma(X_1), \ldots, \gamma(X_n)) \in sat(\lfloor t \rfloor, \varphi, \overline{X})\}$ is a regular subset of $T(F^{(n)})$ (we recall that $F^{(n)} = F \times \{0, 1\}^n$ and that $\rho((f, w)) = \rho(f)$ for $f \in F$ and $w \in \{0, 1\}^n$). The mapping $t * \gamma \mapsto \overrightarrow{\#}\gamma$ is an affine mapping : $T(F^{(n)}) \to \mathcal{N}^n$ (see Section 3.1.6 for definitions). To see this we note that, if $t = f(t_1, \ldots, t_k)$ and $t * \gamma = (f, w)(t_1 * \gamma_1, \ldots, t_k * \gamma_k)$, then we have $\overrightarrow{\#}\gamma = w + \overrightarrow{\#}\gamma_1 + \cdots + \overrightarrow{\#}\gamma_k$. The result follows then from Corollary 3.26.

If L is a VR-equational set of graphs or, more generally by Theorem 7.36, the image of TREES under a CMS-transduction, then $L = \tau(T(F))$ for a finite functional signature F and an MS-transduction τ that is parameterless and noncopying (Proposition 7.27(2) and Lemma 7.35). By the Backwards Translation Theorem (Theorem 7.10), an MS formula $\tau^{\#}(\varphi)$ can be constructed from τ and φ such that, for every $t \in T(F)$, we have $sat(\lfloor t \rfloor, \tau^{\#}(\varphi), \overline{X}) = sat(\tau(t), \varphi, \overline{X})$ if $\tau(t)$ is defined, and $sat(\lfloor t \rfloor, \tau^{\#}(\varphi), \overline{X}) = \emptyset$ if $\tau(t)$ is undefined. Hence $\overrightarrow{\#}\, sat(L, \varphi, \overline{X}) = \overrightarrow{\#}\, sat(T(F), \tau^{\#}(\varphi), \overline{X})$. The result follows then from the first case. Since all steps of the proof are effective, a description of this set can be constructed from φ and a specification of L by a grammar, a VR equation system or the definition scheme of a CMS-transduction. ∎

As an example of the use of this theorem, let L be a VR-equational set of graphs and $\varphi(X, Y)$ be a formula expressing that $X \cap Y = \emptyset$ and every vertex in X is adjacent to every vertex of Y. Then $\overrightarrow{\#}\, sat(L, \varphi, (X, Y))$ is the set of all $(m, n) \in \mathcal{N}^2$ such that $core(G)$ has a subgraph isomorphic to $K_{m,n}$ for some $G \in L$. Since the class of semi-linear sets is (effectively) closed under intersection (Theorem 6.1 of [GinSpa]), it is therefore decidable whether $\overrightarrow{\#}\, sat(L, \varphi, (X, Y)) \cap \{(n, n) \mid n \in \mathcal{N}\}$ is finite, i.e., whether there exists $n \in \mathcal{N}$ such that every graph of L is without $K_{n,n}$, see Theorem 4.51. Hence, by Theorem 4.51, it is decidable whether L is HR-equational. (The implication (3) $\Longrightarrow$ (1) of Theorem 4.51 will be proved in Section 9.4.3.)

Let us finally raise a question:

Question 7.43 Is it true that if a set L of $\mathcal{R}$-structures satisfies the *semi-linearity property*, i.e., is such that $\overrightarrow{\#}\, sat(L, \varphi, \overline{X})$ is semi-linear for every CMS formula φ, then its image under a CMS-transduction satisfies also this property?

7.3 Graph transductions using incidence graphs

Monadic second-order formulas can express graph properties by means of two representations of a graph G by relational structures denoted respectively by $\lfloor G \rfloor$ and $\lceil G \rceil$ (cf. Definition 5.17). Hence there are four types of monadic second-order graph transductions since there are two possibilities for the inputs and the outputs.

The notations $MS_{i,j}$ and $C_rMS_{i,j}$ for $i, j \in [2]$ (already defined in Section 1.8.2) refer to these different types where $i = 1$ (resp. $j = 1$) if and only if the representation $\lfloor \cdot \rfloor$ is used for the input (resp. the output). The same notations are used for transductions that transform p-graphs, s-graphs, terms or words.

A $C_rMS_{i,j}$-*transduction* is a transformation of objects, defined by a C_rMS-transduction of the appropriate type, transforming structures that faithfully represent the objects. For $i = 1$ and $j = 1$ we assume that the objects are labeled p-graphs (including simple graphs, terms and words) and for $i = 2$ and $j = 2$ that they are labeled s-graphs (including graphs, terms and words).[20] For example, if we say

---

[20] We identify a word w over a finite alphabet A with the vertex-labeled simple graph $G(w)$ defined in Example 5.2(2), and a term t over a finite signature F with the labeled simple graph $Syn(t)$ defined in Definition 2.14, cf. the paragraph before Corollary 5.23.

that τ is a $C_r\mathrm{MS}_{1,2}$-transduction, this will mean that it is a set of pairs (G,H) such that G is a p-graph represented by $\lfloor G \rfloor_C$ (for some finite C) and H is an s-graph represented by $\lceil H \rceil_D$ (for some finite D), defined by a $C_r\mathrm{MS}$-definition scheme. Note that a $\mathrm{CMS}_{1,1}$-transduction is a monadic second-order p-graph transduction. If necessary, we will make precise the relational signatures used for representing the considered objects.

Let us give more details about the relational structure[21] $\lceil G \rceil := \lfloor Inc(G) \rfloor$, for a graph G. If G is directed, then $Inc(G) := \langle V_G \cup E_G, in_{1G}, in_{2G} \rangle$, where $in_{1G} := \{(e,x) \mid e \in E_G, x \text{ is the tail of } e\}$ and $in_{2G} := \{(e,x) \mid e \in E_G, x \text{ is the head of } e\}$. Hence, $Inc(G)$ is a simple bipartite directed graph where each edge is labeled by 1 or 2. If G is undirected, then in_{1G} and in_{2G} are replaced by the single binary relation $in_G := \{(e,x) \mid e \in E_G, x \text{ is an end vertex of } e\}$. If G is an s-graph, then $Inc(G)$ is the simple s-graph with $Inc(G)^\circ := Inc(G^\circ)$ and $slab_{Inc(G)} := slab_G$. If L is a set of s-graphs, we let $Inc(L) := \{Inc(G) \mid G \in L\}$.

For a simple (possibly labeled) s-graph G of type $\tau(G) \subseteq C$, we have defined in Example 5.2(2) the faithful representation $\lfloor G \rfloor_C$ as the structure obtained from $\lfloor G^\circ \rfloor$ by adding the unary relations $(lab_{aG})_{a\in C}$, where lab_{aG} is the set of a-sources of G (a singleton or the empty set). For an arbitrary s-graph G of type $\tau(G) \subseteq C$, we have defined in Definition 5.17 the faithful representation $\lceil G \rceil_C$ in such a way that it equals $\lfloor Inc(G) \rfloor_C$.[22] If $\tau(G) = C$, then $\lceil G \rceil_C$ is also denoted $\lceil G \rceil$. Moreover, if L is a set of s-graphs of bounded type, then $\lceil L \rceil$ denotes the set $\{\lceil G \rceil_D \mid G \in L\} = \lfloor Inc(L) \rfloor$, where D is the smallest set of port labels such that $\tau(G) \subseteq D$ for every $G \in L$. Note that if $D \subseteq C$, then the representations $\lceil G \rceil_D$ and $\lceil G \rceil_C$ of an s-graph such that $\tau(G) \subseteq D$ are strongly FO-equivalent.

Inclusions between the four classes of monadic second-order graph transductions, and examples showing proper inclusions and incomparability results have been discussed in Section 1.8.2. For the reader's convenience, we reproduce Figure 1.6 of Section 1.8.2.

Some of these inclusions follow from Theorem 7.14 and the obvious fact that the transformation of $\lceil G \rceil$ into $\lfloor G \rfloor$ is an MS-transduction (and even an FO-transduction), in other words that the identity on simple graphs is an $\mathrm{MS}_{2,1}$-transduction. On certain classes $\mathcal{C}$ of simple graphs the identity : $\mathcal{C} \to \mathcal{C}$ is also an $\mathrm{MS}_{1,2}$-transduction. This means that the transformation of $\lfloor G \rfloor$ into $\lceil G \rceil$ for $G \in \mathcal{C}$ is an MS-transduction, and it follows that certain of the proper inclusions of Figure 7.1 become equalities. Let $\mathcal{C}$ be such a class, and i be 1 or 2. Then:

---

[21] All variants defined in Section 5.2.5 are strongly FO-equivalent (Definition 7.9). Hence the results to be stated below also hold for them. We refer the reader to Definition 5.17 for the extension of these definitions to the representation of labeled graphs.

[22] Note that in the undirected (unlabeled) case, the signature $\mathcal{R}^u_{m,C}$ of $\lceil G \rceil_C$ is the signature $\mathcal{R}_{s,C}$ in which *edg* is changed into *in*. Similarly, in the directed (unlabeled) case, $\mathcal{R}^d_{m,C}$ is $\mathcal{R}_{s,C,[\emptyset,\{1,2\}]}$ with the same change.

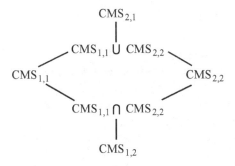

Figure 7.1 The different classes of monadic second-order graph transductions.

(1) every $C_r MS_{2,i}$-transduction with domain included in $\mathcal{C}$ is (effectively) a $C_r MS_{1,i}$-transduction;
(2) every $C_r MS_{i,1}$-transduction with image included in $\mathcal{C}$ is (effectively) a $C_r MS_{i,2}$-transduction.

We have seen in Example 7.8 that the class $\mathcal{C}$ of forests is such a class, and hence so is the class TREES. A similar proof shows that, for any finite functional signature F and any finite alphabet A, the sets of terms $T(F)$ and the sets of words A^* are such classes, cf. Corollary 5.23. Thus, the four types of transductions of terms (or words) are all the same. There are only two types of transductions from trees (or terms, or words) to graphs, because in this case and for each $i = 1$ or 2, the $C_r MS_{1,i}$- and the $C_r MS_{2,i}$-transductions are the same, with effective translations between the definition schemes of different types. In Section 9.4.1 (Theorem 9.44) we will prove that for every class $\mathcal{C}$ of simple and uniformly sparse (labeled) graphs, the identity : $\mathcal{C} \to \mathcal{C}$ is an $MS_{1,2}$-transduction, cf. Theorem 5.22. Implications (1) and (2) are false if $\mathcal{C}$ is the class of simple graphs.

Example 7.44 (Edge subdivision) If G is a simple directed graph, we let $Sub(G)$ be the simple directed graph such that

$$V_{Sub(G)} \;\; := \;\; V_G \cup E_G,$$
$$E_{Sub(G)} \;\; := \;\; \{(u,e),(e,v) \mid (e,u) \in in_{1G} \text{ and } (e,v) \in in_{2G}\}.$$

This transformation consists in replacing every directed edge by a directed path of length 2. It is an $MS_{2,2}$-transduction because the transformation of $\lceil G \rceil$ into $\lceil Sub(G) \rceil$ is a (3-copying and domain-extending) MS-transduction: a vertex v of G is made into a vertex $(v,1)$ of $Sub(G)$, and an edge e of G linking u to v is made into a vertex $(e,1)$ and also into edges $(e,2)$ and $(e,3)$ linking respectively $(u,1)$ to $(e,1)$ and $(e,1)$ to $(v,1)$. Its definition scheme is $\mathcal{D} := \langle \chi, (\delta_i)_{i \in [3]}, (\theta_w)_{w \in \mathcal{R} \otimes [3]} \rangle$ of type $\mathcal{R} \to \mathcal{R}$ where $\mathcal{R} = \{in_1, in_2\}$, with formulas as follows:

χ expresses that G is simple,

$\delta_1(x_1)$ is *True*,

$\delta_2(x_1)$ and $\delta_3(x_1)$ express that x_1 is an edge,

$\theta_{in_1,2,1}(x_1,x_2)$ expresses that x_1 is an edge with tail x_2,

$\theta_{in_1,3,1}(x_1,x_2)$ expresses that x_1 is an edge and $x_2 = x_1$,

$\theta_{in_2,2,1}(x_1,x_2)$ expresses that x_1 is an edge and $x_2 = x_1$,

$\theta_{in_2,3,1}(x_1,x_2)$ expresses that x_1 is an edge with head x_2,

and all other formulas $\theta_{in_i,j,k}$ are *False*.

This transduction has been used in Example 1.47 as an example of an $\mathrm{MS}_{2,2}$-transduction that transforms simple graphs into simple graphs and is not an $\mathrm{MS}_{1,1}$-transduction. $\qquad\square$

Example 7.45 (Graph minors) We will prove that the graph transduction $\lhd^{-1}$ that transforms a graph into its minors is an $\mathrm{MS}_{2,2}$-transduction. We recall from Definition 2.15 that a minor of a concrete graph G is a graph of the form $H = ((G-(Z\cup Y))/\approx_Y)_X$, where $X\subseteq V_G$, $Y\subseteq E_G$, $Z\subseteq V_G\cup E_G$ and the following conditions hold:

M_1 : $Y\subseteq E_{G'}$ where $G':=G-Z$ is the subgraph of G obtained by deleting all vertices and edges of Z, and all edges incident with vertices of Z;

M_2 : X is a cross-section of the equivalence relation $\approx_Y$ on $V_{G'}(=V_G-Z)$ such that $x\approx_Y y$ if and only if x and y are linked by a path with all its edges in Y.

The set Z specifies the vertices and edges to be deleted, Y is the set of edges to be contracted and X is a set of vertices of G chosen to be the vertex set of H. (Vertices of G get fused to become vertices of H; the set X selects, from a set of fused vertices, the one that is kept as a vertex of H.) We denote H by $G[X,Y,Z]$. There exists a noncopying definition scheme $\mathcal{D}$ with three parameters such that $\widehat{\mathcal{D}}(\lceil G\rceil,(X,Y,Z)) = \lceil G[X,Y,Z]\rceil$ for every triple of sets (X,Y,Z) satisfying conditions M_1 and M_2. It is thus clear that $\widehat{\mathcal{D}}(\lceil G\rceil)$ is the set of concrete minors of G. The construction of $\mathcal{D}$ uses Property P2 of Proposition 5.11 and Example 7.3(1).

Hence, we have proved that the transduction $\lhd^{-1}$ from (abstract) graphs to their (abstract) minors is an $\mathrm{MS}_{2,2}$-transduction. It follows that for every graph H there exists an MS_2 sentence $\overline{\mathrm{MINOR}_H}$ such that a graph G has a minor isomorphic to H if and only if $\lceil G\rceil \models \overline{\mathrm{MINOR}_H}$. This sentence is $\exists X,Y,Z.\beta^{\mathcal{D}}$ where β has for models the structures isomorphic to $\lceil H\rceil$ and $\beta^{\mathcal{D}}$ is constructed by Backwards Translation (Theorem 7.10). In Corollary 1.14 we have defined a similar sentence MINOR_H not using edge set quantifications, for H simple and loop-free. $\qquad\square$

Example 7.46 (Definition of orientations) Proposition 9.46 will show that the mapping und^{-1} that associates with an undirected graph all its orientations is an $\mathrm{MS}_{2,2}$-transduction. It is not a $\mathrm{CMS}_{1,1}$-transduction, because otherwise, by Corollary 7.38(2) and since cliques have clique-width 2, tournaments would have bounded clique-width, and we know that this is not the case by Propositions 2.114(3) and 7.39(1). □

7.4 The Equationality Theorem for the HR algebra

Our objective is now to establish a characterization of the HR-equational sets similar to that of the VR-equational sets given by the Equationality Theorem in Section 7.2. This characterization will use monadic second-order transductions that construct incidence graphs. However, we first state some direct consequences of the Equationality Theorem for the VR algebra and of some results of Chapter 2.

Theorem 7.47 Let τ be a graph transduction.

(1) If τ is a $CMS_{2,2}$-transduction, then there exists a computable function $g_\tau : \mathcal{N} \to \mathcal{N}$ such that, for every graph G and every graph H in $\tau(G)$, we have $twd(H) \leq g_\tau(twd(G))$.

(2) If τ is a $CMS_{1,2}$-transduction, then there exists a computable function g'_τ such that $twd(H) \leq g'_\tau(cwd(G))$ for every simple graph G and every graph H in $\tau(G)$.

(3) If τ is a $CMS_{2,1}$-transduction, then there exists a computable function g''_τ such that $cwd(H) \leq g''_\tau(twd(G))$ for every graph G and every simple graph H in $\tau(G)$.

(4) These three results extend to labeled graphs and to path-width and linear clique-width instead of, respectively, tree-width and clique-width.

Proof: (1) Let τ be a $CMS_{2,2}$-transduction that transforms graphs. Then the graph transduction $\tau' := \{(Inc(G), Inc(H)) \mid (G,H) \in \tau\}$ is a $CMS_{1,1}$-transduction (defined by the same definition scheme as τ). Let $Inc(H) \in \tau'(Inc(G))$ and $twd(G) = k$ (without loss of generality we assume $k \geq 2$). Then $twd(Inc(G)) = k$ by Example 2.56(5), hence $cwd(Inc(G)) \leq 2^{4k+4} + k + 1$ by Proposition 2.114(1.1) (we have two edge labels in $Inc(G)$ if G is directed). Hence $cwd(Inc(H)) \leq f_{\tau'}(2^{4k+4} + k + 1)$, where $f_{\tau'}$ is the function of Corollary 7.38(2). Since $Inc(G)$ is bipartite with bipartition (V_G, E_G) and each $e \in E_G$ has degree 2 in $Inc(G)$, this graph is without $K_{3,3}$. Hence we have $twd(H) \leq 12 \cdot f_{\tau'}(2^{4k+4} + k + 1) - 1$ by Proposition 2.115(2). This defines the function g_τ.

The extension to labeled graphs, and especially to edge labels (because vertex labels do not modify tree-width and clique-width) is similar. We get $twd(H) \leq f'_{\tau'}(k)$ for some computable function $f'_{\tau'}$ depending on the (fixed) sets of vertex and edge labels. (The definition scheme of the transduction τ takes into account all relations of the input and output relational structures.)

(2) The proof is similar with a $CMS_{1,1}$-transduction τ' that transforms G into $Inc(H)$. We have $twd(H) \leq 12 \cdot f_{\tau'}(cwd(G)) - 1$.

(3) Here τ' transforms $Inc(G)$ into H and we have $cwd(H) \leq f_{\tau'}(2^{4k+4} + k + 1)$, where $k = twd(G)$.

(4) The proofs of (2) and (3) extend to labeled graphs as that of (1). All these proofs extend to path-width and linear clique-width for labeled graphs. That is, to take an example, if τ is as in (2), then, for some fixed function h_τ, we have $pwd(H) \leq h_\tau(lcwd(G))$ for every simple graph G and every H in $\tau(G)$. The proofs use Theorems 2.114(1.3) and 7.41(2), and Proposition 2.115(3). ∎

These facts are useful for showing quickly that certain sets of graphs have bounded or unbounded clique-width or tree-width, but they do not give precise upper- or lower-bounds. Furthermore, they do not yield a characterization of the HR-equational sets of graphs similar to that of Theorem 7.36 for the VR-equational ones.

The next proposition is similar to Proposition 7.30. It shows that, for each finite set C of source labels, the mapping *val* (defined in Section 2.3.1, Definition 2.32) that associates an s-graph with every term in $T(F_C^{HR})$, is an $MS_{1,2}$-transduction (and hence also an $MS_{2,2}$-transduction, cf. Section 7.3). It is actually not needed for the proof of the Equationality Theorem for the HR algebra but it is interesting anyway. It gives a short proof of the Weak Recognizability Theorem for the HR algebra, cf. Remark 7.31(1). It also gives a fixed-parameter linear algorithm that computes $\tau(G)$, for a given parameter-invariant $CMS_{2,2}$-transduction τ and a given labeled graph G, with parameter $(twd(G), \mathcal{D})$, where $\mathcal{D}$ is the definition scheme for τ, cf. Remark 7.31(2).

Proposition 7.48 For each finite set C of source labels, the mapping $\lfloor t \rfloor \mapsto \lceil val(t) \rceil_C$ with domain $\lfloor T(F_C^{HR}) \rfloor$ is a parameterless MS-transduction of type $\mathcal{R}_{F_C^{HR}} \to \mathcal{R}_{m,C}$. If $D \subseteq C$, then the mapping $\lfloor t \rfloor \mapsto \lceil val(t) \rceil_D$ for $\tau(val(t)) \subseteq D$ (or for $\tau(val(t)) = D$) is also a parameterless MS-transduction, of type $\mathcal{R}_{F_C^{HR}} \to \mathcal{R}_{m,D}$. These results extend to labeled graphs.

Proof: Most of the work has been done in Section 2.3.2 in Definition 2.40 and Proposition 2.41. This proposition constructs $val(t)$ as the quotient of a concrete s-graph $Exp(t)$, called the expansion of t, by an equivalence relation on its vertices denoted by $\approx_t$ that is defined as the equivalence generated by a binary relation R_t on $V_{Exp(t)}$. We will denote by $\lceil Exp(t) \rceil_C + R_t$ the relational structure obtained by adding the binary relation R_t to $\lceil Exp(t) \rceil_C$. We recall that the vertices of $Exp(t)$ are the pairs (u, a), where u is a node of t and a belongs to the type of u (i.e., the type of $val(t/u)$), and hence to the finite set C. Its edges are the nodes of t labeled by the constant symbols that specify edges. It is clear from Definition 2.40 that the mapping $\lfloor t \rfloor \mapsto \lceil Exp(t) \rceil_C + R_t$, for $t \in T(F_C^{HR})$, is a parameterless $(|C| + 1)$-copying MS-transduction, which we denote μ_1. For writing the domain formulas of its definition scheme, we need, for each a in C, an MS formula that expresses that the type of a node u of t (cf. Definition 2.35) contains a. Such a formula is easy to write; it is similar to the one used in the proof of Proposition 7.30 for expressing logically the mapping $port_t$.

We will denote by μ_2 the parameter-invariant transduction (cf. Example 7.3(1)) that transforms $\lceil Exp(t) \rceil_C + R_t$ into the structure $\lceil Exp(t)/\approx_t \rceil_C$ representing the quotient of $Exp(t)$ by the equivalence relation generated by R_t. By Proposition 2.41, $val(t)$ is isomorphic to $Exp(t)/\approx_t$. Hence, the mapping $\lfloor t \rfloor \mapsto \lceil val(t) \rceil_C$ is equal to $\mu_1 \cdot \mu_2$, hence is a parameter-invariant MS-transduction.

In Section 2.3.2, we have also defined concrete quotient graphs $Exp(t)/\approx_t$ denoted by $cval(t)$ and $gval(t)$. It is easy to check that their representations can also be defined from $\lfloor t \rfloor$ by MS-transductions: the only difference with the above construction concerns μ_2 that must be modified so as to select particular cross-sections of $\approx_t$. Since these cross-sections are MS-expressible without the use of parameters, μ_2 and $\mu_1 \cdot \mu_2$ are parameterless MS-transductions. We omit the easy technical details.

The second statement is proved as in the proof of Proposition 7.30. $\blacksquare$

Corollary 7.49 A set of graphs has bounded tree-width (resp. bounded path-width) if and only if it is included in the image of TREES (resp. of PATHS) under a $CMS_{1,2}$-transduction.

Proof: For tree-width, the equivalence follows from Proposition 7.48, Theorem 7.47(2), Proposition 7.27(2) and the characterization of graphs of tree-width at most k as the values of terms over finite subsignatures of $T(F^{HR})$ (Theorem 2.83). For path-width it follows from Proposition 7.48, Theorem 7.47(4), Proposition 7.27(1) and the characterization of graphs of path-width at most k as the values of slim terms over finite subsignatures of $T(F^{HR})$ (Proposition 2.85). $\blacksquare$

This result has several variants: one can replace $CMS_{1,2}$-transductions by $MS_{1,2}$-transductions and the sets TREES and PATHS by equivalent sets by using Proposition 7.27.

The Equationality Theorem for the algebra $\mathbb{JS}$

We start with a result that is similar to Theorem 7.34. In its proof we will actually use Theorem 7.34, coding s-graphs as vertex-labeled incidence graphs (without sources or ports).

If G is a (K, Λ)-labeled s-graph of type included in C, we let G_C be the $(C \cup K, \Lambda)$-labeled graph (without sources) obtained by attaching a as additional label to the a-source of G, for each $a \in \tau(G)$.[23] It is clear that $Inc(G_C) = Inc(G)_C$. Moreover, $\lceil G_C \rceil = \lceil G \rceil_C$ and if G is simple, $\lfloor G_C \rfloor = \lfloor G \rfloor_C$. If L is a set of (K, Λ)-labeled s-graphs of bounded type, we let L_* be the set of $(D \cup K, \Lambda)$-labeled graphs G_D for $G \in L$, where D is the smallest set of source labels such that $\tau(G) \subseteq D$ for every

---

[23] Formally, we would have to use $\overline{C} \cup K$ where $\overline{C}$ is a copy of C disjoint with the set of all source labels (cf. Remark 6.51), but we prefer to disregard this technicality here. Note also that if $C \subseteq D$, then G_C and G_D are the same labeled graph, but with different sets of vertex labels.

$G \in L$. Clearly, $Inc(L_*) = Inc(L)_*$, $\lceil L_* \rceil = \lceil L \rceil$ and if all s-graphs in L are simple, $\lfloor L_* \rfloor = \lfloor L \rfloor$.

Theorem 7.50 Let F be a finite functional signature, let K and Λ be finite sets of vertex and edge labels and let D be a finite set of source labels. Let $\sigma : T(F) \to \mathcal{JS}_{[K,\Lambda]}[D]$ be a $\mathrm{CMS}_{1,2}$-transduction, given by a definition scheme of type $\mathcal{R}_F \to \mathcal{R}_{\mathrm{m},D,[K,\Lambda]}$.

(1) One can construct a finite set C' of source labels that includes D and an invertible MS-transduction $\mu : T(F) \to T(F^{\mathrm{HR}}_{C',[K,\Lambda]})$ such that $\sigma = \mu \cdot val$. If σ is parameterless, then so is μ.

(2) One can construct a regular language $M \subseteq T(F^{\mathrm{HR}}_{C',[K,\Lambda]} - \{\varnothing\}) \cup \{\varnothing\}$ such that $\sigma(T(F)) = val(M)$. The set $\sigma(T(F))$ is HR-equational.

Furthermore, if $Dom(\sigma) \subseteq Slim(F)$, then there exists $k \geq 1$ such that $\mu(T(F)) \subseteq Slim_k(F^{\mathrm{HR}}_{C',[K,\Lambda]})$ and M is a set of k-slim terms.[24]

Proof: The proof of Assertion (2), from Assertion (1), is entirely similar to the one of Theorem 7.34. Hence, it remains to prove Assertion (1). We first consider the case where $D = \varnothing$, i.e., $\sigma(T(F))$ is a set of labeled graphs (without sources). Moreover, we assume that the graphs in $\sigma(T(F))$ are directed and that the definition scheme has type $\mathcal{R}_F \to \mathcal{R}^{\mathrm{d}}_{\mathrm{m},[K,\Lambda]}$. The proof extends without any difficulty to undirected graphs.

The relation $\tau := \{(t, Inc(G)) \mid (t, G) \in \sigma\}$ is a $\mathrm{CMS}_{1,1}$-transduction $: T(F) \to \mathcal{GP}_{[K \cup \Lambda, [2]]}[\varnothing]$. Hence, by Theorem 7.34, we obtain a finite set C of source labels and an invertible MS-transduction $\mu_1 : T(F) \to T(F^{\mathrm{VR}}_{C,[K \cup \Lambda, [2]]})$ such that $\tau = \mu_1 \cdot val$. We will construct an invertible and parameterless MS-transduction μ_2 that transforms every term $t \in T(F^{\mathrm{VR}}_{C,[K \cup \Lambda, [2]]})$ such that $cval(t) = Inc(G)$ into a term t' in $T(F^{\mathrm{HR}}_{C',[K,\Lambda]})$ that evaluates to G, for some finite C'. Then the required invertible MS-transduction is $\mu := \mu_1 \cdot \mu_2$. To guarantee that μ_2 is invertible, we will construct it as the composition of a monadic second-order relabeling and a linear, strict and nonerasing second-order substitution (Proposition 7.22(1), Example 7.3(4)).

Let t and G be as above. Since $Inc(G)$ is without $K_{3,3}$ (cf. the proof of Theorem 7.47), we can use the construction of Proposition 2.115, that defines a tree-decomposition of $Inc(G)$ of width at most $12 \cdot |C| - 1$ whose tree is the syntactic tree $Syn(t)$ of t. (We recall that $V_{Inc(G)} = Occ_0(t)$ is a set of leaves of t, hence a subset of $N_t = N_{Syn(t)}$ which is the domain of the relational structure $\lfloor t \rfloor$.) This construction defines a mapping $f : N_t \to \mathcal{P}(V_G \cup E_G)$ such that:

(a) $(Syn(t), f)$ is a tree-decomposition of $Inc(G)$, each box of which has at most $12 \cdot |C|$ elements;

(b) each $e \in E_G$ is a leaf u of $Syn(t)$, $f(u) = \{u\}$ and the box $f(v)$, where v is the father of u contains e and its end vertices.

---

[24] For the definition of k-slim terms see the paragraphs before Theorem 7.41.

Property (b) is guaranteed by Lemma 2.116(2). It follows that, if we delete from the boxes $f(u)$ the edges of G, then we have a tree-decomposition of G. In order to have a rich tree-decomposition of G, we must ensure that each $e \in E_G$ belongs to a unique box. To obtain this, we restrict f into f' as follows:

$$f'(u) := (f(u) \cap V_G) \cup \{e \in E_G \mid u \text{ is the father of } e\}.$$

It is clear that $(Syn(t), f')$ is a rich tree-decomposition of G.

The construction of Proposition 2.115 (which we do not reproduce here) is based on subsets $Pt(u,a)$ and $Ext(u,a)$ of $V_{Inc(G)}$, with $u \in N_t$ and $a \in C$. It follows from the proof of Proposition 7.30 that, for $x \in V_{Inc(G)}$, the properties $x \in Pt(u,a)$ and $x \in Ext(u,a)$ are MS-expressible in $\lfloor t \rfloor$. Together with the definition of f', this shows that there exists an MS formula $\varphi(u,x)$ such that

$$\lfloor t \rfloor \models \varphi(u,x) \text{ if and only if } u \in N_t, x \in V_G \cup E_G \text{ and } x \in f'(u).$$

Hence, roughly speaking, we have a rich tree-decomposition of G of bounded width, with underlying tree $Syn(t)$, and that is specified in $\lfloor t \rfloor$ by MS formulas (cf. Example 5.2(4)).

Definition 2.80 shows how to transform such a decomposition into a term t' over $F^{HR}_{C',[K,\Lambda]}$ for some set C' of bounded size (in fact, C' is any set of cardinality $12 \cdot |C|$). By combining this transformation with the previous one, we will actually define G by a term t'' over a finite derived signature H of $F^{HR}_{C',[K,\Lambda]}$ such that the transformation of t into t'' is a monadic second-order relabeling. Then $t' := \theta_H(t'')$ evaluates to G, where θ_H is the second-order substitution corresponding to H (by Proposition 2.126). This will give the desired result.

The construction of Definition 2.80 uses a vertex coloring $\gamma : V_G \to C'$ of the considered graph, here G, such that two vertices of the same box have different colors (Lemma 2.78). Since the membership of two vertices in a same box is MS-expressible in $\lfloor t \rfloor$, one can use $|C'|$ set variables to specify such a coloring. In order to do that in a unique way, we take the *first* such coloring for an appropriate linear order defined as follows. Since $V_G \subseteq N_t$, the elements of V_G are linearly ordered by the natural order on $N_t = [|t|] \subseteq \mathcal{N}$ (which is MS-expressible in $\lfloor t \rfloor$ because it corresponds to the usual preorder tree traversal of $Syn(t)$, cf. the proof of Proposition 5.30). We also fix a linear order on C' so that the vertex colorings are ordered lexicographically and this is also MS-expressible.[25]

Then, Definition 2.80 builds from γ and $(Syn(t), f')$ a term t'' over a derived signature H of $F^{HR}_{C',[K,\Lambda]}$ having the same syntactic tree as t (by neglecting node

---

[25] This is, in fact, a particular case of the more general result that for every MS-transduction $\nu : T(F) \to STR(\mathcal{R})$ (for some finite functional signature F and some relational signature $\mathcal{R}$) there exists a parameterless MS-transduction $\nu' : T(F) \to STR(\mathcal{R})$ such that $Dom(\nu') = Dom(\nu)$ and $\nu' \subseteq \nu$ (i.e., ν' computes one element of $\nu(t)$, for every $t \in Dom(\nu)$). A linear order on parameter assignments can be defined, and ν' uses the first such assignment that satisfies the precondition.

labels) and that evaluates to G. These derived operations form a finite set, depending only on C, because they have arity at most 2 and because the graphs $H(u)$ that occur in Equation (2.3) of Definition 2.80, have at most two edges by Property (b) above and the definition of f' (cf. Remark 2.84(2)).[26] Since the decomposition f' of G is MS-expressible in $\lfloor t \rfloor$, it follows that t can be transformed into t'' by a parameterless monadic second-order relabeling. It is clear from Equation (2.3) that the second-order substitution θ_H is linear, strict and nonerasing.

If $Dom(\sigma) \subseteq Slim(F)$, then $\mu_1(T(F))$ is a set of slim terms by Theorem 7.34. It is clear that μ_2 transforms slim terms into $2m$-slim terms, where m is the maximal size of a term with variables that defines a derived operation in the set H.

We now consider the case where D is nonempty, but we treat each s-graph G of type included in D as a graph G_D without sources. Clearly, the relation $\tau := \{(t, Inc(G_D)) \mid (t, G) \in \sigma\}$ is a $CMS_{1,1}$-transduction: $T(F) \to \mathcal{GP}_{[D \cup K \cup \Lambda, [2]]}[\varnothing]$. The transduction μ_2 should now transform every term $t \in T(F^{\mathrm{VR}}_{C,[D \cup K \cup \Lambda, [2]]})$ such that $cval(t) = Inc(G_D)$ into a term t' in $T(F^{\mathrm{HR}}_{C',[K,\Lambda]})$ that evaluates to G, where C' includes D. The proof is as in the case of graphs without sources: the rich tree-decomposition $(Syn(t), f')$ of G has to be defined so that the sources of G are all in the root box. This can be realized by adding the sources of G to *each* box of f'. Then each box of the resulting rich tree-decomposition has at most $12 \cdot |C| + |D|$ elements. Since the sources are identified in G_D by vertex labels, the monadic second-order formula $\varphi(u, x)$ intended to specify f' can ensure this additional condition. As in the first case, the transformation of t into t'' is a parameterless monadic second-order relabeling and the transformation from t'' to t' is a linear, strict and nonerasing second-order substitution. ∎

Theorem 7.51 (Equationality Theorem for the HR algebra) Let L be a set of possibly labeled s-graphs of bounded type. The following conditions are equivalent:
(1) L is HR-equational;
(2) $Inc(L)_*$ is VR-equational;
(3) L is the image of TREES under an $MS_{1,2}$-transduction;[27]
(4) L is the image of TREES under a $CMS_{1,2}$-transduction.

Moreover, L is linear HR-equational if and only if it is the image of PATHS under an $MS_{1,2}$-transduction (or, a $CMS_{1,2}$-transduction) if and only if $Inc(L)_*$ is linear VR-equational.

Proof: (2) $\Longleftrightarrow$ (3) $\Longleftrightarrow$ (4) by the Equationality Theorem for the VR algebra (Theorem 7.36), because $\lfloor Inc(L)_* \rfloor = \lfloor Inc(L) \rfloor = \lceil L \rceil$. The "linear case" (i.e., the corresponding part of the last statement of the theorem) follows from Theorem 7.41(1).

---

[26] In the notation $H(u)$, the letter H does not refer to the derived signature.

[27] More precisely, this means (in the case where the s-graphs in L are (K, Λ)-labeled) that there is an MS-transduction τ of type $\mathcal{R}_m \to \mathcal{R}_{m,D,[K,\Lambda]}$ such that $\lceil L \rceil = \tau(\lfloor \mathrm{TREES} \rfloor)$, where D is the smallest set of source labels such that $\tau(G) \subseteq D$ for every $G \in L$.

$(1) \Longrightarrow (3)$ is immediate from Proposition 7.48 and Proposition 7.27(2). The linear case can be shown similar to the first part of the proof of Theorem 7.41(1). Instead, (the nonlinear case of) $(1) \Longrightarrow (2)$ can be shown without the use of Proposition 7.48, as follows.

Let L be an HR-equational set of directed (K, Λ)-labeled graphs (without sources). Then $Inc(L)$ is also HR-equational: if in an equation system that defines L, we replace everywhere $\overrightarrow{\mathbf{ab}}_\lambda$ by $fg_c(\overrightarrow{\mathbf{ca}}_1 /\!\!/ \overrightarrow{\mathbf{cb}}_2 /\!\!/ \mathbf{c}_\lambda)$ and $\mathbf{a}_\lambda^\ell$ by $fg_c(\overrightarrow{\mathbf{ca}}_1 /\!\!/ \overrightarrow{\mathbf{ca}}_2 /\!\!/ \mathbf{c}_\lambda)$, then we obtain an equation system that defines $Inc(L)$. The subscripts 1 and 2 in $\overrightarrow{\mathbf{ca}}_1$ and $\overrightarrow{\mathbf{cb}}_2$ represent the edge labels 1 and 2 that distinguish the two types of edges (cf. in_1 and in_2) of $Inc(G)$ when G is directed. Since $Inc(L)$ is a set of simple graphs, it is also VR-equational by Theorem 4.49. The proof is similar for undirected graphs.

We now consider graphs with sources. Let D be the smallest set of source labels such that $\tau(G) \subseteq D$ for every s-graph $G \in L$. If G is such an s-graph, then $G_D = fg_D(G /\!\!/ \mathbf{a}_a /\!\!/ \cdots)$, where $/\!\!/$ is applied to all constant symbols $\mathbf{a}_a$ for $a \in \tau(G)$. Each a-source of G is made into an internal vertex of the graph G_D labeled by a. It follows that if L is a homogenous HR-equational set of type D, then $L_* = fg_D(L /\!\!/ \mathbf{a}_a /\!\!/ \cdots)$, hence is an HR-equational set of graphs (without sources). The same holds for every HR-equational set because such a set is a finite union of homogenous HR-equational sets (cf. the paragraph before Definition 4.12). Thus, if L is HR-equational, then so are L_* and $Inc(L)_*$ since this last set is equal to $Inc(L_*)$. Hence $Inc(L)_*$ is VR-equational, which proves that (1) implies (2).

$(4) \Longrightarrow (1)$ is immediate from Proposition 7.27(2) and Theorem 7.50. In the linear case, the proof is similar to the second part of the proof of Theorem 7.41(1) (in particular, that proof also holds if $M \subseteq Slim_k(F')$, taking $|t_i| > k$). ∎

The following corollary is similar to Corollary 7.37.

Corollary 7.52 The image of an HR-equational set of s-graphs under a $CMS_{2,2}$-transduction is HR-equational. Furthermore, the image of an HR-equational set of s-graphs under a $CMS_{2,1}$-transduction is VR-equational and the image of a VR-equational set of p-graphs under a $CMS_{1,2}$-transduction is HR-equational. Similar statements hold for labeled s- and p-graphs, and for the linear equational sets.

Proof: The proof is similar to the one of Corollary 7.37. The first statement is a consequence of the equivalence of (1) and (4) of Theorem 7.51, and the fact that, by Theorem 7.14 and the definitions, $\tau \cdot \theta$ is a $CMS_{1,2}$-transduction if τ is a $CMS_{1,2}$-transduction and θ is a $CMS_{2,2}$-transduction. For the remaining statements one additionally uses the equivalence of (1) and (3) of Theorem 7.36 (and Theorem 7.41(1) for the linear case), together with the more general fact that if τ is a $CMS_{1,i}$-transduction and θ is a $CMS_{i,j}$-transduction, then $\tau \cdot \theta$ is a $CMS_{1,j}$-transduction. ∎

Since the identity is a $CMS_{2,1}$-transduction on simple graphs, we obtain from this corollary that every linear HR-equational set of simple graphs is linear VR-equational: the linear version of Theorem 4.49.

As observed in Section 7.3, it is proved in the main theorem of Section 9.4.1 (Theorem 9.44) that the identity on a set of simple and uniformly sparse labeled graphs is an $MS_{1,2}$-transduction. Using this, we get from Corollary 7.52 that a set of simple labeled graphs is HR-equational if and only if it is VR-equational and has bounded tree-width (see the proof of Theorem 4.51 and Section 9.4.3). Furthermore, it is decidable whether a VR-equational set is HR-equational (see the discussion after Theorem 7.42).

Here is another example of the use of Corollary 7.52. We recall that the *line graph* $Line(G)$ of a graph G has vertex set E_G and an (undirected) edge between e and e' in $E_G = V_{Line(G)}$ if and only if e and e' have an end vertex in common.

Corollary 7.53 The set of line graphs of an HR-equational set of graphs is VR-equational.

Proof: The transduction that transforms $\lceil G \rceil$ into $\lfloor Line(G) \rfloor$ is an FO-transduction, and so the graph transduction $Line$ is an $MS_{2,1}$-transduction (see Example 1.45). Hence, if L is an HR-equational set of graphs, then $Line(L)$ is VR-equational by Corollary 7.52. ∎

From Theorem 7.51 and Corollary 7.52 one can prove that Theorem 7.47 and Corollary 7.49 actually hold for s-graphs and p-graphs, rather than just graphs. These results correspond to Assertions (1) and (2) of Corollary 7.38 and (2) and (3) of Theorem 7.41.

Theorem 7.42, the Semi-Linearity Theorem for images of the set TREES under CMS-transductions (and in particular for VR-equational sets), has an obvious counterpart for HR-equational sets, where one can count edges and not only vertices.

7.5 Decidability of monadic second-order satisfiability problems

One of the consequences of the Recognizability Theorem is that the equational sets of graphs have decidable monadic second-order satisfiability problems (Section 5.4, Theorem 5.80). In this section we use monadic second-order transductions to reduce monadic second-order satisfiability problems for sets of graphs or relational structures to similar problems for related sets. We obtain in particular (by using difficult results of graph theory) that bounded tree-width and bounded clique-width are necessary conditions for the decidability of certain monadic second-order satisfiability problems.

We first recall a definition already given in Section 5.1.6. Let $\mathcal{L}$ be a logical language, and C be a set of structures. The *$\mathcal{L}$-satisfiability problem for C* consists in deciding whether a given sentence in $\mathcal{L}$ is true for some structure in C.

(For sets $\mathcal{L}$ closed under negation, there is an equivalent formulation in terms of $\mathcal{L}$-theories.) For having an easy comparison with what follows, we restate the results of Section 5.4.

The CMS-satisfiability problems of the set of simple labeled graphs of clique-width at most k, for each fixed k, and, more generally, of every VR-equational set of labeled graphs are decidable. So are the CMS_2-satisfiability problems of the set of labeled graphs of tree-width at most k, for each fixed k, and, more generally, of every HR-equational set of labeled graphs.

We will give alternative proofs of these results by using the following theorem.

Theorem 7.54 (Reduction Theorem) Let $r \in \mathcal{N}$, let τ be a $C_r MS$-transduction $: STR(\mathcal{R}) \to STR(\mathcal{R}')$ and let $\mathcal{C} \subseteq STR(\mathcal{R})$. If $\mathcal{C}$ has a decidable $C_r MS$-satisfiability problem, then so has $\tau(\mathcal{C})$.

Proof: Let $\mathcal{C}$ be a subset of $STR(\mathcal{R})$ and A be an algorithm that decides its $C_r MS$-satisfiability problem. Let τ be a $C_r MS$-transduction with parameters $Y_1, \ldots, Y_q$ and φ be a sentence in $C_r MS(\mathcal{R}', \emptyset)$. Then $S' \models \varphi$ for some $S' \in \tau(S)$ such that $S \in \mathcal{C}$ if and only if $S \models \exists Y_1, \ldots, Y_q . \tau^{\#}(\varphi)$ for some S in $\mathcal{C}$. We use here the Backwards Translation Theorem (Theorem 7.10). Its proof gives an algorithm that constructs $\tau^{\#}(\varphi)$ from φ and a definition scheme of τ. Algorithm A can decide if $\exists Y_1, \ldots, Y_q . \tau^{\#}(\varphi)$ is satisfiable in $\mathcal{C}$. Hence, the set $\tau(\mathcal{C})$ has a decidable $C_r MS$-satisfiability problem. ∎

In the particular case where $r = 0$, we obtain a statement relative to MS-satisfiability problems and MS-transductions. By using this theorem, Proposition 7.30 and the classical result that the monadic second-order satisfiability problem for regular languages is decidable (cf. Theorem 1.16), we get immediately that the CMS-satisfiability problem of a VR-equational set of (K, Λ)-labeled graphs, and in particular of the set of (K, Λ)-labeled graphs of clique-width bounded by any fixed k, is decidable. The proof is similar for the CMS_2-satisfiability problem of an HR-equational set of (K, Λ)-labeled graphs, and in particular, for the set of those of tree-width bounded by any fixed k. The following theorem gives some kind of converse statements.

Theorem 7.55
(1) If a set of labeled graphs has a decidable MS_2-satisfiability problem, then it has bounded tree-width.
(2) If a set of simple labeled graphs has a decidable $C_2 MS$-satisfiability problem, then it has bounded clique-width.

Proof: (1) We have shown in Example 7.45 that the transduction $\lhd^{-1}$ that associates with a graph the set of its minors is an $MS_{2,2}$-transduction. If $\mathcal{C}$ has a decidable MS_2-satisfiability problem, then so has the set $\lhd^{-1}(\mathcal{C})$ of all minors of the graphs in $\mathcal{C}$ by the Reduction Theorem (Theorem 7.54). Hence, by Theorem 5.6, the square grids in $\lhd^{-1}(\mathcal{C})$ have bounded size. So, there exists $n \in \mathcal{N}$ such that the graphs of $\mathcal{C}$

do not contain $G_{n \times n}$ as a minor. Since $G_{n \times n}$ is planar, this implies that C has bounded tree-width by the second assertion of Proposition 2.61.

(2) The proof is in [CouOum]. It follows the same scheme but is more complicated. It is based on the notion of *vertex-minor inclusion* instead of that of minor inclusion. The mapping that transforms a simple undirected graph into the set of its vertex-minors is shown to be a C_2MS-transduction, and the proof uses in a crucial way the even cardinality set predicate. For this reason, the hypothesis that the considered set has a decidable C_2MS-satisfiability problem cannot be replaced by the decidability of its MS-satisfiability problem.

The result of [CouOum] is first proved for bipartite and undirected graphs. (This special case concentrates all the difficulties). It is extended first to simple $(K,\emptyset)$-labeled, bipartite and undirected graphs, and then to simple (K,Λ)-labeled graphs by a slight modification of the encoding B_Λ used in Section 6.2.3. This extension is based on the Reduction Theorem and Corollary 7.38(2). ∎

This theorem leaves open several questions.

Can one weaken the hypotheses of Theorem 7.55?

Question 7.56 Is it true that if a set of simple labeled graphs has a decidable MS-satisfiability problem, then it has bounded clique-width? □

This condition is strictly weaker than the decidability of the C_2MS-satisfiability problem by the following proposition.

Proposition 7.57 Let A be a unary relation symbol. There exists a subset of $STR(\{A\})$ that has a decidable MS-satisfiability problem and an undecidable C_2MS-satisfiability problem.

Proof: An element of $STR(\{A\})$ is a pair $S = \langle D_S, A_S \rangle$, where $A_S \subseteq D_S$. For $i, j \in \mathcal{N}$, we let $S_{i,j}$ be the structure S in $STR(\{A\})$ such that $|D_S| = i+j$ and $|A_S| = i$. For each $i \geq 1$, we let φ_i be the C_2MS sentence such that $MOD(\varphi_i) = \{S_{i,j} \mid j \text{ is odd}\}$.

Let $M_1, M_2, \ldots, M_i, \ldots$ be an effective enumeration of all deterministic Turing machines, where each machine is given with an initial configuration. We let C be the set of structures $S_{i,j}$ such that $i \geq 1$ and either $j = 2k \geq 2i$ and after k steps the computation of M_i is not terminated, or $j = 2k + 1 \geq 2i + 1$ and the computation of M_i terminates in k' steps for some $k' \leq k$. Membership in the set C is decidable. Since the computation of M_i terminates if and only if $MOD(\varphi_i) \cap C \neq \emptyset$, the C_2MS-satisfiability problem of C is undecidable.

We now prove that the MS-satisfiability problem for C is decidable. The result of [Cou90b] recalled in Section 5.2.6 (above Lemma 5.27) entails that every sentence φ in $MS(\{A\}, \emptyset)$ is equivalent (in every $\{A\}$-structure S) to a fixed finite disjunction of conditions of the four possible forms, for p, q in $\mathcal{N}$:

$$C_{1,p,q}: \quad |A_S| = p \quad \wedge \quad |D_S - A_S| = q,$$
$$C_{2,p,q}: \quad |A_S| > p \quad \wedge \quad |D_S - A_S| = q,$$
$$C_{3,p,q}: \quad |A_S| = p \quad \wedge \quad |D_S - A_S| > q,$$
$$C_{4,p,q}: \quad |A_S| > p \quad \wedge \quad |D_S - A_S| > q.$$

We prove that, for each (i, p, q), one can decide if $C_{i,p,q}$ is satisfied by some S in $\mathcal{C}$.

Case of $C_{1,p,q}$: The structure $S_{p,q}$ is the only one that satisfies this condition. Hence $C_{1,p,q}$ is satisfied by some structure in $\mathcal{C}$ if and only if $S_{p,q}$ belongs to $\mathcal{C}$. We have noted that membership in $\mathcal{C}$ is decidable, hence this fact is decidable.

Case of $C_{2,p,q}$: A structure $S_{i,j}$ satisfies this condition if and only if $i > p$ and $j = q \geq 2i$. Hence $C_{2,p,q}$ is satisfied by some structure in $\mathcal{C}$ if and only if $S_{i,q}$ belongs to $\mathcal{C}$ for some i such that $p < i \leq q/2$. This is decidable since membership in $\mathcal{C}$ is decidable.

Case of $C_{3,p,q}$: If $p = 0$, this condition cannot hold for any structure in $\mathcal{C}$. If $p > 0$, $q' \geq p$ and $2q' > q$, then one of the two structures $S_{p,2q'}$ and $S_{p,2q'+1}$ belongs to $\mathcal{C}$ and satisfies $C_{3,p,q}$. Hence $C_{3,p,q}$ is satisfied by some structure in $\mathcal{C}$ if and only if $p \neq 0$. (But for given p, we cannot decide if this is by a structure of the form $S_{p,2q'}$ or of the form $S_{p,2q'+1}$.)

Case of $C_{4,p,q}$: As in the previous case, such a condition is always satisfied by $S_{p+1,r}$ for some large enough r.

This proves that the satisfiability in $\mathcal{C}$ of any given MS sentence is decidable, which concludes the proof. ∎

A related open question concerns extensions of Theorem 7.55 to sets of graphs that have decidable $\mathcal{L}$-satisfiability problems for fragments $\mathcal{L}$ of CMS or of CMS_2 that are defined by limitations on quantifications. To make one precise question out of this remark, we ask:

Question 7.58 Is it true that if a set of labeled graphs has a decidable $\mathcal{L}$-satisfiability problem, where $\mathcal{L}$ is the set of existential MS_2-sentences, i.e., of MS_2-sentences written with a sequence of existential set quantifications followed by a first-order formula, then it has bounded tree-width? ☐

Can one extend Theorem 7.55 to relational structures?

The answer is positive for its first assertion by Theorem 9.19. The corresponding extension of the second one is open (cf. Question 9.20).

Question 7.59 Is it true that if a set $\mathcal{C} \subseteq STR(\mathcal{R})$ has a decidable CMS-satisfiability problem, then $\mathcal{C} \subseteq \tau(\text{TREES})$ for some MS-transduction τ from TREES to $STR(\mathcal{R})$? ☐

Could we have a necessary and sufficient condition in Theorem 7.55?

Having bounded tree-width or clique-width is only a necessary condition for sets of graphs to have a decidable monadic second-order satisfiability problem of a certain type. Strong structural conditions like to be a set of paths (equivalently a set of words), do not imply that a set of graphs satisfying such conditions has a decidable MS-satisfiability problem. From the equivalence of monadic second-order logic and finite automata on words, we get that a language $L \subseteq \{0,1\}^*$ has a decidable MS-satisfiability problem if and only if the following problem is decidable:

Given a finite automaton $\mathscr{A}$, decide if $L \cap L(\mathscr{A}) = \emptyset$.

The class of languages for which this problem is decidable contains in particular all languages of the form $\tau(\text{TREES})$, where τ is a monadic second-order transduction, because these languages are those that are HR-equational (equivalently VR-equational) by Theorem 7.51. They will be studied in Section 8.9.

Can one still get decidable satisfiability problems for extensions of monadic second-order logic?

We can ask for "small extensions" of monadic second-order logic such that the corresponding satisfiability problem remains decidable, say at least on trees. We will present three negative results in this direction.

We let CMS+Eq be the extension of CMS logic with the binary set predicate $EqCard(X,Y)$ meaning that X and Y have the same cardinality. If $P \subseteq \mathcal{N}$ is a decidable set of integers, we let CMS+P be the extension of CMS logic with the set predicate $Card_P(X)$, meaning that the cardinality of X belongs to the set of integers P. In the following proposition, we will consider $P = P_{23} := \{n \in \mathcal{N} \mid n = 2^k \text{ or } n = 3^k \text{ for some } k \geq 1\}$.

For $n > 0$, we let D_n be the structure $\langle [n], \leq \rangle$ (it is nothing but $\lfloor a^n \rfloor^\ell$ without the relation lab_a, cf. Example 5.2(1)). Clearly the CMS-satisfiability problem for $\mathcal{D} := \{D_n \mid n \geq 1\}$ is decidable.

Proposition 7.60 If $\mathcal{L}$ is MS+Eq or CMS+P_{23}, then the $\mathcal{L}$-satisfiability problem for $\mathcal{D}$ is undecidable.

Proof: We let U_1, U_2, U_3 be unary relation symbols and W be a ternary symbol. We let $\mathcal{C}$ be the set of $\{\leq, U_1, U_2, U_3, W\}$-structures S that expand the structures D_n with subsets A, B, C interpreting respectively U_1, U_2, U_3 and a ternary relation R interpreting W, in such a way that R is a subset of $A \times B \times C$ that defines a bijection $f_R : A \times B \to C$. Such a bijection is called a *pairing function*.

We let $\alpha(S)$ denote the integer $\min\{|A|, |B|\}$. There exists a parameterless non-copying MS-transduction τ that transforms S in $\mathcal{C}$ into a directed rectangular grid $G_{|A| \times |B|}$ with vertex set C. Its adjacency relation is defined (for $c, c' \in C$) by:

$c \to c'$ if and only if there exist $a, a' \in A$ and $b, b' \in B$ such that $c = f_R(a, b)$ and $c' = f_R(a', b')$ and, either $a = a'$ and b' is the successor of b in B, or $b = b'$ and a' is the successor of a in A. For every sentence $\varphi \in MS(\{edg\}, \emptyset)$, the sentence $\tau^{\#}(\varphi)$ obtained from φ by the Backwards Translation Theorem (Theorem 7.10) belongs to $MS(\{\leq, U_1, U_2, U_3, W\}, \emptyset)$.

Let $\mathcal{L}$ be an extension of monadic second-order logic such as MS+Eq or CMS+P_{23}. Assume that $\theta_A(x_1)$, $\theta_B(x_1)$, $\theta_C(x_1)$ and $\theta_R(x_1, x_2, x_3)$ are formulas in $\mathcal{L}(\{\leq\}, \{x_1, x_2, x_3\})$ satisfying the following, for every $n > 1$:

> if we define $A_n := \{a \in [n] \mid D_n \models \theta_A(a)\}$ and, similarly, B_n, C_n and R_n, then the $\{\leq, U_1, U_2, U_3, W\}$-structure S_n that we obtain by expanding D_n by A_n, B_n, C_n, R_n (interpreting respectively U_1, U_2, U_3, W) belongs to $\mathcal{C}$.

(This means that each R_n defines a pairing function : $A_n \times B_n \to C_n$.)

Then, we define $\mathcal{C}' \subseteq \mathcal{C}$ as the set of structures S_n and we have:

$$\tau(S_n) \models \varphi \qquad \text{if and only if}$$
$$S_n \models \tau^{\#}(\varphi) \qquad \text{if and only if}$$
$$D_n \models \tau^{\#}(\varphi)[\lambda x_1 \cdot \theta_A / U_1, \lambda x_1 \cdot \theta_B / U_2, \lambda x_1 \cdot \theta_C / U_3, \lambda x_1, x_2, x_3 \cdot \theta_R / W].$$

The last sentence belongs to $\mathcal{L}(\{\leq\}, \emptyset)$. It follows that the MS-satisfiability problem for $\tau(\mathcal{C}')$ reduces to the $\mathcal{L}$-satisfiability problem for $\mathcal{D}$. If the integers $\alpha(S_n)$ are unbounded, then the monadic second-order satisfiability problem for $\tau(\mathcal{C}')$ is undecidable (by Theorem 5.6 and the Reduction Theorem[28]), and so is the $\mathcal{L}$-satisfiability problem for $\mathcal{D}$.

We will prove that this is actually the case if $\mathcal{L} = $ MS+Eq and if $\mathcal{L} = $ CMS+P_{23}. In each case, we will construct formulas $\theta_A(x_1)$, $\theta_B(x_1)$, $\theta_C(x_1)$, $\theta_R(x_1, x_2, x_3) \in \mathcal{L}(\{\leq\}, \{x_1, x_2, x_3\})$ that define pairing functions on unbounded sets.

The construction for MS+Eq: Let $x \in [n]$. The interval $[x]$ $(= [1, x])$ can be defined in D_n as $\{y \mid y \leq x\}$. If X is a subset of $[x]$ with maximal element x, we define $I(X)$ as the family of intervals of the ordered set D_n that consists of $[y]$, where $y := \min(X)$, and of the intervals $]y', y]$, where y' and y are consecutive elements of X. (We define $]y', y]$ as $\{z \mid y' < z \leq y\}$.) Clearly $I(X)$ consists of $|X|$ pairwise disjoint intervals and $[x]$ is their union.

For $x, y, z \in [n]$ the condition $z = x + y$ is equivalent to

$$\exists Y (\text{``}[z] = Y \cup [x]\text{''} \wedge EqCard(Y, [y])),$$

---

[28] To obtain a version of Theorem 5.6 for rectangular grids, we apply the Reduction Theorem to the MS-transduction that transforms the rectangular grid $G_{n \times k}$ into the square grid $G_{m \times m}$ where $m = \min\{n, k\}$, cf. (1.7) in the proof of Proposition 5.14.

and the condition $z = x \cdot y$ is equivalent to

$$\exists X (\text{``}z = \max(X)\text{''} \wedge EqCard(X,[x]) \wedge \forall Z(\text{``}Z \in I(X)\text{''} \Rightarrow EqCard(Z,[y]))),$$

where "$[z] = Y \cup [x]$", "$z = \max(X)$" and "$Z \in I(X)$" are easily expressible in terms of x, z, X, Y and Z by monadic second-order formulas.

In each structure D_n such that $n \geq 4$, an MS+Eq formula can determine the maximal element $a > 1$ such that $a^2 \leq n$, because, by the above observation, an MS+Eq formula can express that there exists z in the domain of D_n such that $z = a \cdot a$. Then one can define A_n and B_n to be $[a]$ and C_n to be $[a^2]$, and one can construct the formulas θ_A, θ_B and θ_C that characterize them.

The relation R_n defined by $R_n(x,y,z)$ if and only if $z = x + (y-1) \cdot a$ for $x, y \in [a]$ and $z \in [a^2]$ defines a pairing function : $A_n \times B_n \to C_n$. It is definable by an MS+Eq formula θ_R since the sum and the product are definable by such formulas. We omit the details. We have $\alpha(S_n) = \lfloor \sqrt{n} \rfloor$ for each n, hence the integers $\alpha(S_n)$ are unbounded. This completes the proof.

The construction for CMS+P_{23}: We will now build pairing functions in D_n by making use of the set predicate $Card_{P_{23}}(X)$. For each pair of integers (p,q) with $p \geq 2$ and $q \geq 1$, we define

$$A'(p) := \{2^k \mid 2 \leq k \leq p\},$$
$$B'(q) := \{3^k \mid 1 \leq k \leq q\},$$
$$C'(p,q) := \{2^k 3^m \mid 2 \leq k \leq p, 1 \leq m \leq q\}.$$

When we say that "x is a power of 2 or 3", we exclude the case $x = 1$. We first observe that, for $x \in D_n$:

x is a power of 2 if and only if $Card_{P_{23}}([x]) \wedge Even([x])$ and [29]

x is a power of 3 if and only if $x > 1 \wedge Card_{P_{23}}([x]) \wedge \neg Even([x])$.

Our objective is to define in D_n by a formula of CMS+P_{23} the pairing function $z = x \cdot y$ for $x \in A'(p)$ and $y \in B'(q)$, where $2^p 3^q \leq n$. We will use the following observations. If $k, k', k'' \geq 1$, then:

(a) $2^k + 2^{k'}$ is a power of 2 if and only if $k = k'$;
(b) $3^k + 3^{k'} + 3^{k''}$ is a power of 3 if and only if $k = k' = k''$.

Let us consider the following conditions on $Y \subseteq \mathcal{N}$ and $z \in \mathcal{N}$:

(i) $z = \max(Y)$ and $|Y|$ is a power of 3;
(ii) for every $Z \in I(Y)$ the integer $|Z|$ is a power of 2;

[29] *Even(X)* means that $|X|$ is even (hence it is the same as $Card_{0,2}(X)$).

(iii) for every two distinct $Z, Z' \in I(Y)$, the integer $|Z \cup Z'|$ is a power of 2.

From observation (a) these facts imply that all intervals of $I(Y)$ have the same cardinality and, finally, that $z = \min(Y) \cdot |Y|$, $\min(Y)$ is a power of 2 and $|Y|$ is a power of 3. If $z = 2^k 3^m$ with $k \geq 1$ and $m \geq 1$, then there exists a (unique) set Y as above and 2^k is its least element. We let $\varphi(z, x)$ be the CMS+P_{23} formula expressing that there exists a set Y with least element x that satisfies conditions (i), (ii) and (iii).

We now consider the following conditions similar to (i), (ii) and (iii):

(i′) $z = \max(Y)$, $|Y|$ is a power of 2 and $|Y| \geq 4$;
(ii′) for every $Z \in I(Y)$, the integer $|Z|$ is a power of 3;
(iii′) for every three pairwise distinct $Z, Z', Z'' \in I(Y)$, the integer $|Z \cup Z' \cup Z''|$ is a power of 3.

Since $I(Y)$ contains more than three elements, one can apply observation (b) to get that $z = 2^k 3^m$ with $k \geq 2$, $m \geq 1$, $|Y| = 2^k$ and $3^m = \min(Y)$. Conversely, if $z = 2^k 3^m$, with $k \geq 2$ and $m \geq 1$, there exists such a set Y satisfying (i′)–(iii′). One obtains a CMS+P_{23} formula $\varphi'(z, y)$ expressing these properties with $y = \min(Y)$.

Hence, the CMS+P_{23} formula $\varphi(z, x) \wedge \varphi'(z, y)$ expresses that $z = x \cdot y$ with $x = 2^k$ for some $k \geq 2$ and $y = 3^m$ for some $m \geq 1$.

If $n \geq 12$, then there exist $a, b \in [n]$ such that:

(C) $a = 2^p$ for some $p \geq 2$, $b = 3^q$ for some $q \geq 1$, $ab \leq n$ and $b < a$.

Furthermore, there is a unique such pair (a, b) such that (b, a) is lexicographically maximal and satisfies Condition (C). This pair is of the form $(2^{p_n}, 3^{q_n})$. From the above constructions, we can see that the elements 2^{p_n}, 3^{q_n} and $2^{p_n} \cdot 3^{q_n}$ can be defined in D_n by CMS+P_{23} formulas. So can the sets $A_n := A'(p_n)$, $B_n := B'(q_n)$, $C_n := C'(p_n, q_n)$ and the ternary relation R_n representing $z = x \cdot y$ for $x \in A_n$ and $y \in B_n$. This defines the structures S_n. Letting $r := \lfloor \log(n)/4 \rfloor$, we can see that the pair $(a, b) = (2^{2^r}, 3^r)$ satisfies condition (C). Hence, for a maximal pair (as defined above), we have $q_n \geq r$. It follows that $\alpha(S_n) \geq \lfloor \log(n)/4 \rfloor$, hence the numbers $\alpha(S_n)$ are unbounded. This completes the proof. ∎

This proof suggests one more question.

Question 7.61 Is the (CMS+*Prime*)-satisfiability problem for $\mathcal{D}$ decidable, where *Prime* is the set of prime numbers? [30] □

As final similar case, we define MS+*Auto* as the extension of MS logic with the set predicate *Auto*(X) meaning that the structure induced by X has a nontrivial automorphism. We get that the set of structures $T_n := \langle [n], suc \rangle$ (where *suc* is the successor relation) has an undecidable (MS+*Auto*)-satisfiability problem. The basic observation is that if X is the union of two intervals $[p, q]$ and $[r, s]$ with $r \geq q + 2$, then $T_n[X]$

---

[30] It is proved in [Bes] that the (MS + P)-satisfiability problem is undecidable on words if P is not semi-linear.

has a nontrivial automorphism if and only if $q - p = s - r$. By using it, one can build an MS+*Auto* formula $\varphi(x, y, z)$ to express that x and y are different prime numbers and that $z = x \cdot y$. It is then easy to build pairing functions as in the previous two constructions.

7.6 Questions about logical characterizations of recognizability

Recognizability is equivalent to MS-definability for languages over a finite alphabet or a finite functional signature. It is equivalent to CMS-definability for sets of rooted trees and for sets of terms over an associative and commutative signature. We have proved these results in Chapter 5 (see Section 5.5). On the other hand, we have proved that there are uncountably many HR- and VR-recognizable sets (cf. Section 4.2.3), so that there is no hope to have such characterizations for them. The following conjecture has been made in [Cou91]:

Conjecture 7.62 Every HR-recognizable set of graphs of bounded tree-width is CMS_2-definable.

Together with the Recognizability Theorem for the HR algebra (Theorem 5.68), it would imply that a set of graphs of bounded tree-width is HR-recognizable if and only if it is CMS_2-definable. For simple graphs of bounded tree-width, CMS_2-definability is equivalent to CMS-definability as we will prove in Section 9.4 (cf. Theorem 5.22); we proved that for forests in Example 7.8. Hence, the conjecture implies that a set of simple graphs of bounded tree-width is HR-recognizable if and only if it is CMS-definable. Concerning the algorithmic applications, if the conjecture is true, then no logical language strictly more expressive than counting monadic second-order logic can have fixed-parameter model-checking algorithms based on automata, like those obtained in Chapter 6, with respect to tree-width as parameter.

Since it is hard to prove or disprove this conjecture, relativized versions have been considered. Let us say that the conjecture holds for a set of graphs $\mathcal{C}$ if all its HR-recognizable subsets of bounded tree-width are CMS-definable. The following relativizations have been established: to graphs of tree-width at most 2 ([Cou91]), of tree-width at most 3 and, for each k, to k-connected graphs of tree-width at most k ([Kal]). The cases of graphs of path-width at most k and of all graphs (of tree-width at most k) have been announced in [Kab] and [Lap] respectively, but the full proofs have not been published. The conjecture is still open.

One can raise a similar question for VR-recognizability, about which we make no conjecture.

Question 7.63 Is it true that every VR-recognizable set of simple graphs of bounded clique-width is CMS-definable? □

A positive answer would imply a positive answer to Conjecture 7.62 relativized to the class of simple graphs. Let us prove this implication: if L is an HR-recognizable set of simple graphs of bounded tree-width, then, by Theorem 4.59 (proved in [CouWei]), it is VR-recognizable because it is without $K_{n,n}$ for some n, hence it is CMS-definable (assuming positive the answer to the question) and thus CMS_2-definable.

All above-cited articles use (essentially) the same technique that we now describe. The binary signature $F^{HR}_{[k]}$ has a unique binary operation $/\!/$ which is associative and commutative. Hence, the homomorphism $val : T(F^{HR}_{[k]}) \to \mathbb{JS}$ factorizes through $\mathbb{T}_{AC}(F^{HR}_{[k]})$; we will denote by val_{AC} the corresponding homomorphism $: \mathbb{T}_{AC}(F^{HR}_{[k]}) \to \mathbb{JS}$. The objective is to apply Theorem 5.88 of Section 5.5. Note that we drop $H := \{/\!/\}$ from the notation of Definition 5.85, i.e., we write $\mathbb{T}_{AC}(F^{HR}_{[k]})$ for $\mathbb{T}_{AC}(F^{HR}_{[k]}, H)$. Moreover, we will denote the domain of this algebra by $T_{AC}(F^{HR}_{[k]})$ instead of $T(F^{HR}_{[k]})/\equiv_{AC(H)}$, and refer to its elements as terms (they are, in fact, equivalence classes of terms with respect to the equivalence $\equiv_{AC(H)}$); see Section 5.5 for the representation of these "terms" by relational structures.

Let $\mathcal{C}$ be a set of graphs, k be a nonnegative integer and $\mathcal{C}_k$ be the set of graphs in $\mathcal{C}$ of tree-width at most k. Assume that there exists an integer $f(k) \geq k$ and a $CMS_{2,1}$-transduction τ_k such that for every graph G in $\mathcal{C}_k$:

(1) the set $\tau_k(G)$ contains at least one term in $T_{AC}(F^{HR}_{[f(k)]})$;
(2) $val_{AC}(t) = G$ for every term t in the set $\tau_k(G)$.

If $L \subseteq \mathcal{C}_k$ is HR-recognizable, then it is also recognizable in the algebra $\mathbb{JS}^{gen}[[k]]$ (cf. the proof of Corollary 5.69). Hence the set $val^{-1}_{AC}(L)$ is recognizable in $\mathbb{T}_{AC}(F^{HR}_{[f(k)]})$ (cf. Definition 5.85), and so it is CMS-definable by Theorem 5.88. Then the set of graphs L is $\tau^{-1}_k(val^{-1}_{AC}(L))$ and is CMS-definable by the Backwards Translation Theorem (Theorem 7.10).

The key point is thus the construction of τ_k. We make two observations. First, unless $\mathcal{C}$ is finite or in some other way very particular (for example, the set PATHS), we cannot replace $T_{AC}(F^{HR}_{[f(k)]})$ by $T(F^{HR}_{[f(k)]})$ (and val_{AC} by val) in the above proof method. Consider the set $\mathcal{C}$ of graphs without edges. We have proved in Proposition 7.28 that the rooted trees generated from $\mathcal{C}$ by a fixed CMS-transduction have a bounded number of nonleaf nodes. Hence, Conditions (1) and (2) cannot be satisfied with $T(F^{HR}_{[f(k)]})$ instead of $T_{AC}(F^{HR}_{[f(k)]})$, and even for $k = 0$.

Our second observation is that the transduction τ_k cannot be val^{-1}_{AC}, i.e., it cannot associate with a graph G all the terms in $T_{AC}(F^{HR}_{[f(k)]})$ that evaluate to it. There is a trivial reason: the set $val^{-1}_{AC}(G)$ is infinite, because of Equalities (3), (5)–(7), (9) and (12)–(17) of Proposition 2.48 (Equality (3) alone implies that). For a less trivial counter-example, we consider again the series-parallel graphs, generated by e and the operations $/\!/$ and $\bullet$: the set of different terms modulo the associativity and commutativity of $/\!/$ that denote the simple series-parallel graphs with n vertices is finite but not of cardinality $O(n)$. Hence, no monadic second-order transduction τ_k can construct all such (equivalence classes of) terms from a given series-parallel graph G. Thus, a transduction τ_k satisfying Conditions (1) and (2) must select particular terms that evaluate to the input graph. Such transductions are constructed in [Cou91] and in [Kal].

Instead of a transduction τ_k that constructs terms, one could think of a transduction that constructs for every graph in C_k a structure, as described in Example 5.2(4), that represents a tree-decomposition of G of width at most $f(k)$. (We will do that in Proposition 9.56 for k-chordal graphs.) Such a construction is equivalent to the first one because the transformations between quotient-terms (in $T_{AC}(F^{HR}_{[f(k)]})$) and the corresponding tree-decompositions are monadic second-order transductions, see Definitions 2.75 and 2.80 and the proof of Theorem 7.50.

Here is an auxiliary question, where the associative and commutative operation of $F^{VR}_{[m]}$ is $\oplus$:

Question 7.64 Is it true that for each k there is an integer m and a $CMS_{2,1}$-transduction that associates with every simple graph of clique-width at most k a term in $T_{AC}(F^{VR}_{[m]})$ that evaluates to it? $\square$

A positive answer would not solve Question 7.63 but would perhaps lead to a solution.

7.7 References

The Equationality Theorems for the VR and HR algebras have been proved first (in a different way) in [EngOos97] and in [CouEng]. Proofs similar to the one given for the case of the VR algebra and that extend to the algebra of relational structures are in [Cou92] and [BluCou06].

The linear HR-equational and VR-equational sets are those that are generated by linear context-free hyperedge replacement and vertex replacement graph grammars respectively, where such a grammar is called linear if, as usual, the right-hand side of each of its rewriting rules has at most one occurrence of a nonterminal symbol (see [*Eng97] for a survey). For an example of a linear HR grammar, see Example 4.28.

Monadic second-order transductions from terms to terms are characterized in [BloEng00] and [EngMan99], in terms of attribute grammars and of macro tree transducers, respectively. The HR-equational sets of words and terms are the same as the VR-equational ones. For words, they can be characterized in terms of deterministic tree-walking transducers ([EngHey91]). For terms, they can be characterized in terms of attribute grammars ([BloEng00]) or of macro tree transducers ([EngMan00]). These results will be presented in Chapter 8.

Theorem 7.55(1) is proved in a different way in [See91]. In this article, Seese asks whether a set of graphs that has a decidable MS-theory is the image of a set of trees under an MS-transduction. This question is solved affirmatively (but partially) in [CouOum], see Theorem 7.55(2).

The Reduction Theorem (Theorem 7.54) is a classical tool introduced by Rabin in [Rab65] and [Rab77] under the name of "interpretation of theories" for proving that certain first-order theories are decidable or undecidable by reduction to known cases.

Certain graph transformations that are not monadic second-order transductions satisfy a weak Backwards Translation Theorem. One example is the *unfolding* of rooted, directed and acyclic (i.e., without circuits) graphs that transforms a graph G into the finite tree of its finite directed paths originating from a specified vertex r designated by some label (a unary relation in $\lfloor G \rfloor$ denoting a singleton set). This tree, denoted by $Unf(G,r)$, is called the unfolding of G from r. Since the unfolding is not of linear size increase, it is not a monadic second-order transduction. However, there exists a mapping $\varphi \to \widehat{\varphi}$ from CMS sentences to CMS sentences such that, for every (G,r):

$$Unf(G,r) \models \varphi \text{ if and only if } \lfloor G \rfloor \models \widehat{\varphi}.$$

This result is proved in [CouWal] for all finite and countable directed graphs: their unfoldings are finite or countably infinite trees. However, it holds only for sentences, and not for arbitrary formulas. To see this point, consider a finite graph G with circuits: the bijection described in Remark 7.13(1) between the assignments that satisfy a formula in a structure T obtained from S by a monadic second-order transduction and those that satisfy an associated formula in S cannot exist by an obvious comparison of cardinalities (just take $\varphi(X)$ equal to *True*). But even if G is acyclic, no such bijection can exist.

The definitions by monadic second-order transductions of orientations of graphs (Example 7.46) and of linear orders of their vertex sets (Example 7.3(3)) will be studied in Chapter 9 (Propositions 9.42(1), 9.43 and 9.46); more results can be found in [Cou95a] and [Cou96a].

Theorem 8.1 of [HliSee] states for infinite trees a result similar to Proposition 7.57. The proof of this proposition is based on the proof sketch given in that article.

The comparison of sets of structures via monadic second-order transductions (cf. Section 7.1.7) has been initiated in [BluCou10] where the relation $\sqsubseteq$ between sets of incidence graphs is characterized.

8

Transductions of terms and words

As explained in the introduction of Section 1.7, there are no appropriate finite-state automata or finite-state transducers that work "directly" on graphs. Thus, for graphs, the role of automata and transducers is taken over by monadic second-order logic: instead of being accepted by an automaton, a set of graphs is defined by a monadic second-order sentence; and instead of being computed by a transducer, a graph transduction is defined by a definition scheme consisting of monadic second-order formulas. With respect to automata, the original motivation for this approach was Theorem 1.16 (cf. Theorem 5.82): for terms and words monadic second-order logic and finite-state automata have the same expressive power.

The aim of this chapter is to show that, in a certain sense, the automata-theoretic characterization of monadic-second order logic for sets of terms and words (Theorem 1.16) can be generalized to transductions. This means of course, that the automata should produce output, i.e., that they are transducers. We will concentrate on transductions that are partial functions, and in particular on deterministic devices, i.e., parameterless definition schemes and deterministic transducers. For these, we will show that monadic second-order transductions of words correspond to two-way finite-state transducers, and that monadic second-order transductions of terms correspond to (compositions of) tree-walking transducers. These transducers are well known from Formal Language Theory.

A two-way finite-state transducer is a finite-state automaton with a two-way read-only input tape and a one-way output tape. The input word is placed on the input tape between endmarkers, and the transducer has a reading head that is positioned over a cell of the input tape, at each moment of time. In one computation step, it reads the symbol in that cell, moves its head by one cell to the left or to the right (or keeps it at the same cell), writes some symbols on the output tape, and goes into another state. Such a transducer is said to be a one-way finite-state transducer if it never moves to the left.

A tree-walking transducer is very similar to a two-way finite-state transducer, but is more complicated as both its input and output are trees (or more precisely, syntactic trees of terms). It is a finite-state device, of which the reading head is positioned over a node of the input tree, at each moment of time; this node will be called the "current node." In one computation step, it reads the label of that node and either does not produce output, or produces one labeled node of the output tree. In the first case, it moves its head up to the father or down to one of the sons of the current node (or lets it stay at the same node), and goes into another state. In the second case, the computation of the transducer branches into k parallel computations, where k is the number of sons of the produced output node. Each of these computations then moves its head to the father or one of the sons of the current node (or lets it stay where it is), and goes into another state. Different computations can move to different input nodes and go into different states; in other words, the parallel computations behave independently (and will continue to behave independently in the following steps). The task of the i-th computation is to output the tree rooted at the i-th son of the produced output node. Thus, the output tree is generated in a top-down fashion, with several configurations of the transducer at some of the leaves of the output generated sofar. Each of these configurations represents a parallel branch of computation, which ends when an output node is produced that has no sons (i.e., $k = 0$). It should be noted that a tree-walking transducer can detect whether or not the current node has a father (i.e., whether it is the root), which son it is of its father, and how many sons it has itself. We also note, for readers familiar with alternation, that the branching behavior of a nondeterministic tree-walking transducer is similar to that of an alternating automaton.

We will identify words with terms over a signature consisting of unary function symbols and the constant symbol ϵ. A word $a_1 a_2 \cdots a_n$ is identified with (the syntactic tree of) the term $a_1(a_2(\cdots a_n(\epsilon)\cdots))$. A tree-walking transducer that translates such unary terms (trees) into unary terms (trees) is, apart from notational matters, the same as a two-way finite-state transducer. The symbol ϵ acts as a right endmarker; there is no left endmarker, but the transducer can detect the left end of the word because it is the root of the tree. Walking up and down on the input tree corresponds to the two-way (left and right) motion on the input tape. Since the output tree is unary, the transducer never branches, and so the output is produced as on a one-way output tape.

One might expect or hope that MS-transductions of words correspond to the more usual ones, called rational transductions, that are defined by one-way finite-state transducers (see, e.g., [*Ber]). However, these two classes of transductions have different properties. As observed in the introduction of Section 1.7, the class of rational transductions is closed under composition and inverse, and rational transductions preserve regularity and context-freeness of languages. On the other hand, the class of MS-transductions of words is closed under composition (by Theorem 7.14) but not under inverse. Moreover, inverse MS-transductions of words preserve regularity of languages (by Corollary 7.12) but MS-transductions do not (see Remark 7.23). MS-transductions do not preserve context-freeness either, but (by Corollary 7.37) they do preserve VR-equationality (of which context-freeness is a, proper, special case). These statements will be made more precise in Examples 8.2 and 8.3 and in Section 8.9.

Through the Weak Recognizability Theorem (Section 5.3.9, Corollary 5.69), a finite-state automaton on terms can be viewed as the implementation of a specification of a set of graphs by an MS sentence (i.e., of an MS-definable set of graphs). For describing terms, the implementation formalism corresponds exactly to the specification formalism, in the sense that they have the same expressive power.

This chapter shows how to implement MS-transductions of terms and words by finite-state transducers. In the case of words, the two-way finite-state transducers correspond exactly to the MS-transductions, in the above sense (Theorem 8.10). For terms, MS-transductions can be implemented by the compositions of two tree-walking transducers, the first of which never moves up (Theorem 8.16). These composed transducers correspond exactly to the MS-transductions when restricted to transductions of linear size increase (Theorem 8.17). For terms we will discuss other implementation devices: the tree-walking pushdown tree transducer (Theorem 8.13), the multi bottom-up tree-to-word transducer (Theorem 8.21) and the macro tree transducer (Theorem 8.22). After the automata-theoretic characterization of MS-transductions of terms, we obtain from Theorems 7.34 and 7.50 that each of the above tree transducers can be viewed as the implementation of a specification of a graph transduction by a monadic second-order definition scheme (Theorem 8.18). Moreover, we obtain from Theorem 7.36 (the Equationality Theorem) automata-theoretic characterizations of the VR-equational, equivalently HR-equational, sets of terms and words (Theorems 8.27 and 8.28).

8.1 Terminology

We review some definitions and notation, and we introduce some new terminology, to be used in this chapter.

For transductions τ and τ', their composition is $\tau \cdot \tau' = \{(S,T) \mid (S,U) \in \tau$ and $(U,T) \in \tau'$ for some $U\}$ (see Definition 7.1). For classes X and X' of transductions, $X \cdot X' = \{\tau \cdot \tau' \mid \tau \in X, \tau' \in X'\}$.

8.1.1 Terms and words

In this chapter, all *alphabets are finite*. Moreover, all *functional signatures are finite and one-sorted*, and all *relational signatures are without constant symbols*. A term will also be called a tree; this should not lead to confusion, because in this chapter we will not consider trees other than syntactic trees of terms. Finally, all *unary functional signatures have exactly one constant symbol*, which is denoted ϵ. A term over a unary functional signature will be called a unary term (or unary tree).

For a functional signature F, we use the relational signature $\mathcal{R}_F := \{son_i \mid i \in [\rho(F)]\} \cup \{lab_f \mid f \in F\} \cup \{rt\}$ where $\rho(rt) = 1$. A term t in $T(F)$ is faithfully represented by the $\mathcal{R}_F$-structure $\lfloor t \rfloor := \langle N_t, (son_{it})_{i \in [\rho(F)]}, (lab_{ft})_{f \in F}, rt_t \rangle$ where N_t is the set of nodes of the syntactic tree of t, $lab_{f,t} = Occ(t,f)$, $son_{i,t}$ is the set of pairs (u,v) such that v is the i-th son of u and $rt_t = \{root_t\}$. Without rt, this definition of $\mathcal{R}_F$ and $\lfloor t \rfloor$ is the same as in the beginning of Example 5.2(3) and the beginning of Section 6.3.3. Note that we have taken the rt relation to be of arity 1, to avoid relational signatures with constants. In fact (as observed in Example 5.2(3)), the relation rt is superfluous, because it is FO-expressible in terms of the others: $rt(x) \iff \neg \bigvee_{i \in [\rho(F)]} \exists y.son_i(y,x)$. Thus, whenever we define a monadic second-order transduction to $T(F)$, we need not (and usually will not) specify the relation formulas $\theta_{rt,i}$ that define the rt relation in the output structure. The rt relation is needed, however, because we will use Application 5.61 (see Sections 8.1.3 and 8.4).

In this chapter *we view words as a special case of terms*, viz. unary terms. In this way we avoid the duplication of definitions and proofs. For an alphabet A, there is a natural bijection $\mu_A : A^* \to T(U_A)$, where U_A is the unary functional signature $A \cup \{\epsilon\}$ ($\epsilon \notin A$) with $\rho(a) = 1$ for every $a \in A$ and $\rho(\epsilon) = 0$: the word $a_1 a_2 \cdots a_n$ corresponds to the term $\mu_A(a_1 a_2 \cdots a_n) = a_1(a_2(\cdots a_n(\epsilon) \cdots))$. In fact, μ_A is the inverse of the mapping $val_{\mathbb{W}_{left}(A)}$, see Definition 2.7. A language $L \subseteq A^*$ is regular if and only if the set of unary terms $\mu_A(L) \subseteq T(U_A)$ is regular (see Remark 3.20(4), the discussion after Proposition 3.23, and Remark 3.51(2)).

A word $w \in A^*$ is faithfully represented by the $\mathcal{R}_{U_A}$-structure $\lfloor w \rfloor := \lfloor \mu_A(w) \rfloor$ (cf. Example 5.2(3)). In this chapter it is important that the empty word ε is represented by a structure with a singleton as domain rather than by the empty structure. The reason is that an MS-transduction transforms the empty structure into itself, whereas it can transform a singleton structure into any other structure (Proposition 7.17). For MS-transductions, the definition of $\lfloor w \rfloor$ in Example 5.2(1) is strongly equivalent to the above definition, provided $\lfloor \varepsilon \rfloor$ is defined to have a singleton as domain (see Definition 7.9).

By Propositions 5.29 and 5.30, CMS-transductions taking terms or words as input are also MS-transductions. Hence, without loss of generality, we will only consider MS-transductions of terms and words. We will mainly consider deterministic devices, i.e., parameterless definition schemes and deterministic transducers; to stress this, we will also use "deterministic" instead of "parameterless." We recall that parameterless MS-transductions are partial functions. We denote by DMSOW the class of parameterless MS-transductions of words (where the "D" stands for "deterministic" whence "parameterless"), by DMSOT the class of parameterless MS-transductions of terms, and by DMSOTW those from terms to words. Thus, DMSOT is the class of all transductions $\tau : T(F) \to T(H)$ such that $\tau' := \{(\lfloor s \rfloor, \lfloor t \rfloor) \mid (s,t) \in \tau\}$ (or more precisely, $[\tau']_{iso}$) is a parameterless MS-transduction from $STR(\mathcal{R}_F)$ to $STR(\mathcal{R}_H)$, and similarly for the other two classes (cf. Section 7.1.3).[1] It follows from Theorem 7.14 that DMSOW and DMSOT are closed under composition.

---

[1] We note that every parameter-invariant MS-transduction of terms and/or words is a parameterless MS-transduction, cf. Footnote 25 in the proof of Theorem 7.50.

A term $t \in T(F)$ is a special kind of word over the alphabet F (cf. Definition 2.2). If w_t denotes the term t viewed as a word, then $\lfloor t \rfloor$ and $\lfloor w_t \rfloor$ are different structures. However, the mapping $t \mapsto w_t$ is a noncopying parameterless MS-transduction. Thus, by the closure of the class of MS-transductions under composition, we have that DMSOT $\subseteq$ DMSOTW.

As observed in Definition 7.2 (and in Fact 1.37), every MS-transduction τ (of structures) is of *linear size increase*, i.e., there is an integer k such that $|D_T| \leq k \cdot |D_S|$ for every $(S,T) \in \tau$. For terms and words this implies $|t| \leq k \cdot |s|$ for every $(s,t) \in \tau$, i.e., the size of the output is linear in the size of the input. An arbitrary transduction τ *has finite images* if $\tau(S) = \{T \mid (S,T) \in \tau\}$ is finite (up to isomorphism) for every input structure S. Since, for given n and $\mathcal{R}$, there are only finitely many $\mathcal{R}$-structures T such that $|D_T| \leq n$ (up to isomorphism), every transduction of linear size increase (of type $\mathcal{R}' \to \mathcal{R}$), in particular every MS-transduction, has finite images.

8.1.2 Automata

Recall from Definition 3.46 in Section 3.3.1 that, given a functional signature F, a (bottom-up, ε-free) *automaton* over F (or F-*automaton*) is a tuple $\mathscr{A} = \langle F, Q, \delta, Acc \rangle$ where Q is a set of states, $Acc \subseteq Q$ is the set of accepting states, and δ is a set of transition rules of the form $f[q_1,\ldots,q_k] \to q$ with $f \in F_k$ and $q, q_1,\ldots,q_k \in Q$ (if $k = 0$, the transition rule is also written $f \to q$). An automaton without accepting states is called a *semi-automaton*. Recall from Definition 3.49 that $\mathscr{A}$ is *complete and deterministic* if for every $f \in F_k$ and $q_1,\ldots,q_k \in Q$ there is exactly one transition rule $f[q_1,\ldots,q_k] \to q$ in δ (with $q \in Q$). This definition is based on viewing $\mathscr{A}$ as a bottom-up device.

In Section 8.5 we will also view $\mathscr{A}$ as running *top-down*. In that case, intuitively, Acc is the set of initial states (because $\mathscr{A}$ starts at the root of the input tree), and a transition rule $f[q_1,\ldots,q_k] \to q$ in δ is used from right to left: if $\mathscr{A}$ is in state q at node u with label f, then $\mathscr{A}$ branches and moves to the i-th son of u in state q_i, for each $i \in [k]$. With this view in mind, we say that $\mathscr{A}$ is *top-down complete and deterministic* if Acc is a singleton and for every $q \in Q$ and $f \in F_k$ with $k \neq 0$ there is exactly one transition rule $f[q_1,\ldots,q_k] \to q$ in δ (with $q_1,\ldots,q_k \in Q$).

Let $\mathscr{A} = \langle F, Q, \delta \rangle$ be a complete and deterministic semi-automaton. For $f \in F_k$ and $q_1,\ldots,q_k \in Q$ we denote by $\delta_f(q_1,\ldots,q_k)$ the unique state q such that $f[q_1,\ldots,q_k] \to q$ is a transition rule in δ. For $s \in T(F)$ and $u \in N_s$, *the state in which $\mathscr{A}$ reaches u* is the state $run_{\mathscr{A},s}(u)$, where $run_{\mathscr{A},s}$ is the unique run of $\mathscr{A}$ on s, cf. Definition 3.49. This state can also be defined recursively, as follows: if $u \in Occ(s,f)$ with $\rho(f) = k$, then $run_{\mathscr{A},s}(u) = \delta_f(run_{\mathscr{A},s}(u_1),\ldots,run_{\mathscr{A},s}(u_k))$, where $u_1,\ldots,u_k$ are the sons of u.

Now let $\mathscr{A} = \langle F, Q, \delta, Acc \rangle$ be top-down complete and deterministic. For $s \in T(F)$ and $u \in N_s$ we define $td\text{-}run_{\mathscr{A},s}(u) \in Q$, *the state in which $\mathscr{A}$ reaches u top-down*, recursively as follows: $td\text{-}run_{\mathscr{A},s}(root_s)$ is the state q such that $Acc = \{q\}$, and if $u \in Occ(s,f)$ with $\rho(f) = k \neq 0$ and with sons $u_1,\ldots,u_k$, then, for every $i \in [k]$, *td-run*$_{\mathscr{A},s}(u_i)$ is the state q_i such that $f[q_1,\ldots,q_k] \to q$ is the unique rule in δ with $q = td\text{-}run_{\mathscr{A},s}(u)$. Note that $\mathscr{A}$ need not have a run on s, but if it has one, then *td-run*$_{\mathscr{A},s}$ is the unique run of $\mathscr{A}$ on s.

8.1.3 Logic

Let F be a functional signature. A tree-walking transducer implementing an MS-transduction of terms will need to know at each node u of a tree $s \in T(F)$ which MS formulas $\varphi(x_1)$ are true at u, and in some cases even which formulas $\varphi(x_1,x_2)$ are true at u and some other node. To handle this formally, we define the following theories, similar to, but different from those in the proof of the Equationality Theorem for the VR algebra (more precisely, the proof of Theorem 7.34). For the notation used see Application 5.61.

Let F and $q \in \mathcal{N}$ be fixed. The integer q will be specified later. For $n \in \mathcal{N}$, let $L_n := \widehat{\mathrm{MS}^q}(\mathcal{R}_F, \{x_1,\ldots,x_n\})$ and $L'_n := \widehat{\mathrm{MS}^q}(\mathcal{R}_{F \cup \{w\}}, \{x_1,\ldots,x_n\})$, where w is the special variable used for contexts. Recall from Definition 5.59 and Section 5.6 that L_n and L'_n are finite. For $s \in T(F)$ and $u \in N_s$, we define

$$Th_s^{\downarrow}(u) := \{\varphi \in L_1 \mid \lfloor s \rfloor / u \models \varphi(u)\} \;\;= Th(s, \downarrow u, q, 0, u),$$
$$Th_s^{\uparrow}(u) := \{\varphi \in L'_1 \mid \lfloor s \rfloor \uparrow u \models \varphi(u)\} = Th(s, \uparrow u, q, 0, u), \text{ and}$$
$$Th_s(u) := \{\varphi \in L_1 \mid \lfloor s \rfloor \models \varphi(u)\} \;\;\;\;\;\; = Th(s, \downarrow root_s, q, 0, u),$$

and for $s \in T(F)$ and $x, u \in N_s$, we define

$$Th_s^{\downarrow}(x,u) := \{\varphi \in L_2 \mid \lfloor s \rfloor / u \models \varphi(x,u)\} \quad \text{provided } x \in N_s / u,$$
$$Th_s^{\uparrow}(x,u) := \{\varphi \in L'_2 \mid \lfloor s \rfloor \uparrow u \models \varphi(x,u)\} \quad \text{provided } x \in N_s \uparrow u, \text{ and}$$
$$Th_s(x,u) := \{\varphi \in L_2 \mid \lfloor s \rfloor \models \varphi(x,u)\},$$

and so $Th_s^{\downarrow}(x,u) = Th(s, \downarrow u, q, 0, (x,u))$, and similarly for the other two. Note that if $m \leq n$ then the set of formulas $\mathrm{MS}^q(\mathcal{R}_F, \{x_1,\ldots,x_m\})$ is included in $\mathrm{MS}^q(\mathcal{R}_F, \{x_1,\ldots,x_n\})$, and similarly for $\mathcal{R}_{F \cup \{w\}}$. Hence, e.g., $Th_s^{\downarrow}(u)$ contains all sentences $\widehat{\varphi}$ such that $\varphi \in \mathrm{MS}^q(\mathcal{R}_F, \emptyset)$ and $\lfloor s \rfloor / u \models \varphi$.

The theory $Th_s^{\uparrow}(root_s)$ is independent of s because the $\mathcal{R}_{F \cup \{w\}}$-structure $S := \lfloor s \rfloor \uparrow root_s$ is the same for all s: it has the singleton domain $D_S = \{1\}$ with $lab_w S = rt_S = D_S$, the other relations being empty. We will denote this theory by $Th_{rt}^{\uparrow}$ and call it the *root theory*.

We now use the Splitting Theorem for derived operations (Theorem 5.57 and Corollary 5.60), and in particular Application 5.61, to obtain relationships between the above theories.

Let u be a node of $s \in T(F)$ with label $f \in F_k$ and sons $u_1,\ldots,u_k$. By Equation (5.6) in Application 5.61, there is a mapping $\delta_f : \mathcal{P}(L_1)^k \to \mathcal{P}(L_1)$ (independent from s

and u), such that

$$Th_s^\downarrow(u) = \delta_f(Th_s^\downarrow(u_1), \ldots, Th_s^\downarrow(u_k)).^2$$

Let us express this by saying that

$$Th_s^\downarrow(u) \text{ can be determined from } f \text{ and } Th_s^\downarrow(u_1), \ldots, Th_s^\downarrow(u_k)$$

(implicitly by a fixed mapping, δ_f in this case). Similarly, now using $Z_{f^\uparrow}$ of Equation (5.5) in Application 5.61, there is a mapping $Z_f : \mathcal{P}(L_1') \times \mathcal{P}(L_1)^k \to \mathcal{P}(L_1)$, such that

$$Th_s(u) = Z_f(Th_s^\uparrow(u), Th_s^\downarrow(u_1), \ldots, Th_s^\downarrow(u_k)).$$

In other words, $Th_s(u)$ can be determined from f, $Th_s^\uparrow(u)$, and all theories $Th_s^\downarrow(u_j)$, $j \in [k]$. Moreover, by a similar argument (cf. Equation (5.7) in Application 5.61), for every $i \in [k]$ there is a mapping $Z_{f,i} : \mathcal{P}(L_1') \times \mathcal{P}(L_1)^{k-1} \to \mathcal{P}(L_1')$, such that

$$Th_s^\uparrow(u_i) = Z_{f,i}(Th_s^\uparrow(u), Th_s^\downarrow(u_1), \ldots, Th_s^\downarrow(u_{i-1}), Th_s^\downarrow(u_{i+1}), \ldots, Th_s^\downarrow(u_k)),$$

i.e., $Th_s^\uparrow(u_i)$ can be determined from f, $Th_s^\uparrow(u)$, and all theories $Th_s^\downarrow(u_j)$ with $j \in [k]$, $j \neq i$.

Similar statements are valid when two nodes x and u are considered. For instance, if $x \in N_s \uparrow u$, then $Th_s(x, u)$ can be determined from f, $Th_s^\uparrow(x, u)$, and $Th_s^\downarrow(u_1), \ldots, Th_s^\downarrow(u_k)$. As another example, if $x \in N_s/u_i$, then $Th_s^\downarrow(x, u)$ can be determined from f, $Th_s^\downarrow(x, u_i)$, and all theories $Th_s^\downarrow(u_j)$ with $j \neq i$; moreover, for every $j \neq i$, $Th_s^\uparrow(x, u_j)$ can be determined from f, $Th_s^\uparrow(u)$, $Th_s^\downarrow(x, u_i)$, and all theories $Th_s^\downarrow(u_m)$ with $m \neq i, j$.

We note here that the mappings δ_f considered above can be viewed as the transition rules of a complete and deterministic finite semi-automaton $\mathcal{A} = \langle F, Q, \delta \rangle$ with $Q = \mathcal{P}(L_1)$. Obviously, $run_{\mathcal{A},s}(u) = Th_s^\downarrow(u)$ for every $s \in T(F)$ and $u \in N_s$. Hence, if $\varphi \in L_0$ has quantifier-height $\leq q$ and Acc consists of all subsets of L_1 that contain $\widehat{\varphi}$, then $\bigcup_{q \in Acc} L(\mathcal{A}, q) = \{s \in T(F) \mid s \models \varphi\}$ (see Corollary 5.67).

We observe, as in Step 1 of the proof of Theorem 7.34, that for every $\Phi \subseteq L_1$ there exists an MS formula $\eta_\Phi^\downarrow(x_1)$ such that $\lfloor s \rfloor \models \eta_\Phi^\downarrow(u)$ if and only if $Th_s^\downarrow(u) = \Phi$, and a similar formula $\eta_\Phi^\uparrow(x_1)$ exists expressing that $Th_s^\uparrow(u) = \Phi$. The formula $\eta_\Phi^\downarrow(x_1)$ is

$$\exists X \left[\forall y (y \in X \Leftrightarrow y \leq x_1) \wedge \bigwedge_{\varphi \in \Phi} \varphi'(X, x_1) \wedge \bigwedge_{\varphi \in L_1 - \Phi} \neg \varphi'(X, x_1) \right],$$

where $y \leq x_1$ is a formula expressing that y is a descendant of x_1 and $\varphi'(X, x_1)$ is obtained from $\varphi(x_1) \upharpoonright X$, the relativization of $\varphi(x_1)$ to X (cf. Section 5.2.1), by the

---

[2] To be precise, $\delta_f(\Phi_1, \ldots, \Phi_k) = Z_{f^*, q, 0, (1, 0, \ldots, 0)}(\Phi_1', \ldots, \Phi_k')$, where $\Phi_i \subseteq L_1$ and $\Phi_i' = \Phi_i \cap L_0$.

replacement of each atomic formula $rt(z)$ by $z = x_1$. Similarly, the formula $\eta_\Phi^\uparrow(x_1)$ is

$$\exists X \left[\forall y (y \in X \Leftrightarrow \neg(y < x_1)) \wedge \bigwedge_{\varphi \in \Phi} \varphi''(X, x_1) \wedge \bigwedge_{\varphi \in L_1' - \Phi} \neg\varphi''(X, x_1) \right],$$

where $\varphi''(X, x_1)$ is obtained from $\varphi(x_1) \upharpoonright X$ by the replacement of each atomic formula $lab_w(z)$ by $z = x_1$ and each atomic formula $lab_f(z)$ (where $f \in F$) by $z \neq x_1 \wedge lab_f(z)$.

8.2 Tree-walking transducers

To facilitate the proof of the implementation of MS-transductions of terms by tree-walking transducers, we will consider a hybrid device: a tree-walking transducer that uses MS tests and MS jumps.[3] More precisely, it uses monadic second-order logic for two purposes: for a given input term s, (1) it can use a formula $\varphi(x_1)$ to test a *global* property[4] of the current node x_1 of s, and (2) it can use a formula $\psi(x_1, x_2)$ to describe conditions implying a *jump*[5] in s from the current node x_1 to the next current node x_2. This hybrid transducer will allow us to present the implementation in several steps. To stress the role of determinism, we also consider nondeterministic tree-walking transducers, and prove some of the results on transducers also for the nondeterministic case.

An *MS tree-walking transducer* (abbreviated *ms-twt*) is a 5-tuple $M = (F, H, Q, q_{in}, R)$, where F and H are functional signatures of input and output symbols, respectively, Q is a finite set of states, $q_{in} \in Q$ is the initial state, and R is a finite set of rules. A rule is of the form $\langle q, \varphi \rangle \to \zeta$ such that $q \in Q$, $\varphi \in \mathrm{MS}(\mathcal{R}_F, \{x_1\})$, and ζ equals

(1) $\langle q', \psi \rangle$, or
(2) $h(\langle q_1, \psi_1 \rangle, \ldots, \langle q_k, \psi_k \rangle)$,

where

(i) $q' \in Q$ and $\psi \in \mathrm{MS}(\mathcal{R}_F, \{x_1, x_2\})$
   such that $\lfloor s \rfloor \models \forall x_1 [\varphi(x_1) \Rightarrow \exists! x_2. \psi(x_1, x_2)]$ for every $s \in T(F)$, or
(ii) $k \geq 0$, $h \in H_k$, $q_1, \ldots, q_k \in Q$ and $\psi_1, \ldots, \psi_k \in \mathrm{MS}(\mathcal{R}_F, \{x_1, x_2\})$
   such that $\lfloor s \rfloor \models \forall x_1 [\varphi(x_1) \Rightarrow \exists! x_2. \psi_i(x_1, x_2)]$ for every $s \in T(F)$, $i \in [k]$.

---

[3] This idea is taken from [EngHoo01], where the hybrid MS two-way finite-state transducer was introduced.
[4] A *global* property of a node u of s is a property of the pair (s, u), such as u is the rightmost leaf of s, or all descendants of u that are not leaves have two sons. A *local* property of u is a property of the subgraph of $Syn(s)$ induced by u, its father and its sons (without the labels of the father and the sons), such as u does not have label f, or u is the second son of its father.
[5] The transducer *jumps* from node u to node v by moving its reading head from u directly to v, where v need not be the father or a son of u (or u itself).

As usual the quantifier $\exists!$ stands for "there exists exacly one." Rules of type (1) will be called *epsilon-output* rules.

The above conditions on $\varphi, \psi, \psi_1, \ldots, \psi_k$ are decidable by the Recognizability Theorem for terms (Corollary 5.67) and the decidability of the emptiness of a regular language (cf. Corollary 3.24). Thus, the definition of MS tree-walking transducer is effective.

The ms-twt M is *deterministic* if for every two distinct rules $\langle q_1, \varphi_1 \rangle \rightarrow \zeta_1$ and $\langle q_2, \varphi_2 \rangle \rightarrow \zeta_2$ in R, if $q_1 = q_2$ then φ_1 and φ_2 are mutually exclusive on $T(F)$, i.e., $\lfloor s \rfloor \models \neg \exists x_1 [\varphi_1(x_1) \wedge \varphi_2(x_1)]$ for every $s \in T(F)$. Again, this is a decidable condition.

To define the computations of the MS tree-walking tree transducer M we will use regular grammars (Definition 3.17). For every input term $s \in T(F)$ we define a regular grammar $G_{M,s} := \langle H, X_{M,s}, R_{M,s} \rangle$ over the output signature H. The set $X_{M,s}$ of nonterminals consists of all pairs $\langle q, u \rangle$ with $q \in Q$ and $u \in N_s$ (i.e., u is a node of s). Such a pair is called a *configuration* of M on s. If $\langle q, \varphi \rangle \rightarrow \zeta$ is a rule in R, then $R_{M,s}$ contains a rule $\langle q, u \rangle \rightarrow \zeta'$ for every $u \in N_s$ such that $\lfloor s \rfloor \models \varphi(u)$, where ζ' equals

(1) $\langle q', v \rangle$

  and v is the unique node such that $\lfloor s \rfloor \models \psi(u, v)$, or
(2) $h(\langle q_1, v_1 \rangle, \ldots, \langle q_k, v_k \rangle)$

  and v_i is the unique node such that $\lfloor s \rfloor \models \psi_i(u, v_i)$, for every $i \in [k]$,

respectively (according to the two possible forms of ζ mentioned above). The computations of M on input s are the derivation sequences of the grammar $G_{M,s}$. There may be infinitely many such sequences. The *transduction computed by* M, denoted τ_M, is defined as

$$\tau_M := \{(s, t) \in T(F) \times T(H) \mid t \in L(G_{M,s}, \langle q_{\text{in}}, root_s \rangle)\}.$$

In other words, (s, t) is in τ_M if t is in the language generated by $G_{M,s}$ from the initial configuration $\langle q_{\text{in}}, root_s \rangle$.

If M is deterministic then distinct rules of $G_{M,s}$ have distinct left-hand sides. This implies that M has exactly one (possibly infinite) maximal computation on s starting with $\langle q_{\text{in}}, root_s \rangle$, to which we will refer as *the* computation of M on s. Consequently, $L(G_{M,s}, \langle q_{\text{in}}, root_s \rangle)$ is either empty or a singleton, and hence τ_M is a (partial) function.

The MS tree-walking transducer M is *single-use* if it satisfies the following property: for every $s \in T(F)$, $\xi \in T(H, X_{M,s})$, $t \in T(H)$, and $\langle q, u \rangle \in X_{M,s}$, if $\langle q_0, root_s \rangle \Rightarrow^*_{G_{M,s}} \xi \Rightarrow^*_{G_{M,s}} t$ then $\langle q, u \rangle$ has at most one occurrence in the term ξ. Note that every ms-twt with a unary output signature is single-use. The class of transductions computed by single-use deterministic MS tree-walking transducers is denoted $\text{DTWT}^{\text{MS}}_{\text{su}}$. We will prove in Theorem 8.6 that it equals DMSOT. It follows from the proof that the single-use property is decidable.

The transducer M is an *MS tree-walking tree-to-word transducer* if H is unary (with the constant symbol ϵ). If $B = H - \{\epsilon\}$, then M computes the tree-to-word transduction $\{(s,w) \in T(F) \times B^* \mid (s,\mu_B(w)) \in \tau_M\}$. In this case, one may view M as having a one-way output tape on which the output word is written. The class of transductions computed by deterministic MS tree-walking tree-to-word transducers is denoted DTWTW^{MS}.

The transducer M is an *MS two-way finite-state transducer* if both F and H are unary. If $A = F - \{\epsilon\}$ and $B = H - \{\epsilon\}$, then M computes the transduction of words $\{(v,w) \in A^* \times B^* \mid (\mu_A(v),\mu_B(w)) \in \tau_M\}$.

Finally, M is an *MS tree-walking automaton* if $H = \{\epsilon\}$, with $\rho(\epsilon) = 0$. In that case the *language recognized by* M is defined as $L(M) := \{s \in T(F) \mid (s,\epsilon) \in \tau_M\}$.

An MS tree-walking transducer is jumping rather than walking; moreover, it can do global rather than local tests (cf. the introduction of this section). We say that M is *nonjumping* if the only formulas $\psi(x_1,x_2)$ that are used in the right-hand sides of rules are $x_1 = x_2$ (stay where you are), $son(x_2,x_1)$ (move to the father),[6] and $son_i(x_1,x_2)$ for $i \in [\rho(F)]$ (move to the i-th son). These formulas will also be denoted *stay*, *up*, and *down$_i$*, respectively. We say that M is *local* if it is nonjumping and the only formulas φ that are used in the left-hand sides of rules are boolean combinations of $lab_f(x_1)$ for $f \in F$ (is the label of the current node f ?), $\exists y.son_i(y,x_1)$ for $i \in [\rho(F)]$ (is it the i-th son of its father?) and $rt(x_1)$ (is it the root ?) which can be viewed as an abbreviation of $\neg \bigvee_{i \in [\rho(F)]} \exists y.son_i(y,x_1)$. In our terminology and notation, locality will be indicated by dropping "MS." Thus, a local MS tree-walking transducer is also called a *tree-walking transducer* (*twt* for short), and the class of transductions computed by the deterministic ones is denoted DTWT (and DTWT$_{su}$ for single-use ones). The class of transductions computed by deterministic tree-walking tree-to-word transducers is denoted DTWTW, and the class of transductions (of words) computed by deterministic two-way finite-state transducers is denoted 2DGSM (where "2gsm" stands for "two-way generalized sequential machine," a well-known name for the two-way finite-state transducer).

In the remainder of this section we consider examples of MS-transductions of words and terms that can be computed by tree-walking transducers.

Example 8.1 The *yield* of a term $t \in T(F)$ is the (nonempty) word $yd(t) \in F_0^*$ defined recursively as follows: $yd(f) = f$ for $f \in F_0$, and $yd(f(t_1,\ldots,t_k)) = yd(t_1) \cdots yd(t_k)$ for $f \in F_k$ with $k > 0$ and $t_1,\ldots,t_k \in T(F)$. The mapping yd is a 2-copying parameterless MS-transduction in DMSOTW: the positions of the output word $yd(t)$ are the leaves of the input tree t, plus one extra position (e.g., the second copy of the last leaf). Each (first copy of a) leaf is linked to (the first copy of) the next leaf, in the left-to-right order of the nodes of t, and the first copy of the last leaf is linked to its second copy (which gets label ϵ).

The mapping yd can also be computed by a (local) deterministic tree-walking tree-to-word transducer $M = (F,H,Q,q_{in},R)$. The transducer walks depth-first

---

[6] Note that $son(x_2,x_1)$ abbreviates $\bigvee_{i \in [\rho(F)]} son_i(x_2,x_1)$.

left-to-right through the input tree and outputs the labels of the leaves; it outputs ϵ when it returns to the root. Thus, yd is in the class DTWTW. To consider a concrete example, let A be a finite set. We take $F := \{f\} \cup A$ with $\rho(f) := 2$ and $\rho(a) := 0$ for $a \in A$, and $H := A \cup \{\epsilon\}$ with $\rho(a) := 1$ for $a \in A$ and $\rho(\epsilon) := 0$. The set of states of M is $Q := \{\downarrow, \uparrow_1, \uparrow_2\}$ and $q_{in} := \downarrow$. The rules in R are the following, for $a \in A$ and $i \in \{1,2\}$ (and $br_i(x_1)$ abbreviates $\exists y . son_i(y, x_1)$):

$$
\begin{aligned}
\langle \downarrow, lab_f(x_1) \rangle & \rightarrow \langle \downarrow, down_1 \rangle, \\
\langle \downarrow, lab_a(x_1) \wedge br_i(x_1) \rangle & \rightarrow a(\langle \uparrow_i, up \rangle), \\
\langle \uparrow_1, True \rangle & \rightarrow \langle \downarrow, down_2 \rangle, \\
\langle \uparrow_2, br_i(x_1) \rangle & \rightarrow \langle \uparrow_i, up \rangle, \\
\langle \uparrow_2, rt(x_1) \rangle & \rightarrow \epsilon, \\
\langle \downarrow, lab_a(x_1) \wedge rt(x_1) \rangle & \rightarrow a(\langle \uparrow_2, stay \rangle).
\end{aligned}
$$

Let us also consider a slightly more complicated example: let $H' := A \cup \{\epsilon\}$ with $\rho(a) := 2$ for $a \in A$ and $\rho(\epsilon) := 0$, and define $\tau : T(F) \rightarrow T(H')$ as follows. For every tree $t \in T(F)$, $\tau(t)$ is the unique tree in $T(H')$ such that the sequence of labels of the nodes of any path from the root to a leaf equals $yd(t)\epsilon$ (hence, all leaves are at the same depth in $\tau(t)$). For instance, if $t := f(f(a,b),c)$, then $\tau(t) = a(b(c(\epsilon,\epsilon),c(\epsilon,\epsilon)),b(c(\epsilon,\epsilon),c(\epsilon,\epsilon)))$. The mapping τ is computed by the deterministic tree-walking transducer M' that is obtained from M by changing the right-hand side of the second rule into $a(\langle \uparrow_i, up \rangle, \langle \uparrow_i, up \rangle)$, and similarly the one of the last rule into $a(\langle \uparrow_2, stay \rangle, \langle \uparrow_2, stay \rangle)$. It is clear that M' is not single-use. The mapping τ is not in DMSOT, because it is not of linear size increase (cf. Section 8.1.1): if $|yd(t)| = n$ (and hence $|t| = 2n - 1$) then $|\tau(t)| = 2^{n+1} - 1$.

As an additional example, let $H'' := F \cup \{\sigma, c, d\}$, where σ is binary and c, d are unary. Consider the mapping $t \mapsto \sigma(c^n(t), d^n(t))$ from $T(F)$ to $T(H'')$ where $n = |yd(t)|$. It can be computed by a single-use deterministic tree-walking transducer M'' that starts its computation with the rule $\langle q_{in}, True \rangle \rightarrow \sigma(\langle \downarrow^c, stay \rangle, \langle \downarrow^d, stay \rangle)$, thus branching into two computations. In each branch, M'' walks depth-first left-to-right through t and outputs the symbol c (or d) at each leaf. After returning to the root, it outputs the input tree t (by the rules $\langle q_{id}, lab_f(x_1) \rangle \rightarrow f(\langle q_{id}, down_1 \rangle, \langle q_{id}, down_2 \rangle)$ and $\langle q_{id}, lab_a(x_1) \rangle \rightarrow a$ for $a \in A$).

Example 8.2 In Remark 7.23 an MS-transduction of words is discussed that does not preserve regularity of languages. It transforms the word $a_1 a_2 a_3 \cdots a_{2n}$ into the word $a_1 a_3 \cdots a_{2n-1} a_2 a_4 \cdots a_{2n}$. It is clear that this transduction can be computed by a deterministic two-way finite-state transducer: the transducer walks to the right and outputs all letters at odd positions; then it returns to the beginning of the input (without producing output) and walks again to the right, this time writing the letters at even positions to the output tape.

As a similar, slightly more complicated example, consider the mapping τ on words that associates with $w \in \{a,b,c,d\}^*$ the word $\tau(w) := a^k b^l c^m d^n$, where $k = |w|_a$ (the number of occurrences of a in w), $l = |w|_b$, $m = |w|_c$, and $n = |w|_d$. It is not difficult to see that τ is a parameterless domain-preserving MS-transduction, i.e., that there is a parameterless domain-preserving definition scheme $\mathcal{D} = \langle \chi, \delta, (\theta_R)_{R \in \mathcal{R}_{U_A}} \rangle$ such that $\widehat{\mathcal{D}}(\lfloor \mu_A(w) \rfloor)$ is (isomorphic to) $\lfloor \mu_A(\tau(w)) \rfloor$ for every $w \in A^*$, where $A := \{a,b,c,d\}$. In fact, both χ and δ are *True*, and for every $f \in U_A = A \cup \{\epsilon\}$ we have $\theta_{lab_f} = lab_f(x_1)$ (which means that every node keeps its label). The details of the formula $\theta_{son_1}(x_1, x_2)$ are left to the reader: it should link each occurrence of a in w to the next occurrence of a, unless it is the last occurrence of a, in which case it should be linked to the first occurrence of b; and similarly for b and c, for c and d, and for d and ϵ.

It should also be clear that τ can be computed by a deterministic two-way finite-state transducer. The transducer scans the input word four times (for instance from left to right and from right to left, and then again). At the first scan it outputs all occurrences of a, and at the next scans all occurrences of b, c, and d; finally it outputs ϵ.

Thus, τ is in both DMSOW and 2DGSM. Let L be the regular language $(abcd)^*$. Obviously, $\tau(L) = \{a^n b^n c^n d^n \mid n \geq 0\}$. Hence, τ does not preserve regularity or context-freeness of languages (see the discussion in the introduction of this chapter). We know from Corollary 7.12 that τ^{-1} preserves regularity. Now consider the context-free language $L' = \{a^k b^k c^m d^m \mid k, m \geq 0\}$. Then $\tau^{-1}(L')$ is not context-free; in fact, if it would be context-free then its intersection with the regular language $R = (ac)^* b^* d^*$ would also be context-free, but it is clear that $\tau^{-1}(L') \cap R = \{(ac)^n b^n d^n \mid n \geq 0\}$, a classical noncontext-free language. Thus, inverse MS-transductions of words do not preserve context-freeness of languages.

Example 8.3 A transduction of words $\tau : A^* \to B^*$ is a *rational transduction* if it can be expressed as follows:[7] $\tau(w) = h'(h^{-1}(w) \cap L)$ for every $w \in A^*$, where $L \subseteq C^*$ is a regular language over some alphabet C, and $h : C^* \to A^*$ and $h' : C^* \to B^*$ are homomorphisms (for equivalent characterizations see the book by Sakarovitch [*Sak]). In other words, $\tau = h^{-1} \cdot \mathrm{Id}_L \cdot h'$, where Id_L is the identity on L. It is well known (and the proof is straightforward) that all rational transductions can be computed by two-way, even one-way, finite-state transducers. In fact, the rational transductions are exactly those computed by one-way finite-state transducers (see Chapter 4 of [*Sak]).

It is clear that all homomorphisms are parameterless MS-transductions (the definition scheme for h' is k-copying, where k is the maximal length of a word $h'(c)$ with $c \in C$, cf. Example 7.3(4)). Hence, since L is MS-definable by Theorem 5.15, the partial function $\mathrm{Id}_L \cdot h'$ is a parameterless MS-transduction by the Restriction Theorem (Theorem 7.16). However, inverse homomorphisms are not MS-transductions, because, in general, they do not have finite images (whereas MS-transductions do,

---

[7] This definition is due to Nivat [Niv]; see also [*Ber].

see Section 8.1.1). Thus, the class of MS-transductions of words is not closed under inverse, cf. the discussion in the introduction of this chapter. If the homomorphism h is nonerasing (i.e., $h(c) \neq \varepsilon$ for all $c \in C$), then h^{-1} *is* a noncopying MS-transduction (cf. Example 7.3(4)): its definition scheme uses a parameter X intended to denote the set $\{u_1, \ldots, u_n\} \subseteq Pos(\mu_A(w))$, where u_i is the first position of a subword $h(c_i)$ of the input word w, such that $w = h(c_1) \cdots h(c_n)$ for $c_1, \ldots, c_n \in C$; additionally, it uses parameters X_c for $c \in C$, that form the partition of $\{u_1, \ldots, u_n\}$ such that X_c denotes $\{u_i \mid c_i = c\}$; in other words, the parameters represent a guess of a word $v = c_1 \cdots c_n \in C^*$ such that $h(v) = w$; the output word v is constructed from the positions in the set denoted by X, and such a position is given the label c if it belongs to the set denoted by X_c. Thus, in this case the rational transduction τ is an MS-transduction, by Theorem 7.14. Every rational transduction τ that has finite images, can be expressed as above with h nonerasing (see the book by Berstel [*Ber, Exercise 7.2, page 87], or the one by Autebert and Boasson [*AutBoa, Chapter 1]). Consequently, a rational transduction is an MS-transduction if and only if it has finite images.

Example 8.4 Let $\widetilde{w}$ be the mirror image of the word w. The mapping $\tau(w) := w\widetilde{w}$ is in 2DGSM: the transducer first walks from left to right over the input word, and then back from right to left; it outputs each letter that it meets on its way. If it does *not* produce output during the walk from left to right, it computes the mapping $\tau'(w) := \widetilde{w}$. Both τ and τ' are deterministic MS-transductions (2-copying and noncopying, respectively). Note that neither τ nor τ' is a rational transduction; τ' preserves regularity and context-freeness, but τ does not. For every $k \geq 2$, the mapping $\tau_k(w) := w^k$ is in 2DGSM: the transducer walks k times from left to right over the input word, writing it to the output tape (each time walking back from right to left without producing output). Moreover, τ_k is a k-copying parameterless MS-transduction (in DMSOW); it can be expressed as $copy_k \cdot \mu$, where μ is a noncopying parameterless MS-transduction, cf. Proposition 7.26.

As a more complicated example, consider the alphabet $B := \{a_1, a_2, b\}$ and the mapping $\sigma : B^* \to B^*$ defined by

$$\sigma(w_0 b w_1 b w_2 \cdots b w_n) := w_0 w_0 b w_1 w_1 b w_2 w_2 \cdots b w_n w_n,$$

where $w_i \in \{a_1, a_2\}^*$. The mapping σ is a 2-copying parameterless MS-transduction in DMSOW. It is also in 2DGSM: the transducer scans each subword w_i three times and outputs that word when walking to the right. Now let F be the functional signature $\{f\} \cup B$, where f has arity 2 and the elements of B have arity 0. Then the mapping $\sigma' := yd \cdot \sigma : T(F) \to B^*$ is in DMSOTW (cf. Example 8.1). It is also in DTWTW: a tree-walking tree-to-word transducer can walk from one b-labeled leaf to the next one, or to the previous one.

Example 8.5 The second-order substitutions, defined in Section 2.6.1 (just before Proposition 2.126), and discussed in Example 7.3(4), are often called *tree* *homomorphisms.* Every second-order substitution can be computed by a (local) deterministic tree-walking transducer of which the rules do not use *up*; for instance, if the derived operation h of arity 3 is defined by the term $f(x_3, g(x_1))$, then the transducer has rules $\langle q_{in}, lab_h(x_1) \rangle \to f(\langle q_{in}, down_3 \rangle, \langle q, stay \rangle)$ and $\langle q, True \rangle \to g(\langle q_{in}, down_1 \rangle)$, where q is a "new" state. Moreover, if the second-order substitution is linear (as in the above instance), then the transducer is single-use. Thus, every linear second-order substitution is in DTWT_{su}. It is also a parameterless MS-transduction in DMSOT, cf. Example 7.3(4). As discussed in Example 7.3(4), not all second-order substitutions are MS-transductions. This shows again that tree-walking transducers have more expressive power than MS-transductions of terms.

8.3 The basic characterization

We start by characterizing the MS-transductions of terms by MS tree-walking transducers that are single-use. Intuitively, the single-use restriction means that the tree-walking transducer can output at most k copies of each input node, where k is its number of states: due to determinism, if the transducer visits an input node twice in the same state, then it repeats its computation and hence does not halt and produces no output. This integer k corresponds to the copying number of a definition scheme.

Theorem 8.6 The classes DMSOT and DTWT_{su}^{MS} are equal.

Proof: We first prove the inclusion $\text{DMSOT} \subseteq \text{DTWT}_{su}^{MS}$. Let $\tau : T(F) \to T(H)$ be a deterministic MS-transduction of terms, i.e., $\{(\lfloor s \rfloor, \lfloor t \rfloor) \mid (s,t) \in \tau\}$ is a parameterless MS-transduction from $STR(\mathcal{R}_F)$ to $STR(\mathcal{R}_H)$. Let $\mathcal{D} = \langle \chi, (\delta_i)_{i \in [k]}, (\theta_w)_{w \in \mathcal{R}_H \otimes [k]} \rangle$ be a parameterless definition scheme of type $\mathcal{R}_F \to \mathcal{R}_H$ defining τ, hence $\widehat{\mathcal{D}}(\lfloor s \rfloor) = \lfloor \tau(s) \rfloor$ for every $s \in T(F)$. We define an MS tree-walking transducer $M := (F, H, Q, q_{in}, R)$ that computes τ, as follows. First, its set of states is $Q := \{q_{in}, q_1, \ldots, q_k\}$. Intuitively, for $i \in [k]$, the state q_i represents the i-th copy of the input structure. More precisely, if M is in configuration $\langle q_i, u \rangle$ on the input structure $\lfloor s \rfloor$, then it will output the node (u,i) of the output structure $\lfloor \tau(s) \rfloor$, cf. Definition 7.2. It remains to define the set R. We first define the rules for q_{in}. Intuitively, M starts its computation by jumping to the node of the input tree that will represent the root of the output tree. For every $i \in [k]$, M has the rule

$$\langle q_{in}, \chi \wedge \exists x [\delta_i(x) \wedge \theta_{rt,i}(x)] \rangle \to \langle q_i, \delta_i(x_2) \wedge \theta_{rt,i}(x_2) \rangle.$$

Now suppose that M is in configuration $\langle q_i, u \rangle$. Then it outputs the node (u,i) of the output tree and jumps to the nodes of the input tree that will represent the sons of (u,i). To do this, it has to determine the label of the node (u,i), and the input nodes for its sons. For every $h \in H_m$ and every $j_1, \ldots, j_m \in [k]$, M has the rule $\langle q_i, \varphi \rangle \to h(\langle q_{j_1}, \psi_1 \rangle, \ldots, \langle q_{j_m}, \psi_m \rangle)$, where φ is

$$\chi \wedge \delta_i(x_1) \wedge \theta_{lab_h, i}(x_1) \wedge \bigwedge_{n \in [m]} \exists y [\delta_{j_n}(y) \wedge \theta_{son_n, i, j_n}(x_1, y)],$$

and ψ_n is $\delta_{j_n}(x_2) \wedge \theta_{son_n, i, j_n}(x_1, x_2)$.

It should be clear that M computes τ, and that M is deterministic. Moreover M is single-use, because the configurations of its computation (except the first one) are in bijection with the nodes of the output tree.

For the inclusion $\mathrm{DTWT}_{su}^{MS} \subseteq \mathrm{DMSOT}$, let $M = (F, H, Q, q_{in}, R)$ be a single-use deterministic MS tree-walking transducer. We first consider the special case where M has no epsilon-output rules. Without loss of generality, we assume that $Q = [k]$ for some $k \geq 1$.

Since M is deterministic and single-use, each node of the output tree (for a given input tree s) is produced by a unique configuration $\langle q, u \rangle$ of M, with $q \in [k]$ and $u \in N_s$. Hence a k-copying definition scheme for τ_M can be constructed that uses (u, q) for that node. To simplify the construction we will actually show that τ_M is the composition $\tau_1 \cdot \tau_2$ of two parameterless MS-transductions, of which the first is k-copying and the second is noncopying. This gives the desired result by Theorem 7.14.

The first MS-transduction τ_1 transforms the input tree s into the "computation space" of M on s, which is a structure S in $STR^c(\mathcal{R}_H)$ that encodes the regular grammar $G_{M,s} = \langle H, X_{M,s}, R_{M,s} \rangle$ and the initial configuration $\langle q_{in}, root_s \rangle$. The structure S has the domain $\{(u, q) \mid \langle q, u \rangle \in X_{M,s}\}$, and it has the unary relation $rt_S = \{(root_s, q_{in})\}$. Let $\langle q, u \rangle \to h(\langle q_1, v_1 \rangle, \ldots, \langle q_m, v_m \rangle)$ be a rule in $R_{M,s}$. Then (u, q) is in lab_{hS}, and $((u, q), (v_i, q_i))$ is in son_{iS}, for every $i \in [m]$. Note that S represents an $(H, [\rho(H)])$-labeled directed graph with a designated node (the "root").

This MS-transduction τ_1 is defined by the k-copying parameterless definition scheme $\mathcal{D}_1 := \langle \chi, (\delta_i)_{i \in [k]}, (\theta_w)_{w \in \mathcal{R}_H \otimes [k]} \rangle$ of type $\mathcal{R}_F \to \mathcal{R}_H$. The precondition χ expresses that the input structure represents a term in $T(F)$, see Corollary 5.12. For every $q, q' \in [k]$, δ_q is *True*, $\theta_{rt, q}$ is $rt(x_1)$ if $q = q_{in}$ and *False* otherwise, $\theta_{lab_h, q}$ is the disjunction of all formulas φ such that $\langle q, \varphi \rangle \to h(\langle q_1, \psi_1 \rangle, \ldots, \langle q_m, \psi_m \rangle)$ is a rule of M, and $\theta_{son_i, q, q'}$ is the disjunction of all formulas $\varphi \wedge \psi_i$ such that $\langle q, \varphi \rangle \to h(\langle q_1, \psi_1 \rangle, \ldots, \langle q_m, \psi_m \rangle)$ is a rule of M with $i \in [m]$ and $q_i = q'$.

The second MS-transduction τ_2 will be obtained by the Restriction Theorem from the mapping τ_2' that transforms the computation space into the subgraph induced by all vertices that are reachable from the "root" of the graph. Thus, τ_2 is defined by the noncopying parameterless definition scheme $\mathcal{D}_2 := \langle True, \delta, (\theta_R)_{R \in \mathcal{R}_H} \rangle$ of type $\mathcal{R}_H \to \mathcal{R}_H$ such that δ is the formula $\exists y (rt(y) \wedge \mathrm{TC}[(\lambda u, v \cdot son(u, v)); y, x_1])$, θ_{lab_h} is $lab_h(x_1)$, θ_{son_i} is $son_i(x_1, x_2)$, and θ_{rt} is $rt(x_1)$.[8]

---

[8] For the formula $son(u, v)$ see Footnote 6, and for the transitive closure construction TC see Section 5.2.2.

It remains to check that the resulting graph is a tree (i.e., that s is in the domain of τ_M). Since M is single-use, it suffices to check that (1) there is no circuit (otherwise the computation of M on s does not halt), and (2) every vertex has a label in H (otherwise the computation of M on s aborts). This is expressed by the sentence $\beta := \forall x, y(\text{TC}[(\lambda u, v \cdot son(u, v)); x, y] \Rightarrow \neg son(y, x)) \wedge \forall x \bigvee_{h \in H} lab_h(x)$. Thus, $\tau_2 := \tau_2' \cap (\text{MOD}(\textit{True}) \times \text{MOD}(\beta))$ is a noncopying parameterless MS-transduction by Theorem 7.16.

Finally, we consider a single-use deterministic MS tree-walking transducer $M = (F, H, Q, q_{in}, R)$ with epsilon-output rules. Let the deterministic ms-twt $M' := (F, H \cup \{\odot\}, Q, q_{in}, R')$ be obtained from M by changing every epsilon-output rule $\langle q, \varphi \rangle \rightarrow \langle q', \psi \rangle$ of M into the output producing rule $\langle q, \varphi \rangle \rightarrow \odot(\langle q', \psi \rangle)$ of M', where $\odot$ is a new output symbol of arity 1. Obviously M' is still single-use, and thus we know that $\tau_{M'}$ is in DMSOT. It is also obvious that $\tau_M = \tau_{M'} \cdot \tau_\odot$, where $\tau_\odot : T(H \cup \{\odot\}) \rightarrow T(H)$ is the linear second-order substitution that replaces $\odot(x_1)$ by x_1. By Example 7.3(4), $\tau_\odot$ is a parameterless MS-transduction, and hence so is τ_M by Theorem 7.14. ∎

The subscript "su" cannot be dropped from Theorem 8.6, because DTWT contains transductions that are not of linear size increase, cf. Example 8.1. If the output signature is unary, then it *can* be dropped, because every ms-twt with a unary output signature is single-use. Thus: DMSOTW = DTWT^{MS}.

We observe that the construction of a parameterless definition scheme for an MS tree-walking transducer M, as described in the above proof, can also be carried out in the case where M is not single-use. Then, instead of a tree, the output structure is a directed graph G without circuit (i.e., a tree with shared subtrees). Unfolding G produces the output tree of M. This also shows that *it is decidable whether or not M is single-use*: M is single-use if and only if the indegree of every vertex of G is at most 1, for every output G. Since this is an MS-expressible property, the set of all input terms such that G does not satisfy this property is regular by Corollaries 7.12 and 5.67, and can be checked for emptiness.

8.4 From jumping to walking

Our aim will now be to "get rid of" the MS tests and MS jumps of our hybrid tree-walking transducer. It turns out that MS jumps can be simulated by tree-walking transducers, provided these transducers are still allowed to use MS tests. However, MS tests cannot be handled by tree-walking transducers; thus, for the simulation of tests other means have to be found. We therefore start with the "removal" of jumps.

Theorem 8.7 For every MS tree-walking transducer an equivalent nonjumping MS tree-walking transducer can be constructed. Determinism and the single-use restriction are preserved.

Proof: Let M be an ms-twt with input signature F. The problem is to simulate by a walk a jump of M defined by a formula $\psi \in \mathrm{MS}(\mathcal{R}_F, \{x_1, x_2\})$ in the right-hand side of a rule of M. We may assume that $\lfloor s \rfloor \models \forall x_1 \exists! x_2. \psi(x_1, x_2)$ for every $s \in T(F)$. If this is not the case, we replace ψ by $(\varphi \wedge \psi) \vee (\neg \varphi \wedge x_1 = x_2)$, where $\varphi \in \mathrm{MS}(\mathcal{R}_F, \{x_1\})$ is the formula in the left-hand side of the rule.

We will prove that for every such formula ψ a nonjumping deterministic ms-twt $M_\psi = (F, \emptyset, Q, q_{\mathrm{in}}, R)$ can be constructed that has only epsilon-output rules, and that has a state q_{f} such that for every $s \in T(F)$ and $y, z \in N_s$, $\lfloor s \rfloor \models \psi(y, z)$ if and only if $\langle q_{\mathrm{in}}, y \rangle \Rightarrow_G^* \langle q_{\mathrm{f}}, z \rangle$, where G is the regular grammar $G_{M_\psi, s}$ (cf. Section 8.2). It is easy to see that, using these transducers M_ψ as "subroutines," M can be turned into an equivalent nonjumping ms-twt. Since every M_ψ is deterministic, determinism and the single-use restriction will be preserved. Note that since M_ψ has epsilon-output rules only, it is essentially a tree-walking automaton rather than a tree-walking transducer.

So, let $\psi \in \mathrm{MS}(\mathcal{R}_F, \{x_1, x_2\})$ such that $\lfloor s \rfloor \models \forall x_1 \exists! x_2. \psi(x_1, x_2)$ for every $s \in T(F)$. Let $q := \max\{qh(\psi), m\} + 2$, where m is the quantifier-height of the formula expressing that x_1 is a descendant of x_2. For this q (and F), we will use the definitions in Section 8.1.3.

The transducer $M_\psi := (F, \emptyset, Q, q_{\mathrm{in}}, R)$ to be constructed, should walk from node y to node z if and only if $\lfloor s \rfloor \models \psi(y, z)$. It will do this in a special way: it walks along the shortest (undirected) path in s from y to z, and, roughly speaking, at each node u on this path it keeps track, in its state at that node, of the (finite) theory $Th_s^\downarrow(y, u)$ or $Th_s^\uparrow(y, u)$ depending on whether or not y is a descendant of u.

More precisely, M_ψ first walks from y up to the least common ancestor $lca(y, z)$ of y and z, and at each node u on that path (except y) it is in state $\uparrow_{i, \Phi}$ such that y is a descendant of the i-th son u_i of u and $\Phi = Th_s^\downarrow(y, u_i)$. Then, M_ψ walks from $lca(y, z)$ down to z, and at each node u on that path (except $lca(y, z)$) it is in state $\downarrow_\Phi$ such that $\Phi = Th_s^\uparrow(y, u)$.

Thus, the set Q of states of M_ψ consists of $q_{\mathrm{in}}, q_{\mathrm{f}}, \uparrow_{i, \Phi}$ and $\downarrow_\Phi$, for all $i \in [\rho(F)]$ and $\Phi \subseteq L_2 \cup L_2'$. The right-hand sides of the rules in R are of one of the three forms $\langle \uparrow_{i, \Phi}, up \rangle$, $\langle \downarrow_\Phi, down_j \rangle$, or $\langle q_{\mathrm{f}}, stay \rangle$. Recall that up, $down_j$ and $stay$ denote the formulas $son(x_2, x_1)$, $son_j(x_1, x_2)$, and $x_1 = x_2$, respectively.

For every node u that M_ψ visits, it uses a test of the form

$$\eta_\Phi^\uparrow(x_1) \wedge \exists y_1[son_1(x_1, y_1) \wedge \eta_{\Phi_1}^\downarrow(y_1)] \wedge \cdots \wedge \exists y_k[son_k(x_1, y_k) \wedge \eta_{\Phi_k}^\downarrow(y_k)]$$

to obtain the sets $Th_s^\uparrow(u)$ and $Th_s^\downarrow(u_1), \ldots, Th_s^\downarrow(u_k)$, where $u_1, \ldots, u_k$ are the sons of u (see the end of Section 8.1.3). It should now be clear from Section 8.1.3 that from these sets, and the state $\uparrow_{i, \Phi}$ or $\downarrow_\Phi$ at u, and the label of u (which can of course be obtained by a test $lab_f(x_1)$), M_ψ can determine the set $Th_s(y, u)$. If ψ is in that set (or more precisely, if $\widehat{\psi} \in Th_s(y, u)$), then $u = z$ and M_ψ halts in state q_{f}. Now assume that it is not. Then M_ψ should decide, deterministically, where to go next. Suppose first that M_ψ is in state $\uparrow_{i, \Phi}$. If for some $j \neq i$ the formula $\psi_j(x_1, x_2)$, defined as $\exists v, w[son_j(x_2, v) \wedge w \leq v \wedge \psi(x_1, w)]$, is in $Th_s(y, u)$, then $u = lca(y, z)$ and M_ψ moves down to the j-th son of u; otherwise, M_ψ moves up to the father of u. In the former case the new state is $\downarrow_{\Phi'}$, where Φ' can be determined (according to Section 8.1.3) from Φ, $Th_s^\uparrow(u)$, and all $Th_s^\downarrow(u_m)$ with $m \neq i, j$. In the latter case the new state is $\uparrow_{i', \Phi'}$, where i' can be obtained by a test $\exists v. son_{i'}(v, x_1)$, and Φ' can be determined from Φ and all $Th_s^\downarrow(u_j)$ with $j \neq i$. Suppose now that M_ψ is in state $\downarrow_\Phi$. This case is similar to, but easier than the previous one: if the formula $\psi_j(x_1, x_2)$ is in $Th_s(y, u)$ (and that must be so for exactly one $j \in [k]$), then M_ψ moves down to the j-th son of u in the state $\downarrow_{\Phi'}$, where Φ' can now be determined from Φ and all $Th_s^\downarrow(u_i)$ with $i \neq j$.

It remains to explain how M_ψ starts its walk, in state q_{in}. It first tests whether $y = z$, by checking if the rule $\langle q_{in}, \psi(x_1, x_1) \rangle \rightarrow \langle q_f, stay \rangle$ is applicable. If that is not the case, then M_ψ uses the test formulas $\psi_j(x_1, x_1)$, i.e., $\exists v, w[son_j(x_1, v) \wedge w \leq v \wedge \psi(x_1, w)]$. If such a formula is true, then M_ψ moves down to the j-th son of y in state $\downarrow_\Phi$, where Φ can be determined from $Th_s^\uparrow(y, y)$ and all $Th_s^\downarrow(y_i)$ with $i \neq j$ (where y_i is the i-th son of y). Otherwise, M_ψ moves up to the father of y in state $\uparrow_{i, \Phi}$, where y is the i-th son of its father and $\Phi := Th_s^\downarrow(y, y)$. Note that $Th_s^\uparrow(y, y) = \{\varphi(x_1, x_2) \mid \varphi(x_1, x_1) \in Th_s^\uparrow(y)\}$ and $Th_s^\downarrow(y, y) = \{\varphi(x_1, x_2) \mid \varphi(x_1, x_1) \in Th_s^\downarrow(y)\}$, and that $Th_s^\uparrow(y)$ and $Th_s^\downarrow(y)$ can be obtained by a test.

This ends the description of the nonjumping ms-twt M_ψ. Due to the condition on ψ, which says that for every y there is a unique z such that $\lfloor s \rfloor \models \psi(y, z)$, the transducer M_ψ is deterministic. ∎

8.5 From global to local tests

We first show that MS tests can be simulated by tree-walking transducers in the special case where the input signature is unary. For this we need the fact that a two-way automaton can keep track of the state of a deterministic one-way automaton. This is expressed in the next lemma, in an informal, but hopefully clear way. We use the terminology and notation from Section 8.1.2.

Lemma 8.8 Let M be an MS tree-walking transducer with a unary input signature F, and let $\mathscr{A}$ be a complete and deterministic finite F-semi-automaton. Then an MS tree-walking transducer M' can be constructed that stepwise simulates[9] M and keeps track of the state of $\mathscr{A}$; in its nonlocal rules, M' uses the same MS formulas as M. The same holds if $\mathscr{A}$ is a top-down complete and deterministic finite F-automaton. Determinism and the single-use restriction are preserved in the construction of M' from M.

---

[9] This means (still informally speaking) that every computation of M' can be mapped to a computation of M, by a fixed mapping from the set of states of M' to the set of states of M, provided the intermediate steps that are needed to determine the state of $\mathscr{A}$ are first deleted from the computation of M'; moreover, every computation of M is obtained in this way.

Proof: Let $\mathscr{A} = \langle F, Q, \delta \rangle$ be a complete and deterministic finite semi-automaton. The ms-twt M' should stepwise simulate M and, for every input tree $s \in T(F)$ and current node $u \in N_s$, keep track (in its finite state at u) of $p := run_{\mathscr{A},s}(u)$. In fact, M' will behave in exactly the same way as M, except that after each "simulation step" it uses local epsilon-output rules to compute the new value of p, deterministically.

We will view p as a programming variable, to which assignments can be made. To initialize the variable p, M' first walks down to the unique leaf of s (its ϵ-labeled node), and then walks back to the root, simulating $\mathscr{A}$; note that it can recognize the root by the local test $rt(x_1)$. Then M' sets $p := run_{\mathscr{A},s}(root_s)$, and starts the simulation of M. Suppose that, simulating a step of M, M' moves up to the father of the current node. To update p it simply tests the label of the father, say f, and sets $p := \delta_f(p)$.

Suppose now that M' moves down to the son of the current node u. This is the difficult case. Let f be the label of u. If the label of the son u_1 is ϵ, then M' simply sets $p := \delta_\epsilon$. If it is not, then the new p must be in the set $C := \{q \in Q \mid \delta_f(q) = p\}$ (where C stands for "candidates"). If C is a singleton, the problem is solved. Now suppose it is not a singleton. Then M' walks down from u_1 and computes for each descendant u' of u_1, and each $q \in C$, the set $C(q, u')$ of all states $q' \in Q$ such that, when started at u' in state q', $\mathscr{A}$ reaches u_1 in state q. Note that, for fixed u', the sets $C(q, u')$ are pairwise disjoint. It is easy to see that M' can compute these sets, just by testing the labels of the nodes and using δ. In fact, $C(q, u_1) := \{q\}$ and if u'' is the son of u', then $C(q, u'') := \{q'' \in Q \mid \delta_f(q'') \in C(q, u')\}$, where f is the label of u'.

Now there are two cases: *either* M' arrives at a descendant u' such that exactly one of the sets $C(q, u')$, say $C(\bar{q}, u')$, is nonempty, whereas all the others are empty, *or* M' arrives at the leaf u' of s, in which case there is a unique $\bar{q}$ such that $\delta_\epsilon \in C(\bar{q}, u')$. In both cases, $\bar{q} = run_{\mathscr{A},s}(u_1)$ and M' sets $p := \bar{q}$. But now M' should return to u_1. To this purpose, M' moves up to the father u'' of u' and picks two states $q_1, q_2 \in Q$ that are in two distinct sets $C(q, u'')$ (obviously, when walking down, M' can also store in its state the candidate sets of the father of the current node). Now M' starts the simulation of two incarnations $\mathscr{A}_1$ and $\mathscr{A}_2$ of the automaton $\mathscr{A}$, one in state q_1 and the other in state q_2 at u''. Then M' walks up, simulating both $\mathscr{A}_1$ and $\mathscr{A}_2$. At the very moment where $\mathscr{A}_1$ and $\mathscr{A}_2$ are in the same state (which is the previous value of p), M' is back on node u, and moves down to u_1. This ends the description of M'.[10]

Finally, let $\mathscr{A}$ be a top-down complete and deterministic finite automaton, with $Acc := \{q_0\}$. This time, M' should keep track of $p := td\text{-}run_{\mathscr{A},s}(u)$. The proof is basically the same as the one above, interchanging the up and down moves, and interchanging the roles of the root and the leaf of s. Initially, M' just sets $p := q_0$. When M moves down from u, M' sets $p := q$, where $f[q] \to p$ is the unique rule in δ with right-hand side p and f is the label of u. Now the difficult case is when M

---

[10] To summarize, apart from the finite control, a state of M' contains (at most) a state of M, three states of $\mathscr{A}$, and two functions from Q to $\mathcal{P}(Q)$.

moves up to u from its son u_1. The computation of the new p is similar to the one above, with M' first walking up and computing the candidate sets for the ancestors of u, possibly reaching the root, then walking down to u_1 and finally moving up to u. ∎

Theorem 8.9 For every MS tree-walking transducer with a unary input signature an equivalent (local) tree-walking transducer can be constructed. Determinism and the single-use restriction are preserved.

Proof: Let M be an ms-twt with unary input signature F. By Theorem 8.7 we may assume that M is nonjumping. The remaining problem is how to simulate a global test of M, defined by a formula $\varphi \in \mathrm{MS}(\mathcal{R}_F, \{x_1\})$ in the left-hand side of a rule of M, by an ordinary walk that uses local tests only.

So, let $\varphi \in \mathrm{MS}(\mathcal{R}_F, \{x_1\})$ and let q be the quantifier-height of φ. It suffices to prove that M can be changed into a transducer M' that stepwise simulates M and, moreover, for input tree $s \in T(F)$ and current node $u \in N_s$, additionally keeps track (in its finite state at u) of the theories $p_1 := Th_s^{\downarrow}(u_1)$ (if u has a son u_1) and $P := Th_s^{\uparrow}(u)$. Recall from Section 8.1.3 that there is a mapping Z_f such that $Th_s(u) = Z_f(Th_s^{\uparrow}(u), Th_s^{\downarrow}(u_1), \ldots, Th_s^{\downarrow}(u_k))$, where u has sons $u_1, \ldots, u_k$ (with $k = 1$ or $k = 0$) and label f. Hence, in M', the global test φ can be replaced by the local test consisting of the disjunction of all formulas $lab_f(x_1)$ such that $\rho(f) = 1$ and $\widehat{\varphi} \in Z_f(P, p_1)$, and the formula $lab_\epsilon(x_1)$ if $\widehat{\varphi} \in Z_\epsilon(P)$.

The construction of M' is possible by Lemma 8.8. Recall from Section 8.1.3 that there is a complete and deterministic finite F-semi-automaton $\mathscr{A}$ such that for every $s \in T(F)$ and $u \in N_s$, $run_{\mathscr{A},s}(u) = Th_s^{\downarrow}(u)$. It is easy to construct from $\mathscr{A}$ a complete and deterministic finite F-semi-automaton $\mathscr{A}_1$ such that $run_{\mathscr{A}_1,s}(u) = \langle lab_s(u), Th_s^{\downarrow}(u_1)\rangle$ if u has a son u_1 (where $lab_s(u)$ is the label of u). Thus, by Lemma 8.8, there is an ms-twt M_1 that stepwise simulates M and keeps track of $Th_s^{\downarrow}(u_1)$. It should also be clear from Section 8.1.3, and from the fact that F is unary, that there is a top-down complete and deterministic finite F-automaton $\mathscr{A}_2$ such that $td\text{-}run_{\mathscr{A}_2,s}(u) = Th_s^{\uparrow}(u)$ for every $s \in T(F)$ and $u \in N_s$. In fact, $\mathscr{A}_2$ has the set of states $\mathcal{P}(L_1')$ and the unique initial state $Th_{rt}^{\uparrow}$ (the root theory, defined in Section 8.1.3); the transition rules of $\mathscr{A}_2$ are obtained from the mappings $Z_{f,1} : \mathcal{P}(L_1') \to \mathcal{P}(L_1')$ discussed in Section 8.1.3. Again by Lemma 8.8, applied to M_1 and $\mathscr{A}_2$, there is an mstwt M' that stepwise simulates M and keeps track of both $Th_s^{\downarrow}(u_1)$ and $Th_s^{\uparrow}(u)$. ∎

The next theorem (which is the main result of [EngHoo01]) is now immediate from Theorems 8.6 and 8.9. It shows that the monadic second-order transductions of words are exactly those computed by two-way finite-state transducers (see Examples 8.2 and 8.4). This generalizes the monadic case of Theorem 1.16 to transductions.

Theorem 8.10 The classes of transductions of words DMSOW and 2DGSM are equal. □

In general, it is not possible to "remove" the MS tests from an MS tree-walking transducer, not even when it is an MS tree-walking automaton. This result is proved by Bojańczyk and Colcombet in [BojCol].

Theorem 8.11 The class DTWTW^{MS} contains a transduction that is not in the class DTWTW.[11]

Proof: Every regular tree language L can be recognized by a deterministic MS tree-walking automaton. In fact, the automaton need not walk at all: if the MS sentence φ expresses L (by Theorem 5.15), then the rule $\langle q_{in}, \varphi \rangle \to \epsilon$ suffices. However, not every regular tree language can be recognized by a tree-walking automaton [BojCol].   ∎

Thus, to "remove" MS tests in general, a more powerful type of transducer is needed. One such transducer can be obtained by equipping the nonjumping MS tree-walking transducer with a restricted pushdown. This extended transducer can only push a symbol on the pushdown when it moves down to a son, and it has to pop the top symbol off the pushdown when it moves to the father. In other words, the movements of the pushdown are synchronized with the movements of the reading head. Initially (at the root) the pushdown contains one symbol, and hence, at each moment, the number of symbols on the pushdown equals the *depth* of the current node u, i.e., its number of ancestors (including itself).[12] Each cell of the pushdown corresponds to an ancestor of u, and the content of the cell provides information about that ancestor; in particular, the top of the pushdown corresponds to u itself, and the bottom to the root.

Formally, an *MS tree-walking pushdown transducer* (abbreviated *p-ms-twt*) is a tuple $M = (F, H, Q, q_{in}, \Gamma, \gamma_{in}, R)$, where F, H, Q, and q_{in} are the same as for an ms-twt, Γ is an alphabet of pushdown symbols, γ_{in} is the initial pushdown symbol, and R is a finite set of rules. A rule is of the form $\langle q, \varphi, \gamma \rangle \to \zeta$ with $q \in Q$, $\varphi \in MS(\mathcal{R}_F, \{x_1\})$, $\gamma \in \Gamma$, and either $\zeta \in I$ or $\zeta = h(\zeta_1, \ldots, \zeta_k)$ with $h \in H_k$ and $\zeta_1, \ldots, \zeta_k \in I$, where I is the set of all triples $\langle q', \psi, \beta \rangle$ with $q' \in Q$, $\psi \in MS(\mathcal{R}_F, \{x_1, x_2\})$, and $\beta \in \Gamma^*$, and one of the following three cases holds:

(a) $\psi = up$, $\beta = \varepsilon$, and $\lfloor s \rfloor \models \forall x_1 [\varphi(x_1) \Rightarrow \exists y . son(y, x_1)]$ for every $s \in T(F)$; or
(b) $\psi = stay$ and $\beta \in \Gamma$; or
(c) there exists $i \in [\rho(F)]$ such that $\psi = down_i$, $\beta \in \Gamma^2$, and
$\lfloor s \rfloor \models \forall x_1 [\varphi(x_1) \Rightarrow \exists y . son_i(x_1, y)]$ for every $s \in T(F)$.

The transducer M is *deterministic* if for every two distinct rules $\langle q, \varphi_1, \gamma \rangle \to \zeta_1$ and $\langle q, \varphi_2, \gamma \rangle \to \zeta_2$ in R, φ_1 and φ_2 are mutually exclusive on $T(F)$.

---

[11] More strongly, it contains one that is not even in TWTW, the class of transductions computed by nondeterministic tree-walking tree-to-word transducers.

[12] To be precise, the depth of the root is one, and the depth of a son of a node u is the depth of u plus one.

Just as we did for an ms-twt, we define for every input term $s \in T(F)$ a regular grammar $G_{M,s} := \langle H, X_{M,s}, R_{M,s} \rangle$. The set $X_{M,s}$ of configurations of M on s now consists of all triples $\langle q, u, \pi \rangle$ with $q \in Q$, $u \in N_s$, and $\pi \in \Gamma^*$ such that the length of π equals the depth of u. If $\langle q, \varphi, \gamma \rangle \to \zeta$ is a rule in R, then $R_{M,s}$ contains all rules $\langle q, u, \pi \gamma \rangle \to \zeta'$ for every $u \in N_s$ such that $\lfloor s \rfloor \models \varphi(u)$ and every $\pi \in \Gamma^*$ such that $|\pi \gamma|$ equals the depth of u, where ζ' is defined as follows:

(1) if $\zeta = \langle q', \psi, \beta \rangle$, then

$$\zeta' := \langle q', v, \pi \beta \rangle$$

and v is the unique node such that $\lfloor s \rfloor \models \psi(u, v)$;

(2) if $\zeta = h(\langle q_1, \psi_1, \beta_1 \rangle, \ldots, \langle q_k, \psi_k, \beta_k \rangle)$, then

$$\zeta' := h(\langle q_1, v_1, \pi \beta_1 \rangle, \ldots, \langle q_k, v_k, \pi \beta_k \rangle)$$

and v_i is the unique node such that $\lfloor s \rfloor \models \psi_i(u, v_i)$, for every $i \in [k]$.

The *transduction computed by M* is

$$\tau_M := \{(s, t) \in T(F) \times T(H) \mid t \in L(G_{M,s}, \langle q_{\text{in}}, root_s, \gamma_{\text{in}} \rangle)\}.$$

Note that, by Theorem 8.7, for every ms-twt there is an equivalent p-ms-twt (with $\Gamma = \{\gamma_{\text{in}}\}$).

As for an ms-twt, an MS tree-walking pushdown transducer is a *tree-to-word* transducer if its output signature is unary. Locality of an MS tree-walking pushdown transducer is defined in the same way as for an ms-twt. By P-DTWT we denote the class of transductions computed by deterministic tree-walking pushdown transducers (i.e., by local deterministic MS tree-walking pushdown transducers), and by P-DTWTW the corresponding class in the tree-to-word case.

Theorem 8.12 For every MS tree-walking pushdown transducer an equivalent (local) tree-walking pushdown transducer can be constructed. Determinism is preserved.

Proof: Let M be a p-ms-twt with input signature F. As in Theorem 8.9, the problem is how to simulate a global test, defined by a formula in the left-hand side of a rule of M, by a walk that uses local tests only. Let $\varphi \in \text{MS}(\mathcal{R}_F, \{x_1\})$ be a formula that defines a global test and let q be the quantifier-height of φ. As in the proof of Theorem 8.9, it suffices to prove that M can be changed into a p-ms-twt M' that stepwise simulates M and, moreover, for input tree $s \in T(F)$ and current node $u \in N_s$, additionally keeps track of $p_1 := Th_s^{\downarrow}(u_1), \ldots, p_k := Th_s^{\downarrow}(u_k)$ (if u has k sons $u_1, \ldots, u_k$) and of $P := Th_s^{\uparrow}(u)$.

We first show how M' can compute the theories $p_1, \ldots, p_k$ (they will be recomputed each time M' visits u). Recall again from Section 8.1.3 that there is a complete and deterministic finite semi-automaton $\mathcal{A} = \langle F, Q, \delta \rangle$ such that for every $s \in T(F)$ and $u \in N_s$, $run_{\mathcal{A},s}(u) = Th_s^{\downarrow}(u)$. Thus, $p_i = run_{\mathcal{A},s}(u_i)$ for every $i \in [k]$. Recall from

Section 8.1.2 that $run_{\mathscr{A},s}(u)$ can be defined recursively. Thus, to compute p_i, M' implements this recursion on its pushdown in the obvious way: it executes a depth-first search of the subtree with root u_i and computes $run_{\mathscr{A},s}(u')$ for every descendant u' of u_i, in a bottom-up fashion. To this end, it uses additional pushdown symbols of the form $(b,q_1\cdots q_n)$ with $b \in \{0,1\}$ and $q_1,\ldots,q_n \in Q$. The boolean b just indicates whether or not the corresponding node u' is u_i (this is because M' needs to know when it has returned to u_i). The states $q_1,\ldots,q_n$ are the states of $\mathscr{A}$ in which it reaches the first n sons of u'. Thus, after having computed $p_1,\ldots,p_{i-1}$ (and having stored them in its state), M' calls a subroutine that moves down to the i-th son of u in state $\downarrow$ and pushes the symbol $(1,\varepsilon)$, where ε is the empty sequence of states of $\mathscr{A}$. The subroutine then applies the following rules, for every $f \in F$ with $\rho(f) = k$, $b \in \{0,1\}$, and $q_1,\ldots,q_n \in Q$ with $n \leq k$ (using the states $\downarrow$ and $\uparrow_{b,q}$ for every $b \in \{0,1\}$ and $q \in Q$):

(1) if $n < k$, then
$$\langle\downarrow,lab_f(x_1),(b,q_1\cdots q_n)\rangle \to \langle\downarrow,down_{n+1},(b,q_1\cdots q_n)(0,\varepsilon)\rangle;$$
(2) if $n = k$ and $\delta_f(q_1,\ldots,q_n) = q$, then
$$\langle\downarrow,lab_f(x_1),(b,q_1\cdots q_n)\rangle \to \langle\uparrow_{b,q},up,\varepsilon\rangle;$$
(3) if $n < k$, then
$$\langle\uparrow_{0,q},lab_f(x_1),(b,q_1\cdots q_n)\rangle \to \langle\downarrow,stay,(b,q_1\cdots q_n q)\rangle.$$

Thus, the subroutine returns to u in some state $\uparrow_{1,q}$, and then M' stores $p_i := q$ in its state.

It remains to show how M' keeps track of $P = Th_s^{\uparrow}(u)$. The value of P is stored on the pushdown. Thus, if M has pushdown alphabet Γ and initial pushdown symbol γ_{in}, then M' uses the pushdown symbols (γ,S) with $\gamma \in \Gamma$ and $S \subseteq L_1'$, and has initial pushdown symbol $(\gamma_{in},Th_{rt}^{\uparrow})$, where $Th_{rt}^{\uparrow}$ is the root theory. Clearly, it suffices to show how P can be computed when M' moves down from node u to the i-th son u_i, simulating a step of M. That is easy: at u, M' has computed $p_1 = Th_s^{\downarrow}(u_1),\ldots,p_k = Th_s^{\downarrow}(u_k)$ (as indicated above), and it has stored $P = Th_s^{\uparrow}(u)$ in the top of the pushdown. From this it can determine $Th_s^{\uparrow}(u_i) = Z_{f,i}(P,p_1,\ldots,p_{i-1},p_{i+1},\ldots,p_k)$ as observed in Section 8.1.3, and push $Th_s^{\uparrow}(u_i)$ as second component of the new top symbol when moving down to u_i.   ∎

Theorems 8.6, 8.7 and 8.12 together show that the monadic second-order transductions of terms can be computed by tree-walking pushdown tree transducers.

Theorem 8.13 The following inclusions hold:

$$\text{DMSOT} \subset \text{P-DTWT} \quad \text{and} \quad \text{DMSOTW} \subset \text{P-DTWTW.} \qquad \square$$

The inclusions are proper because the word transduction $a^n \mapsto a^{2^n}$ is in P-DTWTW: the pushdown can be used to count to 2^n, in binary. This also shows that DTWT^{MS} is properly included in P-DTWT: a deterministic MS tree-walking transducer with set of states Q can visit each node of $a^n(\epsilon)$ at most $|Q|$ times (otherwise it is in a loop) and hence the height of the output tree is at most $|Q| \cdot (n+1)$.

It follows from Theorem 7.29 (based on Proposition 6.54) that every parameterless MS-transduction of terms can be computed in linear time (but usually with a large constant). We present here an alternative, but similar, proof using the construction in the proof of Theorem 8.12. As in the case of Theorem 7.29, the corresponding linear-time algorithm is not directly applicable, due to the large constant.

Theorem 8.14 Every MS-transduction τ in DMSOT can be computed in linear time, that is, for a given input term s the value of $\tau(s)$ can be computed in time $O(|s|)$.

Proof: Let τ be an MS-transduction in DMSOT. By Theorems 8.6 and 8.7, the transduction τ is computed by a single-use deterministic nonjumping MS tree-walking transducer $M = (F, H, Q, q_{\text{in}}, R)$. We first observe that M computes in linear time, assuming that each computation step takes one time unit. More precisely, for every $s \in T(F)$ and $t \in T(H)$, if $\langle q_0, \text{root}_s \rangle = \xi_0 \Rightarrow_{G_{M,s}} \xi_1 \Rightarrow_{G_{M,s}} \xi_2 \Rightarrow_{G_{M,s}} \cdots \Rightarrow_{G_{M,s}} \xi_n = t$, then $n \le |Q| \cdot |s|$. To see this, we show that each configuration $\langle q, u \rangle$ of M is rewritten at most once during this computation. Suppose it is rewritten both in ξ_i and in ξ_j, with $i < j$. Let ξ_i be of the form $\xi_i = w_0 \langle q_1, u_1 \rangle w_1 \cdots \langle q_k, u_k \rangle w_k$, where each $\langle q_l, u_l \rangle$ is a configuration and $w_0, w_1, \ldots, w_k$ do not contain configurations. Then ξ_j is of the form $\xi_j = w_0 \zeta_1 w_1 \cdots \zeta_k w_k$ with $\langle q_l, u_l \rangle \Rightarrow^*_{G_{M,s}} \zeta_l$ for every $l \in [k]$. Let $\langle q, u \rangle = \langle q_m, u_m \rangle$. Since $\langle q, u \rangle$ is rewritten in ξ_i, $\langle q, u \rangle$ does not occur in ζ_m, because otherwise the deterministic transducer M would loop and the computation would be infinite. Thus, $\langle q, u \rangle$ occurs in one of $\zeta_1, \ldots, \zeta_{m-1}, \zeta_{m+1}, \ldots, \zeta_k$. Now consider the computation $\xi_0 \Rightarrow^*_{G_{M,s}} \xi_i \Rightarrow^*_{G_{M,s}} \xi \Rightarrow^*_{G_{M,s}} \xi_j \Rightarrow^*_{G_{M,s}} t$ with $\xi = w_0 \zeta_1 w_1 \cdots \zeta_{m-1} w_{m-1} \langle q_m, u_m \rangle w_m \zeta_{m+1} w_{m+1} \cdots \zeta_k w_k$. It contradicts the fact that M is single-use, because $\langle q, u \rangle$ occurs at least twice in ξ.

Thus, assuming that each computation step takes constant time, $\tau(s)$ can be computed in linear time: on input s, at most $|Q| \cdot |s|$ computation steps of M are executed. However, in each computation step, global tests must be simulated. For each such test (with quantifier-height q), as in the proof of Theorem 8.12, it now suffices to show that M can be stepwise simulated in such a way that, for input tree $s \in T(F)$ and current node $u \in N_s$, the information $p_1 := run_{\mathcal{A}, s}(u_1), \ldots, p_k := run_{\mathcal{A}, s}(u_k)$ (if u has k sons $u_1, \ldots, u_k$) and $P := Th_s^{\uparrow}(u)$ is available, where $\mathcal{A}$ is the complete and deterministic F-semi-automaton such that $run_{\mathcal{A}, s}(u_i) = Th_s^{\downarrow}(u_i)$ (see Section 8.1.3).

Rather than computing the states $p_1, \ldots, p_k$ each time M visits u (as in the proof of Theorem 8.12), we can preprocess the input tree s by first computing the (unique) run $run_{\mathcal{A}, s} : Pos(s) \to Q_{\mathcal{A}}$ of $\mathcal{A}$ on s, where $Q_{\mathcal{A}}$ is the set of states of $\mathcal{A}$. Now $p_1, \ldots, p_k$ are available to M in constant time whenever it visits u. The run of $\mathcal{A}$ on s can be computed in time $O(|s|)$ and so, since M has only finitely many global tests, the total preprocessing time is linear.

The set of formulas P can be computed using the pushdown of a p-ms-twt. As described at the end of the proof of Theorem 8.12, this pushdown can be updated in constant time in each computation step of M (using $p_1, \ldots, p_k$). Thus, P is available whenever M visits u. ∎

Next we use Theorem 8.13 to prove that every MS tree-walking transducer can be simulated by a composition of two tree-walking transducers, the first of which only walks down in the input tree. A deterministic tree-walking transducer is called a *deterministic top-down tree transducer* if it does not use the formula *up* in the right-hand sides of its rules (cf. Example 8.5). The class of transductions computed by deterministic top-down tree transducers is denoted by $\mathrm{DTWT}_\downarrow$.

Theorem 8.15 For every tree-walking pushdown transducer M, two tree-walking transducers M_1 and M_2 can be constructed such that $\tau_M = \tau_{M_1} \cdot \tau_{M_2}$. Moreover, M_1 is a deterministic top-down tree transducer, and if M is deterministic then so is M_2.

Proof: Let $M = (F, H, Q, q_{\mathrm{in}}, \Gamma, \gamma_{\mathrm{in}}, R)$ be a (local) p-twt and enumerate Γ as $\{\gamma_1, \ldots, \gamma_n\}$. Without loss of generality we assume that during the computations of M, the bottom symbol of the pushdown is always equal to γ_{in} (M can store the actual value of the bottom symbol in its state). We also assume without loss of generality that in case (c) of the definition of a p-ms-twt, $\beta = \gamma\gamma'$ for some $\gamma' \in \Gamma$, i.e., the symbol γ' is pushed on top of the pushdown, without changing the previous top symbol γ (if γ has to be changed, one first uses a *stay*-rule). Finally, we assume that $\beta = \gamma$ in case (b) of the definition of a p-ms-twt, i.e., the pushdown is unchanged (if it has to be changed, one first moves up and then moves down again).

The transducer M_1 preprocesses the input tree s in such a way that M_2 can stepwise simulate M on $\tau_{M_1}(s)$, using the structure of $\tau_{M_1}(s)$ to implement the pushdown of M. More precisely, for each node $u \in N_s$ and each pushdown $\pi \in \Gamma^m$, where m is the depth of u, $\tau_{M_1}(s)$ has a node $[u, \pi]$ with label $[f, \gamma]$, such that f is the label of u in s and γ is the top symbol of π. Thus, π will correspond to the second components of the labels of the ancestors of $[u, \pi]$.

We define $M_1 := (F, F \times \Gamma, \Gamma, \gamma_{\mathrm{in}}, R_1)$. Note that M_1 uses the pushdown symbols of M as states. For every $\gamma \in \Gamma$ and $f \in F$ of arity k, the arity of $[f, \gamma]$ is nk, and R_1 contains the rule $\langle \gamma, lab_f(x_1) \rangle \to \zeta_{\gamma, f}$ where $\zeta_{\gamma, f}$ is:

$$[f, \gamma](\langle \gamma_1, down_1 \rangle, \ldots, \langle \gamma_n, down_1 \rangle, \ldots, \langle \gamma_1, down_k \rangle, \ldots, \langle \gamma_n, down_k \rangle).$$

This rule changes each son u_i of the current node u into n sons that will receive the additional labels $\gamma_1, \ldots, \gamma_n$ respectively. Note that after application of the rule, the j-th copy of u_i is the m-th son of the output node with $m = n(i-1) + j$.

Finally we define $M_2 := (F \times \Gamma, H, Q, q_{\mathrm{in}}, R_2)$ as follows. Let $\langle q, \varphi, \gamma \rangle \to \zeta$ be a rule of M. It is simulated by the rule

$$\langle q, \varphi' \wedge \bigvee_{f \in F} lab_{[f, \gamma]}(x_1) \rangle \to \zeta'$$

in R_2, where:

- φ' is obtained from φ by changing every subformula $lab_f(x_1)$ into the formula $\bigvee_{\gamma \in \Gamma} lab_{[f,\gamma]}(x_1)$, and every subformula $\exists y.son_i(y,x_1)$ into the formula $\bigvee_{j \in [n]} \exists y.son_{n(i-1)+j}(y,x_1)$; and
- ζ' is obtained from ζ by changing every triple $\langle q',up,\varepsilon \rangle$ into $\langle q',up \rangle$, every triple $\langle q',stay,\gamma \rangle$ into $\langle q',stay \rangle$, and every triple $\langle q',down_i,\gamma\gamma_j \rangle$ into $\langle q',down_{n(i-1)+j} \rangle$.

It is clear that $\tau_M(s) = \tau_{M_2}(\tau_{M_1}(s))$ for every $s \in T(F)$. ∎

Thus, P-DTWT $\subseteq$ DTWT$_\downarrow$ · DTWT. It is straightforward to show that this is, in fact, an equality (see also Theorem 8.22).

From Theorems 8.13 and 8.15 we obtain the next result.

Theorem 8.16 The class DMSOT is included in the class DTWT$_\downarrow$ · DTWT. $\square$

This theorem can be strengthened to a characterization of DMSOT in terms of tree-walking transducers, which can be viewed as a generalization of Theorem 1.16 to transductions. We omit the complicated proof (see [EngMan03a], as explained in Section 8.10). Recall from Section 8.1.1 that, trivially, every MS-transduction is of linear size increase. The class DTWT$_\downarrow$ · DTWT contains transductions that are *not* of linear size increase, because even DTWT$_\downarrow$ does, cf. Examples 8.5 and 7.3(4). The characterization is that DMSOT and DTWT$_\downarrow$ · DTWT have the same expressive power with respect to transductions of linear size increase. In other words, the monadic second-order transductions of terms are exactly the transductions of linear size increase that are computed by the composition of two tree-walking tree transducers, the first of which is a top-down tree transducer.

Theorem 8.17 A transduction τ of terms is in the class DMSOT if and only if (1) τ is in the class DTWT$_\downarrow$ · DTWT and (2) τ is of linear size increase. $\square$

Of course, as a special case, this result also holds for term-to-word transductions, i.e., DMSOTW equals the class of transductions of linear size increase in DTWT$_\downarrow$ · DTWTW.

It is also shown in [EngMan03a], by the same complicated proof, that it is decidable for $\tau \in$ DTWT$_\downarrow$ · DTWT (given as a composition of two tree-walking transducers) whether or not τ is of linear size increase, i.e., whether or not it is in DMSOT; and if so, a definition scheme for τ can be constructed.

We now show that every monadic second-order transduction of graphs of bounded clique-width can be computed, on the level of terms over F^{VR}, by a tree-walking pushdown transducer (or by the composition of two tree-walking transducers, the first of which is a top-down tree transducer). This can be viewed as a generalization of the Weak Recognizability Theorem to transductions.

Theorem 8.18 Let D_1, D_2 and C_1 be finite sets of port labels with $D_1 \subseteq C_1$, and let $\tau : \mathcal{GP}[D_1] \to \mathcal{GP}[D_2]$ be a parameterless monadic second-order transduction of p-graphs, given by a definition scheme of type $\mathcal{R}_{s,D_1} \to \mathcal{R}_{s,D_2}$. One can construct a finite set of port labels C_2 that includes D_2 and a deterministic tree-walking pushdown transducer M with input signature $F_{C_1}^{VR}$ and output signature $F_{C_2}^{VR}$ such that $\tau(val(t)) = val(\tau_M(t))$ for every $t \in T(F_{C_1}^{VR})$ with $\pi(val(t)) \subseteq D_1$. A similar result holds for labeled p-graphs.

Proof: By Proposition 7.30 and Theorem 7.14, the mapping $val \cdot \tau$ is a parameterless monadic second-order transduction : $T(F_{C_1}^{VR}) \to \mathcal{GP}[D_2]$. Hence, by Theorem 7.34 one can construct C_2 and a parameterless (invertible) MS-transduction $\mu : T(F_{C_1}^{VR}) \to T(F_{C_2}^{VR})$ such that $val \cdot \tau = \mu \cdot val$. By Theorem 8.13, μ can be computed by a deterministic tree-walking pushdown transducer M.   ∎

Remark 8.19 (1) This result also holds for parameter-invariant monadic second-order transductions of p-graphs (rather than parameterless ones). In that case the term transduction μ is not parameterless (and in general not even parameter-invariant), but can be replaced by a parameterless MS-transduction $\mu' \subseteq \mu$ such that $Dom(\mu') = Dom(\mu)$, see Footnote 25 in the proof of Theorem 7.50.

(2) The result also holds for $CMS_{i,j}$-transductions (Section 7.3), for each i and j, additionally using Proposition 7.48 and Theorem 7.50. For instance, if $\tau : \mathcal{GP}[D_1] \to \mathcal{JS}[D_2]$ is a parameterless $CMS_{1,2}$-transduction, then the transducer M has input signature $F_{C_1}^{VR}$ and output signature $F_{C_2}^{HR}$.

(3) Theorem 8.18 can also be used as an alternative way of evaluating a parameterless (and even parameter-invariant) monadic second-order graph transduction τ, using Theorem 8.14 instead of Theorem 7.29, see Remark 7.31(2). The input graph is parsed, the resulting term t_1 is translated into a term t_2 for the output graph using Theorem 8.14, and the value of t_2 is computed (cf. Proposition 2.45). The advantage of this method is that further transformations of the output graph need not parse that graph, because t_2 is still available.   □

8.6 Multi bottom-up tree-to-word transducers

In this section we present an alternative tree transducer model for DMSOTW. It can be viewed as an automaton on terms with infinitely many states and transition rules. Thus, it operates in a bottom-up fashion. Instead of computing one word translation of the input tree, it computes a finite number of translations simultaneously (which is indicated by "multi"). At each node u of the input tree s, the finitely many translations of the subtree s/u form, together with the finite state of the transducer, the state of the infinite automaton. The translations at u are concatenations of letters of the output alphabet and the translations at the sons $u_1,\ldots,u_k$ of u. The words computed by the transducer are *not* viewed as unary trees.

Formally, a (complete and deterministic) *multi bottom-up tree-to-word transducer* (abbreviated *mbott*) is a tuple $M = (F, B, Q, \mathcal{X}, x_{\text{out}}, \delta, Acc)$, where F is a functional signature of input symbols, B is an alphabet of output symbols, Q is a finite set of states, $Acc \subseteq Q$ is the set of accepting states, $\mathcal{X}$ is a finite set of variables, $x_{\text{out}} \in \mathcal{X}$ is the output variable and δ is a finite set of transition rules. A transition rule is of the form $f[q_1, \ldots, q_k] \to [q, \alpha]$, where $k \geq 0, f \in F_k, q_1, \ldots, q_k, q \in Q$ and α is a mapping $: \mathcal{X} \to ((\mathcal{X} \times [k]) \cup B)^*$. For every $q_1, \ldots, q_k$ and every $f \in F_k$ there is exactly one pair $[q, \alpha]$ such that $f[q_1, \ldots, q_k] \to [q, \alpha]$ is in δ (which means that M is complete and deterministic). We will denote $[q, \alpha]$ by $\delta_f(q_1, \ldots, q_k)$.

A mapping $\alpha : \mathcal{X} \to ((\mathcal{X} \times [k]) \cup B)^*$ can be viewed as a set of *assignments* $x := \alpha(x)$ for $x \in \mathcal{X}$, where $\alpha(x)$ is of the form $w_0 \langle x_1, i_1 \rangle w_1 \cdots \langle x_n, i_n \rangle w_n$ with $n \geq 0$, $w_j \in B^*$ for $j \in [0, n]$, $x_j \in \mathcal{X}$ and $i_j \in [k]$ for $j \in [n]$. If the above rule is used at a node u with sons $u_1, \ldots, u_k$, then the pair $\langle x, i \rangle$ stands for the variable x of the i-th son (and x for the one of the father).

To define the runs of the mbott M we use the (complete and deterministic) infinite F-automaton $\mathscr{A}_M = \langle F, Q_\infty, \delta_\infty, Acc_\infty \rangle$. The set Q_∞ of states of $\mathscr{A}_M$ consists of all pairs $\langle q, \beta \rangle$ such that $q \in Q$ and β is a *valuation*, i.e., a mapping $: \mathcal{X} \to B^*$. Such a pair is called a *configuration* of M; intuitively, $\beta(x)$ is the current value of the variable x. If $f[q_1, \ldots, q_k] \to [q, \alpha]$ is a transition rule in δ and $\beta_1, \ldots, \beta_k$ are valuations, then δ_∞ contains the transition rule $f[\langle q_1, \beta_1 \rangle, \ldots, \langle q_k, \beta_k \rangle] \to \langle q, \beta \rangle$, where β is the valuation defined as follows: if $x := w_0 \langle x_1, i_1 \rangle w_1 \cdots \langle x_n, i_n \rangle w_n$ is an assignment of α, then $\beta(x) = w_0 \beta_{i_1}(x_1) w_1 \cdots \beta_{i_n}(x_n) w_n$. The set Acc_∞ of accepting states consists of all configurations $\langle q, \beta \rangle$ with $q \in Acc$. The runs of M are the runs of $\mathscr{A}_M$. The *transduction computed by M*, denoted τ_M, is defined as

$$\tau_M := \bigcup_{q \in Acc} \{(s, \beta(x_{\text{out}})) \in T(F) \times B^* \mid s \in L(\mathscr{A}_M, \langle q, \beta \rangle)\}.$$

Note that $L(\mathscr{A}_M)$ is the domain of τ_M; obviously, it is a regular language (because the values of the variables do not influence the runs of $\mathscr{A}_M$).

The mbott M is a *one-way multi word transducer* if F is unary. Then M computes the word transduction $\{(v, w) \in A^* \times B^* \mid (\mu_A(v), w) \in \tau_M\}$, where $A = F - \{\epsilon\}$. It can be viewed as having a one-way input tape, on which it walks from right to left.

The mbott M is *linear* if every set α of assignments is linear, i.e., if each $\langle x, i \rangle$ occurs at most once in $\alpha(x_1) \cdots \alpha(x_m)$, where $\mathcal{X} = \{x_1, \ldots, x_m\}$. Intuitively, such a transducer is "single-use" in the sense that, for a run of M, the value of each variable in a configuration of that run is used at most once as a subword of the output word. To formally define the single-use property, we need some more notation.

Let $s \in T(F)$. For a node $u \in N_s$ we denote by $run_{M,s}(u)$ the state in which M reaches u: the first component of $run_{\mathscr{A}_M,s}(u)$. By α_u we denote the set of assignments that is used by M when reaching u: the second component of $\delta_f(run_{M,s}(u_1), \ldots, run_{M,s}(u_k))$, where u has label f and sons $u_1, \ldots, u_k$. We now define a binary relation $\gamma(u)$ on the set of variables $\mathcal{X}$. Intuitively, (x,y) is in $\gamma(u)$ if the current values of x and y both contain as a subword the value of a variable z at some proper descendant of u; we say that x and y are *linked* and that $\gamma(u)$ is a *linking relation*. Formally, $\gamma(u)$ is the set of all (x,y) such that $\alpha_u(x)$ has an occurrence $\langle x',i \rangle$ and $\alpha_u(y)$ has an occurrence $\langle y',i \rangle$ (with the same $i \in [k]$) and either $x' = y'$ or $(x',y') \in \gamma(u_i)$. We now define an mbott to be single-use by forbidding it to use a variable (of a son) twice or to use two linked variables (of a son) in the right-hand side of an assignment.

The mbott M is *single-use* if for every $s \in T(F)$, $u \in N_s$ and $x \in \mathcal{X}$,

(1) each $\langle y,i \rangle$ has at most one occurrence in $\alpha_u(x)$, and
(2) if $\langle y,i \rangle$ and $\langle z,i \rangle$ (with the same $i \in [k]$ and $y \neq z$) both occur in $\alpha_u(x)$, then $(y,z) \notin \gamma(u_i)$.

It is clear that the transduction computed by a single-use mbott M is of linear size increase. By $\text{DMBOT}_{\text{lsi}}$ we will denote the class of transductions of linear size increase that are computed by mbotts, by DMBOT_{su} the class of transductions computed by single-use mbotts, and by $\text{DMBOT}_{\text{lin}}$ the class of transductions : $T(F) \to B^*$ of the form $\{(s,w) \mid (\#(s),w) \in \tau_M\}$, where M is a linear mbott with input signature $F \cup \{\#\}$ and $\#$ is a special unary function symbol not in F (which, intuitively, allows M to recognize "the end of the input"). For the one-way multi word transducers the corresponding classes are denoted $1\text{DMWT}_{\text{lsi}}$, 1DMWT_{su} and $1\text{DMWT}_{\text{lin}}$.

Example 8.20 We construct a single-use mbott $M = (F,B,Q,\mathcal{X},x_{\text{out}},\delta,Acc)$ that computes the transduction $\sigma' : T(F) \to B^*$ from Example 8.4, where $B = \{a_1,a_2,b\}$ and $F = \{f\} \cup B$ with $f \in F_2$ and $a_1,a_2,b \in F_0$. For every $s \in T(F)$, if $yd(s) = w_0bw_1bw_2 \cdots bw_n$, with $w_i \in \{a_1,a_2\}^*$, then $\sigma'(s) := w_0w_0bw_1w_1bw_2w_2 \cdots bw_nw_n$.

The mbott M has states b and a, both accepting; when M reaches node u of s, it is in state b if u has a descendant with label b, and in state a otherwise. It has six variables: l_1, l_2, m, r_1, r_2 and x_{out}. If $yd(s/u) = w_0bw_1bw_2 \cdots bw_n$, then the value of both l_1 and l_2 is w_0, the value of m is $bw_1w_1bw_2w_2 \cdots b$, the value of both r_1 and r_2 is w_n, and the value of x_{out} is $\sigma'(s/u)$. The transition rules of M are as follows:

$$b \to [b,\alpha_b], \quad a_i \to [a,\alpha_{a_i}] \text{ for } i = 1,2,$$
$$f[a,a] \to [a,\alpha_{a,a}], \quad f[a,b] \to [b,\alpha_{a,b}],$$
$$f[b,a] \to [b,\alpha_{b,a}] \text{ and } f[b,b] \to [b,\alpha_{b,b}].$$

The set of assignments α_b is $l_1 := \varepsilon$, $l_2 := \varepsilon$, $m := b$, $r_1 := \varepsilon$, $r_2 := \varepsilon$ and $x_{\text{out}} := b$, and the set of assignments α_{a_1} is $l_1 := a_1$, $l_2 := a_1$, $m := \varepsilon$, $r_1 := a_1$, $r_2 := a_1$ and $x_{\text{out}} := a_1a_1$. The set of assignments $\alpha_{b,b}$ is as follows:

$$l_1 := \langle l_1,1 \rangle, \quad l_2 := \langle l_2,1 \rangle,$$
$$m := \langle m,1 \rangle \langle r_1,1 \rangle \langle l_1,2 \rangle \langle r_2,1 \rangle \langle l_2,2 \rangle \langle m,2 \rangle,$$
$$r_1 := \langle r_1,2 \rangle, \quad r_2 := \langle r_2,2 \rangle,$$
$$x_{\text{out}} := \langle l_1,1 \rangle \langle l_2,1 \rangle \alpha_{b,b}(m) \langle r_1,2 \rangle \langle r_2,2 \rangle.$$

The other sets of assignments are similar.

If the assignments to x_{out} would not be there, M would be linear. Since x_{out} is not used in the right-hand sides of assignments, this implies that M is single-use. We note that this way of handling x_{out} can be used to show that $DMBOT_{lin} \subseteq DMBOT_{su}$: a linear mbott realizing the same transduction as M has assignments $x_{out} := \varepsilon$, except in the additional transition rules $\#[a] \to [a, \alpha_{\#,a}]$ and $\#[b] \to [b, \alpha_{\#,b}]$: in the first rule $x_{out} := \langle l_1, 1 \rangle \langle l_2, 1 \rangle$, and in the second rule $x_{out} := \langle l_1, 1 \rangle \langle l_2, 1 \rangle \langle m, 1 \rangle \langle r_1, 1 \rangle \langle r_2, 1 \rangle$. $\qquad\square$

In the next theorem we present the characterization of MS tree-to-word transductions by single-use mbotts. By restricting to unary input signatures, the MS transductions of words are characterized by single-use one-way multi word transducers (as an alternative to Theorem 8.10).

Theorem 8.21 The classes DMSOTW and $DMBOT_{su}$ are equal, and so are the classes DMSOW and $1DMWT_{su}$.

Proof: We first show the inclusion $DTWTW^{MS} \subseteq DMBOT_{su}$, cf. Theorem 8.6. Let $M = (F, H, Q, q_{in}, R)$ be a nonjumping deterministic MS tree-walking tree-to-word transducer (cf. Theorem 8.7), and let $B := H - \{\epsilon\}$. Let q be the maximal quantifier-height of the tests φ in the left-hand sides of the rules of M, and let $\mathscr{A} = \langle F, Q_{\mathscr{A}}, \delta_{\mathscr{A}} \rangle$ be the complete and deterministic finite F-semi-automaton $\mathscr{A}$ such that for every $s \in T(F)$ and $u \in N_s$, $run_{\mathscr{A},s}(u) = Th_s^{\downarrow}(u)$ (cf. the proofs of Theorems 8.9 and 8.12). Note that $Q_{\mathscr{A}} = \mathcal{P}(L_1)$.

We will construct a single-use mbott $M' = (F, B, Q', \mathcal{X}, x_{out}, \delta, Acc)$ that computes the same tree-to-word transduction as M. To ensure that M' is single-use, it computes the linking relation $\gamma(u)$ in its state. The set of states Q' of M' is defined to consist of all triples $\langle p, \sigma, \gamma \rangle$ such that $p \in Q_{\mathscr{A}}$, σ is a mapping $: Q \times \mathcal{P}(L_1') \to Q \cup \{@, \bot\}$ (called a *transition table* of M) and $\gamma \subseteq \mathcal{X} \times \mathcal{X}$. Intuitively, when M' reaches node $u \in N_s$, the transition table σ means the following: if $\sigma(q, Th_s^{\uparrow}(u)) = q' \in Q$ then the transducer M, when started in state q at node u, has a computation on s/u that ends by moving up to the father of u in state q'; moreover, if $\sigma(q, Th_s^{\uparrow}(u)) = @$ then the computation of M does not leave s/u and is successful, and if $\sigma(q, Th_s^{\uparrow}(u)) = \bot$ then the computation of M does not leave s/u and is unsuccessful (i.e., aborts or is infinite). The set Acc of accepting states is defined to consist of all triples $\langle p, \sigma, \gamma \rangle$ such that $\sigma(q_{in}, Th_{rt}^{\uparrow}) = @$, where $Th_{rt}^{\uparrow}$ is the root theory.

The set of variables $\mathcal{X}$ of M' is defined to be $Q \times \mathcal{P}(L_1')$. Intuitively, when M' reaches node $u \in N_s$, the value of the variable $\langle q, Th_s^{\uparrow}(u) \rangle$ is the output word in B^* that is produced by M during the above-mentioned computation (if $\sigma(q, Th_s^{\uparrow}(u)) \neq \bot$; otherwise the value is ε). The output variable x_{out} is defined to be $\langle q_{in}, Th_{rt}^{\uparrow} \rangle$.

It remains to define the transition rules of M'. Consider a left-hand side $f[\langle p_1, \sigma_1, \gamma_1 \rangle, \ldots, \langle p_k, \sigma_k, \gamma_k \rangle]$ of a rule of M'. Intuitively, M' has reached (simultaneously) the sons $u_1, \ldots, u_k$ of a node $u \in N_s$ (for some input tree s), it knows $p_1 := Th_s^{\downarrow}(u_1), \ldots, p_k := Th_s^{\downarrow}(u_k)$ and it knows the transition tables $\sigma_1, \ldots, \sigma_k$ of

$u_1, \ldots, u_k$. It additionally knows the linking relations $\gamma_1 = \gamma(u_1), \ldots, \gamma_k = \gamma(u_k)$ and it knows the following: for $P := Th_s^\uparrow(u_i)$, if the variables $\langle q_1, P \rangle$ and $\langle q_2, P \rangle$ are linked at u_i, i.e., $(\langle q_1, P \rangle, \langle q_2, P \rangle) \in \gamma_i$, then the corresponding computations of M on s/u_i both contain the same subcomputation corresponding to a variable $\langle q', P' \rangle$ at a proper descendant of u. These facts can be formally proved by bottom-up induction on u; we omit the details. Note that M' does *not* know $Th_s^\uparrow(u)$, but is prepared for every possible $Th_s^\uparrow(u)$.

We will construct the right-hand side $[\langle p, \sigma, \gamma \rangle, \alpha]$ of the rule. First, we define p to be $\delta_{\mathscr{A}}(p_1, \ldots, p_k)$, i.e., M' computes $Th_s^\downarrow(u)$ in the first component of its state. Let us now define $\sigma(q, P)$ and $\alpha(\langle q, P \rangle)$ for given $q \in Q$ and $P \subseteq L_1'$. We let $Th := Z_f(P, p_1, \ldots, p_k)$ and, for $i \in [k]$, $Th_i := Z_{f,i}(P, p_1, \ldots, p_{i-1}, p_{i+1}, \ldots, p_k)$, cf. Section 8.1.3; thus, if $P = Th_s^\uparrow(u)$, then $Th = Th_s(u)$ and $Th_i = Th_s^\uparrow(u_i)$. We now start a simulation of the transducer M, starting in state q. During this simulation, we keep track of the current state of M (initialized to q) and of the current prefix of $\alpha(\langle q, P \rangle)$ constructed so far (initialized to the empty word ε). Moreover, we keep track of the current position of M: the father or the i-th son (initialized to the father). If M is at the father and the current state is q', then the applicable rules of M are of one of the three forms $\langle q', \varphi \rangle \to \langle q'', \psi \rangle$, or $\langle q', \varphi \rangle \to b(\langle q'', \psi \rangle)$, or $\langle q', \varphi \rangle \to \epsilon$, such that $\widehat{\varphi}$ is in Th (where ψ is *stay*, *up*, or *down$_i$*). If there is no such rule, or there are two or more, then we *abort* the simulation, i.e., we end the simulation and define $\sigma(q, P)$ to be $\perp$ and $\alpha(\langle q, P \rangle)$ to be ε. Now assume that there is exactly one such rule. If it is of the third form, we end the simulation and define $\sigma(q, P)$ to be @ and $\alpha(\langle q, P \rangle)$ to be the current prefix. If the rule is of the first or second form, then we change the current state to q'', and we add b to the current prefix in the second case. Moreover, if ψ is *down$_i$*, we change the current position to the i-th son; if ψ is *up*, we end the simulation and define $\sigma(q, P)$ to be the current state and $\alpha(\langle q, P \rangle)$ to be the current prefix. Now let (our simulation of) M be at the i-th son in state q'. If the current prefix has an occurrence of a variable $\langle \langle \bar{q}, Th_i \rangle, i \rangle$ of the i-th son such that $(\langle \bar{q}, Th_i \rangle, \langle q', Th_i \rangle) \in \gamma_i$, then we abort the simulation, as above, because the computation of M on s/u contains a certain subcomputation on s/u_i twice, and hence is infinite. Otherwise, let $q'' := \sigma_i(q', Th_i)$. If $q'' = \perp$, then we abort the simulation, as above. If $q'' = @$, then the simulation ends and we define $\sigma(q, P)$ to be @ and $\alpha(\langle q, P \rangle)$ to be the current prefix to which $\langle \langle q', Th_i \rangle, i \rangle$ is added: the variable $\langle q', Th_i \rangle$ at the i-th son. If $q'' \in Q$, then we change the current state to q'' and the current position to the father, and we add $\langle \langle q', Th_i \rangle, i \rangle$ to the current prefix. The only remaining problem is that the simulation can be infinite. Since M is deterministic, this can only happen when there is a repetition of the pair consisting of the current state and the current position. In that case we abort the simulation, as above.

Finally, we define γ in the same way as we defined $\gamma(u)$: γ is the set of all $(x, y) \in \mathcal{X} \times \mathcal{X}$ such that $\alpha(x)$ has an occurrence $\langle x', i \rangle$ and $\alpha(y)$ has an occurrence $\langle y', i \rangle$ (with the same $i \in [k]$) and either $x' = y'$ or $(x', y') \in \gamma_i$. It ensures that M' computes $\gamma(u)$ in the third component of its state.

This ends the construction of M'. It should be clear from this construction, and the given intuitive comments, that M' "simulates" M and thus computes the same transduction as M. To show that M' is single-use, we first check Condition (1) in the definition of single-use (given in the second paragraph before Example 8.20). If, in the above construction of $\alpha(\langle q, P \rangle)$, the variable $\langle \langle q', Th_i \rangle, i \rangle$ of the i-th son would be added for the second time to the current prefix, then the pair consisting of the current state (q') and the current position (the i-th son) repeats, and thus $\alpha(\langle q, P \rangle) = \varepsilon$. Condition (2) holds by the construction of M'.

For the proof of the inclusion $\mathrm{DMBOT_{su}} \subseteq \mathrm{DTWTW^{MS}}$, consider a single-use mbott $M = (F, B, Q, \mathcal{X}, x_{\mathrm{out}}, \delta, Acc)$. For $s \in T(F)$ and $u \in N_s$, we define the set of variables $var(u) \subseteq \mathcal{X}$ by top-down induction on u, as follows: $var(root_s)$ is $\{x_{\mathrm{out}}\}$ and if u has sons $u_1, \ldots, u_k$, then $var(u_i)$ consists of all $y \in \mathcal{X}$ such that $\langle y, i \rangle$ has an occurrence in $\alpha_u(x)$ for some $x \in var(u)$. Intuitively, $var(u)$ is the set of variables of which the value at u is used to compute the output word. It follows from the definitions (with a proof by top-down induction on u) that $(x, y) \notin \gamma(u)$ for distinct variables $x, y \in var(u)$. Hence, again by the definitions, for each $y \in var(u_i)$ there is a unique $x \in var(u)$ such that $\langle y, i \rangle$ occurs in $\alpha_u(x)$. We construct a nonjumping deterministic MS tree-walking tree-to-word transducer $M' := (F, H, Q', q_{\mathrm{in}}, R)$ that uses its MS tests to compute α_u and $var(u)$ at each node u. It can do this because for each $q \in Q$ an MS formula expressing that $run_{M,s}(u) = q$ can be constructed as in the proof of Theorem 5.15 (since M has a unique run on s). Then, for each set of assignments α, an MS formula can be constructed expressing that $\alpha_u = \alpha$, and from that one can construct, for each $\mathcal{Y} \subseteq \mathcal{X}$, an MS formula expressing that $var(u) = \mathcal{Y}$, in a way similar to the proof of Theorem 5.15 (but top-down instead of bottom-up).

The unary output signature H of M' is $B \cup \{\epsilon\}$. In the construction of M' we will assume, without loss of generality, that it has rules of the form $\langle q, \varphi \rangle \to w(\langle q', \psi \rangle)$ for arbitrary $w \in B^*$, meaning that w is written to the output; using auxiliary states such a rule can easily be changed into a finite number of official rules, each of them writing one letter to the output. The set of states Q' of M' is $\mathcal{X} \times (\{\downarrow\} \cup \{\uparrow_i \mid i \in \rho(F)\})$ with $q_{\mathrm{in}} := \langle x_{\mathrm{out}}, \downarrow \rangle$. Intuitively, if M' is at node u in state $\langle x, \downarrow \rangle$, then it starts writing the value of the variable x at u to the output; if it is in state $\langle x, \uparrow_i \rangle$ at node u, then it has just finished writing the value of x at the i-th son of u to the output. It is always the case that $x \in var(u)$; in particular, if u is the root, then $x = x_{\mathrm{out}}$. We now describe the rules of M', in an informal way.

Let M' be at node u in state $\langle x, \downarrow \rangle$ and let $\alpha_u(x) = w_0 \langle x_1, i_1 \rangle w_1 \cdots \langle x_n, i_n \rangle w_n$. If $n > 0$, then M' outputs w_0 and moves down to the i_1-th son in state $\langle x_1, \downarrow \rangle$. If $n = 0$ and u is not the root, then M' outputs w_0 and moves up in state $\langle x, \uparrow_i \rangle$ to its father (of which it is the i-th son). If $n = 0$ and u is the root, then M' outputs w_0 and halts (using a rule $\langle \langle x, \downarrow \rangle, \varphi \rangle \to \epsilon$).

Let M' be at node u in state $\langle y, \uparrow_i \rangle$. Then M' determines the unique $x \in var(u)$ such that $\langle y, i \rangle$ occurs in $\alpha_u(x)$. Let $\alpha_u(x) = w_0 \langle x_1, i_1 \rangle w_1 \cdots \langle x_n, i_n \rangle w_n$ and let $\langle y, i \rangle = \langle x_j, i_j \rangle$. If $j < n$, then M' outputs w_j and moves down to the i_{j+1}-th son in state $\langle x_{j+1}, \downarrow \rangle$. If $j = n$ and u is not the root, then M' outputs w_n and moves up in state $\langle x, \uparrow_b \rangle$ to its father (of which it is the b-th son). If $j = n$ and u is the root, then M' outputs w_n and halts.

This ends the description of M'. It should be clear that it computes the same transduction as M. ∎

The one-way multi word transducer walks from right to left on its input tape. However, we may as well assume that it walks from left to right, as usual, because if τ is a deterministic MS-transduction of words, then so is $\{(\tilde{v}, w) \mid (v, w) \in \tau\}$, cf. Example 8.4.

As observed in Section 8.1.1, DMSOT is contained in DMSOTW. Hence we obtain from Theorem 8.21 that DMSOT $\subset$ DMBOT$_{\text{su}}$. Thus, all deterministic MS-transductions of terms can be implemented on the single-use mbott, as an alternative to the tree-walking pushdown transducer (Theorem 8.13) or the composition of two tree-walking transducers (Theorem 8.17). In particular, by (the proof of) Theorem 8.18, every monadic second-order transduction of graphs of bounded clique-width can be computed, on the level of terms, by a single-use mbott. It should be noted that the values of the variables of an mbott that computes a tree transduction, need not be subtrees of the output tree (but they are subwords of the output term). A characterization of DMSOT as a subset of DMBOT$_{\text{su}}$ is not known.

The multi bottom-up *tree* transducer was introduced in [FülKV04]. It is the mbott with an output signature (rather than alphabet), such that for every assignment $x := \alpha(x)$ the right-hand side $\alpha(x)$ is a term, where each variable $\langle x, i \rangle$ of a son has arity 0. The linear version of the multi bottom-up tree transducer is studied in [FülKV05] and [EngLM].

This section was inspired by [AluČer] and [AluDan]. In [AluČer] the linear one-way multi word transducer is introduced (called *streaming string transducer*), as a model for the computation of deterministic MS-transductions of words, and it is shown that DMSOW = 1DMWT$_{\text{lin}}$. In [AluDan] a generalization of the linear mbott (called *streaming tree transducer*) is presented, as a model for the computation of the transductions in DMSOTW. It is not clear whether DMSOTW = DMBOT$_{\text{lin}}$. Note that Theorem 8.21 also holds with subscript "lsi" instead of "su," as can be concluded from the main result of [FülKV04], Theorems 7.1 and 7.4 of [EngMan03a], and Theorem 7.7 of [EngMan99] (recall from the paragraph before Example 8.20 that "lsi" stands for linear size increase).

8.7 Attribute grammars and macro tree transducers

The formalisms used in the litterature to prove the results on MS-transductions of terms in Sections 8.3, 8.4 and 8.5 (as opposed to those on MS-transductions of words) differ from the tree-walking transducer and the tree-walking pushdown transducer: instead, they are the *attribute grammar* and the *macro tree transducer*. Here we have proved those results using the tree-walking transducer as a formal model for two reasons: (1) to have uniform proofs for words and terms, and (2) to avoid the rather complicated definitions of attribute grammar and macro tree transducer. In this section we discuss these formalisms, without going into details.

The *attribute grammar* is a compiler construction tool (see, e.g., [*AhoLSU], [*DeraJL] and [*WilMau]) introduced in [Knu]. It is essentially a deterministic tree-walking transducer M of which the states are called attributes. The value of an attribute q at a node u of the input tree is the term computed by M when started at node u in state q, interpreted in some semantic domain (i.e., in an H-algebra, where H is the output signature). The transduction computed by the attribute grammar consists of all pairs (s, v) such that v is the value of the attribute q_{in} at the root of the input tree s. The input tree is restricted to be a parse tree of a given context-free grammar, and the attributes are divided into *synthesized* and *inherited* attributes: states in which the transducer M, roughly speaking, walks down and up the tree, respectively. This division of states is not essential, but allows a specification of the rules of M that is local to the rules of the context-free grammar. The *attributed tree transducer* (see [*FülVog], [Fül]) allows arbitrary trees over some input signature, and does not interpret the output term, but retains synthesized and inherited attributes. The current version of the tree-walking transducer, without the distinction between synthesized and inherited states, is introduced in [MilSV] as a model of XML transformations. In that paper, it is in addition equipped with a finite number of pebbles that it can drop on the nodes of the input tree (which is why it is called the *pebble tree transducer*). We stress here that the attribute grammar, the attributed tree transducer, and the tree-walking transducer are essentially notational variations of the same formalism. The precise formal relationship between the latter two transducers is explained in Section 3.2 of [EngMan03b]. The deterministic tree-walking transducer is slightly more powerful than the deterministic attributed tree transducer, because the latter is assumed to be "noncircular," which means that every attribute has a value at every node, i.e., that the transducer always halts when started at any node in any state.

The tree-walking tree-to-word transducer was introduced in [AhoUll71], as a model of syntax-directed translation. It can be viewed as an attribute grammar of which the values of the attributes are words in the algebra $\mathbb{W}_{left}(A)$ for some alphabet A. The main result of [AhoUll71] is in essence the equality $DTWTW^{MS} = DMBOT_{lsi}$ (cf. the end of the previous section). The multi bottom-up tree-to-word transducer can be viewed as a two-phase process: phase 1 computes the state components of the run of the transducer on the input tree, and phase 2 is an attribute grammar with synthesized attributes only, of which the values are words in the algebra $\mathbb{W}(A)$ (also called a "generalized syntax-directed translation scheme").

To appreciate the relationship of Theorems 8.13, 8.16 and 8.17 to the litterature, it is necessary to discuss the *macro tree transducer* [*Eng80], [CouFra] and [EngVog85], which was introduced as a formal model of denotational semantics. The macro tree transducer is obtained from the top-down tree transducer by letting its states (which can be viewed as recursive function procedures) have parameters of type "output tree"; for a formal definition that is close to the tree-walking transducer, see [EngMan03b]. Here we can explain the deterministic macro tree transducer in a different way, as a generalization of the multi bottom-up tree-to-word transducer: we allow the variables of the mbott to have terms as values, instead of words. To give more details, and to avoid confusion, we will from now on say that the mbott has *registers* rather than variables. Then, a deterministic *macro tree transducer* is a tuple $M = (F, H, Q, \mathcal{X}, x_{\text{out}}, \delta, Acc)$, defined in the same way as the mbott, except that both H (the set of output symbols) and $\mathcal{X}$ (the set of registers) are functional signatures, and x_{out} has arity 0. If the register x has rank n, then its values are terms with variables in $T(H, \{y_1, \ldots, y_n\})$, where we use $y_1, y_2, \ldots$ (not in $\mathcal{X}$) as standard variables, also called parameters. Moreover, a set of assignments α (used in a transition rule) consists of assignments $x := \alpha(x)$ with $\alpha(x) \in T((\mathcal{X} \times [k]) \cup H, \{y_1, \ldots, y_n\})$ if x has arity n, where each $\langle x', i \rangle \in \mathcal{X} \times [k]$ has the same arity as x'. The value $\beta(x)$ of register x at a father node is obtained by applying to $\alpha(x)$ the second-order substitution that interprets each $\langle x', i \rangle$ as the derived function defined by the term with variables $\beta_i(x')$ (the value of register x' at the i-th son).

For readers familiar with the macro tree transducer we note that in the usual formulation of a macro tree transducer, $\mathcal{X}$ is called the set of states, x_{out} the initial state, and Q is the set of states of a "look-ahead automaton" employed by the transducer (so it is actually a macro tree transducer with "regular look-ahead," cf. [EngVog85]). A transition rule $f[q_1, \ldots, q_k] \to [q, \alpha]$ in δ combines the transition rule $f[q_1, \ldots, q_k] \to q$ of the look-ahead automaton with the rules $x := \alpha(x)$ of the macro tree transducer for input symbol f. Note that if all registers have arity 0, we obtain the multi bottom-up *tree* transducer of [FülKV04]. The main result of [FülKV04] is that it has the same power as the deterministic top-down tree transducer with "regular look-ahead."

Let DMT denote the class of tree transductions computed by deterministic macro tree transducers (it is denoted DMT_{OI} in [EngVog85]). The proof of Theorem 8.21 can be generalized to show that $\text{DTWT}^{\text{MS}} \subseteq \text{DMT}$. In fact, the macro tree transducer has the same power as the tree-walking pushdown transducer, as shown for total functions in Theorem 5.16 of [EngVog86].[13] For partial functions we now fill a gap in the litterature, for completeness sake.

Theorem 8.22 The classes DMT, P-DTWT and $\text{DTWT}_\downarrow \cdot \text{DTWT}$ are equal.

Proof: In Theorem 6.18 of [EngVog85] it is shown that every $\tau \in \text{DMT}$ can be written as $\tau = \tau_1 \cdot \tau_2$, where τ_1 is the identity on some regular set R of trees, and $\tau_2 \in \text{DMT}$ is a total function and hence $\tau_2 \in \text{P-DTWT}$ by Theorem 5.16 of [EngVog86]. It should be clear that $\tau_1 \cdot \tau_2$ is in P-DTWT: the p-twt just starts by checking that the input tree is in R (see the proof of Theorem 8.12). This proves that $\text{DMT} \subseteq \text{P-DTWT}$. The inclusion $\text{P-DTWT} \subseteq \text{DTWT}_\downarrow \cdot \text{DTWT}$ is shown in Theorem 8.15. Finally, the inclusion $\text{DTWT}_\downarrow \cdot \text{DTWT} \subseteq \text{DMT}$ follows from Theorem 35 of [EngMan03b] (which

---

[13] In that theorem, one should take the storage type S to be the tree storage type TR (as defined in Definition 3.17 of [EngVog86]).

shows, for $n = 0$, that DTWT $\subseteq$ DMT) and Theorem 7.6(3) of [EngVog85] (which shows that DTWT$_\downarrow \cdot$ DMT $\subseteq$ DMT). ∎

Thus, by Theorem 8.17, DMSOT is the class of deterministic macro tree transductions of linear size increase, and, by Theorem 8.18, every monadic second-order transduction of graphs of bounded clique-width can be computed, on the level of terms, by a deterministic macro tree transducer.

8.8 Nondeterminism

We will denote the classes of transductions computed by nondeterministic devices (in particular, by definition schemes with parameters) by dropping the D from the notation for the corresponding deterministic devices.

We first note that the nondeterminism of MS-definition schemes is different in nature from that of tree-walking transducers. This is so because (the semantics of) a definition scheme uses a global guess of the values assigned to the parameters, whereas a transducer makes a local guess whenever it visits an input node. It can make different guesses at different visits to the same node. Since, for a given input term s, each parameter has $2^{|s|}$ possible values, every transduction τ in MSOT has finite images, i.e., $\tau(s)$ is finite for every input s. This follows also from the fact that τ is of linear size increase, cf. Section 8.1.1.

Proposition 8.23 The classes MSOW and 2GSM are incomparable. Hence, the classes MSOT and TWT$_{\text{su}}^{\text{MS}}$ are incomparable.

Proof: Obviously, two-way finite-state transducers can compute transductions that do not have finite images, for instance, the one that translates the word a (consisting of one symbol a) into the words a^n for all $n \in \mathcal{N}$. In fact, even one-way finite-state transducers can do that, see also Example 8.3. Hence, 2GSM is not included in MSOW.

To show that MSOW is not included in 2GSM, consider the transduction τ of words from $A^* := \{a\}^*$ to $B^* := \{b, c, \#\}^*$ such that for every $n \in \mathcal{N}$, we have $\tau(a^n) := \{w\#w \mid w \in \{b, c\}^*, |w| = n\}$. It is clear that $\tau \in$ MSOT: the definition scheme is 2-copying and has one parameter, which indicates for each input position whether it will be relabeled with b or c. This transduction cannot be computed by a two-way finite-state transducer. Intuitively, that is because the transducer would have to visit each input position twice, but, in general, cannot make the same choice (b or c) during the two visits. Formally, assume that τ is computed by such a transducer M with k states. Choose n such that $2^n > k(n+1)$. Consider the behavior of M on input $\mu_A(a^n)$, i.e., the unary term $a(a(\cdots a(\epsilon) \cdots))$. Clearly, M has $k(n+1)$ configurations on this input tree. Now consider the configuration of M when it has just produced the output node with label $\#$. As there are 2^n possible output words $w\#w$ for a^n, there exist two words

w_1 and w_2 for which this configuration is the same. This means that, on input $\mu_A(a^n)$, M can switch its computation of $w_1\#w_1$ halfway to the computation of $w_2\#w_2$, thus producing output $w_1\#w_2$, a contradiction.　　　　　　　　　　　　　　　　■

Since the nondeterminism of an MS-transduction just consists of choosing the values of the parameters, it can easily be expressed as a (nondeterministic) relabeling of terms, cf. Proposition 7.24(2). Let REL be the class of all relabelings $\pi_{F,n}^{-1}$, where F is a functional signature and $n \in \mathcal{N}$ (cf. the end of Definition 7.20).

Proposition 8.24 The classes MSOT and REL · DMSOT are equal.

Proof: The inclusion MSOT $\subseteq$ REL · DMSOT is immediate from Proposition 7.24(2). The other inclusion follows from the fact that REL $\subseteq$ MSOT and the closure of the class MSOT under composition.　　　　　　　　　　　　■

In view of this proposition, the nondeterministic case is not as interesting as the deterministic case. The results for DMSOT can be turned into results for MSOT by composition with REL. In particular, MSOT = REL · DTWT$_{su}^{MS}$ by Theorem 8.6, MSOT $\subset$ REL · P-DTWT by Theorem 8.13, and MSOT is the class of transductions in REL · DTWT$_\downarrow$ · DTWT of linear size increase by Theorem 8.17 (because $\pi_{F,n}^{-1}$ is size preserving, and $\pi_{F,n}^{-1}(t) \neq \emptyset$ for every $t \in T(F)$).

For MSOW a similar, but slightly more complicated version of Proposition 8.24 holds (see Theorem 19 of [EngHoo01]). Identifying $(f,(0,\dots,0)) \in F^{(n)}$ with $f \in F$, we obtain that $\tau(\varepsilon)$ is a singleton for every word transduction τ in REL · DMSOW, because the constant symbol ϵ in the unary input tree can only be relabeled by $(\epsilon,(0,\dots,0))$, and so REL · DMSOW is a proper subset of MSOW. Thus, for words, a class slightly larger than REL is needed.

8.9 VR-equational sets of terms and words

Having characterized the MS-transductions of terms, we can apply the Equationality Theorem (Theorem 7.36) to obtain an automata-theoretic characterization of the VR-equational sets of terms. A set of terms $L \subseteq T(F)$ is defined to be *VR-equational* if the set of $(F,[\rho(F)])$-labeled graphs $\{Syn(t) \mid t \in L\}$ is VR-equational. A similar definition holds for sets of words, viewed as unary trees. Since $\lfloor t \rfloor = \lfloor Syn(t) \rfloor$ for every term in $T(F)$, we obtain from the Equationality Theorem that L is VR-equational if and only if it is the image of TREES under an MS-transduction from graphs to terms. From Corollary 7.37 it then follows that the class of VR-equational sets of terms is preserved under MS-transductions of terms. In particular, the class of VR-equational sets of words is preserved under MS-transductions of words (see the discussion in the introduction of this chapter).

The *HR-equational* sets of terms (and words) can be defined in the same way. From the Equationality Theorem for the HR algebra (Theorem 7.51) one then obtains that

L is HR-equational if and only if it is the image of TREES under an $\text{MS}_{1,2}$-transduction from graphs to terms. Since for terms and words the $\text{MS}_{1,2}$-transductions and MS-transductions are the same (see Section 7.3), the HR-equational sets of terms or words are the same as the VR-equational ones (as also shown in [EngHey94], for hyperedge replacement and vertex replacement context-free graph grammars).

To characterize the VR-equational sets of terms and words, we first prove an additional characterization of the MS-transductions of terms. Let MSOT-REL denote the class of monadic second-order relabelings of terms, cf. Definition 7.20. As usual, DMSOT-REL denotes the restriction to parameterless definition schemes.

Theorem 8.25 The classes $\text{DTWT}_{\text{su}}^{\text{MS}}$ and $\text{DMSOT-REL} \cdot \text{DTWT}_{\text{su}}$ are equal.

Proof: The inclusion from right to left follows from Theorem 8.6 and the fact that DMSOT is closed under composition. To prove the inclusion in the other direction, let $M = (F, H, Q, q_{\text{in}}, R)$ be a deterministic single-use MS tree-walking transducer. By Theorem 8.7 we may assume that M is nonjumping. Let $\varphi_1, \ldots, \varphi_n \in \text{MS}(\mathcal{R}_F, \{x_1\})$ be the formulas that are used in the left-hand sides of the rules of M. Recall the notation from Definition 6.24(c). We will define a monadic second-order relabeling τ from $T(F)$ to $T(F^{(n)})$ such that $\tau(s) = s * \gamma_s$, where γ_s is the assignment $\{X_1, \ldots, X_n\} \to \mathcal{P}(N_s)$ with $\gamma_s(X_i) = \{u \in N_s \mid \lfloor s \rfloor \models \varphi_i(u)\}$ for every $i \in [n]$. Then M can be changed into a (local) tree-walking transducer M' that receives $\tau(s)$ as input and stepwise simulates M on input s: in the left-hand sides of its rules, it uses the boolean information in the label $(f, (w_1, \ldots, w_n))$ of the current node (with $w_i \in \{0, 1\}$) instead of the tests $\varphi_1, \ldots, \varphi_n$.

To ensure that M' is deterministic, we take τ from $T(F)$ to $T(F')$, where F' is the subset of $F^{(n)}$ consisting of the symbols $(f, (w_1, \ldots, w_n))$ such that for all $i, j \in [n]$, if φ_i and φ_j are mutually exclusive on $T(F)$, then $w_i = 0$ or $w_j = 0$. The parameterless definition scheme of τ has type $\mathcal{R}_F \to \mathcal{R}_{F'}$ and relation formulas $\theta_{lab_{(f,(w_1,\ldots,w_n))}}$ equal to $lab_f(x_1) \wedge (\bigwedge_{i \in [n], w_i=1} \varphi_i) \wedge (\bigwedge_{i \in [n], w_i=0} \neg \varphi_i)$. Note the similarity with Definition 7.21. The tree-walking transducer M' is obtained from M by changing, in the left-hand sides of its rules, each φ_i into the disjunction of all formulas $lab_{(f,(w_1,\ldots,w_n))}(x_1)$ such that $(f, (w_1, \ldots, w_n)) \in F'$ and $w_i = 1$. Obviously, M' is deterministic and single-use. ∎

Theorem 8.26 The following equalities hold:

$$\text{DMSOT} = \text{DMSOT-REL} \cdot \text{DTWT}_{\text{su}} \quad \text{and} \quad \text{MSOT} = \text{MSOT-REL} \cdot \text{DTWT}_{\text{su}}.$$

Proof: The first equality is immediate from Theorems 8.6 and 8.25. The second equality then follows from Proposition 8.24 and the fact that the equality $\text{REL} \cdot \text{DMSOT-REL} = \text{MSOT-REL}$ is an easy special case of that same proposition. ∎

Now we can characterize the VR-equational sets of terms and words. The class of regular sets of terms is denoted REGT. For a class X of transductions, $X(\text{REGT})$

denotes $\{\tau(L) \mid \tau \in X, L \in \text{REGT}\}$, i.e., the images of the regular languages under the transductions from X.

Theorem 8.27 Let L be a set of terms. The language L is VR-equational if and only if it is in $\text{DTWT}_{\text{su}}(\text{REGT})$.

Proof: Let $L \subseteq T(H)$ be VR-equational, where H is a functional signature. By the Equationality Theorem (Theorem 7.36) and by Theorem 7.27(2), there is a functional signature F and an MS-transduction τ, such that $L = \tau(T(F))$. By Theorem 8.26 there is a monadic second-order relabeling ρ and a transduction $\mu \in \text{DTWT}_{\text{su}}$ such that $L = \mu(\rho(T(F)))$. Since $T(F)$ is regular, $\rho(T(F))$ is regular by Proposition 7.22(3). This shows that $L \in \text{DTWT}_{\text{su}}(\text{REGT})$.

Now let $L = \mu(R)$ with $\mu \in \text{DTWT}_{\text{su}}$ and $R \in \text{REGT}$, such that $L \subseteq T(H)$ and $R \subseteq T(F)$. By Theorem 8.6, μ is in MSOT. Since R is MS-definable, we obtain from μ by the Restriction Theorem (Theorem 7.16) an MS-transduction τ such that $L = \tau(T(F))$. Clearly, we may assume that F is a fat signature. By the Equationality Theorem and Theorem 7.27(2), L is VR-equational. ∎

In a similar way, using Theorems 7.41(1) and 7.27(1), it can be shown that a set of terms is *linear* VR-equational if and only if it is the image of a regular set of words under a deterministic single-use tree-walking transducer with a unary input signature (i.e., a word-to-tree transducer).[14]

As an example, the set of terms $\{\sigma(c^n(t), d^n(t)) \mid t \in T(F), n = |yd(t)|\}$ is VR-equational, see Example 8.1. Similarly, an example of a linear VR-equational set of terms is $\{\sigma(c^n(\mu_A(w))), d^n(\mu_A(w))) \mid w \in A^*, n = |w|\}$, where A is any alphabet.

Since deterministic tree-walking transducers with a unary output signature are single-use, Theorem 8.27 immediately gives the next result. We denote by REGW the class of regular word languages.

Theorem 8.28 Let L be a set of words. The language L is VR-equational if and only if it is in $\text{DTWTW}(\text{REGT})$, and it is linear VR-equational if and only if it is in $\text{2DGSM}(\text{REGW})$. □

By Theorems 8.21 and 8.6 these classes are the same as $\text{DMBOT}_{\text{su}}(\text{REGT})$ and $\text{1DMWT}_{\text{su}}(\text{REGW})$, respectively.[15] The equality of $\text{DTWTW}(\text{REGT})$ and $\text{DMBOT}_{\text{su}}(\text{REGT})$ is essentially stated in Corollary 4.11(i) of [EngRS] and due to [AhoUll71] (cf. Section 8.7). Other characterizations of $\text{DTWTW}(\text{REGT})$ can be found in Section 6 of [*Eng97]. A characterization of $\text{TWTW}(\text{REGT})$ as path languages of VR-equational sets of graphs is proved in Theorem 28 of [EngOos96].

---

[14] In the only-if direction one first obtains a transducer with a unary input signature that may have several constant symbols, which can easily be replaced by one with a unary input signature that only has the constant symbol ϵ.

[15] It is straightforward to show that $\text{DMBOT}_{\text{su}}(\text{REGT}) = \text{DMBOT}_{\text{lin}}(\text{REGT})$ and $\text{1DMWT}_{\text{su}}(\text{REGW}) = \text{1DMWT}_{\text{lin}}(\text{REGW})$: a monadic second-order relabeling can add the set $var(u)$ to the label of every node u of the input tree (cf. the second part of the proof of Theorem 8.21); the mbott can then assign the empty word to the variables not in $var(u)$.

Examples of linear VR-equational languages are $\{a^n b^n c^n d^n \mid n \geq 0\}$, $\{w\widetilde{w} \mid w \in A^*\}$ and $\{w^k \mid w \in A^*\}$ for each $k \geq 2$, see Examples 8.2 and 8.4.

Remark 8.29 (1) In Theorems 8.27 and 8.28 one can replace REGT by the class of tree languages $T(F)$ where F is a functional signature (and similarly for REGW). Thus, a set of terms is VR-equational if and only if it is the image (range) of a transduction in DTWT_{su}, and similarly for words. This is proved as follows. Let $M = (F, H, Q, q_{\text{in}}, R)$ be a single-use deterministic tree-walking transducer and let $\mathscr{A} = \langle F, Q_{\mathscr{A}}, \delta_{\mathscr{A}}, Acc_{\mathscr{A}} \rangle$ be a finite automaton. We will construct a single-use deterministic tree-walking transducer M' such that $\tau_{M'}(T(F \times Q_{\mathscr{A}})) = \tau_M(L(\mathscr{A}))$, where $F \times Q_{\mathscr{A}}$ is the functional signature such that each $\langle f, q \rangle \in F \times Q_{\mathscr{A}}$ has the same arity as f. First we define the finite automaton $\mathscr{A}' := \langle F \times Q_{\mathscr{A}}, Q_{\mathscr{A}}, \delta', Acc_{\mathscr{A}} \rangle$, where δ' consists of the rules $\langle f, q \rangle [q_1, \ldots, q_k] \to q$ such that $f[q_1, \ldots, q_k] \to_{\mathscr{A}} q$. Clearly, the terms in $L(\mathscr{A}')$ represent the accepting runs of $\mathscr{A}$ on the terms in $L(\mathscr{A})$. The transducer M' first checks that the input tree s belongs to $L(\mathscr{A}')$, using a depth-first walk verifying that each node with label $\langle f, q \rangle$ has sons with labels $\langle f_1, q_1 \rangle, \ldots, \langle f_k, q_k \rangle$ such that $f[q_1, \ldots, q_k] \to_{\mathscr{A}} q$. Then it stepwise simulates M on the tree obtained from s by changing every label $\langle f, q \rangle$ into f.

It can even be shown rather easily that REGT can be replaced by the singleton class $\{T(F)\}$, where F is any fat signature (cf. Section 7.1.7). We omit the details.

(2) The class of context-free languages is properly contained in the class of VR-equational languages (see Example 4.3(12) and see the discussion in the introduction of this chapter). To show the containment, it is straightforward to transform a context-free grammar into a hyperedge replacement grammar. Alternatively, Theorem 8.28 can be used as follows. Let G be a context-free grammar. It is well known, and easy to see, that the set $P(G)$ of parse trees of G is a regular language. Since $L(G) = yd(P(G))$, where $L(G)$ is the language generated by G from the initial nonterminal, and since the yield mapping yd is in DTWTW (see Example 8.1), $L(G)$ is in DTWTW(REGT) and hence VR-equational. Note that for each $k \geq 2$ the language $\{w^k \mid w \in L(G)\}$ is also VR-equational: a tree-walking transducer can output the yield of the tree k times. The language $\{a^n b^n c^n d^n \mid n \geq 0\}$ of Example 8.2 is a noncontext-free (linear) VR-equational language.

(3) By Theorem 5.80, every VR-equational set L of terms (or words) has a decidable MS-satisfiability problem. As discussed in Section 7.5 (after Question 7.59), this is equivalent to the fact that it is decidable for a finite automaton $\mathscr{A}$ whether $L \cap L(\mathscr{A}) = \emptyset$. The latter problem is, in fact, decidable for every language L in TWT(REGT) in exponential time (see [Eng09]). $\qquad\square$

The use of the Equationality Theorem in the proof of Theorems 8.27 and 8.28 is not essential. The term generating power of hyperedge replacement graph grammars was first investigated in [EngHey92], where it is shown by a direct proof, not involving MS logic, that DTWT(REGT) is the class of sets of terms that can be generated by

HR grammars, if one allows the graph grammars to generate directed graphs without circuits (i.e., trees with shared subtrees), cf. the discussion at the end of Section 8.3. The construction transforms HR grammars and attribute grammars into each other (see Section 8.7). From this construction Theorem 8.27 easily follows. The present proof is given, also in terms of attribute grammars, in [BloEng00] (see Corollary 19). The characterization in Theorem 8.28 of the VR-equational sets of words is shown in [EngHey91] with a direct proof, not involving logic. The construction transforms HR grammars and tree-walking tree-to-word transducers into each other. The word generating power of HR grammars was first investigated in [HabKre] and Chapter V of [*Hab], see also Section 2.5.2 of [*DreKH].

The advantage of tree-walking transducers over VR equation systems is that they can be "programmed." To show that a given language is VR-equational, it is usually easier to describe the working of a tree-walking transducer than to construct a VR equation system.

8.10 References

Words

The two-way finite-state transducer is a particular type of two-tape Turing machine. It was first studied in [AhoUll70]. Lemma 8.8 is a general technique for two-way machines, proved in Lemma 3 of [HopUll] (see also page 212 of [AhoHU]). Using this lemma, it is shown in [ChyJák] that the class 2DGSM is closed under composition.

The incomparability of MSOW and 2GSM (Proposition 8.23) is Corollary 16 of [EngHoo01]. It is shown in [EngHoo07] that a word transduction τ is in the class MSOW if and only if (1) τ is in the class 2GSM · 2DGSM and (2) τ has finite images.[16] Moreover, it is decidable for $\tau \in$ 2GSM · 2DGSM (given as a composition of two two-way finite-state transducers) whether or not τ has finite images, i.e., whether or not it is in MSOW; and if so, a definition scheme for τ can be constructed. And finally, these results even hold for transductions τ that are compositions of arbitrarily many two-way finite-state transducers. It should also be noted that, in the above statements, "has finite images" can be replaced by "is of linear size increase."

A Hennie machine is a two-way finite-state transducer that, like a Turing machine, can write on its input tape. However, it may only visit each cell of its input tape a bounded number of times. It is shown in [EngHoo01] that MSOW (DMSOW) is the class of word transductions computed by nondeterministic (deterministic) Hennie machines.

---

[16] Since, obviously, every rational transduction is in 2GSM, this proves (and strengthens) the result mentioned at the end of Example 8.3: a rational transduction is an MS-transduction if and only if it has finite images.

In [AluDes] it is shown that MSOW equals the class of nondeterministic streaming string transductions, i.e., the word transductions computed by nondeterministic linear one-way multi word transducers.

Terms

As discussed in Section 8.7, the tree-walking transducer is essentially the attribute grammar of [Knu]. The single-use restriction was introduced (for attribute grammars) in [Gan] and [Gie], where it is shown that single-use attributed tree transducers are closed under composition.

The basic characterization $DMSOT = DTWT_{su}^{MS}$ of Theorem 8.6 is one of the main results of [BloEng00]; more precisely, it is proved there, in terms of attribute grammars, that $DMSOT = DMSOT\text{-}REL \cdot DTWT_{su}$, see Theorem 8.26. How to get rid of jumps (Theorem 8.7) is shown in Theorems 8 and 9 of [BloEng97]. The class DMSOT-REL is characterized in terms of finite-valued attribute grammars in Theorem 3.7 of [NevBus] and in Theorem 10 of [BloEng00] (independently), and in terms of compositions of bottom-up and top-down finite-state relabelings in Theorem 4.4 of [EngMan99] (cf. the discussion following Lemma 4.5). This implies Theorem 8.14 (as stated in Corollary 18 of [BloEng00]).

The tree-walking pushdown tree-to-word transducer (also called the *checking tree-pushdown transducer*) was introduced in [EngRS] and the tree-walking pushdown tree transducer (also called the *indexed tree transducer* or RT(P(TR))-transducer) in [EngVog86]. The pushdown of such a transducer can be viewed as a pushdown of (colored) pebbles: the pebbles are placed on the ancestors of the current node. In [EngHS] the tree-walking pushdown tree transducer is generalized to a transducer called I-PTT, which can place the pebbles anywhere on the input tree; it also generalizes the pebble tree transducer of [MilSV]. The decomposition result of Theorem 8.15 is a special case of Lemma 3 of [EngHS] (where the special case allows the first transducer to be deterministic).

The inclusions of Theorems 8.13 and 8.16 are proved in [EngMan99] in the form $DMSOT \subseteq DMT$, cf. Theorem 8.22. The characterization in Theorem 8.17 of DMSOT as the class of transductions in $DTWT_{\downarrow} \cdot DTWT$ of linear size increase, is the main result of [EngMan03a], where it is proved for DMT rather than for $DTWT_{\downarrow} \cdot DTWT$. The proof is for total functions only, but it should be clear from the proof of Theorem 6.18 of [EngVog85] (cf. the proof of Theorem 8.22) that it then also holds for partial functions: note that if a partial function τ is in DMSOT then it can be written as $\tau = \tau_1 \cdot \tau_2$, where τ_1 is the identity on some regular language, and $\tau_2 \in DMSOT$ is a total function (just view the precondition χ of the definition scheme for τ as a separate transduction τ_1, and take the definition scheme for τ_2 to be the one of τ with $\chi := True$). It is announced in [Man] that the if-direction of Theorem 8.17 even holds for transductions τ that are compositions of arbitrarily many deterministic tree-walking transducers (or macro tree transducers). In Theorem 7.1 of [EngMan99], DMSOT is characterized by syntactically restricted classes of macro tree transducers.

No results for MSOT and TWT are known that are similar to those for MSOW and 2GSM. Since it is easy to see that REL $\subseteq$ TWT$_\downarrow$ (the top-down tree transducer visits each node exactly once and guesses its new label), it follows from Proposition 8.24 and Theorem 8.16 that MSOT is included in TWT$_\downarrow \cdot$ DTWT$_\downarrow \cdot$ DTWT. But it is unknown whether MSOT equals the set of transductions of linear size increase in this class.

Characterizations of the class of VR-equational sets of terms that are closely related to Theorem 8.27, are given in Theorem 8.1 of [Dre99] (in terms of top-down tree transducers and hypergraph substitution) and in Theorem 5 of [EngMan00] (in terms of macro tree transducers; see also Corollary 7.3 of [EngMan99]). In [EngMan00] several natural types of term generating HR grammars are investigated.

We finally note that the equivalence problem is decidable for transductions in DMSOT, as proved in [EngMan06]. This result extends to MS-transductions from graphs to terms on VR-equational sets of graphs, but there is no hope to extend it to all graphs. It is shown in [AluDan] that the problem can be decided in exponential space for streaming tree transducers. For words, i.e., for deterministic two-way finite-state transducers, the decidability of the equivalence problem was first proved in [Gur]. It is open whether it is decidable for deterministic tree-walking tree transducers (and hence for deterministic macro tree transducers, cf. [*Eng80]).

9

Relational structures

In the previous chapters, we have used relational structures to represent labeled graphs, with or without sources or ports. Such structures are binary, i.e., their relations have arity at most 2. However, the Recognizability Theorem has been stated and proved for the algebra $\mathbb{STR}$ of all relational structures, whose operations are the disjoint union and the (unary) quantifier-free definable operations. Monadic second-order transductions have been defined, and the Backwards Translation Theorem has also been proved for general relational structures.

In this chapter, we will show how the previous results concerning graphs, in particular the logical characterizations of sets of graphs of bounded tree-width and clique-width, and several algorithmic and decidability results, extend more or less easily to relational structures. Certain extensions raise challenging open problems.

In Section 9.2, we will consider relational structures of bounded tree-width represented by their incidence graphs. This representation will offer the possibility of writing monadic second-order formulas (to be called MS_2 formulas) where quantified variables can denote sets of tuples of the relations. The results about graphs of bounded tree-width and MS_2 formulas will generalize easily.

In Section 9.3, we will define a complexity measure on relational structures that extends clique-width and most of its properties. However, as noted above, several important questions will remain open.

In Section 9.4, we will establish the Sparseness Theorem, which is a difficult theorem about the elimination of quantifications over sets of tuples, hence about cases where MS_2 formulas are no more expressive than MS formulas. Its particular case concerning graphs, which was announced in Theorem 5.22 and Section 7.3 (see also Theorem 4.51), will be the basis of the induction establishing the general result. Its proof already uses relational structures.

Before developing technical notions, we will detail in Section 9.1 two uses of ternary relational structures in the theory of ordered sets (already presented in Section 1.9.2).

9.1 Two types of ternary relational structures related to ordered sets

In this section, we discuss the notions of betweenness and of cyclic ordering that are formalized as ternary relations.

9.1.1 Betweenness

With a finite linear order $\langle D, \leq \rangle$ such that $|D| \geq 3$, we associate the following ternary relation, called its *betweenness relation*:[1]

$$B(x,y,z) :\Longleftrightarrow (x < y < z) \vee (z < y < x),$$

where $u < v$ means "$u \leq v$ and $u \neq v$". We denote it by $B(\leq)$. This relation satisfies the following properties, for all $x,y,z,t \in D$:

(B1) $B(x,y,z) \Rightarrow x \neq y \wedge x \neq z \wedge y \neq z$;
(B2) $B(x,y,z) \Rightarrow B(z,y,x)$;
(B3) $B(x,y,z) \Rightarrow \neg B(y,z,x)$;
(B4) $B(x,y,z) \wedge B(y,z,t) \Rightarrow B(x,y,t) \wedge B(x,z,t)$;
(B5) $B(x,y,z) \wedge B(y,t,z) \Rightarrow B(x,y,t) \wedge B(x,t,z)$;
(B6) $x \neq y \wedge x \neq z \wedge y \neq z \Rightarrow B(x,y,z) \vee B(y,z,x) \vee B(z,x,y)$.

By Property (B3), the three possibilities of Property (B6) are mutually exclusive. We will prove that these properties characterize betweenness relations.

Proposition 9.1 Every ternary relation on D satisfying Conditions (B1)–(B6) is $B(\leq)$ for some linear order $\leq$, hence is a betweenness relation.

Proof: Let B be a ternary relation satisfying properties (B1)–(B6). For every two distinct elements u and v of D, we define $[[u,v]] := \{x \in D \mid B(u,x,v)\}$. We choose a pair $\{a,b\}$ such that $[[a,b]]$ has a maximal cardinality. We first make some observations about a and b.

First, there is no element u satisfying $B(u,a,b)$. Assume on the contrary that we have $B(u,a,b)$. Then (B5) implies that, for every x in $[[a,b]]$, we have $B(u,x,b)$. Hence $[[a,b]] \cup \{a\} \subseteq [[u,b]]$ and $[[a,b]]$ is not of maximal cardinality by (B1). We can exchange a and b in this proof, hence we cannot have $B(u,b,a)$. By (B2), we cannot have $B(a,b,u)$ for any u.

Second, we have $D = \{a,b\} \cup [[a,b]]$. Because if some element u is not in $\{a,b\}$, we must have by (B6): $B(a,u,b) \vee B(u,b,a) \vee B(b,a,u)$. By (B2) and the first observations, we must have $B(a,u,b)$, hence $u \in [[a,b]]$.

---

[1] This notion arises in a natural way in axiomatizations of planar geometry in terms of points, lines and intersections of lines. See, e.g., [CouOli].

Third, we cannot have $B(u,a,v)$. Otherwise, since we have $B(a,u,b)$, hence $B(b,u,a)$, we have $B(b,a,v)$ by (B4), which contradicts our first observations.

By using a, we define a binary relation $\leq$ as follows:

$$x \leq y \quad \text{if and only if} \quad x = y \text{ or } x = a \text{ or } B(a,x,y).$$

This relation is reflexive. Properties (B1), (B2), (B3) and (B5) imply that it is anti-symmetric and transitive. Hence it is a partial order. We now prove that it is linear. Consider x and $y \neq x$. If $x = a$ or $y = a$, we have $x \leq y$ or $y \leq x$ respectively. Otherwise, by (B6) we have $B(a,x,y) \vee B(x,y,a) \vee B(y,a,x)$. The last case is excluded by the third observation and the others give respectively $x \leq y$ and, with (B2), $y \leq x$.

It remains to prove that $B = B(\leq)$. Let x, y and z be such that $x < y < z$. If $x = a$, we have $B(x,y,z)$ (by the definition of $y \leq z$). Otherwise, we cannot have y or z equal to a. Hence, we must have $B(a,x,y)$ and $B(a,y,z)$. Then (B5) gives $B(x,y,z)$. Finally, let us assume $B(x,y,z)$. Since $\leq$ is a linear order, we have one of the six cases: $x < y < z$, $z < y < x, y < x < z$, etc. The first two give $B(\leq)(x,y,z)$ as desired. The last four give $B(y,x,z)$ and $B(x,z,y)$ which are not possible by (B2) and (B3). This completes the proof.

If in this proof, we use b instead of a, then we obtain the opposite linear order, $\leq^{-1}$. ∎

A subset X of D^3 is *consistent for betweenness* if $X \subseteq B$ for some betweenness relation B on D. The problem BETWEENNESS that consists in deciding whether a given set $X \subseteq D^3$ is consistent (for betweenness) is NP-complete ([*GarJoh], Problem MS1; but no proof or reference to a published proof is given). If X is consistent, we define $\widehat{X}$ as the intersection of all betweenness relations that contain X. We say that X is a *partial betweenness relation on D* if $\widehat{X} = X$. It is clear that if $Y = \widehat{X}$, then $\widehat{Y} = \widehat{X}$. Hence, $\widehat{X}$ is a partial betweenness relation, and even the smallest partial betweenness relation that contains X. We will say that it is *generated* by X. The set $\widehat{X}$ satisfies properties (B1)–(B5), but, as we will see in Remark 9.5(2), a set that satisfies these five properties is not necessarily a partial betweenness relation.

If $X \subseteq D^3$, we denote by X^+ the least set $Y \subseteq D^3$ that contains X and satisfies Conditions (B2), (B4) and (B5). We have $X^+ \subseteq \widehat{X}$, and the inclusion may be strict (see Remark 9.5(1) for an example). This fact shows a difference with the case of partial orders that we now discuss. If $Z \subseteq D^2$ is contained in some linear order (we will also say *extended by* instead of "contained in"), we define $\widetilde{Z}$ as the intersection of all linear orders containing Z. We claim that Z is a partial order if and only if $Z = \widetilde{Z}$. Note that this equivalence is not a definition because partial orders are defined independently, as reflexive, antisymmetric and transitive binary relations. It is clear that $\widetilde{Z}$ has these three properties because it is an intersection of relations satisfying them. Hence $Z = \widetilde{Z}$ implies that Z is a partial order. For the other direction, we let Z be a partial order and (x,y) belong to $D^2 - Z$. We must prove that $(x,y) \notin \widetilde{Z}$, hence that there exists a linear order $\leq$ that extends Z in such a way that $y < x$. If $(y,x) \in Z$, then one can take for $\leq$ any linear order that extends Z. If $(y,x) \notin Z$, then the binary relation E on D consisting of (y,x) and the pairs $(u,v) \in Z$ such that $u \neq v$ is acyclic[2] (easy to check). Hence, its reflexive and transitive closure is a partial order, hence is extended by a linear order. Any such linear order has the desired property. This completes the proof.

Hence our definition of partial betweenness is a natural generalization of that of a partial order. However, it does not seem to be first-order definable, as is the notion of a partial order. The following questions are open:

Question 9.2
(1) Is the class of ternary relations that are consistent for betweenness MS-definable?
(2) Is the class of partial betweenness relations MS-definable? □

However, these classes are definable by monadic second-order sentences that allow quantifications over triples and sets of triples of the given ternary relation. Some definitions are needed.

We let T be a ternary relation symbol. We define the *incidence graph* of a $\{T\}$-structure $\langle D,X \rangle$ as the binary relational structure $Inc(\langle D,X \rangle) := \langle D \cup X, in_1, in_2, in_3 \rangle$, where $in_i(t,d)$ holds if and only if $d \in D$, $t \in X$ and d is the i-th element of t.

A class $\mathcal{C}$ of $\{T\}$-structures is *MS₂-definable* if the membership of S in $\mathcal{C}$ is equivalent to $Inc(S) \models \varphi$ for some sentence φ in $MS(\{in_1, in_2, in_3\}, \emptyset)$.

Proposition 9.3 The class of ternary relations that are consistent for betweenness and the class of partial betweenness relations are MS₂-definable.

Proof: We first prove the MS₂-expressibility of consistency (implicitly, in the sequel, consistency for betweenness). Let $X \subseteq D^3$. For $Y \subseteq X$, we define $Z(Y) \subseteq D^2$ as follows:

$$Z(Y) := \{(x,y) \mid (x,y,z) \text{ or } (z,x,y) \text{ belongs to } Y \text{ for some } z \in D\}$$

$$\cup \{(x,y) \mid (y,x,z) \text{ or } (z,y,x) \text{ belongs to } X - Y \text{ for some } z \in D\}.$$

Claim 9.3.1 A set $X \subseteq D^3$ is consistent if and only if the relation $Z(Y)$ is acyclic for some subset Y of X.

Proof: Let $X \subseteq B(\leq)$. We define $Y := \{(x,y,z) \in X \mid x < y < z\}$. Hence, if $(x,y,z) \in X - Y$, then $z < y < x$. Clearly, $Z(Y)$ is included in $<$ and, hence, is acyclic. Conversely, if $Y \subseteq X$ and $Z(Y)$ is acyclic, then $Z(Y)$ is extended by some linear order $\leq$, and hence $X \subseteq B(\leq)$, because if $(x,y,z) \in Y$, then $x < y < z$, and if $(x,y,z) \in X - Y$, then $z < y < x$. □

Since the transitive closure of an MS-definable relation is MS-definable (Section 5.2.2), the condition of this claim is expressible by an MS sentence over the structure $Inc(\langle D,X \rangle)$. Hence consistency is an MS₂-expressible property.

---

[2] A binary relation E on D is *acyclic* if the directed graph $\langle D,E \rangle$ has no circuit.

The following claim entails that partial betweenness is MS_2-expressible. For every triple $t = (x,y,z) \in D^3$, we define $sh(t)$ (read "shift") as (y,z,x) and $sh^*(t)$ as $\{t, sh(t), sh(sh(t))\}$. Property (B3) says that if $t \in B$, then $sh(t) \notin B$. Property (B6) says that if t is a triple of pairwise distinct elements of D, then $B \cap sh^*(t) \neq \emptyset$.

Claim 9.3.2 Let $X \subseteq D^3$ be consistent. We have $X = \widehat{X}$ if and only if X satisfies (B1)–(B5) and the following:

(B7) For every triple t of pairwise distinct elements of D such that $X \cap sh^*(t) = \emptyset$, the set $X \cup \{t'\}$ is consistent for at least two elements t' of $sh^*(t)$.

Proof: As observed above, $\widehat{X}$ satisfies (B1)–(B5). Let B be a betweenness relation such that $X \subseteq B$.

Assume that $X = \widehat{X}$ and let t be a triple of pairwise distinct elements such that $X \cap sh^*(t) = \emptyset$. Since B satisfies (B6) and (B3), we have $B \cap sh^*(t) = \{t'\}$ for some t', hence $X \cup \{t'\}$ is consistent. There exists a betweenness relation $B' \supseteq X$ such that $B' \cap sh^*(t) = \{t''\}$ and $t' \neq t''$, because otherwise $t' \in \widehat{X}$, which would contradict the hypotheses that $X = \widehat{X}$ and $X \cap sh^*(t) = \emptyset$. Hence, $X \cup \{t''\}$ is consistent and (B7) holds.

Let us conversely assume that X satisfies (B1)–(B5) and (B7). Assume that some triple t belongs to $\widehat{X} - X$. We have $X \cap sh^*(t) = \emptyset$ because if $sh(t) \in X$, then t and $sh(t)$ belong to $\widehat{X}$, which contradicts (B3) that holds for $\widehat{X}$; and if $sh^2(t)$ belongs to X, then $sh^2(t)$ and $sh(sh^2(t)) = t$ belong to $\widehat{X}$, which again contradicts (B3). By (B7), the set $X \cup \{t'\}$ is consistent for some triple $t' \in \{sh(t), sh^2(t)\}$. Hence $X \cup \{t'\} \subseteq B'$, where B' is a betweenness relation. The relation B' cannot contain t (because it contains also $sh(t)$ or $sh^2(t)$), which contradicts the assumption that $t \in \widehat{X}$. Hence $\widehat{X} = X$. □

This ends the proof of the proposition. ∎

Example 9.4 (1) We let $D := \{a,b,c,d,e,f\}$. The set $X := \{(a,b,c), (e,d,c), (d,e,f)\}$ is consistent because $X \subseteq B(\leq_0)$, where $a <_0 b <_0 c <_0 d <_0 e <_0 f$.

In order to illustrate the proof of Proposition 9.3, we consider several sets Y. If $Y := \{(a,b,c), (d,e,f)\}$, then $Z(Y) = \{(a,b), (b,c), (d,e), (e,f), (c,d)\}$. Its reflexive and transitive closure is the linear order $\leq_0$. If $Y := \{(d,e,f), (e,d,c)\}$, then $Z(Y)$ contains (d,e) and (e,d), and hence is not acyclic.

Let us now take $Y := \{(a,b,c), (e,d,c)\}$. Then $Z(Y) = \{(a,b), (b,c), (e,d), (d,c), (f,e)\}$. Its reflexive and transitive closure is a partial order that is not linear and has several linear extensions, in particular: $f <_1 a <_1 e <_1 d <_1 b <_1 c$ and $f <_2 e <_2 d <_2 a <_2 b <_2 c$. We have $X \subseteq B(\leq)$ for each of them.

An easy computation yields $X^+ = X \cup X^{-1} \cup \{(c,d,f), (f,d,c), (c,e,f), (f,e,c)\}$ (where X^{-1} is the set of triples (x,y,z) such that $(z,y,x) \in X$). We will show that this set is equal to $\widehat{X}$, the partial betweenness relation generated by X. We have

$X^+ \subseteq \widehat{X}$. It is not hard to see that the linear orders $\leq$ for which $X \subseteq B(\leq)$ are those such that $a < b < c$ or $c < b < a$, and $c < d < e < f$ or $f < e < d < c$. From this observation, we can check that $\widehat{X} = X^+$, and thus, $\overline{X^+} = X^+$. This set satisfies (B1)–(B5), but not (B6). Consider for example $t = (a,b,d)$ not in X^+. We have $sh^*(t) = \{(a,b,d),(b,d,a),(d,a,b)\}$ and $X^+ \cap sh^*(t) = \emptyset$. For each triple $sh^i(t)$, with $i = 0,1,2$, the set $X^+ \cup \{sh^i(t)\}$ is consistent as it is included in $B(\leq_i)$.

(2) Here is another example. We let $D := \{a,b,c,d\}$ and $X := \{(a,b,c),(a,b,d)\}$. The only linear orders with betweenness relations containing X are $a < b < c < d$ and $a < b < d < c$ and the opposite orders. Again we have $X^+ = \widehat{X}$, equal to $X \cup X^{-1}$. The triple $t = (b,c,d)$ is not in $\widehat{X}$ and the triple $(d,b,c) = sh^2(t)$ is not consistent with $\widehat{X}$. Hence, $sh^*(t)$ contains only two triples that are consistent with $\widehat{X}$, namely t and $sh(t) = (c,d,b)$.

Remark 9.5 (1) Let $X \subseteq D^3$ satisfy Condition (B1). The set X^+ satisfies (B2), (B4) and (B5) (by definition) and can be computed in polynomial time in $|D|$. If X^+ does not satisfy (B3), then X is not consistent. If X^+ satisfies (B3), we cannot conclude that X is consistent. Otherwise, we would have a polynomial algorithm for the problem BETWEENNESS that is NP-complete.

(2) Here is an example such that $X^+ \neq \widehat{X}$. We let $D := \{a,b,c,d,e\}$ and $X := \{(a,c,b),(a,d,b),(c,e,d)\}$. Then $X^+ = X \cup X^{-1}$. It is not hard to see that $\widehat{X}$ contains (a,e,b), which is not in X^+. Since X is consistent, X^+ satisfies (B1)–(B5). It does not satisfy (B7), since the sets $X^+ \cup \{(e,b,a)\}$ and $X^+ \cup \{(b,a,e)\}$ are not consistent. $\square$

We now define a notion of width (a variant of the tree-width of Definition 9.12 below) that will yield a fixed-parameter tractable algorithm for the problem BETWEENNESS. For $X \subseteq D^3$, we define $twd^{Inc}(X)$ as the tree-width of the labeled graph[3] $Inc(\langle D,X \rangle)$, and the *size* of $\langle D,X \rangle$ as $|D| + |X|$.

Corollary 9.6 The problem BETWEENNESS is fixed-parameter linear with respect to twd^{Inc}.

Proof: This follows from Theorem 6.4(1) and Proposition 9.3. ∎

We will conclude the notion of betweenness with an open question:

Question 9.7 Can one compute in polynomial time, for each fixed k, the set $\widehat{X}$ for any given set $X \subseteq D^3$, assumed to be consistent and such that $twd^{Inc}(X) \leq k$? Is this even possible by a fixed-parameter tractable algorithm with twd^{Inc} as parameter?

---

[3] Another possibility is to define the tree-width of $\langle D,X \rangle$ as that of the graph with vertex set D and an edge between any two vertices that are components of some triple in X. By Proposition 9.14(1), this notion of tree-width is the same as the one of Definition 9.12. By Proposition 9.14(2), the two notions of width are related by linear inequalities. Hence Corollary 9.6 also holds for this alternative definition.

9.1.2 Cyclic ordering

We now consider in a similar way the notion of cyclic ordering. With a finite linear order $\langle D, \leq \rangle$ such that $|D| \geq 3$, we associate the ternary relation

$$C(x,y,z) :\Longleftrightarrow (x < y < z) \vee (y < z < x) \vee (z < x < y).$$

If $D = \{d_1, \ldots, d_n\}$ with $d_1 < d_2 < \cdots < d_n$ and $d_1, \ldots, d_n$ are points on a circle such that, according to some orientation of the circle, d_{i+1} follows d_i and d_1 follows d_n, then $C(x,y,z)$ expresses that, if one "walks" along the circle according to this orientation by starting at x, then one meets y before z. We denote by $C(\leq)$ the ternary relation associated with $\leq$ in this way. A relation of this form is a *cyclic ordering*. A cyclic ordering C satisfies the following properties, for all x,y,z in its domain D:

(C1) $C(x,y,z) \Rightarrow x \neq y \wedge x \neq z \wedge y \neq z$;
(C2) $C(x,y,z) \Rightarrow C(y,z,x)$;
(C3) $C(x,y,z) \Rightarrow \neg C(x,z,y)$;
(C4) $C(x,y,z) \wedge C(y,t,z) \Rightarrow C(x,y,t) \wedge C(x,t,z)$;
(C5) $x \neq y \wedge x \neq z \wedge y \neq z \Rightarrow C(x,y,z) \vee C(x,z,y)$.

Conversely:

Proposition 9.8 Every ternary relation on D satisfying Conditions (C1)–(C5) is $C(\leq)$ for some linear order $\leq$, hence is a cyclic ordering.

Proof: Let C satisfy (C1)–(C5). We let a be any element of D and we define on D a binary relation $\leq$ in a similar way as in the case of betweenness:

$$x \leq y \text{ if and only if } x = y \text{ or } x = a \text{ or } C(a,x,y).$$

We claim that this relation is a linear order on D such that $C = C(\leq)$. It is reflexive. By (C1), (C2) and (C4), it is antisymmetric and transitive. Hence, it is a partial order. Condition (C5) implies that this partial order is linear. It remains to prove that $C = C(\leq)$.

We first prove that $C(\leq) \subseteq C$. Let x,y,z be such that $x < y < z$. We have $C(a,y,z)$ by the definition of $\leq$, and we want to prove $C(x,y,z)$. If $x = a$, we are done. Otherwise, we have $C(a,x,y)$ and $C(a,y,z)$, hence by Properties (C2) and (C4) we have $C(x,y,z)$, as was to be proved.

For the converse, we let x,y,z be such that $C(x,y,z)$. Since $\leq$ is a linear order and x, y and z are pairwise distinct, we have one of the six possibilities $x < y < z$, $x < z < y$, $y < z < x$, etc. Three of them entail $C(\leq)(x,y,z)$. The three others would yield $C(x,z,y)$, which contradicts Condition (C3) that C was assumed to satisfy. Hence we have $C(\leq)(x,y,z)$. ∎

A subset X of D^3 is *consistent for cyclic ordering* if $X \subseteq C$ for some cyclic ordering C on D. It is proved in [GalMeg] that the problem CYCLIC ORDERING, which consists of deciding if a set $X \subseteq D^3$ is consistent for cyclic ordering, is NP-complete.

As for betweenness, if $X \subseteq D^3$ is consistent for cyclic ordering, we define $\widehat{X}$ as the intersection of all cyclic orderings C on D such that $X \subseteq C$. A *partial cyclic ordering* on a set D is a subset X of D^3 such that $\widehat{X} = X$. If X is consistent (implicitly, in the sequel, for cyclic ordering), then $\widehat{X}$ is the smallest partial cyclic ordering containing X, and we say that it is *generated* by X. We have the same open questions as for betweenness (cf. Question 9.2), and the same positive results (cf. Proposition 9.3 and Corollary 9.6), based on the following proposition:

Proposition 9.9 The class of ternary relations that are consistent for cyclic ordering and the class of partial cyclic orderings are MS_2-definable.

Proof: We first consider consistency. We observe that if X is a subset of $C(\leq)$, then we can partition it as $X_0 \cup X_1 \cup X_2$, where

$$X_0 := \{(x,y,z) \in X \mid x < y < z\},$$
$$X_1 := \{(x,y,z) \in X \mid y < z < x\},$$
$$X_2 := \{(x,y,z) \in X \mid z < x < y\}.$$

Claim 9.9.1 The set X is consistent if and only if it can be partitioned into three sets X_0, X_1, X_2 such that the binary relation $Z(X_0, X_1, X_2)$ associated with them by:

$(u,v) \in Z(X_0, X_1, X_2)$ if and only if for some w:

(u,v,w) or (w,u,v) belongs to X_0, or

(v,w,u) or (w,u,v) belongs to X_1, or

(u,v,w) or (v,w,u) belongs to X_2,

is acyclic.

Proof: Let X be consistent, with $X \subseteq C(\leq)$ for some linear order $\leq$. The binary relation $Z(X_0, X_1, X_2)$ associated as above with X_0, X_1 and X_2 is included in $\leq$, hence, is acyclic. Conversely, if for some partition X_0, X_1, X_2 of X, the relation $Z(X_0, X_1, X_2)$ is acyclic, then there is a linear order $\leq$ extending it, and we have $X \subseteq C(\leq)$. □

The condition of this claim is MS_2-expressible. The second assertion is proved as the corresponding one for betweenness by means of the following observation: for $X \subseteq D^3$ that is consistent, we have $\widehat{X} = X$ if and only if X satisfies Conditions (C1)–(C4) and for every triple (x,y,z) of pairwise distinct elements, if (x,y,z) and (x,z,y) are not in X, then $X \cup \{(x,y,z)\}$ and $X \cup \{(x,z,y)\}$ are consistent. ∎

An embedding of a loop-free graph on a surface yields, for each vertex x, a cyclic ordering C_x of the set E_x of edges incident with this vertex. Conversely, such an embedding can be specified, up to homeomorphism, by a family of cyclic orderings C_x of E_x, for each vertex x. (See [*MohaTho] for the theory of graphs on surfaces.) If, instead of C_x, one is given a subset of E_x^3 for each vertex x, then one gets a *partial specification* (possibly unrealizable) of an embedding of the given graph on some surface. The problem of consistency of cyclic orderings is thus a subproblem of that of checking the realizability of a partial specification of an embedding. This topic is studied in [Dus].

The following notion of cyclic separation is defined from cyclic ordering in the same way that betweenness is defined from linear order. Let D be a set of at least four elements. A *cyclic separation* is a 4-ary relation E on D such that, for some cyclic ordering C on D, we have

$$E(u,v,w,x) \Longleftrightarrow (C(u,v,w) \wedge C(w,x,u)) \vee (C(w,v,u) \wedge C(u,x,w)).$$

This definition means that, for any layout of the elements of D on a circle, if C is the associated cyclic ordering, then the chord linking u and w crosses the chord linking v and x. We leave to the interested reader the task of axiomatizing cyclic separations and of establishing results analogous to Propositions 9.3 and 9.9.

9.2 Relational structures of bounded tree-width

We show how the results about graphs of bounded tree-width and CMS_2 logic extend (easily) to relational structures. We will first define CMS_2 formulas for expressing properties of relational structures. By representing a graph by its incidence graph (the corresponding relational structure is denoted by $\lceil G \rceil$ for a graph G), we can express monadic second-order graph properties by formulas using quantifications on sets of edges. The notation CMS_2 refers to such formulas, to be interpreted in $\lceil G \rceil$. For expressing properties of relational structures, we can act similarly and replace any relational structure by a richer one where the tuples of its relations are elements of the domain. This structure, actually an edge-labeled graph, allows to use quantified variables denoting sets of tuples of the relations of the original structure. As for graphs, we get in this way a stronger expressive power than with "plain" monadic second-order formulas.

We recall from Definition 5.1 that if $\mathcal{R}$ is a relational signature, we denote by $\mathcal{R}_0$ the set of its constant symbols, by $\mathcal{R}_i$ for $i > 0$ the set of its symbols of arity i, and by $\mathcal{R}_+$ the set $\mathcal{R} - \mathcal{R}_0$. The maximal arity of a symbol in $\mathcal{R}$ is $\rho(\mathcal{R})$. A relational signature is binary if $\rho(\mathcal{R}) \leq 2$.

Definition 9.10 (The incidence graph of a relational structure) Let $\mathcal{R}$ be a relational signature. We define $\mathcal{R}^{Inc}$ as the binary relational signature $\mathcal{R} \cup In_{\mathcal{R}}$ where $In_{\mathcal{R}} := \{in_i \mid 1 \leq i \leq \rho(\mathcal{R})\}$. The symbols of $\mathcal{R}$ are all unary in $\mathcal{R}^{Inc}$ and those of $In_{\mathcal{R}}$ are binary.

The *incidence graph* of a concrete $\mathcal{R}$-structure $S = \langle D_S, (R_S)_{R \in \mathcal{R}} \rangle$ is the concrete $\mathcal{R}^{Inc}$-structure $Inc(S)$ defined as

$$\langle D_S \cup T_S, (R_{Inc(S)})_{R \in \mathcal{R}}, in_{1\, Inc(S)}, \ldots, in_{k\, Inc(S)} \rangle,$$

where $k := \rho(\mathcal{R})$ and

$T_S := \{(R, d_1, \ldots, d_{\rho(R)}) \mid R \in \mathcal{R}_+, (d_1, \ldots, d_{\rho(R)}) \in R_S\};$
    if $R \in \mathcal{R}_+$, then
$R_{Inc(S)}(d) :\Longleftrightarrow d = (R, d_1, \ldots, d_{\rho(R)}) \in T_S$ for some $d_1, \ldots, d_{\rho(R)} \in D_S,$
    if $R = c \in \mathcal{R}_0$, then
$R_{Inc(S)}(d) :\Longleftrightarrow d = c_S$, and
    for all $i = 1, \ldots, k$:
$in_{i\, Inc(S)}(d, d') :\Longleftrightarrow d \in T_S, d' \in D_S$ and $d = (R, d_1, \ldots, d_{\rho(R)})$ for some
$R \in \mathcal{R}_+$ and $d_1, \ldots, d_{\rho(R)}$ such that $d' = d_i.$

Note that each relation $in_{i\, Inc(S)}$ is functional. Since $\mathcal{R}^{Inc}$ is a binary signature, the $\mathcal{R}^{Inc}$-structures can be identified with simple, directed, bipartite $(\mathcal{R}, In_{\mathcal{R}})$-labeled graphs (cf. Example 5.2(2)). Thus, we do not distinguish the graph $G := Inc(S)$ from the structure $\lfloor G \rfloor = Inc(S)$.

It is clear that if S and S' are isomorphic concrete structures, then so are $Inc(S)$ and $Inc(S')$. Hence, the operation Inc is well defined for abstract structures too. If $\mathcal{C}$ is a set of (concrete or abstract) structures, then $Inc(\mathcal{C})$ denotes the set of their (concrete or abstract) incidence graphs.

The purpose of introducing T_S in the domain of the structure $Inc(S)$ is to allow variables to denote sets of tuples of the relations of S. However, this is not necessary for unary relations. The pairs (R, d_1) in T_S can be deleted, and we obtain in this way a smaller set, denoted by T'_S, and the same expressive power for monadic second-order formulas. Hence, we define for S as above:

$$T'_S := \{(R, d_1, \ldots, d_{\rho(R)}) \mid R \in \mathcal{R}, \rho(R) > 1, (d_1, \ldots, d_{\rho(R)}) \in R_S\},$$

and

$$Inc'(S) := \langle D_S \cup T'_S, (R_S)_{R \in \mathcal{R}_1}, (R_{Inc(S)})_{R \in \mathcal{R} - \mathcal{R}_1}, in_{1\, Inc'(S)}, \ldots, in_{k\, Inc'(S)} \rangle,$$

which is also an $\mathcal{R}^{Inc}$-structure.

For a fixed relational signature $\mathcal{R}$, the injective mapping $Inc(S) \mapsto Inc'(S)$ is an FO-transduction, and its inverse is a $(|\mathcal{R}_1| + 1)$-copying FO-transduction. Thus, the two representations $Inc(S)$ and $Inc'(S)$ are strongly equivalent, cf. Definition 7.9. All our results will hold for Inc' as well as for Inc. The constructions with Inc will be simpler because unary relations will not have to be considered separately.

The next lemma will prove that relational structures are faithfully represented by their incidence graphs. It is thus natural to use monadic second-order sentences interpreted over incidence graphs to characterize classes of relational structures. A CMS formula over $\mathcal{R}^{Inc}$ will be called a *CMS$_2$ formula* over $\mathcal{R}$. A property of $\mathcal{R}$-structures S is *CMS$_2$-expressible* if it is equivalent to $Inc(S) \models \varphi$ (or to $Inc'(S) \models \varphi$) for some CMS sentence φ. A class of $\mathcal{R}$-structures is *CMS$_2$-definable* if the membership of an $\mathcal{R}$-structure in this class is a CMS$_2$-expressible property. All the terminology related with CMS$_2$ formulas will thus be applied to relational structures *via* their incidence graphs.

Lemma 9.11 Let $\mathcal{R}$ be a relational signature.
(1) A structure U in $STR^c(\mathcal{R}^{Inc})$ is isomorphic to $Inc(S)$ for some concrete $\mathcal{R}$-structure S if and only if the following conditions hold, where $T_U := \{t \in D_U \mid (t,d) \in A_U$ for some $A \in In_{\mathcal{R}}$ and $d \in D_U\}$:

(*I1*) for every $A \in In_{\mathcal{R}}$, the relation A_U is functional : $T_U \to (D_U - T_U)$;
(*I2*) the sets R_U for $R \in \mathcal{R}_+$ form a partition of T_U;
(*I3*) if $t \in R_U$ for $R \in \mathcal{R}_+$, then there exists a pair (t,d) in $in_i{}_U$ for some d if and only if $i \in [\rho(R)]$;
(*I4*) if t and $t' \neq t$ belong to R_U for $R \in \mathcal{R}_+$, then there exist $i \in [\rho(R)]$ and $d, d' \neq d$ such that (t,d) and (t',d') belong to $in_i{}_U$;
(*I5*) if $c \in \mathcal{R}_0$, then the set c_U is a singleton and $c_U \subseteq D_U - T_U$.

(2) The mapping $Inc^{-1} : STR(\mathcal{R}^{Inc}) \to STR(\mathcal{R})$ is a parameterless noncopying FO-transduction.
(3) Two concrete $\mathcal{R}$-structures S and S' are isomorphic if and only if $Inc(S)$ and $Inc(S')$ are isomorphic.
(4) Similar results hold for Inc'.

Proof: (1) Conditions (I1) to (I5) are clearly necessary. Conversely, let us assume that they hold for some $\mathcal{R}^{Inc}$-structure U. We define as follows an $\mathcal{R}$-structure S. Its domain is $D_U - T_U$; for every c in $\mathcal{R}_0$, we let c_S be the unique element of c_U (we use here (I5)) and for every $R \in \mathcal{R}_+$, we let R_S be the set of tuples $(d_1, \ldots, d_{\rho(R)})$ such that, for some $t \in R_U$, we have $(t, d_i) \in in_i{}_U$ for each $i \in [\rho(R)]$. It is easy to check that there is a unique isomorphism between $Inc(S)$ and U that is the identity on $D_U - T_U$. In particular, a tuple $(R, d_1, \ldots, d_{\rho(R)})$ in T_S corresponds by Conditions (I1)–(I4) to a unique element t of T_U.

(2) It is also easy to check that if h is an isomorphism from U to $Inc(S')$ for an $\mathcal{R}$-structure S', then the restriction of h to $D_U - T_U$ is an isomorphism from S to S'. An FO-transduction can construct S from U: Conditions (I1)–(I5) are expressible by a first-order sentence α, that can be taken as the precondition of the definition scheme to be defined. In proving (1), we have given a construction of S from U in $STR^c(\mathcal{R}^{Inc})$ satisfying Conditions (I1)–(I5). We omit the details of the definition scheme.

(3) The if direction is clear from (2).

(4) Easy proofs.   ∎

Tree-decompositions and tree-width of relational structures

We will extend to relational structures the notion of a tree-decomposition defined for graphs in Section 2.4.1. We will also extend to them the main decidability and model-checking results obtained for graphs of bounded tree-width and CMS_2 formulas.

Definition 9.12 (Tree-decompositions of relational structures) A *tree-decomposition* of a concrete $\mathcal{R}$-structure S is a pair (T, f) such that T is a concrete rooted tree and f is a mapping : $N_T \to \mathcal{P}(D_S)$ such that:

(i) $D_S = \bigcup \{ f(u) \mid u \in N_T \}$;

(ii) for every $R \in \mathcal{R}_+$ and $(d_1, \ldots, d_n) \in R_S$, there exists a *box* $f(u)$ containing $d_1, \ldots, d_n$;

(iii) for every $d \in D_S$, the set $f^{-1}(d) := \{ u \in N_T \mid d \in f(u) \}$ is connected in T.

We call Condition (iii) the *connectivity condition*. The *width* of (T, f) is $wd(T, f) := \max\{ | f(u)| \mid u \in N_T \} - 1$ and the *tree-width* of S, denoted by $twd(S)$, is the smallest width of a tree-decomposition of S. Its *path-width*, denoted by $pwd(S)$, is defined in a similar way with the additional condition that T is a directed path. We denote by $TWD(\mathcal{R}, \leq k)$ and by $PWD(\mathcal{R}, \leq k)$ the classes of $\mathcal{R}$-structures of tree-width and, respectively, of path-width at most k.

Remark 9.13 Definition 9.12 states no condition about the constants of the considered structure. In order to have an exact generalization of the notion of tree-decomposition of a graph with sources, we could add to Conditions (i)–(iii) the following:

(iv) all constants of S belong to the root box.

This would be natural because the sources of a graph become constants of the corresponding relational structure. However, we will not need this condition because we will not extend to relational structures the expression of tree-decompositions of graphs by terms in $T(F^{HR})$. Imposing Condition (iv) would restrict the notion of tree-decomposition and yield a different notion of tree-width. If we denote it by $twd'(S)$, then we have

$$twd(S) \leq twd'(S) \leq twd(S) + |\mathcal{R}_0|$$

for every S in $STR(\mathcal{R})$, because every tree-decomposition can be made into one satisfying (iv) by adding the constants to all boxes. It follows that the same subsets of $STR(\mathcal{R})$ have bounded tree-width according to both notions of tree-decomposition.   □

The tree-width and the path-width of a relational structure S are closely related to the tree-widths and path-widths of its incidence graph and of its *adjacency graph*,[4] that is, the simple, loop-free and undirected graph $Adj(S)$ with vertex set equal to D_S and edges between d and $d' \neq d$ whenever d and d' occur in some tuple of S.

We let $twd^{Inc}(S)$ denote the tree-width of the graph $Inc(S)$. (We extend a notation introduced for stating Corollary 9.6.) This value is also the tree-width of $Inc'(S)$ except in the uninteresting case where $\rho(\mathcal{R}) \leq 1$.

Proposition 9.14 Let $\mathcal{R}$ be a relational signature such that $\rho(\mathcal{R}) \geq 1$. For every relational structure $S \in STR^c(\mathcal{R})$, we have:

(1) S and $Adj(S)$ have the same tree-decompositions, and hence we have $twd(S) = twd(Adj(S))$;

(2) $twd^{Inc}(S) - 1 \leq twd(S) \leq \rho(\mathcal{R}) \cdot (twd^{Inc}(S) + 1) - 1$.

The same properties hold for path-width instead of tree-width.

Proof: (1) By the definitions, every tree-decomposition of S is also one of $Adj(S)$. Conversely, if (Z,f) is a tree-decomposition of $Adj(S)$, then every tuple $(d_1, \dots, d_n)$ of a relation R_S induces a clique in $Adj(S)$ and, hence, is contained in some box of (Z,f) by Proposition 2.58. Hence, (Z,f) is also a tree-decomposition of S.

(2) Let (Z_0, f_0) be a tree-decomposition of $S \in STR^c(\mathcal{R})$. We extend it as follows to a tree-decomposition (Z,f) of $Inc(S)$. For each tuple $(R, d_1, \dots, d_n)$ in T_S, i.e., such that $R \in \mathcal{R}_+$ and $(d_1, \dots, d_n) \in R_S$, we select one node u of Z_0 such that $d_1, \dots, d_n \in f_0(u)$ (there exists one by Condition (ii) of Definition 9.12), we add to Z_0 a new node u' that we make into a son of u in Z and we define $f(u') := f_0(u) \cup \{(R, d_1, \dots, d_n)\}$. It is clear that Z is a rooted tree, that $Z_0 \subseteq Z$ and that all nodes of Z that are not in Z_0 are leaves. Each element of T_S belongs to a single box $f(u)$ of (Z,f), such that u is a leaf. The pair (Z,f) is a tree-decomposition of $Inc(S)$. If (Z_0, f_0) has width k, then (Z,f) has width k or $k+1$. This gives the first inequality. If (Z_0, f_0) is a path-decomposition, we construct similarly a path-decomposition (Z,f) of $Inc(S)$: instead of adding u' as new leaf, we insert it between u and its son. Hence, Z is a path.

For proving the second inequality, we let (Z,f) be a tree-decomposition of $Inc(S)$ of width k. We turn it into a tree-decomposition (Z,f') of S by defining, for each $u \in N_Z$, the box $f'(u)$ as

$$(f(u) \cap D_S) \cup \{d \in D_S \mid d \text{ belongs to some tuple in } f(u) \cap T_S\}.$$

This means that, in a box $f(u)$, we replace each tuple $(R, d_1, \dots, d_n)$ by the elements $d_1, \dots, d_n$ of D_S. It follows that $|f'(u)| \leq (k+1) \cdot \rho(\mathcal{R})$ because if $p := |f(u) \cap D_S|$ and $q := |f(u) \cap T_S|$, then $p + q \leq k + 1$ and $|f'(u)| \leq p + q \cdot \rho(\mathcal{R}) \leq (k+1) \cdot \rho(\mathcal{R})$. It remains to check that (Z,f') is a tree-decomposition. We first check the connectivity

---

[4] This graph, augmented with a loop on each vertex, is called the *Gaifman graph* of S in [*Lib04]. It is useful for the study of the expressive power of first-order formulas.

condition. Let d in D_S belong to the tuples $t_1, \ldots, t_p$. We have $f'^{-1}(d) = f^{-1}(d) \cup f^{-1}(t_1) \cup \cdots \cup f^{-1}(t_p)$, where each component of this union is connected. Since for each i there is a box $f(u)$ that contains t_i and d, we have $f^{-1}(d) \cap f^{-1}(t_i) \neq \emptyset$. This implies that $f'^{-1}(d)$ is connected. The other conditions are clearly satisfied. $\blacksquare$

By the first assertion of this proposition, the algorithms that construct tree-decompositions of graphs (cf. Section 6.2.1) can be used to construct tree-decompositions of relational structures. The next results exploit the particular structure of incidence graphs.

Lemma 9.15 Let $\mathcal{R}$ be a relational signature. The identity on $Inc(STR(\mathcal{R}))$ (and on $Inc'(STR(\mathcal{R}))$) is a $(\rho(\mathcal{R}) + 1)$-copying $\mathrm{MS}_{1,2}$-transduction.

Proof: Let $k := \rho(\mathcal{R})$ and $G := Inc(S)$ for S in $STR(\mathcal{R})$. The edges of G are directed and labeled by $in_1, in_2, \ldots$ or in_k, and two edges with the same tail have different labels. Hence, an $\mathrm{MS}_{1,2}$-transduction can specify an edge from x to y that is labeled by in_j as the pair $(x, j+1)$, and a vertex x as the pair $(x, 1)$. Hence, $\lceil G \rceil$ is defined from $\lfloor G \rfloor$ (i.e., from $Inc(S)$ by a convention made in Definition 9.10) by a $(k+1)$-copying MS-transduction. The detailed construction is similar to that for the incidence graphs of forests done in Example 7.8. The proof is almost the same for Inc'. $\blacksquare$

This lemma and Proposition 9.14 yield the following corollary, which extends Corollary 7.49:

Corollary 9.16 A subset $\mathcal{C}$ of $STR(\mathcal{R})$ has bounded tree-width if and only if $Inc(\mathcal{C})$ is included in the image of TREES under an MS-transduction.

Proof: By Proposition 9.14(2), a subset $\mathcal{C}$ of $STR(\mathcal{R})$ has bounded tree-width if and only if $Inc(\mathcal{C})$ has bounded tree-width, hence, by Corollary 7.49, if and only if $Inc(\mathcal{C})$ is included in $\tau(\text{TREES})$ for some $\mathrm{MS}_{1,2}$-transduction τ. By Lemma 9.15, the last equivalence is valid with τ assumed to be an MS-transduction (cf. Section 7.3). $\blacksquare$

Proposition 9.17 For every relational signature $\mathcal{R}$ and integer k, the classes $Inc(TWD(\mathcal{R}, \leq k))$ and $Inc(PWD(\mathcal{R}, \leq k))$ are HR-equational.

Proof: The class of labeled graphs $Inc(STR(\mathcal{R}))$ is characterized by an FO sentence α as we have seen in the proof of Lemma 9.11. There exists a (noncopying) FO-transduction τ such that $Dom(\tau) = Inc(STR(\mathcal{R}))$ and $\tau(Inc(S)) = Adj(S)$ for every $\mathcal{R}$-structure S. Its precondition is α and the other formulas are easy to write. By Proposition 9.14(1), the class $\tau^{-1}(TWD(\leq k)) = \{Inc(S) \mid twd(Adj(S)) \leq k\}$ equals $Inc(TWD(\mathcal{R}, \leq k))$. By the Backwards Translation Theorem (Corollary 7.12) and since the class of graphs $TWD(\leq k)$ is MS-definable (because it is characterized by a finite set of excluded minors, cf. Corollary 2.60(2) and Proposition 5.11), it follows that $Inc(TWD(\mathcal{R}, \leq k))$ is MS-definable. This class has also tree-width at most $k + 1$ by Proposition 9.14(2), hence it is HR-equational by Example 4.3(8) and the Filtering Theorem (Corollary 5.71).

The proof is the same for path-width because the class of graphs $PWD(\leq k)$ is also characterized by a finite set of excluded minors. ■

CMS$_2$-satisfiability and model-checking

Theorem 9.18 For every relational signature $\mathcal{R}$ and every integer k, the class $TWD(\mathcal{R}, \leq k)$ has a decidable CMS$_2$-satisfiability problem.

Proof: The CMS$_2$-satisfiability problem for a set of $\mathcal{R}$-structures $\mathcal{C}$ is, by definition, the CMS-satisfiability problem for its set of incidence graphs $Inc(\mathcal{C})$. By Proposition 9.17, the set $Inc(TWD(\mathcal{R}, \leq k))$ is HR-equational. Hence, by Theorem 5.80(2), its CMS-satisfiability problem is decidable. ■

This result extends Corollary 5.81(2). Next is a converse that extends Theorem 7.55(1).

Theorem 9.19 Let $\mathcal{R}$ be a relational signature. If a subset $\mathcal{C}$ of $STR(\mathcal{R})$ has a decidable MS$_2$-satisfiability problem, then it has bounded tree-width, equivalently, $Inc(\mathcal{C})$ is included in τ(TREES) for some MS-transduction τ.

Proof: Let $\mathcal{C} \subseteq STR(\mathcal{R})$ have a decidable MS$_2$-satisfiability problem. The MS-satisfiability problem for $Inc(\mathcal{C})$ is thus decidable by the definitions. By Lemma 9.15 and the Reduction Theorem (Theorem 7.54), its MS$_2$-satisfiability problem is decidable. Hence, by Theorem 7.55(1), it has bounded tree-width and so has $\mathcal{C}$ by Proposition 9.14(2). ■

If we only assume that $\mathcal{C}$ has a decidable MS-satisfiability problem, we cannot draw any similar conclusion. The following question is open (cf. Question 7.59):

Question 9.20 Is it true that, for every relational signature $\mathcal{R}$, if a set of $\mathcal{R}$-structures has a decidable CMS-satisfiability problem, then it is included in τ(TREES) for some monadic second-order transduction τ?

If $\mathcal{R}$ is binary, then the answer is positive and this follows from Theorem 7.55(2) since simple directed labeled graphs correspond to relational structures over binary signatures (cf. Definition 2.11). It is also positive if the considered set $\mathcal{C}$ is uniformly k-sparse for some k: this is a consequence of a result to be proved in Section 9.4. (Uniform sparsity means that the structures in $\mathcal{C}$ have "few tuples.") The question is open even if $\mathcal{R}$ consists of a single relation of arity 3.

We now consider the model-checking problem, i.e., the problem of checking whether a given sentence φ is true in a given relational structure S. The major concern is to have efficient, or at least fixed-parameter tractable, algorithms for fixed sentences. We recall from Definition 5.3 that the *size* of an $\mathcal{R}$-structure S is defined as

$$\|S\| := \|\mathcal{R}\| + |D_S| + \sum_{R \in \mathcal{R}_+} \rho(R) \cdot |R_S|,$$

hence, it is $\Theta(\|Inc(S)\|)$ for fixed $\mathcal{R}$. The following theorem is a direct application of Theorem 6.4(1) about graphs of bounded tree-width:

Theorem 9.21 The model-checking problem for relational structures and CMS_2 sentences is fixed-parameter linear with parameter $(twd(S), \varphi)$, where S is the input structure and φ is the input sentence.

Proof: Let be given a relational signature $\mathcal{R}$, a concrete $\mathcal{R}$-structure S and a CMS_2 sentence φ. By the definitions, φ is true in S if and only if $Inc(S) \models \varphi$. The structure $Inc(S)$ can be constructed from S in time $O(\|S\|)$. Its size and its tree-width are proportional to those of S (by Proposition 9.14(2) for the tree-width). Hence, Theorem 6.4(1) can be applied to the $(\mathcal{R}, In_{\mathcal{R}})$-labeled graph $Inc(S)$ and the CMS sentence φ, and it yields a fixed-parameter linear algorithm as desired. ∎

The other algorithmic problems considered in Chapter 6 for graphs of bounded tree-width extend in a similar way.

By the proof of Theorem 9.21 and by Proposition 9.14(2) for path-width, the observations made in the introduction of Section 6.3.5 about the model-checking of CMS_2 sentences on graphs of bounded path-width extend to relational structures. In particular, given a structure S, we construct the labeled graph $Inc(S)$: its size and path-width are proportional to those of S. We construct a slim HR-term evaluating to this graph, and from it a VR-term, as explained in the introduction of Section 6.3.5. We use an $F^{VR}_{[\mathcal{R}, In_{\mathcal{R}}]}$-automaton on this term. We obtain a fixed-parameter linear algorithm for parameter $(pwd(S), \varphi)$, similar to that of Theorem 9.21. The advantage is that the corresponding automata are easier to construct.

9.3 Terms denoting relational structures

In Chapter 5 we have defined the many-sorted algebra $\mathbb{STR}$ of relational structures with signature F^{QF}, of which the graph algebra $\mathbb{GP}^t$ defined in Chapter 2 is a subalgebra. From this definition, we can obtain denotations of relational structures by terms. Our objective is to prove that the value mapping from terms (over finite subsignatures of F^{QF}) to relational structures is a monadic second-order transduction. This result will generalize Proposition 7.30 that concerns in a similar way terms over the signature F^{VR} and the graphs they define. We will also discuss the extension to relational structures of several notions and results relative to graphs, in particular clique-width, fixed-parameter model-checking algorithms and decidability of monadic second-order satisfiability problems.

We first review some definitions relative to the many-sorted F^{QF}-algebra $\mathbb{STR}$ defined in Section 5.3.7 (Definition 5.62). Its set of sorts $\mathcal{S}$ is the countable set of relational signatures, and the set $STR(\mathcal{R})$ of (abstract) $\mathcal{R}$-structures is the domain of sort $\mathcal{R}$. The operations of F^{QF} are the disjoint union and the (unary) quantifier-free operations; its constant symbols denote the structures with at most one element. The many-sorted algebra $\mathbb{GP}^t$ of graphs with ports defined in Section 2.6.3 is a subalgebra of $\mathbb{STR}$.

9.3.1 Monadic second-order model-checking problems

If $t \in T(F^{QF})$, then we denote by $Sig(t)$ the subsignature of F^{QF} consisting of all symbols that have occurrences in t, and by $val(t)$ the relational structure that is its value. The following result is an immediate application of the Recognizability Theorem for the algebra $\mathbb{STR}$. It can be seen as a generalization of Theorem 6.3(3).

Theorem 9.22 Let $\mathcal{R}$ be a relational signature. The problem of checking if a sentence $\varphi \in CMS(\mathcal{R}, \emptyset)$ is true in the structure $val(t)$, where t is a term in $T(F^{QF})$, is solvable by a fixed-parameter linear algorithm with parameter $(Sig(t), \varphi)$.

Proof: Let be given a relational signature $\mathcal{R}$ and $t \in T(F^{QF})$. First, we compute $Sig(t)$ and check whether $val(t)$ is actually an $\mathcal{R}$-structure. This can be done in time $O(|t|)$. Together with Assertions (1) and (2) of Proposition 3.76, and Theorem 3.62, the Recognizability Theorem for $\mathbb{STR}$ (Theorem 5.75) yields that the set L of terms $s \in T(Sig(t))$ such that $S := val(s)$ is an $\mathcal{R}$-structure that satisfies φ, is regular. It is recognized by a finite automaton that can be constructed from $Sig(t)$ and φ. The membership of t in L can be checked in time $O(|t|)$. ∎

This algorithm operates in time bounded by $f(Sig(t), \varphi) \cdot |t|$ for some computable function f. We can consider $\mathcal{R}$ as part of the input and we still have a fixed-parameter linear algorithm with a function f depending also on $\mathcal{R}$. In most cases, however, model-checking problems concern relational structures over fixed signatures. Statements and proofs are then slightly simpler.

The algorithm takes as input a term t denoting a relational structure S and not S itself. The *parsing problem*, that is, the construction of a term in $T(F)$ for some given finite subsignature F of F^{QF} that evaluates to a given relational structure has presently no polynomial algorithm that can play the role of the one of [Bod96] (for graphs of bounded tree-width) or of that of [HliOum] (for graphs of bounded clique-width) that we have used in Theorem 6.4. Let us make an optimistic conjecture:

Conjecture 9.23 For every relational signature $\mathcal{R}$ and every finite subsignature F of F^{QF}, one can construct a polynomial-time algorithm doing the following:

for every $\mathcal{R}$-structure S, it either answers that S is not the value of any term in $T(F)$ or outputs a term in $T(F^{QF})$ that evaluates to S. □

This conjecture would imply the following:

For every set C of $\mathcal{R}$-structures that is generated by a finite subsignature F of F^{QF} and every sentence $\varphi \in CMS(\mathcal{R}, \emptyset)$, one can decide in polynomial time if a given structure S belonging to C satisfies φ.

Note that such a polynomial algorithm is not required to check if S belongs to C, and is not necessarily fixed-parameter tractable for the parameter (F, φ).

Proposition 6.8 has established the validity of Conjecture 9.23 for binary signatures (and for $F \subseteq F^{VR}$) since the corresponding structures are the same as simple labeled graphs. The resulting algorithm is even fixed-parameter cubic.

9.3.2 From terms to relational structures by MS-transductions

We will prove that, for every finite subsignature F of F^{QF} the evaluation mapping *val* from terms in $T(F)$ to relational structures is a monadic second-order transduction. Proposition 7.30 proves this result for the special case of graphs defined by terms over F^{VR}.

For $\mathcal{R} \in \mathcal{S}$, we will denote by $F^{QF} \restriction \mathcal{R}$ the finite set of operations of F^{QF} whose sorts are subsets of $\mathcal{R}$.[5] If F is a finite subsignature of F^{QF}, then $F \subseteq F^{QF} \restriction \mathcal{R}$, where $\mathcal{R}$ is the union of all signatures $\mathcal{R}'$ that occur in the types of the symbols of F. Clearly, $\mathcal{R}$ is finite.[6]

We will denote by *cval*(t) the concrete structure defined as follows by a term $t \in T(F^{QF})$. Its domain is a subset of $Occ_0(t)$, the set of occurrences of the constant symbols that denote nonempty structures. For defining *cval*(t) if $t = f(t_1)$, we use the operation on concrete structures specified by the definition scheme of f. We also use the union of disjoint concrete structures. The domain of *cval*(t) may be a proper subset of $Occ_0(t)$ because a QF operation f may delete elements of its input structures. If $t = t_1 \oplus t_2$, then *cval*(t) is the union of the disjoint structures *cval*$(t)/u_1$ and *cval*$(t)/u_2$, where u_1 and u_2 are the two sons of the root of t. They are actually disjoint because their domains are sets of leaves of t that are respectively below u_1 and below u_2. Clearly, *val*(t) is the abstract structure isomorphic to *cval*(t). These definitions generalize in the obvious way those of Section 2.5.2.

If $t \in T(F^{QF} \restriction \mathcal{R})$, then *cval*$(t)$ is an $\mathcal{R}'$-structure for some $\mathcal{R}' \subseteq \mathcal{R}$. We will denote by $T(F^{QF} \restriction \mathcal{R})_{\mathcal{R}'}$ the set of terms that evaluate to an $\mathcal{R}'$-structure. The following proposition generalizes Proposition 7.30:

Proposition 9.24 For each $\mathcal{R} \in \mathcal{S}$ and $\mathcal{R}' \subseteq \mathcal{R}$, the mapping

$$val : T(F^{QF} \restriction \mathcal{R})_{\mathcal{R}'} \to STR(\mathcal{R}')$$

is a parameterless noncopying MS-transduction.

---

[5] To be precise, we select finitely many QFO definition schemes for the finitely many QF operations, cf. Corollary 5.95.

[6] Extending the convention of Example 5.44 and Definition 5.62, we assume that the arity mappings of all relational signatures in $\mathcal{S}$ agree. Hence, the union of two sorts in $\mathcal{S}$ belongs to $\mathcal{S}$.

Proof: Let $\mathcal{R} \in \mathcal{S}$. We let $F := F^{QF} \upharpoonright \mathcal{R}$.

If $t \in T(F)$ and $u \in N_t$, then the concrete structure $cval(t)/u$, which is isomorphic to $val(t/u)$, has a domain that is a possibly proper subset of the set of elements of $Occ_0(t)$ that are below u.

For each $m \geq 1$, we let X_m be the (standard) set of first-order variables $\{x_1, \ldots, x_m\}$, and $Atom(\mathcal{R}, X_m)$ be the set of atomic formulas of $QF(\mathcal{R}, X_m)$, including the trivial formulas $x_i = x_i$ and $a = a$ (for $a \in \mathcal{R}_0$) which will be useful. For a formula $\beta \in QF(\mathcal{R}, X_m)$, we denote by $Var(\beta) \subseteq X_m$ the set of variables that occur in β, and by $Con(\beta) \subseteq \mathcal{R}_0$ the set of constant symbols that occur in β.

Let $t \in T(F)$ and γ be an X_m-assignment in $\lfloor t \rfloor$ such that $\gamma(x_i)$ is a leaf of t for each i. If $u \in N_t$ and $\beta \in QF(\mathcal{R}, X_m)$, then we write:

$(cval(t)/u, \gamma) \models \beta$ if and only if

> (1) $\gamma(x_i) \in D_{cval(t)/u}$ for each i such that $x_i \in Var(\beta)$,
> (2) $Con(\beta)$ is included in the sort of t/u, and
> (3) $(cval(t)/u, \gamma \upharpoonright Var(\beta)) \models \beta$.

It follows that $(cval(t)/u, \gamma) \models x_i = x_i$ if and only if $\gamma(x_i) \in D_{cval(t)/u}$, which implies in particular that $\gamma(x_i) \leq_t u$. Also, $(cval(t)/u, \gamma) \models a = a$ if and only if a is in the sort of t/u.

For t and γ as above and for every $\alpha \in Atom(\mathcal{R}, X_m)$, we define

$$E_\alpha(t, \gamma) := \{u \in N_t \mid (cval(t)/u, \gamma) \models \alpha\}.$$

We will prove that these sets can be defined in $\lfloor t \rfloor$ by a monadic second-order formula. For this purpose, we introduce a set variable Y_α for every $\alpha \in A := Atom(\mathcal{R}, X_m)$, and we let $\mathcal{W}_m$ be the set of these set variables Y_α. We will use the following auxiliary construction. If C is a Boolean set term[7] over $\mathcal{W}_m$ and if $(E_\alpha)_{\alpha \in A}$ is a family of sets, then we denote by $C((E_\alpha)_{\alpha \in A})$ the set defined by C with E_α as value of Y_α.

Claim 9.24.1 For each $\beta \in QF(\mathcal{R}, X_m)$ there exists a Boolean set term C_β over $\mathcal{W}_m$ such that, for all t, γ and u as above, we have

$$(cval(t)/u, \gamma) \models \beta \text{ if and only if } u \in C_\beta((E_\alpha(t, \gamma))_{\alpha \in A}).$$

Proof: We can assume that β is built without disjunction, because if it is not, it can be transformed into an equivalent formula β' in $QF(\mathcal{R}, X_m)$ without disjunctions that has the same free variables and constant symbols. The latter condition is necessary to ensure that $(cval(t)/u, \gamma) \models \beta'$ if and only if $(cval(t)/u, \gamma) \models \beta$, because of the particular meaning of $\models$ that we are using. This equivalence would not hold with, e.g., β equal to $R(x_1, x_2)$ and β' equal to $R(x_1, x_2) \wedge x_3 = x_3$.

---

[7] That is, a term in $T(\{\cup, \cap, ^-\}, \mathcal{W}_m)$, cf. Definition 6.32.

We now define C_β by the following induction:

$$
\begin{array}{lll}
C_\alpha & \text{is} & Y_\alpha \text{ if } \alpha \text{ is atomic,} \\
C_{\beta \wedge \beta'} & \text{is} & C_\beta \cap C_{\beta'}, \\
C_{\neg \beta} & \text{is} & \overline{C_\beta} \cap D_\beta,
\end{array}
$$

where D_β is the intersection of all set variables $Y_{x_i = x_i}$ such that $x_i \in Var(\beta)$ and all set variables $Y_{a=a}$ such that $a \in Con(\beta)$.

It is clear from the definitions that C_β satisfies the required property. $\qquad\square$

The central part of the proof is the following claim:

Claim 9.24.2 There exists a formula Ψ_m in $\mathrm{MS}(\mathcal{R}_F, \mathcal{W}_m \cup X_m)$ such that a $(\mathcal{W}_m \cup X_m)$-assignment γ in $\lfloor t \rfloor$ satisfies Ψ_m if and only if, for every i, $\gamma(x_i)$ is a leaf of t and, for every $\alpha \in A := Atom(\mathcal{R}, X_m)$, we have

$$
\gamma(Y_\alpha) = E_\alpha(t, \gamma \restriction X_m).
$$

Proof: The construction of Ψ_m will be based on the following informally stated fact:

if u is a node of t, then the condition "$u \in E_\alpha(t, \gamma \restriction X_m)$" is equivalent to a Boolean combination of conditions of the form "$u' \in E_{\alpha'}(t, \gamma \restriction X_m)$" for the sons u' of u and atomic formulas $\alpha' \in A$.

This condition will be expressed by a monadic second-order formula $\psi_{f,\alpha}$ in $\mathrm{MS}(\mathcal{R}_F, \mathcal{W}_m \cup X_m \cup \{u\})$, where f is the label of u. And then, the formula Ψ_m will be defined as:

$$
\text{``}x_1 \text{ is a leaf''} \wedge \cdots \wedge \text{``}x_m \text{ is a leaf''}
$$
$$
\wedge \bigwedge\nolimits_{\alpha \in A, f \in F} \forall u(lab_f(u) \Rightarrow (u \in Y_\alpha \Leftrightarrow \psi_{f,\alpha})).
$$

We now describe the formulas $\psi_{f,\alpha}$. There are several cases that depend on f occurring at u.

Case 1: $u \in Occ(t, \varnothing_{\mathcal{R}'})$. We define $\psi_{f,\alpha}$ as *False*, except that $\psi_{f,True}$ is *True*.

Case 2: $u \in Occ(t, \diamondsuit_{B,\mathcal{R}'})$. We recall that $\diamondsuit_{B,\mathcal{R}'}$ denotes the $\mathcal{R}'$-structure with a singleton domain equal to, say, $\{*\}$ and empty relations for all R of positive arity not in B. The element $*$ is the value of all constant symbols.

We define $\psi_{f,\alpha}$ as the conjunction of the equalities $x_i = u$ for each $x_i \in Var(\alpha)$ if $Con(\alpha) \subseteq \mathcal{R}'$ and $\diamondsuit_{B,\mathcal{R}'} \models \alpha(*, \ldots, *)$, and as *False* otherwise.

Case 3: $u \in Occ(t, \oplus)$. The two sons of u will be denoted by u_1 and u_2. We define $\psi_{f,\alpha}$ as

$$
\exists u_1, u_2[son_1(u, u_1) \wedge son_2(u, u_2) \wedge (u_1 \in Y_\alpha \vee u_2 \in Y_\alpha)].
$$

This is correct because an instantiated atomic formula (i.e., given with values to its variables) is valid in the disjoint union of two structures if and only if it is valid in one of them.

Case 4: $u \in Occ(t,f)$, where f is a unary QF operation of type $\mathcal{R}' \to \mathcal{R}''$. The definition of $\psi_{f,\alpha}$ will use Claim 9.24.1. We will write the formulas $\psi_{f,\alpha}$ by using $u \in C_\beta$ as an abbreviation of the corresponding Boolean combination of formulas $u \in Y_\alpha$, where C_β is as in Claim 9.24.1. We let f be defined by the QFO definition scheme $\langle \delta, (\theta_R)_{R \in \mathcal{R}''_+}, (\kappa_{c,d})_{c \in \mathcal{R}'_0, d \in \mathcal{R}''_0} \rangle$. The unique son of its occurrence u will be denoted by u_1. We consider several cases.

Case (4.1): α is $x_i = x_j$. Then $u \in E_\alpha(t,\gamma)$ if and only if $(cval(t)/u_1,\gamma) \models \delta[x_i/x] \wedge x_i = x_j$. Hence, we define $\psi_{f,\alpha}$ as the formula

$$\exists u_1 (son_1(u,u_1) \wedge u_1 \in C_{\delta[x_i/x]} \wedge u_1 \in Y_{x_i=x_j}).$$

Case (4.2): α is $x_i = a$ or $a = x_i$, where $a \in \mathcal{R}_0$. If $a \in \mathcal{R}_0 - \mathcal{R}''_0$, then $\psi_{f,\alpha}$ is *False*. Now let $a \in \mathcal{R}''_0$. We have $u \in E_\alpha(t,\gamma)$ if and only if, for some $c \in \mathcal{R}'_0$, we have $(cval(t)/u_1,\gamma) \models \kappa_{c,a} \wedge x_i = c$. Hence, we define $\psi_{f,\alpha}$ as the formula

$$\exists u_1 [son_1(u,u_1) \wedge \bigvee_{c \in \mathcal{R}'_0} (u_1 \in C_{\kappa_{c,a}} \wedge u_1 \in Y_{x_i=c})].$$

Case (4.3): α is $a = b$, where $a, b \in \mathcal{R}_0$. On the basis of similar observations, we define $\psi_{f,\alpha}$ either as *False* or as the formula

$$\exists u_1 [son_1(u,u_1) \wedge \bigvee_{c,d \in \mathcal{R}'_0} (u_1 \in C_{\kappa_{c,a}} \wedge u_1 \in C_{\kappa_{d,b}} \wedge u_1 \in Y_{c=d})].$$

Case (4.4): α is $R(t_1,\ldots,t_n)$, where $R \in \mathcal{R}''_n$ and the terms t_i are variables or constant symbols. We assume, without loss of generality, that the formula $\theta_R(x_1,\ldots,x_n) \Rightarrow \delta[x_i/x]$ is universally valid. If all terms t_i are variables, then, since we have $u \in E_\alpha(t,\gamma)$ if and only if $(cval(t)/u_1,\gamma) \models \theta_R[t_1/x_1,\ldots,t_n/x_n]$, we can define $\psi_{f,\alpha}$ to be the formula

$$\exists u_1 [son_1(u,u_1) \wedge u_1 \in C_{\theta_R[t_1/x_1,\ldots,t_n/x_n]}].$$

If some (or all) of the terms t_i are constant symbols, then we combine this definition and those used in Cases (4.2) and (4.3). Just to take an example, if t_1 is $a \in \mathcal{R}''_0$ and $t_2,\ldots,t_n$ are variables, then we define $\psi_{f,\alpha}$ as the formula

$$\exists u_1 [son_1(u,u_1) \wedge \bigvee_{c \in \mathcal{R}'_0} (u_1 \in C_{\kappa_{c,a}} \wedge u_1 \in C_{\theta_R[c/x_1,t_2/x_2,\ldots,t_n/x_n]})].$$

The extension to the case where several of the terms t_i are constant symbols is straightforward but lengthy to write. This completes the construction of the formulas $\psi_{f,\alpha}$ and the proof of the claim. $\qquad\square$

We now complete the proof of the proposition. Let $\mathcal{R}' \subseteq \mathcal{R}$. The output structures may have constants, and these constants will be specified by unary relations. Hence, we will define an MS-definition scheme of type $\mathcal{R}_F \to \mathcal{R}'_*$, cf. Section 7.1.2. It will be of the form $\mathcal{D}_* := \langle \chi, \Delta, (\Theta_R)_{R \in \mathcal{R}'_*} \rangle$.

Its precondition χ must verify that the input structure actually represents a term t in $T(F)_{\mathcal{R}'}$, cf. Corollary 5.12. The other formulas are required to work correctly under the assumption that the input structure is of the form (or is isomorphic to) $\lfloor t \rfloor$ for some term t in $T(F)_{\mathcal{R}'}$. We denote by $\overline{\mathcal{W}}_m$ a tuple enumerating the set $\mathcal{W}_m$. We assume that rt is in $\mathcal{R}_F$, with $\rho(rt) = 0$.

The formula Δ, with free variable x_1, is defined as

$$\exists \overline{\mathcal{W}}_1 (\Psi_1 \wedge rt \in Y_{x_1 = x_1}).$$

Note that the tuple $\overline{\mathcal{W}}_1$ does not consist only of $Y_{x_1 = x_1}$.

For each R in $\mathcal{R}'_m$ such that $m \geq 1$, we define the relation formula Θ_R with free variables $x_1, \ldots, x_m$ as

$$\exists \overline{\mathcal{W}}_m (\Psi_m \wedge rt \in Y_{R(x_1, \ldots, x_m)}).$$

If $a \in \mathcal{R}'_0$, then we let Θ_{lab_a} with free variable x_1 be defined as

$$\exists \overline{\mathcal{W}}_1 (\Psi_1 \wedge rt \in Y_{x_1 = a}).$$

It follows from this construction that $\widehat{\mathcal{D}_*}$ associates with every term t in $T(F)_{\mathcal{R}'}$ the relational structure $cval(t)_*$ in $STR^c_*(\mathcal{R}'_*)$.   ∎

Corollary 9.25 Every set of relational structures that is equational in $\mathbb{STR}$ is the image of TREES under an MS-transduction.   □

Its converse, proved in [BluCou06] (Proposition 63 and Theorem 68), gives an Equationality Theorem for relational structures that is analogous to Theorem 7.36. It will be stated in Theorem 9.31.

Proposition 9.24 yields another proof of Theorem 9.22, one that is not based on the proof of the Recognizability Theorem given in Chapter 5, cf. Remark 7.31(1). However, the proof of Proposition 9.24 uses complicated constructions and is thus not usable in practice. Constructions of finite automata generalizing those of Section 6.3.4 are left for future research.

9.3.3 Width notions for relational structures

In Chapter 2 we have defined the graph algebra $\mathbb{GP}$ with signature F^{VR} and the closely related notion of clique-width. In Chapter 7, we have characterized the sets of graphs of bounded clique-width as the subsets of the images of TREES under monadic second-order transductions. In the present section, we aim to generalize the notion of clique-width and the associated results. We want to obtain the following:

(i) a "good" notion of *width* of a relational structure, defined in terms of its generation by a "small" finite subsignature of F^{QF}, that is equivalent to clique-width for graphs;

(ii) such that bounded tree-width implies bounded width; and

(iii) such that the sets of relational structures of bounded width have a logical characterization along the lines of Corollaries 7.38(1), 7.49 and 9.16.

We first present a natural extension to relational structures of the signature F^{VR} of the graph algebra $\mathbb{GP}$. This extension is useful for handling relational structures of bounded path-width (cf. Proposition 9.28 and the remarks on model-checking at the end of Section 9.2), but it does not satisfy the above properties (ii) and (iii). This indicates that it does not have enough operations. A richer signature and an associated satisfactory width notion will be defined in Section 9.3.4.

A first extension of F^{VR} to relational structures

The following definitions are relative to a fixed relational signature $\mathcal{R}$ without constant symbols.

Definition 9.26 (The operations of F^{relVR} on $\mathcal{R}$-structures) Let $\mathcal{A}$ be a countable set of labels. We will construct structures in $STR(\mathcal{R})$ by using at intermediate stages certain "$\mathcal{A}$-labeled" $\mathcal{R}$-structures. For each $a \in \mathcal{A}$, we let lab_a be a unary relation not in $\mathcal{R}$. It will hold the set of elements of a structure labeled by a. We let $Lab(C) := \{lab_a \mid a \in C\}$ for each finite subset C of $\mathcal{A}$.

The labels of the sets C will be attached to the elements of structures and will be used to construct structures similarly as port labels are used to construct graphs. However, an element may have several labels or no label at all, whereas in a p-graph, each vertex has one and only one port label.

We define a many-sorted signature $F^{relVR}_{\mathcal{R}}$, more simply denoted by F^{relVR} since $\mathcal{R}$ is fixed. Each finite subset C of $\mathcal{A}$ is a sort and the corresponding domain is $STR(\mathcal{R} \cup Lab(C))$. The constant symbols of F^{relVR} of sort C are $\mathbf{a}_C$ for each a in C and $\varnothing_C$. Its operations are the disjoint union $\oplus_{C,C'}$ of type $C \times C' \to C \cup C'$ and the following unary operations:

the *tuple-creating operation* $add_{c_1,\ldots,c_k,R,C}$ of type $C \to C$ for every $k > 0, R \in \mathcal{R}_k$ and $c_1, \ldots, c_k \in C$, and

the *label-modifying operation* $mdf_{Z,C,C'}$ of type $C \to C'$ for every $Z \subseteq C \times C'$.

The symbol $\varnothing_C$ denotes the empty $(\mathcal{R} \cup Lab(C))$-structure and $\mathbf{a}_C$ denotes the $(\mathcal{R} \cup Lab(C))$-structure S having a unique element that belongs to $lab_a S$, with all other relations being empty. Hence, $\mathbf{a}_C$ is another notation for $\diamondsuit_{\{lab_a\}, \mathcal{R} \cup Lab(C)}$ (cf. Definition 5.62, Section 5.3.7).

The unary operations are as follows, for S in $STR^c(\mathcal{R} \cup Lab(C))$. We define $add_{c_1,\ldots,c_k,R,C}(S)$ as the concrete $(\mathcal{R} \cup Lab(C))$-structure S' such that

$$D_{S'} := D_S,$$
$$R_{S'} := R_S \cup \{(x_1,\ldots,x_k) \mid x_i \in lab_{c_i}S \text{ for each } i \in [k]\},$$
$$U_{S'} := U_S \text{ for every } U \in (\mathcal{R} - \{R\}) \cup Lab(C).$$

We define $mdf_{Z,C,C'}(S)$ as the concrete $(\mathcal{R} \cup Lab(C'))$-structure S' such that $D_{S'} := D_S$, $R_{S'} := R_S$ for every $R \in \mathcal{R}$ and, for every $a \in C'$ and $x \in D_S$:

$$x \in lab_a S' \text{ if and only if } x \in lab_b S \text{ for some } b \in C \text{ such that } (b,a) \in Z.$$

(If there is no b such that $(b,a) \in Z$, then $mdf_{Z,C,C'}$ deletes the label a.)

These unary operations are quantifier-free and domain preserving (i.e., they are DP-QF operations). The operation $mdf_{Z,C,C'}$ is a quantifier-free relabeling (cf. Definition 7.20, Section 7.1.6).

For each finite set C, we define F_C^{relVR} as the set of operations of F^{relVR} whose sorts (the sorts occurring in their input and output types) are subsets of C.

Every term t in $T(F_C^{\mathrm{relVR}})$ evaluates to a concrete $(\mathcal{R} \cup Lab(C))$-structure $cval(t)$ whose domain is the set $Occ_0(t)$ of *all* occurrences in t of constant symbols not defining empty structures (since the unary operations do not delete elements, each occurrence in $Occ_0(t)$ specifies an element of $D_{cval(t)}$). The definitions and the notation of Section 2.5.2 concerning the concrete graphs defined by terms in $T(F^{\mathrm{VR}})$ extend in a straightforward manner.

Every structure S in $STR(\mathcal{R} \cup Lab(C))$ is isomorphic to $cval(t)$ for some term t in $T(F_{C'}^{\mathrm{relVR}})$, where C' is a finite subset of $\mathcal{A}$ such that $|C'| \leq |D_S| + |C|$: the proof of Proposition 2.104(1) extends easily.

We define the *relational width* $relwd(S)$ of $S \in STR(\mathcal{R} \cup Lab(C))$ as the minimal cardinality of a set C' such that some term in $T(F_{C'}^{\mathrm{relVR}})$ evaluates to (a concrete structure isomorphic to) S. The operations of F^{relVR} and this notion of width have been introduced in Section 5 of [CouER] (but the term "relational width" is not used in this article). The subsets of $STR(\mathcal{R})$ that are equational in the corresponding many-sorted F^{relVR}-algebra, that we will denote by $\mathbb{STR}_{\mathcal{R}}^{\mathrm{rel}}$, are the sets defined by certain grammars, called the *separated handle-rewriting hypergraph grammars* (cf. Theorem 6.1 of [CouER]). This algebra and the corresponding notion of width satisfy Property (i) but not Properties (ii) and (iii) as we will prove. The following proposition and its proof are from [AdlAdl].

Proposition 9.27 Let $\mathcal{R} := \{R\}$ with R ternary. There exists a subset L of $STR(\mathcal{R})$ that has tree-width 3, unbounded relational width and that is the image of TREES under a monadic second-order transduction.

Proof: We let $\mathcal{D}$ be the class of simple, directed and loop-free graphs. For $G \in \mathcal{D}$, we define $f(G)$ in $STR^c(\mathcal{R})$ as follows:

$$D_{f(G)} := V_G \cup E_G,$$
$$R_{f(G)} := \{(x,y,e) \mid e \in E_G, e : x \to_G y\}.$$

It is clear that $twd(f(G)) \leq twd(G) + 1$.

We denote by $\diamond$ the graph with the single vertex $*$ and no edge. We recall that the binary symbol $\overset{\to}{\otimes}$ denotes the complete directed join on graphs (cf. Example 2.108, Section 2.5.4).

Claim 9.27.1 If $G \in \mathcal{D}$ and $relwd(f(G \overset{\to}{\otimes} \diamond)) = k$, then $pwd(G) \leq 2^k - 1$.

Proof: Assume that $S = f(G \overset{\to}{\otimes} \diamond) = cval(t)$ for some $t \in T(F_{[k]}^{\text{relVR}})$. Let $u \in N_t$ and $x \in D_S$ with $x \leq_t u$. For $i \in [k]$, we say that x *has label i at u* in t if $lab_{iU}(x)$ holds for $U := cval(t)/u$ (hence U is the substructure of S induced by the elements of its domain that are below u). We say that x is *distinguished at u* if there exist $w \in N_t$ and $a,b,c \in [k]$ such that:

(i) $u \leq_t w$;
(ii) w is an occurrence of $add_{a,b,c,R}$;
(iii) for each $j \in \{a,b,c\}$ there exists $y \in D_{cval(t)/w}$ such that y has label j at w; and
(iv) there exists $i \in \{a,b,c\}$ such that x has label i at w.

We observe that if x and x' are two vertices of G such that $x,x' \leq_t u$ and x is distinguished at u, then x and x' do not have the same labels at u, i.e., x has some label i at u that x' has not at u or vice-versa. To see this, we let w and a,b,c satisfy Conditions (i)–(iv) above, and assume that x and x' do have the same labels at u. Then, they also have the same labels at w. By Conditions (i)–(iv), $R_{cval(t)/w}$ (and hence R_S) contains either two triples (x,y,z) and (x',y,z), or two triples (y,x,z) and (y,x',z). This fact contradicts the fact that $S = f(G')$ for some graph G'.

We let P be the path in t from the root to the leaf $*$. For every node w on P, we let $g(w)$ be the set of vertices of G that are distinguished at w. We now check that (P,g) is a path-decomposition of G.

For every $x \in V_G$, we let $I(x)$ be the set of nodes w of t at which x is distinguished. It is an interval, i.e., a set of consecutive nodes, of the path in t from the root to the leaf x. For every $(x,y,z) \in R_S$, there is in t an occurrence w of an operation $add_{a,b,c,R}$ that creates this triple, and so, x,y,z are distinguished at w. Hence the set $I(x) \cap I(y)$ is not empty and is an interval. Since for every $x \in V_G$, we have $(x,*,z) \in R_S$ for some z, we have $I(x) \cap I(*) \neq \emptyset$ and so, x belongs to $g(w)$ for some w on P.

If $x \to_G y$, then (x,y,z) belongs to R_S for some z, hence $I(x) \cap I(y) \neq \emptyset$. Since the three intervals $I(x), I(y)$ and $I(*)$ have pairwise nonempty intersections, there is some w in $I(x) \cap I(y) \cap I(*)$, hence $x,y \in g(w)$ and w is on P. The connectivity condition holds since $g^{-1}(x) = I(x) \cap I(*)$ is an interval for every x in V_G. Hence (P,g) is a path-decomposition of G. Since each box $g(w)$ consists of vertices having different sets of labels from $[k]$ at w, we have $|g(w)| \leq 2^k$. Hence $pwd(G) \leq 2^k - 1$. This completes the proof of the claim. $\qquad\square$

For proving the proposition, we let K be the set of rooted trees, so that $K \subseteq \mathcal{D}$, and L be the set $f(K \overset{\rightarrow}{\otimes} \diamond) \subseteq STR(\mathcal{R})$. Since trees have unbounded path-width (cf. Section 2.4.1), it follows from the claim that L has unbounded relational width. We have $twd(K) = 1$, $twd(K \overset{\rightarrow}{\otimes} \diamond) = 2$ and $twd(L) = 3$. It is not hard to see that the mapping $T \mapsto f(T \overset{\rightarrow}{\otimes} \diamond)$ from K to $STR(\mathcal{R})$ is a 2-copying FO-transduction. This completes the proof. ∎

Every set of $\mathcal{R}$-structures that is equational in the algebra $\mathbb{STR}_{\mathcal{R}}^{\mathrm{rel}}$, has bounded relational width (cf. Proposition 4.44). Hence the set L of Proposition 9.27 is not equational in $\mathbb{STR}_{\mathcal{R}}^{\mathrm{rel}}$.

The following proposition, proved in [AdlAdl], establishes another relation between relational width and path-width. We have proved a similar fact in Assertion (1.3) of Proposition 2.114.

Proposition 9.28 The relational width of a structure is at most its path-width + 1.

Proof: We sketch the proof. Let S be an $\mathcal{R}$-structure having a path-decomposition (P,f) of width $k - 1$. Each box has at most k elements. By Propositions 2.67 and 9.14(1), we can assume that it is 1-downwards increasing and that its root box has a single element. By Lemma 2.78, we can define a coloring γ of D_S with k colors, say the elements of $[k]$, such that any two elements in a box have different colors. Let the path P be $d_1 \to \cdots \to d_n$. Hence, $\{d_1,\dots,d_n\}$ is in bijection with, and will be considered as identical to, the domain D_S, cf. Example 5.2(4). For each $i \in [n]$, we let S_i be the substructure of S induced by the set $\{d_1,\dots,d_i\}$. Furthermore, if $d \in D_{S_i} \cap f(d_i)$, we label it in S_i by $\gamma(d)$; this label will be deleted in S_j for some $j > i$ or at the very end.

Then S_{i+1} is easily built from S_i by the operations of $F_{[k]}^{\mathrm{relVR}}$ that use the k labels of $[k]$, called above colors. Informally, the tuples of S are added one by one as one builds successively $S_1,\dots,S_n$. At the end, all colors are deleted. ∎

This fact has consequences for the model-checking of CMS sentences on relational structures of bounded path-width. We have observed after Theorem 9.21 that we can construct path-decompositions in linear time, and from them, terms over $F_{[\mathcal{R},In_\mathcal{R}]}^{\mathrm{VR}}$ with "few" port labels (by Assertion (1.3) of Proposition 2.114) allowing to check CMS_2 sentences. By Proposition 9.28, we can construct simpler terms over F^{relVR} that suffice for checking CMS sentences. Furthermore, the corresponding finite automata, extending those of Section 6.3.4, are "easy" to build. (This point remains to be verified by an implementation extending that of [CouDur11].)

9.3.4 A powerful subsignature of F^{QF} and another width for relational structures

The converse of Corollary 9.25 is proved in [BluCou06] (Proposition 63). It is even proved for a subsignature of F^{QF} that is as powerful as F^{QF} in the sense that it yields all the equational sets of $\mathbb{STR}$: every equation system over F^{QF} can be transformed into an equivalent one over this restricted signature. We will only state this result and the necessary definitions. We refer the reader to [BluCou06] for the rather technical proofs.

In what follows, we only consider relational signatures without constants. In this way, we simplify the presentation and focus on the main aspects. The case of constants will be discussed briefly afterwards (in Remark 9.32(b)).

Definition 9.29 (The signature $F_{\mathcal{R},\mathcal{B}}^{\text{redQF}}$) Let $\mathcal{R}$ and $\mathcal{B}$ be disjoint relational signatures without constants. The set $\mathcal{B}$ will play the role of $Lab(C)$ in the previous definition. We will later require that $\rho(\mathcal{B}) < \rho(\mathcal{R})$.

We let $F_{\mathcal{R},\mathcal{B}}^{\text{redQF}}$ be the "reduced" subsignature of F^{QF} defined as follows. It has the single sort $\mathcal{R} \cup \mathcal{B}$, with domain $STR(\mathcal{R} \cup \mathcal{B})$. Its unique binary operation is the disjoint union $\oplus$. Its constant symbols are $\varnothing$ to denote the empty structure and the symbols $\diamondsuit_{C,\mathcal{R}\cup\mathcal{B}}$ for $C \subseteq \mathcal{R} \cup \mathcal{B}$ to denote the structures with one element (cf. Definition 5.62). It has the following unary operations:

(a) For every $C \subseteq \mathcal{B}$, the operation del_C empties the relations of C, without modifying the domain of the input structure. We distinguish it from an operation fg_C that would generalize the forgetting operation of Definition 7.19 by deleting C from the signature: here, del_C keeps the relations of C in the signature of the output structure.

(b) For every arity preserving mapping $h : \mathcal{R} \cup \mathcal{B} \to \mathcal{R} \cup \mathcal{B}$ that is the identity on $\mathcal{R}$, the operation $relab_h$ replaces every relation R_S of the input structure S by the union of the relations R'_S such that $R = h(R')$. In other words, each tuple $(R, d_1, \ldots, d_n)$ of T_S is replaced by $(h(R), d_1, \ldots, d_n)$. A relation R of arity n has a defining formula θ_R equal to $\bigvee_{R' \in h^{-1}(R)} R'(x_1, \ldots, x_n)$.

(c) For every $T, U \in \mathcal{B}$, every $R \in \mathcal{R} \cup \mathcal{B}$ and every surjective mapping $h : [m] \to [k+l]$, where $k = \rho(T)$, $l = \rho(U)$ and $m = \rho(R)$, we define the quantifier-free operation $add_{T,U,R,h} : STR^c(\mathcal{R}) \to STR^c(\mathcal{R})$ that redefines R by means of the formula θ_R:

$$R(x_1, \ldots, x_m)$$
$$\vee [T(x_{i_1}, \ldots, x_{i_k}) \wedge U(x_{i_{k+1}}, \ldots, x_{i_{k+l}}) \wedge \bigwedge_{j,j' \in [m], h(j)=h(j')} x_j = x_{j'}],$$

where, for each $s \in [k+l]$, i_s is the smallest element of $h^{-1}(s)$. This operation modifies no relations other than R.

All these unary operations are domain preserving.

The intuition is as follows: the relations of $\mathcal{B}$ are "temporary"– they can be modified or deleted. They generalize the port labels used to build graphs of bounded clique-width. They can be used to add tuples to the relations in $\mathcal{R}$. The tuples of a relation in $\mathcal{R}$ cannot be removed. Similarly, the operations of F^{VR} cannot delete edges.

Example 9.30 We clarify these definitions with some examples.

(1) If G is a p-graph of type included in C, represented by the relational structure $\lfloor G \rfloor_C := \langle V_G, edg_G, (lab_aG)_{a \in C} \rangle$, then

$$\overrightarrow{add}_{a,b}(G) = G' \text{ if and only if } add_{lab_a, lab_b, edg, h}(\lfloor G \rfloor_C) = \lfloor G' \rfloor_C,$$

where h is the identity : $[2] \to [2]$. Also, if $h' : C \to C$, then $relab_{h'}(G) = G'$ if and only if $relab_g(\lfloor G \rfloor_C) = \lfloor G' \rfloor_C$ where $g(edg) := edg$ and $g(lab_a) := lab_{h'(a)}$. Hence, F_C^{VR} is a subsignature of $F_{\mathcal{R}_s, Lab(C)}^{\mathrm{redQF}}$.

(2) Let $U \in \mathcal{B}$ and $U' \in \mathcal{R} \cup \mathcal{B} - \{U\}$ be of the same arity. If $h : \mathcal{R} \cup \mathcal{B} \to \mathcal{R} \cup \mathcal{B}$ is such that $h(U) = U'$ and $h(R) = R$ for $R \neq U$, then the operation $relab_h$ applied to a structure S empties U_S and replaces U'_S by $U'_S \cup U_S$.

(3) We let $\mathcal{B}$ contain the binary relation symbol W and $\mathcal{R}$ the 6-ary symbol R. We let f be the unary operation $add_{W,W,R,h}$, where $h : [6] \to [4]$ is such that $h(1) = h(2) = 1$, $h(3) = 2$, $h(4) = 3$, $h(5) = h(6) = 4$. In a structure S where $u, x, y, z \in D_S$ and $(u,x), (y,z)$ belong to W_S, the operation f adds to $R_{f(S)}$ the tuple (u, u, x, y, z, z). Note that u, x, y, z need not be pairwise distinct. $\qquad \square$

Theorem 9.31 (Equationality Theorem for relational structures) Let $\mathcal{R}$ be a relational signature without constant symbols. The following properties of a subset L of $STR(\mathcal{R})$ are equivalent:

(1) L is equational in the algebra $\mathbb{STR}$.

(2) L is the image of TREES under an MS-transduction.

(3) $L = val(M)$ for some regular subset M of $T(F_{\mathcal{R}, \mathcal{B}}^{\mathrm{redQF}})$ and some relational signature $\mathcal{B}$ such that $\rho(\mathcal{B}) < \rho(\mathcal{R})$.

Proof: (1) $\Longrightarrow$ (2) by Proposition 9.24 and the characterization of equational sets as images of regular languages (cf. Corollary 3.19 and Proposition 3.23(3)).

(2) $\Longrightarrow$ (3): Proposition 63 of [BluCou06] establishes this statement. It corresponds to Theorem 7.34(1) with $T(F_{\mathcal{R}, \mathcal{B}}^{\mathrm{redQF}})$ in place of $T(F_{C,[K,\Lambda]}^{\mathrm{VR}})$, and for a relational signature $\mathcal{B}$ such that $\rho(\mathcal{B}) < \rho(\mathcal{R})$. The result follows then with the usual tools like Proposition 7.27.

(3) $\Longrightarrow$ (1): Property (3) implies, with the results of Chapter 3, that L is equational in the subalgebra of $\mathbb{STR}$ generated by $F_{\mathcal{R}, \mathcal{B}}^{\mathrm{redQF}}$. Hence, it is equational in $\mathbb{STR}$. $\qquad \blacksquare$

Remark 9.32 (a) Property (1) allows L to be defined by an equation system written with operations of F^{QF} involving relational structures with constant symbols. (In particular, some unknowns of this system may generate structures with constants.) The equivalence with Property (3) implies that this system can be replaced, for defining L, by one not using such operations.

(b) If a set $L \subseteq STR(\mathcal{R})$, where $\mathcal{R}$ contains constant symbols, is equational in $\mathbb{STR}$, then L_*, the corresponding set of structures without constants, satisfies Property (2) and thus Property (3). In order to get an Equationality Theorem for $\mathcal{R}$-structures, it suffices to prove that, if L_* is equational, then so is L in the algebra $\mathbb{STR}$. We will spare the reader this scholarly exercise.

(c) An alternative to Property (3) is established in [SheDor]: informally, auxiliary relations of arity at most 3 suffice, but not exactly in the sense of Property (3). This article proves that if L satisfies Property (2), then it is $val(M)$ for some regular subset M of $T(F_{nc}^{QF} \upharpoonright (\mathcal{R} \cup \mathcal{B}))$, where the auxiliary set $\mathcal{B}$ consists of relation symbols of arity at most 3.

(d) In Property (3) it can be required additionally that the set M is included in $T(F_{\mathcal{R},\mathcal{B}}^{redQF} - \{\varnothing\}) \cup \{\varnothing\}$. We can eliminate $\varnothing$ from every term of M by a proof similar to the one of Claim 7.34.1, since we have $g(\varnothing) = \varnothing$ for every unary operation g (together with Equalities (1) and (3) of Proposition 2.101). $\qquad\square$

Yet another width?

Definition 9.33 (The width $w_\mathcal{R}$) We will use the following notion of *size* of a relational signature $\mathcal{R}$ without constant symbols that differs slightly from the one defined in Definition 5.3. We let

$$\|\mathcal{R}\| := \sum \{\rho(R) \mid R \in \mathcal{R}\}.$$

We assume that $\rho(\mathcal{R}) > 1$, otherwise, $\mathcal{R}$-structures are of no interest. For every $S \in STR(\mathcal{R})$, we let $w_\mathcal{R}(S)$ be the minimum size of a signature $\mathcal{B}$, disjoint with $\mathcal{R}$, such that $\rho(\mathcal{B}) < \rho(\mathcal{R})$ and S is the value of a term over $F_{\mathcal{R},\mathcal{B}}^{redQF}$.

It is clear that every $\mathcal{R}$-structure S is the value of a term over $F_{\mathcal{R},\mathcal{B}}^{redQF}$ for some large enough set $\mathcal{B}$. For example, if $\rho(\mathcal{R}) = 3$ and $n = |D_S|$, one can construct such a term by using a set $\mathcal{B}$ consisting of n unary relations and $n(n-1)/2$ binary relations. In general, we have $w_\mathcal{R}(S) \leq \rho(\mathcal{R}) \cdot n^{\lceil \rho(\mathcal{R})/2 \rceil}$.

If L is a set of $\mathcal{R}$-structures, we let $w_\mathcal{R}(L) := \max\{w_\mathcal{R}(S) \mid S \in L\}$.

For the following proposition, we recall that $\mathcal{R}_{s,[K,\Lambda]}$ is the relational signature used for representing simple (K,Λ)-labeled graphs (cf. Examples 5.2(2)).

Proposition 9.34

(1) For every simple (K,Λ)-labeled graph G, we have

$$w_{\mathcal{R}_{s,[K,\Lambda]}}(\lfloor G \rfloor) \leq cwd(G) \leq f(w_{\mathcal{R}_{s,[K,\Lambda]}}(\lfloor G \rfloor)),$$

for some fixed computable function f depending on (K,Λ).
(2) For every relational signature $\mathcal{R}$ and every $S, S' \in STR(\mathcal{R})$ such that S is a substructure[8] of S', we have $w_\mathcal{R}(S') \leq w_\mathcal{R}(S)$.

---

[8] We recall from Definition 5.1 that S is a substructure of S' if $D_S \subseteq D_{S'}$ and $R_S = R_{S'} \cap D_S^{\rho(R)}$ for every $R \in \mathcal{R}$. It generalizes the notion of an induced subgraph.

(3) For every relational signature $\mathcal{R}$ and positive integer k, the set of $\mathcal{R}$-structures S such that $w_{\mathcal{R}}(S) \leq k$ is the image of TREES under a monadic second-order transduction. Conversely, if $L \subseteq STR(\mathcal{R})$ and L is included in the image of TREES under a monadic second-order transduction, then $w_{\mathcal{R}}(L)$ is finite.

(4) For every two relational signatures $\mathcal{R}$ and $\mathcal{R}'$ without constant symbols, and every monadic second-order transduction τ of type $\mathcal{R} \to \mathcal{R}'$, there exists a computable function f_τ such that $w_{\mathcal{R}'}(U) \leq f_\tau(w_{\mathcal{R}}(S))$ for every $S \in STR(\mathcal{R})$ and every U in $\tau(S)$.

Proof: (1) If G has clique-width k, then it is the value of a term t in $T(F_{C,[K,\Lambda]}^{VR})$ where $|C| = k$. Since $F_{C,[K,\Lambda]}^{VR}$ is a subsignature of $F_{\mathcal{R}s,[K,\Lambda],Lab(C)}^{redQF}$ (cf. Example 9.30(1)) and $\|Lab(C)\| = |C|$, we get the first inequality. (We may have a strict inequality because the graphs defined by the terms in $T(F_{\mathcal{R}s,[K,\Lambda],Lab(C)}^{redQF})$ may have several labels from C attached to the same vertex. Hence, these terms are not necessarily terms in $T(F_{C,[K,\Lambda]}^{VR})$.)

The second inequality follows from Proposition 9.24 and Corollary 7.38(2). The function f is computable because Proposition 9.24 is effective (and see the last sentence of the proof of Corollary 7.38(2)).

(2) The proof is similar to the proof of the corresponding result for clique-width (Proposition 2.105(1)). Let S and S' be as in the statement. Let S be the value of a term t in $F_{\mathcal{R},\mathcal{B}}^{redQF}$ such that $\|\mathcal{B}\| = w_{\mathcal{R}}(S)$.

The elements of D_S are in bijection with the occurrences of the constant symbols in t different from $\varnothing$ because the unary operations of $F_{\mathcal{R},\mathcal{B}}^{redQF}$ are all domain preserving. Let us replace by $\varnothing$ those occurrences that correspond to the elements of $D_S - D_{S'}$. We obtain a term t' in $T(F_{\mathcal{R},\mathcal{B}}^{redQF})$ that evaluates to the structure $S[D_{S'}] = S'$ (because for every unary operation g in $F_{\mathcal{R},\mathcal{B}}^{redQF}$, every structure U and every set $D \subseteq D_U$, we have $g(T)[D] = g(T[D])$.) Hence $w_{\mathcal{R}}(S') \leq w_{\mathcal{R}}(S)$. If needed, we can eliminate $\varnothing$ from t' (cf. Remark 9.32(d)).

(3) The first assertion follows from Proposition 9.24 and Proposition 7.27. The second one follows from the implication (2) $\Longrightarrow$ (3) of Theorem 9.31.

(4) An immediate consequence of Proposition 9.24 and the usual arguments (cf. the proof of Corollary 7.38(2)). ∎

We get also that $w_{\mathcal{R}}(L)$ is finite if L has bounded tree-width (because $Inc(L)$ is included in the image of TREES under an MS-transduction by Corollary 9.16, hence L is included in the image of TREES under an MS-transduction, hence $w_{\mathcal{R}}(L)$ is finite). Hence, the numerical parameter $w_{\mathcal{R}}(S)$ satisfies the Conditions (i) to (iii) stated at the beginning of Section 9.3.3. However, we give it no name because some equivalent one having better combinatorial and algorithmic properties may be found in the (next) future and names are not that easy to find for mathematical notions, so we prefer to spare them.

9.4 Sparse relational structures

We will prove that for expressing properties of simple graphs that are sparse, i.e., that have "few" edges in a precise sense to be defined, CMS_2 formulas are no more powerful than CMS formulas. This result, which we call the *Sparseness Theorem*, applies in particular to planar graphs, to graphs of bounded tree-width and, more generally, to the graphs that exclude some fixed graph as a minor. It applies also to graphs of bounded degree. These facts have been stated in connection with Theorems 1.44, 1.49, 4.51, 4.59 and 5.22. We will actually prove this theorem for relational structures and not only for graphs. The CMS_2 formulas over relational structures, which can use quantifications on sets of tuples, have been defined in Section 9.2.

All relational signatures will be without constant symbols. The incidence graph of a concrete $\mathcal{R}$-structure is a simple, directed and labeled concrete graph, and it will be identified (as in Section 9.2, Definition 9.10) with the corresponding $\mathcal{R}^{Inc}$-structure.

The proofs of the Sparseness Theorem and all necessary lemmas will consist of constructions of definition schemes for specifying monadic second-order transductions. A definition scheme specifies in the first place a concrete transduction. We will have to construct definition schemes that specify particular concrete transductions, and not only transductions of abstract structures. Hence, in this section and unless otherwise specified, we will adopt the following convention:

All graphs, relational structures and transductions will be concrete.

Our tools concerning transductions, in particular Theorem 7.10 (the Backwards Translation Theorem) and Theorem 7.14 (about compositions of transductions), have been formulated for concrete MS-transductions. The following definitions concern only concrete structures and transductions.

Definition 9.35 (Unions of concrete structures and transductions) Let $\mathcal{R}$, $\mathcal{R}'$ and $\mathcal{R}''$ be relational signatures without constant symbols. If $S \in STR^c(\mathcal{R})$ and $S' \in STR^c(\mathcal{R}')$, then we denote by $S + S'$ the concrete $\mathcal{R} \cup \mathcal{R}'$-structure T such that

$$\begin{aligned}
D_T &:= D_S \cup D_{S'} \quad (D_S \text{ and } D_{S'} \text{ need not be disjoint}), \\
R_T &:= R_S \cup R_{S'} \quad \text{if } R \in \mathcal{R} \cap \mathcal{R}', \\
R_T &:= R_S \quad \text{if } R \in \mathcal{R} - \mathcal{R}', \\
R_T &:= R_{S'} \quad \text{if } R \in \mathcal{R}' - \mathcal{R}.
\end{aligned}$$

If τ and τ' are concrete transductions of respective types $\mathcal{R} \to \mathcal{R}'$ and $\mathcal{R} \to \mathcal{R}''$, then we denote by $\tau + \tau'$ the concrete transduction of type $\mathcal{R} \to \mathcal{R}' \cup \mathcal{R}''$ such that, for every $S \in Dom(\tau) \cap Dom(\tau')$ (its domain), $(\tau + \tau')(S)$ is the set of structures $T + T'$ such that $T \in \tau(S)$ and $T' \in \tau'(S)$.

If τ and τ' are concrete C_rMS-transductions, then $\tau + \tau'$ is also a concrete C_rMS-transduction. We sketch the construction of a definition scheme for $\tau + \tau'$

from definition schemes $\langle \chi, (\delta_i)_{i \in I}, (\theta_w)_{w \in \mathcal{R}' \otimes I} \rangle$ and $\langle \chi', (\delta_i')_{i \in I'}, (\theta_w')_{w \in \mathcal{R}'' \otimes I'} \rangle$ for τ and τ' respectively. Note that $D_T \subseteq D_S \times I$ if $T \in \tau(S)$ and that $D_T \subseteq D_S \times I'$ if $T \in \tau'(S)$. We can assume that the sets of parameters used for τ and for τ' are disjoint. The definition scheme for $\tau + \tau'$ has all of them as parameters and has precondition $\chi \wedge \chi'$. Its domain formulas are δ_i'' for all $i \in I \cup I'$ where

$$
\begin{array}{lll}
\delta_i'' & \text{is} \quad \delta_i \vee \delta_i' & \text{if } i \in I \cap I', \\
\delta_i'' & \text{is} \quad \delta_i & \text{if } i \in I - I', \\
\delta_i'' & \text{is} \quad \delta_i' & \text{if } i \in I' - I.
\end{array}
$$

The relation formulas are easy to write.

We recall from Definitions 9.10 and 5.1 that if S belongs to $STR^c(\mathcal{R})$, then T_S denotes the set of tuples of S, and if $X \subseteq D_S$, then $S[X]$ denotes the (induced) substructure of S with domain X and relations $R_{S[X]} := R_S \cap X^{\rho(R)}$ for every $R \in \mathcal{R}$.

Definition 9.36 (Uniformly k-sparse relational structures) Let k be a positive integer. A relational structure S is k-*sparse* if $|T_S| \leq k \cdot |D_S|$. It is *uniformly k-sparse* if $S[X]$ is k-sparse for every subset X of D_S. We denote by $US_k(\mathcal{R})$ the class of uniformly k-sparse (concrete) $\mathcal{R}$-structures.

A similar notion is defined for graphs in Section 5.2.5 (cf. Theorem 5.22): a graph G is *uniformly k-sparse* if $|E_H| \leq k \cdot |V_H|$ for each of its induced subgraphs H. In Remark 9.39(2) below, we will compare uniform sparsity for labeled graphs and for the associated relational structures. We note immediately that whether a graph G is uniformly k-sparse does not depend on its possible vertex and edge labels. However, the relational structure $\lfloor G \rfloor$ represents these labels and its size depends on them.

A set of graphs or relational structures is *uniformly k-sparse* if its elements are all uniformly k-sparse.

The main result of this section is formulated in the following two theorems:

Theorem 9.37 A class of simple labeled graphs or of $\mathcal{R}$-structures that is uniformly k-sparse is CMS_2-definable if and only if it is CMS-definable. $\qquad\square$

The "if" direction of this theorem is immediate from the Backwards Translation Theorem (Corollary 7.12) and the fact that the inverse of the mapping *Inc* is an FO-transduction of abstract structures (Lemma 9.11(2)). Thus, again by Corollary 7.12, Theorem 9.37 is a consequence of the following more technical statement (cf. Theorem 1.49).

Theorem 9.38 (Sparseness Theorem) The mapping *Inc* is a domain-extending MS-transduction on the following classes:

the class $US_k(\mathcal{R})$ of uniformly k-sparse $\mathcal{R}$-structures, for every $k > 0$ and every relational signature $\mathcal{R}$ without constant symbols; and the classes of simple, directed or undirected, uniformly k-sparse (K, Λ)-labeled graphs, for every $k > 0$ and every pair (K, Λ) of finite sets of labels. $\square$

By Definition 7.2, this means that a domain-extending MS-transduction τ constructs, for every S in $US_k(\mathcal{R})$, one or more structures with domains included in $D_S \times I$ for some finite set I, that are all isomorphic to $Inc(S)$ by an isomorphism that maps (a, i) to a for some fixed element i of I.

The proof of Theorem 9.38 will construct a domain-extending p-copying transduction that will extend the domain D_S of an input structure S by new elements in bijection with the tuples of S. Such a construction is possible only if S has at most $(p-1) \cdot |D_S|$ tuples. Theorem 9.38 shows that, conversely, such a construction is possible for any fixed $k > 0$ and for all structures in $US_k(\mathcal{R})$. We will first establish it for simple, labeled, directed and undirected graphs.[9] The full proof will be given in Section 9.4.2, after Theorem 9.61.

9.4.1 Edge set quantifications in uniformly k-sparse graphs

We will denote by M^{u} the class of loop-free undirected graphs, by M^{d} the class of loop-free directed graphs, and by A^{d} the subclass of M^{d} consisting of the acyclic graphs (those without circuits). We recall from Definition 2.9 that an *orientation* H of a graph $G \in M^{\mathrm{u}}$ is a directed graph such that $und(H) = G$: each edge is given a direction. Note that every graph in M^{u} has an acyclic orientation in A^{d}: order the vertices and let the direction of the edges respect the order.

We will denote by S^{u} the class of simple, loop-free undirected graphs. An orientation of a graph $G \in S^{\mathrm{u}}$ is thus a directed graph H such that $V_H = V_G$, $edg_H \cap edg_H^{-1} = \emptyset$ and $edg_G = edg_H \cup edg_H^{-1}$. We will denote by S_*^{d} the class of orientations of the graphs in S^{u} and by S_k^{d} the subclass of those of indegree at most k. Hence, S_*^{d} is the class of simple, loop-free directed graphs that have no pair of opposite edges. We will denote by US_k the class of graphs in S^{u} that are uniformly k-sparse.

Remark 9.39 (1) In all this section, we will *not* consider an undirected graph G as a directed graph with pairs of opposite edges.

(2) If G is a simple directed graph, then $|edg_G| = |E_G|$. Hence, it is uniformly k-sparse if and only if the structure $\lfloor G \rfloor$ is uniformly k-sparse. If G is undirected, then

$$|E_G| \leq |edg_G| = |E_G| + |E_G - Loops(G)| \leq 2 \cdot |E_G|,$$

where $Loops(G)$ is the set of loops of G. Hence, the structure $\lfloor G \rfloor$ is uniformly $2k$-sparse if G is uniformly k-sparse. Conversely, if $\lfloor G \rfloor$ is uniformly k-sparse, then G is uniformly k-sparse.

---

[9] It was proved for words and terms (viewed as graphs) in Corollary 5.23, and for forests in Example 7.8.

Let us now consider a (K, Λ)-labeled graph G. The tuples of the structure $\lfloor G \rfloor$ are the pairs representing the labeled edges (each edge has a unique label) and unary tuples representing the labels of the vertices. Hence, if G is directed, then

$$|E_G| \leq |T_{\lfloor G \rfloor}| \leq |E_G| + |K| \cdot |V_G|.$$

It follows that $\lfloor G \rfloor$ is uniformly $(k + |K|)$-sparse if G is uniformly k-sparse. If G is undirected, then we have similarly

$$|E_G| \leq |T_{\lfloor G \rfloor}| \leq 2 \cdot |E_G| + |K| \cdot |V_G|.$$

These remarks entail that a set L of simple, directed or undirected, (K, Λ)-labeled graphs is uniformly k-sparse for some k if and only if the set of structures $\lfloor L \rfloor$ is uniformly m-sparse for some m.

(3) The reader will prove easily that for every (K, Λ)-labeled graph G, the relational structure $\lceil G \rceil = \lfloor Inc(G) \rfloor$ is uniformly $(2 + |K|)$-sparse.

(4) We now compare uniform sparsity for graphs to some related notions. The *average degree* of a graph G (possibly with loops and multiple edges) is defined as $2 \cdot |E_G| / |V_G|$. Hence, a set of graphs is uniformly k-sparse if and only if its graphs and their subgraphs have average degree at most $2k$.

The property that a graph is uniformly k-sparse is also related with its arboricity. A graph G (possibly with multiple edges) has *arboricity at most k* if the undirected graph $und(G)$ (or G if it is undirected) is the union of k edge-disjoint forests (i.e., is equal to $G_1 \cup \cdots \cup G_k$, where each graph G_i is a forest and $E_{G_i} \cap E_{G_j} = \emptyset$ if $i \neq j$). The arboricity of G is the smallest such k. This value is thus undefined for graphs with loops. Arboricity has the following characterization: a loop-free graph G has arboricity at most k if and only if $|E_{G[X]}| \leq k \cdot (|X| - 1)$ for every nonempty subset X of V_G ([*Fra], Theorem 6.13). Hence a graph of arboricity k is uniformly k-sparse. Conversely, a loop-free graph that is uniformly k-sparse has arboricity at most $2k$ as one checks easily. This bound is best possible since a graph with $2k$ parallel edges between two vertices has arboricity $2k$ and is uniformly k-sparse. Hence a set of loop-free graphs has bounded arboricity if and only if it is uniformly k-sparse for some k.

(5) It is clear from the definitions that every graph of degree at most $2d$, and every directed graph of indegree at most d, is uniformly d-sparse. Every simple, loop-free, undirected and planar graph is uniformly 3-sparse ([*Die], Corollary 4.2.10). Corollary 2.74(1) shows that a simple, loop-free and undirected graph of tree-width at most k is uniformly k-sparse (cf. the proof of Theorem 4.51). More generally, if G is a simple graph that does not contain K_p as a minor, then it has at most $a \cdot p\sqrt{\log(p)} \cdot |V_G|$ edges for some constant a (see [Thom]). The same holds for its subgraphs, hence G is uniformly $(a \cdot p\sqrt{\log(p)})$-sparse.   □

The following proposition collects properties of uniformly k-sparse graphs that will be useful for our constructions. A *homomorphism* $h : G \to H$, where G and H belong to S^d_*, is a mapping : $V_G \to V_H$ such that $x \to_G y$ implies $h(x) \to_H h(y)$. For each positive integer k, we define $q(k) := 2^{2k(k+1)+1} - 1$.

Proposition 9.40

(1) A graph in S^u is uniformly k-sparse if and only if it has an orientation of indegree at most k.

(2) Every simple directed graph of indegree at most k has a proper vertex $(2k + 1)$-coloring.[10]

(3) For every $k > 0$, there exists a graph $Q(k)$ in S^d_* with vertex set $[q(k)]$ such that for every $G \in S^d_k$ there exists a homomorphism : $G \to Q(k)$.

Proof: (1) This fact is proved in [*Fra], Theorem 6.13. We will prove a generalization of it to hypergraphs in Proposition 9.58.

(2) Let G belong to S^d_k. By induction on $|V_G|$, we will construct a proper vertex $(2k + 1)$-coloring defined as a mapping $\gamma : V_G \to [2k + 1]$. Since the sum of the indegrees of all vertices is equal to that of their outdegrees (and to $|E_G|$), some vertex v has outdegree at most k, hence degree at most $2k$. Since the graph $G - v$ also belongs to S^d_k it has by induction a proper vertex $(2k + 1)$-coloring defined as a mapping γ'. Since the degree of v is at most $2k$, γ' can be extended into a proper vertex $(2k + 1)$-coloring of G.

(3) This fact is proved in [NešSV], Theorem 10. For completeness we reproduce its proof. Let us fix $k > 0$ and define $p := 2k(k+1)+1$. We construct as follows a directed graph $Q(k)$ with $q(k) = 2^p - 1$ vertices. Its vertices are the tuples $(i; a_1, \ldots, a_{i-1})$ for $i \in [p]$ and $a_s \in \{0, 1\}$ for all $s \in [i - 1]$. If $a = (i; a_1, \ldots, a_{i-1})$ and $b = (j; b_1, \ldots, b_{j-1})$ with $j < i$, there is an edge from a to b if $a_j = 1$ and one from b to a if $a_j = 0$. (There is no edge between $(i; a_1, \ldots, a_{i-1})$ and $(i; b_1, \ldots, b_{i-1})$.)

Let G belong to S^d_k. We first show that it has a proper vertex p-coloring γ such that, if $x \to_G y \to_G z$, then $\gamma(x) \neq \gamma(z)$: consider the graph G' consisting of G augmented with the edges from x to z for all triples (x, y, z) as above. This graph has indegree at most $k(k + 1)$, hence, by (2), it has a proper vertex p-coloring γ, which is the desired one for G.

We now define a homomorphism h from G to $Q(k)$. If $x \in V_G$ we define $h(x) := (i; a_1, \ldots, a_{i-1})$, where $i := \gamma(x)$, and, for $1 \leq j < i$, $a_j := 0$ if and only if there is an edge $y \to_G x$ for some vertex y such that $\gamma(y) = j$.

We now check that h is indeed a homomorphism. Let x and y be adjacent vertices of G such that $j = \gamma(y) < i = \gamma(x)$. Let $h(x) = a = (i; a_1, \ldots, a_{i-1})$ and $h(y) = b = (j; b_1, \ldots, b_{j-1})$. If $y \to_G x$, then $a_j = 0$, hence we have $b \to_{Q(k)} a$. If $x \to_G y$, then

---

10 *Proper* means that two adjacent vertices have different colors.

there is no vertex z such that $z \to_G x$ and $\gamma(z) = j$, because γ is a proper coloring of G'. Hence, $a_j = 1$ and we have $a \to_{Q(k)} b$.

The vertices of $Q(k)$ can be encoded by the integers in $[q(k)]$, which gives the stated result.   ∎

In the next definition and until the end of this section, we denote by $\mathcal{X}_n$ any n-tuple of distinct set variables $(X_1, \ldots, X_n)$.

Definition 9.41 (Orientations and orderings defined by MS formulas) Let $\mathcal{C}$ be a set of simple, loop-free undirected graphs. A pair of MS formulas $(\chi(\mathcal{X}_n), \theta(\mathcal{X}_n, x_1, x_2))$ *defines orientations of every graph in $\mathcal{C}$* (or *orients the graphs of $\mathcal{C}$*) if, for every graph G in $\mathcal{C}$,

$$\lfloor G \rfloor \models (\exists X_1, \ldots, X_n. \chi) \wedge \forall X_1, \ldots, X_n(\chi \Rightarrow \widehat{\theta}),$$

where $\widehat{\theta}$ is the formula with free variables in $\mathcal{X}_n$ defined as

$$\forall x, y[edg(x,y) \Leftrightarrow (\theta(x,y) \vee \theta(y,x))] \wedge \forall x, y[\theta(x,y) \Rightarrow \neg\theta(y,x)].$$

These conditions mean that, for every tuple $(U_1, \ldots, U_n)$ of sets of vertices satisfying χ, if $W := \{(u,v) \mid u, v \in V_G \text{ and } \lfloor G \rfloor \models \theta(U_1, \ldots, U_n, u, v)\}$, then the directed graph $H := \langle V_G, W \rangle$ is an orientation of G. Furthermore, for each graph G in $\mathcal{C}$ there is such an n-tuple.

Equivalently, they mean that the domain-preserving MS-transduction with definition scheme $\langle \chi, True, \theta_{edg} \rangle$, where θ_{edg} (the formula that specifies the relation edg_H of an output graph H) is θ, associates with every graph $G \in \mathcal{C}$ one or more orientations of G. The variables in $\mathcal{X}_n$ are its parameters. If (χ, θ) defines orientations of the graphs in $\mathcal{C}$, then $(\widehat{\theta}, \theta)$ does the same, and for the graphs of a possibly larger class. However, it is frequently clearer to use χ instead of $\widehat{\theta}$.

Similarly, if $\mathcal{C}$ is a set of directed or undirected graphs, a pair of formulas (χ, θ) as above *defines partial orders on V_G for every graph G in $\mathcal{C}$* (or *partially orders the graphs of $\mathcal{C}$*) if the same conditions hold, where $\widehat{\theta}$ is defined here as

$$\forall x. \theta(x,x) \wedge \forall x, y[(\theta(x,y) \wedge \theta(y,x)) \Rightarrow x = y]$$
$$\wedge \forall x, y, z[(\theta(x,y) \wedge \theta(y,z)) \Rightarrow \theta(x,z)].$$

In this case, the set $W := \{(u,v) \mid u, v \in V_G \text{ and } \lfloor G \rfloor \models \theta(U_1, \ldots, U_n, u, v)\}$ is a partial order on V_G. The domain-preserving MS-transduction with definition scheme $\langle \chi, True, \theta_{edg}, \theta_{\leq} \rangle$ such that θ_{edg} is $edg(x_1, x_2)$ and $\theta_{\leq}$ is θ, associates with $\lfloor G \rfloor$ for $G \in \mathcal{C}$ one or more structures of the form $\langle V_G, edg_G, \leq \rangle$, where $\leq$ is a partial order on V_G. In Section 5.2.6 (Definition 5.28), we had a similar definition for specifying linear orders by MS formulas.

The integer $q(k)$ in the following statement is defined before Proposition 9.40.

Proposition 9.42

(1) For each $k > 0$, there exists a pair of FO formulas $(\chi(\mathcal{X}_{q(k)}), \theta(\mathcal{X}_{q(k)}, x_1, x_2))$ that defines orientations of indegree at most k of every graph in US_k.
(2) For each k, the class of graphs US_k is MS-definable.

Proof: (1) By Proposition 9.40(1), a graph G in S^u is uniformly k-sparse if and only if it has an orientation of indegree at most k. We will specify such an orientation by means of a vertex coloring, based on the third assertion of Proposition 9.40 of which we will use the corresponding notions. A k-*good coloring* of G is a mapping $\gamma : V_G \to [q(k)]$ such that:

(i) if u and v are adjacent in G, then $\gamma(u)$ and $\gamma(v)$ are adjacent in $Q(k)$;
(ii) the orientation H of G defined by: $u \to_H v$ if and only if $\gamma(u) \to_{Q(k)} \gamma(v)$, has indegree at most k.

By Proposition 9.40(3), a graph G has an orientation of indegree at most k if and only if it has a k-good coloring: if G has an orientation $H \in S_k^d$, then the homomorphism from H to $Q(k)$ is a k-good coloring. A coloring $\gamma : V_G \to [q(k)]$ can be described by a $q(k)$-tuple of subsets of V_G, $(U_1, \ldots, U_{q(k)}) := (\gamma^{-1}(1), \ldots, \gamma^{-1}(q(k)))$, which defines a partition of V_G (some of its sets may be empty). The formulas χ and θ to be constructed have free variables $X_1, \ldots, X_{q(k)}$ and are easy to write. The formula χ expresses, for every graph G in S^u, that

- $U_1, \ldots, U_{q(k)}$ define a partition of V_G; and
- the mapping $\gamma : V_G \to [q(k)]$ such that $\gamma(u) := i$ for every $i \in [q(k)]$ and $u \in U_i$, is a k-good coloring.

The formula $\theta(x_1, x_2)$ is $edg(x_1, x_2) \wedge \bigvee_{i \to_{Q(k)} j}(x_1 \in X_i \wedge x_2 \in X_j)$, expressing that $\{x_1, x_2\}$ is an edge such that $\gamma(x_1) \to_{Q(k)} \gamma(x_2)$.

(2) It follows from (1) that US_k is defined with respect to the class S^u by the MS sentence $\exists \mathcal{X}_{q(k)} \cdot \chi$.[11]   ∎

For every vertex x of a directed graph G, we define $Adj_G^-(x)$ as the set $\{y \mid y \to_G x\}$.

Proposition 9.43 For each $k > 0$, there exists a pair of MS formulas $(\chi(\mathcal{X}_n), \theta(\mathcal{X}_n, x_1, x_2))$ that defines, for every graph G in S_k^d, partial orders on V_G that are linear on $Adj_G^-(x)$ for each $x \in V_G$.

Before giving the quite complicated proof, we show that the last two propositions yield the assertions of the Sparseness Theorem (Theorem 9.38) relative to graphs and to binary relational structures, which we restate as follows:

---

[11] The definition of US_k is not monadic second-order expressible, but the characterization of Proposition 9.40(1) yields the result. The class of graphs G that are 2-*sparse*, i.e., for which $|E_G| \leq 2 \cdot |V_G|$, is not MS-definable. The proof technique of Proposition 5.13 can be adapted to prove this fact.

Theorem 9.44 The mapping *Inc* is a domain-extending MS-transduction on the classes $US_k(\mathcal{R})$, for every $k > 0$ and every binary relational signature $\mathcal{R}$, and on the classes of simple, directed or undirected, uniformly k-sparse (K, Λ)-labeled graphs, for every $k > 0$ and every pair (K, Λ) of finite sets of labels.

Proof: *The special case of undirected graphs*: For making the proof more readable, we first prove that *Inc* is a domain-extending MS-transduction on the class US_k. Thus, we have to construct a domain-extending MS-transduction τ that transforms each uniformly k-sparse graph G in S^u into $Inc(G)$ (more precisely, into a graph isomorphic to $Inc(G)$). By Propositions 9.42(1) and 9.43 there exists a domain-preserving MS-transduction τ_1 that transforms G into one or more structures $\langle V_H, edg_H, \leq \rangle$ such that $H \in S_k^d$ is an orientation of G and $\leq$ is a partial order on V_H ($= V_G$) that is linear on each set $Adj_H^-(x)$. Its parameters are the set variables of the involved formulas χ and θ, where those of Proposition 9.43 must be taken disjoint from those of Proposition 9.42(1). It remains to construct a parameterless domain-extending MS-transduction τ_2 that transforms this structure into $Inc(G) = \langle V_G \cup E_G, in_G \rangle$. Then we can define the parameter-invariant MS-transduction τ to be $\tau_1 \cdot \tau_2$ by Theorem 7.14 (on the composition of MS-transductions).

For each $x \in V_H$, we enumerate $Adj_H^-(x)$ as $\{p_1(x), p_2(x), \ldots, p_l(x)\}$ with $l \leq k$ and $p_1(x) < p_2(x) < \cdots < p_l(x)$. An edge $p_i(x) \rightarrow_H x$ (which is the edge between x and $p_i(x)$ in E_G) is then defined in $Inc(G)$ as the pair (x, i). Hence we can construct a $(k + 1)$-copying MS-transduction τ_2 that associates with $(H, \leq)$ given as $\langle V_G, edg_H, \leq \rangle$ the structure $T = \langle D_T, in_T \rangle$ such that

$$D_T := (V_G \times \{0\}) \cup \{(x, i) \mid x \in V_G, 1 \leq i \leq |Adj_H^-(x)|\} \subseteq V_G \times [0, k],$$
$$in_T := \{((x, i), (y, 0)) \mid (x, i) \in D_T, i \geq 1, y = p_i(x)\}$$
$$\cup \{((x, i), (x, 0)) \mid (x, i) \in D_T, i \geq 1\}.$$

It is convenient here to define D_T as a subset of $V_G \times [0, k]$ rather than of the usual set $V_G \times [k + 1]$. The set $V_G \times \{0\}$ is the copy of V_G in D_T. It is clear that T is isomorphic to $Inc(G)$, by an isomorphism that maps $(x, 0)$ to x (and that maps (x, i) to the edge between x and $p_i(x)$ for $i \geq 1$). It is straightforward to write a $(k + 1)$-copying domain-extending definition scheme $\mathcal{D}$ such that $\widehat{\mathcal{D}}(\langle V_G, edg_H, \leq \rangle) = T$. Hence, *Inc* is a domain-extending MS-transduction on the class US_k.

The case of binary relational structures: Let $S = \langle D_S, (R_S)_{R \in \mathcal{R}} \rangle \in US_k(\mathcal{R})$ and let $G := Adj(S) \in S^u$ be its adjacency graph (with $V_G = D_S$, cf. Proposition 9.14). Since $|E_G| \leq |T_S|$, the graph G is in US_k. Thus, by Propositions 9.42(1) and 9.43 there exists a domain-preserving MS-transduction $\overline{\tau}_1$ that transforms S into one or more structures $\langle D_S, (R_S)_{R \in \mathcal{R}}, edg_H, \leq \rangle$, where the structure $\langle V_H, edg_H, \leq \rangle$ satisfies the same conditions as in the above special case, with $V_H = V_G = D_S$.[12] For each $x \in D_S$, we define the enumeration $p_1(x), \ldots, p_l(x)$ of $Adj_H^-(x)$ as above.

Our goal is now to define a parameterless domain-extending MS-transduction $\bar{\tau}_2$ that transforms each such structure $\langle D_S, (R_S)_{R \in \mathcal{R}}, edg_H, \leq \rangle$ into $Inc(S)$. Clearly, it suffices to construct a structure U isomorphic to $Inc'(S)$ (the simplified version of $Inc(S)$, cf. Definition 9.10) by a domain-extending MS-transduction: by composing that transduction with the domain-preserving FO-transduction that transforms $Inc'(S)$ into $Inc(S)$ we get the result.

We define U with domain included in $D_S \times (\{0\} \cup ([2k+1] \times \mathcal{R}_2))$.

(1) Every element x of D_S corresponds to $(x, 0)$ in D_U.
(2) If $(x, x) \in R_S$, for $R \in \mathcal{R}_2$, then the triple (R, x, x) in T_S corresponds to $(x, (1, R))$ in D_U.

We now consider $(x, y) \in R_S$ with $x \neq y$, for $R \in \mathcal{R}_2$.

(3) If $x = p_i(y)$, then the triple (R, x, y) in T_S corresponds to $(y, (1 + i, R))$ in D_U.
(4) If $y = p_i(x)$, then (R, x, y) corresponds to $(x, (1 + k + i, R))$ in D_U.

In cases (2) and (4), we define $R_U((x, a))$ to hold (with a equal to $(1, R)$ or to $(1 + k + i, R)$); in case (3), we define $R_U((y, a))$ to hold (with a equal to $(1 + i, R)$). For a unary relation R in $\mathcal{R}_1$, we need only make sure that $R_U((x, 0))$ holds if and only if $R_S(x)$ holds.

It is easy to write the relation formulas that define in_1 and in_2. Note that

$$in_{1U} := \{((x, (1, R)), (x, 0)) \mid (x, (1, R)) \in D_U\}$$
$$\cup \{((y, (1 + i, R)), (x, 0)) \mid (y, (1 + i, R)) \in D_U, x = p_i(y)\}$$
$$\cup \{((x, (1 + k + i, R)), (x, 0)) \mid (x, (1 + k + i, R)) \in D_U\},$$

and similarly for in_{2U}.

The final transduction $\bar{\tau} := \bar{\tau}_1 \cdot \bar{\tau}_2$ is parameter-invariant.

The case of labeled graphs: For directed (K, Λ)-labeled graphs, we can use the construction of the previous case with Λ in place of $\mathcal{R}_2$ and K in place of $\mathcal{R}_1$. For undirected (K, Λ)-labeled graphs, the construction is essentially the same except that we need only use $[k+1]$ instead of $[2k+1]$. We omit the easy details. ∎

The proof of Proposition 9.43, which will be given after Proposition 9.54, will use a number of definitions and results of independent interest.

In what follows we will not distinguish between a simple directed labeled graph G and the corresponding structure $\lfloor G \rfloor$. In particular, for an arbitrary labeled graph G, we will not distinguish between the graph $Inc(G)$ and the structure $\lceil G \rceil$.

---

[12] To be precise, $\bar{\tau}_1 := Id_{\mathcal{R}} + Adj \cdot \tau_1$, where $Id_{\mathcal{R}}$ is the identity on $STR^c(\mathcal{R})$ and τ_1 is defined above (and for $+$ see Definition 9.35). Note that, without loss of generality, we assume that $edg \notin \mathcal{R}$.

Definition 9.45 (Normal and spanning forests) Let F_0 be a forest and R be a subset of V_{F_0} that consists of one vertex from each connected component of F_0. By Definition 2.13, $F := (F_0, R)$ is a rooted forest. The corresponding descendant relation is a partial order denoted by $\leq_F$; its set of maximal elements is R, also denoted by $roots_F$.

If, furthermore, F_0 is a subgraph of an undirected graph G, we say that F is a *rooted forest in G*. Such a forest is *spanning* if $V_{F_0} = V_G$. It is *normal*[13] (without being necessarily spanning) if every two vertices of F that are adjacent in G are comparable with respect to $\leq_F$. If F is normal and spanning, then we let $G(F)$ be the orientation of G defined as follows: an edge linking x and $y \neq x$ is directed from x to y if $y <_F x$ and from y to x if $x <_F y$.

Proposition 9.46 The transduction $und^{-1} : M^u \to M^d$ is an $MS_{2,2}$-transduction, and so is the transduction $und^{-1} \cap (M^u \times A^d)$ that defines the acyclic orientations of every graph in M^u.[14]

Proof: We first construct an $MS_{2,2}$-transduction $\alpha : M^u \to M^d$ such that $\alpha \subseteq und^{-1}$ and $M^u \subseteq Dom(\alpha)$. It will define one or more orientations of each undirected graph G. To be precise, it will define all orientations $G(F)$ where F is a normal spanning forest in G.

We let $\mathcal{D}$ be the domain-preserving definition scheme $\langle \chi, True, \theta_{in_1}, \theta_{in_2} \rangle$ with two parameters such that:

(1) $Inc(G) \models \chi(X, U)$ if and only if $X \subseteq E_G$, $U \subseteq V_G$ and $F := (F_0, U)$ is a normal spanning forest in G, where F_0 is the subgraph of G with edge set X and vertex set V_G;

(2) for every X, U satisfying (1) and every $e \in E_G$ with two end vertices x and y, if $y <_F x$, then:

    (a) $Inc(G) \models \theta_{in_1}(X, U, e, z)$ if and only if $z = x$; and

    (b) $Inc(G) \models \theta_{in_2}(X, U, e, z)$ if and only if $z = y$;

(3) for every X, U satisfying (1), if $Inc(G) \models \theta_{in_i}(X, U, w, z)$ for $i = 1$ or 2, then w is an edge of G (and z is an end vertex of w).

These conditions can easily be expressed in monadic second-order logic; for Condition (1) see Example 5.18.

Since every graph has a normal spanning forest (such a forest can be constructed by a depth-first traversal of each connected component), the transduction $\alpha := \widehat{\mathcal{D}}$ defines one or more orientations for all graphs in M^u. In the orientations constructed

---

[13] This notion extends that of a normal tree defined in [*Die]. See also Definition 2.13 and Example 2.56(6).

[14] Here, we can define *all* orientations of any given graph by an $MS_{2,2}$-transduction. The construction differs from that of Proposition 9.42(1) in which we defined one or more orientations by an $MS_{1,1}$-transduction. It is not possible to define all orientations of every simple undirected graph by a single $MS_{1,1}$-transduction, see Example 7.46.

by α, any two parallel edges are oriented in the same direction, but even for simple graphs, α does not produce all orientations. An additional transduction can reverse the orientations of an arbitrary set of edges, hence yield all orientations.

There exists a domain-preserving $MS_{2,2}$-transduction $\beta : M^d \to M^d$ that is equal to *und* $\cdot$ *und*$^{-1}$, i.e., that associates with every loop-free directed graph G the set of all orientations of *und*(G). This transduction uses a parameter intended to take for values any set of edges: the edges whose direction is to be reversed. Hence $\alpha \cdot \beta$ is the desired transduction. It uses three parameters.

The second assertion is an application of the Restriction Theorem (Theorem 7.16) since the class A^d is MS_2-definable with respect to M^d. ∎

Remark 9.47 This proposition yields the particular case of Proposition 9.43, where $k = 2$:

There is a pair (χ, θ) of MS formulas that defines, for every $G \in S_2^d$ (the set of orientations of indegree at most 2 of the simple undirected graphs), one or more partial orders on its vertex set that are linear on each set $Adj_G^-(x)$.

We sketch the proof. Consider $G \in S_2^d$ and let H be the undirected graph having the same vertices as G and having an edge e_x between u and $v \neq u$ for each x such that $u \to_G x$ and $v \to_G x$. The graph $Inc(H)$ is easily constructed from G by an MS-transduction: it uses a copy, say $(x, 2)$, of such a vertex x as the element of $Inc(H)$ representing the edge e_x of H linking u and v. By Proposition 9.46, an MS-transduction can define from $Inc(H)$ all (incidence graphs of) acyclic orientations of H. This can also be done by an MS-transduction taking G as input. The reflexive and transitive closure of each acyclic orientation (of H) yields an appropriate partial order on V_G. □

The following notion of (undirected) hypergraph will be used as a technical tool.[15] In such a hypergraph, a hyperedge is a nonempty set of vertices. The main application in this section will be to the hypergraph associated with a graph G in S_*^d: each nonempty set $Adj_G^-(x)$, for $x \in V_G$, is defined to be a hyperedge. In Definitions 9.53 and 9.57 below, we will define orientations of hypergraphs.

Definition 9.48 (Hypergraphs and forests in hypergraphs) (a) A *hypergraph* is a triple $H = \langle V_H, E_H, vert_H \rangle$ such that $V_H \cap E_H = \emptyset$ and $vert_H(e)$ is a nonempty subset of V_H for each e in E_H. The elements of E_H are the *hyperedges* of H. Those of V_H (respectively of $vert_H(e)$) are the *vertices* of H (respectively of e). The *rank* of a hyperedge e is the integer $|vert_H(e)|$. The *maximal rank* of H (respectively its *minimal rank*) is the maximal (respectively the minimal) rank of its hyperedges. We denote by $\mathcal{H}_m$ the set of hypergraphs of maximal rank at most m. An *m-hypergraph*

---

[15] We have used hypergraphs of a different type in Section 4.1.5.

is a hypergraph with all hyperedges of rank m. We say that H is *simple* if $vert_H(e) \neq vert_H(e')$ for $e \neq e'$.

The *incidence graph* of H is the simple directed bipartite graph $Inc(H) := \langle V_H \cup E_H, in_H \rangle$ such that $in_H := \{(e,v) \in E_H \times V_H \mid v \in vert_H(e)\}$. We will use it as a relational structure for representing H faithfully. This representation allows quantifications over sets of hyperedges (viewed as a structure it would be natural to denote it by $\lceil H \rceil$).

The following alternative, weaker representation, does not. For each positive integer m, we let edg^m be an m-ary relation symbol. If H has maximal rank m, we define $\lfloor H \rfloor := \langle V_H, edg_H^1, \ldots, edg_H^m \rangle$, where $edg_H^p := \{(x_1, \ldots, x_p) \mid vert_H(e) = \{x_1, \ldots, x_p\}$ for some $e \in E_H$ of rank $p\}$ for each p. This definition yields a faithful representation of the simple hypergraphs of maximal rank m.

(b) Let H be a hypergraph of minimal rank at least 2. A *forest in H*, or an *H-forest*, is a rooted forest F defined as a triple $\langle N_F, son_F, roots_F \rangle$ (where N_F is its set of nodes and $roots_F$ its set of roots) such that:

(i) $N_F \subseteq E_H$;
(ii) if e' is a son of e, then $|vert_H(e) \cap vert_H(e')| = 1$;
(iii) if $e, e' \in N_F$, $e \neq e'$ and $vert_H(e) \cap vert_H(e') \neq \emptyset$, then either e and e' are adjacent in F or they are two sons of some hyperedge e'' such that $vert_H(e'') \cap vert_H(e) = vert_H(e'') \cap vert_H(e') = vert_H(e) \cap vert_H(e').$[16]

Note that the binary relation son_F can be determined (in a unique way) from the sets N_F and $roots_F$ and the mapping $vert_H$.[17] We will do this by a monadic second-order formula in the proof of Proposition 9.50.

(c) Let F be an H-forest. We define $V(F)$ as the set of vertices of the hyperedges of H that belong to N_F. For $e \in N_F$, we let

$$
V(F,e) := \begin{cases} vert_H(e) & \text{if } e \text{ is a root of } F, \\ vert_H(e) - vert_H(e') & \text{if } e' \text{ is the father of } e. \end{cases}
$$

Hence $\{V(F,e) \mid e \in N_F\}$ is a partition of $V(F)$. We let $\preceq_F$ be the quasi-order on $V(F)$ defined as follows:

$x \preceq_F y$ if and only if $x \in V(F,e)$ and $y \in V(F,e')$ for some e, e' such that $e \leq_F e'$.

We write $x \prec_F y$ if $x \preceq_F y$ and we do not have $y \preceq_F x$; we let $x \equiv_F y$ if $x \preceq_F y \preceq_F x$, equivalently, if x and y belong to $V(F,e)$ for some e.

---

[16] By (ii), this implies that $|vert_H(e) \cap vert_H(e')| = 1$.

[17] The relation son_F can be determined in a top-down way on the trees of F. If e is a root, then e' is a son of e if and only if $e' \neq e$ and $vert_H(e') \cap vert_H(e) \neq \emptyset$. If e has sons $e_1, \ldots, e_n$, then e' is a son of e_i if and only if $e' \notin \{e, e_1, \ldots, e_n\}$ and $vert_H(e') \cap vert_H(e_i) \neq \emptyset$.

(d) An H-forest F is *spanning* if $vert_H(e) \cap V(F) \neq \emptyset$ for every $e \in E_H$. It is *normal* if, for every e in E_H that satisfies $|vert_H(e) \cap V(F)| \geq 2$, there exist x, y in $vert_H(e) \cap V(F)$ such that $x \neq y$ and $x \preceq_F y$.

Lemma 9.49 Every hypergraph H of minimal rank at least 2 has a normal and spanning H-forest.

Proof: We adapt the classical depth-first traversal algorithm for graphs. Let F be a rooted forest. A linear order $\leq$ on its set of nodes N_F is *depth-first* if, for all e, e', f in N_F:

(i) if $e' \leq_F e$, then $e \leq e'$;
(ii) if e and f are incomparable with respect to $\leq_F$, $e < f$ and $e' \leq_F e$, then $e' < f$.

The *maximal branch* of F is the path from a root to the node that is maximal with respect to $\leq$. By (i), that node is a leaf of F.

We will construct a finite increasing sequence of H-forests $F_1 \subseteq F_2 \subseteq \cdots \subseteq F_i \subseteq \cdots$ satisfying the following conditions:

(1) $roots_{F_1} \subseteq roots_{F_2} \subseteq \cdots \subseteq roots_{F_i} \subseteq \cdots$;
(2) $N_{F_1} = \{e_1\}$ and $N_{F_{i+1}} = N_{F_i} \cup \{e_{i+1}\}$ for some sequence $e_1, e_2, \ldots, e_i, \ldots$ of pairwise distinct hyperedges;
(3) the linear order $(e_1, \ldots, e_i)$ is depth-first in F_i.

This sequence of H-forests has length at most $|E_H|$ by (2) and we take its last element as the result of the construction. Note that the maximal branch of F_i is the path in F_i from a root to e_i.

We now describe the construction of the sequence $(F_i)_{i \geq 1}$. We define e_1 as any hyperedge and let F_1 consist only of it. Assuming $F_1, \ldots, F_i$ already constructed, we define F_{i+1} as follows:

Case 1: There exists $e \in E_H - N_{F_i}$ such that $|vert_H(e) \cap V(F_i)| = 1$. We let $j \leq i$ be the largest index such that $|vert_H(e) \cap V(F_i)| = |vert_H(e) \cap V(F_i, e_j)| = 1$ for some $e \in E_H - N_{F_i}$. We let e_{i+1} be one of these hyperedges e and define a forest F_{i+1} extending F_i by letting $N_{F_{i+1}}$ be $N_{F_i} \cup \{e_{i+1}\}$ and e_{i+1} a son of e_j. In this case, we let $roots_{F_{i+1}} := roots_{F_i}$.

Case 2: There is no hyperedge e satisfying Case 1 and there exists $e \in E_H - N_{F_i}$ such that $vert_H(e) \cap V(F_i) = \emptyset$. We take any of these hyperedges e as hyperedge e_{i+1} and define F_{i+1} by letting $N_{F_{i+1}} := N_{F_i} \cup \{e_{i+1}\}$ and $roots_{F_{i+1}} := roots_{F_i} \cup \{e_{i+1}\}$.

Case 3: Cases 1 and 2 are not applicable. This case terminates the construction. We return F_i as the desired forest F.

As an example, Figure 9.1 shows a hypergraph with hyperedges all of rank 3 (they are drawn as triangles). The algorithm constructs a normal and spanning forest $F = F_{11}$ with $N_F = \{1, 2, \ldots, 11\}$ (where e_i is denoted by i). It has roots 1 and 11, and

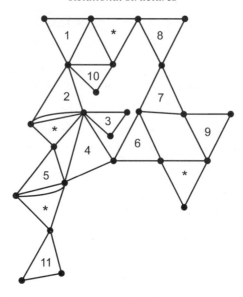

Figure 9.1 A hypergraph H with all hyperedges of rank 3.

directed edges $1 \to 2$, $2 \to 3$, $2 \to 4$, etc. (see Figure 9.2). The hyperedges marked *
in Figure 9.1 are not in N_F.

Going back to the general case, it is clear that each F_i is an H-forest, and that the
last one, say F_m, is spanning (otherwise, F_m could be extended by an application of
Case 1 or 2). We denote by $Touch(F_i)$ the set of hyperedges e of H not in N_{F_i} such
that $|vert_H(e) \cap V(F_i)| = 1$. We now prove that each forest of the sequence is normal.

Claim 9.49.1 For every $i \geq 1$:

(a) $(e_1, \ldots, e_i)$ is depth-first in F_i;
(b) F_i is a normal H-forest;
(c) if $e \in Touch(F_i)$, then the vertex in $vert_H(e) \cap V(F_i)$ belongs to $V(F_i, f)$ for some
    hyperedge f on the maximal branch of F_i.

Proof: We use induction on i. Properties (a), (b) and (c) are trivially true for F_1.

We first assume that F_{i+1} is constructed by Case 1. We let e_{i+1} and e_j be as in the
description of Case 1.

Property (c) that holds for F_i implies that e_j is on the maximal branch of F_i, and
this implies Property (a) for F_{i+1}.

We now prove that F_{i+1} is normal, i.e., satisfies Property (b). Let e in E_H be such
that $|vert_H(e) \cap V(F_{i+1})| \geq 2$. If two vertices of $vert_H(e)$ belong to $V(F_i)$, then the
conclusion holds because, by the induction hypothesis, F_i is normal. If e has only
one vertex in $V(F_i)$, say y, then it has another one, say x, in $vert_H(e_{i+1}) - V(F_i)$;
by Property (c) for F_i, $y \in V(F_i, f)$ for some f on its maximal branch, and f is

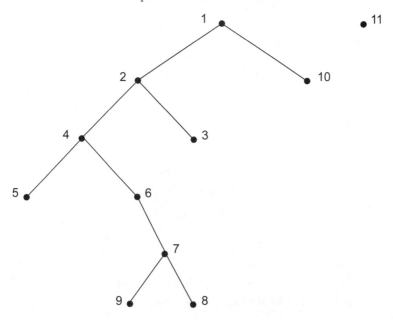

Figure 9.2 A normal and spanning forest in H.

above e_j by the choice of j, and thus $x \preceq_{F_{i+1}} y$. Otherwise, e has two vertices x,y in $vert_H(e_{i+1}) - V(F_i)$ and $x \equiv_{F_{i+1}} y$. This completes the proof that F_{i+1} satisfies Property (b).

It remains to prove that F_{i+1} satisfies Property (c). Let $e \in Touch(F_{i+1})$. If the vertex in $vert_H(e) \cap V(F_{i+1})$ belongs to $V(F_i)$, then, as above, by Property (c) for F_i, this vertex is in $V(F_i,f)$ for some f on its maximal branch, and f is above e_j. Hence, f is on the maximal branch of F_{i+1}. Otherwise, it belongs to $vert_H(e_{i+1}) - V(F_i)$, and we have the desired conclusion because e_{i+1} is on the maximal branch of F_{i+1}.

This completes the proof when F_{i+1} is constructed by Case 1.

We now consider F_{i+1} constructed by Case 2. Property (a) is clearly true from the definitions. The proof of Property (b) is the same as for Case 1, except that the hyperedge e cannot have only one vertex in $V(F_i)$, because otherwise Case 2 would not have been applicable for defining F_{i+1}. Property (c) is also clear because $Touch(F_i)$ is empty: if $e \in Touch(F_{i+1})$, the vertex in $vert_H(e) \cap V(F_{i+1})$ belongs to $vert_H(e_{i+1})$, and e_{i+1} is on the maximal branch of F_{i+1}.

This completes the proof of the claim.   □

Thus F_m, the last forest of the sequence, is normal by Property (b) of the claim, and the proof of the lemma is complete. ∎

We now prove that the notion of a normal and spanning H-forest can be formalized in monadic second-order logic.

Proposition 9.50 There exists a pair of MS formulas $(\chi(Y_1, Y_2), \theta(Y_1, Y_2, x_1, x_2))$ over the signature $\{in\}$ such that for every hypergraph H of minimal rank at least 2, and all sets $N, R \subseteq E_H$:

(i) $Inc(H) \models \chi(N, R)$ if and only if there exists a normal and spanning H-forest F such that $N_F = N$ and $roots_F = R$;

(ii) for every (N, R) satisfying $\chi(N, R)$ and every $e, e' \in N$, we have $Inc(H) \models \theta(N, R, e, e')$ if and only if $e \leq_F e'$ where F is the H-forest such that $N_F = N$ and $roots_F = R$.[18]

Proof: Our first objective is to construct a formula $\theta(Y_1, Y_2, x_1, x_2)$ such that, if F is an H-forest in a hypergraph H of minimal rank at least 2, then, for every $e, e' \in N_F$,

$$Inc(H) \models \theta(N_F, roots_F, e, e') \text{ if and only if } e \leq_F e'.$$

If H is a hypergraph, we say that $X \subseteq E_H$ is *connected* if the graph $G(X)$ with vertex set X and edge set $\{(e, e') \mid vert_H(e) \cap vert_H(e') \neq \emptyset\}$ is connected. With this notion, we can characterize certain paths in an H-forest F. For $e \in N_F$, the nodes of the unique path in F from a root to e form the set $P(e) \subseteq N_F$ characterized as follows:

$P(e)$ is the minimal connected subset of E_H that contains e and some root,

where minimality is understood with respect to set inclusion.[19] By using these paths, we can characterize the partial order $\leq_F$ by

$$e \leq_F e' \text{ if and only if } P(e') \subseteq P(e).$$

From these observations, the formula θ is easy to construct. We can use it to construct similar formulas that express that e is a son of e' in F, that $x \in V(F, e)$ and that $x \preceq_F y$ for $x, y \in V_H$. Note that e is a son of e' in F if and only if $e <_F e'$ and $vert_H(e) \cap vert_H(e') \neq \emptyset$.

Using these formulas it is also easy to write the formula χ. It expresses, for arbitrary subsets N and R of E_H, that the tuple $\langle N, son, R \rangle$ is a normal and spanning H-forest, where *son* is the set of all $(e', e) \in N \times N$ such that $e' \neq e$, $Inc(H) \models \theta(N, R, e, e')$ and $vert_H(e) \cap vert_H(e') \neq \emptyset$. ∎

Definition 9.51 (Splittings of hypergraphs) A *splitting* of a hypergraph H is a pair (H_1, H_2) of hypergraphs related to H as follows:

(a) $V_{H_1} = V_{H_2} = V_H$;

---

[18] As observed in Definition 9.48(b), there is at most one H-forest F such that $N_F = N$ and $roots_F = R$.

[19] Clearly, $P(e)$ is connected by Definition 9.48(b)(ii). If X is a connected subset of E_H that contains e and some root r, then there is a path π in $G(X)$ from r to e. By Definition 9.48(b)(iii), the path π can contain nodes of F that are not in $P(e)$. If f is such a node, then π has a subpath of the form $h_1 - h_2 - \cdots - f - \cdots - h_q$ such that h_1 and h_q are in $P(e)$ and $h_2, \ldots, f, \ldots, h_q$ are sons of h_1 that, except h_q, are not in $P(e)$. All nodes of $P(e)$ are on the path π, hence in X.

(b) $E_{H_1} \cup E_{H_2} = E_H$;

(c) if $e \in E_{H_1} \cap E_{H_2}$, then $vert_H(e) = vert_{H_1}(e) \cup vert_{H_2}(e)$ and $vert_{H_1}(e) \cap vert_{H_2}(e) = \emptyset$;

(d) if $e \in E_{H_1} - E_{H_2}$, then $vert_H(e) = vert_{H_1}(e)$;

(e) if $e \in E_{H_2} - E_{H_1}$, then $vert_H(e) = vert_{H_2}(e)$.

A splitting (H_1, H_2) of H will be represented by the relational structure $S := \langle V_H \cup E_H, in_S^1, in_S^2 \rangle$ such that $in_S^i := \{(e,v) \mid e \in E_{H_i}, v \in vert_{H_i}(e)\}$. We denote it by $S(H_1, H_2)$ if we need to refer to H_1 and H_2. Note that the domain of S is that of $Inc(H)$ and that in_S^1 and in_S^2 form a partition of in_H. We have $Inc(H_i) = \langle V_H \cup E_i, in_S^i \rangle$, where $E_i := \{e \in E_H \mid in_S^i(e,u)$ for some $u \in V_H\}$. Hence, $Inc(H_1)$ and $Inc(H_2)$ can be obtained from $S(H_1, H_2)$ by FO-transductions.

Let $\mathcal{C}$ be a set of hypergraphs. A triple of MS formulas $(\chi(\mathcal{X}_n), \theta_1(\mathcal{X}_n, x_1, x_2), \theta_2(\mathcal{X}_n, x_1, x_2))$, over the relational signature $\{in\}$, *defines splittings of every hypergraph in $\mathcal{C}$* if, for every hypergraph H in $\mathcal{C}$,

$$Inc(H) \models (\exists X_1, \ldots, X_n. \chi) \wedge \forall X_1, \ldots, X_n (\chi \Rightarrow \theta),$$

where θ is the formula with free variables in $\mathcal{X}_n$ defined as

$$\forall y, x[in(y,x) \Leftrightarrow (\theta_1(y,x) \vee \theta_2(y,x))] \wedge \forall y, x. \neg(\theta_1(y,x) \wedge \theta_2(y,x)).$$

These conditions mean that for every tuple $(U_1, \ldots, U_n)$ of subsets of $V_H \cup E_H$ satisfying χ in $Inc(H)$, if $W_i := \{(u,v) \mid u, v \in V_H \cup E_H, Inc(H) \models \theta_i(U_1, \ldots, U_n, u, v)\}$, then the structure $\langle V_H \cup E_H, W_1, W_2 \rangle$ represents a splitting of H. Furthermore, for each hypergraph H in $\mathcal{C}$ there is such an n-tuple.

Equivalently, they mean that the domain-preserving MS-transduction with definition scheme $\langle \chi, True, \theta_1, \theta_2 \rangle$ associates with every incidence graph $Inc(H)$, where H is a hypergraph in $\mathcal{C}$, one or more structures $S(H_1, H_2)$ representing some of its splittings.

Recall that $\mathcal{H}_m$ denotes the set of hypergraphs of maximal rank at most m.

Lemma 9.52 For each $m \geq 2$, there exists a triple of MS formulas $(\chi(\mathcal{X}_{m+2}), \theta_1(\mathcal{X}_{m+2}, x_1, x_2), \theta_2(\mathcal{X}_{m+2}, x_1, x_2))$ that defines splittings of every hypergraph in $\mathcal{H}_m$ into pairs of hypergraphs in $\mathcal{H}_{m-1}$.

Proof: We have to show that there is a domain-preserving MS-transduction that associates with $Inc(H)$, for every hypergraph H of maximal rank at most m, one or more structures $S(H_1, H_2)$ representing splittings of H into hypergraphs H_1, H_2 of maximal rank at most $m - 1$.

Let H be of maximal rank at most m. We will put in H_1 all hyperedges of rank 1. We continue the construction for H' obtained from H by deleting the hyperedges of rank 1.

Let F be a normal and spanning H'-forest (Lemma 9.49) and let $\overline{U} = (U_1, \ldots, U_m)$ be a partition of $V(F)$ such that for every $e \in N_F$, distinct vertices of $V(F,e)$ belong to distinct sets U_i. We refine the quasi-order $\preceq_F$ on the set $V(F)$ defined in Definition 9.48(c) into the partial order $\preceq_{F,\overline{U}}$ defined as follows:

$$x \preceq_{F,\overline{U}} y \text{ if and only if}$$

$$\text{either } x \prec_F y, \text{ or } x \equiv_F y \text{ and } x \in U_i, y \in U_j \text{ with } i \leq j.$$

For every $e \in E_{H'}$, we let $Min(e)$ be the set of minimal elements of $vert_{H'}(e) \cap V(F)$ with respect to $\preceq_{F,\overline{U}}$; this set is not empty since $vert_{H'}(e) \cap V(F) \neq \emptyset$ (because F is spanning).

Claim 9.52.1 For every $e \in E_{H'}$, the set $Min(e)$ is a proper subset of $vert_{H'}(e)$.

Proof: Let $e \in E_{H'}$, so $|vert_{H'}(e)| \geq 2$. If $vert_{H'}(e) \cap V(F) = \{x\}$, then $Min(e) = \{x\}$ and $Min(e)$ is a proper subset of $vert_{H'}(e)$. Otherwise, since F is normal, the set $vert_{H'}(e) \cap V(F)$ contains two elements x and y such that $x \preceq_F y$. If $x \prec_F y$, then $y \notin Min(e)$. If $x \equiv_F y$, then either $x \prec_{F,\overline{U}} y$ and $y \notin Min(e)$, or $y \prec_{F,\overline{U}} x$ and $x \notin Min(e)$. In all cases, $Min(e)$ is a proper subset of $vert_{H'}(e)$. $\quad\square$

To complete the proof of the lemma, we construct a domain-preserving MS-transduction that has a definition scheme $\mathcal{D} := \langle \chi, True, \theta_1, \theta_2 \rangle$ with parameters $X_1, \ldots, X_{m+2}$. The formulas of $\mathcal{D}$ are constructed using the pair of formulas of Proposition 9.50. The precondition χ expresses that these parameters are intended to take values as follows, for input $Inc(H)$:

X_1, X_2 will determine a normal and spanning H'-forest F such that N_F is the value of X_1 and $roots_F$ is the value of X_2; the sets $X_3, \ldots, X_{m+2}$ will define a partition $\overline{U}$ associated with F as described above.

The partial order $\preceq_{F,\overline{U}}$ can be expressed by an MS formula $\alpha(\mathcal{X}_{m+2}, x, y)$ from which we can build a formula $\alpha'(\mathcal{X}_{m+2}, e, x)$ expressing, for $F, \overline{U}$ defined by $X_1, \ldots, X_{m+2}$, that $e \in E_{H'}$ and $x \in Min(e)$. One obtains in this way a splitting (H_1', H_2) of H' such that $E_{H_1'} = E_{H_2} = E_{H'}$ and, for $e \in E_{H'}$:

$$vert_{H_1'}(e) := Min(e), \text{ and}$$
$$vert_{H_2}(e) := vert_{H'}(e) - Min(e).$$

We turn it into a splitting (H_1, H_2) of H such that H_1 is obtained by adding to H_1' the hyperedges of rank 1 of H. By the claim, if e has rank $p \in [2, m]$, then $|Min(e)| \leq p - 1$ and $|vert_H(e) - Min(e)| \leq p - 1$. It follows that H_1 and H_2 have maximal ranks at most $m - 1$.

We define the MS formulas θ_1 and θ_2 with free variables in $\mathcal{X}_{m+2} \cup \{x_1, x_2\}$ as follows:

$$\theta_1 \quad \text{is} \quad \alpha'(\mathcal{X}_{m+2}, x_1, x_2) \vee \forall y (in(x_1, y) \Leftrightarrow y = x_2), \quad \text{and}$$

$$\theta_2 \quad \text{is} \quad in(x_1, x_2) \wedge \neg \theta_1.$$

These formulas define respectively the relations in_S^1 and in_S^2 of $S := S(H_1, H_2)$. Hence, we have an MS-transduction transforming $Inc(H)$ into one or more structures $S(H_1, H_2)$ as desired. ∎

An undirected graph is a hypergraph of maximal rank at most 2. Splitting it is nothing but choosing an orientation for each of its edges that are not loops. The construction of Lemma 9.52 is thus a generalization of the construction of the transduction α in the proof of Proposition 9.46.

Definition 9.53 (Orientations of hypergraphs) Let $H = \langle V_H, E_H, vert_H \rangle$ be a hypergraph. An *orientation* of H is a pair $(H, \leq)$, where $\leq$ is a mapping that associates with every e in E_H a linear order $\leq_e$ on its set of vertices $vert_H(e)$. If $\overline{H}$ is an orientation of H, we let $\sqsubseteq_{\overline{H}}$ be the binary relation on V_H defined as the reflexive and transitive closure of the union of the relations $\leq_e$ for all $e \in E_H$. We say that $\overline{H}$ is *acyclic* if $\sqsubseteq_{\overline{H}}$ is antisymmetric, i.e., if it is a partial order. This is equivalent to requiring that the linear orders $\leq_e$ have a common linear extension, and also, to the acyclicity of the directed loop-free graph G such that $V_G = V_H$ and $edg_G = \bigcup \{(x,y) \mid x <_e y$ for some e in $E_H\}$.

An orientation $\overline{H}$ of H can be identified with the tuple $\langle V_H, E_H, vert_{\overline{H}} \rangle$ such that, for every $e \in E_H$, $vert_{\overline{H}}(e)$ is the enumeration of $vert_H(e)$ in increasing order with respect to $<_e$. It can be faithfully represented by the following relational structure that can be considered as a refinement of $Inc(H)$. We define

$$Inc(\overline{H}) := \langle V_H \cup E_H, order_{\overline{H}} \rangle,$$

where

$$order_{\overline{H}} := \{(e, x, y) \mid x, y \in vert_H(e), x \leq_e y\}.^{20}$$

This ternary relation can replace in_H because $(e, x, x) \in order_{\overline{H}}$ if and only if $x \in vert_H(e)$.

We will say that $\overline{H}$ as above is *simple* if $vert_{\overline{H}}(e) \neq vert_{\overline{H}}(e')$ for $e \neq e'$. If H has maximal rank m, then we define $\lfloor \overline{H} \rfloor := \langle V_H, edg_{\overline{H}}^1, \ldots, edg_{\overline{H}}^m \rangle$ where, for each p, we let $edg_{\overline{H}}^p := \{(x_1, \ldots, x_p) \mid vert_{\overline{H}}(e) = (x_1, \ldots, x_p)$ for some $e \in E_H$ of rank $p\}$. It is

---

[20] For an undirected graph H, which is a hypergraph of maximal rank at most 2, the notion of an orientation $\overline{H}$ and its incidence graph $Inc(\overline{H})$ were defined earlier, in a slightly different way. It should be clear, however, that this difference can be disregarded: structures $\langle V \cup E, in_1, in_2 \rangle$ and $\langle V \cup E, order \rangle$, such that $order = \{(e, x, y) \mid (e, x) \in in_1, (e, y) \in in_2\}$ and $in_1 = \{(e, x) \mid (e, x, y) \in order$ for some $y \in V\}$ (and similarly for in_2), can be transformed into each other by domain-preserving FO-transductions.

a faithful representation of simple orientations of hypergraphs of maximal rank m. These definitions extend those of Definition 9.48(a).

As for splittings, we define how orientations can be determined by formulas. A pair of MS formulas $(\mu(\mathcal{X}_n), \varphi(\mathcal{X}_n, x_1, x_2, x_3))$, over the relational signature $\{in\}$, *defines orientations of every hypergraph in* a set C if, for every hypergraph $H \in C$,

$$Inc(H) \models (\exists X_1, \ldots, X_n \cdot \mu) \wedge \forall X_1, \ldots, X_n (\chi \Rightarrow \widehat{\varphi}),$$

where $\widehat{\varphi}$ is the formula with free variables in $\mathcal{X}_n$ defined as

$$\forall e, x \cdot \varphi(e, x, x) \wedge \forall e, x, y (\varphi(e, x, y) \wedge \varphi(e, y, x) \Rightarrow x = y)$$
$$\wedge \forall e, x, y, z (\varphi(e, x, y) \wedge \varphi(e, y, z) \Rightarrow \varphi(e, x, z))$$
$$\wedge \forall e, x, y [(in(e, x) \wedge in(e, y)) \Leftrightarrow (\varphi(e, x, y) \vee \varphi(e, y, x))].$$

This means that, for every n-tuple $(U_1, \ldots, U_n)$ of subsets of $V_H \cup E_H$ satisfying μ in $Inc(H)$, if $W := \{(e, u, v) \mid e, u, v \in V_H \cup E_H, Inc(H) \models \varphi(U_1, \ldots, U_n, e, u, v)\}$, then the structure $\langle V_H \cup E_H, W \rangle$ represents an orientation of H, and that there is at least one such n-tuple in $Inc(H)$. Equivalently, it means that the domain-preserving MS-transduction with definition scheme $\langle \mu, True, \varphi \rangle$ associates with every incidence graph $Inc(H)$, where H is a hypergraph in C, one or more incidence structures $Inc(\overline{H})$ representing some orientations $\overline{H}$ of H.

We will also consider similar formulas that define all orientations of H in C, and all its acyclic orientations.

The next result generalizes Proposition 9.46.

Proposition 9.54 For each integer $m \geq 2$:
(1) there exists a pair of MS formulas $(\mu(\mathcal{X}_n), \varphi(\mathcal{X}_n, x_1, x_2, x_3))$ that defines orientations of every hypergraph in $\mathcal{H}_m$, i.e., there exists a domain-preserving MS-transduction that defines one or more orientations of every hypergraph of maximal rank at most m, where hypergraphs and their orientations are represented by their incidence structures;[21]
(2) there exist domain-preserving MS-transductions that define all orientations and all acyclic orientations of every hypergraph of maximal rank at most m, where again, hypergraphs and their orientations are represented by their incidence structures.

Proof: (1) The construction is by induction on m. Since $\mathcal{H}_2$ is the set of undirected graphs, the case $m = 2$ follows easily from Proposition 9.46. The corresponding formulas use two set variables.

---

[21] Generalizing the terminology of Section 5.2.5 in an obvious way, such a transduction could be called an $MS_{2,2}$-transduction.

For $m > 2$, we will construct $(\mu(\mathcal{X}_n), \varphi(\mathcal{X}_n, x_1, x_2, x_3))$ by using the corresponding pair of formulas for the case $m - 1$.

We first give the idea. Let H belong to $\mathcal{H}_m$, and be given by its incidence graph $Inc(H)$. By Lemma 9.52, a structure $S(H_1, H_2)$ representing a splitting (H_1, H_2) of H with H_1 and H_2 in $\mathcal{H}_{m-1}$, is defined by monadic second-order formulas using an $(m + 2)$-tuple $\mathcal{U}$ of parameters. We can use on $Inc(H_1)$ and $Inc(H_2)$ (definable in $S := S(H_1, H_2)$ by first-order formulas) the formulas μ', φ' obtained by induction, that define orientations of H_1 and H_2. By combining these orientations, which we define below by the local linear orders $\leq_{1,e}$ and $\leq_{2,e}$, we can define an orientation of H as follows. For $e \in E_H$ and $x, y \in vert_H(e)$, we let

$$x \leq_e y \text{ if and only if } \quad x, y \in vert_{H_1}(e) \text{ and } x \leq_{1,e} y \text{ in } H_1, \text{ or}$$
$$x, y \in vert_{H_2}(e) \text{ and } x \leq_{2,e} y \text{ in } H_2, \text{ or}$$
$$x \in vert_{H_1}(e) \text{ and } y \in vert_{H_2}(e).$$

We now explain how $\mathcal{X}$, μ and φ can be constructed. We will use pairwise disjoint tuples $\mathcal{U}$, $\mathcal{Y}$ and $\mathcal{Z}$ of set variables (parameters). We let τ be the domain-preserving MS-transduction with definition scheme $\langle \chi, True, \theta_1, \theta_2 \rangle$, where the triple $(\chi(\mathcal{U}), \theta_1(\mathcal{U}, x_1, x_2), \theta_2(\mathcal{U}, x_1, x_2))$ defines splittings $S(H_1, H_2)$ of every hypergraph H in $\mathcal{H}_m$ by Lemma 9.52. For $i = 1, 2$ we denote by τ_i the composition of τ with the parameterless noncopying FO-transduction that transforms $S(H_1, H_2)$ into H_i.

We let $(\mu'(\mathcal{Y}), \varphi'(\mathcal{Y}, x_1, x_2, x_3))$ be a pair of formulas that defines orientations of every hypergraph in $\mathcal{H}_{m-1}$; such formulas do exist by the induction hypothesis. We let μ_1 and φ_1 be the backwards translations of μ' and φ' relative to τ_1, cf. Section 7.1.4; they define an orientation of H_1. We let $\mathcal{Z}$ be a disjoint copy of $\mathcal{Y}$, $\mu''(\mathcal{Z})$ and $\varphi''(\mathcal{Z}, x_1, x_2, x_3)$ be obtained from μ' and φ' by the corresponding substitution of variables and finally, we let μ_2 and φ_2 be the backwards translations of μ'' and φ'' relative to τ_2; they define an orientation of H_2.

We will use the tuple of parameters $\mathcal{X}_n := \mathcal{U} \cdot \mathcal{Y} \cdot \mathcal{Z}$ and the precondition μ defined as $\chi(\mathcal{U}) \wedge \mu_1(\mathcal{Y}) \wedge \mu_2(\mathcal{Z})$. The following formula φ expresses the above definition of $\leq_e$ in terms of $\leq_{1,e}$ and $\leq_{2,e}$. We define it as

$$\varphi_1(\mathcal{Y}, x_1, x_2, x_3) \vee \varphi_2(\mathcal{Z}, x_1, x_2, x_3) \vee [\varphi_1(\mathcal{Y}, x_1, x_2, x_2) \wedge \varphi_2(\mathcal{Z}, x_1, x_3, x_3)],$$

so that we have the desired pair (μ, φ).

(2) For each m, there is an MS-transduction β_m that transforms $Inc(\overline{H})$ into the incidence graphs representing all the orientations of H. Its construction generalizes the construction of the $MS_{2,2}$-transduction β that defines $und \cdot und^{-1}$ (in the proof of Proposition 9.46). Its definition uses $m! - 1$ parameters to specify, for each $e \in E_H$, if and how its linear order $\leq_e$ on $vert_H(e)$ must be permuted into another linear order.

The construction of all acyclic orientations follows then by the Restriction Theorem (Theorem 7.16), because the property of an orientation to be acyclic is MS-expressible (due to the fact that the acyclicity of a directed graph is MS-expressible). We omit the routine details. ∎

Proof of Proposition 9.43: We will denote by ω_m the transduction that constructs all acyclic orientations of the hypergraphs of maximal rank at most m, see Proposition 9.54(2). Note that every hypergraph has at least one acyclic orientation: take any linear order of all its vertices.

Let G be a directed graph in S_k^d, i.e., such that each set $Adj_G^-(x)$ defined as $\{y \mid y \to_G x\}$ has at most k elements. We let H be the hypergraph of maximal rank at most k such that

$$V_H := V_G \times \{1\},$$
$$E_H := \{(x,2) \mid x \in V_G, Adj_G^-(x) \neq \emptyset\},$$
$$vert_H((x,2)) := \{(y,1) \mid y \in Adj_G^-(x)\} \text{ for every } (x,2) \in E_H.$$

We will identify V_H and V_G in what follows. The mapping γ that transforms every graph $G \in S_k^d$ into $Inc(H)$ is clearly a parameterless 2-copying FO-transduction. (We recall that for simple graphs G, we identify $\lfloor G \rfloor$ and G.) The MS-transduction $\gamma \cdot \omega_k$ transforms such a graph G into $Inc(\overline{H})$, where $\overline{H}$ is an acyclic orientation of H. It uses a (large!) tuple of parameters, say $\mathcal{X}$; let $\chi(\mathcal{X})$ be its precondition.

We let $\nu(x_1,x_2)$ be the MS formula that defines, in every structure $Inc(\overline{H})$ representing an oriented hypergraph, the binary relation $\sqsubseteq_{\overline{H}}$ on V_H; this relation is a partial order if and only if $\overline{H}$ is acyclic.

We let $\theta(\mathcal{X},x_1,x_2)$ be the backwards translation of $\nu(x_1,x_2)$ relative to $\gamma \cdot \omega_k$, cf. Section 7.1.4. The pair (χ,θ) satisfies the required conditions: for every $\mathcal{X}$-assignment in $G \in S_k^d$ that satisfies χ, the formula θ defines in G a partial order on V_G that induces an orientation $\overline{H}$ of the corresponding hypergraph H, and we recall that $V_H = V_G$. Hence, this partial order is linear on each set $Adj_G^-(x)$ by the definition of H. ∎

The proof of Proposition 9.43 is now complete, and thus Theorem 9.44 (the Sparseness Theorem for graphs and binary structures) is also proved.

Example 9.55 Here is a set $\mathcal{C}$ of simple directed graphs that is not uniformly k-sparse for any k, but on which the mapping *Inc* is an FO-transduction.[22]

For each $n \geq 1$, we let G_n be the graph with vertex set consisting of the elements of $[n]$ and of its subsets of cardinality 2. Its edges are $i \to j$, $i \to \{i,j\}$ and $j \to \{i,j\}$ for $1 \leq i < j \leq n$. These graphs are 3-sparse, but not uniformly k-sparse because the subgraphs of G_n induced by $[n]$ are not. It is easy to see that *Inc* is a domain-extending 4-copying FO-transduction on the class $\mathcal{C} := \{G_n \mid n \geq 1\}$: a vertex x is represented by $(x,1)$, an edge $i \to j$ by $(\{i,j\},2)$, an edge $i \to \{i,j\}$ by $(\{i,j\},3)$ and an edge $j \to \{i,j\}$ by $(\{i,j\},4)$, where in all these cases $i < j$. □

---

[22] This example answers Problem 5.1 of [Blu10].

In the next section, we will extend Theorem 9.44 to relational structures that are not binary, in order to prove the Sparseness Theorem (Theorems 9.37 and 9.38). Before doing so, we apply some techniques developed in this section to the definition of optimal tree-decompositions of chordal graphs by MS-transductions.

We recall from Example 5.2(4) that every graph G has tree-decompositions (T, f), and even optimal ones, such that $N_T = V_G$. In such a case, the relational structure $S_1(G, T, f) := \langle V_G, edg_G, son_T, box_f \rangle$ represents simultaneously G and (T, f) (with $(u, x) \in box_f$ if and only if $x \in f(u)$). We have observed that, in some cases, we can even omit the relation box_f and use $S_2(G, T, f) := \langle V_G, edg_G, son_T \rangle$ because box_f (i.e., the mapping f) can be determined from edg_G and son_T by a monadic second-order formula. This is the case if (T, f) is a normal tree-decomposition (cf. Example 2.56(6)). We now show that optimal normal tree-decompositions of a chordal graph G can be defined by an MS-transduction that takes $\lceil G \rceil$ as input, as already discussed in Example 5.18 (with a slightly different proof). For k-chordal graphs, $\lfloor G \rfloor$ can be taken as input.

Proposition 9.56

(1) There exists an MS-transduction that associates with $Inc(G)$, for every chordal graph G, one or more structures of the form $S_1(G, T, f)$ such that (T, f) is a normal tree-decomposition of G that is optimal.

(2) For each $k \geq 2$, there exists an MS-transduction that associates with every k-chordal graph G one or more structures of the form $S_1(G, T, f)$ such that (T, f) is a normal tree-decomposition of G that is optimal.

Proof: (1) We recall from Proposition 2.72 that a nonempty, connected, simple, loop-free and undirected graph G is chordal if and only if it has a simplicial orientation, i.e., an acyclic orientation H such that for every vertex u, the set $Adj_H^-(u)$ is a clique in G. According to the proof of (3) $\Longrightarrow$ (4) of Proposition 2.72, from such an orientation H, the binary relation on $V_H = V_G$ defined as

$$\{(w, u) \in edg_H \mid \text{ there is no directed path in } H \text{ of length at least 2 from } w \text{ to } u\},$$

is son_T for some rooted tree T that is a normal spanning tree of H and also of G. Following the proof of (4) $\Longrightarrow$ (1) of Proposition 2.72, we define also, for every $u \in N_T = V_G$,

$$f(u) := \{u\} \cup \{w \in V_G \mid (u, w) \in edg_G \text{ and } (w, u) \in son_T^*\},$$

and we obtain a normal tree-decomposition of G that is optimal because every box is a clique.

We let $S_1(G, T, f)$ be the corresponding relational structure (it depends on the orientation H of G). By Proposition 9.46 there exists an MS-transduction τ that associates with $Inc(G)$, for every undirected graph G, all its orientations. That an orientation H is simplicial is MS-expressible. Hence, by the Restriction Theorem (Theorem 7.16), there exists an MS-transduction τ' that constructs (from $Inc(G)$) the simplicial orientations of G, whenever there exist some, which is the case if G is chordal. From a simplicial orientation of G, the relations son_T and box_f are definable by monadic second-order formulas: this is clear from their definitions above. We get in this way the desired transduction.

(2) The transduction τ' defined above in (1) takes as input $Inc(G)$, and not G itself, in order to be able to specify all orientations of G. If G is k-chordal, its simplicial orientations are of indegree at most $k-1$ and, by Proposition 9.42(1), they can be specified by an MS-transduction using $q(k-1)$ parameters. This gives the stated result for k-chordal graphs.   ∎

Hence optimal tree-decompositions (T,f) of k-chordal graphs G (for fixed k) can be specified by structures $S_1(G,T,f)$ that can be defined "inside" $\lfloor G \rfloor$. As already said in Section 7.6, the extension of this result to graphs of tree-width at most k is an open question.

9.4.2 Uniformly k-sparse relational structures

We recall from Definition 9.36 that a relational structure S is uniformly k-sparse if $|T_{S[X]}| \leq k \cdot |X|$ for every subset X of D_S, where, by Definition 9.10, T_S is the set of tuples of a structure S. Similarly, a hypergraph H is *k-sparse* if $|E_H| \leq k \cdot |V_H|$, and it is *uniformly k-sparse* if, for every $X \subseteq V_H$, the induced hypergraph $H[X]$ is k-sparse (the hyperedges of $H[X]$ are those of H with all their vertices in X). For $k \geq 1$ and $m \geq 2$, we let $US_{k,m}$ denote the class of uniformly k-sparse simple m-hypergraphs. Hence, we have $US_{k,2} = US_k$.

In this section we prove the Sparseness Theorem (Theorem 9.38). A large part of the proof consists in showing that the mapping Inc is a domain-extending MS-transduction on the class $US_{k,m}$ (Theorem 9.61). Before giving an overview of the proof, we need a definition and a proposition.

Definition 9.57 (Semi-orientations of hypergraphs) A *semi-orientation* of a hypergraph H is a pair $\overrightarrow{H} = (H,hd)$, where hd is a mapping $E_H \rightarrow V_H$ such that $hd(e) \in vert_H(e)$ for every $e \in E_H$. We call $hd(e)$ the *head* of e. The *indegree* of a vertex v in $\overrightarrow{H}$ is the number $indeg_{\overrightarrow{H}}(v)$ of hyperedges whose head is v. The *indegree* of $\overrightarrow{H}$ is the maximal indegree of its vertices.

We let $Dir(\overrightarrow{H})$ be the simple loop-free directed graph D such that $V_D := V_H$ and $edg_D := \{(x,y) \mid x \in vert_H(e) \text{ and } y = hd(e) \neq x \text{ for some } e \in E_H\}$. Clearly, the indegree of $Dir(\overrightarrow{H})$ is at most $m-1$ times the indegree of $\overrightarrow{H}$.

If H has maximal rank m, then we define the relational structure $\lfloor \overrightarrow{H} \rfloor$ similar to $\lfloor H \rfloor$ (cf. Definition 9.48(a)) by using, for each $p \in [m]$, the p-ary relation $edg_{\overrightarrow{H}}^p$ such that

$(x_1, \ldots, x_p) \in edg^p_{\overrightarrow{H}}$ if and only if

$vert_H(e) = \{x_1, \ldots, x_p\}$ for some $e \in E_H$ of rank p such that $x_p = hd(e)$,

instead of edg^p_H. It is a faithful representation of semi-orientations of simple hypergraphs of maximal rank m.

We prove two combinatorial properties of semi-orientations; the first generalizes Proposition 9.40(1).

Proposition 9.58 Let $k \geq 1$ and $m \geq 2$.

(1) A hypergraph is uniformly k-sparse if and only if it has a semi-orientation of indegree at most k.
(2) A hypergraph H of maximal rank at most m having a semi-orientation $\overrightarrow{H}$ of indegree at most k has a semi-orientation $\overrightarrow{K}$ of indegree at most mk^2 such that $Dir(\overrightarrow{K}) \in S^d_*$, whence $Dir(\overrightarrow{K}) \in S^d_{k'}$ where $k' := (m-1)mk^2$.

Proof: (1) The "if" part is clear since for every hypergraph H with a semi-orientation $\overrightarrow{H}$, we have

$$|E_H| = \sum_{v \in V_H} indeg_{\overrightarrow{H}}(v).$$

"Only if." We generalize the proof of this result for the particular case of undirected graphs that is given in Theorem 6.13 of [*Fra]. We have already used this case in the proof of Proposition 9.42. We first observe that, by the previous equality, if a hypergraph H is k-sparse and has a semi-orientation $\overrightarrow{H}$, then $\sum_{v \in V_H} indeg_{\overrightarrow{H}}(v) \leq k \cdot |V_H|$, so that we must have $indeg_{\overrightarrow{H}}(w) < k$ for some w if, for some v, we have $indeg_{\overrightarrow{H}}(v) > k$.

Consider H uniformly k-sparse. Let $\overrightarrow{H} = (H, hd)$ be any semi-orientation of H. We say that a vertex v is *bad* if $indeg_{\overrightarrow{H}}(v) > k$. We let the *badness* of $\overrightarrow{H}$ be the sum of the indegrees $indeg_{\overrightarrow{H}}(v)$ such that v is bad. We are looking for a semi-orientation of badness 0.

Let the chosen semi-orientation $\overrightarrow{H}$ have positive badness: we will transform it into a semi-orientation (H, hd') of smaller badness. Let v be a bad vertex. Let X be the set of vertices u such that there is a directed path in $Dir(\overrightarrow{H})$ from u to v. Then, the induced hypergraph $H[X]$ is k-sparse. Note that if $hd(e) \in X$, then $vert_H(e) \subseteq X$, hence the indegree of a vertex in $H[X]$ with respect to the semi-orientation of $H[X]$ induced by hd is the same as its indegree in $\overrightarrow{H}$. By the initial observation, $H[X]$ has a vertex w such that $indeg_{\overrightarrow{H}}(w) < k$. Hence, there exists a sequence of hyperedges $e_1, \ldots, e_n$ and a directed path in $Dir(\overrightarrow{H})$ of the form $(v_0, v_1, \ldots, v_n)$ with $v_0 = w$ and $v_n = v$ such that $hd(e_i) = v_i$ and $v_{i-1} \in vert_H(e_i)$ for each $i \in [n]$. Clearly, $e_1, \ldots, e_n$ are pairwise distinct, because if $e_i = e_j$, then $v_i = v_j$. We now define hd' on E_H from *hd* as follows:

$$hd'(e_i) := v_{i-1} \quad \text{for } i \in [n],$$
$$hd'(e) := hd(e) \quad \text{for } e \in E_H - \{e_1, \ldots, e_n\}.$$

It is clear that in $\overrightarrow{K} := (H, hd')$ we have:

$$indeg_{\overrightarrow{K}}(v) = indeg_{\overrightarrow{H}}(v) - 1,$$
$$indeg_{\overrightarrow{K}}(x) = indeg_{\overrightarrow{H}}(x) \text{ for } x \in V_H - \{v, w\},$$
$$indeg_{\overrightarrow{K}}(w) = indeg_{\overrightarrow{H}}(w) + 1 \leq k.$$

Hence, the badness of $\overrightarrow{K}$ is equal to the badness of $\overrightarrow{H}$ minus 1, or minus $(k+1)$ if v is no longer bad. By repeating this step one obtains a semi-orientation of H that has badness 0 as desired.

(2) Let H be a hypergraph of maximal rank at most m and $\overrightarrow{H} = (H, hd)$ be a semi-orientation of H. A vertex v is *bad* for $\overrightarrow{H}$ (in a different sense as in (1)) if $v \to w$ and $w \to v$ in $Dir(\overrightarrow{H})$ for some w. We assume that in $\overrightarrow{H}$:

(i) every bad vertex has indegree at most k;
(ii) the other vertices have indegree at most mk^2.

We will modify hd into hd' such that (i) and (ii) still hold for $\overrightarrow{K} := (H, hd')$ and such that $\overrightarrow{K}$ has fewer bad vertices than $\overrightarrow{H}$. By repeating this step, starting from a semi-orientation of indegree at most k, we will end up with a semi-orientation $\overrightarrow{K}$ of H without bad vertices and with indegree at most mk^2 as desired. It will follow that $Dir(\overrightarrow{K}) \in S_{k'}^d$ where $k' := (m-1)mk^2$.

We do that as follows. Let v be a bad vertex. Let $e_1, \ldots, e_l$, where $l \leq k$, be the hyperedges e in $\overrightarrow{H}$ such that $hd(e) = v$, and let $X := (vert_H(e_1) \cup \cdots \cup vert_H(e_l)) - \{v\}$. Hence $|X| \leq k(m-1)$. Let Y be the set of hyperedges e such that $v \in vert_H(e)$ and $hd(e) \in X$. The cardinality of Y is at most $k^2(m-1)$ since every such $hd(e)$ is bad, hence of indegree at most k. We transform hd into hd' by letting $hd'(e) := v$ for each $e \in Y$, and $hd'(e) := hd(e)$ for $e \in E_H - Y$. For $\overrightarrow{K} := (H, hd')$ we have:

$$indeg_{\overrightarrow{K}}(v) \leq k + k^2(m-1) \leq mk^2,$$
$$indeg_{\overrightarrow{K}}(x) \leq indeg_{\overrightarrow{H}}(x) \text{ for } x \in V_H - \{v\}.$$

In $Dir(\overrightarrow{K})$, there are no two opposite edges $v \to x$ and $x \to v$ for any $x \in V_H - \{v\}$, i.e., v is not bad in $\overrightarrow{K}$. Furthermore, if $x, y \in V_H - \{v\}$ and $x \to y$ in $Dir(\overrightarrow{K})$, then this edge "comes from" a hyperedge not in Y, hence was present in $Dir(\overrightarrow{H})$. It follows that bad vertices in $\overrightarrow{K}$ were already bad in $\overrightarrow{H}$, hence the number of bad vertices has decreased by at least one. Properties (i) and (ii) still hold in $\overrightarrow{K}$. ∎

Overview of the proof of Theorem 9.38: First, we reduce the general case to that of simple m-hypergraphs, i.e., such that all hyperedges have the same rank m and no two hyperedges have the same sets of vertices. The proof for these hypergraphs will be by induction on m. Since a simple 2-hypergraph is a simple loop-free undirected graph, the case of $m = 2$ is stated in Theorem 9.44 (cf. the first part of its proof).

For the induction step of the proof, we consider a uniformly k-sparse simple m-hypergraph H. In each hyperedge we select a vertex called its head. We obtain thus a semi-orientation $\vec{H}$ of H and the corresponding graph $Dir(\vec{H})$ with directed edges from each vertex of a hyperedge to its head. By Proposition 9.58, we can do that in such a way that the graph $Dir(\vec{H})$ has bounded indegree and no pairs of opposite edges. Since $Dir(\vec{H})$ is uniformly p-sparse for some p, its orientation, whence also the semi-orientation $\vec{H}$ of H, can be defined from $\lfloor H \rfloor$ by monadic second-order formulas using a vertex coloring of H obtained by Proposition 9.40(3), as in the proof of Proposition 9.42(1).

If $\vec{H}$ is a semi-orientation of H, we let $\partial(\vec{H})$ be its *derived hypergraph*: its vertices are the edges of $Dir(\vec{H})$, and its hyperedges are those of H with the following sets of vertices: if e is a hyperedge of H such that $vert_H(e) = \{x_1,\ldots,x_{m-1},y\}$ with head y, then its set of vertices in $\partial(\vec{H})$ is defined as $\{(x_1,y),\ldots,(x_{m-1},y)\}$, so that $\partial(\vec{H})$ is an $(m-1)$-hypergraph.

Since $Dir(\vec{H})$ has bounded indegree, its incidence graph $Inc(Dir(\vec{H}))$ can be constructed from $\lfloor H \rfloor$ by an MS-transduction (we use here Theorem 9.44) and so can be $\lfloor \partial(\vec{H}) \rfloor$. Since the $(m-1)$-hypergraph $\partial(\vec{H})$ is uniformly $2k$-sparse, the induction hypothesis implies that $Inc(\partial(\vec{H}))$ can be constructed from $\lfloor \partial(\vec{H}) \rfloor$ by an MS-transduction. Hence, it can also be constructed from $\lfloor H \rfloor$ by such a transduction. An MS-transduction can construct $Inc(H)$ from $Inc(\partial(\vec{H}))$, which will conclude the proof. $\qquad\square$

Figure 9.3 shows a semi-orientation $\vec{H}$ of a 4-hypergraph H with hyperedges U, X, Y and Z represented by quadrangles. Their respective heads $0, 5, 0$ and 5 are shown by small triangles. Figure 9.4 shows the directed graph $Dir(\vec{H})$ and Figure 9.5 shows the derived 3-hypergraph $\partial(\vec{H})$ with the hyperedges represented by triangles. More comments will be given in the proof of Theorem 9.61.

For a simple m-hypergraph H, we will omit the relations $edg_H^1,\ldots,edg_H^{m-1}$ of $\lfloor H \rfloor$, because they are empty. Thus $\lfloor H \rfloor = \langle V_H, edg_H^m \rangle$. Similarly for a semi-orientation $\vec{H}$ of H, we write $\lfloor \vec{H} \rfloor = \langle V_H, edg_{\vec{H}}^m \rangle$.

Proposition 9.59 Let $k \geq 1$ and $m \geq 2$. There exists an MS-transduction that associates with $\lfloor H \rfloor$, for every H in $US_{k,m}$, one or more structures $\lfloor \vec{H} \rfloor$ where $\vec{H}$ is a semi-orientation of H of indegree at most mk^2.

Proof: For $m = 2$, we have the desired transduction by Proposition 9.42(1), and even with k instead of mk^2.

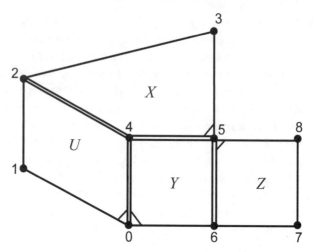

Figure 9.3 A semi-oriented hypergraph $\overrightarrow{H}$.

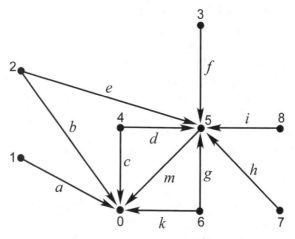

Figure 9.4 The graph $Dir(\overrightarrow{H})$.

We now consider the general case. As in the proof of Proposition 9.42(1), we will specify the required semi-orientation by a vertex coloring of H. By Proposition 9.58, every hypergraph H in $US_{k,m}$ has a semi-orientation $\overrightarrow{H}$ of indegree at most mk^2 such that $D := Dir(\overrightarrow{H}) \in S_{k'}^d$ where $k' := (m-1)mk^2$. By Proposition 9.40(3), there exists a directed graph Q with no pairs of opposite edges and with vertex set $[q]$, where $q := q(k')$, such that there is a homomorphism $\gamma : D \to Q$. Since $V_D = V_H$, γ is a mapping from V_H to $[q]$. If $x \in vert_H(e)$ and $y = hd(e) \neq x$, then $x \to_D y$ and so $\gamma(x) \to_Q \gamma(y)$. Hence, γ is a good coloring, in the following sense. A *good coloring* of a hypergraph H is a mapping $\gamma : V_H \to [q]$ such that

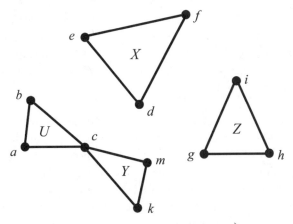

Figure 9.5 The derived 3-hypergraph $\partial(\overrightarrow{H})$.

(i) each set $vert_H(e)$ contains a (unique) vertex x such that $\gamma(y) \to_Q \gamma(x)$ for every other vertex y of $vert_H(e)$;[23]

(ii) the semi-orientation $\overrightarrow{H} := (H, hd)$, where $hd(e)$ is the unique vertex determined by (i), has indegree at most mk^2.

As in the proof of Proposition 9.42(1), a coloring γ can be described by a partition $X_1, \ldots, X_q$ of V_H, such that $X_i = \gamma^{-1}(i)$ for each i. It is straightforward to express Condition (i) by an MS formula with free variables $X_1, \ldots, X_q$. The relation $edg_{\overrightarrow{H}}^m$ is MS-definable from edg_H^m and the sets $X_1, \ldots, X_q$, because we have, for all $x_1, \ldots, x_m$ in V_H,

$$(x_1, \ldots, x_m) \in edg_{\overrightarrow{H}}^m \text{ if and only if}$$

$$(x_1, \ldots, x_m) \in edg_H^m \text{ and } \gamma(x_i) \to_Q \gamma(x_m) \text{ for every } i \in [m-1].$$

Moreover, it is easy to express by an MS formula that the semi-orientation $\overrightarrow{H}$ represented by $\langle V_H, edg_{\overrightarrow{H}}^m \rangle$ has indegree at most mk^2 (using a subformula that expresses that two sets of vertices $\{x_1, \ldots, x_{m-1}\}$ and $\{y_1, \ldots, y_{m-1}\}$ are distinct). Altogether, it can be expressed by an MS formula that $X_1, \ldots, X_q$ is a partition that defines a good coloring, and this formula forms the precondition of the definition scheme of the desired (domain-preserving) monadic second-order transduction, whereas the MS formula for $edg_{\overrightarrow{H}}^m$ is its relation formula. $\blacksquare$

To prove the Sparseness Theorem for simple m-hypergraphs, we will have to compose functional transductions that are domain-extending MS-transductions on classes of structures (as defined in Definition 7.2). Thus, we strengthen Theorem 7.14(3) as follows:

[23] The vertex x is unique because Q has no pairs of opposite edges.

Lemma 9.60 Let $C_1 \subseteq STR^c(\mathcal{R}_1)$ and $C_2 \subseteq STR^c(\mathcal{R}_2)$ be classes of structures, let $f_1 : STR^c(\mathcal{R}_1) \to STR^c(\mathcal{R}_2)$ be a domain-extending MS-transduction on C_1 such that $f_1(C_1) \subseteq C_2$, and let $f_2 : STR^c(\mathcal{R}_2) \to STR^c(\mathcal{R}_3)$ be a domain-extending MS-transduction on C_2. Then $f_1 \cdot f_2 : STR^c(\mathcal{R}_1) \to STR^c(\mathcal{R}_3)$ is a domain-extending MS-transduction on C_1.

Proof: Let τ_i be a concrete domain-extending k_i-copying MS-transduction that realizes f_i (as explained in Definition 7.2). We may assume that, in the definition scheme of τ_1, the domain D_S of an input structure $S \in C_1$ is represented by the copy $D_S \times \{1\}$ in every output structure T (i.e., the first domain formula is *True* and there is an isomorphism from T to $f_1(S)$ that maps $(a, 1)$ to a for every a in D_S). Similarly, we may assume that in the definition scheme of τ_2 the domain D_T of an input structure $T \in C_2$ is represented by the copy $D_T \times \{1\}$ in every output structure U, and hence D_S is represented by $(D_S \times \{1\}) \times \{1\}$ in U.[24] The composition $\tau_1 \cdot \tau_2$ is, in general, *not* a concrete MS-transduction. However, as shown in Case 3 of the proof of Theorem 7.14, it *is* a concrete $k_1 k_2$-copying MS-transduction provided the set $(D_S \times [k_1]) \times [k_2]$ (which contains D_U) is "identified" with the set $D_S \times [k_1 k_2]$ (where, in particular, $((a, 1), 1)$ is identified with $(a, 1)$). Clearly, this identification turns $\tau_1 \cdot \tau_2$ into a domain-preserving MS-transduction in which D_S is represented by the copy $D_S \times \{1\}$. Thus, $f_1 \cdot f_2$ is a domain-preserving MS-transduction on C_1, realized by $\tau_1 \cdot \tau_2$.[25] ∎

Theorem 9.61 Let $k \geq 1$ and $m \geq 2$. The mapping *Inc* that transforms $\lfloor H \rfloor$ into $Inc(H)$ is a domain-extending MS-transduction on $US_{k,m}$.

Proof: The proof is by induction on m, simultaneously for all k. The case $m = 2$ concerns graphs (in S^u) and is stated in Theorem 9.44.

We now consider $m \geq 3$. Let $s\text{-}US_{k,m}$ be the class of semi-orientations of indegree at most mk^2 of hypergraphs in $US_{k,m}$. By Proposition 9.59 (and Theorem 7.14), it suffices to prove that the mapping μ that transforms $\lfloor \overrightarrow{H} \rfloor$ into $Inc(H)$, where $\overrightarrow{H}$ is a semi-orientation of H, is a domain-extending MS-transduction on $s\text{-}US_{k,m}$. In fact, if τ_1 is the domain-preserving MS-transduction that transforms $\lfloor H \rfloor$ into $\lfloor \overrightarrow{H} \rfloor$, by Proposition 9.59, then *Inc* is $\tau_1 \cdot \mu$ on $US_{k,m}$.

Let $\overrightarrow{H} = (H, hd)$ be in $s\text{-}US_{k,m}$, i.e., H is in $US_{k,m}$ and $\overrightarrow{H}$ has indegree at most mk^2. Let $D := Dir(\overrightarrow{H})$. We let $\partial(\overrightarrow{H})$ be the simple $(m-1)$-hypergraph such that

---

[24] Let h_1 be the isomorphism from T to $f_1(S)$ that maps $(a, 1)$ to a. Similarly, let h_2 be the isomorphism from U to $f_2(T)$ that maps $(b, 1)$ to b and in particular $((a, 1), 1)$ to $(a, 1)$. The isomorphism h_1 can be extended to an isomorphism h'_1 from $f_2(T)$ to $f_2(f_1(S))$; thus, it also maps $(a, 1)$ to a. Hence, $h_2 \cdot h'_1$ is an isomorphism from U to $f_2(f_1(S))$ that maps $((a, 1), 1)$ to a.

[25] In the proof of Theorem 7.14, the "identification" is denoted θ. It is an isomorphism from U to the corresponding structure $\overline{\theta}(U)$ and it maps $((a, 1), 1)$ to $(a, 1)$. Then $\tau_1 \cdot \tau_2 \cdot \overline{\theta}$ is a concrete MS-transduction that transforms S into $\overline{\theta}(U)$, and $\theta^{-1} \cdot h_2 \cdot h'_1$ is an isomorphism from $\overline{\theta}(U)$ to $f_2(f_1(S))$ that maps $(a, 1)$ to a. So, $\tau_1 \cdot \tau_2 \cdot \overline{\theta}$ realizes $f_1 \cdot f_2$.

$$V_{\partial(\overrightarrow{H})} := E_D = edg_D \subseteq V_H \times V_H,$$

$$E_{\partial(\overrightarrow{H})} := E_H,$$

$$vert_{\partial(\overrightarrow{H})}(e) := \{(x_1,y),\ldots,(x_{m-1},y)\} \subseteq edg_D$$

if $vert_H(e) = \{x_1,\ldots,x_{m-1},y\}$ and $y = hd(e)$, i.e., if $(x_1,\ldots,x_{m-1},y) \in edg_{\overrightarrow{H}}^m$.

Let us observe that for every $e \in E_{\partial(\overrightarrow{H})}$, we have

$$\bigcup\{vert_D(d) \mid d \in vert_{\partial(\overrightarrow{H})}(e)\} = vert_H(e). \tag{9.1}$$

This is immediate from the definition of $\partial(\overrightarrow{H})$, because $vert_D((x_i,y)) = \{x_i,y\}$.

Let us look again at Figures 9.3–9.5. Figure 9.3 shows a semi-orientation $\overrightarrow{H}$ of a 4-hypergraph H with vertices $0,1,\ldots,8$ and hyperedges U,X,Y,Z with respective heads $0,5,0,5$. Figure 9.4 shows the directed graph $D = Dir(\overrightarrow{H})$. Its edge c from 4 to 0 "comes from" each of the two hyperedges U and Y. This fact is also visible in the 3-hypergraph $\partial(\overrightarrow{H})$ of Figure 9.5 because the edge c is a vertex common to the hyperedges U and Y. The vertices of $\partial(\overrightarrow{H})$ are the edges of D. The hyperedge U has rank 4 in H and rank 3 in $\partial(\overrightarrow{H})$; its vertices are the edges $a = (1,0)$, $b = (2,0)$ and $c = (4,0)$ of D.

Claim 9.61.1 The $(m-1)$-hypergraph $\partial(\overrightarrow{H})$ is uniformly $2k$-sparse.

Proof: Let $X \subseteq V_{\partial(\overrightarrow{H})} = E_D$ and $Y := \bigcup\{vert_D(d) \mid d \in X\} \subseteq V_H$. We have $E_{\partial(\overrightarrow{H})} = E_H$ but these hyperedges have different sets of vertices in $\partial(\overrightarrow{H})$ and in H. It follows from the definition of $\partial(\overrightarrow{H})$ that $E_{\partial(\overrightarrow{H})[X]} \subseteq E_{H[Y]}$: by Equality (9.1), if $vert_{\partial(\overrightarrow{H})}(e) \subseteq X$, then $vert_H(e) \subseteq Y$. Since H is uniformly k-sparse, we have $|E_{H[Y]}| \leq k \cdot |Y|$ and so $|E_{\partial(\overrightarrow{H})[X]}| \leq k \cdot |Y| \leq 2k \cdot |X|$, since we have $|Y| \leq 2 \cdot |X|$. Hence $\partial(\overrightarrow{H})$ is uniformly $2k$-sparse, which completes the proof of the claim. $\square$

We now consider the definability of $D = Dir(\overrightarrow{H})$ and of $\partial(\overrightarrow{H})$ from $\lfloor \overrightarrow{H} \rfloor$ by MS-transductions. We have $\lfloor \overrightarrow{H} \rfloor + Inc(D) = \langle V_H \cup E_D, edg_{\overrightarrow{H}}^m, in_{1,D}, in_{2,D} \rangle$ since $V_D = V_H$. (We use here the union of concrete structures, cf. Definition 9.35.) We have also $\lfloor \overrightarrow{H} \rfloor + Inc(D) + \lfloor \partial(\overrightarrow{H}) \rfloor = \langle V_H \cup E_D, edg_{\overrightarrow{H}}^m, in_{1,D}, in_{2,D}, edg_{\partial(\overrightarrow{H})}^{m-1} \rangle$ since we have $V_{\partial(\overrightarrow{H})} = E_D$. Note that if S is a structure isomorphic to $\langle V_H \cup E_D, edg_{\overrightarrow{H}}^m, in_{1,D}, in_{2,D} \rangle$, then the subset of its domain corresponding to E_D is the set of elements e of D_S such that $(e,x) \in in_{1,S}$ for some x. It follows that a concrete transduction intended to receive input $\lfloor \overrightarrow{H} \rfloor + Inc(D)$ can be described in terms of the sets V_H and E_D and the relations $edg_{\overrightarrow{H}}^m$, $in_{1,D}$ and $in_{2,D}$. For an input structure S isomorphic to $\lfloor \overrightarrow{H} \rfloor + Inc(D)$, it will output structures isomorphic to those obtained for input $\lfloor \overrightarrow{H} \rfloor + Inc(D)$. The same holds for $\lfloor \overrightarrow{H} \rfloor + Inc(D) + \lfloor \partial(\overrightarrow{H}) \rfloor$ and the similar structures defined below.

The constructions of the next claim depend on the fixed integers k and m.

Claim 9.61.2 The mapping τ that transforms the structure $\lfloor \vec{H} \rfloor$ into the structure $\lfloor \vec{H} \rfloor + Inc(D) + \lfloor \partial(\vec{H}) \rfloor$, is a domain-extending MS-transduction on $s\text{-}US_{k,m}$.

Proof: We have the following MS-transductions:

τ_2, domain-preserving, that transforms $\lfloor \vec{H} \rfloor$ into D: its existence is clear from the definition of D and

τ_3, domain-extending, that transforms D into $Inc(D)$, by Theorem 9.44 because $D = Dir(\vec{H})$ is of indegree at most $k' := (m-1)mk^2$ and hence is uniformly k'-sparse.

The definition schemes of these two MS-transductions can be combined, by Definition 9.35, into one for $\tau_4 := Id_1 + \tau_2 \cdot \tau_3$ that transforms $\lfloor \vec{H} \rfloor$ into a structure isomorphic to $\lfloor \vec{H} \rfloor + Inc(D)$, where Id_1 is the identity transduction such that $Id_1(\lfloor \vec{H} \rfloor) = \lfloor \vec{H} \rfloor$.

The structure $\lfloor \partial(\vec{H}) \rfloor = \langle E_D, edg^{m-1}_{\partial(\vec{H})} \rangle$ is definable from $\lfloor \vec{H} \rfloor + Inc(D)$ by an MS-transduction τ_5 because we have

$(e_1, \ldots, e_{m-1}) \in edg^{m-1}_{\partial(\vec{H})}$ if and only if

there are vertices $x_1, \ldots, x_{m-1}, y \in V_H$ such that $e_i : x_i \rightarrow_D y$ and $(x_1, \ldots, x_{m-1}, y) \in edg^m_{\vec{H}}$.

Finally, $\lfloor \vec{H} \rfloor$ is transformed into $\lfloor \vec{H} \rfloor + Inc(D) + \lfloor \partial(\vec{H}) \rfloor$ by the domain-extending MS-transduction $\tau := \tau_4 + \tau_4 \cdot \tau_5 = \tau_4 \cdot (Id_2 + \tau_5)$ where Id_2 is the identity transduction such that $Id_2(\lfloor \vec{H} \rfloor + Inc(D)) = \lfloor \vec{H} \rfloor + Inc(D)$. This completes the proof of Claim 9.61.2. $\qquad\square$

We now complete the proof of the theorem. Our objective is to construct an MS-transduction μ that transforms $\lfloor \vec{H} \rfloor$ into $Inc(H)$ for every $\vec{H}$ in $s\text{-}US_{k,m}$.

By the first claim we have $\partial(\vec{H}) \in US_{2k,m-1}$. Hence, by induction, the mapping μ_1 that transforms $\lfloor \partial(\vec{H}) \rfloor$ into $Inc(\partial(\vec{H})) = \langle E_D \cup E_H, in_{\partial(\vec{H})} \rangle$, is a domain-extending MS-transduction. The domain-extending MS-transduction $\mu_2 := \rho_2 + \rho_3 \cdot \mu_1$ transforms

$$\lfloor \vec{H} \rfloor + Inc(D) + \lfloor \partial(\vec{H}) \rfloor \text{ into } Inc(D) + Inc(\partial(\vec{H})), \text{ where}$$
$$\rho_2(\lfloor \vec{H} \rfloor + Inc(D) + \lfloor \partial(\vec{H}) \rfloor) = Inc(D), \text{ and}$$
$$\rho_3(\lfloor \vec{H} \rfloor + Inc(D) + \lfloor \partial(\vec{H}) \rfloor) = \lfloor \partial(\vec{H}) \rfloor.$$

The transductions ρ_2 and ρ_3 just eliminate components of their input structures.

By the second claim, a domain-extending MS-transduction τ transforms $\lfloor \vec{H} \rfloor$ into $\lfloor \vec{H} \rfloor + Inc(D) + \lfloor \partial(\vec{H}) \rfloor$. Hence, by Lemma 9.60, $\tau \cdot \mu_2$ is a domain-extending MS-transduction that transforms $\lfloor \vec{H} \rfloor$ into

$$Inc(D) + Inc(\partial(\vec{H})) = \langle V_H \cup E_D \cup E_H, in_{1,D}, in_{2,D}, in_{\partial(\vec{H})} \rangle,$$

where $in_{i,D} \subseteq E_D \times V_H$ and $in_{\partial(\vec{H})} \subseteq E_H \times E_D$.

There exists a noncopying MS-transduction μ_3 that transforms the structure $Inc(D) + Inc(\partial(\overrightarrow{H}))$ into $Inc(H) = \langle V_H \cup E_H, in_H \rangle$. The domain formula of μ_3 need only eliminate from the domain of $Inc(D) + Inc(\partial(\overrightarrow{H}))$ the vertices of $\partial(\overrightarrow{H})$, i.e., the elements of E_D. By Equality (9.1), its relation formula $\theta_{in}(e,x)$ that defines in_H, can be written so as to express that $(e,d) \in in_{\partial(\overrightarrow{H})}$ and $(d,x) \in in_{1,D} \cup in_{2,D}$ for some d.

Hence, $Inc(H)$ is definable from $\lfloor \overrightarrow{H} \rfloor$ by the domain-extending MS-transduction μ defined as $\tau \cdot \mu_2 \cdot \mu_3$, so, going back to the elementary transductions used in this proof and in that of Claim 9.61.2, we have

$$Inc = \tau_1 \cdot (Id_1 + \tau_2 \cdot \tau_3) \cdot (Id_2 + \tau_5) \cdot [\rho_2 + (\rho_3 \cdot \mu_1)] \cdot \mu_3$$

on the class $US_{k,m}$.

The validity of the global construction follows from the observations made during the construction of the different transductions in the above expression. ∎

Proof of Theorem 9.38: For $\mathcal{R} = \emptyset$ the result is trivial. If $\mathcal{R} = \{R_1,\ldots,R_p\}$ with $p \geq 1$, then every structure $S \in US_k(\mathcal{R})$ is the union $S = S_1 + \cdots + S_p$ of the structures $S_i \in US_k(\{R_i\})$ such that $S_i := \langle D_S, R_{iS} \rangle$ (cf. Definition 9.35 for unions of concrete structures). If we have domain-extending MS-transductions τ_i that transform S_i into $Inc(S_i)$ for each i, then we can combine (cf. Definition 9.35) their definition schemes into one for a domain-extending MS-transduction τ that transforms S into $Inc(S)$. In particular, if τ_i is $(1 + k_i)$-copying, then τ is $(1 + k_1 + \cdots + k_p)$-copying.

We now consider a structure S in $US_k(\{R\})$ such that R is n-ary. For convenience, we will identify each tuple $(R,x_1,\ldots,x_n)$ in T_S with the corresponding tuple $(x_1,\ldots,x_n)$ in R_S. The structure S can be written $S = S_1 + \cdots + S_n$, where, for every $m \in [n]$, S_m is the structure with domain D_S and relation R_{S_m} consisting of the tuples of R_S having exactly m pairwise distinct elements. As above, it suffices to define, for each $m \in [n]$, a domain-extending MS-transduction τ'_m that transforms S_m into $Inc(S_m)$, and then to take their sum. The case $m = 1$ is easy, with a 2-copying MS-transduction τ'_1 (cf. our comparison of Inc and Inc' in Definition 9.10). It remains to consider the case where $1 < m \leq n$ and to construct τ'_m.

For an n-ary R, we let $US_{k,m}(\{R\})$ be the class of structures $S := \langle D_S, R_S \rangle$ in $US_k(\{R\})$ such that, for every $(x_1,\ldots,x_n) \in R_S$, the set $\{x_i \mid i \in [n]\}$ has cardinality m. For $S \in US_{k,m}(\{R\})$, we let H be the simple m-hypergraph such that $V_H := D_S$ and E_H is the set of sets $\{x_i \mid i \in [n]\}$ such that $(x_1,\ldots,x_n) \in R_S$ with $vert_H(e) := e$ for every $e \in E_H$. (Note that each e is defined as a subset of D_S.)

A domain-preserving MS-transduction can transform $S = \langle D_S, R_S \rangle$ into the structure $\lfloor H \rfloor := \langle D_S, edg_H^m \rangle$, and a domain-extending MS-transduction can transform $\lfloor H \rfloor$ into the structure $Inc(H) = \langle D_S \cup E_H, in_H \rangle$, by Theorem 9.61, since S, and hence H, is uniformly k-sparse. Thus, the mapping that transforms S into the structure $\langle D_S \cup E_H, in_H, R_S \rangle$, which is $S + Inc(H)$, is a domain-extending MS-transduction μ_1 on the class $US_{k,m}(\{R\})$.

We now show that the structure $S + Inc(H)$ can be transformed into a structure U isomorphic to $Inc(S) = \langle D_S \cup T_S, in_{1\,Inc(S)}, \ldots, in_{n\,Inc(S)} \rangle$ by an m^n-copying MS-transduction μ_2 that "preserves" the elements of D_S. We define U with domain included in $(D_S \times \{0\}) \cup (E_H \times [m]^{[n]})$ where $[m]^{[n]}$ is the set of all mappings : $[n] \to [m]$. The set $(D_S \times \{0\})$ is the copy of D_S in U, and T_S is defined as a subset of $E_H \times [m]^{[n]}$.

By applying Proposition 9.43 to the directed graph $\langle D_S \cup E_H, in_H^{-1} \rangle$ (which is in S_m^d), we obtain a pair of MS formulas $(\chi(\mathcal{X}_v), \theta(\mathcal{X}_v, x_1, x_2))$ that defines a partial order $\leq$ on $D_S \cup E_H$ that is linear on each set $vert_H(e)$, for $e \in E_H$. The MS-transduction μ_2 has the set of parameters $\mathcal{X}_v$ and the precondition χ. It uses the formula θ, i.e., the partial order $\leq$, in its domain and relation formulas. For $(e, \pi) \in E_H \times [m]^{[n]}$ and $i \in [n]$, we let $\pi_i(e)$ denote the $\pi(i)$-th element of the set $vert_H(e)$ with respect to $\leq$. The domain formulas of μ_2 define T_S as the set of pairs (e, π) such that $(\pi_1(e), \ldots, \pi_n(e)) \in R_S$, and its relation formulas define $in_{i\,Inc(S)}$, for $i \in [n]$, as the set of all pairs $((e, \pi), (x, 0))$ such that $x = \pi_i(e)$.

By an argument similar to the proof of Lemma 9.60, $\tau'_m := \mu_1 \cdot \mu_2$ is a domain-preserving MS-transduction that transforms S into a structure isomorphic to $Inc(S)$, and hence the mapping Inc is a domain-preserving MS-transduction on the class $US_{k,m}(\{R\})$. ∎

This completes the proof of the Sparseness Theorem, one of the main results of this book.

9.4.3 Consequences

The Sparseness Theorem (Theorem 9.38) has several consequences. We first state those concerning graphs; they follow from Theorem 9.44 proved in Section 9.4.1.

Theorem 9.62 Let $k, r \in \mathcal{N}$ and $i, j \in [2]$. Let $\mathcal{C}$ be a set of uniformly k-sparse, simple, labeled graphs, and $\mathcal{D} \subseteq \mathcal{C}$.
(1) A transduction : $\mathcal{C} \to \mathcal{C}$ is a C_rMS-transduction if and only if it is a C_rMS$_{i,j}$-transduction.[26]
(2) $\mathcal{D}$ is C_rMS-definable with respect to $\mathcal{C}$ if and only if it is C_rMS$_2$-definable with respect to $\mathcal{C}$.
(3) $\mathcal{D}$ has bounded clique-width if and only if it has bounded tree-width.
(4) $\mathcal{D}$ is VR-equational if and only if it is HR-equational.
(5) $\mathcal{D}$ has a decidable C_rMS-satisfiability problem if and only if it has a decidable C_rMS$_2$-satisfiability problem, and decidability implies that $\mathcal{D}$ has bounded tree-width.

Proof: (1) Follows from Theorem 7.14 (about the composition of CMS-transductions); see the discussion in Section 7.3 (before Example 7.44).

---

[26] Transductions here are abstract, not concrete (as postulated at the beginning of Section 9.4).

(2) Follows from Corollary 7.12 (of the Backwards Translation Theorem). It was already stated in Theorem 5.22 (for $r = 0$). The "only if" direction was stated in Theorem 1.44. The "if" direction holds in general and is stated in Proposition 5.20(3).

(3) Follows from Assertions (2) and (3) of Theorem 7.47. The "if" direction holds in general and is stated in Proposition 2.114(1).

(4) Follows from (1) and from the Equationality Theorems (Theorems 7.36 and 7.51), cf. Corollary 7.52. Since the graphs in $\mathcal{D}$ do not have ports or sources, these theorems say that $\mathcal{D}$ is VR-equational (HR-equational) if and only if $\lfloor \mathcal{D} \rfloor$ (respectively, $\lceil \mathcal{D} \rceil$) is the image of $\lfloor \text{TREES} \rfloor$ under an MS-transduction. The "if" direction holds in general and is stated in Theorem 4.49. A result stronger than the "only if" direction was stated in Theorem 4.51: it holds without "uniformly."

(5) Follows from the Reduction Theorem (Theorem 7.54, in Section 7.5). The second part follows from Theorem 7.55(1).　　　　　　　　　　　　　　　■

We have some similar statements with the same proofs[27] for relational structures.

Theorem 9.63 Let $k, r \in \mathcal{N}$ and $i, j \in [2]$. Let $\mathcal{R}$ be a relational signature, $\mathcal{C}$ be a set of uniformly k-sparse $\mathcal{R}$-structures and $\mathcal{D} \subseteq \mathcal{C}$. Assertions (1), (2) and (5) of Theorem 9.62 hold for $\mathcal{C}$ and $\mathcal{D}$.

Note that Assertion (2) is Theorem 9.37.

9.5 References

Theorem 3.7 of [FisMak] claims without proof and erroneously that a set of relational structures of bounded tree-width has bounded relational width, cf. Proposition 9.27. The proof of this proposition simplifies the proof of the same result given in Theorem 7.5 of [CouER]. We use a method from [AdlAdl] where a similar statement is proved. The proof of Proposition 9.28 is due to [AdlAdl].

Orientations defined by monadic second-order formulas have been studied in [Cou95a]. The main result of Section 9.4 is based on [Cou03], where Theorem 9.37 is stated for graphs and relational structures that are countable. However, the proof of Theorem 1.4 in [Cou03] is incorrect. The error has been noted by Blumensath, who gives a correct proof (with a different construction) for infinite relational structures of all cardinalities in [Blu10].

---

[27] See Theorem 9.19 for the second part of Assertion (5).

Conclusion and open problems

We will leave it to the researchers of the next ten years and beyond to decide which of the results presented in this book are the most important. Here we will only indicate what are, in our opinion, the main open problems and research directions, among those that are closely related with the topics of this book.

Algorithmic applications

This topic, studied in Chapter 6, is the most difficult to discuss because of the large number of new articles published every year. Hence, the following comments have a good chance of becoming rapidly obsolete.

The algorithmic consequences of the Recognizability Theorem depend crucially on algorithms that construct graph decompositions of the relevant types or, equivalently, that construct terms over the signatures F^{HR}, F^{VR} which evaluate to the input graphs or relational structures. The efficiency of such algorithms, either for all graphs, but perhaps more promisingly for particular classes of graphs, is a bottleneck for applications.

Another difficulty is due to the sizes of the automata that result from the "compilation" of monadic second-order formulas. Their "cosmological" sizes (cf. [FriGro04], [StoMey], [Wey]) make their constructions intractable for general formulas. However, for certain formulas arising from concrete problems of hardware and software verification they remain manageable, as reported by Klarlund *et al.*, who developed software for that purpose called MONA (cf. [Hen+], [BasKla]). Soguet [Sog] tried using MONA for graphs of bounded tree-width and clique-width, but even for basic graph properties such as connectivity or 3-colorability, the automata become too large to be constructed as soon as one wishes to handle graphs of clique-width more than 3. A more promising perspective presented in [CouDur11] consists of using "fly-automata," i.e., finite deterministic automata whose transitions are defined by programs and not stored in fixed (huge) tables. Hence, an automaton with, say, 2^{100} transition rules specified as a "fly-automaton" will not be computed: for checking a term of size 1000, only the necessary 1000 transitions will be computed. More experiments must be done to verify whether this idea is usable for significant graph properties and for graphs of not too small tree-width or clique-width.

The Recognizability Theorem entails that every MS-expressible graph problem is solvable in polynomial time on graphs of clique-width at most k for each fixed k; more precisely, by a fixed-parameter cubic algorithm. However, some problems that are *not* expressible by MS-formulas also have polynomial algorithms on the same graphs (see [Wan]). The corresponding algorithms are not fixed-parameter tractable with respect to clique-width. The question is:

Does there exist a common logical description for such problems?

Fixed-parameter tractable algorithms for first-order model-checking problems on certain classes of graphs of unbounded tree-width have been constructed: see Grohe's survey [*Gro].[1] The graphs of these classes can be covered by sets of induced sub-graphs of bounded tree-width satisfying certain overlapping constraints. These results use, among other tools, the Recognizability Theorem, which is applied to monadic second-order formulas that are derived from the first-order formula of interest.

*Can one give a kind of algebraic and/or logical description of these graph classes? Can one extend the results surveyed in [*Gro] to fragments of monadic second-order logic that are more expressive than first-order logic?*

Algorithms for enumeration problems and constructions of labeling schemes that answer monadic second-order queries have been briefly cited in Section 6.4.4. Several of these algorithms are based on the Recognizability Theorem.

Can one extend them to these classes of graphs and to first-order formulas?

Such an extension is developed in [CouGK] for labeling schemes intended to solve certain first-order queries.

Algebraic descriptions of graphs and relational structures

Two graph algebras were studied in Chapter 2. The axiomatization of their equational properties leaves, for both of them, open questions which are presented in Sections 2.3.3 and 2.5.3. More important problems are left open for the algebra of relational structures studied in Chapters 5 and 9. The main one is certainly the parsing problem (see Section 9.3.1, Conjecture 9.23), because it is a bottleneck for the construction of fixed-parameter tractable model-checking algorithms for monadic second-order sentences on relational structures.

These algebras yield straightforward extensions to sets of graphs and relational structures of basic notions of Formal Language Theory: context-free and recognizable languages. These extensions are based on equation systems having least solutions and

---

[1] The most powerful result that subsumes all previous ones is in [DvoKT].

on finite congruences. A lot of work has also been done on graph and hypergraph grammars, which do not always yield equational sets, at least with respect to the algebras of Chapter 2 (see [*Roz], [*Eng97]; cf. the last sections of Chapters 1, 2 and 4). These works leave open many questions.

The operations of the algebras of graphs and relational structures $\mathbb{GP}$, $\mathbb{JS}$ and $\mathbb{STR}$ have been selected because of their good "compatibility" with monadic second-order logic: the Recognizability Theorems (Theorems 5.68 and 5.75) witness this "compatibility." However, the operation of unfolding that transforms a directed graph into a finite or infinite tree is also, in a certain sense, "compatible" with monadic second-order logic (as discussed in Section 7.7.) It yields concise descriptions of finite graphs (and also of countable graphs). Can one integrate it as an operation into the present algebraic and logical framework? (It is already widely used for describing countable graphs, see the survey [*BluCL].)

Another question, discussed in Section 9.3.4, is the definition of a "width" parameter for relational structures. The one of Definition 9.33 generalizes clique-width and is robust with respect to monadic second-order transductions (the property of having bounded width for a set of R-structures is preserved under monadic second-order transductions) but one more property would be desirable:

   that the set of R-structures of width at most k be CMS-definable for every R and k.

We recall that tree-width and path-width satisfy such a property, by the excluded minor characterizations of $TWD(\leq k)$ and $PWD(\leq k)$. We have no monadic second-order characterization of the class $CWD(\leq k)$ (as observed in Footnote 28 just before Corollary 5.12). However, the equivalent notion of rank-width (defined in Section 6.2.3, Definition 6.6) enjoys such a property: the class of simple loop-free undirected graphs of rank-width at most k is characterized by finitely many excluded vertex-minors [Oum05], which yields, by the results of [CouOum], the C_2MS-definability of this class.

Graph structure and logic

Many constructions of monadic second-order formulas use graph theoretical properties. The paradigmatic example is the characterization of planar graphs by excluded minors (cf. Chapters 1 and 5) from which a monadic second-order sentence characterizing them can be constructed. The same holds for graphs of tree-width or path-width bounded by a given integer. Other types of excluded configurations are also useful, for example the *vertex-minors* mentioned above, which are essential for the proof of Theorem 7.55(2). This notion is not presented in this (already thick) book: see [Oum05] for its relation with rank-width and [CouOum] for its logical expression. It yields a monadic second-order characterization of *circle graphs*, cf. [Cou08a].

Many important graph classes are characterized by excluded induced subgraphs. In most cases (cographs are an exception), the excluded induced subgraphs form infinite families. We have seen that for chordal graphs (Proposition 2.72). If such a family is monadic second-order definable, then the class of graphs that it characterizes is also monadic second-order definable. This is the case for *interval graphs* and *comparability graphs*, cf. [Cou06a]. The infinite sets of forbidden induced subgraphs that characterize these two classes are VR-equational. Is there a general theorem behind this observation?

Colorability results are also crucial in the proof of the Sparseness Theorem in Section 9.4 about the elimination of edge set quantifications in the monadic second-order expression of graph properties. Hence, deep results of graph theory are useful for constructions of monadic second-order formulas. For example, the Strong Perfect Graph Theorem ([*ChuRST]) entails a monadic second-order expression of perfectness, cf. Example 6.43(3).

In the other direction, certain graph theoretical notions like the decomposition of a graph in 3-blocks, and its modular and split decompositions, can be expressed by monadic second-order transductions (cf. [Cou96a, Cou06b]). We ran out of time to cover this aspect, but we wish to do so in a future edition of the book.

These various interactions between logic (especially monadic second-order logic) and graph theory deserve further investigations.[2] Expressing graph properties using monadic second-order logic is useful in view of algorithmic applications but also raises interesting graph theoretical questions. Particular types of graph decompositions have been defined for proving results like the Strong Perfect Graph Theorem [*ChuRST] and to describe the structure of graphs that exclude a fixed graph as a minor (e.g., clique-sum decompositions, see the surveys [*KawMoh] and [*Gro]). How do they fit with monadic second-order logic?

The well-studied notion of tree-decomposition yields a fascinating question, raised in [Cou91] and and discussed in Section 7.6 (and in Proposition 9.56):

Does there exist, for each k, a function f and a monadic second-order transduction that defines, for every graph of tree-width at most k, a tree-decomposition of this graph of width at most $f(k)$?

A positive answer would characterize the recognizable sets of graphs of bounded tree-width as those that are CMS_2-definable (and of bounded tree-width), see Conjecture 7.62. A negative answer to this conjecture would also be interesting: it would show the existence of model-checking problems that are not monadic second-order expressible but that can be solved by means of finite automata on terms, hence that are fixed-parameter linear with respect to tree-width.

---

[2] Monadic second-order logic is also useful for the definition, the computation and the evaluation (for particular values of their variables) of numerous *graph polynomials*: see [*CouMR] and [Cou08b].

Decidability questions

Among the questions discussed in Section 7.5, the following one (Question 7.56) raised by Seese in [See91] (in a different but equivalent way) is still open, although a partial solution has been given in [CouOum], see Theorem 7.55(2):

Is it true that a set of simple graphs must have bounded clique-width if it has a decidable monadic second-order satisfiability problem?

In this respect, relational structures are much simpler to handle through their incidence graphs rather than directly. This is clear from the constructions of Section 9.2, which reduce the MS_2-satisfiability problem for relational structures to a similar problem for graphs, so that we get an easy extension of Theorem 7.55(1) in Theorem 9.19. But we have no such tool to answer the following question (Question 9.20), which we repeat:

Is it true that if a subset of $STR(\mathcal{R})$ has a decidable CMS-satisfiability problem, then it is included in τ (TREES) for some monadic second-order transduction τ?

MS-transductions of terms and tree transducers

By their very definition, MS-transductions of terms are of linear size increase. Tree-walking transducers and their compositions are far more powerful than MS-transductions of terms (cf. Theorem 8.17), but still have nice properties such as the decidability of type-checking (see [EngMan03b] and [EngHS]) and of the finiteness of their output languages (see [DreEng]). It would be of interest to generalize the definition of MS-transduction in such a way that it captures a larger class of tree transducer transformations, thus allowing the logical specification of such transducers. One could first think of tree transductions of polynomial size increase (see, e.g., the notion of a first-order query in Definition 1.26 of [*Imm]).

The precise relationship between nondeterministic (compositions of) tree transducers and MS-transductions of terms (of which the definition schemes use parameters) has not yet been investigated. For instance, the following problem is open:

Is it decidable whether or not a (nondeterministic) tree-walking transducer is an MS-transduction?

For deterministic tree-walking transducers the problem is decidable [EngMan03a], but its time-complexity is unknown.

The web page www.labri.fr/perso/courcell/TheBook.html will maintain reference updates, new results answering the open questions and errata.

References[1]

Books, book chapters and survey articles

[*AbiHV]   S. Abiteboul, R. Hull and V. Vianu, *Foundations of Databases*, Addison-Wesley, 1995. (74, 78)

[*AhoLSU]   A. Aho, M. Lam, R. Sethi and J. Ullman, *Compilers: Principles, Techniques and Tools*, 2nd edition, Pearson, 2007. (2, 26, 94, 427, 611)

[*ArnNiw]   A. Arnold and D. Niwinski, *Rudiments of μ-Calculus*, North-Holland, 2001. (2, 193, 220, 259)

[*AutBoa]   J.-M. Autebert and L. Boasson, *Transductions Rationnelles: Application aux Langages Algébriques*, Editions Masson, 1988. (590)

[*BaaNip]   F. Baader and T. Nipkow, *Term Rewriting and All That*, Cambridge University Press, 1999. (117, 412)

[*Ber]   J. Berstel, *Transductions and Context-Free Languages*, Teubner Studienbücher, 1979. (58, 552, 579, 589, 590)

[*BluCL]   A. Blumensath, T. Colcombet and C. Löding, Logical theories and compatible operations, in: [*FluGräW], pp. 73–106. (688)

[*Bod93]   H. Bodlaender, A tourist guide through treewidth, *Acta Cybernetica* **11** (1993), 1–22. (51, 121, 185)

[*Bod98]   H. Bodlaender, A partial k-arboretum of graphs with bounded treewidth, *Theoret. Comput. Sci.* **209** (1998), 1–45. (51, 121, 125, 127, 143, 185)

[*Bod06]   H. Bodlaender, Treewidth: characterizations, applications, and computations, in: *Proceedings of WG 2006, 32nd International Workshop, Graph-Theoretic Concepts in Computer Science*, Lecture Notes in Computer Science 4271, Springer, 2006, pp. 1–14. (433, 504)

---

[1] The references are organized in two parts: the first part (with reference labels starting with *) lists books, book chapters and survey articles. The second lists research articles and dissertations. For each reference, we indicate in parentheses the pages where it is cited.

[*BörGG]   E. Börger, E. Grädel and Y. Gurevich, *The Classical Decision Problem*, Springer, 1997. (374)

[*BradMan]   A. Bradley and Z. Manna, *The Calculus of Computation: Decision Procedures with Applications to Verification*, Springer, 2007. (374)

[*BranLS]   A. Brandstädt, V. Le and J. Spinrad, *Graph Classes: A Survey*, SIAM Monographs on Discrete Mathematics and Applications 3, 1999. (134, 160)

[*ChuRST]   M. Chudnovsky, N. Robertson, P. Seymour and R. Thomas, Progress on perfect graphs, *Mathematical Programming, Ser. B* **97** (2003), 405–422. (12, 689)

[*Com+]   H. Comon *et al.*, *Tree Automata Techniques and Applications*, available at: http://tata.gforge.inria.fr/. (37, 58, 78, 83, 180, 188, 221, 225, 227, 249, 255, 259, 426)

[*Cou83]   B. Courcelle, Fundamental properties of infinite trees, *Theoret. Comput. Sci.* **25** (1983), 95–169. (83, 187, 259)

[*Cou86]   B. Courcelle, Equivalences and transformations of regular systems: applications to recursive program schemes and grammars, *Theoret. Comput. Sci.* **42** (1986), 1–122. (259)

[*Cou90a]   B. Courcelle, Recursive applicative program schemes, Chapter 9 of *Handbook of Theoretical Computer Science, Vol. B: Formal Models and Semantics*, J. Van Leeuwen ed., Elsevier, 1990, pp. 459–492. (1 , 259)

[*Cou96b]   B. Courcelle, Basic notions of universal algebra for language theory and graph grammars, *Theoret. Comput. Sci.* **163** (1996), 1–54. (259, 412, 426)

[*Cou97]   B. Courcelle, The expression of graph properties and graph transformations in monadic second-order logic, Chapter 5 of [*Roz], pp. 313–400. (78, 259)

[*CouMR]   B. Courcelle, J. Makowsky and U. Rotics, On the fixed parameter complexity of graph enumeration problems definable in monadic second-order logic, *Discrete Appl. Math.* **108** (2001), 23–52. (504, 689)

[*Cre]   S. Crespi-Reghizzi, *Formal Languages and Compilation*, Springer, 2009. (2, 26, 94, 427)

[*DeraJL]   P. Deransart, M. Jourdan and B. Lorho, *Attribute Grammars*, Lecture Notes in Computer Science 323, Springer, 1988. (611)

[*DersJou]   N. Dershowitz and J.-P. Jouannaud, Rewriting systems, Chapter 6 of *Handbook of Theoretical Computer Science, Vol. B: Formal Models and Semantics*, J. Van Leeuwen ed., Elsevier, 1990, pp. 243–320. (412)

[*DiekRoz]   V. Diekert and G. Rozenberg eds., *The Book of Traces*, World Scientific, 1995. (1, 2, 425)

[*Die]   R. Diestel, *Graph Theory*, 4th edition, Springer, 2010; see: http://diestel-graph-theory.com/. (12, 21, 32, 43, 51, 78, 92, 98, 125, 128, 134, 305, 351, 654, 660)

[*DowFel]   R. Downey and M. Fellows, *Parameterized Complexity*, Springer, 1999. (4, 5, 51, 53, 54, 78, 98, 121, 244, 289, 428, 429, 503)

[*Dre06]   F. Drewes, *Grammatical Picture Generation: A Tree-Based Approach*, Springer, 2006. (219, 259)

[*DreKH]   F. Drewes, H.-J. Kreowski and A. Habel, Hyperedge replacement graph grammars, Chapter 2 of [*Roz], pp. 95–162. (2, 8, 49, 78, 313, 434, 435, 504, 618)

[*EbbFlu]   H. Ebbinghaus and J. Flum, *Finite Model Theory*, Springer, 1999. (57, 425)

[*EhrHR]   A. Ehrenfeucht, T. Harju and G. Rozenberg, *The Theory of 2-Structures*, World Scientific, 1999. (162, 187)

[*Eil]   S. Eilenberg, *Automata, Languages and Machines, Vol A.*, Academic Press, 1974. (30, 79, 221, 229, 235, 255)

[*Eng80]   J. Engelfriet, Some open questions and recent results on tree transducers and tree languages, in: *Formal Language Theory: Perspectives and Open Problems*, R. Book ed., Academic Press, 1980. (611, 620)

[*Eng94]   J. Engelfriet, Graph grammars and tree transducers, in: *Proceedings CAAP 1994 (Colloquium on Trees in Algebra and Programming)*, Lecture Notes in Computer Science 787, Springer, 1994, pp. 15–36. (58, 313)

[*Eng97]   J. Engelfriet, Context-free graph grammars, Chapter 3 of *Handbook of Formal Languages, Vol. 3: Beyond words*, G. Rozenberg and A. Salomaa eds., Springer, 1997, pp. 125–213. (46, 78, 144, 576, 616, 688)

[*EngRoz]   J. Engelfriet and G. Rozenberg, Node replacement graph grammars, Chapter 1 of [*Roz], pp. 1–94. (8, 46, 78, 144, 186, 187, 313, 504)

[*FlaSed]   P. Flajolet and R. Sedgewick, *Analytic Combinatorics*, Cambridge University Press, 2009. (136, 173, 220)

[*FluGräW] J. Flum, E. Grädel and T. Wilke eds., *Logic and Automata: History and Perspectives*, Amsterdam University Press, 2008. (79)

[*FluGro]   J. Flum and M. Grohe, *Parameterized Complexity Theory*, Springer, 2006. (51, 53, 78, 121, 322, 331, 428, 429, 433, 503)

[*Fra]   A. Frank, Connectivity and network flows, in: *Handbook of Combinatorics, Vol. 1*, R. Graham, M. Grötschel and L. Lovász eds., Elsevier, 1997, pp. 111–178. (654, 655, 675)

[*FülVog]   Z. Fülöp and H. Vogler, *Syntax-Directed Semantics: Formal Models based on Tree Transducers*, EATCS Monographs on Theoretical Computer Science, Springer, 1998. (611)

[*GarJoh]   M. Garey and D. Johnson, *Computers and Intractability*, Freeman, 1979. (53, 76, 429, 623)

[*GecSte]   F. Gécseg and M. Steinby, Tree languages, Chapter 1 of *Handbook of Formal Languages, Vol. 3: Beyond Words*, G. Rozenberg and A. Salomaa eds., Springer, 1997, pp. 1–68. (37, 58, 78, 83, 188, 221, 225, 227, 249, 255, 259)

[*GiaRes]   D. Giammarresi and A. Restivo, Two-dimensional languages, Chapter 4 of *Handbook of Formal Languages, Vol. 3: Beyond words*, G. Rozenberg and A. Salomaa eds., Springer, 1997, pp. 215–267. (259, 343)

[*Gol]   M. Golumbic, *Algorithmic Graph Theory and Perfect Graphs*, 2nd edition, Annals of Discrete Mathematics 57, Elsevier, 2004. (132, 134, 477)

[*GotHOS]   G. Gottlob, P. Hliněný, S. Oum and D. Seese, Width parameters beyond tree-width and their applications, *Computer J.* **51** (2008), 326–362. (436)

[*GräTW]   E. Grädel, W. Thomas and T. Wilke, *Automata, Logics, and Infinite Games: A Guide to Current Research*, Lecture Notes in Computer Science 2500, Springer, 2002. (79)

[*Gro]   M. Grohe, Logic, graphs, and algorithms, in: [*FluGräW], pp. 357–422. (11, 78, 504, 687, 689)

[*Hab]   A. Habel, *Hyperedge Replacement: Grammars and Languages*, Lecture Notes in Computer Science 643, Springer, 1992. (49, 313, 618)

[*Har]   M. Harrison, *Introduction to Formal Language Theory*, Addison-Wesley, 1978. (198, 206, 212, 552)

[*Hed]   S. Hedetniemi, References on partial k-trees, *Discrete Appl. Math.* **54** (1994), 281–290. (4, 503)

[*Imm]   N. Immermann, *Descriptive Complexity*, Springer, 1999. (425, 690)

[*JenTof]   T. Jensen and B. Toft, *Graph Coloring Problems*, Wiley-Interscience, 1995. (477)

[*KamLM]   M. Kamiński, V. Lozin and M. Milanič, Recent developments on graphs of bounded clique-width, *Discrete Appl. Math.* **157** (2009), 2747–2761. (185)

[*KawMoh]   K.-I. Kawarabayashi and B. Mohar, Some recent progress and applications in graph minor theory, *Graphs Combin.* **23** (2007), 1–46. (689)

[*Kre]   S. Kreutzer, Algorithmic meta-theorems, in: *Finite and Algorithmic Model Theory*, J. Esparza, C. Michaux and C. Steinhorn eds., Cambridge University Press, 2011, pp. 177–270. (78, 504)

[*Lib04]   L. Libkin, *Elements of Finite Model Theory*, Springer, 2004. (57, 316, 322, 331, 354, 425, 633)

[*Lib06]   L. Libkin, Logic for unranked trees: an overview, *Logical Meth. Comput. Sci.* **2** (2006), 1–31. (259, 412, 426)

[*Mak]   J. Makowsky, Algorithmic uses of the Feferman–Vaught Theorem, *Ann. Pure Appl. Logic* **126** (2004), 159–213. (358, 425, 504)

[*MohaTho]   B. Mohar and C. Thomassen, *Graphs on Surfaces*, Johns Hopkins University Press, 2001. (32, 43, 78, 98, 629)

[*Möhr]   R. Möhring, Computationally tractable classes of ordered sets, in: *Algorithms and Order*, I. Rival ed., Kluwer, 1989, pp. 105–194. (21)

[*MöhRad]   R. Möhring and F. Radermacher, Substitution decomposition for discrete structures and connections with combinatorial optimization, in: *Algebraic and Combinatorial Methods in Operations Research*, Annals of Discrete Mathematics 19, Elsevier, 1984, pp. 257–355. (162, 187)

[*Roz]   G. Rozenberg ed., *Handbook of Graph Grammars and Computing by Graph Transformations, Vol. 1: Foundations*, World Scientific, 1997. (78, 80, 277, 688)

[*Sak]   J. Sakarovitch, *Elements of Automata Theory*, Cambridge University Press, 2009. (30, 58, 79, 221, 229, 235, 236, 287, 509, 589)

[*Tho90]   W. Thomas, Automata on infinite objects, in: *Handbook of Theoretical Computer Science, Vol. B: Formal Models and Semantics*, J. Van Leeuwen ed., Elsevier, 1990, pp. 133–192. (79)

[*Tho97a]   W. Thomas, Languages, automata, and logic, Chapter 7 of *Handbook of Formal Languages, Vol. 3: Beyond Words*, G. Rozenberg and A. Salomaa eds., Springer, 1997, pp. 389–455. (2, 79, 317)

[*Tut]   W. Tutte, *Graph Theory*, Addison-Wesley, 1984. (11)

[*Wec]   W. Wechler, *Universal Algebra for Computer Scientists*, Springer, 1992. (187, 412)

[*WilMau]   R. Wilhelm and D. Maurer, *Compiler Design*, Addison-Wesley, 1995. (611)

Articles and dissertations

[AceEI]   L. Aceto, Z. Esik and A. Ingolfsdottir, A fully equational proof of Parikh's Theorem, *Theoret. Inform. and Appl.* **36** (2002), 129–153. (206, 221)

[AdlAdl]   H. Adler and I. Adler, A note on clique-width and tree-width for structures, preprint, 2008, arXiv:0806.0103v2. (644, 646, 685)

[AhoHU]   A. Aho, J. Hopcroft and J. Ullman, A general theory of translation, *Math. Syst. Theory* **3** (1969), 193–221. (618)

[AhoUll70]   A. Aho and J. Ullman, A characterization of two-way deterministic classes of languages, *J. Comput. Syst. Sci.* **4** (1970), 523–538. (618)

[AhoUll71]   A. Aho and J. Ullman, Translations on a context-free grammar, *Inform. Control* **19** (1971), 439–475. (611, 616)

[AluČer]   R. Alur and P. Černý, Expressiveness of streaming string transducers, in: *Proceedings FSTTCS 2010*, K. Lodaya and M. Mahayan eds., Leibniz International Proceedings in Informatics (LIPIcs) Vol. 8, 2010, pp. 1–12. (610)

[AluDan]   R. Alur and L. D'Antoni, Streaming tree transducers, article in preparation arXiv:1104.2599v1, 2011. (610, 620)

[AluDes]   R. Alur and J. V. Deshmukh, Nondeterministic streaming string transducers, in: *Proceedings ICALP 2011, Part II*, Lecture Notes in Computer Science 6756, Springer, 2011, pp. 1–20. (619)

[ArnCP]   S. Arnborg, D. Corneil and A. Proskurowski, Complexity of finding embeddings in a k-tree, *SIAM J. Algebraic and Discrete Methods* **8** (1987), 277–284. (433, 434)

[ArnCPS]   S. Arnborg, B. Courcelle, A. Proskurowski and D. Seese, An algebraic theory of graph reduction, *J. ACM* **40** (1993), 1134–1164. (143, 433, 501)

[ArnLS]   S. Arnborg, J. Lagergren and D. Seese, Easy problems for tree-decomposable graphs, *J. Algorithms* **12** (1991), 308–340. (502)

[ArnPC]   S. Arnborg, A. Proskurowski and D. Corneil, Forbidden minors characterization of partial 3-trees, *Discrete Math.* **80** (1990), 1–19. (128)

[BacBod06]   E. Bachoore and H. Bodlaender, A branch and bound algorithm for exact, upper, and lower bounds on treewidth, in: *Proceedings of Algorithmic Aspects in Information and Management, Second International Conference*, Lecture Notes in Computer Science 4041, Springer, 2006, pp. 255–266. (433, 504)

[BacBod07]   E. Bachoore and H. Bodlaender, Weighted treewidth: algorithmic techniques and results, in: *Proceedings ISAAC'07*, T. Tokuyama ed., Lecture Notes in Computer Science 4835, Springer, 2007, pp. 893–903. (433, 504)

[Bag]   G. Bagan, MSO queries on tree decomposable structures are computable with linear delay, in: *Proceedings 20th International Workshop Computer Science Logic*, Z. Ésik ed., Lecture Notes in Computer Science 4207, Springer, 2006, pp. 167–181. (500)

[Bar]   D. W. Barnette, 2-Connected spanning subgraphs of planar 3-connected graphs, *J. Comb. Theory B* **61** (1994), 210–216. (356)

[BasKla]   D. Basin and N. Klarlund, Hardware verification using monadic second-order logic, in: *Computer Aided Verification, 7th International Conference*, Lecture Notes in Computer Science 939, Springer, 1995, pp. 31–41. (223, 427, 504, 686)

[BauCou]   M. Bauderon and B. Courcelle, Graph expressions and graph rewritings, *Math. Syst. Theory* **20** (1987), 87–127. (118, 186, 313)

[BeeEKM]   C. Beeri, A. Eyal, S. Kamenkovich and T. Milo, Querying business processes with BP-QL, *Inform. Syst.* **33**(6) (2008), 477–507. (504)

[Bes]   A. Bes, Extensions of monadic second-order logic with cardinality predicates, Paris-Est University, 2011, article in preparation. (573)

[BloEng97] R. Bloem and J. Engelfriet, Monadic second order logic and node relations on graphs and trees, in: *Structures in Logic and Computer Science*, J. Mycielski, G. Rozenberg and A. Salomaa eds., Lecture Notes in Computer Science 1261, Springer, 1997, pp. 144–161. (619)

[BloEng00] R. Bloem and J. Engelfriet, A comparison of tree transductions defined by monadic second order logic and by attribute grammars, *J. Comput. Syst. Sci.* **61** (2000), 1–50. (576, 618, 619)

[Blu10]   A. Blumensath, Guarded second-order logic, spanning trees, and network flows, *Logical Meth. Comput. Sci.* **6** (2010), Paper 4. (672, 685)

[BluCou06] A. Blumensath and B. Courcelle, Recognizability, hypergraph operations, and logical types, *Inform. Comput.* **204** (2006), 853–919. (313, 416, 503, 531, 576, 642, 646, 647, 648)

[BluCou10] A. Blumensath and B. Courcelle, On the monadic second-order transduction hierarchy, *Logical Meth. Comput. Sci.* **6** (2:2) (2010), 1–28. (533, 577)

[BluCou11] A. Blumensath and B. Courcelle, Monadic second-order definable orderings, 2011, submitted for publication in *Logical Meth. Comput. Sci.* (426)

[Bod89]   H. Bodlaender, Achromatic number is NP-complete for cographs and interval graphs, *Inf. Process. Lett.* **31** (1989), 135–138. (476)

[Bod96]   H. Bodlaender, A linear-time algorithm for finding tree-decompositions of small treewidth, *SIAM J. Comput.* **25** (1996), 1305–1317. (54, 55, 433, 434, 436, 637)

[BodFKKT] H. Bodlaender, F. Fomin, A. Koster, D. Kratsch and D. Thilikos, On exact algorithms for treewidth, in: *Proceedings 14th ESA*, Y. Azar and T. Erlebach eds., Lecture Notes in Computer Science 4168, Springer, pp. 672–683. (433)

[BodGHK]   H. Bodlaender, J. Gilbert, H. Hafsteinsson and T. Kloks, Approximating treewidth, pathwidth, frontsize, and shortest elimination tree, *J. Algorithms* **18** (1995), 238–255. (127, 185, 434)

[BodKlo]   H. Bodlaender and T. Kloks, Efficient and constructive algorithms for the pathwidth and treewidth of graphs, *J. Algorithms* **21** (1996), 358–402. (434, 435)

[BodKos10] H. Bodlaender and A. Koster, Treewidth computations I. Upper bounds, *Inform. Comput.* **208** (2010), 259–275. (504)

[BodKos11] H. Bodlaender and A. Koster, Treewidth computations II. Lower bounds, *Inform. Comput.* **209** (2011), 1103–1119. (504)

[BodvAnt]   H. Bodlaender and B. Van Antwerpen-de Fluiter, Reduction algorithms for graphs of small treewidth, *Inform. Comput.* **167** (2001), 86–119. (501)

[BojCol]   M. Bojańczyk and T. Colcombet, Tree-walking automata cannot be determinized, *Theoret. Comput. Sci.* **350** (2006), 164–173. (598)

[BonTal]   I. Boneva and J.-M. Talbot, Automata and logics for unranked and unordered trees, in: *Conference on Rewriting Techniques and Applications*, Lecture Notes in Computer Science 3467, Springer, 2005, pp. 500–515. (259, 412, 426)

[Bor]   R. Borie, Generation of polynomial-time algorithms for some optimization problems on tree-decomposable graphs, *Algorithmica* **14** (1995), 123–137. (504)

[BorPT]   R. Borie, R. Parker and C. Tovey, Automatic generation of linear-time algorithms from predicate calculus descriptions of problems on recursively constructed graph families, *Algorithmica* **7** (1992), 555–581. (504)

[BosDW]   F. Bossut, M. Dauchet and B. Warin, A Kleene Theorem for a class of planar acyclic graphs, *Inform. Comput.* **117** (1995), 251–265. (259)

[BouKMT]   V. Bouchitté, D. Kratsch, H. Müller and I. Todinca, On tree-width approximations, *Discrete Appl. Math.* **136** (2004), 183–196. (434)

[BraDLM]   A. Brandstädt, F. Dragan, H.-O. Le and R. Mosca, New graph classes of bounded clique-width, *Theory Comput. Syst.* **38** (2005), 623–645. (12, 185, 439)

[BraELL]   A. Brandstädt, J. Engelfriet, H.-O. Le and V. Lozin, Clique-width for 4-vertex forbidden subgraphs, *Theory Comput. Syst.* **39** (2006), 561–590. (12, 185, 439)

[Büc]   J. Büchi, Weak second-order arithmetic and finite automata, *Zeitschrift für Mathematische Logik und Grundlagen der Mathematik* **6** (1960), 66–92. (79)

[BuiTV]   B.-M. Bui-Xuan, J. Telle and M. Vatshelle, Boolean-width of graphs, in: *Parameterized and Exact Computation, 4th International Workshop (IWPEC)*, Lecture Notes in Computer Science 5917, Springer, 2009, pp. 61–74. (504)

[CarDF]   R. Carrasco, J. Daciuk and M. Forcada, An implementation of deterministic tree automata minimization, in: *Implementation and Application of Automata, 12th International Conference*, Lecture Notes in Computer Science 4783, Springer, 2007, pp. 122–129. (238)

[ChoSch]   N. Chomsky and M. Schützenberger, The algebraic theory of context-free languages, in: *Computer Programming and Formal Systems*, P. Braffort and D. Hirschberg eds., North-Holland, 1963. (7, 17, 79)

[ChuCLSV]   M. Chudnovsky, G. Cornuéjols, X. Liu, P. Seymour and K. Vušković, Recognizing Berge graphs, *Combinatorica* **25** (2005), 143–186. (478)

[ChuRST]   M. Chudnovsky, N. Robertson, P. Seymour and R. Thomas, The strong perfect graph theorem, *Ann. Math.* **164** (2006), 51–229. (12, 478)

[ChyJák]   M. Chytil and V. Jákl, Serial composition of 2-way finite-state transducers and simple programs on strings, in: *Proceedings ICALP'77*, A. Salomaa and M. Steinby eds., Lecture Notes in Computer Science 52, Springer, 1977, pp. 135–147. (618)

[CorHLRR]   D. Corneil, M. Habib, J.-M. Lanlignel, B. Reed and U. Rotics, Polynomial time recognition of clique-width ≤ 3 graphs, in: *LATIN 2000*, Lecture Notes in Computer Science 1776, Springer, 2000, pp. 126–134. (49, 435)

[CorPS]   D. Corneil, Y. Perl and L. Stewart, A linear time recognition algorithm for cographs, *SIAM J. Comput.* **14** (1985), 926–934. (435)

[CorRot]   D. Corneil and U. Rotics, On the relationship between clique-width and treewidth, *SIAM J. Comput.* **34** (2005), 825–847. (143, 164, 165, 186)

[Cou87]   B. Courcelle, An axiomatic definition of context-free rewriting and its application to NLC graph grammars, *Theoret. Comput. Sci.* **55** (1987), 141–181. (8, 17, 313, 425)

[Cou89a]   B. Courcelle, On recognizable sets and tree automata, in: *Resolution of Equations in Algebraic Structures*, M. Nivat and H. Ait-Kaci eds., Academic Press, 1989, pp. 93–126. (259, 412, 426)

[Cou90b]   B. Courcelle, The monadic second-order logic of graphs I: recognizable sets of finite graphs, *Inform. Comput.* **85** (1990), 12–75. (291, 354, 425, 568)

[Cou91]   B. Courcelle, The monadic second-order logic of graphs V: on closing the gap between definability and recognizability, *Theoret. Comput. Sci.* **80** (1991), 153–202. (574, 576, 689)

[Cou92]   B. Courcelle, The monadic second order logic of graphs VII: graphs as relational structures, *Theoret. Comput. Sci.* **101** (1992), 3–33. (576)

[Cou93]   B. Courcelle, Graph grammars, monadic second-order logic and the theory of graph minors, in: *Graph Structure Theory*, N. Robertson and P. Seymour eds., Contemporary Mathematics 147, American Mathematical Society, 1993, pp. 565–590. (186)

[Cou94a]   B. Courcelle, Recognizable sets of graphs: equivalent definitions and closure properties, *Math. Struct. Comput. Sci.* **4** (1994), 1–32. (249, 313)

[Cou94b]   B. Courcelle, The monadic second-order logic of graphs VI: on several representations of graphs by relational structures, *Discrete Appl. Math.* **54** (1994), 117–149; Erratum **63** (1995), 199–200. (425)

700   *References*

[Cou95a]   B. Courcelle, The monadic second-order logic of graphs VIII:
             orientations, *Ann. Pure Appl. Logic* **72** (1995), 103–143. (169, 552,
             577, 685)

[Cou95b]   B. Courcelle, Structural properties of context-free sets of graphs gen-
             erated by vertex replacement, *Inform. Comput.* **116** (1995), 275–293.
             (313)

[Cou96a]   B. Courcelle, The monadic second-order logic of graphs X: linear
             orderings, *Theoret. Comput. Sci.* **160** (1996), 87–143. (12, 187, 425,
             577, 689)

[Cou99]   B. Courcelle, The monadic second-order logic of graphs XI: hierarchi-
             cal decompositions of connected graphs, *Theoret. Comput. Sci.* **224**
             (1999), 35–58. (12, 129)

[Cou00]   B. Courcelle, The monadic second-order logic of graphs XII: pla-
             nar graphs and planar maps, *Theoret. Comput. Sci.* **237** (2000),
             1–32. (12)

[Cou03]   B. Courcelle, The monadic second-order logic of graphs XIV: uni-
             formly sparse graphs and edge set quantifications, *Theoret. Comput.
             Sci.* **299** (2003), 1–36. (425, 685)

[Cou04]   B. Courcelle, Clique-width of countable graphs: a compactness
             property, *Discrete Math.* **276** (2004), 127–148. (436)

[Cou06a]   B. Courcelle, The monadic second-order logic of graphs XV: on a
             conjecture by D. Seese, *J. Appl. Logic* **4** (2006), 79–114. (12, 337, 689)

[Cou06b]   B. Courcelle, The monadic second-order logic of graphs XVI: canoni-
             cal graph decompositions, *Logical Meth. Comput. Sci.* **2** (2006), 1–46.
             (12, 689)

[Cou08a]   B. Courcelle, Circle graphs and monadic second-order logic, *J. Appl.
             Logic* **6** (2008), 416–442. (12, 688)

[Cou08b]   B. Courcelle, A multivariate interlace polynomial and its computation
             for graphs of bounded clique-width, *Electr. J. Combin.* **15**(1) (2008),
             R69. (689)

[Cou09]   B. Courcelle, Linear delay enumeration and monadic second-order
             logic, *Discrete Appl. Math.* **157** (2009), 2675–2700. (500)

[Cou10]   B. Courcelle, On the model-checking of monadic second-order for-
             mulas with edge set quantifications, 2010, *Discrete Appl. Math.* **160**
             (2012), 866–887. (307, 479, 492, 504)

[CouDur10]   B. Courcelle and I. Durand, Verifying monadic second-order graph
             properties with tree automata, in: *3rd European Lisp Symposium*,
             2010, Lisbon, informal proceedings C. Rhodes ed., pp. 7–21.
             Available at: www.labri.fr/perso/courcell/ArticlesEnCours/BCDurand
             LISP.pdf. (504)

[CouDur11] B. Courcelle and I. Durand, Automata for the verification of monadic second-order graph properties, in preparation, 2011. (475, 476, 478, 504, 646, 686)

[CouEng] B. Courcelle and J. Engelfriet, A logical characterization of hypergraph languages generated by hyperedge replacement grammars, *Math. Syst. Theory* **28** (1995), 515–552. (576)

[CouER] B. Courcelle, J. Engelfriet and G. Rozenberg, Handle-rewriting hyper-graph grammars, *J. Comput. Syst. Sci.* **46** (1993), 218–270. (186, 644, 685)

[CouFra] B. Courcelle and P. Franchi-Zannettacci, Attribute grammars and recursive program schemes I, II, *Theoret. Comput. Sci.* **17** (1982), 163–191, 235–257. (611)

[CouGK] B. Courcelle, C. Gavoille and M. Kanté, Compact labelings for efficient first-order model-checking, *J. Comb. Opt.* **21** (2011), 19–46. (501, 687)

[CouGKT] B. Courcelle, C. Gavoille, M. Kanté and A. Twigg, Optimal labelling for connectivity checking in planar networks with obstacles, submitted, LaBRI, 2008. Available at: http://hal.archives-ouvertes.fr/hal-00367746/fr. (501)

[CouKan] B. Courcelle and M. Kanté, Graph operations characterizing rank-width and balanced graph expressions, in: *Proceedings of WG 2007*, Lecture Notes in Computer Science 4769, Springer, 2007, pp. 66–75. (500)

[CouLag] B. Courcelle and J. Lagergren, Equivalent definitions of recognizability for sets of graphs of bounded tree-width, *Math. Struct. Comput. Sci.* **6** (1996), 141–165. (292, 313)

[CouMak] B. Courcelle and J. Makowsky, Fusion in relational structures and the verification of monadic second-order properties, *Math. Struct. Comput. Sci.* **12** (2002), 203–235. (313)

[CouMakR] B. Courcelle, J. Makowsky and U. Rotics, Linear time solvable optimization problems on graphs of bounded clique-width, *Theory Comput. Syst.* **33** (2000), 125–150. (502, 504)

[CouMos] B. Courcelle and M. Mosbah, Monadic second-order evaluations on tree-decomposable graphs, *Theory Comput. Syst.* **109** (1993), 49–82. (425, 500, 502)

[CouOla] B. Courcelle and S. Olariu, Upper bounds to the clique-width of graphs, *Discrete Appl. Math.* **101** (2000), 77–114. (143, 148, 159, 160, 164, 170, 186, 187)

[CouOli] B. Courcelle and F. Olive, Une axiomatisation au premier ordre des arrangements de pseudo-droites euclidiennes, *Annales de l'Institut Fourier* **49** (1999), 883–903. (622)

References

[CouOum] B. Courcelle and S. Oum, Vertex-minors, monadic second-order logic and a conjecture by Seese, *J. Comb. Theory B* **97** (2007), 91–126. (58, 353, 437, 504, 568, 577, 688, 690)

[CouSén] B. Courcelle and G. Sénizergues, The obstructions of a minor-closed set of graphs defined by a context-free grammar, *Discrete Math.* **182** (1998), 29–51. (314)

[CouTwi] B. Courcelle and A. Twigg, Constrained-path labellings on graphs of bounded clique-width, *Theory Comput. Syst.* **47** (2010), 531–567. (186, 501)

[CouVan] B. Courcelle and R. Vanicat, Query efficient implementation of graphs of bounded clique-width, *Discrete Appl. Math.* **131** (2003), 129–150. (501)

[CouWal] B. Courcelle and I. Walukiewicz, Monadic second-order logic, graph coverings and unfoldings of transition systems, *Ann. Pure Appl. Logic* **92** (1998), 35–62. (577)

[CouWei] B. Courcelle and P. Weil, The recognizability of sets of graphs is a robust property, *Theoret. Comput. Sci.* **342** (2005), 173–228. (187, 308, 312, 313, 426, 575)

[CrePau] C. Crespelle and C. Paul, Fully dynamic recognition algorithm and certificate for directed cographs, *Discrete Appl. Math.* **154** (2006), 1722–1741. (314)

[Din] M. Dinneen, Too many minor order obstructions, *J. UCS* **3** (1997), 1199–1206. (286)

[Don] J. Doner, Tree acceptors and some of their applications, *J. Comput. Syst. Sci.* **4** (1970), 406–451; Erratum, **5** (1971), 453. (45, 79)

[Dre99] F. Drewes, A characterization of the sets of hypertrees generated by hyperedge-replacement graph grammars, *Theory Comput. Syst.* **32** (1999), 159–208. (620)

[DreEng] F. Drewes and J. Engelfriet, Decidability of the finiteness of ranges of tree transductions, *Inform. Comput.* **145** (1998), 1–50. (58, 690)

[Dur] I. Durand, Autowrite: a tool for term rewrite systems and tree automata, *Electr. Notes Theoret. Comput. Sci.* **124** (2005), 29–49. (469)

[DurGra] A. Durand and E. Grandjean, First-order queries on structures of bounded degree are computable with constant delay, *ACM Trans. Comput. Logic* **8** (2007), article 21. (53)

[Dus] V. Dussaux, *Spécifications partielles de dessins de graphes: Etude logique et combinatoire*, Doctoral dissertation, Bordeaux 1 University, 2002. (629)

[DvoKT] Z. Dvorak, D. Kral and R. Thomas, Deciding first-order properties for sparse graphs, in: *Proceedings of the 51st annual IEEE Symposium on Foundations of Computer Science*, 2010, pp. 133–142. (687)

[ElbJT]   M. Elberfeld, A. Jakoby and T. Tantau, Logspace versions of the theorems of Bodlaender and Courcelle, in: *Proceedings of the 51st annual IEEE Symposium on Foundations of Computer Science*, 2010, pp. 143–152. (502)

[Elg]   C. Elgot, Decision problems of finite automata design and related arithmetics, *Trans. Amer. Math. Soc.* **98** (1961), 21–52. (79)

[Eng91]   J. Engelfriet, A regular characterization of graph languages definable in monadic second-order logic, *Theoret. Comput. Sci.* **88** (1991), 139–150. (503)

[Eng09]   J. Engelfriet, The time complexity of typechecking tree-walking tree transducers, *Acta Informatica* **46** (2009), 139–154. (617)

[EngHey91]   J. Engelfriet and L. Heyker, The string generating power of context-free hypergraph grammars, *J. Comput. Syst. Sci.* **43** (1991), 328–360. (576, 618)

[EngHey92]   J. Engelfriet and L. Heyker, Context-free hypergraph grammars have the same term-generating power as attribute grammars, *Acta Informatica* **29** (1992), 161–210. (617)

[EngHey94]   J. Engelfriet and L. Heyker, Hypergraph languages of bounded degree, *J. Comput. Syst. Sci.* **48** (1994), 58–89. (615)

[EngHoo01]   J. Engelfriet and H.J. Hoogeboom, MSO definable string transductions and two-way finite state transducers, *ACM Trans. Comput. Logic* **2** (2001), 216–254. (585, 597, 614, 618)

[EngHoo07]   J. Engelfriet and H.J. Hoogeboom, Finitary compositions of two-way finite-state transductions, *Fundamenta Informaticae* **80** (2007), 111–123. (618)

[EngHS]   J. Engelfriet, H.J. Hoogeboom and B. Samwel, XML transformation by tree-walking transducers with invisible pebbles, in: *Proceedings 26th PODS*, L. Libkin ed., ACM Press, 2007, pp. 63–72. (619, 690)

[EngLM]   J. Engelfriet, E. Lilin and A. Maletti, Extended multi bottom-up tree transducers: composition and decomposition, *Acta Informatica* **46** (2009), 561–590. (610)

[EngMan99]   J. Engelfriet and S. Maneth, Macro tree transducers, attribute grammars, and MSO definable tree translations, *Inform. Comput.* **154** (1999), 34–91. (58, 576, 610, 619, 620)

[EngMan00]   J. Engelfriet and S. Maneth, Tree languages generated by context-free graph grammars, in: *Proceedings TAGT '98*, H. Ehrig, G. Engels, H.-J. Kreowski and G. Rozenberg eds., Lecture Notes in Computer Science 1764, Springer, 2000, pp. 15–29. (576, 620)

[EngMan03a] J. Engelfriet and S. Maneth, Macro tree translations of linear size increase are MSO definable, *SIAM J. Comput.* **32** (2003), 950–1006. (603, 610, 619, 690)

704 *References*

[EngMan03b] J. Engelfriet and S. Maneth, A comparison of pebble tree transducers with macro tree transducers, *Acta Informatica* **39** (2003), 613–698. (611, 612, 690)

[EngMan06] J. Engelfriet and S. Maneth, The equivalence problem for deterministic MSO tree transducers is decidable, *Inf. Process. Lett.* **100** (2006), 206–212. (620)

[EngOos96] J. Engelfriet and V. Van Oostrom, Regular description of context-free graph languages, *J. Comput. Syst. Sci.* **53** (1996), 556–574. (616)

[EngOos97] J. Engelfriet and V. Van Oostrom, Logical description of context-free graph languages, *J. Comput. Syst. Sci.* **55** (1997), 489–503. (576)

[EngRS] J. Engelfriet, G. Rozenberg and G. Slutzki, Tree transducers, L systems, and two-way machines, *J. Comput. Syst. Sci.* **20** (1980), 150–202. (616, 619)

[EngVog85] J. Engelfriet and H. Vogler, Macro tree transducers, *J. Comput. Syst. Sci.* **31** (1985), 71–146. (611, 612, 613, 619)

[EngVog86] J. Engelfriet and H. Vogler, Pushdown machines for the macro tree transducer, *Theoret. Comput. Sci.* **42** (1986), 251–368; Erratum **48** (1986), 339. (612, 619)

[EspGW] W. Espelage, F. Gurski and E. Wanke, Deciding clique-width for graphs of bounded tree-width, *J. Graph Algorithms Appl.* **7** (2003), 141–180. (439)

[FeiHL] U. Feige, M. Hajiaghayi and J. Lee, Improved approximation algorithms for minimum weight vertex separators, *SIAM J. Comput.* **38** (2008), 629–657. (434)

[FelRRS] M. Fellows, F. Rosamond, U. Rotics and S. Szeider, Clique-width is NP-complete, *SIAM J. Discrete Math.* **23** (2009), 909–939. (49, 149, 435)

[FerRR] G. Fertin, A. Raspaud and A. Roychowdhury, On the oriented chromatic number of grids, *Inf. Process. Lett.* **85** (2003), 261–266. (57)

[FisMak] E. Fischer and J. Makowsky, On spectra of sentences of monadic second-order logic with counting, *J. Symbolic Logic* **69** (2004), 617–640. (685)

[FluFG] J. Flum, M. Frick and M. Grohe, Query evaluation via tree-decompositions, *J. ACM* **49** (2002), 716–752. (496, 497, 498)

[FriGro04] M. Frick and M. Grohe, The complexity of first-order and monadic second-order logic revisited, *Ann. Pure Appl. Logic* **130** (2004), 3–31. (55, 439, 686)

[Fül] Z. Fülöp, On attributed tree transducers, *Acta Cybernetica* **5** (1981), 261–279. (611)

[FülKV04] Z. Fülöp, A. Kühnemann and H. Vogler, A bottom-up characterization of deterministic top-down tree transducers with regular look-ahead, *Inf. Process. Lett.* **91** (2004), 57–67. (610, 612)

[FülKV05]   Z. Fülöp, A. Kühnemann and H. Vogler, Linear deterministic multi bottom-up tree transducers, *Theoret. Comput. Sci.* **347** (2005), 276–287. (610)

[Gal]   T. Gallai, Transitiv orientierbare Graphen, *Acta Math. Acad. Sci. Hungar.* **18** (1967), 25–66; English trans. F. Maffray and M. Preissmann, in: *Perfect Graphs*, J. Ramirez Alfonsin and B. Reed eds., Wiley, 2001, pp. 25–66. (11, 12)

[GalMeg]   Z. Galil and N. Meggido, Cyclic ordering is NP-complete, *Theoret. Comput. Sci.* **5** (1977), 179–182. (77, 628)

[GanHli]   R. Ganian and P. Hliněný, On parse trees and Myhill-Nerode-type tools for handling graphs of bounded rank-width, *Discrete Appl. Math.* **158** (2010), 851–867. (504)

[Gan]   H. Ganzinger, Increasing modularity and language-independency in automatically generated compilers, *Sci. Comput. Program.* **3** (1983), 223–278. (619)

[GanRub]   T. Ganzow and S. Rubin, Order-invariant MSO is stronger than Counting MSO in the finite, in: *Proceedings STACS'08*, Leibniz International Proceedings in Informatics (LIPIcs), Vol. 1, 2008, pp. 313–324. (425)

[GarSha]   S. Garfunkel and H. Shank, On the undecidability of finite planar cubic graphs, *J. Symbolic Logic* **37** (1972), 595–597. (425)

[Gav]   F. Gavril, Algorithms for minimum coloring, maximum clique, minimum covering by cliques, and maximum independent set of a chordal graph, *SIAM J. Comput.* **1** (1972), 180–187. (132)

[Gie]   R. Giegerich, Composition and evaluation of attribute coupled grammars, *Acta Informatica* **25** (1988), 355–423. (619)

[GinRic]   S. Ginsburg and H. Rice, Two families of languages related to ALGOL, *J. ACM* **9** (1962), 350–371. (7, 17, 79)

[GinSpa]   S. Ginsburg and E. Spanier, Bounded ALGOL-like languages, *Trans. Amer. Math. Soc.* **113** (1964), 333–368. (555)

[GolRot]   M. Golumbic and U. Rotics, On the clique-width of some perfect graph classes, *Int. J. Found. Comput. Sci.* **11** (2000), 423–443. (160, 186)

[Gon]   G. Gonthier, A computer-checked proof of the Four Colour Theorem, Microsoft Research Cambridge, research report, 2005. (56)

[GotLS]   G. Gottlob, N. Leone and F. Scarcello, Hypertree decompositions and tractable queries, *J. Comput. Syst. Sci.* **64** (2002), 579–627. (78)

[GotPWa]   G. Gottlob, R. Pichler and F. Wei, Bounded treewidth as a key to tractability of knowledge representation and reasoning, *Artif. Intell.* **174** (2010), 105–132. (504)

[GotPWb]   G. Gottlob, R. Pichler and F. Wei, Monadic datalog over finite structures of bounded treewidth, *ACM Trans. Comput. Logic* **12**(1) (2010), article 3. (504)

[GräHO]   E. Grädel, C. Hirsch and M. Otto, Back and forth between guarded and
             modal logics, *ACM Trans. Comput. Logic* **3** (2002), 418–463. (425)

[Gur]   E. Gurari, The equivalence problem for deterministic two-way sequen-
             tial transducers is decidable, *SIAM J. Comput.* **11** (1982), 448–452.
             (620)

[GurWan]   F. Gurski and E. Wanke, The tree-width of clique-width bounded
             graphs without $K_{n,n}$, in: *Proc. Workshop on Graphs, WG'2000*,
             Lecture Notes in Computer Science 1938, Springer, 2000, pp.
             196–205. (166)

[HabKre]   A. Habel and H.-J. Kreowski, Some structural aspects of hyper-
             graph languages generated by hyperedge replacement, in: *Proceedings
             STACS'87*, Lecture Notes in Computer Science 247, Springer, 1987,
             pp. 207–219. (618)

[Hen+]   J. Henriksen *et al.*, Mona: monadic second-order logic in practice,
             in: *Tools and Algorithms for the Construction and Analysis of Sys-
             tems*, Lecture Notes in Computer Science 1019, Springer, 1995, pp.
             89–110. (223, 427, 458, 504, 686)

[Her]   H. Herre, Unentscheidbarkeit in der Graphen Theorie, *Bull. Academ.
             Polonaise des Sciences, Série Sc. Math. Astr. Phys.* **XXI-3** (1973),
             201–208. (425)

[HliOum]   P. Hliněný and S. Oum, Finding branch-decompositions and rank-
             decompositions, *SIAM J. Comput.* **38** (2008), 1012–1032. (55, 436,
             438, 504, 637)

[HliSee]   P. Hliněný and D. Seese, Trees, grids and MSO decidability: from
             graphs to matroids, *Theoret. Comput. Sci.* **351** (2006), 372–393. (577)

[HopUll]   J. Hopcroft and J. Ullman, An approach to a unified theory of automata,
             Bell System Tech. J. **46** (1967), 1793–1829. (618)

[Joha]   O. Johansson, Clique-decomposition, NLC-decomposition and mod-
             ular decomposition: relationships and results for random graphs,
             Congressus Numerantium **132** (1998), 39–60. (158)

[John]   D. Johnson, The (16th) NP-completeness column: an ongoing guide,
             J. Algorithms **6**(3), (1985) 434–451. (3)

[Kab]   V. Kabanets, Recognizability equals definability for partial k-paths,
             in: *Proceedings ICALP'97*, Lecture Notes in Computer Science 1256,
             Springer, 1997, pp. 805–815. (574)

[Kal]   D. Kaller, Definability equals recognizability of partial 3-trees and k-
             connected partial k-trees, *Algorithmica* **27** (2000), 348–381. (574, 576)

[KanRao]   M. Kanté and M. Rao, The rank-width of edge-colored graphs,
             Proceedings of CAI 2011, 2011, arXiv:0709.1433v4. (504)

[Kaw]   K. Kawarabayashi, Hadwiger's conjecture is decidable, Result
             announced at the Graph Minor workshop, Banff, October 2008. (57)

[Kep]   S. Kepser, Querying linguistic treebanks with monadic second-order logic in linear time, *J. Logic Lang. Inform.* **13** (2004), 457–470. (504)

[Kla]   N. Klarlund, Mona & Fido: the logic-automaton connection in practice, in: *Computer Science Logic 1997*, Lecture Notes in Computer Science 1414, Springer, 1998, pp. 311–326. (223, 427, 458, 504)

[KneLan]   J. Kneis and A. Langer, A practical approach to Courcelle's Theorem, *Electr. Notes Theoret. Comput. Sci.* **251** (2009), 65–81. (504)

[Knu]   D. Knuth, Semantics of context-free languages, *Math. Syst. Theory* **2** (1968), 127–145; Correction **5** (1971), 95–96. (611, 619)

[KreTaz]   S. Kreutzer and S. Tazari, Lower bounds for the complexity of monadic second-order logic, in: *Proceedings of the 25th Annual IEEE Symposium on Logic in Computer Science*, 2010, pp. 189–198. (502)

[Lag]   J. Lagergren, Upper bounds on the size of obstructions and intertwines, *J. Comb. Theory B* **73** (1998), 7–40. (128, 286)

[LanWel]   K.-J. Lange and E. Welzl, String grammars with disconnecting or a basic root of the difficulty in graph grammar parsing, *Discrete Appl. Math.* **16** (1987), 17–30. (434)

[Lap]   D. Lapoire, Recognizability equals monadic second-order definability for sets of graphs of bounded tree-width, in: *Proceedings STACS'98*, Lecture Notes in Computer Science 1373, Springer, 1998, pp. 618–628. (574)

[LasNS]   J. Lassez, V. Nguyen and L. Sonenberg, Fixed point theorems and semantics: a folk tale, *Inf. Process. Lett.* **14** (1982), 112–116. (193)

[Lau]   C. Lautemann, The complexity of graph languages generated by hyperedge replacement, *Acta Informatica* **27** (1989), 399–421. (434)

[LeuVW]   J. Leung, O. Vornberger and J. Witthoff, On some variants of the bandwidth minimization problem, *SIAM J. Comput.* **13** (1984), 650–667. (266, 434)

[LodWei]   K. Lodaya and P. Weil, Series-parallel languages and the bounded-width property, *Theoret. Comput. Sci.* **237** (2000), 347–380. (187)

[Loz]   V. Lozin, Minimal classes of graphs of unbounded clique-width, *Annals of Combinatorics*, to appear. (186)

[MakMar]   J. Makowsky and J. Marino, Tree-width and the monadic quantifier hierarchy, *Theoret. Comput. Sci.* **303** (2003), 157–170. (502)

[MakPnu]   J. Makowsky and Y. Pnueli, Arity and alternation in second-order logic, *Ann. Pure Appl. Logic* **78** (1996), 189–202; Erratum **92** (1998), 215. (331)

[MakRot]   J. Makowsky and U. Rotics, On the classes of graphs with few P_4's, *Int. J. Found. Comput. Sci.* **10** (1999), 329–348. (185, 439)

[Man]   S. Maneth, The macro tree transducer hierarchy collapses for functions of linear size increase, in: *Proceedings FSTTCS 2003*, P. Pandya and J. Radhakrishnan eds., Lecture Notes in Computer Science 2914, Springer, 2003, pp. 326–337. (619)

[McDRee] C. McDiarmid and B. Reed, Channel assignment on graphs of bounded tree-width, *Discrete Math.* **273** (2003), 183–192. (504)

[MezWri] J. Mezei and J. Wright, Algebraic automata and context-free sets, *Inform. Control* **11** (1967), 3–29. (2, 17, 24, 36, 79, 259)

[MilSV] T. Milo, D. Suciu and V. Vianu, Typechecking for XML transformers, *J. Comput. Syst. Sci.* **66** (2003), 66–97. (611, 619)

[MooSpie] L. Moonen and F. Spieksma, Exact algorithms for a loading problem with bounded clique width, *Inform. J. Comput.* **18** (2006), 455–465. (504)

[NešSV] J. Nešetřil, E. Sopena and L. Vignal, *T*-preserving homomorphisms of oriented graphs, *Comment. Math. Univ. Carolinae* **38** (1997), 125–136. (655)

[NevBus] F. Neven and J. Van den Bussche, Expressiveness of structured document query languages based on attribute grammars, *J. ACM* **49** (2002), 56–100. (619)

[Niv] M. Nivat, Transductions des langages de Chomsky, *Annales de l'Institut Fourier* **18** (1968), 339–456. (58, 509, 589)

[Oum05] S. Oum, Rank-width and vertex-minors, *J. Comb. Theory B* **95** (2005), 79–100. (12, 504, 688)

[Oum08] S. Oum, Approximating rank-width and clique-width quickly, *ACM Trans. Algorithms* **5** (2008), article 10. (504)

[Oum09] S. Oum, Computing rank-width exactly, *Inf. Process. Lett.* **109** (13) (2009), 745–748. (504)

[OumSey] S. Oum and P. Seymour, Approximating clique-width and branch-width, *J. Comb. Theory B* **96** (2006), 514–528. (55, 435, 436, 504, 637)

[Pil] D. Pilling, Commutative regular equations and Parikh's Theorem, *J. London Math. Soc.* **6** (1973), 663–666. (206)

[Rab65] M. Rabin, A simple method for undecidability proofs, in: *Proc. of the 1964 International Congress for Logic*, North-Holland, 1965, pp. 58–68. (577)

[Rab77] M. Rabin, Decidable theories, in: *Handbook of Mathematical Logic*, J. Barwise ed., North-Holland, 1977, pp. 595–629. (577)

[Rao] M. Rao, MSOL partitioning problems on graphs of bounded treewidth and clique-width, *Theoret. Comput. Sci.* **377** (2007), 260–267. (502)

[Ree] B. Reed, Finding approximate separators and computing tree width quickly, in: *Proceedings 24th Annual ACM Symposium on Theory of Computing*, ACM, 1992, pp. 221–228. (433)

[RobSanST] N. Robertson, D. Sanders, P. Seymour and R. Thomas, The four-colour theorem, *J. Comb. Theory B* **70** (1997), 2–44. (56)

[RobSey90] N. Robertson and P. Seymour, Graph minors IV: tree-width and well-quasi-ordering, *J. Comb. Theory B* **48** (1990), 227–254. (51, 128)

[RobSey03] N. Robertson and P. Seymour, Graph minors XVI: excluding a non-planar graph, *J. Comb. Theory B* **89** (2003), 43–76. (11)

[RobSey04] N. Robertson and P. Seymour, Graph minors XX: Wagner's Conjecture, *J. Comb. Theory B* **92** (2004), 325–357. (98)

[RobST] N. Robertson, P. Seymour and R. Thomas, Hadwiger's conjecture for K_6-free graphs, *Combinatorica* **13** (1993), 279–361. (57)

[See91] D. Seese, The structure of the models of decidable monadic theories of graphs, *Ann. Pure Appl. Logic* **53** (1991), 169–195. (57, 58, 425, 502, 576, 690)

[See96] D. Seese, Linear time computable problems and first-order descriptions, *Math. Struct. Comput. Sci.* **6** (1996), 505–526. (53)

[SheDor] S. Shelah and M. Doron, Relational structures constructible by quantifier-free definable operations, *J. Symbolic Logic* **72** (2007), 1283–1298. (649)

[Sog] D. Soguet, *Génération automatique d'algorithmes linéaires*, Doctoral dissertation, University Paris-Sud, July 2008. (504, 686)

[Sop] E. Sopena, The chromatic number of oriented graphs, *J. Graph Theory* **25** (1997), 191–205. (57)

[StoMey] L. Stockmeyer and A. Meyer, Cosmological lower bound on the circuit complexity of a small problem in logic, *J. ACM* **49** (2002), 753–784. (423, 424, 686)

[TakUK] A. Takahashi, S. Ueno and Y. Kajitani, Minimal acyclic forbidden minors for the family of graphs with bounded path-width, *Discrete Math.* **127** (1994), 293–304. (127, 128)

[ThaWri] J. Thatcher and J. Wright, Generalized finite automata theory with an application to a decision problem of second-order logic, *Math. Syst. Theory* **2** (1968), 57–82. (45, 79)

[Tho82] W. Thomas, Classifying regular events in symbolic logic, *J. Comput. Syst. Sci.* **25** (1982), 360–376. (344)

[Tho97b] W. Thomas, Ehrenfeucht games, the composition method, and the monadic theory of ordinal words, in: *Structures in Logic and Computer Science*, J. Mycielski, G. Rozenberg and A. Salomaa eds., Lecture Notes in Computer Science 1261, Springer, 1997, pp. 118–143. (358)

[Thom] A. Thomason, The extremal function for complete minors, *J. Comb. Theory B* **81** (2001), 318–338. (654)

[Thop] M. Thorup, All structured programs have small tree-width and good register allocation, *Inform. Comput.* **142** (1998), 159–181. (434, 504)

[Tra] B. Trakhtenbrot, Finite automata and the logic of monadic predicates, *Doklady Akademii Nauk SSSR* **140** (1961), 326–329. (79)

[VEBK]   F. Van den Eijkhof, H. Bodlaender and A. Koster, Safe reduction rules for weighted treewidth, *Algorithmica* **47** (2007), 139–158. (433, 504)

[VNMDB]   M. Van den Nest, A. Miyake, W. Dür and H. Briegel, Universal resources for measurement-based quantum computing, *Phys. Rev. Lett.* **97**, 150504 (2006), *arXiv.quant-ph/0604010*. (504)

[Wan]   E. Wanke, k-NLC graphs and polynomial algorithms, *Discrete Appl. Math.* **54** (1994), 251–266. (186, 502, 687)

[Wey]   M. Weyer, Decidability of S1S and S2S, in: *Automata, Logics, and Infinite Games*, Lecture Notes in Computer Science 2500, Springer, 2002, pp. 207–230. (423, 686)

[Wil]   R. Willard, Hereditary undecidability of some theories of finite structures, *J. Symbolic Logic* **59** (1994), 1254–1262. (425)

Index of notation

Sets and sequences

We denote by $\mathcal{P}(A)$ the set of subsets of a set A (its powerset), and by $\mathcal{P}_f(A)$ its set of finite subsets. The difference of two sets A and B is the set $A - B := \{a \in A \mid a \notin B\}$. The empty set is $\emptyset$.

We denote by $Seq(A)$ the set of finite sequences of elements of a set A, and by $s[i]$ the i-th element of $s \in Seq(A)$. The empty sequence is $()$. If A is an alphabet, we denote $Seq(A)$ also by A^*. Its elements are then called words, which are sequences of letters (or symbols). The empty word is also denoted by ε.

We denote by $|A|$ the cardinality of a set A, and also by $|s|$ the length of a sequence s (in particular, of a word). The cardinality of a set A is also denoted by $Card(A)$ in certain cases, for better readability of formulas. We denote by $|s|_a$ the number of occurrences of $a \in A$ in a sequence $s \in Seq(A)$.

Integers

We denote by $\mathcal{Z}$ the set of integers, by $\mathcal{N}$ the set of nonnegative ones and by $\mathcal{N}_+$ the set of positive ones. For $n, m \in \mathcal{Z}$, we let $[n, m] := \{i \in \mathcal{Z} \mid n \leq i \leq m\}$ and $[m] := [1, m]$. We have $[m] = \emptyset$ if $m \leq 0$ and $[n, m] = \emptyset$ if $m < n$.

If $p, q \in \mathcal{N}$, $q \geq 2$, we let $\mathrm{mod}_q(p)$ be the unique integer r in $[0, q - 1]$ such that $p \equiv r \pmod{q}$. If $n \in \mathcal{N}$, then $\exp(n)$ denotes 2^n. All logarithms are in base 2. The function $\exp : \mathcal{N}^2 \to \mathcal{N}$ is defined by $\exp(0, n) = n$ and $\exp(d + 1, n) = 2^{\exp(d, n)}$; thus, $\exp(1, n) = \exp(n)$.

Binary relations and functions

If $R \subseteq A \times A$ is a binary relation on a set A, then R^* denotes its reflexive and transitive closure, R^+ its transitive closure and R^{-1} its inverse $\{(x, y) \mid (y, x) \in R\}$. The identity relation $\{(x, x) \mid x \in A\}$ is denoted by Id_A, or just by Id when A is clear from the context.

If $R \subseteq A \times B$ and $S \subseteq B \times C$ are two binary relations, then $R \cdot S$ denotes their composition, i.e., the relation $\{(x,z) \in A \times C \mid (x,y) \in R$ and $(y,z) \in S$ for some $y \in B\}$. If R and S are functional, i.e., if they define partial functions $f : A \to B$ and $g : B \to C$ respectively, then $R \cdot S$ defines the partial function $g \circ f : A \to C$; in that case we denote $R \cdot S$ also by $S \circ R$. We denote by $\bigcirc_{i \in I} f_i$ the composition in any order of functions f_i that commute pairwise (i.e., such that $f_i \circ f_j = f_j \circ f_i$ for all $i,j \in I$).

The domain of a binary relation $R \subseteq A \times B$ is denoted by $Dom(R) \subseteq A$, and its image is $R(A) \subseteq B$. Thus, $Dom(R) = \{a \in A \mid (a,b) \in R$ for some $b \in B\} = R^{-1}(B)$. For a subset C of A, $R(C) = \{b \in B \mid (a,b) \in R$ for some $a \in C\}$.

The restriction of a mapping $f : A \to B$ to a subset C of A is denoted by $f \restriction C$. Two mappings $f : A \to B$ and $g : A' \to B'$ *agree* if $f(a) = g(a)$ for every $a \in A \cap A'$, i.e., if $f \restriction (A \cap A') = g \restriction (A \cap A')$. We denote by $f \cup g$ their common extension into a mapping $: A \cup A' \to B \cup B'$.

We denote by $[A \to A]_f$ the set of mappings $: A \to A$ that are the identity outside of a finite subset of A and by $Perm_f(A)$ the subset of those that are permutations, i.e., bijections $: A \to A$.

Other symbols

Notation that is self-explanatory (e.g., $Loops(G)$ for the set of loops of a graph G), or that is used in a single section, is not listed. The order below is conceptual: concepts that are close mathematically are put together as closely as possible. General concepts are given before the more technical ones. Symbols are followed by short explanations and the page numbers where they are defined.

Terms and rooted trees (Sections 2.1 and 2.6.1, Definitions 2.13 and 2.14)

Algebras (Sections 2.1 and 2.6.1)

Graphs (Section 2.2)

Graph algebras: graphs with sources (Section 2.3)

Graph algebras: graphs with ports (Section 2.5)

Equational and recognizable sets; automata (Chapters 3, 4 and 6)

Relational structures (Chapters 5, 7 and 9)

Logic (Chapters 5 and 6)

Standard logical notation is not reviewed.

Monadic second-order transductions (Chapters 7 and 9)

Index[1]

---

[1] This index refers to sections, definitions, theorems etc. by using the following abbreviations: a = application, c = corollary, d = definition, e = example, ℓ = lemma, p = proposition, r = remark, s = section and t = theorem.